A Brief Guide to **Getting the Most** from This Book

1 Read the Book

Feature	Description	Benefit
Section-Opening Scenarios	Every section opens with a scenario presenting a unique application of algebra in your life outside the classroom.	Realizing that algebra is everywhere will help motivate your learning. **(See page 106.)**
Detailed Worked-Out Examples	Examples are clearly written and provide step-by-step solutions. No steps are omitted, and each step is thoroughly explained to the right of the mathematics.	The blue annotations will help you understand the solutions by providing the reason why every algebraic step is true. **(See page 111.)**
Applications Using Real-World Data	Interesting applications from nearly every discipline, supported by up-to-date real-world data, are included in every section.	Ever wondered how you'll use algebra? This feature will show you how algebra can solve real problems. **(See page 265.)**
Great Question!	Answers to students' questions offer suggestions for problem solving, point out common errors to avoid, and provide informal hints and suggestions.	By seeing common mistakes, you'll be able to avoid them. This feature should help you not to feel anxious or threatened when asking questions in class. **(See page 109.)**
Brief Reviews	NEW to this edition. Brief Reviews cover skills you already learned but may have forgotten.	Having these refresher boxes easily accessible will help ease anxiety about skills you may have forgotten. **(See page 478.)**
Achieving Success	NEW to this edition. Achieving Success boxes offer strategies for persistence and success in college mathematics courses.	Follow these suggestions to help achieve your full academic potential in college mathematics. **(See page 166.)**
Explanatory Voice Balloons	Voice balloons help to demystify algebra. They translate mathematical language into plain English, clarify problem-solving procedures, and present alternative ways of understanding.	Does math ever look foreign to you? This feature often translates math into everyday English. **(See page 201.)**
Learning Objectives	Every section begins with a list of objectives. Each objective is restated in the margin where the objective is covered.	The objectives focus your reading by emphasizing what is most important and where to find it. **(See page 124.)**
Technology	The screens displayed in the technology boxes show how graphing utilities verify and visualize algebraic results.	Even if you are not using a graphing utility in the course, this feature will help you understand different approaches to problem solving. **(See page 110.)**

2 Work the Problems

Feature	Description	Benefit
Check Point Examples	Each example is followed by a matched problem, called a Check Point, that offers you the opportunity to work a similar exercise. The answers to the Check Points are provided in the answer section.	You learn best by doing. You'll solidify your understanding of worked examples if you try a similar problem right away to be sure you understand what you've just read. **(See page 288.)**
Concept and Vocabulary Checks	These short-answer questions, mainly fill-in-the-blank and true/false items, assess your understanding of the definitions and concepts presented in each section.	It is difficult to learn algebra without knowing its special language. These exercises test your understanding of the vocabulary and concepts. **(See page 229.)**
Extensive and Varied Exercise Sets	An abundant collection of exercises is included in an Exercise Set at the end of each section. Exercises are organized within several categories. Your instructor will usually provide guidance on which exercises to work. The exercises in the first category, Practice Exercises, follow the same order as the section's worked examples.	The parallel order of the Practice Exercises lets you refer to the worked examples and use them as models for solving these problems. **(See page 406.)**
Practice Plus Problems	This category of exercises contains more challenging problems that often require you to combine several skills or concepts.	It is important to dig in and develop your problem-solving skills. Practice Plus Exercises provide you with ample opportunity to do so. **(See page 407.)**
Retaining the Concepts	NEW to this edition. Beginning with Chapter 2, each Exercise Set contains review exercises under the header "Retaining the Concepts."	These exercises improve your understanding of the topics and help maintain mastery of the material. **(See page 234.)**
Preview Problems	Each Exercise Set concludes with three problems to help you prepare for the next section.	These exercises let you review previously covered material that you'll need to be successful for the forthcoming section. Some of these problems will get you thinking about concepts you'll soon encounter. **(See page 312.)**

3 Review for Quizzes and Tests

Feature	Description	Benefit
Mid-Chapter Check Points	At approximately the midway point in the chapter, an integrated set of review exercises allows you to review the skills and concepts you learned separately over several sections.	By combining exercises from the first half of the chapter, the Mid-Chapter Check Points give a comprehensive review before you move on to the material in the remainder of the chapter. **(See page 281.)**
Chapter Review Grids	Each chapter contains a review chart that summarizes the definitions and concepts in every section of the chapter. Examples that illustrate these key concepts are also referenced in the chart.	Review this chart and you'll know the most important material in the chapter! **(See page 454.)**
Chapter Review Exercises	A comprehensive collection of review exercises for each of the chapter's sections follows the grid.	Practice makes perfect. These exercises contain the most significant problems for each of the chapter's sections. **(See page 209.)**
Chapter Tests	Each chapter contains a practice test with approximately 25 problems that cover the important concepts in the chapter. Take the practice test, check your answers, and then watch the Chapter Test Prep Videos to see worked-out solutions for any exercises you miss.	You can use the chapter test to determine whether you have mastered the material covered in the chapter. **(See page 213.)**
Chapter Test Prep Videos	These videos contain worked-out solutions to every exercise in each chapter test and can be found in MyMathLab and on YouTube.	The videos let you review any exercises you miss on the chapter test.
Objective Videos	NEW to this edition. These fresh, interactive videos walk you through the concepts from every objective of the text.	The videos provide you with active learning at your own pace.
Cumulative Review Exercises	Beginning with Chapter 2, each chapter concludes with a comprehensive collection of mixed cumulative review exercises. These exercises combine problems from previous chapters and the present chapter, providing an ongoing cumulative review.	Ever forget what you've learned? These exercises ensure that you are not forgetting anything as you move forward. **(See page 461.)**

COLLEGE ALGEBRA
ESSENTIALS

COLLEGE ALGEBRA
ESSENTIALS
ANNOTATED INSTRUCTOR'S EDITION

Robert Blitzer

Miami Dade College

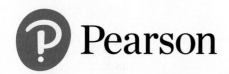

Director, Portfolio Management: Anne Kelly
Courseware Portfolio Manager: Dawn Murrin
Portfolio Management Administrator: Joseph Colella
Content Producer: Kathleen A. Manley
Managing Producer: Karen Wernholm
Producer: Erica Lange
Manager, Courseware QA: Mary Durnwald
Manager, Content Development: Kristina Evans
Product Marketing Manager: Jennifer Edwards
Marketing Assistant: Jennifer Myers

Executive Marketing Manager: Peggy Lucas
Marketing Assistant: Adiranna Valencia
Senior Author Support/Technology Specialist: Joe Vetere
Production Coordination: Francesca Monaco/codeMantra
Text Design and Composition: codeMantra
Illustrations: Scientific Illustrators
Photo Research and Permission Clearance: Cenveo Publisher Services
Cover Design: Studio Montage
Cover Image: Ray_of_Light/Shutterstock

This work is solely for the use of instructors and administrators for the purpose of teaching courses and assessing student learning. Unauthorized dissemination, publication or sale of the work, in whole or in part (including posting on the internet) will destroy the integrity of the work and is strictly prohibited.

Library of Congress Cataloging-in-Publication Data

Names: Blitzer, Robert.
Title: College algebra essentials.
Description: Fifth edition. | Hoboken, NJ : Pearson, [2018] | Includes index.
Identifiers: LCCN 2016030695 | ISBN 9780134469294 ISBN 9780134470191 (Annotated Instructor's Edition)
Subjects: LCSH: Algebra—Textbooks.
Classification: LCC QA152.3.B645 2018 | DDC 512.9—dc23
LC record available at https://lccn.loc.gov/2016030695

ISBN 13: 978-0-13-447019-1
ISBN 10: 0-13-447019-2
ISBN 13: 978-0-13-446929-4
ISBN 10: 0-13-446929-1

CONTENTS

v

PREFACE

I've written *College Algebra Essentials*, **Fifth Edition**, to help diverse students, with different backgrounds and future goals, to succeed. The book has three fundamental goals:

1. To help students acquire a solid foundation in algebra, preparing them for other courses such as calculus, business calculus, and finite mathematics.

2. To show students how algebra can model and solve authentic real-world problems.

3. To enable students to develop problem-solving skills, while fostering critical thinking, within an interesting setting.

One major obstacle in the way of achieving these goals is the fact that very few students actually read their textbook. This has been a regular source of frustration for me and for my colleagues in the classroom. Anecdotal evidence gathered over years highlights two basic reasons that students do not take advantage of their textbook:

- "I'll never use this information."
- "I can't follow the explanations."

I've written every page of the Fifth Edition with the intent of eliminating these two objections. The ideas and tools I've used to do so are described for the student in "A Brief Guide to Getting the Most from This Book," which appears at the front of the book.

A Note on the Essentials Version of College Algebra

- *College Algebra Essentials*, **Fifth Edition**, is a concise version of the **Seventh Edition** of *College Algebra*. The essentials version differs from the Seventh Edition only in terms of length. Chapter 6 (Matrices and Determinants), Chapter 7 (Conic Sections), and Chapter 8 (Sequences, Induction, and Probability) have been eliminated. The essentials version provides a lighter, less expensive alternative to the Seventh Edition for instructors who do not cover the topics in Chapters 6, 7, and 8.

What's New in the Fifth Edition?

- **New Applications and Real-World Data.** The Fifth Edition contains 55 worked-out examples and exercises based on new data sets, and 29 examples and exercises based on data updated from the Fourth Edition. Many of the new applications involve topics relevant to college students, including student-loan debt (Chapter P, Mid-Chapter Check Point, Exercise 31), grade inflation (Exercise Set 1.2, Exercises 97–98), median earnings, by final degree earned (Exercise Set 1.3, Exercises 3–4), excuses for not meeting deadlines (Chapter 1 Summary,

Exercise 36), political orientation of college freshmen (Chapter 2 Summary, Exercise 53), sleep hours of college students (Exercise Set 5.1, Exercise 74), and the number of hours college students study per week, by major (Exercise Set 5.2, Exercises 33–34).

- **Brief Reviews.** Beginning with Chapter 1, the Brief Review boxes that appear throughout the book summarize mathematical skills, many of which are course prerequisites, that students have learned, but which many students need to review. This feature appears whenever a particular skill is first needed and eliminates the need for you to reteach that skill. For more detail, students are referred to the appropriate section and objective in a previous chapter where the topic is fully developed.

- **Achieving Success.** The Achieving Success boxes, appearing at the end of many sections, offer strategies for persistence and success in college mathematics courses.

- **Retaining the Concepts.** Beginning with Chapter 2, Section 2.1, each Exercise Set contains three review exercises under the header "Retaining the Concepts." These exercises are intended for students to review previously covered objectives in order to improve their understanding of the topics and to help maintain their mastery of the material. If students are not certain how to solve a review exercise, they can turn to the section and worked example given in parentheses at the end of each exercise. The Fifth Edition contains 78 new exercises in the "Retaining the Concepts" category.

- **New Blitzer Bonus Videos with Assessment.** Many of the Blitzer Bonus features throughout the textbook have been turned into animated videos that are built into the MyMathLab course. These videos help students make visual connections to algebra and the world around them. Assignable exercises have been created within the MyMathLab course to assess conceptual understanding and mastery. These videos and exercises can be turned into a media assignment within the Blitzer MyMathLab course.

- **Updated Learning Guide.** Organized by the textbook's learning objectives, this updated Learning Guide helps students learn how to make the most of their textbook for test preparation. Projects are now included to give students an opportunity to discover and reinforce the concepts in an active learning environment and are ideal for group work in class.

- **Updated Graphing Calculator Screens.** All screens have been updated using the TI-84 Plus C.

What Content and Organizational Changes Have Been Made to the Fifth Edition?

- **Section P.1 (Algebraic Expressions, Mathematical Models, and Real Numbers)** follows an example on the cost of attending college (Example 2) with a new Blitzer Bonus, "Is College Worthwhile?"

- **Section P.6 (Rational Expressions)** uses the least common denominator to combine rational expressions with different denominators, including expressions having no common factors in their denominators.

- **Section 1.1 (Graphing and Graphing Utilities)** contains a new example of a graph with more than one x-intercept (Example 5(d)).

- **Section 1.4 (Complex Numbers)** includes a new example on dividing complex numbers where the numerator is of the form bi (Example 3). (This is then followed by an example picked up from the Fourth Edition where the numerator is of the form $a + bi$.)

- **Section 1.5 (Quadratic Equations)** provides a step-by-step procedure for solving quadratic equations by completing the square. This procedure forms the framework for the solutions in Examples 4 and 5.

- **Section 1.5 (Quadratic Equations)** contains an example on the quadratic formula (Example 6) where the formula is used to solve a quadratic equation with rational solutions, an equation that students can also solve by factoring.

- **Section 1.5 (Quadratic Equations)** has a new application of the Pythagorean Theorem (Example 11) involving HDTV screens. The example is followed by a new Blitzer Bonus, "Screen Math."

- **Section 1.6 (Other Types of Equations)** includes an example on solving an equation quadratic in form (Example 8),
$$(x^2 - 5)^2 + 3(x^2 - 5) - 10 = 0,$$
where u is a binomial ($u = x^2 - 5$).

- **Section 2.2 (More on Functions and Their Graphs)** contains a new discussion on graphs with three forms of symmetry (Examples 2 and 3) before presenting even and odd functions. A new example (Example 4) addresses identifying even or odd functions from graphs.

- **Section 2.3 (Linear Functions and Slope)** includes a new Blitzer Bonus, "Slope and Applauding Together."

- **Section 2.7 (Inverse Functions)** replaces an example on finding the inverse of $f(x) = \frac{5}{x} + 4$ with an example on finding the inverse of $f(x) = \frac{x + 2}{x - 3}$ (Example 4), a function with two occurrences of x.

- **Section 3.5 (Rational Functions and Their Graphs)** opens with a discussion of college students and video games. This is revisited in a new example (Example 9, "Putting the Video-Game Player Inside the Game") involving the Oculus Rift, a virtual reality headset that enables users to experience video games as immersive three-dimensional environments.

- **Section 5.1 (Systems of Linear Equations in Two Variables)** contains a new discussion on problems involving mixtures, important for many STEM students. A new example (Example 8) illustrates the procedure for solving a mixture problem.

What Familiar Features Have Been Retained in the Fifth Edition?

- **Learning Objectives.** Learning objectives, framed in the context of a student question (What am I supposed to learn?), are clearly stated at the beginning of each section. These objectives help students recognize and focus on the section's most important ideas. The objectives are restated in the margin at their point of use.

- **Chapter-Opening and Section-Opening Scenarios.** Every chapter and every section open with a scenario presenting a unique application of mathematics in students' lives outside the classroom. These scenarios are revisited in the course of the chapter or section in an example, discussion, or exercise.

- **Innovative Applications.** A wide variety of interesting applications, supported by up-to-date, real-world data, are included in every section.

- **Detailed Worked-Out Examples.** Each example is titled, making the purpose of the example clear. Examples are clearly written and provide students with detailed step-by-step solutions. No steps are omitted and each step is thoroughly explained to the right of the mathematics.

- **Explanatory Voice Balloons.** Voice balloons are used in a variety of ways to demystify mathematics. They translate algebraic ideas into everyday English, help clarify problem-solving procedures, present alternative ways of understanding concepts, and connect problem solving to concepts students have already learned.

- **Check Point Examples.** Each example is followed by a similar matched problem, called a Check Point, offering students the opportunity to test their understanding of the example by working a similar exercise. The answers to the Check Points are provided in the answer section.

- **Concept and Vocabulary Checks.** This feature offers short-answer exercises, mainly fill-in-the-blank and true/false items, that assess students' understanding of the definitions and concepts presented in each section. The Concept and Vocabulary Checks appear as separate features preceding the Exercise Sets.

- **Extensive and Varied Exercise Sets.** An abundant collection of exercises is included in an Exercise Set at the end of each section. Exercises are organized within nine category types: Practice Exercises, Practice Plus Exercises, Application Exercises, Explaining the Concepts, Technology Exercises, Critical Thinking Exercises, Group Exercises, Retaining the Concepts, and Preview Exercises. This format makes it easy to create well-rounded homework assignments. The order of the Practice Exercises is exactly the same as the order of the section's worked examples. This parallel order enables students to refer to the titled examples and their detailed explanations to achieve success working the Practice Exercises.

- **Practice Plus Problems.** This category of exercises contains more challenging practice problems that often require students to combine several skills or concepts. With an average of ten Practice Plus problems per Exercise Set, instructors are provided with the option of creating assignments that take Practice Exercises to a more challenging level.

- **Mid-Chapter Check Points.** At approximately the midway point in each chapter, an integrated set of Review Exercises allows students to review and assimilate the skills and concepts they learned separately over several sections.

- **Graphing and Functions.** Graphing is introduced in Chapter 1 and functions are introduced in Chapter 2, with an integrated graphing functional approach emphasized throughout the book. Graphs and functions that model data appear in nearly every section and Exercise Set. Examples and exercises use graphs of functions to explore relationships between data and to provide ways of visualizing a problem's solution. Because functions are the core of this course, students are repeatedly shown how functions relate to equations and graphs.

- **Integration of Technology Using Graphic and Numerical Approaches to Problems.** Side-by-side features in the technology boxes connect algebraic solutions to graphic and numerical approaches to problems. Although the use of graphing utilities is optional, students can use the explanatory voice balloons to understand different approaches to problems even if they are not using a graphing utility in the course.

- **Great Question!** This feature presents a variety of study tips in the context of students' questions. Answers to questions offer suggestions for problem solving, point out common errors to avoid, and provide informal hints and suggestions. As a secondary benefit, this feature should help students not to feel anxious or threatened when asking questions in class.

- **Chapter Summaries.** Each chapter contains a review chart that summarizes the definitions and concepts in every section of the chapter. Examples that illustrate these key concepts are also referenced in the chart.

- **End-of-Chapter Materials.** A comprehensive collection of Review Exercises for each of the chapter's sections follows the Summary. This is followed by a Chapter Test that enables students to test their understanding of the material covered in the chapter. Beginning with Chapter 2, each chapter concludes with a comprehensive collection of mixed Cumulative Review Exercises.

- **Blitzer Bonuses.** These enrichment essays provide historical, interdisciplinary, and otherwise interesting connections to the algebra under study, showing students that math is an interesting and dynamic discipline.

- **Discovery.** Discovery boxes, found throughout the text, encourage students to further explore algebraic concepts. These explorations are optional and their omission does not interfere with the continuity of the topic under consideration.

I hope that my passion for teaching, as well as my respect for the diversity of students I have taught and learned from over the years, is apparent throughout this new edition. By connecting algebra to the whole spectrum of learning, it is my intent to show students that their world is profoundly mathematical, and indeed, π is in the sky.

Robert Blitzer

Acknowledgments

An enormous benefit of authoring a successful series is the broad-based feedback I receive from the students, dedicated users, and reviewers. Every change to this edition is the result of their thoughtful comments and suggestions. I would like to express my appreciation to all the reviewers, whose collective insights form the backbone of this revision. In particular, I would like to thank the following people for reviewing *College Algebra*, *Algebra and Trigonometry*, *Precalculus*, and *Trigonometry*.

Karol Albus, South Plains College

Kayoko Yates Barnhill, Clark College

Timothy Beaver, Isothermal Community College

Jaromir Becan, University of Texas-San Antonio

Imad Benjelloun, Delaware Valley College

Lloyd Best, Pacific Union College

David Bramlett, Jackson State University

Natasha Brewley-Corbin, Georgia Gwinnett College

Denise Brown, Collin College-Spring Creek Campus

David Britz, Raritan Valley Community College

Bill Burgin, Gaston College

Jennifer Cabaniss, Central Texas College

Jimmy Chang, St. Petersburg College

Teresa Chasing Hawk, University of South Dakota

Diana Colt, University of Minnesota-Duluth

Shannon Cornell, Amarillo College

Wendy Davidson, Georgia Perimeter College-Newton

Donna Densmore, Bossier Parish Community College

Disa Enegren, Rose State College

Keith A. Erickson, Georgia Gwinnett College

Nancy Fisher, University of Alabama

Donna Gerken, Miami Dade College

Cynthia Glickman, Community College of Southern Nevada

Sudhir Kumar Goel, Valdosta State University

Donald Gordon, Manatee Community College

David L. Gross, University of Connecticut

Jason W. Groves, South Plains College

Joel K. Haack, University of Northern Iowa

Jeremy Haefner, University of Colorado

Joyce Hague, University of Wisconsin at River Falls

Mike Hall, University of Mississippi

Mahshid Hassani, Hillsborough Community College

Tom Hayes, Montana State University

Christopher N. Hay-Jahans, University of South Dakota

Angela Heiden, St. Clair Community College

Celeste Hernandez, Richland College

Alysmarie Hodges, Eastfield College

Amanda Hood, Copiah-Lincoln Community College

Jo Beth Horney, South Plains College

Heidi Howard, Florida State College at Jacksonville-South Campus

Winfield A. Ihlow, SUNY College at Oswego

Nancy Raye Johnson, Manatee Community College

Dennine Larue, Fairmont State University

Mary Leesburg, Manatee Community College

Christine Heinecke Lehman, Purdue University North Central

Alexander Levichev, Boston University

Zongzhu Lin, Kansas State University

Benjamin Marlin, Northwestern Oklahoma State University

Marilyn Massey, Collin County Community College

Yvelyne McCarthy-Germaine, University of New Orleans

David McMann, Eastfield College

Owen Mertens, Missouri State University-Springfield

James Miller, West Virginia University

Martha Nega, Georgia Perimeter College-Decatur

Shahla Peterman, University of Missouri-St. Louis

Debra A. Pharo, Northwestern Michigan College

Gloria Phoenix, North Carolina Agricultural and Technical State University

Katherine Pinzon, Georgia Gwinnett College

David Platt, Front Range Community College

Juha Pohjanpelto, Oregon State University

Brooke Quinlan, Hillsborough Community College

Janice Rech, University of Nebraska at Omaha

Joseph W. Rody, Arizona State University

Behnaz Rouhani, Georgia Perimeter College-Dunwoody

Judith Salmon, Fitchburg State University

Michael Schramm, Indian River State College

Cynthia Schultz, Illinois Valley Community College

Pat Shelton, North Carolina Agricultural and Technical State University

Jed Soifer, Atlantic Cape Community College

Caroline Spillman, Georgia Perimeter College-Clarkston

Jonathan Stadler, Capital University

Franotis R. Stallworth, Gwinnett Technical College

John David Stark, Central Alabama Community College

Chris Stump, Bethel College

Scott Sykes, University of West Georgia

Richard Townsend, North Carolina Central University

Pamela Trim, Southwest Tennessee Community College

Chris Turner, Arkansas State University

Richard E. Van Lommel, California State University-Sacramento

Dan Van Peursem, University of South Dakota

Philip Van Veldhuizen, University of Nevada at Reno

Jeffrey Weaver, Baton Rouge Community College

Amanda Wheeler, Amarillo College

David White, The Victoria College

Tracy Wienckowski, University of Buffalo

Additional acknowledgments are extended to Dan Miller and Kelly Barber for preparing the solutions manuals; Brad Davis for preparing the answer section, serving as accuracy checker, and writing the new learning guide; the codeMantra formatting team for the book's brilliant paging; Brian Morris and Kevin Morris at Scientific Illustrators for superbly illustrating the book; Francesca Monaco, project manager; and Kathleen Manley, production editor, whose collective talents kept every aspect of this complex project moving through its many stages.

I would like to thank my editor at Pearson, Dawn Murrin, who, with the assistance of Joseph Colella, guided and coordinated the book from manuscript through production. Finally, thanks to Peggy Lucas and Jennifer Edwards for their innovative marketing efforts and to the entire Pearson sales force for their confidence and enthusiasm about the book.

Robert Blitzer

Get the Most Out of

MyMathLab is the leading online homework, tutorial, and assessment program for teaching and learning mathematics, built around Pearson's best-selling content. MyMathLab helps students and instructors improve results; it provides engaging experiences and personalized learning for each student so learning can happen in any environment. Plus, it offers flexible and time-saving course management features to allow instructors to easily manage their classes while remaining in complete control, regardless of course format.

Preparedness

MyMathLab course solutions offer a complete College Algebra or Precalculus course with integrated review of select topics from developmental algebra. These courses help remediate students "just-in-time" and help with student retention of important concepts, ultimately boosting student success.

- Students begin each chapter by completing a Skills Check assignment to pinpoint which developmental topics, if any, they need to review.
- Students who demonstrate mastery of the review topics will move straight into the College Algebra content.
- A personalized review homework assignment will provide extra support for the students who need it.
- Additional review materials (worksheets, videos, and more) are available in an Integrated Review section at the start of each chapter in MyMathLab.

MyMathLab with Integrated Review are appropriate for students who struggle with pre-requisite skills and for co-requisite course models. These Integrated Review MyMathLab courses are available for a variety of College Algebra and Precalculus programs, as well as a variety of other disciplines.

Used by more than 37 million students worldwide, MyMathLab delivers consistent, measurable gains in student learning outcomes, retention, and subsequent course success.

www.mymathlab.com

MyMathLab Online Course
for *College Algebra Essentials* by Robert Blitzer

(access code required)

NEW! Video Program

These fresh, interactive videos walk you through the concepts from every objective of the text. The videos provide an active learning environment where students can work at their own pace.

Your Turn!
Choose the option that best answers the question.

Perform the indicated operation, writing the result in standard form:
$$(-4 - 8i) - (-7 + 2i)$$

 a. $-3 - 10i$
 b. $-11 - 6i$
 c. $-11 + 6i$

NEW! Guided Visualizations

These HTML-based, interactive figures help students visualize the concepts through directed explorations and purposeful manipulation. They encourage active learning, critical thinking, and conceptual learning. They are compatible with iPad and tablet devices.

The Guided Visualizations are located in the Multimedia Library and can be assigned as homework with correlating assessment exercises. Additional Exploratory Exercises are available to help students think more conceptually about the figures and provide an excellent framework for group projects or lecture discussion.

NEW! Workspace Assignments

Students can now show their work like never before! Workspace Assignments allow students to work through an exercise step-by-step, and show their mathematical reasoning as they progress. Students receive immediate feedback after they complete each step, and helpful hints and videos offer guidance when they need it. When accessed via a mobile device, Workspace exercises use handwriting recognition software that allows students to naturally write out their answers. Each student's work is automatically graded and captured in the MyMathLab gradebook so instructors can easily pinpoint exactly where they need to focus their instruction.

www.mymathlab.com

Resources for Success

Instructor Resources

Additional resources can be downloaded from **www.mymathlab.com** or **www.pearsonhighered.com** or hardcopy resources can be ordered from your sales representative.

Annotated Instructor's Edition

Shorter answers are on the page beside the exercises. Longer answers are in the back of the text.

Instructor's Solutions Manual

Fully worked solutions to all textbook exercises.

PowerPoint® Lecture Slides

Fully editable lecture slides that correlate to the textbook.

Mini Lecture Notes

Additional examples and helpful teaching tips for each section.

TestGen®

Enables instructors to build, edit, print, and administer tests using a computerized bank of algorithmic questions developed to cover all the objectives of the text.

Student Resources

Additional resources to help student success are available to be packaged with the Blitzer textbook and MyMathLab access code.

Objective Level Videos

These fresh, interactive videos walk students through the concepts from every objective of the text. The videos provide an active learning environment where students can work at their own pace.

Chapter Test Prep Videos

Students can watch instructors work through step-by-step solutions to all the Chapter Test exercises from the textbook. These are available in MyMathLab and on YouTube.

Student Solutions Manual

Fully worked solutions to odd-numbered exercises and available to be packaged with the textbook.

Learning Guide

This note-taking guide is organized by objective and begins each chapter with an engaging application, providing additional examples and exercises for students to work through for a greater conceptual understanding and mastery of mathematical topics. New to this edition: classroom projects are included with each chapter providing students the opportunity to work collaboratively for stronger conceptual understanding. The Learning Guide is available in PDF and customizable Word file formats in MyMathLab. It can also be packaged with the textbook and MyMathLab access code.

MathTalk Videos

Engaging videos connect mathematics to real-life events and interesting applications. These fun, instructional videos show students that math is relevant to their daily lives and are assignable in MyMathLab. Assignable exercises are available in MyMathLab for these videos to help students retain valuable information presented in the videos.

www.mymathlab.com

TO THE STUDENT

The bar graph shows some of the qualities that students say make a great teacher. It was my goal to incorporate each of these qualities throughout the pages of this book.

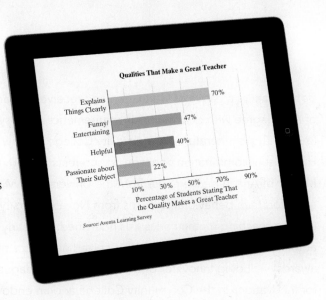

Explains Things Clearly

I understand that your primary purpose in reading *College Algebra Essentials* is to acquire a solid understanding of the required topics in your algebra course. In order to achieve this goal, I've carefully explained each topic. Important definitions and procedures are set off in boxes, and worked-out examples that present solutions in a step-by-step manner appear in every section. Each example is followed by a similar matched problem, called a Check Point, for you to try so that you can actively participate in the learning process as you read the book. (Answers to all Check Points appear in the back of the book.)

Funny & Entertaining

Who says that an algebra textbook can't be entertaining? From our unusual cover to the photos in the chapter and section openers, prepare to expect the unexpected. I hope some of the book's enrichment essays, called Blitzer Bonuses, will put a smile on your face from time to time.

Helpful

I designed the book's features to help you acquire knowledge of college algebra, as well as to show you how algebra can solve authentic problems that apply to your life. These helpful features include:

- **Explanatory Voice Balloons:** Voice balloons are used in a variety of ways to make math less intimidating. They translate algebraic language into everyday English, help clarify problem-solving procedures, present alternative ways of understanding concepts, and connect new concepts to concepts you have already learned.

- **Great Question!:** The book's Great Question! boxes are based on questions students ask in class. The answers to these questions give suggestions for problem solving, point out common errors to avoid, and provide informal hints and suggestions.

- **Achieving Success:** The book's Achieving Success boxes give you helpful strategies for success in learning algebra, as well as suggestions that can be applied for achieving your full academic potential in future college coursework.

- **Chapter Summaries:** Each chapter contains a review chart that summarizes the definitions and concepts in every section of the chapter. Examples from the chapter that illustrate these key concepts are also referenced in the chart. Review these summaries and you'll know the most important material in the chapter!

Passionate about the Subject

I passionately believe that no other discipline comes close to math in offering a more extensive set of tools for application and development of your mind. I wrote the book in Point Reyes National Seashore, 40 miles north of San Francisco. The park consists of 75,000 acres with miles of pristine surf-washed beaches, forested ridges, and bays bordered by white cliffs. It was my hope to convey the beauty and excitement of mathematics using nature's unspoiled beauty as a source of inspiration and creativity. Enjoy the pages that follow as you empower yourself with the algebra needed to succeed in college, your career, and your life.

Regards,

Bob
Robert Blitzer

ABOUT THE AUTHOR

Bob Blitzer is a native of Manhattan and received a Bachelor of Arts degree with dual majors in mathematics and psychology (minor: English literature) from the City College of New York. His unusual combination of academic interests led him toward a Master of Arts in mathematics from the University of Miami and a doctorate in behavioral sciences from Nova University. Bob's love for teaching mathematics was nourished for nearly 30 years at Miami Dade College, where he received numerous teaching awards, including Innovator of the Year from the League for Innovations in the Community College and an endowed chair based on excellence in the classroom. In addition to *College Algebra Essentials*, Bob has written textbooks covering developmental mathematics, introductory algebra, intermediate algebra, trigonometry, algebra and trigonometry, precalculus, and liberal arts mathematics, all published by Pearson. When not secluded in his Northern California writer's cabin, Bob can be found hiking the beaches and trails of Point Reyes National Seashore and tending to the chores required by his beloved entourage of horses, chickens, and irritable roosters.

APPLICATIONS INDEX

COLLEGE ALGEBRA
ESSENTIALS

Prerequisites:
Fundamental Concepts of Algebra

CHAPTER
P

What can algebra possibly have to tell me about

- the skyrocketing cost of a college education?
- student-loan debt?
- my workouts?
- the effects of alcohol?
- the meaning of the national debt that is nearly $19 trillion?
- time dilation on a futuristic high-speed journey to a nearby star?
- racial bias?
- ethnic diversity in the United States?
- the widening imbalance between numbers of women and men on college campuses?

This chapter reviews fundamental concepts of algebra that are prerequisites for the study of college algebra. Throughout the chapter, you will see how the special language of algebra describes your world.

HERE'S WHERE YOU'LL FIND THESE APPLICATIONS:

College costs: Section P.1, Example 2; Exercise Set P.1, Exercises 131–132

Student-loan debt: Mid-Chapter Check Point, Exercise 31

Workouts: Exercise Set P.1, Exercises 129–130

The effects of alcohol: Blitzer Bonus beginning on page 15

The national debt: Section P.2, Example 12

Time dilation: Blitzer Bonus on page 47

Racial bias: Exercise Set P.4, Exercises 91–92

U.S. ethnic diversity: Chapter P Review, Exercise 23

College gender imbalance: Chapter P Test, Exercise 32.

Section P.1

Algebraic Expressions, Mathematical Models, and Real Numbers

What am I supposed to learn?

After studying this section, you should be able to:

1. Evaluate algebraic expressions.
2. Use mathematical models.
3. Find the intersection of two sets.
4. Find the union of two sets.
5. Recognize subsets of the real numbers.
6. Use inequality symbols.
7. Evaluate absolute value.
8. Use absolute value to express distance.
9. Identify properties of the real numbers.
10. Simplify algebraic expressions.

How would your lifestyle change if a gallon of gas cost $9.15? Or if the price of a staple such as milk was $15? That's how much those products would cost if their prices had increased at the same rate college tuition has increased since 1980. (*Source*: Center for College Affordability and Productivity) In this section, you will learn how the special language of algebra describes your world, including the skyrocketing cost of a college education.

Algebraic Expressions

Algebra uses letters, such as x and y, to represent numbers. If a letter is used to represent various numbers, it is called a **variable**. For example, imagine that you are basking in the sun on the beach. We can let x represent the number of minutes that you can stay in the sun without burning with no sunscreen. With a number 6 sunscreen, exposure time without burning is six times as long, or 6 times x. This can be written $6 \cdot x$, but it is usually expressed as $6x$. Placing a number and a letter next to one another indicates multiplication.

Notice that $6x$ combines the number 6 and the variable x using the operation of multiplication. A combination of variables and numbers using the operations of addition, subtraction, multiplication, or division, as well as powers or roots, is called an **algebraic expression**. Here are some examples of algebraic expressions:

$$x + 6, \quad x - 6, \quad 6x, \quad \frac{x}{6}, \quad 3x + 5, \quad x^2 - 3, \quad \sqrt{x} + 7.$$

Many algebraic expressions involve *exponents*. For example, the algebraic expression

$$4x^2 + 330x + 3310$$

approximates the average cost of tuition and fees at public U.S. colleges for the school year ending x years after 2000. The expression x^2 means $x \cdot x$ and is read "x to the second power" or "x squared." The exponent, 2, indicates that the base, x, appears as a factor two times.

Exponential Notation

If n is a counting number (1, 2, 3, and so on),

Exponent or Power

$$b^n = \underbrace{b \cdot b \cdot b \cdot \cdots \cdot b}_{b \text{ appears as a factor } n \text{ times.}}$$

Base

b^n is read "the nth power of b" or "b to the nth power." Thus, the nth power of b is defined as the product of n factors of b. The expression b^n is called an **exponential expression**. Furthermore, $b^1 = b$.

For example,

$$8^2 = 8 \cdot 8 = 64, \quad 5^3 = 5 \cdot 5 \cdot 5 = 125, \quad \text{and} \quad 2^4 = 2 \cdot 2 \cdot 2 \cdot 2 = 16.$$

① Evaluate algebraic expressions.

Evaluating Algebraic Expressions

Evaluating an algebraic expression means to find the value of the expression for a given value of the variable.

Many algebraic expressions involve more than one operation. Evaluating an algebraic expression without a calculator involves carefully applying the following order of operations agreement:

The Order of Operations Agreement

1. Perform operations within the innermost parentheses and work outward. If the algebraic expression involves a fraction, treat the numerator and the denominator as if they were each enclosed in parentheses.
2. Evaluate all exponential expressions.
3. Perform multiplications and divisions **as they occur**, working **from left to right**.
4. Perform additions and subtractions **as they occur**, working **from left to right**.

EXAMPLE 1 Evaluating an Algebraic Expression

Evaluate $7 + 5(x - 4)^3$ for $x = 6$.

SOLUTION

$$7 + 5(x - 4)^3 = 7 + 5(6 - 4)^3 \quad \text{Replace } x \text{ with 6.}$$
$$= 7 + 5(2)^3 \qquad \text{First work inside parentheses: } 6 - 4 = 2.$$
$$= 7 + 5(8) \qquad \text{Evaluate the exponential expression:} \\ 2^3 = 2 \cdot 2 \cdot 2 = 8.$$
$$= 7 + 40 \qquad \text{Multiply: } 5(8) = 40.$$
$$= 47 \qquad\quad \text{Add.} \qquad\qquad \bullet\bullet\bullet$$

✓ Check Point **1** Evaluate $8 + 6(x - 3)^2$ for $x = 13$. 608

② Use mathematical models.

Formulas and Mathematical Models

An **equation** is formed when an equal sign is placed between two algebraic expressions. One aim of algebra is to provide a compact, symbolic description of the world. These descriptions involve the use of *formulas*. A **formula** is an equation that uses variables to express a relationship between two or more quantities.

Here are two examples of formulas related to heart rate and exercise.

Couch-Potato Exercise

$$H = \frac{1}{5}(220 - a)$$

| Heart rate, in beats per minute, | is | $\frac{1}{5}$ of | the difference between 220 and your age. |

Working It

$$H = \frac{9}{10}(220 - a)$$

| Heart rate, in beats per minute, | is | $\frac{9}{10}$ of | the difference between 220 and your age. |

The process of finding formulas to describe real-world phenomena is called **mathematical modeling**. Such formulas, together with the meaning assigned to the variables, are called **mathematical models**. We often say that these formulas model, or describe, the relationships among the variables.

EXAMPLE 2 Modeling the Cost of Attending a Public College

The bar graph in **Figure P.1** shows the average cost of tuition and fees for public four-year colleges, adjusted for inflation. The formula

$$T = 4x^2 + 330x + 3310$$

models the average cost of tuition and fees, T, for public U.S. colleges for the school year ending x years after 2000.

a. Use the formula to find the average cost of tuition and fees at public U.S. colleges for the school year ending in 2010.

b. By how much does the formula underestimate or overestimate the actual cost shown in **Figure P.1**?

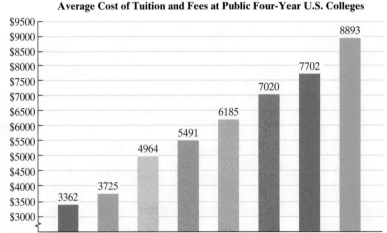

Average Cost of Tuition and Fees at Public Four-Year U.S. Colleges

FIGURE P.1
Source: The College Board

SOLUTION

a. Because 2010 is 10 years after 2000, we substitute 10 for x in the given formula. Then we use the order of operations to find T, the average cost of tuition and fees for the school year ending in 2010.

$T = 4x^2 + 330x + 3310$	This is the given mathematical model.
$T = 4(10)^2 + 330(10) + 3310$	Replace each occurrence of x with 10.
$T = 4(100) + 330(10) + 3310$	Evaluate the exponential expression: $10^2 = 10 \cdot 10 = 100$.
$T = 400 + 3300 + 3310$	Multiply from left to right: $4(100) = 400$ and $330(10) = 3300$.
$T = 7010$	Add.

The formula indicates that for the school year ending in 2010, the average cost of tuition and fees at public U.S. colleges was $7010.

b. **Figure P.1** shows that the average cost of tuition and fees for the school year ending in 2010 was $7020.

The cost obtained from the formula, $7010, underestimates the actual data value by $7020 − $7010, or by $10.

● ● ●

Blitzer Bonus ‖ Is College Worthwhile?

"Questions have intensified about whether going to college is worthwhile," says *Education Pays*, released by the College Board Advocacy & Policy Center. "For the typical student, the investment pays off very well over the course of a lifetime, even considering the expense."

Among the findings in *Education Pays*:

- Mean (average) full-time earnings with a bachelor's degree in 2014 were $62,504, which is $27,768 more than high school graduates.

- Compared with a high school graduate, a four-year college graduate who enrolled in a public university at age 18 will break even by age 33. The college graduate will have earned enough by then to compensate for being out of the labor force for four years and for borrowing enough to pay tuition and fees, shown in **Figure P.1**.

✓ Check Point 2

a. Use the formula $T = 4x^2 + 330x + 3310$, described in Example 2, to find the average cost of tuition and fees at public U.S. colleges for the school year ending in 2014. $8714

b. By how much does the formula underestimate or overestimate the actual cost shown in **Figure P.1**? underestimates by $179

Sometimes a mathematical model gives an estimate that is not a good approximation or is extended to include values of the variable that do not make sense. In these cases, we say that **model breakdown** has occurred. For example, it is not likely that the formula in Example 2 would give a good estimate of tuition and fees in 2050 because it is too far in the future. Thus, model breakdown would occur.

Sets

Before we describe the set of real numbers, let's be sure you are familiar with some basic ideas about sets. A **set** is a collection of objects whose contents can be clearly determined. The objects in a set are called the **elements** of the set. For example, the set of numbers used for counting can be represented by

$$\{1, 2, 3, 4, 5, \ldots\}.$$

The braces, { }, indicate that we are representing a set. This form of representation, called the **roster method**, uses commas to separate the elements of the set. The symbol consisting of three dots after the 5, called an *ellipsis*, indicates that there is no final element and that the listing goes on forever.

A set can also be written in **set-builder notation**. In this notation, the elements of the set are described but not listed. Here is an example:

$$\{x \mid x \text{ is a counting number less than 6}\}.$$

The set of all x such that x is a counting number less than 6.

The same set written using the roster method is

$$\{1, 2, 3, 4, 5\}.$$

If A and B are sets, we can form a new set consisting of all elements that are in both A and B. This set is called the *intersection* of the two sets.

 Find the intersection of two sets.

Definition of the Intersection of Sets

The **intersection** of sets A and B, written $A \cap B$, is the set of elements common to both set A **and** set B. This definition can be expressed in set-builder notation as follows:

$$A \cap B = \{x \mid x \text{ is an element of } A \text{ AND } x \text{ is an element of } B\}.$$

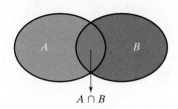

$A \cap B$

FIGURE P.2 Picturing the intersection of two sets

Figure P.2 shows a useful way of picturing the intersection of sets A and B. The figure indicates that $A \cap B$ contains those elements that belong to both A and B at the same time.

EXAMPLE 3 Finding the Intersection of Two Sets

Find the intersection: $\{7, 8, 9, 10, 11\} \cap \{6, 8, 10, 12\}$.

SOLUTION

The elements common to $\{7, 8, 9, 10, 11\}$ and $\{6, 8, 10, 12\}$ are 8 and 10. Thus,

$$\{7, 8, 9, 10, 11\} \cap \{6, 8, 10, 12\} = \{8, 10\}. \qquad \bullet \bullet \bullet$$

⊘ Check Point **3** Find the intersection: $\{3, 4, 5, 6, 7\} \cap \{3, 7, 8, 9\}$. [3, 7]

If a set has no elements, it is called the **empty set**, or the **null set**, and is represented by the symbol \varnothing (the Greek letter phi). Here is an example that shows how the empty set can result when finding the intersection of two sets:

$$\{2, 4, 6\} \cap \{3, 5, 7\} = \varnothing.$$

These sets have no common elements.

Their intersection has no elements and is the empty set.

4 Find the union of two sets.

Another set that we can form from sets A and B consists of elements that are in A or B or in both sets. This set is called the *union* of the two sets.

Definition of the Union of Sets

The **union** of sets A and B, written $A \cup B$, is the set of elements that are members of set A **or** of set B or of both sets. This definition can be expressed in set-builder notation as follows:

$$A \cup B = \{x \mid x \text{ is an element of } A \text{ OR } x \text{ is an element of } B\}.$$

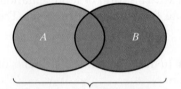

$A \cup B$

FIGURE P.3 Picturing the union of two sets

Figure P.3 shows a useful way of picturing the union of sets A and B. The figure indicates that $A \cup B$ is formed by joining the sets together.

We can find the union of set A and set B by listing the elements of set A. Then we include any elements of set B that have not already been listed. Enclose all elements that are listed with braces. This shows that the union of two sets is also a set.

GREAT QUESTION!

How can I use the words *union* and *intersection* to help me distinguish between these two operations?

Union, as in a marriage union, suggests joining things, or uniting them. Intersection, as in the intersection of two crossing streets, brings to mind the area common to both, suggesting things that overlap.

EXAMPLE 4 Finding the Union of Two Sets

Find the union: $\{7, 8, 9, 10, 11\} \cup \{6, 8, 10, 12\}$.

SOLUTION

To find $\{7, 8, 9, 10, 11\} \cup \{6, 8, 10, 12\}$, start by listing all the elements from the first set, namely, 7, 8, 9, 10, and 11. Now list all the elements from the second set that are not in the first set, namely, 6 and 12. The union is the set consisting of all these elements. Thus,

$$\{7, 8, 9, 10, 11\} \cup \{6, 8, 10, 12\} = \{6, 7, 8, 9, 10, 11, 12\}.$$

Although 8 and 10 appear in both sets,

do not list 8 and 10 twice.

$\bullet \bullet \bullet$

⊘ Check Point **4** Find the union: $\{3, 4, 5, 6, 7\} \cup \{3, 7, 8, 9\}$. [3, 4, 5, 6, 7, 8, 9]

 Recognize subsets of the real numbers.

The Set of Real Numbers

The sets that make up the real numbers are summarized in **Table P.1**. We refer to these sets as **subsets** of the real numbers, meaning that all elements in each subset are also elements in the set of real numbers.

Table P.1 Important Subsets of the Real Numbers

Name/Symbol	Description	Examples	
Natural numbers \mathbb{N}	$\{1, 2, 3, 4, 5, \ldots\}$ These are the numbers that we use for counting.	$2, 3, 5, 17$	
Whole numbers \mathbb{W}	$\{0, 1, 2, 3, 4, 5, \ldots\}$ The set of whole numbers includes 0 and the natural numbers.	$0, 2, 3, 5, 17$	
Integers \mathbb{Z}	$\{\ldots, -5, -4, -3, -2, -1, 0, 1, 2, 3, 4, 5, \ldots\}$ The set of integers includes the negatives of the natural numbers and the whole numbers.	$-17, -5, -3, -2, 0, 2, 3, 5, 17$	
Rational numbers \mathbb{Q}	$\left\{\dfrac{a}{b} \;\middle	\; a \text{ and } b \text{ are integers and } b \neq 0\right\}$ This means that b is not equal to zero. The set of rational numbers is the set of all numbers that can be expressed as a quotient of two integers, with the denominator not 0. Rational numbers can be expressed as terminating or repeating decimals.	$-17 = \frac{-17}{1}, -5 = \frac{-5}{1}, -3, -2,$ $0, 2, 3, 5, 17,$ $\frac{2}{5} = 0.4,$ $\frac{-2}{3} = -0.6666\ldots = -0.\overline{6}$
Irrational numbers \mathbb{I}	The set of irrational numbers is the set of all numbers whose decimal representations are neither terminating nor repeating. Irrational numbers cannot be expressed as a quotient of integers.	$\sqrt{2} \approx 1.414214$ $-\sqrt{3} \approx -1.73205$ $\pi \approx 3.142$ $-\frac{\pi}{2} \approx -1.571$	

Notice the use of the symbol \approx in the examples of irrational numbers. The symbol means "is approximately equal to." Thus,

$$\sqrt{2} \approx 1.414214.$$

We can verify that this is only an approximation by multiplying 1.414214 by itself. The product is very close to, but not exactly, 2:

$$1.414214 \times 1.414214 = 2.000001237796.$$

Not all square roots are irrational. For example, $\sqrt{25} = 5$ because $5^2 = 5 \cdot 5 = 25$. Thus, $\sqrt{25}$ is a natural number, a whole number, an integer, and a rational number $\left(\sqrt{25} = \frac{5}{1}\right)$.

The set of *real numbers* is formed by taking the union of the sets of rational numbers and irrational numbers. Thus, every real number is either rational or irrational, as shown in **Figure P.4**.

TECHNOLOGY

A calculator with a square root key gives a decimal approximation for $\sqrt{2}$, not the exact value.

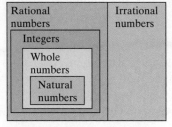

FIGURE P.4 Every real number is either rational or irrational.

Real Numbers

The set of **real numbers** is the set of numbers that are either rational or irrational:

$$\{x \mid x \text{ is rational or } x \text{ is irrational}\}.$$

The symbol \mathbb{R} is used to represent the set of real numbers. Thus,

$$\mathbb{R} = \{x \mid x \text{ is rational}\} \cup \{x \mid x \text{ is irrational}\}.$$

EXAMPLE 5 Recognizing Subsets of the Real Numbers

Consider the following set of numbers:

$$\left\{-7, -\frac{3}{4}, 0, 0.\overline{6}, \sqrt{5}, \pi, 7.3, \sqrt{81}\right\}.$$

List the numbers in the set that are

 a. natural numbers. **b.** whole numbers. **c.** integers.

 d. rational numbers. **e.** irrational numbers. **f.** real numbers.

SOLUTION

a. Natural numbers: The natural numbers are the numbers used for counting. The only natural number in the set $\left\{-7, -\frac{3}{4}, 0, 0.\overline{6}, \sqrt{5}, \pi, 7.3, \sqrt{81}\right\}$ is $\sqrt{81}$ because $\sqrt{81} = 9$. (9 multiplied by itself, or 9^2, is 81.)

b. Whole numbers: The whole numbers consist of the natural numbers and 0. The elements of the set $\left\{-7, -\frac{3}{4}, 0, 0.\overline{6}, \sqrt{5}, \pi, 7.3, \sqrt{81}\right\}$ that are whole numbers are 0 and $\sqrt{81}$.

c. Integers: The integers consist of the natural numbers, 0, and the negatives of the natural numbers. The elements of the set $\left\{-7, -\frac{3}{4}, 0, 0.\overline{6}, \sqrt{5}, \pi, 7.3, \sqrt{81}\right\}$ that are integers are $\sqrt{81}$, 0, and -7.

d. Rational numbers: All numbers in the set $\left\{-7, -\frac{3}{4}, 0, 0.\overline{6}, \sqrt{5}, \pi, 7.3, \sqrt{81}\right\}$ that can be expressed as the quotient of integers are rational numbers. These include $-7\left(-7 = \frac{-7}{1}\right)$, $-\frac{3}{4}, 0\left(0 = \frac{0}{1}\right)$, and $\sqrt{81}\left(\sqrt{81} = \frac{9}{1}\right)$. Furthermore, all numbers in the set that are terminating or repeating decimals are also rational numbers. These include $0.\overline{6}$ and 7.3.

e. Irrational numbers: The irrational numbers in the set $\left\{-7, -\frac{3}{4}, 0, 0.\overline{6}, \sqrt{5}, \pi, 7.3, \sqrt{81}\right\}$ are $\sqrt{5}\left(\sqrt{5} \approx 2.236\right)$ and $\pi(\pi \approx 3.14)$. Both $\sqrt{5}$ and π are only approximately equal to 2.236 and 3.14, respectively. In decimal form, $\sqrt{5}$ and π neither terminate nor have blocks of repeating digits.

f. Real numbers: All the numbers in the given set $\left\{-7, -\frac{3}{4}, 0, 0.\overline{6}, \sqrt{5}, \pi, 7.3, \sqrt{81}\right\}$ are real numbers. • • •

⊘ Check Point **5** Consider the following set of numbers:

$$\left\{-9, -1.3, 0, 0.\overline{3}, \frac{\pi}{2}, \sqrt{9}, \sqrt{10}\right\}.$$

List the numbers in the set that are

 a. natural numbers **b.** whole numbers **c.** integers.

 d. rational numbers **e.** irrational numbers **f.** real numbers.

a. $\sqrt{9}$ **b.** $0, \sqrt{9}$ **c.** $-9, 0, \sqrt{9}$

d. $-9, -1.3, 0, 0.\overline{3}, \sqrt{9}$ **e.** $\frac{\pi}{2}, \sqrt{10}$

f. $-9, -1.3, 0, 0.\overline{3}, \frac{\pi}{2}, \sqrt{9}, \sqrt{10}$

The Real Number Line

The **real number line** is a graph used to represent the set of real numbers. An arbitrary point, called the **origin**, is labeled 0. Select a point to the right of 0 and label it 1. The distance from 0 to 1 is called the **unit distance**. Numbers to the right of the origin are **positive** and numbers to the left of the origin are **negative**. The real number line is shown in **Figure P.5**.

FIGURE P.5 The real number line

GREAT QUESTION!

How did you locate $\sqrt{2}$ as a precise point on the number line in Figure P.6?

We used a right triangle with two legs of length 1. The remaining side has a length measuring $\sqrt{2}$.

We'll have lots more to say about right triangles later in the book.

Real numbers are **graphed** on a number line by placing a dot at the correct location for each number. The integers are easiest to locate. In **Figure P.6**, we've graphed six rational numbers and three irrational numbers on a real number line.

FIGURE P.6 Graphing numbers on a real number line

Every real number corresponds to a point on the number line and every point on the number line corresponds to a real number. We say that there is a **one-to-one correspondence** between all the real numbers and all points on a real number line.

⑥ Use inequality symbols.

Ordering the Real Numbers

On the real number line, the real numbers increase from left to right. The lesser of two real numbers is the one farther to the left on a number line. The greater of two real numbers is the one farther to the right on a number line.

Look at the number line in **Figure P.7**. The integers -4 and -1 are graphed.

FIGURE P.7

Observe that -4 is to the left of -1 on the number line. This means that -4 is less than -1.

$$-4 < -1$$

> −4 is less than −1 because −4 is to the left of −1 on the number line.

In **Figure P.7**, we can also observe that -1 is to the right of -4 on the number line. This means that -1 is greater than -4.

$$-1 > -4$$

> −1 is greater than −4 because −1 is to the right of −4 on the number line.

The symbols $<$ and $>$ are called **inequality symbols**. These symbols always point to the lesser of the two real numbers when the inequality statement is true.

> −4 is less than −1.

$$-4 < -1$$

The symbol points to -4, the lesser number.

> −1 is greater than −4.

$$-1 > -4$$

The symbol still points to -4, the lesser number.

The symbols $<$ and $>$ may be combined with an equal sign, as shown in the following table:

Symbols	Meaning	Examples	Explanation
$a \le b$	a is less than or equal to b.	$2 \le 9$ $9 \le 9$	Because $2 < 9$ Because $9 = 9$
$b \ge a$	b is greater than or equal to a.	$9 \ge 2$ $2 \ge 2$	Because $9 > 2$ Because $2 = 2$

> This inequality is true if either the $<$ part or the $=$ part is true.

> This inequality is true if either the $>$ part or the $=$ part is true.

 Evaluate absolute value.

FIGURE P.8 Absolute value as the distance from 0

Absolute Value

The **absolute value** of a real number a, denoted by $|a|$, is the distance from 0 to a on the number line. **This distance is always taken to be nonnegative.** For example, the real number line in **Figure P.8** shows that

$$|-3| = 3 \quad \text{and} \quad |5| = 5.$$

The absolute value of -3 is 3 because -3 is 3 units from 0 on the number line. The absolute value of 5 is 5 because 5 is 5 units from 0 on the number line. The absolute value of a positive real number or 0 is the number itself. The absolute value of a negative real number, such as -3, is the number without the negative sign.

We can define the absolute value of the real number x without referring to a number line. The algebraic definition of the absolute value of x is given as follows:

Definition of Absolute Value

$$|x| = \begin{cases} x & \text{if } x \geq 0 \\ -x & \text{if } x < 0 \end{cases}$$

If x is nonnegative (that is, $x \geq 0$), the absolute value of x is the number itself. For example,

$$|5| = 5 \qquad |\pi| = \pi \qquad \left|\frac{1}{3}\right| = \frac{1}{3} \qquad |0| = 0.$$

> Zero is the only number whose absolute value is 0.

If x is a negative number (that is, $x < 0$), the absolute value of x is the opposite of x. This makes the absolute value positive. For example,

$$|-3| = -(-3) = 3 \qquad |-\pi| = -(-\pi) = \pi \qquad \left|-\frac{1}{3}\right| = -\left(-\frac{1}{3}\right) = \frac{1}{3}.$$

> This middle step is usually omitted.

Observe that **the absolute value of any nonzero number is always positive.**

EXAMPLE 6 Evaluating Absolute Value

Rewrite each expression without absolute value bars:

a. $|\sqrt{3} - 1|$ **b.** $|2 - \pi|$ **c.** $\dfrac{|x|}{x}$ if $x < 0$.

SOLUTION

a. Because $\sqrt{3} \approx 1.7$, the number inside the absolute value bars, $\sqrt{3} - 1$, is positive. The absolute value of a positive number is the number itself. Thus,

$$|\sqrt{3} - 1| = \sqrt{3} - 1.$$

b. Because $\pi \approx 3.14$, the number inside the absolute value bars, $2 - \pi$, is negative. The absolute value of x when $x < 0$ is $-x$. Thus,

$$|2 - \pi| = -(2 - \pi) = \pi - 2.$$

c. If $x < 0$, then $|x| = -x$. Thus,

$$\frac{|x|}{x} = \frac{-x}{x} = -1.$$

• • •

🖉 **Check Point 6** Rewrite each expression without absolute value bars:

a. $|1 - \sqrt{2}|$ $\sqrt{2} - 1$ **b.** $|\pi - 3|$ $\pi - 3$ **c.** $\dfrac{|x|}{x}$ if $x > 0$. 1

Listed below are several basic properties of absolute value. Each of these properties can be derived from the definition of absolute value.

> **Properties of Absolute Value**
>
> For all real numbers a and b,
>
> **1.** $|a| \geq 0$ **2.** $|-a| = |a|$ **3.** $a \leq |a|$
>
> **4.** $|ab| = |a||b|$ **5.** $\left|\dfrac{a}{b}\right| = \dfrac{|a|}{|b|}, \quad b \neq 0$
>
> **6.** $|a + b| \leq |a| + |b|$ (called the triangle inequality)

⑧ Use absolute value to express distance.

Distance between Points on a Real Number Line

Absolute value is used to find the distance between two points on a real number line. If a and b are any real numbers, the **distance between a and b** is the absolute value of their difference. For example, the distance between 4 and 10 is 6. Using absolute value, we find this distance in one of two ways:

$$|10 - 4| = |6| = 6 \qquad \text{or} \qquad |4 - 10| = |-6| = 6.$$

> The distance between 4 and 10 on the real number line is 6.

Notice that we obtain the same distance regardless of the order in which we subtract.

> **Distance between Two Points on the Real Number Line**
>
> If a and b are any two points on a real number line, then the distance between a and b is given by
>
> $$|a - b| \quad \text{or} \quad |b - a|,$$
>
> where $|a - b| = |b - a|$.

EXAMPLE 7 Distance between Two Points on a Number Line

Find the distance between -5 and 3 on the real number line.

SOLUTION

Because the distance between a and b is given by $|a - b|$, the distance between -5 and 3 is

$$|-5 - 3| = |-8| = 8.$$

> $a = -5$ $b = 3$

FIGURE P.9 The distance between -5 and 3 is 8.

Figure P.9 verifies that there are 8 units between -5 and 3 on the real number line. We obtain the same distance if we reverse the order of the subtraction:

$$|3 - (-5)| = |8| = 8. \qquad \bullet\bullet\bullet$$

✓ **Check Point 7** Find the distance between -4 and 5 on the real number line. ₉

⑨ Identify properties of the real numbers.

Properties of Real Numbers and Algebraic Expressions

When you use your calculator to add two real numbers, you can enter them in any order. The fact that two real numbers can be added in any order is called the **commutative property of addition**. You probably use this property, as well as other

properties of real numbers listed in **Table P.2**, without giving it much thought. The properties of the real numbers are especially useful when working with algebraic expressions. For each property listed in **Table P.2**, a, b, and c represent real numbers, variables, or algebraic expressions.

Table P.2 Properties of the Real Numbers

Name	Meaning	Examples
Commutative Property of Addition	Changing order when adding does not affect the sum. $a + b = b + a$	• $13 + 7 = 7 + 13$ • $13x + 7 = 7 + 13x$
Commutative Property of Multiplication	Changing order when multiplying does not affect the product. $ab = ba$	• $\sqrt{2} \cdot \sqrt{5} = \sqrt{5} \cdot \sqrt{2}$ • $x \cdot 6 = 6x$
Associative Property of Addition	Changing grouping when adding does not affect the sum. $(a + b) + c = a + (b + c)$	• $3 + (8 + x) = (3 + 8) + x$ $= 11 + x$
Associative Property of Multiplication	Changing grouping when multiplying does not affect the product. $(ab)c = a(bc)$	• $-2(3x) = (-2 \cdot 3)x = -6x$
Distributive Property of Multiplication over Addition	Multiplication distributes over addition. $a \cdot (b + c) = a \cdot b + a \cdot c$	• $7(4 + \sqrt{3}) = 7 \cdot 4 + 7 \cdot \sqrt{3}$ $= 28 + 7\sqrt{3}$ • $5(3x + 7) = 5 \cdot 3x + 5 \cdot 7$ $= 15x + 35$
Identity Property of Addition	Zero can be deleted from a sum. $a + 0 = a$ $0 + a = a$	• $\sqrt{3} + 0 = \sqrt{3}$ • $0 + 6x = 6x$
Identity Property of Multiplication	One can be deleted from a product. $a \cdot 1 = a$ $1 \cdot a = a$	• $1 \cdot \pi = \pi$ • $13x \cdot 1 = 13x$
Inverse Property of Addition	The sum of a real number and its additive inverse gives 0, the additive identity. $a + (-a) = 0$ $(-a) + a = 0$	• $\sqrt{5} + (-\sqrt{5}) = 0$ • $-\pi + \pi = 0$ • $6x + (-6x) = 0$ • $(-4y) + 4y = 0$
Inverse Property of Multiplication	The product of a nonzero real number and its multiplicative inverse gives 1, the multiplicative identity. $a \cdot \dfrac{1}{a} = 1, \quad a \neq 0$ $\dfrac{1}{a} \cdot a = 1, \quad a \neq 0$	• $7 \cdot \dfrac{1}{7} = 1$ • $\left(\dfrac{1}{x - 3}\right)(x - 3) = 1, \quad x \neq 3$

The properties of the real numbers in **Table P.2** apply to the operations of addition and multiplication. Subtraction and division are defined in terms of addition and multiplication.

Do the commutative and associative properties work for subtraction and division?

No. Subtraction and division are not commutative operations.

$$a - b \neq b - a \qquad \frac{a}{b} \neq \frac{b}{a}$$

Furthermore, subtraction and division are not associative operations.

$$(a - b) - c \neq a - (b - c)$$
$$(a \div b) \div c \neq a \div (b \div c)$$

Verify each of these four statements using $a = 10, b = 5$, and $c = 2$.

Definitions of Subtraction and Division

Let a and b represent real numbers.

Subtraction: $a - b = a + (-b)$
We call $-b$ the **additive inverse** or **opposite** of b.

Division: $a \div b = a \cdot \frac{1}{b}$, where $b \neq 0$
We call $\frac{1}{b}$ the **multiplicative inverse** or **reciprocal** of b. The quotient of a and b, $a \div b$, can be written in the form $\frac{a}{b}$, where a is the **numerator** and b is the **denominator** of the fraction.

Because subtraction is defined in terms of adding an inverse, the distributive property can be applied to subtraction:

$$a(b - c) = ab - ac$$
$$(b - c)a = ba - ca.$$

For example,

$$4(2x - 5) = 4 \cdot 2x - 4 \cdot 5 = 8x - 20.$$

⑩ Simplify algebraic expressions.

Simplifying Algebraic Expressions

The **terms** of an algebraic expression are those parts that are separated by addition. For example, consider the algebraic expression

$$7x - 9y + z - 3,$$

which can be expressed as

$$7x + (-9y) + z + (-3).$$

This expression contains four terms, namely, $7x$, $-9y$, z, and -3.

The numerical part of a term is called its **coefficient**. In the term $7x$, the 7 is the coefficient. If a term containing one or more variables is written without a coefficient, the coefficient is understood to be 1. Thus, z means $1z$. If a term is a constant, its coefficient is that constant. Thus, the coefficient of the constant term -3 is -3.

$$7x + (-9y) + z + (-3)$$

| Coefficient is 7. | Coefficient is −9. | Coefficient is 1; z means $1z$. | Coefficient is −3. |

The parts of each term that are multiplied are called the **factors** of the term. The **factors** of the term $7x$ are 7 and x.

Like terms are terms that have exactly the same variable factors. For example, $3x$ and $7x$ are like terms. The distributive property in the form

$$ba + ca = (b + c)a$$

enables us to add or subtract like terms. For example,

$$3x + 7x = (3 + 7)x = 10x$$
$$7y^2 - y^2 = 7y^2 - 1y^2 = (7 - 1)y^2 = 6y^2.$$

This process is called **combining like terms**.

What is the bottom line for combining like terms?

To combine like terms mentally, add or subtract the coefficients of the terms. Use this result as the coefficient of the terms' variable factor(s).

An algebraic expression is **simplified** when parentheses have been removed and like terms have been combined.

EXAMPLE 8 Simplifying an Algebraic Expression

Simplify: $6(2x^2 + 4x) + 10(4x^2 + 3x)$.

SOLUTION

$$6(2x^2 + 4x) + 10(4x^2 + 3x)$$

$$= 6 \cdot 2x^2 + 6 \cdot 4x + 10 \cdot 4x^2 + 10 \cdot 3x \qquad \text{Use the distributive property to remove the parentheses.}$$

> $52x^2$ and $54x$ are not like terms. They contain different variable factors, x^2 and x, and cannot be combined.

$$= 12x^2 + 24x + 40x^2 + 30x \qquad\qquad \text{Multiply.}$$

$$= (12x^2 + 40x^2) + (24x + 30x) \qquad \text{Group like terms.}$$

$$= 52x^2 + 54x \qquad\qquad\qquad\qquad \text{Combine like terms.} \quad \bullet\bullet\bullet$$

 Check Point 8 Simplify: $7(4x^2 + 3x) + 2(5x^2 + x)$. $38x^2 + 23x$

Properties of Negatives

The distributive property can be extended to cover more than two terms within parentheses. For example,

> This sign represents subtraction.

> This sign tells us that the number is negative.

$$-3(4x - 2y + 6) = -3 \cdot 4x - (-3) \cdot 2y - 3 \cdot 6$$

$$= -12x - (-6y) - 18$$

$$= -12x + 6y - 18.$$

The voice balloons illustrate that negative signs can appear side by side. They can represent the operation of subtraction or the fact that a real number is negative. Here is a list of properties of negatives and how they are applied to algebraic expressions:

Properties of Negatives

Let a and b represent real numbers, variables, or algebraic expressions.

Property	Examples
1. $(-1)a = -a$	$(-1)4xy = -4xy$
2. $-(-a) = a$	$-(-6y) = 6y$
3. $(-a)b = -ab$	$(-7)4xy = -7 \cdot 4xy = -28xy$
4. $a(-b) = -ab$	$5x(-3y) = -5x \cdot 3y = -15xy$
5. $-(a + b) = -a - b$	$-(7x + 6y) = -7x - 6y$
6. $-(a - b) = -a + b$	$-(3x - 7y) = -3x + 7y$
$\quad = b - a$	$\quad = 7y - 3x$

It is not uncommon to see algebraic expressions with parentheses preceded by a negative sign or subtraction. Properties 5 and 6 in the box, $-(a + b) = -a - b$ and $-(a - b) = -a + b$, are related to this situation. An expression of the form $-(a + b)$ can be simplified as follows:

$$-(a + b) = -1(a + b) = (-1)a + (-1)b = -a + (-b) = -a - b.$$

Do you see a fast way to obtain the simplified expression on the right in the preceding equation? **If a negative sign or a subtraction symbol appears outside parentheses, drop the parentheses and change the sign of every term within the parentheses.** For example,

$$-(3x^2 - 7x - 4) = -3x^2 + 7x + 4.$$

EXAMPLE 9 Simplifying an Algebraic Expression

Simplify: $8x + 2[5 - (x - 3)]$.

SOLUTION

$$8x + 2[5 - (x - 3)]$$

$= 8x + 2[5 - x + 3]$ — Drop parentheses and change the sign of each term in parentheses: $-(x - 3) = -x + 3$.

$= 8x + 2[8 - x]$ — Simplify inside brackets: $5 + 3 = 8$.

$= 8x + 16 - 2x$ — Apply the distributive property:

$2[8 - x] = 2 \cdot 8 - 2x = 16 - 2x.$

$= (8x - 2x) + 16$ — Group like terms.

$= (8 - 2)x + 16$ — Apply the distributive property.

$= 6x + 16$ — Simplify. •••

⌀ Check Point **9** Simplify: $6 + 4[7 - (x - 2)]$. $42 - 4x$

Blitzer Bonus ‖ Using Algebra to Measure Blood-Alcohol Concentration

The amount of alcohol in a person's blood is known as blood-alcohol concentration (BAC), measured in grams of alcohol per deciliter of blood. A BAC of 0.08, meaning 0.08%, indicates that a person has 8 parts alcohol per 10,000 parts blood. In every state in the United States, it is illegal to drive with a BAC of 0.08 or higher.

How Do I Measure My Blood-Alcohol Concentration?

Here's a formula that models BAC for a person who weighs w pounds and who has n drinks* per hour.

$$\text{BAC} = \frac{600n}{w(0.6n + 169)}$$

Number of drinks consumed in an hour

Blood-alcohol concentration

Body weight, in pounds

*A drink can be a 12-ounce can of beer, a 5-ounce glass of wine, or a 1.5-ounce shot of liquor. Each contains approximately 14 grams, or $\frac{1}{2}$ ounce, of alcohol.

Blood-alcohol concentration can be used to quantify the meaning of "tipsy."

BAC	Effects on Behavior
0.05	Feeling of well-being; mild release of inhibitions; absence of observable effects
0.08	Feeling of relaxation; mild sedation; exaggeration of emotions and behavior; slight impairment of motor skills; increase in reaction time
0.12	Muscle control and speech impaired; difficulty performing motor skills; uncoordinated behavior
0.15	Euphoria; major impairment of physical and mental functions; irresponsible behavior; some difficulty standing, walking, and talking
0.35	Surgical anesthesia; lethal dosage for a small percentage of people
0.40	Lethal dosage for 50% of people; severe circulatory and respiratory depression; alcohol poisoning/overdose

Source: National Clearinghouse for Alcohol and Drug Information

(continues on next page)

Keeping in mind the meaning of "tipsy," we can use our model to compare blood-alcohol concentrations of a 120-pound person and a 200-pound person for various numbers of drinks.

We determined each BAC using a calculator, rounding to three decimal places.

Blood-Alcohol Concentrations of a 120-Pound Person

$$\text{BAC} = \frac{600n}{120(0.6n + 169)}$$

n (number of drinks per hour)	1	2	3	4	5	6	7	8	9	10
BAC (blood-alcohol concentration)	0.029	0.059	0.088	0.117	0.145	0.174	0.202	0.230	0.258	0.286

Illegal to drive

Blood-Alcohol Concentrations of a 200-Pound Person

$$\text{BAC} = \frac{600n}{200(0.6n + 169)}$$

n (number of drinks per hour)	1	2	3	4	5	6	7	8	9	10
BAC (blood-alcohol concentration)	0.018	0.035	0.053	0.070	0.087	0.104	0.121	0.138	0.155	0.171

Illegal to drive

Like all mathematical models, the formula for BAC gives approximate rather than exact values. There are other variables that influence blood-alcohol concentration that are not contained in the model. These include the rate at which an individual's body processes alcohol, how quickly one drinks, sex, age, physical condition, and the amount of food eaten prior to drinking.

CONCEPT AND VOCABULARY CHECK

Fill in each blank so that the resulting statement is true.

1. A combination of numbers, variables, and operation symbols is called an algebraic __expression__.

2. If n is a counting number, b^n, read __b to the nth power__, indicates that there are n factors of b. The number b is called the __base__ and the number n is called the __exponent__.

3. An equation that expresses a relationship between two or more variables, such as $H = \frac{9}{10}(220 - a)$, is called a/an __formula__. The process of finding such equations to describe real-world phenomena is called mathematical __modeling__. Such equations, together with the meaning assigned to the variables, are called mathematical __models__.

4. The set of elements common to both set A and set B is called the __intersection__ of sets A and B, and is symbolized by __$A \cap B$__.

5. The set of elements that are members of set A or set B or of both sets is called the __union__ of sets A and B and is symbolized by __$A \cup B$__.

6. The set $\{1, 2, 3, 4, 5, \ldots\}$ is called the set of __natural__ numbers.

7. The set $\{0, 1, 2, 3, 4, 5, \ldots\}$ is called the set of __whole__ numbers

8. The set $\{\ldots, -4, -3, -2, -1, 0, 1, 2, 3, 4, \ldots\}$ is called the set of __integers__.

9. The set of numbers in the form $\frac{a}{b}$, where a and b belong to the set in Exercise 8 and $b \neq 0$, is called the set of __rational__ numbers.

10. The set of numbers whose decimal representations are neither terminating nor repeating is called the set of __irrational__ numbers.

11. Every real number is either a/an __rational__ number or a/an __irrational__ number.

12. The notation $|x|$ is read the __absolute value__ of x. If $x \geq 0$, then $|x| = $ __x__. If $x < 0$, then $|x| = $ __$-x$__.

13. The commutative properties state that $a + b = $ __$b + a$__ and $ab = $ __ba__.

14. The associative properties state that $(a + b) + c = $ __$a + (b + c)$__ and __$(ab)c$__ $= a(bc)$.

15. The distributive property states that $a(b + c) = $ __$ab + ac$__.

16. $a + (-a) = $ __0__: The sum of a real number and its additive __inverse__ is __0__, the additive __identity__.

17. $a \cdot \frac{1}{a} = 1$, $a \neq 0$: The product of a nonzero real number and its multiplicative __inverse__ is __1__, the multiplicative __identity__.

18. An algebraic expression is __simplified__ when parentheses have been removed and like terms have been combined.

19. $-(-a) = $ __a__.

EXERCISE SET P.1

Practice Exercises

In Exercises 1–16, evaluate each algebraic expression for the given value or values of the variable(s).

1. $7 + 5x$, for $x = 10$ 57
2. $8 + 6x$, for $x = 5$ 38
3. $6x - y$, for $x = 3$ and $y = 8$ 10
4. $8x - y$, for $x = 3$ and $y = 4$ 20
5. $x^2 + 3x$, for $x = 8$ 88
6. $x^2 + 5x$, for $x = 6$ 66
7. $x^2 - 6x + 3$, for $x = 7$ 10
8. $x^2 - 7x + 4$, for $x = 8$ 12
9. $4 + 5(x - 7)^3$, for $x = 9$ 44
10. $6 + 5(x - 6)^3$, for $x = 8$ 46
11. $x^2 - 3(x - y)$, for $x = 8$ and $y = 2$ 46
12. $x^2 - 4(x - y)$, for $x = 8$ and $y = 3$ 44
13. $\dfrac{5(x + 2)}{2x - 14}$, for $x = 10$ 10
14. $\dfrac{7(x - 3)}{2x - 16}$, for $x = 9$ 21
15. $\dfrac{2x + 3y}{x + 1}$, for $x = -2$ and $y = 4$ -8
16. $\dfrac{2x + y}{xy - 2x}$, for $x = -2$ and $y = 4$ 0

The formula

$$C = \frac{5}{9}(F - 32)$$

expresses the relationship between Fahrenheit temperature, F, and Celsius temperature, C. In Exercises 17–18, use the formula to convert the given Fahrenheit temperature to its equivalent temperature on the Celsius scale.

17. $50°F$ 10°C
18. $86°F$ 30°C

A football was kicked vertically upward from a height of 4 feet with an initial speed of 60 feet per second. The formula

$$h = 4 + 60t - 16t^2$$

describes the ball's height above the ground, h, in feet, t seconds after it was kicked. Use this formula to solve Exercises 19–20.

19. What was the ball's height 2 seconds after it was kicked? *
20. What was the ball's height 3 seconds after it was kicked? *

In Exercises 21–28, find the intersection of the sets.

21. $\{1, 2, 3, 4\} \cap \{2, 4, 5\}$ {2, 4}
22. $\{1, 3, 7\} \cap \{2, 3, 8\}$ {3}
23. $\{s, e, t\} \cap \{t, e, s\}$ {s, e, t}
24. $\{r, e, a, l\} \cap \{l, e, a, r\}$ *
25. $\{1, 3, 5, 7\} \cap \{2, 4, 6, 8, 10\}$ ∅
26. $\{0, 1, 3, 5\} \cap \{-5, -3, -1\}$ ∅
27. $\{a, b, c, d\} \cap \varnothing$ ∅
28. $\{w, y, z\} \cap \varnothing$ ∅

In Exercises 29–34, find the union of the sets.

29. $\{1, 2, 3, 4\} \cup \{2, 4, 5\}$ {1, 2, 3, 4, 5}
30. $\{1, 3, 7, 8\} \cup \{2, 3, 8\}$ *
31. $\{1, 3, 5, 7\} \cup \{2, 4, 6, 8, 10\}$ *
32. $\{0, 1, 3, 5\} \cup \{2, 4, 6\}$ *
33. $\{a, e, i, o, u\} \cup \varnothing$ {a, e, i, o, u}
34. $\{e, m, p, t, y\} \cup \varnothing$ *

*See Answers to Selected Exercises.

In Exercises 35–38, list all numbers from the given set that are **a.** *natural numbers,* **b.** *whole numbers,* **c.** *integers,* **d.** *rational numbers,* **e.** *irrational numbers,* **f.** *real numbers.*

35. $\left\{-9, -\frac{4}{5}, 0, 0.25, \sqrt{3}, 9.2, \sqrt{100}\right\}$ *
36. $\left\{-7, -0.\overline{6}, 0, \sqrt{49}, \sqrt{50}\right\}$ *
37. $\left\{-11, -\frac{5}{6}, 0, 0.75, \sqrt{5}, \pi, \sqrt{64}\right\}$ *
38. $\left\{-5, -0.\overline{3}, 0, \sqrt{2}, \sqrt{4}\right\}$ *

39. Give an example of a whole number that is not a natural number. 0
40. Give an example of a rational number that is not an integer. *
41. Give an example of a number that is an integer, a whole number, and a natural number. 2 is an example.
42. Give an example of a number that is a rational number, an integer, and a real number. −2 is an example.

Determine whether each statement in Exercises 43–50 is true or false.

43. $-13 \le -2$ true
44. $-6 > 2$ false
45. $4 \ge -7$ true
46. $-13 < -5$ true
47. $-\pi \ge -\pi$ true
48. $-3 > -13$ true
49. $0 \ge -6$ true
50. $0 \ge -13$ true

In Exercises 51–60, rewrite each expression without absolute value bars.

51. $|300|$ 300
52. $|-203|$ 203
53. $|12 - \pi|$ $12 - \pi$
54. $|7 - \pi|$ $7 - \pi$
55. $|\sqrt{2} - 5|$ $5 - \sqrt{2}$
56. $|\sqrt{5} - 13|$ $13 - \sqrt{5}$
57. $\dfrac{-3}{|-3|}$ -1
58. $\dfrac{-7}{|-7|}$ -1
59. $\||-3| - |-7|\|$ 4
60. $\||-5| - |-13|\|$ 8

In Exercises 61–66, evaluate each algebraic expression for $x = 2$ and $y = -5$.

61. $|x + y|$ 3
62. $|x - y|$ 7
63. $|x| + |y|$ 7
64. $|x| - |y|$ -3
65. $\dfrac{y}{|y|}$ -1
66. $\dfrac{|x|}{x} + \dfrac{|y|}{y}$ 0

In Exercises 67–74, express the distance between the given numbers using absolute value. Then find the distance by evaluating the absolute value expression.

67. 2 and 17 $|17 - 2|$; 15
68. 4 and 15 $|15 - 4|$; 11
69. −2 and 5 $|5 - (-2)|$; 7
70. −6 and 8 $|8 - (-6)|$; 14
71. −19 and −4 $|-4 - (-19)|$; 15
72. −26 and −3 *
73. −3.6 and −1.4 $|-1.4 - (-3.6)|$; 2.2
74. −5.4 and −1.2 *

In Exercises 75–84, state the name of the property illustrated.

75. $6 + (-4) = (-4) + 6$ commutative property of addition
76. $11 \cdot (7 + 4) = 11 \cdot 7 + 11 \cdot 4$ distributive property
77. $6 + (2 + 7) = (6 + 2) + 7$ associative property of addition
78. $6 \cdot (2 \cdot 3) = 6 \cdot (3 \cdot 2)$ commutative property of multiplication
79. $(2 + 3) + (4 + 5) = (4 + 5) + (2 + 3)$ *
80. $7 \cdot (11 \cdot 8) = (11 \cdot 8) \cdot 7$ commutative property of multiplication
81. $2(-8 + 6) = -16 + 12$ distributive property
82. $-8(3 + 11) = -24 + (-88)$ distributive property

83. $\dfrac{1}{(x+3)}(x+3) = 1, x \neq -3$ *inverse property of multiplication*

84. $(x+4) + [-(x+4)] = 0$ *inverse property of addition*

In Exercises 85–96, simplify each algebraic expression.

85. $5(3x+4) - 4$ $15x + 16$

86. $2(5x+4) - 3$ $10x + 5$

87. $5(3x-2) + 12x$ $27x - 10$

88. $2(5x-1) + 14x$ *

89. $7(3y-5) + 2(4y+3)$ $29y - 29$

90. $4(2y-6) + 3(5y+10)$ $23y + 6$

91. $5(3y-2) - (7y+2)$ $8y - 12$

92. $4(5y-3) - (6y+3)$ $14y - 15$

93. $7 - 4[3 - (4y-5)]$ $16y - 25$

94. $6 - 5[8 - (2y-4)]$ $10y - 54$

95. $18x^2 + 4 - [6(x^2-2) + 5]$ $12x^2 + 11$

96. $14x^2 + 5 - [7(x^2-2) + 4]$ $7x^2 + 15$

In Exercises 97–102, write each algebraic expression without parentheses.

97. $-(-14x)$ $14x$

98. $-(-17y)$ $17y$

99. $-(2x - 3y - 6)$ $-2x + 3y + 6$

100. $-(5x - 13y - 1)$ *

101. $\frac{1}{3}(3x) + [(4y) + (-4y)]$ x

102. $\frac{1}{2}(2y) + [(-7x) + 7x]$ y

Practice Plus

In Exercises 103–110, insert either $<$, $>$, or $=$ in the shaded area to make a true statement.

103. $|-6|$ ___ $|-3|$ $>$

104. $|-20|$ ___ $|-50|$ $<$

105. $\left|\dfrac{3}{5}\right|$ ___ $|-0.6|$ $=$

106. $\left|\dfrac{5}{2}\right|$ ___ $|-2.5|$ $=$

107. $\dfrac{30}{40} - \dfrac{3}{4}$ ___ $\dfrac{14}{15} \cdot \dfrac{15}{14}$ $<$

108. $\dfrac{17}{18} \cdot \dfrac{18}{17}$ ___ $\dfrac{50}{60} - \dfrac{5}{6}$ $>$

109. $\dfrac{8}{13} \div \dfrac{8}{13}$ ___ $|-1|$ $=$

110. $|-2|$ ___ $\dfrac{4}{17} \div \dfrac{4}{17}$ $>$

In Exercises 111–120, use the order of operations to simplify each expression.

111. $8^2 - 16 \div 2^2 \cdot 4 - 3$ 45

112. $10^2 - 100 \div 5^2 \cdot 2 - 3$ 89

113. $\dfrac{5 \cdot 2 - 3^2}{[3^2 - (-2)]^2}$ $\frac{1}{121}$

114. $\dfrac{10 \div 2 + 3 \cdot 4}{(12 - 3 \cdot 2)^2}$ $\frac{17}{36}$

115. $8 - 3[-2(2-5) - 4(8-6)]$ 14

116. $8 - 3[-2(5-7) - 5(4-2)]$ 26

117. $\dfrac{2(-2) - 4(-3)}{5 - 8}$ $-\frac{8}{3}$

118. $\dfrac{6(-4) - 5(-3)}{9 - 10}$ 9

119. $\dfrac{(5-6)^2 - 2|3-7|}{89 - 3 \cdot 5^2}$ $-\frac{1}{2}$

120. $\dfrac{12 \div 3 \cdot 5|2^2 + 3^2|}{7 + 3 - 6^2}$ -10

In Exercises 121–128, write each English phrase as an algebraic expression. Then simplify the expression. Let x represent the number.

121. A number decreased by the sum of the number and four *

122. A number decreased by the difference between eight and the number $x - (8 - x); 2x - 8$

123. Six times the product of negative five and a number *

124. Ten times the product of negative four and a number *

125. The difference between the product of five and a number and twice the number $5x - 2x; 3x$

126. The difference between the product of six and a number and negative two times the number $6x - (-2x); 8x$

127. The difference between eight times a number and six more than three times the number $8x - (3x + 6); 5x - 6$

128. Eight decreased by three times the sum of a number and six $8 - 3(x + 6); -3x - 10$

Application Exercises

The maximum heart rate, in beats per minute, that you should achieve during exercise is 220 minus your age:

$$220 - a.$$

This algebraic expression gives maximum heart rate in terms of age, a.

The following bar graph shows the target heart rate ranges for four types of exercise goals. The lower and upper limits of these ranges are fractions of the maximum heart rate, $220 - a$. Exercises 129–130 are based on the information in the graph.

Target Heart Rate Ranges for Exercise Goals

Fraction of Maximum Heart Rate, $220 - a$

129. If your exercise goal is to improve cardiovascular conditioning, the graph shows the following range for target heart rate, H, in beats per minute:

Lower limit of range $H = \dfrac{7}{10}(220 - a)$

Upper limit of range $H = \dfrac{4}{5}(220 - a).$

a. What is the lower limit of the heart rate range, in beats per minute, for a 20-year-old with this exercise goal? *

b. What is the upper limit of the heart rate range, in beats per minute, for a 20-year-old with this exercise goal? *

130. If your exercise goal is to improve overall health, the graph shows the following range for target heart rate, H, in beats per minute:

Lower limit of range $H = \dfrac{1}{2}(220 - a)$

Upper limit of range $H = \dfrac{3}{5}(220 - a).$

a. What is the lower limit of the heart rate range, in beats per minute, for a 30-year-old with this exercise goal? *

b. What is the upper limit of the heart rate range, in beats per minute, for a 30-year-old with this exercise goal? *

The bar graph shows the average cost of tuition and fees at private four-year colleges in the United States.

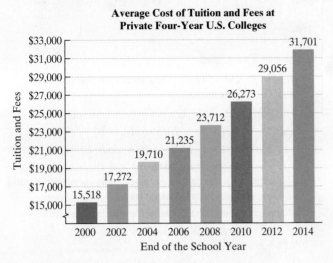

Average Cost of Tuition and Fees at Private Four-Year U.S. Colleges

Source: The College Board

The formula

$$T = 21x^2 + 862x + 15,552$$

models the average cost of tuition and fees, T, at private U.S. colleges for the school year ending x years after 2000. Use this information to solve Exercises 131–132.

131. a. Use the formula to find the average cost of tuition and fees at private U.S. colleges for the school year ending in 2014. $31,736

 b. By how much does the formula underestimate or overestimate the actual cost shown by the graph for the school year ending in 2014? overestimates by $35

 c. Use the formula to project the average cost of tuition and fees at private U.S. colleges for the school year ending in 2020. $41,192

132. a. Use the formula to find the average cost of tuition and fees at private U.S. colleges for the school year ending in 2012. $28,920

 b. By how much does the formula underestimate or overestimate the actual cost shown by the graph for the school year ending in 2012? underestimates by $136

 c. Use the formula to project the average cost of tuition and fees at private U.S. colleges for the school year ending in 2022. $44,680

133. You had $10,000 to invest. You put x dollars in a safe, government-insured certificate of deposit paying 5% per year. You invested the remainder of the money in noninsured corporate bonds paying 12% per year. Your total interest earned at the end of the year is given by the algebraic expression

$$0.05x + 0.12(10,000 - x).$$

 a. Simplify the algebraic expression. $1200 - 0.07x$

 b. Use each form of the algebraic expression to determine your total interest earned at the end of the year if you invested $6000 in the safe, government-insured certificate of deposit. $780

134. It takes you 50 minutes to get to campus. You spend t minutes walking to the bus stop and the rest of the time riding the bus. Your walking rate is 0.06 mile per minute and the bus travels at a rate of 0.5 mile per minute. The total distance walking and traveling by bus is given by the algebraic expression

$$0.06t + 0.5(50 - t).$$

 a. Simplify the algebraic expression. $25 - 0.44t$

 b. Use each form of the algebraic expression to determine the total distance that you travel if you spend 20 minutes walking to the bus stop. 16.2 miles

135. Read the Blitzer Bonus beginning on page 15. Use the formula

$$\text{BAC} = \frac{600n}{w(0.6n + 169)}$$

and replace w with your body weight. Using this formula and a calculator, compute your BAC for integers from $n = 1$ to $n = 10$. Round to three decimal places. According to this model, how many drinks can you consume in an hour without exceeding the legal measure of drunk driving? Answers will vary.

Explaining the Concepts 136–143. Answers will vary.

ACHIEVING SUCCESS

An effective way to understand something is to explain it to someone else. You can do this by using the Explaining the Concepts exercises that ask you to respond with verbal or written explanations. Speaking or writing about a new concept uses a different part of your brain than thinking about the concept. Explaining new ideas verbally will quickly reveal any gaps in your understanding. It will also help you to remember new concepts for longer periods of time.

136. What is an algebraic expression? Give an example with your explanation.

137. If n is a natural number, what does b^n mean? Give an example with your explanation.

138. What does it mean when we say that a formula models real-world phenomena?

139. What is the intersection of sets A and B?

140. What is the union of sets A and B?

141. How do the whole numbers differ from the natural numbers?

142. Can a real number be both rational and irrational? Explain your answer.

143. If you are given two real numbers, explain how to determine which is the lesser.

Critical Thinking Exercises

Make Sense? *In Exercises 144–147, determine whether each statement makes sense or does not make sense, and explain your reasoning.*

144. My mathematical model describes the data for tuition and fees at public four-year colleges for the past ten years extremely well, so it will serve as an accurate prediction for the cost of public colleges in 2050. does not make sense

145. A model that describes the average cost of tuition and fees at private U.S. colleges for the school year ending x years after 2000 cannot be used to estimate the cost of private education for the school year ending in 2000. *

146. The humor in this cartoon is based on the fact that the football will never be hiked.　makes sense

Foxtrot © 2003, 2009 by Bill Amend/Used by permission of Universal Uclick. All rights reserved.

147. Just as the commutative properties change groupings, the associative properties change order.　does not make sense

In Exercises 148–155, determine whether each statement is true or false. If the statement is false, make the necessary change(s) to produce a true statement.　Changes to false statements will vary.

148. Every rational number is an integer.　false

149. Some whole numbers are not integers.　false

150. Some rational numbers are not positive.　true

151. Irrational numbers cannot be negative.　false

152. The term x has no coefficient.　false

153. $5 + 3(x - 4) = 8(x - 4) = 8x - 32$　false

154. $-x - x = -x + (-x) = 0$　false

155. $x - 0.02(x + 200) = 0.98x - 4$　true

In Exercises 156–158, insert either $<$ or $>$ in the shaded area between the numbers to make the statement true.

156. $\sqrt{2}$ ▢ 1.5　$<$

157. $-\pi$ ▢ -3.5　$>$

158. $-\dfrac{3.14}{2}$ ▢ $-\dfrac{\pi}{2}$　$>$

Preview Exercises

Exercises 159–161 will help you prepare for the material covered in the next section.

159. In parts (a) and (b), complete each statement.
 a. $b^4 \cdot b^3 = (b \cdot b \cdot b \cdot b)(b \cdot b \cdot b) = b^?$　b^7
 b. $b^5 \cdot b^5 = (b \cdot b \cdot b \cdot b \cdot b)(b \cdot b \cdot b \cdot b \cdot b) = b^?$　b^{10}
 c. Generalizing from parts (a) and (b), what should be done with the exponents when multiplying exponential expressions with the same base?　Add exponents.

160. In parts (a) and (b), complete each statement.
 a. $\dfrac{b^7}{b^3} = \dfrac{b \cdot b \cdot b \cdot b \cdot b \cdot b \cdot b}{b \cdot b \cdot b} = b^?$　b^4
 b. $\dfrac{b^8}{b^2} = \dfrac{b \cdot b \cdot b \cdot b \cdot b \cdot b \cdot b \cdot b}{b \cdot b} = b^?$　b^6
 c. Generalizing from parts (a) and (b), what should be done with the exponents when dividing exponential expressions with the same base?　Subtract exponents.

161. If 6.2 is multiplied by 10^3, what does this multiplication do to the decimal point in 6.2?　It moves the decimal point 3 places to the right.

Section P.2　Exponents and Scientific Notation

Bigger than the biggest thing ever and then some. Much bigger than that in fact, really amazingly immense, a totally stunning size, real 'wow, that's big', time … Gigantic multiplied by colossal multiplied by staggeringly huge is the sort of concept we're trying to get across here.

Douglas Adams, *The Restaurant at the End of the Universe*

What am I supposed to learn?

After studying this section, you should be able to:

1. Use the product rule.
2. Use the quotient rule.
3. Use the zero-exponent rule.
4. Use the negative-exponent rule.
5. Use the power rule.
6. Find the power of a product.
7. Find the power of a quotient.
8. Simplify exponential expressions.
9. Use scientific notation.

Although Adams's description may not quite apply to this $18.9 trillion national debt, exponents can be used to explore the meaning of this "staggeringly huge" number. In this section, you will learn to use exponents to provide a way of putting large and small numbers in perspective.

The Product and Quotient Rules

We have seen that exponents are used to indicate repeated multiplication. Now consider the multiplication of two exponential expressions, such as $b^4 \cdot b^3$. We are multiplying 4 factors of b and 3 factors of b. We have a total of 7 factors of b:

4 factors of b 3 factors of b

$$b^4 \cdot b^3 = (b \cdot b \cdot b \cdot b)(b \cdot b \cdot b) = b^7.$$

Total: 7 factors of b

The product is exactly the same if we add the exponents:

$$b^4 \cdot b^3 = b^{4+3} = b^7.$$

The fact that $b^4 \cdot b^3 = b^7$ suggests the following rule:

① Use the product rule.

The Product Rule

$$b^m \cdot b^n = b^{m+n}$$

When multiplying exponential expressions with the same base, add the exponents. Use this sum as the exponent of the common base.

EXAMPLE 1 Using the Product Rule

Multiply each expression using the product rule:

 a. $2^2 \cdot 2^3$ **b.** $(6x^4y^3)(5x^2y^7)$.

SOLUTION

 a. $2^2 \cdot 2^3 = 2^{2+3} = 2^5$ or 32 $2^5 = 2 \cdot 2 \cdot 2 \cdot 2 \cdot 2 = 32$

 b. $(6x^4y^3)(5x^2y^7)$

 $= 6 \cdot 5 \cdot x^4 \cdot x^2 \cdot y^3 \cdot y^7$ Use the associative and commutative properties. This step can be done mentally.

 $= 30x^{4+2}y^{3+7}$

 $= 30x^6y^{10}$ • • •

⊘ **Check Point 1** Multiply each expression using the product rule:

 a. $3^3 \cdot 3^2$ 3^5 or 243 **b.** $(4x^3y^4)(10x^2y^6)$. $40x^5y^{10}$

② Use the quotient rule.

Now, consider the division of two exponential expressions, such as the quotient of b^7 and b^3. We are dividing 7 factors of b by 3 factors of b.

$$\frac{b^7}{b^3} = \frac{b \cdot b \cdot b \cdot b \cdot b \cdot b \cdot b}{b \cdot b \cdot b} = \boxed{\frac{b \cdot b \cdot b}{b \cdot b \cdot b}} \cdot b \cdot b \cdot b \cdot b = 1 \cdot b \cdot b \cdot b \cdot b = b^4$$

This factor is equal to 1.

The quotient is exactly the same if we subtract the exponents:

$$\frac{b^7}{b^3} = b^{7-3} = b^4.$$

This suggests the following rule:

> **The Quotient Rule**
>
> $$\frac{b^m}{b^n} = b^{m-n}, \quad b \neq 0$$
>
> When dividing exponential expressions with the same nonzero base, subtract the exponent in the denominator from the exponent in the numerator. Use this difference as the exponent of the common base.

EXAMPLE 2 Using the Quotient Rule

Divide each expression using the quotient rule:

a. $\dfrac{(-2)^7}{(-2)^4}$
b. $\dfrac{30x^{12}y^9}{5x^3y^7}$.

SOLUTION

a. $\dfrac{(-2)^7}{(-2)^4} = (-2)^{7-4} = (-2)^3$ or -8 ⟵ $(-2)^3 = (-2)(-2)(-2) = -8$

b. $\dfrac{30x^{12}y^9}{5x^3y^7} = \dfrac{30}{5} \cdot \dfrac{x^{12}}{x^3} \cdot \dfrac{y^9}{y^7} = 6x^{12-3}y^{9-7} = 6x^9y^2$

• • •

✓ **Check Point 2** Divide each expression using the quotient rule:

a. $\dfrac{(-3)^6}{(-3)^3}$ $(-3)^3$ or -27
b. $\dfrac{27x^{14}y^8}{3x^3y^5}$. $9x^{11}y^3$

③ Use the zero-exponent rule.

Zero as an Exponent

A nonzero base can be raised to the 0 power. The quotient rule can be used to help determine what zero as an exponent should mean. Consider the quotient of b^4 and b^4, where b is not zero. We can determine this quotient in two ways.

$$\frac{b^4}{b^4} = 1 \qquad\qquad \frac{b^4}{b^4} = b^{4-4} = b^0$$

Any nonzero expression divided by itself is 1.

Use the quotient rule and subtract exponents.

This means that b^0 must equal 1.

> **The Zero-Exponent Rule**
>
> If b is any real number other than 0,
>
> $$b^0 = 1.$$

Here are examples involving simplification using the zero-exponent rule:

$$8^0 = 1, \qquad (-6)^0 = 1, \qquad -6^0 = -1, \qquad (5x)^0 = 1, \qquad 5x^0 = 5.$$

Because there are no parentheses, only 6 is raised to the 0 power: $-6^0 = -(6^0) = -1$.

Because there are no parentheses, only x is raised to the 0 power: $5x^0 = 5 \cdot 1 = 5$.

④ Use the negative-exponent rule.

Negative Integers as Exponents

A nonzero base can be raised to a negative power. The quotient rule can be used to help determine what a negative integer as an exponent should mean. Consider the quotient of b^3 and b^5, where b is not zero. We can determine this quotient in two ways.

$$\frac{b^3}{b^5} = \frac{\cancel{b} \cdot \cancel{b} \cdot \cancel{b}}{\cancel{b} \cdot \cancel{b} \cdot \cancel{b} \cdot b \cdot b} = \frac{1}{b^2} \qquad\qquad \frac{b^3}{b^5} = b^{3-5} = b^{-2}$$

> After dividing common factors, we have two factors of b in the denominator.

> Use the quotient rule and subtract exponents.

Notice that $\dfrac{b^3}{b^5}$ equals both b^{-2} and $\dfrac{1}{b^2}$. This means that b^{-2} must equal $\dfrac{1}{b^2}$. This example is a special case of the **negative-exponent rule**.

The Negative-Exponent Rule

If b is any real number other than 0 and n is a natural number, then

$$b^{-n} = \frac{1}{b^n}.$$

EXAMPLE 3 Using the Negative-Exponent Rule

Use the negative-exponent rule to write each expression with a positive exponent. Simplify, if possible:

a. 9^{-2} **b.** $(-2)^{-5}$ **c.** $\dfrac{1}{6^{-2}}$ **d.** $7x^{-5}y^2.$

SOLUTION

a. $9^{-2} = \dfrac{1}{9^2} = \dfrac{1}{81}$

b. $(-2)^{-5} = \dfrac{1}{(-2)^5} = \dfrac{1}{(-2)(-2)(-2)(-2)(-2)} = \dfrac{1}{-32} = -\dfrac{1}{32}$

> Only the sign of the exponent, −5, changes. The base, −2, does not change sign.

c. $\dfrac{1}{6^{-2}} = \dfrac{1}{\dfrac{1}{6^2}} = 1 \cdot \dfrac{6^2}{1} = 6^2 = 36$

d. $7x^{-5}y^2 = 7 \cdot \dfrac{1}{x^5} \cdot y^2 = \dfrac{7y^2}{x^5}$

• • •

✓ Check Point **3** Use the negative-exponent rule to write each expression with a positive exponent. Simplify, if possible:

a. 5^{-2} $\frac{1}{25}$ **b.** $(-3)^{-3}$ $-\frac{1}{27}$ **c.** $\dfrac{1}{4^{-2}}$ 16 **d.** $3x^{-6}y^4.$ $\frac{3y^4}{x^6}$

In Example 3 and Check Point 3, did you notice that

$$\frac{1}{6^{-2}} = 6^2 \qquad \text{and} \qquad \frac{1}{4^{-2}} = 4^2?$$

In general, if a negative exponent appears in a denominator, an expression can be written with a positive exponent using

$$\frac{1}{b^{-n}} = b^n.$$

> **Negative Exponents in Numerators and Denominators**
>
> If b is any real number other than 0 and n is a natural number, then
>
> $$b^{-n} = \frac{1}{b^n} \quad \text{and} \quad \frac{1}{b^{-n}} = b^n.$$
>
> When a negative number appears as an exponent, switch the position of the base (from numerator to denominator or from denominator to numerator) and make the exponent positive. The sign of the base does not change.

⑤ Use the power rule.

The Power Rule for Exponents (Powers to Powers)

The next property of exponents applies when an exponential expression is raised to a power. Here is an example:

$$(b^2)^4.$$

The exponential expression b^2 is raised to the fourth power.

There are four factors of b^2. Thus,

$$(b^2)^4 = b^2 \cdot b^2 \cdot b^2 \cdot b^2 = b^{2+2+2+2} = b^8.$$

Add exponents when multiplying with the same base.

We can obtain the answer, b^8, by multiplying the original exponents:

$$(b^2)^4 = b^{2\cdot4} = b^8.$$

This suggests the following rule:

> **The Power Rule (Powers to Powers)**
>
> $$(b^m)^n = b^{mn}$$
>
> When an exponential expression is raised to a power, multiply the exponents. Place the product of the exponents on the base and remove the parentheses.

EXAMPLE 4 Using the Power Rule (Powers to Powers)

Simplify each expression using the power rule:

 a. $(2^2)^3$ **b.** $(y^5)^{-3}$ **c.** $(b^{-4})^{-2}.$

SOLUTION

 a. $(2^2)^3 = 2^{2\cdot3} = 2^6$ or 64 **b.** $(y^5)^{-3} = y^{5(-3)} = y^{-15} = \dfrac{1}{y^{15}}$

 c. $(b^{-4})^{-2} = b^{(-4)(-2)} = b^8$ • • •

✓ Check Point **4** Simplify each expression using the power rule:

 a. $(3^3)^2$ 3^6 or 729 **b.** $(y^7)^{-2}$ $\frac{1}{y^{14}}$ **c.** $(b^{-3})^{-4}.$ b^{12}

⑥ Find the power of a product.

The Products-to-Powers Rule for Exponents

The next property of exponents applies when we are raising a product to a power. Here is an example:

$$(2x)^4.$$

> The product $2x$ is raised to the fourth power.

There are four factors of $2x$. Thus,

$$(2x)^4 = 2x \cdot 2x \cdot 2x \cdot 2x = 2 \cdot 2 \cdot 2 \cdot 2 \cdot x \cdot x \cdot x \cdot x = 2^4 x^4.$$

We can obtain the answer, $2^4 x^4$, by raising each factor within the parentheses to the fourth power:

$$(2x)^4 = 2^4 x^4.$$

The fact that $(2x)^4 = 2^4 x^4$ suggests the following rule:

Products to Powers

$$(ab)^n = a^n b^n$$

When a product is raised to a power, raise each factor to that power.

EXAMPLE 5 Raising a Product to a Power

Simplify: $(-2y^2)^4$.

SOLUTION

$$
\begin{aligned}
(-2y^2)^4 &= (-2)^4 (y^2)^4 && \text{Raise each factor to the fourth power.} \\
&= (-2)^4 \, y^{2 \cdot 4} && \text{To raise an exponential expression to a power, multiply} \\
& && \text{exponents: } (b^m)^n = b^{mn}. \\
&= 16y^8 && \text{Simplify: } (-2)^4 = (-2)(-2)(-2)(-2) = 16.
\end{aligned}
$$

●●●

✓ Check Point **5** Simplify: $(-4x)^3$. $-64x^3$

The rule for raising a product to a power can be extended to cover three or more factors. For example,

$$(-2xy)^3 = (-2)^3 x^3 y^3 = -8x^3 y^3.$$

7 Find the power of a quotient.

The Quotients-to-Powers Rule for Exponents

The following rule is used to raise a quotient to a power:

Quotients to Powers

If b is a nonzero real number, then

$$\left(\frac{a}{b}\right)^n = \frac{a^n}{b^n}.$$

When a quotient is raised to a power, raise the numerator to that power and divide by the denominator to that power.

EXAMPLE 6 Raising Quotients to Powers

Simplify by raising each quotient to the given power:

a. $\left(-\dfrac{3}{x}\right)^4$ **b.** $\left(\dfrac{x^2}{4}\right)^3$.

SOLUTION

a. $\left(-\dfrac{3}{x}\right)^4 = \dfrac{(-3)^4}{x^4} = \dfrac{(-3)(-3)(-3)(-3)}{x^4} = \dfrac{81}{x^4}$

b. $\left(\dfrac{x^2}{4}\right)^3 = \dfrac{(x^2)^3}{4^3} = \dfrac{x^{2\cdot3}}{4\cdot4\cdot4} = \dfrac{x^6}{64}$ $\bullet\bullet\bullet$

🖊 **Check Point 6** Simplify:

a. $\left(-\dfrac{2}{y}\right)^5$ $-\dfrac{32}{y^5}$ b. $\left(\dfrac{x^5}{3}\right)^3$. $\dfrac{x^{15}}{27}$

⑧ Simplify exponential expressions.

Simplifying Exponential Expressions

Properties of exponents are used to simplify exponential expressions. An exponential expression is **simplified** when

- No parentheses appear.
- No powers are raised to powers.
- Each base occurs only once.
- No negative or zero exponents appear.

Simplifying Exponential Expressions	
	Example
1. If necessary, remove parentheses by using $(ab)^n = a^n b^n$ or $\left(\dfrac{a}{b}\right)^n = \dfrac{a^n}{b^n}$.	$(xy)^3 = x^3 y^3$
2. If necessary, simplify powers to powers by using $(b^m)^n = b^{mn}$.	$(x^4)^3 = x^{4\cdot3} = x^{12}$
3. If necessary, be sure that each base appears only once by using $b^m \cdot b^n = b^{m+n}$ or $\dfrac{b^m}{b^n} = b^{m-n}$.	$x^4 \cdot x^3 = x^{4+3} = x^7$
4. If necessary, rewrite exponential expressions with zero powers as 1 ($b^0 = 1$). Furthermore, write the answer with positive exponents by using $b^{-n} = \dfrac{1}{b^n}$ or $\dfrac{1}{b^{-n}} = b^n$.	$\dfrac{x^5}{x^8} = x^{5-8} = x^{-3} = \dfrac{1}{x^3}$

The following example shows how to simplify exponential expressions. Throughout the example, assume that no variable in a denominator is equal to zero.

EXAMPLE 7 Simplifying Exponential Expressions

Simplify:

a. $(-3x^4y^5)^3$ b. $(-7xy^4)(-2x^5y^6)$ c. $\dfrac{-35x^2y^4}{5x^6y^{-8}}$ d. $\left(\dfrac{4x^2}{y}\right)^{-3}$.

SOLUTION

a. $(-3x^4y^5)^3 = (-3)^3(x^4)^3(y^5)^3$ Raise each factor inside the parentheses to the third power.

$= (-3)^3x^{4\cdot3}y^{5\cdot3}$ Multiply the exponents when raising powers to powers.

$= -27x^{12}y^{15}$ $(-3)^3 = (-3)(-3)(-3) = -27$

b. $(-7xy^4)(-2x^5y^6) = (-7)(-2)xx^5y^4y^6$ Group factors with the same base.

$= 14x^{1+5}y^{4+6}$ When multiplying expressions with the same base, add the exponents.

$= 14x^6y^{10}$ Simplify.

c. $\dfrac{-35x^2y^4}{5x^6y^{-8}} = \left(\dfrac{-35}{5}\right)\left(\dfrac{x^2}{x^6}\right)\left(\dfrac{y^4}{y^{-8}}\right)$ Group factors with the same base.

$= -7x^{2-6}y^{4-(-8)}$ When dividing expressions with the same base, subtract the exponents.

$= -7x^{-4}y^{12}$ Simplify. Notice that $4 - (-8) = 4 + 8 = 12$.

$= \dfrac{-7y^{12}}{x^4}$ Write as a fraction and move the base with the negative exponent, x^{-4}, to the other side of the fraction bar and make the negative exponent positive.

d. $\left(\dfrac{4x^2}{y}\right)^{-3} = \dfrac{(4x^2)^{-3}}{y^{-3}}$ Raise the numerator and the denominator to the -3 power.

$= \dfrac{4^{-3}(x^2)^{-3}}{y^{-3}}$ Raise each factor inside the parentheses to the -3 power.

$= \dfrac{4^{-3}x^{-6}}{y^{-3}}$ Multiply the exponents when raising a power to a power: $(x^2)^{-3} = x^{2(-3)} = x^{-6}$.

$= \dfrac{y^3}{4^3x^6}$ Move each base with a negative exponent to the other side of the fraction bar and make each negative exponent positive.

$= \dfrac{y^3}{64x^6}$ $4^3 = 4\cdot4\cdot4 = 64$ • • •

⊘ Check Point **7** Simplify:

a. $(2x^3y^6)^4$ $16x^{12}y^{24}$ **b.** $(-6x^2y^5)(3xy^3)$ $-18x^3y^8$

c. $\dfrac{100x^{12}y^2}{20x^{16}y^{-4}}$ $\dfrac{5y^6}{x^4}$ **d.** $\left(\dfrac{5x}{y^4}\right)^{-2}$. $\dfrac{y^8}{25x^2}$

GREAT QUESTION!

Simplifying exponential expressions seems to involve lots of steps. Are there common errors I can avoid along the way?

Yes. Here's a list. The first column has the correct simplification. The second column contains common errors you should try to avoid.

Correct	Incorrect	Description of Error
$b^3 \cdot b^4 = b^7$	$b^3 \cdot b^4 = b^{12}$	The exponents should be added, not multiplied.
$3^2 \cdot 3^4 = 3^6$	$3^2 \cdot 3^4 = 9^6$	The common base should be retained, not multiplied.
$\dfrac{5^{16}}{5^4} = 5^{12}$	$\dfrac{5^{16}}{5^4} = 5^4$	The exponents should be subtracted, not divided.
$(4a)^3 = 64a^3$	$(4a)^3 = 4a^3$	Both factors should be cubed.
$b^{-n} = \dfrac{1}{b^n}$	$b^{-n} = -\dfrac{1}{b^n}$	Only the exponent should change sign.
$(a + b)^{-1} = \dfrac{1}{a + b}$	$(a + b)^{-1} = \dfrac{1}{a} + \dfrac{1}{b}$	The exponent applies to the entire expression $a + b$.

⑨ Use scientific notation.

Scientific Notation

Earth is a 4.5-billion-year-old ball of rock orbiting the Sun. Because a billion is 10^9 (see **Table P.3**), the age of our world can be expressed as

$$4.5 \times 10^9.$$

The number 4.5×10^9 is written in a form called *scientific notation.*

Table P.3 Names of Large Numbers

10^2	hundred
10^3	thousand
10^6	million
10^9	billion
10^{12}	trillion
10^{15}	quadrillion
10^{18}	quintillion
10^{21}	sextillion
10^{24}	septillion
10^{27}	octillion
10^{30}	nonillion
10^{100}	googol
10^{googol}	googolplex

Scientific Notation

A number is written in **scientific notation** when it is expressed in the form

$$a \times 10^n,$$

where the absolute value of a is greater than or equal to 1 and less than 10 ($1 \leq |a| < 10$), and n is an integer.

It is customary to use the multiplication symbol, \times, rather than a dot, when writing a number in scientific notation.

Converting from Scientific to Decimal Notation

Here are two examples of numbers in scientific notation:

$$6.4 \times 10^5 \quad \text{means} \quad 640,000.$$

$$2.17 \times 10^{-3} \quad \text{means} \quad 0.00217.$$

Do you see that the number with the positive exponent is relatively large and the number with the negative exponent is relatively small?

We can use n, the exponent on the 10 in $a \times 10^n$, to change a number in scientific notation to decimal notation. If n is **positive**, move the decimal point in a to the **right** n places. If n is **negative**, move the decimal point in a to the **left** $|n|$ places.

EXAMPLE 8 Converting from Scientific to Decimal Notation

Write each number in decimal notation:

 a. 6.2×10^7 **b.** -6.2×10^7 **c.** 2.019×10^{-3} **d.** -2.019×10^{-3}.

SOLUTION

In each case, we use the exponent on the 10 to determine how far to move the decimal point and in which direction. In parts (a) and (b), the exponent is positive, so we move the decimal point to the right. In parts (c) and (d), the exponent is negative, so we move the decimal point to the left.

a. $6.2 \times 10^7 = 62,000,000$ **b.** $-6.2 \times 10^7 = -62,000,000$

$n = 7$ Move the decimal point 7 places to the right. $n = 7$ Move the decimal point 7 places to the right.

c. $2.019 \times 10^{-3} = 0.002019$ **d.** $-2.019 \times 10^{-3} = -0.002019$

$n = -3$ Move the decimal point $|-3|$ places, or 3 places, to the left. $n = -3$ Move the decimal point $|-3|$ places, or 3 places, to the left.

 • • •

✓ Check Point **8** Write each number in decimal notation:

 a. -2.6×10^9 $-2,600,000,000$ **b.** 3.017×10^{-6}. 0.000003017

Converting from Decimal to Scientific Notation

To convert from decimal notation to scientific notation, we reverse the procedure of Example 8.

Converting from Decimal to Scientific Notation

Write the number in the form $a \times 10^n$.

- Determine a, the numerical factor. Move the decimal point in the given number to obtain a number whose absolute value is between 1 and 10, including 1.
- Determine n, the exponent on 10^n. The absolute value of n is the number of places the decimal point was moved. The exponent n is positive if the decimal point was moved to the left, negative if the decimal point was moved to the right, and 0 if the decimal point was not moved.

EXAMPLE 9 Converting from Decimal Notation to Scientific Notation

Write each number in scientific notation:

 a. $34,970,000,000,000$ **b.** $-34,970,000,000,000$
 c. 0.0000000000802 **d.** -0.0000000000802.

SOLUTION

 a. $34,970,000,000,000 = 3.497 \times 10^{13}$

Move the decimal point to get a number whose absolute value is between 1 and 10. The decimal point was moved 13 places to the left, so $n = 13$.

 b. $-34,970,000,000,000 = -3.497 \times 10^{13}$

TECHNOLOGY

You can use your calculator's
$\boxed{\text{EE}}$ (enter exponent) or $\boxed{\text{EXP}}$
key to convert from decimal to
scientific notation. Here is how it's
done for 0.0000000000802.

Many Scientific Calculators

Keystrokes

.0000000000802 $\boxed{\text{EE}}$ $\boxed{=}$

Display

$8.02 - 11$

Many Graphing Calculators

Use the mode setting for scientific
notation.

Keystrokes

.0000000000802 $\boxed{\text{ENTER}}$

Display

$8.02\text{E} - 11$

c. $0.0000000000802 = 8.02 \times 10^{-11}$

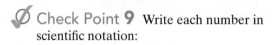

Move the decimal point to get a number whose absolute value is between 1 and 10.

The decimal point was moved 11 places to the right, so $n = -11$.

d. $-0.0000000000802 = -8.02 \times 10^{-11}$ •••

☑ **Check Point 9** Write each number in scientific notation:

a. $5,210,000,000$ 5.21×10^9 **b.** $-0.00000006893.$ -6.893×10^{-8}

GREAT QUESTION!

**In scientific notation, which
numbers have positive
exponents and which have
negative exponents?**

If the absolute value of a
number is greater than 10, it
will have a positive exponent
in scientific notation. If the
absolute value of a number is
less than 1, it will have a negative
exponent in scientific notation.

EXAMPLE 10 Expressing the U.S. Population in Scientific Notation

As of January 2016, the population of the United States was approximately 322 million. Express the population in scientific notation.

SOLUTION

Because a million is 10^6, the 2016 population can be expressed as

$$322 \times 10^6.$$

This factor is not between 1 and 10, so the number is not in scientific notation.

The voice balloon indicates that we need to convert 322 to scientific notation.

$$322 \times 10^6 = (3.22 \times 10^2) \times 10^6 = 3.22 \times 10^{2+6} = 3.22 \times 10^8$$

$322 = 3.22 \times 10^2$

In scientific notation, the population is 3.22×10^8. •••

☑ **Check Point 10** Express 410×10^7 in scientific notation. 4.1×10^9

Computations with Scientific Notation

Properties of exponents are used to perform computations with numbers that are expressed in scientific notation.

TECHNOLOGY

$(6.1 \times 10^5)(4 \times 10^{-9})$
On a Calculator:

Many Scientific Calculators

6.1 $\boxed{\text{EE}}$ 5 $\boxed{\times}$ 4 $\boxed{\text{EE}}$ 9 $\boxed{+/-}$ $\boxed{=}$

Display

$2.44 - 03$

Many Graphing Calculators

6.1 $\boxed{\text{EE}}$ 5 $\boxed{\times}$ 4 $\boxed{\text{EE}}$ $\boxed{(-)}$ 9 $\boxed{\text{ENTER}}$

Display (in scientific notation
mode)

$2.44\text{E} - 3$

EXAMPLE 11 Computations with Scientific Notation

Perform the indicated computations, writing the answers in scientific notation:

a. $(6.1 \times 10^5)(4 \times 10^{-9})$ **b.** $\dfrac{1.8 \times 10^4}{3 \times 10^{-2}}$.

SOLUTION

a. $(6.1 \times 10^5)(4 \times 10^{-9})$

$= (6.1 \times 4) \times (10^5 \times 10^{-9})$ Regroup factors.

$= 24.4 \times 10^{5+(-9)}$ Add the exponents on 10 and multiply the other parts.

$= 24.4 \times 10^{-4}$ Simplify.

$= (2.44 \times 10^1) \times 10^{-4}$ Convert 24.4 to scientific notation: $24.4 = 2.44 \times 10^1$.

$= 2.44 \times 10^{-3}$ $10^1 \times 10^{-4} = 10^{1+(-4)} = 10^{-3}$

b. $\dfrac{1.8 \times 10^4}{3 \times 10^{-2}} = \left(\dfrac{1.8}{3}\right) \times \left(\dfrac{10^4}{10^{-2}}\right)$ Regroup factors.

$= 0.6 \times 10^{4-(-2)}$ Subtract the exponents on 10 and divide the other parts.

$= 0.6 \times 10^6$ Simplify: $4 - (-2) = 4 + 2 = 6$.

$= (6 \times 10^{-1}) \times 10^6$ Convert 0.6 to scientific notation: $0.6 = 6 \times 10^{-1}$.

$= 6 \times 10^5$ $10^{-1} \times 10^6 = 10^{-1+6} = 10^5$ •••

⌀ Check Point **11** Perform the indicated computations, writing the answers in scientific notation:

a. $(7.1 \times 10^5)(5 \times 10^{-7})$ 3.55×10^{-1} **b.** $\dfrac{1.2 \times 10^6}{3 \times 10^{-3}}.$ 4×10^8

Applications: Putting Numbers in Perspective

Due to tax cuts and spending increases, the United States began accumulating large deficits in the 1980s. To finance the deficit, the government had borrowed $18.9 trillion as of January 2016. The graph in **Figure P.10** shows the national debt increasing over time.

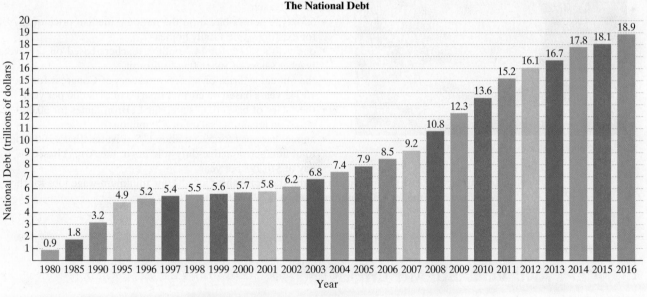

FIGURE P.10
Source: Office of Management and Budget

Example 12 shows how we can use scientific notation to comprehend the meaning of a number such as 18.9 trillion.

EXAMPLE 12 The National Debt

As of January 2016, the national debt was $18.9 trillion, or 18.9×10^{12} dollars. At that time, the U.S. population was approximately 322,000,000 (322 million), or 3.22×10^8. If the national debt was evenly divided among every individual in the United States, how much would each citizen have to pay?

Here is the keystroke sequence for solving Example 12 using a calculator:

18.9 EE 12 \div 3.22 EE 8.

The quotient is displayed by pressing $=$ on a scientific calculator or ENTER on a graphing calculator. The answer can be displayed in scientific or decimal notation. Consult your manual.

SOLUTION

The amount each citizen must pay is the total debt, 18.9×10^{12} dollars, divided by the number of citizens, 3.22×10^8.

$$\frac{18.9 \times 10^{12}}{3.22 \times 10^8} = \left(\frac{18.9}{3.22}\right) \times \left(\frac{10^{12}}{10^8}\right)$$

$$\approx 5.87 \times 10^{12-8}$$

$$= 5.87 \times 10^4$$

$$= 58,700$$

Every U.S. citizen would have to pay approximately $58,700 to the federal government to pay off the national debt. • • •

 Check Point **12** In 2015, there were 680,000 police officers in the United States with yearly wages totaling 4.08×10^{10}. If these wages were evenly divided among all police officers, find the mean, or average, salary of a U.S. police officer. (Source: Bureau of Justice Statistics) $60,000

Blitzer Bonus || Seven Ways to Spend $1 Trillion

Image © photobank.kiev.ua, 2009

Confronting a national debt of $18.9 trillion starts with grasping just how colossal $1 trillion ($1 \times 10^{12}$) actually is. To help you wrap your head around this mind-boggling number, and to put the national debt in further perspective, consider what $1 trillion will buy:

- 40,816,326 new cars based on an average sticker price of $24,500 each
- 5,574,136 homes based on the national median price of $179,400 for existing single-family homes
- one year's salary for 14.7 million teachers based on the average teacher salary of $68,000 in California
- the annual salaries of all 535 members of Congress for the next 10,742 years based on current salaries of $174,000 per year
- the salary of basketball superstar LeBron James for 50,000 years based on an annual salary of $20 million
- annual base pay for 59.5 million U.S. privates (that's 100 times the total number of active-duty soldiers in the Army) based on basic pay of $16,794 per year
- salaries to hire all 2.8 million residents of the state of Kansas in full-time minimum-wage jobs for the next 23 years based on the federal minimum wage of $7.25 per hour

Source: Kiplinger.com

CONCEPT AND VOCABULARY CHECK

Fill in each blank so that the resulting statement is true.

1. The product rule for exponents states that $b^m \cdot b^n = \underline{b^{m+n}}$. When multiplying exponential expressions with the same base, __add__ the exponents.

2. The quotient rule for exponents states that $\dfrac{b^m}{b^n} = \underline{b^{m-n}}$, $b \neq 0$. When dividing exponential expressions with the same nonzero base, __subtract__ the exponents.

3. If $b \neq 0$, then $b^0 = \underline{1}$.

4. The negative-exponent rule states that $b^{-n} = \underline{\frac{1}{b^n}}$, $b \neq 0$.

5. True or false: $5^{-2} = -5^2$ __false__

6. Negative exponents in denominators can be evaluated using $\dfrac{1}{b^{-n}} = \underline{b^n}$, $b \neq 0$.

7. True or false: $\dfrac{1}{8^{-2}} = 8^2$ __true__

8. A positive number is written in scientific notation when it is expressed in the form $a \times 10^n$, where a is __a number greater than or equal to 1 and less than 10__ and n is a/an __integer__.

9. True or false: 4×10^3 is written in scientific notation. __true__

10. True or false: 40×10^2 is written in scientific notation. __false__

EXERCISE SET P.2

Practice Exercises

Evaluate each exponential expression in Exercises 1–22.

1. $5^2 \cdot 2$ $\quad 50$ **2.** $6^2 \cdot 2$ $\quad 72$

3. $(-2)^6$ $\quad 64$ **4.** $(-2)^4$ $\quad 16$

5. -2^6 $\quad -64$ **6.** -2^4 $\quad -16$

7. $(-3)^0$ $\quad 1$ **8.** $(-9)^0$ $\quad 1$

9. -3^0 $\quad -1$ **10.** -9^0 $\quad -1$

11. 4^{-3} $\quad \frac{1}{64}$ **12.** 2^{-6} $\quad \frac{1}{64}$

13. $2^2 \cdot 2^3$ $\quad 32$ **14.** $3^3 \cdot 3^2$ $\quad 243$

15. $(2^2)^3$ $\quad 64$ **16.** $(3^3)^2$ $\quad 729$

17. $\dfrac{2^8}{2^4}$ $\quad 16$ **18.** $\dfrac{3^8}{3^4}$ $\quad 81$

19. $3^{-3} \cdot 3$ $\quad \frac{1}{9}$ **20.** $2^{-3} \cdot 2$ $\quad \frac{1}{4}$

21. $\dfrac{2^3}{2^7}$ $\quad \frac{1}{16}$ **22.** $\dfrac{3^4}{3^7}$ $\quad \frac{1}{27}$

Simplify each exponential expression in Exercises 23–64.

23. $x^{-2}y$ $\quad \frac{y}{x^2}$ **24.** xy^{-3} $\quad \frac{x}{y^3}$

25. $x^0 y^5$ $\quad y^5$ **26.** $x^7 y^0$ $\quad x^7$

27. $x^3 \cdot x^7$ $\quad x^{10}$ **28.** $x^{11} \cdot x^5$ $\quad x^{16}$

29. $x^{-5} \cdot x^{10}$ $\quad x^5$ **30.** $x^{-6} \cdot x^{12}$ $\quad x^6$

31. $(x^3)^7$ $\quad x^{21}$ **32.** $(x^{11})^5$ $\quad x^{55}$

33. $(x^{-5})^3$ $\quad \frac{1}{x^{15}}$ **34.** $(x^{-6})^4$ $\quad \frac{1}{x^{24}}$

35. $\dfrac{x^{14}}{x^7}$ $\quad x^7$ **36.** $\dfrac{x^{30}}{x^{10}}$ $\quad x^{20}$

37. $\dfrac{x^{14}}{x^{-7}}$ $\quad x^{21}$ **38.** $\dfrac{x^{30}}{x^{-10}}$ $\quad x^{40}$

39. $(8x^3)^2$ $\quad 64x^6$ **40.** $(6x^4)^2$ $\quad 36x^8$

41. $\left(-\dfrac{4}{x}\right)^3$ $\quad -\dfrac{64}{x^3}$ **42.** $\left(-\dfrac{6}{y}\right)^3$ $\quad -\dfrac{216}{y^3}$

43. $(-3x^2y^5)^2$ $\quad 9x^4y^{10}$ **44.** $(-3x^4y^6)^3$ $\quad -27x^{12}y^{18}$

45. $(3x^4)(2x^7)$ $\quad 6x^{11}$ **46.** $(11x^5)(9x^{12})$ $\quad 99x^{17}$

47. $(-9x^3y)(-2x^6y^4)$ $\quad 18x^9y^5$ **48.** $(-5x^4y)(-6x^7y^{11})$ $\quad *$

49. $\dfrac{8x^{20}}{2x^4}$ $\quad 4x^{16}$ **50.** $\dfrac{20x^{24}}{10x^6}$ $\quad 2x^{18}$

51. $\dfrac{25a^{13}b^4}{-5a^2b^3}$ $\quad -5a^{11}b$ **52.** $\dfrac{35a^{14}b^6}{-7a^7b^3}$ $\quad -5a^7b^3$

53. $\dfrac{14b^7}{7b^{14}}$ $\quad \frac{2}{b^7}$ **54.** $\dfrac{20b^{10}}{10b^{20}}$ $\quad \frac{2}{b^{10}}$

55. $(4x^3)^{-2}$ $\quad \frac{1}{16x^6}$ **56.** $(10x^2)^{-3}$ $\quad \frac{1}{1000x^6}$

57. $\dfrac{24x^3y^5}{32x^7y^{-9}}$ $\quad \frac{3y^{14}}{4x^4}$ **58.** $\dfrac{10x^4y^9}{30x^{12}y^{-3}}$ $\quad \frac{y^{12}}{3x^8}$

59. $\left(\dfrac{5x^3}{y}\right)^{-2}$ $\quad \frac{y^2}{25x^6}$ **60.** $\left(\dfrac{3x^4}{y}\right)^{-3}$ $\quad \frac{y^3}{27x^{12}}$

61. $\left(\dfrac{-15a^4b^2}{5a^{10}b^{-3}}\right)^3$ $\quad -\dfrac{27b^{15}}{a^{18}}$ **62.** $\left(\dfrac{-30a^{14}b^8}{10a^{17}b^{-2}}\right)^3$ $\quad -\dfrac{27b^{30}}{a^9}$

*See Answers to Selected Exercises.

63. $\left(\dfrac{3a^{-5}b^2}{12a^3b^{-4}}\right)^0$ $\quad 1$ **64.** $\left(\dfrac{4a^{-5}b^3}{12a^3b^{-5}}\right)^0$ $\quad 1$

In Exercises 65–76, write each number in decimal notation without the use of exponents.

65. 3.8×10^2 $\quad 380$ **66.** 9.2×10^2 $\quad 920$

67. 6×10^{-4} $\quad 0.0006$ **68.** 7×10^{-5} $\quad 0.00007$

69. -7.16×10^6 $\quad -7,160,000$ **70.** -8.17×10^6 $\quad -8,170,000$

71. 7.9×10^{-1} $\quad 0.79$ **72.** 6.8×10^{-1} $\quad 0.68$

73. -4.15×10^{-3} $\quad -0.00415$ **74.** -3.14×10^{-3} $\quad -0.00314$

75. -6.00001×10^{10} $\quad *$ **76.** -7.00001×10^{10} $\quad *$

In Exercises 77–86, write each number in scientific notation.

77. $32,000$ $\quad 3.2 \times 10^4$ **78.** $64,000$ $\quad 6.4 \times 10^4$

79. $638,000,000,000,000,000$ $\quad 6.38 \times 10^{17}$

80. $579,000,000,000,000,000$ $\quad 5.79 \times 10^{17}$

81. -5716 $\quad -5.716 \times 10^3$ **82.** -3829 $\quad -3.829 \times 10^3$

83. 0.0027 $\quad 2.7 \times 10^{-3}$ **84.** 0.0083 $\quad 8.3 \times 10^{-3}$

85. -0.00000000504 $\quad -5.04 \times 10^{-9}$

86. -0.00000000405 $\quad -4.05 \times 10^{-9}$

In Exercises 87–106, perform the indicated computations. Write the answers in scientific notation. If necessary, round the decimal factor in your scientific notation answer to two decimal places.

87. $(3 \times 10^4)(2.1 \times 10^3)$ $\quad *$ **88.** $(2 \times 10^4)(4.1 \times 10^3)$ $\quad *$

89. $(1.6 \times 10^{15})(4 \times 10^{-11})$ $\quad *$ **90.** $(1.4 \times 10^{15})(3 \times 10^{-11})$ $\quad *$

91. $(6.1 \times 10^{-8})(2 \times 10^{-4})$ $\quad *$ **92.** $(5.1 \times 10^{-8})(3 \times 10^{-4})$ $\quad *$

93. $(4.3 \times 10^8)(6.2 \times 10^4)$ $\quad *$ **94.** $(8.2 \times 10^8)(4.6 \times 10^4)$ $\quad *$

95. $\dfrac{8.4 \times 10^8}{4 \times 10^5}$ $\quad 2.1 \times 10^3$ **96.** $\dfrac{6.9 \times 10^8}{3 \times 10^5}$ $\quad 2.3 \times 10^3$

97. $\dfrac{3.6 \times 10^4}{9 \times 10^{-2}}$ $\quad 4 \times 10^5$ **98.** $\dfrac{1.2 \times 10^4}{2 \times 10^{-2}}$ $\quad 6 \times 10^5$

99. $\dfrac{4.8 \times 10^{-2}}{2.4 \times 10^6}$ $\quad 2 \times 10^{-8}$ **100.** $\dfrac{7.5 \times 10^{-2}}{2.5 \times 10^6}$ $\quad 3 \times 10^{-8}$

101. $\dfrac{2.4 \times 10^{-2}}{4.8 \times 10^{-6}}$ $\quad 5 \times 10^3$ **102.** $\dfrac{1.5 \times 10^{-2}}{3 \times 10^{-6}}$ $\quad 5 \times 10^3$

103. $\dfrac{480,000,000,000}{0.00012}$ $\quad 4 \times 10^{15}$ **104.** $\dfrac{282,000,000,000}{0.00141}$ $\quad 2 \times 10^{14}$

105. $\dfrac{0.00072 \times 0.003}{0.00024}$ $\quad 9 \times 10^{-3}$ **106.** $\dfrac{66,000 \times 0.001}{0.003 \times 0.002}$ $\quad 1.1 \times 10^7$

Practice Plus

In Exercises 107–114, simplify each exponential expression. Assume that variables represent nonzero real numbers.

107. $\dfrac{(x^{-2}y)^{-3}}{(x^2y^{-1})^3}$ $\quad 1$ **108.** $\dfrac{(xy^{-2})^{-2}}{(x^{-2}y)^{-3}}$ $\quad \frac{y^7}{x^8}$

109. $(2x^{-3}yz^{-6})(2x)^{-5}$ $\quad \frac{y}{16x^8z^6}$ **110.** $(3x^{-4}yz^{-7})(3x)^{-3}$ $\quad \frac{y}{9x^7z^7}$

111. $\left(\dfrac{x^3y^4z^5}{x^{-3}y^{-4}z^{-5}}\right)^{-2}$ $\quad \frac{1}{x^{12}y^{16}z^{20}}$ **112.** $\left(\dfrac{x^4y^5z^6}{x^{-4}y^{-5}z^{-6}}\right)^{-4}$ $\quad \frac{1}{x^{32}y^{40}z^{48}}$

113. $\dfrac{(2^{-1}x^{-2}y^{-1})^{-2}(2x^{-4}y^3)^{-2}(16x^{-3}y^3)^0}{(2x^{-3}y^{-5})^2}$ $\dfrac{x^{18}y^6}{4}$

114. $\dfrac{(2^{-1}x^{-3}y^{-1})^{-2}(2x^{-6}y^4)^{-2}(9x^3y^{-3})^0}{(2x^{-4}y^{-6})^2}$ $\dfrac{x^{26}y^6}{4}$

Application Exercises

The bar graph shows the total amount Americans paid in federal taxes, in trillions of dollars, and the U.S. population, in millions, from 2012 through 2015. Exercises 115–116 are based on the numbers displayed by the graph.

Federal Taxes and the United States Population

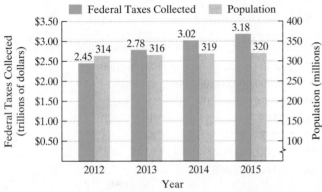

Sources: Internal Revenue Service and U.S. Census Bureau

115. a. In 2015, the United States government collected $3.18 trillion in taxes. Express this number in scientific notation. 3.18×10^{12}

 b. In 2015, the population of the United States was approximately 320 million. Express this number in scientific notation. 3.20×10^8

 c. Use your scientific notation answers from parts (a) and (b) to answer this question: If the total 2015 tax collections were evenly divided among all Americans, how much would each citizen pay? Express the answer in decimal notation, rounded to the nearest dollar. $9938

116. a. In 2014, the United States government collected $3.02 trillion in taxes. Express this number in scientific notation. 3.02×10^{12}

 b. In 2014, the population of the United States was approximately 319 million. Express this number in scientific notation. 3.19×10^8

 c. Use your scientific notation answers from parts (a) and (b) to answer this question: If the total 2014 tax collections were evenly divided among all Americans, how much would each citizen pay? Express the answer in decimal notation, rounded to the nearest dollar. $9467

We have seen that the 2016 U.S. national debt was $18.9 trillion. In Exercises 117–118, you will use scientific notation to put a number like 18.9 trillion in perspective.

117. a. Express 18.9 trillion in scientific notation. 1.89×10^{13}

 b. Four years of tuition, fees, and room and board at a public U.S. college cost approximately $60,000. Express this number in scientific notation. 6×10^4

 c. Use your answers from parts (a) and (b) to determine how many Americans could receive a free college education for $18.9 trillion. 3.15×10^8 or 315,000,000

118. a. Express 18.9 trillion in scientific notation. 1.89×10^{13}

 b. Each year, Americans spend $254 billion on summer vacations. Express this number in scientific notation. 2.54×10^{11}

 c. Use your answers from parts (a) and (b) to determine how many years Americans can have free summer vacations for $18.9 trillion. approximately 74 years

119. In 2012, the United States government spent more than it had collected in taxes, resulting in a budget deficit of $1.09 trillion.

 a. Express 1.09 trillion in scientific notation. 1.09×10^{12}

 b. There are approximately 32,000,000 seconds in a year. Write this number in scienfic notation. 3.2×10^7

 c. Use your answers from parts (a) and (b) to determine approximately how many years is 1.09 trillion seconds. (Note: 1.09 trillion seconds would take us back in time to a period when Neanderthals were using stones to make tools.) 34,062.5 years

120. Refer to the Blitzer Bonus on page 32. Use scientific notation to verify any two of the bulleted items on ways to spend $1 trillion. Answers will vary.

Explaining the Concepts 121–128. Answers will vary.

121. Describe what it means to raise a number to a power. In your description, include a discussion of the difference between -5^2 and $(-5)^2$.

122. Explain the product rule for exponents. Use $2^3 \cdot 2^5$ in your explanation.

123. Explain the power rule for exponents. Use $(3^2)^4$ in your explanation.

124. Explain the quotient rule for exponents. Use $\dfrac{5^8}{5^2}$ in your explanation.

125. Why is $(-3x^2)(2x^{-5})$ not simplified? What must be done to simplify the expression?

126. How do you know if a number is written in scientific notation?

127. Explain how to convert from scientific to decimal notation and give an example.

128. Explain how to convert from decimal to scientific notation and give an example.

Critical Thinking Exercises

Make Sense? *In Exercises 129–132, determine whether each statement makes sense or does not make sense, and explain your reasoning.*

129. There are many exponential expressions that are equal to $36x^{12}$, such as $(6x^6)^2$, $(6x^3)(6x^9)$, $36(x^3)^9$, and $6^2(x^2)^6$. *

130. If 5^{-2} is raised to the third power, the result is a number between 0 and 1. makes sense

131. The population of Colorado is approximately 4.6×10^{12}. *

132. I just finished reading a book that contained approximately 1.04×10^5 words. makes sense

In Exercises 133–140, determine whether each statement is true or false. If the statement is false, make the necessary change(s) to produce a true statement. Changes to false statements will vary.

133. $4^{-2} < 4^{-3}$ false

134. $5^{-2} > 2^{-5}$ true

135. $(-2)^4 = 2^{-4}$ false

136. $5^2 \cdot 5^{-2} > 2^5 \cdot 2^{-5}$ false

137. $534.7 = 5.347 \times 10^3$ false

138. $\dfrac{8 \times 10^{30}}{4 \times 10^{-5}} = 2 \times 10^{25}$ false

139. $(7 \times 10^5) + (2 \times 10^{-3}) = 9 \times 10^2$ false

140. $(4 \times 10^3) + (3 \times 10^2) = 4.3 \times 10^3$ true

141. The mad Dr. Frankenstein has gathered enough bits and pieces (so to speak) for $2^{-1} + 2^{-2}$ of his creature-to-be. Write a fraction that represents the amount of his creature that must still be obtained. $\frac{1}{4}$

142. If $b^A = MN$, $b^C = M$, and $b^D = N$, what is the relationship among A, C, and D? $A = C + D$

143. Our hearts beat approximately 70 times per minute. Express in scientific notation how many times the heart beats over a lifetime of 80 years. Round the decimal factor in your scientific notation answer to two decimal places. *

Group Exercise Answers will vary.

144. Putting Numbers into Perspective. A large number can be put into perspective by comparing it with another number. For example, we put the $18.9 trillion national debt in perspective (Example 12) by comparing this number to the number of U.S. citizens.

For this project, each group member should consult an almanac, a newspaper, or the Internet to find a number greater than one million. Explain to other members of the group the context in which the large number is used. Express the number in scientific notation. Then put the number into perspective by comparing it with another number.

Preview Exercises

Exercises 145–147 will help you prepare for the material covered in the next section.

145. a. Find $\sqrt{16} \cdot \sqrt{4}$. 8

b. Find $\sqrt{16 \cdot 4}$. 8

c. Based on your answers to parts (a) and (b), what can you conclude? $\sqrt{16} \cdot \sqrt{4} = \sqrt{16 \cdot 4}$

146. a. Use a calculator to approximate $\sqrt{300}$ to two decimal places. 17.32

b. Use a calculator to approximate $10\sqrt{3}$ to two decimal places. 17.32

c. Based on your answers to parts (a) and (b), what can you conclude? $\sqrt{300} = 10\sqrt{3}$

147. a. Simplify: $21x + 10x$. $31x$

b. Simplify: $21\sqrt{2} + 10\sqrt{2}$. $31\sqrt{2}$

Section P.3 Radicals and Rational Exponents

What am I supposed to learn?

After studying this section, you should be able to:

① Evaluate square roots.

② Simplify expressions of the form $\sqrt{a^2}$.

③ Use the product rule to simplify square roots.

④ Use the quotient rule to simplify square roots.

⑤ Add and subtract square roots.

⑥ Rationalize denominators.

⑦ Evaluate and perform operations with higher roots.

⑧ Understand and use rational exponents.

This photograph shows mathematical models used by Albert Einstein at a lecture on relativity. Notice the radicals that appear in many of the formulas. Among these models, there is one describing how an astronaut in a moving spaceship ages more slowly than friends who remain on Earth. No description of your world can be complete without roots and radicals. In this section, in addition to reviewing the basics of radical expressions and the use of rational exponents to indicate radicals, you will see how radicals model time dilation for a futuristic high-speed trip to a nearby star.

① Evaluate square roots.

Square Roots

From our earlier work with exponents, we are aware that the square of both 5 and -5 is 25:

$$5^2 = 25 \quad \text{and} \quad (-5)^2 = 25.$$

The reverse operation of squaring a number is finding the *square root* of the number. For example,

- One square root of 25 is 5 because $5^2 = 25$.
- Another square root of 25 is -5 because $(-5)^2 = 25$.

In general, **if $b^2 = a$, then b is a square root of a**.

The symbol $\sqrt{\ }$ is used to denote the *nonnegative* or *principal square root* of a number. For example,

- $\sqrt{25} = 5$ because $5^2 = 25$ and 5 is positive.
- $\sqrt{100} = 10$ because $10^2 = 100$ and 10 is positive.

The symbol $\sqrt{\ }$ that we use to denote the principal square root is called a **radical sign**. The number under the radical sign is called the **radicand**. Together we refer to the radical sign and its radicand as a **radical expression**.

Definition of the Principal Square Root

If a is a nonnegative real number, the nonnegative number b such that $b^2 = a$, denoted by $b = \sqrt{a}$, is the **principal square root** of a.

The symbol $-\sqrt{\ }$ is used to denote the negative square root of a number. For example,

- $-\sqrt{25} = -5$ because $(-5)^2 = 25$ and -5 is negative.
- $-\sqrt{100} = -10$ because $(-10)^2 = 100$ and -10 is negative.

EXAMPLE 1 Evaluating Square Roots

Evaluate:

 a. $\sqrt{64}$ **b.** $-\sqrt{49}$ **c.** $\sqrt{\dfrac{1}{4}}$ **d.** $\sqrt{9 + 16}$ **e.** $\sqrt{9} + \sqrt{16}$.

SOLUTION

a. $\sqrt{64} = 8$ The principal square root of 64 is 8. Check: $8^2 = 64$.

b. $-\sqrt{49} = -7$ The negative square root of 49 is -7. Check: $(-7)^2 = 49$.

c. $\sqrt{\dfrac{1}{4}} = \dfrac{1}{2}$ The principal square root of $\frac{1}{4}$ is $\frac{1}{2}$. Check: $\left(\frac{1}{2}\right)^2 = \frac{1}{4}$.

d. $\sqrt{9 + 16} = \sqrt{25}$ First simplify the expression under the radical sign.
$\qquad\qquad\quad = 5$ Then take the principal square root of 25, which is 5.

e. $\sqrt{9} + \sqrt{16} = 3 + 4$ $\sqrt{9} = 3$ because $3^2 = 9$. $\sqrt{16} = 4$ because $4^2 = 16$.
$\qquad\qquad\qquad = 7$

• • •

GREAT QUESTION!

Is $\sqrt{a + b}$ equal to $\sqrt{a} + \sqrt{b}$?

No. In Example 1, parts (d) and (e), observe that $\sqrt{9 + 16}$ is not equal to $\sqrt{9} + \sqrt{16}$. In general,

$$\sqrt{a + b} \neq \sqrt{a} + \sqrt{b}$$

and

$$\sqrt{a - b} \neq \sqrt{a} - \sqrt{b}.$$

 Check Point **1** Evaluate:

a. $\sqrt{81}$ 9

b. $-\sqrt{9}$ −3

c. $\sqrt{\dfrac{1}{25}}$ $\frac{1}{5}$

d. $\sqrt{36 + 64}$ 10

e. $\sqrt{36} + \sqrt{64}$. 14

A number that is the square of a rational number is called a **perfect square**. All the radicands in Example 1 and Check Point 1 are perfect squares.

- 64 is a perfect square because $64 = 8^2$. Thus, $\sqrt{64} = 8$.

- $\dfrac{1}{4}$ is a perfect square because $\dfrac{1}{4} = \left(\dfrac{1}{2}\right)^2$. Thus, $\sqrt{\dfrac{1}{4}} = \dfrac{1}{2}$.

Let's see what happens to the radical expression \sqrt{x} if x is a negative number. Is the square root of a negative number a real number? For example, consider $\sqrt{-25}$. Is there a real number whose square is −25? No. Thus, $\sqrt{-25}$ is not a real number. In general, **a square root of a negative number is not a real number**.

If a number a is nonnegative ($a \geq 0$), then $(\sqrt{a})^2 = a$. For example,

$$(\sqrt{2})^2 = 2, \quad (\sqrt{3})^2 = 3, \quad (\sqrt{4})^2 = 4, \quad \text{and} \quad (\sqrt{5})^2 = 5.$$

② Simplify expressions of the form $\sqrt{a^2}$.

Simplifying Expressions of the Form $\sqrt{a^2}$

You may think that $\sqrt{a^2} = a$. However, this is not necessarily true. Consider the following examples:

$$\sqrt{4^2} = \sqrt{16} = 4$$
$$\sqrt{(-4)^2} = \sqrt{16} = 4.$$

> The result is not −4, but rather the absolute value of −4, or 4.

Here is a rule for simplifying expressions of the form $\sqrt{a^2}$:

Simplifying $\sqrt{a^2}$

For any real number a,

$$\sqrt{a^2} = |a|.$$

In words, the principal square root of a^2 is the absolute value of a.

For example, $\sqrt{6^2} = |6| = 6$ and $\sqrt{(-6)^2} = |-6| = 6$.

③ Use the product rule to simplify square roots.

The Product Rule for Square Roots

A rule for multiplying square roots can be generalized by comparing $\sqrt{25} \cdot \sqrt{4}$ and $\sqrt{25 \cdot 4}$. Notice that

$$\sqrt{25} \cdot \sqrt{4} = 5 \cdot 2 = 10 \quad \text{and} \quad \sqrt{25 \cdot 4} = \sqrt{100} = 10.$$

Because we obtain 10 in both situations, the original radical expressions must be equal. That is,

$$\sqrt{25} \cdot \sqrt{4} = \sqrt{25 \cdot 4}.$$

This result is a special case of the **product rule for square roots** that can be generalized as follows:

The Product Rule for Square Roots

If a and b represent nonnegative real numbers, then

$$\sqrt{ab} = \sqrt{a} \cdot \sqrt{b} \quad \text{and} \quad \sqrt{a} \cdot \sqrt{b} = \sqrt{ab}.$$

> The square root of a product is the product of the square roots.

> The product of two square roots is the square root of the product of the radicands.

A square root is **simplified** when its radicand has no factors other than 1 that are perfect squares. For example, $\sqrt{500}$ is not simplified because it can be expressed as $\sqrt{100 \cdot 5}$ and 100 is a perfect square. Example 2 shows how the product rule is used to remove from the square root any perfect squares that occur as factors.

> **EXAMPLE 2** Using the Product Rule to Simplify Square Roots

Simplify:
 a. $\sqrt{500}$ **b.** $\sqrt{6x} \cdot \sqrt{3x}$.

SOLUTION

 a. $\sqrt{500} = \sqrt{100 \cdot 5}$ Factor 500. 100 is the greatest perfect square factor.

$\qquad\qquad = \sqrt{100}\sqrt{5}$ Use the product rule: $\sqrt{ab} = \sqrt{a} \cdot \sqrt{b}$.

$\qquad\qquad = 10\sqrt{5}$ Write $\sqrt{100}$ as 10. We read $10\sqrt{5}$ as "ten times the square root of 5."

 b. We can simplify $\sqrt{6x} \cdot \sqrt{3x}$ using the product rule only if $6x$ and $3x$ represent nonnegative real numbers. Thus, $x \geq 0$.

$\sqrt{6x} \cdot \sqrt{3x} = \sqrt{6x \cdot 3x}$ Use the product rule: $\sqrt{a}\sqrt{b} = \sqrt{ab}$.

$\qquad\qquad = \sqrt{18x^2}$ Multiply in the radicand.

$\qquad\qquad = \sqrt{9x^2 \cdot 2}$ Factor 18. 9 is the greatest perfect square factor.

$\qquad\qquad = \sqrt{9x^2}\sqrt{2}$ Use the product rule: $\sqrt{ab} = \sqrt{a} \cdot \sqrt{b}$.

$\qquad\qquad = \sqrt{9}\sqrt{x^2}\sqrt{2}$ Use the product rule to write $\sqrt{9x^2}$ as the product of two square roots.

$\qquad\qquad = 3x\sqrt{2}$ $\sqrt{x^2} = |x| = x$ because $x \geq 0$. ● ● ●

GREAT QUESTION!

When simplifying square roots, what happens if I use a perfect square factor that isn't the greatest perfect square factor possible?

You'll need to simplify even further. For example, consider the following factorization:
$$\sqrt{500} = \sqrt{25 \cdot 20} = \sqrt{25}\sqrt{20} = 5\sqrt{20}.$$

> 25 is a perfect square factor of 500, but not the greatest perfect square factor.

Because 20 contains a perfect square factor, 4, the simplification is not complete.
$$5\sqrt{20} = 5\sqrt{4 \cdot 5} = 5\sqrt{4}\sqrt{5} = 5 \cdot 2\sqrt{5} = 10\sqrt{5}$$

Although the result checks with our simplification using $\sqrt{500} = \sqrt{100 \cdot 5}$, **more work is required when the greatest perfect square factor is not used.**

✓ **Check Point 2** Simplify:
 a. $\sqrt{75}$ $5\sqrt{3}$ **b.** $\sqrt{5x} \cdot \sqrt{10x}$. $5x\sqrt{2}$

④ Use the quotient rule to simplify square roots.

The Quotient Rule for Square Roots

Another property for square roots involves division.

> **The Quotient Rule for Square Roots**
>
> If a and b represent nonnegative real numbers and $b \neq 0$, then
> $$\sqrt{\frac{a}{b}} = \frac{\sqrt{a}}{\sqrt{b}} \quad \text{and} \quad \frac{\sqrt{a}}{\sqrt{b}} = \sqrt{\frac{a}{b}}.$$
>
> The square root of a quotient is the quotient of the square roots.
>
> The quotient of two square roots is the square root of the quotient of the radicands.

EXAMPLE 3 Using the Quotient Rule to Simplify Square Roots

Simplify:

a. $\sqrt{\dfrac{100}{9}}$ **b.** $\dfrac{\sqrt{48x^3}}{\sqrt{6x}}$.

SOLUTION

a. $\sqrt{\dfrac{100}{9}} = \dfrac{\sqrt{100}}{\sqrt{9}} = \dfrac{10}{3}$

b. We can simplify the quotient of $\sqrt{48x^3}$ and $\sqrt{6x}$ using the quotient rule only if $48x^3$ and $6x$ represent nonnegative real numbers and $6x \ne 0$. Thus, $x > 0$.

$$\dfrac{\sqrt{48x^3}}{\sqrt{6x}} = \sqrt{\dfrac{48x^3}{6x}} = \sqrt{8x^2} = \sqrt{4x^2}\sqrt{2} = \sqrt{4}\sqrt{x^2}\sqrt{2} = 2x\sqrt{2}$$

$\sqrt{x^2} = |x| = x$ because $x > 0$. • • •

✓ Check Point **3** Simplify:

a. $\sqrt{\dfrac{25}{16}}$ $\frac{5}{4}$ **b.** $\dfrac{\sqrt{150x^3}}{\sqrt{2x}}$. $5x\sqrt{3}$

⑤ Add and subtract square roots.

GREAT QUESTION!

Should like radicals remind me of like terms?

Yes. Adding or subtracting like radicals is similar to adding or subtracting like terms:

$$7x + 6x = 13x$$

and

$$7\sqrt{11} + 6\sqrt{11} = 13\sqrt{11}.$$

Adding and Subtracting Square Roots

Two or more square roots can be combined using the distributive property provided that they have the same radicand. Such radicals are called **like radicals**. For example,

$$7\sqrt{11} + 6\sqrt{11} = (7 + 6)\sqrt{11} = 13\sqrt{11}.$$

7 square roots of 11 plus 6 square roots of 11 result in 13 square roots of 11.

EXAMPLE 4 Adding and Subtracting Like Radicals

Add or subtract as indicated:

a. $7\sqrt{2} + 5\sqrt{2}$ **b.** $\sqrt{5x} - 7\sqrt{5x}$.

SOLUTION

a. $7\sqrt{2} + 5\sqrt{2} = (7 + 5)\sqrt{2}$ Apply the distributive property.

$= 12\sqrt{2}$ Simplify.

b. $\sqrt{5x} - 7\sqrt{5x} = 1\sqrt{5x} - 7\sqrt{5x}$ Write $\sqrt{5x}$ as $1\sqrt{5x}$.

$= (1 - 7)\sqrt{5x}$ Apply the distributive property.

$= -6\sqrt{5x}$ Simplify. • • •

✓ Check Point **4** Add or subtract as indicated:

a. $8\sqrt{13} + 9\sqrt{13}$ $17\sqrt{13}$ **b.** $\sqrt{17x} - 20\sqrt{17x}$. $-19\sqrt{17x}$

In some cases, radicals can be combined once they have been simplified. For example, to add $\sqrt{2}$ and $\sqrt{8}$, we can write $\sqrt{8}$ as $\sqrt{4 \cdot 2}$ because 4 is a perfect square factor of 8.

$$\sqrt{2} + \sqrt{8} = \sqrt{2} + \sqrt{4 \cdot 2} = 1\sqrt{2} + 2\sqrt{2} = (1 + 2)\sqrt{2} = 3\sqrt{2}$$

EXAMPLE 5 Combining Radicals That First Require Simplification

Add or subtract as indicated:

a. $7\sqrt{3} + \sqrt{12}$ b. $4\sqrt{50x} - 6\sqrt{32x}$.

SOLUTION

a. $7\sqrt{3} + \sqrt{12}$

$= 7\sqrt{3} + \sqrt{4 \cdot 3}$ Split 12 into two factors such that one is a perfect square.

$= 7\sqrt{3} + 2\sqrt{3}$ $\sqrt{4 \cdot 3} = \sqrt{4}\sqrt{3} = 2\sqrt{3}$

$= (7 + 2)\sqrt{3}$ Apply the distributive property. You will find that this step is usually done mentally.

$= 9\sqrt{3}$ Simplify.

b. $4\sqrt{50x} - 6\sqrt{32x}$

$= 4\sqrt{25 \cdot 2x} - 6\sqrt{16 \cdot 2x}$ 25 is the greatest perfect square factor of 50x and 16 is the greatest perfect square factor of 32x.

$= 4 \cdot 5\sqrt{2x} - 6 \cdot 4\sqrt{2x}$ $\sqrt{25 \cdot 2x} = \sqrt{25}\sqrt{2x} = 5\sqrt{2x}$ and $\sqrt{16 \cdot 2x} = \sqrt{16}\sqrt{2x} = 4\sqrt{2x}$.

$= 20\sqrt{2x} - 24\sqrt{2x}$ Multiply: $4 \cdot 5 = 20$ and $6 \cdot 4 = 24$.

$= (20 - 24)\sqrt{2x}$ Apply the distributive property.

$= -4\sqrt{2x}$ Simplify. ● ● ●

⊘ **Check Point 5** Add or subtract as indicated:

a. $5\sqrt{27} + \sqrt{12}$ $17\sqrt{3}$ b. $6\sqrt{18x} - 4\sqrt{8x}$. $10\sqrt{2x}$

⑥ Rationalize denominators.

Rationalizing Denominators

The calculator screen in **Figure P.11** shows approximate values for $\dfrac{1}{\sqrt{3}}$ and $\dfrac{\sqrt{3}}{3}$. The two approximations are the same. This is not a coincidence:

$$\frac{1}{\sqrt{3}} = \frac{1}{\sqrt{3}} \cdot \boxed{\frac{\sqrt{3}}{\sqrt{3}}} = \frac{\sqrt{3}}{\sqrt{9}} = \frac{\sqrt{3}}{3}.$$

Any number divided by itself is 1. Multiplication by 1 does not change the value of $\dfrac{1}{\sqrt{3}}$.

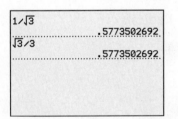

FIGURE P.11 The calculator screen shows approximate values for $\dfrac{1}{\sqrt{3}}$ and $\dfrac{\sqrt{3}}{3}$.

This process involves rewriting a radical expression as an equivalent expression in which the denominator no longer contains any radicals. The process is called **rationalizing the denominator**. If the denominator consists of the square root of a natural number that is not a perfect square, **multiply the numerator and the denominator by the smallest number that produces the square root of a perfect square in the denominator**.

EXAMPLE 6 Rationalizing Denominators

Rationalize the denominator:

a. $\dfrac{15}{\sqrt{6}}$ b. $\dfrac{12}{\sqrt{8}}$.

GREAT QUESTION!

What exactly does rationalizing a denominator do to an irrational number in the denominator?

Rationalizing a numerical denominator makes that denominator a rational number.

SOLUTION

a. If we multiply the numerator and the denominator of $\dfrac{15}{\sqrt{6}}$ by $\sqrt{6}$, the denominator becomes $\sqrt{6} \cdot \sqrt{6} = \sqrt{36} = 6$. Therefore, we multiply by 1, choosing $\dfrac{\sqrt{6}}{\sqrt{6}}$ for 1.

$$\frac{15}{\sqrt{6}} = \frac{15}{\sqrt{6}} \cdot \frac{\sqrt{6}}{\sqrt{6}} = \frac{15\sqrt{6}}{\sqrt{36}} = \frac{15\sqrt{6}}{6} = \frac{5\sqrt{6}}{2}$$

> Multiply by 1.

> Simplify: $\frac{15}{6} = \frac{15 \div 3}{6 \div 3} = \frac{5}{2}$.

b. The *smallest* number that will produce the square root of a perfect square in the denominator of $\dfrac{12}{\sqrt{8}}$ is $\sqrt{2}$, because $\sqrt{8} \cdot \sqrt{2} = \sqrt{16} = 4$. We multiply by 1, choosing $\dfrac{\sqrt{2}}{\sqrt{2}}$ for 1.

$$\frac{12}{\sqrt{8}} = \frac{12}{\sqrt{8}} \cdot \frac{\sqrt{2}}{\sqrt{2}} = \frac{12\sqrt{2}}{\sqrt{16}} = \frac{12\sqrt{2}}{4} = 3\sqrt{2}$$

...

⊘ **Check Point 6** Rationalize the denominator:

a. $\dfrac{5}{\sqrt{3}}$ $\dfrac{5\sqrt{3}}{3}$ **b.** $\dfrac{6}{\sqrt{12}}$. $\sqrt{3}$

Radical expressions that involve the sum and difference of the same two terms are called **conjugates**. Thus,

$$\sqrt{a} + \sqrt{b} \quad \text{and} \quad \sqrt{a} - \sqrt{b}$$

are conjugates. Conjugates are used to rationalize denominators because the product of such a pair contains no radicals:

$$(\sqrt{a} + \sqrt{b})(\sqrt{a} - \sqrt{b})$$

> Multiply each term of $\sqrt{a} - \sqrt{b}$ by each term of $\sqrt{a} + \sqrt{b}$.

$$= \sqrt{a}(\sqrt{a} - \sqrt{b}) + \sqrt{b}(\sqrt{a} - \sqrt{b})$$

> Distribute \sqrt{a} over $\sqrt{a} - \sqrt{b}$.

> Distribute \sqrt{b} over $\sqrt{a} - \sqrt{b}$.

$$= \sqrt{a} \cdot \sqrt{a} - \sqrt{a} \cdot \sqrt{b} + \sqrt{b} \cdot \sqrt{a} - \sqrt{b} \cdot \sqrt{b}$$

$$= (\sqrt{a})^2 - \sqrt{ab} + \sqrt{ab} - (\sqrt{b})^2$$

> $-\sqrt{ab} + \sqrt{ab} = 0$

$$= (\sqrt{a})^2 - (\sqrt{b})^2$$

$$= a - b.$$

> **Multiplying Conjugates**
>
> $$(\sqrt{a} + \sqrt{b})(\sqrt{a} - \sqrt{b}) = (\sqrt{a})^2 - (\sqrt{b})^2 = a - b$$

How can we rationalize a denominator if the denominator contains two terms with one or more square roots? **Multiply the numerator and the denominator by the conjugate of the denominator.** Here are three examples of such expressions:

- $\dfrac{7}{5 + \sqrt{3}}$

- $\dfrac{8}{3\sqrt{2} - 4}$

- $\dfrac{h}{\sqrt{x + h} - \sqrt{x}}$

> The conjugate of the denominator is $5 - \sqrt{3}$.

> The conjugate of the denominator is $3\sqrt{2} + 4$.

> The conjugate of the denominator is $\sqrt{x + h} + \sqrt{x}$.

The product of the denominator and its conjugate is found using the formula

$$\left(\sqrt{a} + \sqrt{b}\right)\left(\sqrt{a} - \sqrt{b}\right) = \left(\sqrt{a}\right)^2 - \left(\sqrt{b}\right)^2 = a - b.$$

The simplified product will not contain a radical.

EXAMPLE 7 Rationalizing a Denominator Containing Two Terms

Rationalize the denominator: $\dfrac{7}{5 + \sqrt{3}}$.

SOLUTION

The conjugate of the denominator is $5 - \sqrt{3}$. If we multiply the numerator and denominator by $5 - \sqrt{3}$, the simplified denominator will not contain a radical. Therefore, we multiply by 1, choosing $\dfrac{5 - \sqrt{3}}{5 - \sqrt{3}}$ for 1.

$$\frac{7}{5 + \sqrt{3}} = \frac{7}{5 + \sqrt{3}} \cdot \frac{5 - \sqrt{3}}{5 - \sqrt{3}} = \frac{7(5 - \sqrt{3})}{5^2 - (\sqrt{3})^2} = \frac{7(5 - \sqrt{3})}{25 - 3}$$

> **Multiply by 1.** $(\sqrt{a} + \sqrt{b})(\sqrt{a} - \sqrt{b})$
> $= (\sqrt{a})^2 - (\sqrt{b})^2$

$$= \frac{7(5 - \sqrt{3})}{22} \text{ or } \frac{35 - 7\sqrt{3}}{22}$$

> In either form of the answer, there is no radical in the denominator.

● ● ●

✓ **Check Point 7** Rationalize the denominator: $\dfrac{8}{4 + \sqrt{5}}$. $\dfrac{8(4 - \sqrt{5})}{11}$ or $\dfrac{32 - 8\sqrt{5}}{11}$

⑦ Evaluate and perform operations with higher roots.

Other Kinds of Roots

We define the **principal nth root** of a real number a, symbolized by $\sqrt[n]{a}$, as follows:

Definition of the Principal nth Root of a Real Number
$$\sqrt[n]{a} = b \text{ means that } b^n = a.$$

If n, the **index**, is even, then a is nonnegative ($a \geq 0$) and b is also nonnegative ($b \geq 0$). If n is odd, a and b can be any real numbers.

For example,

$$\sqrt[3]{64} = 4 \text{ because } 4^3 = 64 \quad \text{and} \quad \sqrt[5]{-32} = -2 \text{ because } (-2)^5 = -32.$$

The same vocabulary that we learned for square roots applies to nth roots. The symbol $\sqrt[n]{}$ is called a **radical** and the expression under the radical is called the **radicand**.

A number that is the nth power of a rational number is called a **perfect nth power**. For example, 8 is a perfect third power, or perfect cube, because $8 = 2^3$. Thus, $\sqrt[3]{8} = \sqrt[3]{2^3} = 2$. In general, one of the following rules can be used to find the nth root of a perfect nth power:

Finding nth Roots of Perfect nth Powers

If n is odd, $\sqrt[n]{a^n} = a$.
If n is even, $\sqrt[n]{a^n} = |a|$.

For example,

$$\sqrt[3]{(-2)^3} = -2 \qquad \text{and} \qquad \sqrt[4]{(-2)^4} = |-2| = 2.$$

Absolute value is not needed with odd roots, but is necessary with even roots.

The Product and Quotient Rules for Other Roots

The product and quotient rules apply to cube roots, fourth roots, and all higher roots.

The Product and Quotient Rules for nth Roots

For all real numbers a and b, where the indicated roots represent real numbers,

$$\sqrt[n]{ab} = \sqrt[n]{a} \cdot \sqrt[n]{b} \qquad \text{and} \qquad \sqrt[n]{a} \cdot \sqrt[n]{b} = \sqrt[n]{ab}$$

The nth root of a product is the product of the nth roots.

The product of two nth roots is the nth root of the product of the radicands.

$$\sqrt[n]{\frac{a}{b}} = \frac{\sqrt[n]{a}}{\sqrt[n]{b}}, b \neq 0 \qquad \text{and} \qquad \frac{\sqrt[n]{a}}{\sqrt[n]{b}} = \sqrt[n]{\frac{a}{b}}, b \neq 0.$$

The nth root of a quotient is the quotient of the nth roots.

The quotient of two nth roots is the nth root of the quotient of the radicands.

EXAMPLE 8 Simplifying, Multiplying, and Dividing Higher Roots

Simplify: **a.** $\sqrt[3]{24}$ **b.** $\sqrt[4]{8} \cdot \sqrt[4]{4}$ **c.** $\sqrt[4]{\dfrac{81}{16}}$.

SOLUTION

a. $\sqrt[3]{24} = \sqrt[3]{8 \cdot 3}$ Find the greatest perfect cube that is a factor of 24. $2^3 = 8$, so 8 is a perfect cube and is the greatest perfect cube factor of 24.

$\qquad = \sqrt[3]{8} \cdot \sqrt[3]{3}$ $\sqrt[n]{ab} = \sqrt[n]{a} \cdot \sqrt[n]{b}$

$\qquad = 2\sqrt[3]{3}$ $\sqrt[3]{8} = 2$

b. $\sqrt[4]{8} \cdot \sqrt[4]{4} = \sqrt[4]{8 \cdot 4}$ $\sqrt[n]{a} \cdot \sqrt[n]{b} = \sqrt[n]{ab}$

$\qquad = \sqrt[4]{32}$ Find the greatest perfect fourth power that is a factor of 32.

$\qquad = \sqrt[4]{16 \cdot 2}$ $2^4 = 16$, so 16 is a perfect fourth power and is the greatest perfect fourth power that is a factor of 32.

$\qquad = \sqrt[4]{16} \cdot \sqrt[4]{2}$ $\sqrt[n]{ab} = \sqrt[n]{a} \cdot \sqrt[n]{b}$

$\qquad = 2\sqrt[4]{2}$ $\sqrt[4]{16} = 2$

c. $\sqrt[4]{\dfrac{81}{16}} = \dfrac{\sqrt[4]{81}}{\sqrt[4]{16}}$ $\quad \sqrt[n]{\dfrac{a}{b}} = \dfrac{\sqrt[n]{a}}{\sqrt[n]{b}}$

$\qquad\qquad = \dfrac{3}{2}$ $\quad \sqrt[4]{81} = 3$ because $3^4 = 81$ and $\sqrt[4]{16} = 2$ because $2^4 = 16.$ • • •

✓ Check Point **8** Simplify:

a. $\sqrt[3]{40}$ $\quad 2\sqrt[3]{5}$ **b.** $\sqrt[5]{8} \cdot \sqrt[5]{8}$ $\quad 2\sqrt[5]{2}$ **c.** $\sqrt[3]{\dfrac{125}{27}}.$ $\quad \frac{5}{3}$

We have seen that adding and subtracting square roots often involves simplifying terms. The same idea applies to adding and subtracting higher roots.

EXAMPLE 9 Combining Cube Roots

Subtract: $5\sqrt[3]{16} - 11\sqrt[3]{2}.$

SOLUTION

$\qquad 5\sqrt[3]{16} - 11\sqrt[3]{2}$

$\qquad = 5\sqrt[3]{8 \cdot 2} - 11\sqrt[3]{2}$ \quad Factor 16. 8 is the greatest perfect cube factor: $2^3 = 8$ and $\sqrt[3]{8} = 2.$

$\qquad = 5 \cdot 2\sqrt[3]{2} - 11\sqrt[3]{2}$ $\quad \sqrt[3]{8 \cdot 2} = \sqrt[3]{8}\sqrt[3]{2} = 2\sqrt[3]{2}$

$\qquad = 10\sqrt[3]{2} - 11\sqrt[3]{2}$ \quad Multiply: $5 \cdot 2 = 10.$

$\qquad = (10 - 11)\sqrt[3]{2}$ \quad Apply the distributive property.

$\qquad = -1\sqrt[3]{2}$ or $-\sqrt[3]{2}$ \quad Simplify. • • •

✓ Check Point **9** Subtract: $3\sqrt[3]{81} - 4\sqrt[3]{3}.$ $\quad 5\sqrt[3]{3}$

⑧ Understand and use rational exponents.

Rational Exponents

We define rational exponents so that their properties are the same as the properties for integer exponents. For example, we know that exponents are multiplied when an exponential expression is raised to a power. For this to be true,

$$\left(7^{\frac{1}{2}}\right)^2 = 7^{\frac{1}{2} \cdot 2} = 7^1 = 7.$$

We also know that

$$\left(\sqrt{7}\right)^2 = \sqrt{7} \cdot \sqrt{7} = \sqrt{49} = 7.$$

Can you see that the square of both $7^{\frac{1}{2}}$ and $\sqrt{7}$ is 7? It is reasonable to conclude that

$$7^{\frac{1}{2}} \quad \text{means} \quad \sqrt{7}.$$

We can generalize the fact that $7^{\frac{1}{2}}$ means $\sqrt{7}$ with the following definition:

The Definition of $a^{\frac{1}{n}}$

If $\sqrt[n]{a}$ represents a real number, where $n \geq 2$ is an integer, then

$$a^{\frac{1}{n}} = \sqrt[n]{a}.$$

The denominator of the rational exponent is the radical's index.

Furthermore,

$$a^{-\frac{1}{n}} = \dfrac{1}{a^{\frac{1}{n}}} = \dfrac{1}{\sqrt[n]{a}}, \quad a \neq 0.$$

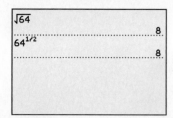
EXAMPLE 10 Using the Definition of $a^{\frac{1}{n}}$

Simplify:

a. $64^{\frac{1}{2}}$ **b.** $125^{\frac{1}{3}}$ **c.** $-16^{\frac{1}{4}}$ **d.** $(-27)^{\frac{1}{3}}$ **e.** $64^{-\frac{1}{3}}$.

SOLUTION

a. $64^{\frac{1}{2}} = \sqrt{64} = 8$

b. $125^{\frac{1}{3}} = \sqrt[3]{125} = 5$

> The denominator is the index.

c. $-16^{\frac{1}{4}} = -(\sqrt[4]{16}) = -2$

> The base is 16 and the negative sign is not affected by the exponent.

d. $(-27)^{\frac{1}{3}} = \sqrt[3]{-27} = -3$

> Parentheses show that the base is −27 and that the negative sign is affected by the exponent.

e. $64^{-\frac{1}{3}} = \dfrac{1}{64^{\frac{1}{3}}} = \dfrac{1}{\sqrt[3]{64}} = \dfrac{1}{4}$

• • •

⊘ **Check Point 10** Simplify:

a. $25^{\frac{1}{2}}$ 5 **b.** $8^{\frac{1}{3}}$ 2 **c.** $-81^{\frac{1}{4}}$ −3 **d.** $(-8)^{\frac{1}{3}}$ −2 **e.** $27^{-\frac{1}{3}}$. $\frac{1}{3}$

In Example 10 and Check Point 10, each rational exponent had a numerator of 1. If the numerator is some other integer, we still want to multiply exponents when raising a power to a power. For this reason,

$$a^{\frac{2}{3}} = \left(a^{\frac{1}{3}}\right)^2 \quad \text{and} \quad a^{\frac{2}{3}} = \left(a^2\right)^{\frac{1}{3}}.$$

> This means $(\sqrt[3]{a})^2$. This means $\sqrt[3]{a^2}$.

Thus,

$$a^{\frac{2}{3}} = \left(\sqrt[3]{a}\right)^2 = \sqrt[3]{a^2}.$$

Do you see that the denominator, 3, of the rational exponent is the same as the index of the radical? The numerator, 2, of the rational exponent serves as an exponent in each of the two radical forms. We generalize these ideas with the following definition:

The Definition of $a^{\frac{m}{n}}$

If $\sqrt[n]{a}$ represents a real number and $\dfrac{m}{n}$ is a positive rational number, $n \geq 2$, then

$$a^{\frac{m}{n}} = \left(\sqrt[n]{a}\right)^m.$$

Also,

$$a^{\frac{m}{n}} = \sqrt[n]{a^m}.$$

Furthermore, if $a^{-\frac{m}{n}}$ is a nonzero real number, then

$$a^{-\frac{m}{n}} = \dfrac{1}{a^{\frac{m}{n}}}.$$

The first form of the definition of $a^{\frac{m}{n}}$, shown again below, involves taking the root first. This form is often preferable because smaller numbers are involved. Notice that the rational exponent consists of two parts, indicated by the following voice balloons:

> The numerator is the exponent.

$$a^{\frac{m}{n}} = \left(\sqrt[n]{a}\right)^m.$$

> The denominator is the radical's index.

EXAMPLE 11 Using the Definition of $a^{\frac{m}{n}}$

Simplify:

a. $27^{\frac{2}{3}}$ **b.** $9^{\frac{3}{2}}$ **c.** $81^{-\frac{3}{4}}$.

SOLUTION

a. $27^{\frac{2}{3}} = \left(\sqrt[3]{27}\right)^2 = 3^2 = 9$

b. $9^{\frac{3}{2}} = \left(\sqrt{9}\right)^3 = 3^3 = 27$

c. $81^{-\frac{3}{4}} = \dfrac{1}{81^{\frac{3}{4}}} = \dfrac{1}{\left(\sqrt[4]{81}\right)^3} = \dfrac{1}{3^3} = \dfrac{1}{27}$

● ● ●

✓ Check Point **11** Simplify:

a. $27^{\frac{4}{3}}$ 81 **b.** $4^{\frac{3}{2}}$ 8 **c.** $32^{-\frac{2}{5}}$. $\frac{1}{4}$

Properties of exponents can be applied to expressions containing rational exponents.

EXAMPLE 12 Simplifying Expressions with Rational Exponents

Simplify using properties of exponents:

a. $\left(5x^{\frac{1}{2}}\right)\left(7x^{\frac{3}{4}}\right)$ **b.** $\dfrac{32x^{\frac{5}{3}}}{16x^{\frac{3}{4}}}$.

SOLUTION

a. $\left(5x^{\frac{1}{2}}\right)\left(7x^{\frac{3}{4}}\right) = 5 \cdot 7x^{\frac{1}{2}} \cdot x^{\frac{3}{4}}$ Group numerical factors and group variable factors with the same base.

$= 35x^{\frac{1}{2}+\frac{3}{4}}$ When multiplying expressions with the same base, add the exponents.

$= 35x^{\frac{5}{4}}$ $\frac{1}{2} + \frac{3}{4} = \frac{2}{4} + \frac{3}{4} = \frac{5}{4}$

b. $\dfrac{32x^{\frac{5}{3}}}{16x^{\frac{3}{4}}} = \left(\dfrac{32}{16}\right)\left(\dfrac{x^{\frac{5}{3}}}{x^{\frac{3}{4}}}\right)$ Group numerical factors and group variable factors with the same base.

$= 2x^{\frac{5}{3}-\frac{3}{4}}$ When dividing expressions with the same base, subtract the exponents.

$= 2x^{\frac{11}{12}}$ $\frac{5}{3} - \frac{3}{4} = \frac{20}{12} - \frac{9}{12} = \frac{11}{12}$

● ● ●

✓ Check Point **12** Simplify using properties of exponents:

a. $\left(2x^{\frac{4}{3}}\right)\left(5x^{\frac{8}{3}}\right)$ $10x^4$ **b.** $\dfrac{20x^4}{5x^{\frac{3}{2}}}$. $4x^{\frac{5}{2}}$

Rational exponents are sometimes useful for simplifying radicals by reducing the index.

EXAMPLE 13 Reducing the Index of a Radical

Simplify: $\sqrt[9]{x^3}$.

SOLUTION

$$\sqrt[9]{x^3} = x^{\frac{3}{9}} = x^{\frac{1}{3}} = \sqrt[3]{x}$$

$\bullet\bullet\bullet$

☑ **Check Point 13** Simplify: $\sqrt[6]{x^3}$. \sqrt{x}

Blitzer Bonus ‖ A Radical Idea: Time Is Relative

The Persistence of Memory (1931), Salvador Dali:. © 2011 MOMA/ARS.

What does travel in space have to do with radicals? Imagine that in the future we will be able to travel at velocities approaching the speed of light (approximately 186,000 miles per second). According to Einstein's theory of special relativity, time would pass more quickly on Earth than it would in the moving spaceship. The special-relativity equation

$$R_a = R_f\sqrt{1 - \left(\frac{v}{c}\right)^2}$$

gives the aging rate of an astronaut, R_a, relative to the aging rate of a friend, R_f, on Earth. In this formula, v is the astronaut's speed and c is the speed of light. As the astronaut's speed approaches the speed of light, we can substitute c for v.

$$R_a = R_f\sqrt{1 - \left(\frac{v}{c}\right)^2} \qquad \text{Einstein's equation gives the aging rate of an astronaut, } R_a, \text{ relative to the aging rate of a friend, } R_f, \text{ on Earth.}$$

$$R_a = R_f\sqrt{1 - \left(\frac{c}{c}\right)^2} \qquad \text{The velocity, } v, \text{ is approaching the speed of light, } c, \text{ so let } v = c.$$

$$= R_f\sqrt{1 - 1} \qquad \left(\frac{c}{c}\right)^2 = 1^2 = 1 \cdot 1 = 1$$

$$= R_f\sqrt{0} \qquad \text{Simplify the radicand: } 1 - 1 = 0.$$

$$= R_f \cdot 0 \qquad \sqrt{0} = 0$$

$$= 0 \qquad \text{Multiply: } R_f \cdot 0 = 0.$$

Close to the speed of light, the astronaut's aging rate, R_a, relative to a friend, R_f, on Earth is nearly 0. What does this mean? As we age here on Earth, the space traveler would barely get older. The space traveler would return to an unknown futuristic world in which friends and loved ones would be long gone.

CONCEPT AND VOCABULARY CHECK

Fill in each blank so that the resulting statement is true.

1. The symbol $\sqrt{}$ is used to denote the nonnegative, or __principal__, square root of a number.

2. $\sqrt{64} = 8$ because $\underline{8^2} = 64$.

3. $\sqrt{a^2} = \underline{|a|}$

4. The product rule for square roots states that if a and b are nonnegative, then $\sqrt{ab} = \underline{\sqrt{a}\cdot\sqrt{b}}$.

5. The quotient rule for square roots states that if a and b are nonnegative and $b \neq 0$, then $\sqrt{\dfrac{a}{b}} = \underline{\dfrac{\sqrt{a}}{\sqrt{b}}}$.

6. $8\sqrt{3} + 10\sqrt{3} = \underline{18\sqrt{3}}$

7. $\sqrt{3} + \sqrt{75} = \sqrt{3} + \sqrt{25\cdot 3} = \sqrt{3} + \underline{5}\,\sqrt{3} = \underline{6\sqrt{3}}$

8. The conjugate of $7 + \sqrt{3}$ is $\underline{7 - \sqrt{3}}$.

9. We rationalize the denominator of $\dfrac{5}{\sqrt{10} - \sqrt{2}}$ by multiplying the numerator and denominator by $\underline{\sqrt{10} + \sqrt{2}}$.

10. In the expression $\sqrt[3]{64}$, the number 3 is called the __index__ and the number 64 is called the __radicand__.

11. $\sqrt[5]{-32} = -2$ because $\underline{(-2)^5} = -32$.

12. If n is odd, $\sqrt[n]{a^n} = \underline{a}$.
 If n is even, $\sqrt[n]{a^n} = \underline{|a|}$.

13. $a^{\frac{1}{n}} = \underline{\sqrt[n]{a}}$

14. $16^{\frac{3}{4}} = \left(\sqrt[4]{16}\right)^3 = (\underline{2})^3 = \underline{8}$

EXERCISE SET P.3

Practice Exercises

Evaluate each expression in Exercises 1–12, or indicate that the root is not a real number.

1. $\sqrt{36}$ 6
2. $\sqrt{25}$ 5
3. $-\sqrt{36}$ −6
4. $-\sqrt{25}$ −5
5. $\sqrt{-36}$ not real
6. $\sqrt{-25}$ not real
7. $\sqrt{25 - 16}$ 3
8. $\sqrt{144 + 25}$ 13
9. $\sqrt{25} - \sqrt{16}$ 1
10. $\sqrt{144} + \sqrt{25}$ 17
11. $\sqrt{(-13)^2}$ 13
12. $\sqrt{(-17)^2}$ 17

Use the product rule to simplify the expressions in Exercises 13–22. In Exercises 17–22, assume that variables represent nonnegative real numbers.

13. $\sqrt{50}$ $5\sqrt{2}$
14. $\sqrt{27}$ $3\sqrt{3}$
15. $\sqrt{45x^2}$ $3|x|\sqrt{5}$
16. $\sqrt{125x^2}$ $5|x|\sqrt{5}$
17. $\sqrt{2x}\cdot\sqrt{6x}$ $2x\sqrt{3}$
18. $\sqrt{10x}\cdot\sqrt{8x}$ $4x\sqrt{5}$
19. $\sqrt{x^3}$ $x\sqrt{x}$
20. $\sqrt{y^3}$ $y\sqrt{y}$
21. $\sqrt{2x^2}\cdot\sqrt{6x}$ $2x\sqrt{3x}$
22. $\sqrt{6x}\cdot\sqrt{3x^2}$ $3x\sqrt{2x}$

Use the quotient rule to simplify the expressions in Exercises 23–32. Assume that $x > 0$.

23. $\sqrt{\dfrac{1}{81}}$ $\frac{1}{9}$
24. $\sqrt{\dfrac{1}{49}}$ $\frac{1}{7}$
25. $\sqrt{\dfrac{49}{16}}$ $\frac{7}{4}$
26. $\sqrt{\dfrac{121}{9}}$ $\frac{11}{3}$
27. $\dfrac{\sqrt{48x^3}}{\sqrt{3x}}$ $4x$
28. $\dfrac{\sqrt{72x^3}}{\sqrt{8x}}$ $3x$
29. $\dfrac{\sqrt{150x^4}}{\sqrt{3x}}$ $5x\sqrt{2x}$
30. $\dfrac{\sqrt{24x^4}}{\sqrt{3x}}$ $2x\sqrt{2x}$
31. $\dfrac{\sqrt{200x^3}}{\sqrt{10x^{-1}}}$ $2x^2\sqrt{5}$
32. $\dfrac{\sqrt{500x^3}}{\sqrt{10x^{-1}}}$ $5x^2\sqrt{2}$

In Exercises 33–44, add or subtract terms whenever possible.

33. $7\sqrt{3} + 6\sqrt{3}$ $13\sqrt{3}$
34. $8\sqrt{5} + 11\sqrt{5}$ $19\sqrt{5}$
35. $6\sqrt{17x} - 8\sqrt{17x}$ $-2\sqrt{17x}$
36. $4\sqrt{13x} - 6\sqrt{13x}$ $-2\sqrt{13x}$

37. $\sqrt{8} + 3\sqrt{2}$ $5\sqrt{2}$
38. $\sqrt{20} + 6\sqrt{5}$ $8\sqrt{5}$
39. $\sqrt{50x} - \sqrt{8x}$ $3\sqrt{2x}$
40. $\sqrt{63x} - \sqrt{28x}$ $\sqrt{7x}$
41. $3\sqrt{18} + 5\sqrt{50}$ $34\sqrt{2}$
42. $4\sqrt{12} - 2\sqrt{75}$ $-2\sqrt{3}$
43. $3\sqrt{8} - \sqrt{32} + 3\sqrt{72} - \sqrt{75}$ $20\sqrt{2} - 5\sqrt{3}$
44. $3\sqrt{54} - 2\sqrt{24} - \sqrt{96} + 4\sqrt{63}$ $\sqrt{6} + 12\sqrt{7}$

In Exercises 45–54, rationalize the denominator.

45. $\dfrac{1}{\sqrt{7}}$ $\frac{\sqrt{7}}{7}$
46. $\dfrac{2}{\sqrt{10}}$ $\frac{\sqrt{10}}{5}$
47. $\dfrac{\sqrt{2}}{\sqrt{5}}$ $\frac{\sqrt{10}}{5}$
48. $\dfrac{\sqrt{7}}{\sqrt{3}}$ $\frac{\sqrt{21}}{3}$
49. $\dfrac{13}{3 + \sqrt{11}}$ $\frac{13(3 - \sqrt{11})}{-2}$
50. $\dfrac{3}{3 + \sqrt{7}}$ $\frac{3(3 - \sqrt{7})}{2}$
51. $\dfrac{7}{\sqrt{5} - 2}$ $7(\sqrt{5} + 2)$
52. $\dfrac{5}{\sqrt{3} - 1}$ $\frac{5(\sqrt{3} + 1)}{2}$
53. $\dfrac{6}{\sqrt{5} + \sqrt{3}}$ $3(\sqrt{5} - \sqrt{3})$
54. $\dfrac{11}{\sqrt{7} - \sqrt{3}}$ $\frac{11(\sqrt{7} + \sqrt{3})}{4}$

Evaluate each expression in Exercises 55–66, or indicate that the root is not a real number.

55. $\sqrt[3]{125}$ 5
56. $\sqrt[3]{8}$ 2
57. $\sqrt[3]{-8}$ −2
58. $\sqrt[3]{-125}$ −5
59. $\sqrt[4]{-16}$ not real
60. $\sqrt[4]{-81}$ not real
61. $\sqrt[4]{(-3)^4}$ 3
62. $\sqrt[4]{(-2)^4}$ 2
63. $\sqrt[5]{(-3)^5}$ −3
64. $\sqrt[5]{(-2)^5}$ −2
65. $\sqrt[5]{-\frac{1}{32}}$ $-\frac{1}{2}$
66. $\sqrt[6]{\frac{1}{64}}$ $\frac{1}{2}$

Simplify the radical expressions in Exercises 67–74 if possible.

67. $\sqrt[3]{32}$ $2\sqrt[3]{4}$
68. $\sqrt[3]{150}$ $\sqrt[3]{150}$
69. $\sqrt[3]{x^4}$ $x\sqrt[3]{x}$
70. $\sqrt[3]{x^5}$ $x\sqrt[3]{x^2}$
71. $\sqrt[3]{9}\cdot\sqrt[3]{6}$ $3\sqrt[3]{2}$
72. $\sqrt[3]{12}\cdot\sqrt[3]{4}$ $2\sqrt[3]{6}$
73. $\dfrac{\sqrt[5]{64x^6}}{\sqrt[5]{2x}}$ $2x$
74. $\dfrac{\sqrt[4]{162x^5}}{\sqrt[4]{2x}}$ $3x$

In Exercises 75–82, add or subtract terms whenever possible.

75. $4\sqrt[5]{2} + 3\sqrt[5]{2}$ $\ 7\sqrt[5]{2}$

76. $6\sqrt[5]{3} + 2\sqrt[5]{3}$ $\ 8\sqrt[5]{3}$

77. $5\sqrt[3]{16} + \sqrt[3]{54}$ $\ 13\sqrt[3]{2}$

78. $3\sqrt[3]{24} + \sqrt[3]{81}$ $\ 9\sqrt[3]{3}$

79. $\sqrt[3]{54xy^3} - y\sqrt[3]{128x}$ $\ -y\sqrt[3]{2x}$

80. $\sqrt[3]{24xy^3} - y\sqrt[3]{81x}$ $\ -y\sqrt[3]{3x}$

81. $\sqrt{2} + \sqrt[3]{8}$ $\ \sqrt{2} + 2$

82. $\sqrt{3} + \sqrt[3]{15}$ $\ \sqrt{3} + \sqrt[3]{15}$

In Exercises 83–90, evaluate each expression without using a calculator.

83. $36^{\frac{1}{2}}$ $\ 6$

84. $121^{\frac{1}{2}}$ $\ 11$

85. $8^{\frac{1}{3}}$ $\ 2$

86. $27^{\frac{1}{3}}$ $\ 3$

87. $125^{\frac{2}{3}}$ $\ 25$

88. $8^{\frac{2}{3}}$ $\ 4$

89. $32^{-\frac{4}{5}}$ $\ \frac{1}{16}$

90. $16^{-\frac{5}{2}}$ $\ \frac{1}{1024}$

In Exercises 91–100, simplify using properties of exponents.

91. $\left(7x^{\frac{1}{3}}\right)\left(2x^{\frac{1}{4}}\right)$ $\ 14x^{7/12}$

92. $\left(3x^{\frac{2}{3}}\right)\left(4x^{\frac{3}{4}}\right)$ $\ 12x^{17/12}$

93. $\dfrac{20x^{\frac{1}{2}}}{5x^{\frac{1}{4}}}$ $\ 4x^{1/4}$

94. $\dfrac{72x^{\frac{3}{4}}}{9x^{\frac{1}{3}}}$ $\ 8x^{5/12}$

95. $\left(x^{\frac{2}{3}}\right)^3$ $\ x^2$

96. $\left(x^{\frac{4}{5}}\right)^5$ $\ x^4$

97. $(25x^4y^6)^{\frac{1}{2}}$ $\ 5x^2|y|^3$

98. $(125x^9y^6)^{\frac{1}{3}}$ $\ 5x^3y^2$

99. $\dfrac{\left(3y^{\frac{1}{4}}\right)^3}{y^{\frac{1}{12}}}$ $\ 27y^{2/3}$

100. $\dfrac{\left(2y^{\frac{1}{5}}\right)^4}{y^{\frac{3}{10}}}$ $\ 16y^{1/2}$

In Exercises 101–108, simplify by reducing the index of the radical.

101. $\sqrt[4]{5^2}$ $\ \sqrt{5}$

102. $\sqrt[4]{7^2}$ $\ \sqrt{7}$

103. $\sqrt[3]{x^6}$ $\ x^2$

104. $\sqrt[4]{x^{12}}$ $\ |x^3|$

105. $\sqrt[6]{x^4}$ $\ \sqrt[3]{x^2}$

106. $\sqrt[9]{x^6}$ $\ \sqrt[3]{x^2}$

107. $\sqrt[9]{x^6y^3}$ $\ \sqrt[3]{x^2y}$

108. $\sqrt[12]{x^4y^8}$ $\ \sqrt[3]{x|y|^2}$

Practice Plus

In Exercises 109–110, evaluate each expression.

109. $\sqrt[3]{\sqrt[4]{16} + \sqrt{625}}$ $\ 3$

110. $\sqrt[3]{\sqrt{\sqrt{169} + \sqrt{9}} + \sqrt{\sqrt[3]{1000} + \sqrt[3]{216}}}$ $\ 2$

In Exercises 111–114, simplify each expression. Assume that all variables represent positive numbers.

111. $(49x^{-2}y^4)^{-\frac{1}{2}}\left(xy^{\frac{1}{2}}\right)$ $\ \dfrac{x^2}{7y^{3/2}}$

112. $(8x^{-6}y^3)^{\frac{1}{3}}\left(x^{\frac{5}{6}}y^{-\frac{1}{3}}\right)^6$ $\ \dfrac{2x^3}{y}$

113. $\left(\dfrac{x^{-\frac{5}{4}}y^{\frac{1}{3}}}{x^{-\frac{3}{4}}}\right)^{-6}$ $\ \dfrac{x^3}{y^2}$

114. $\left(\dfrac{x^{\frac{1}{2}}y^{-\frac{7}{4}}}{y^{-\frac{5}{4}}}\right)^{-4}$ $\ \dfrac{y^2}{x^2}$

Application Exercises

115. The popular comic strip *FoxTrot* follows the off-the-wall lives of the Fox family. Youngest son Jason is forever obsessed by his love of math. In the math-themed strip shown at the top of the next column, Jason shares his opinion in a coded message about the mathematical abilities of his sister Paige.

Solve problems A through Z in the left panel. Then decode Jason Fox's message involving his opinion about the mathematical abilities of his sister Paige shown on the first line. Paige Fox is bad at math.

Hints: Here is the solution for problem C and partial solutions for problems Q and U.

$$C = \sin \frac{\pi}{2} = \sin 90° = 1$$

$$Q = \int_0^2 9x^2\,dx = 3x^3 \Big|_0^2 = 3 \cdot 2^3 - 3 \cdot 0^3 = \underline{}$$

$$U = -3\cos \pi = -3\cos 180° = -3(-1) = \underline{}$$

Note: The comic strip *FoxTrot* is now printed in more than one thousand newspapers. What made cartoonist Bill Amend, a college physics major, put math in the comic? "I always try to use math in the strip to make the joke accessible to anyone," he said. "But if you understand math, hopefully you'll like it that much more!" We highly recommend the math humor in Amend's *FoxTrot* collection *Math, Science, and Unix Underpants* (Andrews McMeel Publishing, 2009).

116. America is getting older. The graph shows the projected elderly U.S. population for ages 65–84 and for ages 85 and older.

Projected Elderly United States Population

Source: U.S. Census Bureau

The formula $E = 5.8\sqrt{x} + 56.4$ models the projected number of elderly Americans ages 65–84, E, in millions, x years after 2020.

a. Use the formula to find the projected increase in the number of Americans ages 65–84, in millions, from 2030 to 2060. Express this difference in simplified radical form. $5.8\sqrt{10}$

b. Use a calculator and write your answer in part (a) to the nearest tenth. Does this rounded decimal overestimate or underestimate the difference in the projected data shown by the bar graph on the previous page? By how much? 18.3; underestimates by 5.8 million

117. The early Greeks believed that the most pleasing of all rectangles were **golden rectangles**, whose ratio of width to height is

$$\frac{w}{h} = \frac{2}{\sqrt{5} - 1}.$$

The Parthenon at Athens fits into a golden rectangle once the triangular pediment is reconstructed.

Rationalize the denominator of the golden ratio. Then use a calculator and find the ratio of width to height, correct to the nearest hundredth, in golden rectangles. *

118. Use Einstein's special-relativity equation

$$R_a = R_f\sqrt{1 - \left(\frac{v}{c}\right)^2},$$

described in the Blitzer Bonus on page 47, to solve this exercise. You are moving at 90% of the speed of light. Substitute $0.9c$ for v, your velocity, in the equation. What is your aging rate, correct to two decimal places, relative to a friend on Earth? If you are gone for 44 weeks, approximately how many weeks have passed for your friend? $0.44R_f$; 100 weeks

The perimeter, P, of a rectangle with length l and width w is given by the formula $P = 2l + 2w$. The area, A, is given by the formula $A = lw$. In Exercises 119–120, use these formulas to find the perimeter and area of each rectangle. Express answers in simplified radical form. Remember that perimeter is measured in linear units, such as feet or meters, and area is measured in square units, such as square feet, ft^2, or square meters, m^2.

119. $P = 18\sqrt{5}$ ft; $A = 100$ ft² **120.** $P = 24\sqrt{5}$ ft; $A = 160$ ft²

Explaining the Concepts 121–128. Answers will vary.

121. Explain how to simplify $\sqrt{10} \cdot \sqrt{5}$.

122. Explain how to add $\sqrt{3} + \sqrt{12}$.

*See Answers to Selected Exercises.

123. Describe what it means to rationalize a denominator. Use both $\dfrac{1}{\sqrt{5}}$ and $\dfrac{1}{5 + \sqrt{5}}$ in your explanation.

124. What difference is there in simplifying $\sqrt[3]{(-5)^3}$ and $\sqrt[4]{(-5)^4}$?

125. What does $a^{\frac{m}{n}}$ mean?

126. Describe the kinds of numbers that have rational fifth roots.

127. Why must a and b represent nonnegative numbers when we write $\sqrt{a} \cdot \sqrt{b} = \sqrt{ab}$? Is it necessary to use this restriction in the case of $\sqrt[3]{a} \cdot \sqrt[3]{b} = \sqrt[3]{ab}$? Explain.

128. Read the Blitzer Bonus on page 47. The future is now: You have the opportunity to explore the cosmos in a starship traveling near the speed of light. The experience will enable you to understand the mysteries of the universe in deeply personal ways, transporting you to unimagined levels of knowing and being. The downside: You return from your two-year journey to a futuristic world in which friends and loved ones are long gone. Do you explore space or stay here on Earth? What are the reasons for your choice?

Critical Thinking Exercises

Make Sense? *In Exercises 129–132, determine whether each statement makes sense or does not make sense, and explain your reasoning.*

129. The joke in this Peanuts cartoon would be more effective if Woodstock had rationalized the denominator correctly in the last frame. does not make sense

Peanuts © 1978 Peanuts Worldwide LLC. Used by permission of Universal Uclick. All rights reserved.

130. Using my calculator, I determined that $6^7 = 279{,}936$, so 6 must be a seventh root of 279,936. makes sense

131. I simplified the terms of $2\sqrt{20} + 4\sqrt{75}$, and then I was able to add the like radicals. does not make sense

132. When I use the definition for $a^{\frac{m}{n}}$, I usually prefer to first raise a to the m power because smaller numbers are involved. does not make sense

In Exercises 133–136, determine whether each statement is true or false. If the statement is false, make the necessary change(s) to produce a true statement. Changes to false statements will vary.

133. $7^{\frac{1}{2}} \cdot 7^{\frac{1}{2}} = 49$ false **134.** $8^{-\frac{1}{3}} = -2$ false

135. The cube root of -8 is not a real number. false

136. $\dfrac{\sqrt{20}}{8} = \dfrac{\sqrt{10}}{4}$ false

In Exercises 137–138, fill in each box to make the statement true.

137. $\left(5 + \sqrt{\Box}\right)\left(5 - \sqrt{\Box}\right) = 22$ 3 in each box

138. $\sqrt{\Box}\, x = 5x^7$ first box: 25; second box: 14

139. Find the exact value of $\sqrt{13 + \sqrt{2} + \dfrac{7}{3 + \sqrt{2}}}$ without the use of a calculator. 4

140. Place the correct symbol, $>$ or $<$, in the shaded area between the given numbers. *Do not use a calculator.* Then check your result with a calculator.

a. $3^{\frac{1}{2}}$ ▨ $3^{\frac{1}{3}}$ $>$

b. $\sqrt{7} + \sqrt{18}$ ▨ $\sqrt{7 + 18}$ $>$

141. a. A mathematics professor recently purchased a birthday cake for her son with the inscription

$$\text{Happy}\left(2^{\frac{5}{2}} \cdot 2^{\frac{3}{4}} \div 2^{\frac{1}{4}}\right)\text{th Birthday.}$$

How old is the son? 8

b. The birthday boy, excited by the inscription on the cake, tried to wolf down the whole thing. Professor Mom, concerned about the possible metamorphosis of her son into a blimp, exclaimed, "Hold on! It is your birthday, so why not take $\dfrac{8^{-\frac{4}{3}} + 2^{-2}}{16^{-\frac{3}{4}} + 2^{-1}}$ of the cake? I'll eat half of what's left over." How much of the cake did the professor eat? $\frac{1}{4}$

Preview Exercises

Exercises 142–144 will help you prepare for the material covered in the next section.

142. Multiply: $(2x^3y^2)(5x^4y^7)$. $10x^7y^9$

143. Use the distributive property to multiply:

$$2x^4(8x^4 + 3x). \quad 16x^8 + 6x^5$$

144. Simplify and express the answer in descending powers of x:

$$2x(x^2 + 4x + 5) + 3(x^2 + 4x + 5).$$

$$2x^3 + 11x^2 + 22x + 15$$

Section P.4 Polynomials

What am I supposed to learn?

After studying this section, you should be able to:

1. Understand the vocabulary of polynomials.
2. Add and subtract polynomials.
3. Multiply polynomials.
4. Use FOIL in polynomial multiplication.
5. Use special products in polynomial multiplication.
6. Perform operations with polynomials in several variables.

Can that be Axl, your author's yellow lab, sharing a special moment with a baby chick? And if it is (it is), what possible relevance can this have to polynomials? An answer is promised before you reach the Exercise Set. For now, we open the section by defining and describing polynomials.

How We Define Polynomials

More than 2 million people have tested their racial prejudice using an online version of the Implicit Association Test. Most groups' average scores fall between "slight" and "moderate" bias, but the differences among groups, by age and by political identification, are intriguing.

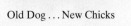

Old Dog . . . New Chicks

In this section's Exercise Set (Exercises 91 and 92), you will be working with models that measure bias:

$$S = 0.3x^3 - 2.8x^2 + 6.7x + 30$$
$$S = -0.03x^3 + 0.2x^2 + 2.3x + 24.$$

In each model, S represents the score on the Implicit Association Test. (Higher scores indicate stronger bias.) In the first model (see Exercise 91), x represents age group. In the second model (see Exercise 92), x represents political identification.

The algebraic expressions that appear on the right sides of the models are examples of *polynomials*. A **polynomial** is a single term or the sum of two or more terms containing variables with whole-number exponents. These particular polynomials each contain four terms. Equations containing polynomials are used in such diverse areas as science, business, medicine, psychology, and sociology. In this section, we review basic ideas about polynomials and their operations.

① Understand the vocabulary of polynomials.

How We Describe Polynomials

Consider the polynomial

$$7x^3 - 9x^2 + 13x - 6.$$

We can express $7x^3 - 9x^2 + 13x - 6$ as

$$7x^3 + (-9x^2) + 13x + (-6).$$

The polynomial contains four terms. It is customary to write the terms in the order of descending powers of the variable. This is the **standard form** of a polynomial.

Some polynomials contain only one variable. Each term of such a polynomial in x is of the form ax^n. If $a \neq 0$, the **degree** of ax^n is n. For example, the degree of the term $7x^3$ is 3.

The Degree of ax^n

If $a \neq 0$, the degree of ax^n is n. The degree of a nonzero constant is 0. The constant 0 has no defined degree.

Here is an example of a polynomial and the degree of each of its four terms:

$$6x^4 - 3x^3 - 2x - 5.$$

| degree 4 | degree 3 | degree 1 | degree of nonzero constant: 0 |

Notice that the exponent on x for the term $2x$, meaning $2x^1$, is understood to be 1. For this reason, the degree of $2x$ is 1. You can think of -5 as $-5x^0$; thus, its degree is 0.

A polynomial is simplified when it contains no grouping symbols and no like terms. A simplified polynomial that has exactly one term is called a **monomial**. A **binomial** is a simplified polynomial that has two terms. A **trinomial** is a simplified polynomial with three terms. Simplified polynomials with four or more terms have no special names.

The **degree of a polynomial** is the greatest degree of all the terms of the polynomial. For example, $4x^2 + 3x$ is a binomial of degree 2 because the degree of the first term is 2, and the degree of the other term is less than 2. Also, $7x^5 - 2x^2 + 4$ is a trinomial of degree 5 because the degree of the first term is 5, and the degrees of the other terms are less than 5.

Up to now, we have used x to represent the variable in a polynomial. However, any letter can be used. For example,

- $7x^5 - 3x^3 + 8$ is a polynomial (in x) of degree 5. Because there are three terms, the polynomial is a trinomial.
- $6y^3 + 4y^2 - y + 3$ is a polynomial (in y) of degree 3. Because there are four terms, the polynomial has no special name.
- $z^7 + \sqrt{2}$ is a polynomial (in z) of degree 7. Because there are two terms, the polynomial is a binomial.

We can tie together the threads of our discussion with the formal definition of a polynomial in one variable. In this definition, the coefficients of the terms are represented by a_n (read "a sub n"), a_{n-1} (read "a sub n minus 1"), a_{n-2}, and so on. The small letters to the lower right of each a are called **subscripts** and are *not exponents*. Subscripts are used to distinguish one constant from another when a large and undetermined number of such constants are needed.

GREAT QUESTION!

Why doesn't the constant 0 have a degree?

We can express 0 in many ways, including $0x$, $0x^2$, and $0x^3$. It is impossible to assign a single exponent on the variable. This is why 0 has no defined degree.

Definition of a Polynomial in x

A **polynomial in x** is an algebraic expression of the form

$$a_n x^n + a_{n-1} x^{n-1} + a_{n-2} x^{n-2} + \cdots + a_1 x + a_0,$$

where $a_n, a_{n-1}, a_{n-2}, \ldots, a_1$, and a_0 are real numbers, $a_n \neq 0$, and n is a nonnegative integer. The polynomial is of **degree n**, a_n is the **leading coefficient**, and a_0 is the **constant term**.

② Add and subtract polynomials.

Can I use a vertical format to add and subtract polynomials?

Yes. Arrange like terms in columns and combine vertically:

$$7x^3 - 8x^2 + 9x - 6$$
$$\underline{-2x^3 + 6x^2 + 3x - 9}$$
$$5x^3 - 2x^2 + 12x - 15$$

The like terms can be combined by adding their coefficients and keeping the same variable factor.

③ Multiply polynomials.

Because monomials with the same base and different exponents can be multiplied, can they also be added?

No. Don't confuse adding and multiplying monomials.

Addition:

$$5x^4 + 6x^4 = 11x^4$$

Multiplication:

$$(5x^4)(6x^4) = (5 \cdot 6)(x^4 \cdot x^4)$$
$$= 30x^{4+4}$$
$$= 30x^8$$

Only like terms can be added or subtracted, but unlike terms may be multiplied.

Addition:

$5x^4 + 3x^2$ cannot be simplified.

Multiplication:

$$(5x^4)(3x^2) = (5 \cdot 3)(x^4 \cdot x^2)$$
$$= 15x^{4+2}$$
$$= 15x^6$$

Adding and Subtracting Polynomials

Polynomials are added and subtracted by combining like terms. For example, we can combine the monomials $-9x^3$ and $13x^3$ using addition as follows:

$$-9x^3 + 13x^3 = (-9 + 13)x^3 = 4x^3.$$

These like terms both contain x to the third power.	Add coefficients and keep the same variable factor, x^3.

EXAMPLE 1 Adding and Subtracting Polynomials

Perform the indicated operations and simplify:

a. $(-9x^3 + 7x^2 - 5x + 3) + (13x^3 + 2x^2 - 8x - 6)$

b. $(7x^3 - 8x^2 + 9x - 6) - (2x^3 - 6x^2 - 3x + 9).$

SOLUTION

a. $(-9x^3 + 7x^2 - 5x + 3) + (13x^3 + 2x^2 - 8x - 6)$

$= (-9x^3 + 13x^3) + (7x^2 + 2x^2) + (-5x - 8x) + (3 - 6)$ Group like terms.

$= 4x^3 + 9x^2 + (-13x) + (-3)$ Combine like terms.

$= 4x^3 + 9x^2 - 13x - 3$ Simplify.

b. $(7x^3 - 8x^2 + 9x - 6) - (2x^3 - 6x^2 - 3x + 9)$ Rewrite subtraction as addition of the additive inverse.

Change the sign of each coefficient.

$= (7x^3 - 8x^2 + 9x - 6) + (-2x^3 + 6x^2 + 3x - 9)$

$= (7x^3 - 2x^3) + (-8x^2 + 6x^2) + (9x + 3x) + (-6 - 9)$ Group like terms.

$= 5x^3 + (-2x^2) + 12x + (-15)$ Combine like terms.

$= 5x^3 - 2x^2 + 12x - 15$ Simplify. • • •

 Check Point **1** Perform the indicated operations and simplify:

a. $(-17x^3 + 4x^2 - 11x - 5) + (16x^3 - 3x^2 + 3x - 15)$ $-x^3 + x^2 - 8x - 20$

b. $(13x^3 - 9x^2 - 7x + 1) - (-7x^3 + 2x^2 - 5x + 9).$ $20x^3 - 11x^2 - 2x - 8$

Multiplying Polynomials

The product of two monomials is obtained by using properties of exponents. For example,

$$(-8x^6)(5x^3) = -8 \cdot 5x^{6+3} = -40x^9.$$

Multiply coefficients and add exponents.

Furthermore, we can use the distributive property to multiply a monomial and a polynomial that is not a monomial. For example,

$$3x^4(2x^3 - 7x + 3) = 3x^4 \cdot 2x^3 - 3x^4 \cdot 7x + 3x^4 \cdot 3 = 6x^7 - 21x^5 + 9x^4.$$

Monomial Trinomial

How do we multiply two polynomials if neither is a monomial? For example, consider

$$(2x + 3)(x^2 + 4x + 5).$$

Binomial Trinomial

One way to perform $(2x + 3)(x^2 + 4x + 5)$ is to distribute $2x$ throughout the trinomial

$$2x(x^2 + 4x + 5)$$

and 3 throughout the trinomial

$$3(x^2 + 4x + 5).$$

Then combine the like terms that result.

> **Multiplying Polynomials When Neither Is a Monomial**
>
> Multiply each term of one polynomial by each term of the other polynomial. Then combine like terms.

EXAMPLE 2 Multiplying a Binomial and a Trinomial

Multiply: $(2x + 3)(x^2 + 4x + 5)$.

SOLUTION

$(2x + 3)(x^2 + 4x + 5)$	
$= 2x(x^2 + 4x + 5) + 3(x^2 + 4x + 5)$	Multiply the trinomial by each term of the binomial.
$= 2x \cdot x^2 + 2x \cdot 4x + 2x \cdot 5 + 3x^2 + 3 \cdot 4x + 3 \cdot 5$	Use the distributive property.
$= 2x^3 + 8x^2 + 10x + 3x^2 + 12x + 15$	Multiply monomials: Multiply coefficients and add exponents.
$= 2x^3 + 11x^2 + 22x + 15$	Combine like terms: $8x^2 + 3x^2 = 11x^2$ and $10x + 12x = 22x$. • • •

Another method for performing the multiplication is to use a vertical format similar to that used for multiplying whole numbers.

$$
\begin{array}{r}
x^2 + 4x + 5 \\
2x + 3 \\
\hline
3x^2 + 12x + 15 \quad \scriptstyle 3(x^2 + 4x + 5) \\
2x^3 + 8x^2 + 10x \quad \scriptstyle 2x(x^2 + 4x + 5) \\
\hline
2x^3 + 11x^2 + 22x + 15
\end{array}
$$

Write like terms in the same column.

Combine like terms.

✓ **Check Point 2** Multiply: $(5x - 2)(3x^2 - 5x + 4)$. $\scriptstyle 15x^3 - 31x^2 + 30x - 8$

④ Use FOIL in polynomial multiplication.

The Product of Two Binomials: FOIL

Frequently, we need to find the product of two binomials. One way to perform this multiplication is to distribute each term in the first binomial through the second binomial. For example, we can find the product of the binomials $3x + 2$ and $4x + 5$ as follows:

$$
\begin{aligned}
(3x + 2)(4x + 5) &= 3x(4x + 5) + 2(4x + 5) \\
&= 3x(4x) + 3x(5) + 2(4x) + 2(5) \\
&= 12x^2 + 15x + 8x + 10.
\end{aligned}
$$

Distribute $3x$ over $4x + 5$. Distribute 2 over $4x + 5$.

We'll combine these like terms later. For now, our interest is in how to obtain each of these four terms.

We can also find the product of $3x + 2$ and $4x + 5$ using a method called FOIL, which is based on our preceding work. Any two binomials can be quickly multiplied

by using the FOIL method, in which **F** represents the product of the **first** terms in each binomial, **O** represents the product of the **outside** terms, **I** represents the product of the **inside** terms, and **L** represents the product of the **last**, or second, terms in each binomial. For example, we can use the FOIL method to find the product of the binomials $3x + 2$ and $4x + 5$ as follows:

In general, here's how to use the FOIL method to find the product of $ax + b$ and $cx + d$:

Using the FOIL Method to Multiply Binomials

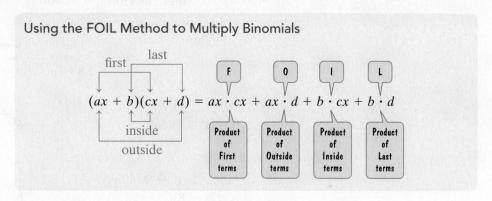

EXAMPLE 3 Using the FOIL Method

Multiply: $(3x + 4)(5x - 3)$.

SOLUTION

$$(3x + 4)(5x - 3) = 3x \cdot 5x + 3x(-3) + 4 \cdot 5x + 4(-3)$$
$$= 15x^2 - 9x + 20x - 12$$
$$= 15x^2 + 11x - 12 \qquad \text{Combine like terms.} \qquad \bullet\bullet\bullet$$

\checkmark **Check Point 3** Multiply: $(7x - 5)(4x - 3)$. $28x^2 - 41x + 15$

5 Use special products in polynomial multiplication.

Multiplying the Sum and Difference of Two Terms

We can use the FOIL method to multiply $A + B$ and $A - B$ as follows:

$$(A + B)(A - B) = A^2 - AB + AB - B^2 = A^2 - B^2.$$

Notice that the outside and inside products have a sum of 0 and the terms cancel. The FOIL multiplication provides us with a quick rule for multiplying the sum and difference of two terms, referred to as a *special-product formula*.

The Product of the Sum and Difference of Two Terms

$$(A + B)(A - B) = A^2 - B^2$$

| The product of the sum and the difference of the same two terms | is | the square of the first term minus the square of the second term. |

EXAMPLE 4 Finding the Product of the Sum and Difference of Two Terms

Multiply:

 a. $(4y + 3)(4y - 3)$ **b.** $(5a^4 + 6)(5a^4 - 6)$.

SOLUTION

Use the special-product formula shown.

$$(A + B)(A - B) \quad = \quad A^2 \quad - \quad B^2$$

| First term squared | − | Second term squared | = | Product |

 a. $(4y + 3)(4y - 3) \;=\; (4y)^2 \;-\; 3^2 \;=\; 16y^2 - 9$

 b. $(5a^4 + 6)(5a^4 - 6) \;=\; (5a^4)^2 \;-\; 6^2 \;=\; 25a^8 - 36$ • • •

✓ Check Point **4** Multiply:

 a. $(7x + 8)(7x - 8)$ $49x^2 - 64$ **b.** $(2y^3 - 5)(2y^3 + 5)$. $4y^6 - 25$

The Square of a Binomial

Let us find $(A + B)^2$, the square of a binomial sum. To do so, we begin with the FOIL method and look for a general rule.

 F **O** **I** **L**

$$(A + B)^2 = (A + B)(A + B) = A \cdot A + A \cdot B + A \cdot B + B \cdot B$$
$$= A^2 + 2AB + B^2$$

This result implies the following rule, which is another example of a special-product formula:

The Square of a Binomial Sum

$$(A + B)^2 \quad = \quad A^2 \quad + \quad 2AB \quad + \quad B^2$$

| The square of a binomial sum | is | first term squared | plus | 2 times the product of the terms | plus | last term squared. |

EXAMPLE 5 Finding the Square of a Binomial Sum

Multiply:

a. $(x + 3)^2$ b. $(3x + 7)^2$.

SOLUTION

Use the special-product formula shown.

$$(A + B)^2 = A^2 + 2AB + B^2$$

	(First Term)2	+	2 · Product of the Terms	+	(Last Term)2	= Product
a. $(x + 3)^2 =$	x^2	+	$2 \cdot x \cdot 3$	+	3^2	$= x^2 + 6x + 9$
b. $(3x + 7)^2 =$	$(3x)^2$	+	$2(3x)(7)$	+	7^2	$= 9x^2 + 42x + 49$

• • •

✓ **Check Point 5** Multiply:

a. $(x + 10)^2$ $x^2 + 20x + 100$ b. $(5x + 4)^2$. $25x^2 + 40x + 16$

A similar pattern occurs for $(A - B)^2$, the square of a binomial difference. Using the FOIL method on $(A - B)^2$, we obtain the following rule:

The Square of a Binomial Difference

EXAMPLE 6 Finding the Square of a Binomial Difference

Multiply:

a. $(x - 4)^2$ b. $(5y - 6)^2$.

SOLUTION

Use the special-product formula shown.

$$(A - B)^2 = A^2 - 2AB + B^2$$

	(First Term)2	−	2 · Product of the Terms	+	(Last Term)2	= Product
a. $(x - 4)^2 =$	x^2	−	$2 \cdot x \cdot 4$	+	4^2	$= x^2 - 8x + 16$
b. $(5y - 6)^2 =$	$(5y)^2$	−	$2(5y)(6)$	+	6^2	$= 25y^2 - 60y + 36$

• • •

✓ **Check Point 6** Multiply:

a. $(x - 9)^2$ $x^2 - 18x + 81$ b. $(7x - 3)^2$. $49x^2 - 42x + 9$

Special Products

There are several products that occur so frequently that it's convenient to memorize the form, or pattern, of these formulas.

Special Products

Let A and B represent real numbers, variables, or algebraic expressions.

Special Product	Example
Sum and Difference of Two Terms	
$(A + B)(A - B) = A^2 - B^2$	$(2x + 3)(2x - 3) = (2x)^2 - 3^2$ $\qquad\qquad\qquad\quad = 4x^2 - 9$
Squaring a Binomial	
$(A + B)^2 = A^2 + 2AB + B^2$	$(y + 5)^2 = y^2 + 2 \cdot y \cdot 5 + 5^2$ $\qquad\quad = y^2 + 10y + 25$
$(A - B)^2 = A^2 - 2AB + B^2$	$(3x - 4)^2$ $= (3x)^2 - 2 \cdot 3x \cdot 4 + 4^2$ $= 9x^2 - 24x + 16$
Cubing a Binomial	
$(A + B)^3 = A^3 + 3A^2B + 3AB^2 + B^3$	$(x + 4)^3$ $= x^3 + 3x^2(4) + 3x(4)^2 + 4^3$ $= x^3 + 12x^2 + 48x + 64$
$(A - B)^3 = A^3 - 3A^2B + 3AB^2 - B^3$	$(x - 2)^3$ $= x^3 - 3x^2(2) + 3x(2)^2 - 2^3$ $= x^3 - 6x^2 + 12x - 8$

6 Perform operations with polynomials in several variables.

Polynomials in Several Variables

A **polynomial in two variables**, x and y, contains the sum of one or more monomials in the form $ax^n y^m$. The constant, a, is the **coefficient**. The exponents, n and m, represent whole numbers. The **degree** of the monomial $ax^n y^m$ is $n + m$.

Here is an example of a polynomial in two variables:

The coefficients are 7, −17, 1, −6, and 9.

$$7x^2y^3 \quad - \quad 17x^4y^2 \quad + \quad xy \quad - \quad 6y^2 \quad + \quad 9.$$

| Degree of monomial: $2 + 3 = 5$ | Degree of monomial: $4 + 2 = 6$ | Degree of monomial $(1x^1y^1)$: $1 + 1 = 2$ | Degree of monomial $(-6x^0y^2)$: $0 + 2 = 2$ | Degree of monomial $(9x^0y^0)$: $0 + 0 = 0$ |

The **degree of a polynomial in two variables** is the highest degree of all its terms. For the preceding polynomial, the degree is 6.

Polynomials containing two or more variables can be added, subtracted, and multiplied just like polynomials that contain only one variable. For example, we can add the monomials $-7xy^2$ and $13xy^2$ as follows:

$$-7xy^2 + 13xy^2 = (-7 + 13)xy^2 = 6xy^2.$$

These like terms both contain the variable factors x and y^2.

Add coefficients and keep the same variable factors, xy^2.

EXAMPLE 7 Subtracting Polynomials in Two Variables

Subtract:
$$(5x^3 - 9x^2y + 3xy^2 - 4) - (3x^3 - 6x^2y - 2xy^2 + 3).$$

SOLUTION

$(5x^3 - 9x^2y + 3xy^2 - 4) - (3x^3 - 6x^2y - 2xy^2 + 3)$

Change the sign of each coefficient.

$= (5x^3 - 9x^2y + 3xy^2 - 4) + (-3x^3 + 6x^2y + 2xy^2 - 3)$ Add the opposite of the polynomial being subtracted.

$= (5x^3 - 3x^3) + (-9x^2y + 6x^2y) + (3xy^2 + 2xy^2) + (-4 - 3)$ Group like terms.

$= 2x^3 - 3x^2y + 5xy^2 - 7$ Combine like terms by adding coefficients and keeping the same variable factors.

• • •

⌀ Check Point **7** Subtract: $(x^3 - 4x^2y + 5xy^2 - y^3) - (x^3 - 6x^2y + y^3).$

$2x^2y + 5xy^2 - 2y^3$

EXAMPLE 8 Multiplying Polynomials in Two Variables

Multiply:

 a. $(x + 4y)(3x - 5y)$ **b.** $(5x + 3y)^2.$

SOLUTION

We will perform the multiplication in part (a) using the FOIL method. We will multiply in part (b) using the formula for the square of a binomial sum, $(A + B)^2$.

 a. $(x + 4y)(3x - 5y)$ Multiply these binomials using the FOIL method.

| F | O | I | L |

$= (x)(3x) + (x)(-5y) + (4y)(3x) + (4y)(-5y)$

$= 3x^2 - 5xy + 12xy - 20y^2$

$= 3x^2 + 7xy - 20y^2$ Combine like terms.

$(A + B)^2 = A^2 + 2 \cdot A \cdot B + B^2$

 b. $(5x + 3y)^2 = (5x)^2 + 2(5x)(3y) + (3y)^2$

$= 25x^2 + 30xy + 9y^2$

• • •

⌀ Check Point **8** Multiply:

 a. $(7x - 6y)(3x - y)$ $21x^2 - 25xy + 6y^2$ **b.** $(2x + 4y)^2.$ $4x^2 + 16xy + 16y^2$

Blitzer Bonus || Labrador Retrievers and Polynomial Multiplication

The color of a Labrador retriever is determined by its pair of genes. A single gene is inherited at random from each parent. The black-fur gene, B, is dominant. The yellow-fur gene, Y, is recessive. This means that labs with at least one black-fur gene (BB or BY) have black coats. Only labs with two yellow-fur genes (YY) have yellow coats.

Axl, your author's yellow lab, inherited his genetic makeup from two black BY parents.

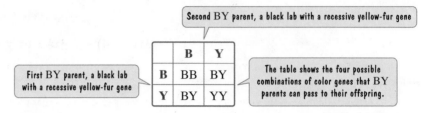

Second BY parent, a black lab with a recessive yellow-fur gene

First BY parent, a black lab with a recessive yellow-fur gene

	B	**Y**
B	BB	BY
Y	BY	YY

The table shows the four possible combinations of color genes that BY parents can pass to their offspring.

Because YY is one of four possible outcomes, the probability that a yellow lab like Axl will be the offspring of these black parents is $\frac{1}{4}$.

The probabilities suggested by the table can be modeled by the expression $\left(\frac{1}{2}B + \frac{1}{2}Y\right)^2$.

$$\left(\frac{1}{2}B + \frac{1}{2}Y\right)^2 = \left(\frac{1}{2}B\right)^2 + 2\left(\frac{1}{2}B\right)\left(\frac{1}{2}Y\right) + \left(\frac{1}{2}Y\right)^2$$

$$= \frac{1}{4}BB + \frac{1}{2}BY + \frac{1}{4}YY$$

The probability of a black lab with two dominant black genes is $\frac{1}{4}$.

The probability of a black lab with a recessive yellow gene is $\frac{1}{2}$.

The probability of a yellow lab with two recessive yellow genes is $\frac{1}{4}$.

CONCEPT AND VOCABULARY CHECK

Fill in each blank so that the resulting statement is true.

1. A polynomial is a single term or the sum of two or more terms containing variables with exponents that are ___whole___ numbers.

2. It is customary to write the terms of a polynomial in the order of descending powers of the variable. This is called the ___standard___ form of a polynomial.

3. A simplified polynomial that has exactly one term is called a/an ___monomial___.

4. A simplified polynomial that has two terms is called a/an ___binomial___.

5. A simplified polynomial that has three terms is called a/an ___trinomial___.

6. If $a \neq 0$, the degree of ax^n is ___n___.

7. Polynomials are added by combining ___like___ terms.

8. To multiply $7x^3(4x^5 - 8x^2 + 6)$, use the ___distributive___ property to multiply each term of the trinomial ___$4x^5 - 8x^2 + 6$___ by the monomial ___$7x^3$___.

9. To multiply $(5x + 3)(x^2 + 8x + 7)$, begin by multiplying each term of $x^2 + 8x + 7$ by ___$5x$___. Then multiply each term of $x^2 + 8x + 7$ by ___3___. Then combine ___like___ terms.

10. When using the FOIL method to find $(x + 7)(3x + 5)$, the product of the first terms is ___$3x^2$___, the product of the outside terms is ___$5x$___, the product of the inside terms is ___$21x$___, and the product of the last terms is ___35___.

11. $(A + B)(A - B) = $ ___$A^2 - B^2$___. The product of the sum and difference of the same two terms is the square of the first term ___minus___ the square of the second term.

12. $(A + B)^2 = $ ___$A^2 + 2AB + B^2$___. The square of a binomial sum is the first term ___squared___ plus 2 times the ___product of the terms___ plus the last term ___squared___.

13. $(A - B)^2 = $ ___$A^2 - 2AB + B^2$___. The square of a binomial difference is the first term squared ___minus___ 2 times the ___product of the terms___ ___plus___ the last term squared.

plus or minus?

14. If $a \neq 0$, the degree of ax^ny^m is ___$n + m$___.

EXERCISE SET P.4

Practice Exercises

In Exercises 1–4, is the algebraic expression a polynomial? If it is, write the polynomial in standard form.

1. $2x + 3x^2 - 5$ yes; $3x^2 + 2x - 5$ 2. $2x + 3x^{-1} - 5$ no

3. $\dfrac{2x + 3}{x}$ no 4. $x^2 - x^3 + x^4 - 5$ yes*

In Exercises 5–8, find the degree of the polynomial.

5. $3x^2 - 5x + 4$ 2 6. $-4x^3 + 7x^2 - 11$ 3

7. $x^2 - 4x^3 + 9x - 12x^4 + 63$ 4 8. $x^2 - 8x^3 + 15x^4 + 91$ 4

In Exercises 9–14, perform the indicated operations. Write the resulting polynomial in standard form and indicate its degree.

9. $(-6x^3 + 5x^2 - 8x + 9) + (17x^3 + 2x^2 - 4x - 13)$ *

10. $(-7x^3 + 6x^2 - 11x + 13) + (19x^3 - 11x^2 + 7x - 17)$ *

11. $(17x^3 - 5x^2 + 4x - 3) - (5x^3 - 9x^2 - 8x + 11)$ *

12. $(18x^4 - 2x^3 - 7x + 8) - (9x^4 - 6x^3 - 5x + 7)$ *

13. $(5x^2 - 7x - 8) + (2x^2 - 3x + 7) - (x^2 - 4x - 3)$ *

14. $(8x^2 + 7x - 5) - (3x^2 - 4x) - (-6x^3 - 5x^2 + 3)$ *

In Exercises 15–58, find each product.

15. $(x + 1)(x^2 - x + 1)$ $x^3 + 1$ 16. $(x + 5)(x^2 - 5x + 25)$ *

17. $(2x - 3)(x^2 - 3x + 5)$ * 18. $(2x - 1)(x^2 - 4x + 3)$ *

19. $(x + 7)(x + 3)$ * 20. $(x + 8)(x + 5)$ *

21. $(x - 5)(x + 3)$ * 22. $(x - 1)(x + 2)$ *

23. $(3x + 5)(2x + 1)$ * 24. $(7x + 4)(3x + 1)$ *

25. $(2x - 3)(5x + 3)$ * 26. $(2x - 5)(7x + 2)$ *

27. $(5x^2 - 4)(3x^2 - 7)$ * 28. $(7x^2 - 2)(3x^2 - 5)$ *

29. $(8x^3 + 3)(x^2 - 5)$ * 30. $(7x^3 + 5)(x^2 - 2)$ *

31. $(x + 3)(x - 3)$ $x^2 - 9$ 32. $(x + 5)(x - 5)$ $x^2 - 25$

33. $(3x + 2)(3x - 2)$ $9x^2 - 4$ 34. $(2x + 5)(2x - 5)$ $4x^2 - 25$

35. $(5 - 7x)(5 + 7x)$ $25 - 49x^2$ 36. $(4 - 3x)(4 + 3x)$ $16 - 9x^2$

37. $(4x^2 + 5x)(4x^2 - 5x)$ * 38. $(3x^2 + 4x)(3x^2 - 4x)$ *

39. $(1 - y^5)(1 + y^5)$ $1 - y^{10}$ 40. $(2 - y^5)(2 + y^5)$ $4 - y^{10}$

41. $(x + 2)^2$ $x^2 + 4x + 4$ 42. $(x + 5)^2$ $x^2 + 10x + 25$

43. $(2x + 3)^2$ $4x^2 + 12x + 9$ 44. $(3x + 2)^2$ $9x^2 + 12x + 4$

45. $(x - 3)^2$ $x^2 - 6x + 9$ 46. $(x - 4)^2$ $x^2 - 8x + 16$

47. $(4x^2 - 1)^2$ $16x^4 - 8x^2 + 1$ 48. $(5x^2 - 3)^2$ $25x^4 - 30x^2 + 9$

49. $(7 - 2x)^2$ $4x^2 - 28x + 49$ 50. $(9 - 5x)^2$ $25x^2 - 90x + 81$

51. $(x + 1)^3$ $x^3 + 3x^2 + 3x + 1$ 52. $(x + 2)^3$ *

53. $(2x + 3)^3$ * 54. $(3x + 4)^3$ *

55. $(x - 3)^3$ $x^3 - 9x^2 + 27x - 27$ 56. $(x - 1)^3$ $x^3 - 3x^2 + 3x - 1$

57. $(3x - 4)^3$ * 58. $(2x - 3)^3$ *

In Exercises 59–66, perform the indicated operations. Indicate the degree of the resulting polynomial.

59. $(5x^2y - 3xy) + (2x^2y - xy)$ $7x^2y - 4xy; 3$

60. $(-2x^2y + xy) + (4x^2y + 7xy)$ $2x^2y + 8xy; 3$

*See Answers to Selected Exercises.

61. $(4x^2y + 8xy + 11) + (-2x^2y + 5xy + 2)$ *

62. $(7x^4y^2 - 5x^2y^2 + 3xy) + (-18x^4y^2 - 6x^2y^2 - xy)$ *

63. $(x^3 + 7xy - 5y^2) - (6x^3 - xy + 4y^2)$ *

64. $(x^4 - 7xy - 5y^3) - (6x^4 - 3xy + 4y^3)$ *

65. $(3x^4y^2 + 5x^3y - 3y) - (2x^4y^2 - 3x^3y - 4y + 6x)$ *

66. $(5x^4y^2 + 6x^3y - 7y) - (3x^4y^2 - 5x^3y - 6y + 8x)$ *

In Exercises 67–82, find each product.

67. $(x + 5y)(7x + 3y)$ * 68. $(x + 9y)(6x + 7y)$ *

69. $(x - 3y)(2x + 7y)$ * 70. $(3x - y)(2x + 5y)$ *

71. $(3xy - 1)(5xy + 2)$ * 72. $(7x^2y + 1)(2x^2y - 3)$ *

73. $(7x + 5y)^2$ $49x^2 + 70xy + 25y^2$

74. $(9x + 7y)^2$ $81x^2 + 126xy + 49y^2$

75. $(x^2y^2 - 3)^2$ $x^4y^4 - 6x^2y^2 + 9$

76. $(x^2y^2 - 5)^2$ $x^4y^4 - 10x^2y^2 + 25$

77. $(x - y)(x^2 + xy + y^2)$ * 78. $(x + y)(x^2 - xy + y^2)$ *

79. $(3x + 5y)(3x - 5y)$ * 80. $(7x + 3y)(7x - 3y)$ *

81. $(7xy^2 - 10y)(7xy^2 + 10y)$ * 82. $(3xy^2 - 4y)(3xy^2 + 4y)$ *

Practice Plus

In Exercises 83–90, perform the indicated operation or operations.

83. $(3x + 4y)^2 - (3x - 4y)^2$ $48xy$

84. $(5x + 2y)^2 - (5x - 2y)^2$ $40xy$

85. $(5x - 7)(3x - 2) - (4x - 5)(6x - 1)$ $-9x^2 + 3x + 9$

86. $(3x + 5)(2x - 9) - (7x - 2)(x - 1)$ $-x^2 - 8x - 47$

87. $(2x + 5)(2x - 5)(4x^2 + 25)$ $16x^4 - 625$

88. $(3x + 4)(3x - 4)(9x^2 + 16)$ $81x^4 - 256$

89. $\dfrac{(2x - 7)^5}{(2x - 7)^3}$ $4x^2 - 28x + 49$ 90. $\dfrac{(5x - 3)^6}{(5x - 3)^4}$ $25x^2 - 30x + 9$

Application Exercises

91. The bar graph shows the differences among age groups on the Implicit Association Test that measures levels of racial prejudice. Higher scores indicate stronger bias.

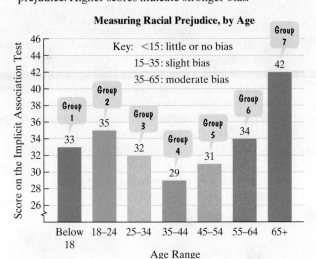

Measuring Racial Prejudice, by Age

Key: <15: little or no bias
15–35: slight bias
35–65: moderate bias

Source: The Race Implicit Association Test on the Project Implicit Demonstration Website

(Refer to the graph shown at the bottom of the previous page.)

a. The data can be described by the following polynomial model of degree 3:

$$S = 0.2x^3 - 1.5x^2 + 3.4x + 25$$
$$+ (0.1x^3 - 1.3x^2 + 3.3x + 5).$$

In this polynomial model, S represents the score on the Implicit Association Test for age group x. Simplify the model. $S = 0.3x^3 - 2.8x^2 + 6.7x + 30$

b. Use the simplified form of the model from part (a) to find the score on the Implicit Association Test for the group in the 45–54 age range. How well does the model describe the score displayed by the bar graph? 31: It's the same.

92. The bar graph shows the differences among political identification groups on the Implicit Association Test that measure levels of racial prejudice. Higher scores indicate stronger bias.

Measuring Racial Prejudice, by Political Identification

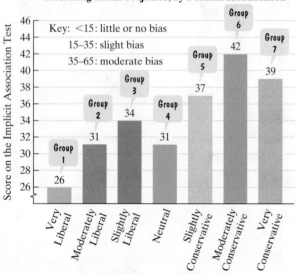

Political Identification

Source: The Race Implicit Association Test on the Project Implicit Demonstration Website

a. The data can be described by the following polynomial model of degree 3:

$$S = -0.02x^3 + 0.4x^2 + 1.2x + 22$$
$$+ (-0.01x^3 - 0.2x^2 + 1.1x + 2).$$

In this polynomial model, S represents the score on the Implicit Association Test for political identification group x. Simplify the model. $S = -0.03x^3 + 0.2x^2 + 2.3x + 24$

b. Use the simplified form of the model from part (a) to find the score on the Implicit Association Test for the slightly conservative political identification group. Does the model underestimate or overestimate the score displayed by the bar graph? By how much? 36.75; underestimates by 0.25

The volume, V, of a rectangular solid with length l, width w, and height h is given by the formula $V = lwh$. In Exercises 93–94, use this formula to write a polynomial in standard form that models, or represents, the volume of the open box.

93.

$8 - 2x$

$10 - 2x$

x

$4x^3 - 36x^2 + 80x$

94.

$8 - 2x$

$5 - 2x$

x

$4x^3 - 26x^2 + 40x$

In Exercises 95–96, write a polynomial in standard form that models, or represents, the area of the shaded region.

95.

$6x + 22$

$x + 9$

$x + 5$

$x + 3$ $x + 1$

96.

$x + 4$ $4x + 10$

$x + 2$

$x + 3$ $x + 1$

Explaining the Concepts 97–102. Answers will vary.

97. What is a polynomial in x?

98. Explain how to subtract polynomials.

99. Explain how to multiply two binomials using the FOIL method. Give an example with your explanation.

100. Explain how to find the product of the sum and difference of two terms. Give an example with your explanation.

101. Explain how to square a binomial difference. Give an example with your explanation.

102. Explain how to find the degree of a polynomial in two variables.

Critical Thinking Exercises

Make Sense? *In Exercises 103–106, determine whether each statement makes sense or does not make sense, and explain your reasoning.*

103. Knowing the difference between factors and terms is important: In $(3x^2y)^2$, I can distribute the exponent 2 on each factor, but in $(3x^2 + y)^2$, I cannot do the same thing on each term. makes sense

104. I used the FOIL method to find the product of $x + 5$ and $x^2 + 2x + 1$. does not make sense

105. Many English words have prefixes with meanings similar to those used to describe polynomials, such as *monologue, binocular,* and *tricuspid.* makes sense

106. Special-product formulas have patterns that make their multiplications quicker than using the FOIL method. *

In Exercises 107–110, determine whether each statement is true or false. If the statement is false, make the necessary change(s) to produce a true statement. Changes to false statements will vary.

107. $(3x^3 + 2)(3x^3 - 2) = 9x^9 - 4$ false

108. $(x - 5)^2 = x^2 - 5x + 25$ false

109. $(x + 1)^2 = x^2 + 1$ false

110. Suppose a square garden has an area represented by $9x^2$ square feet. If one side is made 7 feet longer and the other side is made 2 feet shorter, then the trinomial that models the area of the larger garden is $9x^2 + 15x - 14$ square feet. true

In Exercises 111–113, perform the indicated operations.

111. $[(7x + 5) + 4y][(7x + 5) - 4y]$ $49x^2 + 70x + 25 - 16y^2$

112. $[(3x + y) + 1]^2$ $9x^2 + 6xy + y^2 + 6x + 2y + 1$

113. $(x^n + 2)(x^n - 2) - (x^n - 3)^2$ $6x^n - 13$

114. Express the area of the plane figure shown as a polynomial in standard form. $x^2 + 2x$

Preview Exercises

Exercises 115–117 will help you prepare for the material covered in the next section. In each exercise, replace the boxed question mark with an integer that results in the given product. Some trial and error may be necessary.

115. $(x + 3)(x + \boxed{?}) = x^2 + 7x + 12$ 4

116. $(x - \boxed{?})(x - 12) = x^2 - 14x + 24$ 2

117. $(4x + 1)(2x - \boxed{?}) = 8x^2 - 10x - 3$ 3

CHAPTER P Mid-Chapter Check Point

WHAT YOU KNOW: We defined the real numbers $[\{x \mid x \text{ is rational}\} \cup \{x \mid x \text{ is irrational}\}]$ and graphed them as points on a number line. We reviewed the basic rules of algebra, using these properties to simplify algebraic expressions. We expanded our knowledge of exponents to include exponents other than natural numbers:

$$b^0 = 1; \quad b^{-n} = \frac{1}{b^n}; \quad \frac{1}{b^{-n}} = b^n; \quad b^{\frac{1}{n}} = \sqrt[n]{b};$$

$$b^{\frac{m}{n}} = \left(\sqrt[n]{b}\right)^m = \sqrt[n]{b^m}; \quad b^{-\frac{m}{n}} = \frac{1}{b^{\frac{m}{n}}}.$$

We used properties of exponents to simplify exponential expressions and properties of radicals to simplify radical expressions. Finally, we performed operations with polynomials. We used a number of fast methods for finding products of polynomials, including the FOIL method for multiplying binomials, a special-product formula for the product of the sum and difference of two terms $[(A + B)(A - B) = A^2 - B^2]$, and special-product formulas for squaring binomials $[(A + B)^2 = A^2 + 2AB + B^2;$ $(A - B)^2 = A^2 - 2AB + B^2]$.

In Exercises 1–25, simplify the given expression or perform the indicated operation (and simplify, if possible), whichever is appropriate.

1. $(3x + 5)(4x - 7)$ *

2. $(3x + 5) - (4x - 7)$ $-x + 12$

3. $\sqrt{6} + 9\sqrt{6}$ $10\sqrt{6}$

4. $3\sqrt{12} - \sqrt{27}$ $3\sqrt{3}$

5. $7x + 3[9 - (2x - 6)]$ *

6. $(8x - 3)^2$ $64x^2 - 48x + 9$

7. $\left(x^{\frac{1}{3}}y^{-\frac{1}{2}}\right)^6$ $\frac{x^2}{y^3}$

8. $\left(\frac{2}{7}\right)^0 - 32^{-\frac{2}{5}}$ $\frac{3}{4}$

9. $(2x - 5) - (x^2 - 3x + 1)$ $-x^2 + 5x - 6$

10. $(2x - 5)(x^2 - 3x + 1)$ *

11. $x^3 + x^3 - x^3 \cdot x^3$ $-x^6 + 2x^3$

12. $(9a - 10b)(2a + b)$ *

13. $\{a, c, d, e\} \cup \{c, d, f, h\}$ *

14. $\{a, c, d, e\} \cap \{c, d, f, h\}$ $\{c, d\}$

15. $(3x^2y^3 - xy + 4y^2) - (-2x^2y^3 - 3xy + 5y^2)$ *

16. $\frac{24x^2y^{13}}{-2x^5y^{-2}}$ $-\frac{12y^{15}}{x^3}$

17. $\left(\frac{1}{3}x^{-5}y^4\right)(18x^{-2}y^{-1})$ $\frac{6y^3}{x^7}$

18. $\sqrt[12]{x^4}$ $\sqrt[3]{x}$

19. $\frac{24 \times 10^3}{2 \times 10^6}$ (Express the answer in scientific notation.) 1.2×10^{-2}

20. $\frac{\sqrt[3]{32}}{\sqrt[3]{2}}$ $2\sqrt[3]{2}$

21. $(x^3 + 2)(x^3 - 2)$ $x^6 - 4$

22. $(x^2 + 2)^2$ $x^4 + 4x^2 + 4$

23. $\sqrt{50} \cdot \sqrt{6}$ $10\sqrt{3}$

24. $\frac{11}{7 - \sqrt{3}}$ $\frac{77 + 11\sqrt{3}}{46}$

25. $\frac{11}{\sqrt{3}}$ $\frac{11\sqrt{3}}{3}$

26. List all the rational numbers in this set:

$$\left\{-11, -\frac{3}{7}, 0, 0.45, \sqrt{23}, \sqrt{25}\right\}.$$ $-11, -\frac{3}{7}, 0, 0.45, \sqrt{25}$

In Exercises 27–28, rewrite each expression without absolute value bars.

27. $|2 - \sqrt{13}|$ $\sqrt{13} - 2$

28. $x^2|x|$ if $x < 0$ $-x^3$

29. If the population of the United States is approximately 3.0×10^8 and each person produces about 4.6 pounds of garbage per day, express the total number of pounds of garbage produced in the United States in one day in scientific notation. 1.38×10^9 pounds

30. A human brain contains 3×10^{10} neurons and a gorilla brain contains 7.5×10^9 neurons. How many times as many neurons are in the brain of a human as in the brain of a gorilla? 4

31. College students are graduating with the highest debt burden in history. The bar graph shows the mean, or average, student-loan debt in the United States for five selected graduating years from 2001 through 2013.

Mean Student-Loan Debt in the U.S.

Source: Pew Research Center

Here are two mathematical models for the data shown by the graph. In each formula, D represents mean student-loan debt, in dollars, x years after 2000.

Model 1 $D = 1188x + 16,218$

Model 2 $D = 46x^2 + 541x + 17,650$

a. Which model better describes the data for 2001? Model 1

b. Does the polynomial model of degree 2 underestimate or overestimate mean student-loan debt for 2013? By how much? underestimates by $593

*See Answers to Selected Exercises.

Section P.5 — Factoring Polynomials

What am I supposed to learn?

After studying this section, you should be able to:

1. Factor out the greatest common factor of a polynomial.
2. Factor by grouping.
3. Factor trinomials.
4. Factor the difference of squares.
5. Factor perfect square trinomials.
6. Factor the sum or difference of two cubes.
7. Use a general strategy for factoring polynomials.
8. Factor algebraic expressions containing fractional and negative exponents.

A two-year-old boy is asked, "Do you have a brother?" He answers, "Yes." "What is your brother's name?" "Tom." Asked if Tom has a brother, the two-year-old replies, "No." The child can go in the direction from self to brother, but he cannot reverse this direction and move from brother back to self.

As our intellects develop, we learn to reverse the direction of our thinking. Reversibility of thought is found throughout algebra. For example, we can multiply polynomials and show that

$$5x(2x + 3) = 10x^2 + 15x.$$

We can also reverse this process and express the resulting polynomial as

$$10x^2 + 15x = 5x(2x + 3).$$

Factoring a polynomial expressed as the sum of monomials means finding an equivalent expression that is a product.

Factoring $10x^2 + 15x$

Sum of monomials | Equivalent expression that is a product

$$10x^2 + 15x = 5x(2x + 3)$$

The factors of $10x^2 + 15x$ are $5x$ and $2x + 3$.

In this section, we will be **factoring over the set of integers**, meaning that the coefficients in the factors are integers. Polynomials that cannot be factored using integer coefficients are called **irreducible over the integers**, or **prime**.

The goal in factoring a polynomial is to use one or more factoring techniques until each of the polynomial's factors, except possibly for a monomial factor, is prime or irreducible. In this situation, the polynomial is said to be **factored completely**.

We will now discuss basic techniques for factoring polynomials.

① Factor out the greatest common factor of a polynomial.

Common Factors

In any factoring problem, the first step is to look for the *greatest common factor*. The **greatest common factor**, abbreviated GCF, is an expression of the highest degree that divides each term of the polynomial. The distributive property in the reverse direction

$$ab + ac = a(b + c)$$

can be used to factor out the greatest common factor.

EXAMPLE 1 Factoring Out the Greatest Common Factor

Factor:

a. $18x^3 + 27x^2$ **b.** $x^2(x + 3) + 5(x + 3)$.

GREAT QUESTION!

Is there a rule that can help me determine the greatest common factor?

Yes. The variable part of the greatest common factor always contains the *smallest* power of a variable or algebraic expression that appears in all terms of the polynomial.

SOLUTION

a. First, determine the greatest common factor.

9 is the greatest integer that divides 18 and 27.

$$18x^3 + 27x^2$$

x^2 is the greatest expression that divides x^3 and x^2.

The GCF of the two terms of the polynomial is $9x^2$.

$$18x^3 + 27x^2$$
$$= 9x^2(2x) + 9x^2(3) \qquad \text{Express each term as the product of the GCF and its other factor.}$$
$$= 9x^2(2x + 3) \qquad \text{Factor out the GCF.}$$

b. In this situation, the greatest common factor is the common binomial factor $(x + 3)$. We factor out this common factor as follows:

$$x^2(x + 3) + 5(x + 3) = (x + 3)(x^2 + 5). \quad \text{Factor out the common binomial factor.}$$

• • •

 Check Point **1** Factor:

a. $10x^3 - 4x^2$ $\quad 2x^2(5x - 2)$ **b.** $2x(x - 7) + 3(x - 7)$. $\quad (x - 7)(2x + 3)$

2 Factor by grouping.

Factoring by Grouping

Some polynomials have only a greatest common factor of 1. However, by a suitable grouping of the terms, it still may be possible to factor. This process, called **factoring by grouping**, is illustrated in Example 2.

EXAMPLE 2 Factoring by Grouping

Factor: $x^3 + 4x^2 + 3x + 12$.

SOLUTION

There is no factor other than 1 common to all terms. However, we can group terms that have a common factor:

$$\boxed{x^3 + 4x^2} + \boxed{3x + 12}.$$

Common factor is x^2. Common factor is 3.

We now factor the given polynomial as follows:

$$x^3 + 4x^2 + 3x + 12$$
$$= (x^3 + 4x^2) + (3x + 12) \qquad \text{Group terms with common factors.}$$
$$= x^2(x + 4) + 3(x + 4) \qquad \text{Factor out the greatest common factor from the grouped terms. The remaining two terms have } x + 4 \text{ as a common binomial factor.}$$
$$= (x + 4)(x^2 + 3). \qquad \text{Factor out the GCF, } x + 4.$$

Thus, $x^3 + 4x^2 + 3x + 12 = (x + 4)(x^2 + 3)$. Check the factorization by multiplying the right side of the equation using the FOIL method. Because the factorization is correct, you should obtain the original polynomial.

• • •

DISCOVERY

In Example 2, group the terms as follows:

$$(x^3 + 3x) + (4x^2 + 12).$$

Factor out the greatest common factor from each group and complete the factoring process. Describe what happens. What can you conclude?

 Check Point **2** Factor: $x^3 + 5x^2 - 2x - 10$. $\quad (x + 5)(x^2 - 2)$

③ Factor trinomials.

Factoring Trinomials

To factor a trinomial of the form $ax^2 + bx + c$, a little trial and error may be necessary.

A Strategy for Factoring $ax^2 + bx + c$

Assume, for the moment, that there is no greatest common factor.

1. Find two **First** terms whose product is ax^2:

$$(\square x + \quad)(\square x + \quad) = ax^2 + bx + c.$$

2. Find two **Last** terms whose product is c:

$$(\square x + \square)(\square x + \square) = ax^2 + bx + c.$$

3. By trial and error, perform steps 1 and 2 until the sum of the **Outside** product and **Inside** product is bx:

$$(\square x + \square)(\square x + \square) = ax^2 + bx + c.$$

Sum of O + I

If no such combination exists, the polynomial is prime.

GREAT QUESTION!

Should I feel discouraged if it takes me a while to get the correct factorization?

The *error* part of the factoring strategy plays an important role in the process. If you do not get the correct factorization the first time, this is not a bad thing. This error is often helpful in leading you to the correct factorization.

EXAMPLE 3 Factoring a Trinomial Whose Leading Coefficient Is 1

Factor: $x^2 + 6x + 8$.

SOLUTION

Step 1 Find two First terms whose product is x^2.

$$x^2 + 6x + 8 = (x \quad)(x \quad)$$

Step 2 Find two Last terms whose product is 8.

Factors of 8	8, 1	4, 2	−8, −1	−4, −2

Step 3 Try various combinations of these factors. The correct factorization of $x^2 + 6x + 8$ is the one in which the sum of the **Outside** and **Inside** products is equal to $6x$. Here is a list of the possible factorizations:

Possible Factorizations of $x^2 + 6x + 8$	Sum of Outside and Inside Products (Should Equal 6x)
$(x + 8)(x + 1)$	$x + 8x = 9x$
$(x + 4)(x + 2)$	$2x + 4x = 6x$ ← This is the required middle term.
$(x − 8)(x − 1)$	$−x − 8x = −9x$
$(x − 4)(x − 2)$	$−2x − 4x = −6x$

Thus, $x^2 + 6x + 8 = (x + 4)(x + 2)$ or $(x + 2)(x + 4)$. • • •

In factoring a trinomial of the form $x^2 + bx + c$, you can speed things up by listing the factors of c and then finding their sums. We are interested in a sum of b. For example, in factoring $x^2 + 6x + 8$, we are interested in the factors of 8 whose sum is 6.

Factors of 8	8, 1	4, 2	−8, −1	−4, −2
Sum of Factors	9	6	−9	−6

This is the desired sum.

Thus, $x^2 + 6x + 8 = (x + 4)(x + 2)$.

✓ Check Point **3** Factor: $x^2 + 13x + 40$. $(x + 8)(x + 5)$ or $(x + 5)(x + 8)$

EXAMPLE 4 Factoring a Trinomial Whose Leading Coefficient Is 1

Factor: $x^2 + 3x - 18$.

SOLUTION

Step 1 Find two First terms whose product is x^2.

$$x^2 + 3x - 18 = (x \qquad)(x \qquad)$$

To find the second term of each factor, we must find two integers whose product is -18 and whose sum is 3.

Step 2 Find two Last terms whose product is -18.

Factors of -18	$18, -1$	$-18, 1$	$9, -2$	$-9, 2$	$6, -3$	$-6, 3$

Step 3 Try various combinations of these factors. We are looking for the pair of factors whose sum is 3.

Factors of -18	$18, -1$	$-18, 1$	$9, -2$	$-9, 2$	$6, -3$	$-6, 3$
Sum of Factors	17	-17	7	-7	3	-3

This is the desired sum.

Thus, $x^2 + 3x - 18 = (x + 6)(x - 3)$ or $(x - 3)(x + 6)$. • • •

GREAT QUESTION!

Is there a way to eliminate some of the combinations of factors for a trinomial whose leading coefficient is 1?

Yes. To factor $x^2 + bx + c$ when c is positive, find two numbers with the same sign as the middle term.

$$x^2 + 6x + 8 = (x + 4)(x + 2) \qquad x^2 - 5x + 6 = (x - 3)(x - 2)$$

Same signs Same signs

To factor $x^2 + bx + c$ when c is negative, find two numbers with opposite signs whose sum is the coefficient of the middle term.

$$x^2 + 3x - 18 = (x + 6)(x - 3) \qquad x^2 - 2x - 99 = (x + 9)(x - 11)$$

Negative Opposite signs Negative Opposite signs

✓ Check Point **4** Factor: $x^2 - 5x - 14$. $(x - 7)(x + 2)$ or $(x + 2)(x - 7)$

EXAMPLE 5 Factoring a Trinomial Whose Leading Coefficient Is Not 1

Factor: $8x^2 - 10x - 3$.

GREAT QUESTION!

When factoring trinomials, must I list every possible factorization before getting the correct one?

With practice, you will find that it is not necessary to list every possible factorization of the trinomial. As you practice factoring, you will be able to narrow down the list of possible factors to just a few. When it comes to factoring, practice makes perfect.

SOLUTION

Step 1 Find two First terms whose product is $8x^2$.

$$8x^2 - 10x - 3 \stackrel{?}{=} (8x \quad)(x \quad)$$
$$8x^2 - 10x - 3 \stackrel{?}{=} (4x \quad)(2x \quad)$$

Step 2 Find two Last terms whose product is -3. The possible factorizations are $1(-3)$ and $-1(3)$.

Step 3 Try various combinations of these factors. The correct factorization of $8x^2 - 10x - 3$ is the one in which the sum of the **O**utside and **I**nside products is equal to $-10x$. Here is a list of the possible factorizations:

Possible Factorizations of $8x^2 - 10x - 3$	Sum of Outside and Inside Products (Should Equal $-10x$)
$(8x + 1)(x - 3)$	$-24x + x = -23x$
$(8x - 3)(x + 1)$	$8x - 3x = 5x$
$(8x - 1)(x + 3)$	$24x - x = 23x$
$(8x + 3)(x - 1)$	$-8x + 3x = -5x$
$(4x + 1)(2x - 3)$	$-12x + 2x = -10x$
$(4x - 3)(2x + 1)$	$4x - 6x = -2x$
$(4x - 1)(2x + 3)$	$12x - 2x = 10x$
$(4x + 3)(2x - 1)$	$-4x + 6x = 2x$

These four factorizations use $(8x \quad)(x \quad)$ with $1(-3)$ and $-1(3)$ as factorizations of -3.

These four factorizations use $(4x \quad)(2x \quad)$ with $1(-3)$ and $-1(3)$ as factorizations of -3.

This is the required middle term.

Thus, $8x^2 - 10x - 3 = (4x + 1)(2x - 3)$ or $(2x - 3)(4x + 1)$.

Use FOIL multiplication to check either of these factorizations. • • •

☑ **Check Point 5** Factor: $6x^2 + 19x - 7$. $(3x - 1)(2x + 7)$ or $(2x + 7)(3x - 1)$

GREAT QUESTION!

I zone out reading your long lists of possible factorizations. Are there any rules for shortening these lists?

Here are some suggestions for reducing the list of possible factorizations for $ax^2 + bx + c$:

1. If b is relatively small, avoid the larger factors of a.

2. If c is positive, the signs in both binomial factors must match the sign of b.

3. If the trinomial has no common factor, no binomial factor can have a common factor.

4. Reversing the signs in the binomial factors changes the sign of bx, the middle term.

EXAMPLE 6 Factoring a Trinomial in Two Variables

Factor: $2x^2 - 7xy + 3y^2$.

SOLUTION

Step 1 Find two First terms whose product is $2x^2$.

$$2x^2 - 7xy + 3y^2 = (2x \quad)(x \quad)$$

Step 2 Find two Last terms whose product is $3y^2$. The possible factorizations are $(y)(3y)$ and $(-y)(-3y)$.

Step 3 Try various combinations of these factors. The correct factorization of $2x^2 - 7xy + 3y^2$ is the one in which the sum of the **O**utside and **I**nside products is equal to $-7xy$. Here is a list of possible factorizations:

Possible Factorizations of $2x^2 - 7xy + 3y^2$	Sum of Outside and Inside Products (Should Equal $-7xy$)
$(2x + 3y)(x + y)$	$2xy + 3xy = 5xy$
$(2x + y)(x + 3y)$	$6xy + xy = 7xy$
$(2x - 3y)(x - y)$	$-2xy - 3xy = -5xy$
$(2x - y)(x - 3y)$	$-6xy - xy = -7xy$

This is the required middle term.

Thus,

$$2x^2 - 7xy + 3y^2 = (2x - y)(x - 3y) \quad \text{or} \quad (x - 3y)(2x - y).$$

Use FOIL multiplication to check either of these factorizations. • • •

 Check Point **6** Factor: $3x^2 - 13xy + 4y^2$. $(3x - y)(x - 4y)$ or $(x - 4y)(3x - y)$

4 Factor the difference of squares.

Factoring the Difference of Two Squares

A method for factoring the difference of two squares is obtained by reversing the special product for the sum and difference of two terms.

> ### The Difference of Two Squares
>
> If A and B are real numbers, variables, or algebraic expressions, then
>
> $$A^2 - B^2 = (A + B)(A - B).$$
>
> In words: The difference of the squares of two terms factors as the product of a sum and a difference of those terms.

EXAMPLE 7 Factoring the Difference of Two Squares

Factor: **a.** $x^2 - 4$ **b.** $81x^2 - 49$.

SOLUTION

We must express each term as the square of some monomial. Then we use the formula for factoring $A^2 - B^2$.

a. $x^2 - 4 \;=\; x^2 - 2^2 \;=\; (x + 2)(x - 2)$

$$A^2 \;-\; B^2 \;=\; (A \;+\; B)\,(A \;-\; B)$$

b. $81x^2 - 49 = (9x)^2 - 7^2 = (9x + 7)(9x - 7)$ • • •

 Check Point **7** Factor:

a. $x^2 - 81$ $(x + 9)(x - 9)$ **b.** $36x^2 - 25$. $(6x + 5)(6x - 5)$

We have seen that a polynomial is factored completely when it is written as the product of prime polynomials. To be sure that you have factored completely, check to see whether any factors with more than one term in the factored polynomial can be factored further. If so, continue factoring.

GREAT QUESTION!

Why isn't factoring $x^4 - 81$ as $(x^2 + 9)(x^2 - 9)$ a complete factorization?

The second factor, $x^2 - 9$, is itself a difference of two squares and can be factored.

EXAMPLE 8 A Repeated Factorization

Factor completely: $x^4 - 81$.

SOLUTION

$$
\begin{aligned}
x^4 - 81 &= (x^2)^2 - 9^2 \\
&= (x^2 + 9)(x^2 - 9) \\
&= (x^2 + 9)(x^2 - 3^2) \\
&= (x^2 + 9)(x + 3)(x - 3)
\end{aligned}
$$

Express as the difference of two squares.

The factors are the sum and the difference of the expressions being squared.

The factor $x^2 - 9$ is the difference of two squares and can be factored.

The factors of $x^2 - 9$ are the sum and the difference of the expressions being squared. • • •

⊘ Check Point **8** Factor completely: $81x^4 - 16.$ $(9x^2 + 4)(3x + 2)(3x - 2)$

⑤ Factor perfect square trinomials.

Factoring Perfect Square Trinomials

Our next factoring technique is obtained by reversing the special products for squaring binomials. The trinomials that are factored using this technique are called **perfect square trinomials**.

> **Factoring Perfect Square Trinomials**
>
> Let A and B be real numbers, variables, or algebraic expressions.
>
> **1.** $A^2 + 2AB + B^2 = (A + B)^2$ **2.** $A^2 - 2AB + B^2 = (A - B)^2$
>
> Same sign Same sign

The two items in the box show that perfect square trinomials, $A^2 + 2AB + B^2$ and $A^2 - 2AB + B^2$, come in two forms: one in which the coefficient of the middle term is positive and one in which the coefficient of the middle term is negative. Here's how to recognize a perfect square trinomial:

1. The first and last terms are squares of monomials or integers.

2. The middle term is twice the product of the expressions being squared in the first and last terms.

> **EXAMPLE 9** Factoring Perfect Square Trinomials
>
> Factor: **a.** $x^2 + 6x + 9$ **b.** $25x^2 - 60x + 36.$

SOLUTION

a. $x^2 + 6x + 9 = x^2 + 2 \cdot x \cdot 3 + 3^2 = (x + 3)^2$ The middle term has a positive sign.

$$A^2 \ + \ 2AB \ + \ B^2 \ = \ (A \ + \ B)^2$$

b. We suspect that $25x^2 - 60x + 36$ is a perfect square trinomial because $25x^2 = (5x)^2$ and $36 = 6^2$. The middle term can be expressed as twice the product of $5x$ and 6.

$$25x^2 - 60x + 36 = (5x)^2 - 2 \cdot 5x \cdot 6 + 6^2 = (5x - 6)^2$$

$$A^2 \ - \ 2AB \ + \ B^2 \ = \ (A \ - \ B)^2$$

• • •

⊘ Check Point **9** Factor:

a. $x^2 + 14x + 49$ $(x + 7)^2$ **b.** $16x^2 - 56x + 49.$ $(4x - 7)^2$

6 Factor the sum or difference of two cubes.

Factoring the Sum or Difference of Two Cubes

We can use the following formulas to factor the sum or the difference of two cubes:

GREAT QUESTION!

A Cube of SOAP

The formulas for factoring $A^3 + B^3$ and $A^3 - B^3$ are difficult to remember and easy to confuse. Can you help me out?

When factoring sums or differences of cubes, observe the sign patterns shown by the voice balloons in the box. The word *SOAP* is a way to remember these patterns:

S O A P.

| Same signs | Opposite signs | Always Positive |

> **Factoring the Sum or Difference of Two Cubes**
>
> **1.** Factoring the Sum of Two Cubes
>
> $$A^3 + B^3 = (A + B)(A^2 - AB + B^2)$$
>
> Same signs Opposite signs Always positive
>
> **2.** Factoring the Difference of Two Cubes
>
> $$A^3 - B^3 = (A - B)(A^2 + AB + B^2)$$
>
> Same signs Opposite signs Always positive

EXAMPLE 10 Factoring Sums and Differences of Two Cubes

Factor: **a.** $x^3 + 8$ **b.** $64x^3 - 125$.

SOLUTION

a. To factor $x^3 + 8$, we must express each term as the cube of some monomial. Then we use the formula for factoring $A^3 + B^3$.

$$x^3 + 8 = x^3 + 2^3 = (x + 2)(x^2 - x \cdot 2 + 2^2) = (x + 2)(x^2 - 2x + 4)$$

$$A^3 + B^3 = (A + B)(A^2 - AB + B^2)$$

b. To factor $64x^3 - 125$, we must express each term as the cube of some monomial. Then we use the formula for factoring $A^3 - B^3$.

$$64x^3 - 125 = (4x)^3 - 5^3 = (4x - 5)[(4x)^2 + (4x)(5) + 5^2]$$

$$A^3 - B^3 = (A - B)(A^2 + AB + B^2)$$

$$= (4x - 5)(16x^2 + 20x + 25)$$ • • •

☑ Check Point **10** Factor:

a. $x^3 + 1$ $(x + 1)(x^2 - x + 1)$ **b.** $125x^3 - 8$. $(5x - 2)(25x^2 + 10x + 4)$

7 Use a general strategy for factoring polynomials.

A Strategy for Factoring Polynomials

It is important to practice factoring a wide variety of polynomials so that you can quickly select the appropriate technique. The polynomial is factored completely when all its polynomial factors, except possibly for monomial factors, are prime. Because of the commutative property, the order of the factors does not matter.

A Strategy for Factoring a Polynomial

1. If there is a common factor, factor out the GCF.

2. Determine the number of terms in the polynomial and try factoring as follows:

 a. If there are two terms, can the binomial be factored by using one of the following special forms?

 Difference of two squares: $A^2 - B^2 = (A + B)(A - B)$

 Sum of two cubes: $A^3 + B^3 = (A + B)(A^2 - AB + B^2)$

 Difference of two cubes: $A^3 - B^3 = (A - B)(A^2 + AB + B^2)$

 b. If there are three terms, is the trinomial a perfect square trinomial? If so, factor by using one of the following special forms:

 $$A^2 + 2AB + B^2 = (A + B)^2$$
 $$A^2 - 2AB + B^2 = (A - B)^2.$$

 If the trinomial is not a perfect square trinomial, try factoring by trial and error.

 c. If there are four or more terms, try factoring by grouping.

3. Check to see if any factors with more than one term in the factored polynomial can be factored further. If so, factor completely.

EXAMPLE 11 Factoring a Polynomial

Factor: $2x^3 + 8x^2 + 8x$.

SOLUTION

Step 1 If there is a common factor, factor out the GCF. Because $2x$ is common to all terms, we factor it out.

$$2x^3 + 8x^2 + 8x = 2x(x^2 + 4x + 4) \quad \text{Factor out the GCF.}$$

Step 2 Determine the number of terms and factor accordingly. The factor $x^2 + 4x + 4$ has three terms and is a perfect square trinomial. We factor using $A^2 + 2AB + B^2 = (A + B)^2$.

$$2x^3 + 8x^2 + 8x = 2x(x^2 + 4x + 4)$$
$$= 2x(x^2 + 2 \cdot x \cdot 2 + 2^2)$$

$$\underbrace{\qquad}_{A^2 \ + \ 2AB \ + \ B^2}$$

$$= 2x(x + 2)^2 \qquad\qquad A^2 + 2AB + B^2 = (A + B)^2$$

Step 3 Check to see if factors can be factored further. In this problem, they cannot. Thus,

$$2x^3 + 8x^2 + 8x = 2x(x + 2)^2. \qquad \bullet\bullet\bullet$$

✓ Check Point **11** Factor: $3x^3 - 30x^2 + 75x$. $3x(x - 5)^2$

EXAMPLE 12 Factoring a Polynomial

Factor: $x^2 - 25a^2 + 8x + 16$.

SOLUTION

Step 1 If there is a common factor, factor out the GCF. Other than 1 or −1, there is no common factor.

Step 2 Determine the number of terms and factor accordingly. There are four terms. We try factoring by grouping. It can be shown that grouping into two groups of two terms does not result in a common binomial factor. Let's try grouping as a difference of squares.

$$x^2 - 25a^2 + 8x + 16$$

$$= (x^2 + 8x + 16) - 25a^2 \qquad \text{Rearrange terms and group as a perfect square trinomial minus } 25a^2 \text{ to obtain a difference of squares.}$$

$$= (x + 4)^2 - (5a)^2 \qquad \text{Factor the perfect square trinomial.}$$

$$= (x + 4 + 5a)(x + 4 - 5a) \qquad \text{Factor the difference of squares. The factors are the sum and difference of the expressions being squared.}$$

Step 3 Check to see if factors can be factored further. In this case, they cannot, so we have factored completely. • • •

⊘ **Check Point 12** Factor: $x^2 - 36a^2 + 20x + 100.$ $(x + 10 + 6a)(x + 10 - 6a)$

8 Factor algebraic expressions containing fractional and negative exponents.

Factoring Algebraic Expressions Containing Fractional and Negative Exponents

Although expressions containing fractional and negative exponents are not polynomials, they can be simplified using factoring techniques.

EXAMPLE 13 Factoring Involving Fractional and Negative Exponents

Factor and simplify: $x(x + 1)^{-\frac{3}{4}} + (x + 1)^{\frac{1}{4}}.$

SOLUTION

The greatest common factor of $x(x + 1)^{-\frac{3}{4}} + (x + 1)^{\frac{1}{4}}$ is $x + 1$ with the *smaller exponent* in the two terms. Thus, the greatest common factor is $(x + 1)^{-\frac{3}{4}}$.

$$x(x + 1)^{-\frac{3}{4}} + (x + 1)^{\frac{1}{4}}$$

$$= (x + 1)^{-\frac{3}{4}}x + (x + 1)^{-\frac{3}{4}}(x + 1)^{\frac{4}{4}} \qquad \text{Express each term as the product of the greatest common factor and its other factor. Note that } (x + 1)^{-\frac{3}{4}}(x + 1)^{\frac{4}{4}} = (x + 1)^{-\frac{3}{4}+\frac{4}{4}} = (x + 1)^{\frac{1}{4}}.$$

$$= (x + 1)^{-\frac{3}{4}}x + (x + 1)^{-\frac{3}{4}}(x + 1) \qquad \text{Simplify: } (x + 1)^{\frac{4}{4}} = (x + 1).$$

$$= (x + 1)^{-\frac{3}{4}}[x + (x + 1)] \qquad \text{Factor out the greatest common factor.}$$

$$= \frac{2x + 1}{(x + 1)^{\frac{3}{4}}} \qquad b^{-n} = \frac{1}{b^n} \qquad \text{• • •}$$

⊘ **Check Point 13** Factor and simplify: $x(x - 1)^{-\frac{1}{2}} + (x - 1)^{\frac{1}{2}}.$ $\frac{2x - 1}{(x - 1)^{1/2}}$

ACHIEVING SUCCESS

In the next section, you will see how mathematical models involving quotients of polynomials describe environmental issues. Factoring is an essential skill for working with such models. **Success in mathematics cannot be achieved without a complete understanding of factoring.** Be sure to work all the assigned exercises in Exercise Set P.5 so that you can apply each of the factoring techniques discussed in this section. The more deeply you force your brain to think about factoring by working many exercises, the better will be your chances of achieving success in future math courses.

CONCEPT AND VOCABULARY CHECK

Here is a list of the factoring techniques that we have discussed.

a. Factoring out the GCF

b. Factoring by grouping

c. Factoring trinomials by trial and error

d. Factoring the difference of two squares

$$A^2 - B^2 = (A + B)(A - B)$$

e. Factoring perfect square trinomials

$$A^2 + 2AB + B^2 = (A + B)^2$$
$$A^2 - 2AB + B^2 = (A - B)^2$$

f. Factoring the sum of two cubes

$$A^3 + B^3 = (A + B)(A^2 - AB + B^2)$$

g. Factoring the difference of two cubes

$$A^3 - B^3 = (A - B)(A^2 + AB + B^2)$$

Fill in each blank by writing the letter of the technique (a through g) for factoring the polynomial.

1. $16x^2 - 25$ ___d___

2. $27x^3 - 1$ ___g___

3. $x^2 + 7x + xy + 7y$ ___b___

4. $4x^2 + 8x + 3$ ___c___

5. $9x^2 + 24x + 16$ ___e___

6. $5x^2 + 10x$ ___a___

7. $x^3 + 1000$ ___f___

8. The algebraic expression $(x + 1)^{\frac{1}{2}} - \frac{1}{3}(x + 1)^{\frac{3}{2}}$ can be factored using $\underline{(x + 1)^{\frac{1}{2}}}$ as the greatest common factor.

EXERCISE SET P.5

Practice Exercises

In Exercises 1–10, factor out the greatest common factor.

1. $18x + 27$ $9(2x + 3)$

2. $16x - 24$ $8(2x - 3)$

3. $3x^2 + 6x$ $3x(x + 2)$

4. $4x^2 - 8x$ $4x(x - 2)$

5. $9x^4 - 18x^3 + 27x^2$ *

6. $6x^4 - 18x^3 + 12x^2$ *

7. $x(x + 5) + 3(x + 5)$ *

8. $x(2x + 1) + 4(2x + 1)$ *

9. $x^2(x - 3) + 12(x - 3)$ *

10. $x^2(2x + 5) + 17(2x + 5)$ *

In Exercises 11–16, factor by grouping.

11. $x^3 - 2x^2 + 5x - 10$ *

12. $x^3 - 3x^2 + 4x - 12$ *

13. $x^3 - x^2 + 2x - 2$ *

14. $x^3 + 6x^2 - 2x - 12$ *

15. $3x^3 - 2x^2 - 6x + 4$ *

16. $x^3 - x^2 - 5x + 5$ *

In Exercises 17–38, factor each trinomial, or state that the trinomial is prime.

17. $x^2 + 5x + 6$ $(x + 2)(x + 3)$

18. $x^2 + 8x + 15$ $(x + 3)(x + 5)$

19. $x^2 - 2x - 15$ $(x - 5)(x + 3)$

20. $x^2 - 4x - 5$ $(x - 5)(x + 1)$

21. $x^2 - 8x + 15$ $(x - 5)(x - 3)$

22. $x^2 - 14x + 45$ $(x - 5)(x - 9)$

23. $3x^2 - x - 2$ $(3x + 2)(x - 1)$

24. $2x^2 + 5x - 3$ $(2x - 1)(x + 3)$

25. $3x^2 - 25x - 28$ *

26. $3x^2 - 2x - 5$ $(3x - 5)(x + 1)$

27. $6x^2 - 11x + 4$ *

28. $6x^2 - 17x + 12$ *

29. $4x^2 + 16x + 15$ *

30. $8x^2 + 33x + 4$ *

31. $9x^2 - 9x + 2$ $(3x - 2)(3x - 1)$

32. $9x^2 + 5x - 4$ $(9x - 4)(x + 1)$

33. $20x^2 + 27x - 8$ *

34. $15x^2 - 19x + 6$ *

35. $2x^2 + 3xy + y^2$ *

36. $3x^2 + 4xy + y^2$ *

37. $6x^2 - 5xy - 6y^2$ *

38. $6x^2 - 7xy - 5y^2$ *

In Exercises 39–48, factor the difference of two squares.

39. $x^2 - 100$ $(x + 10)(x - 10)$

40. $x^2 - 144$ $(x + 12)(x - 12)$

41. $36x^2 - 49$ $(6x + 7)(6x - 7)$

42. $64x^2 - 81$ $(8x + 9)(8x - 9)$

43. $9x^2 - 25y^2$ *

44. $36x^2 - 49y^2$ *

45. $x^4 - 16$ *

46. $x^4 - 1$ *

47. $16x^4 - 81$ *

48. $81x^4 - 1$ *

In Exercises 49–56, factor each perfect square trinomial.

49. $x^2 + 2x + 1$ $(x + 1)^2$

50. $x^2 + 4x + 4$ $(x + 2)^2$

51. $x^2 - 14x + 49$ $(x - 7)^2$

52. $x^2 - 10x + 25$ $(x - 5)^2$

53. $4x^2 + 4x + 1$ $(2x + 1)^2$

54. $25x^2 + 10x + 1$ $(5x + 1)^2$

55. $9x^2 - 6x + 1$ $(3x - 1)^2$

56. $64x^2 - 16x + 1$ $(8x - 1)^2$

In Exercises 57–64, factor using the formula for the sum or difference of two cubes.

57. $x^3 + 27$ *

58. $x^3 + 64$ *

59. $x^3 - 64$ *

60. $x^3 - 27$ *

61. $8x^3 - 1$ *

62. $27x^3 - 1$ *

63. $64x^3 + 27$ *

64. $8x^3 + 125$ *

*See Answers to Selected Exercises.

In Exercises 65–92, factor completely, or state that the polynomial is prime.

65. $3x^3 - 3x$ $3x(x + 1)(x - 1)$ **66.** $5x^3 - 45x$ $5x(x + 3)(x - 3)$

67. $4x^2 - 4x - 24$ * **68.** $6x^2 - 18x - 60$ *

69. $2x^4 - 162$ * **70.** $7x^4 - 7$ *

71. $x^3 + 2x^2 - 9x - 18$ * **72.** $x^3 + 3x^2 - 25x - 75$ *

73. $2x^2 - 2x - 112$ * **74.** $6x^2 - 6x - 12$ *

75. $x^3 - 4x$ $x(x + 2)(x - 2)$ **76.** $9x^3 - 9x$ $9x(x + 1)(x - 1)$

77. $x^2 + 64$ prime **78.** $x^2 + 36$ prime

79. $x^3 + 2x^2 - 4x - 8$ * **80.** $x^3 + 2x^2 - x - 2$ *

81. $y^5 - 81y$ * **82.** $y^5 - 16y$ *

83. $20y^4 - 45y^2$ * **84.** $48y^4 - 3y^2$ *

85. $x^2 - 12x + 36 - 49y^2$ * **86.** $x^2 - 10x + 25 - 36y^2$ *

87. $9b^2x - 16y - 16x + 9b^2y$ $(x + y)(3b + 4)(3b - 4)$

88. $16a^2x - 25y - 25x + 16a^2y$ $(x + y)(4a + 5)(4a - 5)$

89. $x^2y - 16y + 32 - 2x^2$ $(y - 2)(x + 4)(x - 4)$

90. $12x^2y - 27y - 4x^2 + 9$ $(3y - 1)(2x + 3)(2x - 3)$

91. $2x^3 - 8a^2x + 24x^2 + 72x$ $2x(x + 6 + 2a)(x + 6 - 2a)$

92. $2x^3 - 98a^2x + 28x^2 + 98x$ $2x(x + 7 + 7a)(x + 7 - 7a)$

In Exercises 93–102, factor and simplify each algebraic expression.

93. $x^{\frac{3}{2}} - x^{\frac{1}{2}}$ $x^{1/2}(x - 1)$ **94.** $x^{\frac{3}{4}} - x^{\frac{1}{4}}$ $x^{1/4}(x^{1/2} - 1)$

95. $4x^{-\frac{2}{3}} + 8x^{\frac{1}{3}}$ * **96.** $12x^{-\frac{3}{4}} + 6x^{\frac{1}{4}}$ *

97. $(x + 3)^{\frac{1}{2}} - (x + 3)^{\frac{3}{2}}$ * **98.** $(x^2 + 4)^{\frac{3}{2}} + (x^2 + 4)^{\frac{7}{2}}$ *

99. $(x + 5)^{-\frac{1}{2}} - (x + 5)^{-\frac{3}{2}}$ $\dfrac{x + 4}{(x + 5)^{3/2}}$

100. $(x^2 + 3)^{-\frac{2}{3}} + (x^2 + 3)^{-\frac{5}{3}}$ $\dfrac{x^2 + 4}{(x^2 + 3)^{5/3}}$

101. $(4x - 1)^{\frac{1}{2}} - \frac{1}{3}(4x - 1)^{\frac{3}{2}}$ $-\dfrac{4(4x - 1)^{1/2}(x - 1)}{3}$

102. $-8(4x + 3)^{-2} + 10(5x + 1)(4x + 3)^{-1}$ *

Practice Plus

In Exercises 103–114, factor completely.

103. $10x^2(x + 1) - 7x(x + 1) - 6(x + 1)$ *

104. $12x^2(x - 1) - 4x(x - 1) - 5(x - 1)$ *

105. $6x^4 + 35x^2 - 6$ * **106.** $7x^4 + 34x^2 - 5$ *

107. $y^7 + y$ * **108.** $(y + 1)^3 + 1$ *

109. $x^4 - 5x^2y^2 + 4y^4$ * **110.** $x^4 - 10x^2y^2 + 9y^4$ *

111. $(x - y)^4 - 4(x - y)^2$ $(x - y)^2(x - y + 2)(x - y - 2)$

112. $(x + y)^4 - 100(x + y)^2$ $(x + y)^2(x + y + 10)(x + y - 10)$

113. $2x^2 - 7xy^2 + 3y^4$ * **114.** $3x^2 + 5xy^2 + 2y^4$ *

Application Exercises

115. Your computer store is having an incredible sale. The price on one model is reduced by 40%. Then the sale price is reduced by another 40%. If x is the computer's original price, the sale price can be modeled by

$$(x - 0.4x) - 0.4(x - 0.4x).$$

a. Factor out $(x - 0.4x)$ from each term. Then simplify the resulting expression. $(x - 0.4x)(1 - 0.4) = 0.36x$

b. Use the simplified expression from part (a) to answer these questions. With a 40% reduction followed by a 40% reduction, is the computer selling at 20% of its original price? If not, at what percentage of the original price is it selling? no; 36%

116. Your local electronics store is having an end-of-the-year sale. The price on a plasma television had been reduced by 30%. Now the sale price is reduced by another 30%. If x is the television's original price, the sale price can be modeled by

$$(x - 0.3x) - 0.3(x - 0.3x).$$

a. Factor out $(x - 0.3x)$ from each term. Then simplify the resulting expression. $(x - 0.3x)(1 - 0.3) = 0.49x$

b. Use the simplified expression from part (a) to answer these questions. With a 30% reduction followed by a 30% reduction, is the television selling at 40% of its original price? If not, at what percentage of the original price is it selling? no; 49%

In Exercises 117–120,

a. *Write an expression for the area of the shaded region.*

b. *Write the expression in factored form.*

117. *

118. *

119. *

120.

a. $x^2 + xy + xy + y^2 = x^2 + 2xy + y^2$

b. $(x + y)^2$

In Exercises 121–122, find the formula for the volume of the region outside the smaller rectangular solid and inside the larger rectangular solid. Then express the volume in factored form.

121.

$4a^3 - 4ab^2$
$= 4a(a + b)(a - b)$

122.

$3a^3 - 3ab^2$
$= 3a(a + b)(a - b)$

Explaining the Concepts 123–127. Answers will vary.

123. Using an example, explain how to factor out the greatest common factor of a polynomial.

124. Suppose that a polynomial contains four terms. Explain how to use factoring by grouping to factor the polynomial.

125. Explain how to factor $3x^2 + 10x + 8$.

126. Explain how to factor the difference of two squares. Provide an example with your explanation.

127. What is a perfect square trinomial and how is it factored?

128. Explain how to factor $x^3 + 1$. 128–129. Answers will vary.

129. What does it mean to factor completely?

Critical Thinking Exercises

Make Sense? *In Exercises 130–133, determine whether each statement makes sense or does not make sense, and explain your reasoning.*

130. Although $20x^3$ appears in both $20x^3 + 8x^2$ and $20x^3 + 10x$, I'll need to factor $20x^3$ in different ways to obtain each polynomial's factorization. makes sense

131. You grouped the polynomial's terms using different groupings than I did, yet we both obtained the same factorization. *

132. I factored $4x^2 - 100$ completely and obtained $(2x + 10)(2x - 10)$. does not make sense

133. First factoring out the greatest common factor makes it easier for me to determine how to factor the remaining factor, assuming that it is not prime. makes sense

In Exercises 134–137, determine whether each statement is true or false. If the statement is false, make the necessary change(s) to produce a true statement. Changes to false statements will vary.

134. $x^4 - 16$ is factored completely as $(x^2 + 4)(x^2 - 4)$. false

135. The trinomial $x^2 - 4x - 4$ is a prime polynomial. true

136. $x^2 + 36 = (x + 6)^2$ false

137. $x^3 - 64 = (x + 4)(x^2 + 4x - 16)$ false

In Exercises 138–141, factor completely.

138. $x^{2n} + 6x^n + 8$ $(x^n + 4)(x^n + 2)$

139. $-x^2 - 4x + 5$ $-(x + 5)(x - 1)$

140. $x^4 - y^4 - 2x^3y + 2xy^3$ $(x - y)^3(x + y)$

141. $(x - 5)^{-\frac{1}{2}}(x + 5)^{-\frac{1}{2}} - (x + 5)^{\frac{1}{2}}(x - 5)^{-\frac{3}{2}}$ *

In Exercises 142–143, find all integers b so that the trinomial can be factored.

142. $x^2 + bx + 15$ $8, -8, 16, -16$ **143.** $x^2 + 4x + b$ *

Preview Exercises

Exercises 144–146 will help you prepare for the material covered in the next section.

144. Factor the numerator and the denominator. Then simplify by dividing out the common factor in the numerator and the denominator.

$$\frac{x^2 + 6x + 5}{x^2 - 25} \quad \frac{(x + 5)(x + 1)}{(x + 5)(x - 5)} = \frac{x + 1}{x - 5}$$

In Exercises 145–146, perform the indicated operation. Where possible, reduce the answer to its lowest terms.

145. $\frac{5}{4} \cdot \frac{8}{15}$ $\frac{2}{3}$ **146.** $\frac{1}{2} + \frac{2}{3}$ $\frac{7}{6}$

Section P.6 Rational Expressions

What am I supposed to learn?

After studying this section, you should be able to:

① Specify numbers that must be excluded from the domain of a rational expression.

② Simplify rational expressions.

③ Multiply rational expressions.

④ Divide rational expressions.

⑤ Add and subtract rational expressions.

⑥ Simplify complex rational expressions.

How do we describe the costs of reducing environmental pollution? We often use algebraic expressions involving quotients of polynomials. For example, the algebraic expression

$$\frac{250x}{100 - x}$$

describes the cost, in millions of dollars, to remove x percent of the pollutants that are discharged into a river. Removing a modest percentage of pollutants, say 40%, is far less costly than removing a substantially greater percentage, such as 95%. We see this by evaluating the algebraic expression for $x = 40$ and $x = 95$.

Evaluating $\dfrac{250x}{100 - x}$ for

$x = 40$: $x = 95$:

Cost is $\dfrac{250(40)}{100 - 40} \approx 167.$ Cost is $\dfrac{250(95)}{100 - 95} = 4750.$

The cost increases from approximately \$167 million to a possibly prohibitive \$4750 million, or \$4.75 billion. Costs spiral upward as the percentage of removed pollutants increases.

DISCOVERY

What happens if you try substituting 100 for x in

$$\frac{250x}{100 - x}?$$

What does this tell you about the cost of cleaning up all of the river's pollutants?

 Specify numbers that must be excluded from the domain of a rational expression.

Many algebraic expressions that describe costs of environmental projects are examples of *rational expressions*. First we will define rational expressions. Then we will review how to perform operations with such expressions.

Rational Expressions

A **rational expression** is the quotient of two polynomials. Some examples are

$$\frac{x - 2}{4}, \quad \frac{4}{x - 2}, \quad \frac{x}{x^2 - 1}, \quad \text{and} \quad \frac{x^2 + 1}{x^2 + 2x - 3}.$$

The set of real numbers for which an algebraic expression is defined is the **domain** of the expression. Because rational expressions indicate division and division by zero is undefined, **we must exclude numbers from a rational expression's domain that make the denominator zero.**

EXAMPLE 1 Excluding Numbers from the Domain

Find all the numbers that must be excluded from the domain of each rational expression:

a. $\dfrac{4}{x - 2}$ **b.** $\dfrac{x}{x^2 - 1}$ **c.** $\dfrac{9x}{x^2 + 3x - 18}.$

SOLUTION

To determine the numbers that must be excluded from each domain, examine the denominators.

a. $\dfrac{4}{x - 2}$ **b.** $\dfrac{x}{x^2 - 1} = \dfrac{x}{(x + 1)(x - 1)}$ **c.** $\dfrac{9x}{x^2 + 3x - 18} = \dfrac{9x}{(x + 6)(x - 3)}$

This denominator would equal zero if $x = 2$. This factor would equal zero if $x = -1$. This factor would equal zero if $x = 1$. This factor would equal zero if $x = -6$. This factor would equal zero if $x = 3$.

For the rational expression in part (a), we must exclude 2 from the domain. For the rational expression in part (b), we must exclude both -1 and 1 from the domain. For the rational expression in part (c), we must exclude both -6 and 3 from the domain. These excluded numbers are often written to the right of a rational expression:

$$\frac{4}{x - 2}, x \neq 2 \qquad \frac{x}{x^2 - 1}, x \neq -1, x \neq 1 \qquad \frac{9x}{x^2 + 3x - 18}, x \neq -6, x \neq 3. \quad \bullet\bullet\bullet$$

✓ Check Point **1** Find all the numbers that must be excluded from the domain of each rational expression:

a. $\dfrac{7}{x + 5}$ -5 **b.** $\dfrac{x}{x^2 - 36}$ $6, -6$ **c.** $\dfrac{7x}{x^2 - 5x - 14}.$ $-2, 7$

 Simplify rational expressions.

Simplifying Rational Expressions

A rational expression is **simplified** if its numerator and denominator have no common factors other than 1 or -1. The following procedure can be used to simplify rational expressions:

Simplifying Rational Expressions

1. Factor the numerator and the denominator completely.
2. Divide both the numerator and the denominator by any common factors.

EXAMPLE 2 Simplifying Rational Expressions

Simplify:

a. $\dfrac{x^3 + x^2}{x + 1}$ **b.** $\dfrac{x^2 + 6x + 5}{x^2 - 25}$.

SOLUTION

a. $\dfrac{x^3 + x^2}{x + 1} = \dfrac{x^2(x + 1)}{x + 1}$ Factor the numerator. Because the denominator is $x + 1$, $x \neq -1$.

$= \dfrac{x^2(\overset{1}{\cancel{x + 1}})}{\underset{1}{\cancel{x + 1}}}$ Divide out the common factor, $x + 1$.

$= x^2, x \neq -1$ Denominators of 1 need not be written because $\frac{a}{1} = a$.

b. $\dfrac{x^2 + 6x + 5}{x^2 - 25} = \dfrac{(x + 5)(x + 1)}{(x + 5)(x - 5)}$ Factor the numerator and denominator. Because the denominator is $(x + 5)(x - 5)$, $x \neq -5$ and $x \neq 5$.

$= \dfrac{(\overset{1}{\cancel{x + 5}})(x + 1)}{(\underset{1}{\cancel{x + 5}})(x - 5)}$ Divide out the common factor, $x + 5$.

$= \dfrac{x + 1}{x - 5}, \quad x \neq -5, \quad x \neq 5$ • • •

✓ **Check Point 2** Simplify:

a. $\dfrac{x^3 + 3x^2}{x + 3}$ $x^2, x \neq -3$ **b.** $\dfrac{x^2 - 1}{x^2 + 2x + 1}$. $\frac{x - 1}{x + 1}, x \neq -1$

③ Multiply rational expressions.

Multiplying Rational Expressions

The product of two rational expressions is the product of their numerators divided by the product of their denominators. Here is a step-by-step procedure for multiplying rational expressions:

Multiplying Rational Expressions

1. Factor all numerators and denominators completely.
2. Divide numerators and denominators by common factors.
3. Multiply the remaining factors in the numerators and multiply the remaining factors in the denominators.

EXAMPLE 3 Multiplying Rational Expressions

Multiply: $\dfrac{x - 7}{x - 1} \cdot \dfrac{x^2 - 1}{3x - 21}$.

SOLUTION

$$\frac{x-7}{x-1} \cdot \frac{x^2-1}{3x-21}$$

This is the given multiplication problem.

$$= \frac{x-7}{x-1} \cdot \frac{(x+1)(x-1)}{3(x-7)}$$

Factor as many numerators and denominators as possible. Because the denominators have factors of $x-1$ and $x-7$, $x \neq 1$ and $x \neq 7$.

$$= \frac{x - \overset{1}{\cancel{7}}}{\cancel{x}-\underset{1}{\cancel{1}}} \cdot \frac{(x+1)(\overset{1}{\cancel{x}-1})}{3(\cancel{x}-\underset{1}{\cancel{7}})}$$

Divide numerators and denominators by common factors.

$$= \frac{x+1}{3}, x \neq 1, x \neq 7$$

Multiply the remaining factors in the numerators and denominators.

> These excluded numbers from the domain must also be excluded from the simplified expression's domain.

⊘ Check Point **3** Multiply:

$$\frac{x+3}{x^2-4} \cdot \frac{x^2-x-6}{x^2+6x+9}. \qquad \frac{x-3}{(x-2)(x+3)}, x \neq 2, x \neq -2, x \neq -3$$

4 Divide rational expressions.

Dividing Rational Expressions

The quotient of two rational expressions is the product of the first expression and the multiplicative inverse, or reciprocal, of the second expression. The reciprocal is found by interchanging the numerator and the denominator. Thus, **we find the quotient of two rational expressions by inverting the divisor and multiplying.**

EXAMPLE 4 Dividing Rational Expressions

Divide: $\dfrac{x^2-2x-8}{x^2-9} \div \dfrac{x-4}{x+3}$.

> This divisor cannot be zero, so $x \neq 4$. But also $x+3 \neq 0$, so $x \neq -3$. Thus, neither the numerator nor the denominator of the divisor can be zero.

SOLUTION

$$\frac{x^2-2x-8}{x^2-9} \div \frac{x-4}{x+3}$$

This is the given division problem.

$$= \frac{x^2-2x-8}{x^2-9} \cdot \frac{x+3}{x-4}$$

Invert the divisor and multiply.

$$= \frac{(x-4)(x+2)}{(x+3)(x-3)} \cdot \frac{x+3}{x-4}$$

Factor as many numerators and denominators as possible. For nonzero denominators, $x \neq -3$, $x \neq 3$, and $x \neq 4$.

$$= \frac{(\overset{1}{\cancel{x-4}})(x+2)}{(\cancel{x+3})(x-3)} \cdot \frac{(\overset{1}{\cancel{x+3}})}{(\cancel{x-4})}$$

Divide numerators and denominators by common factors.

$$= \frac{x+2}{x-3}, x \neq -3, x \neq 3, x \neq 4$$

Multiply the remaining factors in the numerators and in the denominators.

⊘ Check Point **4** Divide:

$$\frac{x^2-2x+1}{x^3+x} \div \frac{x^2+x-2}{3x^2+3}. \qquad \frac{3(x-1)}{x(x+2)}, x \neq 1, x \neq 0, x \neq -2$$

 Add and subtract rational expressions.

Adding and Subtracting Rational Expressions with the Same Denominator

We add or subtract rational expressions with the same denominator by (1) adding or subtracting the numerators, (2) placing this result over the common denominator, and (3) simplifying, if possible.

EXAMPLE 5 Subtracting Rational Expressions with the Same Denominator

Subtract: $\dfrac{5x + 1}{x^2 - 9} - \dfrac{4x - 2}{x^2 - 9}$.

SOLUTION

GREAT QUESTION!

When subtracting a numerator containing more than one term, do I have to insert parentheses like you did in Example 5?

Yes. When a numerator is being subtracted, we must **subtract every term in that expression.** Parentheses indicate that every term inside is being subtracted.

> Don't forget the parentheses.

$$\frac{5x + 1}{x^2 - 9} - \frac{4x - 2}{x^2 - 9} = \frac{5x + 1 - (4x - 2)}{x^2 - 9}$$

Subtract numerators and include parentheses to indicate that both terms are subtracted. Place this difference over the common denominator.

$$= \frac{5x + 1 - 4x + 2}{x^2 - 9}$$

Remove parentheses and then change the sign of each term in parentheses.

$$= \frac{x + 3}{x^2 - 9}$$

Combine like terms.

$$= \frac{\overset{1}{x + 3}}{\underset{1}{(x + 3)}(x - 3)}$$

Factor and simplify ($x \neq -3$ and $x \neq 3$).

$$= \frac{1}{x - 3}, x \neq -3, x \neq 3$$

• • •

✓ **Check Point 5** Subtract: $\dfrac{x}{x + 1} - \dfrac{3x + 2}{x + 1}$. $-2, x \neq -1$

Adding and Subtracting Rational Expressions with Different Denominators

We can gain insight into adding rational expressions with different denominators by looking closely at what we do when adding fractions with different denominators. For example, suppose that we want to add $\frac{1}{2}$ and $\frac{2}{3}$. We must first write the fractions with the same denominator. We look for the smallest number that contains both 2 and 3 as factors. This number, 6, is then used as the *least common denominator*, or LCD.

The **least common denominator**, or LCD, of several rational expressions is a polynomial consisting of the product of all prime factors in the denominators, with each factor raised to the greatest power of its occurrence in any denominator.

Finding the Least Common Denominator

1. Factor each denominator completely.

2. List the factors of the first denominator.

3. Add to the list in step 2 any factors of the second denominator that do not appear in the list.

4. Form the product of each different factor from the list in step 3. This product is the least common denominator.

EXAMPLE 6 Finding the Least Common Denominator

Find the LCD of

$$\frac{x + 2}{2x - 3} \quad \text{and} \quad \frac{4}{x + 3}.$$

SOLUTION

Step 1 Factor each denominator completely.

$$2x - 3 = 1(2x - 3)$$

$$x + 3 = 1(x + 3)$$

Factors are 1 and 2x − 3. $\dfrac{x + 2}{2x - 3}$ $\dfrac{4}{x + 3}$ Factors are 1 and x + 3.

Step 2 List the factors of the first denominator.

$$1, 2x - 3$$

Step 3 Add any unlisted factors from the second denominator. One factor of the second denominator is already in our list. That factor is 1. However, the other factor, $x + 3$, is not listed in step 2. We add $x + 3$ to the list. We have

$$1, 2x - 3, x + 3.$$

Step 4 The least common denominator is the product of all factors in the final list. Thus

$$1(2x - 3)(x + 3)$$

or $(2x - 3)(x + 3)$ is the least common denominator of $\dfrac{x + 2}{2x - 3}$ and $\dfrac{4}{x + 3}$. • • •

⬭ Check Point **6** Find the LCD of

$$\frac{3}{x + 1} \quad \text{and} \quad \frac{5}{x - 1}. \quad (x + 1)(x - 1)$$

In Example 6 and Check Point 6, the denominators do not have any common factors other than 1 and −1. Thus, the LCD is simply the product of the denominators. However, in Example 7, the denominators both have a factor of $x + 3$.

EXAMPLE 7 Finding the Least Common Denominator

Find the LCD of

$$\frac{7}{5x^2 + 15x} \quad \text{and} \quad \frac{9}{x^2 + 6x + 9}.$$

SOLUTION

Step 1 Factor each denominator completely.

$$5x^2 + 15x = 5x(x + 3)$$

$$x^2 + 6x + 9 = (x + 3)^2 \quad \text{or} \quad (x + 3)(x + 3)$$

Factors are 5, x, and x + 3. $\dfrac{7}{5x^2 + 15x}$ $\dfrac{9}{x^2 + 6x + 9}$ Factors are x + 3 and x + 3.

Step 2 List the factors of the first denominator.

$$5, x, x + 3$$

Step 3 Add any unlisted factors from the second denominator. One factor of $x^2 + 6x + 9$ is already in our list. That factor is $x + 3$. However, the other factor of $x + 3$ is not listed in step 2. We add a second factor of $x + 3$ to the list. We have

$$5, x, x + 3, x + 3.$$

Step 4 The least common denominator is the product of all factors in the final list. Thus,

$$5x(x + 3)(x + 3) \quad \text{or} \quad 5x(x + 3)^2$$

is the least common denominator. • • •

In Example 7, if you were to multiply the denominators as given, you would get a common denominator, but not the LCD. While any common denominator can be used to add or subtract rational expressions, using the LCD greatly reduces the amount of work involved.

⊘ Check Point **7** Find the least common denominator of

$$\frac{3}{x^2 - 6x + 9} \quad \text{and} \quad \frac{7}{x^2 - 9}. \quad {\scriptstyle (x - 3)(x - 3)(x + 3) \text{ or } (x - 3)^2(x + 3)}$$

Finding the least common denominator for two (or more) rational expressions is the first step needed to add or subtract the expressions.

> **Adding and Subtracting Rational Expressions That Have Different Denominators**
>
> 1. Find the LCD of the rational expressions.
> 2. Rewrite each rational expression as an equivalent expression whose denominator is the LCD. To do so, multiply the numerator and the denominator of each rational expression by any factor(s) needed to convert the denominator into the LCD.
> 3. Add or subtract numerators, placing the resulting expression over the LCD.
> 4. If possible, simplify the resulting rational expression.

EXAMPLE 8 Subtracting Rational Expressions with Different Denominators

Subtract: $\dfrac{x + 2}{2x - 3} - \dfrac{4}{x + 3}$.

SOLUTION

Step 1 Find the least common denominator. In Example 6, we found that the LCD for these rational expressions is $(2x - 3)(x + 3)$.

Step 2 Write equivalent expressions with the LCD as denominators. We must rewrite each rational expression with a denominator of $(2x - 3)(x + 3)$. We do so by multiplying both the numerator and the denominator of each rational expression by any factor needed to convert the expression's denominator into the LCD.

$$\frac{x + 2}{2x - 3} \cdot \frac{x + 3}{x + 3} = \frac{(x + 2)(x + 3)}{(2x - 3)(x + 3)} \qquad \frac{4}{x + 3} \cdot \frac{2x - 3}{2x - 3} = \frac{4(2x - 3)}{(2x - 3)(x + 3)}$$

> Multiply the numerator and denominator by $x + 3$ to get $(2x - 3)(x + 3)$, the LCD.

> Multiply the numerator and denominator by $2x - 3$ to get $(2x - 3)(x + 3)$, the LCD.

Because $\dfrac{x + 3}{x + 3} = 1$ and $\dfrac{2x - 3}{2x - 3} = 1$, we are not changing the value of either rational expression, only its appearance.

Now we are ready to perform the indicated subtraction.

$$\frac{x + 2}{2x - 3} - \frac{4}{x + 3} \qquad \text{This is the given problem.}$$

$$= \frac{(x + 2)(x + 3)}{(2x - 3)(x + 3)} - \frac{4(2x - 3)}{(x + 3)(2x - 3)} \qquad \begin{array}{l} \text{Multiply each numerator and denominator} \\ \text{by the extra factor required to form} \\ (2x - 3)(x + 3), \text{ the LCD.} \end{array}$$

Step 3 Subtract numerators, putting this difference over the LCD.

$$= \frac{(x + 2)(x + 3) - 4(2x - 3)}{(2x - 3)(x + 3)}$$

$$= \frac{x^2 + 5x + 6 - (8x - 12)}{(2x - 3)(x + 3)}$$ Multiply in the numerator using FOIL and the distributive property.

$$= \frac{x^2 + 5x + 6 - 8x + 12}{(2x - 3)(x + 3)}$$ Remove parentheses and then change the sign of each term in parentheses.

$$= \frac{x^2 - 3x + 18}{(2x - 3)(x + 3)}, x \neq \frac{3}{2}, x \neq -3$$ Combine like terms in the numerator.

Step 4 If necessary, simplify. Because the numerator is prime, no further simplification is possible. •••

\oslash Check Point **8** Add: $\dfrac{x}{x - 3} + \dfrac{x - 1}{x + 3}$.

EXAMPLE 9 Adding Rational Expressions with Different Denominators

Add: $\dfrac{x + 3}{x^2 + x - 2} + \dfrac{2}{x^2 - 1}$.

SOLUTION

Step 1 Find the least common denominator. Start by factoring the denominators.

$$x^2 + x - 2 = (x + 2)(x - 1)$$
$$x^2 - 1 = (x + 1)(x - 1)$$

The factors of the first denominator are $x + 2$ and $x - 1$. The only factor from the second denominator that is not listed is $x + 1$. Thus, the least common denominator is

$$(x + 2)(x - 1)(x + 1).$$

Step 2 Write equivalent expressions with the LCD as denominators. We must rewrite each rational expression with a denominator of $(x + 2)(x - 1)(x + 1)$. We do so by multiplying both the numerator and the denominator of each rational expression by any factor(s) needed to convert the expression's denominator into the LCD.

$$\frac{x + 3}{(x + 2)(x - 1)} \cdot \frac{x + 1}{x + 1} = \frac{(x + 3)(x + 1)}{(x + 2)(x - 1)(x + 1)} \qquad \frac{2}{(x + 1)(x - 1)} \cdot \frac{x + 2}{x + 2} = \frac{2(x + 2)}{(x + 2)(x - 1)(x + 1)}$$

Multiply the numerator and denominator by $x + 1$ to get $(x + 2)(x - 1)(x + 1)$, the **LCD**. Multiply the numerator and denominator by $x + 2$ to get $(x + 2)(x - 1)(x + 1)$, the **LCD**.

Because $\dfrac{x + 1}{x + 1} = 1$ and $\dfrac{x + 2}{x + 2} = 1$, we are not changing the value of either rational expression, only its appearance.

Now we are ready to perform the indicated addition.

$$\frac{x + 3}{x^2 + x - 2} + \frac{2}{x^2 - 1}$$ This is the given problem.

$$= \frac{x + 3}{(x + 2)(x - 1)} + \frac{2}{(x + 1)(x - 1)}$$ Factor the denominators. The LCD is $(x + 2)(x - 1)(x + 1)$.

$$= \frac{(x + 3)(x + 1)}{(x + 2)(x - 1)(x + 1)} + \frac{2(x + 2)}{(x + 2)(x - 1)(x + 1)}$$ Rewrite as equivalent expressions with the LCD.

Step 3 **Add numerators, putting this sum over the LCD.**

$$= \frac{(x+3)(x+1) + 2(x+2)}{(x+2)(x-1)(x+1)}$$

$$= \frac{x^2 + 4x + 3 + 2x + 4}{(x+2)(x-1)(x+1)}$$

Perform the multiplications in the numerator.

$$= \frac{x^2 + 6x + 7}{(x+2)(x-1)(x+1)}, x \neq -2, x \neq 1, x \neq -1$$

Combine like terms in the numerator: $4x + 2x = 6x$ and $3 + 4 = 7$.

Step 4 **If necessary, simplify.** Because the numerator is prime, no further simplification is possible.

● ● ●

✓ Check Point **9** Subtract: $\dfrac{x}{x^2 - 10x + 25} - \dfrac{x-4}{2x - 10}$. $\frac{-x^2 + 11x - 20}{2(x-5)^2}, x \neq 5$

GREAT QUESTION!

How can I quickly tell that $x^2 + 6x + 7$ is prime?

The only positive factors of 7 are 1 and 7. Their sum is 8, not 6.

⑥ Simplify complex rational expressions.

Complex Rational Expressions

Complex rational expressions, also called **complex fractions**, have numerators or denominators containing one or more rational expressions. Here are two examples of such expressions:

$$\dfrac{1 + \dfrac{1}{x}}{1 - \dfrac{1}{x}}$$

Separate rational expressions occur in the numerator and the denominator.

$$\dfrac{\dfrac{1}{x+h} - \dfrac{1}{x}}{h}.$$

Separate rational expressions occur in the numerator.

One method for simplifying a complex rational expression is to combine its numerator into a single expression and combine its denominator into a single expression. Then perform the division by inverting the denominator and multiplying.

EXAMPLE 10 Simplifying a Complex Rational Expression

Simplify: $\dfrac{1 + \dfrac{1}{x}}{1 - \dfrac{1}{x}}$.

SOLUTION

Step 1 **Add to get a single rational expression in the numerator.**

$$1 + \frac{1}{x} = \frac{1}{1} + \frac{1}{x} = \frac{1 \cdot x}{1 \cdot x} + \frac{1}{x} = \frac{x}{x} + \frac{1}{x} = \frac{x+1}{x}$$

The LCD is $1 \cdot x$, or x.

Step 2 **Subtract to get a single rational expression in the denominator.**

$$1 - \frac{1}{x} = \frac{1}{1} - \frac{1}{x} = \frac{1 \cdot x}{1 \cdot x} - \frac{1}{x} = \frac{x}{x} - \frac{1}{x} = \frac{x-1}{x}$$

The LCD is $1 \cdot x$, or x.

Step 3 **Perform the division indicated by the main fraction bar: Invert and multiply. If possible, simplify.**

$$\frac{1 + \dfrac{1}{x}}{1 - \dfrac{1}{x}} = \frac{\dfrac{x+1}{x}}{\dfrac{x-1}{x}} = \frac{x+1}{x} \cdot \frac{x}{x-1} = \frac{x+1}{\overset{1}{\cancel{x}}} \cdot \frac{\overset{1}{\cancel{x}}}{x-1} = \frac{x+1}{x-1}$$

Invert and multiply.

● ● ●

✓ Check Point **10** Simplify: $\dfrac{\dfrac{1}{x} - \dfrac{3}{2}}{\dfrac{1}{x} + \dfrac{3}{4}}$. $\frac{2(2-3x)}{4+3x}, x \neq 0, x \neq -\frac{4}{3}$

A second method for simplifying a complex rational expression is to find the least common denominator of all the rational expressions in its numerator and denominator. Then multiply each term in its numerator and denominator by this least common denominator. Because we are multiplying by a form of 1, we will obtain an equivalent expression that does not contain fractions in its numerator or denominator. Here we use this method to simplify the complex rational expression in Example 10.

$$\dfrac{1 + \dfrac{1}{x}}{1 - \dfrac{1}{x}} = \dfrac{\left(1 + \dfrac{1}{x}\right)}{\left(1 - \dfrac{1}{x}\right)} \cdot \dfrac{x}{x}$$

The least common denominator of all the rational expressions is x. Multiply the numerator and denominator by x. Because $\dfrac{x}{x} = 1$, we are not changing the complex fraction ($x \neq 0$).

$$= \dfrac{1 \cdot x + \dfrac{1}{x} \cdot x}{1 \cdot x - \dfrac{1}{x} \cdot x}$$

Use the distributive property. Be sure to distribute x to every term.

$$= \dfrac{x + 1}{x - 1}, x \neq 0, x \neq 1$$

Multiply. The complex rational expression is now simplified.

EXAMPLE 11 Simplifying a Complex Rational Expression

Simplify: $\dfrac{\dfrac{1}{x + h} - \dfrac{1}{x}}{h}$.

SOLUTION

We will use the method of multiplying each of the three terms, $\dfrac{1}{x + h}$, $\dfrac{1}{x}$, and h, by the least common denominator. The least common denominator is $x(x + h)$.

$$\dfrac{\dfrac{1}{x + h} - \dfrac{1}{x}}{h}$$

$$= \dfrac{\left(\dfrac{1}{x + h} - \dfrac{1}{x}\right)x(x + h)}{hx(x + h)}$$

Multiply the numerator and denominator by $x(x + h)$, $h \neq 0$, $x \neq 0$, $x \neq -h$.

$$= \dfrac{\dfrac{1}{x + h} \cdot x(x + h) - \dfrac{1}{x} \cdot x(x + h)}{hx(x + h)}$$

Use the distributive property in the numerator.

$$= \dfrac{x - (x + h)}{hx(x + h)}$$

Simplify: $\dfrac{1}{x + h} \cdot x(x + h) = x$ and $\dfrac{1}{x} \cdot x(x + h) = x + h$.

$$= \dfrac{x - x - h}{hx(x + h)}$$

Subtract in the numerator. Remove parentheses and change the sign of each term in parentheses.

$$= \frac{\overset{1}{-\cancel{h}}}{\underset{1}{\cancel{h}x(x + h)}}$$

Simplify: $x - x - h = -h$.

$$= -\frac{1}{x(x + h)}, h \neq 0, x \neq 0, x \neq -h$$ Divide the numerator and denominator by h. ● ● ●

\checkmark **Check Point 11** Simplify: $\dfrac{\dfrac{1}{x + 7} - \dfrac{1}{x}}{7}$. $-\frac{1}{x(x+7)}, x \neq 0, x \neq -7$

CONCEPT AND VOCABULARY CHECK

Fill in each blank so that the resulting statement is true.

1. A rational expression is the quotient of two ___polynomials___.

2. The set of real numbers for which a rational expression is defined is the ___domain___ of the expression. We must exclude all numbers from this set that make the denominator of the rational expression ___0___.

3. We simplify a rational expression by ___factoring___ the numerator and the denominator completely. Then we divide the numerator and the denominator by any ___common factors___.

4. $\dfrac{x}{5} \cdot \dfrac{x}{3} = \dfrac{\frac{x^2}{15}}{\rule{2cm}{0.4pt}}$

5. $\dfrac{x}{5} \div \dfrac{x}{3} = \dfrac{\frac{3}{5}}{\rule{2cm}{0.4pt}}, x \neq 0$

6. $\dfrac{x^2}{3} - \dfrac{x - 4}{3} = \dfrac{\frac{x^2 - x + 4}{3}}{\rule{2cm}{0.4pt}}$

7. Consider the following subtraction problem:

$$\frac{x - 1}{x^2 + x - 6} - \frac{x - 2}{x^2 + 4x + 3}.$$

The factors of the first denominator are ___$x + 3$ and $x - 2$___. The factors of the second denominator are ___$x + 3$ and $x + 1$___.

The LCD is ___$(x + 3)(x - 2)(x + 1)$___.

8. An equivalent expression for $\dfrac{3x + 2}{x - 5}$ with a denominator of $(3x + 4)(x - 5)$ can be obtained by multiplying the numerator and denominator by ___$3x + 4$___.

9. A rational expression whose numerator or denominator or both contain rational expressions is called a/an ___complex___ rational expression or a/an ___complex___ fraction.

10. $\dfrac{\dfrac{1}{x + 3} - \dfrac{1}{x}}{3} = \dfrac{x(x + 3)}{x(x + 3)} \cdot \dfrac{\left(\dfrac{1}{x + 3} - \dfrac{1}{x}\right)}{3} = \dfrac{x - (x + 3)}{3x(x + 3)}$

$$= \frac{\overset{-3}{\rule{1cm}{0.4pt}}}{3x(x + 3)}$$

$$= -\frac{\overset{1}{\rule{1cm}{0.4pt}}}{x(x + 3)}$$

EXERCISE SET P.6

Practice Exercises

In Exercises 1–6, find all numbers that must be excluded from the domain of each rational expression.

1. $\dfrac{7}{x - 3}$ 3

2. $\dfrac{13}{x + 9}$ −9

3. $\dfrac{x + 5}{x^2 - 25}$ 5, −5

4. $\dfrac{x + 7}{x^2 - 49}$ 7, −7

5. $\dfrac{x - 1}{x^2 + 11x + 10}$ −1, −10

6. $\dfrac{x - 3}{x^2 + 4x - 45}$ −9, 5

In Exercises 7–14, simplify each rational expression. Find all numbers that must be excluded from the domain of the simplified rational expression.

7. $\dfrac{3x - 9}{x^2 - 6x + 9}$ $\frac{3}{x - 3}, x \neq 3$

8. $\dfrac{4x - 8}{x^2 - 4x + 4}$ *

9. $\dfrac{x^2 - 12x + 36}{4x - 24}$ $\frac{x - 6}{4}, x \neq 6$

10. $\dfrac{x^2 - 8x + 16}{3x - 12}$ *

11. $\dfrac{y^2 + 7y - 18}{y^2 - 3y + 2}$ $\frac{y + 9}{y - 1}, y \neq 1, 2$

12. $\dfrac{y^2 - 4y - 5}{y^2 + 5y + 4}$ *

13. $\dfrac{x^2 + 12x + 36}{x^2 - 36}$ $\frac{x + 6}{x - 6}, x \neq 6, -6$

14. $\dfrac{x^2 - 14x + 49}{x^2 - 49}$ *

In Exercises 15–32, multiply or divide as indicated.

15. $\dfrac{x - 2}{3x + 9} \cdot \dfrac{2x + 6}{2x - 4}$ $\frac{1}{3}, x \neq 2, -3$

16. $\dfrac{6x + 9}{3x - 15} \cdot \dfrac{x - 5}{4x + 6}$ *

17. $\dfrac{x^2 - 9}{x^2} \cdot \dfrac{x^2 - 3x}{x^2 + x - 12}$ *

18. $\dfrac{x^2 - 4}{x^2 - 4x + 4} \cdot \dfrac{2x - 4}{x + 2}$ *

19. $\dfrac{x^2 - 5x + 6}{x^2 - 2x - 3} \cdot \dfrac{x^2 - 1}{x^2 - 4}$ $\frac{x - 1}{x + 2}, x \neq -2, -1, 2, 3$

20. $\dfrac{x^2 + 5x + 6}{x^2 + x - 6} \cdot \dfrac{x^2 - 9}{x^2 - x - 6}$ $\frac{x + 3}{x - 2}, x \neq -3, -2, 2, 3$

21. $\dfrac{x^3 - 8}{x^2 - 4} \cdot \dfrac{x + 2}{3x}$ *

22. $\dfrac{x^2 + 6x + 9}{x^3 + 27} \cdot \dfrac{1}{x + 3}$ *

*See Answers to Selected Exercises.

23. $\dfrac{x+1}{3} \div \dfrac{3x+3}{7}$ $\frac{7}{9}, x \neq -1$ **24.** $\dfrac{x+5}{7} \div \dfrac{4x+20}{9}$ *

25. $\dfrac{x^2-4}{x} \div \dfrac{x+2}{x-2}$ * **26.** $\dfrac{x^2-4}{x-2} \div \dfrac{x+2}{4x-8}$ *

27. $\dfrac{4x^2+10}{x-3} \div \dfrac{6x^2+15}{x^2-9}$ $\frac{2(x+3)}{3}, x \neq 3, -3$

28. $\dfrac{x^2+x}{x^2-4} \div \dfrac{x^2-1}{x^2+5x+6}$ $\frac{x(x+3)}{(x-2)(x-1)}, x \neq 2, 1, -1, -2, -3$

29. $\dfrac{x^2-25}{2x-2} \div \dfrac{x^2+10x+25}{x^2+4x-5}$ $\frac{x-5}{2}, x \neq 1, -5$

30. $\dfrac{x^2-4}{x^2+3x-10} \div \dfrac{x^2+5x+6}{x^2+8x+15}$ $1, x \neq 2, -2, -3, -5$

31. $\dfrac{x^2+x-12}{x^2+x-30} \cdot \dfrac{x^2+5x+6}{x^2-2x-3} \div \dfrac{x+3}{x^2+7x+6}$ *

32. $\dfrac{x^3-25x}{4x^2} \cdot \dfrac{2x^2-2}{x^2-6x+5} \div \dfrac{x^2+5x}{7x+7}$ *

In Exercises 33–58, add or subtract as indicated.

33. $\dfrac{4x+1}{6x+5} + \dfrac{8x+9}{6x+5}$ $2, x \neq -\frac{5}{6}$ **34.** $\dfrac{3x+2}{3x+4} + \dfrac{3x+6}{3x+4}$ *

35. $\dfrac{x^2-2x}{x^2+3x} + \dfrac{x^2+x}{x^2+3x}$ $\frac{2x-1}{x+3}, x \neq 0, -3$

36. $\dfrac{x^2-4x}{x^2-x-6} + \dfrac{4x-4}{x^2-x-6}$ $\frac{x-2}{x-3}, x \neq -2, 3$

37. $\dfrac{4x-10}{x-2} - \dfrac{x-4}{x-2}$ $3, x \neq 2$ **38.** $\dfrac{2x+3}{3x-6} - \dfrac{3-x}{3x-6}$ *

39. $\dfrac{x^2+3x}{x^2+x-12} - \dfrac{x^2-12}{x^2+x-12}$ $\frac{3}{x-3}, x \neq 3, -4$

40. $\dfrac{x^2-4x}{x^2-x-6} - \dfrac{x-6}{x^2-x-6}$ $\frac{x-2}{x+2}, x \neq -2, 3$

41. $\dfrac{3}{x+4} + \dfrac{6}{x+5}$ * **42.** $\dfrac{8}{x-2} + \dfrac{2}{x-3}$ *

43. $\dfrac{3}{x+1} - \dfrac{3}{x}$ $-\frac{3}{x(x+1)}, x \neq -1, 0$ **44.** $\dfrac{4}{x} - \dfrac{3}{x+3}$ *

45. $\dfrac{2x}{x+2} + \dfrac{x+2}{x-2}$ * **46.** $\dfrac{3x}{x-3} - \dfrac{x+4}{x+2}$ *

47. $\dfrac{x+5}{x-5} + \dfrac{x-5}{x+5}$ * **48.** $\dfrac{x+3}{x-3} + \dfrac{x-3}{x+3}$ *

49. $\dfrac{3}{2x+4} + \dfrac{2}{3x+6}$ * **50.** $\dfrac{5}{2x+8} + \dfrac{7}{3x+12}$ *

51. $\dfrac{4}{x^2+6x+9} + \dfrac{4}{x+3}$ * **52.** $\dfrac{3}{5x+2} + \dfrac{5x}{25x^2-4}$ *

53. $\dfrac{3x}{x^2+3x-10} - \dfrac{2x}{x^2+x-6}$ *

54. $\dfrac{x}{x^2-2x-24} - \dfrac{x}{x^2-7x+6}$ *

55. $\dfrac{x+3}{x^2-1} - \dfrac{x+2}{x-1}$ * **56.** $\dfrac{x+5}{x^2-4} - \dfrac{x+1}{x-2}$ *

57. $\dfrac{4x^2+x-6}{x^2+3x+2} - \dfrac{3x}{x+1} + \dfrac{5}{x+2}$ $\frac{x-1}{x+2}, x \neq -2, -1$

58. $\dfrac{6x^2+17x-40}{x^2+x-20} + \dfrac{3}{x-4} - \dfrac{5x}{x+5}$ *

In Exercises 59–72, simplify each complex rational expression.

59. $\dfrac{\frac{x}{3}-1}{x-3}$ $\frac{1}{3}, x \neq 3$ **60.** $\dfrac{\frac{x}{4}-1}{x-4}$ $\frac{1}{4}, x \neq 4$

61. $\dfrac{1+\frac{1}{x}}{3-\frac{1}{x}}$ $\frac{x+1}{3x-1}, x \neq 0, \frac{1}{3}$ **62.** $\dfrac{8+\frac{1}{x}}{4-\frac{1}{x}}$ $\frac{8x+1}{4x-1}, x \neq 0, \frac{1}{4}$

63. $\dfrac{\frac{1}{x}+\frac{1}{y}}{x+y}$ $\frac{1}{xy}, x \neq 0, y \neq 0, x \neq -y$

64. $\dfrac{1-\frac{1}{x}}{xy}$ $\frac{x-1}{x^2y}, x \neq 0, y \neq 0$

65. $\dfrac{x-\frac{x}{x+3}}{x+2}$ $\frac{x}{x+3}, x \neq -2, -3$ **66.** $\dfrac{x-3}{x-\frac{3}{x-2}}$ *

67. $\dfrac{\frac{3}{x-2}-\frac{4}{x+2}}{\frac{7}{x^2-4}}$ $-\frac{x-14}{7}, x \neq -2, 2$ **68.** $\dfrac{\frac{x}{x-2}+1}{\frac{3}{x^2-4}+1}$ *

69. $\dfrac{\frac{1}{x+1}}{\frac{1}{x^2-2x-3}+\frac{1}{x-3}}$ $\frac{x-3}{x+2}, x \neq -2, -1, 3$

70. $\dfrac{\frac{6}{x^2+2x-15}-\frac{1}{x-3}}{\frac{1}{x+5}+1}$ $-\frac{x-1}{(x+6)(x-3)}, x \neq -6, -5, 3$

71. $\dfrac{\frac{1}{(x+h)^2}-\frac{1}{x^2}}{h}$ * **72.** $\dfrac{\frac{x+h}{x+h+1}-\frac{x}{x+1}}{h}$ *

Practice Plus

In Exercises 73–80, perform the indicated operations. Simplify the result, if possible.

73. $\left(\dfrac{2x+3}{x+1} \cdot \dfrac{x^2+4x-5}{2x^2+x-3}\right) - \dfrac{2}{x+2}$ $\frac{x^2+5x+8}{(x+2)(x+1)}$

74. $\dfrac{1}{x^2-2x-8} \div \left(\dfrac{1}{x-4} - \dfrac{1}{x+2}\right)$ $\frac{1}{6}$

75. $\left(2-\dfrac{6}{x+1}\right)\left(1+\dfrac{3}{x-2}\right)$ 2

76. $\left(4-\dfrac{3}{x+2}\right)\left(1+\dfrac{5}{x-1}\right)$ $\frac{(4x+5)(x+4)}{(x+2)(x-1)}$

77. $\dfrac{y^{-1}-(y+5)^{-1}}{5}$ $\frac{1}{y(y+5)}$ **78.** $\dfrac{y^{-1}-(y+2)^{-1}}{2}$ $\frac{1}{y(y+2)}$

79. $\left(\dfrac{1}{a^3-b^3} \cdot \dfrac{ac+ad-bc-bd}{1}\right) - \dfrac{c-d}{a^2+ab+b^2}$ *

80. $\dfrac{ab}{a^2+ab+b^2} + \left(\dfrac{ac-ad-bc+bd}{ac-ad+bc-bd} \div \dfrac{a^3-b^3}{a^3+b^3}\right)$ *

Application Exercises

81. The rational expression

$$\frac{130x}{100 - x}$$

describes the cost, in millions of dollars, to inoculate x percent of the population against a particular strain of flu.

a. Evaluate the expression for $x = 40, x = 80$, and $x = 90$. Describe the meaning of each evaluation in terms of percentage inoculated and cost. *

b. For what value of x is the expression undefined? $x = 100$

c. What happens to the cost as x approaches 100%? How can you interpret this observation? *

82. The average rate on a round-trip commute having a one-way distance d is given by the complex rational expression

$$\frac{2d}{\dfrac{d}{r_1} + \dfrac{d}{r_2}},$$

in which r_1 and r_2 are the average rates on the outgoing and return trips, respectively. Simplify the expression. Then find your average rate if you drive to campus averaging 40 miles per hour and return home on the same route averaging 30 miles per hour. Explain why the answer is not 35 miles per hour. *

83. The bar graph shows the estimated number of calories per day needed to maintain energy balance for various gender and age groups for moderately active lifestyles. (Moderately active means a lifestyle that includes physical activity equivalent to walking 1.5 to 3 miles per day at 3 to 4 miles per hour, in addition to the light physical activity associated with typical day-to-day life.)

Calories Needed to Maintain Energy Balance for Moderately Active Lifestyles

Source: U.S.D.A.

a. The mathematical model

$$W = -66x^2 + 526x + 1030$$

describes the number of calories needed per day, W, by women in age group x with moderately active lifestyles. According to the model, how many calories per day are needed by women between the ages of 19 and 30, inclusive, with this lifestyle? Does this underestimate or overestimate the number shown by the graph? By how much? 2078; underestimates by 22 calories

b. The mathematical model

$$M = -120x^2 + 998x + 590$$

describes the number of calories needed per day, M, by men in age group x with moderately active lifestyles. According to the model, how many calories per day are needed by men between the ages of 19 and 30, inclusive, with this lifestyle? Does this underestimate or overestimate the number shown by the graph? By how much? 2662; underestimates by 38 calories

c. Write a simplified rational expression that describes the ratio of the number of calories needed per day by women in age group x to the number of calories needed per day by men in age group x for people with moderately active lifestyles. *

In Exercises 84–85, express the perimeter of each rectangle as a single rational expression.

84. **85.**

$$\frac{4x^2 + 14x}{(x + 3)(x + 4)} \qquad \frac{4x^2 + 22x}{(x + 5)(x + 6)}$$

Explaining the Concepts 86–97. Answers will vary.

86. What is a rational expression?

87. Explain how to determine which numbers must be excluded from the domain of a rational expression.

88. Explain how to simplify a rational expression.

89. Explain how to multiply rational expressions.

90. Explain how to divide rational expressions.

91. Explain how to add or subtract rational expressions with the same denominators.

92. Explain how to add rational expressions with different denominators. Use $\dfrac{3}{x + 5} + \dfrac{7}{(x + 2)(x + 5)}$ in your explanation.

93. Explain how to find the least common denominator for denominators of $x^2 - 100$ and $x^2 - 20x + 100$.

94. Describe two ways to simplify $\dfrac{\dfrac{3}{x} + \dfrac{2}{x^2}}{\dfrac{1}{x^2} + \dfrac{2}{x}}$.

Explain the error in Exercises 95–97. Then rewrite the right side of the equation to correct the error that now exists.

95. $\dfrac{1}{a} + \dfrac{1}{b} = \dfrac{1}{a + b}$ **96.** $\dfrac{1}{x} + 7 = \dfrac{1}{x + 7}$

97. $\dfrac{a}{x} + \dfrac{a}{b} = \dfrac{a}{x + b}$

Critical Thinking Exercises

Make Sense? *In Exercises 98–101, determine whether each statement makes sense or does not make sense, and explain your reasoning.*

98. I evaluated $\dfrac{3x - 3}{4x(x - 1)}$ for $x = 1$ and obtained 0. *

99. The rational expressions

$$\frac{7}{14x} \quad \text{and} \quad \frac{7}{14 + x}$$

can both be simplified by dividing each numerator and each denominator by 7. does not make sense

100. When performing the division

$$\frac{7x}{x + 3} \div \frac{(x + 3)^2}{x - 5},$$

I began by dividing the numerator and the denominator by the common factor, $x + 3$. does not make sense

101. I subtracted $\frac{3x - 5}{x - 1}$ from $\frac{x - 3}{x - 1}$ and obtained a constant. *

In Exercises 102–105, determine whether each statement is true or false. If the statement is false, make the necessary change(s) to produce a true statement. Changes to false statements will vary.

102. $\frac{x^2 - 25}{x - 5} = x - 5$ false

103. The expression $\frac{-3y - 6}{y + 2}$ simplifies to the consecutive integer that follows -4. true

104. $\frac{2x - 1}{x - 7} + \frac{3x - 1}{x - 7} - \frac{5x - 2}{x - 7} = 0$ true

105. $6 + \frac{1}{x} = \frac{7}{x}$ false

In Exercises 106–108, perform the indicated operations.

106. $\frac{1}{x^n - 1} - \frac{1}{x^n + 1} - \frac{1}{x^{2n} - 1}$ $\frac{1}{x^{2n} - 1}$

107. $\left(1 - \frac{1}{x}\right)\left(1 - \frac{1}{x + 1}\right)\left(1 - \frac{1}{x + 2}\right)\left(1 - \frac{1}{x + 3}\right)$ $\frac{x - 1}{x + 3}$

108. $(x - y)^{-1} + (x - y)^{-2}$ *

109. In one short sentence, five words or less, explain what

$$\frac{\dfrac{1}{x} + \dfrac{1}{x^2} + \dfrac{1}{x^3}}{\dfrac{1}{x^4} + \dfrac{1}{x^5} + \dfrac{1}{x^6}}$$

does to each number x. It cubes x.

Preview Exercises

Exercises 110–112 will help you prepare for the material covered in the first section of the next chapter.

110. If $y = 4 - x^2$, find the value of y that corresponds to values of x for each integer starting with -3 and ending with 3. *

111. If $y = 1 - x^2$, find the value of y that corresponds to values of x for each integer starting with -3 and ending with 3. *

112. If $y = |x + 1|$, find the value of y that corresponds to values of x for each integer starting with -4 and ending with 2. 3; 2; 1; 0; 1; 2; 3

CHAPTER P Summary, Review, and Test

SUMMARY: BASIC FORMULAS

Definition of Absolute Value

$$|x| = \begin{cases} x & \text{if } x \geq 0 \\ -x & \text{if } x < 0 \end{cases}$$

Distance between Points a and b on a Number Line

$$|a - b| \quad \text{or} \quad |b - a|$$

Properties of Real Numbers

Commutative	$a + b = b + a$
	$ab = ba$
Associative	$(a + b) + c = a + (b + c)$
	$(ab)c = a(bc)$
Distributive	$a(b + c) = ab + ac$
Identity	$a + 0 = a$
	$a \cdot 1 = a$
Inverse	$a + (-a) = 0$
	$a \cdot \dfrac{1}{a} = 1, a \neq 0$

Properties of Exponents

$$b^{-n} = \frac{1}{b^n}, \quad b^0 = 1, \quad b^m \cdot b^n = b^{m+n},$$

$$(b^m)^n = b^{mn}, \quad \frac{b^m}{b^n} = b^{m-n}, \quad (ab)^n = a^n b^n, \quad \left(\frac{a}{b}\right)^n = \frac{a^n}{b^n}$$

Product and Quotient Rules for nth Roots

$$\sqrt[n]{ab} = \sqrt[n]{a} \cdot \sqrt[n]{b}, \quad \sqrt[n]{\frac{a}{b}} = \frac{\sqrt[n]{a}}{\sqrt[n]{b}}$$

Rational Exponents

$$a^{\frac{1}{n}} = \sqrt[n]{a}, \quad a^{-\frac{1}{n}} = \frac{1}{a^{\frac{1}{n}}} = \frac{1}{\sqrt[n]{a}},$$

$$a^{\frac{m}{n}} = (\sqrt[n]{a})^m = \sqrt[n]{a^m}, \quad a^{-\frac{m}{n}} = \frac{1}{a^{\frac{m}{n}}}$$

Special Products

$$(A + B)(A - B) = A^2 - B^2$$

$$(A + B)^2 = A^2 + 2AB + B^2$$

$$(A - B)^2 = A^2 - 2AB + B^2$$

$$(A + B)^3 = A^3 + 3A^2B + 3AB^2 + B^3$$

$$(A - B)^3 = A^3 - 3A^2B + 3AB^2 - B^3$$

Factoring Formulas

$$A^2 - B^2 = (A + B)(A - B)$$

$$A^2 + 2AB + B^2 = (A + B)^2$$

$$A^2 - 2AB + B^2 = (A - B)^2$$

$$A^3 + B^3 = (A + B)(A^2 - AB + B^2)$$

$$A^3 - B^3 = (A - B)(A^2 + AB + B^2)$$

REVIEW EXERCISES

You can use these review exercises, like the review exercises at the end of each chapter, to test your understanding of the chapter's topics. However, you can also use these exercises as a prerequisite test to check your mastery of the fundamental algebra skills needed in this book.

P.1

In Exercises 1–2, evaluate each algebraic expression for the given value or values of the variable(s).

1. $3 + 6(x - 2)^3$ for $x = 4$ 51
2. $x^2 - 5(x - y)$ for $x = 6$ and $y = 2$ 16
3. You are riding along an expressway traveling x miles per hour. The formula

$$S = 0.015x^2 + x + 10$$

models the recommended safe distance, S, in feet, between your car and other cars on the expressway. What is the recommended safe distance when your speed is 60 miles per hour? 124 feet

In Exercises 4–7, let $A = \{a, b, c\}, B = \{a, c, d, e\}$, and $C = \{a, d, f, g\}$. Find the indicated set.

4. $A \cap B$ $\{a, c\}$
5. $A \cup B$ $\{a, b, c, d, e\}$
6. $A \cup C$ $\{a, b, c, d, f, g\}$
7. $C \cap A$ $\{a\}$
8. Consider the set:

$$\left\{ -17, -\tfrac{9}{13}, 0, 0.75, \sqrt{2}, \pi, \sqrt{81} \right\}.$$

List all numbers from the set that are **a.** natural numbers, **b.** whole numbers, **c.** integers, **d.** rational numbers, **e.** irrational numbers, **f.** real numbers. *

In Exercises 9–11, rewrite each expression without absolute value bars.

9. $|-103|$ 103
10. $|\sqrt{2} - 1|$ $\sqrt{2} - 1$
11. $|3 - \sqrt{17}|$ $\sqrt{17} - 3$

12. Express the distance between the numbers -17 and 4 using absolute value. Then evaluate the absolute value. *

In Exercises 13–18, state the name of the property illustrated.

13. $3 + 17 = 17 + 3$ commutative property of addition
14. $(6 \cdot 3) \cdot 9 = 6 \cdot (3 \cdot 9)$ associative property of multiplication
15. $\sqrt{3}(\sqrt{5} + \sqrt{3}) = \sqrt{15} + 3$ distributive property
16. $(6 \cdot 9) \cdot 2 = 2 \cdot (6 \cdot 9)$ commutative property of multiplication
17. $\sqrt{3}(\sqrt{5} + \sqrt{3}) = (\sqrt{5} + \sqrt{3})\sqrt{3}$ *
18. $(3 \cdot 7) + (4 \cdot 7) = (4 \cdot 7) + (3 \cdot 7)$ *

In Exercises 19–22, simplify each algebraic expression.

19. $5(2x - 3) + 7x$ $17x - 15$

20. $\tfrac{1}{5}(5x) + [(3y) + (-3y)] - (-x)$ $2x$
21. $3(4y - 5) - (7y + 2)$ $5y - 17$
22. $8 - 2[3 - (5x - 1)]$ $10x$
23. The diversity index, from 0 (no diversity) to 100, measures the chance that two randomly selected people are a different race or ethnicity. The diversity index in the United States varies widely from region to region, from as high as 81 in Hawaii to as low as 11 in Vermont. The bar graph shows the national diversity index for the United States for four years in the period from 1980 through 2010.

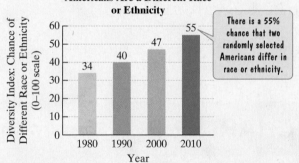

Chance That Two Randomly Selected Americans Are a Different Race or Ethnicity

There is a 55% chance that two randomly selected Americans differ in race or ethnicity.

Source: *USA Today*

The data in the graph can be modeled by the formula

$$D = 0.005x^2 + 0.55x + 34,$$

where D is the national diversity index in the United States x years after 1980. According to the formula, what was the U.S. diversity index in 2010? How does this compare with the index displayed by the bar graph? 55; It's the same.

P.2

Evaluate each exponential expression in Exercises 24–27.

24. $(-3)^3(-2)^2$ -108
25. $2^{-4} + 4^{-1}$ $\tfrac{5}{16}$
26. $5^{-3} \cdot 5$ $\tfrac{1}{25}$
27. $\dfrac{3^3}{3^6}$ $\tfrac{1}{27}$

Simplify each exponential expression in Exercises 28–31.

28. $(-2x^4y^3)^3$ $-8x^{12}y^9$
29. $(-5x^3y^2)(-2x^{-11}y^{-2})$ $\dfrac{10}{x^8}$
30. $(2x^3)^{-4}$ $\dfrac{1}{16x^{12}}$
31. $\dfrac{7x^5y^6}{28x^{15}y^{-2}}$ $\dfrac{y^8}{4x^{10}}$

In Exercises 32–33, write each number in decimal notation.

32. 3.74×10^4 37,400
33. 7.45×10^{-5} 0.0000745

In Exercises 34–35, write each number in scientific notation.

34. 3,590,000 3.59×10^6
35. 0.00725 7.25×10^{-3}

*See Answers to Selected Exercises.

In Exercises 36–37, perform the indicated operation and write the answer in decimal notation.

36. $(3 \times 10^3)(1.3 \times 10^2)$ * **37.** $\dfrac{6.9 \times 10^3}{3 \times 10^5}$ 0.023

In 2009, the United States government spent more than it had collected in taxes, resulting in a budget deficit of $1.35 trillion. In Exercises 38–40, you will use scientific notation to put a number like 1.35 trillion in perspective. Use 10^{12} for 1 trillion.

38. Express 1.35 trillion in scientific notation. 1.35×10^{12}
39. There are approximately 32,000,000 seconds in a year. Express this number in scientific notation. 3.2×10^7
40. Use your scientific notation answers from Exercises 38 and 39 to answer this question: How many years is 1.35 trillion seconds? Round to the nearest year. 42,188 years

P.3

Use the product rule to simplify the expressions in Exercises 41–44. In Exercises 43–44, assume that variables represent nonnegative real numbers.

41. $\sqrt{300}$ $10\sqrt{3}$ **42.** $\sqrt{12x^2}$ $2|x|\sqrt{3}$
43. $\sqrt{10x} \cdot \sqrt{2x}$ $2x\sqrt{5}$ **44.** $\sqrt{r^3}$ $r\sqrt{r}$

Use the quotient rule to simplify the expressions in Exercises 45–46.

45. $\sqrt{\dfrac{121}{4}}$ $\frac{11}{2}$ **46.** $\dfrac{\sqrt{96x^3}}{\sqrt{2x}}$ (Assume that $x > 0$.) *

In Exercises 47–49, add or subtract terms whenever possible.

47. $7\sqrt{5} + 13\sqrt{5}$ $20\sqrt{5}$
48. $2\sqrt{50} + 3\sqrt{8}$ $16\sqrt{2}$
49. $4\sqrt{72} - 2\sqrt{48}$ $24\sqrt{2} - 8\sqrt{3}$

In Exercises 50–53, rationalize the denominator.

50. $\dfrac{30}{\sqrt{5}}$ $6\sqrt{5}$ **51.** $\dfrac{\sqrt{2}}{\sqrt{3}}$ $\frac{\sqrt{6}}{3}$
52. $\dfrac{5}{6 + \sqrt{3}}$ $\frac{5(6-\sqrt{3})}{33}$ **53.** $\dfrac{14}{\sqrt{7} - \sqrt{5}}$ $7(\sqrt{7} + \sqrt{5})$

Evaluate each expression in Exercises 54–57 or indicate that the root is not a real number.

54. $\sqrt[3]{125}$ 5 **55.** $\sqrt[5]{-32}$ -2
56. $\sqrt[4]{-125}$ not real **57.** $\sqrt[4]{(-5)^4}$ 5

Simplify the radical expressions in Exercises 58–62.

58. $\sqrt[3]{81}$ $3\sqrt[3]{3}$ **59.** $\sqrt[3]{y^5}$ $y\sqrt[3]{y^2}$
60. $\sqrt[4]{8} \cdot \sqrt[4]{10}$ $2\sqrt[4]{5}$ **61.** $4\sqrt[3]{16} + 5\sqrt[3]{2}$ $13\sqrt[3]{2}$
62. $\dfrac{\sqrt[4]{32x^5}}{\sqrt[4]{16x}}$ (Assume that $x > 0$.) $x\sqrt[4]{2}$

In Exercises 63–68, evaluate each expression.

63. $16^{\frac{1}{2}}$ 4 **64.** $25^{-\frac{1}{2}}$ $\frac{1}{5}$ **65.** $125^{\frac{1}{3}}$ 5
66. $27^{-\frac{1}{3}}$ $\frac{1}{3}$ **67.** $64^{\frac{2}{3}}$ 16 **68.** $27^{-\frac{4}{3}}$ $\frac{1}{81}$

In Exercises 69–71, simplify using properties of exponents.

69. $\left(5x^{\frac{2}{3}}\right)\left(4x^{\frac{1}{4}}\right)$ $20x^{11/12}$ **70.** $\dfrac{15x^{\frac{3}{4}}}{5x^{\frac{1}{2}}}$ $3x^{1/4}$
71. $(125x^6)^{\frac{2}{3}}$ $25x^4$
72. Simplify by reducing the index of the radical: $\sqrt[6]{y^3}$. \sqrt{y}

P.4

In Exercises 73–74, perform the indicated operations. Write the resulting polynomial in standard form and indicate its degree.

73. $(-6x^3 + 7x^2 - 9x + 3) + (14x^3 + 3x^2 - 11x - 7)$ *
74. $(13x^4 - 8x^3 + 2x^2) - (5x^4 - 3x^3 + 2x^2 - 6)$ *

In Exercises 75–81, find each product.

75. $(3x - 2)(4x^2 + 3x - 5)$ * **76.** $(3x - 5)(2x + 1)$ *
77. $(4x + 5)(4x - 5)$ $16x^2 - 25$ **78.** $(2x + 5)^2$ $4x^2 + 20x + 25$
79. $(3x - 4)^2$ $9x^2 - 24x + 16$ **80.** $(2x + 1)^3$ *
81. $(5x - 2)^3$ $125x^3 - 150x^2 + 60x - 8$

In Exercises 82–83, perform the indicated operations. Indicate the degree of the resulting polynomial.

82. $(7x^2 - 8xy + y^2) + (-8x^2 - 9xy - 4y^2)$ *
83. $(13x^3y^2 - 5x^2y - 9x^2) - (-11x^3y^2 - 6x^2y + 3x^2 - 4)$ *

In Exercises 84–88, find each product.

84. $(x + 7y)(3x - 5y)$ * **85.** $(3x - 5y)^2$ *
86. $(3x^2 + 2y)^2$ * **87.** $(7x + 4y)(7x - 4y)$ *
88. $(a - b)(a^2 + ab + b^2)$ $a^3 - b^3$

P.5

In Exercises 89–105, factor completely, or state that the polynomial is prime.

89. $15x^3 + 3x^2$ $3x^2(5x + 1)$ **90.** $x^2 - 11x + 28$ *
91. $15x^2 - x - 2$ * **92.** $64 - x^2$ $(8 + x)(8 - x)$
93. $x^2 + 16$ prime **94.** $3x^4 - 9x^3 - 30x^2$ *
95. $20x^7 - 36x^3$ $4x^3(5x^4 - 9)$ **96.** $x^3 - 3x^2 - 9x + 27$ *
97. $16x^2 - 40x + 25$ $(4x - 5)^2$ **98.** $x^4 - 16$ *
99. $y^3 - 8$ $(y - 2)(y^2 + 2y + 4)$ **100.** $x^3 + 64$ $(x + 4)(x^2 - 4x + 16)$
101. $3x^4 - 12x^2$ * **102.** $27x^3 - 125$ *
103. $x^5 - x$ * **104.** $x^3 + 5x^2 - 2x - 10$ *
105. $x^2 + 18x + 81 - y^2$ $(x + 9 + y)(x + 9 - y)$

In Exercises 106–108, factor and simplify each algebraic expression.

106. $16x^{-\frac{3}{4}} + 32x^{\frac{1}{4}}$ $\dfrac{16(1 + 2x)}{x^{3/4}}$

107. $(x^2 - 4)(x^2 + 3)^{\frac{1}{2}} - (x^2 - 4)^2(x^2 + 3)^{\frac{3}{2}}$

108. $12x^{-\frac{1}{2}} + 6x^{-\frac{3}{2}}$ $\dfrac{6(2x + 1)}{x^{3/2}}$

P.6

In Exercises 109–111, simplify each rational expression. Also, list all numbers that must be excluded from the domain.

109. $\dfrac{x^3 + 2x^2}{x + 2}$ * **110.** $\dfrac{x^2 + 3x - 18}{x^2 - 36}$ * **111.** $\dfrac{x^2 + 2x}{x^2 + 4x + 4}$ *

In Exercises 112–114, multiply or divide as indicated.

112. $\dfrac{x^2 + 6x + 9}{x^2 - 4} \cdot \dfrac{x + 3}{x - 2}$ * **113.** $\dfrac{6x + 2}{x^2 - 1} \div \dfrac{3x^2 + x}{x - 1}$ *

114. $\dfrac{x^2 - 5x - 24}{x^2 - x - 12} \div \dfrac{x^2 - 10x + 16}{x^2 + x - 6}$ *

In Exercises 115–118, add or subtract as indicated.

115. $\dfrac{2x - 7}{x^2 - 9} - \dfrac{x - 10}{x^2 - 9}$ * **116.** $\dfrac{3x}{x + 2} + \dfrac{x}{x - 2}$ *

117. $\dfrac{x}{x^2 - 9} + \dfrac{x - 1}{x^2 - 5x + 6}$ *

118. $\dfrac{4x - 1}{2x^2 + 5x - 3} - \dfrac{x + 3}{6x^2 + x - 2}$ *

In Exercises 119–121, simplify each complex rational expression.

119. $\dfrac{\frac{1}{x} - \frac{1}{2}}{\frac{1}{3} - \frac{x}{6}}$ * **120.** $\dfrac{3 + \frac{12}{x}}{1 - \frac{16}{x^2}}$ * **121.** $\dfrac{3 - \frac{1}{x + 3}}{3 + \frac{1}{x + 3}}$ *

CHAPTER P TEST

In Exercises 1–18, simplify the given expression or perform the indicated operation (and simplify, if possible), whichever is appropriate.

1. $5(2x^2 - 6x) - (4x^2 - 3x)$ $6x^2 - 27x$

2. $7 + 2[3(x + 1) - 2(3x - 1)]$ $-6x + 17$

3. $\{1, 2, 5\} \cap \{5, a\}$ $\{5\}$

4. $\{1, 2, 5\} \cup \{5, a\}$ $\{1, 2, 5, a\}$

5. $(2x^2y^3 - xy + y^2) - (-4x^2y^3 - 5xy - y^2)$ *

6. $\dfrac{30x^3y^4}{6x^9y^{-4}}$ $\dfrac{5y^8}{x^6}$

7. $\sqrt{6r}\sqrt{3r}$ (Assume that $r \ge 0$.) $3r\sqrt{2}$

8. $4\sqrt{50} - 3\sqrt{18}$ $11\sqrt{2}$ **9.** $\dfrac{3}{5 + \sqrt{2}}$ $\dfrac{3(5 - \sqrt{2})}{23}$

10. $\sqrt[3]{16x^4}$ $2x\sqrt[3]{2x}$ **11.** $\dfrac{x^2 + 2x - 3}{x^2 - 3x + 2}$ *

12. $\dfrac{5 \times 10^{-6}}{20 \times 10^{-8}}$ (Express the answer in scientific notation.) *

13. $(2x - 5)(x^2 - 4x + 3)$ * **14.** $(5x + 3y)^2$ *

15. $\dfrac{2x + 8}{x - 3} \div \dfrac{x^2 + 5x + 4}{x^2 - 9}$ * **16.** $\dfrac{x}{x + 3} + \dfrac{5}{x - 3}$ *

17. $\dfrac{2x + 3}{x^2 - 7x + 12} - \dfrac{2}{x - 3}$ * **18.** $\dfrac{\frac{1}{x} - \frac{1}{3}}{\frac{1}{x}}$ $\dfrac{3 - x}{3}, x \neq 0$

In Exercises 19–24, factor completely, or state that the polynomial is prime.

19. $x^2 - 9x + 18$ $(x - 3)(x - 6)$ **20.** $x^3 + 2x^2 + 3x + 6$ *

21. $25x^2 - 9$ $(5x + 3)(5x - 3)$ **22.** $36x^2 - 84x + 49$ $(6x - 7)^2$

23. $y^3 - 125$ $(y - 5)(y^2 + 5y + 25)$ **24.** $x^2 + 10x + 25 - 9y^2$ *

25. Factor and simplify:

$$x(x + 3)^{-\frac{3}{5}} + (x + 3)^{\frac{2}{5}}. \quad \frac{2x + 3}{(x + 3)^{3/5}}$$

26. List all the rational numbers in this set:

$$\left\{-7, -\tfrac{4}{5}, 0, 0.25, \sqrt{3}, \sqrt{4}, \tfrac{22}{7}, \pi\right\}. \;*$$

In Exercises 27–28, state the name of the property illustrated.

27. $3(2 + 5) = 3(5 + 2)$ commutative property of addition

28. $6(7 + 4) = 6 \cdot 7 + 6 \cdot 4$ distributive property

29. Express in scientific notation: 0.00076. 7.6×10^{-4}

30. Evaluate: $27^{-\frac{5}{3}}$. $\frac{1}{243}$

31. In 2007, world population was approximately 6.6×10^9. By some projections, world population will double by 2040. Express the population at that time in scientific notation. *

32. Big (Lack of) Men on Campus In 2007, 135 women received bachelor's degrees for every 100 men. According to the U.S. Department of Education, that gender imbalance has widened, as shown by the bar graph.

Percentage of Bachelor's Degrees Awarded to United States Men and Women

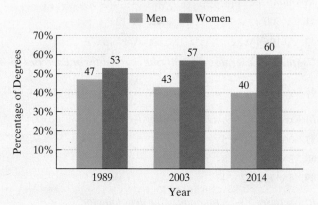

Source: U.S. Department of Education

The data for bachelor's degrees can be described by the following mathematical models:

a. According to the first formula, what percentage of bachelor's degrees were awarded to men in 2003? Does this underestimate or overestimate the actual percent shown by the bar graph? By how much?
43.08%; overestimates by 0.08%

b. Use the given formulas to write a new formula with a rational expression that models the ratio of the percentage of bachelor's degrees received by men to the percentage received by women n years after 1989. Name this new mathematical model R, for ratio. $R = \dfrac{-0.28n + 47}{0.28n + 53}$

c. Use the formula for R to find the ratio of bachelor's degrees received by men to degrees received by women in 2014. According to the model, how many women received bachelor's degrees for every two men in 2014? How well does this describe the data shown by the graph? $\frac{2}{3}$;
3 women for every 2 men; It describes the data exactly.

Equations and Inequalities

The belief that humor and laughter can have positive benefits on our lives is not new. The Bible tells us, "A merry heart doeth good like a medicine, but a broken spirit drieth the bones." (Proverbs 17:22)

Some random humor factoids:

- The average adult laughs 15 times each day. (Newhouse News Service)
- Forty-six percent of people who are telling a joke laugh more than the people they are telling it to. (*U.S. News and World Report*)
- Eighty percent of adult laughter does not occur in response to jokes or funny situations. (*Independent*)
- Algebra can be used to model the influence that humor plays in our responses to negative life events. (Bob Blitzer, *College Algebra*)

That last tidbit that your author threw into the list is true. Based on our sense of humor, there is actually a formula that predicts how we will respond to difficult life events.

Formulas can be used to explain what is happening in the present and to make predictions about what might occur in the future. In this chapter, you will learn to use formulas in new ways that will help you to recognize patterns, logic, and order in a world that can appear chaotic to the untrained eye.

> **HERE'S WHERE YOU'LL FIND THIS APPLICATION:**
> A mathematical model that includes sense of humor as a variable is developed in Example 9 of Section 1.2.

MATT GROENING

Section 1.1 Graphs and Graphing Utilities

The beginning of the seventeenth century was a time of innovative ideas and enormous intellectual progress in Europe. English theatergoers enjoyed a succession of exciting new plays by Shakespeare. William Harvey proposed the radical notion that the heart was a pump for blood rather than the center of emotion. Galileo, with his newfangled invention called the telescope, supported the theory of Polish astronomer Copernicus that the Sun, not the Earth, was the center of the solar system. Monteverdi was writing the world's first grand operas. French mathematicians Pascal and Fermat invented a new field of mathematics called probability theory.

Into this arena of intellectual electricity stepped French aristocrat René Descartes (1596–1650). Descartes (pronounced "day cart"), propelled by the creativity surrounding him, developed a new branch of mathematics that brought together algebra and geometry in a unified way— a way that visualized numbers as points on a graph, equations as geometric figures, and geometric figures as equations. This new branch of mathematics, called *analytic geometry*, established Descartes as one of the founders of modern thought and among the most original mathematicians and philosophers of any age. We begin this section by looking at Descartes's deceptively simple idea, called the **rectangular coordinate system** or (in his honor) the **Cartesian coordinate system**.

① Plot points in the rectangular coordinate system.

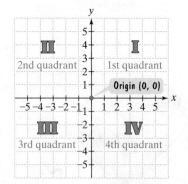

FIGURE 1.1 The rectangular coordinate system

GREAT QUESTION!

What's the significance of the word "ordered" when describing a pair of real numbers?

The phrase *ordered pair* is used because order is important. The order in which coordinates appear makes a difference in a point's location. This is illustrated in **Figure 1.2**.

Points and Ordered Pairs

Descartes used two number lines that intersect at right angles at their zero points, as shown in **Figure 1.1**. The horizontal number line is the **x-axis**. The vertical number line is the **y-axis**. The point of intersection of these axes is their zero points, called the **origin**. Positive numbers are shown to the right and above the origin. Negative numbers are shown to the left and below the origin. The axes divide the plane into four quarters, called **quadrants**. The points located on the axes are not in any quadrant.

Each point in the rectangular coordinate system corresponds to an **ordered pair** of real numbers, (x, y). Examples of such pairs are $(-5, 3)$ and $(3, -5)$. The first number in each pair, called the **x-coordinate**, denotes the distance and direction from the origin along the x-axis. The second number in each pair, called the **y-coordinate**, denotes vertical distance and direction along a line parallel to the y-axis or along the y-axis itself.

Figure 1.2 shows how we **plot**, or locate, the points corresponding to the ordered pairs $(-5, 3)$ and $(3, -5)$. We plot $(-5, 3)$ by going 5 units from 0 to the left along the x-axis. Then we go 3 units up parallel to the y-axis. We plot $(3, -5)$ by going 3 units from 0 to the right along the x-axis and 5 units down parallel to the y-axis. The phrase "the points corresponding to the ordered pairs $(-5, 3)$ and $(3, -5)$" is often abbreviated as "the points $(-5, 3)$ and $(3, -5)$."

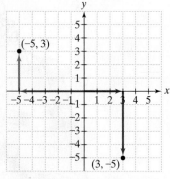

FIGURE 1.2 Plotting $(-5, 3)$ and $(3, -5)$

EXAMPLE 1 Plotting Points in the Rectangular Coordinate System

Plot the points: $A(-3, 5)$, $B(2, -4)$, $C(5, 0)$, $D(-5, -3)$, $E(0, 4)$, and $F(0, 0)$.

SOLUTION

See **Figure 1.3**. We move from the origin and plot the points in the following way:

$A(-3, 5)$:	3 units left, 5 units up
$B(2, -4)$:	2 units right, 4 units down
$C(5, 0)$:	5 units right, 0 units up or down
$D(-5, -3)$:	5 units left, 3 units down
$E(0, 4)$:	0 units right or left, 4 units up
$F(0, 0)$:	0 units right or left, 0 units up or down

Notice that the origin is represented by (0, 0).

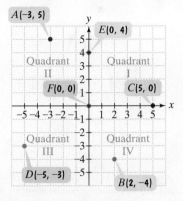

FIGURE 1.3 Plotting points

● ● ●

 Check Point 1 Plot the points: $A(-2, 4)$, $B(4, -2)$, $C(-3, 0)$, and $D(0, -3)$. *

2 Graph equations in the rectangular coordinate system.

Graphs of Equations

A relationship between two quantities can be expressed as an **equation in two variables**, such as

$$y = 4 - x^2.$$

A **solution of an equation in two variables**, x and y, is an ordered pair of real numbers with the following property: When the x-coordinate is substituted for x and the y-coordinate is substituted for y in the equation, we obtain a true statement. For example, consider the equation $y = 4 - x^2$ and the ordered pair $(3, -5)$. When 3 is substituted for x and -5 is substituted for y, we obtain the statement $-5 = 4 - 3^2$, or $-5 = 4 - 9$, or $-5 = -5$. Because this statement is true, the ordered pair $(3, -5)$ is a solution of the equation $y = 4 - x^2$. We also say that $(3, -5)$ **satisfies** the equation.

We can generate as many ordered-pair solutions as desired to $y = 4 - x^2$ by substituting numbers for x and then finding the corresponding values for y. For example, suppose we let $x = 3$:

The **graph of an equation in two variables** is the set of all points whose coordinates satisfy the equation. One method for graphing such equations is the **point-plotting method**. First, we find several ordered pairs that are solutions of the equation. Next, we plot these ordered pairs as points in the rectangular coordinate system. Finally, we connect the points with a smooth curve or line. This often gives us a picture of all ordered pairs that satisfy the equation.

EXAMPLE 2 Graphing an Equation Using the Point-Plotting Method

Graph $y = 4 - x^2$. Select integers for x, starting with -3 and ending with 3.

SOLUTION

For each value of x, we find the corresponding value for y.

FIGURE 1.4 The graph of $y = 4 - x^2$

We selected integers from -3 to 3, inclusive, to include three negative numbers, 0, and three positive numbers. We also wanted to keep the resulting computations for y relatively simple.

Start with x. Compute y. Form the ordered pair (x, y).

x	$y = 4 - x^2$	Ordered Pair (x, y)
-3	$y = 4 - (-3)^2 = 4 - 9 = -5$	$(-3, -5)$
-2	$y = 4 - (-2)^2 = 4 - 4 = 0$	$(-2, 0)$
-1	$y = 4 - (-1)^2 = 4 - 1 = 3$	$(-1, 3)$
0	$y = 4 - 0^2 = 4 - 0 = 4$	$(0, 4)$
1	$y = 4 - 1^2 = 4 - 1 = 3$	$(1, 3)$
2	$y = 4 - 2^2 = 4 - 4 = 0$	$(2, 0)$
3	$y = 4 - 3^2 = 4 - 9 = -5$	$(3, -5)$

Now we plot the seven points and join them with a smooth curve, as shown in **Figure 1.4**. The graph of $y = 4 - x^2$ is a curve where the part of the graph to the right of the y-axis is a reflection of the part to the left of it and vice versa. The arrows on the left and the right of the curve indicate that it extends indefinitely in both directions. ● ● ●

Check Point 2 Graph $y = 4 - x$. Select integers for x, starting with -3 and ending with 3. *

A Brief Review • Absolute Value

Absolute value of x → $|x| = \begin{cases} x & \text{if } x \geq 0 \\ -x & \text{if } x < 0 \end{cases}$

If x is nonnegative $(x \geq 0)$, the absolute value of x is the number itself. If x is negative $(x < 0)$, the absolute value of x is the opposite of x, making the absolute value positive. For more detail, see Section P.1, Objective 7.

EXAMPLE 3 Graphing an Equation Using the Point-Plotting Method

Graph $y = |x|$. Select integers for x, starting with -3 and ending with 3.

SOLUTION

For each value of x, we find the corresponding value for y.

| x | $y = |x|$ | Ordered Pair (x, y) |
|---|---|---|
| -3 | $y = |-3| = 3$ | $(-3, 3)$ |
| -2 | $y = |-2| = 2$ | $(-2, 2)$ |
| -1 | $y = |-1| = 1$ | $(-1, 1)$ |
| 0 | $y = |0| = 0$ | $(0, 0)$ |
| 1 | $y = |1| = 1$ | $(1, 1)$ |
| 2 | $y = |2| = 2$ | $(2, 2)$ |
| 3 | $y = |3| = 3$ | $(3, 3)$ |

*See Answers to Selected Exercises.

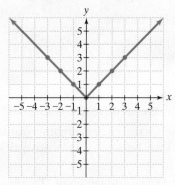

FIGURE 1.5 The graph of $y = |x|$

We plot the points and connect them, resulting in the graph shown in **Figure 1.5**. The graph is V-shaped and centered at the origin. For every point (x, y) on the graph, the point $(-x, y)$ is also on the graph. This shows that the absolute value of a positive number is the same as the absolute value of its opposite. ●●●

⊘ **Check Point 3** Graph $y = |x + 1|$. Select integers for x, starting with -4 and ending with 2. *

GREAT QUESTION!

Are there methods other than point plotting to graph equations in two variables?

Yes. In Chapters 2 and 3, we will be studying graphs of equations in two variables in which

$$y = \text{a polynomial in } x.$$

Do not be concerned that we have not yet learned techniques, other than plotting points, for graphing such equations. As you solve some of the equations in this chapter, we will display graphs simply to enhance your visual understanding of your work. For now, think of graphs of first-degree polynomials as lines, and graphs of second-degree polynomials as symmetric U-shaped bowls that open upward or downward.

③ Interpret information about a graphing utility's viewing rectangle or table.

Graphing Equations and Creating Tables Using a Graphing Utility

Graphing calculators and graphing software packages for computers are referred to as **graphing utilities** or graphers. A graphing utility is a powerful tool that quickly generates the graph of an equation in two variables. **Figures 1.6(a)** and **1.6(b)** show two such graphs for the equations in Examples 2 and 3.

FIGURE 1.6(a) The graph of $y = 4 - x^2$

FIGURE 1.6(b) The graph of $y = |x|$

GREAT QUESTION!

I'm not using a graphing calculator, so should I skip this part of the section?

Even if you are not using a graphing utility in the course, read this part of the section. Knowing about viewing rectangles will enable you to understand the graphs that we display in the technology boxes throughout the book.

What differences do you notice between these graphs and the graphs that we drew by hand? They do seem a bit "jittery." Arrows do not appear on the left and right ends of the graphs. Furthermore, numbers are not given along the axes. For both graphs in **Figure 1.6**, the x-axis extends from -10 to 10 and the y-axis also extends from -10 to 10. The distance represented by each consecutive tick mark is one unit. We say that the **viewing rectangle**, or the **viewing window**, is $[-10, 10, 1]$ by $[-10, 10, 1]$.

Xmin	Xmax	Xscl		Ymin	Ymax	Yscl
$[-10,$	$10,$	$1]$	by	$[-10,$	$10,$	$1]$
The minimum x-value along the x-axis is -10.	The maximum x-value along the x-axis is 10.	Distance between consecutive tick marks on the x-axis is one unit.		The minimum y-value along the y-axis is -10.	The maximum y-value along the y-axis is 10.	Distance between consecutive tick marks on the y-axis is one unit.

To graph an equation in x and y using a graphing utility, enter the equation and specify the size of the viewing rectangle. The size of the viewing rectangle sets minimum and maximum values for both the x- and y-axes. Enter these values, as well as the values representing the distances between consecutive tick marks, on the respective axes. The $[-10, 10, 1]$ by $[-10, 10, 1]$ viewing rectangle used in **Figure 1.6** is called the **standard viewing rectangle**.

*See Answers to Selected Exercises.

Understanding the Viewing Rectangle

What is the meaning of a $[-2, 3, 0.5]$ by $[-10, 20, 5]$ viewing rectangle?

SOLUTION

We begin with $[-2, 3, 0.5]$, which describes the x-axis. The minimum x-value is -2 and the maximum x-value is 3. The distance between consecutive tick marks is 0.5.

Next, consider $[-10, 20, 5]$, which describes the y-axis. The minimum y-value is -10 and the maximum y-value is 20. The distance between consecutive tick marks is 5.

Figure 1.7 illustrates a $[-2, 3, 0.5]$ by $[-10, 20, 5]$ viewing rectangle. To make things clearer, we've placed numbers by each tick mark. These numbers do not appear on the axes when you use a graphing utility to graph an equation. •••

FIGURE 1.7 A $[-2, 3, 0.5]$ by $[-10, 20, 5]$ viewing rectangle

The minimum x-value is -100, the maximum x-value is 100, and the distance between consecutive tick marks is 50. The minimum y-value is -100, the maximum y-value is 100, and the distance between consecutive tick marks is 10. See Answers to Selected Exercises.

⊘ Check Point **4** What is the meaning of a $[-100, 100, 50]$ by $[-100, 100, 10]$ viewing rectangle? Create a figure like the one in **Figure 1.7** that illustrates this viewing rectangle.

On many graphing utilities, the display screen is five-eighths as high as it is wide. By using a square setting, you can equally space the x and y tick marks. (This does not occur in the standard viewing rectangle.) Graphing utilities can also *zoom in* and *zoom out*. When you zoom in, you see a smaller portion of the graph, but you do so in greater detail. When you zoom out, you see a larger portion of the graph. Thus, zooming out may help you to develop a better understanding of the overall character of the graph. With practice, you will become more comfortable with graphing equations in two variables using your graphing utility. You will also develop a better sense of the size of the viewing rectangle that will reveal needed information about a particular graph.

Graphing utilities can also be used to create tables showing solutions of equations in two variables. Use the Table Setup function to choose the starting value of x and to input the increment, or change, between the consecutive x-values. The corresponding y-values are calculated based on the equation(s) in two variables in the $\boxed{Y=}$ screen. In **Figure 1.8**, we used a TI-84 Plus C to create a table for $y = 4 - x^2$ and $y = |x|$, the equations in Examples 2 and 3.

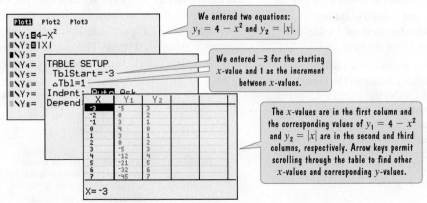

FIGURE 1.8 Creating a table for $y_1 = 4 - x^2$ and $y_2 = |x|$

4 Use a graph to determine intercepts.

Intercepts

An **x-intercept** of a graph is the x-coordinate of a point where the graph intersects the x-axis. For example, look at the graph of $y = 4 - x^2$ in **Figure 1.9** at the top of the next page. The graph crosses the x-axis at $(-2, 0)$ and $(2, 0)$. Thus, the x-intercepts are -2 and 2. **The y-coordinate corresponding to an x-intercept is always zero.**

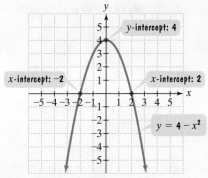

FIGURE 1.9 Intercepts of $y = 4 - x^2$

A **y-intercept** of a graph is the y-coordinate of a point where the graph intersects the y-axis. The graph of $y = 4 - x^2$ in **Figure 1.9** shows that the graph crosses the y-axis at $(0, 4)$. Thus, the y-intercept is 4. **The x-coordinate corresponding to a y-intercept is always zero.**

GREAT QUESTION!

Are single numbers the only way to represent intercepts? Can ordered pairs also be used?

Mathematicians tend to use two ways to describe intercepts. Did you notice that we are using single numbers? If a is an x-intercept of a graph, then the graph passes through the point $(a, 0)$. If b is a y-intercept of a graph, then the graph passes through the point $(0, b)$.

Some books state that the x-intercept is the *point* $(a, 0)$ and the x-intercept is *at a* on the x-axis. Similarly, the y-intercept is the *point* $(0, b)$ and the y-intercept is *at b* on the y-axis. In these descriptions, the intercepts are the actual points where the graph intersects the axes.

Although we'll describe intercepts as single numbers, we'll immediately state the point on the x- or y-axis that the graph passes through. Here's the important thing to keep in mind:

x-intercept: The corresponding value of y is 0.

y-intercept: The corresponding value of x is 0.

EXAMPLE 5 Identifying Intercepts

Identify the x- and y-intercepts.

a. b. c. d.

SOLUTION

a. The graph crosses the x-axis at $(-1, 0)$. Thus, the x-intercept is -1. The graph crosses the y-axis at $(0, 2)$. Thus, the y-intercept is 2.

b. The graph crosses the x-axis at $(3, 0)$, so the x-intercept is 3. This vertical line does not cross the y-axis. Thus, there is no y-intercept.

c. This graph crosses the x- and y-axes at the same point, the origin. Because the graph crosses both axes at $(0, 0)$, the x-intercept is 0 and the y-intercept is 0.

d. The graph crosses the x-axis at $(-2, 0)$ and $(2, 0)$. Thus, the x-intercepts are -2 and 2. The graph crosses the y-axis at $(0, -3)$. Thus, the y-intercept is -3. •••

⊘ Check Point **5** Identify the x- and y-intercepts.

a. b. c. d.

a. x-intercept: -3; y-intercept: 5 **b.** no x-intercept; y-intercept: 4 **c.** x-intercept: 0; y-intercept: 0 **d.** x-intercepts: -1 and 1; y-intercept: 3

Figure 1.10 illustrates that a graph may have no intercepts or several intercepts.

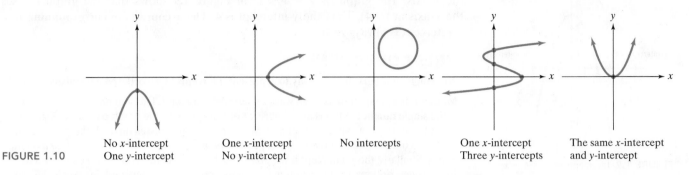

FIGURE 1.10

No *x*-intercept
One *y*-intercept

One *x*-intercept
No *y*-intercept

No intercepts

One *x*-intercept
Three *y*-intercepts

The same *x*-intercept
and *y*-intercept

⑤ Interpret information given
 by graphs.

Interpreting Information Given by Graphs

Line graphs are often used to illustrate trends over time. Some measure of time, such as months or years, frequently appears on the horizontal axis. Amounts are generally listed on the vertical axis. Points are drawn to represent the given information. The graph is formed by connecting the points with line segments.

A line graph displays information in the first quadrant of a rectangular coordinate system. By identifying points on line graphs and their coordinates, you can interpret specific information given by the graph.

A Brief Review • Mathematical Models

The process of finding formulas to describe real-world phenomena is called mathematical modeling. Such formulas, together with the meaning assigned to the variables, are called mathematical models. For more detail, see Section P.1, Objective 2.

EXAMPLE 6 Age at Marriage and the Probability of Divorce

Divorce rates are considerably higher for couples who marry in their teens. The line graphs in **Figure 1.11** show the percentages of marriages ending in divorce based on the wife's age at marriage.

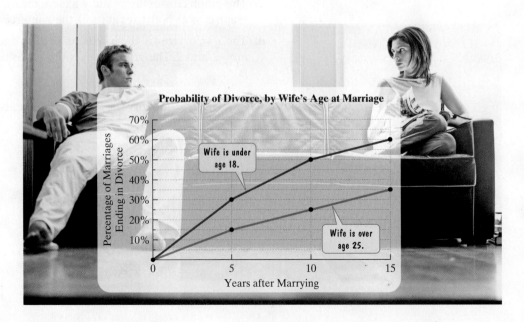

FIGURE 1.11
Source: B. E. Pruitt et al., *Human Sexuality*, Prentice Hall, 2007.

Probability of Divorce, by Wife's Age at Marriage

Wife is under age 18.

Wife is over age 25.

Percentage of Marriages Ending in Divorce

Years after Marrying

Here are two mathematical models that approximate the data displayed by the line graphs:

Wife is under 18 at time of marriage.	Wife is over 25 at time of marriage.
$d = 4n + 5$	$d = 2.3n + 1.5.$

In each model, the variable n is the number of years after marriage and the variable d is the percentage of marriages ending in divorce.

a. Use the appropriate formula to determine the percentage of marriages ending in divorce after 10 years when the wife is over 25 at the time of marriage.

b. Use the appropriate line graph in **Figure 1.11** to determine the percentage of marriages ending in divorce after 10 years when the wife is over 25 at the time of marriage.

c. Does the value given by the mathematical model underestimate or overestimate the actual percentage of marriages ending in divorce after 10 years as shown by the graph? By how much?

SOLUTION

a. Because the wife is over 25 at the time of marriage, we use the formula on the right, $d = 2.3n + 1.5$. To find the percentage of marriages ending in divorce after 10 years, we substitute 10 for n and evaluate the formula.

$$d = 2.3n + 1.5 \qquad \text{This is one of the two given mathematical models.}$$
$$d = 2.3(10) + 1.5 \qquad \text{Replace } n \text{ with 10.}$$
$$d = 23 + 1.5 \qquad \text{Multiply: } 2.3(10) = 23.$$
$$d = 24.5 \qquad \text{Add.}$$

The model indicates that 24.5% of marriages end in divorce after 10 years when the wife is over 25 at the time of marriage.

b. Now let's use the line graph that shows the percentage of marriages ending in divorce when the wife is over 25 at the time of marriage. The graph is shown again in **Figure 1.12**. To find the percentage of marriages ending in divorce after 10 years:

- Locate 10 on the horizontal axis and locate the point above 10.
- Read across to the corresponding percent on the vertical axis.

The actual data displayed by the graph indicate that 25% of these marriages end in divorce after 10 years.

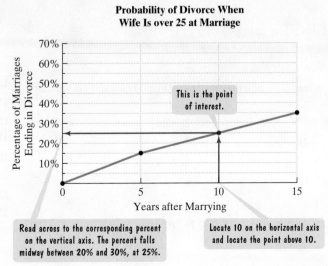

Probability of Divorce When Wife Is over 25 at Marriage

FIGURE 1.12

c. The value obtained by evaluating the mathematical model, 24.5%, is close to, but slightly less than, the actual percentage of divorces, 25.0%. The difference between these percents is 25.0% − 24.5%, or 0.5%. The value given by the mathematical model, 24.5%, underestimates the actual percent, 25%, by only 0.5, providing a fairly accurate description of the data. •••

 Check Point **6**

a. Use the appropriate formula from Example 6 to determine the percentage of marriages ending in divorce after 15 years when the wife is under 18 at the time of marriage. 65%

b. Use the appropriate line graph in **Figure 1.11** on page 100 to determine the percentage of marriages ending in divorce after 15 years when the wife is under 18 at the time of marriage. 60%

c. Does the value given by the mathematical model underestimate or overestimate the actual percentage of marriages ending in divorce after 15 years as shown by the graph? By how much? overestimates by 5

CONCEPT AND VOCABULARY CHECK

Fill in each blank so that the resulting statement is true.

1. In the rectangular coordinate system, the horizontal number line is called the __x-axis__.

2. In the rectangular coordinate system, the vertical number line is called the __y-axis__.

3. In the rectangular coordinate system, the point of intersection of the horizontal axis and the vertical axis is called the __origin__.

4. The axes of the rectangular coordinate system divide the plane into regions, called __quadrants__. There are __four__ of these regions.

5. The first number in an ordered pair such as (8, 3) is called the __x-coordinate__. The second number in such an ordered pair is called the __y-coordinate__.

6. The ordered pair (4, 19) is a/an __solution__ of the equation $y = x^2 + 3$ because when 4 is substituted for x and 19 is substituted for y, we obtain a true statement. We also say that (4, 19) __satisfies__ the equation.

7. The x-coordinate of a point where a graph crosses the x-axis is called a/an __x-intercept__. The y-coordinate of such a point is always __zero__.

8. The y-coordinate of a point where a graph crosses the y-axis is called a/an __y-intercept__. The x-coordinate of such a point is always __zero__.

EXERCISE SET 1.1

Practice Exercises

In Exercises 1–12, plot the given point in a rectangular coordinate system.

1. (1, 4) ∗

2. (2, 5) ∗

3. (−2, 3) ∗

4. (−1, 4) ∗

5. (−3, −5) ∗

6. (−4, −2) ∗

7. (4, −1) ∗

8. (3, −2) ∗

9. (−4, 0) ∗

10. (0, −3) ∗

11. $\left(\frac{7}{2}, -\frac{3}{2}\right)$ ∗

12. $\left(-\frac{5}{2}, \frac{3}{2}\right)$ ∗

Graph each equation in Exercises 13–28. Let x = −3, −2, −1, 0, 1, 2, and 3.

13. $y = x^2 - 2$ ∗

14. $y = x^2 + 2$ ∗

15. $y = x - 2$ ∗

16. $y = x + 2$ ∗

17. $y = 2x + 1$ ∗

18. $y = 2x - 4$ ∗

19. $y = -\frac{1}{2}x$ ∗

20. $y = -\frac{1}{2}x + 2$ ∗

21. $y = 2|x|$ ∗

22. $y = -2|x|$ ∗

23. $y = |x| + 1$ ∗

24. $y = |x| - 1$ ∗

25. $y = 9 - x^2$ ∗

26. $y = -x^2$ ∗

27. $y = x^3$ ∗

28. $y = x^3 - 1$ ∗

∗See Answers to Selected Exercises.

In Exercises 29–32, match the viewing rectangle with the correct figure. Then label the tick marks in the figure to illustrate this viewing rectangle.

29. $[-5, 5, 1]$ by $[-5, 5, 1]$ c **30.** $[-10, 10, 2]$ by $[-4, 4, 2]$ d

31. $[-20, 80, 10]$ by $[-30, 70, 10]$ b

32. $[-40, 40, 20]$ by $[-1000, 1000, 100]$ a

a.

b.

c.

d.

The table of values was generated by a graphing utility with a TABLE feature. Use the table to solve Exercises 33–40.

X	Y₁	Y₂
-3	9	5
-2	4	4
-1	1	3
0	0	2
1	1	1
2	4	0
3	9	-1
4	16	-2
5	25	-3
6	36	-4
7	49	-5

X=-3

33. Which equation corresponds to Y_2 in the table? c

 a. $y_2 = x + 8$ **b.** $y_2 = x - 2$

 c. $y_2 = 2 - x$ **d.** $y_2 = 1 - 2x$

34. Which equation corresponds to Y_1 in the table? b

 a. $y_1 = -3x$ **b.** $y_1 = x^2$

 c. $y_1 = -x^2$ **d.** $y_1 = 2 - x$

35. Does the graph of Y_2 pass through the origin? no

36. Does the graph of Y_1 pass through the origin? yes

37. At which point does the graph of Y_2 cross the *x*-axis? (2, 0)

38. At which point does the graph of Y_2 cross the *y*-axis? (0, 2)

39. At which points do the graphs of Y_1 and Y_2 intersect? *

40. For which values of *x* is $Y_1 = Y_2$? −2 and 1

In Exercises 41–46, use the graph to **a.** *determine the x-intercepts, if any;* **b.** *determine the y-intercepts, if any. For each graph, tick marks along the axes represent one unit each.*

41.

a. 2 b. −4

42.

a. 1 b. 2

43.

a. 1, −2 b. 2

44.

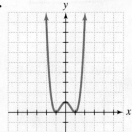

a. 1, −1 b. 1

45.

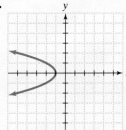

a. −1 b. none

46.

a. none b. 2

Practice Plus

In Exercises 47–50, write each English sentence as an equation in two variables. Then graph the equation.

47. The *y*-value is four more than twice the *x*-value. $y = 2x + 4$*

48. The *y*-value is the difference between four and twice the *x*-value. $y = 4 - 2x$*

49. The *y*-value is three decreased by the square of the *x*-value. *

50. The *y*-value is two more than the square of the *x*-value. *

*See Answers to Selected Exercises.

In Exercises 51–54, graph each equation.

51. $y = 5$ (Let $x = -3, -2, -1, 0, 1, 2,$ and 3.) *

52. $y = -1$ (Let $x = -3, -2, -1, 0, 1, 2,$ and 3.) *

53. $y = \dfrac{1}{x}$ (Let $x = -2, -1, -\dfrac{1}{2}, -\dfrac{1}{3}, \dfrac{1}{3}, \dfrac{1}{2}, 1,$ and 2.) *

54. $y = -\dfrac{1}{x}$ (Let $x = -2, -1, -\dfrac{1}{2}, -\dfrac{1}{3}, \dfrac{1}{3}, \dfrac{1}{2}, 1,$ and 2.) *

Application Exercises

The graphs show the percentage of high school seniors who had ever used alcohol or marijuana.

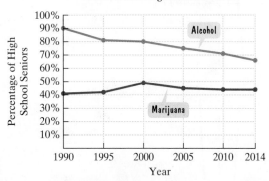

**Alcohol and Marijuana Use by
United States High School Seniors**

Source: University of Michigan Institute for Social Research

The data can be described by the following mathematical models:

Percentage of seniors using alcohol → $A = -0.9n + 88$

Number of years after 1990

Percentage of seniors using marijuana → $M = 0.1n + 43$.

Use this information to solve Exercises 55–56.

55. a. Use the appropriate line graph to estimate the percentage of seniors who used marijuana in 2010. ≈44%

 b. Use the appropriate formula to determine the percentage of seniors who used marijuana in 2010. How does this compare with your estimate in part (a)? 45%*

 c. Use the appropriate line graph to estimate the percentage of seniors who used alcohol in 2010. ≈71%

 d. Use the appropriate formula to determine the percentage of seniors who used alcohol in 2010. How does this compare with your estimate in part (c)? 70%*

 e. For the period from 1990 through 2014, in which year was marijuana use by seniors at a maximum? Estimate the percentage of seniors who used marijuana in that year. *

56. a. Use the appropriate line graph to estimate the percentage of seniors who used alcohol in 2014. ≈66%

 b. Use the appropriate formula to determine the percentage of seniors who used alcohol in 2014. How does this compare with your estimate in part (a)? 66.4%*

 c. Use the appropriate line graph to estimate the percentage of seniors who used marijuana in 2014. ≈44%

d. Use the appropriate formula to determine the percentage of seniors who used marijuana in 2014. How does this compare with your estimate in part (c)? 45.4%*

e. For the period from 1990 through 2014, in which year was alcohol use by seniors at a maximum? What percentage of seniors used alcohol in that year. *

Contrary to popular belief, older people do not need less sleep than younger adults. However, the line graphs show that they awaken more often during the night. The numerous awakenings are one reason why some elderly individuals report that sleep is less restful than it had been in the past. Use the line graphs to solve Exercises 57–60.

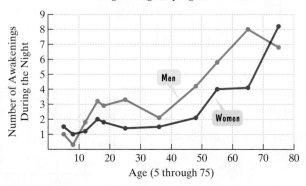

**Average Number of Awakenings
During the Night, by Age and Gender**

Source: Stephen Davis and Joseph Palladino, *Psychology*, 5th Edition, Prentice Hall, 2007.

57. At which age, estimated to the nearest year, do women have the least number of awakenings during the night? What is the average number of awakenings at that age? 8; 1

58. At which age do men have the greatest number of awakenings during the night? What is the average number of awakenings at that age? 65; 8

59. Estimate, to the nearest tenth, the difference between the average number of awakenings during the night for 25-year-old men and 25-year-old women. about 1.9

60. Estimate, to the nearest tenth, the difference between the average number of awakenings during the night for 18-year-old men and 18-year-old women. about 1.1

Explaining the Concepts 61–65. Answers will vary.

61. What is the rectangular coordinate system?

62. Explain how to plot a point in the rectangular coordinate system. Give an example with your explanation.

63. Explain why $(5, -2)$ and $(-2, 5)$ do not represent the same point.

64. Explain how to graph an equation in the rectangular coordinate system.

65. What does a $[-20, 2, 1]$ by $[-4, 5, 0.5]$ viewing rectangle mean?

Technology Exercise

66. Use a graphing utility to verify each of your hand-drawn graphs in Exercises 13–28. Experiment with the settings for the viewing rectangle to make the graph displayed by the graphing utility resemble your hand-drawn graph as much as possible. Answers will vary.

Critical Thinking Exercises

Make Sense? *In Exercises 67–70, determine whether each statement makes sense or does not make sense, and explain your reasoning.*

67. The rectangular coordinate system provides a geometric picture of what an equation in two variables looks like. *

68. There is something wrong with my graphing utility because it is not displaying numbers along the *x*- and *y*-axes. *

69. I used the ordered pairs $(-2, 2)$, $(0, 0)$, and $(2, 2)$ to graph a straight line. does not make sense

70. I used the ordered pairs does not make sense
(time of day, calories that I burned)
to obtain a graph that is a horizontal line.

In Exercises 71–74, determine whether each statement is true or false. If the statement is false, make the necessary change(s) to produce a true statement. Changes to false statements will vary.

71. If the product of a point's coordinates is positive, the point must be in quadrant I. false

72. If a point is on the *x*-axis, it is neither up nor down, so $x = 0$. false

73. If a point is on the *y*-axis, its *x*-coordinate must be 0. true

74. The ordered pair $(2, 5)$ satisfies $3y - 2x = -4$. false

In Exercises 75–78, list the quadrant or quadrants satisfying each condition.

75. $xy > 0$ I, III

76. $\dfrac{y}{x} < 0$ II, IV

77. $x^3 > 0$ and $y^3 < 0$ IV

78. $x^3 < 0$ and $y^3 > 0$ II

In Exercises 79–82, match the story with the correct figure. The figures are labeled (a), (b), (c), and (d).

79. As the blizzard got worse, the snow fell harder and harder. a

80. The snow fell more and more softly. d

81. It snowed hard, but then it stopped. After a short time, the snow started falling softly. b

82. It snowed softly, and then it stopped. After a short time, the snow started falling hard. c

a.

b.

c.

d.
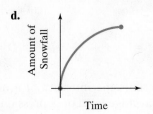

In Exercises 83–86, select the graph that best illustrates each story.

83. An airplane flew from Miami to San Francisco. b

a.

b.

c.

d.

84. At noon, you begin to breathe in. a

a.

b.

c.

d.
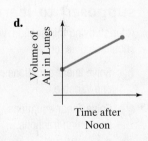

85. Measurements are taken of a person's height from birth to age 100. c

a.

b.

c.

d.
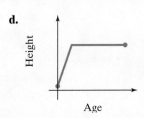

86. You begin your bike ride by riding down a hill. Then you ride up another hill. Finally, you ride along a level surface before coming to a stop. b

a.

b.

c.

d.

Preview Exercises

Exercises 87–89 will help you prepare for the material covered in the next section.

87. If 6 is substituted for x in the equation

$$2(x - 3) - 17 = 13 - 3(x + 2),$$

is the resulting statement true or false? true

88. Multiply and simplify: $12\left(\dfrac{x + 2}{4} - \dfrac{x - 1}{3}\right).$ $-x + 10$

89. Multiply and simplify: $(x - 3)\left(\dfrac{3}{x - 3} + 9\right).$ $9x - 24$

Section 1.2 Linear Equations and Rational Equations

What am I supposed to learn?

After studying this section, you should be able to:

1. Solve linear equations in one variable.
2. Solve linear equations containing fractions.
3. Solve rational equations with variables in the denominators.
4. Recognize identities, conditional equations, and inconsistent equations.
5. Solve applied problems using mathematical models.

FIGURE 1.13
Source: Stephen Davis and Joseph Palladino, *Psychology*, 5th Edition, Prentice Hall, 2007.

The belief that humor and laughter can have positive benefits on our lives is not new. The graphs in **Figure 1.13** indicate that persons with a low sense of humor have higher levels of depression in response to negative life events than those with a high sense of humor. These graphs can be modeled by the following formulas:

Low-Humor Group High-Humor Group

$$D = \dfrac{10}{9}x + \dfrac{53}{9} \qquad D = \dfrac{1}{9}x + \dfrac{26}{9}.$$

In each formula, x represents the intensity of a negative life event (from 1, low, to 10, high) and D is the average level of depression in response to that event.

Suppose that the low-humor group averages a level of depression of 10 in response to a negative life event. We can determine the intensity of that event by substituting 10 for D in the low-humor model, $D = \frac{10}{9}x + \frac{53}{9}$:

$$10 = \frac{10}{9}x + \frac{53}{9}.$$

The two sides of an equation can be reversed. So, we can also express this equation as

$$\frac{10}{9}x + \frac{53}{9} = 10.$$

Notice that the highest exponent on the variable is 1. Such an equation is called a *linear equation in one variable*. In this section, we will study how to solve linear equations.

① Solve linear equations in one variable.

Solving Linear Equations in One Variable

We begin with a general definition of a linear equation in one variable.

> ### Definition of a Linear Equation
>
> A **linear equation in one variable** x is an equation that can be written in the form
>
> $$ax + b = 0,$$
>
> where a and b are real numbers, and $a \neq 0$.

An example of a linear equation in one variable is

$$4x + 12 = 0.$$

Solving an equation in x involves determining all values of x that result in a true statement when substituted into the equation. Such values are **solutions**, or **roots**, of the equation. For example, substitute -3 for x in $4x + 12 = 0$. We obtain

$$4(-3) + 12 = 0, \quad \text{or} \quad -12 + 12 = 0.$$

This simplifies to the true statement $0 = 0$. Thus, -3 is a solution of the equation $4x + 12 = 0$. We also say that -3 **satisfies** the equation $4x + 12 = 0$, because when we substitute -3 for x, a true statement results. The set of all such solutions is called the equation's **solution set**. For example, the solution set of the equation $4x + 12 = 0$ is $\{-3\}$ because -3 is the equation's only solution.

Two or more equations that have the same solution set are called **equivalent equations**. For example, the equations

$$4x + 12 = 0 \quad \text{and} \quad 4x = -12 \quad \text{and} \quad x = -3$$

are equivalent equations because the solution set for each is $\{-3\}$. To solve a linear equation in x, we transform the equation into an equivalent equation one or more times. Our final equivalent equation should be of the form

$$x = \text{a number}.$$

The solution set of this equation is the set consisting of the number.

To generate equivalent equations, we will use the principles in the box at the top of the next page.

Generating Equivalent Equations

An equation can be transformed into an equivalent equation by one or more of the following operations:

Operation	Example
1. Simplify an expression by removing grouping symbols and combining like terms.	$3(x - 6) = 6x - x$ $3x - 18 = 5x$
2. Add (or subtract) the same real number or variable expression on *both* sides of the equation.	$3x - 18 = 5x$ $3x - 18 - 3x = 5x - 3x$ $-18 = 2x$ *Subtract 3x from both sides of the equation.*
3. Multiply (or divide) by the same *nonzero* quantity on *both* sides of the equation.	$-18 = 2x$ $\dfrac{-18}{2} = \dfrac{2x}{2}$ *Divide both sides of the equation by 2.* $-9 = x$
4. Interchange the two sides of the equation.	$-9 = x$ $x = -9$

If you look closely at the equations in the box, you will notice that we have solved the equation $3(x - 6) = 6x - x$. The final equation, $x = -9$, with x isolated on the left side, shows that $\{-9\}$ is the solution set. The idea in solving a linear equation is to get the variable by itself on one side of the equal sign and a number by itself on the other side.

EXAMPLE 1 Solving a Linear Equation

Solve and check: $2x + 3 = 17$.

SOLUTION

Our goal is to obtain an equivalent equation with x isolated on one side and a number on the other side.

$$2x + 3 = 17 \qquad \text{This is the given equation.}$$
$$2x + 3 - 3 = 17 - 3 \qquad \text{Subtract 3 from both sides.}$$
$$2x = 14 \qquad \text{Simplify.}$$
$$\frac{2x}{2} = \frac{14}{2} \qquad \text{Divide both sides by 2.}$$
$$x = 7 \qquad \text{Simplify.}$$

Now we check the proposed solution, 7, by replacing x with 7 in the original equation.

$$2x + 3 = 17 \qquad \text{This is the original equation.}$$
$$2 \cdot 7 + 3 \stackrel{?}{=} 17 \qquad \text{Substitute 7 for } x. \text{ The question mark indicates that we do not yet know if the two sides are equal.}$$
$$14 + 3 \stackrel{?}{=} 17 \qquad \text{Multiply: } 2 \cdot 7 = 14.$$
$$17 = 17 \qquad \text{Add: } 14 + 3 = 17. \qquad \text{This statement is true.}$$

Because the check results in a true statement, we conclude that the solution set of the given equation is $\{7\}$. • • •

⊘ **Check Point 1** Solve and check: $4x + 5 = 29$. {6}

GREAT QUESTION!

What are the differences between what I'm supposed to do with algebraic expressions and algebraic equations?

We simplify algebraic expressions. We solve algebraic equations. Notice the differences between the procedures:

Simplifying an Algebraic Expression

Simplify: $3(x - 7) - (5x - 11)$.

> This is not an equation.
> There is no equal sign.

Solution $3(x - 7) - (5x - 11)$

$= 3x - 21 - 5x + 11$

$= (3x - 5x) + (-21 + 11)$

$= -2x + (-10)$

$= -2x - 10$

> Stop! Further simplification is not possible. Avoid the common error of setting $-2x - 10$ equal to 0.

Solving an Algebraic Equation

Solve: $3(x - 7) - (5x - 11) = 14$.

> This is an equation.
> There is an equal sign.

Solution $3(x - 7) - (5x - 11) = 14$

$3x - 21 - 5x + 11 = 14$

$-2x - 10 = 14$

> Add 10 to both sides.

$-2x - 10 + 10 = 14 + 10$

$-2x = 24$

> Divide both sides by -2.

$\dfrac{-2x}{-2} = \dfrac{24}{-2}$

$x = -12$

The solution set is $\{-12\}$.

Here is a step-by-step procedure for solving a linear equation in one variable. Not all of these steps are necessary to solve every equation.

Solving a Linear Equation

1. Simplify the algebraic expression on each side by removing grouping symbols and combining like terms.
2. Collect all the variable terms on one side and all the numbers, or constant terms, on the other side.
3. Isolate the variable and solve.
4. Check the proposed solution in the original equation.

EXAMPLE 2 Solving a Linear Equation

Solve and check: $2(x - 3) - 17 = 13 - 3(x + 2)$.

SOLUTION

Step 1 Simplify the algebraic expression on each side.

> Do not begin with $13 - 3$.
> Multiplication (the distributive property) is applied before subtraction.

$2(x - 3) - 17 = 13 - 3(x + 2)$ This is the given equation.

$2x - 6 - 17 = 13 - 3x - 6$ Use the distributive property.

$2x - 23 = -3x + 7$ Combine like terms.

Step 2 Collect variable terms on one side and constant terms on the other side. We will collect variable terms on the left by adding $3x$ to both sides. We will collect the numbers on the right by adding 23 to both sides.

$$2x - 23 = -3x + 7 \qquad \text{This is the equation with each side}$$
$$\text{simplified.}$$
$$2x - 23 + 3x = -3x + 7 + 3x \qquad \text{Add } 3x \text{ to both sides.}$$
$$5x - 23 = 7 \qquad \text{Simplify: } 2x + 3x = 5x.$$
$$5x - 23 + 23 = 7 + 23 \qquad \text{Add 23 to both sides.}$$
$$5x = 30 \qquad \text{Simplify.}$$

Step 3 Isolate the variable and solve. We isolate the variable, x, by dividing both sides of $5x = 30$ by 5.

$$\frac{5x}{5} = \frac{30}{5} \qquad \text{Divide both sides by 5.}$$
$$x = 6 \qquad \text{Simplify.}$$

Step 4 Check the proposed solution in the original equation. Substitute 6 for x in the original equation.

$$2(x - 3) - 17 = 13 - 3(x + 2) \qquad \text{This is the original equation.}$$
$$2(6 - 3) - 17 \overset{?}{=} 13 - 3(6 + 2) \qquad \text{Substitute 6 for } x.$$
$$2(3) - 17 \overset{?}{=} 13 - 3(8) \qquad \text{Simplify inside parentheses.}$$
$$6 - 17 \overset{?}{=} 13 - 24 \qquad \text{Multiply.}$$
$$-11 = -11 \qquad \text{Subtract.}$$

DISCOVERY

Solve the equation in Example 2 by collecting terms with the variable on the right and numerical terms on the left. What do you observe?

The true statement $-11 = -11$ verifies that the solution set is $\{6\}$. •••

TECHNOLOGY

Numeric and Graphic Connections

In many algebraic situations, technology provides numeric and visual insights into problem solving. For example, you can use a graphing utility to check the solution of a linear equation, giving numeric and geometric meaning to the solution. Enter each side of the equation separately under y_1 and y_2. Then use the table or the graphs to locate the x-value for which the y-values are the same. This x-value is the solution.

Let's verify our work in Example 2 and show that 6 is the solution of

$$2(x - 3) - 17 = 13 - 3(x + 2).$$

Enter $y_1 = 2(x - 3) - 17$ in the $\boxed{y=}$ screen.

Enter $y_2 = 13 - 3(x + 2)$ in the $\boxed{y=}$ screen.

Numeric Check

Display a table for y_1 and y_2.

When $x = 6$, y_1 and y_2 have the same value, namely −11. This verifies that 6 is the solution of $2(x - 3) - 17 = 13 - 3(x + 2)$.

Graphic Check

Display graphs for y_1 and y_2 and use the intersection feature. The solution is the x-coordinate of the intersection point.

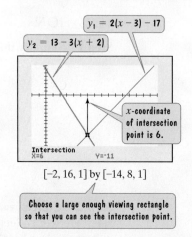

$[-2, 16, 1]$ by $[-14, 8, 1]$

Choose a large enough viewing rectangle so that you can see the intersection point.

⊘ Check Point **2** Solve and check: $4(2x + 1) = 29 + 3(2x - 5)$. {5}

2 Solve linear equations containing fractions.

Linear Equations with Fractions

Equations are easier to solve when they do not contain fractions. How do we remove fractions from an equation? We begin by multiplying both sides of the equation by the least common denominator of any fractions in the equation. The least common denominator is the smallest number that all denominators will divide into. Multiplying every term on both sides of the equation by the least common denominator will eliminate the fractions in the equation. Example 3 shows how we "clear an equation of fractions."

EXAMPLE 3 Solving a Linear Equation Involving Fractions

Solve and check: $\dfrac{x + 2}{4} - \dfrac{x - 1}{3} = 2$.

SOLUTION

The fractional terms have denominators of 4 and 3. The smallest number that is divisible by 4 and 3 is 12. We begin by multiplying both sides of the equation by 12, the least common denominator.

$$\frac{x + 2}{4} - \frac{x - 1}{3} = 2$$ This is the given equation.

$$12\left(\frac{x + 2}{4} - \frac{x - 1}{3}\right) = 12 \cdot 2$$ Multiply both sides by 12.

$$12\left(\frac{x + 2}{4}\right) - 12\left(\frac{x - 1}{3}\right) = 24$$ Use the distributive property and multiply each term on the left by 12.

$$\overset{3}{\cancel{12}}\left(\frac{x + 2}{\underset{1}{\cancel{4}}}\right) - \overset{4}{\cancel{12}}\left(\frac{x - 1}{\underset{1}{\cancel{3}}}\right) = 24$$ Divide out common factors in each multiplication on the left.

$$3(x + 2) - 4(x - 1) = 24$$ The fractions are now cleared.

$$3x + 6 - 4x + 4 = 24$$ Use the distributive property.

$$-x + 10 = 24$$ Combine like terms: $3x - 4x = -x$ and $6 + 4 = 10$.

$$-x + 10 - 10 = 24 - 10$$ Subtract 10 from both sides.

$$-x = 14$$ Simplify.

> **We're not finished. A negative sign should not precede the variable.**

Isolate x by multiplying or dividing both sides of this equation by -1.

$$\frac{-x}{-1} = \frac{14}{-1}$$ Divide both sides by -1.

$$x = -14$$ Simplify.

Check the proposed solution. Substitute -14 for x in the original equation. You should obtain $2 = 2$. This true statement verifies that the solution set is $\{-14\}$. • • •

⊘ Check Point **3** Solve and check: $\dfrac{x - 3}{4} = \dfrac{5}{14} - \dfrac{x + 5}{7}$. {1}

 Solve rational equations with variables in the denominators.

Rational Equations

A **rational equation** is an equation containing one or more rational expressions. In Example 3, we solved a rational equation with constants in the denominators. This rational equation was a linear equation. Now, let's consider a rational equation such as

$$\frac{1}{x} = \frac{1}{5} + \frac{3}{2x}.$$

Can you see how this rational equation differs from the rational equation that we solved earlier? The variable, x, appears in two of the denominators. Although this rational equation is not a linear equation, the solution procedure still involves multiplying each side by the least common denominator. However, we must avoid any values of the variable that make a denominator zero. For example, examine the denominators in the equation

$$\frac{1}{x} = \frac{1}{5} + \frac{3}{2x}.$$

> This denominator would equal zero if $x = 0$.

> This denominator would equal zero if $x = 0$.

We see that x cannot equal zero. With this in mind, let's solve the equation.

A Brief Review • The Least Common Denominator

The least common denominator, or LCD, of several rational expressions is a polynomial consisting of the product of all prime factors in the denominators, with each factor raised to the greatest power of its occurrence in any denominator. For more detail, see Section P.6, Objective 5, Examples 6 and 7.

EXAMPLE 4 Solving a Rational Equation

Solve: $\dfrac{1}{x} = \dfrac{1}{5} + \dfrac{3}{2x}.$

SOLUTION

The denominators are x, 5, and $2x$. The least common denominator is $10x$. We begin by multiplying both sides of the equation by $10x$. We will also write the restriction that x cannot equal zero to the right of the equation.

$$\frac{1}{x} = \frac{1}{5} + \frac{3}{2x}, x \neq 0 \qquad \text{This is the given equation.}$$

$$10x \cdot \frac{1}{x} = 10x\left(\frac{1}{5} + \frac{3}{2x}\right) \qquad \text{Multiply both sides by 10x.}$$

$$10x \cdot \frac{1}{x} = 10x \cdot \frac{1}{5} + 10x \cdot \frac{3}{2x} \qquad \text{Use the distributive property. Be sure to multiply each term by 10x.}$$

$$10\overset{}{x} \cdot \frac{1}{\overset{}{x}} = \overset{2}{10x} \cdot \frac{1}{\underset{1}{5}} + \overset{5}{10x} \cdot \frac{3}{\underset{1}{2x}} \qquad \text{Divide out common factors in the multiplications.}$$

$$10 = 2x + 15 \qquad \text{Complete the multiplications.}$$

Observe that the resulting equation,

$$10 = 2x + 15,$$

is now cleared of fractions. With the variable term, $2x$, already on the right, we will collect constant terms on the left by subtracting 15 from both sides.

$$10 - 15 = 2x + 15 - 15$$ Subtract 15 from both sides.

$$-5 = 2x$$ Simplify.

Finally, we isolate the variable, x, in $-5 = 2x$ by dividing both sides by 2.

$$\frac{-5}{2} = \frac{2x}{2}$$ Divide both sides by 2.

$$-\frac{5}{2} = x$$ Simplify.

We check our solution by substituting $-\frac{5}{2}$ into the original equation or by using a calculator. With a calculator, evaluate each side of the equation for $x = -\frac{5}{2}$, or for $x = -2.5$. Note that the original restriction that $x \neq 0$ is met. The solution set is $\left\{ -\frac{5}{2} \right\}$. • • •

⊘ **Check Point 4** Solve: $\dfrac{5}{2x} = \dfrac{17}{18} - \dfrac{1}{3x}$. [3]

EXAMPLE 5 Solving a Rational Equation

Solve: $\dfrac{x}{x-3} = \dfrac{3}{x-3} + 9$.

SOLUTION

We must avoid any values of the variable x that make a denominator zero.

$$\frac{x}{x-3} = \frac{3}{x-3} + 9$$

These denominators are zero if $x = 3$.

We see that x cannot equal 3. With denominators of $x - 3$, $x - 3$, and 1, the least common denominator is $x - 3$. We multiply both sides of the equation by $x - 3$. We also write the restriction that x cannot equal 3 to the right of the equation.

$$\frac{x}{x-3} = \frac{3}{x-3} + 9, \quad x \neq 3$$ This is the given equation.

$$(x-3) \cdot \frac{x}{x-3} = (x-3)\left(\frac{3}{x-3} + 9\right)$$ Multiply both sides by $x - 3$.

$$(x-3) \cdot \frac{x}{x-3} = (x-3) \cdot \frac{3}{x-3} + (x-3) \cdot 9$$ Use the distributive property.

$$(x\!-\!3) \cdot \frac{x}{x\!-\!3} = (x\!-\!3) \cdot \frac{3}{x\!-\!3} + 9(x-3)$$ Divide out common factors in two of the multiplications.

$$x = 3 + 9(x-3)$$ Simplify.

The resulting equation is cleared of fractions. We now solve for x.

$$x = 3 + 9x - 27$$ Use the distributive property.

$$x = 9x - 24$$ Combine numerical terms.

$$x - 9x = 9x - 24 - 9x$$ Subtract $9x$ from both sides.

$$-8x = -24$$ Simplify.

$$\frac{-8x}{-8} = \frac{-24}{-8}$$ Divide both sides by -8.

$$x = 3$$ Simplify.

The proposed solution, 3, is *not* a solution because of the restriction that $x \neq 3$. There is *no solution to this equation*. The solution set for this equation contains no elements. The solution set is \varnothing, the empty set. • • •

 Check Point **5** Solve: $\dfrac{x}{x-2} = \dfrac{2}{x-2} - \dfrac{2}{3}$. \varnothing

A Brief Review • Factoring Trinomials

Factoring the trinomial $ax^2 + bx + c$ involves finding an equivalent expression that is a product. To factor $ax^2 + bx + c$,

1. Find two **First** terms whose product is ax^2:

$$(\Box x +)(\Box x +) = ax^2 + bx + c.$$

2. Find two **Last** terms whose product is c:

$$(\Box x + \Box)(\Box x + \Box) = ax^2 + bx + c.$$

3. By trial and error, perform steps 1 and 2 until the sum of the **O**utside product and **I**nside product is bx:

$$(\Box x + \Box)(\Box x + \Box) = ax^2 + bx + c.$$
$$\text{I}$$
$$\text{O}$$
$$\text{Sum of O + I}$$

For more detail, see Section P.5, Objective 3.

EXAMPLE 6 Solving a Rational Equation to Determine When Two Equations Are Equal

Consider the equations

$$y_1 = \frac{3}{x+6} + \frac{1}{x-2} \quad \text{and} \quad y_2 = \frac{4}{x^2 + 4x - 12}.$$

Find all values of x for which $y_1 = y_2$.

SOLUTION

Because we are interested in one or more values of x that cause y_1 and y_2 to be equal, we set the expressions that define y_1 and y_2 equal to each other:

$$\frac{3}{x+6} + \frac{1}{x-2} = \frac{4}{x^2 + 4x - 12}.$$

To identify values of x that make denominators zero, let's factor $x^2 + 4x - 12$, the denominator on the right. This factorization, $x^2 + 4x - 12 = (x+6)(x-2)$, is also necessary in identifying the least common denominator.

$$\frac{3}{x+6} + \frac{1}{x-2} = \frac{4}{(x+6)(x-2)}$$

| This denominator is zero if $x = -6$. | This denominator is zero if $x = 2$. | This denominator is zero if $x = -6$ or $x = 2$. |

We see that x cannot equal -6 or 2. The least common denominator is $(x+6)(x-2)$.

$$\frac{3}{x + 6} + \frac{1}{x - 2} = \frac{4}{(x + 6)(x - 2)}, \quad x \neq -6, \quad x \neq 2$$

This is the equation we need to solve with a denominator factored.

$$(x + 6)(x - 2)\left(\frac{3}{x + 6} + \frac{1}{x - 2}\right) = (x + 6)(x - 2) \cdot \frac{4}{(x + 6)(x - 2)}$$

Multiply both sides by $(x + 6)(x - 2)$, the LCD.

$$(x + 6)(x - 2) \cdot \frac{3}{x + 6} + (x + 6)(x - 2) \cdot \frac{1}{x - 2} = (x + 6)(x - 2) \cdot \frac{4}{(x + 6)(x - 2)}$$

Use the distributive property and divide out common factors.

$$3(x - 2) + 1(x + 6) = 4$$

Simplify. This equation is cleared of fractions.

$$3x - 6 + x + 6 = 4$$

Use the distributive property.

$$4x = 4$$

Combine like terms.

$$\frac{4x}{4} = \frac{4}{4}$$

Divide both sides by 4.

$$x = 1$$

Simplify. This is not part of the restriction that $x \neq -6$ and $x \neq 2$.

The value of x for which $y_1 = y_2$ is 1.

Check

Is $y_1 = y_2$ when $x = 1$? We use the given equations

$$y_1 = \frac{3}{x + 6} + \frac{1}{x - 2} \quad \text{and} \quad y_2 = \frac{4}{x^2 + 4x - 12}$$

to answer the question. We can check 1 by hand or with a graphing utility.

Checking by Hand

Substitute 1 for x in y_1 and y_2.

$$y_1 = \frac{3}{1 + 6} + \frac{1}{1 - 2} = \frac{3}{7} + \frac{1}{-1}$$

$$= \frac{3}{7} - 1 = \frac{3}{7} - \frac{7}{7} = -\frac{4}{7}$$

$$y_2 = \frac{4}{1^2 + 4 \cdot 1 - 12} = \frac{4}{1 + 4 - 12}$$

$$= \frac{4}{-7} = -\frac{4}{7}$$

When $x = 1$, y_1 and y_2 have the same value, namely, $-\dfrac{4}{7}$.

Checking with a Graphing Utility

Display a table showing values for y_1 and y_2. Enter the equations as y_1 and y_2, and be careful with parentheses.

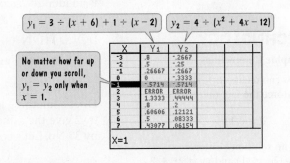

No matter how far up or down you scroll, $y_1 = y_2$ only when $x = 1$.

\varnothing **Check Point 6** Consider the equations

$$y_1 = \frac{1}{x + 4} + \frac{1}{x - 4} \quad \text{and} \quad y_2 = \frac{22}{x^2 - 16}.$$

Find all values of x for which $y_1 = y_2$ and check.

④ Recognize identities, conditional equations, and inconsistent equations.

Types of Equations

Equations can be placed into categories that depend on their solution sets.

An equation that is true for all real numbers for which both sides are defined is called an **identity**. An example of an identity is

$$x + 3 = x + 2 + 1.$$

Every number plus 3 is equal to that number plus 2 plus 1. Therefore, the solution set to this equation is the set of all real numbers, expressed as $\{x \mid x \text{ is a real number}\}$.

GREAT QUESTION!

What's the bottom line on all the vocabulary associated with types of equations?

If you are concerned by the vocabulary of equation types, keep in mind that there are three possible situations. We can state these situations informally as follows:

1. x = a real number

> Conditional equation

2. x = all real numbers

> Identity

3. x = no real numbers.

> Inconsistent equation

Another example of an identity is

$$\frac{2x}{x} = 2.$$

Because division by 0 is undefined, this equation is true for all real number values of x except 0. The solution set is the set of nonzero real numbers, expressed as $\{x \mid x \text{ is a real number and } x \neq 0\}$.

An equation that is not an identity, but that is true for at least one real number, is called a **conditional equation**. The equation $2x + 3 = 17$ is an example of a conditional equation. The equation is not an identity and is true only if x is 7.

An **inconsistent equation** is an equation that is not true for even one real number. An example of an inconsistent equation is

$$x = x + 7.$$

There is no number that is equal to itself plus 7. The equation's solution set is \varnothing, the empty set. Some inconsistent equations are less obvious than this. Consider the equation in Example 5,

$$\frac{x}{x - 3} = \frac{3}{x - 3} + 9.$$

This equation is not true for any real number and has no solution. Thus, it is inconsistent.

If you eliminate the variable while solving a linear equation, a true statement such as $6 = 6$ or a false statement such as $2 = 3$ will be the result. **If a true statement results, the equation is an identity that is true for all real numbers. If a false statement results, the equation is an inconsistent equation with no solution.**

EXAMPLE 7 Categorizing an Equation

Solve and determine whether the equation

$$2(x + 1) = 2x + 3$$

is an identity, a conditional equation, or an inconsistent equation.

TECHNOLOGY

Graphic Connections

How can technology visually reinforce the fact that the equation

$$2(x + 1) = 2x + 3$$

has no solution? Enter $y_1 = 2(x + 1)$ and $y_2 = 2x + 3$. The graphs of y_1 and y_2 appear to be parallel lines with no intersection point. This supports our conclusion that $2(x + 1) = 2x + 3$ is an inconsistent equation with no solution.

$[-5, 2, 1]$ by $[-5, 5, 1]$

SOLUTION

Begin by applying the distributive property on the left side. We obtain

$$2x + 2 = 2x + 3.$$

Does something look strange? Can doubling a number and increasing the product by 2 give the same result as doubling the same number and increasing the product by 3? No. Let's continue solving the equation by subtracting $2x$ from both sides.

$$2x - 2x + 2 = 2x - 2x + 3$$

> Keep reading. 2 = 3 is not the solution.

$$2 = 3$$

The original equation, $2(x + 1) = 2x + 3$, is equivalent to the statement $2 = 3$, which is false for every value of x. The equation is inconsistent and has no solution. The solution set is \varnothing, the empty set.

• • •

✓ **Check Point 7** Solve and determine whether the equation

$$4x - 7 = 4(x - 1) + 3$$

is an identity, a conditional equation, or an inconsistent equation. \varnothing; inconsistent equation

EXAMPLE 8 Categorizing an Equation

Solve and determine whether the equation

$$4x + 6 = 6(x + 1) - 2x$$

is an identity, a conditional equation, or an inconsistent equation.

SOLUTION

$$4x + 6 = 6(x + 1) - 2x$$ This is the given equation.

$$4x + 6 = 6x + 6 - 2x$$ Apply the distributive property on the right side.

$$4x + 6 = 4x + 6$$ Combine like terms on the right side: $6x - 2x = 4x$.

Can you see that the equation $4x + 6 = 4x + 6$ is true for every value of x? Let's continue solving the equation by subtracting $4x$ from both sides.

$$4x - 4x + 6 = 4x - 4x + 6$$

$$6 = 6$$ Keep reading. $6 = 6$ is not the solution.

The original equation is equivalent to the statement $6 = 6$, which is true for every value of x. The equation is an identity, and all real numbers are solutions. You can express this by using the following notation:

$$\{x \mid x \text{ is a real number}\}.$$ • • •

TECHNOLOGY

Numeric Connections

A graphing utility's ⌐TABLE⌐ feature can be used to numerically verify that the solution set of

$$4x + 6 = 6(x + 1) - 2x$$

is the set of all real numbers.

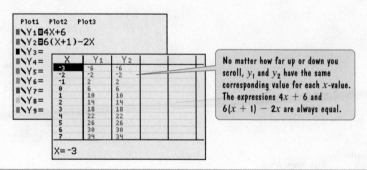

No matter how far up or down you scroll, y_1 and y_2 have the same corresponding value for each x-value. The expressions $4x + 6$ and $6(x + 1) - 2x$ are always equal.

✓ Check Point **8** Solve and determine whether the equation

$$7x + 9 = 9(x + 1) - 2x$$

is an identity, a conditional equation, or an inconsistent equation.

$\{x \mid x \text{ is a real number}\}$; identity

 Solve applied problems using mathematical models.

Sense of Humor and Depression

FIGURE 1.13 (repeated)

Applications

Our next example shows how the procedure for solving linear equations can be used to find the value of a variable in a mathematical model.

EXAMPLE 9 An Application: Responding to Negative Life Events

In the section opener, we introduced line graphs, repeated in **Figure 1.13**, indicating that persons with a low sense of humor have higher levels of depression in response to negative life events than those with a high sense of humor. These graphs can be modeled by the following formulas:

Low-Humor Group

$$D = \frac{10}{9}x + \frac{53}{9}$$

High-Humor Group

$$D = \frac{1}{9}x + \frac{26}{9}.$$

In each formula, x represents the intensity of a negative life event (from 1, low, to 10, high) and D is the average level of depression in response to that event. If the high-humor group averages a level of depression of 3.5, or $\frac{7}{2}$, in response to a negative life event, what is the intensity of that event? How is the solution shown on the red line graph in **Figure 1.13**?

SOLUTION

We are interested in the intensity of a negative life event with an average level of depression of $\frac{7}{2}$ for the high-humor group. We substitute $\frac{7}{2}$ for D in the high-humor model and solve for x, the intensity of the negative life event.

$$D = \frac{1}{9}x + \frac{26}{9}$$ This is the given formula for the high-humor group.

$$\frac{7}{2} = \frac{1}{9}x + \frac{26}{9}$$ Replace D with $\frac{7}{2}$.

$$18 \cdot \frac{7}{2} = 18\left(\frac{1}{9}x + \frac{26}{9}\right)$$ Multiply both sides by 18, the least common denominator.

$$18 \cdot \frac{7}{2} = 18 \cdot \frac{1}{9}x + 18 \cdot \frac{26}{9}$$ Use the distributive property.

$$\overset{9}{\underset{1}{18}} \cdot \frac{7}{2} = \overset{2}{\underset{1}{18}} \cdot \frac{1}{9}x + \overset{2}{\underset{1}{18}} \cdot \frac{26}{9}$$ Divide out common factors in the multiplications.

$$63 = 2x + 52$$ Complete the multiplications. The fractions are now cleared.

$$63 - 52 = 2x + 52 - 52$$ Subtract 52 from both sides to get constants on the left.

$$11 = 2x$$ Simplify.

$$\frac{11}{2} = \frac{2x}{2}$$ Divide both sides by 2.

$$\frac{11}{2} = x$$ Simplify.

FIGURE 1.14

The formula indicates that if the high-humor group averages a level of depression of 3.5 in response to a negative life event, the intensity of that event is $\frac{11}{2}$, or 5.5. This is illustrated by the point (5.5, 3.5) on the line graph for the high-humor group in **Figure 1.14**. • • •

 Check Point **9** Use the model for the low-humor group given in Example 9 to solve this problem. If the low-humor group averages a level of depression of 10 in response to a negative life event, what is the intensity of that event? How is the solution shown on the blue line graph in **Figure 1.13**? 3.7; by the point (3.7, 10)

ACHIEVING SUCCESS

Because concepts in mathematics build on each other, **it is extremely important that you complete all homework assignments**. This requires more than attempting a few of the assigned exercises. When it comes to assigned homework, you need to do four things and to do these things consistently throughout any math course:

1. Attempt to work every assigned problem.
2. Check your answers.
3. Correct your errors.
4. Ask for help with the problems you have attempted, but do not understand.

Having said this, **don't panic at the length of the Exercise Sets**. You are not expected to work all, or even most, of the problems. Your professor will provide guidance on which exercises to work by assigning those problems that are consistent with the goals and objectives of your course.

CONCEPT AND VOCABULARY CHECK

Fill in each blank so that the resulting statement is true.

1. An equation in the form $ax + b = 0$, $a \neq 0$, such as $3x + 17 = 0$, is called a/an __linear__ equation in one variable.

2. Two or more equations that have the same solution set are called __equivalent__ equations.

3. The first step in solving $7 + 3(x - 2) = 2x + 10$ is to __apply the distributive property__.

4. The fractions in the equation

$$\frac{x}{4} = 2 + \frac{x - 3}{3}$$

can be eliminated by multiplying both sides by the __least common denominator__ of $\frac{x}{4}$ and $\frac{x-3}{3}$, which is __12__.

5. We reject any proposed solution of a rational equation that causes a denominator to equal __0__.

6. The first step in solving

$$\frac{4}{x} + \frac{1}{2} = \frac{5}{x}$$

is to multiply both sides by __2x__.

7. The first step in solving

$$\frac{x - 6}{x + 5} = \frac{x - 3}{x + 1}$$

is to multiply both sides by __$(x + 5)(x + 1)$__.

8. The restrictions on the variable in the rational equation

$$\frac{1}{x - 2} - \frac{2}{x + 4} = \frac{2x - 1}{x^2 + 2x - 8}$$

are __$x \neq 2$__ and __$x \neq -4$__.

9. $\dfrac{5}{x + 4} + \dfrac{3}{x + 3} = \dfrac{12x + 9}{(x + 4)(x + 3)}$

$$(x + 4)(x + 3)\left(\frac{5}{x + 4} + \frac{3}{x + 3}\right)$$

$$= (x + 4)(x + 3)\left(\frac{12x + 9}{(x + 4)(x + 3)}\right)$$

The resulting equation cleared of fractions is __$5(x + 3) + 3(x + 4) = 12x + 9$__.

10. An equation that is true for all real numbers for which both sides are defined is called a/an __identity__.

11. An equation that is not true for even one real number is called a/an __inconsistent__ equation.

EXERCISE SET 1.2

In Exercises 1–16, solve and check each linear equation.

1. $7x - 5 = 72$ {11}

2. $6x - 3 = 63$ {11}

3. $11x - (6x - 5) = 40$ {7}

4. $5x - (2x - 10) = 35$ *

5. $2x - 7 = 6 + x$ {13}

6. $3x + 5 = 2x + 13$ {8}

7. $7x + 4 = x + 16$ {2}

8. $13x + 14 = 12x - 5$ *

9. $3(x - 2) + 7 = 2(x + 5)$ {9}

10. $2(x - 1) + 3 = x - 3(x + 1)$ {-1}

11. $3(x - 4) - 4(x - 3) = x + 3 - (x - 2)$ {-5}

12. $2 - (7x + 5) = 13 - 3x$ {-4}

13. $16 = 3(x - 1) - (x - 7)$ {6}

14. $5x - (2x + 2) = x + (3x - 5)$ {3}

15. $25 - [2 + 5y - 3(y + 2)] =$
$-3(2y - 5) - [5(y - 1) - 3y + 3]$ {-2}

16. $45 - [4 - 2y - 4(y + 7)] =$
$-4(1 + 3y) - [4 - 3(y + 2) - 2(2y - 5)]$ $\{-\frac{81}{11}\}$

Exercises 17–30 contain linear equations with constants in denominators. Solve each equation.

17. $\dfrac{x}{3} = \dfrac{x}{2} - 2$ *

18. $\dfrac{x}{5} = \dfrac{x}{6} + 1$ *

19. $20 - \dfrac{x}{3} = \dfrac{x}{2}$ {24}

20. $\dfrac{x}{5} - \dfrac{1}{2} = \dfrac{x}{6}$ *

21. $\dfrac{3x}{5} = \dfrac{2x}{3} + 1$ *

22. $\dfrac{x}{2} = \dfrac{3x}{4} + 5$ {-20}

23. $\dfrac{3x}{5} - x = \dfrac{x}{10} - \dfrac{5}{2}$ {5}

24. $2x - \dfrac{2x}{7} = \dfrac{x}{2} + \dfrac{17}{2}$ {7}

25. $\dfrac{x + 3}{6} = \dfrac{3}{8} + \dfrac{x - 5}{4}$ $\{\frac{33}{2}\}$

26. $\dfrac{x + 1}{4} = \dfrac{1}{6} + \dfrac{2 - x}{3}$ {1}

27. $\dfrac{x}{4} = 2 + \dfrac{x - 3}{3}$ {-12}

28. $5 + \dfrac{x - 2}{3} = \dfrac{x + 3}{8}$ {-19}

29. $\dfrac{x + 1}{3} = 5 - \dfrac{x + 2}{7}$ $\{\frac{46}{5}\}$

30. $\dfrac{3x}{5} - \dfrac{x - 3}{2} = \dfrac{x + 2}{3}$ $\{\frac{25}{7}\}$

*Exercises 31–50 contain rational equations with variables in denominators. For each equation, **a.** write the value or values of the variable that make a denominator zero. These are the restrictions on the variable. **b.** Keeping the restrictions in mind, solve the equation.*

31. $\dfrac{4}{x} = \dfrac{5}{2x} + 3$ a. 0 b. $\{\frac{1}{2}\}$

32. $\dfrac{5}{x} = \dfrac{10}{3x} + 4$ a. 0 b. $\{\frac{5}{12}\}$

33. $\dfrac{2}{x} + 3 = \dfrac{5}{2x} + \dfrac{13}{4}$ a. 0 b. {-2}

34. $\dfrac{7}{2x} - \dfrac{5}{3x} = \dfrac{22}{3}$ a. 0 b. $\{\frac{1}{4}\}$

35. $\dfrac{2}{3x} + \dfrac{1}{4} = \dfrac{11}{6x} - \dfrac{1}{3}$ a. 0 b. {2}

36. $\dfrac{5}{2x} - \dfrac{8}{9} = \dfrac{1}{18} - \dfrac{1}{3x}$ *

37. $\dfrac{x - 2}{2x} + 1 = \dfrac{x + 1}{x}$ *

38. $\dfrac{4}{x} = \dfrac{9}{5} - \dfrac{7x - 4}{5x}$ *

39. $\dfrac{1}{x - 1} + 5 = \dfrac{11}{x - 1}$ *

40. $\dfrac{3}{x + 4} - 7 = \dfrac{-4}{x + 4}$ *

41. $\dfrac{8x}{x + 1} = 4 - \dfrac{8}{x + 1}$ *

42. $\dfrac{2}{x - 2} = \dfrac{x}{x - 2} - 2$ *

43. $\dfrac{3}{2x - 2} + \dfrac{1}{2} = \dfrac{2}{x - 1}$ a. 1 b. {2}

44. $\dfrac{3}{x + 3} = \dfrac{5}{2x + 6} + \dfrac{1}{x - 2}$ a. -3, 2 b. {-8}

45. $\dfrac{3}{x + 2} + \dfrac{2}{x - 2} = \dfrac{8}{(x + 2)(x - 2)}$ a. -2, 2 b. \varnothing

46. $\dfrac{5}{x + 2} + \dfrac{3}{x - 2} = \dfrac{12}{(x + 2)(x - 2)}$ a. -2, 2 b. \varnothing

47. $\dfrac{2}{x + 1} - \dfrac{1}{x - 1} = \dfrac{2x}{x^2 - 1}$ a. -1, 1 b. {-3}

48. $\dfrac{4}{x + 5} + \dfrac{2}{x - 5} = \dfrac{32}{x^2 - 25}$ a. -5, 5 b. {7}

49. $\dfrac{1}{x - 4} - \dfrac{5}{x + 2} = \dfrac{6}{x^2 - 2x - 8}$ a. -2, 4 b. \varnothing

50. $\dfrac{6}{x + 3} - \dfrac{5}{x - 2} = \dfrac{-20}{x^2 + x - 6}$ a. -3, 2 b. {7}

In Exercises 51–56, find all values of x satisfying the given conditions.

51. $y_1 = 5(2x - 8) - 2$, $y_2 = 5(x - 3) + 3$, and $y_1 = y_2$. 6

52. $y_1 = 7(3x - 2) + 5$, $y_2 = 6(2x - 1) + 24$, and $y_1 = y_2$. 3

53. $y_1 = \dfrac{x - 3}{5}$, $y_2 = \dfrac{x - 5}{4}$, and $y_1 - y_2 = 1$. -7

54. $y_1 = \dfrac{x + 1}{4}$, $y_2 = \dfrac{x - 2}{3}$, and $y_1 - y_2 = -4$. 59

55. $y_1 = \dfrac{5}{x + 4}$, $y_2 = \dfrac{3}{x + 3}$, $y_3 = \dfrac{12x + 19}{x^2 + 7x + 12}$, and $y_1 + y_2 = y_3$. 2

56. $y_1 = \dfrac{2x - 1}{x^2 + 2x - 8}$, $y_2 = \dfrac{2}{x + 4}$, $y_3 = \dfrac{1}{x - 2}$, and $y_1 + y_2 = y_3$. 3

In Exercises 57–60, find all values of x such that y = 0.

57. $y = 4[x - (3 - x)] - 7(x + 1)$ 19

58. $y = 2[3x - (4x - 6)] - 5(x - 6)$ 6

59. $y = \dfrac{x + 6}{3x - 12} - \dfrac{5}{x - 4} - \dfrac{2}{3}$ -1

60. $y = \dfrac{1}{5x + 5} - \dfrac{3}{x + 1} + \dfrac{7}{5}$ 1

In Exercises 61–68, solve each equation. Then determine whether the equation is an identity, a conditional equation, or an inconsistent equation.

61. $5x + 9 = 9(x + 1) - 4x$ *

62. $4x + 7 = 7(x + 1) - 3x$ *

63. $3(x + 2) = 7 + 3x$ \varnothing*

64. $4(x + 5) = 21 + 4x$ \varnothing*

65. $10x + 3 = 8x + 3$ {0}*

66. $5x + 7 = 2x + 7$ {0}*

67. $\dfrac{2x}{x - 3} = \dfrac{6}{x - 3} + 4$ \varnothing*

68. $\dfrac{3}{x - 3} = \dfrac{x}{x - 3} + 3$ \varnothing*

The equations in Exercises 69–80 combine the types of equations we have discussed in this section. Solve each equation. Then state whether the equation is an identity, a conditional equation, or an inconsistent equation.

69. $\dfrac{x + 5}{2} - 4 = \dfrac{2x - 1}{3}$ {-7}*

70. $\dfrac{x + 2}{7} = 5 - \dfrac{x + 1}{3}$ $\{\frac{46}{5}\}$*

71. $\dfrac{2}{x - 2} = 3 + \dfrac{x}{x - 2}$ \varnothing*

72. $\dfrac{6}{x + 3} + 2 = \dfrac{-2x}{x + 3}$ \varnothing*

73. $8x - (3x + 2) + 10 = 3x$ {−4}; conditional equation

74. $2(x + 2) + 2x = 4(x + 1)$ {x|x is a real number}; identity

75. $\dfrac{2}{x} + \dfrac{1}{2} = \dfrac{3}{4}$ {8}* **76.** $\dfrac{3}{x} - \dfrac{1}{6} = \dfrac{1}{3}$ {6}*

77. $\dfrac{4}{x - 2} + \dfrac{3}{x + 5} = \dfrac{7}{(x + 5)(x - 2)}$ {−1}*

78. $\dfrac{1}{x - 1} = \dfrac{1}{(2x + 3)(x - 1)} + \dfrac{4}{2x + 3}$ {3}*

79. $\dfrac{4x}{x + 3} - \dfrac{12}{x - 3} = \dfrac{4x^2 + 36}{x^2 - 9}$ ∅; inconsistent equation

80. $\dfrac{4}{x^2 + 3x - 10} - \dfrac{1}{x^2 + x - 6} = \dfrac{3}{x^2 - x - 12}$ $\{\frac{1}{7}\}$*

In Exercises 81–84, use the $\boxed{\text{Y=}}$ screen to write the equation being solved. Then use the table to solve the equation.

81.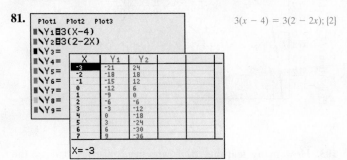
$3(x - 4) = 3(2 - 2x)$; {2}

82.
$3(2x - 5) = 5x + 2$; {17}

83.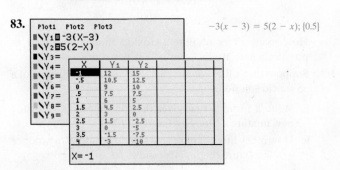
$-3(x - 3) = 5(2 - x)$; {0.5}

84.
$2x - 5 = 4(3x + 1) - 2$; {−0.7}

Practice Plus

85. Evaluate $x^2 - x$ for the value of x satisfying $4(x - 2) + 2 = 4x - 2(2 - x)$. 2

86. Evaluate $x^2 - x$ for the value of x satisfying $2(x - 6) = 3x + 2(2x - 1)$. 6

87. Evaluate $x^2 - (xy - y)$ for x satisfying $\dfrac{3(x + 3)}{5} = 2x + 6$ and y satisfying $-2y - 10 = 5y + 18$. −7

88. Evaluate $x^2 - (xy - y)$ for x satisfying $\dfrac{13x - 6}{4} = 5x + 2$ and y satisfying $5 - y = 7(y + 4) + 1$. −5

In Exercises 89–96, solve each equation.

89. $[(3 + 6)^2 \div 3] \cdot 4 = -54x$ {−2}

90. $2^3 - [4(5 - 3)^3] = -8x$ {3}

91. $5 - 12x = 8 - 7x - [6 \div 3(2 + 5^3) + 5x]$ ∅

92. $2(5x + 58) = 10x + 4(21 \div 3.5 - 11)$ ∅

93. $0.7x + 0.4(20) = 0.5(x + 20)$ {10}

94. $0.5(x + 2) = 0.1 + 3(0.1x + 0.3)$ {0}

95. $4x + 13 - \{2x - [4(x - 3) - 5]\} = 2(x - 6)$ {−2}

96. $-2\{7 - [4 - 2(1 - x) + 3]\} = 10 - [4x - 2(x - 3)]$ $\{\frac{4}{3}\}$

Application Exercises

Grade Inflation. *The bar graph shows the percentage of U.S. college freshmen with an average grade of A in high school.*

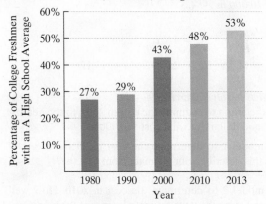

Percentage of U.S. College Freshmen with an Average Grade of A (A− to A+) in High School

Source: Higher Education Research Institute

The data displayed by the bar graph can be described by the mathematical model

$$p = \frac{4x}{5} + 25,$$

where x is the number of years after 1980 and p is the percentage of U.S. college freshmen who had an average grade of A in high school. Use this information to solve Exercises 97–98.

97. a. According to the formula, in 2010, what percentage of U.S. college freshmen had an average grade of A in high school? Does this underestimate or overestimate the percent displayed by the bar graph? By how much?
49%; overestimates by 1%

b. If trends shown by the formula continue, project when 57% of U.S. college freshmen will have had an average grade of A in high school. 2020

98. Refer to the mathematical model and the bar graph on the previous page.

a. According to the formula, in 2000, what percentage of U.S. college freshmen had an average grade of A in high school? Does this underestimate or overestimate the percent displayed by the bar graph? By how much? 41%; underestimates by 2%

b. If trends shown by the formula continue, project when 65% of U.S. college freshmen will have had an average grade of A in high school. 2030

The line graph shows the cost of inflation. What cost $10,000 in 1984 would cost the amount shown by the graph in subsequent years.

The Cost of Inflation

Source: U.S. Bureau of Labor Statistics

Here are two mathematical models for the data shown by the graph. In each formula, C represents the cost x years after 1990 of what cost $10,000 in 1984.

Model 1 $C = 442x + 12,969$

Model 2 $C = 2x^2 + 390x + 13,126$

Use these models to solve Exercises 99–102.

99. a. Use the graph to estimate the cost in 2010, to the nearest thousand dollars, of what cost $10,000 in 1984. $22,000

b. Use model 1 to determine the cost in 2010. How well does this describe your estimate from part (a)? $21,809; reasonably well

c. Use model 2 to determine the cost in 2010. How well does this describe your estimate from part (a)? $21,726; reasonably well

100. a. Use the graph to estimate the cost in 2000, to the nearest thousand dollars, of what cost $10,000 in 1984. $17,000

b. Use model 1 to determine the cost in 2000. How well does this describe your estimate from part (a)? $17,389; reasonably well

c. Use model 2 to determine the cost in 2000. How well does this describe your estimate from part (a)? $17,226; reasonably well

101. Use model 1 to determine in which year the cost will be $26,229 for what cost $10,000 in 1984. 2020

102. Use model 1 to determine in which year the cost will be $25,345 for what cost $10,000 in 1984. 2018

Formulas with rational expressions are often used to model learning. Many of these formulas model the proportion of correct responses in terms of the number of trials of a particular task. One such model, called a learning curve, is

$$P = \frac{0.9x - 0.4}{0.9x + 0.1},$$

where P is the proportion of correct responses after x trials. If P = 0, there are no correct responses. If P = 1, all responses are correct. The graph of the rational formula is shown. Use the formula to solve Exercises 103–104.

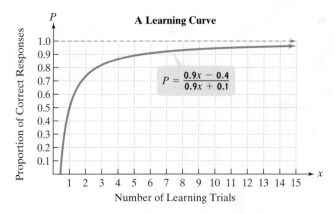

A Learning Curve

103. How many learning trials are necessary for 0.95 of the responses to be correct? Identify your solution as a point on the graph. 11 learning trials; (11, 0.95)

104. How many learning trials are necessary for 0.5 of the responses to be correct? Identify your solution as a point on the graph. 1 learning trial; (1, 0.5)

105. A company wants to increase the 10% peroxide content of its product by adding pure peroxide (100% peroxide). If x liters of pure peroxide are added to 500 liters of its 10% solution, the concentration, C, of the new mixture is given by

$$C = \frac{x + 0.1(500)}{x + 500}.$$

How many liters of pure peroxide should be added to produce a new product that is 28% peroxide? 125 liters

106. Suppose that x liters of pure acid are added to 200 liters of a 35% acid solution.

a. Write a formula that gives the concentration, C, of the new mixture. (*Hint:* See Exercise 105.) *

b. How many liters of pure acid should be added to produce a new mixture that is 74% acid? 300 liters

Explaining the Concepts 107–108. Answers will vary.

107. What is a linear equation in one variable? Give an example of this type of equation.

108. Suppose that you solve $\frac{x}{5} - \frac{x}{2} = 1$ by multiplying both sides by 20 rather than the least common denominator (namely, 10). Describe what happens. If you get the correct solution, why do you think we clear the equation of fractions by multiplying by the *least* common denominator?

109. Suppose you are an algebra teacher grading the following solution on an examination:

$$-3(x - 6) = 2 - x$$
$$-3x - 18 = 2 - x$$
$$-2x - 18 = 2$$
$$-2x = -16$$
$$x = 8.$$

You should note that 8 checks, so the solution set is {8}. The student who worked the problem therefore wants full credit. Can you find any errors in the solution? If full credit is 10 points, how many points should you give the student? Justify your position. **109–115.** Answers will vary.

110. Explain how to find restrictions on the variable in a rational equation.

111. Why should restrictions on the variable in a rational equation be listed before you begin solving the equation?

112. What is an identity? Give an example.

113. What is a conditional equation? Give an example.

114. What is an inconsistent equation? Give an example.

115. Describe the trend shown by the graph in Exercises 103–104 in terms of learning new tasks. What happens initially and what happens as time increases?

Technology Exercises

In Exercises 116–119, use your graphing utility to enter each side of the equation separately under y_1 and y_2. Then use the utility's $\boxed{\text{TABLE}}$ *or* $\boxed{\text{GRAPH}}$ *feature to solve the equation.*

116. $5x + 2(x - 1) = 3x + 10$ {3}*

117. $2x + 3(x - 4) = 4x - 7$ {5}*

118. $\dfrac{x - 3}{5} - 1 = \dfrac{x - 5}{4}$ {-7}*

119. $\dfrac{2x - 1}{3} - \dfrac{x - 5}{6} = \dfrac{x - 3}{4}$ {-5}*

Critical Thinking Exercises

Make Sense? *In Exercises 120–123, determine whether each statement makes sense or does not make sense, and explain your reasoning.*

120. The model $P = -0.18n + 2.1$ describes the number of pay phones, P, in millions, n years after 2000, so I have to solve a linear equation to determine the number of pay phones in 2002. does not make sense

121. Although I can solve $3x + \frac{1}{5} = \frac{1}{4}$ by first subtracting $\frac{1}{5}$ from both sides, I find it easier to begin by multiplying both sides by 20, the least common denominator. makes sense

122. Because I know how to clear an equation of fractions, I decided to clear the equation $0.5x + 8.3 = 12.4$ of decimals by multiplying both sides by 10. makes sense

123. Because $x = x + 5$ is an inconsistent equation, the graphs of $y = x$ and $y = x + 5$ should not intersect. makes sense

In Exercises 124–127, determine whether each statement is true or false. If the statement is false, make the necessary change(s) to produce a true statement. Changes to false statements will vary.

124. The equation $-7x = x$ has no solution. false

125. The equations $\dfrac{x}{x - 4} = \dfrac{4}{x - 4}$ and $x = 4$ are equivalent. *

126. The equations $3y - 1 = 11$ and $3y - 7 = 5$ are equivalent. true

127. If a and b are any real numbers, then $ax + b = 0$ always has one number in its solution set. false

128. If x represents a number, write an English sentence about the number that results in an inconsistent equation.

129. Find b such that $\dfrac{7x + 4}{b} + 13 = x$ has a solution set given by {-6}. 2

130. Find b such that $\dfrac{4x - b}{x - 5} = 3$ has a solution set given by ∅. 20

Preview Exercises

Exercises 131–133 will help you prepare for the material covered in the next section.

131. Jane's salary exceeds Jim's by $150 per week. If x represents Jim's weekly salary, write an algebraic expression that models Jane's weekly salary. $x + 150$

132. A telephone texting plan has a monthly fee of $20 with a charge of $0.05 per text. Write an algebraic expression that models the plan's monthly cost for x text messages. $20 + 0.05x$

133. If the width of a rectangle is represented by x and the length is represented by $x + 200$, write a simplified algebraic expression that models the rectangle's perimeter. $4x + 400$

Section 1.3

Models and Applications

What am I supposed to learn?

After studying this section, you should be able to:

1. Use linear equations to solve problems.

2. Solve a formula for a variable.

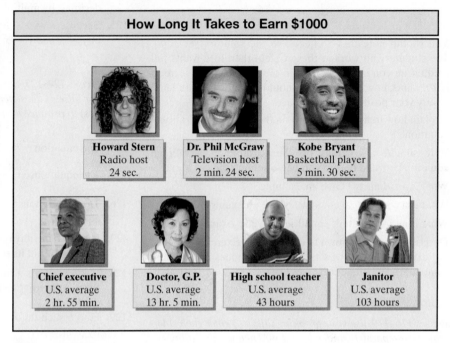

How Long It Takes to Earn $1000

Howard Stern
Radio host
24 sec.

Dr. Phil McGraw
Television host
2 min. 24 sec.

Kobe Bryant
Basketball player
5 min. 30 sec.

Chief executive
U.S. average
2 hr. 55 min.

Doctor, G.P.
U.S. average
13 hr. 5 min.

High school teacher
U.S. average
43 hours

Janitor
U.S. average
103 hours

Source: *Time*

In this section, you'll see examples and exercises focused on how much money Americans earn. These situations illustrate a step-by-step strategy for solving problems. As you become familiar with this strategy, you will learn to solve a wide variety of problems.

1. Use linear equations to solve problems.

Problem Solving with Linear Equations

We have seen that a model is a mathematical representation of a real-world situation. In this section, we will be solving problems that are presented in English. This means that we must obtain models by translating from the ordinary language of English into the language of algebraic equations. To translate, however, we must understand the English prose and be familiar with the forms of algebraic language. Following are some general steps we will use in solving word problems:

GREAT QUESTION!

Why are word problems important?

There is great value in reasoning through the steps for solving a word problem. This value comes from the problem-solving skills that you will attain and is often more important than the specific problem or its solution.

Strategy for Solving Word Problems

Step 1 Read the problem carefully several times until you can state in your own words what is given and what the problem is looking for. Let *x* (or any variable) represent one of the unknown quantities in the problem.

Step 2 If necessary, write expressions for any other unknown quantities in the problem in terms of *x*.

Step 3 Write an equation in *x* that models the verbal conditions of the problem.

Step 4 Solve the equation and answer the problem's question.

Step 5 Check the solution *in the original wording* of the problem, not in the equation obtained from the words.

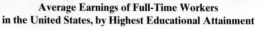

EXAMPLE 1 Education Pays Off

The bar graph in **Figure 1.15** shows average yearly earnings in the United States by highest educational attainment.

FIGURE 1.15
Source: U.S. Census Bureau (2012 data)

The average yearly salary of a man with a bachelor's degree exceeds that of a man with an associate's degree by $25 thousand. The average yearly salary of a man with a master's degree exceeds that of a man with an associate's degree by $45 thousand. Combined, three men with each of these degrees earn $214 thousand. Find the average yearly salary of men with each of these levels of education.

SOLUTION

Step 1 Let x represent one of the unknown quantities. We know something about salaries of men with bachelor's degrees and master's degrees: They exceed the salary of a man with an associate's degree by $25 thousand and $45 thousand, respectively. We will let

x = the average yearly salary of a man with an associate's degree
(in thousands of dollars).

Step 2 Represent other unknown quantities in terms of x. Because a man with a bachelor's degree earns $25 thousand more than a man with an associate's degree, let

$x + 25$ = the average yearly salary of a man with a bachelor's degree.

Because a man with a master's degree earns $45 thousand more than a man with an associate's degree, let

$x + 45$ = the average yearly salary of a man with a master's degree.

Step 3 Write an equation in x that models the conditions. Combined, three men with each of these degrees earn $214 thousand.

Salary: associate's degree	plus	salary: bachelor's degree	plus	salary: master's degree	equals	$214 thousand.
x	$+$	$(x + 25)$	$+$	$(x + 45)$	$=$	214

Step 4 Solve the equation and answer the question.

$$x + (x + 25) + (x + 45) = 214 \qquad \text{This is the equation that models the problem's conditions.}$$

$$3x + 70 = 214 \qquad \text{Remove parentheses, regroup, and combine like terms.}$$

$$3x = 144 \qquad \text{Subtract 70 from both sides.}$$

$$x = 48 \qquad \text{Divide both sides by 3.}$$

Because we isolated the variable in the model and obtained $x = 48$,

average salary with an associate's degree $= x = 48$

average salary with a bachelor's degree $= x + 25 = 48 + 25 = 73$

average salary with a master's degree $= x + 45 = 48 + 45 = 93$.

Men with associate's degrees average $48 thousand per year, men with bachelor's degrees average $73 thousand per year, and men with master's degrees average $93 thousand per year.

Step 5 Check the proposed solution in the original wording of the problem. The problem states that combined, three men with each of these educational attainments earn $214 thousand. Using the salaries we determined in step 4, the sum is

$48 thousand + $73 thousand + $93 thousand, or $214 thousand,

which satisfies the problem's conditions. • • •

✓ Check Point **1** The average yearly salary of a woman with a bachelor's degree exceeds that of a woman with an associate's degree by $14 thousand. The average yearly salary of a woman with a master's degree exceeds that of a woman with an associate's degree by $26 thousand. Combined, three women with each of these educational attainments earn $139 thousand. Find the average yearly salary of women with each of these levels of education. (These salaries are illustrated by the bar graph on the previous page.) associate's degree: $33 thousand; bachelor's degree: $47 thousand; master's degree: $59 thousand

GREAT QUESTION!

Example 1 involves using the word *exceeds* to represent two of the unknown quantities. Can you help me to write algebraic expressions for quantities described using *exceeds*?

Modeling with the word *exceeds* can be a bit tricky. It's helpful to identify the smaller quantity. Then add to this quantity to represent the larger quantity. For example, suppose that Tim's height exceeds Tom's height by a inches. Tom is the shorter person. If Tom's height is represented by x, then Tim's height is represented by $x + a$.

Your author teaching math in 1969

EXAMPLE 2 Modeling Attitudes of College Freshmen

Researchers have surveyed college freshmen every year since 1969. **Figure 1.16** at the top of the next page shows that attitudes about some life goals have changed dramatically over the years. In particular, the freshman class of 2013 was more interested in making money than the freshmen of 1969 had been. In 1969, 42% of first-year college students considered "being well-off financially" essential or very important. For the period from 1969 through 2013, this percentage increased by approximately 0.9 each year. If this trend continues, by which year will all college freshmen consider "being well-off financially" essential or very important?

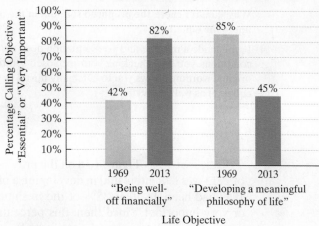

Life Objectives of College Freshmen, 1969–2013

FIGURE 1.16
Source: Higher Education Research Institute

SOLUTION

Step 1 Let *x* represent one of the unknown quantities. We are interested in the year when all college freshmen, or 100% of the freshmen, will consider this life objective essential or very important. Let

x = the number of years after 1969 when all freshmen will consider "being well-off financially" essential or very important.

Step 2 Represent other unknown quantities in terms of *x*. There are no other unknown quantities to find, so we can skip this step.

Step 3 Write an equation in *x* that models the conditions. The problem states that the 1969 percentage increased by approximately 0.9 each year.

The 1969 percentage	increased by	0.9 each year for *x* years	equals	100% of the freshmen.
42	+	0.9*x*	=	100

Step 4 Solve the equation and answer the question.

$$42 + 0.9x = 100$$ This is the equation that models the problem's conditions.

$$42 - 42 + 0.9x = 100 - 42$$ Subtract 42 from both sides.

$$0.9x = 58$$ Simplify.

$$\frac{0.9x}{0.9} = \frac{58}{0.9}$$ Divide both sides by 0.9.

$$x = 64.\overline{4} \approx 64$$ Simplify and round to the nearest whole number.

Using current trends, by approximately 64 years after 1969, or in 2033, all freshmen will consider "being well-off financially" essential or very important.

Step 5 Check the proposed solution in the original wording of the problem. The problem states that all freshmen (100%, represented by 100 using the model) will consider the objective essential or very important. Does this approximately occur if we increase the 1969 percentage, 42, by 0.9 each year for 64 years, our proposed solution?

$$42 + 0.9(64) = 42 + 57.6 = 99.6 \approx 100$$

This verifies that using trends shown in **Figure 1.16**, all first-year college students will consider the objective of being well-off financially essential or very important approximately 64 years after 1969. ● ● ●

GREAT QUESTION!

Why should I check the proposed solution in the original wording of the problem and not in the equation?

If you made a mistake and your equation does not correctly model the problem's conditions, you'll be checking your proposed solution in an incorrect model. Using the original wording allows you to catch any mistakes you may have made in writing the equation as well as solving it.

⊘ Check Point **2** **Figure 1.16** on the previous page shows that the freshman class of 2013 was less interested in developing a philosophy of life than the freshmen of 1969 had been. In 1969, 85% of the freshmen considered this objective essential or very important. Since then, this percentage has decreased by approximately 0.9 each year. If this trend continues, by which year will only 25% of college freshmen consider "developing a meaningful philosophy of life" essential or very important? by 67 years after 1969, or in 2036

EXAMPLE 3 Modeling Options for a Toll

The toll to a bridge costs $7. Commuters who use the bridge frequently have the option of purchasing a monthly discount pass for $30. With the discount pass, the toll is reduced to $4. For how many bridge crossings per month will the total monthly cost without the discount pass be the same as the total monthly cost with the discount pass?

SOLUTION

Step 1 Let x represent one of the unknown quantities. Let

$$x = \text{the number of bridge crossings per month.}$$

Step 2 Represent other unknown quantities in terms of x. There are no other unknown quantities, so we can skip this step.

Step 3 Write an equation in x that models the conditions. The monthly cost without the discount pass is the toll, $7, times the number of bridge crossings per month, x. The monthly cost with the discount pass is the cost of the pass, $30, plus the toll, $4, times the number of bridge crossings per month, x.

The monthly cost without the discount pass	must equal	the monthly cost with the discount pass.
$7x$	$=$	$30 + 4x$

Step 4 Solve the equation and answer the question.

$7x = 30 + 4x$	This is the equation that models the problem's conditions.
$3x = 30$	Subtract $4x$ from both sides.
$x = 10$	Divide both sides by 3.

Because x represents the number of bridge crossings per month, the total monthly cost without the discount pass will be the same as the total monthly cost with the discount pass for 10 bridge crossings per month.

Step 5 Check the proposed solution in the original wording of the problem. The problem states that the monthly cost without the discount pass should be the same as the monthly cost with the discount pass. Let's see if they are the same with 10 bridge crossings per month.

$$\text{Cost without the discount pass} = \$7(10) = \$70$$

$$\text{Cost with the discount pass} = \$30 + \$4(10) = \$30 + \$40 = \$70$$

With 10 bridge crossings per month, both options cost $70 for the month. Thus the proposed solution, 10 bridge crossings, satisfies the problem's conditions. •••

TECHNOLOGY

Numeric and Graphic Connections

We can use a graphing utility to numerically or graphically verify our work in Example 3.

The monthly cost without the discount pass must equal the monthly cost with the discount pass.

$$7x = 30 + 4x$$

Enter $y_1 = 7x$. Enter $y_2 = 30 + 4x$.

Numeric Check

Display a table for y_1 and y_2.

X	Y₁	Y₂
2	14	38
4	28	46
6	42	54
8	56	62
10	70	70
12	84	78
14	98	86
16	112	94
18	126	102
20	140	110
22	154	118
X=10		

When $x = 10$, y_1 and y_2 have the same value, 70. With 10 bridge crossings, costs are the same, $70, for both options.

Graphic Check

Display graphs for y_1 and y_2. Use the intersection feature.

Graphs intersect at (10, 70). With 10 bridge crossings, costs are the same, $70, for both options.

[0, 20, 2] by [0, 150, 5]

☑ Check Point **3** The toll to a bridge costs $5. Commuters who use the bridge frequently have the option of purchasing a monthly discount pass for $40. With the discount pass, the toll is reduced to $3. For how many bridge crossings per month will the total monthly cost without the discount pass be the same as the total monthly cost with the discount pass? 20 bridge crossings per month

EXAMPLE 4 A Price Reduction on a Digital Camera

Your local computer store is having a terrific sale on digital cameras. After a 40% price reduction, you purchase a digital camera for $276. What was the camera's price before the reduction?

SOLUTION

Step 1 Let x represent one of the unknown quantities. We will let

$$x = \text{the price of the digital camera prior to the reduction.}$$

Step 2 Represent other unknown quantities in terms of x. There are no other unknown quantities to find, so we can skip this step.

Step 3 **Write an equation in x that models the conditions.** The camera's original price minus the 40% reduction is the reduced price, $276.

$$x - 0.4x = 276$$

Step 4 **Solve the equation and answer the question.**

$x - 0.4x = 276$ This is the equation that models the problem's conditions.

$0.6x = 276$ Combine like terms: $x - 0.4x = 1x - 0.4x = (1 - 0.4)x = 0.6x$.

$\dfrac{0.6x}{0.6} = \dfrac{276}{0.6}$ Divide both sides by 0.6.

$x = 460$ Simplify: $0.6\,)\overline{276.0}$

The digital camera's price before the reduction was $460.

Step 5 **Check the proposed solution in the original wording of the problem.** The price before the reduction, $460, minus the 40% reduction should equal the reduced price given in the original wording, $276:

$$460 - 40\% \text{ of } 460 = 460 - 0.4(460) = 460 - 184 = 276.$$

This verifies that the digital camera's price before the reduction was $460. •••

○ Check Point **4** After a 30% price reduction, you purchase a new computer for $840. What was the computer's price before the reduction? $1200

Our next example is about simple interest. Simple interest involves interest calculated only on the amount of money that we invest, called the **principal**. The formula $I = Pr$ is used to find the simple interest, I, earned for one year when the principal, P, is invested at an annual interest rate, r. Dual investment problems involve different amounts of money in two or more investments, each paying a different rate.

EXAMPLE 5 Solving a Dual Investment Problem

Your grandmother needs your help. She has $50,000 to invest. Part of this money is to be invested in noninsured bonds paying 15% annual interest. The rest of this money is to be invested in a government-insured certificate of deposit paying 7% annual interest. She told you that she requires $6000 per year in extra income from the combination of these investments. How much money should be placed in each investment?

SOLUTION

Step 1 **Let x represent one of the unknown quantities.** We will let

$x = $ the amount invested in the noninsured bonds at 15%.

Step 2 Represent other unknown quantities in terms of x. The other quantity that we seek is the amount invested at 7% in the certificate of deposit. Because the total amount Grandma has to invest is $50,000 and we already used up x,

$50,000 - x =$ the amount invested in the certificate of deposit at 7%.

Step 3 Write an equation in x that models the conditions. Because Grandma requires $6000 in total interest, the interest for the two investments combined must be $6000. Interest is Pr or rP for each investment.

Step 4 Solve the equation and answer the question.

$0.15x + 0.07(50,000 - x) = 6000$	This is the equation that models the problem's conditions.
$0.15x + 3500 - 0.07x = 6000$	Use the distributive property.
$0.08x + 3500 = 6000$	Combine like terms.
$0.08x = 2500$	Subtract 3500 from both sides.
$\dfrac{0.08x}{0.08} = \dfrac{2500}{0.08}$	Divide both sides by 0.08.
$x = 31,250$	Simplify.

Thus,

the amount invested at 15% = x = 31,250.

the amount invested at 7% = 50,000 − 31,250 = 18,750.

Grandma should invest $31,250 at 15% and $18,750 at 7%.

Step 5 Check the proposed solution in the original wording of the problem. The problem states that the total interest from the dual investments should be $6000. Can Grandma count on $6000 interest? The interest earned on $31,250 at 15% is ($31,250) (0.15), or $4687.50. The interest earned on $18,750 at 7% is ($18,750)(0.07), or $1312.50. The total interest is $4687.50 + $1312.50, or $6000, exactly as it should be. You've made your grandmother happy. (Now if you would just visit her more often . . .) • • •

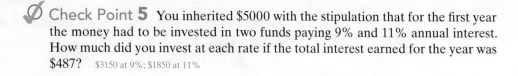

Ⓧ Check Point **5** You inherited $5000 with the stipulation that for the first year the money had to be invested in two funds paying 9% and 11% annual interest. How much did you invest at each rate if the total interest earned for the year was $487? $3150 at 9%; $1850 at 11%

Solving geometry problems usually requires a knowledge of basic geometric ideas and formulas. Formulas for area, perimeter, and volume are given in **Table 1.1**.

Table 1.1 Common Formulas for Area, Perimeter, and Volume

Square	Rectangle	Circle	Triangle	Trapezoid
$A = s^2$	$A = lw$	$A = \pi r^2$	$A = \frac{1}{2}bh$	$A = \frac{1}{2}h(a + b)$
$P = 4s$	$P = 2l + 2w$	$C = 2\pi r$		

Cube	Rectangular Solid	Circular Cylinder	Sphere	Cone
$V = s^3$	$V = lwh$	$V = \pi r^2 h$	$V = \frac{4}{3}\pi r^3$	$V = \frac{1}{3}\pi r^2 h$

We will be using the formula for the perimeter of a rectangle, $P = 2l + 2w$, in our next example. The formula states that a rectangle's perimeter is the sum of twice its length and twice its width.

EXAMPLE 6 Finding the Dimensions of an American Football Field

The length of an American football field is 200 feet more than the width. If the perimeter of the field is 1040 feet, what are its dimensions?

SOLUTION

Step 1 Let x represent one of the unknown quantities. We know something about the length; the length is 200 feet more than the width. We will let

$$x = \text{the width.}$$

Step 2 Represent other unknown quantities in terms of x. Because the length is 200 feet more than the width, we add 200 to the width to represent the length. Thus,

$$x + 200 = \text{the length.}$$

Figure 1.17 illustrates an American football field and its dimensions.

FIGURE 1.17 An American football field

Step 3 Write an equation in x that models the conditions. Because the perimeter of the field is 1040 feet,

Twice the length	plus	twice the width	is	the perimeter.
$2(x + 200)$	$+$	$2x$	$=$	$1040.$

Step 4 **Solve the equation and answer the question.**

$$2(x + 200) + 2x = 1040 \quad \text{This is the equation that models the problem's conditions.}$$

$$2x + 400 + 2x = 1040 \quad \text{Apply the distributive property.}$$

$$4x + 400 = 1040 \quad \text{Combine like terms: } 2x + 2x = 4x.$$

$$4x = 640 \quad \text{Subtract 400 from both sides.}$$

$$x = 160 \quad \text{Divide both sides by 4.}$$

Thus,

$$\text{width} = x = 160.$$

$$\text{length} = x + 200 = 160 + 200 = 360.$$

The dimensions of an American football field are 160 feet by 360 feet. (The 360-foot length is usually described as 120 yards.)

Step 5 **Check the proposed solution in the original wording of the problem.** The perimeter of the football field using the dimensions that we found is

$$2(160 \text{ feet}) + 2(360 \text{ feet}) = 320 \text{ feet} + 720 \text{ feet} = 1040 \text{ feet}.$$

Because the problem's wording tells us that the perimeter is 1040 feet, our dimensions are correct.

• • •

GREAT QUESTION!

Should I draw pictures like Figure 1.17 when solving geometry problems?

When solving word problems, particularly problems involving geometric figures, drawing a picture of the situation is often helpful. Label x on your drawing and, where appropriate, label other parts of the drawing in terms of x.

 Check Point **6** The length of a rectangular basketball court is 44 feet more than the width. If the perimeter of the basketball court is 288 feet, what are its dimensions? 50 feet by 94 feet

 2 Solve a formula for a variable.

Solving a Formula for One of Its Variables

We know that solving an equation is the process of finding the number (or numbers) that make the equation a true statement. All of the equations we have solved contained only one letter, x.

By contrast, formulas contain two or more letters, representing two or more variables. An example is the formula for the perimeter of a rectangle:

$$P = 2l + 2w.$$

We say that this formula is solved for the variable P because P is alone on one side of the equation and the other side does not contain a P.

Solving a formula for a variable means rewriting the formula so that the variable is isolated on one side of the equation. It does not mean obtaining a numerical value for that variable.

To solve a formula for one of its variables, treat that variable as if it were the only variable in the equation. Think of the other variables as if they were numbers. Isolate all terms with the specified variable on one side of the equation and all terms without the specified variable on the other side. Then divide both sides by the same nonzero quantity to get the specified variable alone. The next two examples show how to do this.

EXAMPLE 7 Solving a Formula for a Variable

Solve the formula $P = 2l + 2w$ for l.

SOLUTION

First, isolate $2l$ on the right by subtracting $2w$ from both sides. Then solve for l by dividing both sides by 2.

> We need to isolate l.

$P = 2l + 2w$	This is the given formula.
$P - 2w = 2l + 2w - 2w$	Isolate $2l$ by subtracting $2w$ from both sides.
$P - 2w = 2l$	Simplify.
$\dfrac{P - 2w}{2} = \dfrac{2l}{2}$	Solve for l by dividing both sides by 2.
$\dfrac{P - 2w}{2} = l$	Simplify.

Equivalently,

$$l = \frac{P - 2w}{2}.$$

• • •

⬥ **Check Point 7** Solve the formula $P = 2l + 2w$ for w. $w = \dfrac{P - 2l}{2}$

EXAMPLE 8 Solving a Formula for a Variable That Occurs Twice

The formula

$$A = P + Prt$$

describes the amount, A, that a principal of P dollars is worth after t years when invested at a simple annual interest rate, r. Solve this formula for P.

GREAT QUESTION!

Can I solve $A = P + Prt$ for P by subtracting Prt from both sides and writing

$$A - Prt = P?$$

No. When a formula is solved for a specified variable, that variable must be isolated on one side. The variable P occurs on both sides of

$$A - Prt = P.$$

SOLUTION

Notice that all the terms with P already occur on the right side of the formula.

> We need to isolate P.

$$A = P + Prt$$

We can factor P from the two terms on the right to convert the two occurrences of P into one.

$A = P + Prt$	This is the given formula.
$A = P(1 + rt)$	Factor out P on the right side of the equation.
$\dfrac{A}{1 + rt} = \dfrac{P(1 + rt)}{1 + rt}$	Divide both sides by $1 + rt$.
$\dfrac{A}{1 + rt} = P$	Simplify: $\dfrac{P(1 + rt)}{1(1 + rt)} = \dfrac{P}{1} = P.$

Equivalently,

$$P = \frac{A}{1 + rt}.$$

• • •

⬥ **Check Point 8** Solve the formula $P = C + MC$ for C. $C = \dfrac{P}{1 + M}$

ACHIEVING SUCCESS

Do not expect to solve every word problem immediately. As you read each problem, underline the important parts. It's a good idea to read the problem at least twice. Be persistent, but use the **"Ten Minutes of Frustration" Rule**. If you have exhausted every possible means for solving a problem and you are still bogged down, stop after ten minutes. Put a question mark by the exercise and move on. When you return to class, ask your professor for assistance.

CONCEPT AND VOCABULARY CHECK

Fill in each blank so that the resulting statement is true.

1. According to the U.S. Office of Management and Budget, the 2011 budget for defense exceeded the budget for education by $658.6 billion. If x represents the budget for education, in billions of dollars, the budget for defense can be represented by __$x + 658.6$__.

2. In 2000, 31% of U.S. adults viewed a college education as essential for success. For the period from 2000 through 2010, this percentage increased by approximately 2.4 each year. The percentage of U.S. adults who viewed a college education as essential for success x years after 2000 can be represented by __$31 + 2.4x$__.

3. A text message plan costs $4 per month plus $0.15 per text. The monthly cost for x text messages can be represented by __$4 + 0.15x$__.

4. I purchased a computer after a 15% price reduction. If x represents the computer's original price, the reduced price can be represented by __$x - 0.15x$ or $0.85x$__.

5. The combined yearly interest for x dollars invested at 12% and 30,000 $- x$ dollars invested at 9% is __$0.12x + 0.09(30{,}000 - x)$__.

6. Solving a formula for a variable means rewriting the formula so that the variable is __isolated on one side__.

7. The first step in solving $IR + Ir = E$ for I is to obtain a single occurrence of I by __factoring__ I from the two terms on the left.

EXERCISE SET 1.3

Practice and Application Exercises

How will you spend your average life expectancy of 78 years? The bar graph shows the average number of years you will devote to each of your most time-consuming activities. Exercises 1–2 are based on the data displayed by the graph.

Source: U.S. Bureau of Labor Statistics

1. According to the American Bureau of Labor Statistics, you will devote 37 years to sleeping and watching TV. The number of years sleeping will exceed the number of years watching TV by 19. Over your lifetime, how many years will you spend on each of these activities? TV: 9 years; sleeping: 28 years

2. According to the American Bureau of Labor Statistics, you will devote 32 years to sleeping and eating. The number of years sleeping will exceed the number of years eating by 24. Over your lifetime, how many years will you spend on each of these activities? eating: 4 years; sleeping: 28 years

The bar graph shows median yearly earnings of full-time workers in the United States for people 25 years and over with a college education, by final degree earned. Exercises 3–4 are based on the data displayed by the graph.

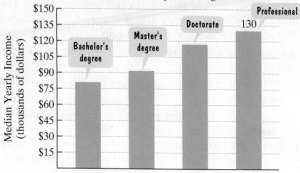

Source: U.S. Census Bureau

(Exercises 3–4 are based on the graph at the bottom-right of the previous page.)

3. The median yearly salary of an American whose final degree is a master's is $70 thousand less than twice that of an American whose final degree is a bachelor's. Combined, two people with each of these educational attainments earn $173 thousand. Find the median yearly salary of Americans with each of these final degrees. Bachelor's: $81 thousand; Master's: $92 thousand

4. The median yearly salary of an American whose final degree is a doctorate is $45 thousand less than twice that of an American whose final degree is a bachelor's. Combined, two people with each of these educational attainments earn $198 thousand. Find the median yearly salary of Americans with each of these final degrees. Bachelor's: $81 thousand; Doctorate: $117 thousand

Despite booming new car sales with their cha-ching sounds, the average age of vehicles on U.S. roads is not going down. The bar graph shows the average price of new cars in the United States and the average age of cars on U.S. roads for two selected years. Exercises 5–6 are based on the information displayed by the graph.

Average Price of New Cars and Average Age of Cars on U.S. Roads

Source: Kelley Blue Book, IHS Automotive/Polk

5. In 2014, the average price of a new car was $37,600. For the period shown, new-car prices increased by approximately $1250 per year. If this trend continues, how many years after 2014 will the price of a new car average $46,350? In which year will this occur? 7 years after 2014; 2021

6. In 2014, the average age of cars on U.S. roads was 11.3 years. For the period shown, this average age increased by approximately 0.2 year per year. If this trend continues, how many years after 2014 will the average age of vehicles on U.S. roads be 12.3 years? In which year will this occur? 5 years after 2014; 2019

7. A new car worth $24,000 is depreciating in value by $3000 per year.
 a. Write a formula that models the car's value, y, in dollars, after x years. $y = 24{,}000 - 3000x$
 b. Use the formula from part (a) to determine after how many years the car's value will be $9000. 5 years
 c. Graph the formula from part (a) in the first quadrant of a rectangular coordinate system. Then show your solution to part (b) on the graph. *

*See Answers to Selected Exercises.

8. A new car worth $45,000 is depreciating in value by $5000 per year.
 a. Write a formula that models the car's value, y, in dollars, after x years. $y = 45{,}000 - 5000x$
 b. Use the formula from part (a) to determine after how many years the car's value will be $10,000. 7 years
 c. Graph the formula from part (a) in the first quadrant of a rectangular coordinate system. Then show your solution to part (b) on the graph. *

9. You are choosing between two health clubs. Club A offers membership for a fee of $40 plus a monthly fee of $25. Club B offers membership for a fee of $15 plus a monthly fee of $30. After how many months will the total cost at each health club be the same? What will be the total cost for each club? after 5 months; $165

10. You need to rent a rug cleaner. Company A will rent the machine you need for $22 plus $6 per hour. Company B will rent the same machine for $28 plus $4 per hour. After how many hours of use will the total amount spent at each company be the same? What will be the total amount spent at each company? 3 hours; $40

11. The bus fare in a city is $1.25. People who use the bus have the option of purchasing a monthly discount pass for $15.00. With the discount pass, the fare is reduced to $0.75. Determine the number of times in a month the bus must be used so that the total monthly cost without the discount pass is the same as the total monthly cost with the discount pass. 30 times

12. A discount pass for a bridge costs $30 per month. The toll for the bridge is normally $5.00, but it is reduced to $3.50 for people who have purchased the discount pass. Determine the number of times in a month the bridge must be crossed so that the total monthly cost without the discount pass is the same as the total monthly cost with the discount pass. 20 times

13. In 2010, there were 13,300 students at college A, with a projected enrollment increase of 1000 students per year. In the same year, there were 26,800 students at college B, with a projected enrollment decline of 500 students per year.
 a. According to these projections, when will the colleges have the same enrollment? What will be the enrollment in each college at that time? 2019; 22,300 students
 b. Use the following table to check your work in part (a) numerically. What equations were entered for Y_1 and Y_2 to obtain this table? $Y_1 = 13{,}300 + 1000x$; $Y_2 = 26{,}800 - 500x$

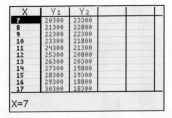

14. In 2000, the population of Greece was 10,600,000, with projections of a population decrease of 28,000 people per year. In the same year, the population of Belgium was 10,200,000, with projections of a population decrease of 12,000 people per year. (Source: United Nations) According to these projections, when will the two countries have the same population? What will be the population at that time? 2025; 9,900,000

15. After a 20% reduction, you purchase a television for $336. What was the television's price before the reduction? $420

16. After a 30% reduction, you purchase a dictionary for $30.80. What was the dictionary's price before the reduction? $44

17. Including 8% sales tax, an inn charges $162 per night. Find the inn's nightly cost before the tax is added. $150

18. Including 5% sales tax, an inn charges $252 per night. Find the inn's nightly cost before the tax is added. $240

Exercises 19–20 involve markup, the amount added to the dealer's cost of an item to arrive at the selling price of that item.

19. The selling price of a refrigerator is $584. If the markup is 25% of the dealer's cost, what is the dealer's cost of the refrigerator? $467.20

20. The selling price of a scientific calculator is $15. If the markup is 25% of the dealer's cost, what is the dealer's cost of the calculator? $12

21. You invested $7000 in two accounts paying 6% and 8% annual interest. If the total interest earned for the year was $520, how much was invested at each rate? *

22. You invested $11,000 in two accounts paying 5% and 8% annual interest. If the total interest earned for the year was $730, how much was invested at each rate? *

23. Things did not go quite as planned. You invested $8000, part of it in stock that paid 12% annual interest. However, the rest of the money suffered a 5% loss. If the total annual income from both investments was $620, how much was invested at each rate? $6000 at 12%; $2000 at 5% loss

24. Things did not go quite as planned. You invested $12,000, part of it in stock that paid 14% annual interest. However, the rest of the money suffered a 6% loss. If the total annual income from both investments was $680, how much was invested at each rate? $7000 at 14%; $5000 at 6% loss

25. A rectangular soccer field is twice as long as it is wide. If the perimeter of the soccer field is 300 yards, what are its dimensions? 50 yd by 100 yd

26. A rectangular swimming pool is three times as long as it is wide. If the perimeter of the pool is 320 feet, what are its dimensions? 40 ft by 120 ft

27. The length of the rectangular tennis court at Wimbledon is 6 feet longer than twice the width. If the court's perimeter is 228 feet, what are the court's dimensions? 36 ft by 78 ft

28. The length of a rectangular pool is 6 meters less than twice the width. If the pool's perimeter is 126 meters, what are its dimensions? 23 m by 40 m

29. The rectangular painting in the figure shown measures 12 inches by 16 inches and is surrounded by a frame of uniform width around the four edges. The perimeter of the rectangle formed by the painting and its frame is 72 inches. Determine the width of the frame. 2 in.

30. The rectangular swimming pool in the figure shown measures 40 feet by 60 feet and is surrounded by a path of uniform width around the four edges. The perimeter of the rectangle formed by the pool and the surrounding path is 248 feet. Determine the width of the path. 6 ft

31. An automobile repair shop charged a customer $448, listing $63 for parts and the remainder for labor. If the cost of labor is $35 per hour, how many hours of labor did it take to repair the car? 11 hours

32. A repair bill on a sailboat came to $1603, including $532 for parts and the remainder for labor. If the cost of labor is $63 per hour, how many hours of labor did it take to repair the sailboat? 17 hours

33. An HMO pamphlet contains the following recommended weight for women: "Give yourself 100 pounds for the first 5 feet plus 5 pounds for every inch over 5 feet tall." Using this description, what height corresponds to a recommended weight of 135 pounds? 5 ft 7 in.

34. A job pays an annual salary of $33,150, which includes a holiday bonus of $750. If paychecks are issued twice a month, what is the gross amount for each paycheck? $1350

35. Answer the question in the following *Peanuts* cartoon strip. (*Note:* You may not use the answer given in the cartoon!) 7 oz

Peanuts © 1978 Peanuts Worldwide LLC. Used by permission of Universal Uclick. All rights reserved.

36. The rate for a particular international person-to-person telephone call is $0.43 for the first minute, $0.32 for each additional minute, and a $2.10 service charge. If the cost of a call is $5.73, how long did the person talk? 11 minutes

In Exercises 37–56, solve each formula for the specified variable. Do you recognize the formula? If so, what does it describe?

37. $A = lw$ for w $w = \dfrac{A}{l}$

38. $D = RT$ for R $R = \dfrac{D}{T}$

39. $A = \frac{1}{2}bh$ for b $b = \dfrac{2A}{h}$

40. $V = \frac{1}{3}Bh$ for B $B = \dfrac{3V}{h}$

41. $I = Prt$ for P $P = \dfrac{I}{rt}$

42. $C = 2\pi r$ for r $r = \dfrac{C}{2\pi}$

43. $E = mc^2$ for m $m = \dfrac{E}{c^2}$

44. $V = \pi r^2 h$ for h $h = \dfrac{V}{\pi r^2}$

45. $T = D + pm$ for p *

46. $P = C + MC$ for M *

47. $A = \frac{1}{2}h(a + b)$ for a *

48. $A = \frac{1}{2}h(a + b)$ for b *

49. $S = P + Prt$ for r *

50. $S = P + Prt$ for t *

51. $B = \dfrac{F}{S - V}$ for S *

52. $S = \dfrac{C}{1 - r}$ for r *

53. $IR + Ir = E$ for I $I = \dfrac{E}{R + r}$

54. $A = 2lw + 2lh + 2wh$ for h $h = \dfrac{A - 2lw}{2l + 2w}$

55. $\dfrac{1}{p} + \dfrac{1}{q} = \dfrac{1}{f}$ for f $f = \dfrac{pq}{p + q}$

56. $\dfrac{1}{R} = \dfrac{1}{R_1} + \dfrac{1}{R_2}$ for R_1 $R_1 = \dfrac{RR_2}{R_2 - R}$

Explaining the Concepts 57–60. Answers will vary.

57. In your own words, describe a step-by-step approach for solving algebraic word problems.

58. Write an original word problem that can be solved using a linear equation. Then solve the problem.

59. Explain what it means to solve a formula for a variable.

60. Did you have difficulties solving some of the problems that were assigned in this Exercise Set? Discuss what you did if this happened to you. Did your course of action enhance your ability to solve algebraic word problems?

Technology Exercises

61. Use a graphing utility to numerically or graphically verify your work in any one exercise from Exercises 9–12. For assistance on how to do this, refer to the Technology box on page 129. Answers will vary.

62. A tennis club offers two payment options. Members can pay a monthly fee of $30 plus $5 per hour for court rental time. The second option has no monthly fee, but court time costs $7.50 per hour.
 a. Write a mathematical model representing total monthly costs for each option for x hours of court rental time. *
 b. Use a graphing utility to graph the two models in a [0, 15, 1] by [0, 120, 20] viewing rectangle. *
 c. Use your utility's trace or intersection feature to determine where the two graphs intersect. Describe what the coordinates of this intersection point represent in practical terms. (12, 90); For 12 hours, the costs are both $90.
 d. Verify part (c) using an algebraic approach by setting the two models equal to one another and determining how many hours one has to rent the court so that the two plans result in identical monthly costs.

Critical Thinking Exercises

Make Sense? *In Exercises 63–66, determine whether each statement makes sense or does not make sense, and explain your reasoning.*

63. By modeling attitudes of college freshmen from 1969 through 2013, I can make precise predictions about the attitudes of the freshman class of 2040. does not make sense

64. I find the hardest part in solving a word problem is writing the equation that models the verbal conditions. makes sense

65. I solved the formula for one of its variables, so now I have a numerical value for that variable. does not make sense

66. After a 35% reduction, a computer's price is $780, so I determined the original price, x, by solving $x - 0.35 = 780$. *

67. At the north campus of a performing arts school, 10% of the students are music majors. At the south campus, 90% of the students are music majors. The campuses are merged into one east campus. If 42% of the 1000 students at the east campus are music majors, how many students did the north and south campuses have before the merger? *

68. The price of a dress is reduced by 40%. When the dress still does not sell, it is reduced by 40% of the reduced price. If the price of the dress after both reductions is $72, what was the original price? $200

69. In a film, the actor Charles Coburn plays an elderly "uncle" character criticized for marrying a woman when he is 3 times her age. He wittily replies, "Ah, but in 20 years time I shall only be twice her age." How old is the "uncle" and the woman? Coburn: 60 years old; woman: 20 years old

70. Suppose that we agree to pay you 8¢ for every problem in this chapter that you solve correctly and fine you 5¢ for every problem done incorrectly. If at the end of 26 problems we do not owe each other any money, how many problems did you solve correctly? 10 correct answers

71. It was wartime when the Ricardos found out Mrs. Ricardo was pregnant. Ricky Ricardo was drafted and made out a will, deciding that $14,000 in a savings account was to be divided between his wife and his child-to-be. Rather strangely, and certainly with gender bias, Ricky stipulated that if the child were a boy, he would get twice the amount of the mother's portion. If it were a girl, the mother would get twice the amount the girl was to receive. We'll never know what Ricky was thinking of, for (as fate would have it) he did not return from war. Mrs. Ricardo gave birth to twins—a boy and a girl. How was the money divided? *

72. A thief steals a number of rare plants from a nursery. On the way out, the thief meets three security guards, one after another. To each security guard, the thief is forced to give one-half the plants that he still has, plus 2 more. Finally, the thief leaves the nursery with 1 lone palm. How many plants were originally stolen? 36 plants

73. Solve for C: $V = C - \dfrac{C - S}{L}N.$ $C = \dfrac{VL - SN}{L - N}$

Group Exercise

74. One of the best ways to learn how to *solve* a word problem in algebra is to *design* word problems of your own. Creating a word problem makes you very aware of precisely how much information is needed to solve the problem. You must also focus on the best way to present information to a reader and on how much information to give. As you write your problem, you gain skills that will help you solve problems created by others.

The group should design five different word problems that can be solved using linear equations. All of the problems should be on different topics. For example, the group should not have more than one problem on simple interest. The group should turn in both the problems and their algebraic solutions. Answers will vary.

Preview Exercises

Exercises 75–77 will help you prepare for the material covered in the next section.

75. Multiply: $(7 - 3x)(-2 - 5x)$. $\quad 15x^2 - 29x - 14^*$

76. Simplify: $\sqrt{18} - \sqrt{8}$. $\quad \sqrt{2}$

77. Rationalize the denominator: $\dfrac{7 + 4\sqrt{2}}{2 - 5\sqrt{2}}$. $\quad \dfrac{54 + 43\sqrt{2}}{46}$

Section 1.4 Complex Numbers

What am I supposed to learn?

After studying this section, you should be able to:

① Add and subtract complex numbers.

② Multiply complex numbers.

③ Divide complex numbers.

④ Perform operations with square roots of negative numbers.

Who is this kid warning us about our eyeballs turning black if we attempt to find the square root of −9? Don't believe what you hear on the street. Although square roots of negative numbers are not real numbers, they do play a significant role in algebra. In this section, we move beyond the real numbers and discuss square roots with negative radicands.

The Imaginary Unit *i*

In the next section, we will study equations whose solutions may involve the square roots of negative numbers. Because the square of a real number is never negative, there is no real number x such that $x^2 = -1$. To provide a setting in which such equations have solutions, mathematicians have invented an expanded system of numbers, the complex numbers. The *imaginary number i*, defined to be a solution of the equation $x^2 = -1$, is the basis of this new number system.

© Roz Chast/The New Yorker Collection/Cartoonbank

The Imaginary Unit *i*

The **imaginary unit *i*** is defined as

$$i = \sqrt{-1}, \quad \text{where } i^2 = -1.$$

Using the imaginary unit i, we can express the square root of any negative number as a real multiple of i. For example,

$$\sqrt{-25} = \sqrt{-1}\sqrt{25} = i\sqrt{25} = 5i.$$

We can check this result by squaring $5i$ and obtaining -25.

$$(5i)^2 = 5^2i^2 = 25(-1) = -25$$

A new system of numbers, called *complex numbers*, is based on adding multiples of i, such as $5i$, to real numbers.

FIGURE 1.18 The complex number system

Complex Numbers and Imaginary Numbers

The set of all numbers in the form

$$a + bi,$$

with real numbers a and b, and i, the imaginary unit, is called the set of **complex numbers**. The real number a is called the **real part** and the real number b is called the **imaginary part** of the complex number $a + bi$. If $b \neq 0$, then the complex number is called an **imaginary number** (**Figure 1.18**). An imaginary number in the form bi is called a **pure imaginary number**.

Here are some examples of complex numbers. Each number can be written in the form $a + bi$.

$$-4 + 6i \qquad\qquad 2i = 0 + 2i \qquad\qquad 3 = 3 + 0i$$

| a, the real part, is −4. | b, the imaginary part, is 6. | a, the real part, is 0. | b, the imaginary part, is 2. | a, the real part, is 3. | b, the imaginary part, is 0. |

Can you see that b, the imaginary part, is not zero in the first two complex numbers? Because $b \neq 0$, these complex numbers are imaginary numbers. Furthermore, the imaginary number $2i$ is a pure imaginary number. By contrast, the imaginary part of the complex number on the right is zero. This complex number is not an imaginary number. The number 3, or $3 + 0i$, is a real number.

A complex number is said to be **simplified** if it is expressed in the **standard form** $a + bi$. If b contains a radical, we usually write i before the radical. For example, we write $7 + 3i\sqrt{5}$ rather than $7 + 3\sqrt{5}i$, which could easily be confused with $7 + 3\sqrt{5i}$.

Expressed in standard form, two complex numbers are equal if and only if their real parts are equal and their imaginary parts are equal.

Equality of Complex Numbers

$a + bi = c + di$ if and only if $a = c$ and $b = d$.

1 Add and subtract complex numbers.

Operations with Complex Numbers

The form of a complex number $a + bi$ is like the binomial $a + bx$. Consequently, we can add, subtract, and multiply complex numbers using the same methods we used for binomials, remembering that $i^2 = -1$.

Adding and Subtracting Complex Numbers

1. $(a + bi) + (c + di) = (a + c) + (b + d)i$

In words, this says that you add complex numbers by adding their real parts, adding their imaginary parts, and expressing the sum as a complex number.

2. $(a + bi) - (c + di) = (a - c) + (b - d)i$

In words, this says that you subtract complex numbers by subtracting their real parts, subtracting their imaginary parts, and expressing the difference as a complex number.

EXAMPLE 1 Adding and Subtracting Complex Numbers

Perform the indicated operations, writing the result in standard form:

 a. $(5 - 11i) + (7 + 4i)$ **b.** $(-5 + i) - (-11 - 6i)$.

SOLUTION

 a. $(5 - 11i) + (7 + 4i)$

$= 5 - 11i + 7 + 4i$	Remove the parentheses.
$= 5 + 7 - 11i + 4i$	Group real and imaginary terms.
$= (5 + 7) + (-11 + 4)i$	Add real parts and add imaginary parts.
$= 12 - 7i$	Simplify.

 b. $(-5 + i) - (-11 - 6i)$

$= -5 + i + 11 + 6i$	Remove the parentheses. Change signs of real and imaginary parts in the complex number being subtracted.
$= -5 + 11 + i + 6i$	Group real and imaginary terms.
$= (-5 + 11) + (1 + 6)i$	Add real parts and add imaginary parts.
$= 6 + 7i$	Simplify.

 • • •

GREAT QUESTION!

Are operations with complex numbers similar to operations with polynomials?

Yes. The following examples, using the same integers as in Example 1, show how operations with complex numbers are just like operations with polynomials. We are simply combining like terms.

 a. $(5 - 11x) + (7 + 4x) = 12 - 7x$

 b. $(-5 + x) - (-11 - 6x) = -5 + x + 11 + 6x$

 $= 6 + 7x$

⊘ **Check Point 1** Perform the indicated operations, writing the result in standard form:

 a. $(5 - 2i) + (3 + 3i)$ $8 + i$ **b.** $(2 + 6i) - (12 - i)$. $-10 + 7i$

Multiplication of complex numbers is performed the same way as multiplication of polynomials, using the distributive property and the FOIL method. After completing the multiplication, we replace any occurrences of i^2 with -1. This idea is illustrated in the next example.

A Brief Review • FOIL

The FOIL method is used to find the product of the binomials $ax + b$ and $cx + d$.

Using the FOIL Method to Multiply Binomials

$$(ax + b)(cx + d) = ax \cdot cx + ax \cdot d + b \cdot cx + b \cdot d$$

For more detail, see Section P.4, Objective 4.

② Multiply complex numbers.

<div style="text-align:center">

EXAMPLE 2 Multiplying Complex Numbers

</div>

Find the products:

 a. $4i(3 - 5i)$ **b.** $(7 - 3i)(-2 - 5i)$.

SOLUTION

a. $4i(3 - 5i)$

$$= 4i \cdot 3 - 4i \cdot 5i \quad \text{Distribute } 4i \text{ throughout the parentheses.}$$
$$= 12i - 20i^2 \quad \text{Multiply.}$$
$$= 12i - 20(-1) \quad \text{Replace } i^2 \text{ with } -1.$$
$$= 20 + 12i \quad \text{Simplify to } 12i + 20 \text{ and write in standard form.}$$

b. $(7 - 3i)(-2 - 5i)$

$$\begin{array}{cccc} \text{F} & \text{O} & \text{I} & \text{L} \end{array}$$

$$= -14 - 35i + 6i + 15i^2 \quad \text{Use the FOIL method.}$$
$$= -14 - 35i + 6i + 15(-1) \quad i^2 = -1$$
$$= -14 - 15 - 35i + 6i \quad \text{Group real and imaginary terms.}$$
$$= -29 - 29i \quad \text{Combine real and imaginary terms.} \quad \bullet\bullet\bullet$$

✓ **Check Point 2** Find the products:

 a. $7i(2 - 9i)$ $63 + 14i$ **b.** $(5 + 4i)(6 - 7i)$. $58 - 11i$

③ Divide complex numbers.

Complex Conjugates and Division

It is possible to multiply imaginary numbers and obtain a real number. This occurs when we multiply $a + bi$ and $a - bi$.

$$\begin{array}{cccc} \text{F} & \text{O} & \text{I} & \text{L} \end{array}$$

$$(a + bi)(a - bi) = a^2 - abi + abi - b^2i^2 \quad \text{Use the FOIL method.}$$
$$= a^2 - b^2(-1) \quad i^2 = -1$$
$$= a^2 + b^2 \quad \text{Notice that this product eliminates } i.$$

For the complex number $a + bi$, we define its *complex conjugate* to be $a - bi$. The multiplication of complex conjugates results in a real number.

Conjugate of a Complex Number

The **complex conjugate** of the number $a + bi$ is $a - bi$, and the complex conjugate of $a - bi$ is $a + bi$. The multiplication of complex conjugates gives a real number.

$$(a + bi)(a - bi) = a^2 + b^2$$
$$(a - bi)(a + bi) = a^2 + b^2$$

Complex conjugates are used to divide complex numbers. The goal of the division procedure is to obtain a real number in the denominator. This real number becomes the denominator of a and b in the quotient $a + bi$. By multiplying the numerator and the denominator of the division by the complex conjugate of the denominator, you will obtain this real number in the denominator.

EXAMPLE 3 Using Complex Conjugates to Divide Complex Numbers

Divide and express the result in standard form: $\dfrac{3i}{4+i}$.

SOLUTION

The complex conjugate of the denominator, $4 + i$, is $4 - i$. Multiplication of both the numerator and the denominator by $4 - i$ will eliminate i from the denominator while maintaining the value of the expression.

$$\frac{3i}{4+i} = \frac{3i}{(4+i)} \cdot \frac{(4-i)}{(4-i)}$$
Multiply the numerator and the denominator by the complex conjugate of the denominator.

$$= \frac{3i \cdot 4 - 3i \cdot i}{4^2 + 1^2}$$
Use the distributive property in the numerator and $(a+bi)(a-bi) = a^2 + b^2$ in the denominator.

$$= \frac{12i - 3i^2}{17}$$
Multiply in the numerator. In the denominator, $4^2 + 1^2 = 16 + 1 = 17$.

$$= \frac{12i - 3(-1)}{17}$$
Replace i^2 with -1 in the numerator.

$$= \frac{12i + 3}{17}$$
Multiply in the numerator.

$$= \frac{12i}{17} + \frac{3}{17}$$
Divide the numerator by 17.

$$= \frac{3}{17} + \frac{12}{17}i$$
Express the answer in standard form.

Observe that the quotient is expressed in the standard form $a + bi$, with $a = \frac{3}{17}$ and $b = \frac{12}{17}$. ● ● ●

 Check Point **3** Divide and express the result in standard form: $\dfrac{5i}{7+i}$. $\frac{1}{10} + \frac{7}{10}i$

EXAMPLE 4 Using Complex Conjugates to Divide Complex Numbers

Divide and express the result in standard form: $\dfrac{7+4i}{2-5i}$.

SOLUTION

The complex conjugate of the denominator, $2 - 5i$, is $2 + 5i$. Multiplication of both the numerator and the denominator by $2 + 5i$ will eliminate i from the denominator while maintaining the value of the expression.

$$\frac{7+4i}{2-5i} = \frac{(7+4i)}{(2-5i)} \cdot \frac{(2+5i)}{(2+5i)}$$
Multiply the numerator and the denominator by the complex conjugate of the denominator.

F O I L

$$= \frac{14 + 35i + 8i + 20i^2}{2^2 + 5^2}$$
Use the FOIL method in the numerator and $(a-bi)(a+bi) = a^2 + b^2$ in the denominator.

$$= \frac{14 + 43i + 20(-1)}{29}$$
In the numerator, combine imaginary terms and replace i^2 with -1. In the denominator, $2^2 + 5^2 = 4 + 25 = 29$.

$$= \frac{-6 + 43i}{29}$$

Combine real terms in the numerator of
$\frac{14 + 43i + 20(-1)}{29}$: $14 + 20(-1) = 14 - 20 = -6$.

$$= -\frac{6}{29} + \frac{43}{29}i$$

Express the answer in standard form.

Observe that the quotient is expressed in the standard form $a + bi$, with $a = -\frac{6}{29}$ and $b = \frac{43}{29}$.

● ● ●

✓ Check Point **4** Divide and express the result in standard form: $\frac{5 + 4i}{4 - i}$. $\frac{16}{17} + \frac{21}{17}i$

④ Perform operations with square roots of negative numbers.

Roots of Negative Numbers

The square of $4i$ and the square of $-4i$ both result in -16:

$$(4i)^2 = 16i^2 = 16(-1) = -16 \qquad (-4i)^2 = 16i^2 = 16(-1) = -16.$$

Consequently, in the complex number system -16 has two square roots, namely, $4i$ and $-4i$. We call $4i$ the **principal square root** of -16.

Principal Square Root of a Negative Number

For any positive real number b, the **principal square root** of the negative number $-b$ is defined by

$$\sqrt{-b} = i\sqrt{b}.$$

GREAT QUESTION!

In the definition $\sqrt{-b} = i\sqrt{b}$, why did you write i in the front? Could I write $\sqrt{-b} = \sqrt{b}i$?

Yes, $\sqrt{-b} = \sqrt{b}i$. However, it's tempting to write $\sqrt{-b} = \sqrt{bi}$, which is incorrect. In order to avoid writing i under a radical, let's agree to write i before any radical.

Consider the multiplication problem

$$5i \cdot 2i = 10i^2 = 10(-1) = -10.$$

This problem can also be given in terms of principal square roots of negative numbers:

$$\sqrt{-25} \cdot \sqrt{-4}.$$

Because the product rule for radicals only applies to real numbers, multiplying radicands is incorrect. **When performing operations with square roots of negative numbers, begin by expressing all square roots in terms of i.** Then perform the indicated operation.

Correct:	**Incorrect:**
$\sqrt{-25} \cdot \sqrt{-4} = i\sqrt{25} \cdot i\sqrt{4}$	$\sqrt{-25} \cdot \sqrt{-4} = \sqrt{(-25)(-4)}$
$= 5i \cdot 2i$	$= \sqrt{100}$
$= 10i^2 = 10(-1) = -10$	$= 10$

EXAMPLE 5 Operations Involving Square Roots of Negative Numbers

Perform the indicated operations and write the result in standard form:

a. $\sqrt{-18} - \sqrt{-8}$ **b.** $\left(-1 + \sqrt{-5}\right)^2$ **c.** $\dfrac{-25 + \sqrt{-50}}{15}$.

SOLUTION

Begin by expressing all square roots of negative numbers in terms of i.

a. $\sqrt{-18} - \sqrt{-8} = i\sqrt{18} - i\sqrt{8} = i\sqrt{9 \cdot 2} - i\sqrt{4 \cdot 2}$

$$= 3i\sqrt{2} - 2i\sqrt{2} = i\sqrt{2}$$

$$(A + B)^2 = A^2 + 2 \cdot A \cdot B + B^2$$

b. $\left(-1 + \sqrt{-5}\right)^2 = \left(-1 + i\sqrt{5}\right)^2 = (-1)^2 + 2(-1)\left(i\sqrt{5}\right) + \left(i\sqrt{5}\right)^2$

$$= 1 - 2i\sqrt{5} + 5i^2$$

$$= 1 - 2i\sqrt{5} + 5(-1)$$

$$= -4 - 2i\sqrt{5}$$

c. $\dfrac{-25 + \sqrt{-50}}{15}$

$= \dfrac{-25 + i\sqrt{50}}{15}$ $\quad \sqrt{-b} = i\sqrt{b}$

$= \dfrac{-25 + 5i\sqrt{2}}{15}$ $\quad \sqrt{50} = \sqrt{25 \cdot 2} = 5\sqrt{2}$

$= \dfrac{-25}{15} + \dfrac{5i\sqrt{2}}{15}$ Write the complex number in standard form.

$= -\dfrac{5}{3} + i\dfrac{\sqrt{2}}{3}$ Simplify by dividing numerators and denominators by 5. ● ● ●

✓ Check Point **5** Perform the indicated operations and write the result in standard form:

a. $\sqrt{-27} + \sqrt{-48}$ **b.** $\left(-2 + \sqrt{-3}\right)^2$ **c.** $\dfrac{-14 + \sqrt{-12}}{2}$.

a. $7i\sqrt{3}$ **b.** $1 - 4i\sqrt{3}$ **c.** $-7 + i\sqrt{3}$

ACHIEVING SUCCESS

Write down all the steps. In this textbook, examples are written that provide step-by-step solutions. No steps are omitted and each step is explained to the right of the mathematics. Some professors are careful to write down every step of a problem as they are doing it in class; others aren't so fastidious. In either case, write down what the professor puts up and when you get home, fill in whatever steps have been omitted (if any). In your math work, including homework and tests, show clear step-by-step solutions. Detailed solutions help organize your thoughts and enhance understanding. Doing too many steps mentally often results in preventable mistakes.

CONCEPT AND VOCABULARY CHECK

Fill in each blank so that the resulting statement is true.

1. The imaginary unit i is defined as $i = \underline{\sqrt{-1}}$, where $i^2 = \underline{-1}$.

2. The set of all numbers in the form $a + bi$ is called the set of $\underline{\text{complex}}$ numbers. If $b \neq 0$, then the number is also called a/an $\underline{\text{imaginary}}$ number. If $b = 0$, then the number is also called a/an $\underline{\text{real}}$ number.

3. $-9i + 3i = \underline{-6i}$

4. $10i - (-4i) = \underline{14i}$

5. Consider the following multiplication problem:

$$(3 + 2i)(6 - 5i).$$

Using the FOIL method, the product of the first terms is $\underline{18}$, the product of the outside terms is $\underline{-15i}$, and the product of the inside terms is $\underline{12i}$. The product of the last terms in terms of i^2 is $\underline{-10i^2}$, which simplifies to $\underline{10}$.

6. The conjugate of $2 - 9i$ is $\underline{2 + 9i}$.

7. The division

$$\frac{7 + 4i}{2 - 5i}$$

is performed by multiplying the numerator and denominator by $\underline{2 + 5i}$.

8. $\sqrt{-20} = \underline{i}\sqrt{20} = \underline{i}\sqrt{4 \cdot 5} = \underline{2i\sqrt{5}}$

EXERCISE SET 1.4

Practice Exercises

In Exercises 1–8, add or subtract as indicated and write the result in standard form.

1. $(7 + 2i) + (1 - 4i)$ $8 - 2i$

2. $(-2 + 6i) + (4 - i)$ $2 + 5i$

3. $(3 + 2i) - (5 - 7i)$ $-2 + 9i$

4. $(-7 + 5i) - (-9 - 11i)$ $2 + 16i$

5. $6 - (-5 + 4i) - (-13 - i)$ $24 - 3i$

6. $7 - (-9 + 2i) - (-17 - i)$ $33 - i$

7. $8i - (14 - 9i)$ $-14 + 17i$

8. $15i - (12 - 11i)$ $-12 + 26i$

In Exercises 9–20, find each product and write the result in standard form.

9. $-3i(7i - 5)$ $21 + 15i$

10. $-8i(2i - 7)$ $16 + 56i$

11. $(-5 + 4i)(3 + i)$ $-19 + 7i$

12. $(-4 - 8i)(3 + i)$ $-4 - 28i$

13. $(7 - 5i)(-2 - 3i)$ $-29 - 11i$

14. $(8 - 4i)(-3 + 9i)$ $12 + 84i$

15. $(3 + 5i)(3 - 5i)$ 34

16. $(2 + 7i)(2 - 7i)$ 53

17. $(-5 + i)(-5 - i)$ 26

18. $(-7 - i)(-7 + i)$ 50

19. $(2 + 3i)^2$ $-5 + 12i$

20. $(5 - 2i)^2$ $21 - 20i$

In Exercises 21–28, divide and express the result in standard form.

21. $\dfrac{2}{3 - i}$ $\frac{3}{5} + \frac{1}{5}i$

22. $\dfrac{3}{4 + i}$ $\frac{12}{17} - \frac{3}{17}i$

23. $\dfrac{2i}{1 + i}$ $1 + i$

24. $\dfrac{5i}{2 - i}$ $-1 + 2i$

25. $\dfrac{8i}{4 - 3i}$ $-\frac{24}{25} + \frac{32}{25}i$

26. $\dfrac{-6i}{3 + 2i}$ $-\frac{12}{13} - \frac{18}{13}i$

27. $\dfrac{2 + 3i}{2 + i}$ $\frac{7}{5} + \frac{4}{5}i$

28. $\dfrac{3 - 4i}{4 + 3i}$ $-i$

In Exercises 29–44, perform the indicated operations and write the result in standard form.

29. $\sqrt{-64} - \sqrt{-25}$ $3i$

30. $\sqrt{-81} - \sqrt{-144}$ $-3i$

31. $5\sqrt{-16} + 3\sqrt{-81}$ $47i$

32. $5\sqrt{-8} + 3\sqrt{-18}$ $19i\sqrt{2}$

33. $\left(-2 + \sqrt{-4}\right)^2$ $-8i$

34. $\left(-5 - \sqrt{-9}\right)^2$ $16 + 30i$

35. $\left(-3 - \sqrt{-7}\right)^2$ $2 + 6i\sqrt{7}$

36. $\left(-2 + \sqrt{-11}\right)^2$ $-7 - 4i\sqrt{11}$

37. $\dfrac{-8 + \sqrt{-32}}{24}$ $-\frac{1}{3} + i\frac{\sqrt{2}}{6}$

38. $\dfrac{-12 + \sqrt{-28}}{32}$ $-\frac{3}{8} + i\frac{\sqrt{7}}{16}$

39. $\dfrac{-6 - \sqrt{-12}}{48}$ $-\frac{1}{8} - i\frac{\sqrt{3}}{24}$

40. $\dfrac{-15 - \sqrt{-18}}{33}$ $-\frac{5}{11} - i\frac{\sqrt{2}}{11}$

41. $\sqrt{-8}\left(\sqrt{-3} - \sqrt{5}\right)$ $-2\sqrt{6} - 2i\sqrt{10}$

42. $\sqrt{-12}\left(\sqrt{-4} - \sqrt{2}\right)$ $-4\sqrt{3} - 2i\sqrt{6}$

43. $\left(3\sqrt{-5}\right)\left(-4\sqrt{-12}\right)$ $24\sqrt{15}$

44. $\left(3\sqrt{-7}\right)\left(2\sqrt{-8}\right)$ $-12\sqrt{14}$

Practice Plus

In Exercises 45–50, perform the indicated operation(s) and write the result in standard form.

45. $(2 - 3i)(1 - i) - (3 - i)(3 + i)$ $-11 - 5i$

46. $(8 + 9i)(2 - i) - (1 - i)(1 + i)$ $23 + 10i$

47. $(2 + i)^2 - (3 - i)^2$ $-5 + 10i$

48. $(4 - i)^2 - (1 + 2i)^2$ $18 - 12i$

49. $5\sqrt{-16} + 3\sqrt{-81}$ $0 + 47i$ or $47i$

50. $5\sqrt{-8} + 3\sqrt{-18}$ $0 + 19i\sqrt{2}$ or $19i\sqrt{2}$

51. Evaluate $x^2 - 2x + 2$ for $x = 1 + i$. 0

52. Evaluate $x^2 - 2x + 5$ for $x = 1 - 2i$. 0

53. Evaluate $\dfrac{x^2 + 19}{2 - x}$ for $x = 3i$. $\frac{20}{13} + \frac{30}{13}i$

54. Evaluate $\dfrac{x^2 + 11}{3 - x}$ for $x = 4i$. $-\frac{3}{5} - \frac{4}{5}i$

Application Exercises

Complex numbers are used in electronics to describe the current in an electric circuit. Ohm's law relates the current in a circuit, I, in amperes, the voltage of the circuit, E, in volts, and the resistance of the circuit, R, in ohms, by the formula $E = IR$. Use this formula to solve Exercises 55–56.

55. Find E, the voltage of a circuit, if $I = (4 - 5i)$ amperes and $R = (3 + 7i)$ ohms. $(47 + 13i)$ volts

56. Find E, the voltage of a circuit, if $I = (2 - 3i)$ amperes and $R = (3 + 5i)$ ohms. $(21 + i)$ volts

57. The mathematician Girolamo Cardano is credited with the first use (in 1545) of negative square roots in solving the now-famous problem, "Find two numbers whose sum is 10 and whose product is 40." Show that the complex numbers $5 + i\sqrt{15}$ and $5 - i\sqrt{15}$ satisfy the conditions of the problem. (Cardano did not use the symbolism $i\sqrt{15}$ or even $\sqrt{-15}$. He wrote R.m 15 for $\sqrt{-15}$, meaning "radix minus 15." He regarded the numbers $5 +$ R.m 15 and $5 -$ R.m 15 as "fictitious" or "ghost numbers," and considered the problem "manifestly impossible." But in a mathematically adventurous spirit, he exclaimed, "Nevertheless, we will operate.") *

Explaining the Concepts 58–66. Answers will vary.

58. What is i?

59. Explain how to add complex numbers. Provide an example with your explanation.

60. Explain how to multiply complex numbers and give an example.

61. What is the complex conjugate of $2 + 3i$? What happens when you multiply this complex number by its complex conjugate?

62. Explain how to divide complex numbers. Provide an example with your explanation.

63. Explain each of the three jokes in the cartoon on page 139.

64. A stand-up comedian uses algebra in some jokes, including one about a telephone recording that announces "You have just reached an imaginary number. Please multiply by i and dial again." Explain the joke.

Explain the error in Exercises 65–66.

65. $\sqrt{-9} + \sqrt{-16} = \sqrt{-25} = i\sqrt{25} = 5i$

66. $\left(\sqrt{-9}\right)^2 = \sqrt{-9} \cdot \sqrt{-9} = \sqrt{81} = 9$

*See Answers to Selected Exercises.

Critical Thinking Exercises

Make Sense? *In Exercises 67–70, determine whether each statement makes sense or does not make sense, and explain your reasoning.*

67. The humor in the cartoon is based on the fact that "rational" and "real" have different meanings in mathematics and in everyday speech. makes sense

68. The word *imaginary* in imaginary numbers tells me that these numbers are undefined. does not make sense

69. By writing the imaginary number $5i$, I can immediately see that 5 is the constant and i is the variable. does not make sense

70. When I add or subtract complex numbers, I am basically combining like terms. makes sense

© 2007 GJ Caulkins

In Exercises 71–74, determine whether each statement is true or false. If the statement is false, make the necessary change(s) to produce a true statement. Changes to false statements will vary.

71. Some irrational numbers are not complex numbers. false

72. $(3 + 7i)(3 - 7i)$ is an imaginary number. false

73. $\dfrac{7 + 3i}{5 + 3i} = \dfrac{7}{5}$ false

74. In the complex number system, $x^2 + y^2$ (the sum of two squares) can be factored as $(x + yi)(x - yi)$. true

In Exercises 75–77, perform the indicated operations and write the result in standard form.

75. $\dfrac{4}{(2 + i)(3 - i)}$ $\frac{14}{25} - \frac{2}{25}i$

76. $\dfrac{1 + i}{1 + 2i} + \dfrac{1 - i}{1 - 2i}$ $\frac{6}{5} + 0i$ or $\frac{6}{5}$

77. $\dfrac{8}{1 + \dfrac{2}{i}}$ $\frac{8}{5} + \frac{16}{5}i$

Preview Exercises

Exercises 78–80 will help you prepare for the material covered in the next section.

78. Factor: $2x^2 + 7x - 4$. $(2x - 1)(x + 4)$

79. Factor: $x^2 - 6x + 9$. $(x - 3)(x - 3)$ or $(x - 3)^2$

80. Evaluate

$$\dfrac{-b - \sqrt{b^2 - 4ac}}{2a}$$

for $a = 2, b = 9$, and $c = -5$. -5

Section 1.5 Quadratic Equations

What am I supposed to learn?

After studying this section, you should be able to:

1. Solve quadratic equations by factoring.
2. Solve quadratic equations by the square root property.
3. Solve quadratic equations by completing the square.
4. Solve quadratic equations using the quadratic formula.
5. Use the discriminant to determine the number and type of solutions.
6. Determine the most efficient method to use when solving a quadratic equation.
7. Solve problems modeled by quadratic equations.

*I'm very well acquainted, too, with matters mathematical,
I understand equations, both simple and quadratical.
About binomial theorem I'm teeming with a lot of news,
With many cheerful facts about the square of
the hypotenuse.*

—Gilbert and Sullivan,
The Pirates of Penzance

Equations quadratical? Cheerful news about the square of the hypotenuse? You've come to the right place. In this section, we study a number of methods for solving *quadratic equations*, equations in which the highest exponent on the variable is 2. (Yes, it's *quadratic* and not *quadratical*, despite the latter's rhyme with mathematical.) We also look at applications of quadratic equations, introducing (cheerfully, of course) the Pythagorean Theorem and the square of the hypotenuse.

We begin by defining a quadratic equation.

Definition of a Quadratic Equation

A **quadratic equation** in x is an equation that can be written in the **general form**

$$ax^2 + bx + c = 0,$$

where a, b, and c are real numbers, with $a \neq 0$. A quadratic equation in x is also called a **second-degree polynomial equation** in x.

1. Solve quadratic equations by factoring.

Solving Quadratic Equations by Factoring

A Brief Review • Factoring

Factoring a polynomial expressed as the sum of monomials means finding an equivalent expression that is a product. The factoring methods used in this section, factoring out the greatest common factor and factoring trinomials, are reviewed in Section P.5, Objectives 1 and 3.

Here is an example of a quadratic equation in general form:

$$x^2 - 7x + 10 = 0.$$

$a = 1$ $b = -7$ $c = 10$

We can factor the left side of this equation. We obtain $(x - 5)(x - 2) = 0$. If a quadratic equation has zero on one side and a factored expression on the other side, it can be solved using the **zero-product principle**.

The Zero-Product Principle

If the product of two algebraic expressions is zero, then at least one of the factors is equal to zero.

$$\text{If } AB = 0, \text{ then } A = 0 \text{ or } B = 0.$$

For example, consider the equation $(x - 5)(x - 2) = 0$. According to the zero-product principle, this product can be zero only if at least one of the factors is zero. We set each individual factor equal to zero and solve each resulting equation for x.

$$(x - 5)(x - 2) = 0$$
$$x - 5 = 0 \quad \text{or} \quad x - 2 = 0$$
$$x = 5 \qquad\qquad x = 2$$

We can check each of these proposed solutions, 5 and 2, in the original quadratic equation, $x^2 - 7x + 10 = 0$. Substitute each one separately for x in the equation.

Check 5:	**Check 2:**
$x^2 - 7x + 10 = 0$	$x^2 - 7x + 10 = 0$
$5^2 - 7 \cdot 5 + 10 \stackrel{?}{=} 0$	$2^2 - 7 \cdot 2 + 10 \stackrel{?}{=} 0$
$25 - 35 + 10 \stackrel{?}{=} 0$	$4 - 14 + 10 \stackrel{?}{=} 0$
$0 = 0, \quad$ true	$0 = 0, \quad$ true

The resulting true statements indicate that the solutions are 2 and 5. The solution set is $\{2, 5\}$. Note that with a quadratic equation, we can have two solutions, compared to the conditional linear equation that had one.

Solving a Quadratic Equation by Factoring

1. If necessary, rewrite the equation in the general form $ax^2 + bx + c = 0$, moving all nonzero terms to one side, thereby obtaining zero on the other side.
2. Factor completely.
3. Apply the zero-product principle, setting each factor containing a variable equal to zero.
4. Solve the equations in step 3.
5. Check the solutions in the original equation.

EXAMPLE 1 Solving Quadratic Equations by Factoring

Solve by factoring:

 a. $4x^2 - 2x = 0$ **b.** $2x^2 + 7x = 4$.

SOLUTION

a. We begin with $4x^2 - 2x = 0$.

Step 1 Move all nonzero terms to one side and obtain zero on the other side. All nonzero terms are already on the left and zero is on the other side, so we can skip this step.

Step 2 Factor. We factor out $2x$ from the two terms on the left side.

$$4x^2 - 2x = 0 \quad \text{\small This is the given equation.}$$
$$2x(2x - 1) = 0 \quad \text{\small Factor.}$$

Steps 3 and 4 **Set each factor equal to zero and solve the resulting equations.**
We apply the zero-product principle to $2x(2x - 1) = 0$.

$$2x = 0 \quad \text{or} \quad 2x - 1 = 0$$
$$x = 0 \qquad\qquad 2x = 1$$
$$x = \tfrac{1}{2}$$

Step 5 **Check the solutions in the original equation.**

Check 0:	**Check $\tfrac{1}{2}$:**
$4x^2 - 2x = 0$	$4x^2 - 2x = 0$
$4 \cdot 0^2 - 2 \cdot 0 \stackrel{?}{=} 0$	$4\left(\tfrac{1}{2}\right)^2 - 2\left(\tfrac{1}{2}\right) \stackrel{?}{=} 0$
$0 - 0 \stackrel{?}{=} 0$	$4\left(\tfrac{1}{4}\right) - 2\left(\tfrac{1}{2}\right) \stackrel{?}{=} 0$
$0 = 0, \quad$ true	$1 - 1 \stackrel{?}{=} 0$
	$0 = 0, \quad$ true

The solution set is $\left\{0, \tfrac{1}{2}\right\}$.

b. Next, we solve $2x^2 + 7x = 4$.

Step 1 **Move all nonzero terms to one side and obtain zero on the other side.** Subtract 4 from both sides and write the equation in general form.

$$2x^2 + 7x = 4 \qquad \text{This is the given equation.}$$
$$2x^2 + 7x - 4 = 4 - 4 \qquad \text{Subtract 4 from both sides.}$$
$$2x^2 + 7x - 4 = 0 \qquad \text{Simplify.}$$

Step 2 **Factor.**

$$2x^2 + 7x - 4 = 0$$
$$(2x - 1)(x + 4) = 0$$

Steps 3 and 4 **Set each factor equal to zero and solve the resulting equations.**

$$2x - 1 = 0 \quad \text{or} \quad x + 4 = 0$$
$$2x = 1 \qquad\qquad x = -4$$
$$x = \tfrac{1}{2}$$

Step 5 **Check the solutions in the original equation.**

Check $\tfrac{1}{2}$:	**Check -4:**
$2x^2 + 7x = 4$	$2x^2 + 7x = 4$
$2\left(\tfrac{1}{2}\right)^2 + 7\left(\tfrac{1}{2}\right) \stackrel{?}{=} 4$	$2(-4)^2 + 7(-4) \stackrel{?}{=} 4$
$\tfrac{1}{2} + \tfrac{7}{2} \stackrel{?}{=} 4$	$32 + (-28) \stackrel{?}{=} 4$
$4 = 4, \quad$ true	$4 = 4, \quad$ true

The solution set is $\left\{-4, \tfrac{1}{2}\right\}$. • • •

⊘ Check Point **1** Solve by factoring:
a. $3x^2 - 9x = 0$ {0, 3} **b.** $2x^2 + x = 1$. $\left\{-1, \tfrac{1}{2}\right\}$

TECHNOLOGY

Graphic Connections

You can use a graphing utility to check the real solutions of a quadratic equation. **The real solutions of $ax^2 + bx + c = 0$ correspond to the x-intercepts of the graph of $y = ax^2 + bx + c$.** For example, to check the solutions of $2x^2 + 7x = 4$, or $2x^2 + 7x - 4 = 0$, graph $y = 2x^2 + 7x - 4$. The U-shaped, bowl-like graph is shown on the right. Note that it is important to have all nonzero terms on one side of the quadratic equation before entering it into the graphing utility. The x-intercepts are -4 and $\tfrac{1}{2}$, and the graph of $y = 2x^2 + 7x - 4$ passes through $(-4, 0)$ and $\left(\tfrac{1}{2}, 0\right)$. This verifies that $\left\{-4, \tfrac{1}{2}\right\}$ is the solution set of $2x^2 + 7x - 4 = 0$, or equivalently, $2x^2 + 7x = 4$.

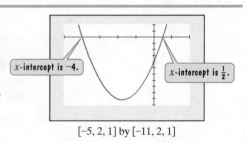

x-intercept is -4.

x-intercept is $\tfrac{1}{2}$.

$[-5, 2, 1]$ by $[-11, 2, 1]$

② Solve quadratic equations by the square root property.

Solving Quadratic Equations by the Square Root Property

Quadratic equations of the form $u^2 = d$, where u is an algebraic expression and d is a nonzero real number, can be solved by the *square root property*. First, isolate the squared expression u^2 on one side of the equation and the number d on the other side. Then take the square root of both sides. Remember, there are two numbers whose square is d. One number is \sqrt{d} and one is $-\sqrt{d}$.

We can use factoring to verify that $u^2 = d$ has these two solutions.

$$u^2 = d \qquad \text{This is the given equation.}$$

$$u^2 - d = 0 \qquad \text{Move all terms to one side and obtain zero on the other side.}$$

$$\left(u + \sqrt{d}\right)\left(u - \sqrt{d}\right) = 0 \qquad \text{Factor.}$$

$$u + \sqrt{d} = 0 \quad \text{or} \quad u - \sqrt{d} = 0 \qquad \text{Set each factor equal to zero.}$$

$$u = -\sqrt{d} \qquad\qquad u = \sqrt{d} \qquad \text{Solve the resulting equations.}$$

Because the solutions differ only in sign, we can write them in abbreviated notation as $u = \pm\sqrt{d}$. We read this as "u equals positive or negative the square root of d" or "u equals plus or minus the square root of d."

Now that we have verified these solutions, we can solve $u^2 = d$ directly by taking square roots. This process is called the **square root property**.

The Square Root Property

If u is an algebraic expression and d is a nonzero number, then $u^2 = d$ has exactly two solutions:

$$\text{If } u^2 = d, \text{ then } u = \sqrt{d} \text{ or } u = -\sqrt{d}.$$

Equivalently,

$$\text{If } u^2 = d, \text{ then } u = \pm\sqrt{d}.$$

Before you can apply the square root property, a squared expression must be isolated on one side of the equation.

EXAMPLE 2 Solving Quadratic Equations by the Square Root Property

Solve by the square root property:

a. $3x^2 - 15 = 0$ **b.** $9x^2 + 25 = 0$ **c.** $(x - 2)^2 = 6$.

SOLUTION

To apply the square root property, we need a squared expression by itself on one side of the equation.

$$3x^2 - 15 = 0 \qquad\qquad 9x^2 + 25 = 0 \qquad\qquad (x - 2)^2 = 6$$

| We want x^2 by itself. | We want x^2 by itself. | The squared expression is by itself. |

a.

$$3x^2 - 15 = 0 \qquad \text{This is the original equation.}$$

$$3x^2 = 15 \qquad \text{Add 15 to both sides.}$$

$$x^2 = 5 \qquad \text{Divide both sides by 3.}$$

$$x = \sqrt{5} \text{ or } x = -\sqrt{5} \qquad \text{Apply the square root property. Equivalently, } x = \pm\sqrt{5}.$$

By checking both proposed solutions in the original equation, we can confirm that the solution set is $\left\{-\sqrt{5}, \sqrt{5}\right\}$ or $\left\{\pm\sqrt{5}\right\}$.

b. $9x^2 + 25 = 0$ This is the original equation.

$\qquad 9x^2 = -25$ Subtract 25 from both sides.

$\qquad x^2 = -\dfrac{25}{9}$ Divide both sides by 9.

$\qquad x = \pm\sqrt{-\dfrac{25}{9}}$ Apply the square root property.

$\qquad x = \pm i\sqrt{\dfrac{25}{9}} = \pm\dfrac{5}{3}i$ Express solutions in terms of i.

Because the equation has an x^2-term and no x-term, we can check both proposed solutions, $\pm\dfrac{5}{3}i$, at once.

Check $\dfrac{5}{3}i$ **and** $-\dfrac{5}{3}i$**:**

$$9x^2 + 25 = 0$$

$$9\left(\pm\dfrac{5}{3}i\right)^2 + 25 \overset{?}{=} 0$$

$$9\left(\dfrac{25}{9}i^2\right) + 25 \overset{?}{=} 0$$

$$25i^2 + 25 \overset{?}{=} 0$$

$\boxed{i^2 = -1}$

$$25(-1) + 25 \overset{?}{=} 0$$

$$0 = 0 \quad \text{true}$$

The solutions are $-\dfrac{5}{3}i$ and $\dfrac{5}{3}i$. The solution set is $\left\{-\dfrac{5}{3}i, \dfrac{5}{3}i\right\}$ or $\left\{\pm\dfrac{5}{3}i\right\}$.

c. $(x - 2)^2 = 6$ This is the original equation.

$\qquad x - 2 = \pm\sqrt{6}$ Apply the square root property.

$\qquad x = 2 \pm \sqrt{6}$ Add 2 to both sides.

By checking both values in the original equation, we can confirm that the solution set is $\left\{2 + \sqrt{6}, 2 - \sqrt{6}\right\}$ or $\left\{2 \pm \sqrt{6}\right\}$. • • •

⊘ **Check Point 2** Solve by the square root property:

a. $3x^2 - 21 = 0$ **b.** $5x^2 + 45 = 0$ **c.** $(x + 5)^2 = 11$.

 a. $\{-\sqrt{7}, \sqrt{7}\}$ **b.** $\{-3i, 3i\}$ **c.** $\{-5 + \sqrt{11}, -5 - \sqrt{11}\}$

TECHNOLOGY

Graphic Connections

The graph of

$$y = 9x^2 + 25$$

has no x-intercepts. This shows that

$$9x^2 + 25 = 0$$

has no real solutions. Example 2(b) algebraically establishes that the solutions are imaginary numbers.

$[-3, 3, 1]$ by $[-5, 100, 5]$

③ Solve quadratic equations by completing the square.

Completing the Square

A Brief Review • Perfect Square Trinomials

Perfect square trinomials, $A^2 + 2AB + B^2$ and $A^2 - 2AB + B^2$, come in two forms: one in which the coefficient of the middle term is positive and one in which the coefficient of the middle term is negative. Perfect square trinomials are factored as follows:

$$A^2 + 2AB + B^2 = (A + B)^2 \qquad A^2 - 2AB + B^2 = (A - B)^2$$

 Same sign Same sign

For more detail, see Section P.5, Objective 5.

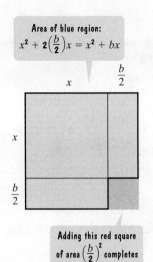

Area of blue region:
$$x^2 + 2\left(\frac{b}{2}\right)x = x^2 + bx$$

Adding this red square of area $\left(\frac{b}{2}\right)^2$ completes the larger square.

FIGURE 1.19 Visualizing completing the square

How do we solve an equation in the form $ax^2 + bx + c = 0$ if the trinomial $ax^2 + bx + c$ cannot be factored? We cannot use the zero-product principle in such a case. However, we can convert the equation into an equivalent equation that can be solved using the square root property. This is accomplished by **completing the square**.

> **Completing the Square**
>
> If $x^2 + bx$ is a binomial, then by adding $\left(\frac{b}{2}\right)^2$, which is the square of half the coefficient of x, a perfect square trinomial will result. That is,
> $$x^2 + bx + \left(\frac{b}{2}\right)^2 = \left(x + \frac{b}{2}\right)^2.$$

Squares and rectangles make it possible to visualize completing the square, as shown in **Figure 1.19**.

EXAMPLE 3 Creating Perfect Square Trinomials by Completing the Square

What term should be added to each binomial so that it becomes a perfect square trinomial? Write and factor the trinomial.

a. $x^2 + 8x$ **b.** $x^2 - 7x$ **c.** $x^2 + \frac{3}{5}x$

SOLUTION

To complete the square, we must add a term to each binomial. The term that should be added is the square of half the coefficient of x.

$x^2 + 8x$ | $x^2 - 7x$ | $x^2 + \frac{3}{5}x$

Add $\left(\frac{8}{2}\right)^2 = 4^2$. Add 16 to complete the square.

Add $\left(\frac{-7}{2}\right)^2$, or $\frac{49}{4}$, to complete the square.

Add $\left(\frac{1}{2} \cdot \frac{3}{5}\right)^2 = \left(\frac{3}{10}\right)^2$. Add $\frac{9}{100}$ to complete the square.

a. The coefficient of the x-term in $x^2 + 8x$ is 8. Half of 8 is 4, and $4^2 = 16$. Add 16. The result is a perfect square trinomial.

$$x^2 + 8x + 16 = (x + 4)^2$$

b. The coefficient of the x-term in $x^2 - 7x$ is -7. Half of -7 is $-\frac{7}{2}$, and $\left(-\frac{7}{2}\right)^2 = \frac{49}{4}$. Add $\frac{49}{4}$. The result is a perfect square trinomial.

$$x^2 - 7x + \frac{49}{4} = \left(x - \frac{7}{2}\right)^2$$

c. The coefficient of the x-term in $x^2 + \frac{3}{5}x$ is $\frac{3}{5}$. Half of $\frac{3}{5}$ is $\frac{1}{2} \cdot \frac{3}{5}$, or $\frac{3}{10}$, and $\left(\frac{3}{10}\right)^2 = \frac{9}{100}$. Add $\frac{9}{100}$. The result is a perfect square trinomial.

$$x^2 + \frac{3}{5}x + \frac{9}{100} = \left(x + \frac{3}{10}\right)^2$$

• • •

GREAT QUESTION!

I'm not accustomed to factoring perfect square trinomials in which fractions are involved. Is there a rule or an observation that can make the factoring easier?

Yes. The constant in the factorization is always half the coefficient of x.

$$x^2 - 7x + \frac{49}{4} = \left(x - \frac{7}{2}\right)^2 \qquad x^2 + \frac{3}{5}x + \frac{9}{100} = \left(x + \frac{3}{10}\right)^2$$

Half the coefficient of x, -7, is $-\frac{7}{2}$. Half the coefficient of x, $\frac{3}{5}$, is $\frac{3}{10}$.

✓ Check Point **3** What term should be added to each binomial so that it becomes a perfect square trinomial? Write and factor the trinomial.

 a. $x^2 + 6x$ **b.** $x^2 - 5x$ **c.** $x^2 + \frac{2}{3}x$

 a. $9; x^2 + 6x + 9 = (x + 3)^2$ **b.** $\frac{25}{4}; x^2 - 5x + \frac{25}{4} = \left(x - \frac{5}{2}\right)^2$ **c.** $\frac{1}{9}; x^2 + \frac{2}{3}x + \frac{1}{9} = \left(x + \frac{1}{3}\right)^2$

We can solve any quadratic equation by completing the square and then applying the square root property.

Solving a Quadratic Equation by Completing the Square

To solve $ax^2 + bx + c = 0$ by completing the square:

1. If a, the leading coefficient, is not 1, divide both sides by a. This makes the coefficient of the x^2-term 1.

2. Isolate the variable terms on one side of the equation and the constant term on the other side of the equation.

3. Complete the square.

 a. Add the square of half the coefficient of x to both sides of the equation.

 b. Factor the resulting perfect square trinomial.

4. Use the square root property and solve for x.

EXAMPLE 4 Solving a Quadratic Equation by Completing the Square

Solve by completing the square: $x^2 - 6x + 4 = 0$.

SOLUTION

Step 1 **Divide both sides by a to make the leading coefficient 1.**

 $x^2 - 6x + 4 = 0$ This is the given equation.
 The leading coefficient, a, is 1,
 so we can skip this step.

Step 2 **Isolate the variable terms on one side of the equation and the constant term on the other side.**

 $x^2 - 6x = -4$ Subtract 4 from both sides.
 Now that $x^2 - 6x$, the binomial,
 is isolated, we can complete the square.

GREAT QUESTION!

When I solve a quadratic equation by completing the square, doesn't this result in a new equation? How do I know that the solutions of this new equation are the same as those of the given equation?

When you complete the square for the binomial expression $x^2 + bx$, you obtain a different polynomial. When you solve a quadratic equation by completing the square, you obtain an equation with the same solution set because the constant needed to complete the square is added to *both sides*.

Step 3(a) Complete the square. Add the square of half the coefficient of x to both sides of the equation.

$$x^2 - 6x + 9 = -4 + 9 \quad \text{Half of } -6 \text{ is } -3 \text{ and } (-3)^2 = 9.$$

Step 3(b) Factor the resulting perfect square trinomial.

$$(x - 3)^2 = 5 \qquad \text{Factor } x^2 - 6x + 9 \text{ using } A^2 - 2AB + B^2 = (A - B)^2.$$
$$\text{Simplify the right side: } -4 + 9 = 5.$$

Step 4 Use the square root property and solve for x.

$$x - 3 = \sqrt{5} \quad \text{or} \quad x - 3 = -\sqrt{5} \qquad \text{Apply the square root property.}$$
$$x = 3 + \sqrt{5} \qquad x = 3 - \sqrt{5} \qquad \text{Add 3 to both sides in each equation.}$$

The solutions are $3 \pm \sqrt{5}$ and the solution set is $\{3 + \sqrt{5}, 3 - \sqrt{5}\}$, or $\{3 \pm \sqrt{5}\}$. • • •

 Check Point **4** Solve by completing the square: $x^2 + 4x - 1 = 0$. $\{-2 \pm \sqrt{5}\}$

EXAMPLE 5 Solving a Quadratic Equation by Completing the Square

Solve by completing the square: $9x^2 - 6x - 4 = 0$.

SOLUTION

Step 1 Divide both sides by a to make the leading coefficient 1.

$$9x^2 - 6x - 4 = 0 \qquad \text{This is the given equation.}$$

$$\frac{9x^2 - 6x - 4}{9} = \frac{0}{9} \qquad \text{Divide both sides by the leading coefficient, 9.}$$

$$\frac{9x^2}{9} - \frac{6x}{9} - \frac{4}{9} = \frac{0}{9} \qquad \text{Divide each term by 9.}$$

$$x^2 - \frac{2}{3}x - \frac{4}{9} = 0 \qquad \text{Simplify.}$$

Step 2 Isolate the variable terms on one side of the equation and the constant term on the other side.

$$x^2 - \frac{2}{3}x = \frac{4}{9} \qquad \text{Add } \tfrac{4}{9} \text{ to both sides to isolate the binomial.}$$

Step 3(a) Complete the square. Add the square of half the coefficient of x to both sides of the equation.

$$x^2 - \frac{2}{3}x + \frac{1}{9} = \frac{4}{9} + \frac{1}{9} \qquad \text{Half of } -\tfrac{2}{3} \text{ is } \tfrac{1}{2}\left(-\tfrac{2}{3}\right) = -\tfrac{2}{6}, \text{ or } -\tfrac{1}{3}, \text{ and } \left(-\tfrac{1}{3}\right)^2 = \tfrac{1}{9}.$$

Step 3(b) Factor the resulting perfect square trinomial.

$$\left(x - \frac{1}{3}\right)^2 = \frac{5}{9} \qquad \text{Factor } x^2 - \tfrac{2}{3}x + \tfrac{1}{9} \text{ using } A^2 - 2AB + B^2 = (A - B)^2.$$
$$\text{Simplify the right side: } \tfrac{4}{9} + \tfrac{1}{9} = \tfrac{5}{9}.$$

●●●

TECHNOLOGY

Graphic Connections

Obtain a decimal approximation for each solution of

$$9x^2 - 6x - 4 = 0,$$

the equation in Example 5.

$$\frac{1 + \sqrt{5}}{3} \approx 1.1; \quad \frac{1 - \sqrt{5}}{3} \approx -0.4$$

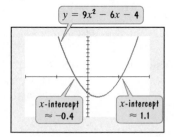

$[-2, 2, 1]$ by $[-10, 10, 1]$

The x-intercepts of $y = 9x^2 - 6x - 4$ verify the solutions.

Step 4 Use the square root property and solve for x. We apply the square root property to $\left(x - \dfrac{1}{3}\right)^2 = \dfrac{5}{9}$.

$$x - \frac{1}{3} = \sqrt{\frac{5}{9}} \quad \text{or} \quad x - \frac{1}{3} = -\sqrt{\frac{5}{9}} \qquad \text{Apply the square root property.}$$

$$x - \frac{1}{3} = \frac{\sqrt{5}}{3} \qquad\qquad x - \frac{1}{3} = -\frac{\sqrt{5}}{3} \qquad \sqrt{\frac{5}{9}} = \frac{\sqrt{5}}{\sqrt{9}} = \frac{\sqrt{5}}{3}$$

$$x = \frac{1}{3} + \frac{\sqrt{5}}{3} \qquad\qquad x = \frac{1}{3} - \frac{\sqrt{5}}{3} \qquad \text{Add } \tfrac{1}{3} \text{ to both sides and solve for } x.$$

$$x = \frac{1 + \sqrt{5}}{3} \qquad\qquad x = \frac{1 - \sqrt{5}}{3} \qquad \text{Express solutions with a common denominator.}$$

The solutions are $\dfrac{1 \pm \sqrt{5}}{3}$ and the solution set is $\left\{\dfrac{1 \pm \sqrt{5}}{3}\right\}$.

☑ **Check Point 5** Solve by completing the square: $2x^2 + 3x - 4 = 0$.

$$\left\{\frac{-3 \pm \sqrt{41}}{4}\right\}$$

Solving Quadratic Equations Using the Quadratic Formula

We can use the method of completing the square to derive a formula that can be used to solve all quadratic equations. The derivation given below also shows a particular quadratic equation, $3x^2 - 2x - 4 = 0$, to specifically illustrate each of the steps.

Deriving the Quadratic Formula

General Form of a Quadratic Equation	Comment	A Specific Example
$ax^2 + bx + c = 0, a > 0$	This is the given equation.	$3x^2 - 2x - 4 = 0$
$x^2 + \dfrac{b}{a}x + \dfrac{c}{a} = 0$	Divide both sides by the coefficient of x^2.	$x^2 - \dfrac{2}{3}x - \dfrac{4}{3} = 0$
$x^2 + \dfrac{b}{a}x = -\dfrac{c}{a}$	Isolate the binomial by adding $-\dfrac{c}{a}$ on both sides of the equation.	$x^2 - \dfrac{2}{3}x = \dfrac{4}{3}$
$x^2 + \dfrac{b}{a}x + \left(\dfrac{b}{2a}\right)^2 = -\dfrac{c}{a} + \left(\dfrac{b}{2a}\right)^2$ (half)²	Complete the square. Add the square of half the coefficient of x to both sides.	$x^2 - \dfrac{2}{3}x + \left(-\dfrac{1}{3}\right)^2 = \dfrac{4}{3} + \left(-\dfrac{1}{3}\right)^2$ (half)²
$x^2 + \dfrac{b}{a}x + \dfrac{b^2}{4a^2} = -\dfrac{c}{a} + \dfrac{b^2}{4a^2}$		$x^2 - \dfrac{2}{3}x + \dfrac{1}{9} = \dfrac{4}{3} + \dfrac{1}{9}$
$\left(x + \dfrac{b}{2a}\right)^2 = -\dfrac{c}{a}\cdot\dfrac{4a}{4a} + \dfrac{b^2}{4a^2}$	Factor on the left side and obtain a common denominator on the right side.	$\left(x - \dfrac{1}{3}\right)^2 = \dfrac{4}{3}\cdot\dfrac{3}{3} + \dfrac{1}{9}$
$\left(x + \dfrac{b}{2a}\right)^2 = \dfrac{-4ac + b^2}{4a^2}$	Add fractions on the right side.	$\left(x - \dfrac{1}{3}\right)^2 = \dfrac{12 + 1}{9}$
$\left(x + \dfrac{b}{2a}\right)^2 = \dfrac{b^2 - 4ac}{4a^2}$		$\left(x - \dfrac{1}{3}\right)^2 = \dfrac{13}{9}$
$x + \dfrac{b}{2a} = \pm\sqrt{\dfrac{b^2 - 4ac}{4a^2}}$	Apply the square root property.	$x - \dfrac{1}{3} = \pm\sqrt{\dfrac{13}{9}}$
$x + \dfrac{b}{2a} = \pm\dfrac{\sqrt{b^2 - 4ac}}{2a}$	Take the square root of the quotient, simplifying the denominator.	$x - \dfrac{1}{3} = \pm\dfrac{\sqrt{13}}{3}$
$x = \dfrac{-b}{2a} \pm \dfrac{\sqrt{b^2 - 4ac}}{2a}$	Solve for x by subtracting $\dfrac{b}{2a}$ from both sides.	$x = \dfrac{1}{3} \pm \dfrac{\sqrt{13}}{3}$
$x = \dfrac{-b \pm \sqrt{b^2 - 4ac}}{2a}$	Combine fractions on the right side.	$x = \dfrac{1 \pm \sqrt{13}}{3}$

The formula shown at the bottom of the left column on the previous page is called the *quadratic formula*. A similar proof shows that the same formula can be used to solve quadratic equations if *a*, the coefficient of the x^2-term, is negative.

④ Solve quadratic equations using the quadratic formula.

The Quadratic Formula

The solutions of a quadratic equation in general form $ax^2 + bx + c = 0$, with $a \neq 0$, are given by the **quadratic formula**:

$$x = \frac{-b \pm \sqrt{b^2 - 4ac}}{2a}.$$

x equals negative *b* plus or minus the square root of $b^2 - 4ac$, all divided by 2*a*.

To use the quadratic formula, rewrite the quadratic equation in general form if necessary. Then determine the numerical values for *a* (the coefficient of the x^2-term), *b* (the coefficient of the *x*-term), and *c* (the constant term). Substitute the values of *a*, *b*, and *c* into the quadratic formula and evaluate the expression. The \pm sign indicates that there are two (not necessarily distinct) solutions of the equation.

EXAMPLE 6 Solving a Quadratic Equation Using the Quadratic Formula

Solve using the quadratic formula: $8x^2 + 2x - 1 = 0$.

SOLUTION

The given equation is in general form. Begin by identifying the values for *a*, *b*, and *c*.

$$8x^2 + 2x - 1 = 0$$

$a = 8 \quad b = 2 \quad c = -1$

Substituting these values into the quadratic formula and simplifying gives the equation's solutions.

$$x = \frac{-b \pm \sqrt{b^2 - 4ac}}{2a}$$
Use the quadratic formula.

$$x = \frac{-2 \pm \sqrt{2^2 - 4(8)(-1)}}{2(8)}$$
Substitute the values for *a*, *b*, and *c*: $a = 8, b = 2,$ and $c = -1$.

$$= \frac{-2 \pm \sqrt{4 - (-32)}}{16}$$
$2^2 - 4(8)(-1) = 4 - (-32)$

$$= \frac{-2 \pm \sqrt{36}}{16}$$
$4 - (-32) = 4 + 32 = 36$

$$= \frac{-2 \pm 6}{16}$$
$\sqrt{36} = 6$

Now we will evaluate this expression in two different ways to obtain the two solutions. On the left, we will *add* 6 to -2. On the right, we will *subtract* 6 from -2.

$$x = \frac{-2 + 6}{16} \quad \text{or} \quad x = \frac{-2 - 6}{16}$$

$$= \frac{4}{16} = \frac{1}{4} \qquad\qquad = \frac{-8}{16} = -\frac{1}{2}$$

The solutions are $-\frac{1}{2}$ and $\frac{1}{4}$, and the solution set is $\left\{-\frac{1}{2}, \frac{1}{4}\right\}$. ● ● ●

TECHNOLOGY

Graphic Connections

The graph of

$$y = 8x^2 + 2x - 1$$

has *x*-intercepts at $-\frac{1}{2}$ and $\frac{1}{4}$. This verifies that $\{-\frac{1}{2}, \frac{1}{4}\}$ is the solution set of the quadratic equation

$$8x^2 + 2x - 1 = 0.$$

$y = 8x^2 + 2x - 1$

x-intercept is $-\frac{1}{2}$. | *x*-intercept is $\frac{1}{4}$.

$[-2, 2, 1]$ by $[-3, 10, 1]$

In Example 6, the solutions of $8x^2 + 2x - 1 = 0$ are rational numbers. This means that the equation can also be solved by factoring. The reason that the solutions are rational numbers is that $b^2 - 4ac$, the radicand in the quadratic formula, is 36, which is a perfect square. If a, b, and c are rational numbers, all quadratic equations for which $b^2 - 4ac$ is a perfect square have rational solutions.

⊘ Check Point **6** Solve using the quadratic formula: $2x^2 + 9x - 5 = 0$. $\{-5, \frac{1}{2}\}$

EXAMPLE 7 Solving a Quadratic Equation Using the Quadratic Formula

Solve using the quadratic formula: $2x^2 - 6x + 1 = 0$.

SOLUTION

The given equation is in general form. Begin by identifying the values for a, b, and c.

$$2x^2 - 6x + 1 = 0$$

$a = 2 \quad b = -6 \quad c = 1$

Substituting these values into the quadratic formula and simplifying gives the equation's solutions.

$$x = \frac{-b \pm \sqrt{b^2 - 4ac}}{2a}$$

Use the quadratic formula.

$$= \frac{-(-6) \pm \sqrt{(-6)^2 - 4(2)(1)}}{2 \cdot 2}$$

Substitute the values for a, b, and c: $a = 2$, $b = -6$, and $c = 1$.

$$= \frac{6 \pm \sqrt{36 - 8}}{4}$$

$-(-6) = 6$, $(-6)^2 = (-6)(-6) = 36$, and $4(2)(1) = 8$.

$$= \frac{6 \pm \sqrt{28}}{4}$$

Complete the subtraction under the radical.

$$= \frac{6 \pm 2\sqrt{7}}{4}$$

$\sqrt{28} = \sqrt{4 \cdot 7} = \sqrt{4}\sqrt{7} = 2\sqrt{7}$

$$= \frac{2(3 \pm \sqrt{7})}{4}$$

Factor out 2 from the numerator.

$$= \frac{3 \pm \sqrt{7}}{2}$$

Divide the numerator and denominator by 2.

The solution set is $\left\{ \dfrac{3 + \sqrt{7}}{2}, \dfrac{3 - \sqrt{7}}{2} \right\}$ or $\left\{ \dfrac{3 \pm \sqrt{7}}{2} \right\}$. ● ● ●

TECHNOLOGY

You can use a graphing utility to verify that the solutions of $2x^2 - 6x + 1 = 0$ are

$\dfrac{3 \pm \sqrt{7}}{2}$. Begin by entering $y_1 = 2x^2 - 6x + 1$ in the $\boxed{Y=}$ screen. Then evaluate this equation at each of the proposed solutions.

```
Y₁((3+√7)/2)
                    0
Y₁((3-√7)/2)
                    0
■
```

In each case, the value is 0, verifying that the solutions satisfy $2x^2 - 6x + 1 = 0$.

In Example 7, the solutions of $2x^2 - 6x + 1 = 0$ are irrational numbers. This means that the equation cannot be solved using factoring. The reason that the solutions are irrational numbers is that $b^2 - 4ac$, the radicand in the quadratic formula, is 28, which is not a perfect square.

⊘ Check Point **7** Solve using the quadratic formula:
$2x^2 + 2x - 1 = 0$. $\left\{ \frac{-1 \pm \sqrt{3}}{2} \right\}$

GREAT QUESTION!

The simplification of the irrational solutions in Example 7 is kind of tricky. Any suggestions to guide the process?

Many students use the quadratic formula correctly until the last step, where they make an error in simplifying the solutions. Be sure to factor the numerator before dividing the numerator and denominator by the greatest common factor:

$$\frac{6 \pm 2\sqrt{7}}{4} = \frac{2(3 \pm \sqrt{7})}{4} = \frac{\overset{1}{2}(3 \pm \sqrt{7})}{\underset{2}{\cancel{4}}} = \frac{3 \pm \sqrt{7}}{2}.$$

Factor first.

Then divide by the GCF.

You cannot divide just one term in the numerator and the denominator by their greatest common factor.

Incorrect!

$$\frac{6 \pm 2\sqrt{7}}{4} = \frac{\overset{3}{6} \pm 2\sqrt{7}}{\underset{2}{4}} = \frac{3 \pm 2\sqrt{7}}{2} \qquad \frac{6 \pm 2\sqrt{7}}{4} = \frac{6 \pm 2\overset{1}{\sqrt{7}}}{\underset{2}{4}} = \frac{6 \pm \sqrt{7}}{2}$$

Can all irrational solutions of quadratic equations be simplified? No. The following solutions cannot be simplified.

$$\frac{5 \pm 2\sqrt{7}}{2}$$

Other than 1, terms in each numerator have no common factor.

$$\frac{-4 \pm 3\sqrt{7}}{2}$$

EXAMPLE 8 Solving a Quadratic Equation Using the Quadratic Formula

Solve using the quadratic formula: $3x^2 - 2x + 4 = 0$.

SOLUTION

The given equation is in general form. Begin by identifying the values for a, b, and c.

$$3x^2 - 2x + 4 = 0$$

$a = 3$ $b = -2$ $c = 4$

GREAT QUESTION!

Should I check irrational and imaginary solutions by substitution in the given quadratic equation?

No. Checking irrational and imaginary solutions can be time-consuming. The solutions given by the quadratic formula are always correct, unless you have made a careless error. Checking for computational errors or errors in simplification is sufficient.

$$x = \frac{-b \pm \sqrt{b^2 - 4ac}}{2a}$$

Use the quadratic formula.

$$= \frac{-(-2) \pm \sqrt{(-2)^2 - 4(3)(4)}}{2(3)}$$

Substitute the values for a, b, and c: $a = 3$, $b = -2$, and $c = 4$.

$$= \frac{2 \pm \sqrt{4 - 48}}{6}$$

$-(-2) = 2$ and $(-2)^2 = (-2)(-2) = 4$.

$$= \frac{2 \pm \sqrt{-44}}{6}$$

Subtract under the radical. Because the number under the radical sign is negative, the solutions will not be real numbers.

$$= \frac{2 \pm 2i\sqrt{11}}{6}$$

$\sqrt{-44} = \sqrt{4(11)(-1)} = 2i\sqrt{11}$

$$= \frac{2(1 \pm i\sqrt{11})}{6}$$

Factor 2 from the numerator.

$$= \frac{1 \pm i\sqrt{11}}{3}$$

Divide the numerator and denominator by 2.

$$= \frac{1}{3} \pm i\frac{\sqrt{11}}{3}$$

Write the complex numbers in standard form.

The solutions are complex conjugates, and the solution set is

$$\left\{\frac{1}{3} + i\frac{\sqrt{11}}{3}, \frac{1}{3} - i\frac{\sqrt{11}}{3}\right\} \text{ or } \left\{\frac{1}{3} \pm i\frac{\sqrt{11}}{3}\right\}. \quad \bullet \bullet \bullet$$

If $ax^2 + bx + c = 0$ has imaginary solutions, the graph of $y = ax^2 + bx + c$ will not have x-intercepts. This is illustrated by the imaginary solutions of $3x^2 - 2x + 4 = 0$ in Example 8 and the graph of $y = 3x^2 - 2x + 4$ in **Figure 1.20**.

In Example 8, the solutions of $3x^2 - 2x + 4 = 0$ are imaginary numbers. This means that the equation cannot be solved using factoring. The reason that the solutions are imaginary numbers is that $b^2 - 4ac$, the radicand in the quadratic formula, is -44, which is negative. Notice, too, that the solutions are complex conjugates.

$y = 3x^2 - 2x + 4$

$[-2, 2, 1]$ by $[-1, 10, 1]$

FIGURE 1.20 This graph has no x-intercepts.

⊘ Check Point **8** Solve using the quadratic formula:

$$x^2 - 2x + 2 = 0. \quad \{1 + i, 1 - i\}$$

⑤ Use the discriminant to determine the number and type of solutions.

The Discriminant

The quantity $b^2 - 4ac$, which appears under the radical sign in the quadratic formula, is called the **discriminant**. **Table 1.2** shows how the discriminant of the quadratic equation $ax^2 + bx + c = 0$ determines the number and type of solutions.

GREAT QUESTION!

Does the discriminant give me the solutions of a quadratic equation?

No. It gives the number and type of solutions, but not the actual solutions.

Table 1.2 The Discriminant and the Kinds of Solutions to $ax^2 + bx + c = 0$

Discriminant $b^2 - 4ac$	Kinds of Solutions to $ax^2 + bx + c = 0$	Graph of $y = ax^2 + bx + c$
$b^2 - 4ac > 0$	**Two unequal real solutions:** If a, b, and c are rational numbers and the discriminant is a perfect square, the solutions are *rational*. If the discriminant is not a perfect square, the solutions are *irrational* conjugates.	Two x-intercepts
$b^2 - 4ac = 0$	**One solution (a repeated solution) that is a real number:** If a, b, and c are rational numbers, the repeated solution is also a rational number.	One x-intercept
$b^2 - 4ac < 0$	**No real solution; two imaginary solutions:** The solutions are complex conjugates.	No x-intercepts

EXAMPLE 9 Using the Discriminant

For each equation, compute the discriminant. Then determine the number and type of solutions:

a. $3x^2 + 4x - 5 = 0$ **b.** $9x^2 - 6x + 1 = 0$ **c.** $3x^2 - 8x + 7 = 0.$

GREAT QUESTION!

Is the square root sign part of the discriminant?

No. The discriminant is $b^2 - 4ac$. It is not $\sqrt{b^2 - 4ac}$, so do not give the discriminant as a radical.

SOLUTION

Begin by identifying the values for a, b, and c in each equation. Then compute $b^2 - 4ac$, the discriminant.

a. $3x^2 + 4x - 5 = 0$

$a = 3$ $b = 4$ $c = -5$

Substitute and compute the discriminant:

$$b^2 - 4ac = 4^2 - 4 \cdot 3(-5) = 16 - (-60) = 16 + 60 = 76.$$

The discriminant, 76, is a positive number that is not a perfect square. Thus, there are two irrational solutions. (These solutions are conjugates of each other.)

b. $9x^2 - 6x + 1 = 0$

$a = 9$ $b = -6$ $c = 1$

Substitute and compute the discriminant:

$$b^2 - 4ac = (-6)^2 - 4 \cdot 9 \cdot 1 = 36 - 36 = 0.$$

The discriminant, 0, shows that there is only one real solution. This real solution is a rational number.

c. $3x^2 - 8x + 7 = 0$

$a = 3$ $b = -8$ $c = 7$

$$b^2 - 4ac = (-8)^2 - 4 \cdot 3 \cdot 7 = 64 - 84 = -20$$

The negative discriminant, -20, shows that there are two imaginary solutions. (These solutions are complex conjugates of each other.) • • •

 Check Point **9** For each equation, compute the discriminant. Then determine the number and type of solutions:

a. $x^2 + 6x + 9 = 0$ **b.** $2x^2 - 7x - 4 = 0$ **c.** $3x^2 - 2x + 4 = 0$.

a. 0; one real solution that is rational

b. 81; two rational solutions

c. -44; two imaginary solutions that are complex conjugates

⑥ Determine the most efficient method to use when solving a quadratic equation.

Determining Which Method to Use

All quadratic equations can be solved by the quadratic formula. However, if an equation is in the form $u^2 = d$, such as $x^2 = 5$ or $(2x + 3)^2 = 8$, it is faster to use the square root property, taking the square root of both sides. If the equation is not in the form $u^2 = d$, write the quadratic equation in general form ($ax^2 + bx + c = 0$). Try to solve the equation by factoring. If $ax^2 + bx + c$ cannot be factored, then solve the quadratic equation by the quadratic formula.

Because we used the method of completing the square to derive the quadratic formula, we no longer need it for solving quadratic equations. However, we will use completing the square later in the book to help graph circles and other kinds of equations.

GREAT QUESTION!

Is factoring the most important technique to use when solving a quadratic equation?

No. Even though you first learned to solve quadratic equations by factoring, most quadratic equations in general form cannot be factored. Be sure to apply the quadratic formula when this occurs.

Table 1.3 summarizes our observations about which technique to use when solving a quadratic equation.

Table 1.3 Determining the Most Efficient Technique to Use when Solving a Quadratic Equation

Description and Form of the Quadratic Equation	Most Efficient Solution Method	Example
$ax^2 + bx + c = 0$ and $ax^2 + bx + c$ can be factored easily.	Factor and use the zero-product principle.	$3x^2 + 5x - 2 = 0$ $(3x - 1)(x + 2) = 0$ $3x - 1 = 0 \quad$ or $\quad x + 2 = 0$ $x = \dfrac{1}{3} \qquad\qquad x = -2$
$ax^2 + bx = 0$ The quadratic equation has no constant term. $(c = 0)$	Factor and use the zero-product principle.	$6x^2 + 9x = 0$ $3x(2x + 3) = 0$ $3x = 0 \quad$ or $\quad 2x + 3 = 0$ $x = 0 \qquad\qquad 2x = -3$ $\qquad\qquad\qquad x = -\dfrac{3}{2}$
$ax^2 + c = 0$ The quadratic equation has no x-term. $(b = 0)$	Solve for x^2 and apply the square root property.	$7x^2 - 4 = 0$ $7x^2 = 4$ $x^2 = \dfrac{4}{7}$ $x = \pm\sqrt{\dfrac{4}{7}}$ $= \pm\dfrac{2}{\sqrt{7}} = \pm\dfrac{2}{\sqrt{7}} \cdot \dfrac{\sqrt{7}}{\sqrt{7}} = \pm\dfrac{2\sqrt{7}}{7}$
$u^2 = d$; u is a first-degree polynomial.	Use the square root property.	$(x + 4)^2 = 5$ $x + 4 = \pm\sqrt{5}$ $x = -4 \pm \sqrt{5}$
$ax^2 + bx + c = 0$ and $ax^2 + bx + c$ cannot be factored or the factoring is too difficult.	Use the quadratic formula: $x = \dfrac{-b \pm \sqrt{b^2 - 4ac}}{2a}$.	$x^2 - 2x - 6 = 0$ $a = 1 \quad b = -2 \quad c = -6$ $x = \dfrac{-(-2) \pm \sqrt{(-2)^2 - 4(1)(-6)}}{2(1)}$ $= \dfrac{2 \pm \sqrt{4 + 24}}{2(1)}$ $= \dfrac{2 \pm \sqrt{28}}{2} = \dfrac{2 \pm \sqrt{4}\sqrt{7}}{2}$ $= \dfrac{2 \pm 2\sqrt{7}}{2} = \dfrac{2\left(1 \pm \sqrt{7}\right)}{2}$ $= 1 \pm \sqrt{7}$

7 Solve problems modeled by quadratic equations.

Applications

EXAMPLE 10 Blood Pressure and Age

The graphs in **Figure 1.21** illustrate that a person's normal systolic blood pressure, measured in millimeters of mercury (mm Hg), depends on his or her age. (The symbol ⌇ on the vertical axis shows there is a break in values between 0 and 100. Thus, the first tick mark on the vertical axis represents a blood pressure of 100 mm Hg.) The formula

$$P = 0.006A^2 - 0.02A + 120$$

models a man's normal systolic pressure, P, at age A.

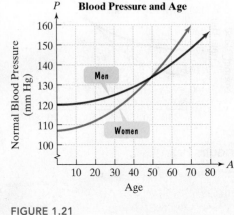

FIGURE 1.21

a. Find the age, to the nearest year, of a man whose normal systolic blood pressure is 125 mm Hg.

b. Use the graphs in **Figure 1.21** to describe the differences between the normal systolic blood pressures of men and women as they age.

SOLUTION

a. We are interested in the age of a man with a normal systolic blood pressure of 125 millimeters of mercury. Thus, we substitute 125 for P in the given formula for men. Then we solve for A, the man's age.

$$P = 0.006A^2 - 0.02A + 120$$ This is the given formula for men.

$$125 = 0.006A^2 - 0.02A + 120$$ Substitute 125 for P.

$$0 = 0.006A^2 - 0.02A - 5$$ Subtract 125 from both sides and write the quadratic equation in general form.

$a = 0.006$ $b = -0.02$ $c = -5$

Because the trinomial on the right side of the equation is prime, we solve using the quadratic formula.

Notice that the variable is A, rather than the usual x.

$$A = \frac{-b \pm \sqrt{b^2 - 4ac}}{2a}$$ Use the quadratic formula.

$$= \frac{-(-0.02) \pm \sqrt{(-0.02)^2 - 4(0.006)(-5)}}{2(0.006)}$$ Substitute the values for a, b, and c: $a = 0.006$, $b = -0.02$, and $c = -5$.

$$= \frac{0.02 \pm \sqrt{0.1204}}{0.012}$$ Use a calculator to simplify the radicand.

$$\approx \frac{0.02 \pm 0.347}{0.012}$$ Use a calculator: $\sqrt{0.1204} \approx 0.347$.

$$A \approx \frac{0.02 + 0.347}{0.012} \quad \text{or} \quad A \approx \frac{0.02 - 0.347}{0.012}$$

$$A \approx 31 \qquad\qquad\qquad A \approx -27$$ Use a calculator and round to the nearest integer.

Reject this solution. Age cannot be negative.

TECHNOLOGY

On most calculators, here is how to approximate

$$\frac{0.02 + \sqrt{0.1204}}{0.012}.$$

Many Scientific Calculators

(.02 + .1204 √)

÷ .012 =

Many Graphing Calculators

(.02 + √ .1204)

÷ .012 ENTER

If your calculator displays an open parenthesis after √ , you'll need to enter another closed parenthesis here. Some calculators require that you press the right arrow key to exit the radical after entering the radicand.

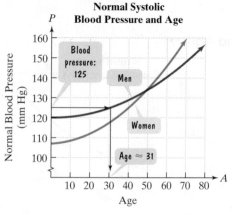

Normal Systolic Blood Pressure and Age

FIGURE 1.22

The positive solution, $A \approx 31$, indicates that 31 is the approximate age of a man whose normal systolic blood pressure is 125 mm Hg. This is illustrated by the black lines with the arrows on the red graph representing men in **Figure 1.22**.

b. Take a second look at the graphs in **Figure 1.22**. Before approximately age 50, the blue graph representing women's normal systolic blood pressure lies below the red graph representing men's normal systolic blood pressure. Thus, up to age 50, women's normal systolic blood pressure is lower than men's, although it is increasing at a faster rate. After age 50, women's normal systolic blood pressure is higher than men's.

⬤⬤⬤

✓ **Check Point 10** The formula $P = 0.01A^2 + 0.05A + 107$ models a woman's normal systolic blood pressure, P, at age A. Use this formula to find the age, to the nearest year, of a woman whose normal systolic blood pressure is 115 mm Hg. Use the blue graph in **Figure 1.22** to verify your solution. approximately 26 years old

In our next example, we will be using the *Pythagorean Theorem* to obtain a mathematical model. The ancient Greek philosopher and mathematician Pythagoras (approximately 582–500 B.C.) founded a school whose motto was "All is number." Pythagoras is best remembered for his work with the **right triangle**, a triangle with one angle measuring 90°. The side opposite the 90° angle is called the **hypotenuse**. The other sides are called **legs**. Pythagoras found that if he constructed squares on each of the legs, as well as a larger square on the hypotenuse, the sum of the areas of the smaller squares is equal to the area of the larger square. This is illustrated in **Figure 1.23**.

This relationship is usually stated in terms of the lengths of the three sides of a right triangle and is called the **Pythagorean Theorem**.

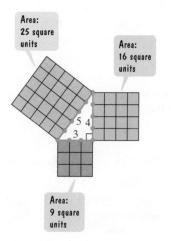

FIGURE 1.23 The area of the large square equals the sum of the areas of the smaller squares.

The Pythagorean Theorem

The sum of the squares of the lengths of the legs of a right triangle equals the square of the length of the hypotenuse.

If the legs have lengths a and b, and the hypotenuse has length c, then

$$a^2 + b^2 = c^2.$$

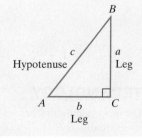

EXAMPLE 11 Screen Math

Did you know that the size of a television screen refers to the length of its diagonal? If the length of the HDTV screen in **Figure 1.24** is 28 inches and its width is 15.7 inches, what is the size of the screen to the nearest inch?

← 28 in. →

15.7 in.

FIGURE 1.24

SOLUTION

Figure 1.24 shows that the length, width, and diagonal of the screen form a right triangle. The diagonal is the hypotenuse of the triangle. We use the Pythagorean Theorem with $a = 28$, $b = 15.7$, and solve for the screen size, c.

$$a^2 + b^2 = c^2$$ This is the Pythagorean Theorem.

$$28^2 + 15.7^2 = c^2$$ Let $a = 28$ and $b = 15.7$.

$$784 + 246.49 = c^2$$ $28^2 = 28 \cdot 28 = 784$ and $15.7^2 = 15.7 \cdot 15.7 = 246.49$.

$$c^2 = 1030.49$$ Add: $784 + 246.49 = 1030.49$.
We also reversed the two sides.

$$c = \sqrt{1030.49} \text{ or } c = -\sqrt{1030.49}$$ Apply the square root property.

$$c \approx 32 \text{ or } c \approx -32$$ Use a calculator and round to the nearest inch.
1030.49 $\boxed{\sqrt{}}$ $\boxed{=}$ or $\boxed{\sqrt{}}$ 1030.49 $\boxed{\text{ENTER}}$

Because c represents the size of the screen, this dimension must be positive. We reject -32. Thus, the screen size of the HDTV is 32 inches. • • •

GREAT QUESTION!

Why did you include a decimal like 15.7 in the screen math example? Squaring 15.7 is awkward. Why not just use 16 inches for the width of the screen?

We wanted to use the actual dimensions for a 32-inch HDTV screen. In the Check Point that follows, we use the exact dimensions of an "old" TV screen (prior to HDTV). A calculator is helpful in squaring each of these dimensions.

FIGURE 1.25

☑ Check Point **11** **Figure 1.25** shows the dimensions of an old TV screen. What is the size of the screen? 32 inches

Blitzer Bonus || Screen Math

That new 32-inch HDTV you want: How much larger than your old 32-incher is it? Actually, based on Example 11 and Check Point 11, it's smaller! **Figure 1.26** compares the screen area of the old 32-inch TV in Check Point 11 with the 32-inch HDTV in Example 11.

Old 32-Inch Screen **HDTV 32-Inch Screen**

Area = length · width Area = length · width
= 25.6 · 19.2 = 28 · 15.7
= 491.52 square inches = 439.6 square inches

FIGURE 1.26

To make sure your HDTV has the same screen area as your old TV, it needs to have a diagonal measure, or screen size, that is 6% larger. Equivalently, take the screen size of the old TV and multiply by 1.06. If you have a 32-inch regular TV, this means the HDTV needs a 34-inch screen ($32 \times 1.06 = 33.92 \approx 34$) if you don't want your new TV picture to be smaller than the old one.

ACHIEVING SUCCESS

A test-taking tip: Go for partial credit. Always show your work. If worse comes to worst, write something down, anything, even if it's a formula that you think might solve a problem or a possible idea or procedure for solving the problem. Here's an example:

Test Question: Solve by the quadratic formula:

$$2x^2 = -4x + 5.$$

Student Solution:

$$2x^2 = -4x + 5$$
$$2x^2 + 4x - 5 = 0$$
$$a = 2 \quad b = 4 \quad c = -5$$

Egad! I forgot the formula.

$$x = \frac{-b \pm \sqrt{?}}{2a} = \frac{-4 \pm \sqrt{?}}{2 \cdot 2}$$

I know I need to find ?.
That will give me the solutions.
Perhaps I can simplify after that.
???

CONCEPT AND VOCABULARY CHECK

Fill in each blank so that the resulting statement is true.

1. An equation that can be written in the general form $ax^2 + bx + c = 0, a \neq 0$, is called a/an ___quadratic___ equation.

2. The zero-product principle states that if $AB = 0$, then ___$A = 0$ or $B = 0$___.

3. The solutions of $ax^2 + bx + c = 0$ correspond to the ___x-intercepts___ for the graph of $y = ax^2 + bx + c$.

4. The square root property states that if $u^2 = d$, then $u = $ ___$\pm\sqrt{d}$___.

5. If $x^2 = 7$, then $x = $ ___$\pm\sqrt{7}$___.

6. To complete the square on $x^2 - 3x$, add ___$\frac{9}{4}$___.

7. To complete the square on $x^2 - \frac{4}{5}x$, add ___$\frac{4}{25}$___.

8. To solve $x^2 + 6x = 7$ by completing the square, add ___9___ to both sides of the equation.

9. To solve $x^2 - \frac{2}{3}x = \frac{4}{9}$ by completing the square, add ___$\frac{1}{9}$___ to both sides of the equation.

10. The solutions of a quadratic equation in the general form $ax^2 + bx + c = 0, a \neq 0$, are given by the quadratic formula $x = $ ___$\frac{-b \pm \sqrt{b^2 - 4ac}}{2a}$___.

11. In order to solve $2x^2 + 9x - 5 = 0$ by the quadratic formula, we use $a = $ ___2___, $b = $ ___9___, and $c = $ ___-5___.

12. In order to solve $x^2 = 4x + 1$ by the quadratic formula, we use $a = $ ___1___, $b = $ ___-4___, and $c = $ ___-1___.

13. $x = \dfrac{-(-4) \pm \sqrt{(-4)^2 - 4(1)(2)}}{2(1)}$ simplifies to $x = $ ___$2 \pm \sqrt{2}$___.

14. $x = \dfrac{-4 \pm \sqrt{4^2 - 4 \cdot 2 \cdot 5}}{2 \cdot 2}$ simplifies to $x = $ ___$\frac{-1 \pm i\sqrt{6}}{2}$___.

15. The discriminant of $ax^2 + bx + c = 0$ is defined by ___$b^2 - 4ac$___.

16. If the discriminant of $ax^2 + bx + c = 0$ is negative, the quadratic equation has ___no___ real solutions.

17. If the discriminant of $ax^2 + bx + c = 0$ is positive, the quadratic equation has ___two___ real solutions.

18. The most efficient technique for solving $(2x + 7)^2 = 25$ is by using ___the square root property___.

19. The most efficient technique for solving $x^2 + 5x - 10 = 0$ is by using ___the quadratic formula___.

20. The most efficient technique for solving $x^2 + 8x + 15 = 0$ is by using ___factoring and the zero-product principle___.

21. A triangle with one angle measuring 90° is called a/an ___right___ triangle. The side opposite the 90° angle is called the ___hypotenuse___. The other sides are called ___legs___.

22. The Pythagorean Theorem states that in any ___right___ triangle, the sum of the squares of the lengths of the ___legs___ equals ___the square of the length of the hypotenuse___.

EXERCISE SET 1.5

Practice Exercises

Solve each equation in Exercises 1–14 by factoring.

1. $x^2 - 3x - 10 = 0$ $\{-2, 5\}$
2. $x^2 - 13x + 36 = 0$ $\{4, 9\}$
3. $x^2 = 8x - 15$ $\{3, 5\}$
4. $x^2 = -11x - 10$ $\{-10, -1\}$
5. $6x^2 + 11x - 10 = 0$ $\{-\frac{5}{2}, \frac{2}{3}\}$
6. $9x^2 + 9x + 2 = 0$ $\{-\frac{2}{3}, -\frac{1}{3}\}$
7. $3x^2 - 2x = 8$ $\{-\frac{4}{3}, 2\}$
8. $4x^2 - 13x = -3$ $\{\frac{1}{4}, 3\}$
9. $3x^2 + 12x = 0$ $\{-4, 0\}$
10. $5x^2 - 20x = 0$ $\{0, 4\}$
11. $2x(x - 3) = 5x^2 - 7x$ $\{0, \frac{1}{3}\}$
12. $16x(x - 2) = 8x - 25$ $\{\frac{5}{4}\}$
13. $7 - 7x = (3x + 2)(x - 1)$ $\{-3, 1\}$
14. $10x - 1 = (2x + 1)^2$ $\{\frac{1}{2}, 1\}$

Solve each equation in Exercises 15–34 by the square root property.

15. $3x^2 = 27$ $\{-3, 3\}$
16. $5x^2 = 45$ $\{-3, 3\}$
17. $5x^2 + 1 = 51$ $\{-\sqrt{10}, \sqrt{10}\}$
18. $3x^2 - 1 = 47$ $\{-4, 4\}$
19. $2x^2 - 5 = -55$ $\{\pm 5i\}$
20. $2x^2 - 7 = -15$ $\{\pm 2i\}$
21. $(x + 2)^2 = 25$ $\{-7, 3\}$
22. $(x - 3)^2 = 36$ $\{-3, 9\}$
23. $3(x - 4)^2 = 15$ $\{4 \pm \sqrt{5}\}$
24. $3(x + 4)^2 = 21$ $\{-4 \pm \sqrt{7}\}$
25. $(x + 3)^2 = -16$ $\{-3 \pm 4i\}$
26. $(x - 1)^2 = -9$ $\{1 \pm 3i\}$
27. $(x - 3)^2 = -5$ $\{3 \pm i\sqrt{5}\}$
28. $(x + 2)^2 = -7$ $\{-2 \pm i\sqrt{7}\}$
29. $(3x + 2)^2 = 9$ $\{-\frac{5}{3}, \frac{1}{3}\}$
30. $(4x - 1)^2 = 16$ $\{-\frac{3}{4}, \frac{5}{4}\}$
31. $(5x - 1)^2 = 7$ *
32. $(8x - 3)^2 = 5$ *
33. $(3x - 4)^2 = 8$ *
34. $(2x + 8)^2 = 27$ *

In Exercises 35–46, determine the constant that should be added to the binomial so that it becomes a perfect square trinomial. Then write and factor the trinomial.

35. $x^2 + 12x$ *
36. $x^2 + 16x$ *
37. $x^2 - 10x$ *
38. $x^2 - 14x$ *
39. $x^2 + 3x$ *
40. $x^2 + 5x$ *
41. $x^2 - 7x$ *
42. $x^2 - 9x$ *
43. $x^2 - \frac{2}{3}x$ *
44. $x^2 + \frac{4}{5}x$ *
45. $x^2 - \frac{1}{3}x$ *
46. $x^2 - \frac{1}{4}x$ *

Solve each equation in Exercises 47–64 by completing the square.

47. $x^2 + 6x = 7$ $\{-7, 1\}$
48. $x^2 + 6x = -8$ $\{-4, -2\}$
49. $x^2 - 2x = 2$ $\{1 \pm \sqrt{3}\}$
50. $x^2 + 4x = 12$ $\{-6, 2\}$
51. $x^2 - 6x - 11 = 0$ $\{3 \pm 2\sqrt{5}\}$
52. $x^2 - 2x - 5 = 0$ $\{1 \pm \sqrt{6}\}$
53. $x^2 + 4x + 1 = 0$ $\{-2 \pm \sqrt{3}\}$
54. $x^2 + 6x - 5 = 0$ $\{-3 \pm \sqrt{14}\}$
55. $x^2 - 5x + 6 = 0$ $\{2, 3\}$
56. $x^2 + 7x - 8 = 0$ $\{-8, 1\}$
57. $x^2 + 3x - 1 = 0$ *
58. $x^2 - 3x - 5 = 0$ *
59. $2x^2 - 7x + 3 = 0$ $\{\frac{1}{2}, 3\}$
60. $2x^2 + 5x - 3 = 0$ $\{-3, \frac{1}{2}\}$
61. $4x^2 - 4x - 1 = 0$ *
62. $2x^2 - 4x - 1 = 0$ *
63. $3x^2 - 2x - 2 = 0$ *
64. $3x^2 - 5x - 10 = 0$ *

Solve each equation in Exercises 65–74 using the quadratic formula.

65. $x^2 + 8x + 15 = 0$ $\{-5, -3\}$
66. $x^2 + 8x + 12 = 0$ $\{-6, -2\}$
67. $x^2 + 5x + 3 = 0$ *
68. $x^2 + 5x + 2 = 0$ *
69. $3x^2 - 3x - 4 = 0$ *
70. $5x^2 + x - 2 = 0$ *
71. $4x^2 = 2x + 7$ *
72. $3x^2 = 6x - 1$ *
73. $x^2 - 6x + 10 = 0$ $\{3 \pm i\}$
74. $x^2 - 2x + 17 = 0$ $\{1 \pm 4i\}$

In Exercises 75–82, compute the discriminant. Then determine the number and type of solutions for the given equation.

75. $x^2 - 4x - 5 = 0$ 36*
76. $4x^2 - 2x + 3 = 0$ −44*
77. $2x^2 - 11x + 3 = 0$ 97*
78. $2x^2 + 11x - 6 = 0$ 169*
79. $x^2 - 2x + 1 = 0$ 0*
80. $3x^2 = 2x - 1$ −8*
81. $x^2 - 3x - 7 = 0$ 37*
82. $3x^2 + 4x - 2 = 0$ 40*

Solve each equation in Exercises 83–108 by the method of your choice.

83. $2x^2 - x = 1$ $\{-\frac{1}{2}, 1\}$
84. $3x^2 - 4x = 4$ $\{-\frac{2}{3}, 2\}$
85. $5x^2 + 2 = 11x$ $\{\frac{1}{5}, 2\}$
86. $5x^2 = 6 - 13x$ $\{-3, \frac{2}{5}\}$
87. $3x^2 = 60$ $\{-2\sqrt{5}, 2\sqrt{5}\}$
88. $2x^2 = 250$ $\{-5\sqrt{5}, 5\sqrt{5}\}$
89. $x^2 - 2x = 1$ $\{1 \pm \sqrt{2}\}$
90. $2x^2 + 3x = 1$ *
91. $(2x + 3)(x + 4) = 1$ *
92. $(2x - 5)(x + 1) = 2$ *
93. $(3x - 4)^2 = 16$ $\{0, \frac{8}{3}\}$
94. $(2x + 7)^2 = 25$ $\{-6, -1\}$
95. $3x^2 - 12x + 12 = 0$ $\{2\}$
96. $9 - 6x + x^2 = 0$ $\{3\}$
97. $4x^2 - 16 = 0$ $\{-2, 2\}$
98. $3x^2 - 27 = 0$ $\{-3, 3\}$
99. $x^2 - 6x + 13 = 0$ $\{3 \pm 2i\}$
100. $x^2 - 4x + 29 = 0$ $\{2 \pm 5i\}$
101. $x^2 = 4x - 7$ $\{2 \pm i\sqrt{3}\}$
102. $5x^2 = 2x - 3$ *
103. $2x^2 - 7x = 0$ $\{0, \frac{7}{2}\}$
104. $2x^2 + 5x = 3$ $\{-3, \frac{1}{2}\}$
105. $\dfrac{1}{x} + \dfrac{1}{x + 2} = \dfrac{1}{3}$ $\{2 \pm \sqrt{10}\}$
106. $\dfrac{1}{x} + \dfrac{1}{x + 3} = \dfrac{1}{4}$ *
107. $\dfrac{2x}{x - 3} + \dfrac{6}{x + 3} = -\dfrac{28}{x^2 - 9}$ $\{-5, -1\}$
108. $\dfrac{3}{x - 3} + \dfrac{5}{x - 4} = \dfrac{x^2 - 20}{x^2 - 7x + 12}$ $\{1, 7\}$

In Exercises 109–114, find the x-intercept(s) of the graph of each equation. Use the x-intercepts to match the equation with its graph. The graphs are shown in $[-10, 10, 1]$ by $[-10, 10, 1]$ viewing rectangles and labeled (a) through (f).

109. $y = x^2 - 4x - 5$ d*
110. $y = x^2 - 6x + 7$ a*
111. $y = -(x + 1)^2 + 4$ f*
112. $y = -(x + 3)^2 + 1$ e*
113. $y = x^2 - 2x + 2$ b*
114. $y = x^2 + 6x + 9$ −3; c

a.

b.

c.

d.

e.

f.

*See Answers to Selected Exercises.

In Exercises 115–122, find all values of x satisfying the given conditions.

115. $y = 2x^2 - 3x$ and $y = 2$. $-\frac{1}{2}$ and 2

116. $y = 5x^2 + 3x$ and $y = 2$. $\frac{2}{5}$ and -1

117. $y_1 = x - 1$, $y_2 = x + 4$, and $y_1 y_2 = 14$. -6 and 3

118. $y_1 = x - 3$, $y_2 = x + 8$, and $y_1 y_2 = -30$. -3 and -2

119. $y_1 = \dfrac{2x}{x + 2}$, $y_2 = \dfrac{3}{x + 4}$, and $y_1 + y_2 = 1$. $\frac{-5 \pm \sqrt{33}}{2}$

120. $y_1 = \dfrac{3}{x - 1}$, $y_2 = \dfrac{8}{x}$, and $y_1 + y_2 = 3$. $\frac{2}{3}$ and 4

121. $y_1 = 2x^2 + 5x - 4$, $y_2 = -x^2 + 15x - 10$, and $y_1 - y_2 = 0$. $\frac{5 \pm \sqrt{7}}{3}$

122. $y_1 = -x^2 + 4x - 2$, $y_2 = -3x^2 + x - 1$, and $y_1 - y_2 = 0$. $\frac{-3 \pm \sqrt{17}}{4}$

Practice Plus

In Exercises 123–124, list all numbers that must be excluded from the domain of each rational expression.

123. $\dfrac{3}{2x^2 + 4x - 9}$ $\frac{-2 \pm \sqrt{22}}{2}$ **124.** $\dfrac{7}{2x^2 - 8x + 5}$ $\frac{4 \pm \sqrt{6}}{2}$

125. When the sum of 6 and twice a positive number is subtracted from the square of the number, 0 results. Find the number. $1 + \sqrt{7}$

126. When the sum of 1 and twice a negative number is subtracted from twice the square of the number, 0 results. Find the number. $\frac{1 - \sqrt{3}}{2}$

In Exercises 127–130, solve each equation by the method of your choice.

127. $\dfrac{1}{x^2 - 3x + 2} = \dfrac{1}{x + 2} + \dfrac{5}{x^2 - 4}$ $\left\{\frac{-1 \pm \sqrt{21}}{2}\right\}$

128. $\dfrac{x - 1}{x - 2} + \dfrac{x}{x - 3} = \dfrac{1}{x^2 - 5x + 6}$ $\left\{\frac{3 \pm \sqrt{5}}{2}\right\}$

129. $\sqrt{2}x^2 + 3x - 2\sqrt{2} = 0$ $\left\{-2\sqrt{2}, \frac{\sqrt{2}}{2}\right\}$

130. $\sqrt{3}x^2 + 6x + 7\sqrt{3} = 0$ $\{-\sqrt{3} \pm 2i\}$

Application Exercises

In a round-robin chess tournament, each player is paired with every other player once. The formula

$$N = \frac{x^2 - x}{2}$$

models the number of chess games, N, that must be played in a round-robin tournament with x chess players. Use this formula to solve Exercises 131–132.

131. In a round-robin chess tournament, 21 games were played. How many players were entered in the tournament? 7

132. In a round-robin chess tournament, 36 games were played. How many players were entered in the tournament? 9

The graph of the formula in Exercises 131–132 is shown. Use the graph to solve Exercises 133–134.

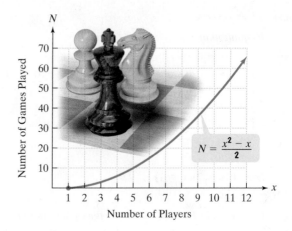

$$N = \frac{x^2 - x}{2}$$

133. Identify your solution to Exercise 131 as a point on the graph. $(7, 21)$

134. Identify your solution to Exercise 132 as a point on the graph. $(9, 36)$

A driver's age has something to do with his or her chance of getting into a fatal car crash. The bar graph shows the number of fatal vehicle crashes per 100 million miles driven for drivers of various age groups. For example, 25-year-old drivers are involved in 4.1 fatal crashes per 100 million miles driven. Thus, when a group of 25-year-old Americans have driven a total of 100 million miles, approximately 4 have been in accidents in which someone died.

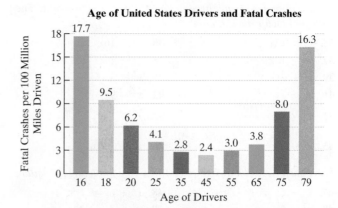

Age of United States Drivers and Fatal Crashes

Source: Insurance Institute for Highway Safety

The number of fatal vehicle crashes per 100 million miles, N, for drivers of age x can be modeled by the formula

$$N = 0.013x^2 - 1.19x + 28.24.$$

Use the formula to solve Exercises 135–136.

135. What age groups are expected to be involved in 3 fatal crashes per 100 million miles driven? How well does the formula model the trend in the actual data shown in the bar graph? 33-year-olds and 58-year olds; reasonably well

136. What age groups are expected to be involved in 10 fatal crashes per 100 million miles driven? How well does the formula model the trend in the actual data shown in the bar graph? 19-year-olds and 72-year-olds; Answers will vary.

Throwing events in track and field include the shot put, the discus throw, the hammer throw, and the javelin throw. The distance that an athlete can achieve depends on the initial velocity of the object thrown and the angle above the horizontal at which the object leaves the hand.

In Exercises 137–138, an athlete whose event is the shot put releases the shot with the same initial velocity, but at different angles.

137. When the shot is released at an angle of 35°, its path can be modeled by the formula

$$y = -0.01x^2 + 0.7x + 6.1,$$

in which x is the shot's horizontal distance, in feet, and y is its height, in feet. This formula is shown by one of the graphs, (a) or (b), in the figure. Use the formula to determine the shot's maximum distance. Use a calculator and round to the nearest tenth of a foot. Which graph, (a) or (b), shows the shot's path? 77.8 ft; (b)

Horizontal Distance
[0, 80, 10] by [0, 40, 10]

138. (Refer to the preceding information and the graphs shown in Exercise 137.) When the shot is released at an angle of 65°, its path can be modeled by the formula

$$y = -0.04x^2 + 2.1x + 6.1,$$

in which x is the shot's horizontal distance, in feet, and y is its height, in feet. This formula is shown by one of the graphs, (a) or (b), in the figure in Exercise 137. Use the formula to determine the shot's maximum distance. Use a calculator and round to the nearest tenth of a foot. Which graph, (a) or (b), shows the shot's path? 55.3 ft; (a)

Use the Pythagorean Theorem and the square root property to solve Exercises 139–143. Express answers in simplified radical form. Then find a decimal approximation to the nearest tenth.

139. A rectangular park is 4 miles long and 2 miles wide. How long is a pedestrian route that runs diagonally across the park? $2\sqrt{5}$ mi, or about 4.5 mi

140. A rectangular park is 6 miles long and 3 miles wide. How long is a pedestrian route that runs diagonally across the park? $3\sqrt{5}$ mi, or about 6.7 mi

141. The base of a 30-foot ladder is 10 feet from a building. If the ladder reaches the flat roof, how tall is the building? *

142. A baseball diamond is actually a square with 90-foot sides. What is the distance from home plate to second base? *

143. a. A wheelchair ramp with a length of 122 inches has a horizontal distance of 120 inches. What is the ramp's vertical distance? 22 in.

b. Construction laws are very specific when it comes to access ramps for the disabled. Every vertical rise of 1 inch requires a horizontal run of 12 inches. Does this ramp satisfy the requirements? no

144. An **isosceles right triangle** has legs that are the same length and acute angles each measuring 45°.

a. Write an expression in terms of a that represents the length of the hypotenuse. $\sqrt{2}a$ or $a\sqrt{2}$

b. Use your result from part (a) to write a sentence that describes the length of the hypotenuse of an isosceles right triangle in terms of the length of a leg. *

145. The length of a rectangular sign is 3 feet longer than the width. If the sign's area is 54 square feet, find its length and width. length: 9 ft; width: 6 ft

146. A rectangular parking lot has a length that is 3 yards greater than the width. The area of the parking lot is 180 square yards. Find the length and the width. 15 yd by 12 yd

147. Each side of a square is lengthened by 3 inches. The area of this new, larger square is 64 square inches. Find the length of a side of the original square. 5 in.

148. Each side of a square is lengthened by 2 inches. The area of this new, larger square is 36 square inches. Find the length of a side of the original square. 4 in.

149. A pool measuring 10 meters by 20 meters is surrounded by a path of uniform width, as shown in the figure below. If the area of the pool and the path combined is 600 square meters, what is the width of the path? 5 m

150. A vacant rectangular lot is being turned into a community vegetable garden measuring 15 meters by 12 meters. A path of uniform width is to surround the garden, as shown in the figure. If the area of the garden and path combined is 378 square meters, find the width of the path. 3 m

151. A machine produces open boxes using square sheets of metal. The figure illustrates that the machine cuts equal-sized squares measuring 2 inches on a side from the corners and then shapes the metal into an open box by turning up the sides. If each box must have a volume of 200 cubic inches, find the length and width of the open box. 10 in.

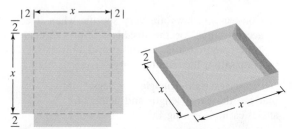

152. A machine produces open boxes using square sheets of metal. The machine cuts equal-sized squares measuring 3 inches on a side from the corners and then shapes the metal into an open box by turning up the sides. If each box must have a volume of 75 cubic inches, find the length and width of the open box. 5 in.

153. A rain gutter is made from sheets of aluminum that are 20 inches wide. As shown in the figure, the edges are turned up to form right angles. Determine the depth of the gutter that will allow a cross-sectional area of 13 square inches. Show that there are two different solutions to the problem. Round to the nearest tenth of an inch. 9.3 in. and 0.7 in.

Flat sheet
20 inches
wide

x $20 - 2x$ x

154. A piece of wire is 8 inches long. The wire is cut into two pieces and then each piece is bent into a square. Find the length of each piece if the sum of the areas of these squares is to be 2 square inches. 4 in. and 4 in.

$\dfrac{x}{4}$ $\dfrac{x}{4}$

$\dfrac{8-x}{4}$ $\dfrac{8-x}{4}$

8 inches

x $8 - x$

cut

Explaining the Concepts 155–163. Answers will vary.

155. What is a quadratic equation?

156. Explain how to solve $x^2 + 6x + 8 = 0$ using factoring and the zero-product principle.

157. Explain how to solve $x^2 + 6x + 8 = 0$ by completing the square.

158. Explain how to solve $x^2 + 6x + 8 = 0$ using the quadratic formula.

159. How is the quadratic formula derived?

160. What is the discriminant and what information does it provide about a quadratic equation?

161. If you are given a quadratic equation, how do you determine which method to use to solve it?

162. Describe the relationship between the real solutions of $ax^2 + bx + c = 0$ and the graph of $y = ax^2 + bx + c$.

163. If a quadratic equation has imaginary solutions, how is this shown on the graph of $y = ax^2 + bx + c$?

Technology Exercises 164–165. Answers will vary.

164. Use a graphing utility and x-intercepts to verify any of the real solutions that you obtained for three of the quadratic equations in Exercises 65–74.

165. Use a graphing utility to graph $y = ax^2 + bx + c$ related to any five of the quadratic equations, $ax^2 + bx + c = 0$, in Exercises 75–82. How does each graph illustrate what you determined algebraically using the discriminant?

Critical Thinking Exercises

Make Sense? *In Exercises 166–169, determine whether each statement makes sense or does not make sense, and explain your reasoning.*

166. Because I want to solve $25x^2 - 169 = 0$ fairly quickly, I'll use the quadratic formula. does not make sense

167. I'm looking at a graph with one x-intercept, so it must be the graph of a linear equation. does not make sense

168. I obtained -17 for the discriminant, so there are two imaginary irrational solutions. does not make sense

169. When I use the square root property to determine the length of a right triangle's side, I don't even bother to list the negative square root. makes sense

In Exercises 170–173, determine whether each statement is true or false. If the statement is false, make the necessary change(s) to produce a true statement. Changes to false statements will vary.

170. The equation $(2x - 3)^2 = 25$ is equivalent to $2x - 3 = 5$. *

171. Any quadratic equation that can be solved by completing the square can be solved by the quadratic formula. true

172. The quadratic formula is developed by applying factoring and the zero-product principle to the quadratic equation $ax^2 + bx + c = 0$. false

173. In using the quadratic formula to solve the quadratic equation $5x^2 = 2x - 7$, we have $a = 5$, $b = 2$, and $c = -7$. false

174. Write a quadratic equation in general form whose solution set is $\{-3, 5\}$. $x^2 - 2x - 15 = 0$

175. Solve for t: $s = -16t^2 + v_0t$. $t = \dfrac{v_0 \pm \sqrt{v_0^2 - 64s}}{32}$

176. A rectangular swimming pool is 12 meters long and 8 meters wide. A tile border of uniform width is to be built around the pool using 120 square meters of tile. The tile is from a discontinued stock (so no additional materials are available) and all 120 square meters are to be used. How wide should the border be? Round to the nearest tenth of a meter. If zoning laws require at least a 2-meter-wide border around the pool, can this be done with the available tile? 2.4 m; yes

Preview Exercises

Exercises 177–179 will help you prepare for the material covered in the next section.

177. Factor completely: $x^3 + x^2 - 4x - 4$. $(x + 1)(x + 2)(x - 2)$

178. Use the special product $(A + B)^2 = A^2 + 2AB + B^2$ to multiply: $\left(\sqrt{x + 4} + 1\right)^2$. $x + 5 + 2\sqrt{x + 4}$

179. If -8 is substituted for x in the equation $5x^{\frac{2}{3}} + 11x^{\frac{1}{3}} + 2 = 0$, is the resulting statement true or false? true

CHAPTER 1 Mid-Chapter Check Point

WHAT YOU KNOW: We used the rectangular coordinate system to represent ordered pairs of real numbers and to graph equations in two variables. We saw that linear equations can be written in the form $ax + b = 0, a \neq 0$, and quadratic equations can be written in the general form $ax^2 + bx + c = 0, a \neq 0$. We solved linear equations. We saw that some equations have no solution, whereas others have all real numbers as solutions. We solved quadratic equations using factoring, the square root property, completing the square, and the quadratic formula. We saw that the discriminant of $ax^2 + bx + c = 0, b^2 - 4ac$, determines the number and type of solutions. We performed operations with complex numbers and used the imaginary unit i ($i = \sqrt{-1}$, where $i^2 = -1$) to represent solutions of quadratic equations with negative discriminants. Only real solutions correspond to x-intercepts. We also solved rational equations by multiplying both sides by the least common denominator and clearing fractions. We developed a strategy for solving a variety of applied problems, using equations to model verbal conditions.

In Exercises 1–12, solve each equation.

1. $-5 + 3(x + 5) = 2(3x - 4)$ *

2. $5x^2 - 2x = 7$ $\{-1, \frac{7}{5}\}$

3. $\dfrac{x - 3}{5} - 1 = \dfrac{x - 5}{4}$ $\{-7\}$

4. $3x^2 - 6x - 2 = 0$ *

5. $4x - 2(1 - x) = 3(2x + 1) - 5$ $\{x | x \text{ is a real number}\}$

6. $5x^2 + 1 = 37$ $\left\{\pm \frac{6\sqrt{5}}{5}\right\}$

7. $x(2x - 3) = -4$ *

8. $\dfrac{3x}{4} - \dfrac{x}{3} + 1 = \dfrac{4x}{5} - \dfrac{3}{20}$ $\{3\}$

9. $(x + 3)^2 = 24$ $\{-3 \pm 2\sqrt{6}\}$

10. $\dfrac{1}{x^2} - \dfrac{4}{x} + 1 = 0$ $\{2 \pm \sqrt{3}\}$

11. $3x + 1 - (x - 5) = 2x - 4$ \varnothing

12. $\dfrac{2x}{x^2 + 6x + 8} = \dfrac{x}{x + 4} - \dfrac{2}{x + 2}$ $\{4\}$

In Exercises 13–17, find the x-intercepts of the graph of each equation.

13. $y = x^2 + 6x + 2$ $-3 - \sqrt{7}$ and $-3 + \sqrt{7}$

14. $y = 4(x + 1) - 3x - (6 - x)$ 1

15. $y = 2x^2 + 26$ no x-intercepts

16. $y = \dfrac{x^2}{3} + \dfrac{x}{2} - \dfrac{2}{3}$ $\frac{-3 - \sqrt{41}}{4}$ and $\frac{-3 + \sqrt{41}}{4}$

17. $y = x^2 - 5x + 8$ no x-intercepts

In Exercises 18–19, find all values of x satisfying the given conditions.

18. $y_1 = 3(2x - 5) - 2(4x + 1), y_2 = -5(x + 3) - 2$, and $y_1 = y_2$. 0

19. $y_1 = 2x + 3, y_2 = x + 2$, and $y_1 y_2 = 10$. -4 and $\frac{1}{2}$

20. Solve by completing the square: $x^2 + 10x - 3 = 0$. *

In Exercises 21–22, without solving the equation, determine the number and type of solutions.

21. $2x^2 + 5x + 4 = 0$ *

22. $10x(x + 4) = 15x - 15$ *

In Exercises 23–25, graph each equation in a rectangular coordinate system.

23. $y = 2x - 1$ *

24. $y = 1 - |x|$ *

25. $y = x^2 + 2$ *

26. Solve for n: $L = a + (n - 1)d$. $n = \frac{L - a}{d} + 1$

27. Solve for l: $A = 2lw + 2lh + 2wh$. $l = \frac{A - 2wh}{2w + 2h}$

28. Solve for f_1: $f = \dfrac{f_1 f_2}{f_1 + f_2}$. $f_1 = -\frac{ff_2}{f - f_2}$ or $f_1 = \frac{ff_2}{f_2 - f}$

29. Although you want to choose a career that fits your interests and abilities, it is good to have an idea of what jobs pay when looking at career options. The bar graph shows the average yearly earnings of full-time employed college graduates with only a bachelor's degree based on their college major.

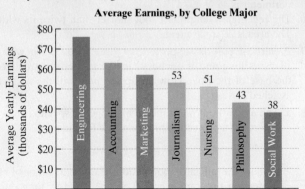

Average Earnings, by College Major

Source: Arthur J. Keown, *Personal Finance*, Pearson.

*See Answers to Selected Exercises.

The average yearly earnings of engineering majors exceeds the earnings of marketing majors by $19 thousand. The average yearly earnings of accounting majors exceeds the earnings of marketing majors by $6 thousand. Combined, the average yearly earnings for these three college majors is $196 thousand. Determine the average yearly earnings, in thousands of dollars, for each of these three college majors. engineering: $76 thousand; accounting: $63 thousand; marketing: $57 thousand

30. The line graph indicates that in 1960, 23% of U.S. taxes came from corporate income tax. For the period from 1960 through 2010, this percentage decreased by approximately 0.28 each year. If this trend continues, by which year will corporations pay zero taxes? Round to the nearest year. by 2042

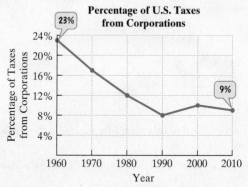

Percentage of U.S. Taxes from Corporations

Source: Office of Management and Budget

31. You invested $25,000 in two accounts paying 8% and 9% annual interest. At the end of the year, the total interest from these investments was $2135. How much was invested at each rate? $11,500 at 8%; $13,500 at 9%

32. The toll to a bridge costs $8. Commuters who use the bridge frequently have the option of purchasing a monthly discount pass for $45. With the discount pass, the toll is reduced to $5. For how many crossings per month will the monthly cost without the discount pass be the same as the monthly cost with the discount pass? 15 bridge crossings per month

33. After a 40% price reduction, you purchase a camcorder for $468. What was the camcorder's price before the reduction? $780

34. You invested $4000. On part of this investment, you earned 4% interest. On the remainder of the investment, you lost 3%. Combining earnings and losses, the annual income from these investments was $55. How much was invested at each rate? $2500 at 4%; $1500 at 3% loss

In Exercises 35–36, find the dimensions of each rectangle.

35. The rectangle's length exceeds twice its width by 5 feet. The perimeter is 46 feet. length: 17 ft; width: 6 ft

36. The rectangle's length is 1 foot shorter than twice its width. The area is 28 square feet. length: 7 ft; width: 4 ft

37. A vertical pole is supported by three wires. Each wire is 13 yards long and is anchored in the ground 5 yards from the base of the pole. How far up the pole will the wires be attached? 12 yd

38. **a.** The alligator, at one time an endangered species, was the subject of a protection program. The formula

$$P = -10x^2 + 475x + 3500$$

models the alligator population, P, after x years of the protection program, where $0 \leq x \leq 12$. How long did it take the population to reach 5990 alligators? 6 years

b. The graph of the formula modeling the alligator population is shown below. Identify your solution from part (a) as a point on the graph. (6, 5990)

Years the Program Is in Effect

39. A substantial percentage of the United States population is foreign-born. The bar graph shows the percentage of foreign-born Americans for selected years from 1920 through 2014.

Percentage of the United States Population That Was Foreign-Born, 1920–2014

Source: U.S. Census Bureau

The percentage, p, of the United States population that was foreign-born x years after 1920 can be modeled by the formula

$$p = 0.004x^2 - 0.35x + 13.9.$$

According to this model, in which year will 19% of the United States population be foreign-born? Round to the nearest year. about 100 yr after 1920, or in 2020

In Exercises 40–45, perform the indicated operations and write the result in standard form.

40. $(6 - 2i) - (7 - i)$ $-1 - i$

41. $3i(2 + i)$ $-3 + 6i$

42. $(1 + i)(4 - 3i)$ $7 + i$

43. $\dfrac{1 + i}{1 - i}$ i

44. $\sqrt{-75} - \sqrt{-12}$ $3i\sqrt{3}$

45. $\left(2 - \sqrt{-3}\right)^2$ $1 - 4i\sqrt{3}$

Section 1.6 | Other Types of Equations

What am I supposed to learn?

After studying this section, you should be able to:

1. Solve polynomial equations by factoring.
2. Solve radical equations.
3. Solve equations with rational exponents.
4. Solve equations that are quadratic in form.
5. Solve equations involving absolute value.
6. Solve problems modeled by equations.

"Bad television is three things: a bullet train to a morally bankrupt youth, a slow spiral into an intellectual void, and of course, a complete blast to watch."

—Dennis Miller

A blast, indeed. Americans spend more of their free time watching TV than any other activity.

What does algebra have to say about this? In particular, what sort of mathematical model can be used to describe the hours per week we watch TV, by annual income? In this

The Real Housewives of Orange County

section (Example 11), we answer the question after studying procedures for solving a variety of different types of equations.

Polynomial Equations

A **polynomial equation** is the result of setting two polynomials equal to each other. The equation is in **general form** if one side is 0 and the polynomial on the other side is in descending powers of the variable. The **degree of a polynomial equation** is the same as the highest degree of any term in the equation. Here are examples of three polynomial equations:

$$3x + 5 = 14 \qquad 2x^2 + 7x = 4 \qquad x^3 + x^2 = 4x + 4.$$

This equation is of degree 1 because 1 is the highest degree.

This equation is of degree 2 because 2 is the highest degree.

This equation is of degree 3 because 3 is the highest degree.

Notice that a polynomial equation of degree 1 is a linear equation. A polynomial equation of degree 2 is a quadratic equation.

Some polynomial equations of degree 3 or higher can be solved by moving all terms to one side, thereby obtaining 0 on the other side. Once the equation is in general form, factor and then set each factor equal to 0.

① Solve polynomial equations by factoring.

EXAMPLE 1 | Solving a Polynomial Equation by Factoring

Solve by factoring: $3x^4 = 27x^2$.

SOLUTION

Step 1 Move all nonzero terms to one side and obtain zero on the other side. Subtract $27x^2$ from both sides.

$$3x^4 = 27x^2 \qquad \text{This is the given equation.}$$
$$3x^4 - 27x^2 = 27x^2 - 27x^2 \qquad \text{Subtract } 27x^2 \text{ from both sides.}$$
$$3x^4 - 27x^2 = 0 \qquad \text{Simplify.}$$

Step 2 Factor. We can factor $3x^2$ from each term.

$$3x^4 - 27x^2 = 0$$
$$3x^2(x^2 - 9) = 0$$

GREAT QUESTION!

Can I solve $3x^4 = 27x^2$ by first dividing both sides by x^2?

No. If you divide both sides of the equation by x^2, you'll lose 0 as a solution. In general, do not divide both sides of an equation by a variable expression because that expression might take on the value 0 and you cannot divide by 0.

Steps 3 and 4 Set each factor equal to zero and solve the resulting equations.

$$3x^2 = 0 \qquad \text{or} \qquad x^2 - 9 = 0$$
$$x^2 = 0 \qquad\qquad\qquad x^2 = 9$$
$$x = \pm\sqrt{0} \qquad\qquad\quad x = \pm\sqrt{9}$$
$$x = 0 \qquad\qquad\qquad\quad x = \pm 3$$

Step 5 Check the solutions in the original equation. Check the three solutions, 0, -3, and 3, by substituting them into the original equation, $3x^4 = 27x^2$. Can you verify that the solution set is $\{-3, 0, 3\}$? •••

✓ **Check Point 1** Solve by factoring: $4x^4 = 12x^2$. $\{-\sqrt{3}, 0, \sqrt{3}\}$

A Brief Review • Factoring by Grouping

Factoring by grouping often involves factoring polynomials with four terms. Group terms with common factors. Factor out the greatest common factor from the grouped terms. Then factor out the common binomial factor. For more detail, see Section P.5, Objective 2.

EXAMPLE 2 Solving a Polynomial Equation by Factoring

Solve by factoring: $x^3 + x^2 = 4x + 4$.

SOLUTION

Step 1 Move all nonzero terms to one side and obtain zero on the other side. Subtract $4x$ and subtract 4 from both sides.

$$x^3 + x^2 = 4x + 4 \qquad\qquad \text{This is the given equation.}$$
$$x^3 + x^2 - 4x - 4 = 4x + 4 - 4x - 4 \quad \text{Subtract } 4x \text{ and } 4 \text{ from both sides.}$$
$$x^3 + x^2 - 4x - 4 = 0 \qquad\qquad \text{Simplify.}$$

Step 2 Factor. Use factoring by grouping. Group terms that have a common factor.

$$\boxed{x^3 + x^2} + \boxed{-4x - 4} = 0$$

Common factor is x^2. Common factor is -4.

$$x^2(x + 1) - 4(x + 1) = 0 \qquad \text{Factor } x^2 \text{ from the first two terms and } -4 \text{ from the last two terms.}$$

$$(x + 1)(x^2 - 4) = 0 \qquad \text{Factor out the common binomial factor, } x + 1, \text{ from each term.}$$

TECHNOLOGY

Numeric Connections

A graphing utility's $\boxed{\text{TABLE}}$ feature can be used to verify numerically that $\{-2, -1, 2\}$ is the solution set of

$$x^3 + x^2 = 4x + 4.$$

Enter Enter
$y_1 = x^3 + x^2$. $y_2 = 4x + 4$.

X	Y1	Y2	
-4	-48	-12	y_1 and y_2 are
-3	-18	-8	equal when
-2	-4	-4	$x = -2$,
-1	0	0	$x = -1$,
0	0	4	and $x = 2$.
1	2	8	
2	12	12	
3	36	16	
4	80	20	
5	150	24	
6	252	28	

X=-4

Steps 3 and 4 Set each factor equal to zero and solve the resulting equations.

$$x + 1 = 0 \qquad \text{or} \qquad x^2 - 4 = 0$$
$$x = -1 \qquad\qquad\qquad x^2 = 4$$
$$x = \pm\sqrt{4} = \pm 2$$

Step 5 Check the solutions in the original equation. Check the three solutions, -1, -2, and 2, by substituting them into $x^3 + x^2 = 4x + 4$, the original equation. Can you verify that the solution set is $\{-2, -1, 2\}$? •••

DISCOVERY

Suggest a method involving intersecting graphs that can be used with a graphing utility to verify that $\{-2, -1, 2\}$ is the solution set of

$$x^3 + x^2 = 4x + 4.$$

Apply this method to verify the solution set.

TECHNOLOGY

Graphic Connections

$[-5, 5, 1]$ by $[-8, 2, 1]$

You can use a graphing utility to check the solutions of $x^3 + x^2 - 4x - 4 = 0$. Graph $y = x^3 + x^2 - 4x - 4$, as shown on the left. The x-intercepts are $-2, -1$, and 2, corresponding to the equation's solutions.

 Check Point 2 Solve by factoring: $2x^3 + 3x^2 = 8x + 12.$ $\left\{-2, -\frac{3}{2}, 2\right\}$

② Solve radical equations.

Radical Equations

A **radical equation** is an equation in which the variable occurs in a square root, cube root, or any higher root. An example of a radical equation is

$$\sqrt{x} = 9.$$

We solve this equation by squaring both sides:

> Squaring both sides eliminates the square root.

$$\left(\sqrt{x}\right)^2 = 9^2$$
$$x = 81.$$

The proposed solution, 81, can be checked in the original equation, $\sqrt{x} = 9$. Because $\sqrt{81} = 9$, the solution is 81 and the solution set is $\{81\}$.

In general, we solve radical equations with square roots by squaring both sides of the equation. We solve radical equations with nth roots by raising both sides of the equation to the nth power. Unfortunately, if n is even, all the solutions of the equation raised to the even power may not be solutions of the original equation. Consider, for example, the equation

$$x = 4.$$

If we square both sides, we obtain

$$x^2 = 16.$$

Solving this equation using the square root property, we obtain

$$x = \pm\sqrt{16} = \pm 4.$$

The new equation $x^2 = 16$ has two solutions, -4 and 4. By contrast, only 4 is a solution of the original equation, $x = 4$. For this reason, **when raising both sides of an equation to an even power, always check proposed solutions in the original equation.**

Here is a general method for solving radical equations with nth roots:

> ### Solving Radical Equations Containing nth Roots
>
> 1. If necessary, arrange terms so that one radical is isolated on one side of the equation.
> 2. Raise both sides of the equation to the nth power to eliminate the isolated nth root.
> 3. Solve the resulting equation. If this equation still contains radicals, repeat steps 1 and 2.
> 4. Check all proposed solutions in the original equation.

Extra solutions may be introduced when you raise both sides of a radical equation to an even power. Such solutions, which are not solutions of the given equation, are called **extraneous solutions** or **extraneous roots**.

EXAMPLE 3 Solving a Radical Equation

Solve: $\sqrt{2x - 1} + 2 = x$.

SOLUTION

Step 1 Isolate a radical on one side. We isolate the radical, $\sqrt{2x - 1}$, by subtracting 2 from both sides.

$$\sqrt{2x - 1} + 2 = x \qquad \text{This is the given equation.}$$
$$\sqrt{2x - 1} = x - 2 \qquad \text{Subtract 2 from both sides.}$$

Step 2 Raise both sides to the nth power. Because n, the index of the radical in the equation $\sqrt{2x - 1} = x - 2$, is 2, we square both sides.

$$\left(\sqrt{2x - 1}\right)^2 = (x - 2)^2$$
$$2x - 1 = x^2 - 4x + 4 \qquad \begin{array}{l}\text{Simplify. Use the formula}\\ (A - B)^2 = A^2 - 2AB + B^2 \text{ on the}\\ \text{right side.}\end{array}$$

Step 3 Solve the resulting equation. Because of the x^2-term, the resulting equation is a quadratic equation. We can obtain 0 on the left side by subtracting $2x$ and adding 1 on both sides.

$$2x - 1 = x^2 - 4x + 4 \qquad \text{The resulting equation is quadratic.}$$
$$0 = x^2 - 6x + 5 \qquad \begin{array}{l}\text{Write in general form, subtracting } 2x\\ \text{and adding 1 on both sides.}\end{array}$$
$$0 = (x - 1)(x - 5) \qquad \text{Factor.}$$
$$x - 1 = 0 \quad \text{or} \quad x - 5 = 0 \qquad \text{Set each factor equal to 0.}$$
$$x = 1 \qquad\qquad x = 5 \qquad \text{Solve the resulting equations.}$$

Step 4 Check the proposed solutions in the original equation.

Check 1:	Check 5:
$\sqrt{2x - 1} + 2 = x$	$\sqrt{2x - 1} + 2 = x$
$\sqrt{2 \cdot 1 - 1} + 2 \overset{?}{=} 1$	$\sqrt{2 \cdot 5 - 1} + 2 \overset{?}{=} 5$
$\sqrt{1} + 2 \overset{?}{=} 1$	$\sqrt{9} + 2 \overset{?}{=} 5$
$1 + 2 \overset{?}{=} 1$	$3 + 2 \overset{?}{=} 5$
$3 = 1$, false	$5 = 5$, true

Thus, 1 is an extraneous solution. The only solution is 5, and the solution set is {5}. • • •

GREAT QUESTION!

Can I square the right side of $\sqrt{2x - 1} = x - 2$ by first squaring x and then squaring 2?

No. Be sure to square *both sides* of an equation. Do *not* square each term.

Correct:
$$\left(\sqrt{2x - 1}\right)^2 = (x - 2)^2$$

Incorrect!
$$\cancel{\left(\sqrt{2x - 1}\right)^2 = x^2 - 2^2}$$

GREAT QUESTION!

If there are two proposed solutions, will one of them always be extraneous?

No. If a radical equation leads to a quadratic equation with two solutions, there will not always be one solution that checks and one that does not. Both solutions of the quadratic equation can satisfy the original radical equation. It is also possible that neither solution of the quadratic equation satisfies the given radical equation.

TECHNOLOGY

Numeric Connections

A graphing utility's TABLE feature provides a numeric check that 1 is not a solution and 5 is a solution of $\sqrt{2x - 1} + 2 = x$.

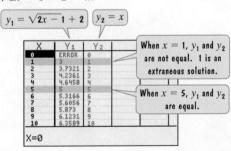

⌀ Check Point **3** Solve: $\sqrt{x + 3} + 3 = x.$ (6)

Solving a radical equation with two or more square root expressions involves isolating a radical, squaring both sides, and then repeating this process. Let's consider an equation containing two square root expressions.

EXAMPLE 4 Solving an Equation That Has Two Radicals

Solve: $\sqrt{3x + 1} - \sqrt{x + 4} = 1.$

SOLUTION

Step 1 Isolate a radical on one side. We can isolate the radical $\sqrt{3x + 1}$ by adding $\sqrt{x + 4}$ to both sides. We obtain

$$\sqrt{3x + 1} = \sqrt{x + 4} + 1.$$

Step 2 Square both sides.

$$\left(\sqrt{3x + 1}\right)^2 = \left(\sqrt{x + 4} + 1\right)^2$$

Squaring the expression on the right side of the equation can be a bit tricky. We have to use the formula

$$(A + B)^2 = A^2 + 2AB + B^2.$$

Focusing on just the right side, here is how the squaring is done:

$$(A + B)^2 \quad = \quad A^2 \; + \; 2 \; \cdot \; A \; \cdot \; B \; + \; B^2$$

$$(\sqrt{x + 4} + 1)^2 = (\sqrt{x + 4})^2 + 2 \cdot \sqrt{x + 4} \cdot 1 + 1^2 = x + 4 + 2\sqrt{x + 4} + 1.$$

Now let's return to squaring both sides.

$$\left(\sqrt{3x + 1}\right)^2 = \left(\sqrt{x + 4} + 1\right)^2 \qquad \text{Square both sides of the equation with an isolated radical.}$$

$$3x + 1 = x + 4 + 2\sqrt{x + 4} + 1 \qquad \left(\sqrt{3x + 1}\right)^2 = 3x + 1; \text{ square the right side using the formula for } (A + B)^2.$$

$$3x + 1 = x + 5 + 2\sqrt{x + 4} \qquad \text{Combine numerical terms on the right side: } 4 + 1 = 5.$$

Can you see that the resulting equation still contains a radical, namely, $\sqrt{x + 4}$? Thus, we need to repeat the first two steps.

Repeat Step 1 Isolate a radical on one side. We isolate $2\sqrt{x + 4}$, the radical term, by subtracting $x + 5$ from both sides. We obtain

$$3x + 1 = x + 5 + 2\sqrt{x + 4} \qquad \text{This is the equation from our last step.}$$

$$2x - 4 = 2\sqrt{x + 4}. \qquad \text{Subtract } x \text{ and subtract 5 from both sides.}$$

Although we can simplify the equation by dividing both sides by 2, this sort of simplification is not always helpful. Thus, we will work with the equation in this form.

Repeat Step 2 Square both sides.

> Be careful in squaring both sides. Use $(A - B)^2 = A^2 - 2AB + B^2$ to square the left side. Use $(AB)^2 = A^2B^2$ to square the right side.

$$(2x - 4)^2 = (2\sqrt{x + 4})^2 \qquad \text{Square both sides.}$$

$$4x^2 - 16x + 16 = 4(x + 4) \qquad \text{Square both 2 and } \sqrt{x + 4} \text{ on the right side.}$$

Step 3 Solve the resulting equation. We solve the quadratic equation $4x^2 - 16x + 16 = 4(x + 4)$ by writing it in general form.

$$4x^2 - 16x + 16 = 4x + 16 \qquad \text{Use the distributive property.}$$
$$4x^2 - 20x = 0 \qquad \text{Subtract } 4x + 16 \text{ from both sides.}$$
$$4x(x - 5) = 0 \qquad \text{Factor.}$$
$$4x = 0 \quad \text{or} \quad x - 5 = 0 \qquad \text{Set each factor equal to zero.}$$
$$x = 0 \qquad\qquad x = 5 \qquad \text{Solve for } x.$$

Step 4 Check the proposed solutions in the original equation.

<div style="display:flex">

Check 0:

$$\sqrt{3x + 1} - \sqrt{x + 4} = 1$$
$$\sqrt{3 \cdot 0 + 1} - \sqrt{0 + 4} \stackrel{?}{=} 1$$
$$\sqrt{1} - \sqrt{4} \stackrel{?}{=} 1$$
$$1 - 2 \stackrel{?}{=} 1$$
$$-1 = 1, \quad \text{false}$$

Check 5:

$$\sqrt{3x + 1} - \sqrt{x + 4} = 1$$
$$\sqrt{3 \cdot 5 + 1} - \sqrt{5 + 4} \stackrel{?}{=} 1$$
$$\sqrt{16} - \sqrt{9} \stackrel{?}{=} 1$$
$$4 - 3 \stackrel{?}{=} 1$$
$$1 = 1, \quad \text{true}$$

</div>

The check indicates that 0 is not a solution. It is an extraneous solution brought about by squaring each side of the equation. The only solution is 5 and the solution set is {5}. • • •

TECHNOLOGY

Graphic Connections

The graph of

$$y = \sqrt{3x + 1} - \sqrt{x + 4} - 1$$

has only one x-intercept at 5. This verifies that the solution set of $\sqrt{3x + 1} - \sqrt{x + 4} = 1$ is {5}.

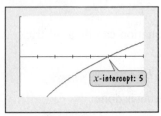

x-intercept: 5

$[0, 7, 1]$ by $[-1, 1, 1]$

GREAT QUESTION!

In Example 4, do I have to use the formula for $(A + B)^2$ in order to square $\sqrt{x + 4} + 1$?

No. You can also use the FOIL method.

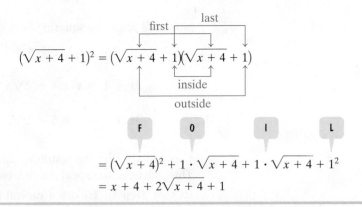

$$= (\sqrt{x + 4})^2 + 1 \cdot \sqrt{x + 4} + 1 \cdot \sqrt{x + 4} + 1^2$$
$$= x + 4 + 2\sqrt{x + 4} + 1$$

✓ **Check Point 4** Solve: $\sqrt{x + 5} - \sqrt{x - 3} = 2$. [4]

③ Solve equations with rational exponents.

Equations with Rational Exponents

We know that expressions with rational exponents represent radicals:

$$a^{\frac{m}{n}} = \left(\sqrt[n]{a}\right)^m = \sqrt[n]{a^m}.$$

For example, the radical equation $3\sqrt[4]{x^3} - 6 = 0$ can be expressed as $3x^{\frac{3}{4}} - 6 = 0$. A radical equation with a rational exponent can be solved by

1. isolating the expression with the rational exponent, and

2. raising both sides of the equation to a power that is the reciprocal of the rational exponent.

Solving Radical Equations of the Form $x^{\frac{m}{n}} = k$

Assume that m and n are positive integers, $\frac{m}{n}$ is in lowest terms, and k is a real number.

1. Isolate the expression with the rational exponent.
2. Raise both sides of the equation to the $\frac{n}{m}$ power.

If m is even:	**If m is odd:**
$x^{\frac{m}{n}} = k$	$x^{\frac{m}{n}} = k$
$\left(x^{\frac{m}{n}}\right)^{\frac{n}{m}} = \pm k^{\frac{n}{m}}$	$\left(x^{\frac{m}{n}}\right)^{\frac{n}{m}} = k^{\frac{n}{m}}$
$x = \pm k^{\frac{n}{m}}$	$x = k^{\frac{n}{m}}$

It is incorrect to insert the \pm symbol when the numerator of the exponent is odd. An odd index has only one root.

3. Check all proposed solutions in the original equation to find out if they are actual solutions or extraneous solutions.

EXAMPLE 5 Solving Equations Involving Rational Exponents

Solve:

a. $3x^{\frac{3}{4}} - 6 = 0$ **b.** $x^{\frac{2}{3}} - \frac{3}{4} = -\frac{1}{2}$.

SOLUTION

a. To solve $3x^{\frac{3}{4}} - 6 = 0$, our goal is to isolate $x^{\frac{3}{4}}$. Then we raise both sides of the equation to the $\frac{4}{3}$ power because $\frac{4}{3}$ is the reciprocal of $\frac{3}{4}$.

$3x^{\frac{3}{4}} - 6 = 0$ This is the given equation; we will isolate $x^{\frac{3}{4}}$.

$3x^{\frac{3}{4}} = 6$ Add 6 to both sides.

$\dfrac{3x^{\frac{3}{4}}}{3} = \dfrac{6}{3}$ Divide both sides by 3.

$x^{\frac{3}{4}} = 2$ Simplify.

$\left(x^{\frac{3}{4}}\right)^{\frac{4}{3}} = 2^{\frac{4}{3}}$ Raise both sides to the $\frac{4}{3}$ power. Because $\frac{m}{n} = \frac{3}{4}$ and m is odd, we do not use the \pm symbol.

$x = 2^{\frac{4}{3}}$ Simplify the left side: $\left(x^{\frac{3}{4}}\right)^{\frac{4}{3}} = x^{\frac{3 \cdot 4}{4 \cdot 3}} = x^{\frac{12}{12}} = x^1 = x.$

The proposed solution is $2^{\frac{4}{3}}$. Complete the solution process by checking this value in the given equation.

$3x^{\frac{3}{4}} - 6 = 0$ This is the original equation.

$3\left(2^{\frac{4}{3}}\right)^{\frac{3}{4}} - 6 \overset{?}{=} 0$ Substitute the proposed solution.

$3 \cdot 2 - 6 \overset{?}{=} 0$ $\left(2^{\frac{4}{3}}\right)^{\frac{3}{4}} = 2^{\frac{4 \cdot 3}{3 \cdot 4}} = 2^{\frac{12}{12}} = 2^1 = 2$

$0 = 0,$ true $3 \cdot 2 - 6 = 6 - 6 = 0$

The solution is $2^{\frac{4}{3}} = \sqrt[3]{2^4} \approx 2.52$. The solution set is $\left\{2^{\frac{4}{3}}\right\}$.

GREAT QUESTION!

Why didn't you eliminate fractions in Example 5(b) by first multiplying both sides by the LCD, 4?

Our goal is to write the equation in the form $x^{\frac{m}{n}} = k$, where the coefficient of the variable term is 1. Multiplying by 4 results in a coefficient of 4 rather than 1.

b. To solve $x^{\frac{2}{3}} - \frac{3}{4} = -\frac{1}{2}$, our goal is to isolate $x^{\frac{2}{3}}$. Then we raise both sides of the equation to the $\frac{3}{2}$ power because $\frac{3}{2}$ is the reciprocal of $\frac{2}{3}$.

$$x^{\frac{2}{3}} - \frac{3}{4} = -\frac{1}{2} \qquad \text{This is the given equation.}$$

$$x^{\frac{2}{3}} = \frac{1}{4} \qquad \text{Add } \tfrac{3}{4} \text{ to both sides: } \tfrac{3}{4} - \tfrac{1}{2} = \tfrac{3}{4} - \tfrac{2}{4} = \tfrac{1}{4}.$$

$$\left(x^{\frac{2}{3}}\right)^{\frac{3}{2}} = \pm\left(\frac{1}{4}\right)^{\frac{3}{2}} \qquad \begin{array}{l}\text{Raise both sides to the } \tfrac{3}{2} \text{ power. Because } \tfrac{m}{n} = \tfrac{2}{3} \\ \text{and } m \text{ is even, the } \pm \text{ symbol is necessary.}\end{array}$$

$$x = \pm\frac{1}{8} \qquad \left(\tfrac{1}{4}\right)^{\frac{3}{2}} = \left(\sqrt{\tfrac{1}{4}}\right)^3 = \left(\tfrac{1}{2}\right)^3 = \tfrac{1}{8}$$

Take a moment to verify that the solution set is $\left\{-\frac{1}{8}, \frac{1}{8}\right\}$. ● ● ●

⊘ Check Point 5 Solve:

a. $5x^{\frac{3}{2}} - 25 = 0$ $\{\sqrt[3]{25}\}$ or $\{5^{2/3}\}$ **b.** $x^{\frac{2}{3}} - 8 = -4$. $\{-8, 8\}$

 Solve equations that are quadratic in form.

GREAT QUESTION!

Do I have to use the letter u in the substitution?

No. Although we selected u, any letter can be used.

Equations That Are Quadratic in Form

Some equations that are not quadratic can be written as quadratic equations using an appropriate substitution. Here are some examples:

Given Equation	Substitution	New Equation
$x^4 - 10x^2 + 9 = 0$ or $(x^2)^2 - 10x^2 + 9 = 0$	$u = x^2$	$u^2 - 10u + 9 = 0$
$5x^{\frac{2}{3}} + 11x^{\frac{1}{3}} + 2 = 0$ or $5\left(x^{\frac{1}{3}}\right)^2 + 11x^{\frac{1}{3}} + 2 = 0$	$u = x^{\frac{1}{3}}$	$5u^2 + 11u + 2 = 0$

An equation that is **quadratic in form** is one that can be expressed as a quadratic equation using an appropriate substitution. Both of the preceding given equations are quadratic in form.

Equations that are quadratic in form contain an expression to a power, the same expression to that power squared, and a constant term. By letting u equal the expression to the power, a quadratic equation in u will result. Now it's easy. Solve this quadratic equation for u. Finally, use your substitution to find the values for the variable in the given equation. Example 6 shows how this is done.

EXAMPLE 6 Solving an Equation Quadratic in Form

Solve: $x^4 - 8x^2 - 9 = 0$.

SOLUTION

Notice that the equation contains an expression to a power, x^2, the same expression to that power squared, x^4 or $(x^2)^2$, and a constant term, -9. We let u equal the expression to the power. Thus,

$$\text{let } u = x^2.$$

TECHNOLOGY

Graphic Connections

The graph of

$$y = x^4 - 8x^2 - 9$$

has x-intercepts at -3 and 3. This verifies that the real solutions of

$$x^4 - 8x^2 - 9 = 0$$

are -3 and 3. The imaginary solutions, $-i$ and i, are not shown as intercepts.

$[-5, 5, 1]$ by $[-25, 10, 5]$

Now we write the given equation as a quadratic equation in u and solve for u.

$x^4 - 8x^2 - 9 = 0$	This is the given equation.
$(x^2)^2 - 8x^2 - 9 = 0$	The given equation contains x^2 and x^2 squared.
$u^2 - 8u - 9 = 0$	Let $u = x^2$. Replace x^2 with u.
$(u - 9)(u + 1) = 0$	Factor.
$u - 9 = 0$ or $u + 1 = 0$	Apply the zero-product principle.
$u = 9$ \qquad $u = -1$	Solve for u.

We're not done! Why not? We were asked to solve for x and we have values for u. We use the original substitution, $u = x^2$, to solve for x. Replace u with x^2 in each equation shown, namely $u = 9$ and $u = -1$.

$u = 9$	$u = -1$	These are the equations in u.
$x^2 = 9$	$x^2 = -1$	Replace u with x^2.
$x = \pm\sqrt{9}$	$x = \pm\sqrt{-1}$	Apply the square root property.
$x = \pm 3$	$x = \pm i$	Simplify.

The solution set is $\{-3, 3, -i, i\}$. The graph in the technology box shows that only the real solutions, -3 and 3, appear as x-intercepts. $\bullet\bullet\bullet$

✓ **Check Point 6** Solve: $x^4 - 5x^2 + 6 = 0$. $\{-\sqrt{3}, -\sqrt{2}, \sqrt{2}, \sqrt{3}\}$

EXAMPLE 7 Solving an Equation Quadratic in Form

Solve: $5x^{\frac{2}{3}} + 11x^{\frac{1}{3}} + 2 = 0$.

SOLUTION

Notice that the equation contains an expression to a power, $x^{\frac{1}{3}}$, the same expression to that power squared, $x^{\frac{2}{3}}$ or $\left(x^{\frac{1}{3}}\right)^2$, and a constant term, 2. We let u equal the expression to the power. Thus,

$$\text{let } u = x^{\frac{1}{3}}.$$

Now we write the given equation as a quadratic equation in u and solve for u.

$5x^{\frac{2}{3}} + 11x^{\frac{1}{3}} + 2 = 0$	This is the given equation.
$5\left(x^{\frac{1}{3}}\right)^2 + 11x^{\frac{1}{3}} + 2 = 0$	The given equation contains $x^{\frac{1}{3}}$ and $x^{\frac{1}{3}}$ squared.
$5u^2 + 11u + 2 = 0$	Let $u = x^{\frac{1}{3}}$.
$(5u + 1)(u + 2) = 0$	Factor.
$5u + 1 = 0$ or $u + 2 = 0$	Set each factor equal to 0.
$5u = -1$ \qquad $u = -2$	Solve for u.
$u = -\dfrac{1}{5}$	

Use the original substitution, $u = x^{\frac{1}{3}}$, to solve for x. Replace u with $x^{\frac{1}{3}}$ in each of the preceding equations, namely $u = -\frac{1}{5}$ and $u = -2$.

$$u = -\frac{1}{5} \qquad u = -2 \qquad \text{These are the equations in } u.$$

$$x^{\frac{1}{3}} = -\frac{1}{5} \qquad x^{\frac{1}{3}} = -2 \qquad \text{Replace } u \text{ with } x^{\frac{1}{3}}.$$

$$\left(x^{\frac{1}{3}}\right)^3 = \left(-\frac{1}{5}\right)^3 \qquad \left(x^{\frac{1}{3}}\right)^3 = (-2)^3 \qquad \text{Solve for } x \text{ by cubing both sides of each equation.}$$

$$x = -\frac{1}{125} \qquad x = -8 \qquad \text{Simplify.}$$

We did not raise both sides of an equation to an even power. A check will show that both -8 and $-\frac{1}{125}$ are solutions of the original equation. The solution set is $\left\{-8, -\frac{1}{125}\right\}$. • • •

✓ Check Point **7** Solve: $3x^{\frac{2}{3}} - 11x^{\frac{1}{3}} - 4 = 0.$ $\left\{-\frac{1}{27}, 64\right\}$

EXAMPLE 8 Solving an Equation Quadratic in Form

Solve: $(x^2 - 5)^2 + 3(x^2 - 5) - 10 = 0.$

SOLUTION

This equation contains $x^2 - 5$ and $x^2 - 5$ squared. We let $u = x^2 - 5.$

$$(x^2 - 5)^2 + 3(x^2 - 5) - 10 = 0 \qquad \text{This is the given equation.}$$

$$u^2 + 3u - 10 = 0 \qquad \text{Let } u = x^2 - 5.$$

$$(u + 5)(u - 2) = 0 \qquad \text{Factor.}$$

$$u + 5 = 0 \quad \text{or} \quad u - 2 = 0 \qquad \text{Set each factor equal to zero.}$$

$$u = -5 \qquad u = 2 \qquad \text{Solve for } u.$$

Use the original substitution, $u = x^2 - 5$, to solve for x. Replace u with $x^2 - 5$ in each of the preceding equations.

$$x^2 - 5 = -5 \quad \text{or} \quad x^2 - 5 = 2 \qquad \text{Replace } u \text{ with } x^2 - 5.$$

$$x^2 = 0 \qquad x^2 = 7 \qquad \text{Solve for } x \text{ by isolating } x^2.$$

$$x = 0 \qquad x = \pm\sqrt{7} \qquad \text{Apply the square root property.}$$

Although we did not raise both sides of an equation to an even power, checking the three proposed solutions in the original equation is a good idea. Do this now and verify that the solutions are $-\sqrt{7}, 0,$ and $\sqrt{7}$, and the solution set is $\left\{-\sqrt{7}, 0, \sqrt{7}\right\}.$ • • •

✓ Check Point **8** Solve: $(x^2 - 4)^2 + (x^2 - 4) - 6 = 0.$ $\left\{-\sqrt{6}, -1, 1, \sqrt{6}\right\}$

DISCOVERY

Solve Example 8 by first simplifying the given equation's left side. Then factor out x^2 and solve the resulting equation. Do you get the same solutions? Which method, substitution or first simplifying, is faster?

⑤ Solve equations involving absolute value.

$|-2| = 2$

$|2| = 2$

FIGURE 1.27

Equations Involving Absolute Value

We have seen that the absolute value of x, denoted $|x|$, describes the distance of x from zero on a number line. Now consider an **absolute value equation**, such as

$$|x| = 2.$$

This means that we must determine real numbers whose distance from the origin on a number line is 2. **Figure 1.27** shows that there are two numbers such that $|x| = 2$, namely, 2 and -2. We write $x = 2$ or $x = -2$. This observation can be generalized as follows:

Rewriting an Absolute Value Equation without Absolute Value Bars

If c is a positive real number and u represents any algebraic expression, then $|u| = c$ is equivalent to $u = c$ or $u = -c.$

Graphic Connections

You can use a graphing utility to verify the solution of an absolute value equation. Consider, for example,

$$|2x - 3| = 11.$$

Graph $y_1 = |2x - 3|$ and $y_2 = 11$. The graphs are shown in a $[-10, 10, 1]$ by $[-1, 15, 1]$ viewing rectangle. The x-coordinates of the intersection points are -4 and 7, verifying that $\{-4, 7\}$ is the solution set.

$[-10, 10, 1]$ by $[-1, 15, 1]$

EXAMPLE 9 Solving an Equation Involving Absolute Value

Solve: $|2x - 3| = 11$.

SOLUTION

$\|2x - 3\| = 11$	This is the given equation.
$2x - 3 = 11$ or $2x - 3 = -11$	Rewrite the equation without absolute value bars: $\|u\| = c$ is equivalent to $u = c$ or $u = -c$.
$2x = 14$ $\qquad\qquad$ $2x = -8$	Add 3 to both sides of each equation.
$x = 7$ $\qquad\qquad$ $x = -4$	Divide both sides of each equation by 2.

Check 7:	**Check −4:**	
$\|2x - 3\| = 11$	$\|2x - 3\| = 11$	This is the original equation.
$\|2(7) - 3\| \overset{?}{=} 11$	$\|2(-4) - 3\| \overset{?}{=} 11$	Substitute the proposed solutions.
$\|14 - 3\| \overset{?}{=} 11$	$\|-8 - 3\| \overset{?}{=} 11$	Perform operations inside the absolute value bars.
$\|11\| \overset{?}{=} 11$	$\|-11\| \overset{?}{=} 11$	
$11 = 11,$ true	$11 = 11,$ true	These true statements indicate that 7 and −4 are solutions.

The solution set is $\{-4, 7\}$. $\bullet\bullet\bullet$

🖉 **Check Point 9** Solve: $|2x - 1| = 5$. $\{-2, 3\}$

EXAMPLE 10 Solving an Equation Involving Absolute Value

Solve: $5|1 - 4x| - 15 = 0$.

SOLUTION

$$5|1 - 4x| - 15 = 0 \qquad \text{This is the given equation.}$$

> We need to isolate $|1 - 4x|$, the absolute value expression.

$5\|1 - 4x\| = 15$	Add 15 to both sides.
$\|1 - 4x\| = 3$	Divide both sides by 5.
$1 - 4x = 3$ or $1 - 4x = -3$	Rewrite $\|u\| = c$ as $u = c$ or $u = -c$.
$-4x = 2$ $\qquad\qquad$ $-4x = -4$	Subtract 1 from both sides of each equation.
$x = -\dfrac{1}{2}$ $\qquad\quad$ $x = 1$	Divide both sides of each equation by −4.

Take a moment to check $-\frac{1}{2}$ and 1, the proposed solutions, in the original equation, $5|1 - 4x| - 15 = 0$. In each case, you should obtain the true statement $0 = 0$. The solution set is $\left\{-\frac{1}{2}, 1\right\}$. These solutions appear as x-intercepts for the graph of $y = 5|1 - 4x| - 15$, as shown in **Figure 1.28**. $\bullet\bullet\bullet$

$[-2, 2, 1]$ by $[-16, 4, 1]$

FIGURE 1.28 The graph of $y = 5|1 - 4x| - 15$

🖉 **Check Point 10** Solve: $4|1 - 2x| - 20 = 0$. $\{-2, 3\}$

The absolute value of a number is never negative. Thus, if u is an algebraic expression and c is a negative number, then $|u| = c$ has no solution. For example,

the equation $|3x - 6| = -2$ has no solution because $|3x - 6|$ cannot be negative. The solution set is \varnothing, the empty set. The graph of $y = |3x - 6| + 2$, shown in **Figure 1.29**, lies above the x-axis and has no x-intercepts.

The absolute value of 0 is 0. Thus, if u is an algebraic expression and $|u| = 0$, the solution is found by solving $u = 0$. For example, the solution of $|x - 2| = 0$ is obtained by solving $x - 2 = 0$. The solution is 2 and the solution set is $\{2\}$.

$[-2, 6, 1]$ by $[-2, 15, 1]$

FIGURE 1.29 An absolute value equation whose graph has no x-intercepts

⑥ Solve problems modeled by equations.

Applications

EXAMPLE 11 Television Viewing, by Annual Income

The graph in **Figure 1.30** shows the average number of hours per week spent watching TV, by annual income. The formula

$$H = -2.3\sqrt{I} + 67.6$$

models weekly television viewing time, H, in hours, by annual income, I, in thousands of dollars. What annual income corresponds to 44.6 hours per week watching TV?

Television Viewing, by Annual Income

FIGURE 1.30
Source: Nielsen Media Research

SOLUTION

Because we are interested in finding the annual income that corresponds to 44.6 hours of weekly viewing time, we substitute 44.6 for H in the given formula. Then we solve for I, the annual income, in thousands of dollars.

$H = -2.3\sqrt{I} + 67.6$	This is the given formula.
$44.6 = -2.3\sqrt{I} + 67.6$	Substitute 44.6 for H.
$-23 = -2.3\sqrt{I}$	Subtract 67.6 from both sides.
$\dfrac{-23}{-2.3} = \dfrac{-2.3\sqrt{I}}{-2.3}$	Divide both sides by -2.3.
$10 = \sqrt{I}$	Simplify: $\dfrac{-23}{-2.3} = 10$ and $\dfrac{-2.3\sqrt{I}}{-2.3} = \sqrt{I}$.
$10^2 = \left(\sqrt{I}\right)^2$	Square both sides.
$100 = I$	Simplify.

The model indicates that an annual income of 100 thousand dollars, or $100,000, corresponds to 44.6 hours of weekly viewing time. ●●●

⊘ Check Point **11** Use the formula in Example 11 to find the annual income that corresponds to 33.1 hours per week watching TV. $225,000

ACHIEVING SUCCESS

Check out the *Learning Guide* that accompanies this textbook.

Benefits of using the *Learning Guide* include:

- It will help you become better organized. This includes organizing your class notes, assigned homework, quizzes, and tests.
- It will enable you to use your textbook more efficiently.
- It will help increase your study skills.
- It will help you prepare for the chapter tests.

Ask your professor about the availability of this textbook supplement.

CONCEPT AND VOCABULARY CHECK

Fill in each blank so that the resulting statement is true.

1. The first step in solving the polynomial equation
$$2x^3 + 3x^2 = 8x + 12$$
is to <u>subtract 8x and subtract 12 from both sides</u>.

2. An equation in which the variable occurs in a square root, cube root, or any higher root is called a/an <u>radical</u> equation.

3. Solutions of a squared equation that are not solutions of the original equation are called <u>extraneous</u> solutions.

4. Consider the equation
$$\sqrt{2x + 1} = x - 7.$$
Squaring the left side and simplifying results in <u>$2x + 1$</u>. Squaring the right side and simplifying results in <u>$x^2 - 14x + 49$</u>.

5. Consider the equation
$$\sqrt{x + 2} = 3 - \sqrt{x - 1}.$$
Squaring the left side and simplifying results in <u>$x + 2$</u>. Squaring the right side and simplifying results in <u>$x + 8 - 6\sqrt{x - 1}$</u>.

6. If $x^{\frac{3}{4}} = 5$, then $x = $ <u>$5^{\frac{4}{3}}$</u>.

7. If $x^{\frac{2}{3}} = 5$, then $x = $ <u>$\pm 5^{\frac{3}{2}}$</u>.

8. We solve $x^4 - 13x^2 + 36 = 0$ by letting $u = $ <u>x^2</u>. We then rewrite the equation in terms of u as <u>$u^2 - 13u + 36 = 0$</u>.

9. We solve $x^{\frac{2}{3}} + 2x^{\frac{1}{3}} - 3 = 0$ by letting $u = $ <u>$x^{\frac{1}{3}}$</u>. We then rewrite the equation in terms of u as <u>$u^2 + 2u - 3 = 0$</u>.

10. If $c > 0$, $|u| = c$ is equivalent to $u = $ <u>c</u> or $u = $ <u>$-c$</u>.

11. $|3x - 1| = 7$ is equivalent to <u>$3x - 1 = 7$</u> or <u>$3x - 1 = -7$</u>.

EXERCISE SET 1.6

Practice Exercises

Solve each polynomial equation in Exercises 1–10 by factoring and then using the zero-product principle.

1. $3x^4 - 48x^2 = 0$ {-4, 0, 4}
2. $5x^4 - 20x^2 = 0$ {-2, 0, 2}
3. $3x^3 + 2x^2 = 12x + 8$ *
4. $4x^3 - 12x^2 = 9x - 27$ *
5. $2x - 3 = 8x^3 - 12x^2$ *
6. $x + 1 = 9x^3 + 9x^2$ *
7. $4y^3 - 2 = y - 8y^2$ $\{-2, -\frac{1}{2}, \frac{1}{2}\}$
8. $9y^3 + 8 = 4y + 18y^2$ *
9. $2x^4 = 16x$ $\{0, 2, -1 \pm i\sqrt{3}\}$
10. $3x^4 = 81x$ *

Solve each radical equation in Exercises 11–30. Check all proposed solutions.

11. $\sqrt{3x + 18} = x$ {6}
12. $\sqrt{20 - 8x} = x$ {2}
13. $\sqrt{x + 3} = x - 3$ {6}
14. $\sqrt{x + 10} = x - 2$ {6}
15. $\sqrt{2x + 13} = x + 7$ {-6}
16. $\sqrt{6x + 1} = x - 1$ {8}
17. $x - \sqrt{2x + 5} = 5$ {10}
18. $x - \sqrt{x + 11} = 1$ {5}
19. $\sqrt{2x + 19} - 8 = x$ {-5}
20. $\sqrt{2x + 15} - 6 = x$ {-3}
21. $\sqrt{3x + 10} = x + 4$ {12}
22. $\sqrt{x - 3} = x - 9$ {9}
23. $\sqrt{x + 8} - \sqrt{x - 4} = 2$ *
24. $\sqrt{x + 5} - \sqrt{x - 3} = 2$ *
25. $\sqrt{x - 5} - \sqrt{x - 8} = 3$ ∅
26. $\sqrt{2x - 3} - \sqrt{x - 2} = 1$ {2, 6}
27. $\sqrt{2x + 3} + \sqrt{x - 2} = 2$ ∅
28. $\sqrt{x + 2} + \sqrt{3x + 7} = 1$ {-2}
29. $\sqrt{3\sqrt{x + 1}} = \sqrt{3x - 5}$ $\{\frac{13 + \sqrt{105}}{6}\}$
30. $\sqrt{1 + 4\sqrt{x}} = 1 + \sqrt{x}$ {0, 4}

Solve each equation with rational exponents in Exercises 31–40. Check all proposed solutions.

31. $x^{\frac{3}{2}} = 8$ {4}
32. $x^{\frac{3}{2}} = 27$ {9}
33. $(x - 4)^{\frac{3}{2}} = 27$ {13}
34. $(x + 5)^{\frac{3}{2}} = 8$ {-1}
35. $6x^{\frac{5}{2}} - 12 = 0$ $\{\sqrt[5]{4}\}$
36. $8x^{\frac{5}{3}} - 24 = 0$ $\{\sqrt[5]{27}\}$
37. $(x - 4)^{\frac{2}{3}} = 16$ {-60, 68}
38. $(x + 5)^{\frac{2}{3}} = 4$ {-13, 3}
39. $(x^2 - x - 4)^{\frac{3}{4}} - 2 = 6$ *
40. $(x^2 - 3x + 3)^{\frac{3}{2}} - 1 = 0$ *

Solve each equation in Exercises 41–60 by making an appropriate substitution.

41. $x^4 - 5x^2 + 4 = 0$ *
42. $x^4 - 13x^2 + 36 = 0$ *
43. $9x^4 = 25x^2 - 16$ $\{-\frac{4}{3}, -1, 1, \frac{4}{3}\}$
44. $4x^4 = 13x^2 - 9$ $\{-\frac{3}{2}, -1, 1, \frac{3}{2}\}$
45. $x - 13\sqrt{x} + 40 = 0$ {25, 64}
46. $2x - 7\sqrt{x} - 30 = 0$ {36}
47. $x^{-2} - x^{-1} - 20 = 0$ $\{-\frac{1}{4}, \frac{1}{5}\}$
48. $x^{-2} - x^{-1} - 6 = 0$ $\{-\frac{1}{2}, \frac{1}{3}\}$
49. $x^{\frac{2}{3}} - x^{\frac{1}{3}} - 6 = 0$ {-8, 27}
50. $2x^{\frac{2}{3}} + 7x^{\frac{1}{3}} - 15 = 0$ *
51. $x^{\frac{3}{2}} - 2x^{\frac{3}{4}} + 1 = 0$ {1}
52. $x^{\frac{2}{5}} + x^{\frac{1}{5}} - 6 = 0$ {-243, 32}
53. $2x - 3x^{\frac{1}{2}} + 1 = 0$ $\{\frac{1}{4}, 1\}$
54. $x + 3x^{\frac{1}{2}} - 4 = 0$ {1}
55. $(x - 5)^2 - 4(x - 5) - 21 = 0$ {2, 12}
56. $(x + 3)^2 + 7(x + 3) - 18 = 0$ {-12, -1}
57. $(x^2 - x)^2 - 14(x^2 - x) + 24 = 0$ {-3, -1, 2, 4}

*See Answers to Selected Exercises.

58. $(x^2 - 2x)^2 - 11(x^2 - 2x) + 24 = 0$ $\{-2, -1, 3, 4\}$

59. $\left(y - \dfrac{8}{y}\right)^2 + 5\left(y - \dfrac{8}{y}\right) - 14 = 0$ $\{-8, -2, 1, 4\}$

60. $\left(y - \dfrac{10}{y}\right)^2 + 6\left(y - \dfrac{10}{y}\right) - 27 = 0$ $\{-10, -2, 1, 5\}$

In Exercises 61–78, solve each absolute value equation or indicate that the equation has no solution.

61. $|x| = 8$ $\{-8, 8\}$

62. $|x| = 6$ $\{-6, 6\}$

63. $|x - 2| = 7$ $\{-5, 9\}$

64. $|x + 1| = 5$ $\{-6, 4\}$

65. $|2x - 1| = 5$ $\{-2, 3\}$

66. $|2x - 3| = 11$ $\{-4, 7\}$

67. $2|3x - 2| = 14$ $\left\{-\frac{5}{3}, 3\right\}$

68. $3|2x - 1| = 21$ $\{-3, 4\}$

69. $7|5x| + 2 = 16$ $\left\{-\frac{2}{5}, \frac{2}{5}\right\}$

70. $7|3x| + 2 = 16$ $\left\{-\frac{2}{3}, \frac{2}{3}\right\}$

71. $2\left|4 - \dfrac{5}{2}x\right| + 6 = 18$ $\left\{-\frac{4}{5}, 4\right\}$

72. $4\left|1 - \dfrac{3}{4}x\right| + 7 = 10$ $\left\{\frac{1}{3}, \frac{7}{3}\right\}$

73. $|x + 1| + 5 = 3$ \varnothing

74. $|x + 1| + 6 = 2$ \varnothing

75. $|2x - 1| + 3 = 3$ $\left\{\frac{1}{2}\right\}$

76. $|3x - 2| + 4 = 4$ $\left\{\frac{2}{3}\right\}$

Hint for Exercises 77–78: Absolute value expressions are equal when the expressions inside the absolute value bars are equal to or opposites of each other.

77. $|3x - 1| = |x + 5|$ $\{-1, 3\}$ **78.** $|2x - 7| = |x + 3|$ $\left\{\frac{4}{3}, 10\right\}$

In Exercises 79–84, find the x-intercepts of the graph of each equation. Then use the x-intercepts to match the equation with its graph. [The graphs are labeled (a) through (f).]

79. $y = \sqrt{x + 2} + \sqrt{x - 1} - 3$ 2; c

80. $y = \sqrt{x - 4} + \sqrt{x + 4} - 4$ 5; a

81. $y = x^{\frac{1}{3}} + 2x^{\frac{1}{6}} - 3$ 1; e

82. $y = x^{-2} - x^{-1} - 6$ $-\frac{1}{2}$ and $\frac{1}{3}$; b

83. $y = (x + 2)^2 - 9(x + 2) + 20$ 2 and 3; f

84. $y = 2(x + 2)^2 + 5(x + 2) - 3$ -5 and $-\frac{3}{2}$; d

a.

[-1, 10, 1] by [-3, 3, 1]

b.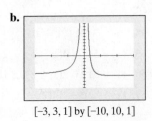

[-3, 3, 1] by [-10, 10, 1]

c.

[-1, 10, 1] by [-4, 4, 1]

d.

[-6, 3, 1] by [-10, 10, 1]

e.

[-1, 10, 1] by [-3, 3, 1]

f.

[-1, 6, 1] by [-1, 10, 1]

In Exercises 85–94, find all values of x satisfying the given conditions.

85. $y = |5 - 4x|$ and $y = 11.$ $-\frac{3}{2}$ and 4

86. $y = |2 - 3x|$ and $y = 13.$ $-\frac{11}{3}$ and 5

87. $y = x + \sqrt{x + 5}$ and $y = 7.$ 4

88. $y = x - \sqrt{x - 2}$ and $y = 4.$ 6

89. $y = 2x^3 + x^2 - 8x + 2$ and $y = 6.$ $-2, -\frac{1}{2}$, and 2

90. $y = x^3 + 4x^2 - x + 6$ and $y = 10.$ $-4, -1$, and 1

91. $y = (x + 4)^{\frac{3}{2}}$ and $y = 8.$ 0

92. $y = (x - 5)^{\frac{3}{2}}$ and $y = 125.$ 30

93. $y_1 = (x^2 - 1)^2$, $y_2 = 2(x^2 - 1)$, and y_1 exceeds y_2 by 3. *

94. $y_1 = 6\left(\dfrac{2x}{x - 3}\right)^2$, $y_2 = 5\left(\dfrac{2x}{x - 3}\right)$, and y_1 exceeds y_2 by 6. *

Practice Plus

In Exercises 95–98, solve each equation.

95. $|x^2 + 2x - 36| = 12$ $\{-8, -6, 4, 6\}$

96. $|x^2 + 6x + 1| = 8$ $\{-7, -3, 1\}$

97. $x(x + 1)^3 - 42(x + 1)^2 = 0$ $\{-7, -1, 6\}$

98. $x(x - 2)^3 - 35(x - 2)^2 = 0$ $\{-5, 2, 7\}$

99. If 5 times a number is decreased by 4, the principal square root of this difference is 2 less than the number. Find the number(s). 8

100. If a number is decreased by 3, the principal square root of this difference is 5 less than the number. Find the number(s). 7

101. Solve for V: $r = \sqrt{\dfrac{3V}{\pi h}}$. $V = \dfrac{\pi r^2 h}{3}$ or $V = \dfrac{1}{3}\pi r^2 h$

102. Solve for A: $r = \sqrt{\dfrac{A}{4\pi}}$. $A = 4\pi r^2$

In Exercises 103–104, list all numbers that must be excluded from the domain of each expression.

103. $\dfrac{|x - 1| - 3}{|x + 2| - 14}$ $-16, 12$ **104.** $\dfrac{x^3 - 2x^2 - 9x + 18}{x^3 + 3x^2 - x - 3}$ $-3, -1, 1$

Application Exercises

A basketball player's hang time is the time spent in the air when shooting a basket. The formula

$$t = \frac{\sqrt{d}}{2}$$

models hang time, t, in seconds, in terms of the vertical distance of a player's jump, d, in feet. Use this formula to solve Exercises 105–106.

105. When Michael Wilson of the Harlem Globetrotters slamdunked a basketball, his hang time for the shot was approximately 1.16 seconds. What was the vertical distance of his jump, rounded to the nearest tenth of a foot? 5.4 ft

106. If hang time for a shot by a professional basketball player is 0.85 second, what is the vertical distance of the jump, rounded to the nearest tenth of a foot? 2.9 ft

Use the graph of the formula for hang time to solve Exercises 107–108.

107. How is your answer to Exercise 105 shown on the graph? *

108. How is your answer to Exercise 106 shown on the graph? *

The graphs show the percentage of jobs in the U.S. labor force held by men and by women from 1970 through 2015. Exercises 109–110 are based on the data displayed by the graphs.

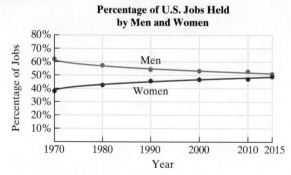

Source: Bureau of Labor Statistics

109. The formula

$$p = 1.6\sqrt{t} + 38$$

models the percentage of jobs in the U.S. labor force, *p*, held by women *t* years after 1970.

a. Use the appropriate graph at the bottom of the previous column to estimate the percentage of jobs in the U.S. labor force held by women in 2010. Give your estimation to the nearest percent. 47% ± 1%

b. Use the mathematical model to determine the percentage of jobs in the U.S. labor force held by women in 2010. Round to the nearest tenth of a percent. 48.1%

c. According to the formula, when will 51% of jobs in the U.S. labor force be held by women? Round to the nearest year. 2036

110. The formula

$$p = -1.6\sqrt{t} + 62$$

models the percentage of jobs in the U.S. labor force, *p*, held by men *t* years after 1970.

a. Use the appropriate graph at the bottom of the previous column to estimate the percentage of jobs in the U.S. labor force held by men in 2010. Give your estimation to the nearest percent. 53% ± 1%

b. Use the mathematical model to determine the percentage of jobs in the U.S. labor force held by men in 2010. Round to the nearest tenth of a percent. 51.9%

c. According to the formula, when will 49% of jobs in the U.S. labor force be held by men? Round to the nearest year. 2036

For each planet in our solar system, its year is the time it takes the planet to revolve once around the Sun. The formula

$$E = 0.2x^{\frac{3}{2}}$$

models the number of Earth days in a planet's year, E, where x is the average distance of the planet from the Sun, in millions of kilometers. Use the equation to solve Exercises 111–112.

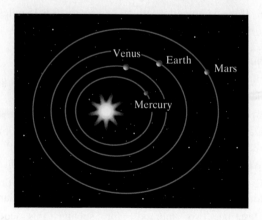

111. We, of course, have 365 Earth days in our year. What is the average distance of Earth from the Sun? Use a calculator and round to the nearest million kilometers. 149 million km

112. There are approximately 88 Earth days in the year of the planet Mercury. What is the average distance of Mercury from the Sun? Use a calculator and round to the nearest million kilometers. 58 million km

Use the Pythagorean Theorem to solve Exercises 113–114.

113. Two vertical poles of lengths 6 feet and 8 feet stand 10 feet apart. A cable reaches from the top of one pole to some point on the ground between the poles and then to the top of the other pole. Where should this point be located to use 18 feet of cable? either 1.2 ft or 7.5 ft from the base of the 6-foot pole

114. Towns *A* and *B* are located 6 miles and 3 miles, respectively, from a major expressway. The point on the expressway closest to town *A* is 12 miles from the point on the expressway closest to town *B*. Two new roads are to be built from *A* to the expressway and then to *B*.

a. Express the combined lengths of the new roads in terms of *x* as shown in the figure. $\sqrt{x^2 + 36} + \sqrt{(12 - x)^2 + 9}$

b. If the combined length of the new roads is 15 miles, what distance does *x* represent? 8 miles

Explaining the Concepts 115–121. Answers will vary.

115. Without actually solving the equation, give a general description of how to solve $x^3 - 5x^2 - x + 5 = 0$.

116. In solving $\sqrt{3x + 4} - \sqrt{2x + 4} = 2$, why is it a good idea to isolate a radical term? What if we don't do this and simply square each side? Describe what happens.

117. What is an extraneous solution to a radical equation?

118. Explain how to recognize an equation that is quadratic in form. Provide two original examples with your explanation.

119. Describe two methods for solving this equation: $x - 5\sqrt{x} + 4 = 0$.

120. Explain how to solve an equation involving absolute value.

121. Explain why the procedure that you explained in Exercise 120 does not apply to the equation $|x - 2| = -3$. What is the solution set for this equation?

Technology Exercises

In Exercises 122–124, use a graphing utility and the graph's x-intercepts to solve each equation. Check by direct substitution. A viewing rectangle is given.

122. $x^3 + 3x^2 - x - 3 = 0$ {−3, −1, 1}
$[-6, 6, 1]$ by $[-6, 6, 1]$

123. $-x^4 + 4x^3 - 4x^2 = 0$ {0, 2}
$[-6, 6, 1]$ by $[-9, 2, 1]$

124. $\sqrt{2x + 13} - x - 5 = 0$ {−2}
$[-5, 5, 1]$ by $[-5, 5, 1]$

Critical Thinking Exercises

Make Sense? *In Exercises 125–128, determine whether each statement makes sense or does not make sense, and explain your reasoning.*

125. When checking a radical equation's proposed solution, I can substitute into the original equation or any equation that is part of the solution process. does not make sense

126. After squaring both sides of a radical equation, the only solution that I obtained was extraneous, so ∅ must be the solution set of the original equation. makes sense

127. The equation $5x^{\frac{2}{3}} + 11x^{\frac{1}{3}} + 2 = 0$ is quadratic in form, but when I reverse the variable terms and obtain $11x^{\frac{1}{3}} + 5x^{\frac{2}{3}} + 2 = 0$, the resulting equation is no longer quadratic in form. does not make sense

128. When I solve an equation that is quadratic in form, it's important to write down the substitution that I am making. *

In Exercises 129–132, determine whether each statement is true or false. If the statement is false, make the necessary change(s) to produce a true statement. Changes to false statements will vary.

129. The first step in solving $\sqrt{x + 6} = x + 2$ is to square both sides, obtaining $x + 6 = x^2 + 4$. false

130. The equations $\sqrt{x + 4} = -5$ and $x + 4 = 25$ have the same solution set. false

131. To solve $x - 9\sqrt{x} + 14 = 0$, we let $\sqrt{u} = x$. false

132. The equation $|x| = -6$ is equivalent to $x = 6$ or $x = -6$. false

133. Solve: $\sqrt{6x - 2} = \sqrt{2x + 3} - \sqrt{4x - 1}$. {$\frac{1}{2}$}

134. Solve without squaring both sides:
$$5 - \frac{2}{x} = \sqrt{5 - \frac{2}{x}}.$$ {$\frac{2}{5}, \frac{1}{2}$}

135. Solve for *x*: $\sqrt[3]{x\sqrt{x}} = 9$. {81}

136. Solve for *x*: $x^{\frac{5}{6}} + x^{\frac{2}{3}} - 2x^{\frac{1}{2}} = 0$. {0, 1}

Preview Exercises

Exercises 137–139 will help you prepare for the material covered in the next section.

137. Is −1 a solution of $3 - 2x \leq 11$? yes

138. Solve: $-2x - 4 = x + 5$. {−3}

139. Solve: $\dfrac{x + 3}{4} = \dfrac{x - 2}{3} + \dfrac{1}{4}$. {14}

Section 1.7

Linear Inequalities and Absolute Value Inequalities

What am I supposed to learn?

After studying this section, you should be able to:

1. Use interval notation.
2. Find intersections and unions of intervals.
3. Solve linear inequalities.
4. Recognize inequalities with no solution or all real numbers as solutions.
5. Solve compound inequalities.
6. Solve absolute value inequalities.

Rent-a-Heap, a car rental company, charges $125 per week plus $0.20 per mile to rent one of their cars. Suppose you are limited by how much money you can spend for the week: You can spend at most $335. If we let x represent the number of miles you drive the heap in a week, we can write an inequality that models the given conditions:

The weekly charge of $125	plus	the charge of $0.20 per mile for x miles	must be less than or equal to	$335.
125	+	0.20x	\leq	335.

Placing an inequality symbol between a polynomial of degee 1 and a constant results in a *linear inequality in one variable.* In this section, we will study how to solve linear inequalities such as $125 + 0.20x \leq 335$. **Solving an inequality** is the process of finding the set of numbers that make the inequality a true statement. These numbers are called the **solutions** of the inequality and we say that they **satisfy** the inequality. The set of all solutions is called the **solution set** of the inequality. Set-builder notation and a new notation, called *interval notation*, are used to represent these solution sets. We begin this section by looking at interval notation.

① Use interval notation.

Interval Notation

Some sets of real numbers can be represented using **interval notation**. Suppose that a and b are two real numbers such that $a < b$.

Interval Notation	Graph
The **open interval** (a, b) represents the set of real numbers between, but not including, a and b. $(a, b) = \{x \mid a < x < b\}$ [x is greater than a $(a < x)$ and x is less than b $(x < b)$.]	The parentheses in the graph and in interval notation indicate that a and b, the endpoints, are excluded from the interval.

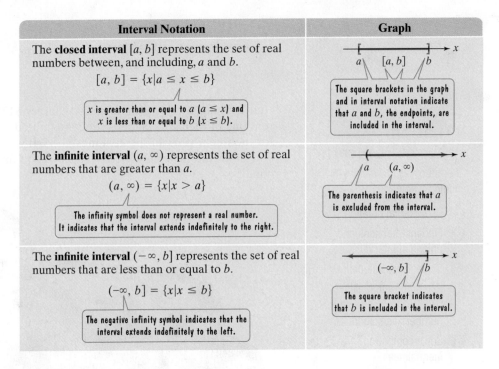

Parentheses and Brackets in Interval Notation

Parentheses indicate endpoints that are not included in an interval. Square brackets indicate endpoints that are included in an interval. Parentheses are always used with ∞ or $-\infty$.

Table 1.4 lists nine possible types of intervals used to describe subsets of real numbers.

Table 1.4 Intervals on the Real Number Line

Let a and b be real numbers such that $a < b$.		
Interval Notation	**Set-Builder Notation**	**Graph**
(a, b)	$\{x \mid a < x < b\}$	
$[a, b]$	$\{x \mid a \leq x \leq b\}$	
$[a, b)$	$\{x \mid a \leq x < b\}$	
$(a, b]$	$\{x \mid a < x \leq b\}$	
(a, ∞)	$\{x \mid x > a\}$	
$[a, \infty)$	$\{x \mid x \geq a\}$	
$(-\infty, b)$	$\{x \mid x < b\}$	
$(-\infty, b]$	$\{x \mid x \leq b\}$	
$(-\infty, \infty)$	$\{x \mid x \text{ is a real number}\}$ or \mathbb{R} (set of real numbers)	

EXAMPLE 1 Using Interval Notation

Express each interval in set-builder notation and graph:

 a. $(-1, 4]$ **b.** $[2.5, 4]$ **c.** $(-4, \infty)$.

SOLUTION

 a. $(-1, 4] = \{x \mid -1 < x \le 4\}$

 b. $[2.5, 4] = \{x \mid 2.5 \le x \le 4\}$

 c. $(-4, \infty) = \{x \mid x > -4\}$

 ● ● ●

✓ **Check Point 1** Express each interval in set-builder notation and graph:

 a. $[-2, 5)$ $\{x \mid -2 \le x < 5\}^*$ **b.** $[1, 3.5]$ $\{x \mid 1 \le x \le 3.5\}^*$ **c.** $(-\infty, -1)$. $\{x \mid x < -1\}^*$

② Find intersections and unions of intervals.

Intersections and Unions of Intervals

In Chapter P, we learned how to find intersections and unions of sets. Recall that $A \cap B$ (A intersection B) is the set of elements common to both set A and set B. By contrast, $A \cup B$ (A union B) is the set of elements in set A or in set B or in both sets.

 Because intervals represent sets, it is possible to find their intersections and unions. Graphs are helpful in this process.

> **Finding Intersections and Unions of Two Intervals**
>
> **1.** Graph each interval on a number line.
> **2. a.** To find the intersection, take the portion of the number line that the two graphs have in common.
> **b.** To find the union, take the portion of the number line representing the total collection of numbers in the two graphs.

EXAMPLE 2 Finding Intersections and Unions of Intervals

Use graphs to find each set:

 a. $(1, 4) \cap [2, 8]$ **b.** $(1, 4) \cup [2, 8]$.

SOLUTION

 a. $(1, 4) \cap [2, 8]$, the intersection of the intervals $(1, 4)$ and $[2, 8]$, consists of the numbers that are in both intervals.

 Graph of $(1, 4)$: $\{x \mid 1 < x < 4\}$

 Graph of $[2, 8]$: $\{x \mid 2 \le x \le 8\}$

 To find $(1, 4) \cap [2, 8]$, take the portion of the number line that the two graphs have in common.

 Numbers in both
 $(1, 4)$ and $[2, 8]$: $\{x \mid 1 < x < 4$ and $2 \le x \le 8\}$

> The numbers common to both intervals
> are those that are greater than or
> equal to 2 and less than 4: [2, 4).

 Thus, $(1, 4) \cap [2, 8] = [2, 4)$.

*See Answers to Selected Exercises.

b. $(1, 4) \cup [2, 8]$, the union of the intervals $(1, 4)$ and $[2, 8]$, consists of the numbers that are in either one interval or the other (or both).

To find $(1, 4) \cup [2, 8]$, take the portion of the number line representing the total collection of numbers in the two graphs.

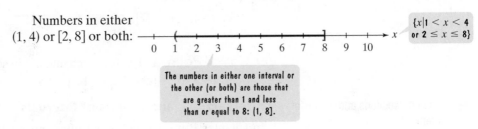

The numbers in either one interval or the other (or both) are those that are greater than 1 and less than or equal to 8: $(1, 8]$.

Thus, $(1, 4) \cup [2, 8] = (1, 8]$.

● ● ●

⌀ **Check Point 2** Use graphs to find each set:
a. $[1, 3] \cap (2, 6)$ $(2, 3]$ **b.** $[1, 3] \cup (2, 6)$. $[1, 6)$

 3 Solve linear inequalities.

Solving Linear Inequalities in One Variable

We know that a linear equation in x can be expressed as $ax + b = 0$. A **linear inequality in x** can be written in one of the following forms:

$$ax + b < 0, \ ax + b \le 0, \ ax + b > 0, \ ax + b \ge 0.$$

In each form, $a \ne 0$.

Back to our question that opened this section: How many miles can you drive your Rent-a-Heap car if you can spend at most $335? We answer the question by solving

$$0.20x + 125 \le 335$$

for x. The solution procedure is nearly identical to that for solving

$$0.20x + 125 = 335.$$

Our goal is to get x by itself on the left side. We do this by subtracting 125 from both sides to isolate $0.20x$:

$0.20x + 125 \le 335$	This is the given inequality.
$0.20x + 125 - 125 \le 335 - 125$	Subtract 125 from both sides.
$0.20x \le 210.$	Simplify.

Finally, we isolate x from $0.20x$ by dividing both sides of the inequality by 0.20:

$\dfrac{0.20x}{0.20} \le \dfrac{210}{0.20}$	Divide both sides by 0.20.
$x \le 1050.$	Simplify.

With at most $335 to spend, you can travel at most 1050 miles.

We started with the inequality $0.20x + 125 \le 335$ and obtained the inequality $x \le 1050$ in the final step. These inequalities have the same solution set, namely $\{x | x \le 1050\}$. Inequalities such as these, with the same solution set, are said to be **equivalent**.

GREAT QUESTION!

What are some common English phrases and sentences that I can model with inequalities?

English phrases such as "at least" and "at most" can be represented by inequalities.

English Sentence	Inequality
x is at least 5.	$x \ge 5$
x is at most 5.	$x \le 5$
x is between 5 and 7.	$5 < x < 7$
x is no more than 5.	$x \le 5$
x is no less than 5.	$x \ge 5$

We isolated x from $0.20x$ by dividing both sides of $0.20x \leq 210$ by 0.20, a positive number. Let's see what happens if we divide both sides of an inequality by a negative number. Consider the inequality $10 < 14$. Divide 10 and 14 by -2:

$$\frac{10}{-2} = -5 \quad \text{and} \quad \frac{14}{-2} = -7.$$

Because -5 lies to the right of -7 on the number line, -5 is greater than -7:

$$-5 > -7.$$

Notice that the direction of the inequality symbol is reversed:

$$
\begin{array}{c}
10 < 14 \\
\uparrow \\
\downarrow \\
-5 > -7.
\end{array}
\qquad
\boxed{\text{Dividing by } -2 \text{ changes the direction of the inequality symbol.}}
$$

In general, **when we multiply or divide both sides of an inequality by a negative number, the direction of the inequality symbol is reversed**. When we reverse the direction of the inequality symbol, we say that we change the *sense* of the inequality.

We can isolate a variable in a linear inequality in the same way we isolate a variable in a linear equation. The following properties are used to create equivalent inequalities.

Properties of Inequalities

Property	The Property in Words	Example
The Addition Property of Inequality If $a < b$, then $a + c < b + c$. If $a < b$, then $a - c < b - c$.	If the same quantity is added to or subtracted from both sides of an inequality, the resulting inequality is equivalent to the original one.	$2x + 3 < 7$ Subtract 3: $2x + 3 - 3 < 7 - 3$. Simplify: $2x < 4$.
The Positive Multiplication Property of Inequality If $a < b$ and c is positive, then $ac < bc$. If $a < b$ and c is positive, then $\dfrac{a}{c} < \dfrac{b}{c}$.	If we multiply or divide both sides of an inequality by the same positive quantity, the resulting inequality is equivalent to the original one.	$2x < 4$ Divide by 2: $\dfrac{2x}{2} < \dfrac{4}{2}$. Simplify: $x < 2$.
The Negative Multiplication Property of Inequality If $a < b$ and c is negative, then $ac > bc$. If $a < b$ and c is negative, then $\dfrac{a}{c} > \dfrac{b}{c}$.	If we multiply or divide both sides of an inequality by the same negative quantity and reverse the direction of the inequality symbol, the resulting inequality is equivalent to the original one.	$-4x < 20$ Divide by -4 and change the sense of the inequality: $\dfrac{-4x}{-4} > \dfrac{20}{-4}$. Simplify: $x > -5$.

EXAMPLE 3 Solving a Linear Inequality

Solve and graph the solution set on a number line:

$$3 - 2x \leq 11.$$

SOLUTION

$3 - 2x \leq 11$	This is the given inequality.
$3 - 2x - 3 \leq 11 - 3$	Subtract 3 from both sides.
$-2x \leq 8$	Simplify.
$\dfrac{-2x}{-2} \geq \dfrac{8}{-2}$	Divide both sides by -2 and change the sense of the inequality.
$x \geq -4$	Simplify.

DISCOVERY

As a partial check, select one number from the solution set of $3 - 2x \leq 11$. Substitute that number into the original inequality. Perform the resulting computations. You should obtain a true statement.

Is it possible to perform a partial check using a number that is not in the solution set? What should happen in this case? Try doing this.

The solution set consists of all real numbers that are greater than or equal to -4, expressed as $\{x \mid x \geq -4\}$ in set-builder notation. The interval notation for this solution set is $[-4, \infty)$. The graph of the solution set is shown as follows:

$$\begin{array}{c} \xleftarrow{\hspace{0.5cm}} \quad [\quad \quad \quad \quad \quad \quad \quad \quad \xrightarrow{\hspace{0.5cm}} x \\ -5 \ -4 \ -3 \ -2 \ -1 \ \ 0 \ \ 1 \ \ 2 \ \ 3 \ \ 4 \ \ 5 \end{array}$$

$\bullet\bullet\bullet$

✓ **Check Point 3** Solve and graph the solution set on a number line:

$$2 - 3x \leq 5. \quad [-1, \infty) \text{ or } \{x \mid x \geq -1\}*$$

EXAMPLE 4 Solving a Linear Inequality

Solve and graph the solution set on a number line:

$$-2x - 4 > x + 5.$$

SOLUTION

Step 1 Simplify each side. Because each side is already simplified, we can skip this step.

Step 2 Collect variable terms on one side and constant terms on the other side. We will collect variable terms of $-2x - 4 > x + 5$ on the left and constant terms on the right.

$-2x - 4 > x + 5$	This is the given inequality.
$-2x - 4 - x > x + 5 - x$	Subtract x from both sides.
$-3x - 4 > 5$	Simplify.
$-3x - 4 + 4 > 5 + 4$	Add 4 to both sides.
$-3x > 9$	Simplify.

Step 3 Isolate the variable and solve. We isolate the variable, x, by dividing both sides by -3. Because we are dividing by a negative number, we must reverse the inequality symbol.

$\dfrac{-3x}{-3} < \dfrac{9}{-3}$	Divide both sides by -3 and change the sense of the inequality.
$x < -3$	Simplify.

GREAT QUESTION!

Do I have to solve the inequality in Example 4 by isolating the variable on the left?

No. You can solve

$$-2x - 4 > x + 5$$

by isolating x on the right side. Add $2x$ to both sides.

$$-2x - 4 + 2x > x + 5 + 2x$$

$$-4 > 3x + 5$$

Now subtract 5 from both sides.

$$-4 - 5 > 3x + 5 - 5$$

$$-9 > 3x$$

Finally, divide both sides by 3.

$$\frac{-9}{3} > \frac{3x}{3}$$

$$-3 > x$$

This last inequality means the same thing as $x < -3$.

Step 4 Express the solution set in set-builder or interval notation and graph the set on a number line. The solution set consists of all real numbers that are less than -3, expressed in set-builder notation as $\{x \mid x < -3\}$. The interval notation for this solution set is $(-\infty, -3)$. The graph of the solution set is shown as follows:

$$\begin{array}{c} \xleftarrow{\hspace{0.5cm}} \quad \quad \quad) \quad \quad \quad \quad \quad \quad \quad \xrightarrow{\hspace{0.5cm}} x \\ -5 \ -4 \ -3 \ -2 \ -1 \ \ 0 \ \ 1 \ \ 2 \ \ 3 \ \ 4 \ \ 5 \end{array}$$

$\bullet\bullet\bullet$

TECHNOLOGY

Numeric and Graphic Connections

You can use a graphing utility to check the solution set of a linear inequality. Enter each side of the inequality separately under y_1 and y_2. Then use the table or the graphs. To use the table, first locate the x-value for which the y-values are the same. Then scroll up or down to locate x values for which y_1 is greater than y_2 or for which y_1 is less than y_2. To use the graphs, locate the intersection point and then find the x-values for which the graph of y_1 lies above the graph of y_2 $(y_1 > y_2)$ or for which the graph of y_1 lies below the graph of y_2 $(y_1 < y_2)$.

Let's verify our work in Example 4 and show that $(-\infty, -3)$ is the solution set of

$$-2x - 4 > x + 5.$$

Enter $y_1 = -2x - 4$ in the $\boxed{y =}$ screen. Enter $y_2 = x + 5$ in the $\boxed{y =}$ screen.

We are looking for values of x for which y_1 is greater than y_2.

Numeric Check	Graphic Check

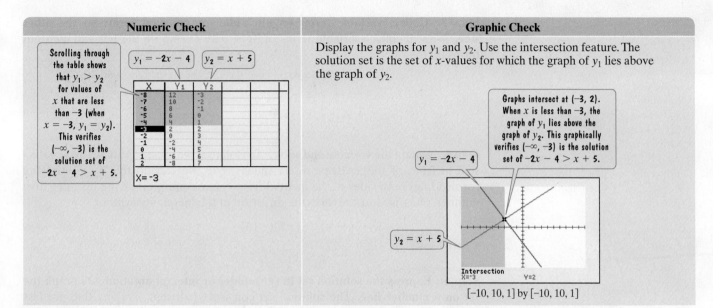

Display the graphs for y_1 and y_2. Use the intersection feature. The solution set is the set of x-values for which the graph of y_1 lies above the graph of y_2.

$[-10, 10, 1]$ by $[-10, 10, 1]$

☑ **Check Point 4** Solve and graph the solution set on a number line:
$3x + 1 > 7x - 15.$ $\{x|x < 4\}$ or $(-\infty, 4)^*$

If an inequality contains fractions with constants in the denominators, begin by multiplying both sides by the least common denominator. This will clear the inequality of fractions.

EXAMPLE 5 Solving a Linear Inequality Containing Fractions

Solve and graph the solution set on a number line:

$$\frac{x + 3}{4} \geq \frac{x - 2}{3} + \frac{1}{4}.$$

SOLUTION

The denominators are 4, 3, and 4. The least common denominator is 12. We begin by multiplying both sides of the inequality by 12.

$\dfrac{x + 3}{4} \geq \dfrac{x - 2}{3} + \dfrac{1}{4}$	This is the given inequality.
$12\left(\dfrac{x + 3}{4}\right) \geq 12\left(\dfrac{x - 2}{3} + \dfrac{1}{4}\right)$	Multiply both sides by 12. Multiplying by a positive number preserves the sense of the inequality.
$\dfrac{12}{1} \cdot \dfrac{x + 3}{4} \geq \dfrac{12}{1} \cdot \dfrac{x - 2}{3} + \dfrac{12}{1} \cdot \dfrac{1}{4}$	Multiply each term by 12. Use the distributive property on the right side.
$\dfrac{\overset{3}{\cancel{12}}}{1} \cdot \dfrac{x + 3}{\underset{1}{\cancel{4}}} \geq \dfrac{\overset{4}{\cancel{12}}}{1} \cdot \dfrac{x - 2}{\underset{1}{\cancel{3}}} + \dfrac{\overset{3}{\cancel{12}}}{1} \cdot \dfrac{1}{\underset{1}{\cancel{4}}}$	Divide out common factors in each multiplication.
$3(x + 3) \geq 4(x - 2) + 3$	The fractions are now cleared.

Now that the fractions have been cleared, we follow the four steps that we used in the previous example.

Step 1 Simplify each side.

$3(x + 3) \geq 4(x - 2) + 3$	This is the inequality with the fractions cleared.
$3x + 9 \geq 4x - 8 + 3$	Use the distributive property.
$3x + 9 \geq 4x - 5$	Simplify.

*See Answers to Selected Exercises.

Step 2 Collect variable terms on one side and constant terms on the other side. We will collect variable terms of $3x + 9 \geq 4x - 5$ on the left and constant terms on the right.

$$3x + 9 - 4x \geq 4x - 5 - 4x \qquad \text{Subtract } 4x \text{ from both sides.}$$
$$-x + 9 \geq -5 \qquad \text{Simplify.}$$
$$-x + 9 - 9 \geq -5 - 9 \qquad \text{Subtract } 9 \text{ from both sides.}$$
$$-x \geq -14 \qquad \text{Simplify.}$$

Step 3 Isolate the variable and solve. To isolate x, we must eliminate the negative sign in front of the x. Because $-x$ means $-1x$, we can do this by multiplying (or dividing) both sides of the inequality by -1. We are multiplying by a negative number. Thus, we must reverse the direction of the inequality symbol.

$$(-1)(-x) \leq (-1)(-14) \qquad \text{Multiply both sides by } -1 \text{ and change the sense of the inequality.}$$
$$x \leq 14 \qquad \text{Simplify.}$$

Step 4 Express the solution set in set-builder or interval notation and graph the set on a number line. The solution set consists of all real numbers that are less than or equal to 14, expressed in set-builder notation as $\{x \mid x \leq 14\}$. The interval notation for this solution set is $(-\infty, 14]$. The graph of the solution set is shown as follows:

•••

✓ Check Point **5** Solve and graph the solution set on a number line:

$$\frac{x - 4}{2} \geq \frac{x - 2}{3} + \frac{5}{6}. \qquad \{x \mid x \geq 13\} \text{ or } [13, \infty)^*$$

④ Recognize inequalities with no solution or all real numbers as solutions.

Inequalities with Unusual Solution Sets

We have seen that some equations have no solution. This is also true for some inequalities. An example of such an inequality is

$$x > x + 1.$$

There is no number that is greater than itself plus 1. This inequality has no solution and its solution set is \varnothing, the empty set.

By contrast, some inequalities are true for all real numbers. An example of such an inequality is

$$x < x + 1.$$

Every real number is less than itself plus 1. The solution set is $\{x \mid x \text{ is a real number}\}$ or \mathbb{R}. In interval notation, the solution set is $(-\infty, \infty)$.

If you attempt to solve an inequality that has no solution, you will eliminate the variable and obtain a false statement, such as $0 > 1$. If you attempt to solve an inequality that is true for all real numbers, you will eliminate the variable and obtain a true statement, such as $0 < 1$.

EXAMPLE 6 Solving Linear Inequalities

Solve each inequality:

 a. $2(x + 4) > 2x + 3$ **b.** $x + 7 \leq x - 2.$

SOLUTION

 a. $2(x + 4) > 2x + 3$ This is the given inequality.
 $2x + 8 > 2x + 3$ Apply the distributive property.
 $2x + 8 - 2x > 2x + 3 - 2x$ Subtract $2x$ from both sides.
 $8 > 3$ Simplify. The statement $8 > 3$ is true.

*See Answers to Selected Exercises.

TECHNOLOGY

Graphic Connections

The graphs of

$$y_1 = 2(x + 4) \text{ and } y_2 = 2x + 3$$

are parallel lines.

[-10, 10, 1] by [-10, 10, 1]

The graph of y_1 is always above the graph of y_2. Every value of x satisfies the inequality $y_1 > y_2$. Thus, the solution set of the inequality

$$2(x + 4) > 2x + 3$$

is $(-\infty, \infty)$.

 Solve compound inequalities.

The inequality $8 > 3$ is true for all values of x. Because this inequality is equivalent to the original inequality, the original inequality is true for all real numbers. The solution set is

$$\{x \mid x \text{ is a real number}\} \text{ or } \mathbb{R} \text{ or } (-\infty, \infty).$$

b.

$x + 7 \leq x - 2$	This is the given inequality.
$x + 7 - x \leq x - 2 - x$	Subtract x from both sides.
$7 \leq -2$	Simplify. The statement $7 \leq -2$ is false.

The inequality $7 \leq -2$ is false for all values of x. Because this inequality is equivalent to the original inequality, the original inequality has no solution. The solution set is \varnothing. • • •

Check Point 6 Solve each inequality:

a. $3(x + 1) > 3x + 2$ $\{x \mid x \text{ is a real number}\}$ or R or $(-\infty, \infty)$

b. $x + 1 \leq x - 1$. \varnothing

Solving Compound Inequalities

We now consider two inequalities such as

$$-3 < 2x + 1 \quad \text{and} \quad 2x + 1 \leq 3,$$

expressed as a **compound inequality**

$$-3 < 2x + 1 \leq 3.$$

The word *and* does not appear when the inequality is written in the shorter form, although intersection is implied. The shorter form enables us to solve both inequalities at once. By performing each operation on all three parts of the inequality, our goal is to **isolate x in the middle**.

> **EXAMPLE 7** Solving a Compound Inequality

Solve and graph the solution set on a number line:

$$-3 < 2x + 1 \leq 3.$$

SOLUTION

We would like to isolate x in the middle. We can do this by first subtracting 1 from all three parts of the compound inequality. Then we isolate x from $2x$ by dividing all three parts of the inequality by 2.

$-3 < 2x + 1 \leq 3$	This is the given inequality.
$-3 - 1 < 2x + 1 - 1 \leq 3 - 1$	Subtract 1 from all three parts.
$-4 < 2x \leq 2$	Simplify.
$\dfrac{-4}{2} < \dfrac{2x}{2} \leq \dfrac{2}{2}$	Divide each part by 2.
$-2 < x \leq 1$	Simplify.

The solution set consists of all real numbers greater than -2 and less than or equal to 1, represented by $\{x \mid -2 < x \leq 1\}$ in set-builder notation and $(-2, 1]$ in interval notation. The graph is shown as follows:

• • •

Check Point 7 Solve and graph the solution set on a number line:

$1 \leq 2x + 3 < 11$. $\{x \mid -1 \leq x < 4\}$ or $[-1, 4)$*

*See Answers to Selected Exercises.

TECHNOLOGY

Numeric and Graphic Connections

Let's verify our work in Example 7 and show that $(-2, 1]$ is the solution set of $-3 < 2x + 1 \le 3$.

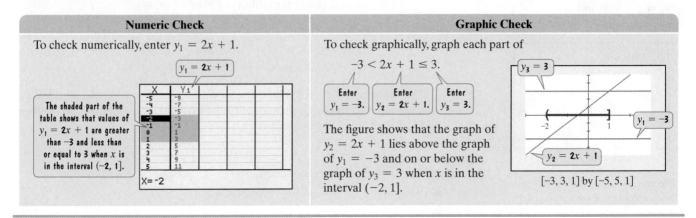

Numeric Check	Graphic Check
To check numerically, enter $y_1 = 2x + 1$.	To check graphically, graph each part of
The shaded part of the table shows that values of $y_1 = 2x + 1$ are greater than -3 and less than or equal to 3 when x is in the interval $(-2, 1]$.	$-3 < 2x + 1 \le 3.$ Enter $y_1 = -3$. Enter $y_2 = 2x + 1$. Enter $y_3 = 3$. The figure shows that the graph of $y_2 = 2x + 1$ lies above the graph of $y_1 = -3$ and on or below the graph of $y_3 = 3$ when x is in the interval $(-2, 1]$. $[-3, 3, 1]$ by $[-5, 5, 1]$

⑥ Solve absolute value inequalities.

FIGURE 1.31 $|x| < 2$, so $-2 < x < 2$.

FIGURE 1.32 $|x| > 2$, so $x < -2$ or $x > 2$.

Solving Inequalities with Absolute Value

We know that $|x|$ describes the distance of x from zero on a real number line. We can use this geometric interpretation to solve an inequality such as

$$|x| < 2.$$

This means that the distance of x from 0 is *less than* 2, as shown in **Figure 1.31**. The interval shows values of x that lie less than 2 units from 0. Thus, x can lie between -2 and 2. That is, x is greater than -2 and less than 2. We write $(-2, 2)$ or $\{x \mid -2 < x < 2\}$.

Some absolute value inequalities use the "greater than" symbol. For example, $|x| > 2$ means that the distance of x from 0 is *greater than* 2, as shown in **Figure 1.32**. Thus, x can be less than -2 *or* greater than 2. We write $x < -2$ or $x > 2$. This can be expressed in interval notation as $(-\infty, -2) \cup (2, \infty)$.

These observations suggest the following principles for solving inequalities with absolute value.

GREAT QUESTION!

The box on the right shows how to rewrite absolute value inequalities without absolute value bars. Once I've done this, how many inequalities do I need to solve?

In the $|u| < c$ case, we have one compound inequality to solve. In the $|u| > c$ case, we have two separate inequalities to solve.

Solving an Absolute Value Inequality

If u is an algebraic expression and c is a positive number,

1. The solutions of $|u| < c$ are the numbers that satisfy $-c < u < c$.
2. The solutions of $|u| > c$ are the numbers that satisfy $u < -c$ or $u > c$.

These rules are valid if $<$ is replaced by \le and $>$ is replaced by \ge.

EXAMPLE 8 Solving an Absolute Value Inequality

Solve and graph the solution set on a number line: $|x - 4| < 3$.

SOLUTION

We rewrite the inequality without absolute value bars.

$$|u| < c \quad \text{means} \quad -c < u < c.$$

$$|x - 4| < 3 \quad \text{means} \quad -3 < x - 4 < 3.$$

We solve the compound inequality by adding 4 to all three parts.

$$-3 < x - 4 < 3$$
$$-3 + 4 < x - 4 + 4 < 3 + 4$$
$$1 < x < 7$$

The solution set consists of all real numbers greater than 1 and less than 7, denoted by $\{x \mid 1 < x < 7\}$ or $(1, 7)$. The graph of the solution set is shown as follows:

We can use the rectangular coordinate system to visualize the solution set of

$$|x - 4| < 3.$$

Figure 1.33 shows the graphs of $y_1 = |x - 4|$ and $y_2 = 3$. The solution set of $|x - 4| < 3$ consists of all values of x for which the blue graph of y_1 lies below the red graph of y_2. These x-values make up the interval $(1, 7)$, which is the solution set.

FIGURE 1.33 The solution set of $|x - 4| < 3$ is $(1, 7)$.

⊘ **Check Point 8** Solve and graph the solution set on a number line: $|x - 2| < 5$. $\{x \mid -3 < x < 7\}$ or $(-3, 7)$*

EXAMPLE 9 Solving an Absolute Value Inequality

Solve and graph the solution set on a number line: $-2|3x + 5| + 7 \geq -13$.

SOLUTION

$$-2|3x + 5| + 7 \geq -13 \qquad \text{This is the given inequality.}$$

> We need to isolate $|3x + 5|$, the absolute value expression.

$$-2|3x + 5| + 7 - 7 \geq -13 - 7 \qquad \text{Subtract 7 from both sides.}$$

$$-2|3x + 5| \geq -20 \qquad \text{Simplify.}$$

$$\frac{-2|3x + 5|}{-2} \leq \frac{-20}{-2} \qquad \text{Divide both sides by } -2 \text{ and change the sense of the inequality.}$$

$$|3x + 5| \leq 10 \qquad \text{Simplify.}$$

$$-10 \leq 3x + 5 \leq 10 \qquad \text{Rewrite without absolute value bars:} \\ |u| \leq c \text{ means } -c \leq u \leq c.$$

> Now we need to isolate x in the middle.

$$-10 - 5 \leq 3x + 5 - 5 \leq 10 - 5 \qquad \text{Subtract 5 from all three parts.}$$

$$-15 \leq 3x \leq 5 \qquad \text{Simplify.}$$

$$\frac{-15}{3} \leq \frac{3x}{3} \leq \frac{5}{3} \qquad \text{Divide each part by 3.}$$

$$-5 \leq x \leq \frac{5}{3} \qquad \text{Simplify.}$$

The solution set is $\left\{ x \mid -5 \leq x \leq \frac{5}{3} \right\}$ in set-builder notation and $\left[-5, \frac{5}{3} \right]$ in interval notation. The graph is shown as follows:

⊘ **Check Point 9** Solve and graph the solution set on a number line: $-3|5x - 2| + 20 \geq -19$. $\left\{ x \mid -\frac{11}{5} \leq x \leq 3 \right\}$ or $\left[-\frac{11}{5}, 3 \right]$*

*See Answers to Selected Exercises.

EXAMPLE 10 Solving an Absolute Value Inequality

Solve and graph the solution set on a number line: $7 < |5 - 2x|$.

SOLUTION

We begin by expressing the inequality with the absolute value expression on the left side:

$$|5 - 2x| > 7.$$

$c < |u|$ means the same thing as $|u| > c$. In both cases, the inequality symbol points to c.

We rewrite this inequality without absolute value bars.

$|u| > c$ means $u < -c$ or $u > c$.

$|5 - 2x| > 7$ means $5 - 2x < -7$ or $5 - 2x > 7$.

Because $|5 - 2x| > 7$ means $5 - 2x < -7$ or $5 - 2x > 7$, we solve $5 - 2x < -7$ and $5 - 2x > 7$ separately. Then we take the union of their solution sets.

$5 - 2x < -7$ or $5 - 2x > 7$		These are the inequalities without absolute value bars.
$5 - 5 - 2x < -7 - 5$ $5 - 5 - 2x > 7 - 5$		Subtract 5 from both sides.
$-2x < -12$ $-2x > 2$		Simplify.
$\dfrac{-2x}{-2} > \dfrac{-12}{-2}$ $\dfrac{-2x}{-2} < \dfrac{2}{-2}$		Divide both sides by -2 and change the sense of each inequality.
$x > 6$ $x < -1$		Simplify.

GREAT QUESTION!

The graph of the solution set in Example 10 consists of two intervals. When is the graph of the solution set of an absolute value inequality a single interval and when is it the union of two intervals?

If u is a linear expression and $c > 0$, the graph of the solution set for $|u| > c$ will be two intervals whose union cannot be represented as a single interval. The graph of the solution set for $|u| < c$ will be a single interval. Avoid the common error of rewriting $|u| > c$ as $-c < u > c$.

The solution set consists of all numbers that are less than -1 or greater than 6. The solution set is $\{x \mid x < -1 \text{ or } x > 6\}$, or, in interval notation $(-\infty, -1) \cup (6, \infty)$. The graph of the solution set is shown as follows:

⊘ Check Point **10** Solve and graph the solution set on a number line: $18 < |6 - 3x|$. $\{x \mid x < -4 \text{ or } x > 8\}$ or $(-\infty, -4) \cup (8, \infty)$*

Applications

Our next example shows how to use an inequality to select the better deal between two pricing options. We use our strategy for solving word problems, modeling the verbal conditions of the problem with a linear inequality.

EXAMPLE 11 Selecting the Better Deal

Acme Car rental agency charges $4 a day plus $0.15 per mile, whereas Interstate rental agency charges $20 a day and $0.05 per mile. How many miles must be driven to make the daily cost of an Acme rental a better deal than an Interstate rental?

*See Answers to Selected Exercises.

SOLUTION

Step 1 Let x represent one of the unknown quantities. We are looking for the number of miles that must be driven in a day to make Acme the better deal. Thus,

$$\text{let } x = \text{the number of miles driven in a day.}$$

Step 2 Represent other unknown quantities in terms of x. We are not asked to find another quantity, so we can skip this step.

Step 3 Write an inequality in x that models the conditions. Acme is a better deal than Interstate if the daily cost of Acme is less than the daily cost of Interstate.

Step 4 Solve the inequality and answer the question.

$4 + 0.15x < 20 + 0.05x$	This is the inequality that models the verbal conditions.
$4 + 0.15x - 0.05x < 20 + 0.05x - 0.05x$	Subtract 0.05x from both sides.
$4 + 0.1x < 20$	Simplify.
$4 + 0.1x - 4 < 20 - 4$	Subtract 4 from both sides.
$0.1x < 16$	Simplify.
$\dfrac{0.1x}{0.1} < \dfrac{16}{0.1}$	Divide both sides by 0.1.
$x < 160$	Simplify.

Thus, driving fewer than 160 miles per day makes Acme the better deal.

Step 5 Check the proposed solution in the original wording of the problem. One way to do this is to take a mileage less than 160 miles per day to see if Acme is the better deal. Suppose that 150 miles are driven in a day.

$$\text{Cost for Acme} = 4 + 0.15(150) = 26.50$$

$$\text{Cost for Interstate} = 20 + 0.05(150) = 27.50$$

Acme has a lower daily cost, making Acme the better deal. • • •

TECHNOLOGY

Graphic Connections

The graphs of the daily cost models for the car rental agencies

$$y_1 = 4 + 0.15x$$

and $y_2 = 20 + 0.05x$

are shown in a $[0, 300, 10]$ by $[0, 40, 4]$ viewing rectangle. The graphs intersect at $(160, 28)$.

To the left of $x = 160$, the graph of Acme's daily cost lies below that of Interstate's daily cost. This shows that for fewer than 160 miles per day, Acme offers the better deal.

⊘ Check Point **11** A car can be rented from Basic Rental for $260 per week with no extra charge for mileage. Continental charges $80 per week plus 25 cents for each mile driven to rent the same car. How many miles must be driven in a week to make the rental cost for Basic Rental a better deal than Continental's? more than 720 miles per week

ACHIEVING SUCCESS

Use index cards to help learn new terms.

Many of the terms, notations, and formulas used in this book will be new to you. Buy a pack of 3×5 index cards. On each card, list a new vocabulary word, symbol, or title of a formula. On the other side of the card, put the definition or formula. Here are four examples related to equations and inequalities.

Effective Index Cards

Review these cards frequently. Use the cards to quiz yourself and prepare for exams.

CONCEPT AND VOCABULARY CHECK

Fill in each blank so that the resulting statement is true.

1. In interval notation, $[2, 5)$ represents the set of real numbers between ___2___ and ___5___, including ___2___ but not including ___5___.
2. In interval notation, $(-2, \infty)$ represents the set of real numbers __greater than__ -2.
3. In interval notation, $(-\infty, -1]$ represents the set of real numbers __less than or equal to__ -1.
4. The set of elements common to both $(-\infty, 9)$ and $(-\infty, 12)$ is __$(-\infty, 9)$__. This represents the __intersection__ of these intervals.
5. The set of elements in $(-\infty, 9)$ or $(-\infty, 12)$ or in both sets is __$(-\infty, 12)$__. This represents the __union__ of these intervals.
6. The linear inequality $-3x - 4 > 5$ can be solved by first __adding 4__ to both sides and then __dividing__ both sides by __-3__, which changes the __direction__ of the inequality symbol from ___>___ to ___<___.

7. In solving an inequality, if you eliminate the variable and obtain a false statement such as $7 < -2$, the solution set is __\varnothing__.
8. In solving an inequality, if you eliminate the variable and obtain a true statement such as $8 > 3$, the solution set is __$(-\infty, \infty)$__.
9. The way to solve $-7 < 3x - 4 \le 5$ is to isolate x in the __middle__.
10. If $c > 0$, $|u| < c$ is equivalent to __$-c$__ $< u <$ __c__.
11. If $c > 0$, $|u| > c$ is equivalent to $u <$ __$-c$__ or $u >$ __c__.
12. $|x - 7| < 2$ can be rewritten without absolute value bars as __$-2 < x - 7 < 2$__.
13. $|x - 7| > 2$ can be rewritten without absolute value bars as __$x - 7 < -2$ or $x - 7 > 2$__.

EXERCISE SET 1.7

Practice Exercises

In Exercises 1–14, express each interval in set-builder notation and graph the interval on a number line.

1. $(1, 6]$ $\{x | 1 < x \le 6\}$*
2. $(-2, 4]$ $\{x | -2 < x \le 4\}$*
3. $[-5, 2)$ $\{x | -5 \le x < 2\}$*
4. $[-4, 3)$ $\{x | -4 \le x < 3\}$*
5. $[-3, 1]$ $\{x | -3 \le x \le 1\}$*
6. $[-2, 5]$ $\{x | -2 \le x \le 5\}$*
7. $(2, \infty)$ $\{x | x > 2\}$*
8. $(3, \infty)$ $\{x | x > 3\}$*
9. $[-3, \infty)$ $\{x | x \ge -3\}$*
10. $[-5, \infty)$ $\{x | x \ge -5\}$*
11. $(-\infty, 3)$ $\{x | x < 3\}$*
12. $(-\infty, 2)$ $\{x | x < 2\}$*
13. $(-\infty, 5.5)$ $\{x | x < 5.5\}$*
14. $(-\infty, 3.5]$ $\{x | x \le 3.5\}$*

In Exercises 15–26, use graphs to find each set.

15. $(-3, 0) \cap [-1, 2]$ $[-1, 0)$
16. $(-4, 0) \cap [-2, 1]$ $[-2, 0)$
17. $(-3, 0) \cup [-1, 2]$ $(-3, 2]$
18. $(-4, 0) \cup [-2, 1]$ $(-4, 1]$
19. $(-\infty, 5) \cap [1, 8)$ $[1, 5)$
20. $(-\infty, 6) \cap [2, 9)$ $[2, 6)$
21. $(-\infty, 5) \cup [1, 8)$ $(-\infty, 8)$
22. $(-\infty, 6) \cup [2, 9)$ $(-\infty, 9)$
23. $[3, \infty) \cap (6, \infty)$ $(6, \infty)$
24. $[2, \infty) \cap (4, \infty)$ $(4, \infty)$
25. $[3, \infty) \cup (6, \infty)$ $[3, \infty)$
26. $[2, \infty) \cup (4, \infty)$ $[2, \infty)$

*See Answers to Selected Exercises.

In all exercises, other than ∅, use interval notation to express solution sets and graph each solution set on a number line.

In Exercises 27–50, solve each linear inequality.

27. $5x + 11 < 26$ $(-\infty, 3)$*
28. $2x + 5 < 17$ $(-\infty, 6)$*
29. $3x - 7 \geq 13$ $[\frac{20}{3}, \infty)$*
30. $8x - 2 \geq 14$ $[2, \infty)$*
31. $-9x \geq 36$ $(-\infty, -4]$*
32. $-5x \leq 30$ $[-6, \infty)$*
33. $8x - 11 \leq 3x - 13$ *
34. $18x + 45 \leq 12x - 8$ *
35. $4(x + 1) + 2 \geq 3x + 6$ $[0, \infty)$*
36. $8x + 3 > 3(2x + 1) + x + 5$ $(5, \infty)$*
37. $2x - 11 < -3(x + 2)$ *
38. $-4(x + 2) > 3x + 20$ *
39. $1 - (x + 3) \geq 4 - 2x$ *
40. $5(3 - x) \leq 3x - 1$ *
41. $\frac{x}{4} - \frac{3}{2} \leq \frac{x}{2} + 1$ $[-10, \infty)$*
42. $\frac{3x}{10} + 1 \geq \frac{1}{5} - \frac{x}{10}$ $[-2, \infty)$
43. $1 - \frac{x}{2} > 4$ $(-\infty, -6)$*
44. $7 - \frac{4}{5}x < \frac{3}{5}$ $(8, \infty)$*
45. $\frac{x - 4}{6} \geq \frac{x - 2}{9} + \frac{5}{18}$ *
46. $\frac{4x - 3}{6} + 2 \geq \frac{2x - 1}{12}$ *
47. $4(3x - 2) - 3x < 3(1 + 3x) - 7$ $(-\infty, \infty)$*
48. $3(x - 8) - 2(10 - x) > 5(x - 1)$ ∅
49. $5(x - 2) - 3(x + 4) \geq 2x - 20$ ∅
50. $6(x - 1) - (4 - x) \geq 7x - 8$ ∅

In Exercises 51–58, solve each compound inequality.

51. $6 < x + 3 < 8$ $(3, 5)$
52. $7 < x + 5 < 11$ $(2, 6)$
53. $-3 \leq x - 2 < 1$ $[-1, 3)$
54. $-6 < x - 4 \leq 1$ $(-2, 5]$
55. $-11 < 2x - 1 \leq -5$ *
56. $3 \leq 4x - 3 < 19$ *
57. $-3 \leq \frac{2}{3}x - 5 < -1$ $[3, 6)$
58. $-6 \leq \frac{1}{2}x - 4 < -3$ *

In Exercises 59–94, solve each absolute value inequality.

59. $|x| < 3$ $(-3, 3)$
60. $|x| < 5$ $(-5, 5)$
61. $|x - 1| \leq 2$ $[-1, 3]$
62. $|x + 3| \leq 4$ $[-7, 1]$
63. $|2x - 6| < 8$ $(-1, 7)$
64. $|3x + 5| < 17$ $(-\frac{22}{3}, 4)$
65. $|2(x - 1) + 4| \leq 8$ $[-5, 3]$
66. $|3(x - 1) + 2| \leq 20$ *
67. $\left|\frac{2x + 6}{3}\right| < 2$ $(-6, 0)$
68. $\left|\frac{3(x - 1)}{4}\right| < 6$ $(-7, 9)$
69. $|x| > 3$ $(-\infty, -3) \cup (3, \infty)$
70. $|x| > 5$ $(-\infty, -5) \cup (5, \infty)$
71. $|x - 1| \geq 2$ *
72. $|x + 3| \geq 4$ *
73. $|3x - 8| > 7$ *
74. $|5x - 2| > 13$ *
75. $\left|\frac{2x + 2}{4}\right| \geq 2$ *
76. $\left|\frac{3x - 3}{9}\right| \geq 1$ *
77. $\left|3 - \frac{2}{3}x\right| > 5$ *
78. $\left|3 - \frac{3}{4}x\right| > 9$ *
79. $3|x - 1| + 2 \geq 8$ *
80. $5|2x + 1| - 3 \geq 9$ *
81. $-2|x - 4| \geq -4$ $[2, 6]$
82. $-3|x + 7| \geq -27$ $[-16, 2]$
83. $-4|1 - x| < -16$ *
84. $-2|5 - x| < -6$ *
85. $3 \leq |2x - 1|$ *
86. $9 \leq |4x + 7|$ *
87. $5 > |4 - x|$ $(-1, 9)$
88. $2 > |11 - x|$ $(9, 13)$
89. $1 < |2 - 3x|$ *
90. $4 < |2 - x|$ *
91. $12 < \left|-2x + \frac{6}{7}\right| + \frac{3}{7}$ *
92. $1 < \left|x - \frac{11}{3}\right| + \frac{7}{3}$ *
93. $4 + \left|3 - \frac{x}{3}\right| \geq 9$ *
94. $\left|2 - \frac{x}{2}\right| - 1 \leq 1$ $[0, 8]$

In Exercises 95–102, use interval notation to represent all values of x satisfying the given conditions.

95. $y_1 = \frac{x}{2} + 3$, $y_2 = \frac{x}{3} + \frac{5}{2}$, and $y_1 \leq y_2$. $(-\infty, -3]$
96. $y_1 = \frac{2}{3}(6x - 9) + 4$, $y_2 = 5x + 1$, and $y_1 > y_2$. $(-\infty, -3)$
97. $y = 1 - (x + 3) + 2x$ and y is at least 4. $[6, \infty)$
98. $y = 2x - 11 + 3(x + 2)$ and y is at most 0. $(-\infty, 1]$
99. $y = |3x - 4| + 2$ and $y < 8$. $(-\frac{2}{3}, \frac{10}{3})$
100. $y = |2x - 5| + 1$ and $y > 9$. $(-\infty, -\frac{3}{2}) \cup (\frac{13}{2}, \infty)$
101. $y = 7 - \left|\frac{x}{2} + 2\right|$ and y is at most 4. $(-\infty, -10] \cup [2, \infty)$
102. $y = 8 - |5x + 3|$ and y is at least 6. $[-1, -\frac{1}{5}]$

Practice Plus

In Exercises 103–104, use the graph of $y = |4 - x|$ to solve each inequality.

103. $|4 - x| < 5$ $(-1, 9)$
104. $|4 - x| \geq 5$ $(-\infty, -1] \cup [9, \infty)$

In Exercises 105–106, use the table to solve each inequality.

105. $-2 \leq 5x + 3 < 13$ $[-1, 2)$

106. $-3 < 2x - 5 \leq 3$ $(1, 4]$

107. When 3 times a number is subtracted from 4, the absolute value of the difference is at least 5. Use interval notation to express the set of all numbers that satisfy this condition. *
108. When 4 times a number is subtracted from 5, the absolute value of the difference is at most 13. Use interval notation to express the set of all numbers that satisfy this condition. *

Application Exercises

The graphs show that the three components of love, namely, passion, intimacy, and commitment, progress differently over time. Passion peaks early in a relationship and then declines. By contrast, intimacy and commitment build gradually. Use the graphs to solve Exercises 109–116.

Source: R. J. Sternberg. A Triangular Theory of Love, *Psychological Review*, 93, 119–135.

109. Use interval notation to write an inequality that expresses for which years in a relationship intimacy is greater than commitment. $(0, 4)$

110. Use interval notation to write an inequality that expresses for which years in a relationship passion is greater than or equal to intimacy. $[0, 5]$

111. What is the relationship between passion and intimacy on the interval $[5, 7)$? intimacy \geq passion*

112. What is the relationship between intimacy and commitment on the interval $[4, 7)$? commitment \geq intimacy*

113. What is the relationship between passion and commitment for $\{x \mid 6 < x < 8\}$? commitment $>$ passion*

114. What is the relationship between passion and commitment for $\{x \mid 7 < x < 9\}$? commitment $>$ passion*

115. What is the maximum level of intensity for passion? After how many years in a relationship does this occur? 9; 3 yr

116. After approximately how many years do levels of intensity for commitment exceed the maximum level of intensity for passion? after approximately $5\frac{1}{2}$ years

In more U.S. marriages, spouses have different faiths. The bar graph shows the percentage of households with an interfaith marriage in 1988 and 2012. Also shown is the percentage of households in which a person of faith is married to someone with no religion.

Source: General Social Survey, University of Chicago

The formula

$$I = \frac{1}{4}x + 26$$

models the percentage of U.S. households with an interfaith marriage, I, x years after 1988. The formula

$$N = \frac{1}{4}x + 6$$

models the percentage of U.S. households in which a person of faith is married to someone with no religion, N, x years after 1988.

Use these models to solve Exercises 117–118.

117. **a.** In which years will more than 33% of U.S. households have an interfaith marriage? years after 2016

 b. In which years will more than 14% of U.S. households have a person of faith married to someone with no religion? years after 2020

 c. Based on your answers to parts (a) and (b), in which years will more than 33% of households have an interfaith marriage and more than 14% have a faith/no religion marriage? years after 2020

 d. Based on your answers to parts (a) and (b), in which years will more than 33% of households have an interfaith marriage or more than 14% have a faith/no religion marriage? years after 2016

118. **a.** In which years will more than 34% of U.S. households have an interfaith marriage? years after 2020

 b. In which years will more than 15% of U.S. households have a person of faith married to someone with no religion? years after 2024

 c. Based on your answers to parts (a) and (b), in which years will more than 34% of households have an interfaith marriage and more than 15% have a faith/no religion marriage? years after 2024

 d. Based on your answers to parts (a) and (b), in which years will more than 34% of households have an interfaith marriage or more than 15% have a faith/no religion marriage? years after 2020

119. The formula for converting Fahrenheit temperature, F, to Celsius temperature, C, is

$$C = \frac{5}{9}(F - 32).$$

If Celsius temperature ranges from 15° to 35°, inclusive, what is the range for the Fahrenheit temperature? Use interval notation to express this range. $[59°\text{F}, 95°\text{F}]$

120. The formula for converting Celsius temperature, C, to Fahrenheit temperature, F, is

$$F = \frac{9}{5}C + 32.$$

If Fahrenheit temperature ranges from 41° to 50°, inclusive, what is the range for Celsius temperature? Use interval notation to express this range. $[5°\text{C}, 10°\text{C}]$

121. If a coin is tossed 100 times, we would expect approximately 50 of the outcomes to be heads. It can be demonstrated that a coin is unfair if h, the number of outcomes that result in heads, satisfies $\left| \frac{h - 50}{5} \right| \geq 1.645$. Describe the number of outcomes that determine an unfair coin that is tossed 100 times. $h \leq 41$ or $h \geq 59$

In Exercises 122–133, use the strategy for solving word problems, modeling the verbal conditions of the problem with a linear inequality.

122. A truck can be rented from Basic Rental for $50 per day plus $0.20 per mile. Continental charges $20 per day plus $0.50 per mile to rent the same truck. How many miles must be driven in a day to make the rental cost for Basic Rental a better deal than Continental's? more than 100 miles*

123. You are choosing between two texting plans. Plan A has a monthly fee of $15 with a charge of $0.08 per text. Plan B has a monthly fee of $3 with a charge of $0.12 per text. How many text messages in a month make plan A the better deal? more than 300 texts*

124. A city commission has proposed two tax bills. The first bill requires that a homeowner pay $1800 plus 3% of the assessed home value in taxes. The second bill requires taxes of $200 plus 8% of the assessed home value. What price range of home assessment would make the first bill a better deal? greater than $32,000*

125. A local bank charges $8 per month plus 5¢ per check. The credit union charges $2 per month plus 8¢ per check. How many checks should be written each month to make the credit union a better deal? 199 checks or fewer*

126. A company manufactures and sells blank audiocassette tapes. The weekly fixed cost is $10,000 and it costs $0.40 to produce each tape. The selling price is $2.00 per tape. How many tapes must be produced and sold each week for the company to generate a profit? more than 6250 tapes*

127. A company manufactures and sells personalized stationery. The weekly fixed cost is $3000 and it costs $3.00 to produce each package of stationery. The selling price is $5.50 per package. How many packages of stationery must be produced and sold each week for the company to generate a profit? more than 1200 packages*

128. An elevator at a construction site has a maximum capacity of 2800 pounds. If the elevator operator weighs 265 pounds and each cement bag weighs 65 pounds, how many bags of cement can be safely lifted on the elevator in one trip? *

129. An elevator at a construction site has a maximum capacity of 3000 pounds. If the elevator operator weighs 245 pounds and each cement bag weighs 95 pounds, how many bags of cement can be safely lifted on the elevator in one trip? *

130. To earn an A in a course, you must have a final average of at least 90%. On the first four examinations, you have grades of 86%, 88%, 92%, and 84%. If the final examination counts as two grades, what must you get on the final to earn an A in the course? at least 95%*

131. On two examinations, you have grades of 86 and 88. There is an optional final examination, which counts as one grade. You decide to take the final in order to get a course grade of A, meaning a final average of at least 90.

 a. What must you get on the final to earn an A in the course? at least a 96*

 b. By taking the final, if you do poorly, you might risk the B that you have in the course based on the first two exam grades. If your final average is less than 80, you will lose your B in the course. Describe the grades on the final that will cause this to happen. less than 66*

132. Parts for an automobile repair cost $175. The mechanic charges $34 per hour. If you receive an estimate for at least $226 and at most $294 for fixing the car, what is the time interval that the mechanic will be working on the job? *

133. The toll to a bridge is $3.00. A three-month pass costs $7.50 and reduces the toll to $0.50. A six-month pass costs $30 and permits crossing the bridge for no additional fee. How many crossings per three-month period does it take for the three-month pass to be the best deal? more than 3 and fewer than 15

Explaining the Concepts 134–141. Answers will vary.

134. When graphing the solutions of an inequality, what does a parenthesis signify? What does a square bracket signify?

135. Describe ways in which solving a linear inequality is similar to solving a linear equation.

136. Describe ways in which solving a linear inequality is different than solving a linear equation.

137. What is a compound inequality and how is it solved?

138. Describe how to solve an absolute value inequality involving the symbol $<$. Give an example.

139. Describe how to solve an absolute value inequality involving the symbol $>$. Give an example.

140. Explain why $|x| < -4$ has no solution.

141. Describe the solution set of $|x| > -4$.

Technology Exercises

In Exercises 142–143, solve each inequality using a graphing utility. Graph each side separately. Then determine the values of x for which the graph for the left side lies above the graph for the right side.

142. $-3(x - 6) > 2x - 2$ $(-\infty, 4)$*

143. $-2(x + 4) > 6x + 16$ $(-\infty, -3)$*

144. Use a graphing utility's ⌷TABLE⌷ feature to verify your work in Exercises 142–143. *

145. A bank offers two checking account plans. Plan A has a base service charge of $4.00 per month plus 10¢ per check. Plan B charges a base service charge of $2.00 per month plus 15¢ per check.

 a. Write models for the total monthly costs for each plan if x checks are written. $C = 4 + 0.10x; C = 2 + 0.15x$

 b. Use a graphing utility to graph the models in the same $[0, 50, 10]$ by $[0, 10, 1]$ viewing rectangle. *

 c. Use the graphs (and the intersection feature) to determine for what number of checks per month plan A will be better than plan B. 41 or more checks

 d. Verify the result of part (c) algebraically by solving an inequality. $\{x | x > 40\}$ or $(40, \infty)$

Critical Thinking Exercises

Make Sense? *In Exercises 146–149, determine whether each statement makes sense or does not make sense, and explain your reasoning.*

146. I prefer interval notation over set-builder notation because it takes less space to write solution sets. makes sense

147. I can check inequalities by substituting 0 for the variable: When 0 belongs to the solution set, I should obtain a true statement, and when 0 does not belong to the solution set, I should obtain a false statement. makes sense

148. In an inequality such as $5x + 4 < 8x - 5$, I can avoid division by a negative number depending on which side I collect the variable terms and on which side I collect the constant terms. makes sense

149. I'll win the contest if I can complete the crossword puzzle in 20 minutes plus or minus 5 minutes, so my winning time, x, is modeled by $|x - 20| \leq 5$. makes sense

In Exercises 150–153, determine whether each statement is true or false. If the statement is false, make the necessary change(s) to produce a true statement. Changes to false statements will vary.

150. $(-\infty, -1] \cap [-4, \infty) = [-4, -1]$ true

151. $(-\infty, 3) \cup (-\infty, -2) = (-\infty, -2)$ false

152. The inequality $3x > 6$ is equivalent to $2 > x$. false

153. All irrational numbers satisfy $|x - 4| > 0$. true

154. What's wrong with this argument? Suppose x and y represent two real numbers, where $x > y$.

$2 > 1$	This is a true statement.
$2(y - x) > 1(y - x)$	Multiply both sides by $y - x$.
$2y - 2x > y - x$	Use the distributive property.
$y - 2x > -x$	Subtract y from both sides.
$y > x$	Add $2x$ to both sides.

The final inequality, $y > x$, is impossible because we were initially given $x > y$. *

155. Write an absolute value inequality for which the interval shown is the solution. a. $|x - 4| < 3$ b. $|x - 4| \geq 3$

Solutions lie within 3 units of 4.

a.

b.

Group Exercise Answers will vary.

156. Each group member should research one situation that provides two different pricing options. These can involve areas such as public transportation options (with or without discount passes), cellphone plans, long-distance telephone plans, or anything of interest. Be sure to bring in all the details for each option. At a second group meeting, select the two pricing situations that are most interesting and relevant. Using each situation, write a word problem about selecting the better of the two options. The word problem should be one that can be solved using a linear inequality. The group should turn in the two problems and their solutions.

Preview Exercises

Exercises 157–159 will help you prepare for the material covered in the first section of the next chapter.

157. Here are two sets of ordered pairs:

$$\text{set 1: } \{(1, 5), (2, 5)\}$$
$$\text{set 2: } \{(5, 1), (5, 2)\}.$$

In which set is each x-coordinate paired with only one y-coordinate? set 1

158. Graph $y = 2x$ and $y = 2x + 4$ in the same rectangular coordinate system. Select integers for x, starting with -2 and ending with 2. *

159. Use the following graph to solve this exercise.

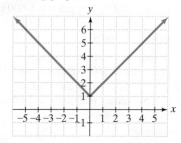

a. What is the y-coordinate when the x-coordinate is 2? 3

b. What are the x-coordinates when the y-coordinate is 4? -3 and 3

c. Describe the x-coordinates of all points on the graph. *

d. Describe the y-coordinates of all points on the graph. *

CHAPTER 1 Summary, Review, and Test

SUMMARY

DEFINITIONS AND CONCEPTS

EXAMPLES

1.1 Graphs and Graphing Utilities

a. The rectangular coordinate system consists of a horizontal number line, the x-axis, and a vertical number line, the y-axis, intersecting at their zero points, the origin. Each point in the system corresponds to an ordered pair of real numbers (x, y). The first number in the pair is the x-coordinate; the second number is the y-coordinate. See Figure 1.1 on page 94. Ex. 1, p. 95

b. An ordered pair is a solution of an equation in two variables if replacing the variables by the corresponding coordinates results in a true statement. The ordered pair is said to satisfy the equation. The graph of the equation is the set of all points whose coordinates satisfy the equation. One method for graphing an equation is to plot ordered-pair solutions and connect them with a smooth curve or line. Ex. 2, p. 96; Ex. 3, p. 96

DEFINITIONS AND CONCEPTS	**EXAMPLES**
c. An x-intercept of a graph is the x-coordinate of a point where the graph intersects the x-axis. The y-coordinate corresponding to an x-intercept is always zero. A y-intercept of a graph is the y-coordinate of a point where the graph intersects the y-axis. The x-coordinate corresponding to a y-intercept is always zero.	Ex. 5, p. 99

1.2 Linear Equations and Rational Equations

a. A linear equation in one variable x can be written in the form $ax + b = 0, a \neq 0$.	
b. The procedure for solving a linear equation is given in the box on page 109.	Ex. 1, p. 108; Ex. 2, p. 109
c. If an equation contains fractions, begin by multiplying both sides by the least common denominator, thereby clearing fractions.	Ex. 3, p. 111
d. A rational equation is an equation containing one or more rational expressions. If an equation contains rational expressions with variable denominators, avoid in the solution set any values of the variable that make a denominator zero.	Ex. 4, p. 112; Ex. 5, p. 113; Ex. 6, p. 114
e. An identity is an equation that is true for all real numbers for which both sides are defined. When solving an identity, the variable is eliminated and a true statement, such as $3 = 3$, results. A conditional equation is not an identity and is true for at least one real number. An inconsistent equation is an equation that is not true for even one real number. A false statement, such as $3 = 7$, results when solving such an equation, whose solution set is \varnothing, the empty set.	Ex. 7, p. 116; Ex. 8, p. 117

1.3 Models and Applications

a. A five-step procedure for solving word problems using equations that model verbal conditions is given in the box on page 124.	Ex. 1, p. 125; Ex. 2, p. 126; Ex. 3, p. 128; Ex. 4, p. 129; Ex. 5, p. 130; Ex. 6, p. 132
b. Solving a formula for a variable means rewriting the formula so that the specified variable is isolated on one side of the equation.	Ex. 7, p. 134; Ex. 8, p. 134

1.4 Complex Numbers

a. The imaginary unit i is defined as $$i = \sqrt{-1}, \text{ where } i^2 = -1.$$ The set of numbers in the form $a + bi$ is called the set of complex numbers; a is the real part and b is the imaginary part. If $b = 0$, the complex number is a real number. If $b \neq 0$, the complex number is an imaginary number. Complex numbers in the form bi are called pure imaginary numbers.	
b. Rules for adding and subtracting complex numbers are given in the box on page 140.	Ex. 1, p. 141
c. To multiply complex numbers, multiply as if they are polynomials. After completing the multiplication, replace i^2 with -1 and simplify.	Ex. 2, p. 142
d. The complex conjugate of $a + bi$ is $a - bi$ and vice versa. The multiplication of complex conjugates gives a real number: $$(a + bi)(a - bi) = a^2 + b^2.$$	
e. To divide complex numbers, multiply the numerator and the denominator by the complex conjugate of the denominator.	Ex. 3, p. 143; Ex. 4, p. 143
f. When performing operations with square roots of negative numbers, begin by expressing all square roots in terms of i. The principal square root of $-b$ is defined by $$\sqrt{-b} = i\sqrt{b}.$$	Ex. 5, p. 145

DEFINITIONS AND CONCEPTS	EXAMPLES

1.5 Quadratic Equations

a. A quadratic equation in x can be written in the general form $ax^2 + bx + c = 0, a \neq 0$.

b. The procedure for solving a quadratic equation by factoring and the zero-product principle is given in the box on page 149.	Ex. 1, p. 149
c. The procedure for solving a quadratic equation by the square root property is given in the box on page 151. If $u^2 = d$, then $u = \pm\sqrt{d}$.	Ex. 2, p. 151
d. All quadratic equations can be solved by completing the square. Isolate the binomial with the two variable terms on one side of the equation. If the coefficient of the x^2-term is not one, divide each side of the equation by this coefficient. Then add the square of half the coefficient of x to both sides. Factor the resulting perfect square trinomial. Use the square root property and solve for x.	Ex. 4, p. 154; Ex. 5, p. 155
e. All quadratic equations can be solved by the quadratic formula $$x = \frac{-b \pm \sqrt{b^2 - 4ac}}{2a}.$$ The formula is derived by completing the square of the equation $ax^2 + bx + c = 0$.	Ex. 6, p. 157; Ex. 7, p. 158; Ex. 8, p. 159
f. The discriminant, $b^2 - 4ac$, indicates the number and type of solutions to the quadratic equation $ax^2 + bx + c = 0$, shown in Table 1.2 on page 160.	Ex. 9, p. 160
g. Table 1.3 on page 162 shows the most efficient technique to use when solving a quadratic equation.	Ex. 10, p. 163; Ex. 11, p. 164

1.6 Other Types of Equations

a. Some polynomial equations of degree 3 or greater can be solved by moving all terms to one side, obtaining zero on the other side, factoring, and using the zero-product principle. Factoring by grouping is often used.	Ex. 1, p. 173; Ex. 2, p. 174		
b. A radical equation is an equation in which the variable occurs in a square root, a cube root, and so on. A radical equation can be solved by isolating the radical and raising both sides of the equation to a power equal to the radical's index. When raising both sides to an even power, check all proposed solutions in the original equation. Eliminate extraneous solutions from the solution set.	Ex. 3, p. 176; Ex. 4, p. 177		
c. A radical equation with a rational exponent can be solved by isolating the expression with the rational exponent and raising both sides of the equation to a power that is the reciprocal of the rational exponent. See the details in the box on page 179.	Ex. 5, p. 179		
d. An equation is quadratic in form if it can be written in the form $au^2 + bu + c = 0$, where u is an algebraic expression and $a \neq 0$. Solve for u and use the substitution that resulted in this equation to find the values for the variable in the given equation.	Ex. 6, p. 180; Ex. 7, p. 181; Ex. 8, p. 182		
e. Absolute value equations in the form $	u	= c, c > 0$, can be solved by rewriting the equation without absolute value bars: $u = c$ or $u = -c$.	Ex. 9, p. 183; Ex. 10, p. 183

1.7 Linear Inequalities and Absolute Value Inequalities

a. Solution sets of inequalities are expressed using set-builder notation and interval notation. In interval notation, parentheses indicate endpoints that are not included in an interval. Square brackets indicate endpoints that are included in an interval. See Table 1.4 on page 190.	Ex. 1, p. 191
b. A procedure for finding intersections and unions of intervals is given in the box on page 191.	Ex. 2, p. 191
c. A linear inequality in one variable x can be expressed as $ax + b \leq 0, ax + b < 0, ax + b \geq 0$, or $ax + b > 0, a \neq 0$.	
d. A linear inequality is solved using a procedure similar to solving a linear equation. However, when multiplying or dividing by a negative number, change the sense of the inequality.	Ex. 3, p. 193; Ex. 4, p. 194; Ex. 5, p. 195
e. A compound inequality with three parts can be solved by isolating the variable in the middle.	Ex. 7, p. 197

DEFINITIONS AND CONCEPTS

f. Inequalities involving absolute value can be solved by rewriting the inequalities without absolute value bars. If c is a positive number,

 1. The solutions of $|u| < c$ are the numbers that satisfy $-c < u < c$.

 2. The solutions of $|u| > c$ are the numbers that satisfy $u < -c$ or $u > c$.

EXAMPLES

Ex. 8, p. 198;
Ex. 9, p. 199;
Ex. 10, p. 200

REVIEW EXERCISES

1.1

Graph each equation in Exercises 1–4. Let $x = -3, -2, -1, 0, 1, 2,$ and 3.

1. $y = 2x - 2$ *

2. $y = x^2 - 3$ *

3. $y = x$ *

4. $y = |x| - 2$ *

5. What does a $[-20, 40, 10]$ by $[-5, 5, 1]$ viewing rectangle mean? Draw axes with tick marks and label the tick marks to illustrate this viewing rectangle. *

In Exercises 6–8, use the graph and determine the x-intercepts if any, and the y-intercepts if any. For each graph, tick marks along the axes represent one unit each.

6. x-intercept: -2; y-intercept: 2

7. x-intercepts: $-2, 2$; y-intercept: -4

8. x-intercept: 5; no y-intercept

Salary after College. In 2010, MonsterCollege surveyed 1250 U.S. college students expecting to graduate in the next several years. Respondents were asked the following question:

What do you think your starting salary will be at your first job after college?

The line graph shows the percentage of college students who anticipated various starting salaries. Use the graph to solve Exercises 9–14.

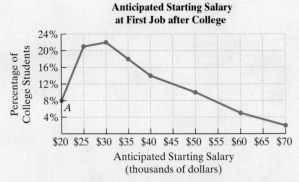

Anticipated Starting Salary at First Job after College

Source: MonsterCollege™

9. What are the coordinates of point A? What does this mean in terms of the information given by the graph? $(20, 8)$*

10. What starting salary was anticipated by the greatest percentage of college students? Estimate the percentage of students who anticipated this salary. $30 thousand; 22%

11. What starting salary was anticipated by the least percentage of college students? Estimate the percentage of students who anticipated this salary. $70 thousand; 2%

12. What starting salaries were anticipated by more than 20% of college students? $25 thousand and $30 thousand

13. Estimate the percentage of students who anticipated a starting salary of $40 thousand. 14%

14. The mathematical model

$$p = -0.01s^2 + 0.8s + 3.7$$

describes the percentage of college students, p, who anticipated a starting salary s, in thousands of dollars. Use this formula to find the percentage of students who anticipated a starting salary of $40 thousand. How does this compare with your estimate in Exercise 13? 19.7%; It's greater than the estimate.

1.2

In Exercises 15–35, solve each equation. Then state whether the equation is an identity, a conditional equation, or an inconsistent equation.

15. $2x - 5 = 7$ {6}; conditional

16. $5x + 20 = 3x$ {-10}; conditional

17. $7(x - 4) = x + 2$ {5}; conditional

18. $1 - 2(6 - x) = 3x + 2$ {-13}; conditional

19. $2(x - 4) + 3(x + 5) = 2x - 2$ {-3}; conditional

20. $2x - 4(5x + 1) = 3x + 17$ {-1}; conditional

21. $7x + 5 = 5(x + 3) + 2x$ ∅; inconsistent

22. $7x + 13 = 2(2x - 5) + 3x + 23$ {x|x is a real number}*

23. $\dfrac{2x}{3} = \dfrac{x}{6} + 1$ {2}; conditional

24. $\dfrac{x}{2} - \dfrac{1}{10} = \dfrac{x}{5} + \dfrac{1}{2}$ {2}; conditional

25. $\dfrac{2x}{3} = 6 - \dfrac{x}{4}$ {$\frac{72}{11}$}; conditional

26. $\dfrac{x}{4} = 2 - \dfrac{x-3}{3}$ {$\frac{36}{7}$}; conditional

27. $\dfrac{3x+1}{3} - \dfrac{13}{2} = \dfrac{1-x}{4}$ {$\frac{77}{15}$}; conditional

28. $\dfrac{9}{4} - \dfrac{1}{2x} = \dfrac{4}{x}$ {2}; conditional

29. $\dfrac{7}{x-5} + 2 = \dfrac{x+2}{x-5}$ ∅; inconsistent

30. $\dfrac{1}{x-1} - \dfrac{1}{x+1} = \dfrac{2}{x^2-1}$ all real numbers except −1 and 1*

31. $\dfrac{5}{x+3} + \dfrac{1}{x-2} = \dfrac{8}{x^2+x-6}$ {$\frac{5}{2}$}; conditional

32. $\dfrac{1}{x+5} = 0$ ∅; inconsistent

33. $\dfrac{4}{x+2} + \dfrac{3}{x} = \dfrac{10}{x^2+2x}$ {$\frac{4}{7}$}; conditional

34. $3 - 5(2x + 1) - 2(x - 4) = 0$ {$\frac{1}{2}$}; conditional

35. $\dfrac{x+2}{x+3} + \dfrac{1}{x^2+2x-3} - 1 = 0$ {2}; conditional

1.3

In Exercises 36–43, use the five-step strategy for solving word problems.

36. The Dog Ate My Calendar. The bar graph shows seven common excuses by college students for not meeting assignment deadlines. The bar heights represent the number of excuses for every 500 excuses that fall into each of these categories.

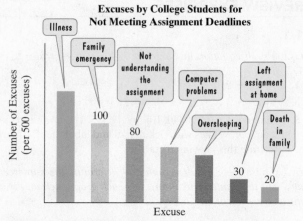

Excuses by College Students for Not Meeting Assignment Deadlines

Source: Roig and Caso, "Lying and Cheating: Fraudulent Excuse Making, Cheating, and Plagiarism," *Journal of Psychology*

For every 500 excuses, the number involving computer problems exceeds the number involving oversleeping by 10. The number involving illness exceeds the number involving oversleeping by 80. Combined, oversleeping, computer problems, and illness account for 270 excuses for not meeting assignment deadlines. For every 500 excuses, determine the number due to oversleeping, computer problems, and illness. oversleeping: 60; computer problems: 70; illness: 140

37. The bar graph shows the average price of a movie ticket for selected years from 1980 through 2013. The graph indicates that in 1980, the average movie ticket price was $2.69. For the period from 1980 through 2013, the price increased by approximately $0.17 per year. If this trend continues, by which year will the average price of a movie ticket be $9.49? by 40 years after 1980, or in 2020

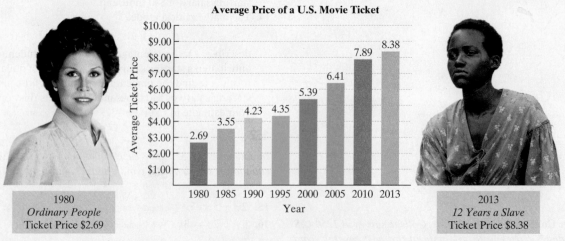

Average Price of a U.S. Movie Ticket

1980
Ordinary People
Ticket Price $2.69

2013
12 Years a Slave
Ticket Price $8.38

Sources: Motion Picture Association of America, National Association of Theater Owners (NATO), and Bureau of Labor Statistics (BLS)

38. You are choosing between two cellphone plans. Data Plan A has a monthly fee of $52 with a charge of $18 per gigabyte (GB). Data Plan B has a monthly fee of $32 with a charge of $22 per GB. For how many GB of data will the costs for the two data plans be the same? 5 GB

39. After a 20% price reduction, a cordless phone sold for $48. What was the phone's price before the reduction? $60

40. A salesperson earns $300 per week plus 5% commission of sales. How much must be sold to earn $800 in a week? $10,000 in sales

41. You invested $9000 in two funds paying 4% and 7% annual interest. At the end of the year, the total interest from these investments was $555. How much was invested at each rate? $2500 at 4%; $6500 at 7%

42. You invested $8000 in two funds paying 2% and 5% annual interest. At the end of the year, the interest from the 5% investment exceeded the interest from the 2% investment by $85. How much money was invested at each rate? *

43. The length of a rectangular field is 6 yards less than triple the width. If the perimeter of the field is 340 yards, what are its dimensions? 44 yd by 126 yd

44. In 2015, there were 14,100 students at college A, with a projected enrollment increase of 1500 students per year. In the same year, there were 41,700 students at college B, with a projected enrollment decline of 800 students per year.

a. Let x represent the number of years after 2015. Write, but do not solve, an equation that can be used to find how many years after 2015 the colleges will have the same enrollment. $14,100 + 1500x = 41,700 - 800x$

b. The following table is based on your equation in part (a). Y_1 represents one side of the equation and Y_2 represents the other side of the equation. Use the table to answer these questions: In which year will the colleges have the same enrollment? What will be the enrollment in each college at that time? 2027; 32,100

X	Y₁	Y₂		
7	24600	36100		
8	26100	35300		
9	27600	34500		
10	29100	33700		
11	30600	32900		
12	32100	32100		
13	33600	31300		
14	35100	30500		
15	36600	29700		
16	38100	28900		
17	39600	28100		

X=7

In Exercises 45–47, solve each formula for the specified variable.

45. $vt + gt^2 = s$ for g $g = \frac{s - vt}{t^2}$ **46.** $T = gr + gvt$ for g *

47. $T = \dfrac{A - P}{Pr}$ for P $P = \dfrac{A}{1 + rT}$

1.4

In Exercises 48–57, perform the indicated operations and write the result in standard form.

48. $(8 - 3i) - (17 - 7i)$ $-9 + 4i$ **49.** $4i(3i - 2)$ $-12 - 8i$

50. $(7 - i)(2 + 3i)$ $17 + 19i$ **51.** $(3 - 4i)^2$ $-7 - 24i$

52. $(7 + 8i)(7 - 8i)$ 113 **53.** $\dfrac{6}{5 + i}$ $\frac{15}{13} - \frac{3}{13}i$

54. $\dfrac{3 + 4i}{4 - 2i}$ $\frac{1}{5} + \frac{11}{10}i$ **55.** $\sqrt{-32} - \sqrt{-18}$ $i\sqrt{2}$

56. $\left(-2 + \sqrt{-100}\right)^2$ $-96 - 40i$ **57.** $\dfrac{4 + \sqrt{-8}}{2}$ $2 + i\sqrt{2}$

1.5

Solve each equation in Exercises 58–59 by factoring.

58. $2x^2 + 15x = 8$ $\{-8, \frac{1}{2}\}$ **59.** $5x^2 + 20x = 0$ $\{-4, 0\}$

Solve each equation in Exercises 60–63 by the square root property.

60. $2x^2 - 3 = 125$ $\{-8, 8\}$ **61.** $\dfrac{x^2}{2} + 5 = -3$ $\{-4i, 4i\}$

62. $(x + 3)^2 = -10$ $\{-3 \pm i\sqrt{10}\}$ **63.** $(3x - 4)^2 = 18$ *

In Exercises 64–65, determine the constant that should be added to the binomial so that it becomes a perfect square trinomial. Then write and factor the trinomial.

64. $x^2 + 20x$ $100; (x + 10)^2$ **65.** $x^2 - 3x$ $\frac{9}{4}; \left(x - \frac{3}{2}\right)^2$

Solve each equation in Exercises 66–67 by completing the square.

66. $x^2 - 12x + 27 = 0$ $\{3, 9\}$

67. $3x^2 - 12x + 11 = 0$ $\left\{2 + \frac{\sqrt{3}}{3}, 2 - \frac{\sqrt{3}}{3}\right\}$

Solve each equation in Exercises 68–70 using the quadratic formula.

68. $x^2 = 2x + 4$ $\{1 \pm \sqrt{5}\}$ **69.** $x^2 - 2x + 19 = 0$ *

70. $2x^2 = 3 - 4x$ $\left\{\frac{-2 \pm \sqrt{10}}{2}\right\}$

In Exercises 71–72, without solving the given quadratic equation, determine the number and type of solutions.

71. $x^2 - 4x + 13 = 0$ * **72.** $9x^2 = 2 - 3x$ *

Solve each equation in Exercises 73–81 by the method of your choice.

73. $2x^2 - 11x + 5 = 0$ $\{\frac{1}{2}, 5\}$ **74.** $(3x + 5)(x - 3) = 5$ *

75. $3x^2 - 7x + 1 = 0$ * **76.** $x^2 - 9 = 0$ $\{-3, 3\}$

77. $(x - 3)^2 - 25 = 0$ $\{-2, 8\}$ **78.** $3x^2 - x + 2 = 0$ *

79. $3x^2 - 10x = 8$ $\{-\frac{2}{3}, 4\}$ **80.** $(x + 2)^2 + 4 = 0$ *

81. $\dfrac{5}{x + 1} + \dfrac{x - 1}{4} = 2$ $\{4 \pm \sqrt{5}\}$

82. The formula $W = 3t^2$ models the weight of a human fetus, W, in grams, after t weeks, where $0 \le t \le 39$. After how many weeks does the fetus weigh 588 grams? 14 weeks

83. One possible reason for the explosion of college tuition involves the decrease in government aid per student. In 2001, higher-education revenues per student averaged $8500. The bar graph shows government aid per U.S. college student from 2005 through 2012. (All figures are adjusted for inflation and expressed in 2012 dollars.)

Higher-Education Government Aid per U.S. College Student

Source: State Higher Education Executive Officers Association

The mathematical model

$$G = -82x^2 + 410x + 7079$$

describes government aid per college student, G, x years after 2005.

a. Does the model underestimate or overestimate aid per student in 2011? By how much? underestimates by $13

b. If we project the model past 2012, determine in which year government aid per student decreased to $4127. 2014

84. An architect is allowed 15 square yards of floor space to add a small bedroom to a house. Because of the room's design in relationship to the existing structure, the width of the rectangular floor must be 7 yards less than two times the length. Find the length and width of the rectangular floor that the architect is permitted. length: 5 yd; width: 3 yd

85. A building casts a shadow that is double the height of the building. If the distance from the end of the shadow to the top of the building is 300 meters, how high is the building? Round to the nearest meter. about 134 m

1.6

Solve each polynomial equation in Exercises 86–87.

86. $2x^4 = 50x^2$ {−5, 0, 5} **87.** $2x^3 - x^2 - 18x + 9 = 0$ *

Solve each radical equation in Exercises 88–89.

88. $\sqrt{2x - 3} + x = 3$ {2} **89.** $\sqrt{x - 4} + \sqrt{x + 1} = 5$ *

Solve the equations with rational exponents in Exercises 90–91.

90. $3x^{\frac{3}{4}} - 24 = 0$ {16} **91.** $(x - 7)^{\frac{2}{3}} = 25$ {−118, 132}

Solve each equation in Exercises 92–93 by making an appropriate substitution.

92. $x^4 - 5x^2 + 4 = 0$ * **93.** $x^{\frac{1}{2}} + 3x^{\frac{1}{4}} - 10 = 0$ {16}

Solve the equations containing absolute value in Exercises 94–95.

94. $|2x + 1| = 7$ {−4, 3} **95.** $2|x - 3| - 6 = 10$ {−5, 11}

Solve each equation in Exercises 96–102 by the method of your choice.

96. $3x^{\frac{4}{3}} - 5x^{\frac{2}{3}} + 2 = 0$ * **97.** $2\sqrt{x - 1} = x$ {2}

98. $|2x - 5| - 3 = 0$ {1, 4} **99.** $x^3 + 2x^2 = 9x + 18$ *

100. $\sqrt{2x + 3} - x = 0$ {3} **101.** $x^3 + 3x^2 - 2x - 6 = 0$ *

102. $-4|x + 1| + 12 = 0$ {−4, 2}

103. Long before the Supreme Court decision legalizing marriage between people of the same sex, a substantial percentage of Americans were in favor of laws prohibiting interracial marriage. Since 1972, the General Social Survey has asked

Do you think that there should be laws against marriages between negroes/blacks/African Americans and whites?

In 1972, 39.3% of the adult U.S. population was in favor of such laws. The bar graph shows the percentage of Americans in favor of legislation prohibiting interracial marriage for five selected years from 1993 through 2002, the last year the data were collected.

Percentage of Americans in Favor of Laws Prohibiting Interracial Marriage

Source: Ben Schott, *Schott's Almanac 2007*, Donnelley and Sons

The formula

$$p = -2.5\sqrt{t} + 17$$

models the percentage of Americans, p, in favor of laws prohibiting interracial marriage t years after 1993. If trends indicated by this model continue, in what year did the percentage favoring such legislation decrease to 7%? by 16 years after 1993, or in 2009

1.7

In Exercises 104–106, express each interval in set-builder notation and graph the interval on a number line.

104. $[-3, 5)$ * **105.** $(-2, \infty)$ * **106.** $(-\infty, 0]$ *

In Exercises 107–110, use graphs to find each set.

107. $(-2, 1] \cap [-1, 3)$ [−1, 1]

108. $(-2, 1] \cup [-1, 3)$ (−2, 3)

109. $[1, 3) \cap (0, 4)$ [1, 3) **110.** $[1, 3) \cup (0, 4)$ (0, 4)

In Exercises 111–121, solve each inequality. Other than \varnothing, use interval notation to express solution sets and graph each solution set on a number line.

111. $-6x + 3 \le 15$ [−2, ∞)*

112. $6x - 9 \ge -4x - 3$ [$\frac{3}{5}$, ∞)*

113. $\frac{x}{3} - \frac{3}{4} - 1 > \frac{x}{2}$ (−∞, −$\frac{21}{2}$)*

114. $6x + 5 > -2(x - 3) - 25$ (−3, ∞)*

115. $3(2x - 1) - 2(x - 4) \ge 7 + 2(3 + 4x)$ (−∞, −2]*

116. $5(x - 2) - 3(x + 4) \ge 2x - 20$ ∅

117. $7 < 2x + 3 \le 9$ (2, 3]*

118. $|2x + 3| \le 15$ [−9, 6]*

119. $\left|\frac{2x + 6}{3}\right| > 2$ (−∞, −6) ∪ (0, ∞)*

120. $|2x + 5| - 7 \ge -6$ (−∞, −3] ∪ [−2, ∞)*

121. $-4|x + 2| + 5 \le -7$ (−∞, −5] ∪ [1, ∞)*

In Exercises 122–123, use interval notation to represent all values of x satisfying the given conditions.

122. $y_1 = -10 - 3(2x + 1)$, $y_2 = 8x + 1$, and $y_1 > y_2$. (−∞, −1)

123. $y = 3 - |2x - 5|$ and y is at least −6. [−2, 7]

124. A car rental agency rents a certain car for $40 per day with unlimited mileage or $24 per day plus $0.20 per mile. How far can a customer drive this car per day for the $24 option to cost no more than the unlimited mileage option? no more than 80 miles per day

125. To receive a B in a course, you must have an average of at least 80% but less than 90% on five exams. Your grades on the first four exams were 95%, 79%, 91%, and 86%. What range of grades on the fifth exam will result in a B for the course? [49%, 99%)

126. A retiree requires an annual income of at least $9000 from an investment paying 7.5% annual interest. How much should the retiree invest to achieve the desired return? at least $120,000

CHAPTER 1 TEST

In Exercises 1–23, solve each equation or inequality. Other than ∅, use interval notation to express solution sets of inequalities and graph these solution sets on a number line.

1. $7(x - 2) = 4(x + 1) - 21$ {−1}

2. $-10 - 3(2x + 1) - 8x - 1 = 0$ {−1}

3. $\dfrac{2x - 3}{4} = \dfrac{x - 4}{2} - \dfrac{x + 1}{4}$ {−6}

4. $\dfrac{2}{x - 3} - \dfrac{4}{x + 3} = \dfrac{8}{x^2 - 9}$ {5}

5. $2x^2 - 3x - 2 = 0$ {−$\frac{1}{2}$, 2}

6. $(3x - 1)^2 = 75$ {$\frac{1 \pm 5\sqrt{3}}{3}$}

7. $(x + 3)^2 + 25 = 0$ {−3 ± 5i}

8. $x(x - 2) = 4$ {1 ± $\sqrt{5}$}

9. $4x^2 = 8x - 5$ {1 ± $\frac{1}{2}i$}

10. $x^3 - 4x^2 - x + 4 = 0$ {−1, 1, 4}

11. $\sqrt{x - 3} + 5 = x$ {7}

12. $\sqrt{8 - 2x} - x = 0$ {2}

13. $\sqrt{x + 4} + \sqrt{x - 1} = 5$ {5}

14. $5x^{\frac{3}{2}} - 10 = 0$ {$\sqrt[3]{4}$}

15. $x^{\frac{2}{3}} - 9x^{\frac{1}{3}} + 8 = 0$ {1, 512}

16. $\left|\dfrac{2}{3}x - 6\right| = 2$ {6, 12}

17. $-3|4x - 7| + 15 = 0$ {$\frac{1}{2}$, 3}

18. $\dfrac{1}{x^2} - \dfrac{4}{x} + 1 = 0$ {2 ± $\sqrt{3}$}

19. $\dfrac{2x}{x^2 + 6x + 8} + \dfrac{2}{x + 2} = \dfrac{x}{x + 4}$ {4}

20. $3(x + 4) \geq 5x - 12$ (−∞, 12]*

21. $\dfrac{x}{6} + \dfrac{1}{8} \leq \dfrac{x}{2} - \dfrac{3}{4}$ [$\frac{21}{8}$, ∞)*

22. $-3 \leq \dfrac{2x + 5}{3} < 6$ [−7, $\frac{13}{2}$)*

23. $|3x + 2| \geq 3$ (−∞, −$\frac{5}{3}$] ∪ [$\frac{1}{3}$, ∞)*

In Exercises 24–25, use interval notation to represent all values of x satisfying the given conditions.

24. $y = 2x - 5$, and y is at least −3 and no more than 7. [1, 6]

25. $y = \left|\dfrac{2 - x}{4}\right|$ and y is at least 1. (−∞, −2] ∪ [6, ∞)

In Exercises 26–27, use graphs to find each set.

26. $[-1, 2) \cup (0, 5]$ [−1, 5]

27. $[-1, 2) \cap (0, 5]$ (0, 2)

In Exercises 28–29, solve each formula for the specified variable.

28. $V = \dfrac{1}{3}lwh$ for h $h = \dfrac{3V}{lw}$

29. $y - y_1 = m(x - x_1)$ for x $x = \dfrac{y - y_1}{m} + x_1$

In Exercises 30–31, graph each equation in a rectangular coordinate system.

30. $y = 2 - |x|$ *

31. $y = x^2 - 4$ *

In Exercises 32–34, perform the indicated operations and write the result in standard form.

32. $(6 - 7i)(2 + 5i)$ 47 + 16i

33. $\dfrac{5}{2 - i}$ 2 + i

34. $2\sqrt{-49} + 3\sqrt{-64}$ 38i

The graphs show the amount being paid in Social Security benefits and the amount going into the system. All data are expressed in billions of dollars. Amounts from 2016 through 2024 are projections.

Social Insecurity: Income and Outflow of the Social Security System

Source: 2004 Social Security Trustees Report

Exercises 35–37 are based on the data shown by the graphs.

35. In 2004, the system's income was $575 billion, projected to increase at an average rate of $43 billion per year. In which year will the system's income be $1177 billion? 2018

36. The data for the system's outflow can be modeled by the formula

$$B = 0.07x^2 + 47.4x + 500,$$

where B represents the amount paid in benefits, in billions of dollars, x years after 2004. According to this model, when will the amount paid in benefits be $1177 billion? Round to the nearest year. 2018

37. How well do your answers to Exercises 35 and 36 model the data shown by the graphs? quite well

38. **Here's Looking at You.** According to University of Texas economist Daniel Hamermesh (*Beauty Pays: Why Attractive People Are More Successful*), strikingly attractive and good-looking men and women can expect to earn an average of $230,000 more in a lifetime than a person who is homely or plain. (Your author feels the need to start affirmative action for the beauty-bereft, consoled by the reality that looks are only one of many things that matter.) The bar graph shows the distribution of looks for American men and women, ranging from homely to strikingly attractive.

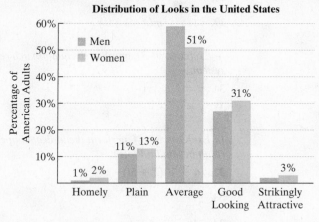

Distribution of Looks in the United States

Source: *Time*, August 22, 2011

The percentage of average-looking men exceeds the percentage of strikingly attractive men by 57. The percentage of good-looking men exceeds the percentage of strikingly attractive men by 25. A total of 88% of American men range between average looking, good-looking, and strikingly attractive. Find the percentage of men who fall within each of these three categories of looks.
average: 59%; good-looking: 27%; strikingly attractive: 2%

39. The costs for two different kinds of heating systems for a small home are given in the following table. After how many years will total costs for solar heating and electric heating be the same? What will be the cost at that time? 26 yr; $33,600

System	Cost to Install	Operating Cost/Year
Solar	$29,700	$150
Electric	$5000	$1100

40. You invested $10,000 in two accounts paying 8% and 10% annual interest. At the end of the year, the total interest from these investments was $940. How much was invested at each rate? $3000 at 8%; $7000 at 10%

41. The length of a rectangular carpet is 4 feet greater than twice its width. If the area is 48 square feet, find the carpet's length and width. length: 12 ft; width: 4 ft

42. A vertical pole is to be supported by a wire that is 26 feet long and anchored 24 feet from the base of the pole. How far up the pole should the wire be attached? 10 ft

43. After a 60% reduction, a jacket sold for $20. What was the jacket's price before the reduction? $50

44. You are choosing between two texting plans. Plan A charges $25 per month for unlimited texting. Plan B has a monthly fee of $13 with a charge of $0.06 per text. How many text messages in a month make plan A the better deal? more than 200 text messages

Functions and Graphs

A vast expanse of open water at the top of our world was once covered with ice. The melting of the Arctic ice caps has forced polar bears to swim as far as 40 miles, causing them to drown in significant numbers. Such deaths were rare in the past.

There is strong scientific consensus that human activities are changing the Earth's climate. Scientists now believe that there is a striking correlation between atmospheric carbon dioxide concentration and global temperature. As both of these variables increase at significant rates, there are warnings of a planetary emergency that threatens to condemn coming generations to a catastrophically diminished future.*

In this chapter, you'll learn to approach our climate crisis mathematically by creating formulas, called functions, that model data for average global temperature and carbon dioxide concentration over time. Understanding the concept of a function will give you a new perspective on many situations, ranging from climate change to using mathematics in a way that is similar to making a movie.

*Sources: Al Gore, *An Inconvenient Truth*, Rodale, 2006; *Time*, April 3, 2006; *Rolling Stone*, September 26, 2013

HERE'S WHERE YOU'LL FIND THESE APPLICATIONS:

A mathematical model involving global warming is developed in Example 9 in Section 2.3.

Using mathematics in a way that is similar to making a movie is discussed in the Blitzer Bonus on page 278.

Section 2.1

Basics of Functions and Their Graphs

Magnified 6000 times, this color-scanned image shows a T-lymphocyte blood cell (green) infected with the HIV virus (red). Depletion of the number of T cells causes destruction of the immune system.

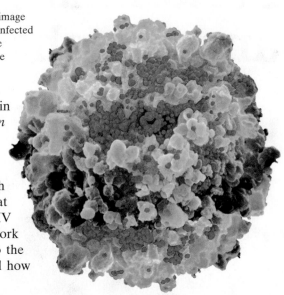

The average number of T cells in a person with HIV is a *function* of time after infection. In this section, you will be introduced to the basics of functions and their graphs. We will analyze the graph of a function using an example that illustrates the progression of HIV and T cell count. Much of our work in this course will be devoted to the important topic of functions and how they model your world.

Relations

Forbes magazine published a list of the highest-paid TV celebrities between June 2013 and June 2014. The results are shown in **Figure 2.1**.

Highest-Paid TV Celebrities between June 2013 and June 2014

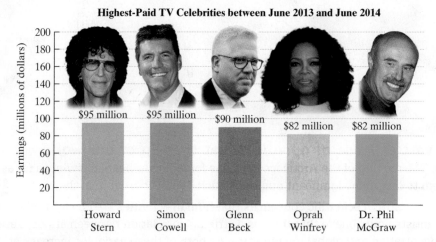

FIGURE 2.1
Source: *Forbes*

1 Find the domain and range of a relation.

The graph indicates a correspondence between a TV celebrity and that person's earnings, in millions of dollars. We can write this correspondence using a set of ordered pairs:

{(Stern, 95), (Cowell, 95), (Beck, 90), (Winfrey, 82), (McGraw, 82)}.

These braces indicate we are representing a set.

The mathematical term for a set of ordered pairs is a *relation*.

> ### Definition of a Relation
>
> A **relation** is any set of ordered pairs. The set of all first components of the ordered pairs is called the **domain** of the relation and the set of all second components is called the **range** of the relation.

EXAMPLE 1 Finding the Domain and Range of a Relation

Find the domain and range of the relation:

{(Stern, 95), (Cowell, 95), (Beck, 90), (Winfrey, 82), (McGraw, 82)}.

SOLUTION

The domain is the set of all first components. Thus, the domain is

{Stern, Cowell, Beck, Winfrey, McGraw}.

The range is the set of all second components. Thus, the range is

{95, 90, 82}.

> Although Stern and Cowell both earned $95 million, it is not necessary to list 95 twice.

> Although Winfrey and McGraw both earned $82 million, it is not necessary to list 82 twice.

● ● ●

⊘ **Check Point 1** Find the domain and range of the relation:

{(0, 9.1), (10, 6.7), (20, 10.7), (30, 13.2), (42, 21.7)}. domain: {0, 10, 20, 30, 42};

range: {9.1, 6.7, 10.7, 13.2, 21.7}

As you worked Check Point 1, did you wonder if there was a rule that assigned the "inputs" in the domain to the "outputs" in the range? For example, for the ordered pair (30, 13.2), how does the output 13.2 depend on the input 30? The ordered pair is based on the data in **Figure 2.2(a)**, which shows the percentage of first-year U.S. college students claiming no religious affiliation for selected years from 1970 through 2012.

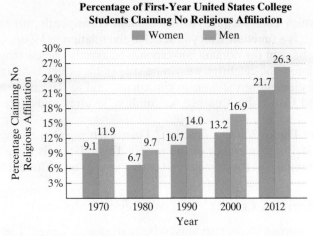

FIGURE 2.2(a) Data for women and men
Source: John Macionis, *Sociology*, Fifteenth Edition, Pearson, 2014.

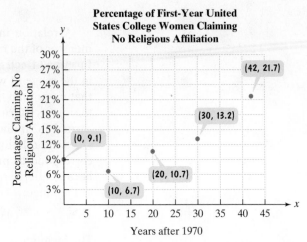

FIGURE 2.2(b) Visually representing the relation for the women's data

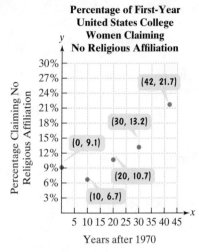

Percentage of First-Year United States College Women Claiming No Religious Affiliation

FIGURE 2.2(b) (repeated)
Visually representing the relation for the women's data

2 Determine whether a relation is a function.

Table 2.1 Highest-Paid TV Celebrities

Celebrity	Earnings (millions of dollars)
Stern	95
Cowell	95
Beck	90
Winfrey	82
McGraw	82

In **Figure 2.2(b)**, we used the data for college women to create the following ordered pairs:

$$\left(\text{years after 1970,} \quad \begin{array}{l} \text{percentage of first-year college} \\ \text{women claiming no religious} \\ \text{affiliation} \end{array} \right).$$

Consider, for example, the ordered pair (30, 13.2).

(30, 13.2)

| 30 years after 1970, or in 2000, | 13.2% of first-year college women claimed no religious affiliation. |

The five points in **Figure 2.2(b)** visually represent the relation formed from the women's data. Another way to visually represent the relation is as follows:

Domain		Range
0	→	9.1
10	→	6.7
20	→	10.7
30	→	13.2
42	→	21.7

Functions

Table 2.1, based on our earlier discussion, shows the highest-paid TV celebrities and their earnings between June 2013 and June 2014, in millions of dollars. We've used this information to define two relations.

Figure 2.3(a) shows a correspondence between celebrities and their earnings. **Figure 2.3(b)** shows a correspondence between earnings and celebrities.

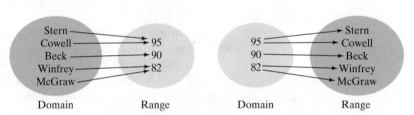

FIGURE 2.3(a) Celebrities correspond to earnings.

FIGURE 2.3(b) Earnings correspond to celebrities.

A relation in which each member of the domain corresponds to exactly one member of the range is a **function**. Can you see that the relation in **Figure 2.3(a)** is a function? Each celebrity in the domain corresponds to exactly one earning amount in the range. If we know the celebrity, we can be sure of his or her earnings. Notice that more than one element in the domain can correspond to the same element in the range: Stern and Cowell both earned $95 million. Winfrey and McGraw both earned $82 million.

Is the relation in **Figure 2.3(b)** a function? Does each member of the domain correspond to precisely one member of the range? This relation is not a function because there are members of the domain that correspond to two different members of the range:

(95, Stern) (95, Cowell) (82, Winfrey) (82, McGraw).

The member of the domain 95 corresponds to both Stern and Cowell. The member of the domain 82 corresponds to both Winfrey and McGraw. If we know that the earnings are $95 million or $82 million, we cannot be sure of the celebrity. Because

a function is a relation in which no two ordered pairs have the same first component and different second components, the ordered pairs (95, Stern) and (95, Cowell) are not ordered pairs of a function. Similarly, (82, Winfrey) and (82, McGraw) are not ordered pairs of a function.

Same first component

(95, Stern) (95, Cowell)

Different second components

Same first component

(82, Winfrey) (82, McGraw)

Different second components

Definition of a Function

A **function** is a correspondence from a first set, called the **domain**, to a second set, called the **range**, such that each element in the domain corresponds to *exactly one* element in the range.

In Check Point 1, we considered a relation that gave a correspondence between years after 1970 and the percentage of first-year college women claiming no religious affiliation. Can you see that this relation is a function?

Each element in the domain

$\{(0, 9.1), (10, 6.7), (20, 10.7), (30, 13.2), (42, 21.7)\}$

corresponds to exactly one element in the range.

However, Example 2 illustrates that not every correspondence between sets is a function.

EXAMPLE 2 Determining Whether a Relation Is a Function

Determine whether each relation is a function:

a. $\{(1, 6), (2, 6), (3, 8), (4, 9)\}$ **b.** $\{(6, 1), (6, 2), (8, 3), (9, 4)\}$.

SOLUTION

We begin by making a figure for each relation that shows the domain and the range (**Figure 2.4**).

Domain Range

FIGURE 2.4(a)

Domain Range

FIGURE 2.4(b)

a. Figure 2.4(a) shows that every element in the domain corresponds to exactly one element in the range. The element 1 in the domain corresponds to the element 6 in the range. Furthermore, 2 corresponds to 6, 3 corresponds to 8, and 4 corresponds to 9. No two ordered pairs in the given relation have the same first component and different second components. Thus, the relation is a function.

b. Figure 2.4(b) shows that 6 corresponds to both 1 and 2. If any element in the domain corresponds to more than one element in the range, the relation is not a function. This relation is not a function; two ordered pairs have the same first component and different second components.

Same first component

(6, 1) (6, 2)

Different second components

GREAT QUESTION!

If I reverse a function's components, will this new relation be a function?

If a relation is a function, reversing the components in each of its ordered pairs may result in a relation that is not a function.

• • •

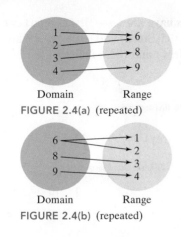

Domain Range
FIGURE 2.4(a) (repeated)

Domain Range
FIGURE 2.4(b) (repeated)

③ Determine whether an equation represents a function.

Look at **Figure 2.4(a)** again. The fact that 1 and 2 in the domain correspond to the same number, 6, in the range does not violate the definition of a function. **A function can have two different first components with the same second component.** By contrast, a relation is not a function when two different ordered pairs have the same first component and different second components. Thus, the relation in **Figure 2.4(b)** is not a function.

⌀ Check Point **2** Determine whether each relation is a function:

a. $\{(1, 2), (3, 4), (5, 6), (5, 8)\}$ **b.** $\{(1, 2), (3, 4), (6, 5), (8, 5)\}$.

 not a function function

Functions as Equations

Functions are usually given in terms of equations rather than as sets of ordered pairs. For example, here is an equation that models the percentage of first-year college women claiming no religious affiliation as a function of time:

$$y = 0.012x^2 - 0.2x + 8.7.$$

The variable x represents the number of years after 1970. The variable y represents the percentage of first-year college women claiming no religious affiliation. The variable y is a function of the variable x. For each value of x, there is one and only one value of y. The variable x is called the **independent variable** because it can be assigned any value from the domain. Thus, x can be assigned any nonnegative integer representing the number of years after 1970. The variable y is called the **dependent variable** because its value depends on x. The percentage claiming no religious affiliation depends on the number of years after 1970. The value of the dependent variable, y, is calculated after selecting a value for the independent variable, x.

We have seen that not every set of ordered pairs defines a function. Similarly, **not all equations with the variables x and y define functions. If an equation is solved for y and more than one value of y can be obtained for a given x, then the equation does not define y as a function of x.**

 EXAMPLE 3 Determining Whether an Equation Represents a Function

Determine whether each equation defines y as a function of x:

 a. $x^2 + y = 4$ **b.** $x^2 + y^2 = 4$.

SOLUTION

Solve each equation for y in terms of x. If two or more values of y can be obtained for a given x, the equation is not a function.

a.

$x^2 + y = 4$	This is the given equation.
$x^2 + y - x^2 = 4 - x^2$	Solve for y by subtracting x^2 from both sides.
$y = 4 - x^2$	Simplify.

From this last equation we can see that for each value of x, there is one and only one value of y. For example, if $x = 1$, then $y = 4 - 1^2 = 3$. The equation defines y as a function of x.

b.

$x^2 + y^2 = 4$	This is the given equation.
$x^2 + y^2 - x^2 = 4 - x^2$	Isolate y^2 by subtracting x^2 from both sides.
$y^2 = 4 - x^2$	Simplify.
$y = \pm\sqrt{4 - x^2}$	Apply the square root property: If $u^2 = d$, then $u = \pm\sqrt{d}$.

The \pm in this last equation shows that for certain values of x (all values between -2 and 2), there are two values of y. For example, if $x = 1$, then $y = \pm\sqrt{4 - 1^2} = \pm\sqrt{3}$. For this reason, the equation does not define y as a function of x. •••

 Check Point **3** Solve each equation for y and then determine whether the equation defines y as a function of x:

a. $2x + y = 6$ \quad $y = 6 - 2x$; function \qquad **b.** $x^2 + y^2 = 1.$ \quad $y = \pm\sqrt{1 - x^2}$; not a function

 Evaluate a function.

Function Notation

If an equation in x and y gives one and only one value of y for each value of x, then the variable y is a function of the variable x. When an equation represents a function, the function is often named by a letter such as f, g, h, F, G, or H. Any letter can be used to name a function. Suppose that f names a function. Think of the domain as the set of the function's inputs and the range as the set of the function's outputs. As shown in **Figure 2.5**, input is represented by x and the output by $f(x)$. The special notation $f(x)$, read "f of x" or "f at x," represents the **value of the function at the number x**.

Let's make this clearer by considering a specific example. We know that the equation

$$y = 0.012x^2 - 0.2x + 8.7$$

defines y as a function of x. We'll name the function f. Now, we can apply our new function notation.

FIGURE 2.5 A "function machine" with inputs and outputs

We read this equation as "f of x equals $0.012x^2 - 0.2x + 8.7$."

Input	Output	Equation
x	$f(x)$	$f(x) = 0.012x^2 - 0.2x + 8.7$

Suppose we are interested in finding $f(30)$, the function's output when the input is 30. To find the value of the function at 30, we substitute 30 for x. We are **evaluating the function** at 30.

$f(x) = 0.012x^2 - 0.2x + 8.7$	This is the given function.
$f(30) = 0.012(30)^2 - 0.2(30) + 8.7$	Replace each occurrence of x with 30.
$\quad\;\; = 0.012(900) - 0.2(30) + 8.7$	Evaluate the exponential expression: $30^2 = 30 \cdot 30 = 900$.
$\quad\;\; = 10.8 - 6 + 8.7$	Perform the multiplications.
$f(30) = 13.5$	Subtract and add from left to right.

The statement $f(30) = 13.5$, read "f of 30 equals 13.5," tells us that the value of the function at 30 is 13.5. When the function's input is 30, its output is 13.5. **Figure 2.6** illustrates the input and output in terms of a function machine.

$$f(30) = 13.5$$

30 years after 1970, or in 2000,	13.5% of first-year college women claimed no religious affiliation.

FIGURE 2.6 A function machine at work

We have seen that in 2000, 13.2% actually claimed nonaffiliation, so our function that models the data slightly overestimates the percent for 2000.

TECHNOLOGY

Graphing utilities can be used to evaluate functions. The screens below show the evaluation of

$$f(x) = 0.012x^2 - 0.2x + 8.7$$

at 30 on a TI-84 Plus C graphing calculator. The function f is named Y_1.

```
Plot1  Plot2  Plot3
∎\Y₁❑0.012X²-0.2X+8.7

∎\Y₂=
∎\Y₃=
∎\Y₄=
∎\Y₅=
∎\Y₆=
∎\Y₇=
```

```
Y₁(30)
                    13.5
```

We used $f(x) = 0.012x^2 - 0.2x + 8.7$ to find $f(30)$. To find other function values, such as $f(40)$ or $f(55)$, substitute the specified input value, 40 or 55, for x in the function's equation.

If a function is named f and x represents the independent variable, the notation $f(x)$ corresponds to the y-value for a given x. Thus,

$$f(x) = 0.012x^2 - 0.2x + 8.7 \quad \text{and} \quad y = 0.012x^2 - 0.2x + 8.7$$

define the same function. This function may be written as

$$y = f(x) = 0.012x^2 - 0.2x + 8.7.$$

EXAMPLE 4 Evaluating a Function

If $f(x) = x^2 + 3x + 5$, evaluate each of the following:

 a. $f(2)$ **b.** $f(x + 3)$ **c.** $f(-x)$.

SOLUTION

We substitute 2, $x + 3$, and $-x$ for x in the equation for f. When replacing x with a variable or an algebraic expression, you might find it helpful to think of the function's equation as

$$f(x) = \boxed{x}^2 + 3\boxed{x} + 5.$$

a. We find $f(2)$ by substituting 2 for x in the equation.

$$f(2) = 2^2 + 3 \cdot 2 + 5 = 4 + 6 + 5 = 15$$

Thus, $f(2) = 15$.

b. We find $f(x + 3)$ by substituting $x + 3$ for x in the equation.

$$f(x + 3) = (x + 3)^2 + 3(x + 3) + 5$$

Equivalently,

$$
\begin{aligned}
f(x + 3) &= (x + 3)^2 + 3(x + 3) + 5 \\
&= x^2 + 6x + 9 + 3x + 9 + 5 && \text{Square } x + 3 \text{ using} \\
&&& (A + B)^2 = A^2 + 2AB + B^2. \\
&&& \text{Distribute 3 throughout the parentheses.} \\
&= x^2 + 9x + 23. && \text{Combine like terms.}
\end{aligned}
$$

DISCOVERY

Using $f(x) = x^2 + 3x + 5$ and the answers in parts (b) and (c):

 1. Is $f(x + 3)$ equal to $f(x) + f(3)$?

 2. Is $f(-x)$ equal to $-f(x)$?

c. We find $f(-x)$ by substituting $-x$ for x in the equation.

$$f(-x) = (-x)^2 + 3(-x) + 5$$

Equivalently,

$$
\begin{aligned}
f(-x) &= (-x)^2 + 3(-x) + 5 \\
&= x^2 - 3x + 5.
\end{aligned}
$$

•••

✓ **Check Point 4** If $f(x) = x^2 - 2x + 7$, evaluate each of the following:

a. $f(-5)$ $_{42}$ **b.** $f(x + 4)$ $_{x^2 + 6x + 15}$ **c.** $f(-x)$. $_{x^2 + 2x + 7}$

⑤ Graph functions by plotting points.

Graphs of Functions

The **graph of a function** is the graph of its ordered pairs. For example, the graph of $f(x) = 2x$ is the set of points (x, y) in the rectangular coordinate system satisfying $y = 2x$. Similarly, the graph of $g(x) = 2x + 4$ is the set of points (x, y) in the rectangular coordinate system satisfying the equation $y = 2x + 4$. In the next example, we graph both of these functions in the same rectangular coordinate system.

EXAMPLE 5 Graphing Functions

Graph the functions $f(x) = 2x$ and $g(x) = 2x + 4$ in the same rectangular coordinate system. Select integers for x, starting with -2 and ending with 2.

SOLUTION

We begin by setting up a partial table of coordinates for each function. Then we plot the five points in each table and connect them, as shown in **Figure 2.7**. The graph of each function is a straight line. Do you see a relationship between the two graphs? The graph of g is the graph of f shifted vertically up by 4 units.

FIGURE 2.7

x	$f(x) = 2x$	(x, y) or $(x, f(x))$
-2	$f(-2) = 2(-2) = -4$	$(-2, -4)$
-1	$f(-1) = 2(-1) = -2$	$(-1, -2)$
0	$f(0) = 2 \cdot 0 = 0$	$(0, 0)$
1	$f(1) = 2 \cdot 1 = 2$	$(1, 2)$
2	$f(2) = 2 \cdot 2 = 4$	$(2, 4)$

x	$g(x) = 2x + 4$	(x, y) or $(x, g(x))$
-2	$g(-2) = 2(-2) + 4 = 0$	$(-2, 0)$
-1	$g(-1) = 2(-1) + 4 = 2$	$(-1, 2)$
0	$g(0) = 2 \cdot 0 + 4 = 4$	$(0, 4)$
1	$g(1) = 2 \cdot 1 + 4 = 6$	$(1, 6)$
2	$g(2) = 2 \cdot 2 + 4 = 8$	$(2, 8)$

Choose x. | Compute $f(x)$ by evaluating f at x. | Form the ordered pair.

Choose x. | Compute $g(x)$ by evaluating g at x. | Form the ordered pair.

● ● ●

The graphs in Example 5 are straight lines. All functions with equations of the form $f(x) = mx + b$ graph as straight lines. Such functions, called **linear functions**, will be discussed in detail in Section 2.3.

TECHNOLOGY

We can use a graphing utility to check the tables and the graphs in Example 5 for the functions

$$f(x) = 2x \qquad \text{and} \qquad g(x) = 2x + 4.$$

Enter $y_1 = 2x$ in the $\boxed{y=}$ screen.

Enter $y_2 = 2x + 4$ in the $\boxed{y=}$ screen.

We entered -2 for the starting x-value and 1 as an increment between x-values to check our tables in Example 1.

Checking Tables

X	Y₁	Y₂
-2	-4	0
-1	-2	2
0	0	4
1	2	6
2	4	8
3	6	10
4	8	12
5	10	14
6	12	16
7	14	18
8	16	20

X=-2

Use the first five ordered pairs (x, y_1) to check the first table.

Use the first five ordered pairs (x, y_2) to check the second table.

Checking Graphs

$y_2 = 2x + 4$

$y_1 = 2x$

$[-6, 6, 1]$ by $[-6, 9, 1]$

We selected this viewing rectangle, or window, to match **Figure 2.7**.

⊘ Check Point **5** Graph the functions $f(x) = 2x$ and $g(x) = 2x - 3$ in the same rectangular coordinate system. Select integers for x, starting with -2 and ending with 2. How is the graph of g related to the graph of f? The graph of g is the graph of f shifted down by 3 units.*

⑥ Use the vertical line test to identify functions.

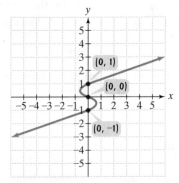

FIGURE 2.8 y is not a function of x because 0 is paired with three values of y, namely, 1, 0, and -1.

The Vertical Line Test

Not every graph in the rectangular coordinate system is the graph of a function. The definition of a function specifies that no value of x can be paired with two or more different values of y. Consequently, if a graph contains two or more different points with the same first coordinate, the graph cannot represent a function. This is illustrated in **Figure 2.8**. Observe that points sharing a common first coordinate are vertically above or below each other.

This observation is the basis of a useful test for determining whether a graph defines y as a function of x. The test is called the **vertical line test**.

The Vertical Line Test for Functions

If any vertical line intersects a graph in more than one point, the graph does not define y as a function of x.

EXAMPLE 6 Using the Vertical Line Test

Use the vertical line test to identify graphs in which y is a function of x.

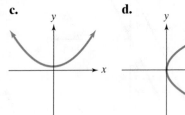

SOLUTION

y is a function of x for the graphs in (b) and (c).

a. y is **not a function** of x. Two values of y correspond to an x-value.

b. y is **a function** of x.

c. y is **a function** of x.

d. y is **not a function** of x. Two values of y correspond to an x-value.

● ● ●

⊘ Check Point **6** Use the vertical line test to identify graphs in which y is a function of x.

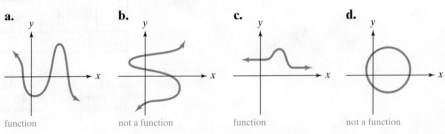

a. function b. not a function c. function d. not a function

*See Answers to Selected Exercises.

 Obtain information about a function from its graph.

Obtaining Information from Graphs

You can obtain information about a function from its graph. At the right or left of a graph, you will find closed dots, open dots, or arrows.

- A closed dot indicates that the graph does not extend beyond this point and the point belongs to the graph.
- An open dot indicates that the graph does not extend beyond this point and the point does not belong to the graph.
- An arrow indicates that the graph extends indefinitely in the direction in which the arrow points.

EXAMPLE 7 Analyzing the Graph of a Function

The human immunodeficiency virus, or HIV, infects and kills helper T cells. Because T cells stimulate the immune system to produce antibodies, their destruction disables the body's defenses against other pathogens. By counting the number of T cells that remain active in the body, the progression of HIV can be monitored. The fewer helper T cells, the more advanced the disease. **Figure 2.9** shows a graph that is used to monitor the average progression of the disease. The average number of T cells, $f(x)$, is a function of time after infection, x.

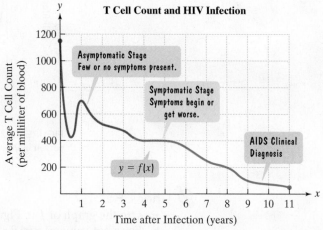

FIGURE 2.9
Source: B. E. Pruitt et al., *Human Sexuality*, Prentice Hall, 2007.

a. Explain why f represents the graph of a function.
b. Use the graph to find and interpret $f(8)$.
c. For what value of x is $f(x) = 350$?
d. Describe the general trend shown by the graph.

SOLUTION

a. No vertical line can be drawn that intersects the graph of f more than once. By the vertical line test, f represents the graph of a function.

b. To find $f(8)$, or f of 8, we locate 8 on the x-axis. **Figure 2.10** shows the point on the graph of f for which 8 is the first coordinate. From this point, we look to the y-axis to find the corresponding y-coordinate. We see that the y-coordinate is 200. Thus,

$$f(8) = 200.$$

When the time after infection is 8 years, the average T cell count is 200 cells per milliliter of blood. (AIDS clinical diagnosis is given at a T cell count of 200 or below.)

FIGURE 2.10 Finding $f(8)$

FIGURE 2.11 Finding x for which $f(x) = 350$

c. To find the value of x for which $f(x) = 350$, we find the approximate location of 350 on the y-axis. **Figure 2.11** shows that there is one point on the graph of f for which 350 is the second coordinate. From this point, we look to the x-axis to find the corresponding x-coordinate. We see that the x-coordinate is 6. Thus,

$$f(x) = 350 \text{ for } x = 6.$$

An average T cell count of 350 occurs 6 years after infection.

d. Figure 2.12 uses voice balloons to describe the general trend shown by the graph.

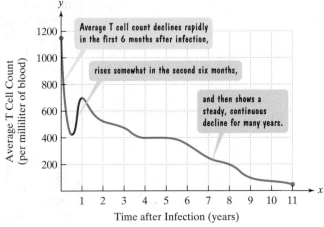

FIGURE 2.12 Describing changing T cell count over time in a person infected with HIV

•••

Check Point 7

a. Use the graph of f in **Figure 2.9** on page 225 to find $f(5)$. 400

b. For what value of x is $f(x) = 100$? 9

c. Estimate the minimum average T cell count during the asymptomatic stage. approximately 425

⑧ Identify the domain and range of a function from its graph.

FIGURE 2.13 Domain and range of f

Identifying Domain and Range from a Function's Graph

A Brief Review • Interval Notation

Square brackets indicate endpoints that are included in an interval. Parentheses indicate endpoints that are not included in an interval. Parentheses are always used with ∞ or $-\infty$. For more detail on interval notation, see Section 1.7, Objective 1.

Figure 2.13 illustrates how the graph of a function is used to determine the function's domain and its range.

Domain: set of inputs

Found on the x-axis

Range: set of outputs

Found on the y-axis

FIGURE 2.14 Domain and range of f

Let's apply these ideas to the graph of the function shown in **Figure 2.14**. To find the domain, look for all the inputs on the x-axis that correspond to points on the graph. Can you see that they extend from -4 to 2, inclusive? The function's domain can be represented as follows:

Using Set-Builder Notation

$$\{x \mid -4 \le x \le 2\}$$

| The set of all x | such that | x is greater than or equal to -4 and less than or equal to 2. |

Using Interval Notation

$$[-4, 2].$$

The square brackets indicate -4 and 2 are included. Note the square brackets on the x-axis in **Figure 2.14**.

To find the range, look for all the outputs on the y-axis that correspond to points on the graph. They extend from 1 to 4, inclusive. The function's range can be represented as follows:

Using Set-Builder Notation

$$\{y \mid 1 \le y \le 4\}$$

| The set of all y | such that | y is greater than or equal to 1 and less than or equal to 4. |

Using Interval Notation

$$[1, 4].$$

The square brackets indicate 1 and 4 are included. Note the square brackets on the y-axis in **Figure 2.14**.

EXAMPLE 8 Identifying the Domain and Range of a Function from Its Graph

Use the graph of each function to identify its domain and its range.

a.

b.

c.

d.

e.

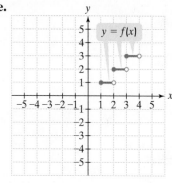

SOLUTION

For the graph of each function, the domain is highlighted in purple on the x-axis and the range is highlighted in green on the y-axis.

a.

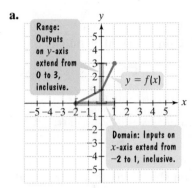

Domain = {x|−2 ≤ x ≤ 1} or [−2, 1]
Range = {y|0 ≤ y ≤ 3} or [0, 3]

b.

Domain = {x|−3 < x ≤ 2} or (−3, 2]
Range = {y|1 < y ≤ 2} or (1, 2]

c.

Domain = {x|−2 ≤ x < 1} or [−2, 1)
Range = {y|1 ≤ y ≤ 5} or [1, 5]

d.

Domain = {x|x ≤ 4} or (−∞, 4]
Range = {y|y ≥ 0} or [0, ∞)

e.

Domain = {x|1 ≤ x < 4} or [1, 4)
Range = {y|y = 1, 2, 3}

• • •

GREAT QUESTION!

The range in Example 8(e) was identified as {y|y = 1, 2, 3}. Why didn't you also use interval notation like you did in the other parts of Example 8?

Interval notation is not appropriate for describing a set of distinct numbers such as {1, 2, 3}. Interval notation, [1, 3], would mean that numbers such as 1.5 and 2.99 are in the range, but they are not. That's why we only used set-builder notation.

☑ Check Point **8** Use the graph of each function to identify its domain and its range.

a.

domain: {x|−2 ≤ x ≤ 1} or [−2, 1];
range: {y|0 ≤ y ≤ 3} or [0, 3]

b.

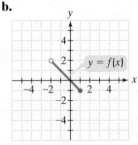

domain: {x|−2 < x ≤ 1} or (−2, 1];
range: {y|−1 ≤ y < 2} or [−1, 2)

c.

domain: {x|−3 ≤ x < 0} or [−3, 0);
range: {y|y = −3, −2, −1}

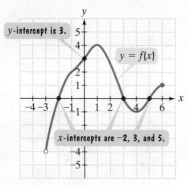

⑨ Identify intercepts from a function's graph.

FIGURE 2.15 Identifying intercepts

Identifying Intercepts from a Function's Graph

Figure 2.15 illustrates how we can identify intercepts from a function's graph. To find the x-intercepts, look for the points at which the graph crosses the x-axis. There are three such points: $(-2, 0)$, $(3, 0)$, and $(5, 0)$. Thus, the x-intercepts are -2, 3, and 5. We express this in function notation by writing $f(-2) = 0$, $f(3) = 0$, and $f(5) = 0$. We say that -2, 3, and 5 are the *zeros of the function*. The **zeros of a function** f are the x-values for which $f(x) = 0$. Thus, the real zeros are the x-intercepts.

To find the y-intercept, look for the point at which the graph crosses the y-axis. This occurs at $(0, 3)$. Thus, the y-intercept is 3. We express this in function notation by writing $f(0) = 3$.

By the definition of a function, for each value of x we can have at most one value for y. What does this mean in terms of intercepts? **A function can have more than one x-intercept but at most one y-intercept.**

ACHIEVING SUCCESS

Analyze the errors you make on quizzes and tests.

For each error, write out the correct solution along with a description of the concept needed to solve the problem correctly. Do your mistakes indicate gaps in understanding concepts or do you at times believe that you are just not a good test taker? Are you repeatedly making the same kinds of mistakes on tests? Keeping track of errors should increase your understanding of the material, resulting in improved test scores.

CONCEPT AND VOCABULARY CHECK

Fill in each blank so that the resulting statement is true.

1. Any set of ordered pairs is called a/an ___relation___. The set of all first components of the ordered pairs is called the ___domain___. The set of all second components of the ordered pairs is called the ___range___.

2. A set of ordered pairs in which each member of the set of first components corresponds to exactly one member of the set of second components is called a/an ___function___.

3. The notation $f(x)$ describes the value of ___f___ at ___x___.

4. True or false: $y = x^2 - 1$ defines y as a function of x. ___true___

5. True or false: $y = \pm\sqrt{x^2 - 1}$ defines y as a function of x. ___false___

6. If $f(x) = x^2 - 5x + 4$, we can find $f(x + 6)$ by replacing each occurrence of ___x___ by ___x + 6___.

7. The graph of a function is the graph of its ___ordered pairs___.

8. If any vertical line intersects a graph ___more than once___, the graph does not define y as a/an ___function___ of x.

9. The shaded set of numbers shown on the x-axis can be expressed in interval notation as ___[0, 3)___. This set represents the function's ___domain___.

10. The shaded set of numbers shown on the y-axis can be expressed in interval notation as ___[1, ∞)___. This set represents the function's ___range___.

11. If the x-intercepts of a function are -1 and 3, then $f(-1) = $ ___0___ and $f(3) = $ ___0___. The x-intercepts, -1 and 3, are called the ___zeros___ of the function.

12. True or false: A function can have more than one y-intercept. ___false___

EXERCISE SET 2.1

Practice Exercises

In Exercises 1–10, determine whether each relation is a function.
Give the domain and range for each relation.

1. $\{(1, 2), (3, 4), (5, 5)\}$ function; {1, 3, 5}; {2, 4, 5}
2. $\{(4, 5), (6, 7), (8, 8)\}$ function; {4, 6, 8}; {5, 7, 8}
3. $\{(3, 4), (3, 5), (4, 4), (4, 5)\}$ not a function; {3, 4}; {4, 5}
4. $\{(5, 6), (5, 7), (6, 6), (6, 7)\}$ not a function; {5, 6}; {6, 7}
5. $\{(3, -2), (5, -2), (7, 1), (4, 9)\}$ function; {3, 4, 5, 7}; {-2, 1, 9}
6. $\{(10, 4), (-2, 4), (-1, 1), (5, 6)\}$ function; {-2, 1, 5, 10}; {1, 4, 6}
7. $\{(-3, -3), (-2, -2), (-1, -1), (0, 0)\}$ function*
8. $\{(-7, -7), (-5, -5), (-3, -3), (0, 0)\}$ function*
9. $\{(1, 4), (1, 5), (1, 6)\}$ not a function; {1}; {4, 5, 6}
10. $\{(4, 1), (5, 1), (6, 1)\}$ function; {4, 5, 6}; {1}

In Exercises 11–26, determine whether each equation defines y as
a function of x.

11. $x + y = 16$ yes
12. $x + y = 25$ yes
13. $x^2 + y = 16$ yes
14. $x^2 + y = 25$ yes
15. $x^2 + y^2 = 16$ no
16. $x^2 + y^2 = 25$ no
17. $x = y^2$ no
18. $4x = y^2$ no
19. $y = \sqrt{x + 4}$ yes
20. $y = -\sqrt{x + 4}$ yes
21. $x + y^3 = 8$ yes
22. $x + y^3 = 27$ yes
23. $xy + 2y = 1$ yes
24. $xy - 5y = 1$ yes
25. $|x| - y = 2$ yes
26. $|x| - y = 5$ yes

In Exercises 27–38, evaluate each function at the given values of
the independent variable and simplify.

27. $f(x) = 4x + 5$
 a. $f(6)$ 29
 b. $f(x + 1)$ 4x + 9 c. $f(-x)$ *
28. $f(x) = 3x + 7$
 a. $f(4)$ 19
 b. $f(x + 1)$ 3x + 10 c. $f(-x)$ *
29. $g(x) = x^2 + 2x + 3$
 a. $g(-1)$ 2
 b. $g(x + 5)$ *
 c. $g(-x)$ *
30. $g(x) = x^2 - 10x - 3$
 a. $g(-1)$ 8
 b. $g(x + 2)$ *
 c. $g(-x)$ *
31. $h(x) = x^4 - x^2 + 1$
 a. $h(2)$ 13
 b. $h(-1)$ 1
 c. $h(-x)$ x⁴ − x² + 1 d. $h(3a)$ 81a⁴ − 9a² + 1
32. $h(x) = x^3 - x + 1$
 a. $h(3)$ 25
 b. $h(-2)$ −5
 c. $h(-x)$ −x³ + x + 1 d. $h(3a)$ 27a³ − 3a + 1
33. $f(r) = \sqrt{r + 6} + 3$
 a. $f(-6)$ 3
 b. $f(10)$ 7
 c. $f(x - 6)$ *
34. $f(r) = \sqrt{25 - r} - 6$
 a. $f(16)$ −3
 b. $f(-24)$ 1
 c. $f(25 - 2x)$ *
35. $f(x) = \dfrac{4x^2 - 1}{x^2}$
 a. $f(2)$ 15/4
 b. $f(-2)$ 15/4
 c. $f(-x)$ 4x²−1/x²

36. $f(x) = \dfrac{4x^3 + 1}{x^3}$
 a. $f(2)$ 33/8
 b. $f(-2)$ 31/8
 c. $f(-x)$ *
37. $f(x) = \dfrac{x}{|x|}$
 a. $f(6)$ 1
 b. $f(-6)$ −1
 c. $f(r^2)$ 1
38. $f(x) = \dfrac{|x + 3|}{x + 3}$
 a. $f(5)$ 1
 b. $f(-5)$ −1
 c. $f(-9 - x)$ *

In Exercises 39–50, graph the given functions, f and g, in the
same rectangular coordinate system. Select integers for x, starting
with −2 and ending with 2. Once you have obtained your graphs,
describe how the graph of g is related to the graph of f.

39. $f(x) = x, g(x) = x + 3$ *
40. $f(x) = x, g(x) = x - 4$ *
41. $f(x) = -2x, g(x) = -2x - 1$ *
42. $f(x) = -2x, g(x) = -2x + 3$ *
43. $f(x) = x^2, g(x) = x^2 + 1$ *
44. $f(x) = x^2, g(x) = x^2 - 2$ *
45. $f(x) = |x|, g(x) = |x| - 2$ *
46. $f(x) = |x|, g(x) = |x| + 1$ *
47. $f(x) = x^3, g(x) = x^3 + 2$ *
48. $f(x) = x^3, g(x) = x^3 - 1$ *
49. $f(x) = 3, g(x) = 5$ *
50. $f(x) = -1, g(x) = 4$ *

In Exercises 51–54, graph the given square root functions, f and g,
in the same rectangular coordinate system. Use the integer
values of x given to the right of each function to obtain ordered
pairs. Because only nonnegative numbers have square roots that
are real numbers, be sure that each graph appears only for values
of x that cause the expression under the radical sign to be greater
than or equal to zero. Once you have obtained your graphs,
describe how the graph of g is related to the graph of f.

51. $f(x) = \sqrt{x}$ $(x = 0, 1, 4, 9)$ and
 $g(x) = \sqrt{x} - 1$ $(x = 0, 1, 4, 9)$ *
52. $f(x) = \sqrt{x}$ $(x = 0, 1, 4, 9)$ and
 $g(x) = \sqrt{x} + 2$ $(x = 0, 1, 4, 9)$ *
53. $f(x) = \sqrt{x}$ $(x = 0, 1, 4, 9)$ and
 $g(x) = \sqrt{x - 1}$ $(x = 1, 2, 5, 10)$ *
54. $f(x) = \sqrt{x}$ $(x = 0, 1, 4, 9)$ and
 $g(x) = \sqrt{x + 2}$ $(x = -2, -1, 2, 7)$ *

In Exercises 55–64, use the vertical line test to identify graphs in
which y is a function of x.

55.

function

56.

function

57.

function

58.

not a function

59.

not a function

60.

not a function

61.

function

62.

not a function

63.

function

64.

function

In Exercises 65–70, use the graph of f to find each indicated function value.

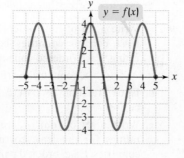

65. $f(-2)$ \quad −4

66. $f(2)$ \quad −4

67. $f(4)$ \quad 4

68. $f(-4)$ \quad 4

69. $f(-3)$ \quad 0

70. $f(-1)$ \quad 0

Use the graph of g to solve Exercises 71–76.

71. Find $g(-4)$. \quad 2

72. Find $g(2)$. \quad −2

73. Find $g(-10)$. \quad 2

74. Find $g(10)$. \quad −2

75. For what value of x is $g(x) = 1$? \quad −2

76. For what value of x is $g(x) = -1$? \quad 1

*In Exercises 77–92, use the graph to determine **a.** the function's domain; **b.** the function's range; **c.** the x-intercepts, if any; **d.** the y-intercept, if any; and **e.** the missing function values, indicated by question marks, below each graph.*

77.

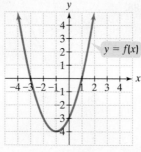

$f(-2) = ?$ \quad $f(2) = ?$ \quad *

78.

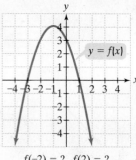

$f(-2) = ?$ \quad $f(2) = ?$ \quad *

79.

$f(-1) = ?$ \quad $f(3) = ?$ \quad *

80.

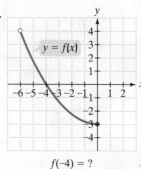

$f(-4) = ?$ \quad $f(3) = ?$ \quad *

81.

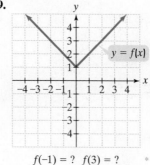

$f(3) = ?$ \quad *

82.

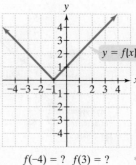

$f(-4) = ?$ \quad *

83.

$f(4) = ?$ \quad *

84.

$f(3) = ?$ \quad *

85.

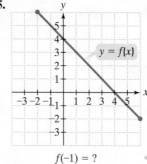

$f(-1) = ?$ \quad *

86.

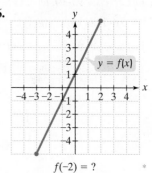

$f(-2) = ?$ \quad *

87.

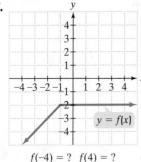

$f(-4) = ?$ $f(4) = ?$ *

88.

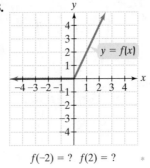

$f(-2) = ?$ $f(2) = ?$ *

89.

a. $(-\infty, \infty)$
b. $(0, \infty)$
c. none
d. 1.5
e. $f(4) = 6$

$f(4) = ?$

90.

a. $(-\infty, 1) \cup (1, \infty)$
b. $(-\infty, 0) \cup (0, \infty)$
c. none
d. -1
e. $f(2) = 1$

$f(2) = ?$

91.

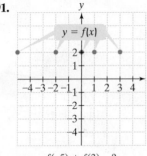

$f(-5) + f(3) = ?$ *

92.

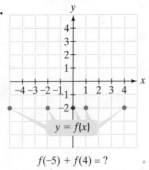

$f(-5) + f(4) = ?$ *

Practice Plus

In Exercises 93–94, let $f(x) = x^2 - x + 4$ *and* $g(x) = 3x - 5$.

93. Find $g(1)$ and $f(g(1))$. *

94. Find $g(-1)$ and $f(g(-1))$. *

In Exercises 95–96, let f and g be defined by the following table:

x	$f(x)$	$g(x)$
-2	6	0
-1	3	4
0	-1	1
1	-4	-3
2	0	-6

95. Find $\sqrt{f(-1) - f(0)} - [g(2)]^2 + f(-2) \div g(2) \cdot g(-1)$. -38

96. Find $|f(1) - f(0)| - [g(1)]^2 + g(1) \div f(-1) \cdot g(2)$. 0

In Exercises 97–98, find $f(-x) - f(x)$ *for the given function f. Then simplify the expression.*

97. $f(x) = x^3 + x - 5$ $-2x^3 - 2x$

98. $f(x) = x^2 - 3x + 7$ $6x$

Application Exercises

The Corruption Perceptions Index uses perceptions of the general public, business people, and risk analysts to rate countries by how likely they are to accept bribes. The ratings are on a scale from 0 to 10, where higher scores represent less corruption. The graph shows the corruption ratings for the world's least corrupt and most corrupt countries. (The rating for the United States is 7.6.) Use the graph to solve Exercises 99–100.

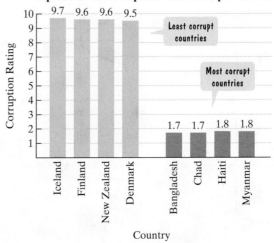

Top Four Least Corrupt and Most Corrupt Countries

Source: Transparency International, *Corruption Perceptions Index*

99. a. Write a set of four ordered pairs in which each of the least corrupt countries corresponds to a corruption rating. Each ordered pair should be in the form
(country, corruption rating). *

b. Is the relation in part (a) a function? Explain your answer. yes*

c. Write a set of four ordered pairs in which corruption ratings for the least corrupt countries correspond to countries. Each ordered pair should be in the form
(corruption rating, country). *

d. Is the relation in part (c) a function? Explain your answer. no*

100. a. Write a set of four ordered pairs in which each of the most corrupt countries corresponds to a corruption rating. Each ordered pair should be in the form
(country, corruption rating). *

b. Is the relation in part (a) a function? Explain your answer. yes*

c. Write a set of four ordered pairs in which corruption ratings for the most corrupt countries correspond to countries. Each ordered pair should be in the form
(corruption rating, country). *

d. Is the relation in part (c) a function? Explain your answer. no*

The bar graph shows your chances of surviving to various ages once you reach 60.

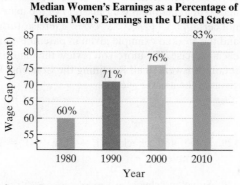

Chances of 60-Year-Olds Surviving to Various Ages

Source: National Center for Health Statistics

The functions

$$f(x) = -2.9x + 286$$

$$\text{and} \quad g(x) = 0.01x^2 - 4.9x + 370$$

model the chance, as a percent, that a 60-year-old will survive to age x. Use this information to solve Exercises 101–102.

101. a. Find and interpret $f(70)$. 83*
 b. Find and interpret $g(70)$. 76*
 c. Which function serves as a better model for the chance of surviving to age 70? *f*

102. a. Find and interpret $f(90)$. 25*
 b. Find and interpret $g(90)$. 10*
 c. Which function serves as a better model for the chance of surviving to age 90? *f*

The wage gap is used to compare the status of women's earnings relative to men's. The wage gap is expressed as a percent and is calculated by dividing the median, or middlemost, annual earnings for women by the median annual earnings for men. The bar graph shows the wage gap for selected years from 1980 through 2010.

Median Women's Earnings as a Percentage of Median Men's Earnings in the United States

Source: Bureau of Labor Statistics

The Graph of a Function Modeling the Data

$$G(x) = -0.01x^2 + x + 60$$

Years after 1980

The function $G(x) = -0.01x^2 + x + 60$ models the wage gap, as a percent, x years after 1980. The graph of function G is shown to the right of the actual data. Use this information to solve Exercises 103–104.

103. a. Find and interpret $G(30)$. Identify this information as a point on the graph of the function. 81; (30, 81)*
 b. Does $G(30)$ overestimate or underestimate the actual data shown by the bar graph? By how much? underestimates by 2%

104. a. Find and interpret $G(10)$. Identify this information as a point on the graph of the function. 69; (10, 69)*
 b. Does $G(10)$ overestimate or underestimate the actual data shown by the bar graph? By how much? underestimates by 2%

In Exercises 105–108, you will be developing functions that model given conditions.

105. A company that manufactures bicycles has a fixed cost of $100,000. It costs $100 to produce each bicycle. The total cost for the company is the sum of its fixed cost and variable costs. Write the total cost, C, as a function of the number of bicycles produced, x. Then find and interpret $C(90)$. *

106. A car was purchased for $22,500. The value of the car decreased by $3200 per year for the first six years. Write a function that describes the value of the car, V, after x years, where $0 \le x \le 6$. Then find and interpret $V(3)$. *

107. You commute to work a distance of 40 miles and return on the same route at the end of the day. Your average rate on the return trip is 30 miles per hour faster than your average rate on the outgoing trip. Write the total time, T, in hours, devoted to your outgoing and return trips as a function of your rate on the outgoing trip, x. Then find and interpret $T(30)$. Hint:

$$\text{Time traveled} = \frac{\text{Distance traveled}}{\text{Rate of travel}}. \quad *$$

108. A chemist working on a flu vaccine needs to mix a 10% sodium-iodine solution with a 60% sodium-iodine solution to obtain a 50-milliliter mixture. Write the amount of sodium iodine in the mixture, S, in milliliters, as a function of the number of milliliters of the 10% solution used, x. Then find and interpret $S(30)$. *

Explaining the Concepts 109–116. Answers will vary.

109. What is a relation? Describe what is meant by its domain and its range.

110. Explain how to determine whether a relation is a function. What is a function?

111. How do you determine if an equation in x and y defines y as a function of x?

112. Does $f(x)$ mean f times x when referring to a function f? If not, what does $f(x)$ mean? Provide an example with your explanation.

113. What is the graph of a function?

114. Explain how the vertical line test is used to determine whether a graph represents a function.

115. Explain how to identify the domain and range of a function from its graph.

116. For people filing a single return, federal income tax is a function of adjusted gross income because for each value of adjusted gross income there is a specific tax to be paid. By contrast, the price of a house is not a function of the lot size on which the house sits because houses on same-sized lots can sell for many different prices.

 a. Describe an everyday situation between variables that is a function.

 b. Describe an everyday situation between variables that is not a function.

Technology Exercise

117. Use a graphing utility to verify any five pairs of graphs that you drew by hand in Exercises 39–54. Answers will vary.

Critical Thinking Exercises

Make Sense? *In Exercises 118–121, determine whether each statement makes sense or does not make sense, and explain your reasoning.*

118. My body temperature is a function of the time of day. *

119. Using $f(x) = 3x + 2$, I found $f(50)$ by applying the distributive property to $(3x + 2)50$. does not make sense

120. I graphed a function showing how paid vacation days depend on the number of years a person works for a company. The domain was the number of paid vacation days. *

121. I graphed a function showing how the average number of annual physician visits depends on a person's age. The domain was the average number of annual physician visits. *

Use the graph of f to determine whether each statement in Exercises 122–125 is true or false.

Graph of f

122. The domain of f is $[-4, -1) \cup (-1, 4]$. false

123. The range of f is $[-2, 2]$. false

124. $f(-1) - f(4) = 2$ true

125. $f(0) = 2.1$ false

126. If $f(x) = 3x + 7$, find $\dfrac{f(a + h) - f(a)}{h}$. 3

127. Give an example of a relation with the following characteristics: The relation is a function containing two ordered pairs. Reversing the components in each ordered pair results in a relation that is not a function. *

128. If $f(x + y) = f(x) + f(y)$ and $f(1) = 3$, find $f(2), f(3)$, and $f(4)$. Is $f(x + y) = f(x) + f(y)$ for all functions? *

Retaining the Concepts

ACHIEVING SUCCESS

According to the Ebbinghaus retention model, you forget 50% of what you learn within one hour. You lose 60% within 24 hours. After 30 days, 70% is gone. Reviewing previously covered topics is an effective way to counteract this phenomenon. From here on, each Exercise Set will contain three review exercises. It is essential to review previously covered topics to improve your understanding of the topics and to help maintain your mastery of the material. If you are not certain how to solve a review exercise, turn to the section and the worked example given in parentheses at the end of each exercise. The more you review the material, the more you retain. Answers to all Retaining the Concepts Exercises are given in the answer section.

129. Solve and check: $-1 + 3(x - 4) = 2x$. (Section 1.2, Example 2) [13]

130. Solve and check: $\dfrac{x - 3}{5} - \dfrac{x - 4}{2} = 5$.

 (Section 1.2, Example 3) [−12]

131. Sharks may be scary, but they are responsible for only three deaths worldwide in 2014. The world's deadliest creatures, ranked by the number of human deaths per year, are mosquitoes, snails, and snakes. The number of deaths by mosquitoes exceeds the number of deaths by snakes by 661 thousand. The number of deaths by snails exceeds the number of deaths by snakes by 106 thousand. Combined, mosquitoes, snails, and snakes result in 1049 thousand (or 1,049,000) human deaths per year. Determine the number of human deaths per year, in thousands, by snakes, mosquitoes, and snails. (Source: World Health Organization) (Section 1.3, Example 1)
 snakes: 94 thousand; mosquitoes: 755 thousand; snails: 200 thousand

Preview Exercises

Exercises 132–134 will help you prepare for the material covered in the next section.

132. The function $C(t) = 20 + 0.40(t - 60)$ describes the monthly cost, $C(t)$, in dollars, for a cellphone plan for t calling minutes, where $t > 60$. Find and interpret $C(100)$. *

133. Use point plotting to graph $f(x) = x + 2$ if $x \le 1$. *

134. Simplify: $2(x + h)^2 + 3(x + h) + 5 - (2x^2 + 3x + 5)$. *

Section 2.2

More on Functions and Their Graphs

It's hard to believe that this gas-guzzler, with its huge fins and overstated design, was available in 1957 for approximately $1800. The line graph in **Figure 2.16** shows the average fuel efficiency, in miles per gallon, of new U.S. passenger cars for selected years from 1955 through 2014. Based on the averages shown by the graph, it's unlikely that this classic 1957 Chevy got more than 15 miles per gallon.

FIGURE 2.16
Source: U.S. Department of Transportation

You are probably familiar with the words used to describe the graph in **Figure 2.16**:

decreased increased minimum maximum

In this section, you will enhance your intuitive understanding of ways of describing graphs by viewing these descriptions from the perspective of functions.

1. Identify intervals on which a function increases, decreases, or is constant.

Increasing and Decreasing Functions

Too late for that flu shot now! It's only 8 A.M. and you're feeling lousy. Your temperature is 101°F. Fascinated by the way that algebra models the world (your author is projecting a bit here), you decide to construct graphs showing your body temperature as a function of the time of day. You decide to let x represent the number of hours after 8 A.M. and $f(x)$ your temperature at time x.

At 8 A.M. your temperature is 101°F and you are not feeling well. However, your temperature starts to decrease. It reaches normal (98.6°F) by 11 A.M. Feeling energized, you construct the graph shown on the right, indicating decreasing temperature for $\{x | 0 < x < 3\}$, or on the interval $(0, 3)$.

Hours after 8 A.M.
Temperature decreases on $(0, 3)$, reaching 98.6° by 11 A.M.

Did creating that first graph drain you of your energy? Your temperature starts to rise after 11 A.M. By 1 P.M., 5 hours after 8 A.M., your temperature reaches 100°F. However, you keep plotting points on your graph. At the right, we can see that your temperature increases for $\{x | 3 < x < 5\}$, or on the interval $(3, 5)$.

The graph of f is decreasing to the left of $x = 3$ and increasing to the right of $x = 3$. Thus, your temperature 3 hours after 8 A.M. was at its lowest point. Your relative minimum temperature was 98.6°.

By 3 P.M., your temperature is no worse than it was at 1 P.M.: It is still 100°F. (Of course, it's no better, either.) Your temperature remained the same, or constant, for $\{x | 5 < x < 7\}$, or on the interval $(5, 7)$.

The time-temperature flu scenario illustrates that a function f is increasing when its graph rises from left to right, decreasing when its graph falls from left to right, and remains constant when it neither rises nor falls. Let's now provide a more precise algebraic description for these intuitive concepts.

Hours after 8 A.M.
Temperature increases on $(3, 5)$.

Hours after 8 A.M.
Temperature remains constant at 100° on $(5, 7)$.

Increasing, Decreasing, and Constant Functions

1. A function is **increasing** on an open interval, I, if $f(x_1) < f(x_2)$ whenever $x_1 < x_2$ for any x_1 and x_2 in the interval.
2. A function is **decreasing** on an open interval, I, if $f(x_1) > f(x_2)$ whenever $x_1 < x_2$ for any x_1 and x_2 in the interval.
3. A function is **constant** on an open interval, I, if $f(x_1) = f(x_2)$ for any x_1 and x_2 in the interval.

Increasing — Rises from left to right

(1) For $x_1 < x_2$ in I,
$f(x_1) < f(x_2)$;
f is increasing on I.

Decreasing — Falls from left to right

(2) For $x_1 < x_2$ in I,
$f(x_1) > f(x_2)$;
f is decreasing on I.

Constant — Horizontal from left to right

(3) For x_1 and x_2 in I,
$f(x_1) = f(x_2)$;
f is constant on I.

EXAMPLE 1 Intervals on Which a Function Increases, Decreases, or Is Constant

State the intervals on which each given function is increasing, decreasing, or constant.

a.

b.

SOLUTION

a. The function is decreasing on the interval $(-\infty, 0)$, increasing on the interval $(0, 2)$, and decreasing on the interval $(2, \infty)$.

b. Although the function's equations are not given, the graph indicates that the function is defined in two pieces. The part of the graph to the left of the *y*-axis shows that the function is constant on the interval $(-\infty, 0)$. The part to the right of the *y*-axis shows that the function is increasing on the interval $(0, \infty)$. • • •

✓ Check Point **1** State the intervals on which the given function is increasing, decreasing, or constant. increasing on $(-\infty, -1)$, decreasing on $(-1, 1)$, increasing on $(1, \infty)$

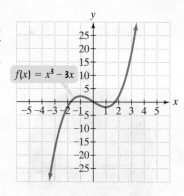

② Use graphs to locate relative maxima or minima.

Relative Maxima and Relative Minima

The points at which a function changes its increasing or decreasing behavior can be used to find any *relative maximum* or *relative minimum* values of the function.

Definitions of Relative Maximum and Relative Minimum

1. A function value $f(a)$ is a **relative maximum** of f if there exists an open interval containing a such that $f(a) > f(x)$ for all $x \neq a$ in the open interval.

2. A function value $f(b)$ is a **relative minimum** of f if there exists an open interval containing b such that $f(b) < f(x)$ for all $x \neq b$ in the open interval.

The word *local* is sometimes used instead of *relative* when describing maxima or minima.

GREAT QUESTION!

What's the difference between where a function has a relative maximum or minimum and the relative maximum or minimum?

- If $f(a)$ is greater than all other values of f near a, then f has a relative maximum at the input value, a. The relative maximum is the output value, $f(a)$. Equivalently, $f(a)$ is a relative maximum *value* of the function.
- If $f(b)$ is less than all other values of f near b, then f has a relative minimum at the input value, b. The relative minimum is the output value, $f(b)$. Equivalently, $f(b)$ is a relative minimum *value* of the function.

FIGURE 2.17 Using a graph to locate where a function has a relative maximum or minimum

If the graph of a function is given, we can often visually locate the number(s) at which the function has a relative maximum or a relative minimum. For example, the graph of f in **Figure 2.17** shows that

- f has a relative maximum at $\dfrac{\pi}{2}$.

 The relative maximum is $f\left(\dfrac{\pi}{2}\right) = 1$.

- f has a relative minimum at $-\dfrac{\pi}{2}$.

 The relative minimum is $f\left(-\dfrac{\pi}{2}\right) = -1$.

3 Test for symmetry.

Tests for Symmetry

Is beauty in the eye of the beholder? Or are there certain objects (or people) that are so well balanced and proportioned that they are universally pleasing to the eye? What constitutes an attractive human face? In **Figure 2.18**, we've drawn lines between paired features. Notice how the features line up almost perfectly. Each half of the face is a mirror image of the other half through the vertical line. If we superimpose a rectangular coordinate system on the attractive face, notice that a point (x, y) on the right has a mirror image at $(-x, y)$ on the left. The attractive face is said to be *symmetric with respect to the y-axis.*

Did you know that graphs of some equations exhibit exactly the kind of symmetry shown by the attractive face in **Figure 2.18**, as well as other kinds of symmetry? The word *symmetry* comes from the Greek *symmetria*, meaning "the same measure." **Figure 2.19** shows three graphs, each with a common form of symmetry. Notice that the graph in **Figure 2.19(a)** shows the y-axis symmetry found in the attractive face.

FIGURE 2.18 To most people, an attractive face is one in which each half is an almost perfect mirror image of the other half.

(a) Symmetric with respect to the y-axis

(b) Symmetric with respect to the x-axis

(c) Symmetric with respect to the origin

FIGURE 2.19 Graphs illustrating the three forms of symmetry

GREAT QUESTION!

The graph in Figure 2.19(b) is not a function. How can it have symmetry?

Good observation! The graph of the relation with x-axis symmetry is not the graph of a function; it fails the vertical line test. Other than $f(x) = 0$ or any graph that lies entirely on the x-axis, graphs with x-axis symmetry do not represent functions: Each point on the graph above the x-axis has a corresponding point below the x-axis with the same x-coordinate and opposite y-coordinate, violating the condition for being a function.

However, even the graphs of relations that are not functions may have symmetry with respect to the y-axis, the x-axis, or the origin. Later in this chapter we will study circles that exhibit all three types of symmetry but are not graphs of functions.

Table 2.2 defines three common forms of symmetry and gives rules to determine if the graph of an equation is symmetric with respect to the y-axis, the x-axis, or the origin.

Table 2.2 Definitions and Tests for Symmetry

Consider an equation in the variables x and y.

Definition of Symmetry	Test for Symmetry
The graph of the equation is **symmetric with respect to the y-axis** if for every point (x, y) on the graph, the point $(-x, y)$ is also on the graph.	Substituting $-x$ for x in the equation results in an equivalent equation.
The graph of the equation is **symmetric with respect to the x-axis** if for every point (x, y) on the graph, the point $(x, -y)$ is also on the graph.	Substituting $-y$ for y in the equation results in an equivalent equation.
The graph of the equation is **symmetric with respect to the origin** if for every point (x, y) on the graph, the point $(-x, -y)$ is also on the graph.	Substituting $-x$ for x and $-y$ for y in the equation results in an equivalent equation.

Not usually the graph of a function (annotation pointing to the x-axis symmetry row)

EXAMPLE 2 Testing for Symmetry

Determine whether the graph of

$$x = y^2 - 1$$

is symmetric with respect to the y-axis, the x-axis, or the origin.

SOLUTION

Test for symmetry with respect to the y-axis. Replace x with $-x$ and see if this results in an equivalent equation.

> These equations are not equivalent.

$x = y^2 - 1$ This is the given equation.

$-x = y^2 - 1$ Replace x with $-x$.

$x = -y^2 + 1$ Multiply both sides by -1 and solve for x.

Replacing x with $-x$ does not give the original equation. Thus, the graph of $x = y^2 - 1$ is not symmetric with respect to the y-axis.

Test for symmetry with respect to the x-axis. Replace y with $-y$ and see if this results in an equivalent equation.

> These equations are equivalent.

$x = y^2 - 1$ This is the given equation.

$x = (-y)^2 - 1$ Replace y with $-y$.

$x = y^2 - 1$ $(-y)^2 = (-y)(-y) = y^2$

Replacing y with $-y$ gives the original equation. Thus, the graph of $x = y^2 - 1$ is symmetric with respect to the x-axis.

Test for symmetry with respect to the origin. Replace x with $-x$ and y with $-y$ and see if this results in an equivalent equation.

$$\begin{array}{ll} x = y^2 - 1 & \text{This is the given equation.} \\ -x = (-y)^2 - 1 & \text{Replace } x \text{ with } -x \text{ and } y \text{ with } -y. \\ -x = y^2 - 1 & (-y)^2 = (-y)(-y) = y^2 \\ x = -y^2 + 1 & \text{Multiply both sides by } -1 \text{ and solve for } x. \end{array}$$

These equations are not equivalent.

Replacing x with $-x$ and y with $-y$ does not give the original equation. Thus, the graph of $x = y^2 - 1$ is not symmetric with respect to the origin.

Our symmetry tests reveal that the graph of $x = y^2 - 1$ is symmetric with respect to the x-axis only.

• • •

The graph of $x = y^2 - 1$ in **Figure 2.20(a)** shows the symmetry with respect to the x-axis that we determined in Example 2. **Figure 2.20(b)** illustrates that the graph fails the vertical line test. Consequently, y is not a function of x. We have seen that if a graph has x-axis symmetry, y is usually not a function of x.

 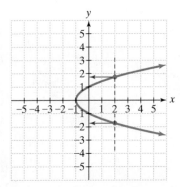

(a) The graph is symmetric with respect to the x-axis.

(b) y is not a function of x. Two values of y correspond to an x-value.

FIGURE 2.20 The graph of $x = y^2 - 1$

✓ Check Point **2** Determine whether the graph of $y = x^2 - 1$ is symmetric with respect to the y-axis, the x-axis, or the origin. symmetric with respect to the y-axis only

EXAMPLE 3 Testing for Symmetry

Determine whether the graph of $y = x^3$ is symmetric with respect to the y-axis, the x-axis, or the origin.

SOLUTION

Test for symmetry with respect to the y-axis. Replace x with $-x$ and see if this results in an equivalent equation.

$$\begin{array}{ll} y = x^3 & \text{This is the given equation.} \\ y = (-x)^3 & \text{Replace } x \text{ with } -x. \\ y = -x^3 & (-x)^3 = (-x)(-x)(-x) = -x^3 \end{array}$$

These equations are not the same.

Replacing x with $-x$ does not give the original equation. Thus, the graph of $y = x^3$ is not symmetric with respect to the y-axis.

Test for symmetry with respect to the x-axis. Replace y with $-y$ and see if this results in an equivalent equation.

These equations are not the same.	$y = x^3$	This is the given equation.
	$-y = x^3$	Replace y with $-y$.
	$y = -x^3$	Multiply both sides by -1 and solve for y.

Replacing y with $-y$ does not give the original equation. Thus, the graph of $y = x^3$ is not symmetric with respect to the x-axis.

Test for symmetry with respect to the origin. Replace x with $-x$ and y with $-y$ and see if this results in an equivalent equation.

These equations are the same.	$y = x^3$	This is the given equation.
	$-y = (-x)^3$	Replace x with $-x$ and y with $-y$.
	$-y = -x^3$	$(-x)^3 = (-x)(-x)(-x) = -x^3$
	$y = x^3$	Multiply both sides by -1 and solve for y.

Replacing x with $-x$ and y with $-y$ gives the original equation. Thus, the graph of $y = x^3$ is symmetric with respect to the origin.

FIGURE 2.21 The graph is symmetric with respect to the origin.

Our symmetry tests reveal that the graph of $y = x^3$ is symmetric with respect to the origin only. The symmetry is shown in **Figure 2.21**. ● ● ●

 Check Point **3** Determine whether the graph of $y^5 = x^3$ is symmetric with respect to the y-axis, the x-axis, or the origin. symmetric with respect to the origin

GREAT QUESTION!

Can the graph of an equation have more than one form of symmetry?

Yes. Consider the graph of $x^2 + y^2 = 25$.

Test for y-axis symmetry.
Replace x with $-x$.
$$x^2 + y^2 = 25 \longleftarrow$$
$$(-x)^2 + y^2 = 25 \quad \text{same}$$
$$x^2 + y^2 = 25 \longleftarrow$$

Test for x-axis symmetry.
Replace y with $-y$.
$$x^2 + y^2 = 25 \longleftarrow$$
$$x^2 + (-y)^2 = 25 \quad \text{same}$$
$$x^2 + y^2 = 25 \longleftarrow$$

Test for origin symmetry.
Replace x with $-x$ and y with $-y$.
$$x^2 + y^2 = 25 \longleftarrow$$
$$(-x)^2 + (-y)^2 = 25 \quad \text{same}$$
$$x^2 + y^2 = 25 \longleftarrow$$

The graph of $x^2 + y^2 = 25$, shown in **Figure 2.22**, is symmetric with respect to the y-axis, the x-axis, and the origin. This "perfectly symmetric" graph is a circle, which we shall study in Section 2.8.

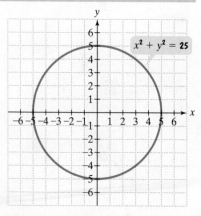

FIGURE 2.22 The graph, a circle, has three kinds of symmetry.

④ Identify even or odd functions and recognize their symmetries.

Even and Odd Functions

We have seen that if a graph is symmetric with respect to the x-axis, it usually fails the vertical line test and is not the graph of a function. However, many functions have graphs that are symmetric with respect to the y-axis or the origin. We give these functions special names.

A function whose graph is symmetric with respect to the y-axis is called an *even function*. If the point (x, y) is on the graph of the function, then the point $(-x, y)$ is also on the graph. Expressing this symmetry in function notation provides the definition of an even function.

Even Functions and Their Symmetries

The function f is an **even function** if

$$f(-x) = f(x) \qquad \text{for all } x \text{ in the domain of } f.$$

The graph of an even function is symmetric with respect to the y-axis.

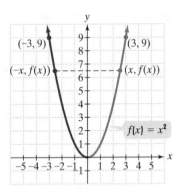

FIGURE 2.23 The graph of $f(x) = x^2$, an even function

An example of an even function is $f(x) = x^2$. The graph, shown in **Figure 2.23**, is symmetric with respect to the y-axis.

Let's show that $f(-x) = f(x)$.

$$f(x) = x^2 \qquad \text{This is the function's equation.}$$

$$f(-x) = (-x)^2 \quad \text{Replace } x \text{ with } -x.$$

$$f(-x) = x^2 \qquad (-x)^2 = (-x)(-x) = x^2$$

Because $f(x) = x^2$ and $f(-x) = x^2$, we see that $f(-x) = f(x)$. This algebraically identifies the function as an even function.

A function whose graph is symmetric with respect to the origin is called an *odd function*. If the point (x, y) is on the graph of the function, then the point $(-x, -y)$ is also on the graph. Expressing this symmetry in function notation provides the definition of an odd function.

Odd Functions and Their Symmetries

The function f is an **odd function** if

$$f(-x) = -f(x) \qquad \text{for all } x \text{ in the domain of } f.$$

The graph of an odd function is symmetric with respect to the origin.

FIGURE 2.24 The graph of $f(x) = x^3$, an odd function

An example of an odd function is $f(x) = x^3$. The graph, shown in **Figure 2.24**, is symmetric with respect to the origin.

Let's show that $f(-x) = -f(x)$.

$$f(x) = x^3 \qquad \text{This is the function's equation.}$$

$$f(-x) = (-x)^3 \quad \text{Replace } x \text{ with } -x.$$

$$f(-x) = -x^3 \qquad (-x)^3 = (-x)(-x)(-x) = -x^3$$

Because $f(x) = x^3$ and $f(-x) = -x^3$, we see that $f(-x) = -f(x)$. This algebraically identifies the function as an odd function.

EXAMPLE 4 Identifying Even or Odd Functions from Graphs

Determine whether each graph is the graph of an even function, an odd function, or a function that is neither even nor odd.

a. **b.** **c.**

SOLUTION

Note that each graph passes the vertical line test and is therefore the graph of a function. In each case, use inspection to determine whether or not there is symmetry. If the graph is symmetric with respect to the y-axis, it is that of an even function. If the graph is symmetric with respect to the origin, it is that of an odd function.

a.

b.

c.

The graph is symmetric with respect to the origin. Therefore, the graph is that of an odd function.

The graph is symmetric with respect to the y-axis. Therefore, the graph is that of an even function.

The graph is neither symmetric with respect to the y-axis nor the origin. Therefore, the graph is that of a function which is neither even nor odd. • • •

✓ Check Point **4** Determine whether each graph is the graph of an even function, an odd function, or a function that is neither even nor odd.

a.

b.

c.

a. even function **b.** neither even nor odd **c.** odd function

In addition to inspecting graphs, we can also use a function's equation and the definitions of *even* and *odd* to determine whether the function is even, odd, or neither.

Identifying Even or Odd Functions from Equations

- **Even function:** $f(-x) = f(x)$
 The right side of the equation of an even function does not change if x is replaced with $-x$.

- **Odd function:** $f(-x) = -f(x)$
 Every term on the right side of the equation of an odd function changes sign if x is replaced with $-x$.

- **Neither even nor odd:** $f(-x) \neq f(x)$ and $f(-x) \neq -f(x)$
 The right side of the equation changes if x is replaced with $-x$, but not every term on the right side changes sign.

EXAMPLE 5 Identifying Even or Odd Functions from Equations

Determine whether each of the following functions is even, odd, or neither. Then determine whether the function's graph is symmetric with respect to the y-axis, the origin, or neither.

 a. $f(x) = x^3 - 6x$ **b.** $g(x) = x^4 - 2x^2$ **c.** $h(x) = x^2 + 2x + 1$

SOLUTION

In each case, replace x with $-x$ and simplify. If the right side of the equation stays the same, the function is even. If every term on the right side changes sign, the function is odd.

a. We use the given function's equation, $f(x) = x^3 - 6x$, to find $f(-x)$.

Use $f(x) = x^3 - 6x$.

> Replace x with $-x$.

$$f(-x) = (-x)^3 - 6(-x) = (-x)(-x)(-x) - 6(-x) = -x^3 + 6x$$

There are two terms on the right side of the given equation, $f(x) = x^3 - 6x$, and each term changed sign when we replaced x with $-x$. Because $f(-x) = -f(x)$, f is an odd function. The graph of f is symmetric with respect to the origin.

b. We use the given function's equation, $g(x) = x^4 - 2x^2$, to find $g(-x)$.

Use $g(x) = x^4 - 2x^2$.

> Replace x with $-x$.

$$g(-x) = (-x)^4 - 2(-x)^2 = (-x)(-x)(-x)(-x) - 2(-x)(-x)$$
$$= x^4 - 2x^2$$

The right side of the equation of the given function, $g(x) = x^4 - 2x^2$, did not change when we replaced x with $-x$. Because $g(-x) = g(x)$, g is an even function. The graph of g is symmetric with respect to the y-axis.

c. We use the given function's equation, $h(x) = x^2 + 2x + 1$, to find $h(-x)$.

Use $h(x) = x^2 + 2x + 1$.

> Replace x with $-x$.

$$h(-x) = (-x)^2 + 2(-x) + 1 = x^2 - 2x + 1$$

The right side of the equation of the given function, $h(x) = x^2 + 2x + 1$, changed when we replaced x with $-x$. Thus, $h(-x) \neq h(x)$, so h is not an even function. The sign of *each* of the three terms in the equation for $h(x)$ did not change when we replaced x with $-x$. Only the second term changed signs. Thus, $h(-x) \neq -h(x)$, so h is not an odd function. We conclude that h is neither an even nor an odd function. The graph of h is neither symmetric with respect to the y-axis nor with respect to the origin.

• • •

GREAT QUESTION!

In the solution on the right, we keep getting $(-x)$ to various powers. Is there a way to avoid having to write $(-x)$ over and over again?

Yes. If you have $(-x)$ to an even power, the negative sign is eliminated.

$$(-x)^{\text{even power}} = x^{\text{even power}}$$

If you have $(-x)$ to an odd power, the result has a negative sign.

$$(-x)^{\text{odd power}} = -x^{\text{odd power}}$$

⊘ **Check Point 5** Determine whether each of the following functions is even, odd, or neither. Then determine whether the function's graph is symmetric with respect to the y-axis, the origin, or neither.

a. even; symmetric with respect to the y-axis
b. odd; symmetric with respect to the origin
c. neither; symmetric to neither the y-axis nor the origin

a. $f(x) = x^2 + 6$ **b.** $g(x) = 7x^3 - x$ **c.** $h(x) = x^5 + 1$

GREAT QUESTION!

Is there a relationship between the exponents in the terms of a function's equation and whether or not the function is even, odd, or neither?

Yes. Consider the functions in Example 5, where the equations all contained polynomials in x.

- $f(x) = x^3 - 6x$: odd function

 > Every term contains x to an odd power.

- $g(x) = x^4 - 2x^2$: even function

 > Every term contains x to an even power.

- $h(x) = x^2 + 2x + 1$: neither an even function nor an odd function

 > There is a combination of even and odd powers.

A constant term has an understood even power of x, x^0. Thus, the function in Check Point 5(c), $h(x) = x^5 + 1$ is a combination of odd and even powers. This function is neither even nor odd.

⑤ Understand and use piecewise functions.

Piecewise Functions

A telephone company offers the following plan:

- $20 per month buys 60 minutes.
- Additional time costs $0.40 per minute.

We can represent this plan mathematically by writing the total monthly cost, C, as a function of the number of calling minutes, t.

$$C(t) = \begin{cases} 20 & \text{if } 0 \le t \le 60 \\ 20 + 0.40(t - 60) & \text{if } t > 60 \end{cases}$$

> The cost is $20 for up to and including 60 calling minutes.

> The cost is $20 plus $0.40 per minute for additional time for more than 60 calling minutes.

$20 for first 60 minutes $0.40 per minute times the number of calling minutes exceeding 60

A function that is defined by two (or more) equations over a specified domain is called a **piecewise function**. Many telephone plans can be represented with piecewise functions. The graph of the piecewise function described above is shown in **Figure 2.25**.

$C(t) = 20 + 0.40(t - 60)$ if $t > 60$

$C(t) = 20$ if $0 \le t \le 60$

FIGURE 2.25

EXAMPLE 6 Evaluating a Piecewise Function

Use the function that describes the telephone plan

$$C(t) = \begin{cases} 20 & \text{if } 0 \le t \le 60 \\ 20 + 0.40(t - 60) & \text{if } t > 60 \end{cases}$$

to find and interpret each of the following:

 a. $C(30)$ **b.** $C(100)$.

SOLUTION

a. To find $C(30)$, we let $t = 30$. Because 30 lies between 0 and 60, we use the first line of the piecewise function.

$$C(t) = 20 \quad \text{This is the function's equation for } 0 \le t \le 60.$$
$$C(30) = 20 \quad \text{Replace } t \text{ with 30. Regardless of this function's}$$
$$\text{input, the constant output is 20.}$$

This means that with 30 calling minutes, the monthly cost is $20. This can be visually represented by the point $(30, 20)$ on the first piece of the graph in **Figure 2.25**.

FIGURE 2.25 (repeated)

a. 20; with 40 calling minutes, the cost is $20.; (40, 20) b. 28; with 80 calling minutes, the cost is $28.; (80, 28)

b. To find $C(100)$, we let $t = 100$. Because 100 is greater than 60, we use the second line of the piecewise function.

$$C(t) = 20 + 0.40(t - 60)$$ This is the function's equation for $t > 60$.
$$C(100) = 20 + 0.40(100 - 60)$$ Replace t with 100.
$$= 20 + 0.40(40)$$ Subtract within parentheses: $100 - 60 = 40$.
$$= 20 + 16$$ Multiply: $0.40(40) = 16$.
$$= 36$$ Add: $20 + 16 = 36$.

Thus, $C(100) = 36$. This means that with 100 calling minutes, the monthly cost is $36. This can be visually represented by the point $(100, 36)$ on the second piece of the graph in **Figure 2.25**. ● ● ●

⊘ **Check Point 6** Use the function in Example 6 to find and interpret each of the following:

 a. $C(40)$ **b.** $C(80)$.

Identify your solutions by points on the graph in **Figure 2.25**.

EXAMPLE 7 Graphing a Piecewise Function

Graph the piecewise function defined by

$$f(x) = \begin{cases} x + 2 & \text{if} \quad x \le 1 \\ 4 & \text{if} \quad x > 1. \end{cases}$$

SOLUTION

We graph f in two parts, using a partial table of coordinates to create each piece. The tables of coordinates and the completed graph are shown in **Figure 2.26**.

GREAT QUESTION!

When I graphed the function in Example 7, here's what I got:

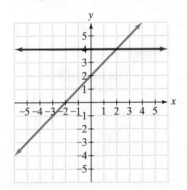

Is my graph ok?

No. You incorrectly ignored the domain for each piece of the function and graphed each equation as if its domain was $(-\infty, \infty)$.

Graph $f(x) = x + 2$ for $x \le 1$.
Domain $= (-\infty, 1]$

x ($x \le 1$)	$f(x) = x + 2$	$(x, f(x))$
1	$f(1) = 1 + 2 = 3$	$(1, 3)$
0	$f(0) = 0 + 2 = 2$	$(0, 2)$
-1	$f(-1) = -1 + 2 = 1$	$(-1, 1)$
-2	$f(-2) = -2 + 2 = 0$	$(-2, 0)$
-3	$f(-3) = -3 + 2 = -1$	$(-3, -1)$
-4	$f(-4) = -4 + 2 = -2$	$(-4, -2)$

Graph $f(x) = 4$ for $x > 1$.
Domain $= (1, \infty)$

x ($x > 1$)	$f(x) = 4$	$(x, f(x))$
1.5	$f(1.5) = 4$	$(1.5, 4)$
2	$f(2) = 4$	$(2, 4)$
3	$f(3) = 4$	$(3, 4)$
4	$f(4) = 4$	$(4, 4)$
5	$f(5) = 4$	$(5, 4)$

The open dot represents a point, $(1, 4)$, not included in the graph. No vertical line intersects the graph more than once. This is the graph of a function.

FIGURE 2.26 Graphing a piecewise function ● ● ●

We can use the graph of the piecewise function in **Figure 2.26** to find the range of f. The range of the blue piece on the left is $\{y \mid y \le 3\}$. The range of the red horizontal piece on the right is $\{y \mid y = 4\}$. Thus, the range of f is

$$\{y \mid y \le 3\} \cup \{y \mid y = 4\}, \text{ or } (-\infty, 3] \cup \{4\}.$$

⊘ **Check Point 7** Graph the piecewise function defined by

$$f(x) = \begin{cases} 3 & \text{if} \quad x \le -1 \\ x - 2 & \text{if} \quad x > -1. \end{cases}$$

*See Answers to Selected Exercises.

Some piecewise functions are called **step functions** because their graphs form discontinuous steps. One such function is called the **greatest integer function**, symbolized by $\text{int}(x)$ or $[\![x]\!]$, where

$$\text{int}(x) = \text{the greatest integer that is less than or equal to } x.$$

For example,

$$\text{int}(1) = 1, \quad \text{int}(1.3) = 1, \quad \text{int}(1.5) = 1, \quad \text{int}(1.9) = 1.$$

> 1 is the greatest integer that is less than or equal to 1, 1.3, 1.5, and 1.9.

Here are some additional examples:

$$\text{int}(2) = 2, \quad \text{int}(2.3) = 2, \quad \text{int}(2.5) = 2, \quad \text{int}(2.9) = 2.$$

> 2 is the greatest integer that is less than or equal to 2, 2.3, 2.5, and 2.9.

Notice how we jumped from 1 to 2 in the function values for $\text{int}(x)$. In particular,

$$\text{If } 1 \leq x < 2, \quad \text{then} \quad \text{int}(x) = 1.$$
$$\text{If } 2 \leq x < 3, \quad \text{then} \quad \text{int}(x) = 2.$$

The graph of $f(x) = \text{int}(x)$ is shown in **Figure 2.27**. The graph of the greatest integer function jumps vertically one unit at each integer. However, the graph is constant between each pair of consecutive integers. The rightmost horizontal step shown in the graph illustrates that

$$\text{If } 5 \leq x < 6, \quad \text{then} \quad \text{int}(x) = 5.$$

In general,

$$\text{If } n \leq x < n + 1, \text{ where } n \text{ is an integer,} \quad \text{then} \quad \text{int}(x) = n.$$

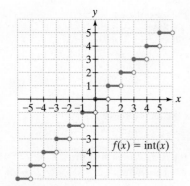

FIGURE 2.27 The graph of the greatest integer function

Find and simplify a function's difference quotient.

Functions and Difference Quotients

In Section 2.4, we will be studying the average rate of change of a function. A ratio, called the *difference quotient*, plays an important role in understanding the rate at which functions change.

> ### Definition of the Difference Quotient of a Function
>
> The expression
>
> $$\frac{f(x + h) - f(x)}{h}$$
>
> for $h \neq 0$ is called the **difference quotient** of the function f.

EXAMPLE 8 Evaluating and Simplifying a Difference Quotient

If $f(x) = 2x^2 - x + 3$, find and simplify each expression:

a. $f(x + h)$ **b.** $\dfrac{f(x + h) - f(x)}{h}, h \neq 0.$

SOLUTION

a. We find $f(x + h)$ by replacing x with $x + h$ each time that x appears in the equation.

$$f(x) = \;\; 2x^2 \quad - \quad x \qquad\quad + 3$$

> Replace x with $x + h$. Replace x with $x + h$. Replace x with $x + h$. Copy the 3. There is no x in this term.

$$f(x + h) = 2(x + h)^2 - (x + h) \qquad + 3$$
$$= 2(x^2 + 2xh + h^2) - x - h + 3$$
$$= 2x^2 + 4xh + 2h^2 - x - h + 3$$

b. Using our result from part (a), we obtain the following:

> This is $f(x + h)$ from part (a).

> This is $f(x)$ from the given equation.

$$\frac{f(x + h) - f(x)}{h} = \frac{\boxed{2x^2 + 4xh + 2h^2 - x - h + 3} - (2x^2 - x + 3)}{h}$$

$$= \frac{2x^2 + 4xh + 2h^2 - x - h + 3 - 2x^2 + x - 3}{h}$$

Remove parentheses and change the sign of each term in the parentheses.

$$= \frac{(2x^2 - 2x^2) + (-x + x) + (3 - 3) + 4xh + 2h^2 - h}{h}$$

Group like terms.

$$= \frac{4xh + 2h^2 - 1h}{h}$$

Simplify.

> We wrote $-h$ as $-1h$ to avoid possible errors in the next factoring step.

$$= \frac{h(4x + 2h - 1)}{h}$$

Factor h from the numerator.

$$= 4x + 2h - 1$$

Divide out identical factors of h in the numerator and denominator.

• • •

✓ **Check Point 8** If $f(x) = -2x^2 + x + 5$, find and simplify each expression:

a. $-2x^2 - 4xh - 2h^2 + x + h + 5$
b. $-4x - 2h + 1, h \neq 0$

a. $f(x + h)$ **b.** $\dfrac{f(x + h) - f(x)}{h}, h \neq 0.$

ACHIEVING SUCCESS

Read ahead. You might find it helpful to use some of your homework time to read (or skim) the section in the textbook that will be covered in your professor's next lecture. Having a clear idea of the new material that will be discussed will help you to understand the class a whole lot better.

CONCEPT AND VOCABULARY CHECK

Fill in each blank so that the resulting statement is true.

1. Assume that f is a function defined on an open interval I and x_1 and x_2 are any elements in the interval I. f is increasing on I if $f(x_1)$ __$< f(x_2)$__ when $x_1 < x_2$. f is decreasing on I if $f(x_1)$ __$> f(x_2)$__ when $x_1 < x_2$. f is constant on I if $f(x_1)$ __$= f(x_2)$__.

2. If $f(a) > f(x)$ in an open interval containing $a, x \neq a$, then the function value $f(a)$ is a relative __maximum__ of f. If $f(b) < f(x)$ in an open interval containing $b, x \neq b$, then the function value $f(b)$ is a relative __minimum__ of f.

3. The graph of an equation is symmetric with respect to the __y-axis__ if substituting $-x$ for x in the equation results in an equivalent equation.

4. The graph of an equation is symmetric with respect to the __x-axis__ if substituting $-y$ for y in the equation results in an equivalent equation.

5. The graph of an equation is symmetric with respect to the __origin__ if substituting $-x$ for x and $-y$ for y in the equation results in an equivalent equation.

6. If f is an even function, then $f(-x) =$ __$f(x)$__. The graph of an even function is symmetric with respect to the __y-axis__.

7. If f is an odd function, then $f(-x) =$ __$-f(x)$__. The graph of an odd function is symmetric with respect to the __origin__.

8. A function defined by two or more equations over a specified domain is called a/an __piecewise__ function.

9. The greatest integer function is defined by $\text{int}(x) =$ the greatest integer that is __less than or equal to x__. For example, $\text{int}(2.5) =$ __2__, $\text{int}(-2.5) =$ __-3__, and $\text{int}(0.5) =$ __0__.

10. The expression

$$\frac{f(x + h) - f(x)}{h}, h \neq 0,$$

is called the __difference quotient__ of the function f. We find this expression by replacing x with __$x + h$__ each time x appears in the function's equation. Then we subtract __$f(x)$__. After simplifying, we factor __h__ from the numerator and divide out identical factors of __h__ in the numerator and denominator.

11. True or false: $f(x + h) = f(x) + h$ __false__

12. True or false: $f(x + h) = f(x) + f(h)$ __false__

EXERCISE SET 2.2

Practice Exercises

In Exercises 1–12, use the graph to determine

 a. intervals on which the function is increasing, if any.
 b. intervals on which the function is decreasing, if any.
 c. intervals on which the function is constant, if any.

1. **2.**

3. **4.**

5. **6.**

7. **8.**

9. **10.**

11. **12.**

In Exercises 13–16, the graph of a function f is given. Use the graph to find each of the following:

 a. The numbers, if any, at which *f* has a relative maximum. What are these relative maxima?

 b. The numbers, if any, at which *f* has a relative minimum. What are these relative minima?

13.

 a. $0; f(0) = 4$
 b. $-3, 3; f(-3) = f(3) = 0$

14.

 a. $0; f(0) = 2$
 b. $-3, 3; f(-3) = f(3) = -1$

15.

 $f(x) = 2x^3 + 3x^2 - 12x + 1$

 a. $-2; f(-2) = 21$
 b. $1; f(1) = -6$

$[-4, 4, 1]$ by $[-15, 25, 5]$

*See Answers to Selected Exercises.

16.
$f(x) = 2x^3 - 15x^2 + 24x + 19$

 a. 1; $f(1) = 30$
 b. 4; $f(4) = 3$

[−2, 6, 1] by [−15, 35, 5]

In Exercises 17–32, determine whether the graph of each equation is symmetric with respect to the y-axis, the x-axis, the origin, more than one of these, or none of these.

17. $y = x^2 + 6$ y-axis
18. $y = x^2 - 2$ y-axis
19. $x = y^2 + 6$ x-axis
20. $x = y^2 - 2$ x-axis
21. $y^2 = x^2 + 6$ all three
22. $y^2 = x^2 - 2$ all three
23. $y = 2x + 3$ none
24. $y = 2x + 5$ none
25. $x^2 - y^3 = 2$ y-axis
26. $x^3 - y^2 = 5$ x-axis
27. $x^2 + y^2 = 100$ all three
28. $x^2 + y^2 = 49$ all three
29. $x^2y^2 + 3xy = 1$ origin
30. $x^2y^2 + 5xy = 2$ origin
31. $y^4 = x^3 + 6$ x-axis
32. $y^5 = x^4 + 2$ y-axis

In Exercises 33–36, use possible symmetry to determine whether each graph is the graph of an even function, an odd function, or a function that is neither even nor odd.

33. even

34. odd

35. odd

36. neither

In Exercises 37–48, determine whether each function is even, odd, or neither. Then determine whether the function's graph is symmetric with respect to the y-axis, the origin, or neither.

37. $f(x) = x^3 + x$ odd*
38. $f(x) = x^3 - x$ odd*
39. $g(x) = x^2 + x$ neither*
40. $g(x) = x^2 - x$ neither*
41. $h(x) = x^2 - x^4$ even*
42. $h(x) = 2x^2 + x^4$ even*
43. $f(x) = x^2 - x^4 + 1$ even; y-axis symmetry
44. $f(x) = 2x^2 + x^4 + 1$ even; y-axis symmetry
45. $f(x) = \frac{1}{5}x^6 - 3x^2$ even*
46. $f(x) = 2x^3 - 6x^5$ odd*
47. $f(x) = x\sqrt{1 - x^2}$ odd*
48. $f(x) = x^2\sqrt{1 - x^2}$ even*

49. Use the graph of f to determine each of the following. Where applicable, use interval notation.

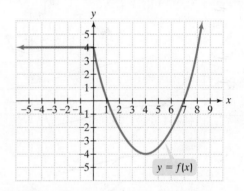

$y = f(x)$

a. the domain of f $(-\infty, \infty)$

b. the range of f $[-4, \infty)$

c. the x-intercepts 1 and 7

d. the y-intercept 4

e. intervals on which f is increasing $(4, \infty)$

f. intervals on which f is decreasing $(0, 4)$

g. intervals on which f is constant $(-\infty, 0)$

h. the number at which f has a relative minimum 4

i. the relative minimum of f −4

j. $f(-3)$ 4

k. the values of x for which $f(x) = -2$ 2 and 6

l. Is f even, odd, or neither? neither

50. Use the graph of f to determine each of the following. Where applicable, use interval notation.

a. the domain of f $(-\infty, \infty)$

b. the range of f $(-\infty, 4]$

c. the x-intercepts -4 and 4

d. the y-intercept 1

e. intervals on which f is increasing $(-\infty, -2)$ or $(0, 3)$

f. intervals on which f is decreasing $(-2, 0)$ or $(3, \infty)$

g. values of x for which $f(x) \le 0$ $(-\infty, -4]$ or $[4, \infty)$

h. the numbers at which f has a relative maximum -2 and 3

i. the relative maxima of f 2 and 4

j. $f(-2)$ 4

k. the values of x for which $f(x) = 0$ -4 and 4

l. Is f even, odd, or neither? neither

51. Use the graph of f to determine each of the following. Where applicable, use interval notation.

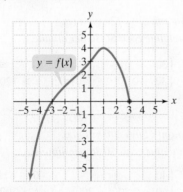

a. the domain of f $(-\infty, 3]$

b. the range of f $(-\infty, 4]$

c. the zeros of f -3 and 3

d. $f(0)$ 3

e. intervals on which f is increasing $(-\infty, 1)$

f. intervals on which f is decreasing $(1, 3)$

g. values of x for which $f(x) \le 0$ $(-\infty, -3]$

h. any relative maxima and the numbers at which they occur Relative maximum of 4 occurs at 1.

i. the value of x for which $f(x) = 4$ 1

j. Is $f(-1)$ positive or negative? positive

52. Use the graph of f to determine each of the following. Where applicable, use interval notation.

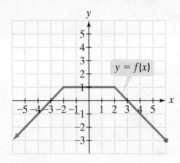

a. the domain of f $(-\infty, 6]$

b. the range of f $(-\infty, 1]$

c. the zeros of f -3 and 3

d. $f(0)$ 1

e. intervals on which f is increasing $(-\infty, -2)$

f. intervals on which f is decreasing $(2, 6)$

g. intervals on which f is constant $(-2, 2)$

h. values of x for which $f(x) > 0$ $(-3, 3)$

i. values of x for which $f(x) = -2$ -5 and 5

j. Is $f(4)$ positive or negative? negative

k. Is f even, odd, or neither? neither

l. Is $f(2)$ a relative maximum? no

In Exercises 53–58, evaluate each piecewise function at the given values of the independent variable.

53. $f(x) = \begin{cases} 3x + 5 & \text{if } x < 0 \\ 4x + 7 & \text{if } x \ge 0 \end{cases}$

 a. $f(-2)$ -1 **b.** $f(0)$ 7 **c.** $f(3)$ 19

54. $f(x) = \begin{cases} 6x - 1 & \text{if } x < 0 \\ 7x + 3 & \text{if } x \ge 0 \end{cases}$

 a. $f(-3)$ -19 **b.** $f(0)$ 3 **c.** $f(4)$ 31

55. $g(x) = \begin{cases} x + 3 & \text{if } x \ge -3 \\ -(x + 3) & \text{if } x < -3 \end{cases}$

 a. $g(0)$ 3 **b.** $g(-6)$ 3 **c.** $g(-3)$ 0

56. $g(x) = \begin{cases} x + 5 & \text{if } x \ge -5 \\ -(x + 5) & \text{if } x < -5 \end{cases}$

 a. $g(0)$ 5 **b.** $g(-6)$ 1 **c.** $g(-5)$ 0

57. $h(x) = \begin{cases} \dfrac{x^2 - 9}{x - 3} & \text{if } x \ne 3 \\ 6 & \text{if } x = 3 \end{cases}$

 a. $h(5)$ 8 **b.** $h(0)$ 3 **c.** $h(3)$ 6

58. $h(x) = \begin{cases} \dfrac{x^2 - 25}{x - 5} & \text{if } x \ne 5 \\ 10 & \text{if } x = 5 \end{cases}$

 a. $h(7)$ 12 **b.** $h(0)$ 5 **c.** $h(5)$ 10

In Exercises 59–70, the domain of each piecewise function is $(-\infty, \infty)$.

a. *Graph each function.*

b. *Use your graph to determine the function's range.*

59. $f(x) = \begin{cases} -x & \text{if } x < 0 \\ x & \text{if } x \geq 0 \end{cases}$ b. $[0, \infty)$*

60. $f(x) = \begin{cases} x & \text{if } x < 0 \\ -x & \text{if } x \geq 0 \end{cases}$ b. $(-\infty, 0]$*

61. $f(x) = \begin{cases} 2x & \text{if } x \leq 0 \\ 2 & \text{if } x > 0 \end{cases}$ b. $(-\infty, 0] \cup \{2\}$*

62. $f(x) = \begin{cases} \frac{1}{2}x & \text{if } x \leq 0 \\ 3 & \text{if } x > 0 \end{cases}$ b. $(-\infty, 0] \cup \{3\}$*

63. $f(x) = \begin{cases} x + 3 & \text{if } x < -2 \\ x - 3 & \text{if } x \geq -2 \end{cases}$ b. $(-\infty, \infty)$*

64. $f(x) = \begin{cases} x + 2 & \text{if } x < -3 \\ x - 2 & \text{if } x \geq -3 \end{cases}$ b. $(-\infty, \infty)$*

65. $f(x) = \begin{cases} 3 & \text{if } x \leq -1 \\ -3 & \text{if } x > -1 \end{cases}$ b. $\{-3, 3\}$*

66. $f(x) = \begin{cases} 4 & \text{if } x \leq -1 \\ -4 & \text{if } x > -1 \end{cases}$ b. $\{-4, 4\}$*

67. $f(x) = \begin{cases} \frac{1}{2}x^2 & \text{if } x < 1 \\ 2x - 1 & \text{if } x \geq 1 \end{cases}$ b. $[0, \infty)$*

68. $f(x) = \begin{cases} -\frac{1}{2}x^2 & \text{if } x < 1 \\ 2x + 1 & \text{if } x \geq 1 \end{cases}$ b. $(-\infty, 0] \cup [3, \infty)$*

69. $f(x) = \begin{cases} 0 & \text{if } x < -4 \\ -x & \text{if } -4 \leq x < 0 \\ x^2 & \text{if } x \geq 0 \end{cases}$ b. $[0, \infty)$*

70. $f(x) = \begin{cases} 0 & \text{if } x < -3 \\ -x & \text{if } -3 \leq x < 0 \\ x^2 - 1 & \text{if } x \geq 0 \end{cases}$ b. $[-1, \infty)$*

In Exercises 71–92, find and simplify the difference quotient

$$\frac{f(x + h) - f(x)}{h}, h \neq 0$$

for the given function.

71. $f(x) = 4x$ 4, $h \neq 0$ **72.** $f(x) = 7x$ 7, $h \neq 0$

73. $f(x) = 3x + 7$ 3, $h \neq 0$ **74.** $f(x) = 6x + 1$ 6, $h \neq 0$

75. $f(x) = x^2$ $2x + h, h \neq 0$ **76.** $f(x) = 2x^2$ $4x + 2h, h \neq 0$

77. $f(x) = x^2 - 4x + 3$ * **78.** $f(x) = x^2 - 5x + 8$ *

79. $f(x) = 2x^2 + x - 1$ * **80.** $f(x) = 3x^2 + x + 5$ *

81. $f(x) = -x^2 + 2x + 4$ * **82.** $f(x) = -x^2 - 3x + 1$ *

83. $f(x) = -2x^2 + 5x + 7$ * **84.** $f(x) = -3x^2 + 2x - 1$ *

85. $f(x) = -2x^2 - x + 3$ * **86.** $f(x) = -3x^2 + x - 1$ *

87. $f(x) = 6$ 0, $h \neq 0$ **88.** $f(x) = 7$ 0, $h \neq 0$

89. $f(x) = \dfrac{1}{x}$ $-\dfrac{1}{x(x + h)}, h \neq 0$ **90.** $f(x) = \dfrac{1}{2x}$ $-\dfrac{1}{2x(x + h)}, h \neq 0$

91. $f(x) = \sqrt{x}$ * **92.** $f(x) = \sqrt{x - 1}$ *

Practice Plus

In Exercises 93–94, let f be defined by the following graph:

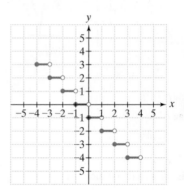

93. Find -18

$$\sqrt{f(-1.5) + f(-0.9)} - [f(\pi)]^2 + f(-3) \div f(1) \cdot f(-\pi).$$

94. Find -3

$$\sqrt{f(-2.5) - f(1.9)} - [f(-\pi)]^2 + f(-3) \div f(1) \cdot f(\pi).$$

A telephone company offers the following plans. Also given are the piecewise functions that model these plans. Use this information to solve Exercises 95–96.

Plan A

- *$30 per month buys 120 minutes.*
- *Additional time costs $0.30 per minute.*

$$C(t) = \begin{cases} 30 & \text{if } 0 \leq t \leq 120 \\ 30 + 0.30(t - 120) & \text{if } t > 120 \end{cases}$$

Plan B

- *$40 per month buys 200 minutes.*
- *Additional time costs $0.30 per minute.*

$$C(t) = \begin{cases} 40 & \text{if } 0 \leq t \leq 200 \\ 40 + 0.30(t - 200) & \text{if } t > 200 \end{cases}$$

95. Simplify the algebraic expression in the second line of the piecewise function for plan A. Then use point-plotting to graph the function. $0.30t - 6$*

96. Simplify the algebraic expression in the second line of the piecewise function for plan B. Then use point-plotting to graph the function. $0.30t - 20$*

In Exercises 97–98, write a piecewise function that models each telephone billing plan. Then graph the function.

97. $50 per month buys 400 minutes. Additional time costs $0.30 per minute. *

98. $60 per month buys 450 minutes. Additional time costs $0.35 per minute. *

Application Exercises

With aging, body fat increases and muscle mass declines. The line graphs show the percent body fat in adult women and men as they age from 25 to 75 years. Use the graphs to solve Exercises 99–106.

Percent Body Fat in Adults

Source: Thompson et al., *The Science of Nutrition*, Benjamin Cummings, 2008

99. State the intervals on which the graph giving the percent body fat in women is increasing and decreasing. *

100. State the intervals on which the graph giving the percent body fat in men is increasing and decreasing. *

101. For what age does the percent body fat in women reach a maximum? What is the percent body fat for that age? *

102. At what age does the percent body fat in men reach a maximum? What is the percent body fat for that age? *

103. Use interval notation to give the domain and the range for the graph of the function for women. *

104. Use interval notation to give the domain and the range for the graph of the function for men. *

105. The function $p(x) = -0.002x^2 + 0.15x + 22.86$ models percent body fat, $p(x)$, where x is the number of years a person's age exceeds 25. Use the graphs to determine whether this model describes percent body fat in women or in men. men

106. The function $p(x) = -0.004x^2 + 0.25x + 33.64$ models percent body fat, $p(x)$, where x is the number of years a person's age exceeds 25. Use the graphs to determine whether this model describes percent body fat in women or in men. women

Here is the Federal Tax Rate Schedule X that specifies the tax owed by a single taxpayer for a recent year.

If Your Taxable Income Is Over	But Not Over	The Tax You Owe Is	Of the Amount Over
$ 0	$ 8500	10%	$ 0
$ 8500	$ 34,500	$ 850.00 + 15%	$ 8500
$ 34,500	$ 83,600	$ 4750.00 + 25%	$ 34,500
$ 83,600	$174,400	$ 17,025.00 + 28%	$ 83,600
$174,400	$379,150	$ 42,449.00 + 33%	$174,400
$379,150	—	$110,016.50 + 35%	$379,150

The preceding tax table can be modeled by a piecewise function, where x represents the taxable income of a single taxpayer and T(x) is the tax owed:

$$T(x) = \begin{cases} 0.10x & \text{if} & 0 < x \le 8500 \\ 850.00 + 0.15(x - 8500) & \text{if} & 8500 < x \le 34{,}500 \\ 4750.00 + 0.25(x - 34{,}500) & \text{if} & 34{,}500 < x \le 83{,}600 \\ 17{,}025.00 + 0.28(x - 83{,}600) & \text{if} & 83{,}600 < x \le 174{,}400 \\ \underline{}\ ? & \text{if} & 174{,}400 < x \le 379{,}150 \\ \underline{}\ ? & \text{if} & x > 379{,}150. \end{cases}$$

Use this information to solve Exercises 107–108.

107. Find and interpret $T(20{,}000)$. 2575*

108. Find and interpret $T(50{,}000)$. 8625*

In Exercises 109–110, refer to the preceding tax table.

109. Find the algebraic expression for the missing piece of $T(x)$ that models tax owed for the domain (174,400, 379,150]. *

110. Find the algebraic expression for the missing piece of $T(x)$ that models tax owed for the domain (379,150, ∞). *

The figure shows the cost of mailing a first-class letter, f(x), as a function of its weight, x, in ounces, for a recent year. Use the graph to solve Exercises 111–114.

Source: Lynn E. Baring, Postmaster, Inverness, CA

111. Find $f(3)$. What does this mean in terms of the variables in this situation? 0.93*

112. Find $f(3.5)$. What does this mean in terms of the variables in this situation? 1.15*

113. What is the cost of mailing a letter that weighs 1.5 ounces? *

114. What is the cost of mailing a letter that weighs 1.8 ounces? *

115. **Furry Finances.** A pet insurance policy has a monthly rate that is a function of the age of the insured dog or cat. For pets whose age does not exceed 4, the monthly cost is $20. The cost then increases by $2 for each successive year of the pet's age.

Age Not Exceeding	Monthly Cost
4	$20
5	$22
6	$24

The cost schedule continues in this manner for ages not exceeding 10. The cost for pets whose ages exceed 10 is $40. Use this information to create a graph that shows the monthly cost of the insurance, $f(x)$, for a pet of age x, where the function's domain is [0, 14]. *

Explaining the Concepts 116–124. Answers will vary.

116. What does it mean if a function f is increasing on an interval?

117. Suppose that a function f whose graph contains no breaks or gaps on (a, c) is increasing on (a, b), decreasing on (b, c), and defined at b. Describe what occurs at $x = b$. What does the function value $f(b)$ represent?

118. Given an equation in x and y, how do you determine if its graph is symmetric with respect to the y-axis?

119. Given an equation in x and y, how do you determine if its graph is symmetric with respect to the x-axis?

120. Given an equation in x and y, how do you determine if its graph is symmetric with respect to the origin?

121. If you are given a function's graph, how do you determine if the function is even, odd, or neither?

122. If you are given a function's equation, how do you determine if the function is even, odd, or neither?

123. What is a piecewise function?

124. Explain how to find the difference quotient of a function f, $\dfrac{f(x + h) - f(x)}{h}$, if an equation for f is given.

Technology Exercises

125. The function

$$f(x) = -0.00002x^3 + 0.008x^2 - 0.3x + 6.95$$

models the number of annual physician visits, $f(x)$, by a person of age x. Graph the function in a $[0, 100, 5]$ by $[0, 40, 2]$ viewing rectangle. What does the shape of the graph indicate about the relationship between one's age and the number of annual physician visits? Use the TABLE or minimum function capability to find the coordinates of the minimum point on the graph of the function. What does this mean? *

In Exercises 126–131, use a graphing utility to graph each function. Use a $[-5, 5, 1]$ by $[-5, 5, 1]$ viewing rectangle. Then find the intervals on which the function is increasing, decreasing, or constant.

126. $f(x) = x^3 - 6x^2 + 9x + 1$ *

127. $g(x) = |4 - x^2|$ *

128. $h(x) = |x - 2| + |x + 2|$ *

129. $f(x) = x^{\frac{1}{3}}(x - 4)$ *

130. $g(x) = x^{\frac{2}{3}}$ *

131. $h(x) = 2 - x^{\frac{2}{5}}$ *

132. a. Graph the functions $f(x) = x^n$ for $n = 2, 4,$ and 6 in a $[-2, 2, 1]$ by $[-1, 3, 1]$ viewing rectangle. *

 b. Graph the functions $f(x) = x^n$ for $n = 1, 3,$ and 5 in a $[-2, 2, 1]$ by $[-2, 2, 1]$ viewing rectangle. *

 c. If n is positive and even, where is the graph of $f(x) = x^n$ increasing and where is it decreasing? *

 d. If n is positive and odd, what can you conclude about the graph of $f(x) = x^n$ in terms of increasing or decreasing behavior? *

 e. Graph all six functions in a $[-1, 3, 1]$ by $[-1, 3, 1]$ viewing rectangle. What do you observe about the graphs in terms of how flat or how steep they are? *

Critical Thinking Exercises

Make Sense? *In Exercises 133–136, determine whether each statement makes sense or does not make sense, and explain your reasoning.*

133. My graph is decreasing on $(-\infty, a)$ and increasing on (a, ∞), so $f(a)$ must be a relative maximum. does not make sense

134. This work by artist Scott Kim (1955–) has the same kind of symmetry as an even function. makes sense

"DYSLEXIA," 1981

135. I graphed

$$f(x) = \begin{cases} 2 & \text{if } x \neq 4 \\ 3 & \text{if } x = 4 \end{cases}$$

and one piece of my graph is a single point. makes sense

136. I noticed that the difference quotient is always zero if $f(x) = c$, where c is any constant. makes sense

137. Sketch the graph of f using the following properties. (More than one correct graph is possible.) f is a piecewise function that is decreasing on $(-\infty, 2), f(2) = 0, f$ is increasing on $(2, \infty)$, and the range of f is $[0, \infty)$.
137–138. Answers will vary.

138. Define a piecewise function on the intervals $(-\infty, 2], (2, 5),$ and $[5, \infty)$ that does not "jump" at 2 or 5 such that one piece is a constant function, another piece is an increasing function, and the third piece is a decreasing function.

139. Suppose that $h(x) = \dfrac{f(x)}{g(x)}$. The function f can be even, odd, or neither. The same is true for the function g.

 a. Under what conditions is h definitely an even function? *

 b. Under what conditions is h definitely an odd function? *

Retaining the Concepts

140. You invested $80,000 in two accounts paying 5% and 7% annual interest. If the total interest earned for the year was $5200, how much was invested at each rate? (Section 1.3, Example 5) $20,000 at 5%; $60,000 at 7%

141. Solve for A: $C = A + Ar$.

 (Section 1.3, Example 8) $A = \dfrac{C}{1 + r}$

142. Solve by the quadratic formula: $5x^2 - 7x + 3 = 0$. *
(Section 1.5, Example 8)

Preview Exercises

Exercises 143–145 will help you prepare for the material covered in the next section.

143. If $(x_1, y_1) = (-3, 1)$ and $(x_2, y_2) = (-2, 4)$, find $\dfrac{y_2 - y_1}{x_2 - x_1}$. *

144. Find the ordered pairs $(\underline{\tfrac{3}{2}}, 0)$ and $(0, \underline{-2})$ satisfying $4x - 3y - 6 = 0$.

145. Solve for y: $3x + 2y - 4 = 0$. $y = \dfrac{-3x + 4}{2}$ or $y = -\dfrac{3}{2}x + 2$

Section 2.3

Linear Functions and Slope

What am I supposed to learn?

After studying this section, you should be able to:

1. Calculate a line's slope.
2. Write the point-slope form of the equation of a line.
3. Write and graph the slope-intercept form of the equation of a line.
4. Graph horizontal or vertical lines.
5. Recognize and use the general form of a line's equation.
6. Use intercepts to graph the general form of a line's equation.
7. Model data with linear functions and make predictions.

Is there a relationship between literacy and child mortality? As the percentage of adult females who are literate increases, does the mortality of children under five decrease? **Figure 2.28** indicates that this is, indeed, the case. Each point in the figure represents one country.

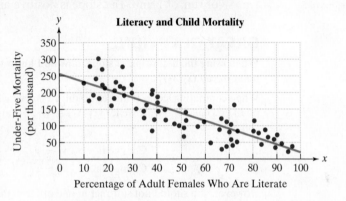

FIGURE 2.28
Source: United Nations

Data presented in a visual form as a set of points is called a **scatter plot**. Also shown in **Figure 2.28** is a line that passes through or near the points. A line that best fits the data points in a scatter plot is called a **regression line**. By writing the equation of this line, we can obtain a model for the data and make predictions about child mortality based on the percentage of literate adult females in a country.

Data often fall on or near a line. In this section, we will use functions to model such data and make predictions. We begin with a discussion of a line's steepness.

1. Calculate a line's slope.

The Slope of a Line

Mathematicians have developed a useful measure of the steepness of a line, called the *slope* of the line. Slope compares the vertical change (the **rise**) to the horizontal change (the **run**) when moving from one fixed point to another along the line. To calculate the slope of a line, we use a ratio that compares the change in y (the rise) to the corresponding change in x (the run).

Definition of Slope

The **slope** of the line through the distinct points (x_1, y_1) and (x_2, y_2) is

$$\frac{\text{Change in } y}{\text{Change in } x} = \frac{\text{Rise}}{\text{Run}} \quad \boxed{\text{Vertical change}} \quad \boxed{\text{Horizontal change}}$$

$$= \frac{y_2 - y_1}{x_2 - x_1},$$

where $x_2 - x_1 \neq 0$.

It is common notation to let the letter m represent the slope of a line. The letter m is used because it is the first letter of the French verb *monter*, meaning "to rise" or "to ascend."

EXAMPLE 1 Using the Definition of Slope

Find the slope of the line passing through each pair of points:

a. $(-3, -1)$ and $(-2, 4)$ **b.** $(-3, 4)$ and $(2, -2)$.

SOLUTION

a. Let $(x_1, y_1) = (-3, -1)$ and $(x_2, y_2) = (-2, 4)$. We obtain the slope as follows:

$$m = \frac{\text{Change in } y}{\text{Change in } x} = \frac{y_2 - y_1}{x_2 - x_1} = \frac{4 - (-1)}{-2 - (-3)} = \frac{4 + 1}{-2 + 3} = \frac{5}{1} = 5.$$

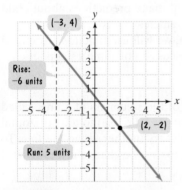

FIGURE 2.29 Visualizing a slope of 5

The situation is illustrated in **Figure 2.29**. The slope of the line is 5. For every vertical change, or rise, of 5 units, there is a corresponding horizontal change, or run, of 1 unit. The slope is positive and the line rises from left to right.

GREAT QUESTION!

When using the definition of slope, how do I know which point to call (x_1, y_1) and which point to call (x_2, y_2)?

When computing slope, it makes no difference which point you call (x_1, y_1) and which point you call (x_2, y_2). If we let $(x_1, y_1) = (-2, 4)$ and $(x_2, y_2) = (-3, -1)$, the slope is still 5:

$$m = \frac{\text{Change in } y}{\text{Change in } x} = \frac{y_2 - y_1}{x_2 - x_1} = \frac{-1 - 4}{-3 - (-2)} = \frac{-5}{-1} = 5.$$

However, you should not subtract in one order in the numerator $(y_2 - y_1)$ and then in a different order in the denominator $(x_1 - x_2)$.

$$\frac{-1 - 4}{-2 - (-3)} = \frac{-5}{1} = -5.$$ Incorrect! The slope is not -5.

b. To find the slope of the line passing through $(-3, 4)$ and $(2, -2)$, we can let $(x_1, y_1) = (-3, 4)$ and $(x_2, y_2) = (2, -2)$. The slope of the line is computed as follows:

$$m = \frac{\text{Change in } y}{\text{Change in } x} = \frac{y_2 - y_1}{x_2 - x_1} = \frac{-2 - 4}{2 - (-3)} = \frac{-6}{5} = -\frac{6}{5}.$$

FIGURE 2.30 Visualizing a slope of $-\frac{6}{5}$

The situation is illustrated in **Figure 2.30**. The slope of the line is $-\frac{6}{5}$. For every vertical change of -6 units (6 units down), there is a corresponding horizontal change of 5 units. The slope is negative and the line falls from left to right. ● ● ●

GREAT QUESTION!

Is it OK to say that a vertical line has no slope?

Always be clear in the way you use language, especially in mathematics. For example, it's not a good idea to say that a line has "no slope." This could mean that the slope is zero or that the slope is undefined.

✅ **Check Point 1** Find the slope of the line passing through each pair of points:
a. $(-3, 4)$ and $(-4, -2)$ 6 **b.** $(4, -2)$ and $(-1, 5)$. $-\frac{7}{5}$

Example 1 illustrates that a line with a positive slope is increasing and a line with a negative slope is decreasing. By contrast, a horizontal line is a constant function and has a slope of zero. A vertical line has no horizontal change, so $x_2 - x_1 = 0$ in the formula for slope. Because we cannot divide by zero, the slope of a vertical line is undefined. This discussion is summarized in **Table 2.3**.

Table 2.3 Possibilities for a Line's Slope

Positive Slope	Negative Slope	Zero Slope	Undefined Slope
$m > 0$	$m < 0$	$m = 0$	m is undefined.
Line rises from left to right.	Line falls from left to right.	Line is horizontal.	Line is vertical.

Blitzer Bonus ‖ Slope and Applauding Together

Using a decibel meter, sociologist Max Atkinson found a piecewise function that models the intensity of a group's applause, $d(t)$, in decibels, over time, t, in seconds.

- Applause starts very fast, reaching a full crescendo of 30 decibels in one second. ($m = 30$)
- Applause remains level at 30 decibels for 5.5 seconds. ($m = 0$)
- Applause trails off by 15 decibels per second for two seconds. ($m = -15$)

These verbal conditions can be modeled by a piecewise function.

$$d(t) = \begin{cases} 30t & 0 \le t \le 1 \\ 30 & 1 < t \le 6.5 \\ -15t + 127.5 & 6.5 < t \le 8.5 \end{cases}$$

Intensity of applause, $d(t)$, in decibels, as a function of time, t, in seconds

The graph of this piecewise function is shown below. The function suggests how strongly people work to coordinate their applause with one another, careful to start clapping at the "right" time and careful to stop when it seems others are stopping as well.

Applauding Together

$m = 0$
Applause remains level for 5.5 seconds.

$m = -15$
Applause trails off fairly fast.

$m = 30$
Applause starts very fast, reaching full crescendo in one second.

Source: *The Sociology Project 2.0*, Pearson, 2016.

② Write the point-slope form of the equation of a line.

The Point-Slope Form of the Equation of a Line

We can use the slope of a line to obtain various forms of the line's equation. For example, consider a nonvertical line that has slope m and that contains the point (x_1, y_1).

FIGURE 2.31 A line passing through (x_1, y_1) with slope m

The line in **Figure 2.31** has slope m and contains the point (x_1, y_1). Let (x, y) represent any other point on the line.

Regardless of where the point (x, y) is located, the steepness of the line in **Figure 2.31** remains the same. Thus, the ratio for the slope stays a constant m. This means that for all points (x, y) along the line

$$m = \frac{\text{Change in } y}{\text{Change in } x} = \frac{y - y_1}{x - x_1}.$$

| We used (x, y) instead of (x_2, y_2) because (x_2, y_2) represents a fixed point and (x, y) is not a fixed point on the line.

We can clear the fraction by multiplying both sides by $x - x_1$, the least common denominator.

$$m = \frac{y - y_1}{x - x_1} \qquad \text{This is the slope of the line in Figure 2.31.}$$

$$m(x - x_1) = \frac{y - y_1}{x - x_1} \cdot (x - x_1) \qquad \text{Multiply both sides by } x - x_1.$$

$$m(x - x_1) = y - y_1 \qquad \text{Simplify: } \frac{y - y_1}{x - x_1} \cdot (x - x_1) = y - y_1.$$

Now, if we reverse the two sides, we obtain the *point-slope form* of the equation of a line.

Point-Slope Form of the Equation of a Line

The **point-slope form of the equation** of a nonvertical line with slope m that passes through the point (x_1, y_1) is

$$y - y_1 = m(x - x_1).$$

GREAT QUESTION!

When using $y - y_1 = m(x - x_1)$, for which variables do I substitute numbers?

When writing the point-slope form of a line's equation, you will never substitute numbers for x and y. You will substitute values for $x_1, y_1,$ and m.

For example, the point-slope form of the equation of the line passing through $(1, 5)$ with slope 2 ($m = 2$) is

$$y - 5 = 2(x - 1).$$

We will soon be expressing the equation of a nonvertical line in function notation. To do so, we need to solve the point-slope form of a line's equation for y. Example 2 illustrates how to isolate y on one side of the equal sign.

EXAMPLE 2 Writing an Equation for a Line in Point-Slope Form

Write an equation in point-slope form for the line with slope 4 that passes through the point $(-1, 3)$. Then solve the equation for y.

SOLUTION

We use the point-slope form of the equation of a line with $m = 4$, $x_1 = -1$, and $y_1 = 3$.

$$y - y_1 = m(x - x_1) \qquad \text{This is the point-slope form of the equation.}$$

$$y - 3 = 4[x - (-1)] \qquad \text{Substitute the given values: } m = 4 \text{ and } (x_1, y_1) = (-1, 3).$$

$$y - 3 = 4(x + 1) \qquad \text{We now have an equation in point-slope form for the given line.}$$

Now we solve this equation for y.

We need to isolate y.

$$y - 3 = 4(x + 1) \qquad \text{This is the point-slope form of the equation.}$$

$$y - 3 = 4x + 4 \qquad \text{Use the distributive property.}$$

$$y = 4x + 7 \qquad \text{Add 3 to both sides.}$$

• • •

⊘ Check Point **2** Write an equation in point-slope form for the line with slope 6 that passes through the point $(2, -5)$. Then solve the equation for y.
$y + 5 = 6(x - 2); y = 6x - 17$

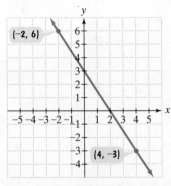

FIGURE 2.32 Write an equation in point-slope form for this line.

EXAMPLE 3 Writing an Equation for a Line in Point-Slope Form

Write an equation in point-slope form for the line passing through the points $(4, -3)$ and $(-2, 6)$. (See **Figure 2.32**.) Then solve the equation for y.

SOLUTION

To use the point-slope form, we need to find the slope. The slope is the change in the y-coordinates divided by the corresponding change in the x-coordinates.

$$m = \frac{6 - (-3)}{-2 - 4} = \frac{9}{-6} = -\frac{3}{2} \quad \begin{array}{l} \text{This is the definition of slope} \\ \text{using } (4, -3) \text{ and } (-2, 6). \end{array}$$

We can take either point on the line to be (x_1, y_1). Let's use $(x_1, y_1) = (4, -3)$. Now, we are ready to write the point-slope form of the equation.

$$y - y_1 = m(x - x_1) \quad \text{This is the point-slope form of the equation.}$$

$$y - (-3) = -\tfrac{3}{2}(x - 4) \quad \text{Substitute: } (x_1, y_1) = (4, -3) \text{ and } m = -\tfrac{3}{2}.$$

$$y + 3 = -\tfrac{3}{2}(x - 4) \quad \text{Simplify.}$$

We now have an equation in point-slope form for the line shown in **Figure 2.32**. Now, we solve this equation for y.

We need to isolate y.

$$y + 3 = -\tfrac{3}{2}(x - 4) \quad \text{This is the point-slope form of the equation.}$$

$$y + 3 = -\tfrac{3}{2}x + 6 \quad \text{Use the distributive property.}$$

$$y = -\tfrac{3}{2}x + 3 \quad \text{Subtract 3 from both sides.} \quad \bullet\bullet\bullet$$

DISCOVERY

You can use either point for (x_1, y_1) when you write a point-slope equation for a line. Rework Example 3 using $(-2, 6)$ for (x_1, y_1). Once you solve for y, you should still obtain

$$y = -\tfrac{3}{2}x + 3.$$

⊘ Check Point **3** Write an equation in point-slope form for the line passing through the points $(-2, -1)$ and $(-1, -6)$. Then solve the equation for y.
$y + 1 = -5(x + 2)$ or $y + 6 = -5(x + 1); y = -5x - 11$

3 Write and graph the slope-intercept form of the equation of a line.

FIGURE 2.33 A line with slope m and y-intercept b

The Slope-Intercept Form of the Equation of a Line

Let's write the point-slope form of the equation of a nonvertical line with slope m and y-intercept b. The line is shown in **Figure 2.33**. Because the y-intercept is b, the line passes through $(0, b)$. We use the point-slope form with $x_1 = 0$ and $y_1 = b$.

$$y - y_1 = m(x - x_1)$$

Let $y_1 = b$. **Let $x_1 = 0$.**

We obtain

$$y - b = m(x - 0).$$

Simplifying on the right side gives us

$$y - b = mx.$$

Finally, we solve for y by adding b to both sides.

$$y = mx + b$$

Thus, if a line's equation is written as $y = mx + b$ with y isolated on one side, the coefficient of x is the line's slope and the constant term is the y-intercept. This form of a line's equation is called the *slope-intercept form* of the line.

> ### Slope-Intercept Form of the Equation of a Line
>
> The **slope-intercept form of the equation** of a nonvertical line with slope m and y-intercept b is
>
> $$y = mx + b.$$

The slope-intercept form of a line's equation, $y = mx + b$, can be expressed in function notation by replacing y with $f(x)$:

$$f(x) = mx + b.$$

We have seen that functions in this form are called **linear functions**. Thus, in the equation of a linear function, the coefficient of x is the line's slope and the constant term is the y-intercept. Here are two examples:

$$y = 2x - 4 \qquad\qquad f(x) = \tfrac{1}{2}x + 2.$$

| The slope is 2. | The y-intercept is −4. | The slope is $\tfrac{1}{2}$. | The y-intercept is 2. |

If a linear function's equation is in slope-intercept form, we can use the y-intercept and the slope to obtain its graph.

GREAT QUESTION!

If the slope is an integer, such as 2, why should I express it as $\tfrac{2}{1}$ for graphing purposes?

Writing the slope, m, as a fraction allows you to identify the rise (the fraction's numerator) and the run (the fraction's denominator).

> ### Graphing $y = mx + b$ Using the Slope and y-Intercept
>
> 1. Plot the point containing the y-intercept on the y-axis. This is the point $(0, b)$.
> 2. Obtain a second point using the slope, m. Write m as a fraction, and use rise over run, starting at the point containing the y-intercept, to plot this point.
> 3. Use a straightedge to draw a line through the two points. Draw arrowheads at the ends of the line to show that the line continues indefinitely in both directions.

EXAMPLE 4 Graphing Using the Slope and y-Intercept

Graph the linear function: $f(x) = -\dfrac{3}{2}x + 2.$

SOLUTION

The equation of the line is in the form $f(x) = mx + b$. We can find the slope, m, by identifying the coefficient of x. We can find the y-intercept, b, by identifying the constant term.

$$f(x) = -\tfrac{3}{2}x + 2$$

| The slope is $-\tfrac{3}{2}$. | The y-intercept is 2. |

Now that we have identified the slope, $-\tfrac{3}{2}$, and the y-intercept, 2, we use the three-step procedure to graph the equation.

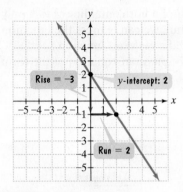

FIGURE 2.34 The graph of $f(x) = -\frac{3}{2}x + 2$

Step 1 Plot the point containing the y-intercept on the y-axis. The y-intercept is 2. We plot $(0, 2)$, shown in **Figure 2.34**.

Step 2 Obtain a second point using the slope, m. Write m as a fraction, and use rise over run, starting at the point containing the y-intercept, to plot this point. The slope, $-\frac{3}{2}$, is already written as a fraction.

$$m = -\frac{3}{2} = \frac{-3}{2} = \frac{\text{Rise}}{\text{Run}}$$

We plot the second point on the line by starting at $(0, 2)$, the first point. Based on the slope, we move 3 units *down* (the rise) and 2 units to the *right* (the run). This puts us at a second point on the line, $(2, -1)$, shown in **Figure 2.34**.

Step 3 Use a straightedge to draw a line through the two points. The graph of the linear function $f(x) = -\frac{3}{2}x + 2$ is shown as a blue line in **Figure 2.34**. •••

 Check Point 4 Graph the linear function: $f(x) = \frac{3}{5}x + 1$. *

4 Graph horizontal or vertical lines.

Equations of Horizontal and Vertical Lines

If a line is horizontal, its slope is zero: $m = 0$. Thus, the equation $y = mx + b$ becomes $y = b$, where b is the y-intercept. All horizontal lines have equations of the form $y = b$.

EXAMPLE 5 Graphing a Horizontal Line

Graph $y = -4$ in the rectangular coordinate system.

SOLUTION

All ordered pairs that are solutions of $y = -4$ have a value of y that is always -4. Any value can be used for x. In the table on the right, we have selected three of the possible values for x: $-2, 0,$ and 3. The table shows that three ordered pairs that are solutions of $y = -4$ are $(-2, -4), (0, -4),$ and $(3, -4)$. Drawing a line that passes through the three points gives the horizontal line shown in **Figure 2.35**.

x	$y = -4$	(x, y)
-2	-4	$(-2, -4)$
0	-4	$(0, -4)$
3	-4	$(3, -4)$

For all choices of x, y is a constant -4.

•••

FIGURE 2.35 The graph of $y = -4$ or $f(x) = -4$

 Check Point 5 Graph $y = 3$ in the rectangular coordinate system. *

Equation of a Horizontal Line

A horizontal line is given by an equation of the form

$$y = b,$$

where b is the y-intercept of the line. The slope of a horizontal line is zero.

Because any vertical line can intersect the graph of a horizontal line $y = b$ only once, a horizontal line is the graph of a function. Thus, we can express the equation $y = b$ as $f(x) = b$. This linear function is often called a **constant function**.

Next, let's see what we can discover about the graph of an equation of the form $x = a$ by looking at an example.

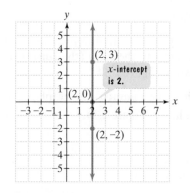

FIGURE 2.36 The graph of $x = 2$

EXAMPLE 6 Graphing a Vertical Line

Graph the linear equation: $x = 2$.

SOLUTION

All ordered pairs that are solutions of $x = 2$ have a value of x that is always 2. Any value can be used for y. In the table on the right, we have selected three of the possible values for y: -2, 0, and 3. The table shows that three ordered pairs that are solutions of $x = 2$ are $(2, -2)$, $(2, 0)$, and $(2, 3)$. Drawing a line that passes through the three points gives the vertical line shown in **Figure 2.36**.

For all choices of y,		
$x = 2$	y	(x, y)
2	-2	$(2, -2)$
2	0	$(2, 0)$
2	3	$(2, 3)$

• • •

> **Equation of a Vertical Line**
>
> A vertical line is given by an equation of the form
> $$x = a,$$
> where a is the x-intercept of the line. The slope of a vertical line is undefined.

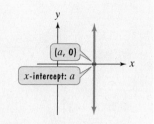

Does a vertical line represent the graph of a linear function? No. Look at the graph of $x = 2$ in **Figure 2.36**. A vertical line drawn through $(2, 0)$ intersects the graph infinitely many times. This shows that infinitely many outputs are associated with the input 2. **No vertical line represents a linear function.**

✓ Check Point **6** Graph the linear equation: $x = -3$. *

⑤ Recognize and use the general form of a line's equation.

The General Form of the Equation of a Line

The vertical line whose equation is $x = 5$ cannot be written in slope-intercept form, $y = mx + b$, because its slope is undefined. However, every line has an equation that can be expressed in the form $Ax + By + C = 0$. For example, $x = 5$ can be expressed as $1x + 0y - 5 = 0$, or $x - 5 = 0$. The equation $Ax + By + C = 0$ is called the *general form* of the equation of a line.

> **General Form of the Equation of a Line**
>
> Every line has an equation that can be written in the **general form**
> $$Ax + By + C = 0,$$
> where A, B, and C are real numbers, and A and B are not both zero.

GREAT QUESTION!

In the general form
$Ax + By + C = 0$, **can I immediately determine that the slope is A and the y-intercept is B?**

No. Avoid this common error. You need to solve $Ax + By + C = 0$ for y before finding the slope and the y-intercept.

If the equation of a nonvertical line is given in general form, it is possible to find the slope, m, and the y-intercept, b, for the line. We solve the equation for y, transforming it into the slope-intercept form $y = mx + b$. In this form, the coefficient of x is the slope of the line and the constant term is its y-intercept.

EXAMPLE 7 Finding the Slope and the y-Intercept

Find the slope and the y-intercept of the line whose equation is $3x + 2y - 4 = 0$.

SOLUTION

The equation is given in general form. We begin by rewriting it in the form $y = mx + b$. We need to solve for y.

$3x + 2y - 4 = 0$	This is the given equation.
$2y = -3x + 4$	Isolate the term containing y by adding $-3x + 4$ to both sides.
$\dfrac{2y}{2} = \dfrac{-3x + 4}{2}$	Divide both sides by 2.
$y = -\dfrac{3}{2}x + 2$	On the right, divide each term in the numerator by 2 to obtain slope-intercept form.

Our goal is to isolate y.

slope y-intercept

The coefficient of x, $-\frac{3}{2}$, is the slope and the constant term, 2, is the y-intercept. This is the form of the equation that we graphed in **Figure 2.34** on page 261. • • •

✓ **Check Point 7** Find the slope and the y-intercept of the line whose equation is $3x + 6y - 12 = 0$. Then use the y-intercept and the slope to graph the equation. slope: $-\frac{1}{2}$; y-intercept: 2*

⑥ Use intercepts to graph the general form of a line's equation.

Using Intercepts to Graph $Ax + By + C = 0$

Example 7 and Check Point 7 illustrate that one way to graph the general form of a line's equation is to convert to slope-intercept form, $y = mx + b$. Then use the slope and the y-intercept to obtain the graph.

A second method for graphing $Ax + By + C = 0$ uses intercepts. This method does not require rewriting the general form in a different form.

Using Intercepts to Graph $Ax + By + C = 0$

1. Find the x-intercept. Let $y = 0$ and solve for x. Plot the point containing the x-intercept on the x-axis.

2. Find the y-intercept. Let $x = 0$ and solve for y. Plot the point containing the y-intercept on the y-axis.

3. Use a straightedge to draw a line through the two points containing the intercepts. Draw arrowheads at the ends of the line to show that the line continues indefinitely in both directions.

As long as none of A, B, and C is zero, the graph of $Ax + By + C = 0$ will have distinct x- and y-intercepts, and this three-step method can be used to graph the equation.

*See Answers to Selected Exercises.

EXAMPLE 8 Using Intercepts to Graph a Linear Equation

Graph using intercepts: $4x - 3y - 6 = 0$.

SOLUTION

Step 1 Find the x-intercept. Let $y = 0$ and solve for x.

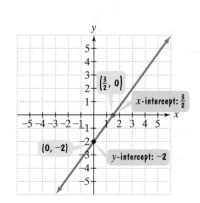

FIGURE 2.37 The graph of $4x - 3y - 6 = 0$

$$4x - 3 \cdot 0 - 6 = 0 \qquad \text{Replace } y \text{ with 0 in } 4x - 3y - 6 = 0.$$
$$4x - 6 = 0 \qquad \text{Simplify.}$$
$$4x = 6 \qquad \text{Add 6 to both sides.}$$
$$x = \frac{6}{4} = \frac{3}{2} \qquad \text{Divide both sides by 4.}$$

The x-intercept is $\frac{3}{2}$, so the line passes through $\left(\frac{3}{2}, 0\right)$ or $(1.5, 0)$, as shown in **Figure 2.37**.

Step 2 Find the y-intercept. Let $x = 0$ and solve for y.

$$4 \cdot 0 - 3y - 6 = 0 \qquad \text{Replace } x \text{ with 0 in } 4x - 3y - 6 = 0.$$
$$-3y - 6 = 0 \qquad \text{Simplify.}$$
$$-3y = 6 \qquad \text{Add 6 to both sides.}$$
$$y = -2 \qquad \text{Divide both sides by } -3.$$

The y-intercept is -2, so the line passes through $(0, -2)$, as shown in **Figure 2.37**.

Step 3 Graph the equation by drawing a line through the two points containing the intercepts. The graph of $4x - 3y - 6 = 0$ is shown in **Figure 2.37**. • • •

⌀ Check Point **8** Graph using intercepts: $3x - 2y - 6 = 0$. *

We've covered a lot of territory. Let's take a moment to summarize the various forms for equations of lines.

Equations of Lines	
1. Point-slope form	$y - y_1 = m(x - x_1)$
2. Slope-intercept form	$y = mx + b$ or $f(x) = mx + b$
3. Horizontal line	$y = b$ or $f(x) = b$
4. Vertical line	$x = a$
5. General form	$Ax + By + C = 0$

⑦ Model data with linear functions and make predictions.

Applications

Linear functions are useful for modeling data that fall on or near a line.

EXAMPLE 9 Modeling Global Warming

The amount of carbon dioxide in the atmosphere, measured in parts per million, has been increasing as a result of the burning of oil and coal. The buildup of gases and particles traps heat and raises the planet's temperature. The bar graph in **Figure 2.38(a)** gives the average atmospheric concentration of carbon dioxide and the average global temperature for six selected years. The data are displayed as a set of six points in a rectangular coordinate system in **Figure 2.38(b)**.

*See Answers to Selected Exercises.

FIGURE 2.38(a)
Source: National Oceanic and Atmospheric Administration

FIGURE 2.38(b)

a. Shown on the scatter plot in **Figure 2.38(b)** is a line that passes through or near the six points. Write the slope-intercept form of this equation using function notation.

b. The preindustrial concentration of atmospheric carbon dioxide was 280 parts per million. The United Nations' Intergovernmental Panel on Climate Change predicts global temperatures will rise between 2°F and 5°F if carbon dioxide concentration doubles from the preindustrial level. Compared to the average global temperature of 57.99°F for 2009, how well does the function from part (a) model this prediction?

SOLUTION

a. The line in **Figure 2.38(b)** passes through (326, 57.06) and (385, 57.99). We start by finding its slope.

$$m = \frac{\text{Change in } y}{\text{Change in } x} = \frac{57.99 - 57.06}{385 - 326} = \frac{0.93}{59} \approx 0.02$$

The slope indicates that for each increase of one part per million in carbon dioxide concentration, the average global temperature is increasing by approximately 0.02°F.

Now we write the line's equation in slope-intercept form.

$y - y_1 = m(x - x_1)$	Begin with the point-slope form.
$y - 57.06 = 0.02(x - 326)$	Either ordered pair can be (x_1, y_1). Let $(x_1, y_1) = (326, 57.06)$. From above, $m \approx 0.02$.
$y - 57.06 = 0.02x - 6.52$	Apply the distributive property: $0.02(326) = 6.52$.
$y = 0.02x + 50.54$	Add 57.06 to both sides and solve for y.

A linear function that models average global temperature, $f(x)$, for an atmospheric carbon dioxide concentration of x parts per million is

TECHNOLOGY

You can use a graphing utility to obtain a model for a scatter plot in which the data points fall on or near a straight line. After entering the data in **Figure 2.38(a)** on the previous page, a graphing utility displays a scatter plot of the data and the regression line, that is, the line that best fits the data.

[310, 390, 10] by [56.8, 58.4, 0.2]

Also displayed is the regression line's equation.

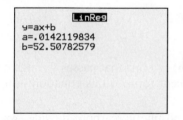

A linear function that models average global temperature, $f(x)$, for an atmospheric carbon dioxide concentration of x parts per million is

$$f(x) = 0.02x + 50.54.$$

b. If carbon dioxide concentration doubles from its preindustrial level of 280 parts per million, which many experts deem very likely, the concentration will reach 280×2, or 560 parts per million. We use the linear function to predict average global temperature at this concentration.

$$f(x) = 0.02x + 50.54 \qquad \text{Use the function from part (a).}$$
$$f(560) = 0.02(560) + 50.54 \qquad \text{Substitute 560 for } x.$$
$$= 11.2 + 50.54 = 61.74$$

Our model projects an average global temperature of 61.74°F for a carbon dioxide concentration of 560 parts per million. Compared to the average global temperature of 57.99° for 2009 shown in **Figure 2.38(a)** on the previous page, this is an increase of

$$61.74°F - 57.99°F = 3.75°F.$$

This is consistent with a rise between 2°F and 5°F as predicted by the Intergovernmental Panel on Climate Change. • • •

✓ Check Point **9** Use the data points (317, 57.04) and (354, 57.64), shown, but not labeled, in **Figure 2.38(b)** on the previous page to obtain a linear function that models average global temperature, $f(x)$, for an atmospheric carbon dioxide concentration of x parts per million. Round m to three decimal places and b to one decimal place. Then use the function to project average global temperature at a concentration of 600 parts per million. $f(x) = 0.016x + 52.0; 61.6°F$

ACHIEVING SUCCESS

Read your lecture notes before starting your homework.

Often homework problems, and later the test problems, are variations of the ones done by your professor in class.

CONCEPT AND VOCABULARY CHECK

Fill in each blank so that the resulting statement is true.

1. Data presented in a visual form as a set of points is called a/an ___scatter plot___. A line that best fits this set of points is called a/an ___regression___ line.

2. The slope, m, of a line through the distinct points (x_1, y_1) and (x_2, y_2) is given by the formula $m = \dfrac{y_2 - y_1}{x_2 - x_1}$.

3. If a line rises from left to right, the line has ___positive___ slope.

4. If a line falls from left to right, the line has ___negative___ slope.

5. The slope of a horizontal line is ___zero___.

6. The slope of a vertical line is ___undefined___.

7. The point-slope form of the equation of a nonvertical line with slope m that passes through the point (x_1, y_1) is ___$y - y_1 = m(x - x_1)$___.

8. The slope-intercept form of the equation of a line is ___$y = mx + b$___, where m represents the ___slope___ and b represents the ___y-intercept___.

9. In order to graph the line whose equation is $y = \dfrac{2}{5}x + 3$, begin by plotting the point ___(0, 3)___.

From this point, we move ___2___ units up (the rise) and ___5___ units to the right (the run).

10. The graph of the equation $y = 3$ is a/an ___horizontal___ line.

11. The graph of the equation $x = -2$ is a/an ___vertical___ line.

12. The equation $Ax + By + C = 0$, where A and B are not both zero, is called the ___general___ form of the equation of a line.

EXERCISE SET 2.3

Practice Exercises

In Exercises 1–10, find the slope of the line passing through each pair of points or state that the slope is undefined. Then indicate whether the line through the points rises, falls, is horizontal, or is vertical.

1. $(4, 7)$ and $(8, 10)$ $\frac{3}{4}$; rises
2. $(2, 1)$ and $(3, 4)$ 3; rises
3. $(-2, 1)$ and $(2, 2)$ $\frac{1}{4}$; rises
4. $(-1, 3)$ and $(2, 4)$ $\frac{1}{3}$; rises
5. $(4, -2)$ and $(3, -2)$ 0*
6. $(4, -1)$ and $(3, -1)$ 0*
7. $(-2, 4)$ and $(-1, -1)$ −5; falls
8. $(6, -4)$ and $(4, -2)$ −1; falls
9. $(5, 3)$ and $(5, -2)$ undefined*
10. $(3, -4)$ and $(3, 5)$ undefined*

In Exercises 11–38, use the given conditions to write an equation for each line in point-slope form and slope-intercept form.
Slope-intercept form is given here.

11. Slope $= 2$, passing through $(3, 5)$ $y = 2x - 1$*
12. Slope $= 4$, passing through $(1, 3)$ $y = 4x - 1$*
13. Slope $= 6$, passing through $(-2, 5)$ $y = 6x + 17$*
14. Slope $= 8$, passing through $(4, -1)$ $y = 8x - 33$*
15. Slope $= -3$, passing through $(-2, -3)$ $y = -3x - 9$*
16. Slope $= -5$, passing through $(-4, -2)$ $y = -5x - 22$*
17. Slope $= -4$, passing through $(-4, 0)$ $y = -4x - 16$*
18. Slope $= -2$, passing through $(0, -3)$ $y = -2x - 3$*
19. Slope $= -1$, passing through $\left(-\frac{1}{2}, -2\right)$ $y = -x - \frac{5}{2}$*
20. Slope $= -1$, passing through $\left(-4, -\frac{1}{4}\right)$ $y = -x - \frac{17}{4}$*
21. Slope $= \frac{1}{2}$, passing through the origin $y = \frac{1}{2}x$*
22. Slope $= \frac{1}{3}$, passing through the origin $y = \frac{1}{3}x$*
23. Slope $= -\frac{2}{3}$, passing through $(6, -2)$ $y = -\frac{2}{3}x + 2$*
24. Slope $= -\frac{3}{5}$, passing through $(10, -4)$ $y = -\frac{3}{5}x + 2$*
25. Passing through $(1, 2)$ and $(5, 10)$ $y = 2x$*
26. Passing through $(3, 5)$ and $(8, 15)$ $y = 2x - 1$*
27. Passing through $(-3, 0)$ and $(0, 3)$ $y = x + 3$*
28. Passing through $(-2, 0)$ and $(0, 2)$ $y = x + 2$*
29. Passing through $(-3, -1)$ and $(2, 4)$ $y = x + 2$*
30. Passing through $(-2, -4)$ and $(1, -1)$ $y = x - 2$*
31. Passing through $(-3, -2)$ and $(3, 6)$ $y = \frac{4}{3}x + 2$*
32. Passing through $(-3, 6)$ and $(3, -2)$ $y = -\frac{4}{3}x + 2$*
33. Passing through $(-3, -1)$ and $(4, -1)$ $y = -1$*
34. Passing through $(-2, -5)$ and $(6, -5)$ $y = -5$*
35. Passing through $(2, 4)$ with x-intercept $= -2$ $y = x + 2$*
36. Passing through $(1, -3)$ with x-intercept $= -1$ $y = -\frac{3}{2}x - \frac{3}{2}$*
37. x-intercept $= -\frac{1}{2}$ and y-intercept $= 4$ $y = 8x + 4$*
38. x-intercept $= 4$ and y-intercept $= -2$ $y = \frac{1}{2}x - 2$*

In Exercises 39–48, give the slope and y-intercept of each line whose equation is given. Then graph the linear function.

39. $y = 2x + 1$ $m = 2; b = 1$*
40. $y = 3x + 2$ $m = 3; b = 2$*
41. $f(x) = -2x + 1$ *
42. $f(x) = -3x + 2$ *
43. $f(x) = \frac{3}{4}x - 2$ *
44. $f(x) = \frac{3}{4}x - 3$ *
45. $y = -\frac{3}{5}x + 7$ *
46. $y = -\frac{2}{5}x + 6$ *
47. $g(x) = -\frac{1}{2}x$ $m = -\frac{1}{2}; b = 0$*
48. $g(x) = -\frac{1}{3}x$ $m = -\frac{1}{3}; b = 0$*

In Exercises 49–58, graph each equation in a rectangular coordinate system.

49. $y = -2$ *
50. $y = 4$ *
51. $x = -3$ *
52. $x = 5$ *
53. $y = 0$ *
54. $x = 0$ *
55. $f(x) = 1$ *
56. $f(x) = 3$ *
57. $3x - 18 = 0$ *
58. $3x + 12 = 0$ *

In Exercises 59–66,

 a. Rewrite the given equation in slope-intercept form.
 b. Give the slope and y-intercept.
 c. Use the slope and y-intercept to graph the linear function.

59. $3x + y - 5 = 0$ *
60. $4x + y - 6 = 0$ *
61. $2x + 3y - 18 = 0$ *
62. $4x + 6y + 12 = 0$ *
63. $8x - 4y - 12 = 0$ *
64. $6x - 5y - 20 = 0$ *
65. $3y - 9 = 0$ a. $y = 3$*
66. $4y + 28 = 0$ a. $y = -7$*

In Exercises 67–72, use intercepts to graph each equation.

67. $6x - 2y - 12 = 0$ *
68. $6x - 9y - 18 = 0$ *
69. $2x + 3y + 6 = 0$ *
70. $3x + 5y + 15 = 0$ *
71. $8x - 2y + 12 = 0$ *
72. $6x - 3y + 15 = 0$ *

Practice Plus

In Exercises 73–76, find the slope of the line passing through each pair of points or state that the slope is undefined. Assume that all variables represent positive real numbers. Then indicate whether the line through the points rises, falls, is horizontal, or is vertical.

73. $(0, a)$ and $(b, 0)$ $m = -\frac{a}{b}$; falls
74. $(-a, 0)$ and $(0, -b)$ *
75. (a, b) and $(a, b + c)$ *
76. $(a - b, c)$ and $(a, a + c)$ *

In Exercises 77–78, give the slope and y-intercept of each line whose equation is given. Assume that $B \neq 0$.

77. $Ax + By = C$ *
78. $Ax = By - C$ *

In Exercises 79–80, find the value of y if the line through the two given points is to have the indicated slope.

79. $(3, y)$ and $(1, 4), m = -3$ −2
80. $(-2, y)$ and $(4, -4), m = \frac{1}{3}$ −6

In Exercises 81–82, graph each linear function.

81. $3x - 4f(x) - 6 = 0$ *
82. $6x - 5f(x) - 20 = 0$ *

83. If one point on a line is $(3, -1)$ and the line's slope is -2, find the y-intercept. 5
84. If one point on a line is $(2, -6)$ and the line's slope is $-\frac{3}{2}$, find the y-intercept. −3

Use the figure to make the lists in Exercises 85–86.

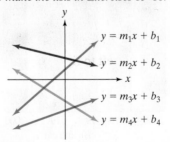

85. List the slopes m_1, m_2, m_3, and m_4 in order of decreasing size.
m_1, m_3, m_2, m_4
86. List the y-intercepts b_1, b_2, b_3, and b_4 in order of decreasing size.
b_2, b_1, b_4, b_3

*See Answers to Selected Exercises.

Application Exercises

Americans are getting married later in life, or not getting married at all. In 2010, more than half of Americans ages 25 through 29 were unmarried. The bar graph shows the percentage of never-married men and women in this age group for four selected years. The data are displayed as two sets of four points each, one scatter plot for the percentage of never-married American men and one for the percentage of never-married American women. Also shown for each scatter plot is a line that passes through or near the four points. Use these lines to solve Exercises 87–88.

Percentage of United States Population Never Married, Ages 25–29

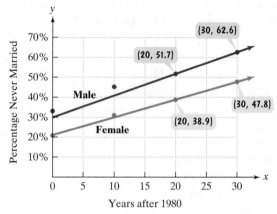

Source: U.S. Census Bureau

87. In this exercise, you will use the blue line for the women shown on the scatter plot to develop a model for the percentage of never-married American females ages 25–29.

　a. Use the two points whose coordinates are shown by the voice balloons to find the point-slope form of the equation of the line that models the percentage of never-married American females ages 25–29, *y*, *x* years after 1980. *

　b. Write the equation from part (a) in slope-intercept form. Use function notation. $f(x) = 0.89x + 21.1$

　c. Use the linear function to predict the percentage of never-married American females, ages 25–29, in 2020. 56.7%

The bar graph gives the life expectancy for American men and women born in six selected years. In Exercises 89–90, you will use the data to obtain models for life expectancy and make predictions about how long American men and women will live in the future.

Life Expectancy in the United States, by Year of Birth

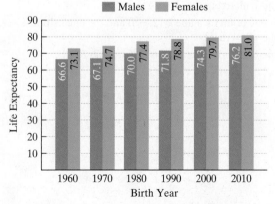

Source: National Center for Health Statistics

88. In this exercise, you will use the red line for the men shown on the scatter plot to develop a model for the percentage of never-married American males ages 25–29.

　a. Use the two points whose coordinates are shown by the voice balloons to find the point-slope form of the equation of the line that models the percentage of never-married American males ages 25–29, *y*, *x* years after 1980. *

　b. Write the equation from part (a) in slope-intercept form. Use function notation. $f(x) = 1.09x + 29.9$

　c. Use the linear function to predict the percentage of never-married American males, ages 25–29, in 2015. 68.05%

89. Use the data for males shown in the bar graph at the bottom of the previous column to solve this exercise.

　a. Let *x* represent the number of birth years after 1960 and let *y* represent male life expectancy. Create a scatter plot that displays the data as a set of six points in a rectangular coordinate system. *

　b. Draw a line through the two points that show male life expectancies for 1980 and 2000. Use the coordinates of these points to write a linear function that models life expectancy, $E(x)$, for American men born *x* years after 1960. *

　c. Use the function from part (b) to project the life expectancy of American men born in 2020. 78.6 yr

90. Use the data for females shown in the bar graph at the bottom of the previous column to solve this exercise.

　a. Let *x* represent the number of birth years after 1960 and let *y* represent female life expectancy. Create a scatter plot that displays the data as a set of six points in a rectangular coordinate system. *

　b. Draw a line through the two points that show female life expectancies for 1970 and 2000. Use the coordinates of these points to write a linear function that models life expectancy, $E(x)$, for American women born *x* years after 1960. Round the slope to two decimal places. *

　c. Use the function from part (b) to project the life expectancy of American women born in 2020. *

91. Shown, again, is the scatter plot that indicates a relationship between the percentage of adult females in a country who are literate and the mortality of children under five. Also shown is a line that passes through or near the points. Find a linear function that models the data by finding the slope-intercept form of the line's equation. Use the function to make a prediction about child mortality based on the percentage of adult females in a country who are literate. *

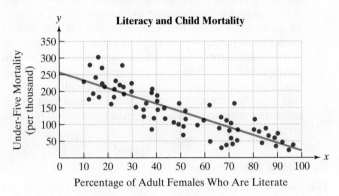

Literacy and Child Mortality

Source: United Nations

92. Just as money doesn't buy happiness for individuals, the two don't necessarily go together for countries either. However, the scatter plot does show a relationship between a country's annual per capita income and the percentage of people in that country who call themselves "happy."

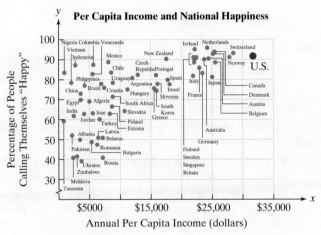

Per Capita Income and National Happiness

Source: Richard Layard, *Happiness: Lessons from a New Science,* Penguin, 2005

Draw a line that fits the data so that the spread of the data points around the line is as small as possible. Use the coordinates of two points along your line to write the slope-intercept form of its equation. Express the equation in function notation and use the linear function to make a prediction about national happiness based on per capita income. Answers will vary.

Explaining the Concepts 93–99. Answers will vary.

93. What is the slope of a line and how is it found?

94. Describe how to write the equation of a line if the coordinates of two points along the line are known.

95. Explain how to derive the slope-intercept form of a line's equation, $y = mx + b$, from the point-slope form

$$y - y_1 = m(x - x_1).$$

96. Explain how to graph the equation $x = 2$. Can this equation be expressed in slope-intercept form? Explain.

97. Explain how to use the general form of a line's equation to find the line's slope and y-intercept.

98. Explain how to use intercepts to graph the general form of a line's equation.

99. Take another look at the scatter plot in Exercise 91. Although there is a relationship between literacy and child mortality, we cannot conclude that increased literacy causes child mortality to decrease. Offer two or more possible explanations for the data in the scatter plot.

Technology Exercises

Use a graphing utility to graph each equation in Exercises 100–103. Then use the TRACE *feature to trace along the line and find the coordinates of two points. Use these points to compute the line's slope. Check your result by using the coefficient of x in the line's equation.*

100. $y = 2x + 4$ $m = 2*$
101. $y = -3x + 6$ $m = -3*$
102. $y = -\frac{1}{2}x - 5$ $m = -\frac{1}{2}*$
103. $y = \frac{3}{4}x - 2$ $m = \frac{3}{4}*$

104. Is there a relationship between wine consumption and deaths from heart disease? The table gives data from 19 developed countries.

| | | | | | | | France |
Country	A	B	C	D	E	F	G
Liters of alcohol from drinking wine, per person per year (x)	2.5	3.9	2.9	2.4	2.9	0.8	9.1
Deaths from heart disease, per 100,000 people per year (y)	211	167	131	191	220	297	71

| | | | | | | | | | | | U.S. |
Country	H	I	J	K	L	M	N	O	P	Q	R	S
(x)	0.8	0.7	7.9	1.8	1.9	0.8	6.5	1.6	5.8	1.3	1.2	2.7
(y)	211	300	107	167	266	227	86	207	115	285	199	172

Source: *New York Times*

a. Use the statistical menu of your graphing utility to enter the 19 ordered pairs of data items shown in the table. *

b. Use the scatter plot capability to draw a scatter plot of the data. *

c. Select the linear regression option. Use your utility to obtain values for a and b for the equation of the regression line, $y = ax + b$. You may also be given a **correlation coefficient**, r. Values of r close to 1 indicate that the points can be described by a linear relationship and the regression line has a positive slope. Values of r close to -1 indicate that the points can be described by a linear relationship and the regression line has a negative slope. Values of r close to 0 indicate no linear relationship between the variables. In this case, a linear model does not accurately describe the data. *

d. Use the appropriate sequence (consult your manual) to graph the regression equation on top of the points in the scatter plot. *

Critical Thinking Exercises

Make Sense? *In Exercises 105–108, determine whether each statement makes sense or does not make sense, and explain your reasoning.*

105. The graph of my linear function at first increased, reached a maximum point, and then decreased. does not make sense

106. A linear function that models tuition and fees at public four-year colleges from 2000 through 2016 has negative slope. *

107. Because the variable m does not appear in $Ax + By + C = 0$, equations in this form make it impossible to determine the line's slope. does not make sense

108. The federal minimum wage was $5.15 per hour from 1997 through 2006, so $f(x) = 5.15$ models the minimum wage, $f(x)$, in dollars, for the domain $\{1997, 1998, 1999, \ldots, 2006\}$. *

In Exercises 109–112, determine whether each statement is true or false. If the statement is false, make the necessary change(s) to produce a true statement. Changes to false statements will vary.

109. The equation $y = mx + b$ shows that no line can have a y-intercept that is numerically equal to its slope. false

110. Every line in the rectangular coordinate system has an equation that can be expressed in slope-intercept form. *

111. The graph of the linear function $5x + 6y - 30 = 0$ is a line passing through the point $(6, 0)$ with slope $-\frac{5}{6}$. true

112. The graph of $x = 7$ in the rectangular coordinate system is the single point $(7, 0)$. false

In Exercises 113–114, find the coefficients that must be placed in each shaded area so that the function's graph will be a line satisfying the specified conditions.

113. ▨ $x +$ ▨ $y - 12 = 0$; x-intercept $= -2$; y-intercept $= 4$ *

114. ▨ $x +$ ▨ $y - 12 = 0$; y-intercept $= -6$; slope $= \frac{1}{2}$ *

115. Prove that the equation of a line passing through $(a, 0)$ and $(0, b)(a \neq 0, b \neq 0)$ can be written in the form $\frac{x}{a} + \frac{y}{b} = 1$. Why is this called the *intercept form* of a line?

116. Excited about the success of celebrity stamps, post office officials were rumored to have put forth a plan to institute two new types of thermometers. On these new scales, $°E$ represents degrees Elvis and $°M$ represents degrees Madonna. If it is known that $40°E = 25°M$, $280°E = 125°M$, and degrees Elvis is linearly related to degrees Madonna, write an equation expressing E in terms of M. $E = 2.4M - 20$

Group Exercise Answers will vary.

117. In Exercises 87–88, we used the data in a bar graph to develop linear functions that modeled the percentage of never-married American females and males, ages 25–29. For this group exercise, you might find it helpful to pattern your work after Exercises 87 and 88. Group members should begin by consulting an almanac, newspaper, magazine, or the Internet to find data that appear to lie approximately on or near a line. Working by hand or using a graphing utility, group members should construct scatter plots for the data that were assembled. If working by hand, draw a line that approximately fits the data in each scatter plot and then write its equation as a function in slope-intercept form. If using a graphing utility, obtain the equation of each regression line. Then use each linear function's equation to make predictions about what might occur in the future. Are there circumstances that might affect the accuracy of the prediction? List some of these circumstances.

Retaining the Concepts

118. In 1994, there were 714 violent crime incidents per 100,000 Americans. For the period from 1994 through 2014, this number decreased by approximately 17 incidents per 100,000 people each year. If this trend continues, by which year will violent crime incidents decrease to 289 per 100,000 people? (Source: FBI)

(Section 1.3, Example 2) by 2019

In Exercises 119–120, solve and graph the solution set on a number line.

119. $\dfrac{x + 3}{4} \geq \dfrac{x - 2}{3} + 1$

(Section 1.7, Example 5) $(-\infty, 5]$*

120. $3|2x + 6| - 9 < 15$

(Section 1.7, Example 9) $(-7, 1)$*

Preview Exercises

Exercises 121–123 will help you prepare for the material covered in the next section.

121. Write the slope-intercept form of the equation of the line passing through $(-3, 1)$ whose slope is the same as the line whose equation is $y = 2x + 1$. $y = 2x + 7$ or $f(x) = 2x + 7$

122. Write an equation in general form of the line passing through $(3, -5)$ whose slope is the negative reciprocal (the reciprocal with the opposite sign) of $-\frac{1}{4}$. $4x - y - 17 = 0$

123. If $f(x) = x^2$, find

$$\frac{f(x_2) - f(x_1)}{x_2 - x_1},$$

where $x_1 = 1$ and $x_2 = 4$. 5

Section 2.4 | More on Slope

What am I supposed to learn?

After studying this section, you should be able to:

1. Find slopes and equations of parallel and perpendicular lines.
2. Interpret slope as rate of change.
3. Find a function's average rate of change.

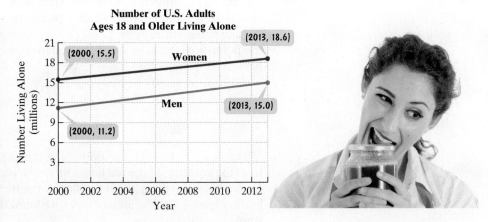

Number of U.S. Adults Ages 18 and Older Living Alone

(2000, 15.5) Women (2013, 18.6)

Men (2013, 15.0)

(2000, 11.2)

FIGURE 2.39
Source: U.S. Census Bureau

A best guess at the future of our nation indicates that the numbers of men and women living alone will increase each year. **Figure 2.39** shows that in 2013, 15.0 million men and 18.6 million women lived alone, an increase over the numbers displayed in the graph for 2000.

Take a second look at **Figure 2.39**. Can you tell that the green graph representing men has a greater slope than the red graph representing women? This indicates a greater rate of change in the number of men living alone than in the number of women living alone over the period from 2000 through 2013. In this section, you will learn to interpret slope as a rate of change. You will also explore the relationships between slopes of parallel and perpendicular lines.

Parallel and Perpendicular Lines

Two nonintersecting lines that lie in the same plane are **parallel**. If two lines do not intersect, the ratio of the vertical change to the horizontal change is the same for both lines. Because two parallel lines have the same "steepness," they must have the same slope.

> **Slope and Parallel Lines**
> 1. If two nonvertical lines are parallel, then they have the same slope.
> 2. If two distinct nonvertical lines have the same slope, then they are parallel.
> 3. Two distinct vertical lines, both with undefined slopes, are parallel.

1. Find slopes and equations of parallel and perpendicular lines.

EXAMPLE 1 Writing Equations of a Line Parallel to a Given Line

Write an equation of the line passing through $(-3, 1)$ and parallel to the line whose equation is $y = 2x + 1$. Express the equation in point-slope form and slope-intercept form.

SOLUTION

The situation is illustrated in **Figure 2.40**. We are looking for the equation of the red line passing through $(-3, 1)$ and parallel to the blue line whose equation is $y = 2x + 1$. How do we obtain the equation of this red line? Notice that the line passes through the point $(-3, 1)$. Using the point-slope form of the line's equation, we have $x_1 = -3$ and $y_1 = 1$.

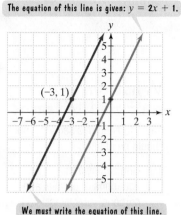

The equation of this line is given: $y = 2x + 1$.

$(-3, 1)$

We must write the equation of this line.

FIGURE 2.40

$$y - y_1 = m(x - x_1)$$

$y_1 = 1$ $x_1 = {}^-3$

With $(x_1, y_1) = (-3, 1)$, the only thing missing from the equation of the red line is m, the slope. Do we know anything about the slope of either line in **Figure 2.40**? The answer is yes; we know the slope of the blue line on the right, whose equation is given.

$$y = 2x + 1$$

The slope of the blue line on the right in **Figure 2.40** is 2.

Parallel lines have the same slope. Because the slope of the blue line is 2, the slope of the red line, the line whose equation we must write, is also 2: $m = 2$. We now have values for x_1, y_1, and m for the red line.

$$y - y_1 = m(x - x_1)$$

$y_1 = 1$ $m = 2$ $x_1 = {}^-3$

The point-slope form of the red line's equation is

$$y - 1 = 2[x - (-3)] \text{ or }$$
$$y - 1 = 2(x + 3).$$

Solving for y, we obtain the slope-intercept form of the equation.

$y - 1 = 2x + 6$ Apply the distributive property.

$y = 2x + 7$ Add 1 to both sides. This is the slope-intercept form, $y = mx + b$, of the equation. Using function notation, the equation is $f(x) = 2x + 7$. • • •

⊘ Check Point **1** Write an equation of the line passing through $(-2, 5)$ and parallel to the line whose equation is $y = 3x + 1$. Express the equation in point-slope form and slope-intercept form. $y - 5 = 3(x + 2); y = 3x + 11 \text{ or } f(x) = 3x + 11$

Two lines that intersect at a right angle (90°) are said to be **perpendicular**, shown in **Figure 2.41**. The relationship between the slopes of perpendicular lines is not as obvious as the relationship between parallel lines. **Figure 2.41** shows line AB, with slope $\frac{c}{d}$. Rotate line AB counterclockwise 90° to the left to obtain line $A'B'$, perpendicular to line AB. The figure indicates that the rise and the run of the new line are reversed from the original line, but the former rise, the new run, is now negative. This means that the slope of the new line is $-\frac{d}{c}$. Notice that the product of the slopes of the two perpendicular lines is -1:

$$\left(\frac{c}{d}\right)\left(-\frac{d}{c}\right) = -1.$$

This relationship holds for all perpendicular lines and is summarized in the box at the top of the next page.

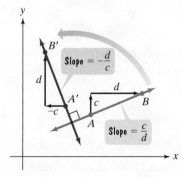

FIGURE 2.41 Slopes of perpendicular lines

Slope and Perpendicular Lines

1. If two nonvertical lines are perpendicular, then the product of their slopes is -1.

2. If the product of the slopes of two lines is -1, then the lines are perpendicular.

3. A horizontal line having zero slope is perpendicular to a vertical line having undefined slope.

An equivalent way of stating this relationship is to say that **one line is perpendicular to another line if its slope is the *negative reciprocal* of the slope of the other line**. For example, if a line has slope 5, any line having slope $-\frac{1}{5}$ is perpendicular to it. Similarly, if a line has slope $-\frac{3}{4}$, any line having slope $\frac{4}{3}$ is perpendicular to it.

> **EXAMPLE 2** Writing Equations of a Line Perpendicular to a Given Line

a. Find the slope of any line that is perpendicular to the line whose equation is $x + 4y - 8 = 0$.

b. Write the equation of the line passing through $(3, -5)$ and perpendicular to the line whose equation is $x + 4y - 8 = 0$. Express the equation in general form.

SOLUTION

a. We begin by writing the equation of the given line, $x + 4y - 8 = 0$, in slope-intercept form. Solve for y.

$$x + 4y - 8 = 0 \qquad \text{This is the given equation.}$$
$$4y = -x + 8 \qquad \text{To isolate the } y\text{-term, subtract } x \text{ and add 8 on both sides.}$$
$$y = -\frac{1}{4}x + 2 \qquad \text{Divide both sides by 4.}$$

Slope is $-\frac{1}{4}$.

The given line has slope $-\frac{1}{4}$. Any line perpendicular to this line has a slope that is the negative reciprocal of $-\frac{1}{4}$. Thus, the slope of any perpendicular line is 4.

b. Let's begin by writing the point-slope form of the perpendicular line's equation. Because the line passes through the point $(3, -5)$, we have $x_1 = 3$ and $y_1 = -5$. In part (a), we determined that the slope of any line perpendicular to $x + 4y - 8 = 0$ is 4, so the slope of this particular perpendicular line must also be 4: $m = 4$.

$$y - y_1 = m(x - x_1)$$

$y_1 = -5$ $m = 4$ $x_1 = 3$

The point-slope form of the perpendicular line's equation is

$$y - (-5) = 4(x - 3) \text{ or}$$
$$y + 5 = 4(x - 3).$$

How can we express this equation, $y + 5 = 4(x - 3)$, in general form $(Ax + By + C = 0)$? We need to obtain zero on one side of the equation. Let's do this and keep A, the coefficient of x, positive.

$y + 5 = 4(x - 3)$	This is the point-slope form of the line's equation.
$y + 5 = 4x - 12$	Apply the distributive property.
$y - y + 5 - 5 = 4x - y - 12 - 5$	To obtain O on the left, subtract y and subtract 5 on both sides.
$0 = 4x - y - 17$	Simplify.

In general form, the equation of the perpendicular line is $4x - y - 17 = 0$.
• • •

⏀ Check Point **2**

a. Find the slope of any line that is perpendicular to the line whose equation is $x + 3y - 12 = 0$. ₃

b. Write the equation of the line passing through $(-2, -6)$ and perpendicular to the line whose equation is $x + 3y - 12 = 0$. Express the equation in general form. $3x - y = 0$

② Interpret slope as rate of change.

Slope as Rate of Change

Slope is defined as the ratio of a change in y to a corresponding change in x. It describes how fast y is changing with respect to x. For a linear function, slope may be interpreted as the rate of change of the dependent variable per unit change in the independent variable.

Our next example shows how slope can be interpreted as a rate of change in an applied situation. When calculating slope in applied problems, keep track of the units in the numerator and the denominator.

EXAMPLE 3 Slope as a Rate of Change

The line graphs for the number of women and men living alone are shown again in **Figure 2.42**. Find the slope of the line segment for the women. Describe what this slope represents.

SOLUTION

We let x represent a year and y the number of women living alone in that year. The two points shown on the line segment for women have the following coordinates:

$(2000, 15.5)$ $(2013, 18.6)$.

In 2000, 15.5 million U.S. women lived alone. In 2013, 18.6 million U.S. women lived alone.

Now we compute the slope:

$$m = \frac{\text{Change in } y}{\text{Change in } x} = \frac{18.6 - 15.5}{2013 - 2000}$$

The unit in the numerator is *million women*.

The unit in the denominator is *year*.

$$= \frac{3.1}{13} \approx \frac{0.24 \text{ million women}}{\text{year}}.$$

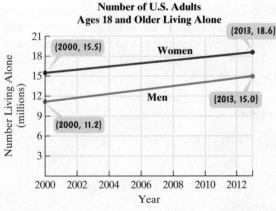

Number of U.S. Adults Ages 18 and Older Living Alone

(2013, 18.6)
(2000, 15.5) — Women
(2013, 15.0)
Men
(2000, 11.2)

Number Living Alone (millions)

Year

FIGURE 2.42

Source: U.S. Census Bureau

The slope indicates that the number of American women living alone increased at a rate of approximately 0.24 million each year for the period from 2000 through 2013. The rate of change is 0.24 million women per year.
• • •

$m \approx 0.29$; The number of men living alone increased at a rate of 0.29 million per year. The rate of change is 0.29 million men per year.

⊘ Check Point 3 Use the ordered pairs in **Figure 2.42** to find the slope of the green line segment for the men. Express the slope correct to two decimal places and describe what it represents.

In Check Point 3, did you find that the slope of the line segment for the men is different from that of the women? The rate of change for the number of men living alone is greater than the rate of change for the number of women living alone. The green line segment representing men in **Figure 2.42** is steeper than the red line segment representing women. If you extend the line segments far enough, the resulting lines will intersect. They are not parallel.

③ Find a function's average rate of change.

The Average Rate of Change of a Function

If the graph of a function is not a straight line, the **average rate of change** between any two points is the slope of the line containing the two points. This line is called a **secant line**. For example, **Figure 2.43** shows the graph of a particular man's height, in inches, as a function of his age, in years. Two points on the graph are labeled: $(13, 57)$ and $(18, 76)$. At age 13, this man was 57 inches tall and at age 18, he was 76 inches tall.

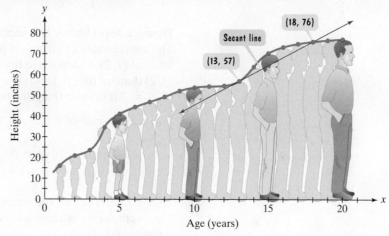

FIGURE 2.43 Height as a function of age

The man's average growth rate between ages 13 and 18 is the slope of the secant line containing $(13, 57)$ and $(18, 76)$:

$$m = \frac{\text{Change in } y}{\text{Change in } x} = \frac{76 - 57}{18 - 13} = \frac{19}{5} = 3\frac{4}{5}.$$

This man's average rate of change, or average growth rate, from age 13 to age 18 was $3\frac{4}{5}$, or 3.8, inches per year.

The Average Rate of Change of a Function

Let $(x_1, f(x_1))$ and $(x_2, f(x_2))$ be distinct points on the graph of a function f. (See **Figure 2.44**.) The **average rate of change of f** from x_1 to x_2 is

$$\frac{f(x_2) - f(x_1)}{x_2 - x_1}.$$

FIGURE 2.44

EXAMPLE 4 Finding the Average Rate of Change

Find the average rate of change of $f(x) = x^2$ from

a. $x_1 = 0$ to $x_2 = 1$ **b.** $x_1 = 1$ to $x_2 = 2$ **c.** $x_1 = -2$ to $x_2 = 0$.

SOLUTION

a. The average rate of change of $f(x) = x^2$ from $x_1 = 0$ to $x_2 = 1$ is

$$\frac{f(x_2) - f(x_1)}{x_2 - x_1} = \frac{f(1) - f(0)}{1 - 0} = \frac{1^2 - 0^2}{1} = 1.$$

Figure 2.45(a) shows the secant line of $f(x) = x^2$ from $x_1 = 0$ to $x_2 = 1$. The average rate of change is positive and the function is increasing on the interval $(0, 1)$.

b. The average rate of change of $f(x) = x^2$ from $x_1 = 1$ to $x_2 = 2$ is

$$\frac{f(x_2) - f(x_1)}{x_2 - x_1} = \frac{f(2) - f(1)}{2 - 1} = \frac{2^2 - 1^2}{1} = 3.$$

Figure 2.45(b) shows the secant line of $f(x) = x^2$ from $x_1 = 1$ to $x_2 = 2$. The average rate of change is positive and the function is increasing on the interval $(1, 2)$. Can you see that the graph rises more steeply on the interval $(1, 2)$ than on $(0, 1)$? This is because the average rate of change from $x_1 = 1$ to $x_2 = 2$ is greater than the average rate of change from $x_1 = 0$ to $x_2 = 1$.

c. The average rate of change of $f(x) = x^2$ from $x_1 = -2$ to $x_2 = 0$ is

$$\frac{f(x_2) - f(x_1)}{x_2 - x_1} = \frac{f(0) - f(-2)}{0 - (-2)} = \frac{0^2 - (-2)^2}{2} = \frac{-4}{2} = -2.$$

Figure 2.45(c) shows the secant line of $f(x) = x^2$ from $x_1 = -2$ to $x_2 = 0$. The average rate of change is negative and the function is decreasing on the interval $(-2, 0)$.

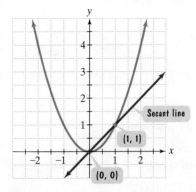

FIGURE 2.45(a) The secant line of $f(x) = x^2$ from $x_1 = 0$ to $x_2 = 1$

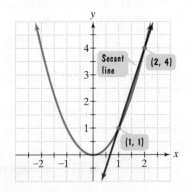

FIGURE 2.45(b) The secant line of $f(x) = x^2$ from $x_1 = 1$ to $x_2 = 2$

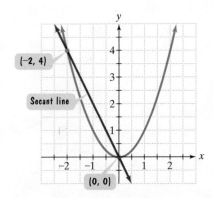

FIGURE 2.45(c) The secant line of $f(x) = x^2$ from $x_1 = -2$ to $x_2 = 0$ ● ● ●

✓ Check Point **4** Find the average rate of change of $f(x) = x^3$ from

a. $x_1 = 0$ to $x_2 = 1$ 1 **b.** $x_1 = 1$ to $x_2 = 2$ 7 **c.** $x_1 = -2$ to $x_2 = 0$. 4

Suppose we are interested in the average rate of change of f from $x_1 = x$ to $x_2 = x + h$. In this case, the average rate of change is

$$\frac{f(x_2) - f(x_1)}{x_2 - x_1} = \frac{f(x + h) - f(x)}{x + h - x} = \frac{f(x + h) - f(x)}{h}.$$

Do you recognize the last expression? It is the difference quotient that you used in Section 2.2. Thus, **the difference quotient gives the average rate of change of a function from x to $x + h$.** In the difference quotient, h is thought of as a number very close to 0. In this way, the average rate of change can be found for a very short interval.

FIGURE 2.46 Concentration of a drug as a function of time

EXAMPLE 5 Finding the Average Rate of Change

When a person receives a drug injected into a muscle, the concentration of the drug in the body, measured in milligrams per 100 milliliters, is a function of the time elapsed after the injection, measured in hours. **Figure 2.46** shows the graph of such a function, where x represents hours after the injection and $f(x)$ is the drug's concentration at time x. Find the average rate of change in the drug's concentration between 3 and 7 hours.

SOLUTION

At 3 hours, the drug's concentration is 0.05 and at 7 hours, the concentration is 0.02. The average rate of change in its concentration between 3 and 7 hours is

$$\frac{f(x_2) - f(x_1)}{x_2 - x_1} = \frac{f(7) - f(3)}{7 - 3} = \frac{0.02 - 0.05}{7 - 3} = \frac{-0.03}{4} = -0.0075.$$

The average rate of change is -0.0075. This means that the drug's concentration is decreasing at an average rate of 0.0075 milligram per 100 milliliters per hour. • • •

GREAT QUESTION!

Can you clarify how you determine the units that you use when describing slope as a rate of change?

Units used to describe x and y tend to "pile up" when expressing the rate of change of y with respect to x. The unit used to express the rate of change of y with respect to x is

the unit used to describe y **per** the unit used to describe x.

In **Figure 2.46**, y, or drug concentration, is described in milligrams per 100 milliliters.

In **Figure 2.46**, x, or time, is described in hours.

In Example 5, the rate of change is described in terms of milligrams per 100 milliliters per hour.

⊘ Check Point **5** Use **Figure 2.46** to find the average rate of change in the drug's concentration between 1 hour and 3 hours. 0.01 mg per 100 mL per hr

Blitzer Bonus || How Calculus Studies Change

Take a rapid sequence of still photographs of a moving scene and project them onto a screen at thirty shots a second or faster. Our eyes see the results as continuous motion. The small difference between one frame and the next cannot be detected by the human visual system. The idea of calculus likewise regards continuous motion as made up of a sequence of still configurations. Calculus masters the mystery of movement by "freezing the frame" of a continuously changing process, instant by instant. For example, **Figure 2.47** shows a male's changing height over intervals of time. Over the period of time from P to D, his average rate of growth is his change in height—that is, his height at time D minus his height at time P—divided by the change in time from P to D. This is the slope of secant line PD.

The secant lines PD, PC, PB, and PA shown in **Figure 2.47** have slopes that show average growth rates for successively shorter periods of time. Calculus makes these time frames so small that they approach a single point—that is, a single instant in time. This point is shown as point P in **Figure 2.47**. The slope of the line that touches the graph at P gives the male's growth rate at one instant in time, P.

FIGURE 2.47 Analyzing continuous growth over intervals of time and at an instant in time

CONCEPT AND VOCABULARY CHECK

Fill in each blank so that the resulting statement is true.

1. If two nonvertical lines are parallel, then they have ___the same___ slope.
2. If two nonvertical lines are perpendicular, then the product of their slopes is ___−1___.
3. Consider the line whose equation is $y = -\frac{1}{3}x + 5$. The slope of any line that is parallel to this line is ___$-\frac{1}{3}$___. The slope of any line that is perpendicular to this line is ___3___.
4. Consider the line whose equation is $2x + y - 6 = 0$. The slope of any line that is parallel to this line is ___−2___. The slope of any line that is perpendicular to this line is ___$\frac{1}{2}$___.

5. The slope of the line through the distinct points (x_1, y_1) and (x_2, y_2) can be interpreted as the rate of change in ___y___ with respect to ___x___.
6. If $(x_1, f(x_1))$ and $(x_2, f(x_2))$ are distinct points on the graph of a function f, the average rate of change of f from x_1 to x_2 is ___$\dfrac{f(x_2) - f(x_1)}{x_2 - x_1}$___.

EXERCISE SET 2.4

Practice Exercises

In Exercises 1–4, write an equation for line L in point-slope form and slope-intercept form.

1.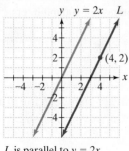

L is parallel to $y = 2x$.

$y - 2 = 2(x - 4);\ y = 2x - 6$

2.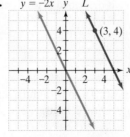

L is parallel to $y = -2x$.

$y - 4 = -2(x - 3);\ y = -2x + 10$

3.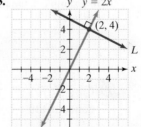

L is perpendicular to $y = 2x$.

$y - 4 = -\frac{1}{2}(x - 2);\ y = -\frac{1}{2}x + 5$

4.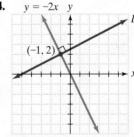

L is perpendicular to $y = -2x$.

$y - 2 = \frac{1}{2}(x + 1);\ y = \frac{1}{2}x + \frac{5}{2}$

In Exercises 5–8, use the given conditions to write an equation for each line in point-slope form and slope-intercept form.

5. Passing through $(-8, -10)$ and parallel to the line whose equation is $y = -4x + 3$ $y + 10 = -4(x + 8); y = -4x - 42$

6. Passing through $(-2, -7)$ and parallel to the line whose equation is $y = -5x + 4$ $y + 7 = -5(x + 2); y = -5x - 17$

7. Passing through $(2, -3)$ and perpendicular to the line whose equation is $y = \frac{1}{5}x + 6$ $y + 3 = -5(x - 2); y = -5x + 7$

8. Passing through $(-4, 2)$ and perpendicular to the line whose equation is $y = \frac{1}{3}x + 7$ $y - 2 = -3(x + 4); y = -3x - 10$

In Exercises 9–12, use the given conditions to write an equation for each line in point-slope form and general form. General form is given here.

9. Passing through $(-2, 2)$ and parallel to the line whose equation is $2x - 3y - 7 = 0$ $2x - 3y + 10 = 0^*$

10. Passing through $(-1, 3)$ and parallel to the line whose equation is $3x - 2y - 5 = 0$ $3x - 2y + 9 = 0^*$

11. Passing through $(4, -7)$ and perpendicular to the line whose equation is $x - 2y - 3 = 0$ $2x + y - 1 = 0^*$

12. Passing through $(5, -9)$ and perpendicular to the line whose equation is $x + 7y - 12 = 0$ $7x - y - 44 = 0^*$

In Exercises 13–18, find the average rate of change of the function from x_1 to x_2.

13. $f(x) = 3x$ from $x_1 = 0$ to $x_2 = 5$ 3

14. $f(x) = 6x$ from $x_1 = 0$ to $x_2 = 4$ 6

15. $f(x) = x^2 + 2x$ from $x_1 = 3$ to $x_2 = 5$ 10

16. $f(x) = x^2 - 2x$ from $x_1 = 3$ to $x_2 = 6$ 7

17. $f(x) = \sqrt{x}$ from $x_1 = 4$ to $x_2 = 9$ $\frac{1}{5}$

18. $f(x) = \sqrt{x}$ from $x_1 = 9$ to $x_2 = 16$ $\frac{1}{7}$

Practice Plus

In Exercises 19–24, write an equation in slope-intercept form of a linear function f whose graph satisfies the given conditions.

19. The graph of f passes through $(-1, 5)$ and is perpendicular to the line whose equation is $x = 6$. $f(x) = 5$

20. The graph of f passes through $(-2, 6)$ and is perpendicular to the line whose equation is $x = -4$. $f(x) = 6$

21. The graph of f passes through $(-6, 4)$ and is perpendicular to the line that has an x-intercept of 2 and a y-intercept of -4. $f(x) = -\frac{1}{2}x + 1$

22. The graph of f passes through $(-5, 6)$ and is perpendicular to the line that has an x-intercept of 3 and a y-intercept of -9. $f(x) = -\frac{1}{3}x + \frac{13}{3}$

23. The graph of f is perpendicular to the line whose equation is $3x - 2y - 4 = 0$ and has the same y-intercept as this line. $f(x) = -\frac{2}{3}x - 2$

24. The graph of f is perpendicular to the line whose equation is $4x - y - 6 = 0$ and has the same y-intercept as this line. $f(x) = -\frac{1}{4}x - 6$

Application Exercises

The bar graph shows that as costs changed over the decades, Americans devoted less of their budget to groceries and more to health care.

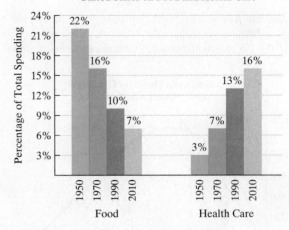

Percentage of Total Spending in the United States on Food and Health Care

Source: *Time*, October 10, 2011

In Exercises 25–26, find a linear function in slope-intercept form that models the given description. Each function should model the percentage of total spending, $p(x)$, by Americans x years after 1950.

25. In 1950, Americans spent 22% of their budget on food. This has decreased at an average rate of approximately 0.25% per year since then. $p(x) = -0.25x + 22$

26. In 1950, Americans spent 3% of their budget on health care. This has increased at an average rate of approximately 0.22% per year since then. $p(x) = 0.22x + 3$

The stated intent of the 1994 "don't ask, don't tell" policy was to reduce the number of discharges of gay men and lesbians from the military. Nearly 14,000 active-duty gay servicemembers were dismissed under the policy, which officially ended in 2011, after 18 years. The line graph shows the number of discharges under "don't ask, don't tell" from 1994 through 2010. Use the data displayed by the graph to solve Exercises 27–28.

Number of Active-Duty Gay Servicemembers Discharged from the Military for Homosexuality

Source: General Accountability Office

(In Exercises 27–28, be sure to refer to the graph at the bottom of the previous page.)

27. Find the average rate of change, rounded to the nearest whole number, from 1994 through 1998. Describe what this means. *

28. Find the average rate of change, rounded to the nearest whole number, from 2001 through 2006. Describe what this means. *

The function $f(x) = 1.1x^3 - 35x^2 + 264x + 557$ models the number of discharges, $f(x)$, under "don't ask, don't tell" x years after 1994. Use this model and its graph, shown below, to solve Exercises 29–30.

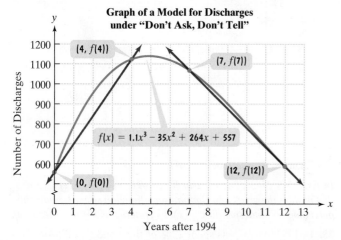

Graph of a Model for Discharges under "Don't Ask, Don't Tell"

$f(x) = 1.1x^3 - 35x^2 + 264x + 557$

Years after 1994

29. a. Find the slope of the secant line, rounded to the nearest whole number, from $x_1 = 0$ to $x_2 = 4$. 142

b. Does the slope from part (a) underestimate or overestimate the average yearly increase that you determined in Exercise 27? By how much? *

30. a. Find the slope of the secant line, rounded to the nearest whole number, from $x_1 = 7$ to $x_2 = 12$. −96

b. Does the slope from part (b) underestimate or overestimate the average yearly decrease that you determined in Exercise 28? By how much? *

Explaining the Concepts 31–36. Answers will vary.

31. If two lines are parallel, describe the relationship between their slopes.

32. If two lines are perpendicular, describe the relationship between their slopes.

33. If you know a point on a line and you know the equation of a line perpendicular to this line, explain how to write the line's equation.

34. A formula in the form $y = mx + b$ models the average retail price, y, of a new car x years after 2000. Would you expect m to be positive, negative, or zero? Explain your answer.

35. What is a secant line?

36. What is the average rate of change of a function?

Technology Exercise

37. a. Why are the lines whose equations are $y = \frac{1}{3}x + 1$ and $y = -3x - 2$ perpendicular? Product of slopes is −1.

b. Use a graphing utility to graph the equations in a $[-10, 10, 1]$ by $[-10, 10, 1]$ viewing rectangle. Do the lines appear to be perpendicular? no *

c. Now use the zoom square feature of your utility. Describe what happens to the graphs. Explain why this is so. *

Critical Thinking Exercises

Make Sense? *In Exercises 38–41, determine whether each statement makes sense or does not make sense, and explain your reasoning.*

38. I computed the slope of one line to be $-\frac{3}{5}$ and the slope of a second line to be $-\frac{5}{3}$, so the lines must be perpendicular. does not make sense

39. I have linear functions that model changes for men and women over the same time period. The functions have the same slope, so their graphs are parallel lines, indicating that the rate of change for men is the same as the rate of change for women. makes sense

40. The graph of my function is not a straight line, so I cannot use slope to analyze its rates of change. does not make sense

41. According to the Blitzer Bonus on page 278, calculus studies change by analyzing slopes of secant lines over successively shorter intervals. makes sense

42. What is the slope of a line that is perpendicular to the line whose equation is $Ax + By + C = 0, A \neq 0$ and $B \neq 0$? *

43. Determine the value of A so that the line whose equation is $Ax + y - 2 = 0$ is perpendicular to the line containing the points $(1, -3)$ and $(-2, 4)$. $-\frac{3}{7}$

Retaining the Concepts

44. Solve and check: $24 + 3(x + 2) = 5(x - 12)$. (Section 1.2, Example 2) [45]

45. After a 30% price reduction, you purchase a television for $980. What was the television's price before the reduction? (Section 1.3, Example 4) $1400

46. Solve: $2x^{\frac{2}{3}} - 5x^{\frac{1}{3}} - 3 = 0$. (Section 1.6, Example 7) $\left\{-\frac{1}{8}, 27\right\}$

Preview Exercises

Exercises 47–49 will help you prepare for the material covered in the next section. In each exercise, graph the functions in parts (a) and (b) in the same rectangular coordinate system.

47. a. Graph $f(x) = |x|$ using the ordered pairs $(-3, f(-3))$, $(-2, f(-2)), (-1, f(-1)), (0, f(0)), (1, f(1)), (2, f(2))$, and $(3, f(3))$. *

b. Subtract 4 from each y-coordinate of the ordered pairs in part (a). Then graph the ordered pairs and connect them with two linear pieces. *

c. Describe the relationship between the graph in part (b) and the graph in part (a). *

48. a. Graph $f(x) = x^2$ using the ordered pairs $(-3, f(-3))$, $(-2, f(-2)), (-1, f(-1)), (0, f(0)), (1, f(1)), (2, f(2))$, and $(3, f(3))$. *

b. Add 2 to each x-coordinate of the ordered pairs in part (a). Then graph the ordered pairs and connect them with a smooth curve. *

c. Describe the relationship between the graph in part (b) and the graph in part (a). *

49. a. Graph $f(x) = x^3$ using the ordered pairs $(-2, f(-2))$, $(-1, f(-1)), (0, f(0)), (1, f(1))$, and $(2, f(2))$. *

b. Replace each x-coordinate of the ordered pairs in part (a) with its opposite, or additive inverse. Then graph the ordered pairs and connect them with a smooth curve. *

c. Describe the relationship between the graph in part (b) and the graph in part (a). *

CHAPTER 2 Mid-Chapter Check Point

WHAT YOU KNOW: We learned that a function is a relation in which no two ordered pairs have the same first component and different second components. We represented functions as equations and used function notation. We graphed functions and applied the vertical line test to identify graphs of functions. We determined the domain and range of a function from its graph, using inputs on the x-axis for the domain and outputs on the y-axis for the range. We used graphs to identify intervals on which functions increase, decrease, or are constant, as well as to locate relative maxima or minima. We determined when graphs of equations are symmetric with respect to the y-axis (no change when $-x$ is substituted for x), the x-axis (no change when $-y$ is substituted for y), and the origin (no change when $-x$ is substituted for x and $-y$ is substituted for y). We identified even functions [$f(-x) = f(x)$: y-axis symmetry] and odd functions [$f(-x) = -f(x)$: origin symmetry]. Finally, we studied linear functions and slope, using slope (change in y divided by change in x) to develop various forms for equations of lines:

Point-slope form	Slope-intercept form	Horizontal line	Vertical line	General form
$y - y_1 = m(x - x_1)$	$y = f(x) = mx + b$	$y = f(x) = b$	$x = a$	$Ax + By + C = 0.$

We saw that parallel lines have the same slope and that perpendicular lines have slopes that are negative reciprocals. For linear functions, slope was interpreted as the rate of change of the dependent variable per unit change in the independent variable. For nonlinear functions, the slope of the secant line between $(x_1, f(x_1))$ and $(x_2, f(x_2))$ described the average rate of change of f from x_1 to x_2: $\dfrac{f(x_2) - f(x_1)}{x_2 - x_1}$.

In Exercises 1–6, determine whether each relation is a function. Give the domain and range for each relation.

1. $\{(2, 6), (1, 4), (2, -6)\}$ no; {1, 2}; {-6, 4, 6}

2. $\{(0, 1), (2, 1), (3, 4)\}$ yes; {0, 2, 3}; {1, 4}

3. *

4. *

5. *

6. *

In Exercises 7–8, determine whether each equation defines y as a function of x.

7. $x^2 + y = 5$ yes

8. $x + y^2 = 5$ no

Use the graph of f to solve Exercises 9–24. Where applicable, use interval notation.

$y = f(x)$

9. Explain why f represents the graph of a function. *
10. Find the domain of f. $(-\infty, \infty)$
11. Find the range of f. $(-\infty, 4]$
12. Find the x-intercept(s). -6 and 2
13. Find the y-intercept. 3
14. Find the interval(s) on which f is increasing. $(-\infty, -2)$
15. Find the interval(s) on which f is decreasing. $(-2, \infty)$
16. At what number does f have a relative maximum? -2
17. What is the relative maximum of f? 4
18. Find $f(-4)$. 3
19. For what value or values of x is $f(x) = -2$? -7 and 3
20. For what value or values of x is $f(x) = 0$? -6 and 2
21. For what values of x is $f(x) > 0$? $(-6, 2)$
22. Is $f(100)$ positive or negative? negative
23. Is f even, odd, or neither? neither
24. Find the average rate of change of f from $x_1 = -4$ to $x_2 = 4$. -1

In Exercises 25–26, determine whether the graph of each equation is symmetric with respect to the y-axis, the x-axis, the origin, more than one of these, or none of these.

25. $x = y^2 + 1$ x-axis **26.** $y = x^3 - 1$ none

*See Answers to Selected Exercises.

In Exercises 27–38, graph each equation in a rectangular coordinate system.

27. $y = -2x$ *

28. $y = -2$ *

29. $x + y = -2$ *

30. $y = \frac{1}{3}x - 2$ *

31. $x = 3.5$ *

32. $4x - 2y = 8$ *

33. $f(x) = x^2 - 4$ *

34. $f(x) = x - 4$ *

35. $f(x) = |x| - 4$ *

36. $5y = -3x$ *

37. $5y = 20$ *

38. $f(x) = \begin{cases} -1 & \text{if} \quad x \le 0 \\ 2x + 1 & \text{if} \quad x > 0 \end{cases}$ *

39. Let $f(x) = -2x^2 + x - 5$.

 a. Find $f(-x)$. Is f even, odd, or neither? neither*

 b. Find $\dfrac{f(x + h) - f(x)}{h}, h \ne 0$. $-4x - 2h + 1, h \ne 0$

40. Let $C(x) = \begin{cases} 30 & \text{if} \quad 0 \le x \le 200 \\ 30 + 0.40(x - 200) & \text{if} \quad x > 200 \end{cases}$.

 a. Find $C(150)$. 30 **b.** Find $C(250)$. 50

In Exercises 41–44, write a function in slope-intercept form whose graph satisfies the given conditions.

41. Slope $= -2$, passing through $(-4, 3)$ $f(x) = -2x - 5$

42. Passing through $(-1, -5)$ and $(2, 1)$ $f(x) = 2x - 3$

43. Passing through $(3, -4)$ and parallel to the line whose equation is $3x - y - 5 = 0$ $f(x) = 3x - 13$

44. Passing through $(-4, -3)$ and perpendicular to the line whose equation is $2x - 5y - 10 = 0$ $f(x) = -\frac{5}{2}x - 13$

45. Determine whether the line through $(2, -4)$ and $(7, 0)$ is parallel to a second line through $(-4, 2)$ and $(1, 6)$. yes

46. Exercise is useful not only in preventing depression, but also as a treatment. The following graphs show the percentage of patients with depression in remission when exercise (brisk walking) was used as a treatment. (The control group that engaged in no exercise had 11% of the patients in remission.)

Exercise and Percentage of Patients with Depression in Remission

Source: *Newsweek*, March 26, 2007

 a. Find the slope of the line passing through the two points shown by the voice balloons. Express the slope as a decimal. 0.16

 b. Use your answer from part (a) to complete this statement: For each minute of brisk walking, the percentage of patients with depression in remission increased by ___0.16___ %. The rate of change is ___0.16___ % per ___minute of brisk walking___.

47. Find the average rate of change of $f(x) = 3x^2 - x$ from $x_1 = -1$ to $x_2 = 2$. 2

Section 2.5 — Transformations of Functions

What am I supposed to learn?

After studying this section, you should be able to:

① Recognize graphs of common functions.

② Use vertical shifts to graph functions.

③ Use horizontal shifts to graph functions.

④ Use reflections to graph functions.

⑤ Use vertical stretching and shrinking to graph functions.

⑥ Use horizontal stretching and shrinking to graph functions.

⑦ Graph functions involving a sequence of transformations.

Have you seen *Terminator 2, The Mask*, or *The Matrix?* These were among the first films to use spectacular effects in which a character or object having one shape was transformed in a fluid fashion into a quite different shape. The name for such a transformation is **morphing**. The effect allows a real actor to be seamlessly transformed into a computer-generated animation. The animation can be made to perform impossible feats before it is morphed back to the conventionally filmed image.

Like transformed movie images, the graph of one function can be turned into the graph of a different function. To do this, we need to rely on a function's equation. Knowing that a graph is a transformation of a familiar graph makes graphing easier.

① Recognize graphs of common functions.

Graphs of Common Functions

Table 2.4 gives names to seven frequently encountered functions in algebra. The table shows each function's graph and lists characteristics of the function. Study the shape of each graph and take a few minutes to verify the function's characteristics from its graph. Knowing these graphs is essential for analyzing their transformations into more complicated graphs.

Table 2.4 Algebra's Common Graphs

Constant Function

- Domain: $(-\infty, \infty)$
- Range: the single number c
- Constant on $(-\infty, \infty)$

- Even function

Identity Function

- Domain: $(-\infty, \infty)$
- Range: $(-\infty, \infty)$
- Increasing on $(-\infty, \infty)$

- Odd function

Absolute Value Function

- Domain: $(-\infty, \infty)$
- Range: $[0, \infty)$
- Decreasing on $(-\infty, 0)$
 and increasing on $(0, \infty)$
- Even function

Standard Quadratic Function

- Domain: $(-\infty, \infty)$
- Range: $[0, \infty)$
- Decreasing on $(-\infty, 0)$
 and increasing on $(0, \infty)$
- Even function

Square Root Function

- Domain: $[0, \infty)$
- Range: $[0, \infty)$
- Increasing on $(0, \infty)$

- Neither even nor odd

Standard Cubic Function

- Domain: $(-\infty, \infty)$
- Range: $(-\infty, \infty)$
- Increasing on $(-\infty, \infty)$

- Odd function

Cube Root Function

- Domain: $(-\infty, \infty)$
- Range: $(-\infty, \infty)$
- Increasing on $(-\infty, \infty)$

- Odd function

DISCOVERY

The study of how changing a function's equation can affect its graph can be explored with a graphing utility. Use your graphing utility to verify the hand-drawn graphs as you read this section.

 Use vertical shifts to graph functions.

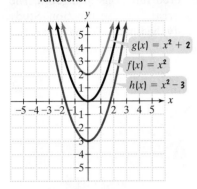

FIGURE 2.48 Vertical shifts

GREAT QUESTION!

If I'm using the graph of a familiar function, how do I actually obtain the graph of a transformation?

To keep track of transformations and obtain their graphs, identify a number of points on the given function's graph. Then analyze what happens to the coordinates of these points with each transformation.

Vertical Shifts

Let's begin by looking at three graphs whose shapes are the same. **Figure 2.48** shows the graphs. The black graph in the middle is the standard quadratic function, $f(x) = x^2$. Now, look at the blue graph on the top. The equation of this graph, $g(x) = x^2 + 2$, adds 2 to the right side of $f(x) = x^2$. The y-coordinate of each point of g is 2 more than the corresponding y-coordinate of each point of f. What effect does this have on the graph of f? It shifts the graph vertically up by 2 units.

The graph of g — $g(x) = x^2 + 2 = f(x) + 2$ — shifts the graph of f up 2 units.

Finally, look at the red graph on the bottom in **Figure 2.48**. The equation of this graph, $h(x) = x^2 - 3$, subtracts 3 from the right side of $f(x) = x^2$. The y-coordinate of each point of h is 3 less than the corresponding y-coordinate of each point of f. What effect does this have on the graph of f? It shifts the graph vertically down by 3 units.

The graph of h — $h(x) = x^2 - 3 = f(x) - 3$ — shifts the graph of f down 3 units.

In general, if c is positive, $y = f(x) + c$ shifts the graph of f upward c units and $y = f(x) - c$ shifts the graph of f downward c units. These are called **vertical shifts** of the graph of f.

Vertical Shifts

Let f be a function and c a positive real number.

- The graph of $y = f(x) + c$ is the graph of $y = f(x)$ shifted c units vertically upward.
- The graph of $y = f(x) - c$ is the graph of $y = f(x)$ shifted c units vertically downward.

EXAMPLE 1 Vertical Shift Downward

Use the graph of $f(x) = |x|$ to obtain the graph of $g(x) = |x| - 4$.

SOLUTION

The graph of $g(x) = |x| - 4$ has the same shape as the graph of $f(x) = |x|$. However, it is shifted down vertically 4 units.

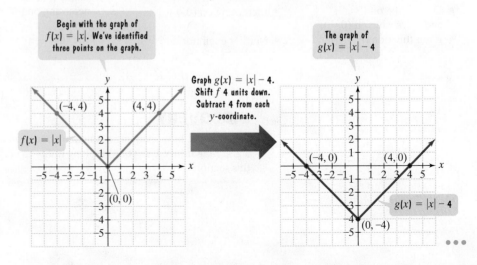

• • •

⌀ Check Point **1** Use the graph of $f(x) = |x|$ to obtain the graph of $g(x) = |x| + 3$. *

Horizontal Shifts

We return to the graph of $f(x) = x^2$, the standard quadratic function. In **Figure 2.49**, the graph of function f is in the middle of the three graphs. By contrast to the vertical shift situation, this time there are graphs to the left and to the right of the graph of f. Look at the blue graph on the right. The equation of this graph, $g(x) = (x - 3)^2$, subtracts 3 from each value of x before squaring it. What effect does this have on the graph of $f(x) = x^2$? It shifts the graph horizontally to the right by 3 units.

$$g(x) = (x - 3)^2 = f(x - 3)$$

| The graph of g | shifts the graph of f 3 units to the right. |

Does it seem strange that *subtracting* 3 in the domain causes a shift of 3 units to the *right*? Perhaps a partial table of coordinates for each function will numerically convince you of this shift.

x	$f(x) = x^2$	x	$g(x) = (x - 3)^2$
-2	$(-2)^2 = 4$	1	$(1 - 3)^2 = (-2)^2 = 4$
-1	$(-1)^2 = 1$	2	$(2 - 3)^2 = (-1)^2 = 1$
0	$0^2 = 0$	3	$(3 - 3)^2 = 0^2 = 0$
1	$1^2 = 1$	4	$(4 - 3)^2 = 1^2 = 1$
2	$2^2 = 4$	5	$(5 - 3)^2 = 2^2 = 4$

Notice that for the values of $f(x)$ and $g(x)$ to be the same, the values of x used in graphing g must each be 3 units greater than those used to graph f. For this reason, the graph of g is the graph of f shifted 3 units to the right.

Now, look at the red graph on the left in **Figure 2.49**. The equation of this graph, $h(x) = (x + 2)^2$, adds 2 to each value of x before squaring it. What effect does this have on the graph of $f(x) = x^2$? It shifts the graph horizontally to the left by 2 units.

$$h(x) = (x + 2)^2 = f(x + 2)$$

| The graph of h | shifts the graph of f 2 units to the left. |

In general, if c is positive, $y = f(x + c)$ shifts the graph of f to the left c units and $y = f(x - c)$ shifts the graph of f to the right c units. These are called **horizontal shifts** of the graph of f.

Horizontal Shifts

Let f be a function and c a positive real number.

- The graph of $y = f(x + c)$ is the graph of $y = f(x)$ shifted to the left c units.
- The graph of $y = f(x - c)$ is the graph of $y = f(x)$ shifted to the right c units.

③ Use horizontal shifts to graph functions.

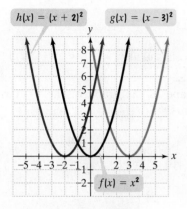

FIGURE 2.49 Horizontal shifts

GREAT QUESTION!

Using my intuition, it seems that $f(x + c)$ should cause a shift to the right and $f(x - c)$ should cause a shift to the left. Is my intuition on target when it comes to these horizontal shifts?

No. On a number line, if x represents a number and c is positive, then $x + c$ lies c units to the right of x and $x - c$ lies c units to the left of x. This orientation does not apply to horizontal shifts: $f(x + c)$ causes a shift of c units to the left and $f(x - c)$ causes a shift of c units to the right.

*See Answers to Selected Exercises.

EXAMPLE 2 Horizontal Shift to the Left

Use the graph of $f(x) = \sqrt{x}$ to obtain the graph of $g(x) = \sqrt{x + 5}$.

SOLUTION

Compare the equations for $f(x) = \sqrt{x}$ and $g(x) = \sqrt{x + 5}$. The equation for g adds 5 to each value of x before taking the square root.

$$y = g(x) = \sqrt{x + 5} = f(x + 5)$$

| The graph of g | shifts the graph of f 5 units to the left. |

The graph of $g(x) = \sqrt{x + 5}$ has the same shape as the graph of $f(x) = \sqrt{x}$. However, it is shifted horizontally to the left 5 units.

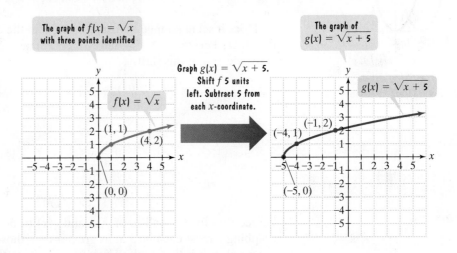

GREAT QUESTION!

What's the difference between $f(x) + c$ and $f(x + c)$?

- $y = f(x) + c$ shifts the graph of $y = f(x)$ c units vertically upward.
- $y = f(x + c)$ shifts the graph of $y = f(x)$ c units horizontally to the left.

There are analogous differences between $f(x) - c$ and $f(x - c)$.

✓ Check Point **2** Use the graph of $f(x) = \sqrt{x}$ to obtain the graph of $g(x) = \sqrt{x - 4}$. *

Some functions can be graphed by combining horizontal and vertical shifts. These functions will be variations of a function whose equation you know how to graph, such as the standard quadratic function, the standard cubic function, the square root function, the cube root function, or the absolute value function.

In our next example, we will use the graph of the standard quadratic function, $f(x) = x^2$, to obtain the graph of $h(x) = (x + 1)^2 - 3$. We will graph three functions:

$$f(x) = x^2 \qquad g(x) = (x + 1)^2 \qquad h(x) = (x + 1)^2 - 3.$$

| Start by graphing the standard quadratic function. | Shift the graph of f horizontally one unit to the left. | Shift the graph of g vertically down 3 units. |

*See Answers to Selected Exercises.

EXAMPLE 3 Combining Horizontal and Vertical Shifts

Use the graph of $f(x) = x^2$ to obtain the graph of $h(x) = (x + 1)^2 - 3$.

SOLUTION

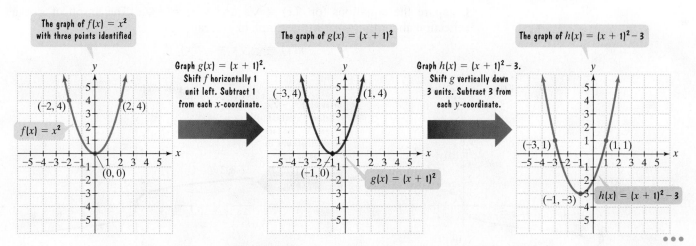

The graph of $f(x) = x^2$ with three points identified

$f(x) = x^2$

$(-2, 4)$ $(2, 4)$

$(0, 0)$

Graph $g(x) = (x + 1)^2$. Shift f horizontally 1 unit left. Subtract 1 from each x-coordinate.

The graph of $g(x) = (x + 1)^2$

$(-3, 4)$ $(1, 4)$

$(-1, 0)$

$g(x) = (x + 1)^2$

Graph $h(x) = (x + 1)^2 - 3$. Shift g vertically down 3 units. Subtract 3 from each y-coordinate.

The graph of $h(x) = (x + 1)^2 - 3$

$(-3, 1)$ $(1, 1)$

$(-1, -3)$

$h(x) = (x + 1)^2 - 3$

DISCOVERY

Work Example 3 by first shifting the graph of $f(x) = x^2$ three units down, graphing $g(x) = x^2 - 3$. Now, shift this graph one unit left to graph $h(x) = (x + 1)^2 - 3$. Did you obtain the last graph shown in the solution of Example 3? What can you conclude?

Check Point **3** Use the graph of $f(x) = \sqrt{x}$ to obtain the graph of $h(x) = \sqrt{x - 1} - 2$. *

④ Use reflections to graph functions.

Reflections of Graphs

This photograph shows a reflection of an old bridge in a Maryland river. This perfect reflection occurs because the surface of the water is absolutely still. A mild breeze rippling the water's surface would distort the reflection.

Is it possible for graphs to have mirror-like qualities? Yes. **Figure 2.50** shows the graphs of $f(x) = x^2$ and $g(x) = -x^2$. The graph of g is a **reflection about the x-axis** of the graph of f. For corresponding values of x, the y-coordinates of g are the opposites of the y-coordinates of f. In general, the graph of $y = -f(x)$ reflects the graph of f about the x-axis. Thus, the graph of g is a reflection of the graph of f about the x-axis because

$$g(x) = -x^2 = -f(x).$$

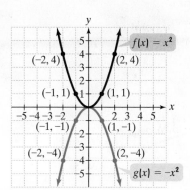

$f(x) = x^2$

$(-2, 4)$ $(2, 4)$

$(-1, 1)$ $(1, 1)$

$(-1, -1)$ $(1, -1)$

$(-2, -4)$ $(2, -4)$

$g(x) = -x^2$

FIGURE 2.50 Reflection about the x-axis

Reflection about the x-Axis

The graph of $y = -f(x)$ is the graph of $y = f(x)$ reflected about the x-axis.

EXAMPLE 4 Reflection about the x-Axis

Use the graph of $f(x) = \sqrt[3]{x}$ to obtain the graph of $g(x) = -\sqrt[3]{x}$.

SOLUTION

Compare the equations for $f(x) = \sqrt[3]{x}$ and $g(x) = -\sqrt[3]{x}$. The graph of g is a reflection about the x-axis of the graph of f because

$$g(x) = -\sqrt[3]{x} = -f(x).$$

Check Point **4** Use the graph of $f(x) = |x|$ to obtain the graph of $g(x) = -|x|$. *

It is also possible to reflect graphs about the y-axis.

Reflection about the y-Axis

The graph of $y = f(-x)$ is the graph of $y = f(x)$ reflected about the y-axis.

For each point (x, y) on the graph of $y = f(x)$, the point $(-x, y)$ is on the graph of $y = f(-x)$.

EXAMPLE 5 Reflection about the y-Axis

Use the graph of $f(x) = \sqrt{x}$ to obtain the graph of $h(x) = \sqrt{-x}$.

SOLUTION

Compare the equations for $f(x) = \sqrt{x}$ and $h(x) = \sqrt{-x}$. The graph of h is a reflection about the y-axis of the graph of f because

$$h(x) = \sqrt{-x} = f(-x).$$

Check Point **5** Use the graph of $f(x) = \sqrt[3]{x}$ to obtain the graph of $h(x) = \sqrt[3]{-x}$. *

⑤ Use vertical stretching and shrinking to graph functions.

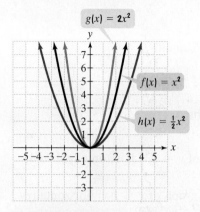

FIGURE 2.51 Vertically stretching and shrinking $f(x) = x^2$

Vertical Stretching and Shrinking

Morphing does much more than move an image horizontally, vertically, or about an axis. An object having one shape is transformed into a different shape. Horizontal shifts, vertical shifts, and reflections do not change the basic shape of a graph. Graphs remain rigid and proportionally the same when they undergo these transformations. How can we shrink and stretch graphs, thereby altering their basic shapes?

Look at the three graphs in **Figure 2.51**. The black graph in the middle is the graph of the standard quadratic function, $f(x) = x^2$. Now, look at the blue graph on the top. The equation of this graph is $g(x) = 2x^2$, or $g(x) = 2f(x)$. Thus, for each x, the y-coordinate of g is two times as large as the corresponding y-coordinate on the graph of f. The result is a narrower graph because the values of y are rising faster. We say that the graph of g is obtained by vertically *stretching* the graph of f. Now, look at the red graph on the bottom. The equation of this graph is $h(x) = \frac{1}{2}x^2$, or $h(x) = \frac{1}{2}f(x)$. Thus, for each x, the y-coordinate of h is one-half as large as the corresponding y-coordinate on the graph of f. The result is a wider graph because the values of y are rising more slowly. We say that the graph of h is obtained by vertically *shrinking* the graph of f.

These observations can be summarized as follows:

Vertically Stretching and Shrinking Graphs

Let f be a function and c a positive real number.

- If $c > 1$, the graph of $y = cf(x)$ is the graph of $y = f(x)$ vertically stretched by multiplying each of its y-coordinates by c.
- If $0 < c < 1$, the graph of $y = cf(x)$ is the graph of $y = f(x)$ vertically shrunk by multiplying each of its y-coordinates by c.

GREAT QUESTION!

Does vertical stretching or shrinking change a graph's shape?

Yes. A vertical stretch moves a function's graph away from the x-axis. A vertical shrink compresses a function's graph toward the x-axis. The other transformations we have discussed (vertical shifts, horizontal shifts, and reflections) change only the position of a function's graph without changing the shape of the basic graph.

EXAMPLE 6 Vertically Shrinking a Graph

Use the graph of $f(x) = x^3$ to obtain the graph of $h(x) = \frac{1}{2}x^3$.

SOLUTION

The graph of $h(x) = \frac{1}{2}x^3$ is obtained by vertically shrinking the graph of $f(x) = x^3$.

● ● ●

⊘ Check Point **6** Use the graph of $f(x) = |x|$ to obtain the graph of $g(x) = 2|x|$. ＊

6 Use horizontal stretching and shrinking to graph functions.

Horizontal Stretching and Shrinking

It is also possible to stretch and shrink graphs horizontally.

GREAT QUESTION!

How does horizontal shrinking or stretching change a graph's shape?

A horizontal shrink compresses a function's graph toward the y-axis. A horizontal stretch moves a function's graph away from the y-axis.

Horizontally Stretching and Shrinking Graphs

Let f be a function and c a positive real number.

- If $c > 1$, the graph of $y = f(cx)$ is the graph of $y = f(x)$ horizontally shrunk by dividing each of its x-coordinates by c.
- If $0 < c < 1$, the graph of $y = f(cx)$ is the graph of $y = f(x)$ horizontally stretched by dividing each of its x-coordinates by c.

Shrinking : $c > 1$ **Stretching : $0 < c < 1$**

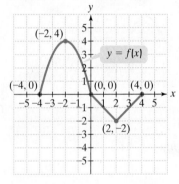

FIGURE 2.52

EXAMPLE 7 Horizontally Stretching and Shrinking a Graph

Use the graph of $y = f(x)$ in **Figure 2.52** to obtain each of the following graphs:

 a. $g(x) = f(2x)$ **b.** $h(x) = f\left(\tfrac{1}{2}x\right)$.

SOLUTION

 a. The graph of $g(x) = f(2x)$ is obtained by horizontally shrinking the graph of $y = f(x)$.

b. The graph of $h(x) = f\left(\frac{1}{2}x\right)$ is obtained by horizontally stretching the graph of $y = f(x)$.

FIGURE 2.53

✓ **Check Point 7** Use the graph of $y = f(x)$ in **Figure 2.53** to obtain each of the following graphs:

a. $g(x) = f(2x)$ * **b.** $h(x) = f\left(\frac{1}{2}x\right)$. *

⑦ Graph functions involving a sequence of transformations.

Sequences of Transformations

Table 2.5 summarizes the procedures for transforming the graph of $y = f(x)$.

Table 2.5 Summary of Transformations

In each case, c represents a positive real number.

To Graph:	Draw the Graph of f and:	Changes in the Equation of $y = f(x)$
Vertical shifts $y = f(x) + c$ $y = f(x) - c$	Raise the graph of f by c units. Lower the graph of f by c units.	c is added to $f(x)$. c is subtracted from $f(x)$.
Horizontal shifts $y = f(x + c)$ $y = f(x - c)$	Shift the graph of f to the left c units. Shift the graph of f to the right c units.	x is replaced with $x + c$. x is replaced with $x - c$.
Reflection about the x-axis $y = -f(x)$	Reflect the graph of f about the x-axis.	$f(x)$ is multiplied by -1.
Reflection about the y-axis $y = f(-x)$	Reflect the graph of f about the y-axis.	x is replaced with $-x$.
Vertical stretching or shrinking $y = cf(x), c > 1$ $y = cf(x), 0 < c < 1$	Multiply each y-coordinate of $y = f(x)$ by c, vertically stretching the graph of f. Multiply each y-coordinate of $y = f(x)$ by c, vertically shrinking the graph of f.	$f(x)$ is multiplied by $c, c > 1$. $f(x)$ is multiplied by $c, 0 < c < 1$.
Horizontal stretching or shrinking $y = f(cx), c > 1$ $y = f(cx), 0 < c < 1$	Divide each x-coordinate of $y = f(x)$ by c, horizontally shrinking the graph of f. Divide each x-coordinate of $y = f(x)$ by c, horizontally stretching the graph of f.	x is replaced with $cx, c > 1$. x is replaced with $cx, 0 < c < 1$.

*See Answers to Selected Exercises.

Order of Transformations

A function involving more than one transformation can be graphed by performing transformations in the following order:

1. Horizontal shifting
2. Stretching or shrinking
3. Reflecting
4. Vertical shifting

EXAMPLE 8 Graphing Using a Sequence of Transformations

Use the graph of $y = f(x)$ given in **Figure 2.52** of Example 7 on page 290, and repeated below, to graph $y = -\frac{1}{2}f(x - 1) + 3$.

SOLUTION

Our graphs will evolve in the following order:

1. **Horizontal shifting:** Graph $y = f(x - 1)$ by shifting the graph of $y = f(x)$ 1 unit to the right.
2. **Shrinking:** Graph $y = \frac{1}{2}f(x - 1)$ by shrinking the previous graph by a factor of $\frac{1}{2}$.
3. **Reflecting:** Graph $y = -\frac{1}{2}f(x - 1)$ by reflecting the previous graph about the x-axis.
4. **Vertical shifting:** Graph $y = -\frac{1}{2}f(x - 1) + 3$ by shifting the previous graph up 3 units.

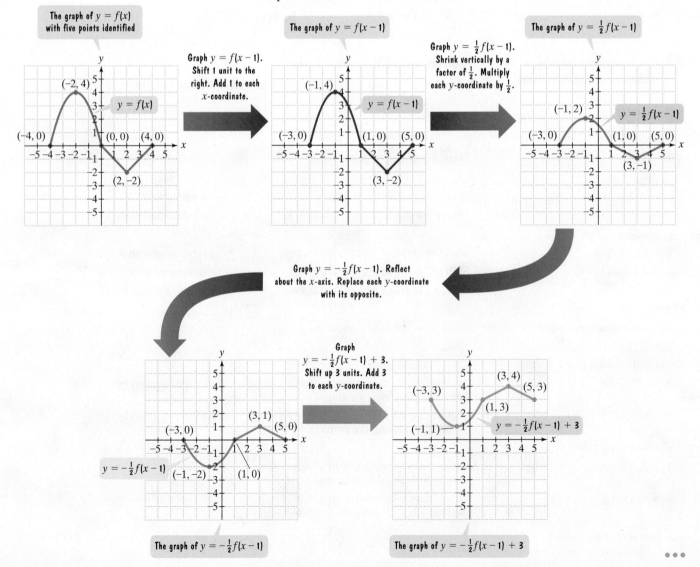

☑ **Check Point 8** Use the graph of $y = f(x)$ given in **Figure 2.53** of Check Point 7 on page 291 to graph $y = -\frac{1}{3}f(x + 1) - 2$. ∗

EXAMPLE 9 Graphing Using a Sequence of Transformations

Use the graph of $f(x) = x^2$ to graph $g(x) = 2(x + 3)^2 - 1$.

SOLUTION

Our graphs will evolve in the following order:

1. Horizontal shifting: Graph $y = (x + 3)^2$ by shifting the graph of $f(x) = x^2$ three units to the left.

2. Stretching: Graph $y = 2(x + 3)^2$ by stretching the previous graph by a factor of 2.

3. Vertical shifting: Graph $g(x) = 2(x + 3)^2 - 1$ by shifting the previous graph down 1 unit.

☑ **Check Point 9** Use the graph of $f(x) = x^2$ to graph $g(x) = 2(x - 1)^2 + 3$. ∗

∗See Answers to Selected Exercises.

ACHIEVING SUCCESS

When using your professor's office hours, show up prepared. If you are having difficulty with a concept or problem, bring your work so that your instructor can determine where you are having trouble. If you miss a lecture, read the appropriate section in the textbook, borrow class notes, and attempt the assigned homework before your office visit. Because this text has an accompanying video lesson for every objective, you might find it helpful to view the video covering the material you missed. It is not realistic to expect your professor to rehash all or part of a class lecture during office hours.

CONCEPT AND VOCABULARY CHECK

Fill in each blank so that the resulting statement is true.

1. The graph of $y = f(x) - 5$ is obtained by a/an _____vertical_____ shift of the graph of $y = f(x)$ ___down___ a distance of 5 units.

2. The graph of $y = f(x - 5)$ is obtained by a/an ___horizontal___ shift of the graph of $y = f(x)$ ___to the right___ a distance of 5 units.

3. The graph of $y = -f(x)$ is the graph of $y = f(x)$ reflected about the ___x-axis___.

4. The graph of $y = f(-x)$ is the graph of $y = f(x)$ reflected about the ___y-axis___.

5. The graph of $y = 5f(x)$ is obtained by a/an ___vertical___ stretch of the graph of $y = f(x)$ by multiplying each of its ___y___-coordinates by 5.

6. The graph of $y = f\left(\frac{1}{5}x\right)$ is obtained by a/an ___horizontal___ stretch of the graph of $y = f(x)$ by multiplying each of its ___x___-coordinates by 5.

7. True or false: The graph of $g(x) = \sqrt{x} + 4$ is the graph of $f(x) = \sqrt{x}$ shifted horizontally to the right by 4 units. ___false___

EXERCISE SET 2.5

Practice Exercises

In Exercises 1–16, use the graph of $y = f(x)$ to graph each function g.

1. $g(x) = f(x) + 1$ *
2. $g(x) = f(x) - 1$ *
3. $g(x) = f(x + 1)$ *
4. $g(x) = f(x - 1)$ *
5. $g(x) = f(x - 1) - 2$ *
6. $g(x) = f(x + 1) + 2$ *
7. $g(x) = f(-x)$ *
8. $g(x) = -f(x)$ *
9. $g(x) = -f(x) + 3$ *
10. $g(x) = f(-x) + 3$ *
11. $g(x) = \frac{1}{2}f(x)$ *
12. $g(x) = 2f(x)$ *
13. $g(x) = f\left(\frac{1}{2}x\right)$ *
14. $g(x) = f(2x)$ *
15. $g(x) = -f\left(\frac{1}{2}x\right) + 1$ *
16. $g(x) = -f(2x) - 1$ *

*See Answers to Selected Exercises.

In Exercises 17–32, use the graph of $y = f(x)$ to graph each function g.

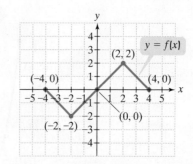

17. $g(x) = f(x) - 1$ *
18. $g(x) = f(x) + 1$ *
19. $g(x) = f(x - 1)$ *
20. $g(x) = f(x + 1)$ *
21. $g(x) = f(x - 1) + 2$ *
22. $g(x) = f(x + 1) - 2$ *
23. $g(x) = -f(x)$ *
24. $g(x) = f(-x)$ *
25. $g(x) = f(-x) + 1$ *
26. $g(x) = -f(x) + 1$ *
27. $g(x) = 2f(x)$ *
28. $g(x) = \frac{1}{2}f(x)$ *
29. $g(x) = f(2x)$ *
30. $g(x) = f\left(\frac{1}{2}x\right)$ *
31. $g(x) = 2f(x + 2) + 1$ *
32. $g(x) = 2f(x + 2) - 1$ *

In Exercises 33–44, use the graph of y = f(x) to graph each function g.

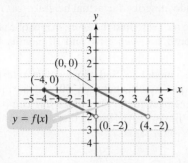

33. $g(x) = f(x) + 2$ *

34. $g(x) = f(x) - 2$ *

35. $g(x) = f(x + 2)$ *

36. $g(x) = f(x - 2)$ *

37. $g(x) = -f(x + 2)$ *

38. $g(x) = -f(x - 2)$ *

39. $g(x) = -\frac{1}{2}f(x + 2)$ *

40. $g(x) = -\frac{1}{2}f(x - 2)$ *

41. $g(x) = -\frac{1}{2}f(x + 2) - 2$ *

42. $g(x) = -\frac{1}{2}f(x - 2) + 2$ *

43. $g(x) = \frac{1}{2}f(2x)$ *

44. $g(x) = 2f\left(\frac{1}{2}x\right)$ *

In Exercises 45–52, use the graph of y = f(x) to graph each function g.

45. $g(x) = f(x - 1) - 1$ *

46. $g(x) = f(x + 1) + 1$ *

47. $g(x) = -f(x - 1) + 1$ *

48. $g(x) = -f(x + 1) - 1$ *

49. $g(x) = 2f\left(\frac{1}{2}x\right)$ *

50. $g(x) = \frac{1}{2}f(2x)$ *

51. $g(x) = \frac{1}{2}f(x + 1)$ *

52. $g(x) = 2f(x - 1)$ *

In Exercises 53–66, begin by graphing the standard quadratic function, $f(x) = x^2$. Then use transformations of this graph to graph the given function.

53. $g(x) = x^2 - 2$ *

54. $g(x) = x^2 - 1$ *

55. $g(x) = (x - 2)^2$ *

56. $g(x) = (x - 1)^2$ *

57. $h(x) = -(x - 2)^2$ *

58. $h(x) = -(x - 1)^2$ *

59. $h(x) = (x - 2)^2 + 1$ *

60. $h(x) = (x - 1)^2 + 2$ *

61. $g(x) = 2(x - 2)^2$ *

62. $g(x) = \frac{1}{2}(x - 1)^2$ *

63. $h(x) = 2(x - 2)^2 - 1$ *

64. $h(x) = \frac{1}{2}(x - 1)^2 - 1$ *

65. $h(x) = -2(x + 1)^2 + 1$

66. $h(x) = -2(x + 2)^2 + 1$ *

In Exercises 67–80, begin by graphing the square root function, $f(x) = \sqrt{x}$. Then use transformations of this graph to graph the given function.

67. $g(x) = \sqrt{x} + 2$ *

68. $g(x) = \sqrt{x} + 1$ *

69. $g(x) = \sqrt{x + 2}$ *

70. $g(x) = \sqrt{x + 1}$ *

71. $h(x) = -\sqrt{x + 2}$ *

72. $h(x) = -\sqrt{x + 1}$ *

73. $h(x) = \sqrt{-x + 2}$ *

74. $h(x) = \sqrt{-x + 1}$ *

75. $g(x) = \frac{1}{2}\sqrt{x + 2}$ *

76. $g(x) = 2\sqrt{x + 1}$ *

77. $h(x) = \sqrt{x + 2} - 2$ *

78. $h(x) = \sqrt{x + 1} - 1$ *

79. $g(x) = 2\sqrt{x + 2} - 2$ *

80. $g(x) = 2\sqrt{x + 1} - 1$ *

In Exercises 81–94, begin by graphing the absolute value function, $f(x) = |x|$. Then use transformations of this graph to graph the given function.

81. $g(x) = |x| + 4$ *

82. $g(x) = |x| + 3$ *

83. $g(x) = |x + 4|$ *

84. $g(x) = |x + 3|$ *

85. $h(x) = |x + 4| - 2$ *

86. $h(x) = |x + 3| - 2$ *

87. $h(x) = -|x + 4|$ *

88. $h(x) = -|x + 3|$ *

89. $g(x) = -|x + 4| + 1$ *

90. $g(x) = -|x + 4| + 2$ *

91. $h(x) = 2|x + 4|$ *

92. $h(x) = 2|x + 3|$ *

93. $g(x) = -2|x + 4| + 1$ *

94. $g(x) = -2|x + 3| + 2$ *

In Exercises 95–106, begin by graphing the standard cubic function, $f(x) = x^3$. Then use transformations of this graph to graph the given function.

95. $g(x) = x^3 - 3$ *

96. $g(x) = x^3 - 2$ *

97. $g(x) = (x - 3)^3$ *

98. $g(x) = (x - 2)^3$ *

99. $h(x) = -x^3$ *

100. $h(x) = -(x - 2)^3$ *

101. $h(x) = \frac{1}{2}x^3$ *

102. $h(x) = \frac{1}{4}x^3$ *

103. $r(x) = (x - 3)^3 + 2$ *

104. $r(x) = (x - 2)^3 + 1$ *

105. $h(x) = \frac{1}{2}(x - 3)^3 - 2$ *

106. $h(x) = \frac{1}{2}(x - 2)^3 - 1$ *

In Exercises 107–118, begin by graphing the cube root function, $f(x) = \sqrt[3]{x}$. Then use transformations of this graph to graph the given function.

107. $g(x) = \sqrt[3]{x} + 2$ *

108. $g(x) = \sqrt[3]{x} - 2$ *

109. $g(x) = \sqrt[3]{x + 2}$ *

110. $g(x) = \sqrt[3]{x - 2}$ *

111. $h(x) = \frac{1}{2}\sqrt[3]{x + 2}$ *

112. $h(x) = \frac{1}{2}\sqrt[3]{x - 2}$ *

113. $r(x) = \frac{1}{2}\sqrt[3]{x + 2} - 2$ *

114. $r(x) = \frac{1}{2}\sqrt[3]{x - 2} + 2$ *

115. $h(x) = -\sqrt[3]{x + 2}$ *

116. $h(x) = -\sqrt[3]{x - 2}$ *

117. $g(x) = \sqrt[3]{-x - 2}$ *

118. $g(x) = \sqrt[3]{-x + 2}$ *

Practice Plus

*In Exercises 119–122, use transformations of the graph of the greatest integer function, $f(x) = \text{int}(x)$, to graph each function. (The graph of $f(x) = \text{int}(x)$ is shown in **Figure 2.27** on page 247.)*

119. $g(x) = 2 \text{ int } (x + 1)$ *

120. $g(x) = 3 \text{ int } (x - 1)$ *

121. $h(x) = \text{int}(-x) + 1$ *

122. $h(x) = \text{int}(-x) - 1$ *

In Exercises 123–126, write a possible equation for the function whose graph is shown. Each graph shows a transformation of a common function.

123.

$[-2, 8, 1]$ by $[-1, 4, 1]$

124.

$[-3, 3, 1]$ by $[-6, 6, 1]$

125.

$[-5, 3, 1]$ by $[-5, 10, 1]$

126.

$[-1, 9, 1]$ by $[-1, 5, 1]$

Application Exercises

127. The function $f(x) = 2.9\sqrt{x} + 20.1$ models the median height, $f(x)$, in inches, of boys who are x months of age. The graph of f is shown.

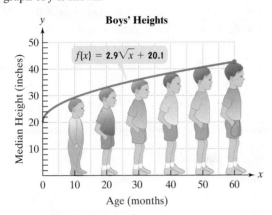

Boys' Heights

$f(x) = 2.9\sqrt{x} + 20.1$

Median Height (inches)

Age (months)

Source: Laura Walther Nathanson, *The Portable Pediatrician for Parents*

a. Describe how the graph can be obtained using transformations of the square root function $f(x) = \sqrt{x}$. *

b. According to the model, what is the median height of boys who are 48 months, or four years, old? Use a calculator and round to the nearest tenth of an inch. The actual median height for boys at 48 months is 40.8 inches. How well does the model describe the actual height? 40.2 in.; very well

c. Use the model to find the average rate of change, in inches per month, between birth and 10 months. Round to the nearest tenth. 0.9 in. per month

d. Use the model to find the average rate of change, in inches per month, between 50 and 60 months. Round to the nearest tenth. How does this compare with your answer in part (c)? How is this difference shown by the graph? 0.2 in. per month*

128. The function $f(x) = 3.1\sqrt{x} + 19$ models the median height, $f(x)$, in inches, of girls who are x months of age. The graph of f is shown.

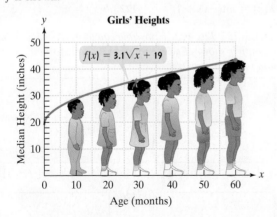

Girls' Heights

$f(x) = 3.1\sqrt{x} + 19$

Median Height (inches)

Age (months)

Source: Laura Walther Nathanson, *The Portable Pediatrician for Parents*

a. Describe how the graph can be obtained using transformations of the square root function $f(x) = \sqrt{x}$. *

b. According to the model, what is the median height of girls who are 48 months, or 4 years, old? Use a calculator and round to the nearest tenth of an inch. The actual median height for girls at 48 months is 40.2 inches. How well does the model describe the actual height? 40.5 in.; very well

c. Use the model to find the average rate of change, in inches per month, between birth and 10 months. Round to the nearest tenth. 1.0 in. per month

d. Use the model to find the average rate of change, in inches per month, between 50 and 60 months. Round to the nearest tenth. How does this compare with your answer in part (c)? How is this difference shown by the graph? 0.2 in. per month*

Explaining the Concepts 129–134. Answers will vary.

129. What must be done to a function's equation so that its graph is shifted vertically upward?

130. What must be done to a function's equation so that its graph is shifted horizontally to the right?

131. What must be done to a function's equation so that its graph is reflected about the x-axis?

132. What must be done to a function's equation so that its graph is reflected about the y-axis?

133. What must be done to a function's equation so that its graph is stretched vertically?

134. What must be done to a function's equation so that its graph is shrunk horizontally?

Technology Exercises

135. a. Use a graphing utility to graph $f(x) = x^2 + 1$. *

 b. Graph $f(x) = x^2 + 1, g(x) = f(2x), h(x) = f(3x)$, and $k(x) = f(4x)$ in the same viewing rectangle. *

 c. Describe the relationship among the graphs of f, g, h, and k, with emphasis on different values of x for points on all four graphs that give the same y-coordinate.

 d. Generalize by describing the relationship between the graph of f and the graph of g, where $g(x) = f(cx)$ for $c > 1$.

 e. Try out your generalization by sketching the graphs of $f(cx)$ for $c = 1, c = 2, c = 3$, and $c = 4$ for a function of your choice. c–e. Answers will vary.

136. a. Use a graphing utility to graph $f(x) = x^2 + 1$. *

 b. Graph $f(x) = x^2 + 1, g(x) = f\left(\frac{1}{2}x\right)$, and $h(x) = f\left(\frac{1}{4}x\right)$ in the same viewing rectangle. *

 c. Describe the relationship among the graphs of f, g, and h, with emphasis on different values of x for points on all three graphs that give the same y-coordinate.

 d. Generalize by describing the relationship between the graph of f and the graph of g, where $g(x) = f(cx)$ for $0 < c < 1$.

 e. Try out your generalization by sketching the graphs of $f(cx)$ for $c = 1$, and $c = \frac{1}{2}$, and $c = \frac{1}{4}$ for a function of your choice. c–e. Answers will vary.

Critical Thinking Exercises

Make Sense? *During the winter, you program your home thermostat so that at midnight, the temperature is 55°. This temperature is maintained until 6 A.M. Then the house begins to warm up so that by 9 A.M. the temperature is 65°. At 6 P.M. the house begins to cool. By 9 P.M., the temperature is again 55°.*

The graph illustrates home temperature, f(t), as a function of hours after midnight, t.

Home Temperature as a Function of Time

In Exercises 137–140, determine whether each statement makes sense or does not make sense, and explain your reasoning. If the statement makes sense, graph the new function on the domain $[0, 24]$. *If the statement does not make sense, correct the function in the statement and graph the corrected function on the domain* $[0, 24]$.

137. I decided to keep the house 5° warmer than before, so I reprogrammed the thermostat to $y = f(t) + 5$. *

138. I decided to keep the house 5° cooler than before, so I reprogrammed the thermostat to $y = f(t) - 5$. *

139. I decided to change the heating schedule to start one hour earlier than before, so I reprogrammed the thermostat to $y = f(t - 1)$. does not make sense

140. I decided to change the heating schedule to start one hour later than before, so I reprogrammed the thermostat to $y = f(t + 1)$. does not make sense

In Exercises 141–144, determine whether each statement is true or false. If the statement is false, make the necessary change(s) to produce a true statement. Changes to false statements will vary.

141. If $f(x) = |x|$ and $g(x) = |x + 3| + 3$, then the graph of g is a translation of the graph of f three units to the right and three units upward. false

142. If $f(x) = -\sqrt{x}$ and $g(x) = \sqrt{-x}$, then f and g have identical graphs. false

143. If $f(x) = x^2$ and $g(x) = 5(x^2 - 2)$, then the graph of g can be obtained from the graph of f by stretching f five units followed by a downward shift of two units. false

144. If $f(x) = x^3$ and $g(x) = -(x - 3)^3 - 4$, then the graph of g can be obtained from the graph of f by moving f three units to the right, reflecting about the x-axis, and then moving the resulting graph down four units. true

In Exercises 145–148, functions f and g are graphed in the same rectangular coordinate system. If g is obtained from f through a sequence of transformations, find an equation for g.

145. $g(x) = -(x + 4)^2$

146. $g(x) = -|x - 5| + 1$

147. $g(x) = -\sqrt{x - 2} + 2$

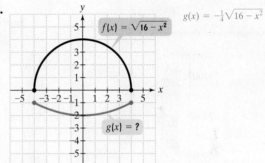

148. $g(x) = -\frac{1}{4}\sqrt{16 - x^2} - 1$

For Exercises 149–152, assume that (a, b) is a point on the graph of f. What is the corresponding point on the graph of each of the following functions?

149. $y = f(-x)$ $(-a, b)$ **150.** $y = 2f(x)$ $(a, 2b)$

151. $y = f(x - 3)$ $(a + 3, b)$ **152.** $y = f(x) - 3$ $(a, b - 3)$

Retaining the Concepts

153. The length of a rectangle exceeds the width by 13 yards. If the perimeter of the rectangle is 82 yards, what are its dimensions?

(Section 1.3, Example 6) length: 27 yd; width: 14 yd

154. Solve: $\sqrt{x + 10} - 4 = x$.

(Section 1.6, Example 3) $\{-1\}$

155. Multiply and write the product in standard form: $(3 - 7i)(5 + 2i)$.

(Section 1.4, Example 2) $29 - 29i$

Preview Exercises

Exercises 156–158 will help you prepare for the material covered in the next section.

In Exercises 156–157, perform the indicated operation or operations.

156. $(2x - 1)(x^2 + x - 2)$ $2x^3 + x^2 - 5x + 2$

157. $(f(x))^2 - 2f(x) + 6$, where $f(x) = 3x - 4$ $9x^2 - 30x + 30$

158. Simplify: $\dfrac{\frac{2}{3}}{\frac{3}{x} - 1}$. $\dfrac{2x}{3 - x}$

Section 2.6 Combinations of Functions; Composite Functions

What am I supposed to learn?

After studying this section, you should be able to:

① Find the domain of a function.

② Combine functions using the algebra of functions, specifying domains.

③ Form composite functions.

④ Determine domains for composite functions.

⑤ Write functions as compositions.

We're born. We die. **Figure 2.54** quantifies these statements by showing the number of births and deaths in the United States from 2000 through 2011.

In this section, we look at these data from the perspective of functions. By considering the yearly change in the U.S. population, you will see that functions can be subtracted using procedures that will remind you of combining algebraic expressions.

Number of Births and Deaths in the United States

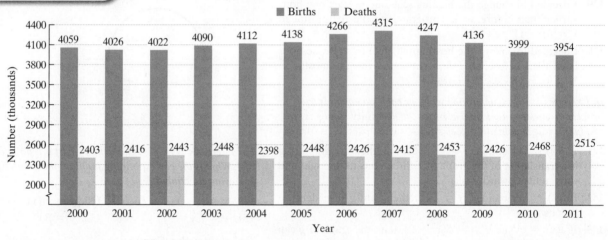

FIGURE 2.54
Source: U.S. Department of Health and Human Services

① Find the domain of a function.

The Domain of a Function

We begin with two functions that model the data in **Figure 2.54**.

$$B(x) = -2.6x^2 + 49x + 3994 \qquad D(x) = -0.6x^2 + 7x + 2412$$

Number of births, $B(x)$, in thousands, x years after **2000**

Number of deaths, $D(x)$, in thousands, x years after **2000**

The years in **Figure 2.54** extend from 2000 through 2011. Because x represents the number of years after 2000,

$$\text{Domain of } B = \{0, 1, 2, 3, \ldots, 11\}$$

and

$$\text{Domain of } D = \{0, 1, 2, 3, \ldots, 11\}.$$

Functions that model data often have their domains explicitly given with the function's equation. However, for most functions, only an equation is given and the domain is not specified. In cases like this, the domain of a function f is the largest set of real numbers for which the value of $f(x)$ is a real number. For example, consider the function

$$f(x) = \frac{1}{x - 3}.$$

Because division by 0 is undefined, the denominator, $x - 3$, cannot be 0. Thus, x cannot equal 3. The domain of the function consists of all real numbers other than 3, represented by

Domain of $f = \{x \mid x$ is a real number and $x \neq 3\}$.

Using interval notation,

Domain of $f = (-\infty, 3) \cup (3, \infty)$.

All real numbers less than 3 or All real numbers greater than 3

Now consider a function involving a square root:

$$g(x) = \sqrt{x - 3}.$$

Because only nonnegative numbers have square roots that are real numbers, the expression under the square root sign, $x - 3$, must be nonnegative. We can use inspection to see that $x - 3 \geq 0$ if $x \geq 3$. The domain of g consists of all real numbers that are greater than or equal to 3:

Domain of $g = \{x \mid x \geq 3\}$ or $[3, \infty)$.

Finding a Function's Domain

If a function f does not model data or verbal conditions, its domain is the largest set of real numbers for which the value of $f(x)$ is a real number. Exclude from a function's domain real numbers that cause division by zero and real numbers that result in an even root, such as a square root, of a negative number.

EXAMPLE 1 Finding the Domain of a Function

Find the domain of each function:

 a. $f(x) = x^2 - 7x$ **b.** $g(x) = \dfrac{3x + 2}{x^2 - 2x - 3}$

 c. $h(x) = \sqrt{3x + 12}$ **d.** $j(x) = \dfrac{3x + 2}{\sqrt{14 - 2x}}.$

SOLUTION

The domain is the set of all real numbers, $(-\infty, \infty)$, unless x appears in a denominator or in an even root, such as a square root.

 a. The function $f(x) = x^2 - 7x$ contains neither division nor a square root. For every real number, x, the algebraic expression $x^2 - 7x$ represents a real number. Thus, the domain of f is the set of all real numbers.

Domain of $f = (-\infty, \infty)$

 b. The function $g(x) = \dfrac{3x + 2}{x^2 - 2x - 3}$ contains division. Because division by 0 is undefined, we must exclude from the domain the values of x that cause the denominator, $x^2 - 2x - 3$, to be 0. We can identify these values by setting $x^2 - 2x - 3$ equal to 0.

GREAT QUESTION!

Despite the voice balloons, the notation for the domain of f on the right is a mouthful! Will you be using set operations with interval notation in this section? What should I already know?

Yes, you'll be seeing the intersection and the union of sets that are expressed in interval notation. Recall that the intersection of sets A and B, written $A \cap B$, is the set of elements common to both set A and set B. When sets A and B are in interval notation, to find the intersection, graph each interval and take the portion of the number line that the two graphs have in common. We will also be using notation involving the union of sets A and B, $A \cup B$, meaning the set of elements in A or in B or in both. For more detail, see Section P.1, Objectives 3 and 4, and Section 1.7, Objective 2.

$$x^2 - 2x - 3 = 0 \quad \text{Set the function's denominator equal to O.}$$
$$(x + 1)(x - 3) = 0 \quad \text{Factor.}$$
$$x + 1 = 0 \quad \text{or} \quad x - 3 = 0 \quad \text{Set each factor equal to O.}$$
$$x = -1 \qquad x = 3 \quad \text{Solve the resulting equations.}$$

We must exclude -1 and 3 from the domain of $g(x) = \dfrac{3x + 2}{x^2 - 2x - 3}$.

$$\text{Domain of } g = (-\infty, -1) \cup (-1, 3) \cup (3, \infty)$$

GREAT QUESTION!

When finding the domain of a function, when do I have to factor?

In parts (a) and (b), observe when to factor and when not to factor a polynomial.

- $f(x) = x^2 - 7x$

- $g(x) = \dfrac{3x + 2}{x^2 - 2x - 3}$

> Do not factor $x^2 - 7x$ and set it equal to zero. No values of x need be excluded from the domain.

> Do factor $x^2 - 2x - 3$ and set it equal to zero. We must exclude values of x that cause this denominator to be zero.

c. The function $h(x) = \sqrt{3x + 12}$ contains an even root. Because only nonnegative numbers have real square roots, the quantity under the radical sign, $3x + 12$, must be greater than or equal to 0.

$$3x + 12 \geq 0 \quad \text{Set the function's radicand greater than or equal to O.}$$
$$3x \geq -12 \quad \text{Subtract 12 from both sides.}$$
$$x \geq -4 \quad \text{Divide both sides by 3. Division by a positive number preserves the sense of the inequality.}$$

The domain of h consists of all real numbers greater than or equal to -4.

$$\text{Domain of } h = [-4, \infty)$$

The domain is highlighted on the x-axis in **Figure 2.55**.

d. The function $j(x) = \dfrac{3x + 2}{\sqrt{14 - 2x}}$ contains both an even root and division. Because only nonnegative numbers have real square roots, the quantity under the radical sign, $14 - 2x$, must be greater than or equal to 0. But wait, there's more! Because division by 0 is undefined, $14 - 2x$ cannot equal 0. Thus, $14 - 2x$ must be strictly greater than 0.

$$14 - 2x > 0 \quad \text{Set the function's radicand greater than O.}$$
$$-2x > -14 \quad \text{Subtract 14 from both sides.}$$
$$x < 7 \quad \text{Divide both sides by } -2. \text{ Division by a negative number reverses the direction of the inequality.}$$

The domain of j consists of all real numbers less than 7.

$$\text{Domain of } j = (-\infty, 7) \qquad \bullet\bullet\bullet$$

(left margin figure)

$h(x) = \sqrt{3x + 12}$

Domain $[-4, \infty)$

$[-10, 10, 1]$ by $[-10, 10, 1]$

FIGURE 2.55

☑ **Check Point 1** Find the domain of each function:

a. $f(x) = x^2 + 3x - 17$ $\quad (-\infty, \infty)$
b. $g(x) = \dfrac{5x}{x^2 - 49}$ $\quad (-\infty, -7) \cup (-7, 7) \cup (7, \infty)$

c. $h(x) = \sqrt{9x - 27}$ $\quad [3, \infty)$
d. $j(x) = \dfrac{5x}{\sqrt{24 - 3x}}$. $\quad (-\infty, 8)$

② Combine functions using the algebra of functions, specifying domains.

The Algebra of Functions

We can combine functions using addition, subtraction, multiplication, and division by performing operations with the algebraic expressions that appear on the right side of the equations. For example, the functions $f(x) = 2x$ and $g(x) = x - 1$ can be combined to form the sum, difference, product, and quotient of f and g. Here's how it's done:

For each function, $f(x) = 2x$ and $g(x) = x - 1$.

Sum: $f + g$

$$(f + g)(x) = f(x) + g(x)$$
$$= 2x + (x - 1) = 3x - 1$$

Difference: $f - g$

$$(f - g)(x) = f(x) - g(x)$$
$$= 2x - (x - 1) = 2x - x + 1 = x + 1$$

Product: fg

$$(fg)(x) = f(x) \cdot g(x)$$
$$= 2x(x - 1) = 2x^2 - 2x$$

Quotient: $\dfrac{f}{g}$

$$\left(\frac{f}{g}\right)(x) = \frac{f(x)}{g(x)} = \frac{2x}{x - 1}, x \neq 1.$$

The domain for each of these functions consists of all real numbers that are common to the domains of f and g. Using D_f to represent the domain of f and D_g to represent the domain of g, the domain for each function is $D_f \cap D_g$. In the case of the quotient function $\dfrac{f(x)}{g(x)}$, we must remember not to divide by 0, so we add the further restriction that $g(x) \neq 0$.

The Algebra of Functions: Sum, Difference, Product, and Quotient of Functions

Let f and g be two functions. The **sum** $f + g$, the **difference** $f - g$, the **product** fg, and the **quotient** $\dfrac{f}{g}$ are functions whose domains are the set of all real numbers common to the domains of f and g $(D_f \cap D_g)$, defined as follows:

1. Sum: $(f + g)(x) = f(x) + g(x)$
2. Difference: $(f - g)(x) = f(x) - g(x)$
3. Product: $(fg)(x) = f(x) \cdot g(x)$
4. Quotient: $\left(\dfrac{f}{g}\right)(x) = \dfrac{f(x)}{g(x)}$, provided $g(x) \neq 0$.

EXAMPLE 2 Combining Functions

Let $f(x) = 2x - 1$ and $g(x) = x^2 + x - 2$. Find each of the following functions:

a. $(f + g)(x)$ **b.** $(f - g)(x)$ **c.** $(fg)(x)$ **d.** $\left(\dfrac{f}{g}\right)(x)$.

Determine the domain for each function.

SOLUTION

a. $(f + g)(x) = f(x) + g(x)$ This is the definition of the sum $f + g$.

$\qquad\qquad = (2x - 1) + (x^2 + x - 2)$ Substitute the given functions.

$\qquad\qquad = x^2 + 3x - 3$ Remove parentheses and combine like terms.

b. $(f - g)(x) = f(x) - g(x)$ This is the definition of the difference $f - g$.

$\qquad\qquad = (2x - 1) - (x^2 + x - 2)$ Substitute the given functions.

$\qquad\qquad = 2x - 1 - x^2 - x + 2$ Remove parentheses and change the sign of each term in the second set of parentheses.

$\qquad\qquad = -x^2 + x + 1$ Combine like terms and arrange terms in descending powers of x.

c. $(fg)(x) = f(x) \cdot g(x)$ This is the definition of the product fg.

$$= (2x - 1)(x^2 + x - 2)$$ Substitute the given functions.

$$= 2x(x^2 + x - 2) - 1(x^2 + x - 2)$$ Multiply each term in the second factor by $2x$ and -1, respectively.

$$= 2x^3 + 2x^2 - 4x - x^2 - x + 2$$ Use the distributive property.

$$= 2x^3 + (2x^2 - x^2) + (-4x - x) + 2$$ Rearrange terms so that like terms are adjacent.

$$= 2x^3 + x^2 - 5x + 2$$ Combine like terms.

d. $\left(\dfrac{f}{g}\right)(x) = \dfrac{f(x)}{g(x)}$ This is the definition of the quotient $\dfrac{f}{g}$.

$$= \dfrac{2x - 1}{x^2 + x - 2}$$ Substitute the given functions. This rational expression cannot be simplified.

Because the equations for f and g do not involve division or contain even roots, the domain of both f and g is the set of all real numbers. Thus, the domain of $f + g, f - g$, and fg is the set of all real numbers, $(-\infty, \infty)$.

The function $\dfrac{f}{g}$ contains division. We must exclude from its domain values of x that cause the denominator, $x^2 + x - 2$, to be 0. Let's identify these values.

$$x^2 + x - 2 = 0 \quad \text{Set the denominator of } \tfrac{f}{g} \text{ equal to 0.}$$

$$(x + 2)(x - 1) = 0 \quad \text{Factor.}$$

$$x + 2 = 0 \quad \text{or} \quad x - 1 = 0 \quad \text{Set each factor equal to 0.}$$

$$x = -2 \qquad\qquad x = 1 \quad \text{Solve the resulting equations.}$$

We must exclude -2 and 1 from the domain of $\dfrac{f}{g}$.

$$\text{Domain of } \dfrac{f}{g} = (-\infty, -2) \cup (-2, 1) \cup (1, \infty)$$ • • •

a. $(f + g)(x) = x^2 + x - 6; (-\infty, \infty)$

b. $(f - g)(x) = -x^2 + x + 4; (-\infty, \infty)$

c. $(fg)(x) = x^3 - 5x^2 - x + 5; (-\infty, \infty)$

d. $\left(\dfrac{f}{g}\right)(x) = \dfrac{x - 5}{x^2 - 1};$
$(-\infty, -1) \cup (-1, 1) \cup (1, \infty)$

✅ Check Point **2** Let $f(x) = x - 5$ and $g(x) = x^2 - 1$. Find each of the following functions:

 a. $(f + g)(x)$ **b.** $(f - g)(x)$ **c.** $(fg)(x)$ **d.** $\left(\dfrac{f}{g}\right)(x)$.

Determine the domain for each function.

EXAMPLE 3 Adding Functions and Determining the Domain

Let $f(x) = \sqrt{x + 3}$ and $g(x) = \sqrt{x - 2}$. Find each of the following:

 a. $(f + g)(x)$ **b.** the domain of $f + g$.

SOLUTION

a. $(f + g)(x) = f(x) + g(x) = \sqrt{x + 3} + \sqrt{x - 2}$

b. The domain of $f + g$ is the set of all real numbers that are common to the domain of f and the domain of g. Thus, we must find the domains of f and g before finding their intersection.

 • $f(x) = \sqrt{x + 3}$ • $g(x) = \sqrt{x - 2}$

$x + 3$ must be nonnegative:
$x + 3 \ge 0$. $D_f = [-3, \infty)$

$x - 2$ must be nonnegative:
$x - 2 \ge 0$. $D_g = [2, \infty)$

FIGURE 2.56 Finding the domain of the sum $f + g$

Now, we can use a number line to determine $D_f \cap D_g$, the domain of $f + g$. **Figure 2.56** shows the domain of f, $[-3, \infty)$, in blue and the domain of g, $[2, \infty)$, in red. Can you see that all real numbers greater than or equal to 2 are common to both domains? This is shown in purple on the number line. Thus, the domain of $f + g$ is $[2, \infty)$. ●●●

TECHNOLOGY

Graphic Connections

The graph on the left is the graph of

$$y = \sqrt{x + 3} + \sqrt{x - 2}$$

in a $[-3, 10, 1]$ by $[0, 8, 1]$ viewing rectangle. The graph reveals what we discovered algebraically in Example 3(b). The domain of this function is $[2, \infty)$.

Domain: $[2, \infty)$

 Check Point **3** Let $f(x) = \sqrt{x - 3}$ and $g(x) = \sqrt{x + 1}$. Find each of the following:

a. $(f + g)(x)$ $(f + g)(x) = \sqrt{x - 3} + \sqrt{x + 1}$ **b.** the domain of $f + g$. $[3, \infty)$

EXAMPLE 4 Applying the Algebra of Functions

We opened the section with functions that model the number of births and deaths in the United States from 2000 through 2011:

$$B(x) = -2.6x^2 + 49x + 3994 \qquad D(x) = -0.6x^2 + 7x + 2412.$$

Number of births, $B(x)$, in thousands, x years after 2000

Number of deaths, $D(x)$, in thousands, x years after 2000

a. Write a function that models the change in U.S. population for the years from 2000 through 2011.

b. Use the function from part (a) to find the change in U.S. population in 2008.

c. Does the result in part (b) overestimate or underestimate the actual population change in 2008 obtained from the data in **Figure 2.54** on page 298? By how much?

SOLUTION

a. The change in population is the number of births minus the number of deaths. Thus, we will find the difference function, $B - D$.

$(B - D)(x)$

$= B(x) - D(x)$

$= (-2.6x^2 + 49x + 3994) - (-0.6x^2 + 7x + 2412)$ Substitute the given functions.

$= -2.6x^2 + 49x + 3994 + 0.6x^2 - 7x - 2412$ Remove parentheses and change the sign of each term in the second set of parentheses.

$= (-2.6x^2 + 0.6x^2) + (49x - 7x) + (3994 - 2412)$ Group like terms.

$= -2x^2 + 42x + 1582$ Combine like terms.

The function

$$(B - D)(x) = -2x^2 + 42x + 1582$$

models the change in U.S. population, in thousands, x years after 2000.

b. Because 2008 is 8 years after 2000, we substitute 8 for x in the difference function $(B - D)(x)$.

$$(B - D)(x) = -2x^2 + 42x + 1582 \qquad \text{Use the difference function } B - D.$$
$$(B - D)(8) = -2(8)^2 + 42(8) + 1582 \qquad \text{Substitute 8 for } x.$$
$$= -2(64) + 42(8) + 1582 \qquad \text{Evaluate the exponential expression: } 8^2 = 64.$$
$$= -128 + 336 + 1582 \qquad \text{Perform the multiplications.}$$
$$= 1790 \qquad \text{Add from left to right.}$$

We see that $(B - D)(8) = 1790$. The model indicates that there was a population increase of 1790 thousand, or approximately 1,790,000 people, in 2008.

c. The data for 2008 in **Figure 2.54** on page 298 show 4247 thousand births and 2453 thousand deaths.

$$\text{population change} = \text{births} - \text{deaths}$$
$$= 4247 - 2453 = 1794$$

The actual population increase was 1794 thousand, or 1,794,000. Our model gave us an increase of 1790 thousand. Thus, the model underestimates the actual increase by $1794 - 1790$, or 4 thousand people. • • •

◯ Check Point **4** Use the birth and death models from Example 4.

a. Write a function that models the total number of births and deaths in the United States for the years from 2000 through 2011. $(B + D)(x) = -3.2x^2 + 56x + 6406$

b. Use the function from part (a) to find the total number of births and deaths in the United States in 2003. 6545.2 thousand

c. Does the result in part (b) overestimate or underestimate the actual number of total births and deaths in 2003 obtained from the data in **Figure 2.54** on page 298? By how much? overestimates by 7.2 thousand

③ Form composite functions.

Composite Functions

There is another way of combining two functions. To help understand this new combination, suppose that your local computer store is having a sale. The models that are on sale cost either $300 less than the regular price or 85% of the regular price. If x represents the computer's regular price, the discounts can be modeled with the following functions:

$$f(x) = x - 300 \qquad g(x) = 0.85x.$$

| The computer is on sale for $300 less than its regular price. | The computer is on sale for 85% of its regular price. |

At the store, you bargain with the salesperson. Eventually, she makes an offer you can't refuse. The sale price will be 85% of the regular price followed by a $300 reduction:

$$0.85x - 300.$$

| 85% of the regular price | followed by a $300 reduction |

In terms of the functions f and g, this offer can be obtained by taking the output of $g(x) = 0.85x$, namely, $0.85x$, and using it as the input of f:

$$f(x) = x - 300$$

> Replace x with $0.85x$, the output of $g(x) = 0.85x$.

$$f(0.85x) = 0.85x - 300.$$

Because $0.85x$ is $g(x)$, we can write this last equation as

$$f(g(x)) = 0.85x - 300.$$

We read this equation as "f of g of x is equal to $0.85x - 300$." We call $f(g(x))$ the **composition of the function f with g**, or a **composite function**. This composite function is written $f \circ g$. Thus,

$$(f \circ g)(x) = f(g(x)) = 0.85x - 300.$$

> This can be read "f of g of x" or "f composed with g of x."

Like all functions, we can evaluate $f \circ g$ for a specified value of x in the function's domain. For example, here's how to find the value of the composite function describing the offer you cannot refuse at 1400:

$$(f \circ g)(x) = 0.85x - 300$$

> Replace x with 1400.

$$(f \circ g)(1400) = 0.85(1400) - 300 = 1190 - 300 = 890.$$

This means that a computer that regularly sells for $1400 is on sale for $890 subject to both discounts. We can use a partial table of coordinates for each of the discount functions, g and f, to verify this result numerically.

Computer's regular price	85% of the regular price		85% of the regular price	$300 reduction
x	$g(x) = 0.85x$		x	$f(x) = x - 300$
1200	1020		1020	720
1300	1105		1105	805
1400	1190		1190	890

Using these tables, we can find $(f \circ g)(1400)$:

$$(f \circ g)(1400) = f(g(1400)) = f(1190) = 890.$$

> The table for g shows that $g(1400) = 1190$. The table for f shows that $f(1190) = 890$.

This verifies that a computer that regularly sells for $1400 is on sale for $890 subject to both discounts.

Before you run out to buy a computer, let's generalize our discussion of the computer's double discount and define the composition of any two functions.

The Composition of Functions

The **composition of the function f with g** is denoted by $f \circ g$ and is defined by the equation

$$(f \circ g)(x) = f(g(x)).$$

The **domain of the composite function $f \circ g$** is the set of all x such that

1. x is in the domain of g and
2. $g(x)$ is in the domain of f.

The composition of f with g, $f \circ g$, is illustrated in **Figure 2.57**.

Step 1 Input x into g. **Step 2** Input $g(x)$ into f.

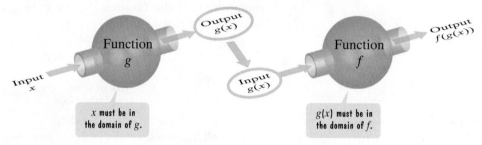

FIGURE 2.57

The figure reinforces the fact that the inside function g in $f(g(x))$ is done first.

> **EXAMPLE 5** Forming Composite Functions

Given $f(x) = 3x - 4$ and $g(x) = x^2 - 2x + 6$, find each of the following:

a. $(f \circ g)(x)$ **b.** $(g \circ f)(x)$ **c.** $(g \circ f)(1)$.

SOLUTION

a. We begin with $(f \circ g)(x)$, the composition of f with g. Because $(f \circ g)(x)$ means $f(g(x))$, we must replace each occurrence of x in the equation for f with $g(x)$.

$$f(x) = 3x - 4 \qquad \text{This is the given equation for } f.$$

Replace x with $g(x)$.

$$(f \circ g)(x) = f(g(x)) = 3g(x) - 4$$
$$= 3(x^2 - 2x + 6) - 4 \qquad \begin{array}{l}\text{Because } g(x) = x^2 - 2x + 6,\\ \text{replace } g(x) \text{ with } x^2 - 2x + 6.\end{array}$$
$$= 3x^2 - 6x + 18 - 4 \qquad \text{Use the distributive property.}$$
$$= 3x^2 - 6x + 14 \qquad \text{Simplify.}$$

Thus, $(f \circ g)(x) = 3x^2 - 6x + 14$.

b. Next, we find $(g \circ f)(x)$, the composition of g with f. Because $(g \circ f)(x)$ means $g(f(x))$, we must replace each occurrence of x in the equation for g with $f(x)$.

$$g(x) = x^2 - 2x + 6 \qquad \text{This is the equation for } g.$$

Replace x with $f(x)$.

$$(g \circ f)(x) = g(f(x)) = (f(x))^2 - 2f(x) + 6$$
$$= (3x - 4)^2 - 2(3x - 4) + 6 \qquad \begin{array}{l}\text{Because } f(x) = 3x - 4,\\ \text{replace } f(x) \text{ with } 3x - 4.\end{array}$$

$$= 9x^2 - 24x + 16 - 6x + 8 + 6 \qquad \text{Use } (A - B)^2 =$$
$$A^2 - 2AB + B^2$$
$$\text{to square } 3x - 4.$$

$$= 9x^2 - 30x + 30 \qquad \text{Simplify:}$$
$$-24x - 6x = -30x$$
$$\text{and } 16 + 8 + 6 = 30.$$

Thus, $(g \circ f)(x) = 9x^2 - 30x + 30$. **Notice that $(f \circ g)(x)$ is not the same function as $(g \circ f)(x)$.**

c. We can use $(g \circ f)(x)$ to find $(g \circ f)(1)$.

$$(g \circ f)(x) = 9x^2 - 30x + 30$$

Replace x with 1.

$$(g \circ f)(1) = 9 \cdot 1^2 - 30 \cdot 1 + 30 = 9 - 30 + 30 = 9$$

It is also possible to find $(g \circ f)(1)$ without determining $(g \circ f)(x)$.

$$(g \circ f)(1) = g(f(1)) = g(-1) = 9$$

First find $f(1)$.	Next find $g(-1)$.
$f(x) = 3x - 4$, so	$g(x) = x^2 - 2x + 6$, so
$f(1) = 3 \cdot 1 - 4 = -1$.	$g(-1) = (-1)^2 - 2(-1) + 6$
	$= 1 + 2 + 6 = 9$.

• • •

a. $(f \circ g)(x) = 10x^2 - 5x + 1$
b. $(g \circ f)(x) = 50x^2 + 115x + 65$

✓ **Check Point 5** Given $f(x) = 5x + 6$ and $g(x) = 2x^2 - x - 1$, find each of the following:

a. $(f \circ g)(x)$ b. $(g \circ f)(x)$ c. $(f \circ g)(-1)$. 16

④ Determine domains for composite functions.

We need to be careful in determining the domain for a composite function.

Excluding Values from the Domain of $(f \circ g)(x) = f(g(x))$

The following values must be excluded from the input x:

• If x is not in the domain of g, it must not be in the domain of $f \circ g$.
• Any x for which $g(x)$ is not in the domain of f must not be in the domain of $f \circ g$.

EXAMPLE 6 Forming a Composite Function and Finding Its Domain

A Brief Review •
Simplifying Complex Fractions

One method for simplifying the complex fraction on the right is to find the least common denominator of all the rational expressions in the numerator and the denominator. Then multiply each term in the numerator and denominator by this least common denominator. The procedures for simplifying complex fractions can be found in Section P.6, Objective 6.

Given $f(x) = \dfrac{2}{x - 1}$ and $g(x) = \dfrac{3}{x}$, find each of the following:

a. $(f \circ g)(x)$ b. the domain of $f \circ g$.

SOLUTION

a. Because $(f \circ g)(x)$ means $f(g(x))$, we must replace x in $f(x) = \dfrac{2}{x - 1}$ with $g(x)$.

$$(f \circ g)(x) = f(g(x)) = \frac{2}{g(x) - 1} = \frac{2}{\dfrac{3}{x} - 1} = \frac{2}{\dfrac{3}{x} - 1} \cdot \frac{x}{x} = \frac{2x}{3 - x}$$

$g(x) = \dfrac{3}{x}$

Simplify the complex fraction by multiplying by $\dfrac{x}{x}$, or 1.

Thus, $(f \circ g)(x) = \dfrac{2x}{3 - x}$.

b. We determine values to exclude from the domain of $(f \circ g)(x)$ in two steps.

Rules for Excluding Numbers from the Domain of $(f \circ g)(x) = f(g(x))$	Applying the Rules to $f(x) = \dfrac{2}{x-1}$ and $g(x) = \dfrac{3}{x}$
If x is not in the domain of g, it must not be in the domain of $f \circ g$.	Because $g(x) = \dfrac{3}{x}$, 0 is not in the domain of g. Thus, 0 must be excluded from the domain of $f \circ g$.
Any x for which $g(x)$ is not in the domain of f must not be in the domain of $f \circ g$.	Because $f(g(x)) = \dfrac{2}{g(x)-1}$, we must exclude from the domain of $f \circ g$ any x for which $g(x) = 1$. $\dfrac{3}{x} = 1$ Set $g(x)$ equal to 1. $3 = x$ Multiply both sides by x. 3 must be excluded from the domain of $f \circ g$.

We see that 0 and 3 must be excluded from the domain of $f \circ g$. The domain of $f \circ g$ is

$$(-\infty, 0) \cup (0, 3) \cup (3, \infty).$$

• • •

 Check Point 6 Given $f(x) = \dfrac{4}{x+2}$ and $g(x) = \dfrac{1}{x}$, find each of the following:

a. $(f \circ g)(x)$ $(f \circ g)(x) = \frac{4x}{1+2x}$ **b.** the domain of $f \circ g$. $\left(-\infty, -\frac{1}{2}\right) \cup \left(-\frac{1}{2}, 0\right) \cup (0, \infty)$

⑤ Write functions as compositions.

Decomposing Functions

When you form a composite function, you "compose" two functions to form a new function. It is also possible to reverse this process. That is, you can "decompose" a given function and express it as a composition of two functions. Although there is more than one way to do this, there is often a "natural" selection that comes to mind first. For example, consider the function h defined by

$$h(x) = (3x^2 - 4x + 1)^5.$$

The function h takes $3x^2 - 4x + 1$ and raises it to the power 5. A natural way to write h as a composition of two functions is to raise the function $g(x) = 3x^2 - 4x + 1$ to the power 5. Thus, if we let

$$f(x) = x^5 \text{ and } g(x) = 3x^2 - 4x + 1, \text{ then}$$
$$(f \circ g)(x) = f(g(x)) = f(3x^2 - 4x + 1) = (3x^2 - 4x + 1)^5.$$

GREAT QUESTION!

Is there a procedure I can use to write a function as a composition?

Yes. Suppose the form of function h is $h(x) = (\text{algebraic expression})^{\text{power}}$. Function h can be expressed as a composition, $f \circ g$, using

$f(x) = x^{\text{power}}$

$g(x) = \text{algebraic expression}$.

EXAMPLE 7 Writing a Function as a Composition

Express $h(x)$ as a composition of two functions:

$$h(x) = \sqrt[3]{x^2 + 1}.$$

SOLUTION

The function h takes $x^2 + 1$ and takes its cube root. A natural way to write h as a composition of two functions is to take the cube root of the function $g(x) = x^2 + 1$. Thus, we let

$$f(x) = \sqrt[3]{x} \text{ and } g(x) = x^2 + 1.$$

We can check this composition by finding $(f \circ g)(x)$. This should give the original function, namely, $h(x) = \sqrt[3]{x^2 + 1}$.

$$(f \circ g)(x) = f(g(x)) = f(x^2 + 1) = \sqrt[3]{x^2 + 1} = h(x)$$ • • •

✓ Check Point **7** Express $h(x)$ as a composition of two functions:

$$h(x) = \sqrt{x^2 + 5}.$$
If $f(x) = \sqrt{x}$ and $g(x) = x^2 + 5$, then $h(x) = (f \circ g)(x)$.

ACHIEVING SUCCESS

Learn from your mistakes. Being human means making mistakes. By finding and understanding your errors, you will become a better math student.

Source of Error	Remedy
Not Understanding a Concept	Review the concept by finding a similar example in your textbook or class notes. Ask your professor questions to help clarify the concept.
Skipping Steps	Show clear step-by-step solutions. Detailed solution procedures help organize your thoughts and enhance understanding. Doing too many steps mentally often results in preventable mistakes.
Carelessness	Write neatly. Not being able to read your own math writing leads to errors. Avoid writing in pen so you won't have to put huge marks through incorrect work.

"You can achieve your goal if you persistently pursue it."

—Cha Sa-Soon, a 68-year-old South Korean woman who passed her country's written driver's-license exam on her 950th try (Source: *Newsweek*)

CONCEPT AND VOCABULARY CHECK

Fill in each blank so that the resulting statement is true.

1. We exclude from a function's domain real numbers that cause division by __zero__.

2. We exclude from a function's domain real numbers that result in a square root of a/an __negative__ number.

3. $(f + g)(x) = $ __$f(x) + g(x)$__

4. $(f - g)(x) = $ __$f(x) - g(x)$__

5. $(fg)(x) = $ __$f(x) \cdot g(x)$__

6. $\dfrac{f}{g}(x) = $ __$\dfrac{f(x)}{g(x)}$__ , provided __$g(x)$__ $\neq 0$

7. The domain of $f(x) = 5x + 7$ consists of all real numbers, represented in interval notation as __$(-\infty, \infty)$__.

8. The domain of $g(x) = \dfrac{3}{x - 2}$ consists of all real numbers except 2, represented in interval notation as $(-\infty, 2) \cup$ __$(2, \infty)$__.

9. The domain of $h(x) = \dfrac{1}{x} + \dfrac{7}{x - 3}$ consists of all real numbers except 0 and 3, represented in interval notation as $(-\infty, 0) \cup$ __$(0, 3)$__ \cup __$(3, \infty)$__.

10. The notation $f \circ g$, called the __composition__ of the function f with g, is defined by $(f \circ g)(x) = $ __$f(g(x))$__.

11. I find $(f \circ g)(x)$ by replacing each occurrence of x in the equation for __f__ with __$g(x)$__.

12. The notation $g \circ f$, called the __composition__ of the function g with f, is defined by $(g \circ f)(x) = $ __$g(f(x))$__.

13. I find $(g \circ f)(x)$ by replacing each occurrence of x in the equation for __g__ with __$f(x)$__.

14. True or false: $f \circ g$ is the same function as $g \circ f$. __false__

15. True or false: $f(g(x)) = f(x) \cdot g(x)$ __false__

16. If $f(g(x)) = \dfrac{3}{g(x) - 4}$ and $g(x) = \dfrac{8}{x}$, then 0 and __2__ must be excluded from the domain of $f \circ g$.

EXERCISE SET 2.6

Practice Exercises

In Exercises 1–30, find the domain of each function.

1. $f(x) = 3(x - 4)$ $(-\infty, \infty)$

2. $f(x) = 2(x + 5)$ $(-\infty, \infty)$

3. $g(x) = \dfrac{3}{x - 4}$ *

4. $g(x) = \dfrac{2}{x + 5}$ *

5. $f(x) = x^2 - 2x - 15$ *

6. $f(x) = x^2 + x - 12$ *

7. $g(x) = \dfrac{3}{x^2 - 2x - 15}$ *

8. $g(x) = \dfrac{2}{x^2 + x - 12}$ *

9. $f(x) = \dfrac{1}{x + 7} + \dfrac{3}{x - 9}$ *

10. $f(x) = \dfrac{1}{x + 8} + \dfrac{3}{x - 10}$ *

11. $g(x) = \dfrac{1}{x^2 + 1} - \dfrac{1}{x^2 - 1}$ $(-\infty, -1) \cup (-1, 1) \cup (1, \infty)$

12. $g(x) = \dfrac{1}{x^2 + 4} - \dfrac{1}{x^2 - 4}$ $(-\infty, -2) \cup (-2, 2) \cup (2, \infty)$

13. $h(x) = \dfrac{4}{\dfrac{3}{x} - 1}$ *

14. $h(x) = \dfrac{5}{\dfrac{4}{x} - 1}$ *

15. $f(x) = \dfrac{1}{\dfrac{4}{x - 1} - 2}$ *

16. $f(x) = \dfrac{1}{\dfrac{4}{x - 2} - 3}$ *

17. $f(x) = \sqrt{x - 3}$ $[3, \infty)$

18. $f(x) = \sqrt{x + 2}$ $[-2, \infty)$

19. $g(x) = \dfrac{1}{\sqrt{x - 3}}$ $(3, \infty)$

20. $g(x) = \dfrac{1}{\sqrt{x + 2}}$ $(-2, \infty)$

21. $g(x) = \sqrt{5x + 35}$ $[-7, \infty)$

22. $g(x) = \sqrt{7x - 70}$ $[10, \infty)$

23. $f(x) = \sqrt{24 - 2x}$ $(-\infty, 12]$

24. $f(x) = \sqrt{84 - 6x}$ $(-\infty, 14]$

25. $h(x) = \sqrt{x - 2} + \sqrt{x + 3}$ $[2, \infty)$

26. $h(x) = \sqrt{x - 3} + \sqrt{x + 4}$ $[3, \infty)$

27. $g(x) = \dfrac{\sqrt{x - 2}}{x - 5}$ *

28. $g(x) = \dfrac{\sqrt{x - 3}}{x - 6}$ *

29. $f(x) = \dfrac{2x + 7}{x^3 - 5x^2 - 4x + 20}$ $(-\infty, -2) \cup (-2, 2) \cup (2, 5) \cup (5, \infty)$

30. $f(x) = \dfrac{7x + 2}{x^3 - 2x^2 - 9x + 18}$ $(-\infty, -3) \cup (-3, 2) \cup (2, 3) \cup (3, \infty)$

In Exercises 31–50, find $f + g$, $f - g$, fg, and $\frac{f}{g}$. Determine the domain for each function.

31. $f(x) = 2x + 3, g(x) = x - 1$ *

32. $f(x) = 3x - 4, g(x) = x + 2$ *

33. $f(x) = x - 5, g(x) = 3x^2$ *

34. $f(x) = x - 6, g(x) = 5x^2$ *

35. $f(x) = 2x^2 - x - 3, g(x) = x + 1$ *

36. $f(x) = 6x^2 - x - 1, g(x) = x - 1$ *

37. $f(x) = 3 - x^2, g(x) = x^2 + 2x - 15$ *

38. $f(x) = 5 - x^2, g(x) = x^2 + 4x - 12$ *

39. $f(x) = \sqrt{x}, g(x) = x - 4$ *

40. $f(x) = \sqrt{x}, g(x) = x - 5$ *

41. $f(x) = 2 + \dfrac{1}{x}, g(x) = \dfrac{1}{x}$ *

42. $f(x) = 6 - \dfrac{1}{x}, g(x) = \dfrac{1}{x}$ *

43. $f(x) = \dfrac{5x + 1}{x^2 - 9}, g(x) = \dfrac{4x - 2}{x^2 - 9}$ *

44. $f(x) = \dfrac{3x + 1}{x^2 - 25}, g(x) = \dfrac{2x - 4}{x^2 - 25}$ *

45. $f(x) = \dfrac{8x}{x - 2}, g(x) = \dfrac{6}{x + 3}$ *

46. $f(x) = \dfrac{9x}{x - 4}, g(x) = \dfrac{7}{x + 8}$ *

47. $f(x) = \sqrt{x + 4}, g(x) = \sqrt{x - 1}$ *

48. $f(x) = \sqrt{x + 6}, g(x) = \sqrt{x - 3}$ *

49. $f(x) = \sqrt{x - 2}, g(x) = \sqrt{2 - x}$ *

50. $f(x) = \sqrt{x - 5}, g(x) = \sqrt{5 - x}$ *

In Exercises 51–66, find

 a. $(f \circ g)(x)$ **b.** $(g \circ f)(x)$ **c.** $(f \circ g)(2)$ **d.** $(g \circ f)(2)$.

51. $f(x) = 2x, g(x) = x + 7$ **a.** $2x + 14$ **b.** $2x + 7$ **c.** 18 **d.** 11

52. $f(x) = 3x, g(x) = x - 5$ **a.** $3x - 15$ **b.** $3x - 5$ **c.** -9 **d.** 1

53. $f(x) = x + 4, g(x) = 2x + 1$ **a.** $2x + 5$ **b.** $2x + 9$ **c.** 9 **d.** 13

54. $f(x) = 5x + 2, g(x) = 3x - 4$ **a.** $15x - 18$ **b.** $15x + 2$ **c.** 12 **d.** 32

55. $f(x) = 4x - 3, g(x) = 5x^2 - 2$ **c.** 69 **d.** 123*

56. $f(x) = 7x + 1, g(x) = 2x^2 - 9$ **c.** -6 **d.** 441*

57. $f(x) = x^2 + 2, g(x) = x^2 - 2$ **c.** 6 **d.** 34*

58. $f(x) = x^2 + 1, g(x) = x^2 - 3$ **c.** 2 **d.** 22*

59. $f(x) = 4 - x, g(x) = 2x^2 + x + 5$ **c.** -11 **d.** 15*

60. $f(x) = 5x - 2, g(x) = -x^2 + 4x - 1$ **c.** 13 **d.** -33*

61. $f(x) = \sqrt{x}, g(x) = x - 1$ **c.** 1 **d.** $\sqrt{2} - 1$*

62. $f(x) = \sqrt{x}, g(x) = x + 2$ **c.** 2 **d.** $\sqrt{2} + 2$*

63. $f(x) = 2x - 3, g(x) = \dfrac{x + 3}{2}$ **a.** x **b.** x **c.** 2 **d.** 2

64. $f(x) = 6x - 3, g(x) = \dfrac{x + 3}{6}$ **a.** x **b.** x **c.** 2 **d.** 2

65. $f(x) = \dfrac{1}{x}, g(x) = \dfrac{1}{x}$ **a.** x **b.** x **c.** 2 **d.** 2

66. $f(x) = \dfrac{2}{x}, g(x) = \dfrac{2}{x}$ **a.** x **b.** x **c.** 2 **d.** 2

In Exercises 67–74, find

 a. $(f \circ g)(x)$ **b.** *the domain of $f \circ g$.*

67. $f(x) = \dfrac{2}{x + 3}, g(x) = \dfrac{1}{x}$ **a.** $(f \circ g)(x) = \dfrac{2x}{1 + 3x}$*

68. $f(x) = \dfrac{5}{x + 4}, g(x) = \dfrac{1}{x}$ **a.** $(f \circ g)(x) = \dfrac{5x}{1 + 4x}$*

69. $f(x) = \dfrac{x}{x + 1}, g(x) = \dfrac{4}{x}$ **a.** $(f \circ g)(x) = \dfrac{4}{4 + x}$*

70. $f(x) = \dfrac{x}{x + 5}, g(x) = \dfrac{6}{x}$ **a.** $(f \circ g)(x) = \dfrac{6}{6 + 5x}$*

71. $f(x) = \sqrt{x}, g(x) = x - 2$ **a.** $(f \circ g)(x) = \sqrt{x - 2}$ **b.** $[2, \infty)$

72. $f(x) = \sqrt{x}, g(x) = x - 3$ **a.** $(f \circ g)(x) = \sqrt{x - 3}$ **b.** $[3, \infty)$

73. $f(x) = x^2 + 4, g(x) = \sqrt{1 - x}$ **a.** $(f \circ g)(x) = 5 - x*$

74. $f(x) = x^2 + 1, g(x) = \sqrt{2 - x}$ **a.** $(f \circ g)(x) = 3 - x*$

In Exercises 75–82, express the given function h as a composition of two functions f and g so that $h(x) = (f \circ g)(x)$.

75. $h(x) = (3x - 1)^4$ *

76. $h(x) = (2x - 5)^3$ *

77. $h(x) = \sqrt[3]{x^2 - 9}$ *

78. $h(x) = \sqrt{5x^2 + 3}$ *

79. $h(x) = |2x - 5|$ *

80. $h(x) = |3x - 4|$ *

81. $h(x) = \dfrac{1}{2x - 3}$ *

82. $h(x) = \dfrac{1}{4x + 5}$ *

Practice Plus

Use the graphs of f and g to solve Exercises 83–90.

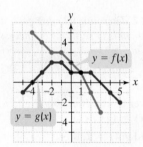

83. Find $(f + g)(-3)$. 5

84. Find $(g - f)(-2)$. −1

85. Find $(fg)(2)$. −1

86. Find $\left(\dfrac{g}{f}\right)(3)$. 0

87. Find the domain of $f + g$. *

88. Find the domain of $\dfrac{f}{g}$. *

89. Graph $f + g$. *

90. Graph $f - g$. *

In Exercises 91–94, use the graphs of f and g to evaluate each composite function.

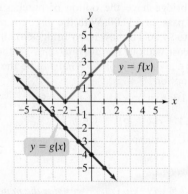

91. $(f \circ g)(-1)$ 1

92. $(f \circ g)(1)$ 3

93. $(g \circ f)(0)$ −6

94. $(g \circ f)(-1)$ −5

In Exercises 95–96, find all values of x satisfying the given conditions.

95. $f(x) = 2x - 5, g(x) = x^2 - 3x + 8$, and $(f \circ g)(x) = 7$. *

96. $f(x) = 1 - 2x, g(x) = 3x^2 + x - 1$, and $(f \circ g)(x) = -5$. *

Application Exercises

The bar graph shows the population of the United States, in millions, for seven selected years.

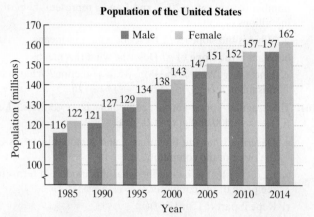

Population of the United States

Source: U.S. Census Bureau

Here are two functions that model the data:

$M(x) = 1.48x + 115.1$ — Male U.S. population, $M(x)$, in millions, x years after 1985

$F(x) = 1.44x + 120.9$ — Female U.S. population, $F(x)$, in millions, x years after 1985

Use the functions to solve Exercises 97–98.

97. a. Write a function that models the total U.S. population for the years shown in the bar graph. $(M + F)(x) = 2.92x + 236$

 b. Use the function from part (a) to find the total U.S. population in 2010. 309 million

 c. How well does the result in part (b) model the actual total U.S. population in 2010 shown by the bar graph? It's the same.

98. a. Write a function that models the difference between the female U.S. population and the male U.S. population for the years shown in the bar graph. $(M - F)(x) = -0.04x + 5.8$

 b. Use the function from part (a) to find how many more women than men there were in the U.S. population in 2010. 4.8 million

 c. Does the result in part (b) overestimate or underestimate the actual difference between the female and male population in 2010 shown by the bar graph? By how much? underestimates by 0.2 million

99. A company that sells radios has yearly fixed costs of $600,000. It costs the company $45 to produce each radio. Each radio will sell for $65. The company's costs and revenue are modeled by the following functions, where x represents the number of radios produced and sold:

$C(x) = 600,000 + 45x$ This function models the company's costs.

$R(x) = 65x$. This function models the company's revenue.

Find and interpret $(R - C)(20,000)$, $(R - C)(30,000)$, and $(R - C)(40,000)$. *

100. A department store has two locations in a city. From 2012 through 2016, the profits for each of the store's two branches are modeled by the functions $f(x) = -0.44x + 13.62$ and $g(x) = 0.51x + 11.14$. In each model, x represents the number of years after 2012, and f and g represent the profit, in millions of dollars.

 a. What is the slope of f? Describe what this means. −0.44*

 b. What is the slope of g? Describe what this means. 0.51*

 c. Find $f + g$. What is the slope of this function? What does this mean? $(f + g)(x) = 0.07x + 24.76; m = 0.07*$

101. The regular price of a computer is x dollars. Let $f(x) = x - 400$ and $g(x) = 0.75x$.

 a. Describe what the functions f and g model in terms of the price of the computer. *

 b. Find $(f \circ g)(x)$ and describe what this models in terms of the price of the computer. $(f \circ g)(x) = 0.75x - 400*$

 c. Repeat part (b) for $(g \circ f)(x)$. $(g \circ f)(x) = 0.75(x - 400)*$

 d. Which composite function models the greater discount on the computer, $f \circ g$ or $g \circ f$? Explain. $f \circ g*$

102. The regular price of a pair of jeans is x dollars. Let $f(x) = x - 5$ and $g(x) = 0.6x$.

 a. Describe what functions f and g model in terms of the price of the jeans. *

 b. Find $(f \circ g)(x)$ and describe what this models in terms of the price of the jeans. $(f \circ g)(x) = 0.6x - 5*$

 c. Repeat part (b) for $(g \circ f)(x)$. $(g \circ f)(x) = 0.6(x - 5)*$

 d. Which composite function models the greater discount on the jeans, $f \circ g$ or $g \circ f$? Explain. $f \circ g*$

Explaining the Concepts 103–107. Answers will vary.

103. If a function is defined by an equation, explain how to find its domain.

104. If equations for f and g are given, explain how to find $f - g$.

105. If equations for two functions are given, explain how to obtain the quotient function and its domain.

106. Describe a procedure for finding $(f \circ g)(x)$. What is the name of this function?

107. Describe the values of x that must be excluded from the domain of $(f \circ g)(x)$.

Technology Exercises

108. Graph $y_1 = x^2 - 2x$, $y_2 = x$, and $y_3 = y_1 \div y_2$ in the same $[-10, 10, 1]$ by $[-10, 10, 1]$ viewing rectangle. Then use the ⟨TRACE⟩ feature to trace along y_3. What happens at $x = 0$? Explain why this occurs. *

109. Graph $y_1 = \sqrt{2 - x}$, $y_2 = \sqrt{x}$, and $y_3 = \sqrt{2 - y_2}$ in the same $[-4, 4, 1]$ by $[0, 2, 1]$ viewing rectangle. If y_1 represents f and y_2 represents g, use the graph of y_3 to find the domain of $f \circ g$. Then verify your observation algebraically. *

Critical Thinking Exercises

Make Sense? *In Exercises 110–113, determine whether each statement makes sense or does not make sense, and explain your reasoning.*

110. I used a function to model data from 1990 through 2015. The independent variable in my model represented the number of years after 1990, so the function's domain was $\{x \mid x = 0, 1, 2, 3, \dots, 25\}$. makes sense

111. I have two functions. Function f models total world population x years after 2000 and function g models population of the world's more-developed regions x years after 2000. I can use $f - g$ to determine the population of the world's less-developed regions for the years in both function's domains. makes sense

112. I must have made a mistake in finding the composite functions $f \circ g$ and $g \circ f$, because I notice that $f \circ g$ is not the same function as $g \circ f$. does not make sense

113. This diagram illustrates that $f(g(x)) = x^2 + 4$. does not make sense

In Exercises 114–117, determine whether each statement is true or false. If the statement is false, make the necessary change(s) to produce a true statement. Changes to false statements will vary.

114. If $f(x) = x^2 - 4$ and $g(x) = \sqrt{x^2 - 4}$, then $(f \circ g)(x) = -x^2$ and $(f \circ g)(5) = -25$. false

115. There can never be two functions f and g, where $f \neq g$, for which $(f \circ g)(x) = (g \circ f)(x)$. false

116. If $f(7) = 5$ and $g(4) = 7$, then $(f \circ g)(4) = 35$. false

117. If $f(x) = \sqrt{x}$ and $g(x) = 2x - 1$, then $(f \circ g)(5) = g(2)$. *

118. Prove that if f and g are even functions, then fg is also an even function. *

119. Define two functions f and g so that $f \circ g = g \circ f$. *

Retaining the Concepts

120. Solve and check: $\dfrac{x - 1}{5} - \dfrac{x + 3}{2} = 1 - \dfrac{x}{4}$.

 (Section 1.2, Example 3) $\{-54\}$

121. The toll to a bridge costs $6.00. Commuters who frequently use the bridge have the option of purchasing a monthly discount pass for $30.00. With the discount pass, the toll is reduced to $4.00. For how many bridge crossings per month will the cost without the discount pass be the same as the cost with the discount pass? What will be the monthly cost for each option?

 (Section 1.3, Example 3) 15 crossings; $90

122. Solve for y: $Ax + By = Cy + D$.

 (Section 1.3, Example 8) $y = \dfrac{D}{A + B - C}$

Preview Exercises

Exercises 123–125 will help you prepare for the material covered in the next section.

123. Consider the function defined by

$$\{(-2, 4), (-1, 1), (1, 1), (2, 4)\}.$$

Reverse the components of each ordered pair and write the resulting relation. Is this relation a function? no*

124. Solve for y: $x = \dfrac{5}{y} + 4$. $y = \dfrac{5}{x - 4}$

125. Solve for y: $x = y^2 - 1, y \geq 0$. $y = \sqrt{x + 1}$

Section 2.7

Inverse Functions

What am I supposed to learn?

After studying this section, you should be able to:

1. Verify inverse functions.
2. Find the inverse of a function.
3. Use the horizontal line test to determine if a function has an inverse function.
4. Use the graph of a one-to-one function to graph its inverse function.
5. Find the inverse of a function and graph both functions on the same axes.

Based on Shakespeare's *Romeo and Juliet*, the film *West Side Story* swept the 1961 Academy Awards with ten Oscars. The top four movies to win the most Oscars are shown in **Table 2.6**.

Table 2.6 Films Winning the Most Oscars

Movie	Year	Number of Academy Awards
Ben-Hur	1960	11
Titanic	1998	11
The Lord of the Rings: The Return of the King	2003	11
West Side Story	1961	10

Source: Russell Ash, *The Top 10 of Everything, 2011*

We can use the information in **Table 2.6** to define a function. Let the domain of the function be the set of four movies shown in the table. Let the range be the number of Academy Awards for each of the respective films. The function can be written as follows:

f: {(*Ben-Hur*, 11), (*Titanic*, 11), (*The Lord of the Rings*, 11), (*West Side Story*, 10)}.

Now let's "undo" f by interchanging the first and second components in each of the ordered pairs. Switching the inputs and outputs of f, we obtain the following relation:

Same first component

Undoing f: {(11, *Ben-Hur*), (11, *Titanic*), (11, *The Lord of the Rings*), (10, *West Side Story*)}.

Different second components

Can you see that this relation is not a function? Three of its ordered pairs have the same first component and different second components. This violates the definition of a function.

If a function f is a set of ordered pairs, (x, y), then the changes produced by f can be "undone" by reversing the components of all the ordered pairs. The resulting relation, (y, x), may or may not be a function. In this section, we will develop these ideas by studying functions whose compositions have a special "undoing" relationship.

Inverse Functions

Here are two functions that describe situations related to the price of a computer, x:

$$f(x) = x - 300 \qquad g(x) = x + 300.$$

Function f subtracts \$300 from the computer's price and function g adds \$300 to the computer's price. Let's see what $f(g(x))$ does. Put $g(x)$ into f:

$$f(x) = x - 300 \qquad \text{This is the given equation for } f.$$

> Replace x with $g(x)$.

$$f(g(x)) = g(x) - 300$$
$$= x + 300 - 300 \qquad \text{Because } g(x) = x + 300,$$
$$\text{replace } g(x) \text{ with } x + 300.$$

> This is the computer's original price.

$$= x.$$

Using $f(x) = x - 300$ and $g(x) = x + 300$, we see that $f(g(x)) = x$. By putting $g(x)$ into f and finding $f(g(x))$, the computer's price, x, went through two changes: the first, an increase; the second, a decrease:

$$x + 300 - 300.$$

The final price of the computer, x, is identical to its starting price, x.

In general, if the changes made to x by a function g are undone by the changes made by a function f, then

$$f(g(x)) = x.$$

Assume, also, that this "undoing" takes place in the other direction:

$$g(f(x)) = x.$$

Under these conditions, we say that each function is the *inverse function* of the other. The fact that g is the inverse of f is expressed by renaming g as f^{-1}, read "f-inverse." For example, the inverse functions

$$f(x) = x - 300 \qquad g(x) = x + 300$$

are usually named as follows:

$$f(x) = x - 300 \qquad f^{-1}(x) = x + 300.$$

We can use partial tables of coordinates for f and f^{-1} to gain numerical insight into the relationship between a function and its inverse function.

Computer's regular price	\$300 reduction	Price with \$300 reduction	\$300 price increase
x	$f(x) = x - 300$	x	$f^{-1}(x) = x + 300$
1200	900	900	1200
1300	1000	1000	1300
1400	1100	1100	1400

Ordered pairs for f:
(1200, 900), (1300, 1000), (1400, 1100)

Ordered pairs for f^{-1}:
(900, 1200), (1000, 1300), (1100, 1400)

The tables illustrate that if a function f is the set of ordered pairs (x, y), then its inverse, f^{-1}, is the set of ordered pairs (y, x). Using these tables, we can see how one function's changes to x are undone by the other function:

$$(f^{-1} \circ f)(1300) = f^{-1}(f(1300)) = f^{-1}(1000) = 1300.$$

> The table for f shows that $f(1300) = 1000.$

> The table for f^{-1} shows that $f^{-1}(1000) = 1300.$

The final price of the computer, \$1300, is identical to its starting price, \$1300.

With these ideas in mind, we present the formal definition of the inverse of a function:

Definition of the Inverse of a Function

Let f and g be two functions such that

$$f(g(x)) = x \qquad \text{for every } x \text{ in the domain of } g$$

and

$$g(f(x)) = x \qquad \text{for every } x \text{ in the domain of } f.$$

The function g is the **inverse of the function f** and is denoted by f^{-1} (read "f-inverse"). Thus, $f(f^{-1}(x)) = x$ and $f^{-1}(f(x)) = x$. The domain of f is equal to the range of f^{-1}, and vice versa.

① Verify inverse functions.

EXAMPLE 1 Verifying Inverse Functions

Show that each function is the inverse of the other:

$$f(x) = 3x + 2 \quad \text{and} \quad g(x) = \frac{x-2}{3}.$$

SOLUTION

To show that f and g are inverses of each other, we must show that $f(g(x)) = x$ and $g(f(x)) = x$. We begin with $f(g(x))$.

$$f(x) = 3x + 2 \quad \text{This is the equation for } f.$$

Replace x with $g(x)$.

$$f(g(x)) = 3g(x) + 2 = 3\left(\frac{x-2}{3}\right) + 2 = (x-2) + 2 = x$$

$$g(x) = \frac{x-2}{3}$$

Next, we find $g(f(x))$.

$$g(x) = \frac{x-2}{3} \quad \text{This is the equation for } g.$$

Replace x with $f(x)$.

$$g(f(x)) = \frac{f(x) - 2}{3} = \frac{(3x+2) - 2}{3} = \frac{3x}{3} = x$$

$$f(x) = 3x + 2$$

Because g is the inverse of f (and vice versa), we can use inverse notation and write

$$f(x) = 3x + 2 \quad \text{and} \quad f^{-1}(x) = \frac{x-2}{3}.$$

Notice how f^{-1} undoes the changes produced by f: f changes x by *multiplying* by 3 and *adding* 2, and f^{-1} undoes this by *subtracting* 2 and *dividing* by 3. This "undoing" process is illustrated in **Figure 2.58**. ●●●

FIGURE 2.58 f^{-1} undoes the changes produced by f.

In the figure: f — Multiply by 3. Add 2. $x \to 3x \to 3x+2$; f^{-1} — Subtract 2. Divide by 3. $x \to 3x \to 3x+2$

✓ **Check Point 1** Show that each function is the inverse of the other:

$$f(x) = 4x - 7 \quad \text{and} \quad g(x) = \frac{x+7}{4}.$$

$$f(g(x)) = 4\left(\frac{x+7}{4}\right) - 7 = x + 7 - 7 = x; \quad g(f(x)) = \frac{(4x-7) + 7}{4} = \frac{4x}{4} = x$$

② Find the inverse of a function.

Finding the Inverse of a Function

The definition of the inverse of a function tells us that the domain of f is equal to the range of f^{-1}, and vice versa. This means that if the function f is the set of ordered pairs (x, y), then the inverse of f is the set of ordered pairs (y, x). If a function is defined by an equation, we can obtain the equation for f^{-1}, the inverse of f, by interchanging the role of x and y in the equation for the function f.

Finding the Inverse of a Function

The equation for the inverse of a function f can be found as follows:

1. Replace $f(x)$ with y in the equation for $f(x)$.
2. Interchange x and y.
3. Solve for y. If this equation does not define y as a function of x, the function f does not have an inverse function and this procedure ends. If this equation does define y as a function of x, the function f has an inverse function.
4. If f has an inverse function, replace y in step 3 by $f^{-1}(x)$. We can verify our result by showing that $f(f^{-1}(x)) = x$ and $f^{-1}(f(x)) = x$.

The procedure for finding a function's inverse uses a *switch-and-solve* strategy. Switch x and y, and then solve for y.

EXAMPLE 2 Finding the Inverse of a Function

Find the inverse of $f(x) = 7x - 5$.

SOLUTION

Step 1 Replace $f(x)$ with y:

$$y = 7x - 5.$$

Step 2 Interchange x and y:

$$x = 7y - 5.$$ This is the inverse function.

Step 3 Solve for y:

$$x + 5 = 7y$$ Add 5 to both sides.

$$\frac{x + 5}{7} = y.$$ Divide both sides by 7.

DISCOVERY

In Example 2, we found that if $f(x) = 7x - 5$, then

$$f^{-1}(x) = \frac{x + 5}{7}.$$

Verify this result by showing that

$$f(f^{-1}(x)) = x$$

and

$$f^{-1}(f(x)) = x.$$

Step 4 Replace y with $f^{-1}(x)$:

$$f^{-1}(x) = \frac{x + 5}{7}.$$ The equation is written with f^{-1} on the left.

Thus, the inverse of $f(x) = 7x - 5$ is $f^{-1}(x) = \dfrac{x + 5}{7}$.

The inverse function, f^{-1}, undoes the changes produced by f. f changes x by multiplying by 7 and subtracting 5. f^{-1} undoes this by adding 5 and dividing by 7. • • •

✓ Check Point **2** Find the inverse of $f(x) = 2x + 7$. $f^{-1}(x) = \frac{x-7}{2}$

EXAMPLE 3 Finding the Inverse of a Function

Find the inverse of $f(x) = x^3 + 1$.

SOLUTION

Step 1 Replace $f(x)$ with y: $y = x^3 + 1$.

Step 2 Interchange x and y: $x = y^3 + 1$.

Step 3 Solve for y:

> Our goal is to isolate y. Because $\sqrt[3]{y^3} = y$, we will take the cube root of both sides of the equation.

$$x - 1 = y^3 \qquad \text{Subtract 1 from both sides.}$$

$$\sqrt[3]{x - 1} = \sqrt[3]{y^3} \qquad \text{Take the cube root on both sides.}$$

$$\sqrt[3]{x - 1} = y. \qquad \text{Simplify.}$$

Step 4 Replace y with $f^{-1}(x)$: $f^{-1}(x) = \sqrt[3]{x - 1}$.

Thus, the inverse of $f(x) = x^3 + 1$ is $f^{-1}(x) = \sqrt[3]{x - 1}$. • • •

☑ **Check Point 3** Find the inverse of $f(x) = 4x^3 - 1$. $f^{-1}(x) = \sqrt[3]{\dfrac{x + 1}{4}}$

EXAMPLE 4 Finding the Inverse of a Function

Find the inverse of $f(x) = \dfrac{x + 2}{x - 3}, x \neq 3$.

SOLUTION

Step 1 Replace $f(x)$ with y:

$$y = \frac{x + 2}{x - 3}.$$

Step 2 Interchange x and y:

$$x = \frac{y + 2}{y - 3}.$$

GREAT QUESTION!

When it comes to finding the inverse of a function, does step 3, solving for y, always involve the most amount of work?

Yes. Solving for y is the most challenging step in the four-step procedure.

Step 3 Solve for y:

$$x = \frac{y + 2}{y - 3} \qquad \text{This is the equation from step 2.}$$

$$x(y - 3) = \left(\frac{y + 2}{y - 3}\right)(y - 3) \qquad \begin{array}{l}\text{Clear fractions. Multiply both sides} \\ \text{by } y - 3, y \neq 3.\end{array}$$

$$x(y - 3) = y + 2 \qquad \text{Simplify: } \left(\dfrac{y + 2}{y - 3}\right)(y - 3) = y + 2.$$

> Our goal is to isolate y.

$$xy - 3x = y + 2 \qquad \text{Use the distributive property.}$$

$$xy - y = 3x + 2 \qquad \begin{array}{l}\text{Collect the terms with } y \text{ on the left side:} \\ \text{Subtract } y \text{ from both sides and add } 3x \\ \text{to both sides.}\end{array}$$

$$y(x - 1) = 3x + 2 \qquad \begin{array}{l}\text{Factor out } y \text{ from } xy - y \text{ to obtain a single} \\ \text{occurrence of } y.\end{array}$$

$$\frac{y(x - 1)}{x - 1} = \frac{3x + 2}{x - 1} \qquad \text{Divide both sides by } x - 1, x \neq 1.$$

$$y = \frac{3x + 2}{x - 1}. \qquad \text{Simplify.}$$

Step 4 **Replace y with $f^{-1}(x)$:**

$$f^{-1}(x) = \frac{3x + 2}{x - 1}.$$

Thus, the inverse of $f(x) = \dfrac{x + 2}{x - 3}$ is $f^{-1}(x) = \dfrac{3x + 2}{x - 1}.$

• • •

◯ **Check Point 4** Find the inverse of $f(x) = \dfrac{x + 1}{x - 5}, x \neq 5.$ $\quad f^{-1}(x) = \frac{5x + 1}{x - 1}$

③ Use the horizontal line test to determine if a function has an inverse function.

The Horizontal Line Test and One-to-One Functions

Let's see what happens if we try to find the inverse of the standard quadratic function, $f(x) = x^2$.

Step 1 **Replace $f(x)$ with y:** $y = x^2$.

Step 2 **Interchange x and y:** $x = y^2$.

Step 3 **Solve for y:** We apply the square root property to solve $y^2 = x$ for y. We obtain

$$y = \pm\sqrt{x}.$$

The \pm in $y = \pm\sqrt{x}$ shows that for certain values of x (all positive real numbers), there are two values of y. Because this equation does not represent y as a function of x, the standard quadratic function $f(x) = x^2$ does not have an inverse function.

We can use a few of the solutions of $y = x^2$ to illustrate numerically that this function does not have an inverse:

A function provides exactly one output for each input. Thus, the ordered pairs in the bottom row do not define a function.

Can we look at the graph of a function and tell if it represents a function with an inverse? Yes. The graph of the standard quadratic function $f(x) = x^2$ is shown in **Figure 2.59**. Four units above the x-axis, a horizontal line is drawn. This line intersects the graph at two of its points, $(-2, 4)$ and $(2, 4)$. Inverse functions have ordered pairs with the coordinates reversed. We just saw what happened when we interchanged x and y. We obtained $(4, -2)$ and $(4, 2)$, and these ordered pairs do not define a function.

If any horizontal line, such as the one in **Figure 2.59**, intersects a graph at two or more points, the set of these points will not define a function when their coordinates are reversed. This suggests the **horizontal line test** for inverse functions.

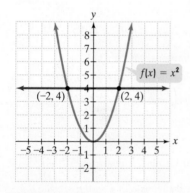

FIGURE 2.59 The horizontal line intersects the graph twice.

DISCOVERY

How might you restrict the domain of $f(x) = x^2$, graphed in **Figure 2.59**, so that the remaining portion of the graph passes the horizontal line test?

The Horizontal Line Test for Inverse Functions

A function f has an inverse that is a function, f^{-1}, if there is no horizontal line that intersects the graph of the function f at more than one point.

EXAMPLE 5 Applying the Horizontal Line Test

Which of the following graphs represent functions that have inverse functions?

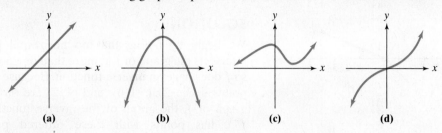

(a) (b) (c) (d)

SOLUTION

Notice that horizontal lines can be drawn in graphs **(b)** and **(c)** that intersect the graphs more than once. These graphs do not pass the horizontal line test. These are not the graphs of functions with inverse functions. By contrast, no horizontal line can be drawn in graphs **(a)** and **(d)** that intersects the graphs more than once. These graphs pass the horizontal line test. Thus, the graphs in parts **(a)** and **(d)** represent functions that have inverse functions.

Has an inverse function No inverse function No inverse function Has an inverse function
(a) (b) (c) (d) • • •

✓ Check Point **5** Which of the following graphs represent functions that have inverse functions? (b) and (c)

(a) (b) (c)

A function passes the horizontal line test when no two different ordered pairs have the same second component. This means that if $x_1 \neq x_2$, then $f(x_1) \neq f(x_2)$. Such a function is called a **one-to-one function**. Thus, **a one-to-one function is a function in which no two different ordered pairs have the same second component. Only one-to-one functions have inverse functions.** Any function that passes the horizontal line test is a one-to-one function. Any one-to-one function has a graph that passes the horizontal line test.

④ Use the graph of a one-to-one function to graph its inverse function.

Graphs of f and f⁻¹

There is a relationship between the graph of a one-to-one function, f, and its inverse, f^{-1}. Because inverse functions have ordered pairs with the coordinates interchanged, if the point (a, b) is on the graph of f, then the point (b, a) is on the graph of f^{-1}. The points (a, b) and (b, a) are symmetric with respect to the line $y = x$. Thus, **the graph of f^{-1} is a reflection of the graph of f about the line $y = x$.** This is illustrated in **Figure 2.60**.

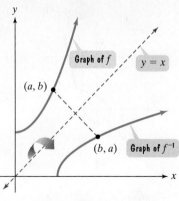

FIGURE 2.60 The graph of f^{-1} is a reflection of the graph of f about $y = x$.

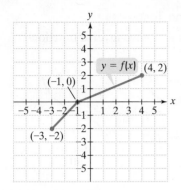

FIGURE 2.61

EXAMPLE 6 Graphing the Inverse Function

Use the graph of f in **Figure 2.61** to draw the graph of its inverse function.

SOLUTION

We begin by noting that no horizontal line intersects the graph of f at more than one point, so f does have an inverse function. Because the points $(-3, -2), (-1, 0)$, and $(4, 2)$ are on the graph of f, the graph of the inverse function, f^{-1}, has points with these ordered pairs reversed. Thus, $(-2, -3), (0, -1)$, and $(2, 4)$ are on the graph of f^{-1}. We can use these points to graph f^{-1}. The graph of f^{-1} is shown in green in **Figure 2.62**. Note that the green graph of f^{-1} is the reflection of the blue graph of f about the line $y = x$.

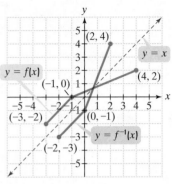

FIGURE 2.62 The graphs of f and f^{-1}

• • •

\oslash Check Point **6** The graph of function f consists of two line segments, one segment from $(-2, -2)$ to $(-1, 0)$ and a second segment from $(-1, 0)$ to $(1, 2)$. Graph f and use the graph to draw the graph of its inverse function. ⁎

⑤ Find the inverse of a function and graph both functions on the same axes.

In our final example, we will first find f^{-1}. Then we will graph f and f^{-1} in the same rectangular coordinate system.

EXAMPLE 7 Finding the Inverse of a Domain-Restricted Function

Find the inverse of $f(x) = x^2 - 1$ if $x \geq 0$. Graph f and f^{-1} in the same rectangular coordinate system.

SOLUTION

The graph of $f(x) = x^2 - 1$ is the graph of the standard quadratic function shifted vertically down 1 unit. **Figure 2.63** shows the function's graph. This graph fails the horizontal line test, so the function $f(x) = x^2 - 1$ does not have an inverse function. By restricting the domain to $x \geq 0$, as given, we obtain a new function whose graph is shown in red in **Figure 2.63**. This red portion of the graph is increasing on the interval $(0, \infty)$ and passes the horizontal line test. This tells us that $f(x) = x^2 - 1$ has an inverse function if we restrict its domain to $x \geq 0$. We use our four-step procedure to find this inverse function. Begin with $f(x) = x^2 - 1, x \geq 0$.

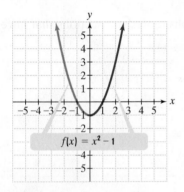

FIGURE 2.63

Step 1 **Replace $f(x)$ with y:** $y = x^2 - 1, x \geq 0$.

Step 2 **Interchange x and y:** $x = y^2 - 1, y \geq 0$.

Step 3 **Solve for y:**

$$x = y^2 - 1, y \geq 0 \quad \text{This is the equation from step 2.}$$

$$x + 1 = y^2 \quad \text{Add 1 to both sides.}$$

$$\sqrt{x + 1} = y \quad \text{Apply the square root property.}$$

> Because $y \geq 0$, take only the principal square root and not the negative square root.

FIGURE 2.64

Step 4 **Replace y with $f^{-1}(x)$:** $f^{-1}(x) = \sqrt{x + 1}$.

Thus, the inverse of $f(x) = x^2 - 1, x \geq 0$, is $f^{-1}(x) = \sqrt{x + 1}$. The graphs of f and f^{-1} are shown in **Figure 2.64**. We obtained the graph of $f^{-1}(x) = \sqrt{x + 1}$ by shifting the graph of the square root function, $y = \sqrt{x}$, horizontally to the left 1 unit. Note that the green graph of f^{-1} is the reflection of the red graph of f about the line $y = x$.

• • •

⁎See Answers to Selected Exercises.

⌀ **Check Point 7** Find the inverse of $f(x) = x^2 + 1$ if $x \geq 0$. Graph f and f^{-1} in the same rectangular coordinate system. $f^{-1}(x) = \sqrt{x-1}$*

ACHIEVING SUCCESS

Organizing and creating your own compact chapter summaries can reinforce what you know and help with the retention of this information. Imagine that your professor will permit two index cards of notes (3 by 5; front and back) on all exams. Organize and create such a two-card summary for the test on this chapter. Begin by determining what information you would find most helpful to include on the cards. Take as long as you need to create the summary. Based on how effective you find this strategy, you may decide to use the technique to help prepare for future exams.

CONCEPT AND VOCABULARY CHECK

Fill in each blank so that the resulting statement is true.

1. The notation f^{-1} means the ___inverse___ of the function f.

2. If the function g is the inverse of the function f, then $f(g(x)) = $ __x__ and $g(f(x)) = $ __x__.

3. A function f has an inverse that is a function if there is no ___horizontal___ line that intersects the graph of f at more than one point. Such a function is called a/an ___one-to-one___ function.

4. The graph of f^{-1} is a reflection of the graph of f about the line whose equation is ___$y = x$___.

EXERCISE SET 2.7

Practice Exercises

In Exercises 1–10, find $f(g(x))$ and $g(f(x))$ and determine whether each pair of functions f and g are inverses of each other.

1. $f(x) = 4x$ and $g(x) = \dfrac{x}{4}$ $f(g(x)) = x; g(f(x)) = x$; inverses

2. $f(x) = 6x$ and $g(x) = \dfrac{x}{6}$ $f(g(x)) = x; g(f(x)) = x$; inverses

3. $f(x) = 3x + 8$ and $g(x) = \dfrac{x-8}{3}$ $f(g(x)) = x; g(f(x)) = x$*

4. $f(x) = 4x + 9$ and $g(x) = \dfrac{x-9}{4}$ $f(g(x)) = x; g(f(x)) = x$*

5. $f(x) = 5x - 9$ and $g(x) = \dfrac{x+5}{9}$ not inverses*

6. $f(x) = 3x - 7$ and $g(x) = \dfrac{x+3}{7}$ not inverses*

7. $f(x) = \dfrac{3}{x-4}$ and $g(x) = \dfrac{3}{x} + 4$ $f(g(x)) = x; g(f(x)) = x$*

8. $f(x) = \dfrac{2}{x-5}$ and $g(x) = \dfrac{2}{x} + 5$ $f(g(x)) = x; g(f(x)) = x$*

9. $f(x) = -x$ and $g(x) = -x$ $f(g(x)) = x; g(f(x)) = x$; inverses

10. $f(x) = \sqrt[3]{x-4}$ and $g(x) = x^3 + 4$ $f(g(x)) = x; g(f(x)) = x$*

The functions in Exercises 11–28 are all one-to-one. For each function,

 a. Find an equation for $f^{-1}(x)$, the inverse function.

 b. Verify that your equation is correct by showing that $f(f^{-1}(x)) = x$ and $f^{-1}(f(x)) = x$.

11. $f(x) = x + 3$ $f^{-1}(x) = x - 3$ **12.** $f(x) = x + 5$ $f^{-1}(x) = x - 5$

13. $f(x) = 2x$ $f^{-1}(x) = \dfrac{x}{2}$ **14.** $f(x) = 4x$ $f^{-1}(x) = \dfrac{x}{4}$

15. $f(x) = 2x + 3$ $f^{-1}(x) = \dfrac{x-3}{2}$ **16.** $f(x) = 3x - 1$ $f^{-1}(x) = \dfrac{x+1}{3}$

17. $f(x) = x^3 + 2$ $f^{-1}(x) = \sqrt[3]{x-2}$

*See Answers to Selected Exercises.

18. $f(x) = x^3 - 1$ $f^{-1}(x) = \sqrt[3]{x+1}$

19. $f(x) = (x + 2)^3$ $f^{-1}(x) = \sqrt[3]{x} - 2$

20. $f(x) = (x - 1)^3$ $f^{-1}(x) = \sqrt[3]{x} + 1$

21. $f(x) = \dfrac{1}{x}$ $f^{-1}(x) = \dfrac{1}{x}$ **22.** $f(x) = \dfrac{2}{x}$ $f^{-1}(x) = \dfrac{2}{x}$

23. $f(x) = \sqrt{x}$ $f^{-1}(x) = x^2, x \geq 0$ **24.** $f(x) = \sqrt[3]{x}$ $f^{-1}(x) = x^3$

25. $f(x) = \dfrac{x+4}{x-2}$ $f^{-1}(x) = \dfrac{2x+4}{x-1}, x \neq 1$

26. $f(x) = \dfrac{x+5}{x-6}$ $f^{-1}(x) = \dfrac{6x+5}{x-1}, x \neq 1$

27. $f(x) = \dfrac{2x+1}{x-3}$ $f^{-1}(x) = \dfrac{3x+1}{x-2}, x \neq 2$

28. $f(x) = \dfrac{2x-3}{x+1}$ $f^{-1}(x) = \dfrac{-x-3}{x-2}, x \neq 2$

Which graphs in Exercises 29–34 represent functions that have inverse functions?

29. *

30. *

31. *

32. *

33. **34.**

In Exercises 35–38, use the graph of f to draw the graph of its inverse function.

35. **36.**

37. **38.**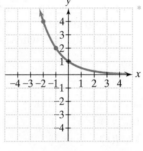

In Exercises 39–52,

 a. Find an equation for $f^{-1}(x)$.

 b. Graph f and f^{-1} in the same rectangular coordinate system.

 c. Use interval notation to give the domain and the range of f and f^{-1}.

39. $f(x) = 2x - 1$ a. $f^{-1}(x) = \dfrac{x+1}{2}$*

40. $f(x) = 2x - 3$ a. $f^{-1}(x) = \dfrac{x+3}{2}$*

41. $f(x) = x^2 - 4, x \geq 0$ a. $f^{-1}(x) = \sqrt{x+4}$*

42. $f(x) = x^2 - 1, x \leq 0$ a. $f^{-1}(x) = -\sqrt{x+1}$*

43. $f(x) = (x-1)^2, x \leq 1$ a. $f^{-1}(x) = -\sqrt{x}+1$*

44. $f(x) = (x-1)^2, x \geq 1$ a. $f^{-1}(x) = \sqrt{x}+1$*

45. $f(x) = x^3 - 1$ a. $f^{-1}(x) = \sqrt[3]{x+1}$*

46. $f(x) = x^3 + 1$ a. $f^{-1}(x) = \sqrt[3]{x-1}$*

47. $f(x) = (x+2)^3$ a. $f^{-1}(x) = \sqrt[3]{x}-2$*

48. $f(x) = (x-2)^3$ a. $f^{-1}(x) = \sqrt[3]{x}+2$*

(Hint for Exercises 49–52: To solve for a variable involving an nth root, raise both sides of the equation to the nth power: $(\sqrt[n]{y})^n = y$.)

49. $f(x) = \sqrt{x-1}$ a. $f^{-1}(x) = x^2 + 1, x \geq 0$*

50. $f(x) = \sqrt{x}+2$ a. $f^{-1}(x) = (x-2)^2, x \geq 2$*

51. $f(x) = \sqrt[3]{x}+1$ a. $f^{-1}(x) = (x-1)^3$*

52. $f(x) = \sqrt[3]{x-1}$ a. $f^{-1}(x) = x^3 + 1$*

Practice Plus

In Exercises 53–58, f and g are defined by the following tables. Use the tables to evaluate each composite function.

x	f(x)
−1	1
0	4
1	5
2	−1

x	g(x)
−1	0
1	1
4	2
10	−1

53. $f(g(1))$ 5 **54.** $f(g(4))$ −1
55. $(g \circ f)(-1)$ 1 **56.** $(g \circ f)(0)$ 2
57. $f^{-1}(g(10))$ 2 **58.** $f^{-1}(g(1))$ −1

In Exercises 59–64, let

$$f(x) = 2x - 5$$
$$g(x) = 4x - 1$$
$$h(x) = x^2 + x + 2.$$

Evaluate the indicated function without finding an equation for the function.

59. $(f \circ g)(0)$ −7 **60.** $(g \circ f)(0)$ −21
61. $f^{-1}(1)$ 3 **62.** $g^{-1}(7)$ 2
63. $g(f[h(1)])$ 11 **64.** $f(g[h(1)])$ 25

Application Exercises

In most societies, women say they prefer to marry men who are older than themselves, whereas men say they prefer women who are younger. The bar graph shows the preferred age difference in a mate in five selected countries. Use this information to solve Exercises 65–66.

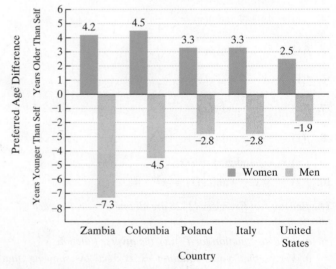

Source: Carole Wade and Carol Tavris, *Psychology*, 6th Edition, Prentice Hall, 2000.

65. a. Consider a function, f, whose domain is the set of the five countries shown in the graph. Let the range be the set of the average number of years women in each of

the respective countries prefer men who are older than themselves. Write the function f as a set of ordered pairs. *

b. Write the relation that is the inverse of f as a set of ordered pairs. Based on these ordered pairs, is f a one-to-one function? Explain your answer. not one-to-one*

66. a. Consider a function, g, whose domain is the set of the five countries shown in the graph. Let the range be the set of negative numbers representing the average number of years men in each of the respective countries prefer women who are younger than themselves. Write the function g as a set of ordered pairs. *

b. Write the relation that is the inverse of g as a set of ordered pairs. Based on these ordered pairs, is g a one-to-one function? Explain your answer. not one-to-one*

67. The graph represents the probability of two people in the same room sharing a birthday as a function of the number of people in the room. Call the function f.

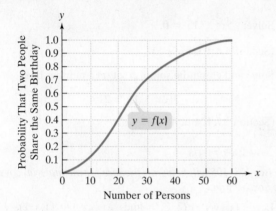

a. Explain why f has an inverse that is a function. *

b. Describe in practical terms the meaning of $f^{-1}(0.25), f^{-1}(0.5),$ and $f^{-1}(0.7)$. *

68. A study of 900 working women in Texas showed that their feelings changed throughout the day. As the graph indicates, the women felt better as time passed, except for a blip (that's slang for relative maximum) at lunchtime.

Average Level of Happiness at Different Times of Day

Source: D. Kahneman et al., "A Survey Method for Characterizing Daily Life Experience," *Science.*

a. Does the graph have an inverse that is a function? Explain your answer. no*

b. Identify two or more times of day when the average happiness level is 3. Express your answers as ordered pairs. 12 P.M. and 7 P.M.; (12, 3) and (19, 3)

c. Do the ordered pairs in part (b) indicate that the graph represents a one-to-one function? Explain your answer. no*

69. The formula

$$y = f(x) = \frac{9}{5}x + 32$$

is used to convert from x degrees Celsius to y degrees Fahrenheit. The formula

$$y = g(x) = \frac{5}{9}(x - 32)$$

is used to convert from x degrees Fahrenheit to y degrees Celsius. Show that f and g are inverse functions. *

Explaining the Concepts 70–75. Answers will vary.

70. Explain how to determine if two functions are inverses of each other.

71. Describe how to find the inverse of a one-to-one function.

72. What is the horizontal line test and what does it indicate?

73. Describe how to use the graph of a one-to-one function to draw the graph of its inverse function.

74. How can a graphing utility be used to visually determine if two functions are inverses of each other?

75. What explanations can you offer for the trends shown by the graph in Exercise 68?

Technology Exercises

In Exercises 76–83, use a graphing utility to graph the function. Use the graph to determine whether the function has an inverse that is a function (that is, whether the function is one-to-one).

76. $f(x) = x^2 - 1$ not one-to-one*

77. $f(x) = \sqrt[3]{2 - x}$ one-to-one*

78. $f(x) = \dfrac{x^3}{2}$ one-to-one*

79. $f(x) = \dfrac{x^4}{4}$ not one-to-one*

80. $f(x) = \text{int}(x - 2)$ not one-to-one*

81. $f(x) = |x - 2|$ not one-to-one*

82. $f(x) = (x - 1)^3$ one-to-one*

83. $f(x) = -\sqrt{16 - x^2}$ not one-to-one*

In Exercises 84–86, use a graphing utility to graph f and g in the same $[-8, 8, 1]$ by $[-5, 5, 1]$ viewing rectangle. In addition, graph the line $y = x$ and visually determine if f and g are inverses.

84. $f(x) = 4x + 4, g(x) = 0.25x - 1$ inverses*

85. $f(x) = \dfrac{1}{x} + 2, g(x) = \dfrac{1}{x - 2}$ inverses*

86. $f(x) = \sqrt[3]{x} - 2, g(x) = (x + 2)^3$ inverses*

Critical Thinking Exercises

Make Sense? *In Exercises 87–90, determine whether each statement makes sense or does not make sense, and explain your reasoning.*

87. I found the inverse of $f(x) = 5x - 4$ in my head: The reverse of multiplying by 5 and subtracting 4 is adding 4 and dividing by 5, so $f^{-1}(x) = \dfrac{x + 4}{5}$. makes sense

88. I'm working with the linear function $f(x) = 3x + 5$ and I do not need to find f^{-1} in order to determine the value of $(f \circ f^{-1})(17)$. makes sense

89. When finding the inverse of a function, I interchange x and y, which reverses the domain and range between the function and its inverse. makes sense

90. I used vertical lines to determine if my graph represents a one-to-one function. does not make sense

In Exercises 91–94, determine whether each statement is true or false. If the statement is false, make the necessary change(s) to produce a true statement. Changes to false statements will vary.

91. The inverse of $\{(1, 4), (2, 7)\}$ is $\{(2, 7), (1, 4)\}$. false

92. The function $f(x) = 5$ is one-to-one. false

93. If $f(x) = 3x$, then $f^{-1}(x) = \dfrac{1}{3x}$. false

94. The domain of f is the same as the range of f^{-1}. true

95. If $f(x) = 3x$ and $g(x) = x + 5$, find $(f \circ g)^{-1}(x)$ and $(g^{-1} \circ f^{-1})(x)$. *

96. Show that
$$f(x) = \frac{3x - 2}{5x - 3}$$
is its own inverse. $(f \circ f)(x) = x$, so f is its own inverse.

97. *Freedom 7* was the spacecraft that carried the first American into space in 1961. Total flight time was 15 minutes and the spacecraft reached a maximum height of 116 miles. Consider a function, s, that expresses *Freedom 7*'s height, $s(t)$, in miles, after t minutes. Is s a one-to-one function? Explain your answer. not one-to-one*

98. If $f(2) = 6$, and f is one-to-one, find x satisfying $8 + f^{-1}(x - 1) = 10$. 7

Group Exercise

99. In Tom Stoppard's play *Arcadia*, the characters dream and talk about mathematics, including ideas involving graphing, composite functions, symmetry, and lack of symmetry in things that are tangled, mysterious, and unpredictable. Group members should read the play. Present a report on the ideas discussed by the characters that are related to concepts that we studied in this chapter. Bring in a copy of the play and read appropriate excerpts. Answers will vary.

Retaining the Concepts

100. Solve by completing the square:
$$2x^2 - 5x + 1 = 0.$$
(Section 1.5, Example 5) $\left\{\dfrac{5 \pm \sqrt{17}}{4}\right\}$

101. Solve: $5x^{\frac{3}{4}} - 15 = 0$.

(Section 1.6, Example 5) $\left\{3^{\frac{4}{3}}\right\}$

102. Solve and graph the solution set on a number line:
$$3|2x - 1| \geq 21.$$
(Section 1.7, Example 10) $(-\infty, -3] \cup [4, \infty)$*

Preview Exercises

Exercises 103–105 will help you prepare for the material covered in the next section.

103. Let $(x_1, y_1) = (7, 2)$ and $(x_2, y_2) = (1, -1)$. Find $\sqrt{(x_2 - x_1)^2 + (y_2 - y_1)^2}$. Express the answer in simplified radical form. $3\sqrt{5}$

104. Use a rectangular coordinate system to graph the circle with center $(1, -1)$ and radius 1. *

105. Solve by completing the square: $y^2 - 6y - 4 = 0$. $\{3 \pm \sqrt{13}\}$

Section 2.8

Distance and Midpoint Formulas; Circles

It's a good idea to know your way around a circle. Clocks, angles, maps, and compasses are based on circles. Circles occur everywhere in nature: in ripples on water, patterns on a butterfly's wings, and cross sections of trees. Some consider the circle to be the most pleasing of all shapes.

The rectangular coordinate system gives us a unique way of knowing a circle. It enables us to translate a circle's geometric definition into an algebraic equation. To do this, we must first develop a formula for the distance between any two points in rectangular coordinates.

The Distance Formula

Using the Pythagorean Theorem, we can find the distance between the two points $P_1(x_1, y_1)$ and $P_2(x_2, y_2)$ in the rectangular coordinate system. The two points are illustrated in **Figure 2.65**.

The distance that we need to find is represented by d and shown in blue. Notice that the distance between the two points on the dashed horizontal line is the absolute value of the difference between the x-coordinates of the points. This distance, $|x_2 - x_1|$, is shown in pink. Similarly, the distance between the two points on the dashed vertical line is the absolute value of the difference between the y-coordinates of the points. This distance, $|y_2 - y_1|$, is also shown in pink.

Because the dashed lines are horizontal and vertical, a right triangle is formed. Thus, we can use the Pythagorean Theorem to find the distance d. Squaring the lengths of the triangle's sides results in positive numbers, so absolute value notation is not necessary.

1. Find the distance between two points.

FIGURE 2.65

$$d^2 = (x_2 - x_1)^2 + (y_2 - y_1)^2$$ Apply the Pythagorean Theorem to the right triangle in Figure 2.65.

$$d = \pm\sqrt{(x_2 - x_1)^2 + (y_2 - y_1)^2}$$ Apply the square root property.

$$d = \sqrt{(x_2 - x_1)^2 + (y_2 - y_1)^2}$$ Because distance is nonnegative, write only the principal square root.

This result is called the **distance formula**.

The Distance Formula

The distance, d, between the points (x_1, y_1) and (x_2, y_2) in the rectangular coordinate system is

$$d = \sqrt{(x_2 - x_1)^2 + (y_2 - y_1)^2}.$$

To compute the distance between two points, find the square of the difference between the x-coordinates plus the square of the difference between the y-coordinates. The principal square root of this sum is the distance.

When using the distance formula, it does not matter which point you call (x_1, y_1) and which you call (x_2, y_2).

EXAMPLE 1 Using the Distance Formula

Find the distance between $(-1, 4)$ and $(3, -2)$. Express the answer in simplified radical form and then round to two decimal places.

SOLUTION

We will let $(x_1, y_1) = (-1, 4)$ and $(x_2, y_2) = (3, -2)$.

$$d = \sqrt{(x_2 - x_1)^2 + (y_2 - y_1)^2}$$ Use the distance formula.

$$= \sqrt{[3 - (-1)]^2 + (-2 - 4)^2}$$ Substitute the given values.

$$= \sqrt{4^2 + (-6)^2}$$ Perform operations inside grouping symbols: $3 - (-1) = 3 + 1 = 4$ and $-2 - 4 = -6$.

$$= \sqrt{16 + 36}$$ Square 4 and -6.

> **Caution: This does not equal $\sqrt{16} + \sqrt{36}$.**

$$= \sqrt{52}$$ Add.

$$= \sqrt{4 \cdot 13} = 2\sqrt{13} \approx 7.21$$ $\sqrt{52} = \sqrt{4 \cdot 13} = \sqrt{4}\sqrt{13} = 2\sqrt{13}$

The distance between the given points is $2\sqrt{13}$ units, or approximately 7.21 units. The situation is illustrated in **Figure 2.66**. • • •

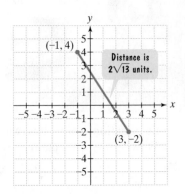

FIGURE 2.66 Finding the distance between two points

A Brief Review • Simplified Radical Form

A square root is simplified when its radicand has no factors other than 1 that are perfect squares. For example, $\sqrt{52}$ is not simplified because it can be expressed as $\sqrt{4 \cdot 13}$ and 4 is a perfect square.

$$\sqrt{52} = \sqrt{4 \cdot 13} = \sqrt{4}\sqrt{13} = 2\sqrt{13}$$

For more detail, see Section P.3, Objective 3.

⊘ **Check Point 1** Find the distance between $(-1, -3)$ and $(2, 3)$. Express the answer in simplified radical form and then round to two decimal places.
$3\sqrt{5} \approx 6.71$

2 Find the midpoint of a line segment.

The Midpoint Formula

The distance formula can be used to derive a formula for finding the midpoint of a line segment between two given points. The formula is given as follows:

The Midpoint Formula

Consider a line segment whose endpoints are (x_1, y_1) and (x_2, y_2). The coordinates of the segment's midpoint are

$$\left(\frac{x_1 + x_2}{2}, \frac{y_1 + y_2}{2} \right).$$

To find the midpoint, take the average of the two x-coordinates and the average of the two y-coordinates.

Do I add or subtract coordinates when applying the midpoint and distance formulas?

The midpoint formula requires finding the *sum* of coordinates. By contrast, the distance formula requires finding the *difference* of coordinates:

Midpoint:
Sum of coordinates

$$\left(\frac{x_1 + x_2}{2}, \frac{y_1 + y_2}{2}\right)$$

Distance:
Difference of coordinates

$$\sqrt{(x_2 - x_1)^2 + (y_2 - y_1)^2}$$

It's easy to confuse the two formulas. Be sure to use addition, not subtraction, when applying the midpoint formula.

EXAMPLE 2 Using the Midpoint Formula

Find the midpoint of the line segment with endpoints $(1, -6)$ and $(-8, -4)$.

SOLUTION

To find the coordinates of the midpoint, we average the coordinates of the endpoints.

$$\text{Midpoint} = \left(\frac{1 + (-8)}{2}, \frac{-6 + (-4)}{2}\right) = \left(\frac{-7}{2}, \frac{-10}{2}\right) = \left(-\frac{7}{2}, -5\right)$$

Average the
x-coordinates.

Average the
y-coordinates.

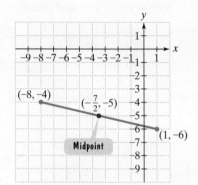

FIGURE 2.67 Finding a line segment's midpoint

Figure 2.67 illustrates that the point $\left(-\frac{7}{2}, -5\right)$ is midway between the points $(1, -6)$ and $(-8, -4)$. • • •

 Check Point **2** Find the midpoint of the line segment with endpoints $(1, 2)$ and $(7, -3)$. $\left(4, -\frac{1}{2}\right)$

Circles

Our goal is to translate a circle's geometric definition into an equation. We begin with this geometric definition.

Definition of a Circle

A **circle** is the set of all points in a plane that are equidistant from a fixed point, called the **center**. The fixed distance from the circle's center to any point on the circle is called the **radius**.

Figure 2.68 is our starting point for obtaining a circle's equation. We've placed the circle into a rectangular coordinate system. The circle's center is (h, k) and its radius is r. We let (x, y) represent the coordinates of any point on the circle.

What does the geometric definition of a circle tell us about the point (x, y) in **Figure 2.68**? The point is on the circle if and only if its distance from the center is r. We can use the distance formula to express this idea algebrgaically:

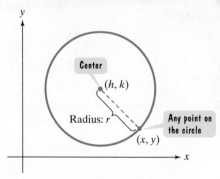

FIGURE 2.68 A circle centered at (h, k) with radius r

The distance between
(x, y) and (h, k) is always r.

$$\sqrt{(x - h)^2 + (y - k)^2} = r.$$

Squaring both sides of $\sqrt{(x-h)^2 + (y-k)^2} = r$ yields the *standard form of the equation of a circle*.

③ Write the standard form of a circle's equation.

The Standard Form of the Equation of a Circle

The **standard form of the equation of a circle** with center (h, k) and radius r is

$$(x-h)^2 + (y-k)^2 = r^2.$$

EXAMPLE 3 Finding the Standard Form of a Circle's Equation

Write the standard form of the equation of the circle with center $(0,0)$ and radius 2. Graph the circle.

SOLUTION

The center is $(0, 0)$. Because the center is represented as (h, k) in the standard form of the equation, $h = 0$ and $k = 0$. The radius is 2, so we will let $r = 2$ in the equation.

$$(x-h)^2 + (y-k)^2 = r^2 \quad \text{This is the standard form of a circle's equation.}$$

$$(x-0)^2 + (y-0)^2 = 2^2 \quad \text{Substitute O for } h, \text{O for } k, \text{ and 2 for } r.$$

$$x^2 + y^2 = 4 \quad \text{Simplify.}$$

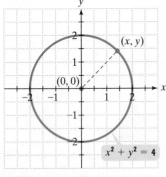

FIGURE 2.69 The graph of $x^2 + y^2 = 4$

The standard form of the equation of the circle is $x^2 + y^2 = 4$. **Figure 2.69** shows the graph. 　　　　●●●

Ⓥ Check Point 3 Write the standard form of the equation of the circle with center $(0,0)$ and radius 4. $x^2 + y^2 = 16$

TECHNOLOGY

$y_1 = \sqrt{4 - x^2}$

$y_2 = -\sqrt{4 - x^2}$

To graph a circle with a graphing utility, first solve the equation for y.

$$x^2 + y^2 = 4$$

$$y^2 = 4 - x^2$$

$$y = \pm\sqrt{4 - x^2} \quad \text{y is not a function of } x.$$

Graph the two equations

$$y_1 = \sqrt{4 - x^2} \quad \text{and} \quad y_2 = -\sqrt{4 - x^2}$$

in the same viewing rectangle. The graph of $y_1 = \sqrt{4 - x^2}$ is the top blue semicircle because y is always positive. The graph of $y_2 = -\sqrt{4 - x^2}$ is the bottom red semicircle because y is always negative. Use a ZOOM SQUARE setting so that the circle looks like a circle. (Many graphing utilities have problems connecting the two semicircles because the segments directly to the left and to the right of the center become nearly vertical.)

Example 3 and Check Point 3 involved circles centered at the origin. The standard form of the equation of all such circles is $x^2 + y^2 = r^2$, where r is the circle's radius. Now, let's consider a circle whose center is not at the origin.

EXAMPLE 4 Finding the Standard Form of a Circle's Equation

Write the standard form of the equation of the circle with center $(-2, 3)$ and radius 4.

SOLUTION

The center is $(-2, 3)$. Because the center is represented as (h, k) in the standard form of the equation, $h = -2$ and $k = 3$. The radius is 4, so we will let $r = 4$ in the equation.

$$(x - h)^2 + (y - k)^2 = r^2$$ This is the standard form of a circle's equation.

$$[x - (-2)]^2 + (y - 3)^2 = 4^2$$ Substitute -2 for h, 3 for k, and 4 for r.

$$(x + 2)^2 + (y - 3)^2 = 16$$ Simplify.

The standard form of the equation of the circle is $(x + 2)^2 + (y - 3)^2 = 16$. • • •

⊘ **Check Point 4** Write the standard form of the equation of the circle with center $(0, -6)$ and radius 10. $x^2 + (y + 6)^2 = 100$

④ Give the center and radius of a circle whose equation is in standard form.

EXAMPLE 5 Using the Standard Form of a Circle's Equation to Graph the Circle

a. Find the center and radius of the circle whose equation is

$$(x - 2)^2 + (y + 4)^2 = 9.$$

b. Graph the equation.
c. Use the graph to identify the relation's domain and range.

SOLUTION

a. We begin by finding the circle's center, (h, k), and its radius, r. We can find the values for h, k, and r by comparing the given equation to the standard form of the equation of a circle, $(x - h)^2 + (y - k)^2 = r^2$.

$$(x - 2)^2 + (y + 4)^2 = 9$$

$$(x - 2)^2 + (y - (-4))^2 = 3^2$$

This is $(x - h)^2$, with $h = 2$. This is $(y - k)^2$, with $k = -4$. This is r^2, with $r = 3$.

We see that $h = 2$, $k = -4$, and $r = 3$. Thus, the circle has center $(h, k) = (2, -4)$ and a radius of 3 units.

b. To graph this circle, first locate the center $(2, -4)$. Because the radius is 3, you can locate at least four points on the circle by going out three units to the right, to the left, up, and down from the center.

The points three units to the right and to the left of $(2, -4)$ are $(5, -4)$ and $(-1, -4)$, respectively. The points three units up and down from $(2, -4)$ are $(2, -1)$ and $(2, -7)$, respectively.

Using these points, we obtain the graph in **Figure 2.70**.

c. The four points that we located on the circle can be used to determine the relation's domain and range. The points $(-1, -4)$ and $(5, -4)$ show that values of x extend from -1 to 5, inclusive:

$$\text{Domain} = [-1, 5].$$

The points $(2, -7)$ and $(2, -1)$ show that values of y extend from -7 to -1, inclusive:

$$\text{Range} = [-7, -1].$$ • • •

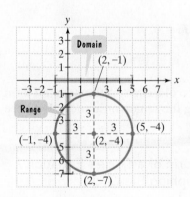

FIGURE 2.70 The graph of $(x - 2)^2 + (y + 4)^2 = 9$

GREAT QUESTION!

What's the bottom line when identifying h and k in the standard form of a circle's equation?

It's easy to make sign errors when finding h and k, the coordinates of a circle's center, (h, k). Keep in mind that h and k are the numbers that *follow the subtraction signs* in a circle's equation:

$$(x - 2)^2 + (y + 4)^2 = 9$$
$$(x - 2)^2 + (y - (-4))^2 = 9.$$

The number *after* the subtraction is 2: $h = 2$.

The number *after* the subtraction is -4: $k = -4$.

a. center: $(-3, 1)$; radius: 2

c. domain: $[-5, -1]$; range: $[-1, 3]$

✓ Check Point 5

a. Find the center and radius of the circle whose equation is
$$(x + 3)^2 + (y - 1)^2 = 4.$$

b. Graph the equation.

c. Use the graph to identify the relation's domain and range.

If we square $x - 2$ and $y + 4$ in the standard form of the equation in Example 5, we obtain another form for the circle's equation.

$$(x - 2)^2 + (y + 4)^2 = 9 \qquad \text{This is the standard form of the equation in Example 5.}$$

$$x^2 - 4x + 4 + y^2 + 8y + 16 = 9 \qquad \text{Square } x - 2 \text{ and } y + 4.$$

$$x^2 + y^2 - 4x + 8y + 20 = 9 \qquad \text{Combine constants and rearrange terms.}$$

$$x^2 + y^2 - 4x + 8y + 11 = 0 \qquad \text{Subtract 9 from both sides.}$$

This result suggests that an equation in the form $x^2 + y^2 + Dx + Ey + F = 0$ can represent a circle. This is called the *general form of the equation of a circle*.

The General Form of the Equation of a Circle

The **general form of the equation of a circle** is
$$x^2 + y^2 + Dx + Ey + F = 0,$$
where D, E, and F are real numbers.

⑤ Convert the general form of a circle's equation to standard form.

We can convert the general form of the equation of a circle to the standard form $(x - h)^2 + (y - k)^2 = r^2$. We do so by completing the square on x and y. Let's see how this is done.

A Brief Review •
Completing the Square

To complete the square, on $x^2 + bx$, take half the coefficient of x. Then square this number. By adding the square of half the coefficient of x, a perfect square trinomial will result. For more detail, see Section 1.5, Objective 3.

EXAMPLE 6 Converting the General Form of a Circle's Equation to Standard Form and Graphing the Circle

Write in standard form and graph: $x^2 + y^2 + 4x - 6y - 23 = 0$.

SOLUTION

Because we plan to complete the square on both x and y, let's rearrange the terms so that x-terms are arranged in descending order, y-terms are arranged in descending order, and the constant term appears on the right.

$$x^2 + y^2 + 4x - 6y - 23 = 0 \qquad \text{This is the given equation.}$$

$$(x^2 + 4x \quad) + (y^2 - 6y \quad) = 23 \qquad \text{Rewrite in anticipation of completing the square.}$$

*See Answers to Selected Exercises.

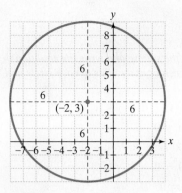

FIGURE 2.71 The graph of $(x + 2)^2 + (y - 3)^2 = 36$

$$(x^2 + 4x + 4) + (y^2 - 6y + 9) = 23 + 4 + 9$$

> Complete the square on x: $\frac{1}{2} \cdot 4 = 2$ and $2^2 = 4$, so add 4 to both sides.
> Complete the square on y: $\frac{1}{2}(-6) = -3$ and $(-3)^2 = 9$, so add 9 to both sides.

> Remember that numbers added on the left side must also be added on the right side.

$$(x + 2)^2 + (y - 3)^2 = 36$$

> Factor on the left and add on the right.

This last equation, $(x + 2)^2 + (y - 3)^2 = 36$, is in standard form. We can identify the circle's center and radius by comparing this equation to the standard form of the equation of a circle, $(x - h)^2 + (y - k)^2 = r^2$.

$$(x + 2)^2 + (y - 3)^2 = 36$$
$$(x - (-2))^2 + (y - 3)^2 = 6^2$$

> This is $(x - h)^2$, with $h = -2$.
> This is $(y - k)^2$, with $k = 3$.
> This is r^2, with $r = 6$.

We use the center, $(h, k) = (-2, 3)$, and the radius, $r = 6$, to graph the circle. The graph is shown in **Figure 2.71**. •••

TECHNOLOGY

To graph $x^2 + y^2 + 4x - 6y - 23 = 0$, the general form of the circle's equation given in Example 6, rewrite the equation as a quadratic equation in y.

$$y^2 - 6y + (x^2 + 4x - 23) = 0$$

Now solve for y using the quadratic formula, with $a = 1$, $b = -6$, and $c = x^2 + 4x - 23$.

$$y = \frac{-b \pm \sqrt{b^2 - 4ac}}{2a} = \frac{-(-6) \pm \sqrt{(-6)^2 - 4 \cdot 1(x^2 + 4x - 23)}}{2 \cdot 1} = \frac{6 \pm \sqrt{36 - 4(x^2 + 4x - 23)}}{2}$$

Because we will enter these equations, there is no need to simplify. Enter

$$y_1 = \frac{6 + \sqrt{36 - 4(x^2 + 4x - 23)}}{2}$$

and

$$y_2 = \frac{6 - \sqrt{36 - 4(x^2 + 4x - 23)}}{2}.$$

Use a ZOOM SQUARE setting. The graph is shown on the right.

✓ Check Point **6** Write in standard form and graph:

$$x^2 + y^2 + 4x - 4y - 1 = 0.$$ $(x + 2)^2 + (y - 2)^2 = 9$*

CONCEPT AND VOCABULARY CHECK

Fill in each blank so that the resulting statement is true.

1. The distance, d, between the points (x_1, y_1) and (x_2, y_2) in the rectangular coordinate system is $d =$ $\underline{\sqrt{(x_2 - x_1)^2 + (y_2 - y_1)^2}}$.

2. The midpoint of a line segment whose endpoints are (x_1, y_1) and (x_2, y_2) is $(\underline{\frac{x_1 + x_2}{2}}, \underline{\frac{y_1 + y_2}{2}})$.

3. The set of all points in a plane that are equidistant from a fixed point is a/an $\underline{\text{circle}}$. The fixed point is called the $\underline{\text{center}}$. The distance from this fixed point to any point on the geometric figure is called the $\underline{\text{radius}}$.

4. The standard form of the equation of a circle with center (h, k) and radius r is $\underline{(x - h)^2 + (y - k)^2 = r^2}$.

5. The equation $x^2 + y^2 + Dx + Ey + F = 0$ is called the $\underline{\text{general}}$ form of the equation of a circle.

6. In the equation $(x^2 + 4x \ \) + (y^2 - 8y \ \) = 5$, we complete the square on x by adding $\underline{4}$ to both sides. We complete the square on y by adding $\underline{16}$ to both sides.

*See Answers to Selected Exercises.

EXERCISE SET 2.8

Practice Exercises

In Exercises 1–18, find the distance between each pair of points. If necessary, express answers in simplified radical form and then round to two decimals places.

1. (2, 3) and (14, 8) 13

2. (5, 1) and (8, 5) 5

3. (4, −1) and (−6, 3) *

4. (2, −3) and (−1, 5) *

5. (0, 0) and (−3, 4) 5

6. (0, 0) and (3, −4) 5

7. (−2, −6) and (3, −4) *

8. (−4, −1) and (2, −3) *

9. (0, −3) and (4, 1) $4\sqrt{2} \approx 5.66$

10. (0, −2) and (4, 3) *

11. (3.5, 8.2) and (−0.5, 6.2) *

12. (2.6, 1.3) and (1.6, −5.7) *

13. $\left(0, -\sqrt{3}\right)$ and $\left(\sqrt{5}, 0\right)$ $2\sqrt{2} \approx 2.83$

14. $\left(0, -\sqrt{2}\right)$ and $\left(\sqrt{7}, 0\right)$ 3

15. $\left(3\sqrt{3}, \sqrt{5}\right)$ and $\left(-\sqrt{3}, 4\sqrt{5}\right)$ $\sqrt{93} \approx 9.64$

16. $\left(2\sqrt{3}, \sqrt{6}\right)$ and $\left(-\sqrt{3}, 5\sqrt{6}\right)$ $\sqrt{123} \approx 11.09$

17. $\left(\dfrac{7}{3}, \dfrac{1}{5}\right)$ and $\left(\dfrac{1}{3}, \dfrac{6}{5}\right)$ $\sqrt{5} \approx 2.24$

18. $\left(-\dfrac{1}{4}, -\dfrac{1}{7}\right)$ and $\left(\dfrac{3}{4}, \dfrac{6}{7}\right)$ *

In Exercises 19–30, find the midpoint of each line segment with the given endpoints.

19. (6, 8) and (2, 4) (4, 6)

20. (10, 4) and (2, 6) (6, 5)

21. (−2, −8) and (−6, −2) *

22. (−4, −7) and (−1, −3) *

23. (−3, −4) and (6, −8) $(\frac{3}{2}, -6)$

24. (−2, −1) and (−8, 6) $(-5, \frac{5}{2})$

25. $\left(-\dfrac{7}{2}, \dfrac{3}{2}\right)$ and $\left(-\dfrac{5}{2}, -\dfrac{11}{2}\right)$ $(-3, -2)$

26. $\left(-\dfrac{2}{5}, \dfrac{7}{15}\right)$ and $\left(-\dfrac{2}{5}, -\dfrac{4}{15}\right)$ $(-\frac{2}{5}, \frac{1}{10})$

27. $\left(8, 3\sqrt{5}\right)$ and $\left(-6, 7\sqrt{5}\right)$ $(1, 5\sqrt{5})$

28. $\left(7\sqrt{3}, -6\right)$ and $\left(3\sqrt{3}, -2\right)$ $(5\sqrt{3}, -4)$

29. $\left(\sqrt{18}, -4\right)$ and $\left(\sqrt{2}, 4\right)$ $(2\sqrt{2}, 0)$

30. $\left(\sqrt{50}, -6\right)$ and $\left(\sqrt{2}, 6\right)$ $(3\sqrt{2}, 0)$

In Exercises 31–40, write the standard form of the equation of the circle with the given center and radius.

31. Center (0, 0), $r = 7$ *

32. Center (0, 0), $r = 8$ *

33. Center (3, 2), $r = 5$ *

34. Center (2, −1), $r = 4$ *

35. Center (−1, 4), $r = 2$

36. Center (−3, 5), $r = 3$

37. Center (−3, −1), $r = \sqrt{3}$ *

38. Center (−5, −3), $r = \sqrt{5}$ *

39. Center (−4, 0), $r = 10$ *

40. Center (−2, 0), $r = 6$ *

In Exercises 41–52, give the center and radius of the circle described by the equation and graph each equation. Use the graph to identify the relation's domain and range.

41. $x^2 + y^2 = 16$ *

42. $x^2 + y^2 = 49$ *

43. $(x - 3)^2 + (y - 1)^2 = 36$ *

44. $(x - 2)^2 + (y - 3)^2 = 16$ *

45. $(x + 3)^2 + (y - 2)^2 = 4$ *

46. $(x + 1)^2 + (y - 4)^2 = 25$ *

47. $(x + 2)^2 + (y + 2)^2 = 4$ *

48. $(x + 4)^2 + (y + 5)^2 = 36$ *

49. $x^2 + (y - 1)^2 = 1$ *

50. $x^2 + (y - 2)^2 = 4$ *

51. $(x + 1)^2 + y^2 = 25$ *

52. $(x + 2)^2 + y^2 = 16$ *

In Exercises 53–64, complete the square and write the equation in standard form. Then give the center and radius of each circle and graph the equation.

53. $x^2 + y^2 + 6x + 2y + 6 = 0$ $(x + 3)^2 + (y + 1)^2 = 4$*

54. $x^2 + y^2 + 8x + 4y + 16 = 0$ $(x + 4)^2 + (y + 2)^2 = 4$*

55. $x^2 + y^2 - 10x - 6y - 30 = 0$ $(x - 5)^2 + (y - 3)^2 = 64$*

56. $x^2 + y^2 - 4x - 12y - 9 = 0$ $(x - 2)^2 + (y - 6)^2 = 49$*

57. $x^2 + y^2 + 8x - 2y - 8 = 0$ $(x + 4)^2 + (y - 1)^2 = 25$*

58. $x^2 + y^2 + 12x - 6y - 4 = 0$ $(x + 6)^2 + (y - 3)^2 = 49$*

59. $x^2 - 2x + y^2 - 15 = 0$ $(x - 1)^2 + y^2 = 16$*

60. $x^2 + y^2 - 6y - 7 = 0$ $x^2 + (y - 3)^2 = 16$*

61. $x^2 + y^2 - x + 2y + 1 = 0$ $(x - \frac{1}{2})^2 + (y + 1)^2 = \frac{1}{4}$*

62. $x^2 + y^2 + x + y - \frac{1}{2} = 0$ $(x + \frac{1}{2})^2 + (y + \frac{1}{2})^2 = 1$*

63. $x^2 + y^2 + 3x - 2y - 1 = 0$ $(x + \frac{3}{2})^2 + (y - 1)^2 = \frac{17}{4}$*

64. $x^2 + y^2 + 3x + 5y + \frac{9}{4} = 0$ $(x + \frac{3}{2})^2 + (y + \frac{5}{2})^2 = \frac{25}{4}$*

Practice Plus

In Exercises 65–66, a line segment through the center of each circle intersects the circle at the points shown.

a. Find the coordinates of the circle's center.

b. Find the radius of the circle.

c. Use your answers from parts (a) and (b) to write the standard form of the circle's equation.

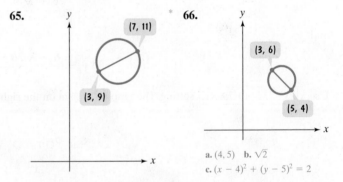

65.

(7, 11)

(3, 9)

66.

(3, 6)

(5, 4)

a. (4, 5) **b.** $\sqrt{2}$

c. $(x - 4)^2 + (y - 5)^2 = 2$

In Exercises 67–70, graph both equations in the same rectangular coordinate system and find all points of intersection. Then show that these ordered pairs satisfy the equations.

67. $x^2 + y^2 = 16$

$x - y = 4$ {(0, −4), (4, 0)}*

68. $x^2 + y^2 = 9$

$x - y = 3$ {(0, −3), (3, 0)}*

69. $(x - 2)^2 + (y + 3)^2 = 4$

$y = x - 3$ {(0, −3), (2, −1)}*

70. $(x - 3)^2 + (y + 1)^2 = 9$

$y = x - 1$ {(0, −1), (3, 2)}*

Application Exercises

The cellphone screen shows coordinates of six cities from a rectangular coordinate system placed on North America by long-distance telephone companies. Each unit in this system represents $\sqrt{0.1}$ *mile.*

In Exercises 71–72, use this information to find the distance, to the nearest mile, between each pair of cities.

Source: Peter H. Dana

71. Boston and San Francisco 2693 miles

72. New Orleans and Houston 328 miles

73. A rectangular coordinate system with coordinates in miles is placed with the origin at the center of Los Angeles. The figure indicates that the University of Southern California is located 2.4 miles west and 2.7 miles south of central Los Angeles. A seismograph on the campus shows that a small earthquake occurred. The quake's epicenter is estimated to be approximately 30 miles from the university. Write the standard form of the equation for the set of points that could be the epicenter of the quake. $(x + 2.4)^2 + (y + 2.7)^2 = 900$

74. The Ferris wheel in the figure has a radius of 68 feet. The clearance between the wheel and the ground is 14 feet. The rectangular coordinate system shown has its origin on the ground directly below the center of the wheel. Use the coordinate system to write the equation of the circular wheel. $x^2 + (y - 82)^2 = 4624$

Explaining the Concepts 75–82. Answers will vary.

75. In your own words, describe how to find the distance between two points in the rectangular coordinate system.

76. In your own words, describe how to find the midpoint of a line segment if its endpoints are known.

77. What is a circle? Without using variables, describe how the definition of a circle can be used to obtain a form of its equation.

78. Give an example of a circle's equation in standard form. Describe how to find the center and radius for this circle.

79. How is the standard form of a circle's equation obtained from its general form?

80. Does $(x - 3)^2 + (y - 5)^2 = 0$ represent the equation of a circle? If not, describe the graph of this equation.

81. Does $(x - 3)^2 + (y - 5)^2 = -25$ represent the equation of a circle? What sort of set is the graph of this equation?

82. Write and solve a problem about the flying time between a pair of cities shown on the cellphone screen for Exercises 71–72. Do not use the pairs in Exercise 71 or Exercise 72. Begin by determining a reasonable average speed, in miles per hour, for a jet flying between the cities.

Technology Exercises

In Exercises 83–85, use a graphing utility to graph each circle whose equation is given. Use a square setting for the viewing window.

83. $x^2 + y^2 = 25$ *

84. $(y + 1)^2 = 36 - (x - 3)^2$ *

85. $x^2 + 10x + y^2 - 4y - 20 = 0$ *

Critical Thinking Exercises

Make Sense? *In Exercises 86–89, determine whether each statement makes sense or does not make sense, and explain your reasoning.*

86. I've noticed that in mathematics, one topic often leads logically to a new topic: makes sense

87. To avoid sign errors when finding h and k, I place parentheses around the numbers that follow the subtraction signs in a circle's equation. makes sense

88. I used the equation $(x + 1)^2 + (y - 5)^2 = -4$ to identify the circle's center and radius. does not make sense

89. My graph of $(x - 2)^2 + (y + 1)^2 = 16$ is my graph of $x^2 + y^2 = 16$ translated two units right and one unit down. *

In Exercises 90–93, determine whether each statement is true or false. If the statement is false, make the necessary change(s) to produce a true statement. Changes to false statements will vary.

90. The equation of the circle whose center is at the origin with radius 16 is $x^2 + y^2 = 16$. false

91. The graph of $(x - 3)^2 + (y + 5)^2 = 36$ is a circle with radius 6 centered at $(-3, 5)$. false

92. The graph of $(x - 4) + (y + 6) = 25$ is a circle with radius 5 centered at $(4, -6)$. false

93. The graph of $(x - 3)^2 + (y + 5)^2 = -36$ is a circle with radius 6 centered at $(3, -5)$. false

94. Show that the points $A(1, 1 + d)$, $B(3, 3 + d)$, and $C(6, 6 + d)$ are collinear (lie along a straight line) by showing that the distance from A to B plus the distance from B to C equals the distance from A to C. $2\sqrt{2} + 3\sqrt{2} = 5\sqrt{2}$

95. Prove the midpoint formula by using the following procedure.

 a. Show that the distance between (x_1, y_1) and $\left(\dfrac{x_1 + x_2}{2}, \dfrac{y_1 + y_2}{2}\right)$ is equal to the distance between (x_2, y_2) and $\left(\dfrac{x_1 + x_2}{2}, \dfrac{y_1 + y_2}{2}\right)$. *

 b. Use the procedure from Exercise 94 and the distances from part (a) to show that the points (x_1, y_1), $\left(\dfrac{x_1 + x_2}{2}, \dfrac{y_1 + y_2}{2}\right)$, and (x_2, y_2) are collinear. *

96. Find the area of the donut-shaped region bounded by the graphs of $(x - 2)^2 + (y + 3)^2 = 25$ and $(x - 2)^2 + (y + 3)^2 = 36$. 11π

97. A **tangent line** to a circle is a line that intersects the circle at exactly one point. The tangent line is perpendicular to the radius of the circle at this point of contact. Write an equation in point-slope form for the line tangent to the circle whose equation is $x^2 + y^2 = 25$ at the point $(3, -4)$. $y + 4 = \frac{3}{4}(x - 3)$

Retaining the Concepts

98. Solve and determine whether the equation

$$7(x - 2) + 5 = 7x - 9$$

is an identity, a conditional equation, or an inconsistent equation. (Section 1.2, Example 8) *

99. Divide and express the result in standard form:

$$\frac{4i + 7}{5 - 2i}.$$

(Section 1.4, Example 4) $\frac{27}{29} + \frac{34}{29}i$

100. Solve and graph the solution set on a number line:

$$-9 \le 4x - 1 < 15.$$

(Section 1.7, Example 7) $[-2, 4)$*

Preview Exercises

Exercises 101–103 will help you prepare for the material covered in the first section of the next chapter.

In Exercises 101–102, solve each quadratic equation by the method of your choice.

101. $0 = -2(x - 3)^2 + 8$ $\{1, 5\}$

102. $-x^2 - 2x + 1 = 0$ $\{-1 \pm \sqrt{2}\}$

103. Use the graph of $f(x) = x^2$ to graph $g(x) = (x + 3)^2 + 1$. *

CHAPTER 2 Summary, Review, and Test

SUMMARY

DEFINITIONS AND CONCEPTS	EXAMPLES
2.1 Basics of Functions and Their Graphs	
a. A relation is any set of ordered pairs. The set of first components is the domain of the relation and the set of second components is the range.	Ex. 1, p. 217
b. A function is a correspondence from a first set, called the domain, to a second set, called the range, such that each element in the domain corresponds to exactly one element in the range. If any element in a relation's domain corresponds to more than one element in the range, the relation is not a function.	Ex. 2, p. 219
c. Functions are usually given in terms of equations involving x and y, in which x is the independent variable and y is the dependent variable. If an equation is solved for y and more than one value of y can be obtained for a given x, then the equation does not define y as a function of x. If an equation defines a function, the value of the function at x, $f(x)$, often replaces y.	Ex. 3, p. 220; Ex. 4, p. 222
d. The graph of a function is the graph of its ordered pairs.	Ex. 5, p. 223

DEFINITIONS AND CONCEPTS	**EXAMPLES**
e. The vertical line test for functions: If any vertical line intersects a graph in more than one point, the graph does not define y as a function of x.	Ex. 6, p. 224
f. The graph of a function can be used to determine the function's domain and its range. To find the domain, look for all the inputs on the x-axis that correspond to points on the graph. To find the range, look for all the outputs on the y-axis that correspond to points on the graph.	Ex. 8, p. 227
g. The zeros of a function, f, are the values of x for which $f(x) = 0$. At the real zeros, the graph of f has x-intercepts. The graph of a function can have more than one x-intercept but at most one y-intercept.	Figure 2.15, p. 229

2.2 More on Functions and Their Graphs

a. A function is increasing on intervals where its graph rises, decreasing on intervals where it falls, and constant on intervals where it neither rises nor falls. Precise definitions are given in the box on page 236.	Ex. 1, p. 237
b. If the graph of a function is given, we can often visually locate the number(s) at which the function has a relative maximum or relative minimum. Precise definitions are given in the box on page 237.	Figure 2.17, p. 238
c. Definitions and tests for symmetry are given in Table 2.2 on page 239. The graph of an equation is: • symmetric with respect to the y-axis if substituting $-x$ for x in the equation gives an equivalent equation. • symmetric with respect to the x-axis if substituting $-y$ for y in the equation gives an equivalent equation. • symmetric with respect to the origin if substituting $-x$ for x and $-y$ for y gives an equivalent equation.	Ex. 2, p. 239; Ex. 3, p. 240
d. The graph of an even function in which $f(-x) = f(x)$ is symmetric with respect to the y-axis. The graph of an odd function in which $f(-x) = -f(x)$ is symmetric with respect to the origin.	Ex. 4, p. 242; Ex. 5, p. 243
e. Piecewise functions are defined by two (or more) equations over a specified domain. Some piecewise functions are called step functions because their graphs form discontinuous steps. An example is $f(x) = \text{int}(x)$, where $\text{int}(x)$ is the greatest integer that is less than or equal to x.	Ex. 6, p. 245; Ex. 7, p. 246; Figure 2.27, p. 247
f. The difference quotient of a function f is $$\frac{f(x+h) - f(x)}{h}, h \neq 0.$$	Ex. 8, p. 247

2.3 Linear Functions and Slope

a. The slope, m, of the line through (x_1, y_1) and (x_2, y_2) is $m = \dfrac{y_2 - y_1}{x_2 - x_1}$.	Ex. 1, p. 256
b. Equations of lines include point-slope form, $y - y_1 = m(x - x_1)$, slope-intercept form, $y = mx + b$, and general form, $Ax + By + C = 0$. The equation of a horizontal line is $y = b$; a vertical line is $x = a$. A vertical line is not a linear function.	Ex. 2, p. 258; Ex. 3, p. 259; Ex. 5, p. 261; Ex. 6, p. 262
c. Linear functions in the form $f(x) = mx + b$ can be graphed using the slope, m, and the y-intercept, b. (See the box on page 260.) Linear equations in the general form $Ax + By + C = 0$ can be solved for y and graphed using the slope and the y-intercept. Intercepts can also be used to graph $Ax + By + C = 0$. (See the box on page 263.)	Ex. 4, p. 260; Ex. 7, p. 263; Ex. 8, p. 264

2.4 More on Slope

a. Parallel lines have equal slopes. Perpendicular lines have slopes that are negative reciprocals.	Ex. 1, p. 271; Ex. 2, p. 273

DEFINITIONS AND CONCEPTS	EXAMPLES

b. The slope of a linear function is the rate of change of the dependent variable per unit change of the independent variable.

Ex. 3, p. 274

c. The average rate of change of f from x_1 to x_2 is

$$\frac{f(x_2) - f(x_1)}{x_2 - x_1}.$$

Ex. 4, p. 276;
Ex. 5, p. 277

2.5 Transformations of Functions

a. Table 2.4 on page 283 shows the graphs of the constant function, $f(x) = c$, the identity function, $f(x) = x$, the absolute value function, $f(x) = |x|$, the standard quadratic function, $f(x) = x^2$, the square root function, $f(x) = \sqrt{x}$, the standard cubic function, $f(x) = x^3$, and the cube root function, $f(x) = \sqrt[3]{x}$. The table also lists characteristics of each function.

b. Table 2.5 on page 291 summarizes how to graph a function using vertical shifts, $y = f(x) \pm c$, horizontal shifts, $y = f(x \pm c)$, reflections about the x-axis, $y = -f(x)$, reflections about the y-axis, $y = f(-x)$, vertical stretching, $y = cf(x), c > 1$, vertical shrinking, $y = cf(x), 0 < c < 1$, horizontal shrinking, $y = f(cx), c > 1$, and horizontal stretching, $y = f(cx), 0 < c < 1$.

Ex. 1, p. 284;
Ex. 2, p. 286;
Ex. 3, p. 287;
Ex. 4, p. 288;
Ex. 5, p. 288;
Ex. 6, p. 289;
Ex. 7, p. 290

c. A function involving more than one transformation can be graphed in the following order: (1) horizontal shifting; (2) stretching or shrinking; (3) reflecting; (4) vertical shifting.

Ex. 8, p. 292;
Ex. 9, p. 293

2.6 Combinations of Functions; Composite Functions

a. If a function f does not model data or verbal conditions, its domain is the largest set of real numbers for which the value of $f(x)$ is a real number. Exclude from a function's domain real numbers that cause division by zero and real numbers that result in an even root, such as a square root, of a negative number.

Ex. 1, p. 299

b. When functions are given as equations, they can be added, subtracted, multiplied, or divided by performing operations with the algebraic expressions that appear on the right side of the equations. Definitions for the sum $f + g$, the difference $f - g$, the product fg, and the quotient $\frac{f}{g}$ functions, with domains $D_f \cap D_g$, and $g(x) \neq 0$ for the quotient function, are given in the box on page 301.

Ex. 2, p. 301;
Ex. 3, p. 302;
Ex. 4, p. 303

c. The composition of functions f and g, $f \circ g$, is defined by $(f \circ g)(x) = f(g(x))$. The domain of the composite function $f \circ g$ is given in the box on page 306. This composite function is obtained by replacing each occurrence of x in the equation for f with $g(x)$.

Ex. 5, p. 306;
Ex. 6, p. 307

d. Expressing a given function as a composition of two functions is called decomposing the given function.

Ex. 7, p. 308

2.7 Inverse Functions

a. If $f(g(x)) = x$ and $g(f(x)) = x$, function g is the inverse of function f, denoted f^{-1} and read "f-inverse." Thus, to show that f and g are inverses of each other, one must show that $f(g(x)) = x$ and $g(f(x)) = x$.

Ex. 1, p. 315

b. The procedure for finding a function's inverse uses a switch-and-solve strategy. Switch x and y, and then solve for y. The procedure is given in the box on page 316.

Ex. 2, p. 316;
Ex. 3, p. 317;
Ex. 4, p. 317

c. The horizontal line test for inverse functions: A function f has an inverse that is a function, f^{-1}, if there is no horizontal line that intersects the graph of the function f at more than one point.

Ex. 5, p. 319

d. A one-to-one function is one in which no two different ordered pairs have the same second component. Only one-to-one functions have inverse functions.

e. If the point (a, b) is on the graph of f, then the point (b, a) is on the graph of f^{-1}. The graph of f^{-1} is a reflection of the graph of f about the line $y = x$.

Ex. 6, p. 320;
Ex. 7, p. 320

DEFINITIONS AND CONCEPTS

EXAMPLES

2.8 Distance and Midpoint Formulas; Circles

a. The distance, d, between the points (x_1, y_1) and (x_2, y_2) is given by $d = \sqrt{(x_2 - x_1)^2 + (y_2 - y_1)^2}$.

Ex. 1, p. 326

b. The midpoint of the line segment whose endpoints are (x_1, y_1) and (x_2, y_2) is the point with coordinates $\left(\dfrac{x_1 + x_2}{2}, \dfrac{y_1 + y_2}{2} \right)$.

Ex. 2, p. 327

c. The standard form of the equation of a circle with center (h, k) and radius r is $(x - h)^2 + (y - k)^2 = r^2$.

Ex. 3, p. 328;
Ex. 4, p. 329;
Ex. 5, p. 329

d. The general form of the equation of a circle is $x^2 + y^2 + Dx + Ey + F = 0$.

e. To convert from the general form to the standard form of a circle's equation, complete the square on x and y.

Ex. 6, p. 330

REVIEW EXERCISES

2.1 and 2.2

In Exercises 1–3, determine whether each relation is a function. Give the domain and range for each relation.

1. $\{(2, 7), (3, 7), (5, 7)\}$ *

2. $\{(1, 10), (2, 500), (13, \pi)\}$ *

3. $\{(12, 13), (14, 15), (12, 19)\}$ not a function; {12, 14}; {13, 15, 19}

In Exercises 4–6, determine whether each equation defines y as a function of x.

4. $2x + y = 8$ yes

5. $3x^2 + y = 14$ yes

6. $2x + y^2 = 6$ no

In Exercises 7–10, evaluate each function at the given values of the independent variable and simplify.

7. $f(x) = 5 - 7x$

 a. $f(4)$ −23 **b.** $f(x + 3)$ −7x − 16 **c.** $f(-x)$ 5 + 7x

8. $g(x) = 3x^2 - 5x + 2$

 a. $g(0)$ 2 **b.** $g(-2)$ 24

 c. $g(x - 1)$ 3x² − 11x + 10 **d.** $g(-x)$ 3x² + 5x + 2

9. $g(x) = \begin{cases} \sqrt{x - 4} & \text{if } x \geq 4 \\ 4 - x & \text{if } x < 4 \end{cases}$

 a. $g(13)$ 3 **b.** $g(0)$ 4 **c.** $g(-3)$ 7

10. $f(x) = \begin{cases} \dfrac{x^2 - 1}{x - 1} & \text{if } x \neq 1 \\ 12 & \text{if } x = 1 \end{cases}$

 a. $f(-2)$ −1 **b.** $f(1)$ 12 **c.** $f(2)$ 3

In Exercises 11–16, use the vertical line test to identify graphs in which y is a function of x.

11.

not a function

12.

function

13.

function

14.

not a function

15.

not a function

16.

function

In Exercises 17–19, use the graph to determine **a.** *the function's domain;* **b.** *the function's range;* **c.** *the x-intercepts, if any;* **d.** *the y-intercept, if there is one;* **e.** *intervals on which the function is increasing, decreasing, or constant; and* **f.** *the missing function values, indicated by question marks, below each graph.*

17.

$f(-2) = ?$ $f(3) = ?$

a. $[-3, 5)$ **b.** $[-5, 0]$
c. −3 **d.** −2
e. increasing: $(-2, 0)$ or $(3, 5)$; decreasing: $(-3, -2)$ or $(0, 3)$
f. $f(-2) = -3; f(3) = -5$

*See Answers to Selected Exercises.

18.

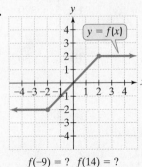

a. $(-\infty, \infty)$ b. $(-\infty, 3]$
c. -2 and 3 d. 3
e. increasing: $(-\infty, 0)$
decreasing: $(0, \infty)$
f. $f(-2) = 0; f(6) = -3$

$f(-2) = ?$ $f(6) = ?$

19.

a. $(-\infty, \infty)$ b. $[-2, 2]$
c. 0 d. 0
e. increasing: $(-2, 2)$; constant:
$(-\infty, -2)$ or $(2, \infty)$
f. $f(-9) = -2; f(14) = 2$

$f(-9) = ?$ $f(14) = ?$

In Exercises 20–21, find each of the following:

a. *The numbers, if any, at which f has a relative maximum. What are these relative maxima?*
b. *The numbers, if any, at which f has a relative minimum. What are these relative minima?*

20. Use the graph in Exercise 17. a. $0; f(0) = -2$*
21. Use the graph in Exercise 18. a. $0; f(0) = 3$ b. none

In Exercises 22–24, determine whether the graph of each equation is symmetric with respect to the y-axis, the x-axis, the origin, more than one of these, or none of these.

22. $y = x^2 + 8$ y-axis
23. $x^2 + y^2 = 17$ all three
24. $x^3 - y^2 = 5$ x-axis

In Exercises 25–27, determine whether each graph is the graph of an even function, an odd function, or a function that is neither even nor odd.

25. odd

26. neither

27. even

In Exercises 28–30, determine whether each function is even, odd, or neither. State each function's symmetry. If you are using a graphing utility, graph the function and verify its possible symmetry.

28. $f(x) = x^3 - 5x$ odd; origin
29. $f(x) = x^4 - 2x^2 + 1$ even; y-axis
30. $f(x) = 2x\sqrt{1 - x^2}$ odd; origin

In Exercises 31–32, the domain of each piecewise function is $(-\infty, \infty)$.

a. *Graph each function.*
b. *Use the graph to determine the function's range.*

31. $f(x) = \begin{cases} 5 & \text{if} \quad x \le -1 \\ -3 & \text{if} \quad x > -1 \end{cases}$ b. $\{-3, 5\}$*

32. $f(x) = \begin{cases} 2x & \text{if} \quad x < 0 \\ -x & \text{if} \quad x \ge 0 \end{cases}$ b. $(-\infty, 0]$*

In Exercises 33–34, find and simplify the difference quotient

$$\frac{f(x + h) - f(x)}{h}, \quad h \ne 0$$

for the given function.

33. $f(x) = 8x - 11$ $8, h \ne 0$
34. $f(x) = -2x^2 + x + 10$ $-4x - 2h + 1, h \ne 0$

35. The graph shows the height, in meters, of an eagle in terms of its time, in seconds, in flight.

a. Is the eagle's height a function of time? Use the graph to explain why or why not. yes*
b. On which interval is the function decreasing? Describe what this means in practical terms. $(3, 12)$*
c. On which intervals is the function constant? What does this mean for each of these intervals? $(0, 3)$ and $(12, 17)$*
d. On which interval is the function increasing? What does this mean? $(17, 30)$*

36. A cargo service charges a flat fee of $5 plus $1.50 for each pound or fraction of a pound. Graph shipping cost, $C(x)$, in dollars, as a function of weight, x, in pounds, for $0 < x \leq 5$. *

2.3 and 2.4

In Exercises 37–40, find the slope of the line passing through each pair of points or state that the slope is undefined. Then indicate whether the line through the points rises, falls, is horizontal, or is vertical.

37. $(3, 2)$ and $(5, 1)$ $-\frac{1}{2}$; falls
38. $(-1, -2)$ and $(-3, -4)$ 1; rises
39. $\left(-3, \frac{1}{4}\right)$ and $\left(6, \frac{1}{4}\right)$ 0; horizontal
40. $(-2, 5)$ and $(-2, 10)$ undefined; vertical

In Exercises 41–44, use the given conditions to write an equation for each line in point-slope form and slope-intercept form.

41. Passing through $(-3, 2)$ with slope -6 $y = -6x - 16$*
42. Passing through $(1, 6)$ and $(-1, 2)$ $y = 2x + 4$*
43. Passing through $(4, -7)$ and parallel to the line whose equation is $3x + y - 9 = 0$ $y + 7 = -3(x - 4); y = -3x + 5$
44. Passing through $(-3, 6)$ and perpendicular to the line whose equation is $y = \frac{1}{3}x + 4$ $y - 6 = -3(x + 3); y = -3x - 3$
45. Write an equation in general form for the line passing through $(-12, -1)$ and perpendicular to the line whose equation is $6x - y - 4 = 0$. $x + 6y + 18 = 0$

In Exercises 46–49, give the slope and y-intercept of each line whose equation is given. Then graph the line.

46. $y = \frac{2}{5}x - 1$ $m = \frac{2}{5}; b = -1$*
47. $f(x) = -4x + 5$ $m = -4; b = 5$*
48. $2x + 3y + 6 = 0$ $m = -\frac{2}{3}; b = -2$*
49. $2y - 8 = 0$ $m = 0; b = 4$*
50. Graph using intercepts: $2x - 5y - 10 = 0$. *
51. Graph: $2x - 10 = 0$. *
52. The bar graph shows the average age at which men in the United States married for the first time from 2009 through 2013. The data are displayed as five points in a scatter plot. Also shown is a line that passes through or near the points.

Men's Average Age of First Marriage

Source: U.S. Census Bureau

a. Use the two points whose coordinates are shown by the voice balloons to find the point-slope form of the equation of the line that models men's average age of first marriage, y, x years after 2008. $y - 28.2 = 0.2(x - 2)$ or $y - 28.6 = 0.2(x - 4)$

b. Write the equation from part (a) in slope-intercept form. Use function notation. $f(x) = 0.2x + 27.8$
c. If trends shown from 2009 through 2013 continue, use the linear function to project men's average age of first marriage in 2020. 30.2 years

53. The graph shows the percentage of college freshmen who were liberal or conservative in 1980 and in 2010.

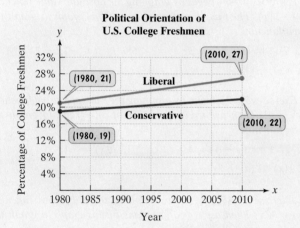

Source: The Higher Education Research Institute

a. Find the slope of the line segment representing liberal college freshmen. 0.2

b. Use your answer from part (a) to complete this statement: For the period shown, the percentage of liberal college freshmen increased each year by approximately _____0.2_____. The rate of change was ___0.2%___ per ___year___.

54. Find the average rate of change of $f(x) = x^2 - 4x$ from $x_1 = 5$ to $x_2 = 9$. 10

2.5

In Exercises 55–59, use the graph of $y = f(x)$ to graph each function g.

55. $g(x) = f(x + 2) + 3$ *
56. $g(x) = \frac{1}{2}f(x - 1)$ *
57. $g(x) = -f(2x)$ *
58. $g(x) = 2f\left(\frac{1}{2}x\right)$ *
59. $g(x) = -f(-x) - 1$ *

In Exercises 60–63, begin by graphing the standard quadratic function, $f(x) = x^2$. Then use transformations of this graph to graph the given function.

60. $g(x) = x^2 + 2$ *
61. $h(x) = (x + 2)^2$ *
62. $r(x) = -(x + 1)^2$ *
63. $y(x) = \frac{1}{2}(x - 1)^2 + 1$ *

In Exercises 64–66, begin by graphing the square root function, $f(x) = \sqrt{x}$. Then use transformations of this graph to graph the given function.

64. $g(x) = \sqrt{x + 3}$ *

65. $h(x) = \sqrt{3 - x}$ *

66. $r(x) = 2\sqrt{x + 2}$ *

In Exercises 67–69, begin by graphing the absolute value function, $f(x) = |x|$. Then use transformations of this graph to graph the given function.

67. $g(x) = |x + 2| - 3$ *

68. $h(x) = -|x - 1| + 1$ *

69. $r(x) = \frac{1}{2}|x + 2|$ *

In Exercises 70–72, begin by graphing the standard cubic function, $f(x) = x^3$. Then use transformations of this graph to graph the given function.

70. $g(x) = \frac{1}{2}(x - 1)^3$ *

71. $h(x) = -(x + 1)^3$ *

72. $r(x) = \frac{1}{4}x^3 - 1$ *

In Exercises 73–75, begin by graphing the cube root function, $f(x) = \sqrt[3]{x}$. Then use transformations of this graph to graph the given function.

73. $g(x) = \sqrt[3]{x + 2} - 1$ *

74. $h(x) = -\sqrt[3]{2x}$ *

75. $r(x) = -2\sqrt[3]{-x}$ *

2.6

In Exercises 76–81, find the domain of each function.

76. $f(x) = x^2 + 6x - 3$ $(-\infty, \infty)$

77. $g(x) = \dfrac{4}{x - 7}$ *

78. $h(x) = \sqrt{8 - 2x}$ $(-\infty, 4]$

79. $f(x) = \dfrac{x}{x^2 + 4x - 21}$ $(-\infty, -7) \cup (-7, 3) \cup (3, \infty)$

80. $g(x) = \dfrac{\sqrt{x - 2}}{x - 5}$ $[2, 5) \cup (5, \infty)$

81. $f(x) = \sqrt{x - 1} + \sqrt{x + 5}$ $[1, \infty)$

In Exercises 82–84, find $f + g$, $f - g$, fg, and $\frac{f}{g}$. Determine the domain for each function.

82. $f(x) = 3x - 1$, $g(x) = x - 5$ *

83. $f(x) = x^2 + x + 1$, $g(x) = x^2 - 1$ *

84. $f(x) = \sqrt{x + 7}$, $g(x) = \sqrt{x - 2}$ *

*In Exercises 85–86, find **a.** $(f \circ g)(x)$; **b.** $(g \circ f)(x)$; **c.** $(f \circ g)(3)$.*

85. $f(x) = x^2 + 3$, $g(x) = 4x - 1$ **c.** 124*

86. $f(x) = \sqrt{x}$, $g(x) = x + 1$ **c.** 2*

*In Exercises 87–88, find **a.** $(f \circ g)(x)$; **b.** the domain of $(f \circ g)$.*

87. $f(x) = \dfrac{x + 1}{x - 2}$, $g(x) = \dfrac{1}{x}$ **a.** $(f \circ g)(x) = \dfrac{1 + x}{1 - 2x}$*

88. $f(x) = \sqrt{x - 1}$, $g(x) = x + 3$ **a.** $(f \circ g)(x) = \sqrt{x + 2}$*

In Exercises 89–90, express the given function h as a composition of two functions f and g so that $h(x) = (f \circ g)(x)$.

89. $h(x) = (x^2 + 2x - 1)^4$ $f(x) = x^4, g(x) = x^2 + 2x - 1$

90. $h(x) = \sqrt[3]{7x + 4}$ $f(x) = \sqrt[3]{x}, g(x) = 7x + 4$

2.7

In Exercises 91–92, find $f(g(x))$ and $g(f(x))$ and determine whether each pair of functions f and g are inverses of each other.

91. $f(x) = \dfrac{3}{5}x + \dfrac{1}{2}$ and $g(x) = \dfrac{5}{3}x - 2$ not inverses*

92. $f(x) = 2 - 5x$ and $g(x) = \dfrac{2 - x}{5}$ inverses*

The functions in Exercises 93–95 are all one-to-one. For each function,

 a. *Find an equation for $f^{-1}(x)$, the inverse function.*

 b. *Verify that your equation is correct by showing that*
 $f(f^{-1}(x)) = x$ *and* $f^{-1}(f(x)) = x$.

93. $f(x) = 4x - 3$ $f^{-1}(x) = \dfrac{x + 3}{4}$

94. $f(x) = 8x^3 + 1$ $f^{-1}(x) = \sqrt[3]{\dfrac{x - 1}{8}}$ or $f^{-1}(x) = \dfrac{\sqrt[3]{x - 1}}{2}$

95. $f(x) = \dfrac{x - 7}{x + 2}$ $f^{-1}(x) = \dfrac{-2x - 7}{x - 1}$

Which graphs in Exercises 96–99 represent functions that have inverse functions?

96. Inverse function exists.

97. Inverse function does not exist.

98. Inverse function exists.

99. Inverse function does not exist.

100. Use the graph of f in the figure shown to draw the graph of its inverse function. *

$$y = f(x)$$

In Exercises 101–102, find an equation for $f^{-1}(x)$. Then graph f and f^{-1} in the same rectangular coordinate system.

101. $f(x) = 1 - x^2, x \geq 0$ * **102.** $f(x) = \sqrt{x} + 1$ *

2.8

In Exercises 103–104, find the distance between each pair of points. If necessary, express answers in simplified radical form and then round to two decimal places.

103. $(-2, 3)$ and $(3, -9)$ 13 **104.** $(-4, 3)$ and $(-2, 5)$ $2\sqrt{2} \approx 2.83$

In Exercises 105–106, find the midpoint of each line segment with the given endpoints.

105. $(2, 6)$ and $(-12, 4)$ $(-5, 5)$
106. $(4, -6)$ and $(-15, 2)$ $\left(-\frac{11}{2}, -2\right)$

In Exercises 107–108, write the standard form of the equation of the circle with the given center and radius.

107. Center $(0, 0)$, $r = 3$ $x^2 + y^2 = 9$
108. Center $(-2, 4)$, $r = 6$ $(x + 2)^2 + (y - 4)^2 = 36$

In Exercises 109–111, give the center and radius of each circle and graph its equation. Use the graph to identify the relation's domain and range.

109. $x^2 + y^2 = 1$ $(h, k) = (0, 0); r = 1$*
110. $(x + 2)^2 + (y - 3)^2 = 9$ $(h, k) = (-2, 3); r = 3$*
111. $x^2 + y^2 - 4x + 2y - 4 = 0$ $(h, k) = (2, -1); r = 3$*

CHAPTER 2 TEST

1. List by letter all relations that are not functions. b, c, d
 a. $\{(7, 5), (8, 5), (9, 5)\}$
 b. $\{(5, 7), (5, 8), (5, 9)\}$
 c.

 d. $x^2 + y^2 = 100$
 e.

2. Use the graph of $y = f(x)$ to solve this exercise.

$$y = f(x)$$

 a. What is $f(4) - f(-3)$? 5
 b. What is the domain of f? $(-5, 6]$
 c. What is the range of f? $[-4, 5]$
 d. On which interval or intervals is f increasing? $(-1, 2)$
 e. On which interval or intervals is f decreasing? *
 f. For what number does f have a relative maximum? What is the relative maximum? $2; f(2) = 5$
 g. For what number does f have a relative minimum? What is the relative minimum? $-1; f(-1) = -4$
 h. What are the x-intercepts? $-4, 1,$ and 5
 i. What is the y-intercept? -3

3. Use the graph of $y = f(x)$ to solve this exercise.

a. What are the zeros of f? −2 and 2

b. Find the value(s) of x for which $f(x) = -1$. −1 and 1

c. Find the value(s) of x for which $f(x) = -2$. 0

d. Is f even, odd, or neither? even

e. Does f have an inverse function? no

f. Is $f(0)$ a relative maximum, a relative minimum, or neither? relative minimum

g. Graph $g(x) = f(x + 1) - 1$. *

h. Graph $h(x) = \frac{1}{2}f(\frac{1}{2}x)$. *

i. Graph $r(x) = -f(-x) + 1$. *

j. Find the average rate of change of f from $x_1 = -2$ to $x_2 = 1$. $-\frac{1}{3}$

In Exercises 4–15, graph each equation in a rectangular coordinate system. If two functions are indicated, graph both in the same system. Then use your graphs to identify each relation's domain and range.

4. $x + y = 4$ *

5. $x^2 + y^2 = 4$ *

6. $f(x) = 4$ *

7. $f(x) = -\frac{1}{3}x + 2$ *

8. $(x + 2)^2 + (y - 1)^2 = 9$ *

9. $f(x) = \begin{cases} 2 & \text{if } x \le 0 \\ -1 & \text{if } x > 0 \end{cases}$ *

10. $x^2 + y^2 + 4x - 6y - 3 = 0$ *

11. $f(x) = |x|$ and $g(x) = \frac{1}{2}|x + 1| - 2$ *

12. $f(x) = x^2$ and $g(x) = -(x - 1)^2 + 4$ *

13. $f(x) = 2x - 4$ and f^{-1} *

14. $f(x) = x^3 - 1$ and f^{-1} *

15. $f(x) = x^2 - 1, x \ge 0$, and f^{-1} *

In Exercises 16–23, let $f(x) = x^2 - x - 4$ and $g(x) = 2x - 6$.

16. Find $f(x - 1)$. *

17. Find $\dfrac{f(x + h) - f(x)}{h}$. *

18. Find $(g - f)(x)$. $(g - f)(x) = -x^2 + 3x - 2$

19. Find $\left(\dfrac{f}{g}\right)(x)$ and its domain. $\left(\dfrac{f}{g}\right)(x) = \dfrac{x^2 - x - 4}{2x - 6}$*

20. Find $(f \circ g)(x)$. *

21. Find $(g \circ f)(x)$. *

22. Find $g(f(-1))$. −10

23. Find $f(-x)$. Is f even, odd, or neither? neither*

24. Determine if the graph of $x^2 + y^3 = 7$ is symmetric with respect to the y-axis, the x-axis, the origin, or none of these. symmetric with respect to the y-axis

In Exercises 25–26, use the given conditions to write an equation for each line in point-slope form and slope-intercept form.

25. Passing through $(2, 1)$ and $(-1, -8)$ $y = 3x - 5$*

26. Passing through $(-4, 6)$ and perpendicular to the line whose equation is $y = -\frac{1}{4}x + 5$ $y - 6 = 4(x + 4); y = 4x + 22$

27. Write an equation in general form for the line passing through $(-7, -10)$ and parallel to the line whose equation is $4x + 2y - 5 = 0$. $2x + y + 24 = 0$

28. In Section 2.4 (Example 3), we worked with data involving the increasing number of U.S. adults ages 18 and older living alone. The bar graph reinforces the fact that one-person households are growing more common. It shows one-person households as a percentage of the U.S. total for five selected years from 1980 through 2012.

One-Person Households as a Percentage of the U.S. Total

Source: U.S. Census Bureau

a. Shown to the right of the bar graph is a scatter plot with a line passing through two of the data points. Use the two points whose coordinates are shown by the voice balloons to write the point-slope form of an equation that models one-person households as a percentage of the U.S. total, y, x years after 1980. $y - 24.6 = 0.12(x - 10)$ or $y - 25.8 = 0.12(x - 20)$

b. Write the equation from part (a) in slope-intercept form. Use function notation. $f(x) = 0.12x + 23.4$

c. If trends shown by the data continue, use the model from part (b) to project one-person households as a percentage of the U.S. total in 2020. 28.2%

29. Find the average rate of change of $f(x) = 3x^2 - 5$ from $x_1 = 6$ to $x_2 = 10$. 48

30. If $g(x) = \begin{cases} \sqrt{x - 3} & \text{if } x \ge 3 \\ 3 - x & \text{if } x < 3 \end{cases}$, find $g(-1)$ and $g(7)$. 4; 2

In Exercises 31–32, find the domain of each function.

31. $f(x) = \dfrac{3}{x + 5} + \dfrac{7}{x - 1}$ $(-\infty, -5) \cup (-5, 1) \cup (1, \infty)$

32. $f(x) = 3\sqrt{x + 5} + 7\sqrt{x - 1}$ $[1, \infty)$

33. If $f(x) = \dfrac{7}{x - 4}$ and $g(x) = \dfrac{2}{x}$, find $(f \circ g)(x)$ and the domain of $f \circ g$. $(f \circ g)(x) = \dfrac{7x}{2 - 4x}; (-\infty, 0) \cup (0, \frac{1}{2}) \cup (\frac{1}{2}, \infty)$

34. Express $h(x) = (2x + 3)^7$ as a composition of two functions f and g so that $h(x) = (f \circ g)(x)$. $f(x) = x^7; g(x) = 2x + 3$

35. Find the length and the midpoint of the line segment whose endpoints are $(2, -2)$ and $(5, 2)$. 5; $(\frac{7}{2}, 0)$ or $(3.5, 0)$

CUMULATIVE REVIEW EXERCISES (CHAPTERS 1–2)

Use the graph of $y = f(x)$ to solve Exercises 1–5.

1. Find the domain and the range of f. [0, 2); [0, 2]

2. For what value(s) of x is $f(x) = 1$? $\frac{1}{2}$ and $\frac{3}{2}$

3. Find the relative maximum. 2

4. Graph $g(x) = f(x - 1) + 1$. *

5. Graph $h(x) = -2f\left(\frac{1}{2}x\right)$. *

In Exercises 6–10, solve each equation or inequality.

6. $(x + 3)(x - 4) = 8$ {−4, 5}

7. $3(4x - 1) = 4 - 6(x - 3)$ $\left\{\frac{25}{18}\right\}$

8. $\sqrt{x} + 2 = x$ {4}

9. $x^{\frac{2}{3}} - x^{\frac{1}{3}} - 6 = 0$ {−8, 27}

10. $\frac{x}{2} - 3 \le \frac{x}{4} + 2$ $(-\infty, 20]$

In Exercises 11–14, graph each equation in a rectangular coordinate system. If two functions are indicated, graph both in the same system. Then use your graphs to identify each relation's domain and range.

11. $3x - 6y - 12 = 0$ *

12. $(x - 2)^2 + (y + 1)^2 = 4$ *

13. $f(x) = \sqrt[3]{x}$ and $g(x) = \sqrt[3]{x - 3} + 4$ *

14. $f(x) = \sqrt{x - 3} + 2$ and f^{-1} *

In Exercises 15–16, let $f(x) = 4 - x^2$ and $g(x) = x + 5$.

15. Find $\dfrac{f(x + h) - f(x)}{h}$ and simplify. $-2x - h, h \ne 0$

16. Find all values of x satisfying $(f \circ g)(x) = 0$. −7 and −3

17. Write equations in point-slope form, slope-intercept form, and general form for the line passing through $(-2, 5)$ and perpendicular to the line whose equation is $y = -\frac{1}{4}x + \frac{1}{3}$. *

18. You invested \$6000 in two accounts paying 7% and 9% annual interest. At the end of the year, the total interest from these investments was \$510. How much was invested at each rate? \$1500 at 7%; \$4500 at 9%

19. For a summer sales job, you are choosing between two pay arrangements: a weekly salary of \$200 plus 5% commission on sales, or a straight 15% commission. For how many dollars of sales will the earnings be the same regardless of the pay arrangement? \$2000

20. The length of a rectangular garden is 2 feet more than twice its width. If 22 feet of fencing is needed to enclose the garden, what are its dimensions? 3 ft by 8 ft

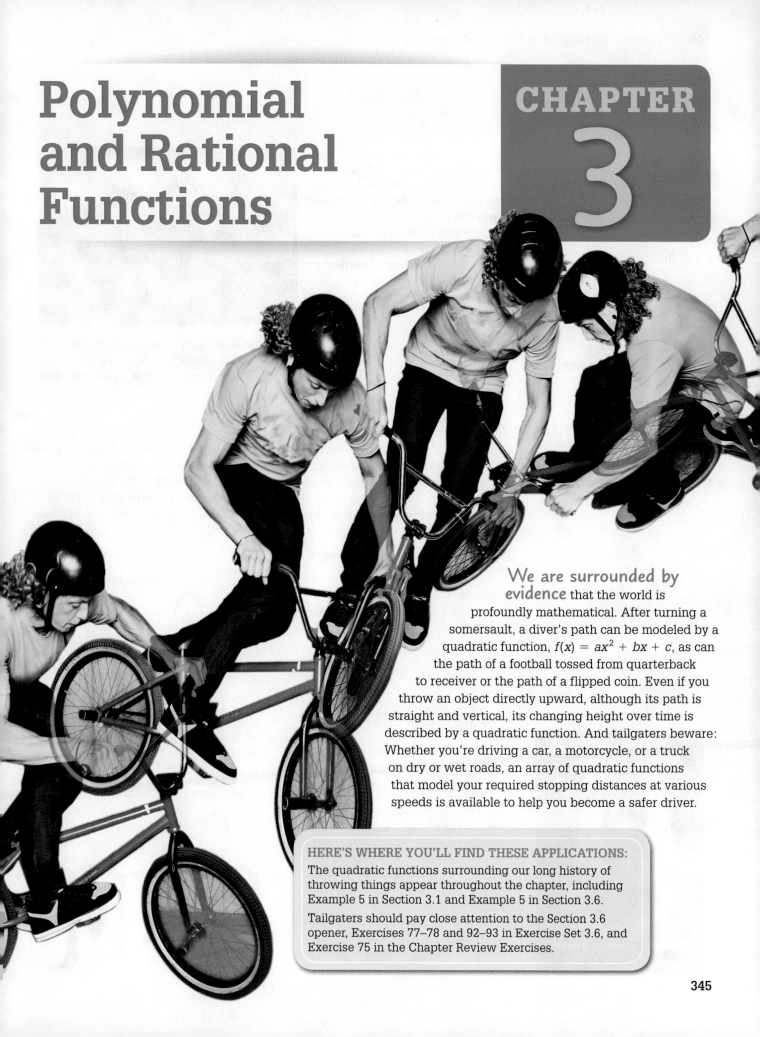

Polynomial and Rational Functions

We are surrounded by evidence that the world is profoundly mathematical. After turning a somersault, a diver's path can be modeled by a quadratic function, $f(x) = ax^2 + bx + c$, as can the path of a football tossed from quarterback to receiver or the path of a flipped coin. Even if you throw an object directly upward, although its path is straight and vertical, its changing height over time is described by a quadratic function. And tailgaters beware: Whether you're driving a car, a motorcycle, or a truck on dry or wet roads, an array of quadratic functions that model your required stopping distances at various speeds is available to help you become a safer driver.

HERE'S WHERE YOU'LL FIND THESE APPLICATIONS:

The quadratic functions surrounding our long history of throwing things appear throughout the chapter, including Example 5 in Section 3.1 and Example 5 in Section 3.6.

Tailgaters should pay close attention to the Section 3.6 opener, Exercises 77–78 and 92–93 in Exercise Set 3.6, and Exercise 75 in the Chapter Review Exercises.

Section 3.1 | Quadratic Functions

1. Recognize characteristics of parabolas.

Many sports involve objects that are thrown, kicked, or hit, and then proceed with no additional force of their own. Such objects are called **projectiles**. Paths of projectiles, as well as their heights over time, can be modeled by *quadratic functions*. A **quadratic function** is any function of the form

$$f(x) = ax^2 + bx + c,$$

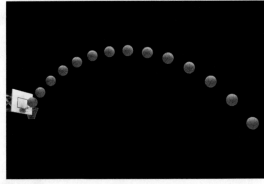

where a, b, and c are real numbers, with $a \neq 0$. A quadratic function is a polynomial function whose greatest exponent is 2. In this section, you will learn to use graphs of quadratic functions to gain a visual understanding of the algebra that describes football, baseball, basketball, the shot put, and other projectile sports.

Graphs of Quadratic Functions

The graph of any quadratic function is called a **parabola**. Parabolas are shaped like bowls or inverted bowls, as shown in **Figure 3.1**. If the coefficient of x^2 (the value of a in $ax^2 + bx + c$) is positive, the parabola opens upward. If the coefficient of x^2 is negative, the parabola opens downward. The **vertex** (or turning point) of the parabola is the lowest point on the graph when it opens upward and the highest point on the graph when it opens downward.

$a > 0$: Parabola opens upward. $a < 0$: Parabola opens downward.

FIGURE 3.1 Characteristics of graphs of quadratic functions

Look at the unusual image of the word *mirror* shown here. The artist, Scott Kim, has created the image so that the two halves of the whole are mirror images of each other. A parabola shares this kind of symmetry, in which a vertical line through the vertex divides the figure in half. Parabolas are symmetric with respect to this line, called the **axis of symmetry**. If a parabola is folded along its axis of symmetry, the two halves match exactly.

② Graph parabolas.

Graphing Quadratic Functions in Standard Form

In our earlier work with transformations, we applied a series of transformations to the graph of $f(x) = x^2$. The graph of this function is a parabola. The vertex for this parabola is $(0, 0)$. In **Figure 3.2(a)**, the graph of $f(x) = ax^2$ for $a > 0$ is shown in black; it opens *upward*. In **Figure 3.2(b)**, the graph of $f(x) = ax^2$ for $a < 0$ is shown in black; it opens *downward*.

Transformations of $f(x) = ax^2$

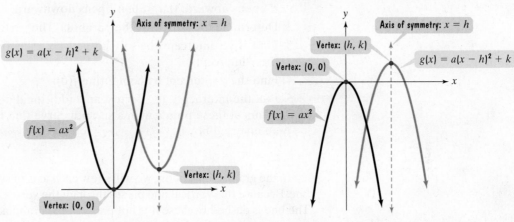

FIGURE 3.2(a) $a > 0$: Parabola opens upward.　　　FIGURE 3.2(b) $a < 0$: Parabola opens downward.

Figure 3.2(a) and **3.2(b)** also show the graph of $g(x) = a(x - h)^2 + k$ in blue. Compare these graphs to those of $f(x) = ax^2$. Observe that h determines a horizontal shift and k determines a vertical shift of the graph of $f(x) = ax^2$:

$$g(x) = a(x - h)^2 + k.$$

If $h > 0$, the graph of $f(x) = ax^2$ is shifted h units to the right.

If $k > 0$, the graph of $y = a(x - h)^2$ is shifted k units up.

Consequently, the vertex $(0, 0)$ on the black graph of $f(x) = ax^2$ moves to the point (h, k) on the blue graph of $g(x) = a(x - h)^2 + k$. The axis of symmetry is the vertical line whose equation is $x = h$.

The form of the expression for g is convenient because it immediately identifies the vertex of the parabola as (h, k). This is the **standard form** of a quadratic function.

The Standard Form of a Quadratic Function

The quadratic function

$$f(x) = a(x - h)^2 + k, \quad a \neq 0$$

is in **standard form**. The graph of f is a parabola whose vertex is the point (h, k). The parabola is symmetric with respect to the line $x = h$. If $a > 0$, the parabola opens upward; if $a < 0$, the parabola opens downward.

The sign of a in $f(x) = a(x - h)^2 + k$ determines whether the parabola opens upward or downward. Furthermore, if $|a|$ is small, the parabola opens more flatly than if $|a|$ is large. Here is a general procedure for graphing parabolas whose equations are in standard form:

Graphing Quadratic Functions with Equations in Standard Form

To graph $f(x) = a(x - h)^2 + k$,

1. Determine whether the parabola opens upward or downward. If $a > 0$, it opens upward. If $a < 0$, it opens downward.
2. Determine the vertex of the parabola. The vertex is (h, k).
3. Find any x-intercepts by solving $f(x) = 0$. The function's real zeros are the x-intercepts.
4. Find the y-intercept by computing $f(0)$.
5. Plot the intercepts, the vertex, and additional points as necessary. Connect these points with a smooth curve that is shaped like a bowl or an inverted bowl.

In the graphs that follow, we will show each axis of symmetry as a dashed vertical line. Because this vertical line passes through the vertex, (h, k), its equation is $x = h$. The line is dashed because it is not part of the parabola.

EXAMPLE 1 Graphing a Quadratic Function in Standard Form

Graph the quadratic function $f(x) = -2(x - 3)^2 + 8$.

SOLUTION

We can graph this function by following the steps in the preceding box. We begin by identifying values for a, h, and k.

Standard form
$$f(x) = a(x - h)^2 + k$$
$$a = -2 \quad h = 3 \quad k = 8$$
Given function
$$f(x) = -2(x - 3)^2 + 8$$

Step 1 Determine how the parabola opens. Note that a, the coefficient of x^2, is -2. Thus, $a < 0$; this negative value tells us that the parabola opens downward.

Step 2 Find the vertex. The vertex of the parabola is at (h, k). Because $h = 3$ and $k = 8$, the parabola has its vertex at $(3, 8)$.

Step 3 Find the x-intercepts by solving $f(x) = 0$. Replace $f(x)$ with 0 in $f(x) = -2(x - 3)^2 + 8$.

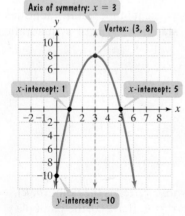

Axis of symmetry: $x = 3$

Vertex: (3, 8)

x-intercept: 1 x-intercept: 5

y-intercept: −10

FIGURE 3.3 The graph of $f(x) = -2(x - 3)^2 + 8$

$0 = -2(x - 3)^2 + 8$	Find x-intercepts, setting $f(x)$ equal to O.
$2(x - 3)^2 = 8$	Solve for x. Add $2(x - 3)^2$ to both sides of the equation.
$(x - 3)^2 = 4$	Divide both sides by 2.
$x - 3 = \sqrt{4} \quad$ or $\quad x - 3 = -\sqrt{4}$	Apply the square root property: If $u^2 = d$, then $u = \sqrt{d}$ or $u = -\sqrt{d}$.
$x - 3 = 2 \qquad\qquad x - 3 = -2$	$\sqrt{4} = 2$
$x = 5 \qquad\qquad\quad x = 1$	Add 3 to both sides in each equation.

The x-intercepts are 1 and 5. The parabola passes through $(1, 0)$ and $(5, 0)$.

Step 4 Find the y-intercept by computing f(0). Replace x with 0 in $f(x) = -2(x - 3)^2 + 8$.

$$f(0) = -2(0 - 3)^2 + 8 = -2(-3)^2 + 8 = -2(9) + 8 = -10$$

The y-intercept is -10. The parabola passes through $(0, -10)$.

Step 5 Graph the parabola. With a vertex at $(3, 8)$, x-intercepts at 5 and 1, and a y-intercept at -10, the graph of f is shown in **Figure 3.3**. The axis of symmetry is the vertical line whose equation is $x = 3$. •••

⊘ **Check Point 1** Graph the quadratic function $f(x) = -(x - 1)^2 + 4$. *

EXAMPLE 2 Graphing a Quadratic Function in Standard Form

Graph the quadratic function $f(x) = (x + 3)^2 + 1$.

SOLUTION

We begin by finding values for a, h, and k.

$$f(x) = a(x - h)^2 + k \quad \text{Standard form of quadratic function}$$
$$f(x) = (x + 3)^2 + 1 \quad \text{Given function}$$
$$f(x) = 1(x - (-3))^2 + 1$$

$$a = 1 \quad h = -3 \quad k = 1$$

Step 1 Determine how the parabola opens. Note that a, the coefficient of x^2, is 1. Thus, $a > 0$; this positive value tells us that the parabola opens upward.

Step 2 Find the vertex. The vertex of the parabola is at (h, k). Because $h = -3$ and $k = 1$, the parabola has its vertex at $(-3, 1)$.

Step 3 Find the x-intercepts by solving f(x) = 0. Replace $f(x)$ with 0 in $f(x) = (x + 3)^2 + 1$. Because the vertex is at $(-3, 1)$, which lies above the x-axis, and the parabola opens upward, it appears that this parabola has no x-intercepts. We can verify this observation algebraically.

$$0 = (x + 3)^2 + 1 \qquad \text{Find possible } x\text{-intercepts, setting } f(x) \text{ equal to 0.}$$

$$-1 = (x + 3)^2 \qquad \text{Solve for } x. \text{ Subtract 1 from both sides.}$$

$$x + 3 = \sqrt{-1} \quad \text{or} \quad x + 3 = -\sqrt{-1} \qquad \text{Apply the square root property.}$$

$$x + 3 = i \qquad\qquad x + 3 = -i \qquad \sqrt{-1} = i$$

$$x = -3 + i \qquad\qquad x = -3 - i \qquad \text{The solutions are } -3 \pm i.$$

Because this equation has no real solutions, the parabola has no x-intercepts.

Step 4 Find the y-intercept by computing f(0). Replace x with 0 in $f(x) = (x + 3)^2 + 1$.

$$f(0) = (0 + 3)^2 + 1 = 3^2 + 1 = 9 + 1 = 10$$

The y-intercept is 10. The parabola passes through $(0, 10)$.

Step 5 Graph the parabola. With a vertex at $(-3, 1)$, no x-intercepts, and a y-intercept at 10, the graph of f is shown in **Figure 3.4**. The axis of symmetry is the vertical line whose equation is $x = -3$. •••

FIGURE 3.4 The graph of $f(x) = (x + 3)^2 + 1$

*See Answers to Selected Exercises.

GREAT QUESTION!

I'm confused about finding h from the equation $f(x) = a(x - h)^2 + k$. Can you help me out?

It's easy to make a sign error when finding h, the x-coordinate of the vertex. In

$$f(x) = a(x - h)^2 + k,$$

h is the number that follows the subtraction sign.

- $f(x) = -2(x - 3)^2 + 8$

 The number after the subtraction is 3: $h = 3$.

- $f(x) = (x + 3)^2 + 1$
 $$= (x - (-3))^2 + 1$$

 The number after the subtraction is -3: $h = -3$.

☑ **Check Point 2** Graph the quadratic function $f(x) = (x - 2)^2 + 1$. *

Graphing Quadratic Functions in the Form $f(x) = ax^2 + bx + c$

Quadratic functions are frequently expressed in the form $f(x) = ax^2 + bx + c$. How can we identify the vertex of a parabola whose equation is in this form? Completing the square provides the answer to this question.

$$f(x) = ax^2 + bx + c$$

$$= a\left(x^2 + \frac{b}{a}x\right) + c \qquad \text{Factor out } a \text{ from } ax^2 + bx.$$

$$= a\left(x^2 + \frac{b}{a}x + \frac{b^2}{4a^2}\right) + c - a\left(\frac{b^2}{4a^2}\right)$$

Complete the square by adding the square of half the coefficient of x.

By completing the square, we added $a \cdot \dfrac{b^2}{4a^2}$. To avoid changing the function's equation, we must subtract this term.

$$= a\left(x + \frac{b}{2a}\right)^2 + c - \frac{b^2}{4a}$$

Write the trinomial as the square of a binomial and simplify the constant term.

Compare this form of the equation with a quadratic function's **standard form**.

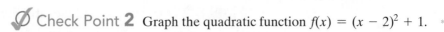

Standard form $f(x) = a(x - h)^2 + k$

$$h = -\frac{b}{2a} \qquad k = c - \frac{b^2}{4a}$$

Equation under discussion $f(x) = a\left(x - \left(-\frac{b}{2a}\right)\right)^2 + c - \frac{b^2}{4a}$

The important part of this observation is that h, the x-coordinate of the vertex, is $-\dfrac{b}{2a}$. The y-coordinate can be found by evaluating the function at $-\dfrac{b}{2a}$.

The Vertex of a Parabola Whose Equation Is $f(x) = ax^2 + bx + c$

Consider the parabola defined by the quadratic function $f(x) = ax^2 + bx + c$. The parabola's vertex is $\left(-\dfrac{b}{2a}, f\left(-\dfrac{b}{2a}\right)\right)$. The x-coordinate is $-\dfrac{b}{2a}$. The y-coordinate is found by substituting the x-coordinate into the parabola's equation and evaluating the function at this value of x.

*See Answers to Selected Exercises.

We can apply our five-step procedure to graph parabolas in the form $f(x) = ax^2 + bx + c$.

Graphing Quadratic Functions with Equations in the Form
$f(x) = ax^2 + bx + c$

To graph $f(x) = ax^2 + bx + c$,

1. Determine whether the parabola opens upward or downward. If $a > 0$, it opens upward. If $a < 0$, it opens downward.

2. Determine the vertex of the parabola. The vertex is $\left(-\dfrac{b}{2a}, f\left(-\dfrac{b}{2a}\right)\right)$.

3. Find any x-intercepts by solving $f(x) = 0$. The real solutions of $ax^2 + bx + c = 0$ are the x-intercepts.

4. Find the y-intercept by computing $f(0)$. Because $f(0) = c$ (the constant term in the function's equation), the y-intercept is c and the parabola passes through $(0, c)$.

5. Plot the intercepts, the vertex, and additional points as necessary. Connect these points with a smooth curve.

EXAMPLE 3 Graphing a Quadratic Function in the Form
$f(x) = ax^2 + bx + c$

Graph the quadratic function $f(x) = -x^2 - 2x + 1$. Use the graph to identify the function's domain and its range.

SOLUTION

Step 1 Determine how the parabola opens. Note that a, the coefficient of x^2, is -1. Thus, $a < 0$; this negative value tells us that the parabola opens downward.

Step 2 Find the vertex. We know that the x-coordinate of the vertex is $x = -\dfrac{b}{2a}$. We identify a, b, and c in $f(x) = ax^2 + bx + c$.

$$f(x) = -x^2 - 2x + 1$$

$\boxed{a = -1}$ $\boxed{b = -2}$ $\boxed{c = 1}$

Substitute the values of a and b into the equation for the x-coordinate:

$$x = -\frac{b}{2a} = -\frac{-2}{2(-1)} = -\left(\frac{-2}{-2}\right) = -1.$$

The x-coordinate of the vertex is -1. We substitute -1 for x in the equation of the function, $f(x) = -x^2 - 2x + 1$, to find the y-coordinate:

$$f(-1) = -(-1)^2 - 2(-1) + 1 = -1 + 2 + 1 = 2.$$

The vertex is at $(-1, 2)$.

Step 3 Find the x-intercepts by solving $f(x) = 0$. Replace $f(x)$ with 0 in $f(x) = -x^2 - 2x + 1$. We obtain $0 = -x^2 - 2x + 1$. This equation cannot be solved by factoring. We will use the quadratic formula to solve it.

$$-x^2 - 2x + 1 = 0$$

$\boxed{a = -1}$ $\boxed{b = -2}$ $\boxed{c = 1}$

We substitute -1 for a, -2 for b, and 1 for c into the quadratic formula.

$$x = \frac{-b \pm \sqrt{b^2 - 4ac}}{2a} = \frac{-(-2) \pm \sqrt{(-2)^2 - 4(-1)(1)}}{2(-1)} = \frac{2 \pm \sqrt{4 - (-4)}}{-2}$$

> To locate the x-intercepts, we need decimal approximations. Thus, there is no need to simplify the radical form of the solutions.

$$x = \frac{2 + \sqrt{8}}{-2} \approx -2.4 \qquad \text{or} \qquad x = \frac{2 - \sqrt{8}}{-2} \approx 0.4$$

The x-intercepts are approximately -2.4 and 0.4. The parabola passes through $(-2.4, 0)$ and $(0.4, 0)$.

Step 4 Find the y-intercept by computing $f(0)$. Replace x with 0 in $f(x) = -x^2 - 2x + 1$.

$$f(0) = -0^2 - 2 \cdot 0 + 1 = 1$$

The y-intercept is 1, which is the constant term in the function's equation. The parabola passes through $(0, 1)$.

Step 5 Graph the parabola. With a vertex at $(-1, 2)$, x-intercepts at approximately -2.4 and 0.4, and a y-intercept at 1, the graph of f is shown in **Figure 3.5(a)**. The axis of symmetry is the vertical line whose equation is $x = -1$.

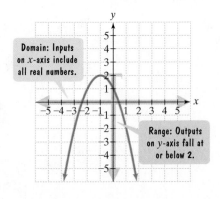

FIGURE 3.5(a) The graph of $f(x) = -x^2 - 2x + 1$

FIGURE 3.5(b) Determining the domain and range of $f(x) = -x^2 - 2x + 1$

GREAT QUESTION!

Are there rules to find domains and ranges of quadratic functions?

Yes. The domain of any quadratic function includes all real numbers. If the vertex is the graph's highest point, the range includes all real numbers at or below the y-coordinate of the vertex. If the vertex is the graph's lowest point, the range includes all real numbers at or above the y-coordinate of the vertex.

Now we are ready to determine the domain and range of $f(x) = -x^2 - 2x + 1$. We can use the parabola, shown again in **Figure 3.5(b)**, to do so. To find the domain, look for all the inputs on the x-axis that correspond to points on the graph. As the graph widens and continues to fall at both ends, can you see that these inputs include all real numbers?

$$\text{Domain of } f \text{ is } \{x \mid x \text{ is a real number}\} \text{ or } (-\infty, \infty).$$

To find the range, look for all the outputs on the y-axis that correspond to points on the graph. **Figure 3.5(b)** shows that the parabola's vertex, $(-1, 2)$, is the highest point on the graph. Because the y-coordinate of the vertex is 2, outputs on the y-axis fall at or below 2.

$$\text{Range of } f \text{ is } \{y \mid y \leq 2\} \text{ or } (-\infty, 2]. \qquad \bullet\bullet\bullet$$

✓ **Check Point 3** Graph the quadratic function $f(x) = -x^2 + 4x + 1$. Use the graph to identify the function's domain and its range. domain: $(-\infty, \infty)$; range: $(-\infty, 5]$*

*See Answers to Selected Exercises.

GREAT QUESTION!

I feel overwhelmed by the amount of information required to graph just one quadratic function. Is there a way I can organize the information and gain a better understanding of the graphing procedure?

You're right: The skills needed to graph a quadratic function combine information from many of the topics we've covered so far. Try organizing the items you need to graph quadratic functions in a table something like this:

Graphing $f(x) = a(x - h)^2 + k$ or $f(x) = ax^2 + bx + c$

1. Opens upward if $a > 0$. Opens downward if $a < 0$.	2. Find the vertex. (h, k) or $\left(-\dfrac{b}{2a}, f\left(-\dfrac{b}{2a}\right)\right)$
3. Find x-intercepts. Solve $f(x) = 0$. Find real solutions.	4. Find the y-intercept. Find $f(0)$.

③ Determine a quadratic function's minimum or maximum value.

Minimum and Maximum Values of Quadratic Functions

Consider the quadratic function $f(x) = ax^2 + bx + c$. If $a > 0$, the parabola opens upward and the vertex is its lowest point. If $a < 0$, the parabola opens downward and the vertex is its highest point. The x-coordinate of the vertex is $-\dfrac{b}{2a}$. Thus, we can find the minimum or maximum value of f by evaluating the quadratic function at $x = -\dfrac{b}{2a}$.

Minimum and Maximum: Quadratic Functions

Consider the quadratic function $f(x) = ax^2 + bx + c$.

1. If $a > 0$, then f has a minimum that occurs at $x = -\dfrac{b}{2a}$. This minimum value is $f\left(-\dfrac{b}{2a}\right)$.

2. If $a < 0$, then f has a maximum that occurs at $x = -\dfrac{b}{2a}$. This maximum value is $f\left(-\dfrac{b}{2a}\right)$.

In each case, the value of x gives the location of the minimum or maximum value. The value of y, or $f\left(-\dfrac{b}{2a}\right)$, gives that minimum or maximum value.

EXAMPLE 4 Obtaining Information about a Quadratic Function from Its Equation

Consider the quadratic function $f(x) = -3x^2 + 6x - 13$.

 a. Determine, without graphing, whether the function has a minimum value or a maximum value.

 b. Find the minimum or maximum value and determine where it occurs.

 c. Identify the function's domain and its range.

SOLUTION

We begin by identifying a, b, and c in the function's equation:

$$f(x) = -3x^2 + 6x - 13.$$

$a = -3$ $b = 6$ $c = -13$

a. Because $a < 0$, the function has a maximum value.

b. The maximum value occurs at

$$x = -\frac{b}{2a} = -\frac{6}{2(-3)} = -\frac{6}{-6} = -(-1) = 1.$$

The maximum value occurs at $x = 1$ and the maximum value of $f(x) = -3x^2 + 6x - 13$ is

$$f(1) = -3 \cdot 1^2 + 6 \cdot 1 - 13 = -3 + 6 - 13 = -10.$$

We see that the maximum is -10 at $x = 1$.

c. Like all quadratic functions, the domain is $(-\infty, \infty)$. Because the function's maximum value is -10, the range includes all real numbers at or below -10. The range is $(-\infty, -10]$. ● ● ●

We can use the graph of $f(x) = -3x^2 + 6x - 13$ to visualize the results of Example 4. **Figure 3.6** shows the graph in a $[-6, 6, 1]$ by $[-50, 20, 10]$ viewing rectangle. The maximum function feature verifies that the function's maximum is -10 at $x = 1$. Notice that x gives the location of the maximum and y gives the maximum value. Notice, too, that the maximum value is -10 and not the ordered pair $(1, -10)$.

Range is $(-\infty, -10]$.

Maximum
X=1 Y=-10

$[-6, 6, 1]$ by $[-50, 20, 10]$

FIGURE 3.6

✓ Check Point **4** Repeat parts (a) through (c) of Example 4 using the quadratic function $f(x) = 4x^2 - 16x + 1000$. **a.** minimum **b.** Minimum is 984 at $x = 2$. **c.** domain: $(-\infty, \infty)$; range: $[984, \infty)$

4 Solve problems involving a quadratic function's minimum or maximum value.

Applications of Quadratic Functions

Many applied problems involve finding the maximum or minimum value of a quadratic function, as well as where this value occurs.

EXAMPLE 5 The Parabolic Path of a Punted Football

Figure 3.7 shows that when a football is kicked, the nearest defensive player is 6 feet from the point of impact with the kicker's foot. The height of the punted football, $f(x)$, in feet, can be modeled by

$$f(x) = -0.01x^2 + 1.18x + 2,$$

where x is the ball's horizontal distance, in feet, from the point of impact with the kicker's foot.

a. What is the maximum height of the punt and how far from the point of impact does this occur?

b. How far must the nearest defensive player, who is 6 feet from the kicker's point of impact, reach to block the punt?

c. If the ball is not blocked by the defensive player, how far down the field will it go before hitting the ground?

d. Graph the function that models the football's parabolic path.

Height of the Punted Football (feet)

Distance from the Point of Impact (feet)

FIGURE 3.7

SOLUTION

a. We begin by identifying the numbers a, b, and c in the function's equation.

$$f(x) = -0.01x^2 + 1.18x + 2$$

| $a = -0.01$ | $b = 1.18$ | $c = 2$ |

Because $a < 0$, the function has a maximum that occurs at $x = -\dfrac{b}{2a}$.

$$x = -\frac{b}{2a} = -\frac{1.18}{2(-0.01)} = -(-59) = 59$$

This means that the maximum height of the punt occurs 59 feet from the kicker's point of impact. The maximum height of the punt is

$$f(59) = -0.01(59)^2 + 1.18(59) + 2 = 36.81,$$

or 36.81 feet.

b. **Figure 3.7** shows that the defensive player is 6 feet from the kicker's point of impact. To block the punt, he must touch the football along its parabolic path. This means that we must find the height of the ball 6 feet from the kicker. Replace x with 6 in the given function, $f(x) = -0.01x^2 + 1.18x + 2$.

$$f(6) = -0.01(6)^2 + 1.18(6) + 2 = -0.36 + 7.08 + 2 = 8.72$$

The defensive player must reach 8.72 feet above the ground to block the punt.

c. Assuming that the ball is not blocked by the defensive player, we are interested in how far down the field it will go before hitting the ground. We are looking for the ball's horizontal distance, x, when its height above the ground, $f(x)$, is 0 feet. To find this x-intercept, replace $f(x)$ with 0 in $f(x) = -0.01x^2 + 1.18x + 2$. We obtain $0 = -0.01x^2 + 1.18x + 2$, or $-0.01x^2 + 1.18x + 2 = 0$. The equation cannot be solved by factoring. We will use the quadratic formula to solve it.

$-0.01x^2 + 1.18x + 2 = 0$

| $a = -0.01$ | $b = 1.18$ | $c = 2$ |

The equation for determining the ball's maximum horizontal distance

Use a calculator to evaluate the radicand.

$$x = \frac{-b \pm \sqrt{b^2 - 4ac}}{2a} = \frac{-1.18 \pm \sqrt{(1.18)^2 - 4(-0.01)(2)}}{2(-0.01)} = \frac{-1.18 \pm \sqrt{1.4724}}{-0.02}$$

$$x = \frac{-1.18 + \sqrt{1.4724}}{-0.02} \quad \text{or} \quad x = \frac{-1.18 - \sqrt{1.4724}}{-0.02}$$

$$x \approx -1.7 \qquad\qquad x \approx 119.7$$

Use a calculator and round to the nearest tenth.

Reject this value. We are interested in the football's height corresponding to horizontal distances from its point of impact onward, or $x \geq 0$.

If the football is not blocked by the defensive player, it will go approximately 119.7 feet down the field before hitting the ground.

d. In terms of graphing the model for the football's parabolic path, $f(x) = -0.01x^2 + 1.18x + 2$, we have already determined the vertex and the approximate x-intercept.

vertex: $(59, 36.81)$

The ball's maximum height, 36.81 feet, occurs at a horizontal distance of 59 feet.

x-intercept: 119.7

The ball's maximum horizontal distance is approximately 119.7 feet.

Height of the Punted Football (feet)

Distance from the
Point of Impact (feet)

FIGURE 3.7 (repeated)

Figure 3.7 indicates that the *y*-intercept is 2, meaning that the ball is kicked from a height of 2 feet. Let's verify this value by replacing *x* with 0 in $f(x) = -0.01x^2 + 1.18x + 2$.

$$f(0) = -0.01 \cdot 0^2 + 1.18 \cdot 0 + 2 = 0 + 0 + 2 = 2$$

Using the vertex, (59, 36.81), the *x*-intercept, 119.7, and the *y*-intercept, 2, the graph of the equation that models the football's parabolic path is shown in **Figure 3.8**. The graph is shown only for $x \geq 0$, indicating horizontal distances that begin at the football's impact with the kicker's foot and end with the ball hitting the ground.

Vertex: (59, 36.81)
The ball's maximum height is 36.81 feet.

$f(x) = -0.01x^2 + 1.18x + 2$

x-intercept: 119.7
The ball hits the ground after traveling a maximum horizontal distance of 119.7 feet.

y-intercept: 2
The ball is kicked from a height of 2 feet.

Football's Height (feet)

Football's Horizontal Distance (feet)

FIGURE 3.8 The parabolic path of a punted football •••

✓ **Check Point 5** An archer's arrow follows a parabolic path. The height of the arrow, *f*(*x*), in feet, can be modeled by

$$f(x) = -0.005x^2 + 2x + 5,$$

where *x* is the arrow's horizontal distance, in feet.

 a. What is the maximum height of the arrow and how far from its release does this occur? maximum height: 205 ft; 200 ft from its release

 b. Find the horizontal distance the arrow travels before it hits the ground. Round to the nearest foot. 402 ft

 c. Graph the function that models the arrow's parabolic path. *

Quadratic functions can also be used to model verbal conditions. Once we have obtained a quadratic function, we can then use the *x*-coordinate of the vertex to determine its maximum or minimum value. Here is a step-by-step strategy for solving these kinds of problems:

> **Strategy for Solving Problems Involving Maximizing or Minimizing Quadratic Functions**
>
> **1.** Read the problem carefully and decide which quantity is to be maximized or minimized.
>
> **2.** Use the conditions of the problem to express the quantity as a function in one variable.
>
> **3.** Rewrite the function in the form $f(x) = ax^2 + bx + c$.
>
> **4.** Calculate $-\dfrac{b}{2a}$. If $a > 0$, f has a minimum at $x = -\dfrac{b}{2a}$. This minimum value is $f\left(-\dfrac{b}{2a}\right)$. If $a < 0$, f has a maximum at $x = -\dfrac{b}{2a}$. This maximum value is $f\left(-\dfrac{b}{2a}\right)$.
>
> **5.** Answer the question posed in the problem.

*See Answers to Selected Exercises.

EXAMPLE 6 Minimizing a Product

Among all pairs of numbers whose difference is 10, find a pair whose product is as small as possible. What is the minimum product?

SOLUTION

Step 1 Decide what must be maximized or minimized. We must minimize the product of two numbers. Calling the numbers x and y, and calling the product P, we must minimize

$$P = xy.$$

Step 2 Express this quantity as a function in one variable. In the formula $P = xy$, P is expressed in terms of two variables, x and y. However, because the difference of the numbers is 10, we can write

$$x - y = 10.$$

We can solve this equation for y in terms of x (or vice versa), substitute the result into $P = xy$, and obtain P as a function of one variable.

$$-y = -x + 10 \quad \text{Subtract } x \text{ from both sides of } x - y = 10.$$
$$y = x - 10 \quad \text{Multiply both sides of the equation by } -1 \text{ and solve for } y.$$

Now we substitute $x - 10$ for y in $P = xy$.

$$P = xy = x(x - 10)$$

Because P is now a function of x, we can write

$$P(x) = x(x - 10)$$

Step 3 Write the function in the form $f(x) = ax^2 + bx + c$. We apply the distributive property to obtain

$$P(x) = x(x - 10) = x^2 - 10x.$$

$$\boxed{a = 1} \quad \boxed{b = -10}$$

Step 4 Calculate $-\dfrac{b}{2a}$. If $a > 0$, the function has a minimum at this value. The voice balloons show that $a = 1$ and $b = -10$.

$$x = -\frac{b}{2a} = -\frac{-10}{2(1)} = -(-5) = 5$$

This means that the product, P, of two numbers whose difference is 10 is a minimum when one of the numbers, x, is 5.

Step 5 Answer the question posed by the problem. The problem asks for the two numbers and the minimum product. We found that one of the numbers, x, is 5. Now we must find the second number, y.

$$y = x - 10 = 5 - 10 = -5$$

The number pair whose difference is 10 and whose product is as small as possible is 5, -5. The minimum product is $5(-5)$, or -25. • • •

Check Point **6** Among all pairs of numbers whose difference is 8, find a pair whose product is as small as possible. What is the minimum product? 4 and -4; -16

TECHNOLOGY

Numeric Connections

The | TABLE | feature of a graphing utility can be used to verify our work in Example 6.

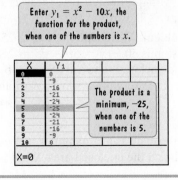

Enter $y_1 = x^2 - 10x$, the function for the product, when one of the numbers is x.

X	Y₁
0	0
1	-9
2	-16
3	-21
4	-24
5	-25
6	-24
7	-21
8	-16
9	-9
10	0

The product is a minimum, -25, when one of the numbers is 5.

X=0

EXAMPLE 7 Maximizing Area

You have 100 yards of fencing to enclose a rectangular region. Find the dimensions of the rectangle that maximize the enclosed area. What is the maximum area?

SOLUTION

Step 1 Decide what must be maximized or minimized. We must maximize area. What we do not know are the rectangle's dimensions, x and y.

Step 2 Express this quantity as a function in one variable. Because we must maximize the area of a rectangle, we have $A = xy$. We need to transform this into a function in which A is represented by one variable. Because you have 100 yards of fencing, the perimeter of the rectangle is 100 yards. This means that

$$2x + 2y = 100.$$

We can solve this equation for y in terms of x, substitute the result into $A = xy$, and obtain A as a function in one variable. We begin by solving for y.

$2y = 100 - 2x$	Subtract 2x from both sides of 2x + 2y = 100.
$y = \dfrac{100 - 2x}{2}$	Divide both sides by 2.
$y = 50 - x$	Divide each term in the numerator by 2.

Now we substitute $50 - x$ for y in $A = xy$.

$$A = xy = x(50 - x)$$

The rectangle and its dimensions are illustrated in **Figure 3.9**. Because A is now a function of x, we can write

$$A(x) = x(50 - x).$$

This function models the area, $A(x)$, of any rectangle whose perimeter is 100 yards in terms of one of its dimensions, x.

Step 3 Write the function in the form $f(x) = ax^2 + bx + c$. We apply the distributive property to obtain

$$A(x) = x(50 - x) = 50x - x^2 = -x^2 + 50x.$$

$a = -1$ $b = 50$

Step 4 Calculate $-\dfrac{b}{2a}$. If $a < 0$, the function has a maximum at this value. The voice balloons show that $a = -1$ and $b = 50$.

$$x = -\frac{b}{2a} = -\frac{50}{2(-1)} = 25$$

This means that the area, $A(x)$, of a rectangle with perimeter 100 yards is a maximum when one of the rectangle's dimensions, x, is 25 yards.

Step 5 Answer the question posed by the problem. We found that $x = 25$. **Figure 3.9** shows that the rectangle's other dimension is $50 - x = 50 - 25 = 25$. The dimensions of the rectangle that maximize the enclosed area are 25 yards by 25 yards. The rectangle that gives the maximum area is actually a square with an area of 25 yards · 25 yards, or 625 square yards. ●●●

FIGURE 3.9 What value of x will maximize the rectangle's area?

TECHNOLOGY

Graphic Connections

The graph of the area function

$$A(x) = x(50 - x)$$

was obtained with a graphing utility using a $[0, 50, 2]$ by $[0, 700, 25]$ viewing rectangle. The maximum function feature verifies that a maximum area of 625 square yards occurs when one of the dimensions is 25 yards.

☑ Check Point **7** You have 120 feet of fencing to enclose a rectangular region. Find the dimensions of the rectangle that maximize the enclosed area. What is the maximum area? dimensions: 30 ft by 30 ft; area: 900 sq ft

Blitzer Bonus || Addressing Stress Parabolically

Moderate stress, high performance

Low stress, low performance

High stress, low performance

Performance

Level of Stress

Stress levels can help or hinder performance. The parabola in **Figure 3.10** serves as a model that shows people under both low stress and high stress perform worse than their moderate-stress counterparts.

FIGURE 3.10
Source: Herbert Benson, *Your Maximum Mind*, Random House, 1987.

CONCEPT AND VOCABULARY CHECK

Fill in each blank so that the resulting statement is true.

1. The quadratic function $f(x) = a(x - h)^2 + k, a \neq 0$, is in ___standard___ form. The graph of f is called a/an ___parabola___ whose vertex is the point ___(h, k)___. The graph opens upward if a ___> 0___ and opens downward if a ___< 0___.

2. Consider the quadratic function $f(x) = ax^2 + bx + c, a \neq 0$. If $a > 0$, then f has a minimum that occurs at $x = $ ___$-\frac{b}{2a}$___. This minimum value is ___$f\left(-\frac{b}{2a}\right)$___. If $a < 0$, then f has a maximum that occurs at $x = $ ___$-\frac{b}{2a}$___. This maximum value is ___$f\left(-\frac{b}{2a}\right)$___.

3. True or false: The graph of $f(x) = (x - 2)^2 + 1$ opens upward. ___true___

4. True or false: The graph of $f(x) = (x + 5)^2 + 3$ has its vertex at $(5, 3)$. ___false___

5. True or false: The y-coordinate of the vertex of $f(x) = 4x^2 - 16x + 300$ is $f(2)$. ___true___

6. The difference between two numbers is 8. If one number is represented by x, the other number can be expressed as ___$x - 8$___. The product of the numbers, $P(x)$, expressed in the form $P(x) = ax^2 + bx + c$, is $P(x) = $ ___$x^2 - 8x$___.

7. The perimeter of a rectangle is 80 feet. If the length of the rectangle is represented by x, its width can be expressed as ___$40 - x$___. The area of the rectangle, $A(x)$, expressed in the form $A(x) = ax^2 + bx + c$, is $A(x) = $ ___$-x^2 + 40x$___.

EXERCISE SET 3.1

Practice Exercises

In Exercises 1–4, the graph of a quadratic function is given. Write the function's equation, selecting from the following options.

$$f(x) = (x + 1)^2 - 1 \qquad g(x) = (x + 1)^2 + 1$$
$$h(x) = (x - 1)^2 + 1 \qquad j(x) = (x - 1)^2 - 1$$

1.

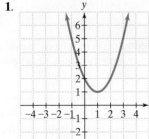

$h(x) = (x - 1)^2 + 1$

2.

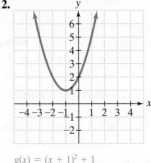

$g(x) = (x + 1)^2 + 1$

3.

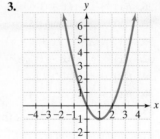

$j(x) = (x - 1)^2 - 1$

4.

$f(x) = (x + 1)^2 - 1$

In Exercises 5–8, the graph of a quadratic function is given. Write the function's equation, selecting from the following options.

$$f(x) = x^2 + 2x + 1 \qquad g(x) = x^2 - 2x + 1$$
$$h(x) = x^2 - 1 \qquad j(x) = -x^2 - 1$$

5.

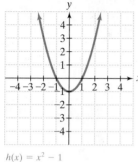

$h(x) = x^2 - 1$

6.

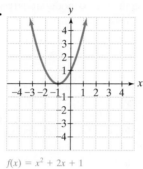

$f(x) = x^2 + 2x + 1$

7.

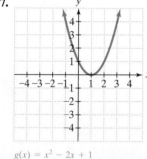

$g(x) = x^2 - 2x + 1$

8.

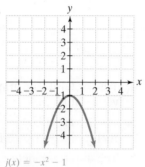

$j(x) = -x^2 - 1$

In Exercises 9–16, find the coordinates of the vertex for the parabola defined by the given quadratic function.

9. $f(x) = 2(x - 3)^2 + 1$ (3, 1) **10.** $f(x) = -3(x - 2)^2 + 12$ *

11. $f(x) = -2(x + 1)^2 + 5$ * **12.** $f(x) = -2(x + 4)^2 - 8$ *

13. $f(x) = 2x^2 - 8x + 3$ (2, -5) **14.** $f(x) = 3x^2 - 12x + 1$ *

15. $f(x) = -x^2 - 2x + 8$ (-1, 9) **16.** $f(x) = -2x^2 + 8x - 1$ *

In Exercises 17–38, use the vertex and intercepts to sketch the graph of each quadratic function. Give the equation of the parabola's axis of symmetry. Use the graph to determine the function's domain and range.

17. $f(x) = (x - 4)^2 - 1$ * **18.** $f(x) = (x - 1)^2 - 2$ *

19. $f(x) = (x - 1)^2 + 2$ * **20.** $f(x) = (x - 3)^2 + 2$ *

21. $y - 1 = (x - 3)^2$ * **22.** $y - 3 = (x - 1)^2$ *

23. $f(x) = 2(x + 2)^2 - 1$ * **24.** $f(x) = \frac{5}{4} - \left(x - \frac{1}{2}\right)^2$ *

25. $f(x) = 4 - (x - 1)^2$ * **26.** $f(x) = 1 - (x - 3)^2$ *

27. $f(x) = x^2 - 2x - 3$ * **28.** $f(x) = x^2 - 2x - 15$ *

29. $f(x) = x^2 + 3x - 10$ * **30.** $f(x) = 2x^2 - 7x - 4$ *

31. $f(x) = 2x - x^2 + 3$ * **32.** $f(x) = 5 - 4x - x^2$ *

33. $f(x) = x^2 + 6x + 3$ * **34.** $f(x) = x^2 + 4x - 1$ *

35. $f(x) = 2x^2 + 4x - 3$ * **36.** $f(x) = 3x^2 - 2x - 4$ *

37. $f(x) = 2x - x^2 - 2$ * **38.** $f(x) = 6 - 4x + x^2$ *

In Exercises 39–44, an equation of a quadratic function is given.

a. *Determine, without graphing, whether the function has a minimum value or a maximum value.*

b. *Find the minimum or maximum value and determine where it occurs.*

c. *Identify the function's domain and its range.*

*See Answers to Selected Exercises.

39. $f(x) = 3x^2 - 12x - 1$ * **40.** $f(x) = 2x^2 - 8x - 3$ *

41. $f(x) = -4x^2 + 8x - 3$ * **42.** $f(x) = -2x^2 - 12x + 3$ *

43. $f(x) = 5x^2 - 5x$ * **44.** $f(x) = 6x^2 - 6x$ *

Practice Plus

In Exercises 45–48, give the domain and the range of each quadratic function whose graph is described.

45. The vertex is $(-1, -2)$ and the parabola opens up. *

46. The vertex is $(-3, -4)$ and the parabola opens down. *

47. Maximum $= -6$ at $x = 10$ domain: $(-\infty, \infty)$; range: $(-\infty, -6]$

48. Minimum $= 18$ at $x = -6$ domain: $(-\infty, \infty)$; range: $[18, \infty)$

In Exercises 49–52, write an equation in standard form of the parabola that has the same shape as the graph of $f(x) = 2x^2$, but with the given point as the vertex.

49. $(5, 3)$ $f(x) = 2(x - 5)^2 + 3$ **50.** $(7, 4)$ $f(x) = 2(x - 7)^2 + 4$

51. $(-10, -5)$ * **52.** $(-8, -6)$ $f(x) = 2(x + 8)^2 - 6$

In Exercises 53–56, write an equation in standard form of the parabola that has the same shape as the graph of $f(x) = 3x^2$ or $g(x) = -3x^2$, but with the given maximum or minimum.

53. Maximum $= 4$ at $x = -2$ * **54.** Maximum $= -7$ at $x = 5$ *

55. Minimum $= 0$ at $x = 11$ * **56.** Minimum $= 0$ at $x = 9$ *

Application Exercises

An athlete whose event is the shot put releases the shot with the same initial velocity but at different angles. The figure shows the parabolic paths for shots released at angles of 35° and 65°. Exercises 57–58 are based on the functions that model the parabolic paths.

57. When the shot whose path is shown by the blue graph is released at an angle of 35°, its height, $f(x)$, in feet, can be modeled by

$$f(x) = -0.01x^2 + 0.7x + 6.1,$$

where x is the shot's horizontal distance, in feet, from its point of release. Use this model to solve parts (a) through (c) and verify your answers using the blue graph.

a. What is the maximum height of the shot and how far from its point of release does this occur? 18.35 ft; 35 ft

b. What is the shot's maximum horizontal distance, to the nearest tenth of a foot, or the distance of the throw? 77.8 ft

c. From what height was the shot released? 6.1 ft

58. When the shot whose path is shown by the red graph on the previous page is released at an angle of 65°, its height, $g(x)$, in feet, can be modeled by

$$g(x) = -0.04x^2 + 2.1x + 6.1,$$

where x is the shot's horizontal distance, in feet, from its point of release. Use this model to solve parts (a) through (c) and verify your answers using the red graph.

a. What is the maximum height, to the nearest tenth of a foot, of the shot and how far from its point of release does this occur? approximately 33.7 ft; 26.25 ft

b. What is the shot's maximum horizontal distance, to the nearest tenth of a foot, or the distance of the throw? 55.3 ft

c. From what height was the shot released? 6.1 ft

59. A ball is thrown upward and outward from a height of 6 feet. The height of the ball, $f(x)$, in feet, can be modeled by

$$f(x) = -0.8x^2 + 2.4x + 6,$$

where x is the ball's horizontal distance, in feet, from where it was thrown.

a. What is the maximum height of the ball and how far from where it was thrown does this occur? 7.8 ft; 1.5 ft

b. How far does the ball travel horizontally before hitting the ground? Round to the nearest tenth of a foot. 4.6 ft

c. Graph the function that models the ball's parabolic path. *

60. A ball is thrown upward and outward from a height of 6 feet. The height of the ball, $f(x)$, in feet, can be modeled by

$$f(x) = -0.8x^2 + 3.2x + 6,$$

where x is the ball's horizontal distance, in feet, from where it was thrown.

a. What is the maximum height of the ball and how far from where it was thrown does this occur? 9.2 ft; 2 ft

b. How far does the ball travel horizontally before hitting the ground? Round to the nearest tenth of a foot. 5.4 ft

c. Graph the function that models the ball's parabolic path. *

61. Among all pairs of numbers whose sum is 16, find a pair whose product is as large as possible. What is the maximum product? 8 and 8; 64

62. Among all pairs of numbers whose sum is 20, find a pair whose product is as large as possible. What is the maximum product? 10 and 10; 100

63. Among all pairs of numbers whose difference is 16, find a pair whose product is as small as possible. What is the minimum product? 8 and −8; −64

64. Among all pairs of numbers whose difference is 24, find a pair whose product is as small as possible. What is the minimum product? 12 and −12; −144

65. You have 600 feet of fencing to enclose a rectangular plot that borders on a river. If you do not fence the side along the river, find the length and width of the plot that will maximize the area. What is the largest area that can be enclosed? length: 300 ft; width: 150 ft; maximum area: 45,000 sq ft

66. You have 200 feet of fencing to enclose a rectangular plot that borders on a river. If you do not fence the side along the river, find the length and width of the plot that will maximize the area. What is the largest area that can be enclosed? length: 100 ft; width: 50 ft; maximum area: 5000 sq ft

67. You have 50 yards of fencing to enclose a rectangular region. Find the dimensions of the rectangle that maximize the enclosed area. What is the maximum area? *

68. You have 80 yards of fencing to enclose a rectangular region. Find the dimensions of the rectangle that maximize the enclosed area. What is the maximum area? *

69. A rectangular playground is to be fenced off and divided in two by another fence parallel to one side of the playground. Six hundred feet of fencing is used. Find the dimensions of the playground that maximize the total enclosed area. What is the maximum area? 150 ft by 100 ft; 15,000 sq ft

70. A rectangular playground is to be fenced off and divided in two by another fence parallel to one side of the playground. Four hundred feet of fencing is used. Find the dimensions of the playground that maximize the total enclosed area. What is the maximum area? 100 ft by $66\frac{2}{3}$ ft; $6666\frac{2}{3}$ sq ft

71. A rain gutter is made from sheets of aluminum that are 20 inches wide by turning up the edges to form right angles. Determine the depth of the gutter that will maximize its cross-sectional area and allow the greatest amount of water to flow. What is the maximum cross-sectional area? 5 in.; 50 sq in.

Flat sheet 20 inches wide

72. A rain gutter is made from sheets of aluminum that are 12 inches wide by turning up the edges to form right angles. Determine the depth of the gutter that will maximize its cross-sectional area and allow the greatest amount of water to flow. What is the maximum cross-sectional area? *

73. Hunky Beef, a local sandwich store, has a fixed weekly cost of $525.00, and variable costs for making a roast beef sandwich are $0.55.

a. Let x represent the number of roast beef sandwiches made and sold each week. Write the weekly cost function, C, for Hunky Beef. (*Hint:* The cost function is the sum of fixed and variable costs.) $C(x) = 0.55x + 525$

b. The function $R(x) = -0.001x^2 + 3x$ describes the money, in dollars, that Hunky Beef takes in each week from the sale of x roast beef sandwiches. Use this revenue function and the cost function from part (a) to write the store's weekly profit function, P. (*Hint:* The profit function is the difference between the revenue and cost functions.) $P(x) = -0.001x^2 + 2.45x - 525$

c. Use the store's profit function to determine the number of roast beef sandwiches it should make and sell each week to maximize profit. What is the maximum weekly profit? 1225 sandwiches; ≈ $975.63

Explaining the Concepts 74–79. Answers will vary.

74. What is a quadratic function?

75. What is a parabola? Describe its shape.

76. Explain how to decide whether a parabola opens upward or downward.

77. Describe how to find a parabola's vertex if its equation is expressed in standard form. Give an example.

78. Describe how to find a parabola's vertex if its equation is in the form $f(x) = ax^2 + bx + c$. Use $f(x) = x^2 - 6x + 8$ as an example.

79. A parabola that opens upward has its vertex at (1, 2). Describe as much as you can about the parabola based on this information. Include in your discussion the number of x-intercepts (if any) for the parabola.

Technology Exercises

80. Use a graphing utility to verify any five of your hand-drawn graphs in Exercises 17–38. Answers will vary.

81. **a.** Use a graphing utility to graph $y = 2x^2 - 82x + 720$ in a standard viewing rectangle. What do you observe? *

b. Find the coordinates of the vertex for the given quadratic function. $(20.5, -120.5)$

c. The answer to part (b) is $(20.5, -120.5)$. Because the leading coefficient, 2, of the given function is positive, the vertex is a minimum point on the graph. Use this fact to help find a viewing rectangle that will give a relatively complete picture of the parabola. With an axis of symmetry at $x = 20.5$, the setting for x should extend past this, so try Xmin = 0 and Xmax = 30. The setting for y should include (and probably go below) the y-coordinate of the graph's minimum y-value, so try Ymin = −130. Experiment with Ymax until your utility shows the parabola's major features. Try Ymax = 750.

d. In general, explain how knowing the coordinates of a parabola's vertex can help determine a reasonable viewing rectangle on a graphing utility for obtaining a complete picture of the parabola. *

In Exercises 82–85, find the vertex for each parabola. Then determine a reasonable viewing rectangle on your graphing utility and use it to graph the quadratic function.

82. $y = -0.25x^2 + 40x$ (80, 1600)*

83. $y = -4x^2 + 20x + 160$ (2.5, 185)*

84. $y = 5x^2 + 40x + 600$ (−4, 520)*

85. $y = 0.01x^2 + 0.6x + 100$ (−30, 91)*

86. The bar graph shows the ratings of *American Idol* from season 1 (2002) through season 12 (2013).

American Idol: Each Season's Champion and Average Number of Viewers

Source: Nielsen

a. Let x represent *American Idol*'s season number and let y represent the average number of viewers, in millions. Use a graphing utility to draw a scatter plot of the data. Explain why a quadratic function is appropriate for modeling these data. *

b. Use the quadratic regression feature to find the quadratic function that best fits the data. Round all numbers to two decimal places. $f(x) = -0.48x^2 + 6.17x + 9.57$

c. Use the model in part (b) to determine the season in which *American Idol* had the greatest number of viewers.

Round to the nearest whole number. According to the model, how many millions of viewers were there in that season? Round to one decimal place. 6; 29.3 million

d. How do the results obtained from the model in part (c) compare with the data displayed by the graph? *

e. Use a graphing utility to draw a scatter plot of the data and graph the quadratic function of best fit on the scatter plot. Can you see why projections based on the graph had the producers of *American Idol* looking for a shake-up? No shake-up was found and the show's final season aired in 2016. *

Critical Thinking Exercises

Make Sense? *In Exercises 87–90, determine whether each statement makes sense or does not make sense, and explain your reasoning.*

87. I must have made an error when graphing this parabola because its axis of symmetry is the *y*-axis. does not make sense

88. I like to think of a parabola's vertex as the point where it intersects its axis of symmetry. makes sense

89. I threw a baseball vertically upward and its path was a parabola. does not make sense

90. **Figure 3.7** on page 354 shows that a linear function provides a better description of the football's path than a quadratic function. does not make sense

In Exercises 91–94, determine whether each statement is true or false. If the statement is false, make the necessary change(s) to produce a true statement. Changes to false statements will vary.

91. No quadratic functions have a range of $(-\infty, \infty)$. true

92. The vertex of the parabola described by $f(x) = 2(x-5)^2 - 1$ is at $(5, 1)$. false

93. The graph of $f(x) = -2(x+4)^2 - 8$ has one *y*-intercept and two *x*-intercepts. false

94. The maximum value of *y* for the quadratic function $f(x) = -x^2 + x + 1$ is 1. false

In Exercises 95–96, find the axis of symmetry for each parabola whose equation is given. Use the axis of symmetry to find a second point on the parabola whose y-coordinate is the same as the given point.

95. $f(x) = 3(x+2)^2 - 5$; $(-1, -2)$ $x = -2; (-3, -2)$

96. $f(x) = (x-3)^2 + 2$; $(6, 11)$ $x = 3; (0, 11)$

In Exercises 97–98, write the equation of each parabola in standard form.

97. Vertex: $(-3, -4)$; The graph passes through the point $(1, 4)$. $f(x) = \frac{1}{2}(x+3)^2 - 4$

98. Vertex: $(-3, -1)$; The graph passes through the point $(-2, -3)$. $f(x) = -2(x+3)^2 - 1$

99. A rancher has 1000 feet of fencing to construct six corrals, as shown in the figure. Find the dimensions that maximize the enclosed area. What is the maximum area?

$x = 166\frac{2}{3}$ ft;
$y = 125$ ft;
$\approx 20{,}833$ ft^2

100. The annual yield per lemon tree is fairly constant at 320 pounds when the number of trees per acre is 50 or fewer. For each additional tree over 50, the annual yield per tree for all trees on the acre decreases by 4 pounds due to overcrowding. Find the number of trees that should be planted on an acre to produce the maximum yield. How many pounds is the maximum yield? 65 trees; 16,900 lb

Group Exercise Answers will vary.

101. Each group member should consult an almanac, newspaper, magazine, or the Internet to find data that initially increase and then decrease, or vice versa, and therefore can be modeled by a quadratic function. Group members should select the two sets of data that are most interesting and relevant. For each data set selected,

 a. Use the quadratic regression feature of a graphing utility to find the quadratic function that best fits the data.

 b. Use the equation of the quadratic function to make a prediction from the data. What circumstances might affect the accuracy of your prediction?

 c. Use the equation of the quadratic function to write and solve a problem involving maximizing or minimizing the function.

Retaining the Concepts

102. Does the equation $3x + y^2 = 10$ define *y* as a function of *x*? (Section 2.1, Example 3) no

103. Use the following graph to solve this exercise.

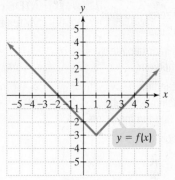

 a. Determine the function's domain. $(-\infty, \infty)$

 b. Determine the function's range. $[-3, \infty)$

 c. What are the *x*-intercepts? -2 and 4

 d. What is the *y*-intercept? -2

 e. Determine $f(-4)$. $f(-4) = 2$
 (Section 2.1, Examples 7 and 8)

104. If $f(x) = 4x^2 - 2x + 7$, find

$$\frac{f(x+h) - f(x)}{h}, h \neq 0,$$

 and simplify. (Section 2.2, Example 8) $8x + 4h - 2$

Preview Exercises

Exercises 105–107 will help you prepare for the material covered in the next section.

105. Factor: $x^3 + 3x^2 - x - 3$. $(x+3)(x+1)(x-1)$

106. If $f(x) = x^3 - 2x - 5$, find $f(2)$ and $f(3)$. Then explain why the continuous graph of f must cross the *x*-axis between 2 and 3. $f(2) = -1; f(3) = 16$*

107. Determine whether $f(x) = x^4 - 2x^2 + 1$ is even, odd, or neither. Describe the symmetry, if any, for the graph of f. even; *y*-axis symmetry

Section 3.2 | Polynomial Functions and Their Graphs

What am I supposed to learn?

After studying this section, you should be able to:

1. Identify polynomial functions.
2. Recognize characteristics of graphs of polynomial functions.
3. Determine end behavior.
4. Use factoring to find zeros of polynomial functions.
5. Identify zeros and their multiplicities.
6. Use the Intermediate Value Theorem.
7. Understand the relationship between degree and turning points.
8. Graph polynomial functions.

In 1980, U.S. doctors diagnosed 41 cases of a rare form of cancer, Kaposi's sarcoma, that involved skin lesions, pneumonia, and severe immunological deficiencies. All cases involved gay men ranging in age from 26 to 51. By the end of 2015, more than one million Americans, straight and gay, male and female, old and young, were infected with the HIV virus.

Modeling AIDS-related data and making predictions about the epidemic's havoc is serious business. **Figure 3.11** shows the number of AIDS cases diagnosed in the United States from 1983 through 2011.

Basketball player Magic Johnson (1959–) tested positive for HIV in 1991.

1. Identify polynomial functions.

Changing circumstances and unforeseen events can result in models for AIDS-related data that are not particularly useful over long periods of time. For example, the function

$$f(x) = -49x^3 + 806x^2 + 3776x + 2503$$

models the number of AIDS cases diagnosed in the United States x years after 1983. The model was obtained using a portion of the data shown in **Figure 3.11**, namely

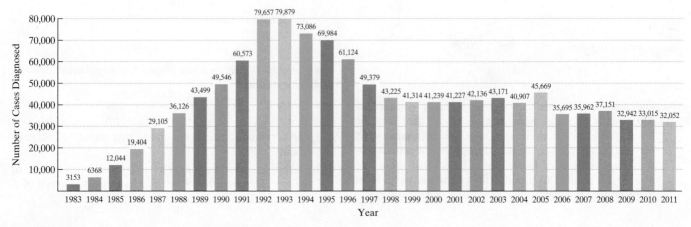

AIDS Cases Diagnosed in the United States, 1983–2011

FIGURE 3.11
Source: Department of Health and Human Services

$$f(x) = -49x^3 + 806x^2 + 3776x + 2503$$

Cases Diagnosed

Years after 1983

[0, 8, 1] by [0, 60,000, 5000]

FIGURE 3.12 The graph of a function modeling the number of AIDS diagnoses from 1983 through 1991

GREAT QUESTION!

What kind of notation is $a_n, a_{n-1}, a_2, a_1, a_0$? I find it confusing.

When defining polynomial functions, we use *subscripted letters*. **Subscripts** are written next to, but slightly below, each a.

a_n ("a sub-n"):
the coefficient of x^n

a_{n-1} ("a sub-$n$$-$1"):
the coefficient of x^{n-1}

a_2 ("a sub-2"):
the coefficient of x^2

a_1 ("a sub-1"):
the coefficient of x

a_0 ("a sub-0"):
the constant term

cases diagnosed from 1983 through 1991, inclusive. **Figure 3.12** shows the graph of f from 1983 through 1991. This function is an example of a *polynomial function of degree 3*.

Definition of a Polynomial Function

Let n be a nonnegative integer and let $a_n, a_{n-1}, \ldots, a_2, a_1, a_0$ be real numbers, with $a_n \neq 0$. The function defined by

$$f(x) = a_n x^n + a_{n-1} x^{n-1} + \cdots + a_2 x^2 + a_1 x + a_0$$

is called a **polynomial function of degree n**. The number a_n, the coefficient of the variable to the highest power, is called the **leading coefficient**.

Polynomial Functions

$$f(x) = -3x^5 + \sqrt{2}x^2 + 5$$

Polynomial function of degree 5

$$\begin{aligned} g(x) &= -3x^4(x-2)(x+3) \\ &= -3x^4(x^2 + x - 6) \\ &= -3x^6 - 3x^5 + 18x^4 \end{aligned}$$

Polynomial function of degree 6

Not Polynomial Functions

$$\begin{aligned} F(x) &= -3\sqrt{x} + \sqrt{2}x^2 + 5 \\ &= -3x^{\frac{1}{2}} + \sqrt{2}x^2 + 5 \end{aligned}$$

The exponent on the variable is not an integer.

$$\begin{aligned} G(x) &= -\frac{3}{x^2} + \sqrt{2}x^2 + 5 \\ &= -3x^{-2} + \sqrt{2}x^2 + 5 \end{aligned}$$

The exponent on the variable is not a nonnegative integer.

A constant function $f(x) = c$, where $c \neq 0$, is a polynomial function of degree 0. A linear function $f(x) = mx + b$, where $m \neq 0$, is a polynomial function of degree 1. A quadratic function $f(x) = ax^2 + bx + c$, where $a \neq 0$, is a polynomial function of degree 2. In this section, we focus on polynomial functions of degree 3 or higher.

② Recognize characteristics of graphs of polynomial functions.

Smooth, Continuous Graphs

Polynomial functions of degree 2 or higher have graphs that are *smooth* and *continuous*. By **smooth**, we mean that the graphs contain only rounded curves with no sharp corners. By **continuous**, we mean that the graphs have no breaks and can be drawn without lifting your pencil from the rectangular coordinate system. These ideas are illustrated in **Figure 3.13**.

Graphs of Polynomial Functions

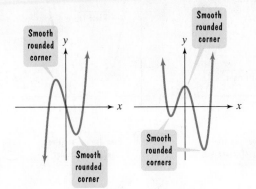

Not Graphs of Polynomial Functions

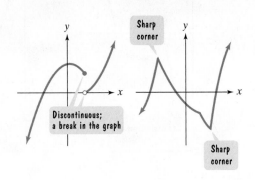

FIGURE 3.13 Recognizing graphs of polynomial functions

3 Determine end behavior.

Years after 1983

[0, 22, 1] by [−10,000, 85,000, 5000]

FIGURE 3.14 By extending the viewing rectangle, we see that y is eventually negative and the function no longer models the number of AIDS cases.

End Behavior of Polynomial Functions

Figure 3.14 shows the graph of the function

$$f(x) = -49x^3 + 806x^2 + 3776x + 2503,$$

which models the number of U.S. AIDS diagnoses from 1983 through 1991. Look at what happens to the graph when we extend the year up through 2005. By year 21 (2004), the values of y are negative and the function no longer models AIDS diagnoses. We've added an arrow to the graph at the far right to emphasize that it continues to decrease without bound. It is this far-right *end behavior* of the graph that makes it inappropriate for modeling AIDS cases into the future.

The behavior of the graph of a function to the far left or the far right is called its **end behavior**. Although the graph of a polynomial function may have intervals where it increases or decreases, the graph will eventually rise or fall without bound as it moves far to the left or far to the right.

How can you determine whether the graph of a polynomial function goes up or down at each end? The end behavior of a polynomial function

$$f(x) = a_n x^n + a_{n-1} x^{n-1} + \cdots + a_1 x + a_0$$

depends upon the leading term $a_n x^n$, because when $|x|$ is large, the other terms are relatively insignificant in size. In particular, the sign of the leading coefficient, a_n, and the degree, n, of the polynomial function reveal its end behavior. In terms of end behavior, only the term of highest degree counts, as summarized by the **Leading Coefficient Test**.

The Leading Coefficient Test

As x increases or decreases without bound, the graph of the polynomial function

$$f(x) = a_n x^n + a_{n-1} x^{n-1} + a_{n-2} x^{n-2} + \cdots + a_1 x + a_0 \quad (a_n \neq 0)$$

eventually rises or falls. In particular,

1. For n odd:

If the leading coefficient is positive, the graph falls to the left and rises to the right. (\swarrow, \nearrow)

$a_n > 0$

If the leading coefficient is negative, the graph rises to the left and falls to the right. (\nwarrow, \searrow)

$a_n < 0$

2. For n even:

If the leading coefficient is positive, the graph rises to the left and rises to the right. (\nwarrow, \nearrow)

$a_n > 0$

If the leading coefficient is negative, the graph falls to the left and falls to the right. (\swarrow, \searrow)

$a_n < 0$

Odd degree; positive leading coefficient

Odd degree; negative leading coefficient

Even degree; positive leading coefficient

Even degree; negative leading coefficient

DISCOVERY

Verify each of the four cases of the Leading Coefficient Test by using a graphing utility to graph $f(x) = x^3$, $f(x) = -x^3$, $f(x) = x^2$, and $f(x) = -x^2$.

GREAT QUESTION!

What's the bottom line on the Leading Coefficient Test?

Odd-degree polynomial functions have graphs with opposite behavior at each end. Even-degree polynomial functions have graphs with the same behavior at each end. Here's a table to help you remember the details:

Leading Term: $a_n x^n$

	Odd n	Even n
$a_n > 0$	↙ ↗	↖ ↗
$a_n < 0$	↖ ↘	↙ ↘

Opposite behavior at each end Same behavior at each end

EXAMPLE 1 Using the Leading Coefficient Test

Use the Leading Coefficient Test to determine the end behavior of the graph of

$$f(x) = x^3 + 3x^2 - x - 3.$$

SOLUTION

We begin by identifying the sign of the leading coefficient and the degree of the polynomial.

$$f(x) = x^3 + 3x^2 - x - 3$$

The leading coefficient, 1, is positive. The degree of the polynomial, 3, is odd.

The degree of the function f is 3, which is odd. Odd-degree polynomial functions have graphs with opposite behavior at each end. The leading coefficient, 1, is positive. Thus, the graph falls to the left and rises to the right (↙, ↗). The graph of f is shown in **Figure 3.15**. •••

Rises right

Falls left

FIGURE 3.15 The graph of $f(x) = x^3 + 3x^2 - x - 3$

☑ **Check Point 1** Use the Leading Coefficient Test to determine the end behavior of the graph of $f(x) = x^4 - 4x^2$. The graph rises to the left and to the right.

EXAMPLE 2 Using the Leading Coefficient Test

Use the Leading Coefficient Test to determine the end behavior of the graph of

$$f(x) = -4x^3(x - 1)^2(x + 5).$$

SOLUTION

Although the equation for f is in factored form, it is not necessary to multiply to determine the degree of the function.

$$f(x) = -4x^3(x - 1)^2(x + 5)$$

Degree of this factor is 3. Degree of this factor is 2. Degree of this factor is 1.

$$f(x) = -4x^3(x - 1)^2(x + 5)$$

The given function (repeated)

When multiplying exponential expressions with the same base, we add the exponents. This means that the degree of f is $3 + 2 + 1$, or 6, which is even. Even-degree polynomial functions have graphs with the same behavior at each end. Without multiplying out, you can see that the leading coefficient is -4, which is negative. Thus, the graph of f falls to the left and falls to the right (\swarrow, \searrow). • • •

⊘ Check Point **2** Use the Leading Coefficient Test to determine the end behavior of the graph of $f(x) = 2x^3(x - 1)(x + 5)$. The graph falls to the left and rises to the right.

EXAMPLE 3 Using the Leading Coefficient Test

Use end behavior to explain why

$$f(x) = -49x^3 + 806x^2 + 3776x + 2503$$

is only an appropriate model for AIDS diagnoses for a limited time period.

SOLUTION

We begin by identifying the sign of the leading coefficient and the degree of the polynomial.

$$f(x) = -49x^3 + 806x^2 + 3776x + 2503$$

The leading coefficient, -49, is negative. The degree of the polynomial, 3, is odd.

The degree of f is 3, which is odd. Odd-degree polynomial functions have graphs with opposite behavior at each end. The leading coefficient, -49, is negative. Thus, the graph rises to the left and falls to the right (\nwarrow, \searrow). The fact that the graph falls to the right indicates that at some point the number of AIDS diagnoses will be negative, an impossibility. If a function has a graph that decreases without bound over time, it will not be capable of modeling nonnegative phenomena over long time periods. Model breakdown will eventually occur. • • •

⊘ Check Point **3** The polynomial function

$$f(x) = -0.27x^3 + 9.2x^2 - 102.9x + 400$$

models the ratio of students to computers in U.S. public schools x years after 1980. Use end behavior to determine whether this function could be an appropriate model for computers in the classroom well into the twenty-first century. Explain your answer. Since n is odd and the leading coefficient is negative, the graph falls to the right. Since the ratio cannot be negative, the model won't be appropriate.

If you use a graphing utility to graph a polynomial function, it is important to select a viewing rectangle that accurately reveals the graph's end behavior. If the viewing rectangle, or window, is too small, it may not accurately show a complete graph with the appropriate end behavior.

EXAMPLE 4 Using the Leading Coefficient Test

The graph of $f(x) = -x^4 + 8x^3 + 4x^2 + 2$ was obtained with a graphing utility using a $[-8, 8, 1]$ by $[-10, 10, 1]$ viewing rectangle. The graph is shown in **Figure 3.16**. Is this a complete graph that shows the end behavior of the function?

$[-8, 8, 1]$ by $[-10, 10, 1]$

FIGURE 3.16

SOLUTION

We begin by identifying the sign of the leading coefficient and the degree of the polynomial.

$$f(x) = -x^4 + 8x^3 + 4x^2 + 2$$

The leading coefficient, −1, is negative. The degree of the polynomial, 4, is even.

The degree of f is 4, which is even. Even-degree polynomial functions have graphs with the same behavior at each end. The leading coefficient, −1, is negative. Thus, the graph should fall to the left and fall to the right (\swarrow, \searrow). The graph in **Figure 3.16** is falling to the left, but it is not falling to the right. Therefore, the graph is not complete enough to show end behavior. A more complete graph of the function is shown in a larger viewing rectangle in **Figure 3.17**.

$[-10, 10, 1]$ by $[-1000, 750, 250]$

FIGURE 3.17

● ● ●

☑ Check Point **4** The graph of $f(x) = x^3 + 13x^2 + 10x - 4$ is shown in a standard viewing rectangle in **Figure 3.18**. Use the Leading Coefficient Test to determine whether this is a complete graph that shows the end behavior of the function. Explain your answer. The graph does not show the end behavior of the function. The graph should fall to the left.

FIGURE 3.18

④ Use factoring to find zeros of polynomial functions.

Zeros of Polynomial Functions

If f is a polynomial function, then the values of x for which $f(x)$ is equal to 0 are called the **zeros** of f. These values of x are the **roots**, or **solutions**, of the polynomial equation $f(x) = 0$. Each real root of the polynomial equation appears as an x-intercept of the graph of the polynomial function.

EXAMPLE 5 Finding Zeros of a Polynomial Function

Find all zeros of $f(x) = x^3 + 3x^2 - x - 3$.

SOLUTION

By definition, the zeros are the values of x for which $f(x)$ is equal to 0. Thus, we set $f(x)$ equal to 0:

$$f(x) = x^3 + 3x^2 - x - 3 = 0.$$

We solve the polynomial equation $x^3 + 3x^2 - x - 3 = 0$ for x as follows:

$x^3 + 3x^2 - x - 3 = 0$	This is the equation needed to find the function's zeros.
$x^2(x + 3) - 1(x + 3) = 0$	Factor x^2 from the first two terms and −1 from the last two terms.
$(x + 3)(x^2 - 1) = 0$	A common factor of $x + 3$ is factored from the expression.
$x + 3 = 0$ or $x^2 - 1 = 0$	Set each factor equal to 0.
$x = -3$ $x^2 = 1$	Solve for x.
$x = \pm 1$	Remember that if $x^2 = d$, then $x = \pm \sqrt{d}$.

The zeros of f are −3, −1, and 1. The graph of f in **Figure 3.19** shows that each zero is an x-intercept. The graph passes through the points $(-3, 0)$, $(-1, 0)$, and $(1, 0)$. ● ● ●

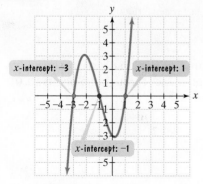

x-intercept: −3 x-intercept: 1

x-intercept: −1

FIGURE 3.19

TECHNOLOGY

Graphic and Numeric Connections

A graphing utility can be used to verify that -3, -1, and 1 are the three real zeros of $f(x) = x^3 + 3x^2 - x - 3$.

Numeric Check

Display a table for the function.

Graphic Check

Display a graph for the function. The x-intercepts indicate that -3, -1, and 1 are the real zeros.

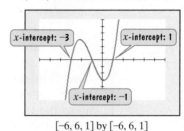

$[-6, 6, 1]$ by $[-6, 6, 1]$

The utility's $\boxed{\text{ZERO}}$ feature on the graph of f also verifies that -3, -1, and 1 are the function's real zeros.

✓ **Check Point 5** Find all zeros of $f(x) = x^3 + 2x^2 - 4x - 8$. -2 and 2

EXAMPLE 6 Finding Zeros of a Polynomial Function

Find all zeros of $f(x) = -x^4 + 4x^3 - 4x^2$.

SOLUTION

We find the zeros of f by setting $f(x)$ equal to 0 and solving the resulting equation.

$$-x^4 + 4x^3 - 4x^2 = 0 \quad \text{We now have a polynomial equation.}$$

$$x^4 - 4x^3 + 4x^2 = 0 \quad \text{Multiply both sides by } -1. \text{ This step is optional.}$$

$$x^2(x^2 - 4x + 4) = 0 \quad \text{Factor out } x^2.$$

$$x^2(x - 2)^2 = 0 \quad \text{Factor completely.}$$

$$x^2 = 0 \quad \text{or} \quad (x - 2)^2 = 0 \quad \text{Set each factor equal to O.}$$

$$x = 0 \qquad\qquad x = 2 \quad \text{Solve for x.}$$

The zeros of $f(x) = -x^4 + 4x^3 - 4x^2$ are 0 and 2. The graph of f, shown in **Figure 3.20**, has x-intercepts at 0 and 2. The graph passes through the points $(0, 0)$ and $(2, 0)$. •••

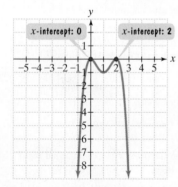

FIGURE 3.20 The zeros of $f(x) = -x^4 + 4x^3 - 4x^2$, namely 0 and 2, are the x-intercepts for the graph of f.

✓ **Check Point 6** Find all zeros of $f(x) = x^4 - 4x^2$. $-2, 0,$ and 2

⑤ Identify zeros and their multiplicities.

Multiplicities of Zeros

We can use the results of factoring to express a polynomial as a product of factors. For instance, in Example 6, we can use our factoring to express the function's equation as follows:

$$f(x) = -x^4 + 4x^3 - 4x^2 = -(x^4 - 4x^3 + 4x^2) = -x^2(x - 2)^2.$$

The factor x occurs twice: $x^2 = x \cdot x$.

The factor $(x - 2)$ occurs twice: $(x - 2)^2 = (x - 2)(x - 2)$.

Notice that each factor occurs twice. In factoring the equation for the polynomial function f, if the same factor $x - r$ occurs k times, but not $k + 1$ times, we call r a **zero with multiplicity k**. For the polynomial function

$$f(x) = -x^2(x - 2)^2,$$

0 and 2 are both zeros with multiplicity 2.

Multiplicity provides another connection between zeros and graphs. The multiplicity of a zero tells us whether the graph of a polynomial function touches the x-axis at the zero and turns around or if the graph crosses the x-axis at the zero. For example, look again at the graph of $f(x) = -x^4 + 4x^3 - 4x^2$ in **Figure 3.20**. Each zero, 0 and 2, is a zero with multiplicity 2. The graph of f touches, but does not cross, the x-axis at each of these zeros of even multiplicity. By contrast, a graph crosses the x-axis at zeros of odd multiplicity.

Multiplicity and x-Intercepts

If r is a zero of **even multiplicity**, then the graph **touches** the x-axis **and turns around** at r. If r is a zero of **odd multiplicity**, then the graph **crosses** the x-axis at r. Regardless of whether the multiplicity of a zero is even or odd, graphs tend to flatten out near zeros with multiplicity greater than one.

If a polynomial function's equation is expressed as a product of linear factors, we can quickly identify zeros and their multiplicities.

EXAMPLE 7 Finding Zeros and Their Multiplicities

Find the zeros of $f(x) = \frac{1}{2}(x + 1)(2x - 3)^2$ and give the multiplicity of each zero. State whether the graph crosses the x-axis or touches the x-axis and turns around at each zero.

SOLUTION

We find the zeros of f by setting $f(x) = \frac{1}{2}(x + 1)(2x - 3)^2$ equal to 0:

$$\frac{1}{2}(x + 1)(2x - 3)^2 = 0.$$

Set each variable factor equal to 0.

$x + 1 = 0$	$2x - 3 = 0$
$x = -1$	$x = \frac{3}{2}$

$$\frac{1}{2}(x + 1)^1(2x - 3)^2 = 0$$

This exponent is 1. Thus, the multiplicity of -1 is **1**.

This exponent is 2. Thus, the multiplicity of $\frac{3}{2}$ is **2**.

-1 is a zero of odd multiplicity. Graph crosses x-axis.

$\frac{3}{2}$ is a zero of even multiplicity. Graph touches x-axis, flattens, and turns around.

$[-3, 3, 1]$ by $[-10, 10, 1]$

FIGURE 3.21 The graph of $f(x) = \frac{1}{2}(x + 1)(2x - 3)^2$

The zeros of $f(x) = \frac{1}{2}(x + 1)(2x - 3)^2$ are -1, with multiplicity 1, and $\frac{3}{2}$, with multiplicity 2. Because the multiplicity of -1 is odd, the graph crosses the x-axis at this zero. Because the multiplicity of $\frac{3}{2}$ is even, the graph touches the x-axis and turns around at this zero. These relationships are illustrated by the graph of f in **Figure 3.21**.

● ● ●

✓ **Check Point 7** Find the zeros of $f(x) = -4\left(x + \frac{1}{2}\right)^2(x - 5)^3$ and give the multiplicity of each zero. State whether the graph crosses the x-axis or touches the x-axis and turns around at each zero. $-\frac{1}{2}$ with multiplicity 2 and 5 with multiplicity 3; touches and turns at $-\frac{1}{2}$ and crosses at 5

⑥ Use the Intermediate Value Theorem.

The Intermediate Value Theorem

The *Intermediate Value Theorem* tells us of the existence of real zeros. The idea behind the theorem is illustrated in **Figure 3.22**. The figure shows that if $(a, f(a))$ lies below the x-axis and $(b, f(b))$ lies above the x-axis, the smooth, continuous graph of the polynomial function f must cross the x-axis at some value c between a and b. This value is a real zero for the function.

These observations are summarized in the **Intermediate Value Theorem**.

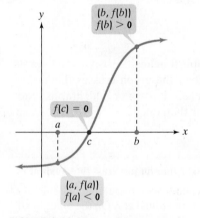

FIGURE 3.22 The graph must cross the x-axis at some value between a and b.

The Intermediate Value Theorem for Polynomial Functions

Let f be a polynomial function with real coefficients. If $f(a)$ and $f(b)$ have opposite signs, then there is at least one value of c between a and b for which $f(c) = 0$. Equivalently, the equation $f(x) = 0$ has at least one real root between a and b.

EXAMPLE 8 Using the Intermediate Value Theorem

Show that the polynomial function $f(x) = x^3 - 2x - 5$ has a real zero between 2 and 3.

SOLUTION

Let us evaluate f at 2 and at 3. If $f(2)$ and $f(3)$ have opposite signs, then there is at least one real zero between 2 and 3. Using $f(x) = x^3 - 2x - 5$, we obtain

$$f(2) = 2^3 - 2 \cdot 2 - 5 = 8 - 4 - 5 = -1$$

$f(2)$ is negative.

and

$$f(3) = 3^3 - 2 \cdot 3 - 5 = 27 - 6 - 5 = 16.$$

$f(3)$ is positive.

[−3, 3, 1] by [−10, 10, 1]

FIGURE 3.23

⑦ Understand the relationship between degree and turning points.

⑧ Graph polynomial functions.

Because $f(2) = -1$ and $f(3) = 16$, the sign change shows that the polynomial function has a real zero between 2 and 3. This zero is actually irrational and is approximated using a graphing utility's ZERO feature as 2.0945515 in **Figure 3.23**.

● ● ●

✓ Check Point **8** Show that the polynomial function $f(x) = 3x^3 - 10x + 9$ has a real zero between −3 and −2. $f(-3) = -42; f(-2) = 5$

Turning Points of Polynomial Functions

The graph of $f(x) = x^5 - 6x^3 + 8x + 1$ is shown in **Figure 3.24**. The graph has four smooth **turning points**.

At each turning point in **Figure 3.24**, the graph changes direction from increasing to decreasing or vice versa. The given equation has 5 as its greatest exponent and is therefore a polynomial function of degree 5. Notice that the graph has four turning points. In general, **if f is a polynomial function of degree n, then the graph of f has at most $n - 1$ turning points.**

Figure 3.24 illustrates that the y-coordinate of each turning point is either a relative maximum or a relative minimum of f. Without the aid of a graphing utility or a knowledge of

FIGURE 3.24 Graph with four turning points

calculus, it is difficult and often impossible to locate turning points of polynomial functions with degrees greater than 2. If necessary, test values can be taken between the x-intercepts to get a general idea of how high the graph rises or how low the graph falls. For the purpose of graphing in this section, a general estimate is sometimes appropriate and necessary.

A Strategy for Graphing Polynomial Functions

Here's a general strategy for graphing a polynomial function. A graphing utility is a valuable complement, but not a necessary component, to this strategy. If you are using a graphing utility, some of the steps listed in the following box will help you to select a viewing rectangle that shows the important parts of the graph.

GREAT QUESTION!

When graphing a polynomial function, how do I determine the location of its turning points?

Without calculus, it is often impossible to give the exact location of turning points. However, you can obtain additional points satisfying the function to estimate how high the graph rises or how low it falls. To obtain these points, use values of x between (and to the left and right of) the x-intercepts.

Graphing a Polynomial Function

$$f(x) = a_n x^n + a_{n-1}x^{n-1} + a_{n-2}x^{n-2} + \cdots + a_1 x + a_0, a_n \neq 0$$

1. Use the Leading Coefficient Test to determine the graph's end behavior.
2. Find x-intercepts by setting $f(x) = 0$ and solving the resulting polynomial equation. If there is an x-intercept at r as a result of $(x - r)^k$ in the complete factorization of $f(x)$, then
 a. If k is even, the graph touches the x-axis at r and turns around.
 b. If k is odd, the graph crosses the x-axis at r.
 c. If $k > 1$, the graph flattens out near $(r, 0)$.
3. Find the y-intercept by computing $f(0)$.
4. Use symmetry, if applicable, to help draw the graph:
 a. y-axis symmetry: $f(-x) = f(x)$
 b. Origin symmetry: $f(-x) = -f(x)$.
5. Use the fact that the maximum number of turning points of the graph is $n - 1$, where n is the degree of the polynomial function, to check whether it is drawn correctly.

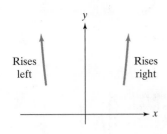

EXAMPLE 9 Graphing a Polynomial Function

Graph: $f(x) = x^4 - 2x^2 + 1$.

SOLUTION

Step 1 **Determine end behavior.** Identify the sign of a_n, the leading coefficient, and the degree, n, of the polynomial function.

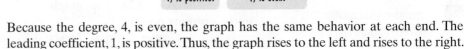

$$f(x) = x^4 - 2x^2 + 1$$

| The leading coefficient, 1, is positive. | The degree of the polynomial function, 4, is even. |

Because the degree, 4, is even, the graph has the same behavior at each end. The leading coefficient, 1, is positive. Thus, the graph rises to the left and rises to the right.

Step 2 **Find x-intercepts (zeros of the function) by setting $f(x) = 0$.**

$$
\begin{array}{ll}
x^4 - 2x^2 + 1 = 0 & \text{Set } f(x) \text{ equal to O.} \\
(x^2 - 1)(x^2 - 1) = 0 & \text{Factor.} \\
(x + 1)(x - 1)(x + 1)(x - 1) = 0 & \text{Factor completely.} \\
(x + 1)^2(x - 1)^2 = 0 & \text{Express the factorization in a more compact form.} \\
(x + 1)^2 = 0 \quad \text{or} \quad (x - 1)^2 = 0 & \text{Set each factorization equal to O.} \\
x = -1 \qquad\qquad x = 1 & \text{Solve for } x.
\end{array}
$$

We see that -1 and 1 are both repeated zeros with multiplicity 2. Because of the even multiplicity, the graph touches the x-axis at -1 and 1 and turns around. Furthermore, the graph tends to flatten out near these zeros with multiplicity greater than one.

Step 3 **Find the y-intercept by computing $f(0)$.** We use $f(x) = x^4 - 2x^2 + 1$ and compute $f(0)$.

$$f(0) = 0^4 - 2 \cdot 0^2 + 1 = 1$$

There is a y-intercept at 1, so the graph passes through $(0, 1)$.

It appears that 1 is a relative maximum, but we need more information to be certain.

Step 4 **Use possible symmetry to help draw the graph.** Our partial graph suggests y-axis symmetry. Let's verify this by finding $f(-x)$.

$$f(x) = x^4 - 2x^2 + 1$$

Replace x with $-x$.

$$f(-x) = (-x)^4 - 2(-x)^2 + 1 = x^4 - 2x^2 + 1$$

Because $f(-x) = f(x)$, the graph of f is symmetric with respect to the y-axis. **Figure 3.25** shows the graph of $f(x) = x^4 - 2x^2 + 1$.

FIGURE 3.25 The graph of $f(x) = x^4 - 2x^2 + 1$

Step 5 Use the fact that the maximum number of turning points of the graph is $n - 1$ to check whether it is drawn correctly. Because $n = 4$, the maximum number of turning points is $4 - 1$, or 3. Because the graph in **Figure 3.25** has three turning points, we have not violated the maximum number possible. Can you see how this verifies that 1 is indeed a relative maximum and $(0, 1)$ is a turning point? If the graph rose above 1 on either side of $x = 0$, it would have to rise above 1 on the other side as well because of symmetry. This would require additional turning points to smoothly curve back to the x-intercepts. The graph already has three turning points, which is the maximum number for a fourth-degree polynomial function. • • •

✓ Check Point **9** Use the five-step strategy to graph $f(x) = x^3 - 3x^2$. *

EXAMPLE 10 Graphing a Polynomial Function

Graph: $f(x) = -2(x - 1)^2(x + 2)$.

SOLUTION

Step 1 Determine end behavior. Identify the sign of a_n, the leading coefficient, and the degree, n, of the polynomial function.

$$f(x) = \underbrace{-2(x - 1)^2(x + 2)}$$

The leading term is $-2x^2 \cdot x$, or $-2x^3$.

> The leading coefficient, -2, is negative.

> The degree of the polynomial function, 3, is odd.

Because the degree, 3, is odd, the graph has opposite behavior at each end. The leading coefficient, -2, is negative. Thus, the graph rises to the left and falls to the right.

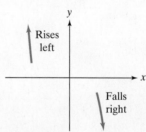

Step 2 Find x-intercepts (zeros of the function) by setting $f(x) = 0$.

$-2(x - 1)^2(x + 2) = 0$	Set $f(x)$ equal to O.
$(x - 1)^2 = 0$ or $x + 2 = 0$	Set each variable factor equal to O.
$x = 1$ $x = -2$	Solve for x.

We see that the zeros are 1 and -2. The multiplicity of 1 is even, so the graph touches the x-axis at 1, flattens, and turns around. The multiplicity of -2 is odd, so the graph crosses the x-axis at -2.

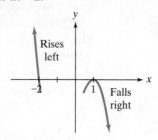

Step 3 Find the y-intercept by computing $f(0)$. We use $f(x) = -2(x - 1)^2(x + 2)$ and compute $f(0)$.

$$f(0) = -2(0 - 1)^2(0 + 2) = -2(1)(2) = -4$$

There is a y-intercept at -4, so the graph passes through $(0, -4)$.

We're not sure about the graph's behavior for x between -2 and 0 or between 0 and 1. Additional points will be helpful. Furthermore, we may need to rescale the y-axis.

Based on the note in the voice balloon, let's evaluate the function at -1.5, -1, -0.5, and 0.5, as well as at -3, 2, and 3.

x	-3	-1.5	-1	-0.5	0.5	2	3
$f(x) = -2(x - 1)^2(x + 2)$	32	-6.25	-8	-6.75	-1.25	-8	-40

In order to accommodate these points, we'll scale the y-axis from -50 to 50, with each tick mark representing 10 units.

Step 4 Use possible symmetry to help draw the graph. Our partial graph illustrates that we have neither y-axis symmetry nor origin symmetry. Using end behavior, intercepts, and the points from our table, **Figure 3.26** shows the graph of $f(x) = -2(x - 1)^2(x + 2)$.

We cannot be sure about the exact location of this turning point. (Yes, your author is certain there's a relative minimum at -1: He took calculus!)

FIGURE 3.26 The graph of $f(x) = -2(x - 1)^2(x + 2)$

Step 5 Use the fact that the maximum number of turning points of the graph is $n - 1$ to check whether it is drawn correctly. The leading term is $-2x^3$, so $n = 3$. The maximum number of turning points is $3 - 1$, or 2. Because the graph in **Figure 3.26** has two turning points, we have not violated the maximum number possible. • • •

✓ Check Point **10** Use the five-step strategy to graph
$f(x) = 2(x + 2)^2(x - 3)$. *

*See Answers to Selected Exercises.

ACHIEVING SUCCESS

Take some time to consider how you are doing in this course. Check your performance by answering the following questions:

- Are you attending all lectures?
- For each hour of class time, are you spending at least two hours outside of class completing all homework assignments, checking answers, correcting errors, and using all resources to get the help that you need?
- Are you reviewing for quizzes and tests?
- Are you reading the textbook? In all college courses, you are responsible for the information in the text, whether or not it is covered in class.
- Are you keeping an organized notebook? Does each page have the appropriate section number from the text on top? Do the pages contain examples your instructor works during lecture and other relevant class notes? Have you included your worked-out homework exercises? Do you keep a special section for graded exams?
- Are you analyzing your mistakes and learning from your errors?
- Are there ways you can improve how you are doing in the course?

CONCEPT AND VOCABULARY CHECK

Fill in each blank so that the resulting statement is true.

1. The degree of the polynomial function $f(x) = -2x^3(x - 1)(x + 5)$ is __5__. The leading coefficient is __−2__.

2. True or false: Some polynomial functions of degree 2 or higher have breaks in their graphs. __false__

3. The behavior of the graph of a polynomial function to the far left or the far right is called its __end__ behavior, which depends upon the __leading__ term.

4. The graph of $f(x) = x^3$ __falls__ to the left and __rises__ to the right.

5. The graph of $f(x) = -x^3$ __rises__ to the left and __falls__ to the right.

6. The graph of $f(x) = x^2$ __rises__ to the left and __rises__ to the right.

7. The graph of $f(x) = -x^2$ __falls__ to the left and __falls__ to the right.

8. True or false: Odd-degree polynomial functions have graphs with opposite behavior at each end. __true__

9. True or false: Even-degree polynomial functions have graphs with the same behavior at each end. __true__

10. Every real zero of a polynomial function appears as a/an __x-intercept__ of the graph.

11. If r is a zero of even multiplicity, then the graph touches the x-axis and __turns around__ at r. If r is a zero of odd multiplicity, then the graph __crosses__ the x-axis at r.

12. If f is a polynomial function and $f(a)$ and $f(b)$ have opposite signs, then there must be at least one value of c between a and b for which $f(c) = $ __0__. This result is called the __Intermediate Value__ Theorem.

13. If f is a polynomial function of degree n, then the graph of f has at most __$n - 1$__ turning points.

EXERCISE SET 3.2

Practice Exercises

In Exercises 1–10, determine which functions are polynomial functions. For those that are, identify the degree.

1. $f(x) = 5x^2 + 6x^3$ yes; 3

2. $f(x) = 7x^2 + 9x^4$ yes; 4

3. $g(x) = 7x^5 - \pi x^3 + \frac{1}{5}x$ *

4. $g(x) = 6x^7 + \pi x^5 + \frac{2}{3}x$ *

5. $h(x) = 7x^3 + 2x^2 + \frac{1}{x}$ no

6. $h(x) = 8x^3 - x^2 + \frac{2}{x}$ no

7. $f(x) = x^{\frac{1}{2}} - 3x^2 + 5$ no

8. $f(x) = x^{\frac{1}{3}} - 4x^2 + 7$ no

9. $f(x) = \frac{x^2 + 7}{x^3}$ no

10. $f(x) = \frac{x^2 + 7}{3}$ yes; 2

*See Answers to Selected Exercises.

In Exercises 11–14, identify which graphs are not those of polynomial functions. 12. and 13. are not polynomial functions.

11.

12.

13.

14.

In Exercises 15–18, use the Leading Coefficient Test to determine the end behavior of the graph of the given polynomial function. Then use this end behavior to match the polynomial function with its graph. [The graphs are labeled (a) through (d).]

15. $f(x) = -x^4 + x^2$ b
16. $f(x) = x^3 - 4x^2$ c
17. $f(x) = (x - 3)^2$ a
18. $f(x) = -x^3 - x^2 + 5x - 3$ d

a.

b.

c.

d.

In Exercises 19–24, use the Leading Coefficient Test to determine the end behavior of the graph of the polynomial function.

19. $f(x) = 5x^3 + 7x^2 - x + 9$ left: falls; right: rises
20. $f(x) = 11x^3 - 6x^2 + x + 3$ left: falls; right: rises
21. $f(x) = 5x^4 + 7x^2 - x + 9$ left: rises; right: rises
22. $f(x) = 11x^4 - 6x^2 + x + 3$ left: rises; right: rises
23. $f(x) = -5x^4 + 7x^2 - x + 9$ left: falls; right: falls
24. $f(x) = -11x^4 - 6x^2 + x + 3$ left: falls; right: falls

In Exercises 25–32, find the zeros for each polynomial function and give the multiplicity for each zero. State whether the graph crosses the x-axis, or touches the x-axis and turns around, at each zero.

25. $f(x) = 2(x - 5)(x + 4)^2$ *
26. $f(x) = 3(x + 5)(x + 2)^2$ *

27. $f(x) = 4(x - 3)(x + 6)^3$ *
28. $f(x) = -3\left(x + \frac{1}{2}\right)(x - 4)^3$ *
29. $f(x) = x^3 - 2x^2 + x$ *
30. $f(x) = x^3 + 4x^2 + 4x$ *
31. $f(x) = x^3 + 7x^2 - 4x - 28$ *
32. $f(x) = x^3 + 5x^2 - 9x - 45$ *

In Exercises 33–40, use the Intermediate Value Theorem to show that each polynomial has a real zero between the given integers.

33. $f(x) = x^3 - x - 1$; between 1 and 2 $f(1) = -1; f(2) = 5$
34. $f(x) = x^3 - 4x^2 + 2$; between 0 and 1 $f(0) = 2; f(1) = -1$
35. $f(x) = 2x^4 - 4x^2 + 1$; between −1 and 0 $f(-1) = -1; f(0) = 1$
36. $f(x) = x^4 + 6x^3 - 18x^2$; between 2 and 3 $f(2) = -8; f(3) = 81$
37. $f(x) = x^3 + x^2 - 2x + 1$; between −3 and −2 *
38. $f(x) = x^5 - x^3 - 1$; between 1 and 2 $f(1) = -1; f(2) = 23$
39. $f(x) = 3x^3 - 10x + 9$; between −3 and −2 *
40. $f(x) = 3x^3 - 8x^2 + x + 2$; between 2 and 3 *

In Exercises 41–64,

a. *Use the Leading Coefficient Test to determine the graph's end behavior.*

b. *Find the x-intercepts. State whether the graph crosses the x-axis, or touches the x-axis and turns around, at each intercept.*

c. *Find the y-intercept.*

d. *Determine whether the graph has y-axis symmetry, origin symmetry, or neither.*

e. *If necessary, find a few additional points and graph the function. Use the maximum number of turning points to check whether it is drawn correctly.*

41. $f(x) = x^3 + 2x^2 - x - 2$ *
42. $f(x) = x^3 + x^2 - 4x - 4$ *
43. $f(x) = x^4 - 9x^2$ *
44. $f(x) = x^4 - x^2$ *
45. $f(x) = -x^4 + 16x^2$ *
46. $f(x) = -x^4 + 4x^2$ *
47. $f(x) = x^4 - 2x^3 + x^2$ *
48. $f(x) = x^4 - 6x^3 + 9x^2$ *
49. $f(x) = -2x^4 + 4x^3$ *
50. $f(x) = -2x^4 + 2x^3$ *
51. $f(x) = 6x^3 - 9x - x^5$ *
52. $f(x) = 6x - x^3 - x^5$ *
53. $f(x) = 3x^2 - x^3$ *
54. $f(x) = \frac{1}{2} - \frac{1}{2}x^4$ *
55. $f(x) = -3(x - 1)^2(x^2 - 4)$ *
56. $f(x) = -2(x - 4)^2(x^2 - 25)$ *
57. $f(x) = x^2(x - 1)^3(x + 2)$ *
58. $f(x) = x^3(x + 2)^2(x + 1)$ *
59. $f(x) = -x^2(x - 1)(x + 3)$ *
60. $f(x) = -x^2(x + 2)(x - 2)$ *
61. $f(x) = -2x^3(x - 1)^2(x + 5)$ *
62. $f(x) = -3x^3(x - 1)^2(x + 3)$ *
63. $f(x) = (x - 2)^2(x + 4)(x - 1)$ *
64. $f(x) = (x + 3)(x + 1)^3(x + 4)$ *

Practice Plus

In Exercises 65–72, complete graphs of polynomial functions whose zeros are integers are shown.

a. Find the zeros and state whether the multiplicity of each zero is even or odd.

b. Write an equation, expressed as the product of factors, of a polynomial function that might have each graph. Use a leading coefficient of 1 or −1, and make the degree of *f* as small as possible.

c. Use both the equation in part (b) and the graph to find the *y*-intercept.

65.

[−5, 5, 1] by [−12, 12, 1]

66.

[−6, 6, 1] by [−40, 40, 10]

67.

[−3, 6, 1] by [−10, 10, 1]

68.

[−3, 3, 1] by [−10, 10, 1]

69.

[−4, 4, 1] by [−40, 4, 4]

70.

[−2, 5, 1] by [−40, 4, 4]

71.

[−3, 3, 1] by [−5, 10, 1]

72.

[−3, 3, 1] by [−5, 10, 1]

Application Exercises

Experts fear that without conservation efforts, tigers could disappear from the wild by 2022. Just one hundred years ago, there were at least 100,000 wild tigers. By 2010, the estimated world tiger population was 3200. The bar graph shows the estimated world tiger population for selected years from 1970 through 2010. Also shown is a polynomial function, with its graph, that models the data. Use this information to solve Exercises 73–74.

Vanishing Tigers

Estimated World Tiger Population

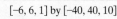

Source: World Wildlife Fund

Graph of a Polynomial Model for the Data

$$f(x) = 0.76x^3 - 30x^2 - 882x + 37,807$$

73. a. Find and interpret $f(40)$. Identify this information as a point on the graph of *f*. 3167; (40, 3167)*

b. Does $f(40)$ overestimate or underestimate the actual data shown by the bar graph? By how much? *

c. Use the Leading Coefficient Test to determine the end behavior to the right for the graph of *f*. Will this function be useful in modeling the world tiger population if conservation efforts to save wild tigers fail? Explain your answer. rises to the right; no*

74. a. Find and interpret $f(10)$. Identify this information as a point on the graph of *f*. 26,747; (10, 26,747)*

b. Does $f(10)$ overestimate or underestimate the actual data shown by the bar graph? By how much? *

c. Use the Leading Coefficient Test to determine the end behavior to the right for the graph of *f*. Might this function be useful in modeling the world tiger population if conservation efforts to save wild tigers are successful? Explain your answer. rises to the right; yes*

75. During a diagnostic evaluation, a 33-year-old woman experienced a panic attack a few minutes after she had been asked to relax her whole body. The graph shows the rapid increase in heart rate during the panic attack.

Heart Rate before and during a Panic Attack

Source: Davis and Palladino, *Psychology*, Fifth Edition, Prentice Hall, 2007.

a. For which time periods during the diagnostic evaluation was the woman's heart rate increasing? *

b. For which time periods during the diagnostic evaluation was the woman's heart rate decreasing? *

c. How many turning points (from increasing to decreasing or from decreasing to increasing) occurred for the woman's heart rate during the first 12 minutes of the diagnostic evaluation? 3

d. Suppose that a polynomial function is used to model the data displayed by the graph using

(time during the evaluation, heart rate).

Use the number of turning points to determine the degree of the polynomial function of best fit. 4

e. For the model in part (d), should the leading coefficient of the polynomial function be positive or negative? Explain your answer. negative*

f. Use the graph to estimate the woman's maximum heart rate during the first 12 minutes of the diagnostic evaluation. After how many minutes did this occur? *

g. Use the graph to estimate the woman's minimum heart rate during the first 12 minutes of the diagnostic evaluation. After how many minutes did this occur? *

76. Volatility at the Pump The graph shows the average price per gallon of gasoline in the United States in January for the period from 2005 through 2011.

Average January Gasoline Price in the United States

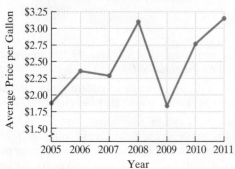

Source: U.S. Energy Information Administration

a. For which years was the average price per gallon in January increasing? 2005 to 2006, 2007 to 2008, and 2009 to 2011

b. For which years was the average price per gallon in January decreasing? 2006 to 2007, and 2008 to 2009

c. How many turning points (from increasing to decreasing or from decreasing to increasing) does the graph have for the period shown? 4

d. Suppose that a polynomial function is used to model the data displayed by the graph using

(number of years after 2005, average January price per gallon).

Use the number of turning points to determine the degree of the polynomial function of best fit. 5

e. For the model in part (d), should the leading coefficient of the polynomial function be positive or negative? Explain your answer. positive*

f. Use the graph to estimate the maximum average January price per gallon. In which year did this occur? *

g. Use the graph to estimate the minimum average January price per gallon. In which year did this occur? *

Explaining the Concepts 77–88. Answers will vary.

77. What is a polynomial function?

78. What do we mean when we describe the graph of a polynomial function as smooth and continuous?

79. What is meant by the end behavior of a polynomial function?

80. Explain how to use the Leading Coefficient Test to determine the end behavior of a polynomial function.

81. Why is a third-degree polynomial function with a negative leading coefficient not appropriate for modeling nonnegative real-world phenomena over a long period of time?

82. What are the zeros of a polynomial function and how are they found?

83. Explain the relationship between the multiplicity of a zero and whether or not the graph crosses or touches the *x*-axis and turns around at that zero.

84. If *f* is a polynomial function, and *f*(*a*) and *f*(*b*) have opposite signs, what must occur between *a* and *b*? If *f*(*a*) and *f*(*b*) have the same sign, does it necessarily mean that this will not occur? Explain your answer.

85. Explain the relationship between the degree of a polynomial function and the number of turning points on its graph.

86. Can the graph of a polynomial function have no *x*-intercepts? Explain.

87. Can the graph of a polynomial function have no *y*-intercept? Explain.

88. Describe a strategy for graphing a polynomial function. In your description, mention intercepts, the polynomial's degree, and turning points.

Technology Exercises 89–93. Answers will vary.

89. Use a graphing utility to verify any five of the graphs that you drew by hand in Exercises 41–64.

Write a polynomial function that imitates the end behavior of each graph in Exercises 90–93. The dashed portions of the graphs indicate that you should focus only on imitating the left and right behavior of the graph and can be flexible about what occurs between the left and right ends. Then use your graphing utility to graph the polynomial function and verify that you imitated the end behavior shown in the given graph.

90.　**91.**

92.　**93.**

In Exercises 94–97, use a graphing utility with a viewing rectangle large enough to show end behavior to graph each polynomial function.

94. $f(x) = x^3 + 13x^2 + 10x - 4$ *
95. $f(x) = -2x^3 + 6x^2 + 3x - 1$ *
96. $f(x) = -x^4 + 8x^3 + 4x^2 + 2$ *
97. $f(x) = -x^5 + 5x^4 - 6x^3 + 2x + 20$ *

In Exercises 98–99, use a graphing utility to graph f and g in the same viewing rectangle. Then use the ZOOM OUT feature to show that f and g have identical end behavior.

98. $f(x) = x^3 - 6x + 1$, $g(x) = x^3$ *
99. $f(x) = -x^4 + 2x^3 - 6x$, $g(x) = -x^4$ *

Critical Thinking Exercises

Make Sense? In Exercises 100–103, determine whether each statement makes sense or does not make sense, and explain your reasoning.

100. When I'm trying to determine end behavior, it's the coefficient of the leading term of a polynomial function that I should inspect. makes sense

101. I graphed $f(x) = (x + 2)^3(x - 4)^2$, and the graph touched the x-axis and turned around at −2. does not make sense

102. I'm graphing a fourth-degree polynomial function with four turning points. does not make sense

103. Although I have not yet learned techniques for finding the x-intercepts of $f(x) = x^3 + 2x^2 - 5x - 6$, I can easily determine the y-intercept. makes sense

In Exercises 104–107, determine whether each statement is true or false. If the statement is false, make the necessary change(s) to produce a true statement. Changes to false statements will vary.

104. If $f(x) = -x^3 + 4x$, then the graph of f falls to the left and falls to the right. false

105. A mathematical model that is a polynomial of degree n whose leading term is $a_n x^n$, n odd and $a_n < 0$, is ideally suited to describe phenomena that have positive values over unlimited periods of time. false

106. There is more than one third-degree polynomial function with the same three x-intercepts. true

107. The graph of a function with origin symmetry can rise to the left and rise to the right. false

Use the descriptions in Exercises 108–109 to write an equation of a polynomial function with the given characteristics. Use a graphing utility to graph your function to see if you are correct. If not, modify the function's equation and repeat this process.

108. Crosses the x-axis at −4, 0, and 3; lies above the x-axis between −4 and 0; lies below the x-axis between 0 and 3. $f(x) = x^3 + x^2 - 12x$

109. Touches the x-axis at 0 and crosses the x-axis at 2; lies below the x-axis between 0 and 2. $f(x) = x^3 - 2x^2$

Retaining the Concepts

110. In 1995, there were 315 death sentences rendered by American juries. For the period from 1995 through 2014, the number of death sentences rendered by juries decreased by approximately 13 per year. If this trend continues, by which year will American juries render 29 death sentences? (Source: Death Penalty Information Center) (Section 1.3, Example 2) by 2017

111. Solve and graph the solution set on a number line:

$$\frac{2x - 3}{4} \geq \frac{3x}{4} + \frac{1}{2}.$$

(Section 1.7, Example 5) $(-\infty, -5]$*

112. Write an equation in point-slope form and slope-intercept form of the line passing through (−10, 3) and (−2, −5). (Section 2.3, Example 3) $y = -x - 7$*

Preview Exercises

Exercises 113–115 will help you prepare for the material covered in the next section.

113. Divide 737 by 21 without using a calculator. Write the answer as

$$\text{quotient} + \frac{\text{remainder}}{\text{divisor}}.$$ $35\frac{2}{21}$

114. Rewrite $4 - 5x - x^2 + 6x^3$ in descending powers of x. *

115. Use

$$\frac{2x^3 - 3x^2 - 11x + 6}{x - 3} = 2x^2 + 3x - 2$$

to factor $2x^3 - 3x^2 - 11x + 6$ completely.
$(x - 3)(2x - 1)(x + 2)$

Section 3.3 Dividing Polynomials; Remainder and Factor Theorems

What am I supposed to learn?

After studying this section, you should be able to:

1. Use long division to divide polynomials.
2. Use synthetic division to divide polynomials.
3. Evaluate a polynomial using the Remainder Theorem.
4. Use the Factor Theorem to solve a polynomial equation.

A moth has moved into your closet. She appeared in your bedroom at night, but somehow her relatively stout body escaped your clutches. Within a few weeks, swarms of moths in your tattered wardrobe suggest that Mama Moth was in the family way. There must be at least 200 critters nesting in every crevice of your clothing.

Two hundred plus moth-tykes from one female moth—is this possible? Indeed it is. The number of eggs, $f(x)$, in a female moth is a function of her abdominal width, x, in millimeters, modeled by

$$f(x) = 14x^3 - 17x^2 - 16x + 34, \quad 1.5 \le x \le 3.5.$$

Because there are 200 moths feasting on your favorite sweaters, Mama's abdominal width can be estimated by finding the solutions of the polynomial equation

$$14x^3 - 17x^2 - 16x + 34 = 200.$$

How can we solve such an equation? You might begin by subtracting 200 from both sides to obtain zero on one side. But then what? The factoring that we used in the previous section will not work in this situation.

In the next section, we will present techniques for solving certain kinds of polynomial equations. These techniques will further enhance your ability to manipulate algebraically the polynomial functions that model your world. Because these techniques are based on understanding polynomial division, in this section we look at two methods for dividing polynomials. (We'll return to Mama Moth's abdominal width in Exercise 51 in the Exercise Set.)

1. Use long division to divide polynomials.

Long Division of Polynomials and the Division Algorithm

In Example 1, we look at division by a polynomial containing more than one term:

$$x + 3 \overline{)x^2 + 10x + 21}.$$

> Divisor has two terms and is a binomial.

> The polynomial dividend has three terms and is a trinomial.

Because this process of dividing a polynomial by a binomial is similar to long division, let's first review long division of whole numbers by dividing 3983 by 26.

> Divisor $26\overline{)3983}$ Dividend

$$\begin{array}{r} 1 \\ 26\overline{)3983} \\ \underline{26}\downarrow \\ 138 \end{array}$$

Divide: $\dfrac{39}{26} = 1$ plus a remainder.

Multiply: $1 \cdot 26 = 26.$

Subtract: $39 - 26 = 13.$

Bring down the next digit in the dividend.

$$26\overline{)3983}$$

15

26
138
130
83

Divide: $\frac{138}{26} = 5$ plus a remainder.

Multiply: $5 \cdot 26 = 130$.

Subtract: $138 - 130 = 8$.

Bring down the next digit in the dividend.

153

$$26\overline{)3983}$$

26
138
130
83
78
5

Divide: $\frac{83}{26} = 3$ plus a remainder.

Multiply: $3 \cdot 26 = 78$.

Subtract: $83 - 78 = 5$.

There are no more digits to bring down so the remainder is 5.

The quotient is 153 and the remainder is 5. This can be written as

Quotient $153\frac{5}{26}$. Remainder Divisor

We see that $26\overline{)3983}^{\,153\frac{5}{26}}$.

This answer can be checked. Multiply the divisor and the quotient. Then add the remainder. If the result is the dividend, the answer is correct. In this case, we have

$$26(153) \quad + \quad 5 \quad = \quad 3978 + 5 \quad = \quad 3983.$$

Divisor Quotient Remainder This is the dividend.

Because we obtained the dividend, the answer to the division problem, $153\frac{5}{26}$, is correct.

When a divisor has more than one term, the four steps used to divide whole numbers—**divide, multiply, subtract, bring down the next term**—form the repetitive procedure for polynomial long division.

EXAMPLE 1 Long Division of Polynomials

Divide $x^2 + 10x + 21$ by $x + 3$.

SOLUTION

The following process illustrates how polynomial division is very similar to division of whole numbers.

$$x + 3\overline{)x^2 + 10x + 21}$$

Arrange the terms of the dividend ($x^2 + 10x + 21$) and the divisor ($x + 3$) in descending powers of x.

$$x + 3\overline{)x^2 + 10x + 21}^{\quad x}$$

Divide x^2 (the first term in the dividend) by x (the first term in the divisor): $\frac{x^2}{x} = x$. Align like terms.

GREAT QUESTION!

You've mentioned more than once that we're dividing by a polynomial with more than one term. Should I use the long-division process when the divisor only has one term?

No. When the divisor is a monomial, you should simply divide each term of the dividend by the monomial and then simplify the resulting expressions.

$$\frac{6x^5 + 9x^4 - 24x^3}{3x^3}$$

$$= \frac{6x^5}{3x^3} + \frac{9x^4}{3x^3} - \frac{24x^3}{3x^3}$$

$$= 2x^2 + 3x - 8$$

This is much simpler than long division.

$$x(x + 3) = x^2 + 3x$$

$$\begin{array}{r} x \\ x + 3\overline{)x^2 + 10x + 21} \\ x^2 + 3x \end{array}$$

Multiply each term in the divisor $(x + 3)$ by x, aligning terms of the product under like terms in the dividend.

$$\begin{array}{r} x \\ x + 3\overline{)x^2 + 10x + 21} \\ \underline{\ominus x^2 \oplus 3x} \\ 7x \end{array}$$

Change signs of the polynomial being subtracted.

Subtract $x^2 + 3x$ from $x^2 + 10x$ by changing the sign of each term in the lower expression and adding.

$$\begin{array}{r} x \\ x + 3\overline{)x^2 + 10x + 21} \\ \underline{x^2 + 3x} \downarrow \\ 7x + 21 \end{array}$$

Bring down 21 from the original dividend and add algebraically to form a new dividend.

$$\begin{array}{r} x + 7 \\ x + 3\overline{)x^2 + 10x + 21} \\ \underline{x^2 + 3x} \\ 7x + 21 \end{array}$$

Find the second term of the quotient. Divide the first term of $7x + 21$ by x, the first term of the divisor: $\dfrac{7x}{x} = 7$.

$$7(x + 3) = 7x + 21$$

$$\begin{array}{r} x + 7 \\ x + 3\overline{)x^2 + 10x + 21} \\ \underline{x^2 + 3x} \\ 7x + 21 \\ \underline{\ominus 7x \oplus 21} \\ 0 \end{array}$$

Remainder

Multiply the divisor $(x + 3)$ by 7, aligning under like terms in the new dividend. Then subtract to obtain the remainder of 0.

The quotient is $x + 7$. Because the remainder is 0, we can conclude that $x + 3$ is a factor of $x^2 + 10x + 21$ and

$$\frac{x^2 + 10x + 21}{x + 3} = x + 7.$$ • • •

⊘ **Check Point 1** Divide $x^2 + 14x + 45$ by $x + 9$. $x + 5$

Before considering additional examples, let's summarize the general procedure for dividing one polynomial by another.

> **Long Division of Polynomials**
>
> 1. Arrange the terms of both the dividend and the divisor in descending powers of any variable.
> 2. **Divide** the first term in the dividend by the first term in the divisor. The result is the first term of the quotient.
> 3. **Multiply** every term in the divisor by the first term in the quotient. Write the resulting product beneath the dividend with like terms lined up.
> 4. **Subtract** the product from the dividend.
> 5. **Bring down** the next term in the original dividend and write it next to the remainder to form a new dividend.
> 6. Use this new expression as the dividend and repeat this process until the remainder can no longer be divided. This will occur when the degree of the remainder (the highest exponent on a variable in the remainder) is less than the degree of the divisor.

In our next long division, we will obtain a nonzero remainder.

EXAMPLE 2 Long Division of Polynomials

Divide $4 - 5x - x^2 + 6x^3$ by $3x - 2$.

SOLUTION

We begin by writing the dividend in descending powers of x.

$$4 - 5x - x^2 + 6x^3 = 6x^3 - x^2 - 5x + 4$$

A Brief Review • Dividing Monomials

When dividing monomials with the same base, subtract the exponent in the denominator from the exponent in the numerator. Use this difference as the exponent of the common base:

$$\frac{6x^3}{3x} = \frac{6}{3}x^{3-1} = 2x^2.$$

For more detail, see Section P.2, Objectives 2 and 8.

Now we divide $3x^2$ by $3x$ to obtain x, multiply x and the divisor, and subtract.

Now we divide $-3x$ by $3x$ to obtain -1, multiply -1 and the divisor, and subtract.

The quotient is $2x^2 + x - 1$ and the remainder is 2. When there is a nonzero remainder, as in this example, list the quotient, plus the remainder above the divisor. Thus,

$$\frac{6x^3 - x^2 - 5x + 4}{3x - 2} = \underbrace{2x^2 + x - 1}_{\text{Quotient}} + \frac{2}{3x - 2}.$$

• • •

An important property of division can be illustrated by clearing fractions in the equation that concluded Example 2. Multiplying both sides of this equation by $3x - 2$ results in the following equation:

$$6x^3 - x^2 - 5x + 4 = (3x - 2)(2x^2 + x - 1) + 2.$$

Dividend Divisor Quotient Remainder

Polynomial long division is checked by multiplying the divisor with the quotient and then adding the remainder. This should give the dividend. The process illustrates the **Division Algorithm**.

The Division Algorithm

If $f(x)$ and $d(x)$ are polynomials, with $d(x) \neq 0$, and the degree of $d(x)$ is less than or equal to the degree of $f(x)$, then there exist unique polynomials $q(x)$ and $r(x)$ such that

$$f(x) \quad = \quad d(x) \quad \cdot \quad q(x) \quad + \quad r(x).$$

Dividend Divisor Quotient Remainder

The remainder, $r(x)$, equals 0 or it is of degree less than the degree of $d(x)$. If $r(x) = 0$, we say that $d(x)$ **divides evenly** into $f(x)$ and that $d(x)$ and $q(x)$ are **factors** of $f(x)$.

☑ **Check Point 2** Divide $7 - 11x - 3x^2 + 2x^3$ by $x - 3$. Express the result in the form quotient, plus remainder divided by divisor. $\quad 2x^2 + 3x - 2 + \dfrac{1}{x - 3}$

If a power of x is missing in either a dividend or a divisor, add that power of x with a coefficient of 0 and then divide. In this way, like terms will be aligned as you carry out the long division.

EXAMPLE 3 Long Division of Polynomials

Divide $6x^4 + 5x^3 + 3x - 5$ by $3x^2 - 2x$.

SOLUTION

We write the dividend, $6x^4 + 5x^3 + 3x - 5$, as $6x^4 + 5x^3 + 0x^2 + 3x - 5$ to keep all like terms aligned.

The division process is finished because the degree of $7x - 5$, which is 1, is less than the degree of the divisor $3x^2 - 2x$, which is 2. The answer is

$$\frac{6x^4 + 5x^3 + 3x - 5}{3x^2 - 2x} = 2x^2 + 3x + 2 + \frac{7x - 5}{3x^2 - 2x}.$$

• • •

☑ **Check Point 3** Divide $2x^4 + 3x^3 - 7x - 10$ by $x^2 - 2x$. $\quad 2x^2 + 7x + 14 + \dfrac{21x - 10}{x^2 - 2x}$

 Use synthetic division to divide polynomials.

Dividing Polynomials Using Synthetic Division

We can use **synthetic division** to divide polynomials if the divisor is of the form $x - c$. This method provides a quotient more quickly than long division. Let's compare the two methods showing $x^3 + 4x^2 - 5x + 5$ divided by $x - 3$.

Notice the relationship between the polynomials in the long division process and the numbers that appear in synthetic division.

These are the coefficients of the dividend $x^3 + 4x^2 - 5x + 5$.

The divisor is $x - 3$. This is 3, or c, in $x - c$.

$$3\overline{\big)\;1\quad 4\quad -5\quad 5}$$
$$3\quad 21\quad 48$$
$$1\quad 7\quad 16\quad 53$$

These are the coefficients of the quotient $x^2 + 7x + 16$.

This is the remainder.

Now let's look at the steps involved in synthetic division.

Synthetic Division

To divide a polynomial by $x - c$:

Example

1. Arrange the polynomial in descending powers, with a 0 coefficient for any missing term.

$$x - 3\overline{\big)\;x^3 + 4x^2 - 5x + 5}$$

2. Write c for the divisor, $x - c$. To the right, write the coefficients of the dividend.

$$3\overline{\big)\;1\quad 4\quad -5\quad 5}$$

3. Write the leading coefficient of the dividend on the bottom row.

$$3\overline{\big)\;1\quad 4\quad -5\quad 5}$$
Bring down 1.
$$1$$

4. Multiply c (in this case, 3) times the value just written on the bottom row. Write the product in the next column in the second row.

$$3\overline{\big)\;1\quad 4\quad -5\quad 5}$$
$$3$$
$$1$$
Multiply by 3: $3 \cdot 1 = 3$.

5. Add the values in this new column, writing the sum in the bottom row.

$$3\overline{\big)\;1\quad 4\quad -5\quad 5}$$
$$3$$ Add.
$$1\quad 7$$

Example

6. Repeat this series of multiplications and additions until all columns are filled in.

7. Use the numbers in the last row to write the quotient, plus the remainder above the divisor. **The degree of the first term of the quotient is one less than the degree of the first term of the dividend.** The final value in this row is the remainder.

<div style="background-color:#ccc; display:inline-block;">EXAMPLE 4</div> Using Synthetic Division

Use synthetic division to divide $5x^3 + 6x + 8$ by $x + 2$.

SOLUTION

The divisor must be in the form $x - c$. Thus, we write $x + 2$ as $x - (-2)$. This means that $c = -2$. Writing a 0 coefficient for the missing x^2-term in the dividend, we can express the division as follows:

$$x - (-2)\overline{)5x^3 + 0x^2 + 6x + 8}.$$

Now we are ready to set up the problem so that we can use synthetic division.

We begin the synthetic division process by bringing down 5. This is followed by a series of multiplications and additions.

6. Multiply: $-2(26) = -52.$

$$\begin{array}{r|rrrr} -2 & 5 & 0 & 6 & 8 \\ & & -10 & 20 & -52 \\ \hline & 5 & -10 & 26 \end{array}$$

Multiply 26 by −2.

7. Add: $8 + (-52) = -44.$

$$\begin{array}{r|rrrr} -2 & 5 & 0 & 6 & 8 \\ & & -10 & 20 & -52 \\ \hline & 5 & -10 & 26 & -44 \end{array}$$ Add.

The numbers in the last row represent the coefficients of the quotient and the remainder. The degree of the first term of the quotient is one less than that of the dividend. Because the degree of the dividend, $5x^3 + 6x + 8$, is 3, the degree of the quotient is 2. This means that the 5 in the last row represents $5x^2$.

$$\begin{array}{r|rrrr} -2 & 5 & 0 & 6 & 8 \\ & & -10 & 20 & -52 \\ \hline & 5 & -10 & 26 & -44 \end{array}$$

The quotient is $5x^2 - 10x + 26$. The remainder is −44.

Thus,

$$x + 2 \overline{)5x^3 + 6x + 8} \quad 5x^2 - 10x + 26 - \frac{44}{x + 2}$$

• • •

✓ **Check Point 4** Use synthetic division to divide $x^3 - 7x - 6$ by $x + 2$.

$x^2 - 2x - 3$

③ Evaluate a polynomial using the Remainder Theorem.

The Remainder Theorem

Let's consider the Division Algorithm when the dividend, $f(x)$, is divided by $x - c$. In this case, the remainder must be a constant because its degree is less than one, the degree of $x - c$.

$$f(x) \quad = \quad (x - c)q(x) \quad + \quad r$$

Dividend Divisor Quotient The remainder, r, is a constant when dividing by $x - c$.

Now let's evaluate f at c.

$f(c) = (c - c)q(c) + r$ Find $f(c)$ by letting $x = c$ in $f(x) = (x - c)q(x) + r$. This will give an expression for r.

$f(c) = 0 \cdot q(c) + r$ $c - c = 0$

$f(c) = r$ $0 \cdot q(c) = 0$ and $0 + r = r$.

What does this last equation mean? If a polynomial is divided by $x - c$, the remainder is the value of the polynomial at c. This result is called the **Remainder Theorem**.

The Remainder Theorem

If the polynomial $f(x)$ is divided by $x - c$, then the remainder is $f(c)$.

Example 5 shows how we can use the Remainder Theorem to evaluate a polynomial function at 2. Rather than substituting 2 for x, we divide the function by $x - 2$. The remainder is $f(2)$.

EXAMPLE 5 Using the Remainder Theorem to Evaluate a Polynomial Function

Given $f(x) = x^3 - 4x^2 + 5x + 3$, use the Remainder Theorem to find $f(2)$.

SOLUTION

By the Remainder Theorem, if $f(x)$ is divided by $x - 2$, then the remainder is $f(2)$. We'll use synthetic division to divide.

$$\underline{2|}\ \begin{array}{rrrr} 1 & -4 & 5 & 3 \\ & 2 & -4 & 2 \\ \hline 1 & -2 & 1 & 5 \end{array}$$

Remainder

The remainder, 5, is the value of $f(2)$. Thus, $f(2) = 5$. We can verify that this is correct by evaluating $f(2)$ directly. Using $f(x) = x^3 - 4x^2 + 5x + 3$, we obtain

$$f(2) = 2^3 - 4 \cdot 2^2 + 5 \cdot 2 + 3 = 8 - 16 + 10 + 3 = 5.$$ • • •

⟡ **Check Point 5** Given $f(x) = 3x^3 + 4x^2 - 5x + 3$, use the Remainder Theorem to find $f(-4)$. $_{-105}$

④ Use the Factor Theorem to solve a polynomial equation.

The Factor Theorem

Let's look again at the Division Algorithm when the divisor is of the form $x - c$.

$$f(x) \quad = \quad (x - c)q(x) \quad + \quad r$$

Dividend Divisor Quotient Constant remainder

By the Remainder Theorem, the remainder r is $f(c)$, so we can substitute $f(c)$ for r:

$$f(x) = (x - c)q(x) + f(c).$$

Notice that if $f(c) = 0$, then

$$f(x) = (x - c)q(x)$$

so that $x - c$ is a factor of $f(x)$. This means that for the polynomial function $f(x)$, if $f(c) = 0$, then $x - c$ is a factor of $f(x)$.

Let's reverse directions and see what happens if $x - c$ is a factor of $f(x)$. This means that

$$f(x) = (x - c)q(x).$$

If we replace x in $f(x) = (x - c)q(x)$ with c, we obtain

$$f(c) = (c - c)q(c) = 0 \cdot q(c) = 0.$$

Thus, if $x - c$ is a factor of $f(x)$, then $f(c) = 0$.

We have proved a result known as the **Factor Theorem**.

The Factor Theorem

Let $f(x)$ be a polynomial.

a. If $f(c) = 0$, then $x - c$ is a factor of $f(x)$.

b. If $x - c$ is a factor of $f(x)$, then $f(c) = 0$.

The example that follows shows how the Factor Theorem can be used to solve a polynomial equation.

EXAMPLE 6 Using the Factor Theorem

Solve the equation $2x^3 - 3x^2 - 11x + 6 = 0$ given that 3 is a zero of $f(x) = 2x^3 - 3x^2 - 11x + 6$.

SOLUTION

We are given that 3 is a zero of $f(x) = 2x^3 - 3x^2 - 11x + 6$. This means that $f(3) = 0$. Because $f(3) = 0$, the Factor Theorem tells us that $x - 3$ is a factor of $f(x)$. We'll use synthetic division to divide $f(x)$ by $x - 3$.

TECHNOLOGY

Graphic Connections

Because the solution set of

$$2x^3 - 3x^2 - 11x + 6 = 0$$

is $\left\{-2, \frac{1}{2}, 3\right\}$, this implies that the polynomial function

$$f(x) = 2x^3 - 3x^2 - 11x + 6$$

has x-intercepts (or zeros) at $-2, \frac{1}{2}$, and 3. This is verified by the graph of f.

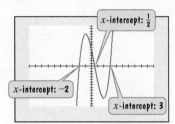

x-intercept: $\frac{1}{2}$

x-intercept: -2

x-intercept: 3

$[-10, 10, 1]$ by $[-15, 15, 1]$

$$\begin{array}{c|cccc} 3 & 2 & -3 & -11 & 6 \\ & & 6 & 9 & -6 \\ \hline & 2 & 3 & -2 & 0 \end{array}$$

The remainder, 0, verifies that $x - 3$ is a factor of $2x^3 - 3x^2 - 11x + 6$.

$$\begin{array}{r} 2x^2 + 3x - 2 \\ x - 3 \overline{\smash{\big)}2x^3 - 3x^2 - 11x + 6} \end{array}$$

Equivalently,

$$2x^3 - 3x^2 - 11x + 6 = (x - 3)(2x^2 + 3x - 2).$$

Now we can solve the polynomial equation.

$2x^3 - 3x^2 - 11x + 6 = 0$	This is the given equation.
$(x - 3)(2x^2 + 3x - 2) = 0$	Factor using the result from the synthetic division.
$(x - 3)(2x - 1)(x + 2) = 0$	Factor the trinomial.
$x - 3 = 0$ or $2x - 1 = 0$ or $x + 2 = 0$	Set each factor equal to O.
$x = 3 \qquad x = \frac{1}{2} \qquad x = -2$	Solve for x.

The solution set is $\left\{-2, \frac{1}{2}, 3\right\}$. • • •

Based on the Factor Theorem, the following statements are useful in solving polynomial equations:

1. If $f(x)$ is divided by $x - c$ and the remainder is zero, then c is a zero of f and c is a root of the polynomial equation $f(x) = 0$.

2. If $f(x)$ is divided by $x - c$ and the remainder is zero, then $x - c$ is a factor of $f(x)$.

⊘ **Check Point 6** Solve the equation $15x^3 + 14x^2 - 3x - 2 = 0$ given that -1 is a zero of $f(x) = 15x^3 + 14x^2 - 3x - 2$. $\left\{-1, -\frac{1}{3}, \frac{2}{5}\right\}$

ACHIEVING SUCCESS

Don't wait too long after class to review your notes. Reading notes while the classroom experience is fresh in your mind will help you to remember what was covered during lecture.

CONCEPT AND VOCABULARY CHECK

Fill in each blank so that the resulting statement is true.

1. Consider the following long division problem:

$$x + 4 \overline{\smash{\big)}6x - 4 + 2x^3}.$$

We begin the division process by rewriting the dividend as $\underline{2x^3 + 0x^2 + 6x - 4}$.

2. Consider the following long division problem:

$$3x - 1 \overline{\smash{\big)}6x^3 + 7x^2 + 12x - 5}.$$

We begin the division process by dividing $\underline{6x^3}$ by $\underline{3x}$. We obtain $\underline{2x^2}$. We write this result above $\underline{7x^2}$ in the dividend.

3. In the following long division problem, the first step has been completed:

$$5x - 2 \overline{)10x^3 + 6x^2 - 9x + 10}.$$
$$2x^2$$

The next step is to multiply __$2x^2$__ and __$5x - 2$__. We obtain __$10x^3 - 4x^2$__. We write this result below __$10x^3 + 6x^2$__.

4. In the following long division problem, the first two steps have been completed:

$$3x - 5 \overline{)6x^2 + 8x - 4}.$$
$$2x$$
$$6x^2 - 10x$$

The next step is to subtract __$6x^2 - 10x$__ from __$6x^2 + 8x$__. We obtain __$18x$__. Then we bring down __-4__ and form the new dividend __$18x - 4$__.

5. In the following long division problem, most of the steps have been completed:

$$2x + 1 \overline{)6x^2 - 7x + 4}.$$
$$3x - 5$$
$$6x^2 + 3x$$
$$-10x + 4$$
$$-10x - 5$$
$$?$$

Completing the step designated by the question mark, we obtain __9__. Thus, the quotient is __$3x - 5$__ and the remainder is __9__. The answer to this long division problem is __$3x - 5 + \dfrac{9}{2x + 1}$__.

6. After performing polynomial long division, the answer may be checked by multiplying the __divisor__ by the __quotient__, and then adding the __remainder__. You should obtain the __dividend__.

7. To divide $x^3 + 5x^2 - 7x + 1$ by $x - 4$ using synthetic division, the first step is to write

$$4 \underline{|\quad} \quad 1 \quad 5 \quad -7 \quad 1.$$

8. To divide $4x^3 - 8x - 2$ by $x + 5$ using synthetic division, the first step is to write

$$-5 \underline{|\quad} \quad 4 \quad 0 \quad -8 \quad -2.$$

9. True or false:

$$-1 \underline{|\quad} \quad 3 \quad -4 \quad 2 \quad -1$$
$$\qquad\qquad -3 \quad 7 \quad -9 \text{ means}$$
$$\qquad\quad 3 \quad -7 \quad 9 \quad -10$$

$$\frac{3x^3 - 4x^2 + 2x - 1}{x + 1} = 3x^2 - 7x + 9 - \frac{10}{x + 1}. \quad \underline{\text{true}}$$

10. The Remainder Theorem states that if the polynomial $f(x)$ is divided by $x - c$, then the remainder is __$f(c)$__.

11. The Factor Theorem states that if f is a polynomial function and $f(c) = 0$, then __$x - c$__ is a factor of $f(x)$.

EXERCISE SET 3.3

Practice Exercises

In Exercises 1–16, divide using long division. State the quotient, $q(x)$, and the remainder, $r(x)$.

1. $(x^2 + 8x + 15) \div (x + 5)$ $x + 3$

2. $(x^2 + 3x - 10) \div (x - 2)$ $x + 5$

3. $(x^3 + 5x^2 + 7x + 2) \div (x + 2)$ $x^2 + 3x + 1$

4. $(x^3 - 2x^2 - 5x + 6) \div (x - 3)$ $x^2 + x - 2$

5. $(6x^3 + 7x^2 + 12x - 5) \div (3x - 1)$ $2x^2 + 3x + 5$

6. $(6x^3 + 17x^2 + 27x + 20) \div (3x + 4)$ $2x^2 + 3x + 5$

7. $(12x^2 + x - 4) \div (3x - 2)$ $4x + 3 + \dfrac{2}{3x - 2}$

8. $(4x^2 - 8x + 6) \div (2x - 1)$ $2x - 3 + \dfrac{3}{2x - 1}$

9. $\dfrac{2x^3 + 7x^2 + 9x - 20}{x + 3}$ *

10. $\dfrac{3x^2 - 2x + 5}{x - 3}$ *

11. $\dfrac{4x^4 - 4x^2 + 6x}{x - 4}$ *

12. $\dfrac{x^4 - 81}{x - 3}$ $x^3 + 3x^2 + 9x + 27$

13. $\dfrac{6x^3 + 13x^2 - 11x - 15}{3x^2 - x - 3}$ *

14. $\dfrac{x^4 + 2x^3 - 4x^2 - 5x - 6}{x^2 + x - 2}$ *

15. $\dfrac{18x^4 + 9x^3 + 3x^2}{3x^2 + 1}$ *

16. $\dfrac{2x^5 - 8x^4 + 2x^3 + x^2}{2x^3 + 1}$ *

In Exercises 17–32, divide using synthetic division.

17. $(2x^2 + x - 10) \div (x - 2)$ *

18. $(x^2 + x - 2) \div (x - 1)$ *

19. $(3x^2 + 7x - 20) \div (x + 5)$ $3x - 8 + \dfrac{20}{x + 5}$

20. $(5x^2 - 12x - 8) \div (x + 3)$ $5x - 27 + \dfrac{73}{x + 3}$

21. $(4x^3 - 3x^2 + 3x - 1) \div (x - 1)$ $4x^2 + x + 4 + \dfrac{3}{x - 1}$

22. $(5x^3 - 6x^2 + 3x + 11) \div (x - 2)$ $5x^2 + 4x + 11 + \dfrac{33}{x - 2}$

23. $(6x^5 - 2x^3 + 4x^2 - 3x + 1) \div (x - 2)$ *

24. $(x^5 + 4x^4 - 3x^2 + 2x + 3) \div (x - 3)$ *

25. $(x^2 - 5x - 5x^3 + x^4) \div (5 + x)$ $x^3 - 10x^2 + 51x - 260 + \dfrac{1300}{x + 5}$

26. $(x^2 - 6x - 6x^3 + x^4) \div (6 + x)$ $x^3 - 12x^2 + 73x - 444 + \dfrac{2664}{x + 6}$

27. $\dfrac{x^5 + x^3 - 2}{x - 1}$ *

28. $\dfrac{x^7 + x^5 - 10x^3 + 12}{x + 2}$ *

29. $\dfrac{x^4 - 256}{x - 4}$ *

30. $\dfrac{x^7 - 128}{x - 2}$ *

31. $\dfrac{2x^5 - 3x^4 + x^3 - x^2 + 2x - 1}{x + 2}$ *

32. $\dfrac{x^5 - 2x^4 - x^3 + 3x^2 - x + 1}{x - 2}$ $x^4 - x^2 + x + 1 + \dfrac{3}{x - 2}$

In Exercises 33–40, use synthetic division and the Remainder Theorem to find the indicated function value.

33. $f(x) = 2x^3 - 11x^2 + 7x - 5$; $f(4)$ -25

34. $f(x) = x^3 - 7x^2 + 5x - 6$; $f(3)$ -27

35. $f(x) = 3x^3 - 7x^2 - 2x + 5$; $f(-3)$ -133

36. $f(x) = 4x^3 + 5x^2 - 6x - 4$; $f(-2)$ -4

37. $f(x) = x^4 + 5x^3 + 5x^2 - 5x - 6$; $f(3)$ 240

38. $f(x) = x^4 - 5x^3 + 5x^2 + 5x - 6$; $f(2)$ 0

39. $f(x) = 2x^4 - 5x^3 - x^2 + 3x + 2$; $f\left(-\dfrac{1}{2}\right)$ 1

40. $f(x) = 6x^4 + 10x^3 + 5x^2 + x + 1$; $f\left(-\dfrac{2}{3}\right)$ $\dfrac{7}{9}$

41. Use synthetic division to divide

$$f(x) = x^3 - 4x^2 + x + 6 \text{ by } x + 1.$$

Use the result to find all zeros of f. $x^2 - 5x + 6; -1, 2, 3$

42. Use synthetic division to divide

$$f(x) = x^3 - 2x^2 - x + 2 \text{ by } x + 1.$$

Use the result to find all zeros of f. $x^2 - 3x + 2; -1, 2, 3$

43. Solve the equation $2x^3 - 5x^2 + x + 2 = 0$ given that 2 is a zero of $f(x) = 2x^3 - 5x^2 + x + 2$. $\left\{-\frac{1}{2}, 1, 2\right\}$

44. Solve the equation $2x^3 - 3x^2 - 11x + 6 = 0$ given that -2 is a zero of $f(x) = 2x^3 - 3x^2 - 11x + 6$. $\left\{-2, \frac{1}{2}, 3\right\}$

45. Solve the equation $12x^3 + 16x^2 - 5x - 3 = 0$ given that $-\frac{3}{2}$ is a root. $\left\{-\frac{3}{2}, -\frac{1}{3}, \frac{1}{2}\right\}$

46. Solve the equation $3x^3 + 7x^2 - 22x - 8 = 0$ given that $-\frac{1}{3}$ is a root. $\left\{-4, -\frac{1}{3}, 2\right\}$

Practice Plus

In Exercises 47–50, use the graph or the table to determine a solution of each equation. Use synthetic division to verify that this number is a solution of the equation. Then solve the polynomial equation.

47. $x^3 + 2x^2 - 5x - 6 = 0$ $2; \{-3, -1, 2\}$

$[0, 4, 1]$ by $[-25, 25, 5]$

48. $2x^3 + x^2 - 13x + 6 = 0$ $-3; \left\{-3, \frac{1}{2}, 2\right\}$

$[-4, 0, 1]$ by $[-25, 25, 5]$

49. $6x^3 - 11x^2 + 6x - 1 = 0$ $1; \left\{\frac{1}{3}, \frac{1}{2}, 1\right\}$

50. $2x^3 + 11x^2 - 7x - 6 = 0$ $1; \left\{-6, -\frac{1}{2}, 1\right\}$

Application Exercises

51. a. Use synthetic division to show that 3 is a solution of the polynomial equation The remainder is 0.

$$14x^3 - 17x^2 - 16x - 177 = 0.$$

b. Use the solution from part (a) to solve this problem. The number of eggs, $f(x)$, in a female moth is a function of her abdominal width, x, in millimeters, modeled by

$$f(x) = 14x^3 - 17x^2 - 16x + 34.$$ 3 mm

What is the abdominal width when there are 211 eggs?

52. a. Use synthetic division to show that 2 is a solution of the polynomial equation The remainder is 0.

$$2h^3 + 14h^2 - 72 = 0.$$

b. Use the solution from part (a) to solve this problem. The width of a rectangular box is twice the height and the length is 7 inches more than the height. If the volume is 72 cubic inches, find the dimensions of the box. 2 in. by 4 in. by 9 in.

In Exercises 53–54, write a polynomial that represents the length of each rectangle.

53.

The width is $x + 0.2$ units. $0.5x^2 - 0.4x + 0.3$

The area is $0.5x^3 - 0.3x^2 + 0.22x + 0.06$ square units.

54.

$8x^2 - 12x + 4$

The width is
$x + \frac{3}{4}$ units.

The area is
$8x^3 - 6x^2 - 5x + 3$
square units.

During the 1980s, the controversial economist Arthur Laffer promoted the idea that tax increases lead to a reduction in government revenue. Called supply-side economics, the theory uses functions such as

$$f(x) = \frac{80x - 8000}{x - 110}, 30 \le x \le 100.$$

This function models the government tax revenue, f(x), in tens of billions of dollars, in terms of the tax rate, x. The graph of the function is shown. It illustrates tax revenue decreasing quite dramatically as the tax rate increases. At a tax rate of (gasp) 100%, the government takes all our money and no one has an incentive to work. With no income earned, zero dollars in tax revenue is generated.

Use function f and its graph to solve Exercises 55–56.

55. a. Find and interpret $f(30)$. Identify the solution as a point on the graph of the function. 70; (30, 70)*

 b. Rewrite the function by using long division to perform

 $$(80x - 8000) \div (x - 110).$$

 Then use this new form of the function to find $f(30)$. Do you obtain the same answer as you did in part (a)? yes*

 c. Is f a polynomial function? Explain your answer. no*

56. a. Find and interpret $f(40)$. Identify the solution as a point on the graph of the function. $68\frac{4}{7}$; $(40, 68\frac{4}{7})$*

 b. Rewrite the function by using long division to perform

 $$(80x - 8000) \div (x - 110).$$

 Then use this new form of the function to find $f(40)$. Do you obtain the same answer as you did in part (a)? yes*

 c. Is f a polynomial function? Explain your answer. no*

Explaining the Concepts 57–64. Answers will vary.

57. Explain how to perform long division of polynomials. Use $2x^3 - 3x^2 - 11x + 7$ divided by $x - 3$ in your explanation.

58. In your own words, state the Division Algorithm.

59. How can the Division Algorithm be used to check the quotient and remainder in a long division problem?

60. Explain how to perform synthetic division. Use the division problem in Exercise 57 to support your explanation.

61. State the Remainder Theorem.

62. Explain how the Remainder Theorem can be used to find $f(-6)$ if $f(x) = x^4 + 7x^3 + 8x^2 + 11x + 5$. What advantage is there to using the Remainder Theorem in this situation rather than evaluating $f(-6)$ directly?

63. How can the Factor Theorem be used to determine if $x - 1$ is a factor of $x^3 - 2x^2 - 11x + 12$?

64. If you know that -2 is a zero of

$$f(x) = x^3 + 7x^2 + 4x - 12,$$

explain how to solve the equation

$$x^3 + 7x^2 + 4x - 12 = 0.$$

Technology Exercise

65. For each equation that you solved in Exercises 43–46, use a graphing utility to graph the polynomial function defined by the left side of the equation. Use end behavior to obtain a complete graph. Then use the graph's x-intercepts to verify your solutions. Answers will vary.

Critical Thinking Exercises

Make Sense? *In Exercises 66–69, determine whether each statement makes sense or does not make sense, and explain your reasoning.*

66. When performing the division $(x^5 + 1) \div (x + 1)$, there's no need for me to follow all the steps involved in polynomial long division because I can work the problem in my head and see that the quotient must be $x^4 + 1$. does not make sense

67. Every time I divide polynomials using synthetic division, I am using a highly condensed form of the long division procedure where omitting the variables and exponents does not involve the loss of any essential data. makes sense

68. The only nongraphic method that I have for evaluating a function at a given value is to substitute that value into the function's equation. does not make sense

69. I found the zeros of function f, but I still need to find the solutions of the equation $f(x) = 0$. does not make sense

In Exercises 70–73, determine whether each statement is true or false. If the statement is false, make the necessary change(s) to produce a true statement. Changes to false statements will vary.

70. If a trinomial in x of degree 6 is divided by a trinomial in x of degree 3, the degree of the quotient is 2. false

71. Synthetic division can be used to find the quotient of $10x^3 - 6x^2 + 4x - 1$ and $x - \frac{1}{2}$. true

72. Any problem that can be done by synthetic division can also be done by the method for long division of polynomials. true

73. If a polynomial long-division problem results in a remainder that is a whole number, then the divisor is a factor of the dividend. false

74. Find k so that $4x + 3$ is a factor of

$$20x^3 + 23x^2 - 10x + k. k = -12$$

75. When $2x^2 - 7x + 9$ is divided by a polynomial, the quotient is $2x - 3$ and the remainder is 3. Find the polynomial. $x - 2$

76. Find the quotient of $x^{3n} + 1$ and $x^n + 1$. $x^{2n} - x^n + 1$

77. Synthetic division is a process for dividing a polynomial by $x - c$. The coefficient of x in the divisor is 1. How might synthetic division be used if you are dividing by $2x - 4$?

78. Use synthetic division to show that 5 is a solution of

$$x^4 - 4x^3 - 9x^2 + 16x + 20 = 0.$$

Then solve the polynomial equation. $\{-2, -1, 2, 5\}$

Retaining the Concepts

79. Use the graph of $y = f(x)$ to graph $y = 2f(x + 1) - 3$. *

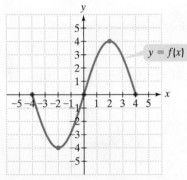

(Section 2.5, Example 8)

80. Given $f(x) = 2x - 3$ and $g(x) = 2x^2 - x + 5$, find each of the following:

a. $(f \circ g)(x)$ $\quad (f \circ g)(x) = 4x^2 - 2x + 7$

b. $(g \circ f)(x)$ $\quad (g \circ f)(x) = 8x^2 - 26x + 26$

c. $(g \circ f)(1)$. $\quad 8$

(Section 2.6, Example 5)

81. Find the inverse of $f(x) = \dfrac{x - 10}{x + 10}$.

(Section 2.7, Example 4) $\quad f^{-1}(x) = \dfrac{-10x - 10}{x - 1}$

Preview Exercises

Exercises 82–84 will help you prepare for the material covered in the next section.

82. Solve: $x^2 + 4x - 1 = 0$. $\quad \{-2 \pm \sqrt{5}\}$

83. Solve: $x^2 + 4x + 6 = 0$. $\quad \{-2 \pm i\sqrt{2}\}$

84. Let $f(x) = a_n(x^4 - 3x^2 - 4)$. If $f(3) = -150$, determine the value of a_n. $\quad -3$

Section 3.4 Zeros of Polynomial Functions

What am I supposed to learn?

After studying this section, you should be able to:

① Use the Rational Zero Theorem to find possible rational zeros.

② Find zeros of a polynomial function.

③ Solve polynomial equations.

④ Use the Linear Factorization Theorem to find polynomials with given zeros.

⑤ Use Descartes's Rule of Signs.

HIERONYMVS CARDANVS.

You stole my formula!

Tartaglia's Secret Formula for One Solution of $x^3 + mx = n$

$$x = \sqrt[3]{\sqrt{\left(\frac{n}{2}\right)^2 + \left(\frac{m}{3}\right)^3} + \frac{n}{2}}$$
$$- \sqrt[3]{\sqrt{\left(\frac{n}{2}\right)^2 + \left(\frac{m}{3}\right)^3} - \frac{n}{2}}$$

Popularizers of mathematics are sharing bizarre stories that are giving math a secure place in popular culture. One episode, able to compete with the wildest fare served up by television reality shows and the tabloids, involves three Italian mathematicians and, of all things, zeros of polynomial functions.

Tartaglia (1499–1557), poor and starving, has found a formula that gives a root for a third-degree polynomial equation. Cardano (1501–1576) begs Tartaglia to reveal the secret formula, wheedling it from him with the promise he will find the impoverished Tartaglia a patron. Then Cardano publishes his famous work *Ars Magna*, in which he presents Tartaglia's formula as his own. Cardano uses his most talented student, Ferrari (1522–1565), who derived a formula for a root of a fourth-degree polynomial equation, to falsely accuse Tartaglia of plagiarism. The dispute becomes violent and Tartaglia is fortunate to escape alive.

The noise from this "You Stole My Formula" episode is quieted by the work of French mathematician Evariste Galois (1811–1832). Galois proved that there is no general formula for finding roots of polynomial equations of degree 5 or higher. There are, however, methods for finding roots. In this section, we study methods for finding zeros of polynomial functions. We begin with a theorem that plays an important role in this process.

GREAT QUESTION!

When finding zeros of polynomial functions, what kinds of numbers can I get?

Here's a quick example that involves all possible kinds of zeros:

$$f(x) = (x + 3)(2x - 1)(x + \sqrt{2})(x - \sqrt{2})(x - 4 + 5i)(x - 4 - 5i).$$

Zeros: $-3,$ $\dfrac{1}{2},$ $-\sqrt{2},$ $\sqrt{2},$ $4 - 5i,$ $4 + 5i$

| Rational zeros | Irrational zeros | Complex imaginary zeros |

| Real zeros | Nonreal zeros |

① Use the Rational Zero Theorem to find possible rational zeros.

The Rational Zero Theorem

The *Rational Zero Theorem* provides us with a tool that we can use to make a list of all possible rational zeros of a polynomial function. Equivalently, the theorem gives all possible rational roots of a polynomial equation. Not every number in the list will be a zero of the function, but every rational zero of the polynomial function will appear somewhere in the list.

> **The Rational Zero Theorem**
>
> If $f(x) = a_n x^n + a_{n-1} x^{n-1} + \cdots + a_1 x + a_0$ has *integer* coefficients and $\dfrac{p}{q}$ (where $\dfrac{p}{q}$ is reduced to lowest terms) is a rational zero of f, then p is a factor of the constant term, a_0, and q is a factor of the leading coefficient, a_n.

You can explore the "why" behind the Rational Zero Theorem in Exercise 92 of Exercise Set 3.4. For now, let's see if we can figure out what the theorem tells us about possible rational zeros. To use the theorem, list all the integers that are factors of the constant term, a_0. Then list all the integers that are factors of the leading coefficient, a_n. Finally, list all possible rational zeros:

$$\text{Possible rational zeros} = \frac{\text{Factors of the constant term}}{\text{Factors of the leading coefficient}}.$$

GREAT QUESTION!

When you list the factors, what does the ± symbol mean?

It's a shorthand way to list two factors at once.

± 1 ± 2 ± 4

| 1 and −1 are factors. | 2 and −2 are factors. | 4 and −4 are factors. |

EXAMPLE 1 Using the Rational Zero Theorem

List all possible rational zeros of $f(x) = -x^4 + 3x^2 + 4$.

SOLUTION

The constant term is 4. We list all of its factors: $\pm 1, \pm 2, \pm 4$. The leading coefficient is -1. Its factors are ± 1.

Factors of the constant term, 4: $\pm 1,$ $\pm 2,$ ± 4

Factors of the leading coefficient, -1: ± 1

Because

$$\text{Possible rational zeros} = \frac{\text{Factors of the constant term}}{\text{Factors of the leading coefficient}},$$

we must take each number in the first row, ± 1, ± 2, ± 4, and divide by each number in the second row, ± 1.

$$\text{Possible rational zeros} = \frac{\text{Factors of 4}}{\text{Factors of } -1} = \frac{\pm 1, \pm 2, \pm 4}{\pm 1} = \pm 1, \quad \pm 2, \quad \pm 4$$

There are six possible rational zeros. The graph of $f(x) = -x^4 + 3x^2 + 4$ is shown in **Figure 3.27**. The x-intercepts are -2 and 2. Thus, -2 and 2 are the actual rational zeros. •••

\oslash **Check Point 1** List all possible rational zeros of $\pm 1, \pm 2, \pm 3, \pm 6$

$$f(x) = x^3 + 2x^2 - 5x - 6.$$

FIGURE 3.27 The graph of $f(x) = -x^4 + 3x^2 + 4$ shows that -2 and 2 are rational zeros.

GREAT QUESTION!

What's the relationship among zeros, roots, and x-intercepts?

The zeros of a function f are the roots, or solutions, of the equation $f(x) = 0$. Furthermore, the real zeros, or real roots, are the x-intercepts of the graph of f.

EXAMPLE 2 Using the Rational Zero Theorem

List all possible rational zeros of $f(x) = 15x^3 + 14x^2 - 3x - 2$.

SOLUTION

The constant term is -2 and the leading coefficient is 15.

$$\text{Possible rational zeros} = \frac{\text{Factors of the constant term, } -2}{\text{Factors of the leading coefficient, } 15} = \frac{\pm 1, \pm 2}{\pm 1, \pm 3, \pm 5, \pm 15}$$

$$= \pm 1, \quad \pm 2, \quad \pm \frac{1}{3}, \quad \pm \frac{2}{3}, \quad \pm \frac{1}{5}, \quad \pm \frac{2}{5}, \quad \pm \frac{1}{15}, \quad \pm \frac{2}{15}$$

| Divide ±1 and ±2 by ±1. | Divide ±1 and ±2 by ±3. | Divide ±1 and ±2 by ±5. | Divide ±1 and ±2 by ±15. |

There are 16 possible rational zeros. The actual solution set of

$$15x^3 + 14x^2 - 3x - 2 = 0$$

is $\left\{-1, -\frac{1}{3}, \frac{2}{5}\right\}$, which contains three of the 16 possible zeros. •••

\oslash **Check Point 2** List all possible rational zeros of

$$f(x) = 4x^5 + 12x^4 - x - 3. \quad \pm 1, \pm 3, \pm \frac{1}{2}, \pm \frac{1}{4}, \pm \frac{3}{2}, \pm \frac{3}{4}$$

② Find zeros of a polynomial function.

How do we determine which (if any) of the possible rational zeros are rational zeros of the polynomial function? To find the first rational zero, we can use a trial-and-error process involving synthetic division: If $f(x)$ is divided by $x - c$ and the remainder is zero, then c is a zero of f. After we identify the first rational zero, we use the result of the synthetic division to factor the original polynomial. Then we set each factor equal to zero to identify any additional rational zeros.

EXAMPLE 3 Finding Zeros of a Polynomial Function

Find all zeros of $f(x) = x^3 + 2x^2 - 5x - 6$.

SOLUTION

We begin by listing all possible rational zeros of $f(x) = x^3 + 2x^2 - 5x - 6$.

Possible rational zeros

$$= \frac{\text{Factors of the constant term, } -6}{\text{Factors of the leading coefficient, } 1} = \frac{\pm 1, \pm 2, \pm 3, \pm 6}{\pm 1} = \pm 1, \pm 2, \pm 3, \pm 6$$

> Divide the eight numbers in the numerator by ± 1.

Now we will use synthetic division to see if we can find a rational zero among the possible rational zeros $\pm 1, \pm 2, \pm 3, \pm 6$. Keep in mind that if $f(x)$ is divided by $x - c$ and the remainder is zero, then c is a zero of f. Let's start by testing 1. If 1 is not a rational zero, then we will test other possible rational zeros.

Test 1.

Possible rational zero

$$\underline{1|} \quad 1 \quad 2 \quad -5 \quad -6$$
$$ \quad 1 \quad 3 \quad -2$$
$$\overline{1 \quad 3 \quad -2 \quad -8}$$

The nonzero remainder shows that 1 is not a zero.

Test 2.

Possible rational zero

$$\underline{2|} \quad 1 \quad 2 \quad -5 \quad -6$$
$$ \quad 2 \quad 8 \quad 6$$
$$\overline{1 \quad 4 \quad 3 \quad 0}$$

The zero remainder shows that 2 is a zero.

The zero remainder tells us that 2 is a zero of the polynomial function $f(x) = x^3 + 2x^2 - 5x - 6$. Equivalently, 2 is a solution, or root, of the polynomial equation $x^3 + 2x^2 - 5x - 6 = 0$. Thus, $x - 2$ is a factor of the polynomial. The first three numbers in the bottom row of the synthetic division on the right, 1, 4, and 3, give the coefficients of the other factor. This factor is $x^2 + 4x + 3$.

$x^3 + 2x^2 - 5x - 6 = 0$ Finding the zeros of $f(x) = x^3 + 2x^2 - 5x - 6$ is the same as finding the roots of this equation.

$(x - 2)(x^2 + 4x + 3) = 0$ Factor using the result from the synthetic division.

$(x - 2)(x + 3)(x + 1) = 0$ Factor completely.

$x - 2 = 0 \quad$ or $\quad x + 3 = 0 \quad$ or $\quad x + 1 = 0$ Set each factor equal to zero.

$x = 2 \qquad\qquad x = -3 \qquad\qquad x = -1$ Solve for x.

The solution set is $\{-3, -1, 2\}$. The zeros of f are $-3, -1,$ and 2. • • •

✓ **Check Point 3** Find all zeros of

$$f(x) = x^3 + 8x^2 + 11x - 20. \quad \text{—5, —4, and 1}$$

Our work in Example 3 involved finding zeros of a third-degree polynomial function. The Rational Zero Theorem is a tool that allows us to rewrite such functions as products of two factors, one linear and one quadratic. Zeros of the quadratic factor are found by factoring, the quadratic formula, or the square root property.

EXAMPLE 4 Finding Zeros of a Polynomial Function

Find all zeros of $f(x) = x^3 + 7x^2 + 11x - 3$.

SOLUTION

We begin by listing all possible rational zeros.

$$\text{Possible rational zeros} = \frac{\text{Factors of the constant term, } -3}{\text{Factors of the leading coefficient, } 1} = \frac{\pm 1, \pm 3}{\pm 1} = \pm 1, \pm 3$$

Now we will use synthetic division to see if we can find a rational zero among the four possible rational zeros, $\pm 1, \pm 3$, of $f(x) = x^3 + 7x^2 + 11x - 3$.

Test 1.	Test −1.	Test 3.	Test −3.

$$
\begin{array}{r|rrrr}
1 & 1 & 7 & 11 & -3 \\
 & & 1 & 8 & 19 \\
\hline
 & 1 & 8 & 19 & 16
\end{array}
\qquad
\begin{array}{r|rrrr}
-1 & 1 & 7 & 11 & -3 \\
 & & -1 & -6 & -5 \\
\hline
 & 1 & 6 & 5 & -8
\end{array}
\qquad
\begin{array}{r|rrrr}
3 & 1 & 7 & 11 & -3 \\
 & & 3 & 30 & 123 \\
\hline
 & 1 & 10 & 41 & 120
\end{array}
\qquad
\begin{array}{r|rrrr}
-3 & 1 & 7 & 11 & -3 \\
 & & -3 & -12 & 3 \\
\hline
 & 1 & 4 & -1 & 0
\end{array}
$$

The zero remainder when testing -3 tells us that -3 is a zero of the polynomial function $f(x) = x^3 + 7x^2 + 11x - 3$. To find all zeros of f, we proceed as follows:

$$x^3 + 7x^2 + 11x - 3 = 0 \qquad \text{Finding the zeros of } f \text{ is the same thing as finding the roots of } f(x) = 0.$$

$$(x + 3)(x^2 + 4x - 1) = 0 \qquad \text{This result is from the last synthetic division, shown above. The first three numbers in the bottom row, 1, 4, and } -1, \text{ give the coefficients of the second factor.}$$

$$x + 3 = 0 \quad \text{or} \quad x^2 + 4x - 1 = 0 \qquad \text{Set each factor equal to 0.}$$

$$x = -3. \qquad \text{Solve the linear equation.}$$

We can use the quadratic formula to solve $x^2 + 4x - 1 = 0$.

$$x = \frac{-b \pm \sqrt{b^2 - 4ac}}{2a} \qquad \text{We use the quadratic formula because } x^2 + 4x - 1 \text{ cannot be factored.}$$

$$= \frac{-4 \pm \sqrt{4^2 - 4(1)(-1)}}{2(1)} \qquad \text{Let } a = 1, b = 4, \text{ and } c = -1.$$

$$= \frac{-4 \pm \sqrt{20}}{2} \qquad \text{Multiply and subtract under the radical: } 4^2 - 4(1)(-1) = 16 - (-4) = 16 + 4 = 20.$$

$$= \frac{-4 \pm 2\sqrt{5}}{2} \qquad \sqrt{20} = \sqrt{4 \cdot 5} = 2\sqrt{5}$$

$$= -2 \pm \sqrt{5} \qquad \text{Divide the numerator and the denominator by 2.}$$

The solution set is $\left\{ -3, -2 - \sqrt{5}, -2 + \sqrt{5} \right\}$. The zeros of $f(x) = x^3 + 7x^2 + 11x - 3$ are $-3, -2 - \sqrt{5}$, and $-2 + \sqrt{5}$. Among these three real zeros, one zero is rational and two are irrational. ● ● ●

☑ **Check Point 4** Find all zeros of $f(x) = x^3 + x^2 - 5x - 2$. $2, \frac{-3 - \sqrt{5}}{2}, \text{ and } \frac{-3 + \sqrt{5}}{2}$

If the degree of a polynomial function or equation is 4 or higher, it is often necessary to find more than one linear factor by synthetic division.

One way to speed up the process of finding the first zero is to graph the function. Any x-intercept is a zero.

③ Solve polynomial equations.

EXAMPLE 5 Solving a Polynomial Equation

Solve: $x^4 - 6x^2 - 8x + 24 = 0$.

SOLUTION

Recall that we refer to the *zeros* of a polynomial function and the *roots* of a polynomial equation. Because we are given an equation, we will use the word "roots," rather than "zeros," in the solution process.

We begin by listing all possible rational roots of $x^4 - 6x^2 - 8x + 24 = 0$.

$$\text{Possible rational roots} = \frac{\text{Factors of the constant term, 24}}{\text{Factors of the leading coefficient, 1}}$$

$$= \frac{\pm 1, \ \pm 2, \ \pm 3, \ \pm 4, \ \pm 6, \ \pm 8, \ \pm 12, \ \pm 24}{\pm 1}$$

$$= \pm 1, \ \pm 2, \ \pm 3, \ \pm 4, \ \pm 6, \ \pm 8, \ \pm 12, \ \pm 24$$

Part of the graph of $f(x) = x^4 - 6x^2 - 8x + 24$ is shown in **Figure 3.28**. Because the x-intercept is 2, we will test 2 by synthetic division and show that it is a root of the given equation. Without the graph, the procedure would be to start the trial-and-error synthetic division with 1 and proceed until a zero remainder is found, as we did in Example 4.

FIGURE 3.28 The graph of $f(x) = x^4 - 6x^2 - 8x + 24$ in a $[-1, 5, 1]$ by $[-2, 10, 1]$ viewing rectangle

$$\begin{array}{r|rrrrr} 2 & 1 & 0 & -6 & -8 & 24 \\ & & 2 & 4 & -4 & -24 \\ \hline & 1 & 2 & -2 & -12 & 0 \end{array}$$

> **Careful!**
> $x^4 - 6x^2 - 8x + 24 = x^4 + 0x^3 - 6x^2 - 8x + 24$

> The zero remainder indicates that 2 is a root of $x^4 - 6x^2 - 8x + 24 = 0$.

Now we can rewrite the given equation in factored form.

$$x^4 - 6x^2 - 8x + 24 = 0 \quad \text{This is the given equation.}$$

$$(x - 2)(x^3 + 2x^2 - 2x - 12) = 0 \quad \text{This is the result obtained from the synthetic division. The first four numbers in the bottom row, 1, 2, } -2, \text{ and } -12, \text{ give the coefficients of the second factor.}$$

$$x - 2 = 0 \quad \text{or} \quad x^3 + 2x^2 - 2x - 12 = 0 \quad \text{Set each factor equal to 0.}$$

We can use the same approach to look for rational roots of the polynomial equation $x^3 + 2x^2 - 2x - 12 = 0$, listing all possible rational roots. Without the graph in **Figure 3.28**, the procedure would be to start testing possible rational roots by trial-and-error synthetic division. However, take a second look at the graph in **Figure 3.28**. Because the graph turns around at 2, this means that 2 is a root of even multiplicity. Thus, 2 must also be a root of $x^3 + 2x^2 - 2x - 12 = 0$, confirmed by the following synthetic division.

$$\begin{array}{r|rrrr} 2 & 1 & 2 & -2 & -12 \\ & & 2 & 8 & 12 \\ \hline & 1 & 4 & 6 & 0 \end{array}$$

> These are the coefficients of $x^3 + 2x^2 - 2x - 12 = 0$.

> The zero remainder indicates that 2 is a root of $x^3 + 2x^2 - 2x - 12 = 0$.

Now we can solve the original equation as follows:

$$x^4 - 6x^2 - 8x + 24 = 0 \quad \text{This is the given equation.}$$

$$(x - 2)(x^3 + 2x^2 - 2x - 12) = 0 \quad \text{This factorization was obtained from the first synthetic division.}$$

$$(x - 2)(x - 2)(x^2 + 4x + 6) = 0 \quad \text{This factorization was obtained from the second synthetic division. The first three numbers in the bottom row, 1, 4, and 6, give the coefficients of the third factor.}$$

$$x - 2 = 0 \quad \text{or} \quad x - 2 = 0 \quad \text{or} \quad x^2 + 4x + 6 = 0 \quad \text{Set each factor equal to 0.}$$

$$x = 2 \qquad\qquad x = 2. \qquad\qquad\qquad \text{Solve the linear equations.}$$

We can use the quadratic formula to solve $x^2 + 4x + 6 = 0$.

$$x = \frac{-b \pm \sqrt{b^2 - 4ac}}{2a}$$ We use the quadratic formula because $x^2 + 4x + 6$ cannot be factored.

$$= \frac{-4 \pm \sqrt{4^2 - 4(1)(6)}}{2(1)}$$ Let $a = 1$, $b = 4$, and $c = 6$.

$$= \frac{-4 \pm \sqrt{-8}}{2}$$ Multiply and subtract under the radical: $4^2 - 4(1)(6) = 16 - 24 = -8$.

$$= \frac{-4 \pm 2i\sqrt{2}}{2}$$ $\sqrt{-8} = \sqrt{4(2)(-1)} = 2i\sqrt{2}$

$$= -2 \pm i\sqrt{2}$$ Simplify.

The solution set of the original equation, $x^4 - 6x^2 - 8x + 24 = 0$, is $\{2, -2 - i\sqrt{2}, -2 + i\sqrt{2}\}$. Note that a graphing utility does not reveal the two imaginary roots. ● ● ●

In Example 5, 2 is a repeated root of the equation with multiplicity 2. Counting this multiple root separately, the fourth-degree equation $x^4 - 6x^2 - 8x + 24 = 0$ has four roots: 2, 2, $-2 + i\sqrt{2}$, and $-2 - i\sqrt{2}$. The equation and its roots illustrate two general properties:

Properties of Roots of Polynomial Equations

1. If a polynomial equation is of degree n, then counting multiple roots separately, the equation has n roots.
2. If $a + bi$ is a root of a polynomial equation with real coefficients ($b \neq 0$), then the imaginary number $a - bi$ is also a root. Imaginary roots, if they exist, occur in conjugate pairs.

 Check Point 5 Solve: $x^4 - 6x^3 + 22x^2 - 30x + 13 = 0$. $\{1, 2 - 3i, 2 + 3i\}$

The Fundamental Theorem of Algebra

The fact that a polynomial equation of degree n has n roots is a consequence of a theorem proved in 1799 by a 22-year-old student named Carl Friedrich Gauss in his doctoral dissertation. His result is called the **Fundamental Theorem of Algebra**.

The Fundamental Theorem of Algebra

If $f(x)$ is a polynomial of degree n, where $n \geq 1$, then the equation $f(x) = 0$ has at least one complex root.

GREAT QUESTION!

Do all polynomial equations have at least one imaginary root?

No. As you read the Fundamental Theorem of Algebra, don't confuse *complex root* with *imaginary root* and conclude that every polynomial equation has at least one imaginary root. Recall that complex numbers, $a + bi$, include both real numbers ($b = 0$) and imaginary numbers ($b \neq 0$).

Suppose, for example, that $f(x) = 0$ represents a polynomial equation of degree n. By the Fundamental Theorem of Algebra, we know that this equation has at least one complex root; we'll call it c_1. By the Factor Theorem, we know that $x - c_1$ is a factor of $f(x)$. Therefore, we obtain

$$(x - c_1)q_1(x) = 0$$ The degree of the polynomial $q_1(x)$ is $n - 1$.

$$x - c_1 = 0 \quad \text{or} \quad q_1(x) = 0.$$ Set each factor equal to O.

If the degree of $q_1(x)$ is at least 1, by the Fundamental Theorem of Algebra, the equation $q_1(x) = 0$ has at least one complex root. We'll call it c_2. The Factor Theorem gives us

$$q_1(x) = 0$$ The degree of $q_1(x)$ is $n - 1$.

$$(x - c_2)q_2(x) = 0$$ The degree of $q_2(x)$ is $n - 2$.

$$x - c_2 = 0 \quad \text{or} \quad q_2(x) = 0.$$ Set each factor equal to O.

Let's see what we have up to this point and then continue the process.

$$f(x) = 0 \quad \text{\small This is the original polynomial equation of degree } n.$$

$$(x - c_1)q_1(x) = 0 \quad \text{\small This is the result from our first application of the} \\ \text{\small Fundamental Theorem.}$$

$$(x - c_1)(x - c_2)q_2(x) = 0 \quad \text{\small This is the result from our second application of the} \\ \text{\small Fundamental Theorem.}$$

By continuing this process, we will obtain the product of n linear factors. Setting each of these linear factors equal to zero results in n complex roots. Thus, if $f(x)$ is a polynomial of degree n, where $n \geq 1$, then $f(x) = 0$ has exactly n roots, where roots are counted according to their multiplicity.

④ Use the Linear Factorization Theorem to find polynomials with given zeros.

The Linear Factorization Theorem

In Example 5, we found that $x^4 - 6x^2 - 8x + 24 = 0$ has $\{2, -2 \pm i\sqrt{2}\}$ as a solution set, where 2 is a repeated root with multiplicity 2. The polynomial can be factored over the complex numbers as follows:

$$f(x) = x^4 - 6x^2 - 8x + 24 \qquad \text{\small These are the four zeros.}$$

$$= [x - (-2 + i\sqrt{2})][x - (-2 - i\sqrt{2})](x - 2)(x - 2).$$

These are the linear factors.

This fourth-degree polynomial has four linear factors. Just as an nth-degree polynomial equation has n roots, an nth-degree polynomial has n linear factors. This is formally stated as the **Linear Factorization Theorem**.

The Linear Factorization Theorem

If $f(x) = a_n x^n + a_{n-1} x^{n-1} + \cdots + a_1 x + a_0$, where $n \geq 1$ and $a_n \neq 0$, then

$$f(x) = a_n(x - c_1)(x - c_2) \cdots (x - c_n),$$

where c_1, c_2, \ldots, c_n are complex numbers (possibly real and not necessarily distinct). In words: An nth-degree polynomial can be expressed as the product of a nonzero constant and n linear factors, where each linear factor has a leading coefficient of 1.

Many of our problems involving polynomial functions and polynomial equations dealt with the process of finding zeros and roots. The Linear Factorization Theorem enables us to reverse this process, finding a polynomial function when the zeros are given.

EXAMPLE 6 Finding a Polynomial Function with Given Zeros

Find a fourth-degree polynomial function $f(x)$ with real coefficients that has $-2, 2$, and i as zeros and such that $f(3) = -150$.

SOLUTION

Because i is a zero and the polynomial has real coefficients, the conjugate, $-i$, must also be a zero. We can now use the Linear Factorization Theorem.

$$f(x) = a_n(x - c_1)(x - c_2)(x - c_3)(x - c_4) \qquad \text{\small This is the linear factorization for a} \\ \text{\small fourth-degree polynomial.}$$

$$= a_n(x + 2)(x - 2)(x - i)(x + i) \qquad \text{\small Use the given zeros:} \\ \text{\small } c_1 = -2, c_2 = 2, c_3 = i, \text{ and, from above,} \\ \text{\small } c_4 = -i.$$

$$= a_n(x^2 - 4)(x^2 + 1) \qquad \text{\small Multiply: } (x - i)(x + i) = x^2 - i^2 = \\ \text{\small } x^2 - (-1) = x^2 + 1.$$

TECHNOLOGY

Graphic Connections

The graph of
$f(x) = -3x^4 + 9x^2 + 12$, shown
in a $[-3, 3, 1]$ by $[-200, 20, 20]$
viewing rectangle, verifies that -2
and 2 are real zeros. By tracing
along the curve, we can check that
$f(3) = -150$.

-2 is a zero. 2 is a zero.

$f(x) = a_n(x^4 - 3x^2 - 4)$	Complete the multiplication for $f(x) = a_n(x^2 - 4)(x^2 + 1)$.
$f(3) = a_n(3^4 - 3 \cdot 3^2 - 4) = -150$	Use the fact that $f(3) = -150$ to find a_n.
$a_n(81 - 27 - 4) = -150$	Solve for a_n.
$50a_n = -150$	Simplify: $81 - 27 - 4 = 50$.
$a_n = -3$	Divide both sides by 50.

Substituting -3 for a_n in the formula for $f(x)$, we obtain

$$f(x) = -3(x^4 - 3x^2 - 4).$$

Equivalently,

$$f(x) = -3x^4 + 9x^2 + 12.$$ • • •

◯✓ **Check Point 6** Find a third-degree polynomial function $f(x)$ with real coefficients that has -3 and i as zeros and such that $f(1) = 8$. $f(x) = x^3 + 3x^2 + x + 3$

⑤ Use Descartes's Rule of Signs.

"An equation can have as many true [positive] roots as it contains changes of sign, from plus to minus or from minus to plus." René Descartes (1596–1650) in *La Géométrie* (1637)

GREAT QUESTION!

Does Descartes's Rule of Signs include both rational and irrational zeros?

Yes. The number of real zeros given by Descartes's Rule of Signs includes rational zeros from a list of possible rational zeros, as well as irrational zeros not on the list. It does not include any imaginary zeros.

Descartes's Rule of Signs

Because an nth-degree polynomial equation might have roots that are imaginary numbers, we should note that such an equation can have *at most n* real roots. **Descartes's Rule of Signs** provides even more specific information about the number of real zeros that a polynomial can have. The rule is based on considering *variations in sign* between consecutive coefficients. For example, the function $f(x) = 3x^7 - 2x^5 - x^4 + 7x^2 + x - 3$ has three sign changes:

$$f(x) = 3x^7 - 2x^5 - x^4 + 7x^2 + x - 3.$$

sign change sign change sign change

Descartes's Rule of Signs

Let $f(x) = a_n x^n + a_{n-1} x^{n-1} + \cdots + a_2 x^2 + a_1 x + a_0$ be a polynomial with real coefficients.

1. The number of *positive real zeros* of f is either
 a. the same as the number of sign changes of $f(x)$

 or

 b. less than the number of sign changes of $f(x)$ by a positive even integer. If $f(x)$ has only one variation in sign, then f has exactly one positive real zero.

2. The number of *negative real zeros* of f is either
 a. the same as the number of sign changes of $f(-x)$

 or

 b. less than the number of sign changes of $f(-x)$ by a positive even integer. If $f(-x)$ has only one variation in sign, then f has exactly one negative real zero.

Table 3.1 illustrates what Descartes's Rule of Signs tells us about the positive real zeros of various polynomial functions.

Table 3.1 Descartes's Rule of Signs and Positive Real Zeros

Polynomial Function	Sign Changes	Conclusion
$f(x) = 3x^7 - 2x^5 - x^4 + 7x^2 + x - 3$ sign change sign change sign change	3	There are 3 positive real zeros. or There is $3 - 2 = 1$ positive real zero.
$f(x) = 4x^5 + 2x^4 - 3x^2 + x + 5$ sign change sign change	2	There are 2 positive real zeros. or There are $2 - 2 = 0$ positive real zeros.
$f(x) = -7x^6 - 5x^4 + x + 9$ sign change	1	There is 1 positive real zero.

EXAMPLE 7 Using Descartes's Rule of Signs

Determine the possible numbers of positive and negative real zeros of $f(x) = x^3 + 2x^2 + 5x + 4$.

SOLUTION

1. To find possibilities for positive real zeros, count the number of sign changes in the equation for $f(x)$. Because all the coefficients are positive, there are no variations in sign. Thus, there are no positive real zeros.

2. To find possibilities for negative real zeros, count the number of sign changes in the equation for $f(-x)$. We obtain this equation by replacing x with $-x$ in the given function.

$$f(x) = x^3 + 2x^2 + 5x + 4$$

Replace x with $-x$.

$$f(-x) = (-x)^3 + 2(-x)^2 + 5(-x) + 4$$
$$= -x^3 + 2x^2 - 5x + 4$$

Now count the sign changes.

$$f(-x) = -x^3 + 2x^2 - 5x + 4$$

sign change sign change sign change

There are three variations in sign. The number of negative real zeros of f is either equal to the number of sign changes, 3, or is less than this number by an even integer. This means that either there are 3 negative real zeros or there is $3 - 2 = 1$ negative real zero. •••

What do the results of Example 7 mean in terms of solving

$$x^3 + 2x^2 + 5x + 4 = 0?$$

Without using Descartes's Rule of Signs, we list the possible rational roots as follows:

Possible rational roots

$$= \frac{\text{Factors of the constant term, 4}}{\text{Factors of the leading coefficient, 1}} = \frac{\pm 1, \pm 2, \pm 4}{\pm 1} = \pm 1, \pm 2, \pm 4.$$

We see that the possible rational roots of $x^3 + 2x^2 + 5x + 4 = 0$ are ± 1, ± 2, and ± 4. However, Descartes's Rule of Signs informed us that $f(x) = x^3 + 2x^2 + 5x + 4$ has no positive real zeros. Thus, the polynomial equation $x^3 + 2x^2 + 5x + 4 = 0$ has no positive real roots. This means that we can eliminate the positive numbers 1, 2, and 4 from our list of possible rational roots. Possible rational roots include only -1, -2, and -4. We can use synthetic division and test the first of these three possible rational roots of $x^3 + 2x^2 + 5x + 4 = 0$ as follows:

Test -1.
$$
\begin{array}{r|rrrr}
-1 & 1 & 2 & 5 & 4 \\
 & & -1 & -1 & -4 \\
\hline
 & 1 & 1 & 4 & 0.
\end{array}
$$

The zero remainder shows that -1 is a root.

By solving the equation $x^3 + 2x^2 + 5x + 4 = 0$, you will find that this equation of degree 3 has three roots. One root is -1 and the other two roots are imaginary numbers in a conjugate pair. Verify this by completing the solution process.

⊘ **Check Point 7** Determine the possible numbers of positive and negative real zeros of $f(x) = x^4 - 14x^3 + 71x^2 - 154x + 120$. 4, 2, or 0 positive zeros; no possible negative zeros

ACHIEVING SUCCESS

Think about finding a tutor.

If you're attending all lectures, taking good class notes, reading the textbook, and doing all the assigned homework, but still having difficulty in a math course, you might want to find a tutor. Many on-campus learning centers and math labs have trained people available to help you. Sometimes a TA who has previously taught the course is available. **Make sure the tutor is both good at math and familiar with the particular course you're taking.** Bring your textbook, class notes, the problems you've done, and information about course policy and tests to each meeting with your tutor. That way he or she can be sure the tutoring sessions address your exact needs.

CONCEPT AND VOCABULARY CHECK

Fill in each blank so that the resulting statement is true.

1. Consider the polynomial function with integer coefficients

 $f(x) = a_n x^n + a_{n-1} x^{n-1} + \cdots + a_1 x + a_0, a_n \neq 0.$

 The Rational Zero Theorem states that if $\dfrac{p}{q}$ is a rational zero of f (where $\dfrac{p}{q}$ is reduced to lowest terms), then p is a factor of ___a_0___ and q is a factor of ___a_n___.

2. True or false: $\dfrac{3}{2}$ is a possible rational zero of $f(x) = 2x^3 + 11x^2 - 7x - 6.$ ___true___

3. True or false: $\dfrac{1}{2}$ is a possible rational zero of $f(x) = 3x^4 - 3x^3 + x^2 - x + 1.$ ___false___

4. If a polynomial equation is of degree n, then counting multiple roots separately, the equation has ___n___ roots.

5. If $a + bi$ is a root of a polynomial equation with real coefficients, $b \neq 0$, then ___$a - bi$___ is also a root of the equation.

6. Consider solving $2x^3 + 11x^2 - 7x - 6 = 0$. The synthetic division shown below indicates that ___-6___ is a root.

 $$
 \begin{array}{r|rrrr}
 -6 & 2 & 11 & -7 & -6 \\
 & & -12 & 6 & 6 \\
 \hline
 & 2 & -1 & -1 & 0
 \end{array}
 $$

 Based on the synthetic division, $2x^3 + 11x^2 - 7x - 6 = 0$ can be written in factored form as ___$(x + 6)(2x^2 - x - 1) = 0$___.

7. The Linear Factorization Theorem states that an nth-degree polynomial can be expressed as the product of a nonzero constant and __n__ linear factors, where each linear factor has a leading coefficient of __1__.

Use Descartes's Rule of Signs to determine whether each statement is true of false.

8. A polynomial function with four sign changes must have four positive real zeros. __false__

9. A polynomial function with one sign change must have one positive real zero. __true__

10. A polynomial function with seven sign changes can have one, three, five, or seven positive real zeros. __true__

EXERCISE SET 3.4

Practice Exercises

In Exercises 1–8, use the Rational Zero Theorem to list all possible rational zeros for each given function.

1. $f(x) = x^3 + x^2 - 4x - 4$ $\pm 1, \pm 2, \pm 4$

2. $f(x) = x^3 + 3x^2 - 6x - 8$ $\pm 1, \pm 2, \pm 4, \pm 8$

3. $f(x) = 3x^4 - 11x^3 - x^2 + 19x + 6$ $\pm 1, \pm 2, \pm 3, \pm 6, \pm\frac{1}{3}, \pm\frac{2}{3}$

4. $f(x) = 2x^4 + 3x^3 - 11x^2 - 9x + 15$ *

5. $f(x) = 4x^4 - x^3 + 5x^2 - 2x - 6$ *

6. $f(x) = 3x^4 - 11x^3 - 3x^2 - 6x + 8$ *

7. $f(x) = x^5 - x^4 - 7x^3 + 7x^2 - 12x - 12$ *

8. $f(x) = 4x^5 - 8x^4 - x + 2$ $\pm 1, \pm 2, \pm\frac{1}{2}, \pm\frac{1}{4}$

In Exercises 9–16,

 a. List all possible rational zeros.

 b. Use synthetic division to test the possible rational zeros and find an actual zero.

 c. Use the quotient from part (b) to find the remaining zeros of the polynomial function.

9. $f(x) = x^3 + x^2 - 4x - 4$ a. $\pm 1, \pm 2, \pm 4$ c. $-2, -1,$ and 2*

10. $f(x) = x^3 - 2x^2 - 11x + 12$ c. $-3, 1,$ and 4*

11. $f(x) = 2x^3 - 3x^2 - 11x + 6$ c. $-2, \frac{1}{2},$ and 3*

12. $f(x) = 2x^3 - 5x^2 + x + 2$ a. $\pm 1, \pm 2, \pm\frac{1}{2}$ c. $-\frac{1}{2}, 1,$ and 2*

13. $f(x) = x^3 + 4x^2 - 3x - 6$ a. $\pm 1, \pm 2, \pm 3, \pm 6$ b. -1*

14. $f(x) = 2x^3 + x^2 - 3x + 1$ a. $\pm 1, \pm\frac{1}{2}$ b. $\frac{1}{2}$*

15. $f(x) = 2x^3 + 6x^2 + 5x + 2$ a. $\pm 1, \pm\frac{1}{2}, \pm 2$ b. -2*

16. $f(x) = x^3 - 4x^2 + 8x - 5$ a. $\pm 1, \pm 5$ b. 1*

In Exercises 17–24,

 a. List all possible rational roots.

 b. Use synthetic division to test the possible rational roots and find an actual root.

 c. Use the quotient from part (b) to find the remaining roots and solve the equation.

17. $x^3 - 2x^2 - 11x + 12 = 0$ c. $\{-3, 1, 4\}$*

18. $x^3 - 2x^2 - 7x - 4 = 0$ a. $\pm 1, \pm 2, \pm 4$ c. $\{-1, 4\}$*

19. $x^3 - 10x - 12 = 0$ b. -2 c. $\{-2, 1 + \sqrt{7}, 1 - \sqrt{7}\}$*

20. $x^3 - 5x^2 + 17x - 13 = 0$ a. $\pm 1, \pm 13$ b. 1 c. $\{1, 2 + 3i, 2 - 3i\}$

21. $6x^3 + 25x^2 - 24x + 5 = 0$ c. $\{-5, \frac{1}{3}, \frac{1}{2}\}$*

22. $2x^3 - 5x^2 - 6x + 4 = 0$ b. $\frac{1}{2}$ c. $\{\frac{1}{2}, 1 + \sqrt{5}, 1 - \sqrt{5}\}$*

23. $x^4 - 2x^3 - 5x^2 + 8x + 4 = 0$ c. $\{-2, 2, 1 + \sqrt{2}, 1 - \sqrt{2}\}$*

24. $x^4 - 2x^2 - 16x - 15 = 0$ c. $\{-1, 3, -1 + 2i, -1 - 2i\}$*

In Exercises 25–32, find an nth-degree polynomial function with real coefficients satisfying the given conditions. If you are using a graphing utility, use it to graph the function and verify the real zeros and the given function value.

25. $n = 3$; 1 and $5i$ are zeros; $f(-1) = -104$ *

26. $n = 3$; 4 and $2i$ are zeros; $f(-1) = -50$ *

27. $n = 3$; -5 and $4 + 3i$ are zeros; $f(2) = 91$ *

28. $n = 3$; 6 and $-5 + 2i$ are zeros; $f(2) = -636$ *

29. $n = 4$; i and $3i$ are zeros; $f(-1) = 20$ $f(x) = x^4 + 10x^2 + 9$

30. $n = 4$; $-2, -\frac{1}{2},$ and i are zeros; $f(1) = 18$ *

31. $n = 4$; $-2, 5,$ and $3 + 2i$ are zeros; $f(1) = -96$ *

32. $n = 4$; $-4, \frac{1}{3},$ and $2 + 3i$ are zeros; $f(1) = 100$ *

In Exercises 33–38, use Descartes's Rule of Signs to determine the possible number of positive and negative real zeros for each given function.

33. $f(x) = x^3 + 2x^2 + 5x + 4$ positive: 0; negative: 3 or 1

34. $f(x) = x^3 + 7x^2 + x + 7$ positive: 0; negative: 3 or 1

35. $f(x) = 5x^3 - 3x^2 + 3x - 1$ positive: 3 or 1; negative: 0

36. $f(x) = -2x^3 + x^2 - x + 7$ positive: 3 or 1; negative: 0

37. $f(x) = 2x^4 - 5x^3 - x^2 - 6x + 4$ positive: 2 or 0; negative: 2 or 0

38. $f(x) = 4x^4 - x^3 + 5x^2 - 2x - 6$ positive: 3 or 1; negative: 1

In Exercises 39–52, find all zeros of the polynomial function or solve the given polynomial equation. Use the Rational Zero Theorem, Descartes's Rule of Signs, and possibly the graph of the polynomial function shown by a graphing utility as an aid in obtaining the first zero or the first root.

39. $f(x) = x^3 - 4x^2 - 7x + 10$ zeros: $-2, 5,$ and 1

40. $f(x) = x^3 + 12x^2 + 21x + 10$ zeros: -1 and -10

41. $2x^3 - x^2 - 9x - 4 = 0$ $\{-\frac{1}{2}, \frac{1 + \sqrt{17}}{2}, \frac{1 - \sqrt{17}}{2}\}$

42. $3x^3 - 8x^2 - 8x + 8 = 0$ $\{\frac{2}{3}, 1 - \sqrt{5}, 1 + \sqrt{5}\}$

43. $f(x) = x^4 - 2x^3 + x^2 + 12x + 8$ zeros: $-1, 2 + 2i, 2 - 2i$

44. $f(x) = x^4 - 4x^3 - x^2 + 14x + 10$ zeros: $-1, 3 + i, 3 - i$

45. $x^4 - 3x^3 - 20x^2 - 24x - 8 = 0$ $\{-1, -2, 3 + \sqrt{13}, 3 - \sqrt{13}\}$

46. $x^4 - x^3 + 2x^2 - 4x - 8 = 0$ $\{-1, 2, 2i, -2i\}$

47. $f(x) = 3x^4 - 11x^3 - x^2 + 19x + 6$ zeros: $-1, 2, -\frac{1}{3}$, and 3

48. $f(x) = 2x^4 + 3x^3 - 11x^2 - 9x + 15$ zeros: $1, -\frac{5}{2}, \sqrt{3}$, and $-\sqrt{3}$

49. $4x^4 - x^3 + 5x^2 - 2x - 6 = 0$ $\{1, -\frac{3}{4}, i\sqrt{2}, -i\sqrt{2}\}$

50. $3x^4 - 11x^3 - 3x^2 - 6x + 8 = 0$ $\{4, \frac{2}{3}, \frac{-1 + i\sqrt{3}}{2}, \frac{-1 - i\sqrt{3}}{2}\}$

51. $2x^5 + 7x^4 - 18x^2 - 8x + 8 = 0$ $\{-2, \frac{1}{2}, \sqrt{2}, -\sqrt{2}\}$

52. $4x^5 + 12x^4 - 41x^3 - 99x^2 + 10x + 24 = 0$ $\{3, -2, -4, \frac{1}{2}, -\frac{1}{2}\}$

Practice Plus

Exercises 53–60 show incomplete graphs of given polynomial functions.

 a. *Find all the zeros of each function.*

 b. *Without using a graphing utility, draw a complete graph of the function.*

53. $f(x) = -x^3 + x^2 + 16x - 16$ **a.** $-4, 1$, and $4*$

$[-5, 0, 1]$ by $[-40, 25, 5]$

54. $f(x) = -x^3 + 3x^2 - 4$ **a.** -1 and $2*$

$[-2, 0, 1]$ by $[-10, 10, 1]$

55. $f(x) = 4x^3 - 8x^2 - 3x + 9$ **a.** -1 and $\frac{3}{2}*$

$[-2, 0, 1]$ by $[-10, 10, 1]$

56. $f(x) = 3x^3 + 2x^2 + 2x - 1$ **a.** $\frac{1}{3}, -\frac{1}{2} \pm i\frac{\sqrt{3}}{2}*$

$[0, 2, \frac{1}{6}]$ by $[-3, 15, 1]$

57. $f(x) = 2x^4 - 3x^3 - 7x^2 - 8x + 6$ **a.** $\frac{1}{2}, 3, -1 \pm i*$

$[0, 1, \frac{1}{4}]$ by $[-10, 10, 1]$

58. $f(x) = 2x^4 + 2x^3 - 22x^2 - 18x + 36$ **a.** $-3, -2, 1, 3*$

$[0, 4, 1]$ by $[-50, 50, 10]$

59. $f(x) = 3x^5 + 2x^4 - 15x^3 - 10x^2 + 12x + 8$

 a. $-2, -1, -\frac{2}{3}, 1$, and $2*$

$[0, 4, 1]$ by $[-20, 25, 5]$

60. $f(x) = -5x^4 + 4x^3 - 19x^2 + 16x + 4$ **a.** $-\frac{1}{5}, 1, \pm 2i*$

$[0, 2, 1]$ by $[-10, 10, 1]$

Application Exercises

A popular model of carry-on luggage has a length that is 10 inches greater than its depth. Airline regulations require that the sum of the length, width, and depth cannot exceed 40 inches. These conditions, with the assumption that this sum is 40 inches, can be modeled by a function that gives the volume of the luggage, V, in cubic inches, in terms of its depth, x, in inches.

Volume	=	depth	·	length	·	width: $40 - (\text{depth} + \text{length})$

$$V(x) = x \cdot (x + 10) \cdot [40 - (x + x + 10)]$$

$$V(x) = x(x + 10)(30 - 2x)$$

Use function V to solve Exercises 61–62.

61. If the volume of the carry-on luggage is 2000 cubic inches, determine two possibilities for its depth. Where necessary, round to the nearest tenth of an inch. 7.8 in. or 10 in.

62. If the volume of the carry-on luggage is 1500 cubic inches, determine two possibilities for its depth. Where necessary, round to the nearest tenth of an inch. 5 in. or 12.2 in.

Use the graph of the function modeling the volume of the carry-on luggage to solve Exercises 63–64.

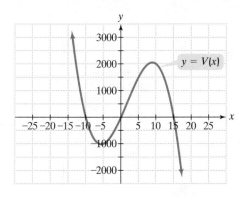

63. a. Identify your answers from Exercise 61 as points on the graph. (7.8, 2000), (10, 2000)

 b. Use the graph to describe a realistic domain, x, for the volume function, where x represents the depth of the carry-on luggage. (0, 15)

64. a. Identify your answers from Exercise 62 as points on the graph. (5, 1500), (12.2, 1500)

 b. Use the graph to describe a realistic domain, x, for the volume function, where x represents the depth of the carry-on luggage. (0, 15)

Explaining the Concepts 65–71. Answers will vary.

65. Describe how to find the possible rational zeros of a polynomial function.

66. How does the linear factorization of $f(x)$, that is,

$$f(x) = a_n(x - c_1)(x - c_2) \cdots (x - c_n),$$

show that a polynomial equation of degree n has n roots?

67. Describe how to use Descartes's Rule of Signs to determine the possible number of positive real zeros of a polynomial function.

68. Describe how to use Descartes's Rule of Signs to determine the possible number of negative roots of a polynomial equation.

69. Why must every polynomial equation with real coefficients of degree 3 have at least one real root?

70. Explain why the equation $x^4 + 6x^2 + 2 = 0$ has no rational roots.

71. Suppose $\frac{3}{4}$ is a root of a polynomial equation. What does this tell us about the leading coefficient and the constant term in the equation?

Technology Exercises

The equations in Exercises 72–75 have real roots that are rational. Use the Rational Zero Theorem to list all possible rational roots. Then graph the polynomial function in the given viewing rectangle to determine which possible rational roots are actual roots of the equation.

72. $2x^3 - 15x^2 + 22x + 15 = 0$; $[-1, 6, 1]$ by $[-50, 50, 10]$ *

73. $6x^3 - 19x^2 + 16x - 4 = 0$; $[0, 2, 1]$ by $[-3, 2, 1]$ $\{\frac{1}{2}, \frac{2}{3}, 2\}$

74. $2x^4 + 7x^3 - 4x^2 - 27x - 18 = 0$; $[-4, 3, 1]$ by $[-45, 45, 15]$ $\{-3, -\frac{3}{2}, -1, 2\}$

75. $4x^4 + 4x^3 + 7x^2 - x - 2 = 0$; $[-2, 2, 1]$ by $[-5, 5, 1]$ $\{\pm\frac{1}{2}\}$

76. Use Descartes's Rule of Signs to determine the possible number of positive and negative real zeros of $f(x) = 3x^4 + 5x^2 + 2$. What does this mean in terms of the graph of f? Verify your result by using a graphing utility to graph f. no real zeros; no x-intercepts

77. Use Descartes's Rule of Signs to determine the possible number of positive and negative real zeros of $f(x) = x^5 - x^4 + x^3 - x^2 + x - 8$. Verify your result by using a graphing utility to graph f. *

78. Write equations for several polynomial functions of odd degree and graph each function. Is it possible for the graph to have no real zeros? Explain. Try doing the same thing for polynomial functions of even degree. Now is it possible to have no real zeros? Answers will vary.

Use a graphing utility to obtain a complete graph for each polynomial function in Exercises 79–82. Then determine the number of real zeros and the number of imaginary zeros for each function.

79. $f(x) = x^3 - 6x - 9$ *

80. $f(x) = 3x^5 - 2x^4 + 6x^3 - 4x^2 - 24x + 16$ *

81. $f(x) = 3x^4 + 4x^3 - 7x^2 - 2x - 3$ *

82. $f(x) = x^6 - 64$ *

Critical Thinking Exercises

Make Sense? *In Exercises 83–86, determine whether each statement makes sense or does not make sense, and explain your reasoning.*

83. I've noticed that $f(-x)$ is used to explore the number of negative real zeros of a polynomial function, as well as to determine whether a function is even, odd, or neither. *

84. By using the quadratic formula, I do not need to bother with synthetic division when solving polynomial equations of degree 3 or higher. does not make sense

85. I'm working with a fourth-degree polynomial function with integer coefficients and zeros at 1 and $3 + \sqrt{5}$. I'm certain that $3 + \sqrt{2}$ cannot also be a zero of this function. *

86. I'm working with the polynomial function $f(x) = x^4 + 3x^2 + 2$ that has four possible rational zeros but no actual rational zeros. makes sense

In Exercises 87–90, determine whether each statement is true or false. If the statement is false, make the necessary change(s) to produce a true statement. Changes to false statements will vary.

87. The equation $x^3 + 5x^2 + 6x + 1 = 0$ has one positive real root. false

88. Descartes's Rule of Signs gives the exact number of positive and negative real roots for a polynomial equation. false

89. Every polynomial equation of degree 3 with real coefficients has at least one real root. true

90. Every polynomial equation of degree n has n distinct solutions. false

91. If the volume of the solid shown in the figure is 208 cubic inches, find the value of x. 3 in.

$x + 2$

3

3

x

$2x + 1$

$x + 5$

92. In this exercise, we lead you through the steps involved in the proof of the Rational Zero Theorem. Consider the polynomial equation Answers will vary.

$$a_n x^n + a_{n-1} x^{n-1} + a_{n-2} x^{n-2} + \cdots + a_1 x + a_0 = 0,$$

and let $\dfrac{p}{q}$ be a rational root reduced to lowest terms.

a. Substitute $\dfrac{p}{q}$ for x in the equation and show that the equation can be written as

$$a_n p^n + a_{n-1} p^{n-1} q + a_{n-2} p^{n-2} q^2 + \cdots + a_1 p q^{n-1} = -a_0 q^n.$$

b. Why is p a factor of the left side of the equation?

c. Because p divides the left side, it must also divide the right side. However, because $\dfrac{p}{q}$ is reduced to lowest terms, p and q have no common factors other than -1 and 1. Because p does divide the right side and has no factors in common with q^n, what can you conclude?

d. Rewrite the equation from part (a) with all terms containing q on the left and the term that does not have a factor of q on the right. Use an argument that parallels parts (b) and (c) to conclude that q is a factor of a_n.

In Exercises 93–96, the graph of a polynomial function is given. What is the smallest degree that each polynomial could have?

93.

3

y

x

94.

3

y

x

95.

y

5

x

96.

y

5

x

97. Explain why a polynomial function of degree 20 cannot cross the x-axis exactly once. Answers will vary.

Retaining the Concepts

98. Consider the equations

$$y_1 = \frac{1}{x - 1} - \frac{1}{x + 1} \text{ and } y_2 = \frac{2}{x^2 - 1}.$$

Find all values of x for which $y_1 = y_2$.
(Section 1.2, Example 6) all real numbers except -1 and 1

99. Write an equation in point-slope form and general form of the line passing through $(-5, 3)$ and perpendicular to the line whose equation is $x + 5y - 7 = 0$.
(Section 2.4, Example 2) $y - 3 = 5(x + 5); 5x - y + 28 = 0$

100. Find the average rate of change of $f(x) = \sqrt{x}$ from $x_1 = 4$ to $x_2 = 9$. (Section 2.4, Example 4) $\dfrac{1}{5}$

Preview Exercises

Exercises 101–103 will help you prepare for the material covered in the next section. Use the graph of function f to solve each exercise.

$y = f(x)$

101. For what values of x is the function undefined? 1 and 2

102. Write the equation of the vertical asymptote, or the vertical line that the graph of f approaches but does not touch. $x = 1$

103. Write the equation of the horizontal asymptote, or the horizontal line that the graph of f approaches but does not touch. $y = 0$

CHAPTER 3 Mid-Chapter Check Point

WHAT YOU KNOW: We graphed quadratic functions using vertices, intercepts, and additional points, as necessary. We learned that the vertex of $f(x) = a(x - h)^2 + k$ is (h, k) and the vertex of $f(x) = ax^2 + bx + c$ is $\left(-\frac{b}{2a}, f\left(-\frac{b}{2a}\right)\right)$. We used the vertex to solve problems that involved minimizing or maximizing quadratic functions. We learned a number of techniques for finding the zeros of a polynomial function f of degree 3 or higher or, equivalently, finding the roots, or solutions, of the equation $f(x) = 0$. For some functions, the zeros were found by factoring $f(x)$. For other functions, we listed possible rational zeros and used synthetic division and the Factor Theorem to determine the zeros. We saw that graphs cross the x-axis at zeros of odd multiplicity and touch the x-axis and turn around at zeros of even multiplicity. We learned to graph polynomial functions using zeros, the Leading Coefficient Test, intercepts, and symmetry. We checked graphs using the fact that a polynomial function of degree n has a graph with at most $n - 1$ turning points. After finding zeros of polynomial functions, we reversed directions by using the Linear Factorization Theorem to find functions with given zeros.

In Exercises 1–4, graph the given quadratic function. Give each function's domain and range.

1. $f(x) = (x - 3)^2 - 4$ *

2. $f(x) = 5 - (x + 2)^2$ *

3. $f(x) = -x^2 - 4x + 5$ *

4. $f(x) = 3x^2 - 6x + 1$ *

In Exercises 5–13, find all zeros of each polynomial function. Then graph the function.

5. $f(x) = (x - 2)^2(x + 1)^3$ zeros: -1 and 2*

6. $f(x) = -(x - 2)^2(x + 1)^2$ zeros: -1 and 2*

7. $f(x) = x^3 - x^2 - 4x + 4$ zeros: $-2, 1,$ and 2*

8. $f(x) = x^4 - 5x^2 + 4$ zeros: $-2, -1, 1,$ and 2*

9. $f(x) = -(x + 1)^6$ zeros: -1*

10. $f(x) = -6x^3 + 7x^2 - 1$ zeros: $-\frac{1}{3}, \frac{1}{2},$ and 1*

11. $f(x) = 2x^3 - 2x$ zeros: $-1, 0,$ and 1*

12. $f(x) = x^3 - 2x^2 + 26x$ zeros: $0, 1 \pm 5i$*

13. $f(x) = -x^3 + 5x^2 - 5x - 3$ zeros: $3, 1 \pm \sqrt{2}$*

In Exercises 14–19, solve each polynomial equation.

14. $x^3 - 3x + 2 = 0$ $\{-2, 1\}$

15. $6x^3 - 11x^2 + 6x - 1 = 0$ $\left\{\frac{1}{3}, \frac{1}{2}, 1\right\}$

16. $(2x + 1)(3x - 2)^3(2x - 7) = 0$ $\left\{-\frac{1}{2}, \frac{2}{3}, \frac{7}{2}\right\}$

17. $2x^3 + 5x^2 - 200x - 500 = 0$ $\left\{-10, -\frac{5}{2}, 10\right\}$

18. $x^4 - x^3 - 11x^2 = x + 12$ $\{-3, 4, \pm i\}$

19. $2x^4 + x^3 - 17x^2 - 4x + 6 = 0$ $\left\{-3, \frac{1}{2}, 1 \pm \sqrt{3}\right\}$

20. A company manufactures and sells bath cabinets. The function

$$P(x) = -x^2 + 150x - 4425$$

models the company's daily profit, $P(x)$, when x cabinets are manufactured and sold per day. How many cabinets should be manufactured and sold per day to maximize the company's profit? What is the maximum daily profit? 75; $1200

21. Among all pairs of numbers whose sum is -18, find a pair whose product is as large as possible. What is the maximum product? -9 and -9; 81

22. The base of a triangle measures 40 inches minus twice the measure of its height. For what measure of the height does the triangle have a maximum area? What is the maximum area? 10 in.; 100 sq in.

In Exercises 23–24, divide, using synthetic division if possible.

23. $(6x^4 - 3x^3 - 11x^2 + 2x + 4) \div (3x^2 - 1)$ *

24. $(2x^4 - 13x^3 + 17x^2 + 18x - 24) \div (x - 4)$ *

In Exercises 25–26, find an nth-degree polynomial function with real coefficients satisfying the given conditions.

25. $n = 3$; 1 and i are zeros; $f(-1) = 8$ *

26. $n = 4$; 2 (with multiplicity 2) and $3i$ are zeros; $f(0) = 36$ *

27. Does $f(x) = x^3 - x - 5$ have a real zero between 1 and 2? *

Section 3.5

Rational Functions and Their Graphs

The current generation of college students grew up playing interactive online games and many continue to play in school. Hundreds of colleges have formed organized gaming teams, many as campus clubs. Enter the Oculus Rift, revolutionizing the way people experience gaming. The Oculus Rift is a virtual reality headset that enables

users to experience video games as immersive three-dimensional environments. Basically, it puts the gamer inside the game.

The cost of manufacturing Oculus Rift headsets can be modeled by rational functions. In this section, you will see that high production levels of the Oculus Rift can keep the price of this amazing invention low, perhaps making this the device that brings home virtual reality to reality.

Rational Functions

Rational functions are quotients of polynomial functions. This means that rational functions can be expressed as

$$f(x) = \frac{p(x)}{q(x)},$$

where p and q are polynomial functions and $q(x) \neq 0$. The **domain** of a rational function is the set of all real numbers except the x-values that make the denominator zero. For example, the domain of the rational function

$$f(x) = \frac{x^2 + 7x + 9}{x(x - 2)(x + 5)}$$

This is $p(x)$.

This is $q(x)$.

is the set of all real numbers except 0, 2, and -5.

① Find the domains of rational functions.

EXAMPLE 1 Finding the Domain of a Rational Function

Find the domain of each rational function:

a. $f(x) = \dfrac{x^2 - 9}{x - 3}$ **b.** $g(x) = \dfrac{x}{x^2 - 9}$ **c.** $h(x) = \dfrac{x + 3}{x^2 + 9}.$

SOLUTION

Rational functions contain division. Because division by 0 is undefined, we must exclude from the domain of each function values of x that cause the polynomial function in the denominator to be 0.

a. The denominator of $f(x) = \dfrac{x^2 - 9}{x - 3}$ is 0 if $x = 3$. Thus, x cannot equal 3.

The domain of f consists of all real numbers except 3. We can express the domain in set-builder or interval notation:

$$\text{Domain of } f = \{x \mid x \neq 3\}$$
$$\text{Domain of } f = (-\infty, 3) \cup (3, \infty).$$

 Use arrow notation.

b. The denominator of $g(x) = \dfrac{x}{x^2 - 9}$ is 0 if $x = -3$ or $x = 3$. Thus, the domain of g consists of all real numbers except -3 and 3. We can express the domain in set-builder or interval notation:

$$\text{Domain of } g = \{x \mid x \neq -3, x \neq 3\}$$
$$\text{Domain of } g = (-\infty, -3) \cup (-3, 3) \cup (3, \infty).$$

c. No real numbers cause the denominator of $h(x) = \dfrac{x+3}{x^2+9}$ to equal 0. The domain of h consists of all real numbers.

$$\text{Domain of } h = (-\infty, \infty) \qquad \bullet\bullet\bullet$$

✓ **Check Point 1** Find the domain of each rational function:

a. $f(x) = \dfrac{x^2 - 25}{x - 5}$ **b.** $g(x) = \dfrac{x}{x^2 - 25}$ **c.** $h(x) = \dfrac{x + 5}{x^2 + 25}$.

The most basic rational function is the **reciprocal function**, defined by $f(x) = \dfrac{1}{x}$. The denominator of the reciprocal function is zero when $x = 0$, so the domain of f is the set of all real numbers except 0.

Let's look at the behavior of f near the excluded value 0. We start by evaluating $f(x)$ to the left of 0.

x approaches 0 from the left.					
x	-1	-0.5	-0.1	-0.01	-0.001
$f(x) = \dfrac{1}{x}$	-1	-2	-10	-100	-1000

Mathematically, we say that "x approaches 0 from the left." From the table and the accompanying graph, it appears that as x approaches 0 from the left, the function values, $f(x)$, decrease without bound. We say that "$f(x)$ approaches negative infinity." We use a special arrow notation to describe this situation symbolically:

$$\text{As } x \to 0^-, f(x) \to -\infty.$$

As x approaches 0 from the left, $f(x)$ approaches negative infinity (that is, the graph falls).

Observe that the minus $(-)$ superscript on the 0 $(x \to 0^-)$ is read "from the left."

Next, we evaluate $f(x)$ to the right of 0.

x approaches 0 from the right.					
x	0.001	0.01	0.1	0.5	1
$f(x) = \dfrac{1}{x}$	1000	100	10	2	1

Mathematically, we say that "x approaches 0 from the right." From the table and the accompanying graph, it appears that as x approaches 0 from the right, the function values, $f(x)$, increase without bound. We say that "$f(x)$ approaches infinity." We again use a special arrow notation to describe this situation symbolically:

$$\text{As } x \to 0^+, f(x) \to \infty.$$

As x approaches 0 from the right, $f(x)$ approaches infinity (that is, the graph rises).

Observe that the plus $(+)$ superscript on the 0 $(x \to 0^+)$ is read "from the right."

Now let's see what happens to the function values of $f(x) = \dfrac{1}{x}$ as x gets farther away from the origin. The following tables suggest what happens to $f(x)$ as x increases or decreases without bound.

x increases without bound:					x decreases without bound:				
x	1	10	100	1000	x	-1	-10	-100	-1000
$f(x) = \dfrac{1}{x}$	1	0.1	0.01	0.001	$f(x) = \dfrac{1}{x}$	-1	-0.1	-0.01	-0.001

It appears that as x increases or decreases without bound, the function values, $f(x)$, are getting progressively closer to 0.

Figure 3.29 illustrates the end behavior of $f(x) = \dfrac{1}{x}$ as x increases or decreases without bound. The graph shows that the function values, $f(x)$, are approaching 0. This means that as x increases or decreases without bound, the graph of f is approaching the horizontal line $y = 0$ (that is, the x-axis). We use arrow notation to describe this situation:

$$\text{As } x \to \infty, f(x) \to 0 \qquad \text{and} \qquad \text{as } x \to -\infty, f(x) \to 0.$$

> As x approaches infinity (that is, increases without bound), $f(x)$ approaches 0.

> As x approaches negative infinity (that is, decreases without bound), $f(x)$ approaches 0.

Thus, as x approaches infinity ($x \to \infty$) or as x approaches negative infinity ($x \to -\infty$), the function values are approaching zero: $f(x) \to 0$.

The graph of the reciprocal function $f(x) = \dfrac{1}{x}$ is shown in **Figure 3.30**. Unlike the graph of a polynomial function, the graph of the reciprocal function has a break and is composed of two branches. Observe that the graph is symmetric with respect to the origin.

GREAT QUESTION!

What is the relationship between x and $\dfrac{1}{x}$ when x is far from 0? What happens if x is close to 0?

If x is far from 0, then $\dfrac{1}{x}$ is close to 0. By contrast, if x is close to 0, then $\dfrac{1}{x}$ is far from 0.

FIGURE 3.29 $f(x)$ approaches 0 as x increases or decreases without bound.

FIGURE 3.30 The graph of the reciprocal function $f(x) = \dfrac{1}{x}$

Blitzer Bonus || The Reciprocal Function as a Knuckle Tattoo

"I got the tattoo because I like the idea of math not being well behaved. That sounds lame and I really don't mean that in some kind of anarchy-type way. I just think that it's kind of nice that something as perfectly functional as math can kink up around the edges."

Kink up around the edges? On the next page, we'll describe the graphic behavior of the reciprocal function using *asymptotes* rather than kink. Asymptotes are lines that graphs approach but never touch. *Asymptote* comes from the Greek word *asymptotos*, meaning "not meeting."

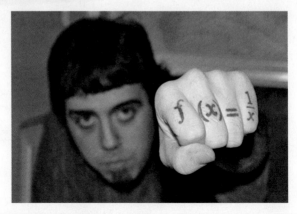

The arrow notation used throughout our discussion of the reciprocal function is summarized in the following box:

Arrow Notation

Symbol	Meaning
$x \to a^+$	x approaches a from the right.
$x \to a^-$	x approaches a from the left.
$x \to \infty$	x approaches infinity; that is, x increases without bound.
$x \to -\infty$	x approaches negative infinity; that is, x decreases without bound.

Another basic rational function is $f(x) = \dfrac{1}{x^2}$. The graph of this even function, with y-axis symmetry and positive function values, is shown in **Figure 3.31**. Like the reciprocal function, the graph has a break and is composed of two distinct branches.

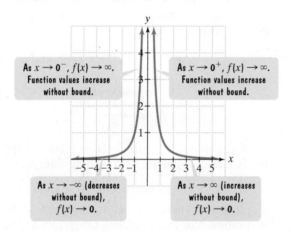

As $x \to 0^-$, $f(x) \to \infty$. Function values increase without bound.

As $x \to 0^+$, $f(x) \to \infty$. Function values increase without bound.

As $x \to -\infty$ (decreases without bound), $f(x) \to 0$.

As $x \to \infty$ (increases without bound), $f(x) \to 0$.

FIGURE 3.31 The graph of $f(x) = \dfrac{1}{x^2}$

③ Identify vertical asymptotes.

Vertical Asymptotes of Rational Functions

Look again at the graph of $f(x) = \dfrac{1}{x^2}$ in **Figure 3.31**. The curve approaches, but does not touch, the y-axis. The y-axis, or $x = 0$, is said to be a *vertical asymptote* of the graph. A rational function may have no vertical asymptotes, one vertical asymptote, or several vertical asymptotes. The graph of a rational function never intersects a vertical asymptote. We will use dashed lines to show asymptotes.

Definition of a Vertical Asymptote

The line $x = a$ is a **vertical asymptote** of the graph of a function f if $f(x)$ increases or decreases without bound as x approaches a.

As $x \to a^+$, $f(x) \to \infty$.

As $x \to a^-$, $f(x) \to \infty$.

As $x \to a^+$, $f(x) \to -\infty$.

As $x \to a^-$, $f(x) \to -\infty$.

Thus, as x approaches a from either the left or the right, $f(x) \to \infty$ or $f(x) \to -\infty$.

If the graph of a rational function has vertical asymptotes, they can be located using the following theorem:

Locating Vertical Asymptotes

If $f(x) = \dfrac{p(x)}{q(x)}$ is a rational function in which $p(x)$ and $q(x)$ have no common factors and a is a zero of $q(x)$, the denominator, then $x = a$ is a vertical asymptote of the graph of f.

EXAMPLE 2 Finding the Vertical Asymptotes of a Rational Function

Find the vertical asymptotes, if any, of the graph of each rational function:

a. $f(x) = \dfrac{x}{x^2 - 9}$ **b.** $g(x) = \dfrac{x+3}{x^2-9}$ **c.** $h(x) = \dfrac{x+3}{x^2+9}$.

SOLUTION

Factoring is usually helpful in identifying zeros of denominators and any common factors in the numerators and denominators.

a. $f(x) = \dfrac{x}{x^2 - 9} = \dfrac{x}{(x+3)(x-3)}$

> This factor is 0 if $x = -3$. This factor is 0 if $x = 3$.

There are no common factors in the numerator and the denominator. The zeros of the denominator are -3 and 3. Thus, the lines $x = -3$ and $x = 3$ are the vertical asymptotes for the graph of f. [See **Figure 3.32(a)**.]

b. We will use factoring to see if there are common factors.

$$g(x) = \frac{x+3}{x^2-9} = \frac{(x+3)}{(x+3)(x-3)} = \frac{1}{x-3}, \text{ provided } x \neq -3$$

> There is a common factor, $x + 3$, so simplify. This denominator is 0 if $x = 3$.

The only zero of the denominator of $g(x)$ in simplified form is 3. Thus, the line $x = 3$ is the only vertical asymptote of the graph of g. [See **Figure 3.32(b)**.]

c. We cannot factor the denominator of $h(x)$ over the real numbers.

$$h(x) = \frac{x+3}{x^2+9}$$

> No real numbers make this denominator 0.

The denominator has no real zeros. Thus, the graph of h has no vertical asymptotes. [See **Figure 3.32(c)**.]

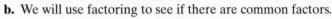

FIGURE 3.32(a) The graph of $f(x) = \dfrac{x}{x^2-9}$ has two vertical asymptotes.

Vertical asymptote: $x = -3$ Vertical asymptote: $x = 3$

FIGURE 3.32(b) The graph of $g(x) = \dfrac{x+3}{x^2-9}$ has one vertical asymptote.

There is a hole in the graph corresponding to $x = -3$.

Vertical asymptote: $x = 3$

FIGURE 3.32(c) The graph of $h(x) = \dfrac{x+3}{x^2+9}$ has no vertical asymptotes.

●●●

☑ **Check Point** **2** Find the vertical asymptotes, if any, of the graph of each rational function:

a. $f(x) = \dfrac{x}{x^2-1}$ **b.** $g(x) = \dfrac{x-1}{x^2-1}$ **c.** $h(x) = \dfrac{x-1}{x^2+1}$.

a. $x = 1, x = -1$ **b.** $x = -1$ **c.** none

TECHNOLOGY

When using a graphing utility to graph a rational function, if the numerator or the denominator contains more than one term, be sure to enclose the polynomials in parentheses. The graph of the rational function

$$f(x) = \frac{x}{x^2 - 9},$$

drawn by hand in **Figure 3.32(a)** on the previous page, is graphed on the right in a [–5, 5, 1] by [–4, 4, 1] viewing rectangle. On some calculators, it may be necessary to press the right arrow key to exit the exponent after entering the exponent 2.

Some graphing utilities incorrectly connect the pieces of the graph, resulting in nearly vertical lines appearing on the screen. If your graphing utility does this, we recommend that you use ⎹DOT⎸ mode. Consult your owner's manual.

GREAT QUESTION!

Do I have to factor to identify a rational function's possible vertical asymptotes or holes?

Yes. It is essential to factor the numerator and the denominator of the rational function.

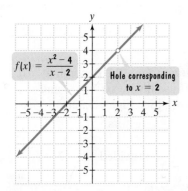

FIGURE 3.33 A graph with a hole corresponding to the denominator's zero

④ Identify horizontal asymptotes.

A value where the denominator of a rational function is zero does not necessarily result in a vertical asymptote. There is a hole corresponding to $x = a$, and not a vertical asymptote, in the graph of a rational function under the following conditions: The value a causes the denominator to be zero, but there is a reduced form of the function's equation in which a does not cause the denominator to be zero. Look back at **Figure 3.32(b)** on the previous page.

Consider, as another example, the function

$$f(x) = \frac{x^2 - 4}{x - 2}.$$

Because the denominator is zero when $x = 2$, the function's domain is all real numbers except 2. However, there is a reduced form of the equation in which 2 does not cause the denominator to be zero:

$$f(x) = \frac{x^2 - 4}{x - 2} = \frac{(x + 2)(x - 2)}{x - 2} = x + 2, \ x \neq 2.$$

Denominator is zero at $x = 2$.

In this reduced form, 2 does not result in a zero denominator.

Figure 3.33 shows that the graph has a hole corresponding to $x = 2$. A graphing utility may not show this feature of the graph.

Horizontal Asymptotes of Rational Functions

Figure 3.30, repeated in the margin, shows the graph of the reciprocal function $f(x) = \dfrac{1}{x}$. As $x \to \infty$ and as $x \to -\infty$, the function values are approaching 0: $f(x) \to 0$. The line $y = 0$ (that is, the x-axis) is a *horizontal asymptote* of the graph. Many, but not all, rational functions have horizontal asymptotes.

FIGURE 3.30 The graph of $f(x) = \dfrac{1}{x}$ (repeated)

Definition of a Horizontal Asymptote

The line $y = b$ is a **horizontal asymptote** of the graph of a function f if $f(x)$ approaches b as x increases or decreases without bound.

As $x \to \infty$, $f(x) \to b$. As $x \to \infty$, $f(x) \to b$. As $x \to \infty$, $f(x) \to b$.

Recall that a rational function may have several vertical asymptotes. By contrast, it can have at most one horizontal asymptote. Although a graph can never intersect a vertical asymptote, it may cross its horizontal asymptote.

If the graph of a rational function has a horizontal asymptote, it can be located using the following theorem:

Locating Horizontal Asymptotes

Let f be the rational function given by

$$f(x) = \frac{a_n x^n + a_{n-1} x^{n-1} + \cdots + a_1 x + a_0}{b_m x^m + b_{m-1} x^{m-1} + \cdots + b_1 x + b_0}, \quad a_n \neq 0, b_m \neq 0.$$

The degree of the numerator is n. The degree of the denominator is m.

1. If $n < m$, the x-axis, or $y = 0$, is the horizontal asymptote of the graph of f.

2. If $n = m$, the line $y = \dfrac{a_n}{b_m}$ is the horizontal asymptote of the graph of f.

3. If $n > m$, the graph of f has no horizontal asymptote.

FIGURE 3.34(a) The horizontal asymptote of the graph is $y = 0$.

EXAMPLE 3 Finding the Horizontal Asymptote of a Rational Function

Find the horizontal asymptote, if there is one, of the graph of each rational function:

a. $f(x) = \dfrac{4x}{2x^2 + 1}$ **b.** $g(x) = \dfrac{4x^2}{2x^2 + 1}$ **c.** $h(x) = \dfrac{4x^3}{2x^2 + 1}.$

SOLUTION

a. $f(x) = \dfrac{4x}{2x^2 + 1}$

The degree of the numerator, 1, is less than the degree of the denominator, 2. Thus, the graph of f has the x-axis as a horizontal asymptote. [See **Figure 3.34(a)**.] The equation of the horizontal asymptote is $y = 0$.

FIGURE 3.34(b) The horizontal asymptote of the graph is $y = 2$.

b. $g(x) = \dfrac{4x^2}{2x^2 + 1}$

The degree of the numerator, 2, is equal to the degree of the denominator, 2. The leading coefficients of the numerator and denominator, 4 and 2, are used to obtain the equation of the horizontal asymptote. The equation of the horizontal asymptote is $y = \frac{4}{2}$ or $y = 2$. [See **Figure 3.34(b)**.]

c. $h(x) = \dfrac{4x^3}{2x^2 + 1}$

The degree of the numerator, 3, is greater than the degree of the denominator, 2. Thus, the graph of h has no horizontal asymptote. [See **Figure 3.34(c)**.] ● ● ●

FIGURE 3.34(c) The graph has no horizontal asymptote.

✓ **Check Point 3** Find the horizontal asymptote, if there is one, of the graph of each rational function:

a. $f(x) = \dfrac{9x^2}{3x^2 + 1}$ **b.** $g(x) = \dfrac{9x}{3x^2 + 1}$ **c.** $h(x) = \dfrac{9x^3}{3x^2 + 1}.$

a. $y = 3$ **b.** $y = 0$ **c.** none

⑤ Use transformations to graph rational functions.

Using Transformations to Graph Rational Functions

Table 3.2 shows the graphs of two rational functions, $f(x) = \dfrac{1}{x}$ and $f(x) = \dfrac{1}{x^2}$. The dashed green lines indicate the asymptotes.

Table 3.2 Graphs of Common Rational Functions

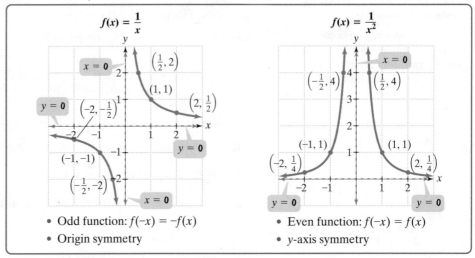

- Odd function: $f(-x) = -f(x)$
- Origin symmetry

- Even function: $f(-x) = f(x)$
- y-axis symmetry

Some rational functions can be graphed using transformations (horizontal shifting, stretching or shrinking, reflecting, vertical shifting) of these two common graphs.

EXAMPLE 4 Using Transformations to Graph a Rational Function

Use the graph of $f(x) = \dfrac{1}{x^2}$ to graph $g(x) = \dfrac{1}{(x-2)^2} + 1$.

SOLUTION

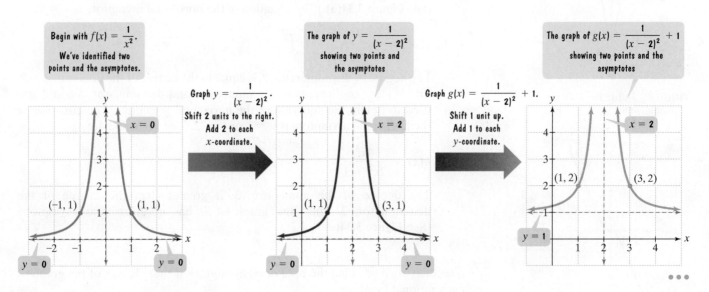

✓ **Check Point 4** Use the graph of $f(x) = \dfrac{1}{x}$ to graph $g(x) = \dfrac{1}{x+2} - 1$. *

*See Answers to Selected Exercises.

⑥ Graph rational functions.

Graphing Rational Functions

Rational functions that are not transformations of $f(x) = \dfrac{1}{x}$ or $f(x) = \dfrac{1}{x^2}$ can be graphed using the following procedure:

> ### Strategy for Graphing a Rational Function
>
> The following strategy can be used to graph
>
> $$f(x) = \frac{p(x)}{q(x)},$$
>
> where p and q are polynomial functions with no common factors.
>
> **1.** Determine whether the graph of f has symmetry.
>
> $$f(-x) = f(x): \qquad y\text{-axis symmetry}$$
> $$f(-x) = -f(x): \qquad \text{origin symmetry}$$
>
> **2.** Find the y-intercept (if there is one) by evaluating $f(0)$.
> **3.** Find the x-intercepts (if there are any) by solving the equation $p(x) = 0$.
> **4.** Find any vertical asymptote(s) by solving the equation $q(x) = 0$.
> **5.** Find the horizontal asymptote (if there is one) using the rule for determining the horizontal asymptote of a rational function.
> **6.** Plot at least one point between and beyond each x-intercept and vertical asymptote.
> **7.** Use the information obtained previously to graph the function between and beyond the vertical asymptotes.

EXAMPLE 5 Graphing a Rational Function

Graph: $f(x) = \dfrac{2x - 1}{x - 1}$.

SOLUTION

Step 1 Determine symmetry.

$$f(-x) = \frac{2(-x) - 1}{-x - 1} = \frac{-2x - 1}{-x - 1} = \frac{2x + 1}{x + 1}$$

Because $f(-x)$ does not equal either $f(x)$ or $-f(x)$, the graph has neither y-axis symmetry nor origin symmetry.

Step 2 Find the y-intercept. Evaluate $f(0)$.

$$f(0) = \frac{2 \cdot 0 - 1}{0 - 1} = \frac{-1}{-1} = 1$$

The y-intercept is 1, so the graph passes through $(0, 1)$.

Step 3 Find x-intercept(s). This is done by solving $p(x) = 0$, where $p(x)$ is the numerator of $f(x)$.

$$2x - 1 = 0 \qquad \text{Set the numerator equal to 0.}$$
$$2x = 1 \qquad \text{Add 1 to both sides.}$$
$$x = \frac{1}{2} \qquad \text{Divide both sides by 2.}$$

The x-intercept is $\frac{1}{2}$, so the graph passes through $\left(\frac{1}{2}, 0\right)$.

Step 4 Find the vertical asymptote(s). Solve $q(x) = 0$, where $q(x)$ is the denominator of $f(x)$, thereby finding zeros of the denominator. (Note that the numerator and denominator of $f(x) = \dfrac{2x - 1}{x - 1}$ have no common factors.)

$$x - 1 = 0 \qquad \text{Set the denominator equal to O.}$$
$$x = 1 \qquad \text{Add 1 to both sides.}$$

The equation of the vertical asymptote is $x = 1$.

Step 5 Find the horizontal asymptote. Because the numerator and denominator of $f(x) = \dfrac{2x - 1}{x - 1}$ have the same degree, 1, the leading coefficients of the numerator and denominator, 2 and 1, respectively, are used to obtain the equation of the horizontal asymptote. The equation is

$$y = \frac{2}{1} = 2.$$

The equation of the horizontal asymptote is $y = 2$.

Step 6 Plot points between and beyond each x-intercept and vertical asymptote. With an x-intercept at $\frac{1}{2}$ and a vertical asymptote at $x = 1$, we evaluate the function at $-2, -1, \dfrac{3}{4}, 2,$ and 4.

x	-2	-1	$\dfrac{3}{4}$	2	4
$f(x) = \dfrac{2x - 1}{x - 1}$	$\dfrac{5}{3}$	$\dfrac{3}{2}$	-2	3	$\dfrac{7}{3}$

Figure 3.35 shows these points, the y-intercept, the x-intercept, and the asymptotes.

Step 7 Graph the function. The graph of $f(x) = \dfrac{2x - 1}{x - 1}$ is shown in **Figure 3.36**.

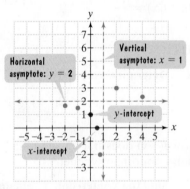

FIGURE 3.35 Preparing to graph the rational function $f(x) = \dfrac{2x - 1}{x - 1}$

FIGURE 3.36 The graph of $f(x) = \dfrac{2x - 1}{x - 1}$ • • •

⊘ **Check Point 5** Graph: $f(x) = \dfrac{3x - 3}{x - 2}$. *

EXAMPLE 6 Graphing a Rational Function

Graph: $f(x) = \dfrac{3x^2}{x^2 - 4}$.

SOLUTION

Step 1 Determine symmetry. $f(-x) = \dfrac{3(-x)^2}{(-x)^2 - 4} = \dfrac{3x^2}{x^2 - 4} = f(x)$: The graph of f is symmetric with respect to the y-axis.

*See Answers to Selected Exercises.

Step 2 Find the y-intercept. $f(0) = \dfrac{3 \cdot 0^2}{0^2 - 4} = \dfrac{0}{-4} = 0$: The y-intercept is 0, so the graph passes through the origin.

Step 3 Find the x-intercept(s). $3x^2 = 0$, so $x = 0$: The x-intercept is 0, verifying that the graph passes through the origin.

Step 4 Find the vertical asymptote(s). Set $q(x) = 0$. (Note that the numerator and denominator of $f(x) = \dfrac{3x^2}{x^2 - 4}$ have no common factors.)

$$x^2 - 4 = 0 \qquad \text{Set the denominator equal to 0.}$$
$$x^2 = 4 \qquad \text{Add 4 to both sides.}$$
$$x = \pm 2 \qquad \text{Use the square root property.}$$

The vertical asymptotes are $x = -2$ and $x = 2$.

Step 5 Find the horizontal asymptote. Because the numerator and denominator of $f(x) = \dfrac{3x^2}{x^2 - 4}$ have the same degree, 2, their leading coefficients, 3 and 1, are used to determine the equation of the horizontal asymptote. The equation is $y = \frac{3}{1} = 3$.

Step 6 Plot points between and beyond each x-intercept and vertical asymptote. With an x-intercept at 0 and vertical asymptotes at $x = -2$ and $x = 2$, we evaluate the function at $-3, -1, 1, 3,$ and 4.

x	−3	−1	1	3	4
$f(x) = \dfrac{3x^2}{x^2 - 4}$	$\dfrac{27}{5}$	−1	−1	$\dfrac{27}{5}$	4

Figure 3.37 shows the points $\left(-3, \dfrac{27}{5}\right), (-1, -1), (1, -1), \left(3, \dfrac{27}{5}\right),$ and $(4, 4),$ the y-intercept, the x-intercept, and the asymptotes.

Step 7 Graph the function. The graph of $f(x) = \dfrac{3x^2}{x^2 - 4}$ is shown in **Figure 3.38**. The y-axis symmetry is now obvious.

GREAT QUESTION!

Do I have to evaluate $f(x)$ at all five of the values shown in the table for Step 6?

No. Because the graph has y-axis symmetry, it is not necessary to evaluate the even function at −3 and again at 3.

$$f(-3) = f(3) = \tfrac{27}{5}$$

This also applies to evaluation at −1 and 1.

TECHNOLOGY

The graph of $y = \dfrac{3x^2}{x^2 - 4}$, generated by a graphing utility, verifies that our hand-drawn graph is correct.

$[-6, 6, 1]$ by $[-6, 6, 1]$

FIGURE 3.37 Preparing to graph $f(x) = \dfrac{3x^2}{x^2 - 4}$

FIGURE 3.38 The graph of $f(x) = \dfrac{3x^2}{x^2 - 4}$

● ● ●

✓ Check Point **6** Graph: $f(x) = \dfrac{2x^2}{x^2 - 9}.$ *

Example 7 illustrates that not every rational function has vertical and horizontal asymptotes.

*See Answers to Selected Exercises.

EXAMPLE 7 Graphing a Rational Function

Graph: $f(x) = \dfrac{x^4}{x^2 + 1}$.

SOLUTION

Step 1 Determine symmetry. $f(-x) = \dfrac{(-x)^4}{(-x)^2 + 1} = \dfrac{x^4}{x^2 + 1} = f(x)$

The graph of f is symmetric with respect to the y-axis.

Step 2 Find the y-intercept. $f(0) = \dfrac{0^4}{0^2 + 1} = \dfrac{0}{1} = 0$: The y-intercept is 0.

Step 3 Find the x-intercept(s). $x^4 = 0$, so $x = 0$: The x-intercept is 0.

Step 4 Find the vertical asymptote(s). Set $q(x) = 0$.

$$x^2 + 1 = 0 \qquad \text{Set the denominator equal to O.}$$
$$x^2 = -1 \qquad \text{Subtract 1 from both sides.}$$

Although this equation has imaginary roots ($x = \pm i$), there are no real roots. Thus, the graph of f has no vertical asymptotes.

Step 5 Find the horizontal asymptote. Because the degree of the numerator, 4, is greater than the degree of the denominator, 2, there is no horizontal asymptote.

Step 6 Plot points between and beyond each x-intercept and vertical asymptote. With an x-intercept at 0 and no vertical asymptotes, let's look at function values at $-2, -1, 1,$ and 2. You can evaluate the function at 1 and 2. Use y-axis symmetry to obtain function values at -1 and -2:

$$f(-1) = f(1) \text{ and } f(-2) = f(2).$$

x				-2	-1	1	2
$f(x) = $		$\dfrac{x^4}{x^2 + 1}$		$\dfrac{16}{5}$	$\dfrac{1}{2}$	$\dfrac{1}{2}$	$\dfrac{16}{5}$

Step 7 Graph the function. Figure 3.39 shows the graph of f using the points obtained from the table and y-axis symmetry. Notice that as x approaches infinity or negative infinity ($x \to \infty$ or $x \to -\infty$), the function values, $f(x)$, are getting larger without bound $[f(x) \to \infty]$. •••

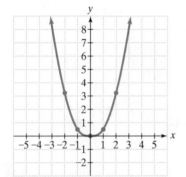

FIGURE 3.39 The graph of $f(x) = \dfrac{x^4}{x^2 + 1}$

✓ Check Point 7 Graph: $f(x) = \dfrac{x^4}{x^2 + 2}$. *

7 Identify slant asymptotes.

Slant Asymptotes

Examine the graph of

$$f(x) = \frac{x^2 + 1}{x - 1},$$

shown in **Figure 3.40**. Note that the degree of the numerator, 2, is greater than the degree of the denominator, 1. Thus, the graph of this function has no horizontal asymptote. However, the graph has a **slant asymptote**, $y = x + 1$.

The graph of a rational function has a slant asymptote if the degree of the numerator is one more than the degree of the denominator. The equation of the slant asymptote can be found by division. For example, to find the slant asymptote for the graph of $f(x) = \dfrac{x^2 + 1}{x - 1}$, divide $x - 1$ into $x^2 + 1$:

FIGURE 3.40 The graph of $f(x) = \dfrac{x^2 + 1}{x - 1}$ with a slant asymptote

$$\begin{array}{r|rrr} 1 & 1 & 0 & 1 \\ & & 1 & 1 \\ \hline & 1 & 1 & 2 \end{array}$$ Remainder

$$1x + 1 + \dfrac{2}{x - 1}$$
$$x - 1 \overline{)x^2 + 0x + 1}.$$

*See Answers to Selected Exercises.

Observe that

$$f(x) = \frac{x^2 + 1}{x - 1} = \underbrace{x + 1}_{} + \frac{2}{x - 1}.$$

> The equation of the slant asymptote is $y = x + 1$.

As $|x| \to \infty$, the value of $\dfrac{2}{x - 1}$ is approximately 0. Thus, when $|x|$ is large, the function is very close to $y = x + 1 + 0$. This means that as $x \to \infty$ or as $x \to -\infty$, the graph of f gets closer and closer to the line whose equation is $y = x + 1$. As shown in **Figure 3.40**, the line $y = x + 1$ is a slant asymptote of the graph.

In general, if $f(x) = \dfrac{p(x)}{q(x)}$, p and q have no common factors, and the degree of p is one greater than the degree of q, find the slant asymptote by dividing $q(x)$ into $p(x)$. The division will take the form

$$\frac{p(x)}{q(x)} = mx + b + \frac{\text{remainder}}{q(x)}.$$

> Slant asymptote:
> $y = mx + b$

The equation of the slant asymptote is obtained by dropping the term with the remainder. Thus, the equation of the slant asymptote is $y = mx + b$.

EXAMPLE 8 Finding the Slant Asymptote of a Rational Function

Find the slant asymptote of $f(x) = \dfrac{x^2 - 4x - 5}{x - 3}$.

SOLUTION

Because the degree of the numerator, 2, is exactly one more than the degree of the denominator, 1, and $x - 3$ is not a factor of $x^2 - 4x - 5$, the graph of f has a slant asymptote. To find the equation of the slant asymptote, divide $x - 3$ into $x^2 - 4x - 5$:

$$\begin{array}{r|rrr} 3 & 1 & -4 & -5 \\ & & 3 & -3 \\ \hline & 1 & -1 & -8 \end{array}$$

> Remainder

$$x - 3 \overline{)x^2 - 4x - 5}\ \ \ 1x - 1 - \frac{8}{x - 3}$$

> Drop the remainder term and you'll have the equation of the slant asymptote.

> Slant asymptote:
> $y = x - 1$

> Vertical asymptote:
> $x = 3$

FIGURE 3.41 The graph of $f(x) = \dfrac{x^2 - 4x - 5}{x - 3}$

The equation of the slant asymptote is $y = x - 1$. Using our strategy for graphing rational functions, the graph of $f(x) = \dfrac{x^2 - 4x - 5}{x - 3}$ is shown in **Figure 3.41**. • • •

⊘ Check Point 8 Find the slant asymptote of $f(x) = \dfrac{2x^2 - 5x + 7}{x - 2}$. $y = 2x - 1$

⑧ Solve applied problems involving rational functions.

Applications

There are numerous examples of asymptotic behavior in functions that model real-world phenomena. Let's consider an example from the business world. The **cost function**, C, for a business is the sum of its fixed and variable costs:

$$C(x) = (\text{fixed cost}) + cx.$$

> Cost per unit times the number of units produced, x

The **average cost** per unit for a company to produce x units is the sum of its fixed and variable costs divided by the number of units produced. The **average cost function** is a rational function that is denoted by \overline{C}. Thus,

> Cost of producing x units: fixed plus variable costs

$$\overline{C}(x) = \frac{(\text{fixed cost}) + cx}{x}.$$

> Number of units produced

EXAMPLE 9 Putting the Video-Game Player Inside the Game

We return to the Oculus Rift, described in the section opener, a virtual reality headset that enables users to experience video games as immersive three-dimensional environments. Suppose the company that manufactures this invention has a fixed monthly cost of $600,000 and that it costs $50 to produce each headset.

 a. Write the cost function, C, of producing x headsets.
 b. Write the average cost function, \overline{C}, of producing x headsets.
 c. Find and interpret $\overline{C}(1000)$, $\overline{C}(10,000)$, and $\overline{C}(100,000)$.
 d. What is the horizontal asymptote for the graph of the average cost function, \overline{C}? Describe what this represents for the company.

SOLUTION

 a. The cost function, C, is the sum of the fixed cost and the variable costs.

$$C(x) = 600,000 + 50x$$

> Fixed cost is $600,000.

> Variable cost: $50 for each headset

 b. The average cost function, \overline{C}, is the sum of fixed and variable costs divided by the number of headsets produced.

$$\overline{C}(x) = \frac{600,000 + 50x}{x} \quad \text{or} \quad \overline{C}(x) = \frac{50x + 600,000}{x}$$

 c. We evaluate

$$\overline{C}(x) = \frac{50x + 600,000}{x}$$

at 1000, 10,000, and 100,000, interpreting the results.

$$\overline{C}(1000) = \frac{50(1000) + 600,000}{1000} = 650$$

The average cost per headset of producing 1000 headsets per month is $650.

$$\overline{C}(10,000) = \frac{50(10,000) + 600,000}{10,000} = 110$$

The average cost per headset of producing 10,000 headsets per month is $110.

$$\overline{C}(100,000) = \frac{50(100,000) + 600,000}{100,000} = 56$$

The average cost per headset of producing 100,000 headsets per month is $56. Notice that with higher production levels, the cost of producing each headset decreases.

d. We developed the average cost function

$$\overline{C}(x) = \frac{50x + 600,000}{x}$$

in which the degree of the numerator, 1, is equal to the degree of the denominator, 1. The leading coefficients of the numerator and denominator, 50 and 1, are used to obtain the equation of the horizontal asymptote. The equation of the horizontal asymptote is

$$y = \frac{50}{1} \quad \text{or} \quad y = 50.$$

The horizontal asymptote is shown in **Figure 3.42**. This means that the more headsets produced each month, the closer the average cost per headset for the company comes to $50. The least possible cost per headset is approaching $50. Competitively low prices take place with high production levels, posing a major problem for small businesses. $\bullet\bullet\bullet$

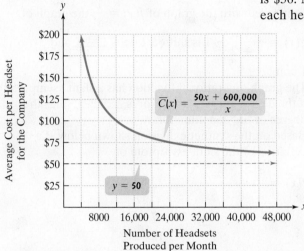

Average Cost per Headset for the Company

Number of Headsets Produced per Month

$\overline{C}(x) = \dfrac{50x + 600,000}{x}$

$y = 50$

FIGURE 3.42

b. $\overline{C}(x) = \dfrac{500,000 + 400x}{x}$

c. $\overline{C}(1000) = 900$: The average cost per wheelchair of producing 1000 wheelchairs per month is $900.; $\overline{C}(10,000) = 450$: The average cost per wheelchair of producing 10,000 wheelchairs per month is $450.; $\overline{C}(100,000) = 405$: The average cost per wheelchair of producing 100,000 wheelchairs per month is $405.

d. $y = 400$; The cost per wheelchair approaches $400 as more wheelchairs are produced.

✓ **Check Point 9** A company is planning to manufacture wheelchairs that are light, fast, and beautiful. The fixed monthly cost will be $500,000 and it will cost $400 to produce each radically innovative chair.

a. Write the cost function, C, of producing x wheelchairs. $C(x) = 500,000 + 400x$

b. Write the average cost function, \overline{C}, of producing x wheelchairs.

c. Find and interpret $\overline{C}(1000)$, $\overline{C}(10,000)$, and $\overline{C}(100,000)$.

d. What is the horizontal asymptote for the graph of the average cost function, \overline{C}? Describe what this represents for the company.

ACHIEVING SUCCESS

Be sure to use the Chapter Test Prep on YouTube for each chapter test. The Chapter Test Prep videos provide step-by-step solutions to every exercise in the test and let you review any exercises you miss.

Are you using any of the other textbook supplements for help and additional study? These include:

• The Student Solutions Manual. This contains fully worked solutions to the odd-numbered section exercises plus all Check Points, Concept and Vocabulary Checks, Review/Preview Exercises, Mid-Chapter Check Points, Chapter Reviews, Chapter Tests, and Cumulative Reviews.

• Objective Videos. These interactive videos highlight important concepts from each objective of the text.

• MyMathLab is a text-specific online course. Math XL is an online homework, tutorial, and assessment system. Ask your instructor whether these are available to you.

CONCEPT AND VOCABULARY CHECK

Fill in each blank so that the resulting statement is true.

1. All rational functions can be expressed as
$$f(x) = \frac{p(x)}{q(x)},$$
where p and q are __polynomial__ functions and $q(x) \neq 0$.

2. True or false: The domain of every rational function is the set of all real numbers. __false__

3. True or false: The graph of the reciprocal function $f(x) = \dfrac{1}{x}$ has a break and is composed of two distinct branches. __true__

4. If the graph of a function f increases or decreases without bound as x approaches a, then the line $x = a$ is a/an __vertical asymptote__ of the graph of f. The equation of such a line for the graph of $f(x) = \dfrac{2}{x + 5}$ is __$x = -5$__.

5. If the graph of a function f approaches b as x increases or decreases without bound, then the line $y = b$ is a/an __horizontal asymptote__ of the graph of f. The equation of such a line for the graph of $y = \dfrac{x - 10}{3x^2 + x + 1}$ is __$y = 0$__. The equation of such a line for the graph of $y = \dfrac{x^2 - 10}{3x^2 + x + 1}$ is __$y = \frac{1}{3}$__.

6. True or false: If the degree of the numerator of a rational function equals the degree of the denominator, then setting y equal to the ratio of the leading coefficients gives the equation of the horizontal asymptote. __true__

7. Compared with the graph of $f(x) = \dfrac{1}{x}$, the graph of $g(x) = \dfrac{1}{x + 2} - 1$ is shifted 2 units __left__ and 1 unit __down__.

8. The graph of a rational function has a slant asymptote if the degree of the numerator is __one more than__ the degree of the denominator.

9. Based on the synthetic division shown below, the equation of the slant asymptote of
$$f(x) = \frac{3x^2 - 7x + 5}{x - 4}$$ is __$y = 3x + 5$__.

$$
\begin{array}{r|rrr}
4 & 3 & -7 & 5 \\
 & & 12 & 20 \\
\hline
 & 3 & 5 & 25
\end{array}
$$

EXERCISE SET 3.5

Practice Exercises

In Exercises 1–8, find the domain of each rational function.

1. $f(x) = \dfrac{5x}{x - 4}$ $\{x \mid x \neq 4\}$

2. $f(x) = \dfrac{7x}{x - 8}$ $\{x \mid x \neq 8\}$

3. $g(x) = \dfrac{3x^2}{(x - 5)(x + 4)}$ $\{x \mid x \neq 5, x \neq -4\}$

4. $g(x) = \dfrac{2x^2}{(x - 2)(x + 6)}$ $\{x \mid x \neq 2, x \neq -6\}$

5. $h(x) = \dfrac{x + 7}{x^2 - 49}$ $\{x \mid x \neq 7, x \neq -7\}$

6. $h(x) = \dfrac{x + 8}{x^2 - 64}$ $\{x \mid x \neq 8, x \neq -8\}$

7. $f(x) = \dfrac{x + 7}{x^2 + 49}$ all real numbers

8. $f(x) = \dfrac{x + 8}{x^2 + 64}$ all real numbers

Use the graph of the rational function in the figure shown to complete each statement in Exercises 9–14.

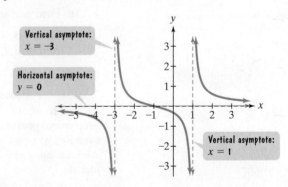

9. As $x \rightarrow -3^-$, $f(x) \rightarrow$ _____. $-\infty$

10. As $x \rightarrow -3^+$, $f(x) \rightarrow$ _____. $+\infty$

11. As $x \rightarrow 1^-$, $f(x) \rightarrow$ _____. $-\infty$

12. As $x \rightarrow 1^+$, $f(x) \rightarrow$ _____. $+\infty$

13. As $x \rightarrow -\infty$, $f(x) \rightarrow$ _____. 0

14. As $x \rightarrow \infty$, $f(x) \rightarrow$ _____. 0

Use the graph of the rational function in the figure shown to complete each statement in Exercises 15–20.

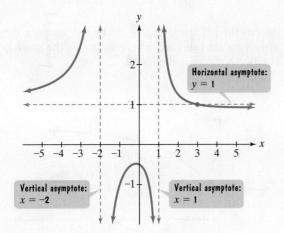

Horizontal asymptote: $y = 1$

Vertical asymptote: $x = -2$

Vertical asymptote: $x = 1$

15. As $x \to 1^+$, $f(x) \to$ _____. $+\infty$
16. As $x \to 1^-$, $f(x) \to$ _____. $-\infty$
17. As $x \to -2^+$, $f(x) \to$ _____. $-\infty$
18. As $x \to -2^-$, $f(x) \to$ _____. $+\infty$
19. As $x \to \infty$, $f(x) \to$ _____. 1
20. As $x \to -\infty$, $f(x) \to$ _____. 1

In Exercises 21–36, find the vertical asymptotes, if any, and the values of x corresponding to holes, if any, of the graph of each rational function.

21. $f(x) = \dfrac{x}{x + 4}$ *

22. $f(x) = \dfrac{x}{x - 3}$ *

23. $g(x) = \dfrac{x + 3}{x(x + 4)}$ *

24. $g(x) = \dfrac{x + 3}{x(x - 3)}$ *

25. $h(x) = \dfrac{x}{x(x + 4)}$ *

26. $h(x) = \dfrac{x}{x(x - 3)}$ *

27. $r(x) = \dfrac{x}{x^2 + 4}$ *

28. $r(x) = \dfrac{x}{x^2 + 3}$ *

29. $f(x) = \dfrac{x^2 - 9}{x - 3}$ *

30. $f(x) = \dfrac{x^2 - 25}{x - 5}$ *

31. $g(x) = \dfrac{x - 3}{x^2 - 9}$ *

32. $g(x) = \dfrac{x - 5}{x^2 - 25}$ *

33. $h(x) = \dfrac{x + 7}{x^2 + 4x - 21}$ *

34. $h(x) = \dfrac{x + 6}{x^2 + 2x - 24}$ *

35. $r(x) = \dfrac{x^2 + 4x - 21}{x + 7}$ *

36. $r(x) = \dfrac{x^2 + 2x - 24}{x + 6}$ *

In Exercises 37–44, find the horizontal asymptote, if there is one, of the graph of each rational function.

37. $f(x) = \dfrac{12x}{3x^2 + 1}$ $y = 0$

38. $f(x) = \dfrac{15x}{3x^2 + 1}$ $y = 0$

39. $g(x) = \dfrac{12x^2}{3x^2 + 1}$ $y = 4$

40. $g(x) = \dfrac{15x^2}{3x^2 + 1}$ $y = 5$

41. $h(x) = \dfrac{12x^3}{3x^2 + 1}$ none

42. $h(x) = \dfrac{15x^3}{3x^2 + 1}$ none

43. $f(x) = \dfrac{-2x + 1}{3x + 5}$ $y = -\frac{2}{3}$

44. $f(x) = \dfrac{-3x + 7}{5x - 2}$ $y = -\frac{3}{5}$

In Exercises 45–56, use transformations of $f(x) = \dfrac{1}{x}$ or $f(x) = \dfrac{1}{x^2}$ to graph each rational function.

45. $g(x) = \dfrac{1}{x - 1}$ *

46. $g(x) = \dfrac{1}{x - 2}$ *

47. $h(x) = \dfrac{1}{x} + 2$ *

48. $h(x) = \dfrac{1}{x} + 1$ *

49. $g(x) = \dfrac{1}{x + 1} - 2$ *

50. $g(x) = \dfrac{1}{x + 2} - 2$ *

51. $g(x) = \dfrac{1}{(x + 2)^2}$ *

52. $g(x) = \dfrac{1}{(x + 1)^2}$ *

53. $h(x) = \dfrac{1}{x^2} - 4$ *

54. $h(x) = \dfrac{1}{x^2} - 3$ *

55. $h(x) = \dfrac{1}{(x - 3)^2} + 1$ *

56. $h(x) = \dfrac{1}{(x - 3)^2} + 2$ *

In Exercises 57–80, follow the seven steps on page 419 to graph each rational function.

57. $f(x) = \dfrac{4x}{x - 2}$ *

58. $f(x) = \dfrac{3x}{x - 1}$ *

59. $f(x) = \dfrac{2x}{x^2 - 4}$ *

60. $f(x) = \dfrac{4x}{x^2 - 1}$ *

61. $f(x) = \dfrac{2x^2}{x^2 - 1}$ *

62. $f(x) = \dfrac{4x^2}{x^2 - 9}$ *

63. $f(x) = \dfrac{-x}{x + 1}$ *

64. $f(x) = \dfrac{-3x}{x + 2}$ *

65. $f(x) = -\dfrac{1}{x^2 - 4}$ *

66. $f(x) = -\dfrac{2}{x^2 - 1}$ *

67. $f(x) = \dfrac{2}{x^2 + x - 2}$ *

68. $f(x) = \dfrac{-2}{x^2 - x - 2}$ *

69. $f(x) = \dfrac{2x^2}{x^2 + 4}$ *

70. $f(x) = \dfrac{4x^2}{x^2 + 1}$ *

71. $f(x) = \dfrac{x + 2}{x^2 + x - 6}$ *

72. $f(x) = \dfrac{x - 4}{x^2 - x - 6}$ *

73. $f(x) = \dfrac{x - 2}{x^2 - 4}$ *

74. $f(x) = \dfrac{x - 3}{x^2 - 9}$ *

75. $f(x) = \dfrac{x^4}{x^2 + 2}$ *

76. $f(x) = \dfrac{2x^4}{x^2 + 1}$ *

77. $f(x) = \dfrac{x^2 + x - 12}{x^2 - 4}$ *

78. $f(x) = \dfrac{x^2}{x^2 + x - 6}$ *

79. $f(x) = \dfrac{3x^2 + x - 4}{2x^2 - 5x}$ *

80. $f(x) = \dfrac{x^2 - 4x + 3}{(x + 1)^2}$ *

*In Exercises 81–88, **a.** Find the slant asymptote of the graph of each rational function and **b.** Follow the seven-step strategy and use the slant asymptote to graph each rational function.*

81. $f(x) = \dfrac{x^2 - 1}{x}$ **a.** $y = x$*

82. $f(x) = \dfrac{x^2 - 4}{x}$ **a.** $y = x$*

83. $f(x) = \dfrac{x^2 + 1}{x}$ **a.** $y = x$*

84. $f(x) = \dfrac{x^2 + 4}{x}$ **a.** $y = x$*

85. $f(x) = \dfrac{x^2 + x - 6}{x - 3}$ *

86. $f(x) = \dfrac{x^2 - x + 1}{x - 1}$ *

87. $f(x) = \dfrac{x^3 + 1}{x^2 + 2x}$ **a.** $y = x - 2$*

88. $f(x) = \dfrac{x^3 - 1}{x^2 - 9}$ **a.** $y = x$*

Practice Plus

In Exercises 89–94, the equation for f is given by the simplified expression that results after performing the indicated operation. Write the equation for f and then graph the function.

89. $\dfrac{5x^2}{x^2 - 4} \cdot \dfrac{x^2 + 4x + 4}{10x^3}$ *

90. $\dfrac{x - 5}{10x - 2} \div \dfrac{x^2 - 10x + 25}{25x^2 - 1}$ *

91. $\dfrac{x}{2x + 6} - \dfrac{9}{x^2 - 9}$ *

92. $\dfrac{2}{x^2 + 3x + 2} - \dfrac{4}{x^2 + 4x + 3}$ *

93. $\dfrac{1 - \dfrac{3}{x + 2}}{1 + \dfrac{1}{x - 2}}$ $f(x) = \dfrac{x - 2}{x + 2}$*

94. $\dfrac{x - \dfrac{1}{x}}{x + \dfrac{1}{x}}$ $f(x) = \dfrac{x^2 - 1}{x^2 + 1}$*

In Exercises 95–98, use long division to rewrite the equation for g in the form

$$quotient + \frac{remainder}{divisor}.$$

Then use this form of the function's equation and transformations of $f(x) = \dfrac{1}{x}$ to graph g.

95. $g(x) = \dfrac{2x + 7}{x + 3}$ *

96. $g(x) = \dfrac{3x + 7}{x + 2}$ *

97. $g(x) = \dfrac{3x - 7}{x - 2}$ *

98. $g(x) = \dfrac{2x - 9}{x - 4}$ *

Application Exercises

99. A company is planning to manufacture mountain bikes. The fixed monthly cost will be $100,000 and it will cost $100 to produce each bicycle.

 a. Write the cost function, C, of producing x mountain bikes. $C(x) = 100x + 100,000$

 b. Write the average cost function, \overline{C}, of producing x mountain bikes. *

 c. Find and interpret $\overline{C}(500)$, $\overline{C}(1000)$, $\overline{C}(2000)$, and $\overline{C}(4000)$. 300; 200; 150; 125*

 d. What is the horizontal asymptote for the graph of the average cost function, \overline{C}? Describe what this means in practical terms. $y = 100$*

100. A company that manufactures running shoes has a fixed monthly cost of $300,000. It costs $30 to produce each pair of shoes.

 a. Write the cost function, C, of producing x pairs of shoes. $C(x) = 30x + 300,000$

 b. Write the average cost function, \overline{C}, of producing x pairs of shoes. *

 c. Find and interpret $\overline{C}(1000)$, $\overline{C}(10,000)$, and $\overline{C}(100,000)$. *

 d. What is the horizontal asymptote for the graph of the average cost function, \overline{C}? Describe what this represents for the company. $y = 30$*

101. The function

$$f(x) = \frac{6.5x^2 - 20.4x + 234}{x^2 + 36}$$

models the pH level, $f(x)$, of the human mouth x minutes after a person eats food containing sugar. The graph of this function is shown in the figure.

a. Use the graph to obtain a reasonable estimate, to the nearest tenth, of the pH level of the human mouth 42 minutes after a person eats food containing sugar. 6.0

b. After eating sugar, when is the pH level the lowest? Use the function's equation to determine the pH level, to the nearest tenth, at this time. after 6 min; about 4.8

c. According to the graph, what is the normal pH level of the human mouth? 6.5

d. What is the equation of the horizontal asymptote associated with this function? Describe what this means in terms of the mouth's pH level over time. $y = 6.5$*

e. Use the graph to describe what happens to the pH level during the first hour. *

102. A drug is injected into a patient and the concentration of the drug in the bloodstream is monitored. The drug's concentration, $C(t)$, in milligrams per liter, after t hours is modeled by

$$C(t) = \frac{5t}{t^2 + 1}.$$

The graph of this rational function, obtained with a graphing utility, is shown in the figure.

[0, 10, 1] by [0, 3, 1]

a. Use the graph to obtain a reasonable estimate of the drug's concentration after 3 hours. ≈ 1.5 mg per liter

b. Use the function's equation displayed in the voice balloon by the graph to determine the drug's concentration after 3 hours. 1.5 mg per liter

c. Use the function's equation to find the horizontal asymptote for the graph. Describe what this means about the drug's concentration in the patient's bloodstream as time increases. $y = 0$*

Among all deaths from a particular disease, the percentage that is smoking related (21–39 cigarettes per day) is a function of the disease's **incidence ratio**. The incidence ratio describes the number of times more likely smokers are than nonsmokers to die from the disease. The following table shows the incidence ratios for heart disease and lung cancer for two age groups.

Incidence Ratios		
	Heart Disease	**Lung Cancer**
Ages 55–64	1.9	10
Ages 65–74	1.7	9

Source: Alexander M. Walker, *Observations and Inference,* Epidemiology Resources Inc., 1991.

For example, the incidence ratio of 9 in the table means that smokers between the ages of 65 and 74 are 9 times more likely than nonsmokers in the same age group to die from lung cancer. The rational function

$$P(x) = \frac{100(x - 1)}{x}$$

models the percentage of smoking-related deaths among all deaths from a disease, P(x), in terms of the disease's incidence ratio, x. The graph of the rational function is shown. Use this function to solve Exercises 103–106.

The Disease's Incidence Ratio:
The number of times more likely smokers are than nonsmokers to die from the disease

103. Find P(10). Describe what this means in terms of the incidence ratio, 10, given in the table. Identify your solution as a point on the graph. *P(10) = 90**

104. Find P(9). Round to the nearest percent. Describe what this means in terms of the incidence ratio, 9, given in the table. Identify your solution as a point on the graph. *P(9) = 89**

105. What is the horizontal asymptote of the graph? Describe what this means about the percentage of deaths caused by smoking with increasing incidence ratios. *y = 100**

106. According to the model and its graph, is there a disease for which all deaths are caused by smoking? Explain your answer. *no **

107. The bar graph shows the amount, in billions of dollars, that the United States government spent on human resources and total budget outlays for six selected years. (Human resources include education, health, Medicare, Social Security, and veterans benefits and services.)

Federal Budget Expenditures on Human Resources

Source: Office of Management and Budget

The function $p(x) = 1.75x^2 - 15.9x + 160$ models the amount, p(x), in billions of dollars, that the United States government spent on human resources x years after 1970. The function $q(x) = 2.1x^2 - 3.5x + 296$ models total budget expenditures, q(x), in billions of dollars, x years after 1970.

a. Use p and q to write a rational function that models the fraction of total budget outlays spent on human resources x years after 1970. *

b. Use the data displayed by the bar graph to find the percentage of federal expenditures spent on human resources in 2010. Round to the nearest percent. *67%*

c. Use the rational function from part (a) to find the percentage of federal expenditures spent on human resources in 2010. Round to the nearest percent. Does this underestimate or overestimate the actual percent that you found in part (b)? By how much? *66%**

d. What is the equation of the horizontal asymptote associated with the rational function in part (a)? If trends modeled by the function continue, what percentage of the federal budget will be spent on human resources over time? Round to the nearest percent. *$y = \frac{1.75}{2.1}$; 83%*

Explaining the Concepts 108–112. Answers will vary.

108. What is a rational function?

109. Use everyday language to describe the graph of a rational function f such that as $x \to -\infty$, $f(x) \to 3$.

110. Use everyday language to describe the behavior of a graph near its vertical asymptote if $f(x) \to \infty$ as $x \to -2^-$ and $f(x) \to -\infty$ as $x \to -2^+$.

111. If you are given the equation of a rational function, explain how to find the vertical asymptotes, if any, of the function's graph.

112. If you are given the equation of a rational function, explain how to find the horizontal asymptote, if there is one, of the function's graph.

113. Describe how to graph a rational function.

114. If you are given the equation of a rational function, how can you tell if the graph has a slant asymptote? If it does, how do you find its equation?

115. Is every rational function a polynomial function? Why or why not? Does a true statement result if the two adjectives *rational* and *polynomial* are reversed? Explain.

116. Although your friend has a family history of heart disease, he smokes, on average, 25 cigarettes per day. He sees the table showing incidence ratios for heart disease (see Exercises 103–106) and feels comfortable that they are less than 2, compared to 9 and 10 for lung cancer. He claims that all family deaths have been from heart disease and decides not to give up smoking. Use the given function and its graph to describe some additional information not given in the table that might influence his decision. 113–116. Answers will vary.

Technology Exercises

117. Use a graphing utility to verify any five of your hand-drawn graphs in Exercises 45–88. Answers will vary.

118. Use a graphing utility to graph $y = \dfrac{1}{x}$, $y = \dfrac{1}{x^3}$, and $\dfrac{1}{x^5}$ in the same viewing rectangle. For odd values of *n*, how does changing *n* affect the graph of $y = \dfrac{1}{x^n}$? *

119. Use a graphing utility to graph $y = \dfrac{1}{x^2}$, $y = \dfrac{1}{x^4}$, and $y = \dfrac{1}{x^6}$ in the same viewing rectangle. For even values of *n*, how does changing *n* affect the graph of $y = \dfrac{1}{x^n}$? *

120. Use a graphing utility to graph

$$f(x) = \frac{x^2 - 4x + 3}{x - 2} \quad \text{and} \quad g(x) = \frac{x^2 - 5x + 6}{x - 2}.$$

What differences do you observe between the graph of *f* and the graph of *g*? How do you account for these differences? *

121. The rational function

$$f(x) = \frac{27{,}725(x - 14)}{x^2 + 9} - 5x$$

models the number of arrests, $f(x)$, per 100,000 drivers, for driving under the influence of alcohol, as a function of a driver's age, *x*.

a. Graph the function in a $[0, 70, 5]$ by $[0, 400, 20]$ viewing rectangle. *

b. Describe the trend shown by the graph. *

c. Use the ZOOM and TRACE features or the maximum function feature of your graphing utility to find the age that corresponds to the greatest number of arrests. How many arrests, per 100,000 drivers, are there for this age group? at age 25, about 356 arrests

Critical Thinking Exercises

Make Sense? *In Exercises 122–125, determine whether each statement makes sense or does not make sense, and explain your reasoning.*

122. I've graphed a rational function that has two vertical asymptotes and two horizontal asymptotes. *

123. My graph of $y = \dfrac{x - 1}{(x - 1)(x - 2)}$ has vertical asymptotes at $x = 1$ and $x = 2$. does not make sense

124. The function

$$f(x) = \frac{1.96x + 3.14}{3.04x + 21.79}$$

models the fraction of nonviolent prisoners in New York State prisons *x* years after 1980. I can conclude from this equation that over time the percentage of nonviolent prisoners will exceed 60%. makes sense

125. As production level increases, the average cost for a company to produce each unit of its product also increases. does not make sense

In Exercises 126–129, determine whether each statement is true or false. If the statement is false, make the necessary change(s) to produce a true statement. Changes to false statements will vary.

126. The graph of a rational function cannot have both a vertical asymptote and a horizontal asymptote. false

127. It is possible to have a rational function whose graph has no *y*-intercept. true

128. The graph of a rational function can have three vertical asymptotes. true

129. The graph of a rational function can never cross a vertical asymptote. true

In Exercises 130–133, write the equation of a rational function $f(x) = \dfrac{p(x)}{q(x)}$ having the indicated properties, in which the degrees of p and q are as small as possible. More than one correct function may be possible. Graph your function using a graphing utility to verify that it has the required properties. 130–133. Answers will vary.

130. *f* has a vertical asymptote given by $x = 3$, a horizontal asymptote $y = 0$, *y*-intercept at -1, and no *x*-intercept.

131. *f* has vertical asymptotes given by $x = -2$ and $x = 2$, a horizontal asymptote $y = 2$, *y*-intercept at $\frac{9}{2}$, *x*-intercepts at -3 and 3, and *y*-axis symmetry.

132. *f* has a vertical asymptote given by $x = 1$, a slant asymptote whose equation is $y = x$, *y*-intercept at 2, and *x*-intercepts at -1 and 2.

133. *f* has no vertical, horizontal, or slant asymptotes, and no *x*-intercepts.

Retaining the Concepts

134. Basic Car Rental charges $20 a day plus $0.10 per mile, whereas Acme Car Rental charges $30 a day plus $0.05 per mile. How many miles must be driven to make the daily cost of a Basic Rental a better deal than an Acme Rental? (Section 1.7, Example 11) less than 200 miles

135. Identify the graphs (a)–(d) in which *y* is a function of *x*.

c.

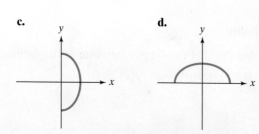

d.

(Section 2.1, Example 6) *y is a function of x in (a) and (d).*

136. Which of the following graphs (a)–(d) represent functions that have an inverse function?

a.

b.

c.

d.

(Section 2.7, Example 5) (b) and (d) have inverse functions.

Preview Exercises

Exercises 137–139 will help you prepare for the material covered in the next section.

137. Solve: $2x^2 + x = 15$. $\left\{-3, \frac{5}{2}\right\}$

138. Solve: $x^3 + x^2 = 4x + 4$. $\{-2, -1, 2\}$

139. Simplify: $\dfrac{x+1}{x+3} - 2$. $\dfrac{-x-5}{x+3}$ or $-\dfrac{x+5}{x+3}$

Section 3.6

Polynomial and Rational Inequalities

What am I supposed to learn?

After studying this section, you should be able to:

1. Solve polynomial inequalities.

2. Solve rational inequalities.

3. Solve problems modeled by polynomial or rational inequalities.

© Warren Miller/The New Yorker Collection/Cartoonbank

Tailgaters beware: If your car is going 35 miles per hour on dry pavement, your required stopping distance is 160 feet, or the width of a football field. At 65 miles per hour, the distance required is 410 feet, or approximately the length of one and one-tenth football fields. **Figure 3.43** shows stopping distances for cars at various speeds on dry roads and on wet roads.

A car's required stopping distance, $f(x)$, in feet, on dry pavement traveling at x miles per hour can be modeled by the quadratic function

$$f(x) = 0.0875x^2 - 0.4x + 66.6.$$

How can we use this function to determine speeds on dry pavement requiring stopping distances that exceed the length of one and one-half football fields, or 540 feet? We must solve the inequality

$$0.0875x^2 - 0.4x + 66.6 > 540.$$

Required stopping distance exceeds 540 feet.

We begin by subtracting 540 from both sides. This will give us zero on the right:

$$0.0875x^2 - 0.4x + 66.6 - 540 > 540 - 540$$
$$0.0875x^2 - 0.4x - 473.4 > 0.$$

Stopping Distances for Cars at Selected Speeds

■ Dry Pavement ■ Wet Pavement

FIGURE 3.43
Source: National Highway Traffic Safety Administration

The form of this inequality is $ax^2 + bx + c > 0$. Such a quadratic inequality is called a *polynomial inequality*.

TECHNOLOGY

We used the statistical menu of a graphing utility and the quadratic regression program to obtain the quadratic function that models stopping distance on dry pavement. After entering the appropriate data from **Figure 3.43** on the previous page, namely,

$(35, 160), (45, 225), (55, 310),$
$(65, 410),$

we obtained the results shown in the screen.

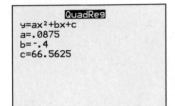

Definition of a Polynomial Inequality

A polynomial inequality is any inequality that can be put into one of the forms

$$f(x) < 0, \quad f(x) > 0, \quad f(x) \leq 0, \quad \text{or} \quad f(x) \geq 0,$$

where f is a polynomial function.

In this section, we establish the basic techniques for solving polynomial inequalities. We will also use these techniques to solve inequalities involving rational functions.

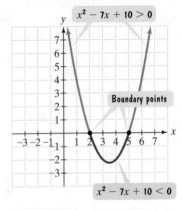

FIGURE 3.44

Solving Polynomial Inequalities

Graphs can help us visualize the solutions of polynomial inequalities. For example, the graph of $f(x) = x^2 - 7x + 10$ is shown in **Figure 3.44**. The x-intercepts, 2 and 5, are **boundary points** between where the graph lies above the x-axis, shown in blue, and where the graph lies below the x-axis, shown in red.

Locating the x-intercepts of a polynomial function, f, is an important step in finding the solution set for polynomial inequalities in the form $f(x) < 0$ or $f(x) > 0$. We use the x-intercepts of f as boundary points that divide the real number line into intervals. On each interval, the graph of f is either above the x-axis $[f(x) > 0]$ or below the x-axis $[f(x) < 0]$. For this reason, x-intercepts play a fundamental role in solving polynomial inequalities. The x-intercepts are found by solving the equation $f(x) = 0$.

① Solve polynomial inequalities.

Procedure for Solving Polynomial Inequalities

1. Express the inequality in the form

$$f(x) < 0 \quad \text{or} \quad f(x) > 0,$$

where f is a polynomial function.
2. Solve the equation $f(x) = 0$. The real solutions are the **boundary points**.
3. Locate these boundary points on a number line, thereby dividing the number line into intervals.
4. Choose one representative number, called a **test value**, within each interval and evaluate f at that number.
 a. If the value of f is positive, then $f(x) > 0$ for all numbers, x, in the interval.
 b. If the value of f is negative, then $f(x) < 0$ for all numbers, x, in the interval.
5. Write the solution set, selecting the interval or intervals that satisfy the given inequality.

This procedure is valid if $<$ is replaced by \leq or $>$ is replaced by \geq. However, if the inequality involves \leq or \geq, include the boundary points [the solutions of $f(x) = 0$] in the solution set.

EXAMPLE 1 Solving a Polynomial Inequality

Solve and graph the solution set on a real number line: $2x^2 + x > 15$.

SOLUTION

Step 1 Express the inequality in the form $f(x) < 0$ or $f(x) > 0$. We begin by rewriting the inequality so that 0 is on the right side.

$$2x^2 + x > 15 \qquad \text{This is the given inequality.}$$
$$2x^2 + x - 15 > 15 - 15 \qquad \text{Subtract 15 from both sides.}$$
$$2x^2 + x - 15 > 0 \qquad \text{Simplify.}$$

This inequality is equivalent to the one we wish to solve. It is in the form $f(x) > 0$, where $f(x) = 2x^2 + x - 15$.

Step 2 Solve the equation $f(x) = 0$. We find the x-intercepts of $f(x) = 2x^2 + x - 15$ by solving the equation $2x^2 + x - 15 = 0$.

$$2x^2 + x - 15 = 0 \qquad \begin{array}{l}\text{This polynomial equation is a}\\\text{quadratic equation.}\end{array}$$
$$(2x - 5)(x + 3) = 0 \qquad \text{Factor.}$$
$$2x - 5 = 0 \quad \text{or} \quad x + 3 = 0 \qquad \text{Set each factor equal to O.}$$
$$x = \tfrac{5}{2} \qquad\qquad x = -3 \qquad \text{Solve for x.}$$

The x-intercepts of f are -3 and $\frac{5}{2}$. We will use these x-intercepts as boundary points on a number line.

Step 3 Locate the boundary points on a number line and separate the line into intervals. The number line with the boundary points is shown as follows:

The boundary points divide the number line into three intervals:

$$(-\infty, -3) \quad \left(-3, \tfrac{5}{2}\right) \quad \left(\tfrac{5}{2}, \infty\right).$$

Step 4 Choose one test value within each interval and evaluate f at that number.

Interval	Test Value	Substitute into $f(x) = 2x^2 + x - 15$	Conclusion
$(-\infty, -3)$	-4	$\begin{aligned}f(-4) &= 2(-4)^2 + (-4) - 15\\ &= 13, \text{positive}\end{aligned}$	$f(x) > 0$ for all x in $(-\infty, -3)$.
$\left(-3, \dfrac{5}{2}\right)$	0	$\begin{aligned}f(0) &= 2 \cdot 0^2 + 0 - 15\\ &= -15, \text{negative}\end{aligned}$	$f(x) < 0$ for all x in $\left(-3, \dfrac{5}{2}\right)$.
$\left(\dfrac{5}{2}, \infty\right)$	3	$\begin{aligned}f(3) &= 2 \cdot 3^2 + 3 - 15\\ &= 6, \text{positive}\end{aligned}$	$f(x) > 0$ for all x in $\left(\dfrac{5}{2}, \infty\right)$.

TECHNOLOGY

Graphic Connections

The graph of $f(x) = 2x^2 + x - 15$ lies above the x-axis, representing $>$, for all x in $(-\infty, -3)$ or $\left(\frac{5}{2}, \infty\right)$.

Step 5 Write the solution set, selecting the interval or intervals that satisfy the given inequality. We are interested in solving $f(x) > 0$, where $f(x) = 2x^2 + x - 15$. Based on our work in step 4, we see that $f(x) > 0$ for all x in $(-\infty, -3)$ or $\left(\frac{5}{2}, \infty\right)$. Thus, the solution set of the given inequality, $2x^2 + x > 15$, or, equivalently, $2x^2 + x - 15 > 0$, is

$$(-\infty, -3) \cup \left(\tfrac{5}{2}, \infty\right) \text{or} \left\{x \mid x < -3 \text{ or } x > \tfrac{5}{2}\right\}.$$

The graph of the solution set on a number line is shown as follows:

● ● ●

⊘ **Check Point 1** Solve and graph the solution set: $x^2 - x > 20$.

$\{x < -4 \text{ or } x > 5\} \text{ or } (-\infty, -4) \cup (5, \infty)^*$

*See Answers to Selected Exercises.

EXAMPLE 2 Solving a Polynomial Inequality

Solve and graph the solution set on a real number line: $4x^2 \leq 1 - 2x$.

SOLUTION

Step 1 Express the inequality in the form $f(x) \leq 0$ or $f(x) \geq 0$. We begin by rewriting the inequality so that 0 is on the right side.

$4x^2 \leq 1 - 2x$	This is the given inequality.
$4x^2 + 2x - 1 \leq 1 - 2x + 2x - 1$	Add $2x$ and subtract 1 on both sides.
$4x^2 + 2x - 1 \leq 0$	Simplify.

This inequality is equivalent to the one we wish to solve. It is in the form $f(x) \leq 0$, where $f(x) = 4x^2 + 2x - 1$.

Step 2 Solve the equation $f(x) = 0$. We will find the x-intercepts of $f(x) = 4x^2 + 2x - 1$ by solving the equation $4x^2 + 2x - 1 = 0$. This equation cannot be solved by factoring. We will use the quadratic formula to solve it.

$$4x^2 + 2x - 1 = 0$$

$$\boxed{a = 4} \quad \boxed{b = 2} \quad \boxed{c = -1}$$

$$x = \frac{-b \pm \sqrt{b^2 - 4ac}}{2a} = \frac{-2 \pm \sqrt{2^2 - 4 \cdot 4(-1)}}{2 \cdot 4} = \frac{-2 \pm \sqrt{4 - (-16)}}{8}$$

$$= \frac{-2 \pm \sqrt{20}}{8} = \frac{-2 \pm \sqrt{4}\sqrt{5}}{8} = \frac{-2 \pm 2\sqrt{5}}{8} = \frac{2(-1 \pm \sqrt{5})}{8} = \frac{-1 \pm \sqrt{5}}{4}$$

$$x = \frac{-1 + \sqrt{5}}{4} \approx 0.3 \qquad x = \frac{-1 - \sqrt{5}}{4} \approx -0.8$$

The x-intercepts of f are $\dfrac{-1 + \sqrt{5}}{4}$ (approximately 0.3) and $\dfrac{-1 - \sqrt{5}}{4}$ (approximately -0.8). We will use these x-intercepts as boundary points on a number line.

Step 3 Locate the boundary points on a number line and separate the line into intervals. The number line with the boundary points is shown as follows:

The boundary points divide the number line into three intervals:

$$\left(-\infty, \frac{-1 - \sqrt{5}}{4} \right) \left(\frac{-1 - \sqrt{5}}{4}, \frac{-1 + \sqrt{5}}{4} \right) \left(\frac{-1 + \sqrt{5}}{4}, \infty \right).$$

Step 4 Choose one test value within each interval and evaluate f at that number.

Interval	Test Value	Substitute into $f(x) = 4x^2 + 2x - 1$	Conclusion
$\left(-\infty, \dfrac{-1 - \sqrt{5}}{4} \right)$	-1	$\begin{aligned} f(-1) &= 4(-1)^2 + 2(-1) - 1 \\ &= 1, \text{positive} \end{aligned}$	$f(x) > 0$ for all x in $\left(-\infty, \dfrac{-1 - \sqrt{5}}{4} \right)$.
$\left(\dfrac{-1 - \sqrt{5}}{4}, \dfrac{-1 + \sqrt{5}}{4} \right)$	0	$\begin{aligned} f(0) &= 4 \cdot 0^2 + 2 \cdot 0 - 1 \\ &= -1, \text{negative} \end{aligned}$	$f(x) < 0$ for all x in $\left(\dfrac{-1 - \sqrt{5}}{4}, \dfrac{-1 + \sqrt{5}}{4} \right)$.
$\left(\dfrac{-1 + \sqrt{5}}{4}, \infty \right)$	1	$\begin{aligned} f(1) &= 4 \cdot 1^2 + 2 \cdot 1 - 1 \\ &= 5, \text{positive} \end{aligned}$	$f(x) > 0$ for all x in $\left(\dfrac{-1 + \sqrt{5}}{4}, \infty \right)$.

TECHNOLOGY

Graphic Connections

The solution set for

$$4x^2 \le 1 - 2x$$

or, equivalently,

$$4x^2 + 2x - 1 \le 0$$

can be verified with a graphing utility. The graph of $f(x) = 4x^2 + 2x - 1$ was obtained using a $[-2, 2, 1]$ by $[-10, 10, 1]$ viewing rectangle. The graph lies on or below the x-axis, representing \le, for all x in

$$\left[\frac{-1 - \sqrt{5}}{4}, \frac{-1 + \sqrt{5}}{4}\right]$$
$$\approx [-0.8, 0.3].$$

Step 5 Write the solution set, selecting the interval or intervals that satisfy the given inequality. We are interested in solving $f(x) \le 0$, where $f(x) = 4x^2 + 2x - 1$. Based on our work in step 4, we see that $f(x) < 0$ for all x in $\left(\frac{-1 - \sqrt{5}}{4}, \frac{-1 + \sqrt{5}}{4}\right)$. However, because the inequality involves \le (less than or *equal to*), we must also include the solutions of $4x^2 + 2x - 1 = 0$, namely, $\frac{-1 - \sqrt{5}}{4}$ and $\frac{-1 + \sqrt{5}}{4}$, in the solution set. Thus, the solution set of the given inequality $4x^2 \le 1 - 2x$, or, equivalently, $4x^2 + 2x - 1 \le 0$, is

$$\left[\frac{-1 - \sqrt{5}}{4}, \frac{-1 + \sqrt{5}}{4}\right] \text{ or } \left\{x \middle| \frac{-1 - \sqrt{5}}{4} \le x \le \frac{-1 + \sqrt{5}}{4}\right\}.$$

The graph of the solution set on a number line is shown as follows:

```
  +---+---+---+---+---+--[+--]+---+---+---+---+--> x
 -5  -4  -3  -2  -1   0   1   2   3   4   5
```
• • •

✓ Check Point 2 Solve and graph the solution set on a real number line: $2x^2 \le -6x - 1$. $\left[\frac{-3 - \sqrt{7}}{2}, \frac{-3 + \sqrt{7}}{2}\right]$ or $\left\{x \middle| \frac{-3 - \sqrt{7}}{2} \le x \le \frac{-3 + \sqrt{7}}{2}\right\}$*

EXAMPLE 3 Solving a Polynomial Inequality

Solve and graph the solution set on a real number line: $x^3 + x^2 \le 4x + 4$.

SOLUTION

Step 1 Express the inequality in the form $f(x) \le 0$ or $f(x) \ge 0$. We begin by rewriting the inequality so that 0 is on the right side.

$x^3 + x^2 \le 4x + 4$	This is the given inequality.
$x^3 + x^2 - 4x - 4 \le 4x + 4 - 4x - 4$	Subtract $4x + 4$ from both sides.
$x^3 + x^2 - 4x - 4 \le 0$	Simplify.

This inequality is equivalent to the one we wish to solve. It is in the form $f(x) \le 0$, where $f(x) = x^3 + x^2 - 4x - 4$.

Step 2 Solve the equation $f(x) = 0$. We find the x-intercepts of $f(x) = x^3 + x^2 - 4x - 4$ by solving the equation $x^3 + x^2 - 4x - 4 = 0$.

$x^3 + x^2 - 4x - 4 = 0$	This polynomial equation is of degree 3.
$x^2(x + 1) - 4(x + 1) = 0$	Factor x^2 from the first two terms and -4 from the last two terms.
$(x + 1)(x^2 - 4) = 0$	A common factor of $x + 1$ is factored from the expression.
$(x + 1)(x + 2)(x - 2) = 0$	Factor completely.
$x + 1 = 0 \quad \text{or} \quad x + 2 = 0 \quad \text{or} \quad x - 2 = 0$	Set each factor equal to 0.
$x = -1 \qquad\qquad x = -2 \qquad\qquad x = 2$	Solve for x.

The x-intercepts of f are -2, -1, and 2. We will use these x-intercepts as boundary points on a number line.

Step 3 Locate the boundary points on a number line and separate the line into intervals. The number line with the boundary points is shown as follows:

```
              -2  -1         2
  +---+---+---•---•---+---+---•---+---+---+--> x
 -5  -4  -3  -2  -1   0   1   2   3   4   5
```

The boundary points divide the number line into four intervals:

$$(-\infty, -2) \quad (-2, -1) \quad (-1, 2) \quad (2, \infty).$$

Step 4 Choose one test value within each interval and evaluate f at that number.

Interval	Test Value	Substitute into $f(x) = x^3 + x^2 - 4x - 4$	Conclusion
$(-\infty, -2)$	-3	$f(-3) = (-3)^3 + (-3)^2 - 4(-3) - 4$ $= -10$, negative	$f(x) < 0$ for all x in $(-\infty, -2)$.
$(-2, -1)$	-1.5	$f(-1.5) = (-1.5)^3 + (-1.5)^2 - 4(-1.5) - 4$ $= 0.875$, positive	$f(x) > 0$ for all x in $(-2, -1)$.
$(-1, 2)$	0	$f(0) = 0^3 + 0^2 - 4 \cdot 0 - 4$ $= -4$, negative	$f(x) < 0$ for all x in $(-1, 2)$.
$(2, \infty)$	3	$f(3) = 3^3 + 3^2 - 4 \cdot 3 - 4$ $= 20$, positive	$f(x) > 0$ for all x in $(2, \infty)$.

Step 5 Write the solution set, selecting the interval or intervals that satisfy the given inequality. We are interested in solving $f(x) \le 0$, where $f(x) = x^3 + x^2 - 4x - 4$. Based on our work in step 4, we see that $f(x) < 0$ for all x in $(-\infty, -2)$ or $(-1, 2)$. However, because the inequality involves \le (less than or *equal to*), we must also include the solutions of $x^3 + x^2 - 4x - 4 = 0$, namely -2, -1, and 2, in the solution set. Thus, the solution set of the given inequality, $x^3 + x^2 \le 4x + 4$, or, equivalently, $x^3 + x^2 - 4x - 4 \le 0$, is

$$(-\infty, -2] \cup [-1, 2]$$

$$\text{or} \quad \{x \mid x \le -2 \text{ or } -1 \le x \le 2\}.$$

The graph of the solution set on a number line is shown as follows:

Check Point 3 Solve and graph the solution set on a real number line: $x^3 + 3x^2 \le x + 3$. $\quad \{x \mid x \le -3 \text{ or } -1 \le x \le 1\} \text{ or } (-\infty, -3] \cup [-1, 1]^*$

Solving Rational Inequalities

A **rational inequality** is any inequality that can be put into one of the forms
$$f(x) < 0, \quad f(x) > 0, \quad f(x) \le 0, \quad \text{or} \quad f(x) \ge 0,$$
where f is a rational function. An example of a rational inequality is

$$\frac{3x + 3}{2x + 4} > 0.$$

This inequality is in the form $f(x) > 0$, where f is the rational function given by

$$f(x) = \frac{3x + 3}{2x + 4}.$$

The graph of f is shown in **Figure 3.45**.

We can find the x-intercept of f by setting the numerator equal to 0:

$$3x + 3 = 0$$
$$3x = -3$$
$$x = -1. \quad \boxed{\text{f has an x-intercept at -1 and passes through $(-1, 0)$.}}$$

② Solve rational inequalities.

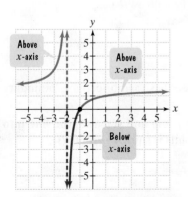

FIGURE 3.45 The graph of $f(x) = \dfrac{3x + 3}{2x + 4}$

*See Answers to Selected Exercises.

We can determine where f is undefined by setting the denominator equal to 0:

$$2x + 4 = 0$$
$$2x = -4$$
$$x = -2.$$

> f is undefined at -2.
> **Figure 3.45** shows that the function's vertical asymptote is $x = -2$.

By setting both the numerator and the denominator of f equal to 0, we obtained -2 and -1. These numbers separate the x-axis into three intervals: $(-\infty, -2)$, $(-2, -1)$, and $(-1, \infty)$. On each interval, the graph of f is either above the x-axis $[f(x) > 0]$ or below the x-axis $[f(x) < 0]$.

Examine the graph in **Figure 3.45** carefully. Can you see that it is above the x-axis for all x in $(-\infty, -2)$ or $(-1, \infty)$, shown in blue? Thus, the solution set of $\dfrac{3x + 3}{2x + 4} > 0$ is $(-\infty, -2) \cup (-1, \infty)$. By contrast, the graph of f lies below the x-axis for all x in $(-2, -1)$, shown in red. Thus, the solution set of $\dfrac{3x + 3}{2x + 4} < 0$ is $(-2, -1)$.

The first step in solving a rational inequality is to bring all terms to one side, obtaining zero on the other side. Then express the rational function on the nonzero side as a single quotient. The second step is to set the numerator and the denominator of the rational function f equal to zero. The solutions of these equations serve as boundary points that separate the real number line into intervals. At this point, the procedure is the same as the one we used for solving polynomial inequalities.

GREAT QUESTION!

Can I begin solving

$$\frac{x + 1}{x + 3} \geq 2$$

by multiplying both sides by $x + 3$?

No. We do not know if $x + 3$ is positive or negative. Thus, we do not know whether or not to change the sense of the inequality.

EXAMPLE 4 Solving a Rational Inequality

Solve and graph the solution set: $\dfrac{x + 1}{x + 3} \geq 2$.

SOLUTION

Step 1 Express the inequality so that one side is zero and the other side is a single quotient. We subtract 2 from both sides to obtain zero on the right.

$$\frac{x + 1}{x + 3} \geq 2 \qquad \text{This is the given inequality.}$$

$$\frac{x + 1}{x + 3} - 2 \geq 0 \qquad \begin{array}{l}\text{Subtract 2 from both sides, obtaining O}\\\text{on the right.}\end{array}$$

$$\frac{x + 1}{x + 3} - \frac{2(x + 3)}{x + 3} \geq 0 \qquad \begin{array}{l}\text{The least common denominator is } x + 3.\\\text{Express 2 in terms of this denominator.}\end{array}$$

$$\frac{x + 1 - 2(x + 3)}{x + 3} \geq 0 \qquad \text{Subtract rational expressions.}$$

$$\frac{x + 1 - 2x - 6}{x + 3} \geq 0 \qquad \text{Apply the distributive property.}$$

$$\frac{-x - 5}{x + 3} \geq 0 \qquad \text{Simplify.}$$

This inequality is equivalent to the one we wish to solve. It is in the form $f(x) \geq 0$, where $f(x) = \dfrac{-x - 5}{x + 3}$.

Step 2 Set the numerator and the denominator of f equal to zero. The real solutions are the boundary points.

$$-x - 5 = 0 \qquad x + 3 = 0 \qquad \begin{array}{l}\text{Set the numerator and denominator equal}\\\text{to O. These are the values that make the}\\\text{previous quotient zero or undefined.}\end{array}$$

$$x = -5 \qquad x = -3 \qquad \text{Solve for } x.$$

We will use these solutions as boundary points on a number line.

Step 3 **Locate the boundary points on a number line and separate the line into intervals.** The number line with the boundary points is shown as follows:

The boundary points divide the number line into three intervals:

$$(-\infty, -5) \quad (-5, -3) \quad (-3, \infty).$$

Step 4 **Choose one test value within each interval and evaluate f at that number.**

Interval	Test Value	Substitute into $f(x) = \dfrac{-x-5}{x+3}$	Conclusion
$(-\infty, -5)$	-6	$f(-6) = \dfrac{-(-6)-5}{-6+3}$ $= -\frac{1}{3}$, negative	$f(x) < 0$ for all x in $(-\infty, -5)$.
$(-5, -3)$	-4	$f(-4) = \dfrac{-(-4)-5}{-4+3}$ $= 1$, positive	$f(x) > 0$ for all x in $(-5, -3)$.
$(-3, \infty)$	0	$f(0) = \dfrac{-0-5}{0+3}$ $= -\frac{5}{3}$, negative	$f(x) < 0$ for all x in $(-3, \infty)$.

Step 5 **Write the solution set, selecting the interval or intervals that satisfy the given inequality.** We are interested in solving $f(x) \geq 0$, where $f(x) = \dfrac{-x-5}{x+3}$. Based on our work in step 4, we see that $f(x) > 0$ for all x in $(-5, -3)$. However, because the inequality involves \geq (greater than or *equal to*), we must also include the solution of $f(x) = 0$, namely, the value that we obtained when we set the numerator of f equal to zero. Thus, we must include -5 in the solution set. The solution set of the given inequality is

$$[-5, -3) \quad \text{or} \quad \{x \mid -5 \leq x < -3\}.$$

> -3 causes the denominator of f to equal zero. It must be excluded from the solution set.

The graph of the solution set on a number line is shown as follows:

 Check Point **4** Solve and graph the solution set: $\dfrac{2x}{x+1} \geq 1$.

$\{x \mid x < -1 \text{ or } x \geq 1\} \text{ or } (-\infty, -1) \cup [1, \infty)$*

Applications

If you throw an object directly upward, although its path is straight and vertical, its changing height over time can be described by a quadratic function.

The Position Function for a Free-Falling Object Near Earth's Surface

An object that is falling or vertically projected into the air has its height above the ground, $s(t)$, in feet, given by

$$s(t) = -16t^2 + v_0 t + s_0,$$

where v_0 is the original velocity (initial velocity) of the object, in feet per second, t is the time that the object is in motion, in seconds, and s_0 is the original height (initial height) of the object, in feet.

GREAT QUESTION!

Which boundary points must I always exclude from the solution set of a rational inequality?

Never include values that cause a rational function's denominator to equal zero. Division by zero is undefined.

TECHNOLOGY

Graphic Connections

The solution set for

$$\frac{x+1}{x+3} \geq 2$$

or, equivalently,

$$\frac{-x-5}{x+3} \geq 0$$

can be verified with a graphing utility. The graph of $f(x) = \dfrac{-x-5}{x+3}$, shown in blue, lies on or above the x-axis, representing \geq, for all x in $[-5, -3)$. We also graphed the vertical asymptote, $x = -3$, to remind us that f is undefined for $x = -3$.

Graph lies on or above the x-axis.

$x = -3$

$x = -5$

$[-8, 8, 1]$ by $[-3, 3, 1]$

3 Solve problems modeled by polynomial or rational inequalities.

In Example 5, we solve a polynomial inequality in a problem about the position of a free-falling object.

EXAMPLE 5 Using the Position Function

A ball is thrown vertically upward from the top of the Leaning Tower of Pisa (190 feet high) with an initial velocity of 96 feet per second (**Figure 3.46**). During which time period will the ball's height exceed that of the tower?

SOLUTION

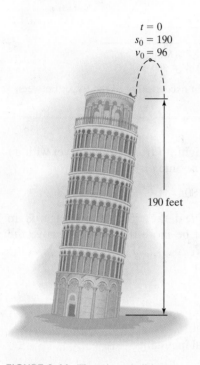

$t = 0$
$s_0 = 190$
$v_0 = 96$

190 feet

FIGURE 3.46 Throwing a ball from 190 feet with a velocity of 96 feet per second

$$s(t) = -16t^2 + v_0 t + s_0 \qquad \text{This is the position function for a free-falling object.}$$

$$s(t) = -16t^2 + 96t + 190 \qquad \text{Because } v_O \text{ (initial velocity)} = 96 \text{ and } s_O \text{ (initial position)} = 190, \text{ substitute these values into the formula.}$$

When will $s(t)$, the ball's height, exceed that of the tower?

$$-16t^2 + 96t + 190 > 190$$

$$-16t^2 + 96t + 190 > 190 \qquad \text{This is the inequality that models the problem's question. We must find } t.$$

$$-16t^2 + 96t > 0 \qquad \text{Subtract 190 from both sides. This inequality is in the form } f(t) > O, \text{ where } f(t) = -16t^2 + 96t.$$

$$-16t^2 + 96t = 0 \qquad \text{Solve the equation } f(t) = O.$$

$$-16t(t - 6) = 0 \qquad \text{Factor.}$$

$$-16t = 0 \quad \text{or} \quad t - 6 = 0 \qquad \text{Set each factor equal to O.}$$

$$t = 0 \qquad\qquad t = 6 \qquad \text{Solve for } t. \text{ The boundary points are O and 6. Locate these values on a number line.}$$

The intervals are $(-\infty, 0)$, $(0, 6)$, and $(6, \infty)$. For our purposes, the mathematical model is useful only from $t = 0$ until the ball hits the ground. (By setting $-16t^2 + 96t + 190$ equal to zero, we find $t \approx 7.57$; the ball hits the ground after approximately 7.57 seconds.) Thus, we use $(0, 6)$ and $(6, 7.57)$ for our intervals.

Interval	Test Value	Substitute into $f(t) = -16t^2 + 96t$	Conclusion
$(0, 6)$	1	$f(1) = -16 \cdot 1^2 + 96 \cdot 1$ $= 80$, positive	$f(t) > 0$ for all t in $(0, 6)$.
$(6, 7.57)$	7	$f(7) = -16 \cdot 7^2 + 96 \cdot 7$ $= -112$, negative	$f(t) < 0$ for all t in $(6, 7.57)$.

We are interested in solving $f(t) > 0$, where $f(t) = -16t^2 + 96t$. We see that $f(t) > 0$ for all t in $(0, 6)$. This means that the ball's height exceeds that of the tower between 0 and 6 seconds.

• • •

TECHNOLOGY

Graphic Connections

The graphs of

$$y_1 = -16x^2 + 96x + 190$$

and

$$y_2 = 190$$

are shown in a

[0, 8, 1] by [0, 360, 36]

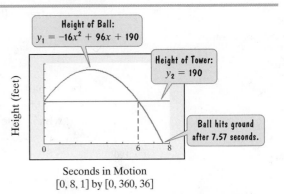

viewing rectangle. The graphs show that the ball's height exceeds that of the tower between 0 and 6 seconds.

 Check Point **5** An object is propelled straight up from ground level with an initial velocity of 80 feet per second. Its height at time t is modeled by

$$s(t) = -16t^2 + 80t,$$

where the height, $s(t)$, is measured in feet and the time, t, is measured in seconds. In which time interval will the object be more than 64 feet above the ground? between 1 and 4 seconds, excluding $t = 1$ and $t = 4$

CONCEPT AND VOCABULARY CHECK

Fill in each blank so that the resulting statement is true.

1. We solve the polynomial inequality $x^2 + 8x + 15 > 0$ by first solving the equation __$x^2 + 8x + 15 = 0$__. The real solutions of this equation, -5 and -3, shown on the number line, are called ___boundary___ points.

2. The points at -5 and -3 shown in Exercise 1 divide the number line into three intervals:
___$(-\infty, -5)$___, ___$(-5, -3)$___, ___$(-3, \infty)$___.

3. True or false: A test value for the leftmost interval on the number line shown in Exercise 1 could be -10. ___true___

4. True or false: A test value for the rightmost interval on the number line shown in Exercise 1 could be 0. ___true___

5. Consider the rational inequality

$$\frac{x-1}{x+2} \geq 0.$$

Setting the numerator and the denominator of $\dfrac{x-1}{x+2}$ equal to zero, we obtain $x = 1$ and $x = -2$. These values are shown as points on the number line. Also shown is information about three test values.

Based on the information shown above, the solution set of $\dfrac{x-1}{x+2} \geq 0$ is ___$(-\infty, -2) \cup [1, \infty)$___.

EXERCISE SET 3.6

Practice Exercises

Solve each polynomial inequality in Exercises 1–42 and graph the solution set on a real number line. Express each solution set in interval notation.

1. $(x - 4)(x + 2) > 0$ *
2. $(x + 3)(x - 5) > 0$ *
3. $(x - 7)(x + 3) \leq 0$ $[-3, 7]$*
4. $(x + 1)(x - 7) \leq 0$ $[-1, 7]$*
5. $x^2 - 5x + 4 > 0$ *
6. $x^2 - 4x + 3 < 0$ $(1, 3)$*
7. $x^2 + 5x + 4 > 0$ *
8. $x^2 + x - 6 > 0$ *
9. $x^2 - 6x + 9 < 0$ ∅*
10. $x^2 - 2x + 1 > 0$ *
11. $3x^2 + 10x - 8 \leq 0$ $\left[-4, \frac{2}{3}\right]$*
12. $9x^2 + 3x - 2 \geq 0$ *
13. $2x^2 + x < 15$ $\left(-3, \frac{5}{2}\right)$*
14. $6x^2 + x > 1$ *
15. $4x^2 + 7x < -3$ $\left(-1, -\frac{3}{4}\right)$*
16. $3x^2 + 16x < -5$ $\left(-5, -\frac{1}{3}\right)$*

*See Answers to Selected Exercises.

17. $5x \leq 2 - 3x^2$ $[-2, \frac{1}{3}]$*
18. $4x^2 + 1 \geq 4x$ $(-\infty, \infty)$*
19. $x^2 - 4x \geq 0$ *
20. $x^2 + 2x < 0$ $(-2, 0)$*
21. $2x^2 + 3x > 0$ *
22. $3x^2 - 5x \leq 0$ $[0, \frac{5}{3}]$*
23. $-x^2 + x \geq 0$ $[0, 1]$*
24. $-x^2 + 2x \geq 0$ $[0, 2]$*
25. $x^2 \leq 4x - 2$ *
26. $x^2 \leq 2x + 2$ *
27. $9x^2 - 6x + 1 < 0$ \varnothing*
28. $4x^2 - 4x + 1 \geq 0$ *
29. $(x - 1)(x - 2)(x - 3) \geq 0$ $[1, 2] \cup [3, \infty)$*
30. $(x + 1)(x + 2)(x + 3) \geq 0$ $[-3, -2] \cup [-1, \infty)$*
31. $x(3 - x)(x - 5) \leq 0$ *
32. $x(4 - x)(x - 6) \leq 0$ *
33. $(2 - x)^2 \left(x - \dfrac{7}{2} \right) < 0$ *
34. $(5 - x)^2 \left(x - \dfrac{13}{2} \right) < 0$ *
35. $x^3 + 2x^2 - x - 2 \geq 0$ *
36. $x^3 + 2x^2 - 4x - 8 \geq 0$ *
37. $x^3 - 3x^2 - 9x + 27 < 0$ *
38. $x^3 + 7x^2 - x - 7 < 0$ *
39. $x^3 + x^2 + 4x + 4 > 0$ *
40. $x^3 - x^2 + 9x - 9 > 0$ *
41. $x^3 \geq 9x^2$ $\{0\} \cup [9, \infty)$*
42. $x^3 \leq 4x^2$ $(-\infty, 4]$*

Solve each rational inequality in Exercises 43–60 and graph the solution set on a real number line. Express each solution set in interval notation.

43. $\dfrac{x - 4}{x + 3} > 0$ $(-\infty, -3) \cup (4, \infty)$*
44. $\dfrac{x + 5}{x - 2} > 0$ *
45. $\dfrac{x + 3}{x + 4} < 0$ $(-4, -3)$*
46. $\dfrac{x + 5}{x + 2} < 0$ $(-5, -2)$*
47. $\dfrac{-x + 2}{x - 4} \geq 0$ $[2, 4)$*
48. $\dfrac{-x - 3}{x + 2} \leq 0$ *
49. $\dfrac{4 - 2x}{3x + 4} \leq 0$ *
50. $\dfrac{3x + 5}{6 - 2x} \geq 0$ $[-\frac{5}{3}, 3)$*
51. $\dfrac{x}{x - 3} > 0$ $(-\infty, 0) \cup (3, \infty)$*
52. $\dfrac{x + 4}{x} > 0$ *
53. $\dfrac{(x + 4)(x - 1)}{x + 2} \leq 0$ *
54. $\dfrac{(x + 3)(x - 2)}{x + 1} \leq 0$ *
55. $\dfrac{x + 1}{x + 3} < 2$ *
56. $\dfrac{x}{x - 1} > 2$ $(1, 2)$*
57. $\dfrac{x + 4}{2x - 1} \leq 3$ *
58. $\dfrac{1}{x - 3} < 1$ *
59. $\dfrac{x - 2}{x + 2} \leq 2$ *
60. $\dfrac{x}{x + 2} \geq 2$ $[-4, -2)$*

Practice Plus

In Exercises 61–64, find the domain of each function.

61. $f(x) = \sqrt{2x^2 - 5x + 2}$ *
62. $f(x) = \dfrac{1}{\sqrt{4x^2 - 9x + 2}}$ *
63. $f(x) = \sqrt{\dfrac{2x}{x + 1} - 1}$ *
64. $f(x) = \sqrt{\dfrac{x}{2x - 1} - 1}$ *

Solve each inequality in Exercises 65–70 and graph the solution set on a real number line.

65. $|x^2 + 2x - 36| > 12$ *
66. $|x^2 + 6x + 1| > 8$ *
67. $\dfrac{3}{x + 3} > \dfrac{3}{x - 2}$ $(-3, 2)$*
68. $\dfrac{1}{x + 1} > \dfrac{2}{x - 1}$ *
69. $\dfrac{x^2 - x - 2}{x^2 - 4x + 3} > 0$ *
70. $\dfrac{x^2 - 3x + 2}{x^2 - 2x - 3} > 0$ *

In Exercises 71–72, use the graph of the polynomial function to solve each inequality.

$f(x) = 2x^3 + 11x^2 - 7x - 6$

$[-7, 3, 1]$ by $[-10, 70, 10]$

71. $2x^3 + 11x^2 \geq 7x + 6$ *
72. $2x^3 + 11x^2 < 7x + 6$ *

In Exercises 73–74, use the graph of the rational function to solve each inequality.

$f(x) = \dfrac{x + 1}{x^2 - 4}$

$[-4, 4, 1]$ by $[-4, 4, 1]$

73. $\dfrac{1}{4(x + 2)} \leq -\dfrac{3}{4(x - 2)}$ *
74. $\dfrac{1}{4(x + 2)} > -\dfrac{3}{4(x - 2)}$ *

Application Exercises

Use the position function

$$s(t) = -16t^2 + v_0 t + s_0$$

$(v_0 = $ initial velocity, $s_0 = $ initial position, $t = $ time$)$

to answer Exercises 75–76.

75. Divers in Acapulco, Mexico, dive headfirst at 8 feet per second from the top of a cliff 87 feet above the Pacific Ocean. During which time period will a diver's height exceed that of the cliff? between 0 and $\frac{1}{2}$ second

76. You throw a ball straight up from a rooftop 160 feet high with an initial velocity of 48 feet per second. During which time period will the ball's height exceed that of the rooftop? *

The functions

$$f(x) = 0.0875x^2 - 0.4x + 66.6$$

Dry pavement

and

Wet pavement

$$g(x) = 0.0875x^2 + 1.9x + 11.6$$

model a car's stopping distance, $f(x)$ or $g(x)$, in feet, traveling at x miles per hour. Function f models stopping distance on dry pavement and function g models stopping distance on wet pavement. The graphs of these functions are shown for

$\{x \mid x \geq 30\}$. *Notice that the figure does not specify which graph is the model for dry roads and which is the model for wet roads. Use this information to solve Exercises 77–78.*

Graphs of Models for Stopping Distances

77. a. Use the given functions at the bottom of the previous page to find the stopping distance on dry pavement and the stopping distance on wet pavement for a car traveling at 35 miles per hour. Round to the nearest foot. dry: 160 ft; wet: 185 ft

 b. Based on your answers to part (a), which rectangular coordinate graph shows stopping distances on dry pavement and which shows stopping distances on wet pavement? dry: (b); wet: (a)

 c. How well do your answers to part (a) model the actual stopping distances shown in **Figure 3.43** on page 431? *

 d. Determine speeds on dry pavement requiring stopping distances that exceed the length of one and one-half football fields, or 540 feet. Round to the nearest mile per hour. How is this shown on the appropriate graph of the models? speeds exceeding 76 mph*

78. a. Use the given functions at the bottom of the previous page to find the stopping distance on dry pavement and the stopping distance on wet pavement for a car traveling at 55 miles per hour. Round to the nearest foot. dry: 309 ft; wet: 381 ft

 b. Based on your answers to part (a), which rectangular coordinate graph shows stopping distances on dry pavement and which shows stopping distances on wet pavement? dry: (b); wet: (a)

 c. How well do your answers to part (a) model the actual stopping distances shown in **Figure 3.43** on page 431? *

 d. Determine speeds on wet pavement requiring stopping distances that exceed the length of one and one-half football fields, or 540 feet. Round to the nearest mile per hour. How is this shown on the appropriate graph of the models? speeds exceeding 68 mph*

79. The perimeter of a rectangle is 50 feet. Describe the possible lengths of a side if the area of the rectangle is not to exceed 114 square feet. $(0, 6] \cup [19, 25)$ in feet

80. The perimeter of a rectangle is 180 feet. Describe the possible lengths of a side if the area of the rectangle is not to exceed 800 square feet. $(0, 10] \cup [80, 90)$ in feet

Explaining the Concepts 81–83. Answers will vary.

81. What is a polynomial inequality?

82. What is a rational inequality?

83. If f is a polynomial or rational function, explain how the graph of f can be used to visualize the solution set of the inequality $f(x) < 0$.

Technology Exercises

84. Use a graphing utility to verify your solution sets to any three of the polynomial inequalities that you solved algebraically in Exercises 1–42. Answers will vary.

85. Use a graphing utility to verify your solution sets to any three of the rational inequalities that you solved algebraically in Exercises 43–60. Answers will vary.

Solve each inequality in Exercises 86–91 using a graphing utility.

86. $x^2 + 3x - 10 > 0$ *

87. $2x^2 + 5x - 3 \leq 0$ *

88. $x^3 + x^2 - 4x - 4 > 0$ *

89. $\dfrac{x - 4}{x - 1} \leq 0$ $(1, 4]$

90. $\dfrac{x + 2}{x - 3} \leq 2$ $(-\infty, 3) \cup [8, \infty)$

91. $\dfrac{1}{x + 1} \leq \dfrac{2}{x + 4}$ *

The graph shows stopping distances for trucks at various speeds on dry roads and on wet roads. Use this information to solve Exercises 92–93.

Stopping Distances for Trucks at Selected Speeds

Source: National Highway Traffic Safety Administration

92. a. Use the statistical menu of your graphing utility and the quadratic regression program to obtain the quadratic function that models a truck's stopping distance, $f(x)$, in feet, on dry pavement traveling at x miles per hour. Round the x-coefficient and the constant term to one decimal place. $f(x) = 0.1125x^2 - 0.1x + 55.9$

 b. Use the function from part (a) to determine speeds on dry pavement requiring stopping distances that exceed 455 feet. Round to the nearest mile per hour. *

93. a. Use the statistical menu of your graphing utility and the quadratic regression program to obtain the quadratic function that models a truck's stopping distance, $f(x)$, in feet, on wet pavement traveling at x miles per hour. Round the x-coefficient and the constant term to one decimal place. $f(x) = 0.1375x^2 + 0.7x + 37.8$

 b. Use the function from part (a) to determine speeds on wet pavement requiring stopping distances that exceed 446 feet. speeds exceeding 52 mph

Critical Thinking Exercises

Make Sense? *In Exercises 94–97, determine whether each statement makes sense or does not make sense, and explain your reasoning.*

94. When solving $f(x) > 0$, where f is a polynomial function, I only pay attention to the sign of f at each test value and not the actual function value. makes sense

95. I'm solving a polynomial inequality that has a value for which the polynomial function is undefined. *

96. Because it takes me longer to come to a stop on a wet road than on a dry road, graph (a) for Exercises 77–78 is the model for stopping distances on wet pavement and graph (b) is the model for stopping distances on dry pavement. *

97. I began the solution of the rational inequality $\frac{x+1}{x+3} \geq 2$ by setting both $x + 1$ and $x + 3$ equal to zero. *

In Exercises 98–101, determine whether each statement is true or false. If the statement is false, make the necessary change(s) to produce a true statement. Changes to false statements will vary.

98. The solution set of $x^2 > 25$ is $(5, \infty)$. false

99. The inequality $\frac{x-2}{x+3} < 2$ can be solved by multiplying both sides by $x + 3$, resulting in the equivalent inequality $x - 2 < 2(x + 3)$. false

100. $(x + 3)(x - 1) \geq 0$ and $\frac{x+3}{x-1} \geq 0$ have the same solution set. false

101. The inequality $\frac{x-2}{x+3} < 2$ can be solved by multiplying both sides by $(x + 3)^2, x \neq -3$, resulting in the equivalent inequality $(x - 2)(x + 3) < 2(x + 3)^2$. true

102. Write a polynomial inequality whose solution set is $[-3, 5]$. sample answer: $x^2 - 2x - 15 \leq 0$

103. Write a rational inequality whose solution set is $(-\infty, -4) \cup [3, \infty)$. sample answer: $\frac{x-3}{x+4} \geq 0$

In Exercises 104–107, use inspection to describe each inequality's solution set. Do not solve any of the inequalities.

104. $(x - 2)^2 > 0$ *

105. $(x - 2)^2 \leq 0$ {2}

106. $(x - 2)^2 < -1$ ∅

107. $\frac{1}{(x-2)^2} > 0$ *

108. The graphing utility screen shows the graph of $y = 4x^2 - 8x + 7$.

[−2, 6, 1] by [−2, 8, 1]

a. Use the graph to describe the solution set of $4x^2 - 8x + 7 > 0$. $(-\infty, \infty)$

b. Use the graph to describe the solution set of $4x^2 - 8x + 7 < 0$. ∅

c. Use an algebraic approach to verify each of your descriptions in parts (a) and (b).

109. The graphing utility screen shows the graph of $y = \sqrt{27 - 3x^2}$. Write and solve a quadratic inequality that explains why the graph only appears for $-3 \leq x \leq 3$. $27 - 3x^2 \geq 0$*

[−5, 5, 1] by [0, 6, 1]

Retaining the Concepts

110. The size of a television screen refers to the length of its diagonal. If the length of an HDTV screen is 28 inches and its width is 15.7 inches, what is the size of the screen, to the nearest inch?

(Section 1.5, Example 11) 32 inches

111. Find the domain of $h(x) = \sqrt{36 - 2x}$.

(Section 2.6, Example 1) $(-\infty, 18]$

112. Use the graph of $y = f(x)$ to graph $y = f(-x) + 3$.

(Section 2.5, Example 8) *

Preview Exercises

Exercises 113–115 will help you prepare for the material covered in the next section.

113. **a.** If $y = kx^2$, find the value of k using $x = 2$ and $y = 64$. *

b. Substitute the value for k into $y = kx^2$ and write the resulting equation. $y = 16x^2$

c. Use the equation from part (b) to find y when $x = 5$. *

114. **a.** If $y = \frac{k}{x}$, find the value of k using $x = 8$ and $y = 12$. *

b. Substitute the value for k into $y = \frac{k}{x}$ and write the resulting equation. $y = \frac{96}{x}$

c. Use the equation from part (b) to find y when $x = 3$. *

115. If $S = \frac{kA}{P}$, find the value of k using $A = 60,000, P = 40$, and $S = 12,000$. 8

Section 3.7 Modeling Using Variation

What am I supposed to learn?

After studying this section, you should be able to:

1. Solve direct variation problems.
2. Solve inverse variation problems.
3. Solve combined variation problems.
4. Solve problems involving joint variation.

Have you ever wondered how telecommunication companies estimate the number of phone calls expected per day between two cities? The formula

$$C = \frac{0.02P_1P_2}{d^2}$$

shows that the daily number of phone calls, C, increases as the populations of the cities, P_1 and P_2, in thousands, increase and decreases as the distance, d, between the cities increases.

Certain formulas occur so frequently in applied situations that they are given special names. Variation formulas show how one quantity changes in relation to other quantities. Quantities can vary *directly*, *inversely*, or *jointly*. In this section, we look at situations that can be modeled by each of these kinds of variation.

1 Solve direct variation problems.

Direct Variation

When you swim underwater, the pressure in your ears depends on the depth at which you are swimming. The formula

$$p = 0.43d$$

describes the water pressure, p, in pounds per square inch, at a depth of d feet. We can use this linear function to determine the pressure in your ears at various depths:

If $d = 20$, $p = 0.43(20) = 8.6$. At a depth of 20 feet, water pressure is 8.6 pounds per square inch.

Doubling the depth doubles the pressure.

If $d = 40$, $p = 0.43(40) = 17.2$. At a depth of 40 feet, water pressure is 17.2 pounds per square inch.

Doubling the depth doubles the pressure.

If $d = 80$, $p = 0.43(80) = 34.4$. At a depth of 80 feet, water pressure is 34.4 pounds per square inch.

The formula $p = 0.43d$ illustrates that water pressure is a constant multiple of your underwater depth. If your depth is doubled, the pressure is doubled; if your depth is tripled, the pressure is tripled; and so on. Because of this, the pressure in your ears is said to **vary directly** as your underwater depth. The **equation of variation** is

$$p = 0.43d.$$

Generalizing our discussion of pressure and depth, we obtain the following statement:

Direct Variation

If a situation is described by an equation in the form

$$y = kx,$$

where k is a nonzero constant, we say that **y varies directly as x** or **y is directly proportional to x**. The number k is called the **constant of variation** or the **constant of proportionality**.

Can you see that **the direct variation equation, $y = kx$, is a special case of the linear function $y = mx + b$?** When $m = k$ and $b = 0, y = mx + b$ becomes $y = kx$. Thus, the slope of a direct variation equation is k, the constant of variation. Because b, the y-intercept, is 0, the graph of a variation equation is a line passing through the origin. This is illustrated in **Figure 3.47**, which shows the graph of $p = 0.43d$: Water pressure varies directly as depth.

FIGURE 3.47 Water pressure at various depths

Problems involving direct variation can be solved using the following procedure. This procedure applies to direct variation problems, as well as to the other kinds of variation problems that we will discuss.

Solving Variation Problems

1. Write an equation that models the given English statement.
2. Substitute the given pair of values into the equation in step 1 and find the value of k, the constant of variation.
3. Substitute the value of k into the equation in step 1.
4. Use the equation from step 3 to answer the problem's question.

EXAMPLE 1 Solving a Direct Variation Problem

The volume of blood, B, in a person's body varies directly as body weight, W. A person who weighs 160 pounds has approximately 5 quarts of blood. Estimate the number of quarts of blood in a person who weighs 200 pounds.

SOLUTION

Step 1 Write an equation. We know that y *varies directly as x* is expressed as

$$y = kx.$$

By changing letters, we can write an equation that models the following English statement: The volume of blood, B, varies directly as body weight, W.

$$B = kW$$

Step 2 Use the given values to find k. A person who weighs 160 pounds has approximately 5 quarts of blood. Substitute 160 for W and 5 for B in the direct variation equation. Then solve for k.

$B = kW$	The volume of blood varies directly as body weight.
$5 = k \cdot 160$	Substitute 160 for W and 5 for B.
$\dfrac{5}{160} = \dfrac{k \cdot 160}{160}$	Divide both sides by 160.
$0.03125 = k$	Express $\frac{5}{160}$, or $\frac{1}{32}$, in decimal form.

Step 3 Substitute the value of k into the equation.

$B = kW$	Use the equation from step 1.
$B = 0.03125W$	Replace k, the constant of variation, with 0.03125.

Step 4 Answer the problem's question. We are interested in estimating the number of quarts of blood in a person who weighs 200 pounds. Substitute 200 for W in $B = 0.03125W$ and solve for B.

$$B = 0.03125W \qquad \text{\small This is the equation from step 3.}$$
$$B = 0.03125(200) \qquad \text{\small Substitute 200 for } W.$$
$$= 6.25 \qquad \text{\small Multiply.}$$

A person who weighs 200 pounds has approximately 6.25 quarts of blood. • • •

⊘ Check Point **1** The number of gallons of water, W, used when taking a shower varies directly as the time, t, in minutes, in the shower. A shower lasting 5 minutes uses 30 gallons of water. How much water is used in a shower lasting 11 minutes? 66 gallons

The direct variation equation $y = kx$ is a linear function. If $k > 0$, then the slope of the line is positive. Consequently, as x increases, y also increases.

A direct variation situation can involve variables to higher powers. For example, y can vary directly as x^2 ($y = kx^2$) or as x^3 ($y = kx^3$).

> **Direct Variation with Powers**
>
> **y varies directly as the nth power of x** if there exists some nonzero constant k such that
>
> $$y = kx^n.$$
>
> We also say that **y is directly proportional to the nth power of x.**

Direct variation with whole number powers is modeled by polynomial functions. In our next example, the graph of the variation equation is the familiar parabola.

EXAMPLE 2 Solving a Direct Variation Problem

The distance, s, that a body falls from rest varies directly as the square of the time, t, of the fall. If skydivers fall 64 feet in 2 seconds, how far will they fall in 4.5 seconds?

SOLUTION

Step 1 Write an equation. We know that y *varies directly as the square of x* is expressed as

$$y = kx^2.$$

By changing letters, we can write an equation that models the following English statement: Distance, s, varies directly as the square of time, t, of the fall.

$$s = kt^2$$

Step 2 Use the given values to find k. Skydivers fall 64 feet in 2 seconds. Substitute 64 for s and 2 for t in the direct variation equation. Then solve for k.

$$s = kt^2 \qquad \text{\small Distance varies directly as the square of time.}$$
$$64 = k \cdot 2^2 \qquad \text{\small Skydivers fall 64 feet in 2 seconds.}$$
$$64 = 4k \qquad \text{\small Simplify: } 2^2 = 4.$$
$$\frac{64}{4} = \frac{4k}{4} \qquad \text{\small Divide both sides by 4.}$$
$$16 = k \qquad \text{\small Simplify.}$$

Distance Fallen by Skydivers over Time

FIGURE 3.48 The graph of $s(t) = 16t^2$

Step 3 Substitute the value of k into the equation.

$s = kt^2$ Use the equation from step 1.

$s = 16t^2$ Replace k, the constant of variation, with 16.

Step 4 Answer the problem's question. How far will the skydivers fall in 4.5 seconds? Substitute 4.5 for t in $s = 16t^2$ and solve for s.

$$s = 16(4.5)^2 = 16(20.25) = 324$$

Thus, in 4.5 seconds, the skydivers will fall 324 feet. •••

We can express the variation equation from Example 2 in function notation, writing

$$s(t) = 16t^2.$$

The distance that a body falls from rest is a function of the time, t, of the fall. The parabola that is the graph of this quadratic function is shown in **Figure 3.48**. The graph increases rapidly from left to right, showing the effects of the acceleration of gravity.

✓ Check Point **2** The weight of a great white shark varies directly as the cube of its length. A great white shark caught off Catalina Island, California, was 15 feet long and weighed 2025 pounds. What was the weight of the 25-foot-long shark in the novel *Jaws*? 9375 pounds

2 Solve inverse variation problems.

FIGURE 3.49

Inverse Variation

The distance from San Francisco to Los Angeles is 420 miles. The time that it takes to drive from San Francisco to Los Angeles depends on the rate at which one drives and is given by

$$\text{Time} = \frac{420}{\text{Rate}}.$$

For example, if you average 30 miles per hour, the time for the drive is

$$\text{Time} = \frac{420}{30} = 14,$$

or 14 hours. If you average 50 miles per hour, the time for the drive is

$$\text{Time} = \frac{420}{50} = 8.4,$$

or 8.4 hours. As your rate (or speed) increases, the time for the trip decreases and vice versa. This is illustrated by the graph in **Figure 3.49**.

We can express the time for the San Francisco–Los Angeles trip using t for time and r for rate:

$$t = \frac{420}{r}.$$

This equation is an example of an **inverse variation** equation. Time, t, **varies inversely** as rate, r. When two quantities vary inversely, one quantity increases as the other decreases and vice versa.

Generalizing, we obtain the following statement:

Inverse Variation

If a situation is described by an equation in the form

$$y = \frac{k}{x},$$

where k is a nonzero constant, we say that **y varies inversely as x** or **y is inversely proportional to x.** The number k is called the **constant of variation**.

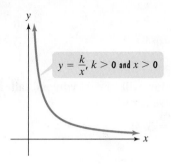

FIGURE 3.50 The graph of the inverse variation equation

Doubling the pressure halves the volume.

Notice that the **inverse variation equation**

$$y = \frac{k}{x}, \quad \text{or} \quad f(x) = \frac{k}{x},$$

is a rational function. For $k > 0$ and $x > 0$, the graph of the function takes on the shape shown in **Figure 3.50**.

We use the same procedure to solve inverse variation problems as we did to solve direct variation problems. Example 3 illustrates this procedure.

EXAMPLE 3 Solving an Inverse Variation Problem

When you use a spray can and press the valve at the top, you decrease the pressure of the gas in the can. This decrease of pressure causes the volume of the gas in the can to increase. Because the gas needs more room than is provided in the can, it expands in spray form through the small hole near the valve. In general, if the temperature is constant, the pressure, P, of a gas in a container varies inversely as the volume, V, of the container. The pressure of a gas sample in a container whose volume is 8 cubic inches is 12 pounds per square inch. If the sample expands to a volume of 22 cubic inches, what is the new pressure of the gas?

SOLUTION

Step 1 Write an equation. We know that y *varies inversely as* x is expressed as

$$y = \frac{k}{x}.$$

By changing letters, we can write an equation that models the following English statement: The pressure, P, of a gas in a container varies inversely as the volume, V.

$$P = \frac{k}{V}$$

Step 2 Use the given values to find k. The pressure of a gas sample in a container whose volume is 8 cubic inches is 12 pounds per square inch. Substitute 12 for P and 8 for V in the inverse variation equation. Then solve for k.

$$P = \frac{k}{V} \qquad \text{Pressure varies inversely as volume.}$$

$$12 = \frac{k}{8} \qquad \text{The pressure in an 8 cubic-inch container is 12 pounds per square inch.}$$

$$12 \cdot 8 = \frac{k}{8} \cdot 8 \qquad \text{Multiply both sides by 8.}$$

$$96 = k \qquad \text{Simplify.}$$

Step 3 Substitute the value of k into the equation.

$$P = \frac{k}{V} \qquad \text{Use the equation from step 1.}$$

$$P = \frac{96}{V} \qquad \text{Replace } k, \text{ the constant of variation, with 96.}$$

Step 4 Answer the problem's question. We need to find the pressure when the volume expands to 22 cubic inches. Substitute 22 for V and solve for P.

$$P = \frac{96}{V} = \frac{96}{22} = \frac{48}{11} = 4\frac{4}{11}$$

When the volume is 22 cubic inches, the pressure of the gas is $4\frac{4}{11}$ pounds per square inch.

● ● ●

 Check Point **3** The length of a violin string varies inversely as the frequency of its vibrations. A violin string 8 inches long vibrates at a frequency of 640 cycles per second. What is the frequency of a 10-inch string? 512 cycles per second

③ Solve combined variation problems.

Combined Variation

In **combined variation**, direct variation and inverse variation occur at the same time. For example, as the advertising budget, A, of a company increases, its monthly sales, S, also increase. Monthly sales vary directly as the advertising budget:

$$S = kA.$$

By contrast, as the price of the company's product, P, increases, its monthly sales, S, decrease. Monthly sales vary inversely as the price of the product:

$$S = \frac{k}{P}.$$

We can combine these two variation equations into one combined equation:

$$S = \frac{kA}{P}.$$

> Monthly sales, S, vary directly as the advertising budget, A, and inversely as the price of the product, P.

The following example illustrates an application of combined variation.

EXAMPLE 4 Solving a Combined Variation Problem

The owners of Rollerblades Plus determine that the monthly sales, S, of their skates vary directly as their advertising budget, A, and inversely as the price of the skates, P. When \$60,000 is spent on advertising and the price of the skates is \$40, the monthly sales are 12,000 pairs of rollerblades.

a. Write an equation of variation that describes this situation.

b. Determine monthly sales if the amount of the advertising budget is increased to \$70,000.

SOLUTION

a. Write an equation.

$$S = \frac{kA}{P}$$

> Translate "sales vary directly as the advertising budget and inversely as the skates' price."

Use the given values to find k.

$$12{,}000 = \frac{k(60{,}000)}{40} \qquad \text{When \$60,000 is spent on advertising} \\ (A = 60{,}000) \text{ and the price is \$40} \\ (P = 40), \text{ monthly sales are 12,000 units} \\ (S = 12{,}000).$$

$$12{,}000 = k \cdot 1500 \qquad \text{Divide 60,000 by 40.}$$

$$\frac{12{,}000}{1500} = \frac{k \cdot 1500}{1500} \qquad \text{Divide both sides of the equation by 1500.}$$

$$8 = k \qquad \text{Simplify.}$$

Therefore, the equation of variation that models monthly sales is

$$S = \frac{8A}{P}. \qquad \text{Substitute 8 for } k \text{ in } S = \frac{kA}{P}.$$

b. The advertising budget is increased to $70,000, so $A = 70,000$. The skates' price is still $40, so $P = 40$.

$$S = \frac{8A}{P}$$ This is the equation from part (a).

$$S = \frac{8(70,000)}{40}$$ Substitute 70,000 for A and 40 for P.

$$S = 14,000$$ Simplify.

With a $70,000 advertising budget and $40 price, the company can expect to sell 14,000 pairs of rollerblades in a month (up from 12,000). •••

✓ Check Point **4** The number of minutes needed to solve an Exercise Set of variation problems varies directly as the number of problems and inversely as the number of people working to solve the problems. It takes 4 people 32 minutes to solve 16 problems. How many minutes will it take 8 people to solve 24 problems? 24 minutes

4 Solve problems involving joint variation.

Joint Variation

Joint variation is a variation in which a variable varies directly as the product of two or more other variables. Thus, the equation $y = kxz$ is read "y varies jointly as x and z."

Joint variation plays a critical role in Isaac Newton's formula for gravitation:

$$F = G\frac{m_1 m_2}{d^2}.$$

The formula states that the force of gravitation, F, between two bodies varies jointly as the product of their masses, m_1 and m_2, and inversely as the square of the distance between them, d. (G is the gravitational constant.) The formula indicates that gravitational force exists between any two objects in the universe, increasing as the distance between the bodies decreases. One practical result is that the pull of the Moon on the oceans is greater on the side of Earth closer to the Moon. This gravitational imbalance is what produces tides.

EXAMPLE 5 Modeling Centrifugal Force

The centrifugal force, C, of a body moving in a circle varies jointly with the radius of the circular path, r, and the body's mass, m, and inversely with the square of the time, t, it takes to move about one full circle. A 6-gram body moving in a circle with radius 100 centimeters at a rate of 1 revolution in 2 seconds has a centrifugal force of 6000 dynes. Find the centrifugal force of an 18-gram body moving in a circle with radius 100 centimeters at a rate of 1 revolution in 3 seconds.

SOLUTION

$$C = \frac{krm}{t^2}$$ Translate "Centrifugal force, C, varies jointly with radius, r, and mass, m, and inversely with the square of time, t."

$$6000 = \frac{k(100)(6)}{2^2}$$ A 6-gram body ($m = 6$) moving in a circle with radius 100 centimeters ($r = 100$) at 1 revolution in 2 seconds ($t = 2$) has a centrifugal force of 6000 dynes ($C = 6000$).

$$6000 = 150k$$ Simplify.

$$40 = k$$ Divide both sides by 150 and solve for k.

$$C = \frac{40rm}{t^2}$$ Substitute 40 for k in the model for centrifugal force.

$$C = \frac{40(100)(18)}{3^2}$$ Find centrifugal force, C, of an 18-gram body ($m = 18$) moving in a circle with radius 100 centimeters ($r = 100$) at 1 revolution in 3 seconds ($t = 3$).

$$= 8000$$ Simplify.

The centrifugal force is 8000 dynes. •••

✓ Check Point **5** The volume of a cone, V, varies jointly as its height, h, and the square of its radius, r. A cone with a radius measuring 6 feet and a height measuring 10 feet has a volume of 120π cubic feet. Find the volume of a cone having a radius of 12 feet and a height of 2 feet. 96π cubic feet

CONCEPT AND VOCABULARY CHECK

Fill in each blank so that the resulting statement is true.

1. y varies directly as x can be modeled by the equation __$y = kx$__, where k is called the __constant of variation__.

2. y varies directly as the nth power of x can be modeled by the equation __$y = kx^n$__.

3. y varies inversely as x can be modeled by the equation __$y = \frac{k}{x}$__.

4. y varies directly as x and inversely as z can be modeled by the equation __$y = \frac{kx}{z}$__.

5. y varies jointly as x and z can be modeled by the equation __$y = kxz$__.

6. In the equation $S = \frac{8A}{P}$, S varies __directly__ as A and __inversely__ as P.

7. In the equation $C = \frac{0.02P_1P_2}{d^2}$, C varies __jointly__ as P_1 and P_2 and __inversely__ as the square of d.

EXERCISE SET 3.7

Practice Exercises

Use the four-step procedure for solving variation problems given on page 445 to solve Exercises 1–10.

1. y varies directly as x. $y = 65$ when $x = 5$. Find y when $x = 12$. 156

2. y varies directly as x. $y = 45$ when $x = 5$. Find y when $x = 13$. 117

3. y varies inversely as x. $y = 12$ when $x = 5$. Find y when $x = 2$. 30

4. y varies inversely as x. $y = 6$ when $x = 3$. Find y when $x = 9$. 2

5. y varies directly as x and inversely as the square of z. $y = 20$ when $x = 50$ and $z = 5$. Find y when $x = 3$ and $z = 6$. $\frac{5}{6}$

6. a varies directly as b and inversely as the square of c. $a = 7$ when $b = 9$ and $c = 6$. Find a when $b = 4$ and $c = 8$. $\frac{7}{4}$

7. y varies jointly as x and z. $y = 25$ when $x = 2$ and $z = 5$. Find y when $x = 8$ and $z = 12$. 240

8. C varies jointly as A and T. $C = 175$ when $A = 2100$ and $T = 4$. Find C when $A = 2400$ and $T = 6$. 300

9. y varies jointly as a and b and inversely as the square root of c. $y = 12$ when $a = 3, b = 2$, and $c = 25$. Find y when $a = 5, b = 3$, and $c = 9$. 50

10. y varies jointly as m and the square of n and inversely as p. $y = 15$ when $m = 2, n = 1$, and $p = 6$. Find y when $m = 3, n = 4$, and $p = 10$. 216

Practice Plus

In Exercises 11–20, write an equation that expresses each relationship. Then solve the equation for y.

11. x varies jointly as y and z. $x = kyz; y = \frac{x}{kz}$

12. x varies jointly as y and the square of z. $x = kyz^2; y = \frac{x}{kz^2}$

13. x varies directly as the cube of z and inversely as y. *

14. x varies directly as the cube root of z and inversely as y. *

15. x varies jointly as y and z and inversely as the square root of w. *

16. x varies jointly as y and z and inversely as the square of w. *

17. x varies jointly as z and the sum of y and w. *

18. x varies jointly as z and the difference between y and w. *

19. x varies directly as z and inversely as the difference between y and w. *

20. x varies directly as z and inversely as the sum of y and w. *

*See Answers to Selected Exercises.

Application Exercises

Use the four-step procedure for solving variation problems given on page 445 to solve Exercises 21–36.

21. An alligator's tail length, *T*, varies directly as its body length, *B*. An alligator with a body length of 4 feet has a tail length of 3.6 feet. What is the tail length of an alligator whose body length is 6 feet? 5.4 ft

|◄——— Body length, *B* ———►|◄——— Tail length, *T* ———►|

22. An object's weight on the Moon, *M*, varies directly as its weight on Earth, *E*. Neil Armstrong, the first person to step on the Moon on July 20, 1969, weighed 360 pounds on Earth (with all of his equipment on) and 60 pounds on the Moon. What is the Moon weight of a person who weighs 186 pounds on Earth? 31 lb

23. The height that a ball bounces varies directly as the height from which it was dropped. A tennis ball dropped from 12 inches bounces 8.4 inches. From what height was the tennis ball dropped if it bounces 56 inches? 80 in.

24. The distance that a spring will stretch varies directly as the force applied to the spring. A force of 12 pounds is needed to stretch a spring 9 inches. What force is required to stretch the spring 15 inches? 20 lb

25. If all men had identical body types, their weight would vary directly as the cube of their height. Shown below is Robert Wadlow, who reached a record height of 8 feet 11 inches (107 inches) before his death at age 22. If a man who is 5 feet 10 inches tall (70 inches) with the same body type as Mr. Wadlow weighs 170 pounds, what was Robert Wadlow's weight shortly before his death? about 607 lb

26. The number of houses that can be served by a water pipe varies directly as the square of the diameter of the pipe. A water pipe that has a 10-centimeter diameter can supply 50 houses.

 a. How many houses can be served by a water pipe that has a 30-centimeter diameter? 450 houses

 b. What size water pipe is needed for a new subdivision of 1250 houses? 50 cm

27. The figure shows that a bicyclist tips the cycle when making a turn. The angle *B*, formed by the vertical direction and the bicycle, is called the banking angle. The banking angle varies inversely as the cycle's turning radius. When the turning radius is 4 feet, the banking angle is 28°. What is the banking angle when the turning radius is 3.5 feet? 32°

28. The water temperature of the Pacific Ocean varies inversely as the water's depth. At a depth of 1000 meters, the water temperature is 4.4° Celsius. What is the water temperature at a depth of 5000 meters? 0.88°C

29. Radiation machines, used to treat tumors, produce an intensity of radiation that varies inversely as the square of the distance from the machine. At 3 meters, the radiation intensity is 62.5 milliroentgens per hour. What is the intensity at a distance of 2.5 meters? 90 milliroentgens per hour

30. The illumination provided by a car's headlight varies inversely as the square of the distance from the headlight. A car's headlight produces an illumination of 3.75 foot-candles at a distance of 40 feet. What is the illumination when the distance is 50 feet? 2.4 foot-candles

31. Body-mass index, or BMI, takes both weight and height into account when assessing whether an individual is underweight or overweight. BMI varies directly as one's weight, in pounds, and inversely as the square of one's height, in inches. In adults, normal values for the BMI are between 20 and 25, inclusive. Values below 20 indicate that an individual is underweight and values above 30 indicate that an individual is obese. A person who weighs 180 pounds and is 5 feet, or 60 inches, tall has a BMI of 35.15. What is the BMI, to the nearest tenth, for a 170-pound person who is 5 feet 10 inches tall? Is this person overweight? 24.4; not overweight

32. One's intelligence quotient, or IQ, varies directly as a person's mental age and inversely as that person's chronological age. A person with a mental age of 25 and a chronological age of 20 has an IQ of 125. What is the chronological age of a person with a mental age of 40 and an IQ of 80? 50

33. The heat loss of a glass window varies jointly as the window's area and the difference between the outside and inside temperatures. A window 3 feet wide by 6 feet long loses 1200 Btu per hour when the temperature outside is 20° colder than the temperature inside. Find the heat loss through a glass window that is 6 feet wide by 9 feet long when the temperature outside is 10° colder than the temperature inside. 1800 Btu per hour

34. Kinetic energy varies jointly as the mass and the square of the velocity. A mass of 8 grams and velocity of 3 centimeters per second has a kinetic energy of 36 ergs. Find the kinetic energy for a mass of 4 grams and velocity of 6 centimeters per second. 72 ergs

35. Sound intensity varies inversely as the square of the distance from the sound source. If you are in a movie theater and you change your seat to one that is twice as far from the speakers, how does the new sound intensity compare to that of your original seat? $\frac{1}{4}$ of what it was originally

36. Many people claim that as they get older, time seems to pass more quickly. Suppose that the perceived length of a period of time is inversely proportional to your age. How long will a year seem to be when you are three times as old as you are now? *

37. The average number of daily phone calls, C, between two cities varies jointly as the product of their populations, P_1 and P_2, and inversely as the square of the distance, d, between them.

 a. Write an equation that expresses this relationship. *

 b. The distance between San Francisco (population: 777,000) and Los Angeles (population: 3,695,000) is 420 miles. If the average number of daily phone calls between the cities is 326,000, find the value of k to two decimal places and write the equation of variation. $k \approx 0.02$; $C = \frac{0.02\,P_1 P_2}{d^2}$

 c. Memphis (population: 650,000) is 400 miles from New Orleans (population: 490,000). Find the average number of daily phone calls, to the nearest whole number, between these cities. 39,813 daily phone calls

38. The force of wind blowing on a window positioned at a right angle to the direction of the wind varies jointly as the area of the window and the square of the wind's speed. It is known that a wind of 30 miles per hour blowing on a window measuring 4 feet by 5 feet exerts a force of 150 pounds. During a storm with winds of 60 miles per hour, should hurricane shutters be placed on a window that measures 3 feet by 4 feet and is capable of withstanding 300 pounds of force? yes*

39. The table shows the values for the current, I, in an electric circuit and the resistance, R, of the circuit.

I (amperes)	0.5	1.0	1.5	2.0	2.5	3.0	4.0	5.0
R (ohms)	12.0	6.0	4.0	3.0	2.4	2.0	1.5	1.2

 a. Graph the ordered pairs in the table of values, with values of I along the x-axis and values of R along the y-axis. Connect the eight points with a smooth curve. *

 b. Does current vary directly or inversely as resistance? Use your graph and explain how you arrived at your answer. *

 c. Write an equation of variation for I and R, using one of the ordered pairs in the table to find the constant of variation. Then use your variation equation to verify the other seven ordered pairs in the table. $R = \frac{6}{I}$

Explaining the Concepts 40–47. Answers will vary.

40. What does it mean if two quantities vary directly?

41. In your own words, explain how to solve a variation problem.

42. What does it mean if two quantities vary inversely?

43. Explain what is meant by combined variation. Give an example with your explanation.

44. Explain what is meant by joint variation. Give an example with your explanation.

In Exercises 45–46, describe in words the variation shown by the given equation.

45. $z = \dfrac{k\sqrt{x}}{y^2}$ **46.** $z = kx^2\sqrt{y}$

47. We have seen that the daily number of phone calls between two cities varies jointly as their populations and inversely as the square of the distance between them. This model, used by telecommunication companies to estimate the line capacities needed among various cities, is called the *gravity model*. Compare the model to Newton's formula for gravitation on page 450 and describe why the name *gravity model* is appropriate.

Technology Exercise

48. Use a graphing utility to graph any three of the variation equations in Exercises 21–30. Then TRACE along each curve and identify the point that corresponds to the problem's solution. Answers will vary.

Critical Thinking Exercises

Make Sense? *In Exercises 49–52, determine whether each statement makes sense or does not make sense, and explain your reasoning.*

49. I'm using an inverse variation equation and I need to determine the value of the dependent variable when the independent variable is zero. does not make sense

50. The graph of this direct variation equation that has a positive constant of variation shows one variable increasing as the other variable decreases. does not make sense

51. When all is said and done, it seems to me that direct variation equations are special kinds of linear functions and inverse variation equations are special kinds of rational functions. *

52. Using the language of variation, I can now state the formula for the area of a trapezoid, $A = \frac{1}{2}h(b_1 + b_2)$, as, "A trapezoid's area varies jointly with its height and the sum of its bases." makes sense

53. In a hurricane, the wind pressure varies directly as the square of the wind velocity. If wind pressure is a measure of a hurricane's destructive capacity, what happens to this destructive power when the wind speed doubles? 4 times greater

54. The illumination from a light source varies inversely as the square of the distance from the light source. If you raise a lamp from 15 inches to 30 inches over your desk, what happens to the illumination? $\frac{1}{4}$ as much

55. The heat generated by a stove element varies directly as the square of the voltage and inversely as the resistance. If the voltage remains constant, what needs to be done to triple the amount of heat generated? *

56. Galileo's telescope brought about revolutionary changes in astronomy. A comparable leap in our ability to observe the universe took place as a result of the Hubble Space Telescope. The space telescope was able to see stars and galaxies whose brightness is $\frac{1}{50}$ of the faintest objects observable using ground-based telescopes. Use the fact that the brightness of a point source, such as a star, varies inversely as the square of its distance from an observer to show that the space telescope was able to see about seven times farther than a ground-based telescope. Answers will vary.

Group Exercise

57. Begin by deciding on a product that interests the group because you are now in charge of advertising this product. Members were told that the demand for the product varies directly as the amount spent on advertising and inversely as the price of the product. However, as more money is spent on advertising, the price of your product rises. Under what conditions would members recommend an increased expense in advertising? Once you've determined what your product is, write formulas for the given conditions and experiment with hypothetical numbers. What other factors might you take into consideration in terms of your recommendation? How do these factors affect the demand for your product?

Retaining the Concepts

58. You invested $20,000 in two accounts paying 7% and 9% annual interest. If the total interest earned for the year is $1550, how much was invested at each rate? (Section 1.3, Example 5) $12,500 at 7%; $7500 at 9%

59. Solve: $\sqrt{x} + \sqrt{x-5} = 5$. (Section 1.6, Example 4) {9}

60. Find the inverse of $f(x) = x^3 + 2$. (Section 2.7, Example 3)
$f^{-1}(x) = \sqrt[3]{x-2}$

Preview Exercises

Exercises 61–63 will help you prepare for the material covered in the first section of the next chapter.

61. Use point plotting to graph $f(x) = 2^x$. Begin by setting up a partial table of coordinates, selecting integers from -3 to 3, inclusive, for x. Because $y = 0$ is a horizontal asymptote, your graph should approach, but never touch, the negative portion of the x-axis. *

In Exercises 62–63, use transformations of your graph from Exercise 61 to graph each function.

62. $g(x) = f(-x) = 2^{-x}$ *
63. $h(x) = f(x) + 1 = 2^x + 1$ *

CHAPTER 3 Summary, Review, and Test

SUMMARY

DEFINITIONS AND CONCEPTS	EXAMPLES

3.1 Quadratic Functions

a. A quadratic function is of the form $f(x) = ax^2 + bx + c, a \neq 0$.

b. The standard form of a quadratic function is $f(x) = a(x - h)^2 + k, a \neq 0$.

c. The graph of a quadratic function is a parabola. The vertex is (h, k) or $\left(-\dfrac{b}{2a}, f\left(-\dfrac{b}{2a}\right)\right)$.

A procedure for graphing a quadratic function in standard form is given in the box on page 348. A procedure for graphing a quadratic function in the form $f(x) = ax^2 + bx + c$ is given in the box on page 351. Ex. 1, p. 348; Ex. 2, p. 349; Ex. 3, p. 351

d. See the box on page 353 for minimum or maximum values of quadratic functions. Ex. 4, p. 353; Ex. 5, p. 354

e. A strategy for solving problems involving maximizing or minimizing quadratic functions is given in the box on page 356. Ex. 6, p. 357; Ex. 7, p. 358

3.2 Polynomial Functions and Their Graphs

a. Polynomial Function of Degree n: $f(x) = a_n x^n + a_{n-1} x^{n-1} + \cdots + a_2 x^2 + a_1 x + a_0, a_n \neq 0$

b. The graphs of polynomial functions are smooth and continuous. Fig. 3.13, p. 365

c. The end behavior of the graph of a polynomial function depends on the leading term, given by the Leading Coefficient Test in the box on page 366. Odd-degree polynomial functions have graphs with opposite behavior at each end. Even-degree polynomial functions have graphs with the same behavior at each end. Ex. 1, p. 367; Ex. 2, p. 367; Ex. 3, p. 368; Ex. 4, p. 368

DEFINITIONS AND CONCEPTS	**EXAMPLES**
d. The values of x for which $f(x)$ is equal to 0 are the zeros of the polynomial function f. These values are the roots, or solutions, of the polynomial equation $f(x) = 0$.	Ex. 5, p. 369; Ex. 6, p. 370
e. If $x - r$ occurs k times in a polynomial function's factorization, r is a repeated zero with multiplicity k. If k is even, the graph touches the x-axis and turns around at r. If k is odd, the graph crosses the x-axis at r.	Ex. 7, p. 371
f. The Intermediate Value Theorem: If f is a polynomial function and $f(a)$ and $f(b)$ have opposite signs, there is at least one value of c between a and b for which $f(c) = 0$.	Ex. 8, p. 372
g. If f is a polynomial of degree n, the graph of f has at most $n - 1$ turning points.	Fig. 3.24, p. 373
h. A strategy for graphing a polynomial function is given in the box on page 373.	Ex. 9, p. 374; Ex. 10, p. 375

3.3 Dividing Polynomials; Remainder and Factor Theorems

a. Long division of polynomials is performed by dividing, multiplying, subtracting, bringing down the next term, and repeating this process until the degree of the remainder is less than the degree of the divisor. The details are given in the box on page 384.	Ex. 1, p. 383; Ex. 2, p. 385; Ex. 3, p. 386
b. The Division Algorithm: $f(x) = d(x)q(x) + r(x)$. The dividend is the product of the divisor and the quotient plus the remainder.	
c. Synthetic division is used to divide a polynomial by $x - c$. The details are given in the box on page 387.	Ex. 4, p. 388
d. The Remainder Theorem: If a polynomial $f(x)$ is divided by $x - c$, then the remainder is $f(c)$.	Ex. 5, p. 390
e. The Factor Theorem: If $x - c$ is a factor of a polynomial function $f(x)$, then c is a zero of f and a root of $f(x) = 0$. If c is a zero of f or a root of $f(x) = 0$, then $x - c$ is a factor of $f(x)$.	Ex. 6, p. 391

3.4 Zeros of Polynomial Functions

a. The Rational Zero Theorem states that the possible rational zeros of a polynomial $\text{function} = \dfrac{\text{Factors of the constant term}}{\text{Factors of the leading coefficient}}$. The theorem is stated in the box on page 396.	Ex. 1, p. 396; Ex. 2, p. 397; Ex. 3, p. 397; Ex. 4, p. 398; Ex. 5, p. 399
b. Number of roots: If $f(x)$ is a polynomial of degree $n \geq 1$, then, counting multiple roots separately, the equation $f(x) = 0$ has n roots.	
c. If $a + bi$ is a root of $f(x) = 0$ with real coefficients, then $a - bi$ is also a root.	
d. The Linear Factorization Theorem: An nth-degree polynomial can be expressed as the product of n linear factors. Thus, $f(x) = a_n(x - c_1)(x - c_2) \cdots (x - c_n)$.	Ex. 6, p. 402
e. Descartes's Rule of Signs: The number of positive real zeros of f equals the number of sign changes of $f(x)$ or is less than that number by an even integer. The number of negative real zeros of f applies a similar statement to $f(-x)$.	Table 3.1, p. 404; Ex. 7, p. 404

3.5 Rational Functions and Their Graphs

a. Rational function: $f(x) = \dfrac{p(x)}{q(x)}$; p and q are polynomial functions and $q(x) \neq 0$. The domain of f is the set of all real numbers excluding values of x that make $q(x)$ zero.	Ex. 1, p. 411
b. Arrow notation is summarized in the box at the top of page 414.	
c. The line $x = a$ is a vertical asymptote of the graph of f if $f(x)$ increases or decreases without bound as x approaches a. Vertical asymptotes are located using the theorem in the box at the bottom of page 414.	Ex. 2, p. 415
d. The line $y = b$ is a horizontal asymptote of the graph of f if $f(x)$ approaches b as x increases or decreases without bound. Horizontal asymptotes are located using the theorem in the box on page 417.	Ex. 3, p. 417

DEFINITIONS AND CONCEPTS

EXAMPLES

e. Table 3.2 on page 418 shows the graphs of $f(x) = \dfrac{1}{x}$ and $f(x) = \dfrac{1}{x^2}$. Some rational functions can be graphed using transformations of these common graphs.

Ex. 4, p. 418

f. A strategy for graphing rational functions is given in the box on page 419.

Ex. 5, p. 419;
Ex. 6, p. 420;
Ex. 7, p. 422

g. The graph of a rational function has a slant asymptote when the degree of the numerator is one more than the degree of the denominator. The equation of the slant asymptote is found using division and dropping the remainder term.

Ex. 8, p. 423

3.6 Polynomial and Rational Inequalities

a. A polynomial inequality can be expressed as $f(x) < 0, f(x) > 0, f(x) \leq 0$, or $f(x) \geq 0$, where f is a polynomial function. A procedure for solving polynomial inequalities is given in the box on page 432.

Ex. 1, p. 432;
Ex. 2, p. 434;
Ex. 3, p. 435

b. A rational inequality can be expressed as $f(x) < 0, f(x) > 0, f(x) \leq 0$, or $f(x) \geq 0$, where f is a rational function. The procedure for solving such inequalities begins with expressing them so that one side is zero and the other side is a single quotient. Find boundary points by setting the numerator and denominator equal to zero. Then follow a procedure similar to that for solving polynomial inequalities.

Ex. 4, p. 437

3.7 Modeling Using Variation

a. A procedure for solving variation problems is given in the box on page 445.

b. **English Statement**	**Equation**	
y varies directly as x. y is directly proportional to x.	$y = kx$	Ex. 1, p. 445
y varies directly as x^n. y is directly proportional to x^n.	$y = kx^n$	Ex. 2, p. 446
y varies inversely as x. y is inversely proportional to x.	$y = \dfrac{k}{x}$	Ex. 3, p. 448; Ex. 4, p. 449
y varies inversely as x^n. y is inversely proportional to x^n.	$y = \dfrac{k}{x^n}$	
y varies jointly as x and z.	$y = kxz$	Ex. 5, p. 450

REVIEW EXERCISES

3.1

In Exercises 1–4, use the vertex and intercepts to sketch the graph of each quadratic function. Give the equation for the parabola's axis of symmetry. Use the graph to determine the function's domain and range.

1. $f(x) = -(x + 1)^2 + 4$ *
2. $f(x) = (x + 4)^2 - 2$ *
3. $f(x) = -x^2 + 2x + 3$ *
4. $f(x) = 2x^2 - 4x - 6$ *

In Exercises 5–6, use the function's equation, and not its graph, to find

 a. *the minimum or maximum value and where it occurs.*

 b. *the function's domain and its range.*

5. $f(x) = -x^2 + 14x - 106$ *
6. $f(x) = 2x^2 + 12x + 703$ *

*See Answers to Selected Exercises.

7. A quarterback tosses a football to a receiver 40 yards downfield. The height of the football, $f(x)$, in feet, can be modeled by

$$f(x) = -0.025x^2 + x + 6,$$

where x is the ball's horizontal distance, in yards, from the quarterback.

 a. What is the ball's maximum height and how far from the quarterback does this occur? 16 ft; 20 yd

 b. From what height did the quarterback toss the football? *

 c. If the football is not blocked by a defensive player nor caught by the receiver, how far down the field will it go before hitting the ground? Round to the nearest tenth of a yard. 45.3 yd

 d. Graph the function that models the football's parabolic path. *

8. A field bordering a straight stream is to be enclosed. The side bordering the stream is not to be fenced. If 1000 yards of fencing material is to be used, what are the dimensions of the largest rectangular field that can be fenced? What is the maximum area? 250 yd by 500 yd; 125,000 sq yd

9. Among all pairs of numbers whose difference is 14, find a pair whose product is as small as possible. What is the minimum product? −7 and 7; −49

3.2

In Exercises 10–13, use the Leading Coefficient Test to determine the end behavior of the graph of the given polynomial function. Then use this end behavior to match the polynomial function with its graph. [The graphs are labeled (a) through (d).]

10. $f(x) = -x^3 + x^2 + 2x$ c **11.** $f(x) = x^6 - 6x^4 + 9x^2$ b

12. $f(x) = x^5 - 5x^3 + 4x$ a **13.** $f(x) = -x^4 + 1$ d

a.

b.

c.

d.

14. The Brazilian Amazon rain forest is the world's largest tropical rain forest, with some of the greatest biodiversity of any region. In 2012, the number of trees cut down in the Amazon dropped to its lowest level in 20 years. The line graph shows the number of square kilometers cleared from 2001 through 2012.

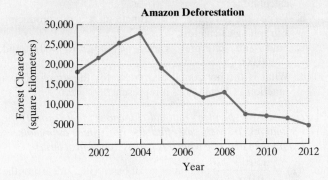

Amazon Deforestation

Source: Brazil's National Institute for Space Research

The data in the line graph can be modeled by the following third- and fourth-degree polynomial functions:

$$f(x) = 81x^3 - 1691x^2 + 8120x + 12{,}465$$
$$g(x) = -15x^4 + 484x^3 - 5155x^2 + 19{,}112x + 2797.$$

Amazon deforestation, in square kilometers, x years after **2000**

a. Use the Leading Coefficient Test to determine the end behavior to the right for the graph of f. rises to the right

b. Assume that the rate at which the Amazon rain forest is being cut down continues to decline. Based on your answer to part (a), will f be useful in modeling Amazon deforestation over an extended period of time? Explain your answer. no*

c. Use the Leading Coefficient Test to determine the end behavior to the right for the graph of g. falls to the right

d. Assume that the rate at which the Amazon rain forest is being cut down continues to decline. Based on your answer to part (c), will g be useful in modeling Amazon deforestation over an extended period of time? Explain your answer. no*

15. The polynomial function

$$f(x) = -0.87x^3 + 0.35x^2 + 81.62x + 7684.94$$

model the number of thefts, $f(x)$, in thousands in the United States x years after 1987. Will this function be useful in modeling the number of thefts over an extended period of time? Explain your answer. no*

In Exercises 16–17, find the zeros for each polynomial function and give the multiplicity of each zero. State whether the graph crosses the x-axis, or touches the x-axis and turns around, at each zero.

16. $f(x) = -2(x - 1)(x + 2)^2(x + 5)^3$ *

17. $f(x) = x^3 - 5x^2 - 25x + 125$ *

18. Show that $f(x) = x^3 - 2x - 1$ has a real zero between 1 and 2. *

In Exercises 19–24,

 a. *Use the Leading Coefficient Test to determine the graph's end behavior.*

 b. *Determine whether the graph has y-axis symmetry, origin symmetry, or neither.*

 c. *Graph the function.*

19. $f(x) = x^3 - x^2 - 9x + 9$ * **20.** $f(x) = 4x - x^3$ *

21. $f(x) = 2x^3 + 3x^2 - 8x - 12$ * **22.** $f(x) = -x^4 + 25x^2$ *

23. $f(x) = -x^4 + 6x^3 - 9x^2$ * **24.** $f(x) = 3x^4 - 15x^3$ *

In Exercises 25–26, graph each polynomial function.

25. $f(x) = 2x^2(x - 1)^3(x + 2)$ *

26. $f(x) = -x^3(x + 4)^2(x - 1)$ *

3.3

In Exercises 27–29, divide using long division.

27. $(4x^3 - 3x^2 - 2x + 1) \div (x + 1)$ $4x^2 - 7x + 5 - \dfrac{4}{x + 1}$

28. $(10x^3 - 26x^2 + 17x - 13) \div (5x - 3)$ *

29. $(4x^4 + 6x^3 + 3x - 1) \div (2x^2 + 1)$ $2x^2 + 3x - 1$

In Exercises 30–31, divide using synthetic division.

30. $(3x^4 + 11x^3 - 20x^2 + 7x + 35) \div (x + 5)$ $3x^3 - 4x^2 + 7$

31. $(3x^4 - 2x^2 - 10x) \div (x - 2)$ $3x^3 + 6x^2 + 10x + 10 + \dfrac{20}{x-2}$

32. Given $f(x) = 2x^3 - 7x^2 + 9x - 3$, use the Remainder Theorem to find $f(-13)$. -5697

33. Use synthetic division to divide $f(x) = 2x^3 + x^2 - 13x + 6$ by $x - 2$. Use the result to find all zeros of f. $2, \frac{1}{2}, -3$

34. Solve the equation $x^3 - 17x + 4 = 0$ given that 4 is a root. $\{4, -2 \pm \sqrt{5}\}$

3.4

In Exercises 35–36, use the Rational Zero Theorem to list all possible rational zeros for each given function.

35. $f(x) = x^4 - 6x^3 + 14x^2 - 14x + 5$ $\pm 1, \pm 5$

36. $f(x) = 3x^5 - 2x^4 - 15x^3 + 10x^2 + 12x - 8$ *

In Exercises 37–38, use Descartes's Rule of Signs to determine the possible number of positive and negative real zeros for each given function.

37. $f(x) = 3x^4 - 2x^3 - 8x + 5$ positive: 2 or 0; negative: 0

38. $f(x) = 2x^5 - 3x^3 - 5x^2 + 3x - 1$ *

39. Use Descartes's Rule of Signs to explain why $2x^4 + 6x^2 + 8 = 0$ has no real roots. *

For Exercises 40–46,

 a. List all possible rational roots or rational zeros.

 b. Use Descartes's Rule of Signs to determine the possible number of positive and negative real roots or real zeros.

 c. Use synthetic division to test the possible rational roots or zeros and find an actual root or zero.

 d. Use the quotient from part (c) to find all the remaining roots or zeros.

40. $f(x) = x^3 + 3x^2 - 4$ a. $\pm 1, \pm 2, \pm 4$ d. $-2, 1$*

41. $f(x) = 6x^3 + x^2 - 4x + 1$ a. $\pm 1, \pm \frac{1}{2}, \pm \frac{1}{3}, \pm \frac{1}{6}$ d. $-1, \frac{1}{3}, \frac{1}{2}$*

42. $8x^3 - 36x^2 + 46x - 15 = 0$ d. $\{\frac{1}{2}, \frac{3}{2}, \frac{5}{2}\}$*

43. $2x^3 + 9x^2 - 7x + 1 = 0$ a. $\pm 1, \pm \frac{1}{2}$ c. $\frac{1}{2}$*

44. $x^4 - x^3 - 7x^2 + x + 6 = 0$ a. $\pm 1, \pm 2, \pm 3, \pm 6$ d. $\{-2, -1, 1, 3\}$*

45. $4x^4 + 7x^2 - 2 = 0$ a. $\pm 1, \pm 2, \pm \frac{1}{2}, \pm \frac{1}{4}$ d. $\{\pm \frac{1}{2}, \pm i\sqrt{2}\}$*

46. $f(x) = 2x^4 + x^3 - 9x^2 - 4x + 4$ a. $\pm 1, \pm 2, \pm 4, \pm \frac{1}{2}$ d. $-2, -1, \frac{1}{2}, 2$*

In Exercises 47–48, find an nth-degree polynomial function with real coefficients satisfying the given conditions. If you are using a graphing utility, graph the function and verify the real zeros and the given function value.

47. $n = 3$; 2 and $2 - 3i$ are zeros; $f(1) = -10$ *

48. $n = 4$; i is a zero; -3 is a zero of multiplicity 2; $f(-1) = 16$ *

In Exercises 49–50, find all the zeros of each polynomial function and write the polynomial as a product of linear factors.

49. $f(x) = 2x^4 + 3x^3 + 3x - 2$ zeros: $-2, \frac{1}{2}, \pm i$*

50. $g(x) = x^4 - 6x^3 + x^2 + 24x + 16$ zeros: $-1, 4$*

In Exercises 51–54, graphs of fifth-degree polynomial functions are shown. In each case, specify the number of real zeros and the number of imaginary zeros. Indicate whether there are any real zeros with multiplicity other than 1.

51.

*

52.

*

53.

*

54.

*

3.5

In Exercises 55–56, use transformations of $f(x) = \dfrac{1}{x}$ or $f(x) = \dfrac{1}{x^2}$ to graph each rational function.

55. $g(x) = \dfrac{1}{(x+2)^2} - 1$ *

56. $h(x) = \dfrac{1}{x-1} + 3$ *

In Exercises 57–64, find the vertical asymptotes, if any, the horizontal asymptote, if one exists, and the slant asymptote, if there is one, of the graph of each rational function. Then graph the rational function.

57. $f(x) = \dfrac{2x}{x^2 - 9}$ *

58. $g(x) = \dfrac{2x - 4}{x + 3}$ *

59. $h(x) = \dfrac{x^2 - 3x - 4}{x^2 - x - 6}$ *

60. $r(x) = \dfrac{x^2 + 4x + 3}{(x+2)^2}$ *

61. $y = \dfrac{x^2}{x + 1}$ *

62. $y = \dfrac{x^2 + 2x - 3}{x - 3}$ *

63. $f(x) = \dfrac{-2x^3}{x^2 + 1}$ *

64. $g(x) = \dfrac{4x^2 - 16x + 16}{2x - 3}$ *

65. A company is planning to manufacture affordable graphing calculators. The fixed monthly cost will be $50,000 and it will cost $25 to produce each calculator.

 a. Write the cost function, C, of producing x graphing calculators. $C(x) = 25x + 50{,}000$

 b. Write the average cost function, \overline{C}, of producing x graphing calculators. *

 c. Find and interpret $\overline{C}(50)$, $\overline{C}(100)$, $\overline{C}(1000)$, and $\overline{C}(100{,}000)$. *

 d. What is the horizontal asymptote for the graph of this function and what does it represent? $y = 25$*

Exercises 66–67 involve rational functions that model the given situations. In each case, find the horizontal asymptote as $x \to \infty$ and then describe what this means in practical terms.

66. $f(x) = \dfrac{150x + 120}{0.05x + 1}$; the number of bass, $f(x)$, after x months in a lake that was stocked with 120 bass $y = 3000$*

67. $P(x) = \dfrac{72{,}900}{100x^2 + 729}$; the percentage, $P(x)$, of people in the United States with x years of education who are unemployed $y = 0^*$

68. The bar graph shows the population of the United States, in millions, for six selected years.

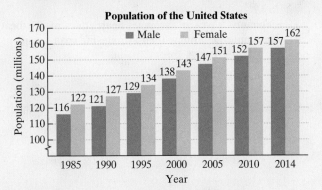

Population of the United States

■ Male ■ Female

Population (millions)

170 · 160 · 150 · 140 · 130 · 120 · 110 · 100

116 122 · 121 127 · 129 134 · 138 143 · 147 151 · 152 157 · 157 162

1985 1990 1995 2000 2005 2010 2014
Year

Source: U.S. Census Bureau

Here are two functions that model the data:

$M(x) = 1.48x + 115.1$ — Male U.S. population, $M(x)$, in millions, x years after 1985

$F(x) = 1.44x + 120.9$. — Female U.S. population, $F(x)$, in millions, x years after 1985

a. Write a function that models the total U.S. population, $P(x)$, in millions, x years after 1985. $P(x) = 2.92x + 236$

b. Write a rational function that models the fraction of men in the U.S. population, $R(x)$, x years after 1985. *

c. What is the equation of the horizontal asymptote associated with the function in part (b)? Round to two decimal places. What does this mean about the percentage of men in the U.S. population over time? $y = 0.51^*$

3.6

In Exercises 69–74, solve each inequality and graph the solution set on a real number line.

69. $2x^2 + 5x - 3 < 0$ $\left(-3, \frac{1}{2}\right)^*$ **70.** $2x^2 + 9x + 4 \geq 0$ *

71. $x^3 + 2x^2 > 3x$ * **72.** $\dfrac{x - 6}{x + 2} > 0$ $(-\infty, -2) \cup (6, \infty)^*$

73. $\dfrac{(x + 1)(x - 2)}{x - 1} \geq 0$ * **74.** $\dfrac{x + 3}{x - 4} \leq 5$ $(-\infty, 4) \cup \left[\frac{23}{4}, \infty\right)^*$

75. The graph shows stopping distances for motorcycles at various speeds on dry roads and on wet roads.

Stopping Distances for Motorcycles at Selected Speeds

■ Dry Pavement ■ Wet Pavement

Stopping Distance (feet)

800 · 700 · 600 · 500 · 400 · 300 · 200 · 100

225 260 · 315 385 · 435 530 · 575 705

35 45 55 65
Speed (miles per hour)

Source: National Highway Traffic Safety Administration

The functions

$$f(x) = 0.125x^2 - 0.8x + 99$$

> Dry pavement

and

> Wet pavement

$$g(x) = 0.125x^2 + 2.3x + 27$$

model a motorcycle's stopping distance, $f(x)$ or $g(x)$, in feet, traveling at x miles per hour. Function f models stopping distance on dry pavement and function g models stopping distance on wet pavement.

a. Use function g to find the stopping distance on wet pavement for a motorcycle traveling at 35 miles per hour. Round to the nearest foot. Does your rounded answer overestimate or underestimate the stopping distance shown by the graph? By how many feet? 261 ft*

b. Use function f to determine speeds on dry pavement requiring stopping distances that exceed 267 feet. *

76. Use the position function

$$s(t) = -16t^2 + v_0t + s_0$$

to solve this problem. A projectile is fired vertically upward from ground level with an initial velocity of 48 feet per second. During which time period will the projectile's height exceed 32 feet? from 1 to 2 seconds

3.7

Solve the variation problems in Exercises 77–82.

77. Many areas of Northern California depend on the snowpack of the Sierra Nevada mountain range for their water supply. The volume of water produced from melting snow varies directly as the volume of snow. Meteorologists have determined that 250 cubic centimeters of snow will melt to 28 cubic centimeters of water. How much water does 1200 cubic centimeters of melting snow produce? 134.4 cm³

78. The distance that a body falls from rest is directly proportional to the square of the time of the fall. If skydivers fall 144 feet in 3 seconds, how far will they fall in 10 seconds? 1600 ft

79. The pitch of a musical tone varies inversely as its wavelength. A tone has a pitch of 660 vibrations per second and a wavelength of 1.6 feet. What is the pitch of a tone that has a wavelength of 2.4 feet? 440 vibrations per second

80. The loudness of a stereo speaker, measured in decibels, varies inversely as the square of your distance from the speaker. When you are 8 feet from the speaker, the loudness is 28 decibels. What is the loudness when you are 4 feet from the speaker? 112 decibels

81. The time required to assemble computers varies directly as the number of computers assembled and inversely as the number of workers. If 30 computers can be assembled by 6 workers in 10 hours, how long would it take 5 workers to assemble 40 computers? 16 hr

82. The volume of a pyramid varies jointly as its height and the area of its base. A pyramid with a height of 15 feet and a base with an area of 35 square feet has a volume of 175 cubic feet. Find the volume of a pyramid with a height of 20 feet and a base with an area of 120 square feet. 800 ft³

83. Heart rates and life spans of most mammals can be modeled using inverse variation. The bar graph shows the average heart rate and the average life span of five mammals.

Heart Rate and Life Span

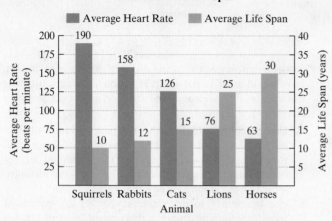

Source: *The Handy Science Answer Book*, Visible Ink Press, 2003.

a. A mammal's average life span, L, in years, varies inversely as its average heart rate, R, in beats per minute. Use the data shown for horses to write the equation that models this relationship. $L = \dfrac{1890}{R}$

b. Is the inverse variation equation in part (a) an exact model or an approximate model for the data shown for lions? an approximate model

c. Elephants have an average heart rate of 27 beats per minute. Determine their average life span. 70 yr

CHAPTER 3 TEST

In Exercises 1–2, use the vertex and intercepts to sketch the graph of each quadratic function. Give the equation for the parabola's axis of symmetry. Use the graph to determine the function's domain and range.

1. $f(x) = (x + 1)^2 + 4$ *

2. $f(x) = x^2 - 2x - 3$ *

3. Determine, without graphing, whether the quadratic function $f(x) = -2x^2 + 12x - 16$ has a minimum value or a maximum value. Then find:

 a. the minimum or maximum value and where it occurs. *

 b. the function's domain and its range. *

4. The function $f(x) = -x^2 + 46x - 360$ models the daily profit, $f(x)$, in hundreds of dollars, for a company that manufactures x computers daily. How many computers should be manufactured each day to maximize profit? What is the maximum daily profit? 23 computers; $16,900

5. Among all pairs of numbers whose sum is 14, find a pair whose product is as large as possible. What is the maximum product? 7 and 7; 49

6. Consider the function $f(x) = x^3 - 5x^2 - 4x + 20$.

 a. Use factoring to find all zeros of f. 5, 2, and −2

 b. Use the Leading Coefficient Test and the zeros of f to graph the function. *

7. Use end behavior to explain why the following graph cannot be the graph of $f(x) = x^5 - x$. Then use intercepts to explain why the graph cannot represent $f(x) = x^5 - x$. *

8. The graph of $f(x) = 6x^3 - 19x^2 + 16x - 4$ is shown in the figure.

 a. Based on the graph of f, find the root of the equation $6x^3 - 19x^2 + 16x - 4 = 0$ that is an integer. 2

 b. Use synthetic division to find the other two roots of $6x^3 - 19x^2 + 16x - 4 = 0$. $\frac{1}{2}, \frac{2}{3}$

9. Use the Rational Zero Theorem to list all possible rational zeros of $f(x) = 2x^3 + 11x^2 - 7x - 6$. $\pm 1, \pm 2, \pm 3, \pm 6, \pm \frac{1}{2}, \pm \frac{3}{2}$

10. Use Descartes's Rule of Signs to determine the possible number of positive and negative real zeros of

$$f(x) = 3x^5 - 2x^4 - 2x^2 + x - 1. \quad *$$

11. Solve: $x^3 + 9x^2 + 16x - 6 = 0$. $\{-3, -3 - \sqrt{11}, -3 + \sqrt{11}\}$

12. Consider the function whose equation is given by $f(x) = 2x^4 - x^3 - 13x^2 + 5x + 15$.

 a. List all possible rational zeros. *

 b. Use the graph of f in the figure shown and synthetic division to find all zeros of the function. $-\sqrt{5}, -1, \frac{3}{2},$ and $\sqrt{5}$

13. Use the graph of $f(x) = x^3 + 3x^2 - 4$ in the figure shown to factor $x^3 + 3x^2 - 4$. $(x-1)(x+2)^2$

14. Find a fourth-degree polynomial function $f(x)$ with real coefficients that has $-1, 1$, and i as zeros and such that $f(3) = 160$. $f(x) = 2x^4 - 2$

15. The figure shows an incomplete graph of $f(x) = -3x^3 - 4x^2 + x + 2$. Find all the zeros of the function. Then draw a complete graph. -1 and $\frac{2}{3}$*

In Exercises 16–21, find the domain of each rational function and graph the function.

16. $f(x) = \dfrac{1}{(x+3)^2}$ *

17. $f(x) = \dfrac{1}{x-1} + 2$ *

18. $f(x) = \dfrac{x}{x^2 - 16}$ *

19. $f(x) = \dfrac{x^2 - 9}{x - 2}$ *

20. $f(x) = \dfrac{x+1}{x^2 + 2x - 3}$ *

21. $f(x) = \dfrac{4x^2}{x^2 + 3}$ *

22. A company is planning to manufacture portable satellite radio players. The fixed monthly cost will be $300,000 and it will cost $10 to produce each player.

 a. Write the average cost function, \overline{C}, of producing x players. *

 b. What is the horizontal asymptote for the graph of this function and what does it represent? $y = 10$*

Solve each inequality in Exercises 23–24 and graph the solution set on a real number line. Express each solution set in interval notation.

23. $x^2 < x + 12$ $(-3, 4)$*

24. $\dfrac{2x + 1}{x - 3} \leq 3$ $(-\infty, 3) \cup [10, \infty)$*

25. The intensity of light received at a source varies inversely as the square of the distance from the source. A particular light has an intensity of 20 foot-candles at 15 feet. What is the light's intensity at 10 feet? 45 foot-candles

CUMULATIVE REVIEW EXERCISES (CHAPTERS 1–3)

Use the graph of $y = f(x)$ to solve Exercises 1–6.

1. Find the domain and the range of f. *

2. Find the zeros and the least possible multiplicity of each zero. -1 and 1; both of multiplicity 2

3. Where does the relative maximum occur? 0

4. Find $(f \circ f)(-1)$. 3

5. Use arrow notation to complete this statement: $f(x) \to \infty$ as _____ or as _____. $x \to -2^+; x \to 2^-$

6. Graph $g(x) = f(x + 2) + 1$. *

In Exercises 7–12, solve each equation or inequality.

7. $|2x - 1| = 3$ $\{-1, 2\}$

8. $3x^2 - 5x + 1 = 0$ $\left\{\frac{5 \pm \sqrt{13}}{6}\right\}$

9. $9 + \dfrac{3}{x} = \dfrac{2}{x^2}$ $\left\{-\frac{2}{3}, \frac{1}{3}\right\}$

10. $x^3 + 2x^2 - 5x - 6 = 0$ $\{-3, -1, 2\}$

11. $|2x - 5| > 3$ $(-\infty, 1) \cup (4, \infty)$

12. $3x^2 > 2x + 5$ $(-\infty, -1) \cup \left(\frac{5}{3}, \infty\right)$

In Exercises 13–18, graph each equation in a rectangular coordinate system. If two functions are given, graph both in the same system.

13. $f(x) = x^3 - 4x^2 - x + 4$ *

14. $f(x) = x^2 + 2x - 8$ *

15. $f(x) = x^2(x - 3)$ *

16. $f(x) = \dfrac{x - 1}{x - 2}$ *

17. $f(x) = |x|$ and $g(x) = -|x| - 1$ *

18. $x^2 + y^2 - 2x + 4y - 4 = 0$ *

In Exercises 19–20, let $f(x) = 2x^2 - x - 1$ and $g(x) = 4x - 1$.

19. Find $(f \circ g)(x)$. $(f \circ g)(x) = 32x^2 - 20x + 2$

20. Find $\dfrac{f(x + h) - f(x)}{h}$. $4x + 2h - 1$

Exponential and Logarithmic Functions

Can I put aside $25,000 when I'm 20 and wind up sitting on half a million dollars by my early fifties? Will population growth lead to a future without comfort or individual choice? Why did I feel I was walking too slowly on my visit to New York City? Are Californians at greater risk from drunk drivers than from earthquakes? What is the difference between earthquakes measuring 6 and 7 on the Richter scale? Can I live forever? And what can possibly be causing merchants at our local shopping mall to grin from ear to ear as they watch the browsers?

The functions that you will be learning about in this chapter will provide you with the mathematics for answering these questions.

You will see how these remarkable functions enable us to predict the future and rediscover the past.

HERE'S WHERE YOU'LL FIND THESE APPLICATIONS:

- You'll be sitting on $500,000 in Example 10 of Section 4.4.
- World population growth: Section 4.5, Example 5
- Population and walking speed: Section 4.5, Check Point 4
- Alcohol and risk of a car accident: Section 4.4, Example 9
- Earthquake intensity: Section 4.2, Example 9
- Immortality: Blitzer Bonus, page 468

We open the chapter with those grinning merchants and the sound of ka-ching!

Section 4.1 Exponential Functions

Just browsing? Take your time. Researchers know, to the dollar, the average amount the typical consumer spends per minute at the shopping mall. And the longer you stay, the more you spend. So if you say you're just browsing, that's just fine with the mall merchants. Browsing is time and, as shown in **Figure 4.1**, time is money.

The data in **Figure 4.1** can be modeled by the function

$$f(x) = 42.2(1.56)^x,$$

where $f(x)$ is the average amount spent, in dollars, at a shopping mall after x hours. Can you see how this function is different from a polynomial function? The variable x is in the exponent. Functions whose equations contain a variable in the exponent are called **exponential functions**. Many real-life situations, including

Mall Browsing Time and Average Amount Spent

FIGURE 4.1
Source: International Council of Shopping Centers Research

population growth, growth of epidemics, radioactive decay, and other changes that involve rapid increase or decrease, can be described using exponential functions.

Definition of the Exponential Function

The **exponential function** f **with base** b is defined by

$$f(x) = b^x \quad \text{or} \quad y = b^x,$$

where b is a positive constant other than 1 ($b > 0$ and $b \neq 1$) and x is any real number.

Here are some examples of exponential functions:

$$f(x) = 2^x \qquad g(x) = 10^x \qquad h(x) = 3^{x+1} \qquad j(x) = \left(\frac{1}{2}\right)^{x-1}.$$

Base is 2. Base is 10. Base is 3. Base is $\frac{1}{2}$.

Each of these functions has a constant base and a variable exponent.

By contrast, the following functions are not exponential functions:

$$F(x) = x^2 \qquad G(x) = 1^x \qquad H(x) = (-1)^x \qquad J(x) = x^x.$$

Variable is the base and not the exponent. The base of an exponential function must be a positive constant other than 1. The base of an exponential function must be positive. Variable is both the base and the exponent.

Why is $G(x) = 1^x$ not classified as an exponential function? The number 1 raised to any power is 1. Thus, the function G can be written as $G(x) = 1$, which is a constant function.

Why is $H(x) = (-1)^x$ not an exponential function? The base of an exponential function must be positive to avoid having to exclude many values of x from the domain that result in nonreal numbers in the range:

$$H(x) = (-1)^x \qquad H\left(\frac{1}{2}\right) = (-1)^{\frac{1}{2}} = \sqrt{-1} = i.$$

> Not an exponential function

> All values of x resulting in even roots of negative numbers produce nonreal numbers.

① Evaluate exponential functions.

You will need a calculator to evaluate exponential expressions. Most scientific calculators have a $\boxed{y^x}$ key. Graphing calculators have a $\boxed{\wedge}$ key. To evaluate expressions of the form b^x, enter the base b, press $\boxed{y^x}$ or $\boxed{\wedge}$, enter the exponent x, and finally press $\boxed{=}$ or $\boxed{\text{ENTER}}$.

EXAMPLE 1 Evaluating an Exponential Function

The exponential function $f(x) = 42.2(1.56)^x$ models the average amount spent, $f(x)$, in dollars, at a shopping mall after x hours. What is the average amount spent, to the nearest dollar, after four hours?

SOLUTION

Because we are interested in the amount spent after four hours, substitute 4 for x and evaluate the function.

$$f(x) = 42.2(1.56)^x \quad \text{This is the given function.}$$
$$f(4) = 42.2(1.56)^4 \quad \text{Substitute 4 for } x.$$

Use a scientific or graphing calculator to evaluate $f(4)$. Press the following keys on your calculator to do this:

Scientific calculator: $42.2 \boxed{\times} 1.56 \boxed{y^x} 4 \boxed{=}$

Graphing calculator: $42.2 \boxed{\times} 1.56 \boxed{\wedge} 4 \boxed{\text{ENTER}}$.

The display should be approximately 249.92566.

$$f(4) = 42.2(1.56)^4 \approx 249.92566 \approx 250$$

Thus, the average amount spent after four hours at a mall is $250. • • •

Evaluating $42.2(1.56)^4$ on a graphing calculator

Ⓥ **Check Point 1** Use the exponential function in Example 1 to find the average amount spent, to the nearest dollar, after three hours at a shopping mall. Does this rounded function value underestimate or overestimate the amount shown in **Figure 4.1**? By how much? approximately $160; overestimates by $11

② Graph exponential functions.

Graphing Exponential Functions

We are familiar with expressions involving b^x, where x is a rational number. For example,

$$b^{1.7} = b^{\frac{17}{10}} = \sqrt[10]{b^{17}} \quad \text{and} \quad b^{1.73} = b^{\frac{173}{100}} = \sqrt[100]{b^{173}}.$$

However, note that the definition of $f(x) = b^x$ includes all real numbers for the domain x. You may wonder what b^x means when x is an irrational number, such as $b^{\sqrt{3}}$ or b^{π}. Using closer and closer approximations for $\sqrt{3}$ ($\sqrt{3} \approx 1.73205$), we can think of $b^{\sqrt{3}}$ as the value that has the successively closer approximations

$$b^{1.7}, b^{1.73}, b^{1.732}, b^{1.73205}, \ldots.$$

In this way, we can graph exponential functions with no holes, or points of discontinuity, at the irrational domain values.

EXAMPLE 2 Graphing an Exponential Function

Graph: $f(x) = 2^x$.

SOLUTION

We begin by setting up a table of coordinates.

> We selected integers from −3 to 3, inclusive, to include three negative numbers, 0, and three positive numbers. We also wanted to keep the resulting computations for y relatively simple.

x	$f(x) = 2^x$
−3	$f(-3) = 2^{-3} = \dfrac{1}{2^3} = \dfrac{1}{8}$
−2	$f(-2) = 2^{-2} = \dfrac{1}{2^2} = \dfrac{1}{4}$
−1	$f(-1) = 2^{-1} = \dfrac{1}{2^1} = \dfrac{1}{2}$
0	$f(0) = 2^0 = 1$
1	$f(1) = 2^1 = 2$
2	$f(2) = 2^2 = 4$
3	$f(3) = 2^3 = 8$

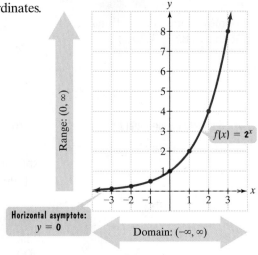

FIGURE 4.2 The graph of $f(x) = 2^x$

We plot these points, connecting them with a continuous curve. **Figure 4.2** shows the graph of $f(x) = 2^x$. Observe that the graph approaches, but never touches, the negative portion of the x-axis. Thus, the x-axis, or $y = 0$, is a horizontal asymptote. The range is the set of all positive real numbers. Although we used integers for x in our table of coordinates, you can use a calculator to find additional points. For example, $f(0.3) = 2^{0.3} \approx 1.231$ and $f(0.95) = 2^{0.95} \approx 1.932$. The points $(0.3, 1.231)$ and $(0.95, 1.932)$ approximately fit the graph. •••

⊘ Check Point **2** Graph: $f(x) = 3^x$. *

EXAMPLE 3 Graphing an Exponential Function

Graph: $g(x) = \left(\dfrac{1}{2}\right)^x$.

SOLUTION

x	$g(x) = \left(\dfrac{1}{2}\right)^x$ or 2^{-x}
−3	$g(-3) = 2^{-(-3)} = 2^3 = 8$
−2	$g(-2) = 2^{-(-2)} = 2^2 = 4$
−1	$g(-1) = 2^{-(-1)} = 2^1 = 2$
0	$g(0) = 2^{-0} = 1$
1	$g(1) = 2^{-1} = \dfrac{1}{2^1} = \dfrac{1}{2}$
2	$g(2) = 2^{-2} = \dfrac{1}{2^2} = \dfrac{1}{4}$
3	$g(3) = 2^{-3} = \dfrac{1}{2^3} = \dfrac{1}{8}$

We begin by setting up a table of coordinates. We compute the function values by noting that

$$g(x) = \left(\frac{1}{2}\right)^x = (2^{-1})^x = 2^{-x}.$$

We plot these points, connecting them with a continuous curve. **Figure 4.3** shows the graph of $g(x) = \left(\frac{1}{2}\right)^x$. This time the graph approaches, but never touches, the *positive* portion of the x-axis. Once again, the x-axis, or $y = 0$, is a horizontal asymptote. The range consists of all positive real numbers.

*See Answers to Selected Exercises.

FIGURE 4.3 The graph of $g(x) = \left(\frac{1}{2}\right)^x$ •••

Do you notice a relationship between the graphs of $f(x) = 2^x$ and $g(x) = \left(\frac{1}{2}\right)^x$ in **Figures 4.2** and **4.3**? The graph of $g(x) = \left(\frac{1}{2}\right)^x$ is the graph of $f(x) = 2^x$ reflected about the y-axis:

$$g(x) = \left(\frac{1}{2}\right)^x = 2^{-x} = f(-x).$$

> Recall that the graph of $y = f(-x)$ is the graph of $y = f(x)$ reflected about the y-axis.

✓ Check Point **3** Graph: $f(x) = \left(\frac{1}{3}\right)^x$. Note that $f(x) = \left(\frac{1}{3}\right)^x = (3^{-1})^x = 3^{-x}$. ∗

Four exponential functions have been graphed in **Figure 4.4**. Compare the black and green graphs, where $b > 1$, to those in blue and red, where $b < 1$. When $b > 1$, the value of y increases as the value of x increases. When $b < 1$, the value of y decreases as the value of x increases. Notice that all four graphs pass through $(0, 1)$.

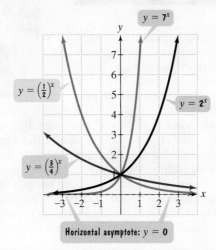

FIGURE 4.4 Graphs of four exponential functions

These graphs illustrate the following general characteristics of exponential functions:

Characteristics of Exponential Functions of the Form $f(x) = b^x$

1. The domain of $f(x) = b^x$ consists of all real numbers: $(-\infty, \infty)$. The range of $f(x) = b^x$ consists of all positive real numbers: $(0, \infty)$.

2. The graphs of all exponential functions of the form $f(x) = b^x$ pass through the point $(0, 1)$ because $f(0) = b^0 = 1(b \neq 0)$. The y-intercept is 1. There is no x-intercept.

3. If $b > 1$, $f(x) = b^x$ has a graph that goes up to the right and is an increasing function. The greater the value of b, the steeper the increase.

4. If $0 < b < 1$, $f(x) = b^x$ has a graph that goes down to the right and is a decreasing function. The smaller the value of b, the steeper the decrease.

5. $f(x) = b^x$ is one-to-one and has an inverse that is a function.

6. The graph of $f(x) = b^x$ approaches, but does not touch, the x-axis. The x-axis, or $y = 0$, is a horizontal asymptote.

Blitzer Bonus || Exponential Growth: The Year Humans Become Immortal

In 2011, *Jeopardy!* aired a three-night match between a personable computer named Watson and the show's two most successful players. The winner: Watson. In the time it took each human contestant to respond to one trivia question, Watson was able to scan the content of one million books. It was also trained to understand the puns and twists of phrases unique to *Jeopardy!* clues.

Watson's remarkable accomplishments can be thought of as a single data point on an exponential curve that models growth in computing power. According to inventor, author, and computer scientist Ray Kurzweil (1948–), computer technology is progressing exponentially, doubling in power each year. What does this mean in terms of the accelerating pace of the graph of $y = 2^x$ that starts slowly and then rockets skyward toward infinity? According to Kurzweil, by 2023, a supercomputer will surpass the brainpower of a human. As progress accelerates exponentially and every hour brings a century's worth of scientific breakthroughs, by 2045, computers will surpass the brainpower equivalent to that of all human brains combined. Here's where it gets exponentially weird: In that year (says Kurzweil), we will be able to scan our consciousness into computers and enter a virtual existence, or swap our bodies for immortal robots. Indefinite life extension will become a reality and people will die only if they choose to.

Transformations of Exponential Functions

The graphs of exponential functions can be translated vertically or horizontally, reflected, stretched, or shrunk. These transformations are summarized in **Table 4.1**.

Table 4.1 Transformations Involving Exponential Functions

In each case, c represents a positive real number.

Transformation	Equation	Description
Vertical translation	$g(x) = b^x + c$	• Shifts the graph of $f(x) = b^x$ upward c units.
	$g(x) = b^x - c$	• Shifts the graph of $f(x) = b^x$ downward c units.
Horizontal translation	$g(x) = b^{x+c}$	• Shifts the graph of $f(x) = b^x$ to the left c units.
	$g(x) = b^{x-c}$	• Shifts the graph of $f(x) = b^x$ to the right c units.
Reflection	$g(x) = -b^x$	• Reflects the graph of $f(x) = b^x$ about the x-axis.
	$g(x) = b^{-x}$	• Reflects the graph of $f(x) = b^x$ about the y-axis.
Vertical stretching or shrinking	$g(x) = cb^x$	• Vertically stretches the graph of $f(x) = b^x$ if $c > 1$.
		• Vertically shrinks the graph of $f(x) = b^x$ if $0 < c < 1$.
Horizontal stretching or shrinking	$g(x) = b^{cx}$	• Horizontally shrinks the graph of $f(x) = b^x$ if $c > 1$.
		• Horizontally stretches the graph of $f(x) = b^x$ if $0 < c < 1$.

EXAMPLE 4 Transformations Involving Exponential Functions

Use the graph of $f(x) = 3^x$ to obtain the graph of $g(x) = 3^{x+1}$.

SOLUTION

The graph of $g(x) = 3^{x+1}$ is the graph of $f(x) = 3^x$ shifted 1 unit to the left.

x	$f(x) = 3^x$
-2	$f(-2) = 3^{-2} = \frac{1}{9}$
-1	$f(-1) = 3^{-1} = \frac{1}{3}$
0	$f(0) = 3^0 = 1$
1	$f(1) = 3^1 = 3$
2	$f(2) = 3^2 = 9$

Begin with a table showing some of the coordinates for f.

Graph $f(x) = 3^x$. We identified three points and the horizontal asymptote.

Graph $g(x) = 3^{x+1}$. Shift f 1 unit left. Subtract 1 from each x-coordinate.

The graph of $g(x) = 3^{x+1}$ with three points and the horizontal asymptote labeled

Horizontal asymptote: $y = 0$

✓ **Check Point 4** Use the graph of $f(x) = 3^x$ to obtain the graph of $g(x) = 3^{x-1}$. *

If an exponential function is translated upward or downward, the horizontal asymptote is shifted by the amount of the vertical shift.

EXAMPLE 5 Transformations Involving Exponential Functions

Use the graph of $f(x) = 2^x$ to obtain the graph of $g(x) = 2^x - 3$.

SOLUTION

The graph of $g(x) = 2^x - 3$ is the graph of $f(x) = 2^x$ shifted down 3 units.

x	$f(x) = 2^x$
-2	$f(-2) = 2^{-2} = \frac{1}{4}$
-1	$f(-1) = 2^{-1} = \frac{1}{2}$
0	$f(0) = 2^0 = 1$
1	$f(1) = 2^1 = 2$
2	$f(2) = 2^2 = 4$

Begin with a table showing some of the coordinates for f.

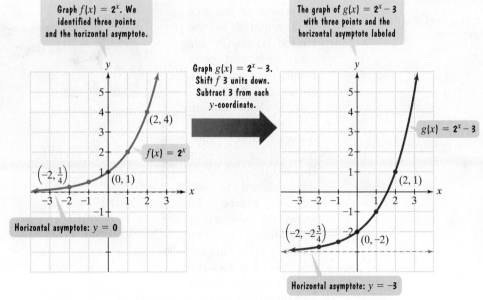

Graph $f(x) = 2^x$. We identified three points and the horizontal asymptote.

Graph $g(x) = 2^x - 3$. Shift f 3 units down. Subtract 3 from each y-coordinate.

The graph of $g(x) = 2^x - 3$ with three points and the horizontal asymptote labeled

Horizontal asymptote: $y = 0$

Horizontal asymptote: $y = -3$

✓ **Check Point 5** Use the graph of $f(x) = 2^x$ to obtain the graph of $g(x) = 2^x + 1$. *

③ Evaluate functions with base *e*.

TECHNOLOGY

Graphic Connections

As $n \to \infty$, the graph of $y = \left(1 + \frac{1}{n}\right)^n$ approaches the graph of $y = e$.

[0, 15, 1] by [0, 3, 1]

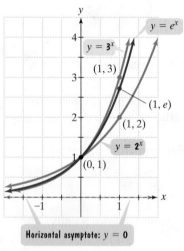

FIGURE 4.5 Graphs of three exponential functions

The Natural Base *e*

An irrational number, symbolized by the letter *e*, appears as the base in many applied exponential functions. The number *e* is defined as the value that $\left(1 + \frac{1}{n}\right)^n$ approaches as *n* gets larger and larger. **Table 4.2** shows values of $\left(1 + \frac{1}{n}\right)^n$ for increasingly large values of *n*. As $n \to \infty$, the approximate value of *e* to nine decimal places is

$$e \approx 2.718281827.$$

The irrational number *e*, approximately 2.72, is called the **natural base**. The function $f(x) = e^x$ is called the **natural exponential function**.

Use a scientific or graphing calculator with an $\boxed{e^x}$ key to evaluate *e* to various powers. For example, to find e^2, press the following keys on most calculators:

Scientific calculator: $2 \boxed{e^x}$

Graphing calculator: $\boxed{e^x} \, 2 \, \boxed{\text{ENTER}}$.

The display should be approximately 7.389.

$$e^2 \approx 7.389$$

The number *e* lies between 2 and 3. Because $2^2 = 4$ and $3^2 = 9$, it makes sense that e^2, approximately 7.389, lies between 4 and 9.

Because $2 < e < 3$, the graph of $y = e^x$ is between the graphs of $y = 2^x$ and $y = 3^x$, shown in **Figure 4.5**.

Table 4.2

n	$\left(1 + \frac{1}{n}\right)^n$
1	2
2	2.25
5	2.48832
10	2.59374246
100	2.704813829
1000	2.716923932
10,000	2.718145927
100,000	2.718268237
1,000,000	2.718280469
1,000,000,000	2.718281827

As $n \to \infty$, $\left(1 + \frac{1}{n}\right)^n \to e$.

In calculus, this is expressed as

$$\lim_{n \to \infty} \left(1 + \frac{1}{n}\right)^n = e.$$

EXAMPLE 6 Gray Wolf Population

Insatiable killer. That's the reputation the gray wolf acquired in the United States in the nineteenth and early twentieth centuries. Although the label was undeserved, an estimated two million wolves were shot, trapped, or poisoned. By 1960, the population was reduced to 800 wolves. **Figure 4.6** shows the rebounding population in two recovery areas after the gray wolf was declared an endangered species and received federal protection.

The exponential function

Gray Wolf Population in Two Recovery Areas for Selected Years

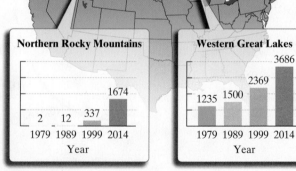

FIGURE 4.6
Source: U.S. Fish and Wildlife Service

$$f(x) = 1145e^{0.0325x}$$

models the gray wolf population of the Western Great Lakes, $f(x)$, x years after 1978.

 a. According to the model, what was the gray wolf population, rounded to the nearest whole number, of the Western Great Lakes in 2014?

 b. Does the model underestimate or overestimate the gray wolf population of the Western Great Lakes in 2014? By how much?

SOLUTION

 a. Because 2014 is 36 years after 1978 (2014 − 1978 = 36), we substitute 36 for x in the given function.

$$f(x) = 1145e^{0.0325x} \qquad \text{This is the given function.}$$
$$f(36) = 1145e^{0.0325(36)} \qquad \text{Substitute 36 fox x.}$$

Perform this computation on your calculator.

This parenthesis is given on some calculators.

The display should be approximately 3689.181571. Thus,

$$f(36) = 1145e^{0.0325(36)} \approx 3689.$$

According to the model, the gray wolf population of the Western Great Lakes in 2014 was approximately 3689 wolves.

 b. **Figure 4.6** shows that the gray wolf population of the Western Great Lakes in 2014 was 3686 wolves. The model in part (a) indicates a population of 3689 wolves. Thus, the model overestimates the population by 3689 − 3686, or by 3 wolves. •••

 ⦶ Check Point **6** Use $f(x) = 1145e^{0.0325x}$, the model described in Example 6, to project the gray wolf population of the Western Great Lakes, rounded to the nearest whole number, in 2017. ≈ 4067

Using exponential functions and projections like those in Example 6 and Check Point 6, in 2011, the U.S. Fish and Wildlife Service removed the Northern Rocky Mountain gray wolf from the endangered species list. Since then, hundreds of wolves have been slaughtered by hunters and trappers. Without protection as an endangered species, free-roaming wolf packs in the United States will be lucky to survive anywhere outside the protection of national parks.

④ Use compound interest formulas.

Compound Interest

We all want a wonderful life with fulfilling work, good health, and loving relationships. And let's be honest: Financial security wouldn't hurt! Achieving this goal depends on understanding how money in savings accounts grows in remarkable ways as a result of *compound interest*. **Compound interest** is interest computed on your original investment as well as on any accumulated interest.

Suppose a sum of money, called the **principal**, P, is invested at an annual percentage rate r, in decimal form, compounded once per year. Because the interest is added to the principal at year's end, the accumulated value, A, is

$$A = P + Pr = P(1 + r).$$

The accumulated amount of money follows this pattern of multiplying the previous principal by $(1 + r)$ for each successive year, as indicated in **Table 4.3**.

Table 4.3

Time in Years	Accumulated Value after Each Compounding
0	$A = P$
1	$A = P(1 + r)$
2	$A = P(1 + r)(1 + r) = P(1 + r)^2$
3	$A = P(1 + r)^2(1 + r) = P(1 + r)^3$
4	$A = P(1 + r)^3(1 + r) = P(1 + r)^4$
\vdots	\vdots
t	$A = P(1 + r)^t$

This formula gives the balance, A, that a principal, P, is worth after t years at interest rate r, compounded once a year.

Most savings institutions have plans in which interest is paid more than once a year. If compound interest is paid twice a year, the compounding period is six months. We say that the interest is **compounded semiannually**. When compound interest is paid four times a year, the compounding period is three months and the interest is said to be **compounded quarterly**. Some plans allow for monthly compounding or daily compounding.

In general, when compound interest is paid n times a year, we say that there are **n compounding periods per year**. **Table 4.4** shows the four most frequently used plans in which interest is paid more than once a year.

Table 4.4 Interest Plans

Name	Number of Compounding Periods per Year	Length of Each Compounding Period
Semiannual Compounding	$n = 2$	6 months
Quarterly Compounding	$n = 4$	3 months
Monthly Compounding	$n = 12$	1 month
Daily Compounding	$n = 365$	1 day

The formula $A = P(1 + r)^t$ can be adjusted to take into account the number of compounding periods in a year. If there are n compounding periods per year, in each time period the interest rate is $\frac{r}{n}$ and there are nt time periods in t years. This results in the following formula for the balance, A, after t years:

$$A = P\left(1 + \frac{r}{n}\right)^{nt}.$$

Some banks use **continuous compounding**, where the number of compounding periods increases infinitely (compounding interest every trillionth of a second, every quadrillionth of a second, etc.). Let's see what happens to the balance, A, as $n \to \infty$.

$$\frac{n}{r} \cdot rt = nt$$

$$A = P\left(1 + \frac{r}{n}\right)^{nt} = P\left[\left(1 + \frac{1}{\frac{n}{r}}\right)^{\frac{n}{r}}\right]^{rt} = P\left[\left(1 + \frac{1}{h}\right)^{h}\right]^{rt} = Pe^{rt}$$

Let $h = \frac{n}{r}$.
As $n \to \infty$, $h \to \infty$.

As $h \to \infty$, by definition
$\left(1 + \frac{1}{h}\right)^{h} \to e$.

We see that the formula for continuous compounding is $A = Pe^{rt}$. Although continuous compounding sounds terrific, it yields only a fraction of a percent more interest over a year than daily compounding.

Formulas for Compound Interest

After t years, the balance, A, in an account with principal P and annual interest rate r (in decimal form) is given by the following formulas:

1. For n compounding periods per year: $A = P\left(1 + \dfrac{r}{n}\right)^{nt}$

2. For continuous compounding: $A = Pe^{rt}$.

EXAMPLE 7 Choosing between Investments

You decide to invest $8000 for 6 years and you have a choice between two accounts. The first pays 7% per year, compounded monthly. The second pays 6.85% per year, compounded continuously. Which is the better investment?

Choosing between Six-Year Investments

Annual Rate	Compounding Period
7%	monthly
6.85%	continuously

SOLUTION

The better investment is the one with the greater balance in the account after 6 years. Let's begin with the account with monthly compounding. We use the compound interest model with $P = 8000$, $r = 7\% = 0.07$, $n = 12$ (monthly compounding means 12 compoundings per year), and $t = 6$.

$$A = P\left(1 + \frac{r}{n}\right)^{nt} = 8000\left(1 + \frac{0.07}{12}\right)^{12 \cdot 6} \approx 12{,}160.84$$

The balance in this account after 6 years is $12,160.84.

For the second investment option, we use the model for continuous compounding with $P = 8000$, $r = 6.85\% = 0.0685$, and $t = 6$.

$$A = Pe^{rt} = 8000e^{0.0685(6)} \approx 12{,}066.60$$

The balance in this account after 6 years is $12,066.60, slightly less than the previous amount. Thus, the better investment is the 7% monthly compounding option. • • •

Ø **Check Point 7** A sum of $10,000 is invested at an annual rate of 8%. Find the balance in the account after 5 years subject to **a.** quarterly compounding and **b.** continuous compounding. a. $14,859.47 b. $14,918.25

ACHIEVING SUCCESS

Avoid coursus interruptus.

Now that you're well into college algebra, don't interrupt the sequence until you have completed all your required math classes. You'll have better results if you take your math courses without a break. If you start, stop, and start again, it's easy to forget what you've learned and lose your momentum.

CONCEPT AND VOCABULARY CHECK

Fill in each blank so that the resulting statement is true.

1. The exponential function f with base b is defined by $f(x) = \underline{b^x}$, $b > 0$ and $b \neq 1$. Using interval notation, the domain of this function is $\underline{(-\infty, \infty)}$ and the range is $\underline{(0, \infty)}$.

2. The graph of the exponential function f with base b approaches, but does not touch, the \underline{x}-axis. This axis, whose equation is $\underline{y = 0}$, is a/an $\underline{\text{horizontal}}$ asymptote.

3. The value that $\left(1 + \dfrac{1}{n}\right)^n$ approaches as n gets larger and larger is the irrational number $\underline{\quad e \quad}$, called the $\underline{\text{natural}}$ base. This irrational number is approximately equal to $\underline{2.72}$.

4. Consider the compound interest formula
$$A = P\left(1 + \frac{r}{n}\right)^{nt}.$$
This formula gives the balance, $\underline{\quad A \quad}$, in an account with principal $\underline{\quad P \quad}$ and annual interest rate $\underline{\quad r \quad}$, in decimal form, subject to compound interest paid $\underline{\quad n \quad}$ times per year.

5. If compound interest is paid twice a year, we say that the interest is compounded $\underline{\text{semiannually}}$. If compound interest is paid four times a year, we say that the interest is compounded $\underline{\text{quarterly}}$. If the number of compounding periods increases infinitely, we call this $\underline{\text{continuous}}$ compounding.

EXERCISE SET 4.1

Practice Exercises

In Exercises 1–10, approximate each number using a calculator. Round your answer to three decimal places.

1. $2^{3.4}$ * 2. $3^{2.4}$ * 3. $3^{\sqrt{5}}$ * 4. $5^{\sqrt{3}}$ * 5. $4^{-1.5}$ *
6. $6^{-1.2}$ * 7. $e^{2.3}$ * 8. $e^{3.4}$ * 9. $e^{-0.95}$ * 10. $e^{-0.75}$ *

In Exercises 11–18, graph each function by making a table of coordinates. If applicable, use a graphing utility to confirm your hand-drawn graph.

11. $f(x) = 4^x$ * 12. $f(x) = 5^x$ * 13. $g(x) = \left(\frac{3}{2}\right)^x$ *
14. $g(x) = \left(\frac{4}{3}\right)^x$ * 15. $h(x) = \left(\frac{1}{2}\right)^x$ * 16. $h(x) = \left(\frac{1}{3}\right)^x$ *
17. $f(x) = (0.6)^x$ * 18. $f(x) = (0.8)^x$ *

In Exercises 19–24, the graph of an exponential function is given. Select the function for each graph from the following options:

$$f(x) = 3^x, g(x) = 3^{x-1}, h(x) = 3^x - 1,$$
$$F(x) = -3^x, G(x) = 3^{-x}, H(x) = -3^{-x}.$$

19.

20.

21.

22.

23.

24.
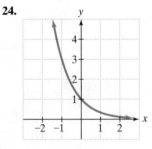

In Exercises 25–34, begin by graphing $f(x) = 2^x$. Then use transformations of this graph to graph the given function. Be sure to graph and give equations of the asymptotes. Use the graphs to determine each function's domain and range. If applicable, use a graphing utility to confirm your hand-drawn graphs.

25. $g(x) = 2^{x+1}$ * 26. $g(x) = 2^{x+2}$ *
27. $g(x) = 2^x - 1$ * 28. $g(x) = 2^x + 2$ *
29. $h(x) = 2^{x+1} - 1$ * 30. $h(x) = 2^{x+2} - 1$ *
31. $g(x) = -2^x$ * 32. $g(x) = 2^{-x}$ *
33. $g(x) = 2 \cdot 2^x$ * 34. $g(x) = \frac{1}{2} \cdot 2^x$ *

The figure shows the graph of $f(x) = e^x$. In Exercises 35–46, use transformations of this graph to graph each function. Be sure to give equations of the asymptotes. Use the graphs to determine each function's domain and range. If applicable, use a graphing utility to confirm your hand-drawn graphs.

35. $g(x) = e^{x-1}$ *

36. $g(x) = e^{x+1}$ *

37. $g(x) = e^x + 2$ *

38. $g(x) = e^x - 1$ *

39. $h(x) = e^{x-1} + 2$ *

40. $h(x) = e^{x+1} - 1$ *

41. $h(x) = e^{-x}$ *

42. $h(x) = -e^x$ *

43. $g(x) = 2e^x$ *

44. $g(x) = \frac{1}{2}e^x$ *

45. $h(x) = e^{2x} + 1$ *

46. $h(x) = e^{\frac{x}{2}} + 2$ *

In Exercises 47–52, graph functions f and g in the same rectangular coordinate system. Graph and give equations of all asymptotes. If applicable, use a graphing utility to confirm your hand-drawn graphs.

47. $f(x) = 3^x$ and $g(x) = 3^{-x}$ *

48. $f(x) = 3^x$ and $g(x) = -3^x$ *

49. $f(x) = 3^x$ and $g(x) = \frac{1}{3} \cdot 3^x$ *

50. $f(x) = 3^x$ and $g(x) = 3 \cdot 3^x$ *

51. $f(x) = \left(\frac{1}{2}\right)^x$ and $g(x) = \left(\frac{1}{2}\right)^{x-1} + 1$ *

52. $f(x) = \left(\frac{1}{2}\right)^x$ and $g(x) = \left(\frac{1}{2}\right)^{x-1} + 2$ *

Use the compound interest formulas $A = P\left(1 + \dfrac{r}{n}\right)^{nt}$ and $A = Pe^{rt}$ to solve Exercises 53–56. Round answers to the nearest cent.

53. Find the accumulated value of an investment of $10,000 for 5 years at an interest rate of 5.5% if the money is **a.** compounded semiannually; **b.** compounded quarterly; **c.** compounded monthly; **d.** compounded continuously. *

54. Find the accumulated value of an investment of $5000 for 10 years at an interest rate of 6.5% if the money is **a.** compounded semiannually; **b.** compounded quarterly; **c.** compounded monthly; **d.** compounded continuously. *

55. Suppose that you have $12,000 to invest. Which investment yields the greater return over 3 years: 7% compounded monthly or 6.85% compounded continuously? 7% monthly

56. Suppose that you have $6000 to invest. Which investment yields the greater return over 4 years: 8.25% compounded quarterly or 8.3% compounded semiannually? 8.25% quarterly

Practice Plus

In Exercises 57–58, graph f and g in the same rectangular coordinate system. Then find the point of intersection of the two graphs.

57. $f(x) = 2^x, g(x) = 2^{-x}$ intersection point: (0, 1)*

58. $f(x) = 2^{x+1}, g(x) = 2^{-x+1}$ intersection point: (0, 2)*

59. Graph $y = 2^x$ and $x = 2^y$ in the same rectangular coordinate system. *

60. Graph $y = 3^x$ and $x = 3^y$ in the same rectangular coordinate system. *

In Exercises 61–64, give the equation of each exponential function whose graph is shown.

61.

$y = 4^x$

62.

$y = 6^x$

63.

$y = -e^x$

64.

$y = e^{-x}$

Application Exercises

Use a calculator with a $\boxed{y^x}$ key or a $\boxed{\wedge}$ key to solve Exercises 65–70.

65. India is currently one of the world's fastest-growing countries. By 2040, the population of India will be larger than the population of China; by 2050, nearly one-third of the world's population will live in these two countries alone. The exponential function $f(x) = 574(1.026)^x$ models the population of India, $f(x)$, in millions, x years after 1974.

a. Substitute 0 for x and, without using a calculator, find India's population in 1974. 574 million

b. Substitute 27 for x and use your calculator to find India's population, to the nearest million, in the year 2001 as modeled by this function. 1148 million

c. Find India's population, to the nearest million, in the year 2028 as predicted by this function. 2295 million

d. Find India's population, to the nearest million, in the year 2055 as predicted by this function. 4590 million

e. What appears to be happening to India's population every 27 years? It appears to double.

66. The 1986 explosion at the Chernobyl nuclear power plant in the former Soviet Union sent about 1000 kilograms of radioactive cesium-137 into the atmosphere. The function $f(x) = 1000(0.5)^{\frac{x}{30}}$ describes the amount, $f(x)$, in kilograms, of cesium-137 remaining in Chernobyl x years after 1986. If even 100 kilograms of cesium-137 remain in Chernobyl's atmosphere, the area is considered unsafe for human habitation. Find $f(80)$ and determine if Chernobyl will be safe for human habitation by 2066. 157.5; no

The formula $S = C(1 + r)^t$ models inflation, where $C =$ the value today, $r =$ the annual inflation rate, and $S =$ the inflated value t years from now. Use this formula to solve Exercises 67–68. Round answers to the nearest dollar.

67. If the inflation rate is 6%, how much will a house now worth $465,000 be worth in 10 years? $832,744

68. If the inflation rate is 3%, how much will a house now worth $510,000 be worth in 5 years? $591,230

69. A decimal approximation for $\sqrt{3}$ is 1.7320508. Use a calculator to find $2^{1.7}$, $2^{1.73}$, $2^{1.732}$, $2^{1.73205}$, and $2^{1.7320508}$. Now find $2^{\sqrt{3}}$. What do you observe? *

70. A decimal approximation for π is 3.141593. Use a calculator to find 2^3, $2^{3.1}$, $2^{3.14}$, $2^{3.141}$, $2^{3.1415}$, $2^{3.14159}$, and $2^{3.141593}$. Now find 2^{π}. What do you observe? *

Use a calculator with an $\boxed{e^x}$ key to solve Exercises 71–76.

The bar graph shows the percentage of U.S. high school seniors who applied to more than three colleges for selected years from 1980 through 2013.

Percentage of U.S. High School Seniors Applying to More Than Three Colleges

- 1980: 32
- 1990: 43
- 1995: 44
- 2000: 51
- 2005: 55
- 2010: 61
- 2013: 69

Source: The Higher Education Research Institute

The data can be modeled by

$$f(x) = x + 31 \quad \text{and} \quad g(x) = 32.7e^{0.0217x},$$

in which $f(x)$ and $g(x)$ represent the percentage of high school seniors who applied to more than three colleges x years after 1980. Use these functions to solve Exercises 71–72. Where necessary, round answers to the nearest percent.

71. a. According to the linear model, what percentage of high school seniors applied to more than three colleges in 2013? 64%

 b. According to the exponential model, what percentage of high school seniors applied to more than three colleges in 2013? 67%

 c. Which function is a better model for the data shown by the bar graph in 2013? exponential model

72. a. According to the linear model, what percentage of high school seniors applied to more than three colleges in 2010? 61%

 b. According to the exponential model, what percentage of high school seniors applied to more than three colleges in 2010? 63%

 c. Which function is a better model for the data shown by the bar graph in 2010? linear model

73. In college, we study large volumes of information—information that, unfortunately, we do not often retain for very long. The function

$$f(x) = 80e^{-0.5x} + 20$$

describes the percentage of information, $f(x)$, that a particular person remembers x weeks after learning the information.

 a. Substitute 0 for x and, without using a calculator, find the percentage of information remembered at the moment it is first learned. 100%

 b. Substitute 1 for x and find the percentage of information that is remembered after 1 week. \approx 68.5%

 c. Find the percentage of information that is remembered after 4 weeks. \approx 30.8%

 d. Find the percentage of information that is remembered after one year (52 weeks). \approx 20%

74. In 1626, Peter Minuit convinced the Wappinger Indians to sell him Manhattan Island for $24. If the Native Americans had put the $24 into a bank account paying 5% interest, how much would the investment have been worth in the year 2010 if interest were compounded

 a. monthly? * **b.** continuously? *

The bar graph shows the percentage of people 25 years of age and older who were college graduates in the United States for eight selected years.

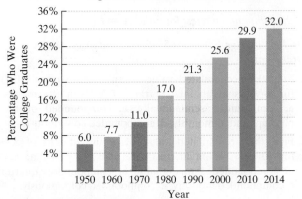

Percentage of College Graduates, Among People Ages 25 and Older, in the United States

- 1950: 6.0
- 1960: 7.7
- 1970: 11.0
- 1980: 17.0
- 1990: 21.3
- 2000: 25.6
- 2010: 29.9
- 2014: 32.0

Source: U.S. Census Bureau

The functions

$$f(x) = 6.43(1.027)^x \quad \text{and} \quad g(x) = \frac{40.9}{1 + 6.6e^{-0.049x}}$$

model the percentage of college graduates, among people ages 25 and older, $f(x)$ or $g(x)$, x years after 1950. Use these functions to solve Exercises 75–76.

75. Which function is a better model for the percentage who were college graduates in 2014? the function g

76. Which function is a better model for the percentage who were college graduates in 1990? the function g

Explaining the Concepts 77–78. Answers will vary.

77. What is an exponential function?

78. What is the natural exponential function?

79. Use a calculator to evaluate $\left(1 + \dfrac{1}{x}\right)^x$ for $x = 10, 100, 1000,$ $10{,}000, 100{,}000,$ and $1{,}000{,}000.$ Describe what happens to the expression as x increases. 79–80. Answers will vary.

80. Describe how you could use the graph of $f(x) = 2^x$ to obtain a decimal approximation for $\sqrt{2}$.

Technology Exercises

81. You have $10,000 to invest. One bank pays 5% interest compounded quarterly and a second bank pays 4.5% interest compounded monthly.

 a. Use the formula for compound interest to write a function for the balance in each bank at any time t. *

 b. Use a graphing utility to graph both functions in an appropriate viewing rectangle. Based on the graphs, which bank offers the better return on your money? *

82. a. Graph $y = e^x$ and $y = 1 + x + \dfrac{x^2}{2}$ in the same viewing rectangle. *

 b. Graph $y = e^x$ and $y = 1 + x + \dfrac{x^2}{2} + \dfrac{x^3}{6}$ in the same viewing rectangle. *

 c. Graph $y = e^x$ and $y = 1 + x + \dfrac{x^2}{2} + \dfrac{x^3}{6} + \dfrac{x^4}{24}$ in the same viewing rectangle. *

 d. Describe what you observe in parts (a)–(c). Try generalizing this observation. Answers will vary.

Critical Thinking Exercises

Make Sense? In Exercises 83–86, determine whether each statement makes sense or does not make sense, and explain your reasoning.

83. My graph of $f(x) = 3 \cdot 2^x$ shows that the horizontal asymptote for f is $x = 3$. does not make sense

84. I'm using a photocopier to reduce an image over and over by 50%, so the exponential function $f(x) = \left(\dfrac{1}{2}\right)^x$ models the new image size, where x is the number of reductions. *

85. Taxing thoughts: I'm looking at data that show the number of pages in the publication that explains the U.S. tax code for selected years from 1945 thorough 2013. A linear function appears to be a better choice than an exponential function for modeling the number of pages in the tax code during this period. does not make sense

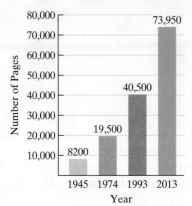

Number of Pages in the Federal Tax Code

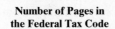

Source: CCH Inc.

86. I use the natural base e when determining how much money I'd have in a bank account that earns compound interest subject to continuous compounding. makes sense

In Exercises 87–90, determine whether each statement is true or false. If the statement is false, make the necessary change(s) to produce a true statement. Changes to false statements will vary.

87. As the number of compounding periods increases on a fixed investment, the amount of money in the account over a fixed interval of time will increase without bound. false

88. The functions $f(x) = 3^{-x}$ and $g(x) = -3^x$ have the same graph. false

89. If $f(x) = 2^x$, then $f(a + b) = f(a) + f(b)$. false

90. The functions $f(x) = \left(\dfrac{1}{3}\right)^x$ and $g(x) = 3^{-x}$ have the same graph. true

91. The graphs labeled (a)–(d) in the figure represent $y = 3^x$, $y = 5^x$, $y = \left(\dfrac{1}{3}\right)^x$, and $y = \left(\dfrac{1}{5}\right)^x$, but not necessarily in that order. Which is which? Describe the process that enables you to make this decision. $y = 3^x$: (d); $y = 5^x$: (c); $y = \left(\dfrac{1}{3}\right)^x$: (a); $y = \left(\dfrac{1}{5}\right)^x$: (b)

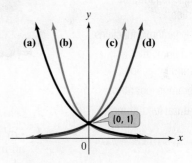

92. Graph $f(x) = 2^x$ and its inverse function in the same rectangular coordinate system. *

93. The *hyperbolic cosine* and *hyperbolic sine* functions are defined by

$$\cosh x = \dfrac{e^x + e^{-x}}{2} \quad \text{and} \quad \sinh x = \dfrac{e^x - e^{-x}}{2}.$$

 a. Show that $\cosh x$ is an even function. *

 b. Show that $\sinh x$ is an odd function. *

 c. Prove that $(\cosh x)^2 - (\sinh x)^2 = 1$. *

Retaining the Concepts

94. Solve for y: $7x + 3y = 18$. (Section 1.3, Example 7) *

95. Find all zeros of $f(x) = x^3 + 5x^2 - 8x + 2$. (Section 3.4, Example 4) $1, -3 \pm \sqrt{11}$

96. Solve and graph the solution set on a number line: $2x^2 + 5x < 12$. (Section 3.6, Example 1) $\left(-4, \dfrac{3}{2}\right)$ *

Preview Exercises

Exercises 97–99 will help you prepare for the material covered in the next section.

97. What problem do you encounter when using the switch-and-solve strategy to find the inverse of $f(x) = 2^x$? (The switch-and-solve strategy is described in the box on page 316.) *

98. 25 to what power gives 5? $(25^? = 5)$ $\frac{1}{2}$

99. Solve: $(x - 3)^2 > 0$. $(-\infty, 3) \cup (3, \infty)$

Section 4.2 | Logarithmic Functions

What am I supposed to learn?

After studying this section, you should be able to:

1. Change from logarithmic to exponential form.
2. Change from exponential to logarithmic form.
3. Evaluate logarithms.
4. Use basic logarithmic properties.
5. Graph logarithmic functions.
6. Find the domain of a logarithmic function.
7. Use common logarithms.
8. Use natural logarithms.

The earthquake that ripped through northern California on October 17, 1989, measured 7.1 on the Richter scale, killed more than 60 people, and injured more than 2400. Shown here is San Francisco's Marina district, where shock waves tossed houses off their foundations and into the street.

A higher measure on the Richter scale is more devastating than it seems because for each increase in one unit on the scale, there is a tenfold increase in the intensity of an earthquake. In this section, our focus is on the inverse of the exponential function, called the logarithmic function. The logarithmic function will help you to understand diverse phenomena, including earthquake intensity, human memory, and the pace of life in large cities.

A Brief Review • Functions and Their Inverses

1. Only one-to-one functions have inverses that are functions. A function, f, has an inverse function, f^{-1}, if there is no horizontal line that intersects the graph of f at more than one point.
2. If a function is one-to-one, its inverse function can be found by interchanging x and y in the function's equation and solving for y.
3. If $f(a) = b$, then $f^{-1}(b) = a$. The domain of f is the range of f^{-1}. The range of f is the domain of f^{-1}.
4. $f(f^{-1}(x)) = x$ and $f^{-1}(f(x)) = x$.
5. The graph of f^{-1} is the reflection of the graph of f about the line $y = x$.

For more detail, see Section 2.7.

The Definition of Logarithmic Functions

No horizontal line can be drawn that intersects the graph of an exponential function at more than one point. This means that the exponential function is one-to-one and has an inverse. Let's use our switch-and-solve strategy from Section 2.7 to find the inverse.

All exponential functions have inverse functions. $f(x) = b^x$

Step 1 Replace $f(x)$ with y: $y = b^x$.

Step 2 Interchange x and y: $x = b^y$.

Step 3 Solve for y: ?

The question mark indicates that we do not have a method for solving $b^y = x$ for y. To isolate the exponent y, a new notation, called *logarithmic notation*, is needed. This notation gives us a way to name the inverse of $f(x) = b^x$. **The inverse function of the exponential function with base b is called the *logarithmic function with base b.***

Definition of the Logarithmic Function

For $x > 0$ and $b > 0, b \neq 1$,

$$y = \log_b x \text{ is equivalent to } b^y = x.$$

The function $f(x) = \log_b x$ is the **logarithmic function with base b**.

The equations

$$y = \log_b x \quad \text{and} \quad b^y = x$$

are different ways of expressing the same thing. The first equation is in **logarithmic form** and the second equivalent equation is in **exponential form**.

Notice that a **logarithm, y, is an exponent.** You should learn the location of the base and exponent in each form.

Location of Base and Exponent in Exponential and Logarithmic Forms

$$\text{Logarithmic Form: } y = \log_b x \qquad \text{Exponential Form: } b^y = x$$

Exponent

Base

Exponent

Base

① Change from logarithmic to exponential form.

EXAMPLE 1 Changing from Logarithmic to Exponential Form

Write each equation in its equivalent exponential form:

a. $2 = \log_5 x$ **b.** $3 = \log_b 64$ **c.** $\log_3 7 = y$.

SOLUTION

We use the fact that $y = \log_b x$ means $b^y = x$.

a. $2 = \log_5 x$ means $5^2 = x$. **b.** $3 = \log_b 64$ means $b^3 = 64$.

Logarithms are exponents.

Logarithms are exponents.

c. $\log_3 7 = y$ or $y = \log_3 7$ means $3^y = 7$. • • •

GREAT QUESTION!

Much of what you've discussed so far involves changing from logarithmic form to the more familiar exponential form. Is there a pattern I can use to help me remember how to do this?

Yes. To change from logarithmic form to exponential form, use this pattern:

$$y = \log_b x \qquad \text{means} \qquad b^y = x.$$

✓ Check Point **1** Write each equation in its equivalent exponential form:

a. $3 = \log_7 x$ $7^3 = x$ **b.** $2 = \log_b 25$ $b^2 = 25$ **c.** $\log_4 26 = y$. $4^y = 26$

② Change from exponential to logarithmic form.

EXAMPLE 2 Changing from Exponential to Logarithmic Form

Write each equation in its equivalent logarithmic form:

a. $12^2 = x$ b. $b^3 = 8$ c. $e^y = 9$.

SOLUTION

We use the fact that $b^y = x$ means $y = \log_b x$.

a. $12^2 = x$ means $2 = \log_{12} x$. b. $b^3 = 8$ means $3 = \log_b 8$.

Exponents are logarithms. Exponents are logarithms.

c. $e^y = 9$ means $y = \log_e 9$. •••

✓ Check Point 2 Write each equation in its equivalent logarithmic form:

a. $2^5 = x$ $5 = \log_2 x$ b. $b^3 = 27$ $3 = \log_b 27$ c. $e^y = 33$. $y = \log_e 33$

③ Evaluate logarithms.

Remembering that logarithms are exponents makes it possible to evaluate some logarithms by inspection. The logarithm of x with base b, $\log_b x$, is the exponent to which b must be raised to get x. For example, suppose we want to evaluate $\log_2 32$. We ask, 2 to what power gives 32? Because $2^5 = 32$, $\log_2 32 = 5$.

EXAMPLE 3 Evaluating Logarithms

Evaluate:

a. $\log_2 16$ b. $\log_7 \frac{1}{49}$ c. $\log_{25} 5$ d. $\log_2 \sqrt[5]{2}$.

SOLUTION

Logarithmic Expression	Question Needed for Evaluation	Logarithmic Expression Evaluated
a. $\log_2 16$	2 to what power gives 16? $2^? = 16$	$\log_2 16 = 4$ because $2^4 = 16$.
b. $\log_7 \frac{1}{49}$	7 to what power gives $\frac{1}{49}$? $7^? = \frac{1}{49}$	$\log_7 \frac{1}{49} = -2$ because $7^{-2} = \frac{1}{7^2} = \frac{1}{49}$.
c. $\log_{25} 5$	25 to what power gives 5? $25^? = 5$	$\log_{25} 5 = \frac{1}{2}$ because $25^{\frac{1}{2}} = \sqrt{25} = 5$.
d. $\log_2 \sqrt[5]{2}$	2 to what power gives $\sqrt[5]{2}$, or $2^{\frac{1}{5}}$? $2^? = \sqrt[5]{2}$	$\log_2 \sqrt[5]{2} = \frac{1}{5}$ because $2^{\frac{1}{5}} = \sqrt[5]{2}$.

•••

✓ Check Point 3 Evaluate:

a. $\log_{10} 100$ 2 b. $\log_5 \frac{1}{125}$ -3 c. $\log_{36} 6$ $\frac{1}{2}$ d. $\log_3 \sqrt[7]{3}$. $\frac{1}{7}$

④ Use basic logarithmic properties.

Basic Logarithmic Properties

Because logarithms are exponents, they have properties that can be verified using properties of exponents.

Basic Logarithmic Properties Involving One

1. $\log_b b = 1$ because 1 is the exponent to which b must be raised to obtain b. ($b^1 = b$)

2. $\log_b 1 = 0$ because 0 is the exponent to which b must be raised to obtain 1. ($b^0 = 1$)

EXAMPLE 4 Using Properties of Logarithms

Evaluate:

 a. $\log_7 7$ **b.** $\log_5 1$.

SOLUTION

 a. Because $\log_b b = 1$, we conclude $\log_7 7 = 1$.

> This means that $7^1 = 7$.

 b. Because $\log_b 1 = 0$, we conclude $\log_5 1 = 0$.

> This means that $5^0 = 1$.

 • • •

⊘ Check Point **4** Evaluate:

 a. $\log_9 9$ 1 **b.** $\log_8 1$. 0

Now that we are familiar with logarithmic notation, let's resume and finish the switch-and-solve strategy for finding the inverse of $f(x) = b^x$.

Step 1 **Replace $f(x)$ with y:** $y = b^x$.

Step 2 **Interchange x and y:** $x = b^y$.

Step 3 **Solve for y:** $y = \log_b x$.

Step 4 **Replace y with $f^{-1}(x)$:** $f^{-1}(x) = \log_b x$.

The completed switch-and-solve strategy illustrates that if $f(x) = b^x$, then $f^{-1}(x) = \log_b x$. The inverse of an exponential function is the logarithmic function with the same base.

In Section 2.7, we saw how inverse functions "undo" one another. In particular,

$$f(f^{-1}(x)) = x \quad \text{and} \quad f^{-1}(f(x)) = x.$$

Applying these relationships to exponential and logarithmic functions, we obtain the following **inverse properties of logarithms**:

Inverse Properties of Logarithms

For $b > 0$ and $b \neq 1$,

$\log_b b^x = x$ The logarithm with base b of b raised to a power equals that power.

$b^{\log_b x} = x$. b raised to the logarithm with base b of a number equals that number.

EXAMPLE 5 Using Inverse Properties of Logarithms

Evaluate:

 a. $\log_4 4^5$ **b.** $6^{\log_6 9}$.

SOLUTION

 a. Because $\log_b b^x = x$, we conclude $\log_4 4^5 = 5$.

 b. Because $b^{\log_b x} = x$, we conclude $6^{\log_6 9} = 9$.

 • • •

⊘ Check Point **5** Evaluate:

 a. $\log_7 7^8$ 8 **b.** $3^{\log_3 17}$. 17

⑤ Graph logarithmic functions.

Graphs of Logarithmic Functions

How do we graph logarithmic functions? We use the fact that a logarithmic function is the inverse of an exponential function. This means that the logarithmic function reverses the coordinates of the exponential function. It also means that the graph of the logarithmic function is a reflection of the graph of the exponential function about the line $y = x$.

The basic properties of logarithms that were listed earlier in this section can be applied to common logarithms.

Properties of Common Logarithms

General Properties	Common Logarithms
1. $\log_b 1 = 0$	**1.** $\log 1 = 0$
2. $\log_b b = 1$	**2.** $\log 10 = 1$
3. $\log_b b^x = x$	**3.** $\log 10^x = x$
4. $b^{\log_b x} = x$	**4.** $10^{\log x} = x$

Inverse properties

The property $\log 10^x = x$ can be used to evaluate common logarithms involving powers of 10. For example,

$$\log 100 = \log 10^2 = 2, \quad \log 1000 = \log 10^3 = 3, \quad \text{and} \quad \log 10^{7.1} = 7.1.$$

EXAMPLE 9 Earthquake Intensity

The magnitude, R, on the Richter scale of an earthquake of intensity I is given by

$$R = \log \frac{I}{I_0},$$

where I_0 is the intensity of a barely felt zero-level earthquake. The earthquake that destroyed San Francisco in 1906 was $10^{8.3}$ times as intense as a zero-level earthquake. What was its magnitude on the Richter scale?

SOLUTION

Because the earthquake was $10^{8.3}$ times as intense as a zero-level earthquake, the intensity, I, is $10^{8.3}I_0$.

$$R = \log \frac{I}{I_0} \qquad \text{This is the formula for magnitude on the Richter scale.}$$

$$R = \log \frac{10^{8.3}I_0}{I_0} \qquad \text{Substitute } 10^{8.3}I_0 \text{ for } I.$$

$$= \log 10^{8.3} \qquad \text{Simplify.}$$

$$= 8.3 \qquad \text{Use the property } \log 10^x = x.$$

San Francisco's 1906 earthquake registered 8.3 on the Richter scale. • • •

⏀ Check Point **9** Use the formula in Example 9 to solve this problem. If an earthquake is 10,000 times as intense as a zero-level quake ($I = 10,000I_0$), what is its magnitude on the Richter scale? 4.0

⑧ Use natural logarithms.

Natural Logarithms

The logarithmic function with base e is called the **natural logarithmic function**. The function $f(x) = \log_e x$ is usually expressed as $f(x) = \ln x$, read "el en of x." A calculator with an ⏍LN⏍ key can be used to evaluate natural logarithms. Keystrokes are identical to those shown for common logarithmic evaluations on page 485.

Like the domain of all logarithmic functions, the domain of the natural logarithmic function $f(x) = \ln x$ is the set of all positive real numbers. Thus, the domain of $f(x) = \ln g(x)$ consists of all x for which $g(x) > 0$.

[–10, 10, 1] by [–10, 10, 1]

FIGURE 4.14 The domain of $f(x) = \ln(3 - x)$ is $(-\infty, 3)$.

[–10, 10, 1] by [–10, 10, 1]

FIGURE 4.15 3 is excluded from the domain of $h(x) = \ln(x - 3)^2$.

EXAMPLE 10 Finding Domains of Natural Logarithmic Functions

Find the domain of each function:

 a. $f(x) = \ln(3 - x)$ **b.** $h(x) = \ln(x - 3)^2$.

SOLUTION

a. The domain of f consists of all x for which $3 - x > 0$. Solving this inequality for x, we obtain $x < 3$. Thus, the domain of f is $\{x \mid x < 3\}$ or $(-\infty, 3)$. This is verified by the graph in **Figure 4.14**.

b. The domain of h consists of all x for which $(x - 3)^2 > 0$. It follows that the domain of h is all real numbers except 3. Thus, the domain of h is $\{x \mid x \neq 3\}$ or $(-\infty, 3) \cup (3, \infty)$. This is shown by the graph in **Figure 4.15**. To make it more obvious that 3 is excluded from the domain, we used a $\boxed{\text{DOT}}$ format.

 • • •

✓ Check Point **10** Find the domain of each function:

 a. $f(x) = \ln(4 - x)$ $(-\infty, 4)$ **b.** $h(x) = \ln x^2$. $(-\infty, 0) \cup (0, \infty)$

The basic properties of logarithms that were listed earlier in this section can be applied to natural logarithms.

Properties of Natural Logarithms

General Properties	**Natural Logarithms**
1. $\log_b 1 = 0$	**1.** $\ln 1 = 0$
2. $\log_b b = 1$	**2.** $\ln e = 1$
3. $\log_b b^x = x$	**3.** $\ln e^x = x$
4. $b^{\log_b x} = x$	**4.** $e^{\ln x} = x$

Inverse properties

Examine the inverse properties, $\ln e^x = x$ and $e^{\ln x} = x$. Can you see how ln and e "undo" one another? For example,

$$\ln e^2 = 2, \quad \ln e^{7x^2} = 7x^2, \quad e^{\ln 2} = 2, \quad \text{and} \quad e^{\ln 7x^2} = 7x^2.$$

EXAMPLE 11 Dangerous Heat: Temperature in an Enclosed Vehicle

When the outside air temperature is anywhere from 72° to 96° Fahrenheit, the temperature in an enclosed vehicle climbs by 43° in the first hour. The bar graph in **Figure 4.16** shows the temperature increase throughout the hour. The function

$$f(x) = 13.4 \ln x - 11.6$$

models the temperature increase, $f(x)$, in degrees Fahrenheit, after x minutes. Use the function to find the temperature increase, to the nearest degree, after 50 minutes. How well does the function model the actual increase shown in **Figure 4.16**?

Temperature Increase in an Enclosed Vehicle

FIGURE 4.16
Source: Professor Jan Null, San Francisco State University

**Temperature Increase
in an Enclosed Vehicle**

FIGURE 4.16 (repeated)

SOLUTION

We find the temperature increase after 50 minutes by substituting 50 for x and evaluating the function at 50.

$$f(x) = 13.4 \ln x - 11.6 \qquad \text{This is the given function.}$$
$$f(50) = 13.4 \ln 50 - 11.6 \qquad \text{Substitute 50 for } x.$$
$$\approx 41 \qquad \qquad \text{Graphing calculator keystrokes:}$$

13.4 ⌊ln⌋ 50 ⌊−⌋ 11.6 ⌊ENTER⌋. On some calculators, a parenthesis is needed after 50.

According to the function, the temperature will increase by approximately 41° after 50 minutes. Because the increase shown in **Figure 4.16** is 41°, the function models the actual increase extremely well. •••

✓ Check Point **11** Use the function in Example 11 to find the temperature increase, to the nearest degree, after 30 minutes. How well does the function model the actual increase shown in **Figure 4.16**? 34°; extremely well

Blitzer Bonus ‖ The Curious Number e

FIGURE 4.17

You will learn more about each curiosity mentioned below if you take calculus.

- The number e was named by the Swiss mathematician Leonhard Euler (1707–1783), who proved that it is the limit as $n \to \infty$ of $\left(1 + \dfrac{1}{n}\right)^n$.
- e features in Euler's remarkable relationship $e^{i\pi} = -1$, in which $i = \sqrt{-1}$.
- The first few decimal places of e are fairly easy to remember:
 $e = 2.7\ 1828\ 1828\ 45\ 90\ 45\ldots$.
- The best approximation of e using numbers less than 1000 is also easy to remember:
 $e \approx \dfrac{878}{323} \approx 2.71826\ldots$.
- Isaac Newton (1642–1727), one of the cofounders of calculus, showed that
 $e^x = 1 + x + \dfrac{x^2}{2!} + \dfrac{x^3}{3!} + \dfrac{x^4}{4!} + \cdots$, from which we obtain $e = 1 + 1 + \dfrac{1}{2!} + \dfrac{1}{3!} + \dfrac{1}{4!} + \cdots$, an infinite sum suitable for calculation because its terms decrease so rapidly. (*Note*: $n!$ (n factorial) is the product of all the consecutive integers from n down to 1: $n! = n(n-1)(n-2)(n-3) \cdot \cdots \cdot 3 \cdot 2 \cdot 1$.)
- The area of the region bounded by $y = \dfrac{1}{x}$, the x-axis, $x = 1$, and $x = t$ (shaded in **Figure 4.17**) is a function of t, designated by $A(t)$. Grégoire de Saint-Vincent, a Belgian Jesuit (1584–1667), spent his entire professional life attempting to find a formula for $A(t)$. With his student, he showed that $A(t) = \ln t$, becoming one of the first mathematicians to make use of the logarithmic function for something other than a computational device.

ACHIEVING SUCCESS

A recent government study cited in *Math: A Rich Heritage* (Globe Fearon Educational Publisher) found this simple fact: **The more college mathematics courses you take, the greater your earning potential will be.** Even jobs that do not require a college degree require mathematical thinking that involves attending to precision, making sense of complex problems, and persevering in solving them. No other discipline comes close to math in offering a more extensive set of tools for application and intellectual development. Take as much math as possible as you continue your journey into higher education.

CONCEPT AND VOCABULARY CHECK

Fill in each blank so that the resulting statement is true.

1. $y = \log_b x$ is equivalent to the exponential form ___$b^y = x$___, $x > 0, b > 0, b \neq 1$.

2. The function $f(x) = \log_b x$ is the ___logarithmic___ function with base ___b___.

3. $\log_b b = $ ___1___

4. $\log_b 1 = $ ___0___

5. $\log_b b^x = $ ___x___

6. $b^{\log_b x} = $ ___x___

7. Using interval notation, the domain of $f(x) = \log_b x$ is ___$(0, \infty)$___ and the range is ___$(-\infty, \infty)$___.

8. The graph of $f(x) = \log_b x$ approaches, but does not touch, the ___y___ -axis. This axis, whose equation is ___$x = 0$___, is a/an ___vertical___ asymptote.

9. The graph of $g(x) = 5 + \log_2 x$ is the graph of $f(x) = \log_2 x$ shifted ___up 5 units___.

10. The graph of $g(x) = \log_3(x + 5)$ is the graph of $f(x) = \log_3 x$ shifted ___to the left 5 units___.

11. The graph of $g(x) = -\log_4 x$ is the graph of $f(x) = \log_4 x$ reflected about the ___x-axis___.

12. The graph of $g(x) = \log_5(-x)$ is the graph of $f(x) = \log_5 x$ reflected about the ___y-axis___.

13. The domain of $g(x) = \log_2(5 - x)$ can be found by solving the inequality ___$5 - x > 0$___.

14. The logarithmic function with base 10 is called the ___common___ logarithmic function. The function $f(x) = \log_{10} x$ is usually expressed as $f(x) = $ ___$\log x$___.

15. The logarithmic function with base e is called the ___natural___ logarithmic function. The function $f(x) = \log_e x$ is usually expressed as $f(x) = $ ___$\ln x$___.

EXERCISE SET 4.2

Practice Exercises

In Exercises 1–8, write each equation in its equivalent exponential form.

1. $4 = \log_2 16$ $2^4 = 16$
2. $6 = \log_2 64$ $2^6 = 64$
3. $2 = \log_3 x$ $3^2 = x$
4. $2 = \log_9 x$ $9^2 = x$
5. $5 = \log_b 32$ $b^5 = 32$
6. $3 = \log_b 27$ $b^3 = 27$
7. $\log_6 216 = y$ $6^y = 216$
8. $\log_5 125 = y$ $5^y = 125$

In Exercises 9–20, write each equation in its equivalent logarithmic form.

9. $2^3 = 8$ *
10. $5^4 = 625$ *
11. $2^{-4} = \frac{1}{16}$ *
12. $5^{-3} = \frac{1}{125}$ *
13. $\sqrt[3]{8} = 2$ *
14. $\sqrt[3]{64} = 4$ *
15. $13^2 = x$ *
16. $15^2 = x$ *
17. $b^3 = 1000$ *
18. $b^3 = 343$ *
19. $7^y = 200$ *
20. $8^y = 300$ *

In Exercises 21–42, evaluate each expression without using a calculator.

21. $\log_4 16$ 2
22. $\log_7 49$ 2
23. $\log_2 64$ 6
24. $\log_3 27$ 3
25. $\log_5 \frac{1}{5}$ −1
26. $\log_6 \frac{1}{6}$ −1
27. $\log_2 \frac{1}{8}$ −3
28. $\log_3 \frac{1}{9}$ −2
29. $\log_7 \sqrt{7}$ $\frac{1}{2}$
30. $\log_6 \sqrt{6}$ $\frac{1}{2}$
31. $\log_2 \frac{1}{\sqrt{2}}$ $-\frac{1}{2}$
32. $\log_3 \frac{1}{\sqrt{3}}$ $-\frac{1}{2}$
33. $\log_{64} 8$ $\frac{1}{2}$
34. $\log_{81} 9$ $\frac{1}{2}$
35. $\log_5 5$ 1
36. $\log_{11} 11$ 1
37. $\log_4 1$ 0
38. $\log_6 1$ 0
39. $\log_5 5^7$ 7
40. $\log_4 4^6$ 6
41. $8^{\log_8 19}$ 19
42. $7^{\log_7 23}$ 23

43. Graph $f(x) = 4^x$ and $g(x) = \log_4 x$ in the same rectangular coordinate system. *

44. Graph $f(x) = 5^x$ and $g(x) = \log_5 x$ in the same rectangular coordinate system. *

45. Graph $f(x) = \left(\frac{1}{2}\right)^x$ and $g(x) = \log_{\frac{1}{2}} x$ in the same rectangular coordinate system. *

46. Graph $f(x) = \left(\frac{1}{4}\right)^x$ and $g(x) = \log_{\frac{1}{4}} x$ in the same rectangular coordinate system. *

In Exercises 47–52, the graph of a logarithmic function is given. Select the function for each graph from the following options:

$$f(x) = \log_3 x, \ g(x) = \log_3(x - 1), \ h(x) = \log_3 x - 1,$$
$$F(x) = -\log_3 x, \ G(x) = \log_3(-x), \ H(x) = 1 - \log_3 x.$$

47. *

48. *

49. *

50. *

51. **52.**

In Exercises 53–58, begin by graphing $f(x) = \log_2 x$. Then use transformations of this graph to graph the given function. What is the vertical asymptote? Use the graphs to determine each function's domain and range.

53. $g(x) = \log_2(x + 1)$ * **54.** $g(x) = \log_2(x + 2)$ *
55. $h(x) = 1 + \log_2 x$ * **56.** $h(x) = 2 + \log_2 x$ *
57. $g(x) = \frac{1}{2}\log_2 x$ * **58.** $g(x) = -2\log_2 x$ *

The figure shows the graph of $f(x) = \log x$. In Exercises 59–64, use transformations of this graph to graph each function. Graph and give equations of the asymptotes. Use the graphs to determine each function's domain and range.

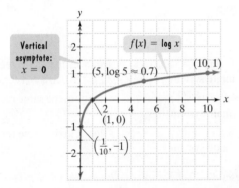

59. $g(x) = \log(x - 1)$ * **60.** $g(x) = \log(x - 2)$ *
61. $h(x) = \log x - 1$ * **62.** $h(x) = \log x - 2$ *
63. $g(x) = 1 - \log x$ * **64.** $g(x) = 2 - \log x$ *

The figure shows the graph of $f(x) = \ln x$. In Exercises 65–74, use transformations of this graph to graph each function. Graph and give equations of the asymptotes. Use the graphs to determine each function's domain and range.

65. $g(x) = \ln(x + 2)$ *
66. $g(x) = \ln(x + 1)$ *
67. $h(x) = \ln(2x)$ *
68. $h(x) = \ln\left(\frac{1}{2}x\right)$ *

69. $g(x) = 2\ln x$ *
70. $g(x) = \frac{1}{2}\ln x$ *
71. $h(x) = -\ln x$ *
72. $h(x) = \ln(-x)$ *
73. $g(x) = 2 - \ln x$ *
74. $g(x) = 1 - \ln x$ *

In Exercises 75–80, find the domain of each logarithmic function.

75. $f(x) = \log_5(x + 4)$ $(-4, \infty)$
76. $f(x) = \log_5(x + 6)$ $(-6, \infty)$
77. $f(x) = \log(2 - x)$ $(-\infty, 2)$
78. $f(x) = \log(7 - x)$ $(-\infty, 7)$
79. $f(x) = \ln(x - 2)^2$ $(-\infty, 2) \cup (2, \infty)$
80. $f(x) = \ln(x - 7)^2$ $(-\infty, 7) \cup (7, \infty)$

In Exercises 81–100, evaluate or simplify each expression without using a calculator.

81. $\log 100$ 2 **82.** $\log 1000$ 3
83. $\log 10^7$ 7 **84.** $\log 10^8$ 8
85. $10^{\log 33}$ 33 **86.** $10^{\log 53}$ 53
87. $\ln 1$ 0 **88.** $\ln e$ 1
89. $\ln e^6$ 6 **90.** $\ln e^7$ 7
91. $\ln\dfrac{1}{e^6}$ -6 **92.** $\ln\dfrac{1}{e^7}$ -7
93. $e^{\ln 125}$ 125 **94.** $e^{\ln 300}$ 300
95. $\ln e^{9x}$ $9x$ **96.** $\ln e^{13x}$ $13x$
97. $e^{\ln 5x^2}$ $5x^2$ **98.** $e^{\ln 7x^2}$ $7x^2$
99. $10^{\log\sqrt{x}}$ \sqrt{x} **100.** $10^{\log\sqrt[3]{x}}$ $\sqrt[3]{x}$

Practice Plus

In Exercises 101–104, write each equation in its equivalent exponential form. Then solve for x.

101. $\log_3(x - 1) = 2$ $3^2 = x - 1; \{10\}$
102. $\log_5(x + 4) = 2$ $5^2 = x + 4; \{21\}$
103. $\log_4 x = -3$ $4^{-3} = x; \{\frac{1}{64}\}$
104. $\log_{64} x = \dfrac{2}{3}$ $64^{2/3} = x; \{16\}$

In Exercises 105–108, evaluate each expression without using a calculator.

105. $\log_3(\log_7 7)$ 0
106. $\log_5(\log_2 32)$ 1
107. $\log_2(\log_3 81)$ 2
108. $\log(\ln e)$ 0

In Exercises 109–112, find the domain of each logarithmic function.

109. $f(x) = \ln(x^2 - x - 2)$ $(-\infty, -1) \cup (2, \infty)$
110. $f(x) = \ln(x^2 - 4x - 12)$ $(-\infty, -2) \cup (6, \infty)$
111. $f(x) = \log\left(\dfrac{x + 1}{x - 5}\right)$ $(-\infty, -1) \cup (5, \infty)$
112. $f(x) = \log\left(\dfrac{x - 2}{x + 5}\right)$ $(-\infty, -5) \cup (2, \infty)$

Application Exercises

The percentage of adult height attained by a girl who is x years old can be modeled by

$$f(x) = 62 + 35 \log(x - 4),$$

where x represents the girl's age (from 5 to 15) and f(x) represents the percentage of her adult height. Use the function to solve Exercises 113–114. Round answers to the nearest tenth of a percent.

113. Approximately what percentage of her adult height has a girl attained at age 13? 95.4%

114. Approximately what percentage of her adult height has a girl attained at age ten? 89.2%

The bar graph shows the average number of hours per week that U.S. wives and husbands engaged in housework in six selected years. Use this information to solve Exercises 115–116.

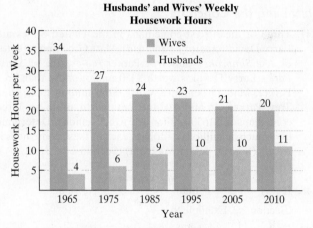

Husbands' and Wives' Weekly Housework Hours

Source: *The Sociology Project 2.0,* 2016, Pearson.

115. The function

$$f(x) = -3.52 \ln x + 34.5$$

models the wives' weekly housework hours, $f(x)$, x years after 1964.

a. Use the function to find the number of weekly housework hours for wives in 2010. Round to the nearest hour. Does this function value overestimate or underestimate the number of hours displayed by the graph? By how much? 21 hours; overestimates by 1 hour

b. Use the function to project wives' weekly housework hours in 2025. Round to the nearest hour. 20 hours

116. The function

$$f(x) = 1.8 \ln x + 3.42$$

models the husbands' weekly housework hours, $f(x)$, x years after 1964.

a. Use the function to find the number of weekly housework hours for husbands in 2010. Round to the nearest hour. Does this function value overestimate or underestimate the number of hours displayed by the graph? By how much? 10 hours; underestimates by 1 hour

b. Use the function to project husbands' weekly housework hours in 2025. Round to the nearest hour. 11 hours

The loudness level of a sound, D, in decibels, is given by the formula

$$D = 10 \log(10^{12} I),$$

where I is the intensity of the sound, in watts per meter2. Decibel levels range from 0, a barely audible sound, to 160, a sound resulting in a ruptured eardrum. (Any exposure to sounds of 130 decibels or higher puts a person at immediate risk for hearing damage.) Use the formula to solve Exercises 117–118.

117. The sound of a blue whale can be heard 500 miles away, reaching an intensity of 6.3×10^6 watts per meter2. Determine the decibel level of this sound. At close range, can the sound of a blue whale rupture the human eardrum? ≈188 db; yes

118. What is the decibel level of a normal conversation, 3.2×10^{-6} watt per meter2? ≈65 db

119. Students in a psychology class took a final examination. As part of an experiment to see how much of the course content they remembered over time, they took equivalent forms of the exam in monthly intervals thereafter. The average score for the group, $f(t)$, after t months was modeled by the function

$$f(t) = 88 - 15 \ln(t + 1), \qquad 0 \le t \le 12.$$

a. What was the average score on the original exam? 88

b. What was the average score after 2 months? 4 months? 6 months? 8 months? 10 months? one year? *

c. Sketch the graph of f (either by hand or with a graphing utility). Describe what the graph indicates in terms of the material retained by the students. *

Explaining the Concepts 120–127. Answers will vary.

120. Describe the relationship between an equation in logarithmic form and an equivalent equation in exponential form.

121. What question can be asked to help evaluate $\log_3 81$?

122. Explain why the logarithm of 1 with base b is 0.

123. Describe the following property using words: $\log_b b^x = x$.

124. Explain how to use the graph of $f(x) = 2^x$ to obtain the graph of $g(x) = \log_2 x$.

125. Explain how to find the domain of a logarithmic function.

126. Logarithmic models are well suited to phenomena in which growth is initially rapid but then begins to level off. Describe something that is changing over time that can be modeled using a logarithmic function.

127. Suppose that a girl is 4 feet 6 inches at age 10. Explain how to use the function in Exercises 113–114 to determine how tall she can expect to be as an adult.

Technology Exercises

In Exercises 128–131, graph f and g in the same viewing rectangle. Then describe the relationship of the graph of g to the graph of f.

128. $f(x) = \ln x, g(x) = \ln(x + 3)$ *

129. $f(x) = \ln x, g(x) = \ln x + 3$ *

130. $f(x) = \log x, g(x) = -\log x$ *

131. $f(x) = \log x, g(x) = \log(x - 2) + 1$ *

132. Students in a mathematics class took a final examination. They took equivalent forms of the exam in monthly intervals thereafter. The average score, $f(t)$, for the group after t months was modeled by the human memory function $f(t) = 75 - 10 \log(t + 1)$, where $0 \le t \le 12$. Use a graphing utility to graph the function. Then determine how many months elapsed before the average score fell below 65. 9 months*

133. In parts (a)–(c), graph f and g in the same viewing rectangle.

a. $f(x) = \ln(3x), g(x) = \ln 3 + \ln x$ *

b. $f(x) = \log(5x^2), g(x) = \log 5 + \log x^2$ *

c. $f(x) = \ln(2x^3), g(x) = \ln 2 + \ln x^3$ *

d. Describe what you observe in parts (a)–(c). Generalize this observation by writing an equivalent expression for $\log_b(MN)$, where $M > 0$ and $N > 0$. *

e. Complete this statement: The logarithm of a product is equal to <u>the sum of the logarithms of its factors</u>.

134. Graph each of the following functions in the same viewing rectangle and then place the functions in order from the one that increases most slowly to the one that increases most rapidly.

$$y = x, y = \sqrt{x}, y = e^x, y = \ln x, y = x^x, y = x^2 \;\; *$$

Critical Thinking Exercises

Make Sense? *In Exercises 135–138, determine whether each statement makes sense or does not make sense, and explain your reasoning.*

135. I've noticed that exponential functions and logarithmic functions exhibit inverse, or opposite, behavior in many ways. For example, a vertical translation shifts an exponential function's horizontal asymptote and a horizontal translation shifts a logarithmic function's vertical asymptote. *

136. I estimate that $\log_8 16$ lies between 1 and 2 because $8^1 = 8$ and $8^2 = 64$. makes sense

137. I can evaluate some common logarithms without having to use a calculator. makes sense

138. An earthquake of magnitude 8 on the Richter scale is twice as intense as an earthquake of magnitude 4. *

In Exercises 139–142, determine whether each statement is true or false. If the statement is false, make the necessary change(s) to produce a true statement. Changes to false statements will vary.

139. $\dfrac{\log_2 8}{\log_2 4} = \dfrac{8}{4}$ false **140.** $\log(-100) = -2$ false

141. The domain of $f(x) = \log_2 x$ is $(-\infty, \infty)$. false

142. $\log_b x$ is the exponent to which b must be raised to obtain x. *

143. Without using a calculator, find the exact value of

$$\frac{\log_3 81 - \log_\pi 1}{\log_{2\sqrt{2}} 8 - \log 0.001}. \quad \tfrac{4}{5} \;\; *$$

144. Without using a calculator, find the exact value of $\log_4[\log_3(\log_2 8)]$. 0

145. Without using a calculator, determine which is the greater number: $\log_4 60$ or $\log_3 40$. $\log_3 40 > \log_4 60$

Group Exercise Answers will vary.

146. This group exercise involves exploring the way we grow. Group members should create a graph for the function that models the percentage of adult height attained by a boy who is x years old, $f(x) = 29 + 48.8 \log(x + 1)$. Let $x = 5, 6, 7, \ldots, 15$, find function values, and connect the resulting points with a smooth curve. Then create a graph for the function that models the percentage of adult height attained by a girl who is x years old, $g(x) = 62 + 35 \log(x - 4)$. Let $x = 5, 6, 7, \ldots, 15$, find function values, and connect the resulting points with a smooth curve. Group members should then discuss similarities and differences in the growth patterns for boys and girls based on the graphs.

Retaining the Concepts

147. Three of the richest comedians in the United States are Larry David (creator of *Seinfeld*), Matt Groening (creator of *The Simpsons*), and Trey Parker (co-creator of *South Park*). Larry David is worth \$450 million more than Trey Parker. Matt Groening is worth \$150 million more than Trey Parker. Combined, the net worth of these three comedians is \$1650 million (or \$16.5 billion). Determine how much, in millions of dollars, each of these comedians is worth. (Source: petamovies.com) (Section 1.3, Example 1)
Parker: \$350 million; David: \$800 million; Groening: \$500 million

148. If $f(x) = mx + b$, find $\dfrac{f(x + h) - f(x)}{h}, h \ne 0$. m
(Section 2.2, Example 8)

149. Find the inverse of $f(x) = x^2 + 4, x \ge 0$. $f^{-1}(x) = \sqrt{x - 4}$
(Section 2.7, Example 7)

Preview Exercises

Exercises 150–152 will help you prepare for the material covered in the next section. In each exercise, evaluate the indicated logarithmic expressions without using a calculator.

150. a. Evaluate: $\log_2 32$. 5
 b. Evaluate: $\log_2 8 + \log_2 4$. 5
 c. What can you conclude about $\log_2 32$, or $\log_2(8 \cdot 4)$? *

151. a. Evaluate: $\log_2 16$. 4
 b. Evaluate: $\log_2 32 - \log_2 2$. 4
 c. What can you conclude about

$$\log_2 16, \text{ or } \log_2\left(\frac{32}{2}\right)? \quad \log_2\left(\tfrac{32}{2}\right) = \log_2 32 - \log_2 2$$

152. a. Evaluate: $\log_3 81$. 4
 b. Evaluate: $2 \log_3 9$. 4
 c. What can you conclude about

$$\log_3 81, \text{ or } \log_3 9^2? \quad \log_3 9^2 = 2 \log_3 9$$

Section 4.3 Properties of Logarithms

What am I supposed to learn?

After studying this section, you should be able to:

1. Use the product rule.
2. Use the quotient rule.
3. Use the power rule.
4. Expand logarithmic expressions.
5. Condense logarithmic expressions.
6. Use the change-of-base property.

1. Use the product rule.

DISCOVERY

We know that log 100,000 = 5. Show that you get the same result by writing 100,000 as 1000 · 100 and then using the product rule. Then verify the product rule by using other numbers whose logarithms are easy to find.

> You can find the proof of the product rule in the appendix.

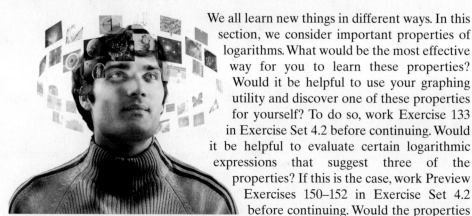

We all learn new things in different ways. In this section, we consider important properties of logarithms. What would be the most effective way for you to learn these properties? Would it be helpful to use your graphing utility and discover one of these properties for yourself? To do so, work Exercise 133 in Exercise Set 4.2 before continuing. Would it be helpful to evaluate certain logarithmic expressions that suggest three of the properties? If this is the case, work Preview Exercises 150–152 in Exercise Set 4.2 before continuing. Would the properties become more meaningful if you could see exactly where they come from? If so, you will find details of the proofs of many of these properties in the appendix. The remainder of our work in this chapter will be based on the properties of logarithms that you learn in this section.

The Product Rule

Properties of exponents correspond to properties of logarithms. For example, when we multiply with the same base, we add exponents:

$$b^m \cdot b^n = b^{m+n}.$$

This property of exponents, coupled with an awareness that a logarithm is an exponent, suggests the following property, called the **product rule**:

The Product Rule

Let b, M, and N be positive real numbers with $b \neq 1$.

$$\log_b(MN) = \log_b M + \log_b N$$

The logarithm of a product is the sum of the logarithms.

When we use the product rule to write a single logarithm as the sum of two logarithms, we say that we are **expanding a logarithmic expression**. For example, we can use the product rule to expand $\ln(7x)$:

$$\ln(7x) = \ln 7 + \ln x.$$

The logarithm of a product is the sum of the logarithms.

EXAMPLE 1 Using the Product Rule

Use the product rule to expand each logarithmic expression:

a. $\log_4(7 \cdot 5)$ **b.** $\log(10x)$.

SOLUTION

a. $\log_4(7 \cdot 5) = \log_4 7 + \log_4 5$ The logarithm of a product is the sum of the logarithms.

b. $\log(10x) = \log 10 + \log x$ The logarithm of a product is the sum of the logarithms. These are common logarithms with base 10 understood.

$= 1 + \log x$ Because $\log_b b = 1$, then $\log 10 = 1$. • • •

 Check Point **1** Use the product rule to expand each logarithmic expression:

a. $\log_6(7 \cdot 11)$ $\log_6 7 + \log_6 11$ **b.** $\log(100x)$. $2 + \log x$

 Use the quotient rule.

The Quotient Rule

When we divide with the same base, we subtract exponents:

$$\frac{b^m}{b^n} = b^{m-n}.$$

This property suggests the following property of logarithms, called the **quotient rule**:

> **The Quotient Rule**
>
> Let b, M, and N be positive real numbers with $b \neq 1$.
>
> $$\log_b\left(\frac{M}{N}\right) = \log_b M - \log_b N$$
>
> The logarithm of a quotient is the difference of the logarithms.

DISCOVERY

We know that $\log_2 16 = 4$. Show that you get the same result by writing 16 as $\dfrac{32}{2}$ and then using the quotient rule. Then verify the quotient rule using other numbers whose logarithms are easy to find.

When we use the quotient rule to write a single logarithm as the difference of two logarithms, we say that we are **expanding a logarithmic expression**. For example, we can use the quotient rule to expand $\log \dfrac{x}{2}$:

$$\log\left(\frac{x}{2}\right) = \log x - \log 2.$$

The logarithm of a quotient is the difference of the logarithms.

EXAMPLE 2 Using the Quotient Rule

Use the quotient rule to expand each logarithmic expression:

a. $\log_7\left(\dfrac{19}{x}\right)$ **b.** $\ln\left(\dfrac{e^3}{7}\right)$.

SOLUTION

a. $\log_7\left(\dfrac{19}{x}\right) = \log_7 19 - \log_7 x$ The logarithm of a quotient is the difference of the logarithms.

b. $\ln\left(\dfrac{e^3}{7}\right) = \ln e^3 - \ln 7$ The logarithm of a quotient is the difference of the logarithms. These are natural logarithms with base e understood.

$= 3 - \ln 7$ Because $\ln e^x = x$, then $\ln e^3 = 3$. • • •

 Check Point **2** Use the quotient rule to expand each logarithmic expression:

a. $\log_8\left(\dfrac{23}{x}\right)$ $\log_8 23 - \log_8 x$ **b.** $\ln\left(\dfrac{e^5}{11}\right)$. $5 - \ln 11$

③ Use the power rule.

The Power Rule

When an exponential expression is raised to a power, we multiply exponents:

$$(b^m)^n = b^{mn}.$$

This property suggests the following property of logarithms, called the **power rule**:

> **The Power Rule**
>
> Let b and M be positive real numbers with $b \neq 1$, and let p be any real number.
>
> $$\log_b M^p = p \log_b M$$
>
> The logarithm of a number with an exponent is the product of the exponent and the logarithm of that number.

When we use the power rule to "pull the exponent to the front," we say that we are **expanding a logarithmic expression**. For example, we can use the power rule to expand $\ln x^2$:

$$\ln x^2 = 2 \ln x.$$

| The logarithm of a number with an exponent | is | the product of the exponent and the logarithm of that number. |

Figure 4.18 shows the graphs of $y = \ln x^2$ and $y = 2 \ln x$ in $[-5, 5, 1]$ by $[-5, 5, 1]$ viewing rectangles. Are $\ln x^2$ and $2 \ln x$ the same? The graphs illustrate that $y = \ln x^2$ and $y = 2 \ln x$ have different domains. The graphs are only the same if $x > 0$. Thus, we should write

$$\ln x^2 = 2 \ln x \text{ for } x > 0.$$

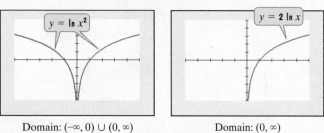

Domain: $(-\infty, 0) \cup (0, \infty)$ Domain: $(0, \infty)$

FIGURE 4.18 $\ln x^2$ and $2 \ln x$ have different domains.

When expanding a logarithmic expression, you might want to determine whether the rewriting has changed the domain of the expression. For the rest of this section, assume that all variables and variable expressions represent positive numbers.

EXAMPLE 3 Using the Power Rule

Use the power rule to expand each logarithmic expression:

a. $\log_5 7^4$ **b.** $\ln \sqrt{x}$ **c.** $\log(4x)^5$.

SOLUTION

a. $\log_5 7^4 = 4 \log_5 7$ The logarithm of a number with an exponent is the exponent times the logarithm of the number.

b. $\ln \sqrt{x} = \ln x^{\frac{1}{2}}$ Rewrite the radical using a rational exponent.

$\quad = \frac{1}{2} \ln x$ Use the power rule to bring the exponent to the front.

c. $\log(4x)^5 = 5 \log(4x)$ We immediately apply the power rule because the entire variable expression, $4x$, is raised to the 5th power. • • •

⊘ Check Point **3** Use the power rule to expand each logarithmic expression:

a. $\log_6 3^9$ $9 \log_6 3$ **b.** $\ln \sqrt[3]{x}$ $\frac{1}{3} \ln x$ **c.** $\log(x + 4)^2$. $2 \log (x + 4)$

④ Expand logarithmic expressions.

Expanding Logarithmic Expressions

It is sometimes necessary to use more than one property of logarithms when you expand a logarithmic expression. Properties for expanding logarithmic expressions are as follows:

Properties for Expanding Logarithmic Expressions

For $M > 0$ and $N > 0$:

1. $\log_b (MN) = \log_b M + \log_b N$ Product rule

2. $\log_b \left(\dfrac{M}{N} \right) = \log_b M - \log_b N$ Quotient rule

3. $\log_b M^p = p \log_b M$ Power rule

GREAT QUESTION!

Are there some common blunders that I can avoid when using properties of logarithms?

The graphs show that

$$\ln(x + 3) \neq \ln x + \ln 3.$$

| $y = \ln x$ shifted 3 units left | $y = \ln x$ shifted ln 3 units up |

In general,

$$\log_b(M + N) \neq \log_b M + \log_b N.$$

$y_1 = \ln (x + 3)$

$y_2 = \ln x + \ln 3$

$[-4, 5, 1]$ by $[-3, 3, 1]$

Try to avoid the following errors:

Incorrect!

$\log_b (M + N) = \log_b M + \log_b N$

$\log_b (M - N) = \log_b M - \log_b N$

$\log_b (M \cdot N) = \log_b M \cdot \log_b N$

$\log_b \left(\dfrac{M}{N} \right) = \dfrac{\log_b M}{\log_b N}$

$\dfrac{\log_b M}{\log_b N} = \log_b M - \log_b N$

$\log_b (MN^p) = p \log_b (MN)$

EXAMPLE 4 Expanding Logarithmic Expressions

Use logarithmic properties to expand each expression as much as possible:

a. $\log_b(x^2 \sqrt{y})$ **b.** $\log_6 \left(\dfrac{\sqrt[3]{x}}{36y^4} \right)$.

SOLUTION

We will have to use two or more of the properties for expanding logarithms in each part of this example.

a. $\log_b (x^2 \sqrt{y}) = \log_b \left(x^2 y^{\frac{1}{2}} \right)$ Use exponential notation.

$= \log_b x^2 + \log_b y^{\frac{1}{2}}$ Use the product rule.

$= 2 \log_b x + \dfrac{1}{2} \log_b y$ Use the power rule.

b. $\log_6\left(\dfrac{\sqrt[3]{x}}{36y^4}\right) = \log_6\left(\dfrac{x^{\frac{1}{3}}}{36y^4}\right)$ Use exponential notation.

$= \log_6 x^{\frac{1}{3}} - \log_6(36y^4)$ Use the quotient rule.

$= \log_6 x^{\frac{1}{3}} - (\log_6 36 + \log_6 y^4)$ Use the product rule on $\log_6(36y^4)$.

$= \dfrac{1}{3}\log_6 x - (\log_6 36 + 4\log_6 y)$ Use the power rule.

$= \dfrac{1}{3}\log_6 x - \log_6 36 - 4\log_6 y$ Apply the distributive property.

$= \dfrac{1}{3}\log_6 x - 2 - 4\log_6 y$ $\log_6 36 = 2$ because 2 is the power to which we must raise 6 to get 36. $(6^2 = 36)$ •••

📝 Check Point **4** Use logarithmic properties to expand each expression as much as possible:

a. $\log_b\left(x^4\sqrt[3]{y}\right)$ $4\log_b x + \frac{1}{3}\log_b y$ **b.** $\log_5\left(\dfrac{\sqrt{x}}{25y^3}\right).$ $\frac{1}{2}\log_5 x - 2 - 3\log_5 y$

⑤ Condense logarithmic expressions.

Condensing Logarithmic Expressions

To **condense a logarithmic expression**, we write the sum or difference of two or more logarithmic expressions as a single logarithmic expression. We use the properties of logarithms to do so.

GREAT QUESTION!

Are the properties listed on the right the same as those in the box on page 496?

Yes. The only difference is that we've reversed the sides in each property from the previous box.

> **Properties for Condensing Logarithmic Expressions**
>
> For $M > 0$ and $N > 0$:
>
> **1.** $\log_b M + \log_b N = \log_b(MN)$ Product rule
>
> **2.** $\log_b M - \log_b N = \log_b\left(\dfrac{M}{N}\right)$ Quotient rule
>
> **3.** $p\log_b M = \log_b M^p$ Power rule

EXAMPLE 5 Condensing Logarithmic Expressions

Write as a single logarithm:

a. $\log_4 2 + \log_4 32$ **b.** $\log(4x - 3) - \log x.$

SOLUTION

a. $\log_4 2 + \log_4 32 = \log_4(2 \cdot 32)$ Use the product rule.

$= \log_4 64$ We now have a single logarithm. However, we can simplify.

$= 3$ $\log_4 64 = 3$ because $4^3 = 64$.

b. $\log(4x - 3) - \log x = \log\left(\dfrac{4x - 3}{x}\right)$ Use the quotient rule. •••

📝 Check Point **5** Write as a single logarithm:

a. $\log 25 + \log 4$ $\log 100 = 2$ **b.** $\log(7x + 6) - \log x.$ $\log\frac{7x + 6}{x}$

Coefficients of logarithms must be 1 before you can condense them using the product and quotient rules. For example, to condense

$$2 \ln x + \ln(x + 1),$$

the coefficient of the first term must be 1. We use the power rule to rewrite the coefficient as an exponent:

> 1. Use the power rule to make the number in front an exponent.

$$2 \ln x + \ln(x + 1) = \ln x^2 + \ln(x + 1) = \ln[x^2(x + 1)].$$

> 2. Use the product rule. The sum of logarithms with coefficients of 1 is the logarithm of the product.

EXAMPLE 6 Condensing Logarithmic Expressions

Write as a single logarithm:

a. $\frac{1}{2} \log x + 4 \log(x - 1)$　　　　　**b.** $3 \ln (x + 7) - \ln x$

c. $4 \log_b x - 2 \log_b 6 - \frac{1}{2} \log_b y.$

SOLUTION

a. $\frac{1}{2} \log x + 4 \log(x - 1)$

$= \log x^{\frac{1}{2}} + \log(x - 1)^4$ 　　Use the power rule so that all coefficients are 1.

$= \log\left[x^{\frac{1}{2}}(x - 1)^4 \right]$ 　　Use the product rule. The condensed form can be expressed as $\log[\sqrt{x}\, (x - 1)^4]$.

b. $3 \ln(x + 7) - \ln x$

$= \ln(x + 7)^3 - \ln x$ 　　Use the power rule so that all coefficients are 1.

$= \ln\left[\dfrac{(x + 7)^3}{x} \right]$ 　　Use the quotient rule.

c. $4 \log_b x - 2 \log_b 6 - \frac{1}{2} \log_b y$

$= \log_b x^4 - \log_b 6^2 - \log_b y^{\frac{1}{2}}$ 　　Use the power rule so that all coefficients are 1.

$= \log_b x^4 - \left(\log_b 36 + \log_b y^{\frac{1}{2}} \right)$ 　　Rewrite as a single subtraction.

$= \log_b x^4 - \log_b\left(36 y^{\frac{1}{2}} \right)$ 　　Use the product rule.

$= \log_b\left(\dfrac{x^4}{36 y^{\frac{1}{2}}} \right)$ or $\log_b\left(\dfrac{x^4}{36 \sqrt{y}} \right)$ 　　Use the quotient rule. 　　• • •

✓ **Check Point 6** Write as a single logarithm:

a. $2 \ln x + \frac{1}{3} \ln(x + 5)$ 　$\ln[x^2 \sqrt[3]{x + 5}]$ 　　**b.** $2 \log(x - 3) - \log x$ 　$\log \dfrac{(x - 3)^2}{x}$

c. $\frac{1}{4} \log_b x - 2 \log_b 5 - 10 \log_b y.$ 　$\log_b \dfrac{\sqrt[4]{x}}{25 y^{10}}$

⑥ Use the change-of-base property.

The Change-of-Base Property

We have seen that all scientific and graphing calculators give the values of both common logarithms (base 10) and natural logarithms (base e). Some graphing calculators also give the value of logarithms with bases other than 10 or e. Whether or not your calculator has this option, you can find a logarithm with any base using the following change-of-base property:

The Change-of-Base Property

For any logarithmic bases a and b, and any positive number M,

$$\log_b M = \frac{\log_a M}{\log_a b}.$$

The logarithm of M with base b is equal to the logarithm of M with any new base divided by the logarithm of b with that new base.

In the change-of-base property, base b is the base of the original logarithm. Base a is a new base that we introduce. Thus, the change-of-base property allows us to change from base b to *any* new base a, as long as the newly introduced base is a positive number not equal to 1.

The change-of-base property is used to write a logarithm in terms of quantities that can be evaluated with a calculator. Because calculators contain keys for common (base 10) and natural (base e) logarithms, we will frequently introduce base 10 or base e.

Change-of-Base Property	Introducing Common Logarithms	Introducing Natural Logarithms
$\log_b M = \dfrac{\log_a M}{\log_a b}$	$\log_b M = \dfrac{\log_{10} M}{\log_{10} b}$	$\log_b M = \dfrac{\log_e M}{\log_e b}$
a is the new introduced base.	10 is the new introduced base.	*e* is the new introduced base.

Using the notations for common logarithms and natural logarithms, we have the following results:

The Change-of-Base Property: Introducing Common and Natural Logarithms

Introducing Common Logarithms

$$\log_b M = \frac{\log M}{\log b}$$

Introducing Natural Logarithms

$$\log_b M = \frac{\ln M}{\ln b}$$

EXAMPLE 7 Changing Base to Common Logarithms

Use common logarithms to evaluate $\log_5 140$.

SOLUTION

Because $\log_b M = \dfrac{\log M}{\log b}$,

$$\log_5 140 = \frac{\log 140}{\log 5}$$

$$\approx 3.07.$$

Use a calculator: 140 [LOG] [÷] 5 [LOG] [=] or [LOG] 140 [÷] [LOG] 5 [ENTER]. On some calculators, parentheses are needed after 140 and 5.

This means that $\log_5 140 \approx 3.07$. •••

✓ Check Point **7** Use common logarithms to evaluate $\log_7 2506$. 4.02

DISCOVERY

Find a reasonable estimate of $\log_5 140$ to the nearest whole number. To what power can you raise 5 in order to get 140? Compare your estimate to the value obtained in Example 7.

EXAMPLE 8 Changing Base to Natural Logarithms

Use natural logarithms to evaluate $\log_5 140$.

SOLUTION

Because $\log_b M = \dfrac{\ln M}{\ln b}$,

$$\log_5 140 = \frac{\ln 140}{\ln 5}$$

$$\approx 3.07.$$ Use a calculator: 140 $\boxed{\text{LN}}$ $\boxed{\div}$ 5 $\boxed{\text{LN}}$ $\boxed{=}$

or $\boxed{\text{LN}}$ 140 $\boxed{\div}$ $\boxed{\text{LN}}$ 5 $\boxed{\text{ENTER}}$. On some calculators, parentheses are needed after 140 and 5.

We have again shown that $\log_5 140 \approx 3.07$. • • •

⊘ Check Point **8** Use natural logarithms to evaluate $\log_7 2506$. 4.02

TECHNOLOGY

We can use the change-of-base property to graph logarithmic functions with bases other than 10 or e on a graphing utility. For example, **Figure 4.19** shows the graphs of

$$y = \log_2 x \quad \text{and} \quad y = \log_{20} x$$

in a $[0, 10, 1]$ by $[-3, 3, 1]$ viewing rectangle. Because $\log_2 x = \dfrac{\ln x}{\ln 2}$ and $\log_{20} x = \dfrac{\ln x}{\ln 20}$, the functions are entered as

$$y_1 = \boxed{\text{LN}}\ x\ \boxed{\div}\ \boxed{\text{LN}}\ 2,$$

and $y_2 = \boxed{\text{LN}}\ x\ \boxed{\div}\ \boxed{\text{LN}}\ 20.$

On some calculators, parentheses are needed after x, **2**, and **20**.

FIGURE 4.19 Using the change-of-base property to graph logarithmic functions

ACHIEVING SUCCESS

Assuming that you have done very well preparing for an exam, **there are certain things you can do that will make you a better test taker**.

- Just before the exam, briefly review the relevant material in the chapter summary.
- Bring everything you need to the exam, including two pencils, an eraser, scratch paper (if permitted), a calculator (if you're allowed to use one), water, and a watch.
- Survey the entire exam quickly to get an idea of its length.
- Read the directions to each problem carefully. Make sure that you have answered the specific question asked.
- Work the easy problems first. Then return to the hard problems you are not sure of. Doing the easy problems first will build your confidence. If you get bogged down on any one problem, you may not be able to complete the exam and receive credit for the questions you can easily answer.
- Attempt every problem. There may be partial credit even if you do not obtain the correct answer.
- Work carefully. Show your step-by-step solutions neatly. Check your work and answers.
- Watch the time. Pace yourself and be aware of when half the time is up. Determine how much of the exam you have completed. This will indicate if you're moving at a good pace or need to speed up. Prepare to spend more time on problems worth more points.
- Never turn in a test early. Use every available minute you are given for the test. If you have extra time, double check your arithmetic and look over your solutions.

CONCEPT AND VOCABULARY CHECK

Fill in each blank so that the resulting statement is true.

1. The product rule for logarithms states that $\log_b(MN) = \underline{\log_b M + \log_b N}$. The logarithm of a product is the $\underline{\text{sum}}$ of the logarithms.

2. The quotient rule for logarithms states that $\log_b\left(\dfrac{M}{N}\right) = \underline{\log_b M - \log_b N}$. The logarithm of a quotient is the $\underline{\text{difference}}$ of the logarithms.

3. The power rule for logarithms states that $\log_b M^p = \underline{p \log_b M}$. The logarithm of a number with an exponent is the $\underline{\text{product}}$ of the exponent and the logarithm of that number.

4. The change-of-base property for logarithms allows us to write logarithms with base b in terms of a new base a. Introducing base a, the property states that
$$\log_b M = \underline{\dfrac{\log_a M}{\log_a b}}.$$

EXERCISE SET 4.3

Practice Exercises

In Exercises 1–40, use properties of logarithms to expand each logarithmic expression as much as possible. Where possible, evaluate logarithmic expressions without using a calculator.

1. $\log_5(7 \cdot 3)$ *
2. $\log_8(13 \cdot 7)$ *
3. $\log_7(7x)$ *
4. $\log_9(9x)$ *
5. $\log(1000x)$ *
6. $\log(10{,}000x)$ *
7. $\log_7\left(\dfrac{7}{x}\right)$ *
8. $\log_9\left(\dfrac{9}{x}\right)$ *
9. $\log\left(\dfrac{x}{100}\right)$ *
10. $\log\left(\dfrac{x}{1000}\right)$ *
11. $\log_4\left(\dfrac{64}{y}\right)$ *
12. $\log_5\left(\dfrac{125}{y}\right)$ *
13. $\ln\left(\dfrac{e^2}{5}\right)$ *
14. $\ln\left(\dfrac{e^4}{8}\right)$ *
15. $\log_b x^3$ $3\log_b x$
16. $\log_b x^7$ $7\log_b x$
17. $\log N^{-6}$ $-6\log N$
18. $\log M^{-8}$ $-8\log M$
19. $\ln\sqrt[5]{x}$ $\frac{1}{5}\ln x$
20. $\ln\sqrt[7]{x}$ $\frac{1}{7}\ln x$
21. $\log_b(x^2 y)$ *
22. $\log_b(xy^3)$ *
23. $\log_4\left(\dfrac{\sqrt{x}}{64}\right)$ *
24. $\log_5\left(\dfrac{\sqrt{x}}{25}\right)$ *
25. $\log_6\left(\dfrac{36}{\sqrt{x+1}}\right)$ *
26. $\log_8\left(\dfrac{64}{\sqrt{x+1}}\right)$ *
27. $\log_b\left(\dfrac{x^2 y}{z^2}\right)$ *
28. $\log_b\left(\dfrac{x^3 y}{z^2}\right)$ *
29. $\log\sqrt{100x}$ *
30. $\ln\sqrt{ex}$ *
31. $\log\sqrt[3]{\dfrac{x}{y}}$ *
32. $\log\sqrt[5]{\dfrac{x}{y}}$ *
33. $\log_b\left(\dfrac{\sqrt{x}y^3}{z^3}\right)$ *
34. $\log_b\left(\dfrac{\sqrt[3]{x}y^4}{z^5}\right)$ *
35. $\log_5\sqrt[3]{\dfrac{x^2 y}{25}}$ *
36. $\log_2\sqrt[5]{\dfrac{xy^4}{16}}$ *
37. $\ln\left[\dfrac{x^3\sqrt{x^2+1}}{(x+1)^4}\right]$ *
38. $\ln\left[\dfrac{x^4\sqrt{x^2+3}}{(x+3)^5}\right]$ *
39. $\log\left[\dfrac{10x^2\sqrt[3]{1-x}}{7(x+1)^2}\right]$ *
40. $\log\left[\dfrac{100x^3\sqrt[3]{5-x}}{3(x+7)^2}\right]$ *

In Exercises 41–70, use properties of logarithms to condense each logarithmic expression. Write the expression as a single logarithm whose coefficient is 1. Where possible, evaluate logarithmic expressions without using a calculator.

41. $\log 5 + \log 2$ 1
42. $\log 250 + \log 4$ 3

43. $\ln x + \ln 7$ $\ln(7x)$
44. $\ln x + \ln 3$ $\ln(3x)$
45. $\log_2 96 - \log_2 3$ 5
46. $\log_3 405 - \log_3 5$ 4
47. $\log(2x+5) - \log x$ *
48. $\log(3x+7) - \log x$ *
49. $\log x + 3\log y$ $\log(xy^3)$
50. $\log x + 7\log y$ $\log(xy^7)$
51. $\frac{1}{2}\ln x + \ln y$ $\ln(y\sqrt{x})$*
52. $\frac{1}{3}\ln x + \ln y$ $\ln(y\sqrt[3]{x})$*
53. $2\log_b x + 3\log_b y$ $\log_b(x^2y^3)$
54. $5\log_b x + 6\log_b y$ $\log_b(x^5y^6)$
55. $5\ln x - 2\ln y$ *
56. $7\ln x - 3\ln y$ *
57. $3\ln x - \frac{1}{3}\ln y$ *
58. $2\ln x - \frac{1}{2}\ln y$ *
59. $4\ln(x+6) - 3\ln x$ *
60. $8\ln(x+9) - 4\ln x$ *
61. $3\ln x + 5\ln y - 6\ln z$ *
62. $4\ln x + 7\ln y - 3\ln z$ *
63. $\frac{1}{2}(\log x + \log y)$ $\log\sqrt{xy}$
64. $\frac{1}{3}(\log_4 x - \log_4 y)$ $\log_4\sqrt[3]{\frac{x}{y}}$
65. $\frac{1}{2}(\log_5 x + \log_5 y) - 2\log_5(x+1)$ *
66. $\frac{1}{3}(\log_4 x - \log_4 y) + 2\log_4(x+1)$ *
67. $\frac{1}{3}[2\ln(x+5) - \ln x - \ln(x^2-4)]$ *
68. $\frac{1}{3}[5\ln(x+6) - \ln x - \ln(x^2-25)]$ *
69. $\log x + \log(x^2-1) - \log 7 - \log(x+1)$ *
70. $\log x + \log(x^2-4) - \log 15 - \log(x+2)$ *

In Exercises 71–78, use common logarithms or natural logarithms and a calculator to evaluate to four decimal places.

71. $\log_5 13$ 1.5937
72. $\log_6 17$ 1.5812
73. $\log_{14} 87.5$ 1.6944
74. $\log_{16} 57.2$ 1.4595
75. $\log_{0.1} 17$ -1.2304
76. $\log_{0.3} 19$ -2.4456
77. $\log_\pi 63$ 3.6193
78. $\log_\pi 400$ 5.2340

In Exercises 79–82, use a graphing utility and the change-of-base property to graph each function.

79. $y = \log_3 x$ *
80. $y = \log_{15} x$ *
81. $y = \log_2(x+2)$ *
82. $y = \log_3(x-2)$ *

Practice Plus

In Exercises 83–88, let $\log_b 2 = A$ and $\log_b 3 = C$. Write each expression in terms of A and C.

83. $\log_b \frac{3}{2}$ $C - A$
84. $\log_b 6$ $A + C$
85. $\log_b 8$ $3A$
86. $\log_b 81$ $4C$
87. $\log_b\sqrt{\dfrac{2}{27}}$ *
88. $\log_b\sqrt{\dfrac{3}{16}}$ *

*See Answers to Selected Exercises.

In Exercises 89–102, determine whether each equation is true or false. Where possible, show work to support your conclusion. If the statement is false, make the necessary change(s) to produce a true statement.

89. $\ln e = 0$ false; $\ln e = 1$

90. $\ln 0 = e$ false*

91. $\log_4(2x^3) = 3\log_4(2x)$ *

92. $\ln(8x^3) = 3\ln(2x)$ true

93. $x \log 10^x = x^2$ true

94. $\ln(x + 1) = \ln x + \ln 1$ *

95. $\ln(5x) + \ln 1 = \ln(5x)$ *

96. $\ln x + \ln(2x) = \ln(3x)$ *

97. $\log(x + 3) - \log(2x) = \dfrac{\log(x + 3)}{\log(2x)}$ false*

98. $\dfrac{\log(x + 2)}{\log(x - 1)} = \log(x + 2) - \log(x - 1)$ false*

99. $\log_6\left(\dfrac{x - 1}{x^2 + 4}\right) = \log_6(x - 1) - \log_6(x^2 + 4)$ true

100. $\log_6[4(x + 1)] = \log_6 4 + \log_6(x + 1)$ true

101. $\log_3 7 = \dfrac{1}{\log_7 3}$ true

102. $e^x = \dfrac{1}{\ln x}$ false*

Application Exercises

103. The loudness level of a sound can be expressed by comparing the sound's intensity to the intensity of a sound barely audible to the human ear. The formula

$$D = 10(\log I - \log I_0)$$

describes the loudness level of a sound, D, in decibels, where I is the intensity of the sound, in watts per meter2, and I_0 is the intensity of a sound barely audible to the human ear.

a. Express the formula so that the expression in parentheses is written as a single logarithm. $D = 10 \log \dfrac{I}{I_0}$

b. Use the form of the formula from part (a) to answer this question: If a sound has an intensity 100 times the intensity of a softer sound, how much larger on the decibel scale is the loudness level of the more intense sound? 20 db

104. The formula

$$t = \frac{1}{c}[\ln A - \ln(A - N)]$$

describes the time, t, in weeks, that it takes to achieve mastery of a portion of a task, where A is the maximum learning possible, N is the portion of the learning that is to be achieved, and c is a constant used to measure an individual's learning style.

a. Express the formula so that the expression in brackets is written as a single logarithm. $t = \dfrac{1}{c} \ln\left(\dfrac{A}{A - N}\right)$

b. The formula is also used to determine how long it will take chimpanzees and apes to master a task. For example, a typical chimpanzee learning sign language can master a maximum of 65 signs. Use the form of the formula from part (a) to answer this question: How many weeks will it take a chimpanzee to master 30 signs if c for that chimp is 0.03? more than $20\frac{1}{2}$ weeks

Explaining the Concepts 105–112. Answers will vary.

105. Describe the product rule for logarithms and give an example.

106. Describe the quotient rule for logarithms and give an example.

107. Describe the power rule for logarithms and give an example.

108. Without showing the details, explain how to condense $\ln x - 2\ln(x + 1)$.

109. Describe the change-of-base property and give an example.

110. Explain how to use your calculator to find $\log_{14} 283$.

111. You overhear a student talking about a property of logarithms in which division becomes subtraction. Explain what the student means by this.

112. Find $\ln 2$ using a calculator. Then calculate each of the following: $1 - \frac{1}{2}$; $1 - \frac{1}{2} + \frac{1}{3}$; $1 - \frac{1}{2} + \frac{1}{3} - \frac{1}{4}$; $1 - \frac{1}{2} + \frac{1}{3} - \frac{1}{4} + \frac{1}{5}$; Describe what you observe.

Technology Exercises

113. **a.** Use a graphing utility (and the change-of-base property) to graph $y = \log_3 x$. *

b. Graph $y = 2 + \log_3 x$, $y = \log_3(x + 2)$, and $y = -\log_3 x$ in the same viewing rectangle as $y = \log_3 x$. Then describe the change or changes that need to be made to the graph of $y = \log_3 x$ to obtain each of these three graphs. *

114. Graph $y = \log x$, $y = \log(10x)$, and $y = \log(0.1x)$ in the same viewing rectangle. Describe the relationship among the three graphs. What logarithmic property accounts for this relationship? *

115. Use a graphing utility and the change-of-base property to graph $y = \log_3 x$, $y = \log_{25} x$, and $y = \log_{100} x$ in the same viewing rectangle.

a. Which graph is on the top in the interval $(0, 1)$? Which is on the bottom? top: $y = \log_{100} x$; bottom: $y = \log_3 x$

b. Which graph is on the top in the interval $(1, \infty)$? Which is on the bottom? top: $y = \log_3 x$; bottom: $y = \log_{100} x$

c. Generalize by writing a statement about which graph is on top, which is on the bottom, and in which intervals, using $y = \log_b x$ where $b > 1$. *

Disprove each statement in Exercises 116–120 by

a. *letting y equal a positive constant of your choice, and*

b. *using a graphing utility to graph the function on each side of the equal sign. The two functions should have different graphs, showing that the equation is not true in general.*

116. $\log(x + y) = \log x + \log y$ 116–120. Answers will vary.

117. $\log\left(\dfrac{x}{y}\right) = \dfrac{\log x}{\log y}$

118. $\ln(x - y) = \ln x - \ln y$

119. $\ln(xy) = (\ln x)(\ln y)$

120. $\dfrac{\ln x}{\ln y} = \ln x - \ln y$

Critical Thinking Exercises

Make Sense? *In Exercises 121–124, determine whether each statement makes sense or does not make sense, and explain your reasoning.*

121. Because I cannot simplify the expression $b^m + b^n$ by adding exponents, there is no property for the logarithm of a sum. *

122. Because logarithms are exponents, the product, quotient, and power rules remind me of properties for operations with exponents. makes sense

123. I can use any positive number other than 1 in the change-of-base property, but the only practical bases are 10 and e because my calculator gives logarithms for these two bases. *

124. I expanded $\log_4 \sqrt{\dfrac{x}{y}}$ by writing the radical using a rational exponent and then applying the quotient rule, obtaining $\dfrac{1}{2}\log_4 x - \log_4 y$. does not make sense

In Exercises 125–128, determine whether each statement is true or false. If the statement is false, make the necessary change(s) to produce a true statement. Changes to false statements will vary.

125. $\ln\sqrt{2} = \dfrac{\ln 2}{2}$ true

126. $\dfrac{\log_7 49}{\log_7 7} = \log_7 49 - \log_7 7$ false

127. $\log_b(x^3 + y^3) = 3\log_b x + 3\log_b y$ false

128. $\log_b(xy)^5 = (\log_b x + \log_b y)^5$ false

129. Use the change-of-base property to prove that

$$\log e = \dfrac{1}{\ln 10}.$$ *

130. If $\log 3 = A$ and $\log 7 = B$, find $\log_7 9$ in terms of A and B. *

131. Write as a single term that does not contain a logarithm:

$$e^{\ln 8x^5 - \ln 2x^2}.$$ $4x^3$

132. If $f(x) = \log_b x$, show that

$$\dfrac{f(x + h) - f(x)}{h} = \log_b\left(1 + \dfrac{h}{x}\right)^{\frac{1}{h}}, h \neq 0.$$ *

133. Use the proof of the product rule in the appendix to prove the quotient rule. *

Retaining the Concepts

134. Given $f(x) = \dfrac{2}{x + 1}$ and $g(x) = \dfrac{1}{x}$, find each of the following:

 a. $(f \circ g)(x)$ $(f \circ g)(x) = \dfrac{2x}{x + 1}$

 b. the domain of $f \circ g$. all real numbers except -1 and 0

 (Section 2.6, Example 6)

135. Use the Leading Coefficient Test to determine the end behavior of the graph of $f(x) = -2x^2(x - 3)^2(x + 5)$.

 (Section 3.2, Example 2) rises left; falls right

136. Graph: $f(x) = \dfrac{4x^2}{x^2 - 9}$.

 (Section 3.5, Example 6) *

Preview Exercises

Exercises 137–139 will help you prepare for the material covered in the next section.

137. Solve for x: $a(x - 2) = b(2x + 3)$. $x = \dfrac{2a + 3b}{a - 2b}$

138. Solve: $x(x - 7) = 3$. $\left\{\dfrac{7 \pm \sqrt{61}}{2}\right\}$

139. Solve: $\dfrac{x + 2}{4x + 3} = \dfrac{1}{x}$. $\{-1, 3\}$

CHAPTER 4 Mid-Chapter Check Point

WHAT YOU KNOW: We evaluated and graphed exponential functions $[f(x) = b^x, b > 0$ and $b \neq 1]$, including the natural exponential function $[f(x) = e^x, e \approx 2.718]$. A function has an inverse that is a function if there is no horizontal line that intersects the function's graph more than once. The exponential function passes this horizontal line test and we called the inverse of the exponential function with base b the logarithmic function with base b. We learned that $y = \log_b x$ is equivalent to $b^y = x$. We evaluated and graphed logarithmic functions, including the common logarithmic function $[f(x) = \log_{10} x$ or $f(x) = \log x]$ and the natural logarithmic function $[f(x) = \log_e x$ or $f(x) = \ln x]$. We learned to use transformations to graph exponential and logarithmic functions. Finally, we used properties of logarithms to expand and condense logarithmic expressions.

In Exercises 1–5, graph f and g in the same rectangular coordinate system. Graph and give equations of all asymptotes. Give each function's domain and range.

1. $f(x) = 2^x$ and $g(x) = 2^x - 3$ *

2. $f(x) = \left(\dfrac{1}{2}\right)^x$ and $g(x) = \left(\dfrac{1}{2}\right)^{x-1}$ *

3. $f(x) = e^x$ and $g(x) = \ln x$ *

4. $f(x) = \log_2 x$ and $g(x) = \log_2(x - 1) + 1$ *

5. $f(x) = \log_{\frac{1}{2}} x$ and $g(x) = -2\log_{\frac{1}{2}} x$ *

In Exercises 6–9, find the domain of each function.

6. $f(x) = \log_3(x + 6)$ $(-6, \infty)$ **7.** $g(x) = \log_3 x + 6$ $(0, \infty)$

8. $h(x) = \log_3(x + 6)^2$ * **9.** $f(x) = 3^{x+6}$ $(-\infty, \infty)$

In Exercises 10–20, evaluate each expression without using a calculator. If evaluation is not possible, state the reason.

10. $\log_2 8 + \log_5 25$ 5 **11.** $\log_3 \dfrac{1}{9}$ -2 **12.** $\log_{100} 10$ $\frac{1}{2}$

13. $\log\sqrt[3]{10}$ $\frac{1}{3}$ **14.** $\log_2(\log_3 81)$ 2 **15.** $\log_3\left(\log_2 \dfrac{1}{8}\right)$ *

16. $6^{\log_6 5}$ 5 **17.** $\ln e^{\sqrt{7}}$ $\sqrt{7}$ **18.** $10^{\log 13}$ 13

19. $\log_{100} 0.1$ $-\frac{1}{2}$ **20.** $\log_\pi \pi^{\sqrt{\pi}}$ $\sqrt{\pi}$

In Exercises 21–22, expand and evaluate numerical terms.

21. $\log\left(\dfrac{\sqrt{xy}}{1000}\right)$ $\frac{1}{2}\log x + \frac{1}{2}\log y - 3$ **22.** $\ln(e^{19}x^{20})$ $19 + 20\ln x$

In Exercises 23–25, write each expression as a single logarithm.

23. $8\log_7 x - \dfrac{1}{3}\log_7 y$ $\log_7\left(\dfrac{x^8}{\sqrt[3]{y}}\right)$ **24.** $7\log_5 x + 2\log_5 x$ $\log_5 x^9$

25. $\dfrac{1}{2}\ln x - 3\ln y - \ln(z - 2)$ $\ln\left[\dfrac{\sqrt{x}}{y^3(z - 2)}\right]$

26. Use the formulas

$$A = P\left(1 + \dfrac{r}{n}\right)^{nt} \quad \text{and} \quad A = Pe^{rt}$$

to solve this exercise. You decide to invest \$8000 for 3 years at an annual rate of 8%. How much more is the return if the interest is compounded continuously than if it is compounded monthly? Round to the nearest dollar. \$8

Section 4.4 Exponential and Logarithmic Equations

1. Use like bases to solve exponential equations.

At age 20, you inherit $30,000. You'd like to put aside $25,000 and eventually have over half a million dollars for early retirement. Is this possible? In Example 10 in this section, you will see how techniques for solving equations with variable exponents provide an answer to this question.

Exponential Equations

An **exponential equation** is an equation containing a variable in an exponent. Examples of exponential equations include

$$2^{3x-8} = 16, \qquad 4^x = 15, \quad \text{and} \quad 40e^{0.6x} = 240.$$

Some exponential equations can be solved by expressing each side of the equation as a power of the same base. All exponential functions are one-to-one—that is, no two different ordered pairs have the same second component. Thus, if b is a positive number other than 1 and $b^M = b^N$, then $M = N$.

> **Solving Exponential Equations by Expressing Each Side as a Power of the Same Base**
>
> $$\text{If} \quad b^M = b^N, \quad \text{then} \quad M = N.$$
>
> Express each side as a power of the same base. Set the exponents equal to each other.
>
> 1. Rewrite the equation in the form $b^M = b^N$.
> 2. Set $M = N$.
> 3. Solve for the variable.

EXAMPLE 1 Solving Exponential Equations

Solve:

a. $2^{3x-8} = 16$ **b.** $27^{x+3} = 9^{x-1}$.

SOLUTION

In each equation, express both sides as a power of the same base. Then set the exponents equal to each other and solve for the variable.

a. Because 16 is 2^4, we express each side of $2^{3x-8} = 16$ in terms of base 2.

$2^{3x-8} = 16$	This is the given equation.
$2^{3x-8} = 2^4$	Write each side as a power of the same base.
$3x - 8 = 4$	If $b^M = b^N$, $b > 0$ and $b \neq 1$, then $M = N$.
$3x = 12$	Add 8 to both sides.
$x = 4$	Divide both sides by 3.

Substituting 4 for x into the original equation produces the true statement $16 = 16$. The solution set is {4}.

b. Because $27 = 3^3$ and $9 = 3^2$, we can express both sides of $27^{x+3} = 9^{x-1}$ in terms of base 3.

$$27^{x+3} = 9^{x-1}$$ This is the given equation.

$$\left(3^3\right)^{x+3} = \left(3^2\right)^{x-1}$$ Write each side as a power of the same base.

$$3^{3(x+3)} = 3^{2(x-1)}$$ When an exponential expression is raised to a power, multiply exponents.

$$3(x + 3) = 2(x - 1)$$ If two powers of the same base are equal, then the exponents are equal.

$$3x + 9 = 2x - 2$$ Apply the distributive property.

$$x + 9 = -2$$ Subtract $2x$ from both sides.

$$x = -11$$ Subtract 9 from both sides.

Substituting -11 for x into the original equation produces $27^{-8} = 9^{-12}$, which simplifies to the true statement $3^{-24} = 3^{-24}$. The solution set is $\{-11\}$. • • •

🗸 **Check Point 1** Solve:

 a. $5^{3x-6} = 125$ {3} **b.** $8^{x+2} = 4^{x-3}$. {−12}

② Use logarithms to solve exponential equations.

Most exponential equations cannot be rewritten so that each side has the same base. Here are two examples:

$$4^x = 15 \qquad\qquad 10^x = 120,000.$$

We cannot rewrite both sides in terms of base 2 or base 4.

We cannot rewrite both sides in terms of base 10.

Logarithms are extremely useful in solving these equations. The solution begins with isolating the exponential expression. Notice that the exponential expression is already isolated in both $4^x = 15$ and $10^x = 120,000$. Then we take the logarithm on both sides. Why can we do this? All logarithmic relations are functions. Thus, if M and N are positive real numbers and $M = N$, then $\log_b M = \log_b N$.

The base that is used when taking the logarithm on both sides of an equation can be any base at all. If the exponential equation involves base 10, as in $10^x = 120,000$, we'll take the common logarithm on both sides. If the exponential equation involves any other base, as in $4^x = 15$, we'll take the natural logarithm on both sides.

Using Logarithms to Solve Exponential Equations

1. Isolate the exponential expression.
2. Take the common logarithm on both sides of the equation for base 10. Take the natural logarithm on both sides of the equation for bases other than 10.
3. Simplify using one of the following properties:
$$\ln b^x = x \ln b \quad \text{or} \quad \ln e^x = x \quad \text{or} \quad \log 10^x = x.$$
4. Solve for the variable.

EXAMPLE 2 Solving Exponential Equations

Solve:

 a. $4^x = 15$ **b.** $10^x = 120,000$.

GREAT QUESTION!

You mentioned that the base used when taking the logarithm on both sides of an equation can be any base at all. What happens if I try to solve $4^x = 15$ by taking the logarithm with base 4 on both sides?

You still get the same solution.

$$4^x = 15$$
$$\log_4 4^x = \log_4 15$$
$$x = \log_4 15 \quad \log_b b^x = x$$
$$x = \frac{\ln 15}{\ln 4} \quad \log_b M = \frac{\ln M}{\ln b}$$

SOLUTION

We will use the natural logarithmic function to solve $4^x = 15$ and the common logarithmic function to solve $10^x = 120{,}000$.

a. Because the exponential expression, 4^x, is already isolated on the left side of $4^x = 15$, we begin by taking the natural logarithm on both sides of the equation.

$4^x = 15$	This is the given equation.
$\ln 4^x = \ln 15$	Take the natural logarithm on both sides.
$x \ln 4 = \ln 15$	Use the power rule and bring the variable exponent to the front: $\ln b^x = x \ln b$.
$x = \dfrac{\ln 15}{\ln 4}$	Solve for x by dividing both sides by $\ln 4$. Equivalently, $x = \log_4 15$.

We now have an exact value for x. We use the exact value for x in the equation's solution set. Thus, the equation's solution is $\dfrac{\ln 15}{\ln 4}$ and the solution set is $\left\{ \dfrac{\ln 15}{\ln 4} \right\}$. We can obtain a decimal approximation by using a calculator: $x \approx 1.95$. Because $4^2 = 16$, it seems reasonable that the solution to $4^x = 15$ is approximately 1.95.

b. Because the exponential expression, 10^x, is already isolated on the left side of $10^x = 120{,}000$, we begin by taking the common logarithm on both sides of the equation.

$10^x = 120{,}000$	This is the given equation.
$\log 10^x = \log 120{,}000$	Take the common logarithm on both sides.
$x = \log 120{,}000$	Use the inverse property $\log 10^x = x$ on the left.

The equation's solution is $\log 120{,}000$ and the solution set is $\{\log 120{,}000\}$. We can obtain a decimal approximation by using a calculator: $x \approx 5.08$. Because $10^5 = 100{,}000$, it seems reasonable that the solution to $10^x = 120{,}000$ is approximately 5.08. ● ● ●

⌀ **Check Point 2** Solve:

a. $5^x = 134$ $\left\{ \dfrac{\ln 134}{\ln 5} \right\}; \approx 3.04$ **b.** $10^x = 8000.$ $\{\log 8000\}; \approx 3.90$

Find each solution set and then use a calculator to obtain a decimal approximation to two decimal places for the solution.

EXAMPLE 3 Solving an Exponential Equation

Solve: $40e^{0.6x} - 3 = 237$.

SOLUTION

We begin by adding 3 to both sides and dividing both sides by 40 to isolate the exponential expression, $e^{0.6x}$. Then we take the natural logarithm on both sides of the equation.

$40e^{0.6x} - 3 = 237$	This is the given equation.
$40e^{0.6x} = 240$	Add 3 to both sides.
$e^{0.6x} = 6$	Isolate the exponential factor by dividing both sides by 40.
$\ln e^{0.6x} = \ln 6$	Take the natural logarithm on both sides.
$0.6x = \ln 6$	Use the inverse property $\ln e^x = x$ on the left.
$x = \dfrac{\ln 6}{0.6} \approx 2.99$	Divide both sides by 0.6 and solve for x.

Thus, the solution of the equation is $\dfrac{\ln 6}{0.6} \approx 2.99$. Try checking this approximate solution in the original equation to verify that $\left\{\dfrac{\ln 6}{0.6}\right\}$ is the solution set. •••

$\left\{\dfrac{\ln 9}{2}\right\}; \approx 1.10$

☑ **Check Point 3** Solve: $7e^{2x} - 5 = 58$. Find the solution set and then use a calculator to obtain a decimal approximation to two decimal places for the solution.

EXAMPLE 4 Solving an Exponential Equation

Solve: $5^{x-2} = 4^{2x+3}$.

SOLUTION

Because each exponential expression is isolated on one side of the equation, we begin by taking the natural logarithm on both sides.

$$5^{x-2} = 4^{2x+3}$$ This is the given equation.

$$\ln 5^{x-2} = \ln 4^{2x+3}$$ Take the natural logarithm on both sides.

> Be sure to insert parentheses around the binomials.

$$(x - 2)\ln 5 = (2x + 3)\ln 4$$ Use the power rule and bring the variable exponents to the front: $\ln b^x = x \ln b$.

> Remember that ln 5 and ln 4 are constants, not variables.

$$x \ln 5 - 2 \ln 5 = 2x \ln 4 + 3 \ln 4$$ Use the distributive property to distribute ln 5 and ln 4 to both terms in parentheses.

$$x \ln 5 - 2x \ln 4 = 2 \ln 5 + 3 \ln 4$$ Collect variable terms involving x on the left by subtracting $2x \ln 4$ and adding $2 \ln 5$ on both sides.

$$x(\ln 5 - 2 \ln 4) = 2 \ln 5 + 3 \ln 4$$ Factor out x from the two terms on the left.

$$x = \frac{2 \ln 5 + 3 \ln 4}{\ln 5 - 2 \ln 4}$$ Isolate x by dividing both sides by $\ln 5 - 2 \ln 4$.

The solution set is $\left\{\dfrac{2 \ln 5 + 3 \ln 4}{\ln 5 - 2 \ln 4}\right\}$. The solution is approximately -6.34. •••

DISCOVERY

Use properties of logarithms to show that the solution in Example 4 can be expressed as

$$\frac{\ln 1600}{\ln\left(\frac{5}{16}\right)}.$$

$\left\{\dfrac{\ln 3 + \ln 7}{2 \ln 3 - \ln 7}\right\}; \approx 12.11$

☑ **Check Point 4** Solve: $3^{2x-1} = 7^{x+1}$. Find the solution set and then use a calculator to obtain a decimal approximation to two decimal places for the solution.

EXAMPLE 5 Solving an Exponential Equation

Solve: $e^{2x} - 4e^x + 3 = 0$.

SOLUTION

The given equation is quadratic in form. If $u = e^x$, the equation can be expressed as $u^2 - 4u + 3 = 0$. Because this equation can be solved by factoring, we factor to isolate the exponential term.

$$e^{2x} - 4e^x + 3 = 0$$ This is the given equation.

$$(e^x - 3)(e^x - 1) = 0$$ Factor on the left. Notice that if $u = e^x$, $u^2 - 4u + 3 = (u - 3)(u - 1)$.

$$e^x - 3 = 0 \quad \text{or} \quad e^x - 1 = 0$$ Set each factor equal to 0.

$$e^x = 3 \qquad\qquad e^x = 1$$ Solve for e^x.

$$\ln e^x = \ln 3 \qquad\quad x = 0$$ Take the natural logarithm on both sides of the first equation. The equation on the right can be solved by inspection.

$$x = \ln 3 \qquad\qquad\qquad$$ $\ln e^x = x$

The solution set is $\{0, \ln 3\}$. The solutions are 0 and $\ln 3$, which is approximately 1.10. •••

TECHNOLOGY

Graphic Connections

Shown below is the graph of $y = e^{2x} - 4e^x + 3$. There are two x-intercepts, one at 0 and one at approximately 1.10. These intercepts verify our algebraic solution.

$[-3, 3, 1]$ by $[-1, 3, 1]$

 Check Point **5** Solve: $e^{2x} - 8e^x + 7 = 0$. Find the solution set and then use a calculator to obtain a decimal approximation to two decimal places, if necessary, for the solutions. {0, ln 7}; ln 7 ≈ 1.95

③ Use the definition of a logarithm to solve logarithmic equations.

Logarithmic Equations

A **logarithmic equation** is an equation containing a variable in a logarithmic expression. Examples of logarithmic equations include

$$\log_4(x + 3) = 2 \quad \text{and} \quad \ln(x + 2) - \ln(4x + 3) = \ln\left(\frac{1}{x}\right).$$

Some logarithmic equations can be expressed in the form $\log_b M = c$. We can solve such equations by rewriting them in exponential form.

Using the Definition of a Logarithm to Solve Logarithmic Equations

1. Express the equation in the form $\log_b M = c$.
2. Use the definition of a logarithm to rewrite the equation in exponential form:

$$\log_b M = c \quad \text{means} \quad b^c = M.$$

 Logarithms are exponents.

3. Solve for the variable.
4. Check proposed solutions in the original equation. Include in the solution set only values for which $M > 0$.

EXAMPLE 6 Solving Logarithmic Equations

Solve:

 a. $\log_4(x + 3) = 2$ **b.** $3\ln(2x) = 12$.

SOLUTION

The form $\log_b M = c$ involves a single logarithm whose coefficient is 1 on one side and a constant on the other side. Equation (a) is already in this form. We will need to divide both sides of equation (b) by 3 to obtain this form.

a. $\log_4(x + 3) = 2$ This is the given equation.

 $4^2 = x + 3$ Rewrite in exponential form: $\log_b M = c$ means $b^c = M$.

 $16 = x + 3$ Square 4.

 $13 = x$ Subtract 3 from both sides.

Check 13:

 $\log_4(x + 3) = 2$ This is the given logarithmic equation.

 $\log_4(13 + 3) \stackrel{?}{=} 2$ Substitute 13 for x.

 $\log_4 16 \stackrel{?}{=} 2$

 $2 = 2$, true $\log_4 16 = 2$ because $4^2 = 16$.

This true statement indicates that the solution set is {13}.

b. $3\ln(2x) = 12$ This is the given equation.

 $\ln(2x) = 4$ Divide both sides by 3.

 $\log_e(2x) = 4$ Rewrite the natural logarithm showing base e. This step is optional.

$$e^4 = 2x \qquad \text{Rewrite in exponential form: } \log_b M = c \text{ means } b^c = M.$$

$$\frac{e^4}{2} = x \qquad \text{Divide both sides by 2.}$$

Check $\dfrac{e^4}{2}$:

$$3\ln(2x) = 12 \qquad \text{This is the given logarithmic equation.}$$

$$3\ln\left[2\left(\frac{e^4}{2}\right)\right] \overset{?}{=} 12 \qquad \text{Substitute } \frac{e^4}{2} \text{ for } x.$$

$$3\ln e^4 \overset{?}{=} 12 \qquad \text{Simplify: } \frac{\cancel{2}}{1}\cdot\frac{e^4}{\cancel{2}} = e^4.$$

$$3\cdot 4 \overset{?}{=} 12 \qquad \text{Because } \ln e^x = x, \text{ we conclude } \ln e^4 = 4.$$

$$12 = 12, \text{ true}$$

This true statement indicates that the solution set is $\left\{\dfrac{e^4}{2}\right\}$. • • •

☑ **Check Point 6** Solve:

a. $\log_2(x - 4) = 3$ {12} **b.** $4\ln(3x) = 8$. $\left\{\frac{e^2}{3}\right\}$

Logarithmic expressions are defined only for logarithms of positive real numbers. **Always check proposed solutions of a logarithmic equation in the original equation. Exclude from the solution set any proposed solution that produces the logarithm of a negative number or the logarithm of 0.**

GREAT QUESTION!

Can a negative number belong to the solution set of a logarithmic equation?

Yes. Here's an example.

$$\log_2(x + 20) = 3 \qquad \text{Solve this equation.}$$

$$2^3 = x + 20 \qquad \text{Rewrite in exponential form.}$$

$$8 = x + 20 \qquad \text{Cube 2.}$$

$$-12 = x \qquad \text{Subtract 20 from both sides.}$$

Check -12:

$$\log_2(-12 + 20) \overset{?}{=} 3 \qquad \text{Substitute } -12 \text{ for } x.$$

$$\log_2 8 \overset{?}{=} 3$$

$$3 = 3, \text{ true} \qquad \log_2 8 = 3 \text{ because } 2^3 = 8.$$

The solution set is $\{-12\}$. Although -12 is negative, it does not produce the logarithm of a negative number in $\log_2(x + 20) = 3$, the given equation. Note that the domain of the expression $\log_2(x + 20)$ is $(-20, \infty)$, which includes negative numbers such as -12.

To rewrite the logarithmic equation $\log_b M = c$ in the equivalent exponential form $b^c = M$, we need a single logarithm whose coefficient is one. It is sometimes necessary to use properties of logarithms to condense logarithms into a single logarithm. In the next example, we use the product rule for logarithms to obtain a single logarithmic expression on the left side.

EXAMPLE 7 Solving a Logarithmic Equation

Solve: $\log_2 x + \log_2(x - 7) = 3$.

SOLUTION

$\log_2 x + \log_2(x - 7) = 3$	This is the given equation.
$\log_2[x(x - 7)] = 3$	Use the product rule to obtain a single logarithm: $\log_b M + \log_b N = \log_b(MN)$.
$2^3 = x(x - 7)$	Rewrite in exponential form: $\log_b M = c$ means $b^c = M$.
$8 = x^2 - 7x$	Evaluate 2^3 on the left and apply the distributive property on the right.
$0 = x^2 - 7x - 8$	Set the equation equal to O.
$0 = (x - 8)(x + 1)$	Factor.
$x - 8 = 0$ or $x + 1 = 0$	Set each factor equal to O.
$x = 8$ $x = -1$	Solve for x.

Check 8:	**Check -1:**
$\log_2 x + \log_2(x - 7) = 3$	$\log_2 x + \log_2(x - 7) = 3$
$\log_2 8 + \log_2(8 - 7) \stackrel{?}{=} 3$	$\log_2(-1) + \log_2(-1 - 7) \stackrel{?}{=} 3$
$\log_2 8 + \log_2 1 \stackrel{?}{=} 3$	The number -1 does not check.
$3 + 0 \stackrel{?}{=} 3$	The domain of a logarithmic function consists
$3 = 3$, true	of positive numbers.

The solution set is {8}. •••

✓ **Check Point 7** Solve: $\log x + \log(x - 3) = 1$. {5}

④ Use the one-to-one property of logarithms to solve logarithmic equations.

Some logarithmic equations can be expressed in the form $\log_b M = \log_b N$, where the bases on both sides of the equation are the same. Because all logarithmic functions are one-to-one, we can conclude that $M = N$.

> **Using the One-to-One Property of Logarithms to Solve Logarithmic Equations**
>
> 1. Express the equation in the form $\log_b M = \log_b N$. This form involves a single logarithm whose coefficient is 1 on each side of the equation.
> 2. Use the one-to-one property to rewrite the equation without logarithms: If $\log_b M = \log_b N$, then $M = N$.
> 3. Solve for the variable.
> 4. Check proposed solutions in the original equation. Include in the solution set only values for which $M > 0$ and $N > 0$.

EXAMPLE 8 Solving a Logarithmic Equation

Solve: $\ln(x + 2) - \ln(4x + 3) = \ln\left(\dfrac{1}{x}\right)$.

SOLUTION

In order to apply the one-to-one property of logarithms, we need a single logarithm whose coefficient is 1 on each side of the equation. The right side is already in this form. We can obtain a single logarithm on the left side by applying the quotient rule.

TECHNOLOGY

Numeric Connections

A graphing utility's [TABLE] feature can be used to verify that {3} is the solution set of

$$\ln(x + 2) - \ln(4x + 3) = \ln\left(\frac{1}{x}\right).$$

$y_1 = \ln(x + 2) - \ln(4x + 3)$

$y_2 = \ln\left(\frac{1}{x}\right)$

y_1 and y_2 are equal when $x = 3$.

5 Solve applied problems involving exponential and logarithmic equations.

A blood alcohol concentration of 0.22 corresponds to near certainty, or a 100% probability, of a car accident.

FIGURE 4.20

$\ln(x + 2) - \ln(4x + 3) = \ln\left(\dfrac{1}{x}\right)$	This is the given equation.
$\ln\left(\dfrac{x + 2}{4x + 3}\right) = \ln\left(\dfrac{1}{x}\right)$	Use the quotient rule to obtain a single logarithm on the left side: $\log_b M - \log_b N = \log_b\left(\dfrac{M}{N}\right).$
$\dfrac{x + 2}{4x + 3} = \dfrac{1}{x}$	Use the one-to-one property: If $\log_b M = \log_b N$, then $M = N.$
$x(4x + 3)\left(\dfrac{x + 2}{4x + 3}\right) = x(4x + 3)\left(\dfrac{1}{x}\right)$	Multiply both sides by $x(4x + 3)$, the LCD.
$x(x + 2) = 4x + 3$	Simplify.
$x^2 + 2x = 4x + 3$	Apply the distributive property.
$x^2 - 2x - 3 = 0$	Subtract $4x + 3$ from both sides and set the equation equal to O.
$(x - 3)(x + 1) = 0$	Factor.
$x - 3 = 0 \quad \text{or} \quad x + 1 = 0$	Set each factor equal to O.
$x = 3 \qquad\qquad x = -1$	Solve for x.

Substituting 3 for x into $\ln(x + 2) - \ln(4x + 3) = \ln\left(\frac{1}{x}\right)$ produces the true statement $\ln\left(\frac{1}{3}\right) = \ln\left(\frac{1}{3}\right)$. However, substituting -1 produces logarithms of negative numbers. Thus, -1 is not a solution. The solution set is {3}. •••

✓ **Check Point 8** Solve: $\ln(x - 3) = \ln(7x - 23) - \ln(x + 1).$ {4, 5}

Applications

Our first applied example provides a mathematical perspective on the old slogan "Alcohol and driving don't mix." In California, where 38% of fatal traffic crashes involve drinking drivers, it is illegal to drive with a blood alcohol concentration of 0.08 or higher. At these levels, drivers may be arrested and charged with driving under the influence.

EXAMPLE 9 Alcohol and Risk of a Car Accident

Medical research indicates that the risk of having a car accident increases exponentially as the concentration of alcohol in the blood increases. The risk is modeled by

$$R = 6e^{12.77x},$$

where x is the blood alcohol concentration and R, given as a percent, is the risk of having a car accident. What blood alcohol concentration corresponds to a 17% risk of a car accident? How is this shown on the graph of R in **Figure 4.20**?

SOLUTION

For a risk of 17%, we let $R = 17$ in the equation and solve for x, the blood alcohol concentration.

$R = 6e^{12.77x}$	This is the given equation.
$6e^{12.77x} = 17$	Substitute 17 for R and (optional) reverse the two sides of the equation.
$e^{12.77x} = \dfrac{17}{6}$	Isolate the exponential factor by dividing both sides by 6.
$\ln e^{12.77x} = \ln\left(\dfrac{17}{6}\right)$	Take the natural logarithm on both sides.
$12.77x = \ln\left(\dfrac{17}{6}\right)$	Use the inverse property $\ln e^x = x$ on the left side.
$x = \dfrac{\ln\left(\dfrac{17}{6}\right)}{12.77} \approx 0.08$	Divide both sides by 12.77.

A blood alcohol concentration of 0.22 corresponds to near certainty, or a 100% probability, of a car accident.

$R = 6e^{12.77x}$

Blood Alcohol Concentration

FIGURE 4.20 (repeated)

For a blood alcohol concentration of 0.08, the risk of a car accident is 17%. This is shown on the graph of R in **Figure 4.20** by the point (0.08, 17) that lies on the blue curve. Take a moment to locate this point on the curve. In many states, it is illegal to drive with a blood alcohol concentration of 0.08. •••

✓ Check Point **9** Use the formula in Example 9 to answer this question: What blood alcohol concentration corresponds to a 7% risk of a car accident? (In many states, drivers under the age of 21 can lose their licenses for driving at this level.) 0.01

Suppose that you inherit $30,000 at age 20. Is it possible to invest $25,000 and have over half a million dollars for early retirement? Our next example illustrates the power of compound interest.

EXAMPLE 10 Revisiting the Formula for Compound Interest

The formula

$$A = P\left(1 + \frac{r}{n}\right)^{nt}$$

describes the accumulated value, A, of a sum of money, P, the principal, after t years at annual percentage rate r (in decimal form) compounded n times a year. How long will it take $25,000 to grow to $500,000 at 9% annual interest compounded monthly?

SOLUTION

$$A = P\left(1 + \frac{r}{n}\right)^{nt}$$

This is the given formula.

$$500,000 = 25,000\left(1 + \frac{0.09}{12}\right)^{12t}$$

A(the desired accumulated value) = $500,000, P(the principal) = $25,000, r(the interest rate) = 9% = 0.09, and $n = 12$ (monthly compounding).

Because of the exponent, 12t, do not distribute 25,000 over the terms in parentheses.

Our goal is to solve the equation for t. Let's reverse the two sides of the equation and then simplify within parentheses.

$$25,000\left(1 + \frac{0.09}{12}\right)^{12t} = 500,000$$

Reverse the two sides of the previous equation.

$$25,000(1 + 0.0075)^{12t} = 500,000$$

Divide within parentheses: $\frac{0.09}{12} = 0.0075$.

$$25,000(1.0075)^{12t} = 500,000$$

Add within parentheses.

$$(1.0075)^{12t} = 20$$

Divide both sides by 25,000.

$$\ln(1.0075)^{12t} = \ln 20$$

Take the natural logarithm on both sides.

$$12t \ln(1.0075) = \ln 20$$

Use the power rule to bring the exponent to the front: $\ln b^x = x \ln b$.

$$t = \frac{\ln 20}{12 \ln 1.0075}$$

Solve for t, dividing both sides by 12 ln 1.0075.

$$\approx 33.4$$

Use a calculator.

After approximately 33.4 years, the $25,000 will grow to an accumulated value of $500,000. If you set aside the money at age 20, you can begin enjoying a life as a half-millionaire at about age 53. •••

✓ Check Point **10** How long, to the nearest tenth of a year, will it take $1000 to grow to $3600 at 8% annual interest compounded quarterly? 16.2 yr

Blitzer Bonus ‖

Playing Doubles: Interest Rates and Doubling Time

One way to calculate what your savings will be worth at some point in the future is to consider doubling time. The following table shows how long it takes for your money to double at different annual interest rates subject to continuous compounding.

Annual Interest Rate	Years to Double
5%	13.9 years
7%	9.9 years
9%	7.7 years
11%	6.3 years

Of course, the first problem is collecting some money to invest. The second problem is finding a reasonably safe investment with a return of 9% or more.

EXAMPLE 11 Revisiting the Model for Height of Children

We have seen that the percentage of adult height attained by a boy who is x years old can be modeled by

$$f(x) = 29 + 48.8 \log(x + 1),$$

where x represents the boy's age (from 5 to 15) and $f(x)$ represents the percentage of his adult height. At what age, rounded to the nearest year, has a boy attained 85% of his adult height?

SOLUTION

To find at what age a boy has attained 85% of his adult height, we substitute 85 for $f(x)$ and solve for x, the boy's age.

$f(x) = 29 + 48.8 \log(x + 1)$	This is the given function.
$85 = 29 + 48.8 \log(x + 1)$	Substitute 85 for $f(x)$.

> Our goal is to isolate $\log(x + 1)$ and then rewrite the equation in exponential form.

$56 = 48.8 \log(x + 1)$	Subtract 29 from both sides.
$\dfrac{56}{48.8} = \log(x + 1)$	Divide both sides by 48.8.
$\dfrac{56}{48.8} = \log_{10}(x + 1)$	Rewrite the common logarithm showing base 10. This step is optional.
$10^{\frac{56}{48.8}} = x + 1$	Rewrite in exponential form.
$10^{\frac{56}{48.8}} - 1 = x$	Subtract 1 from both sides.
$13 \approx x$	Use a calculator: 10 $\boxed{\wedge}$ $\boxed{(}$ 56 $\boxed{\div}$ 48.8 $\boxed{)}$ $\boxed{-}$ 1 $\boxed{\text{ENTER}}$. On some calculators, use the right arrow key to exit the exponent before subtracting 1.

At approximately age 13, a boy has attained 85% of his adult height. This is shown on the graph of the model in **Figure 4.21** by the point $(13, 85)$.

FIGURE 4.21 Graph of a model for the percentage of adult height attained by a boy ● ● ●

✓ Check Point **11** The percentage of adult height attained by a girl who is x years old can be modeled by

$$f(x) = 62 + 35 \log(x - 4),$$

where x represents the girl's age (from 5 to 15) and $f(x)$ represents the percentage of her adult height. At what age has a girl attained 97% of her adult height? 14

ACHIEVING SUCCESS

Many college algebra courses cover only selected sections from the book's remaining chapters. Regardless of the content requirements for your course, **it's never too early to start thinking about a final exam**. Here are some strategies to help you prepare for your final:

- Review your back exams. Be sure you understand any errors that you made. Seek help with any concepts that are still unclear.
- Ask your professor if there are additional materials to help students review for the final. This includes review sheets and final exams from previous semesters.
- Attend any review sessions conducted by your professor or by the math department.
- Use the strategy introduced earlier in the book: Imagine that your professor will permit two 3 by 5 index cards of notes on the final. Organize and create such a two-card summary for the most vital information in the course, including all important formulas. Refer to the chapter summaries in the textbook to prepare your personalized summary.
- For further review, work the relevant exercises in the Cumulative Review exercises at the end of all chapters covered in your course.
- Write your own final exam with detailed solutions for each item. You can use test questions from back exams in mixed order, worked examples from the textbook's chapter summaries, exercises in the Cumulative Reviews, and problems from course handouts. Use your test as a practice final exam.

CONCEPT AND VOCABULARY CHECK

Fill in each blank so that the resulting statement is true.

1. If $b^M = b^N$, then ___$M = N$___.

2. If $2^{4x-1} = 2^7$, then ___$4x - 1$___ $= 7$.

3. If $x \ln 9 = \ln 20$, then $x =$ _____. $\frac{\ln 20}{\ln 9}$

4. If $e^{0.6x} = 6$, then $0.6x =$ ___$\ln 6$___.

5. If $\log_5(x + 1) = 3$, then ___5^3___ $= x + 1$.

6. If $\log_3 x + \log_3(x + 1) = 2$, then \log_3 ___$(x^2 + x)$___ $= 2$.

7. If $\ln\left(\dfrac{7x - 23}{x + 1}\right) = \ln(x - 3)$, then ___$\frac{7x-23}{x+1}$___ $= x - 3$.

8. True or false: $x^4 = 15$ is an exponential equation. ___false___

9. True or false: $4^x = 15$ is an exponential equation. ___true___

10. True or false: -3 is a solution of $\log_5 9 = 2 \log_5 x$. ___false___

11. True or false: -10 is a solution of $\log_5(x + 35) = 2$. ___true___

EXERCISE SET 4.4

Practice Exercises

Solve each exponential equation in Exercises 1–22 by expressing each side as a power of the same base and then equating exponents.

1. $2^x = 64$ {6}
2. $3^x = 81$ {4}
3. $5^x = 125$ {3}
4. $5^x = 625$ {4}
5. $2^{2x-1} = 32$ {3}
6. $3^{2x+1} = 27$ {1}
7. $4^{2x-1} = 64$ {2}
8. $5^{3x-1} = 125$ {$\frac{4}{3}$}
9. $32^x = 8$ {$\frac{3}{5}$}
10. $4^x = 32$ {$\frac{5}{2}$}
11. $9^x = 27$ {$\frac{3}{2}$}
12. $125^x = 625$ {$\frac{4}{3}$}
13. $3^{1-x} = \frac{1}{27}$ {4}
14. $5^{2-x} = \frac{1}{125}$ {5}
15. $6^{\frac{x-3}{4}} = \sqrt{6}$ {5}
16. $7^{\frac{x-2}{6}} = \sqrt{7}$ {5}
17. $4^x = \frac{1}{\sqrt{2}}$ {$-\frac{1}{4}$}
18. $9^x = \frac{1}{\sqrt[3]{3}}$ {$-\frac{1}{6}$}
19. $8^{x+3} = 16^{x-1}$ {13}
20. $8^{1-x} = 4^{x+2}$ {$-\frac{1}{5}$}
21. $e^{x+1} = \frac{1}{e}$ {-2}
22. $e^{x+4} = \frac{1}{e^{2x}}$ {$-\frac{4}{3}$}

Solve each exponential equation in Exercises 23–48. Express the solution set in terms of natural logarithms or common logarithms. Then use a calculator to obtain a decimal approximation, correct to two decimal places, for the solution.

23. $10^x = 3.91$ *
24. $10^x = 8.07$ *
25. $e^x = 5.7$ {ln 5.7}; ≈1.74
26. $e^x = 0.83$ {ln 0.83}; ≈ −0.19
27. $5^x = 17$ *
28. $19^x = 143$ *
29. $5e^x = 23$ {$\ln \frac{23}{5}$}; ≈1.53
30. $9e^x = 107$ {$\ln \frac{107}{9}$}; ≈2.48
31. $3e^{5x} = 1977$ *
32. $4e^{7x} = 10{,}273$ *
33. $e^{1-5x} = 793$ *
34. $e^{1-8x} = 7957$ *
35. $e^{5x-3} - 2 = 10{,}476$ *
36. $e^{4x-5} - 7 = 11{,}243$ *
37. $7^{x+2} = 410$ *
38. $5^{x-3} = 137$ *
39. $7^{0.3x} = 813$ *
40. $3^{\frac{x}{7}} = 0.2$ *
41. $5^{2x+3} = 3^{x-1}$ *
42. $7^{2x+1} = 3^{x+2}$ *
43. $e^{2x} - 3e^x + 2 = 0$ *
44. $e^{2x} - 2e^x - 3 = 0$ *
45. $e^{4x} + 5e^{2x} - 24 = 0$ *
46. $e^{4x} - 3e^{2x} - 18 = 0$ *
47. $3^{2x} + 3^x - 2 = 0$ {0}
48. $2^{2x} + 2^x - 12 = 0$ *

*See Answers to Selected Exercises.

Solve each logarithmic equation in Exercises 49–92. Be sure to reject any value of x that is not in the domain of the original logarithmic expressions. Give the exact answer. Then, where necessary, use a calculator to obtain a decimal approximation, correct to two decimal places, for the solution.

49. $\log_3 x = 4$ {81}

50. $\log_5 x = 3$ {125}

51. $\ln x = 2$ {e^2}; ≈ 7.39

52. $\ln x = 3$ {e^3}; ≈ 20.09

53. $\log_4(x + 5) = 3$ {59}

54. $\log_5(x - 7) = 2$ {32}

55. $\log_2(x + 25) = 4$ {−9}

56. $\log_2(x + 50) = 5$ {−18}

57. $\log_3(x + 4) = -3$ {$\frac{-107}{27}$}

58. $\log_7(x + 2) = -2$ {$-\frac{97}{49}$}

59. $\log_4(3x + 2) = 3$ {$\frac{62}{3}$}

60. $\log_2(4x + 1) = 5$ {$\frac{31}{4}$}

61. $5 \ln(2x) = 20$ *

62. $6 \ln(2x) = 30$ *

63. $6 + 2 \ln x = 5$ *

64. $7 + 3 \ln x = 6$ *

65. $\ln \sqrt{x + 3} = 1$ *

66. $\ln \sqrt{x + 4} = 1$ *

67. $\log_5 x + \log_5(4x - 1) = 1$ {$\frac{5}{4}$}

68. $\log_6(x + 5) + \log_6 x = 2$ {4}

69. $\log_3(x + 6) + \log_3(x + 4) = 1$ {−3}

70. $\log_6(x + 3) + \log_6(x + 4) = 1$ {−1}

71. $\log_2(x + 2) - \log_2(x - 5) = 3$ {6}

72. $\log_4(x + 2) - \log_4(x - 1) = 1$ {2}

73. $2 \log_3(x + 4) = \log_3 9 + 2$ {5}

74. $3 \log_2(x - 1) = 5 - \log_2 4$ {3}

75. $\log_2(x - 6) + \log_2(x - 4) - \log_2 x = 2$ {12}

76. $\log_2(x - 3) + \log_2 x - \log_2(x + 2) = 2$ {8}

77. $\log(x + 4) = \log x + \log 4$ {$\frac{4}{3}$}

78. $\log(5x + 1) = \log(2x + 3) + \log 2$ {5}

79. $\log(3x - 3) = \log(x + 1) + \log 4$ ∅

80. $\log(2x - 1) = \log(x + 3) + \log 3$ ∅

81. $2 \log x = \log 25$ {5}

82. $3 \log x = \log 125$ {5}

83. $\log(x + 4) - \log 2 = \log(5x + 1)$ {$\frac{2}{9}$}

84. $\log(x + 7) - \log 3 = \log(7x + 1)$ {$\frac{1}{5}$}

85. $2 \log x - \log 7 = \log 112$ {28}

86. $\log(x - 2) + \log 5 = \log 100$ {22}

87. $\log x + \log(x + 3) = \log 10$ {2}

88. $\log(x + 3) + \log(x - 2) = \log 14$ {4}

89. $\ln(x - 4) + \ln(x + 1) = \ln(x - 8)$ ∅

90. $\log_2(x - 1) - \log_2(x + 3) = \log_2\left(\dfrac{1}{x}\right)$ {3}

91. $\ln(x - 2) - \ln(x + 3) = \ln(x - 1) - \ln(x + 7)$ {$\frac{11}{3}$}

92. $\ln(x - 5) - \ln(x + 4) = \ln(x - 1) - \ln(x + 2)$ ∅

Practice Plus

In Exercises 93–102, solve each equation.

93. $5^{2x} \cdot 5^{4x} = 125$ {$\frac{1}{2}$}

94. $3^{x+2} \cdot 3^x = 81$ {1}

95. $2|\ln x| - 6 = 0$ {e^3, e^{-3}}

96. $3|\log x| - 6 = 0$ {$100, \frac{1}{100}$}

97. $3^{x^2} = 45$ *

98. $5^{x^2} = 50$ *

99. $\ln(2x + 1) + \ln(x - 3) - 2 \ln x = 0$ {$\frac{5 + \sqrt{37}}{2}$}

100. $\ln 3 - \ln(x + 5) - \ln x = 0$ {$\frac{-5 + \sqrt{37}}{2}$}

101. $5^{x^2 - 12} = 25^{2x}$ {−2, 6}

102. $3^{x^2 - 12} = 9^{2x}$ {−2, 6}

Application Exercises

103. The formula $A = 37.3e^{0.0095t}$ models the population of California, A, in millions, t years after 2010.

 a. What was the population of California in 2010? *

 b. When will the population of California reach 40 million? 2017

104. The formula $A = 25.1e^{0.0187t}$ models the population of Texas, A, in millions, t years after 2010.

 a. What was the population of Texas in 2010? 25.1 million

 b. When will the population of Texas reach 28 million? *

The function $f(x) = 20(0.975)^x$ models the percentage of surface sunlight, $f(x)$, that reaches a depth of x feet beneath the surface of the ocean. The figure shows the graph of this function. Use this information to solve Exercises 105–106.

105. Use the function to determine at what depth, to the nearest foot, there is 1% of surface sunlight. How is this shown on the graph of f? 118 ft; by the point (118, 1)

106. Use the function to determine at what depth, to the nearest foot, there is 3% of surface sunlight. How is this shown on the graph of f? 75 ft; by the point (75, 3)

In Exercises 107–110, complete the table for a savings account subject to n compoundings yearly $\left[A = P\left(1 + \dfrac{r}{n}\right)^{nt} \right]$. *Round answers to one decimal place.*

	Amount Invested	Number of Compounding Periods	Annual Interest Rate	Accumulated Amount	Time t in Years
107.	$12,500	4	5.75%	$20,000	8.2
108.	$7250	12	6.5%	$15,000	11.2
109.	$1000	365	16.8%	$1400	2.0
110.	$5000	365	14.7%	$9000	4.0

In Exercises 111–114, complete the table for a savings account subject to continuous compounding ($A = Pe^{rt}$). Round answers to one decimal place.

	Amount Invested	Annual Interest Rate	Accumulated Amount	Time t in Years
111.	$8000	8%	Double the amount invested	8.7
112.	$8000	20.3%	$12,000	2.0
113.	$2350	15.7%	Triple the amount invested	7.0
114.	$17,425	4.25%	$25,000	8.5

By 2019, nearly $1 out of every $5 spent in the U.S. economy is projected to go for health care. The bar graph shows the percentage of the U.S. gross domestic product (GDP) going toward health care from 2007 through 2014, with a projection for 2019.

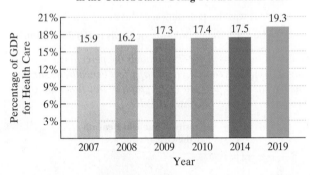

Percentage of the Gross Domestic Product in the United States Going Toward Health Care

Source: Health Affairs (healthaffairs.org)

The data can be modeled by the function $f(x) = 1.2 \ln x + 15.7$, where $f(x)$ is the percentage of the U.S. gross domestic product going toward health care x years after 2006. Use this information to solve Exercises 115–116.

115. a. Use the function to determine the percentage of the U.S. gross domestic product that went toward health care in 2009. Round to the nearest tenth of a percent. Does this underestimate or overestimate the percent displayed by the graph? By how much? 17.0%; underestimates by 0.3%

 b. According to the model, when will 18.5% of the U.S. gross domestic product go toward health care? Round to the nearest year. 2016

116. a. Use the function to determine the percentage of the U.S. gross domestic product that went toward health care in 2008. Round to the nearest tenth of a percent. Does this underestimate or overestimate the percent displayed by the graph? By how much? 16.5%; overestimates by 0.3%

 b. According to the model, when will 18.6% of the U.S. gross domestic product go toward health care? Round to the nearest year. 2017

The function $P(x) = 95 - 30 \log_2 x$ models the percentage, $P(x)$, of students who could recall the important features of a classroom lecture as a function of time, where x represents the number of days that have elapsed since the lecture was given. The figure at the top of the next column shows the graph

of the function. Use this information to solve Exercises 117–118. Round answers to one decimal place.

117. After how many days do only half the students recall the important features of the classroom lecture? (Let $P(x) = 50$ and solve for x.) Locate the point on the graph that conveys this information. 2.8 days; (2.8, 50)

118. After how many days have all students forgotten the important features of the classroom lecture? (Let $P(x) = 0$ and solve for x.) Locate the point on the graph that conveys this information. 9.0 days; (9.0, 0)

The pH scale is used to measure the acidity or alkalinity of a solution. The scale ranges from 0 to 14. A neutral solution, such as pure water, has a pH of 7. An acid solution has a pH less than 7 and an alkaline solution has a pH greater than 7. The lower the pH below 7, the more acidic is the solution. Each whole-number decrease in pH represents a tenfold increase in acidity.

The pH Scale

The pH of a solution is given by

$$pH = -\log x,$$

where x represents the concentration of the hydrogen ions in the solution, in moles per liter. Use the formula to solve Exercises 119–120. Express answers as powers of 10.

119. a. Normal, unpolluted rain has a pH of about 5.6. What is the hydrogen ion concentration? $10^{-5.6}$ mole per liter

b. An environmental concern involves the destructive effects of acid rain. The most acidic rainfall ever had a pH of 2.4. What was the hydrogen ion concentration? *

c. How many times greater is the hydrogen ion concentration of the acidic rainfall in part (b) than the normal rainfall in part (a)? $10^{3.2}$ times greater

120. a. The figure indicates that lemon juice has a pH of 2.3. What is the hydrogen ion concentration? $10^{-2.3}$ mole per liter

b. Stomach acid has a pH that ranges from 1 to 3. What is the hydrogen ion concentration of the most acidic stomach? 10^{-1} mole per liter

c. How many times greater is the hydrogen ion concentration of the acidic stomach in part (b) than the lemon juice in part (a)? $10^{1.3}$ times greater

Explaining the Concepts 121–124. Answers will vary.

121. Explain how to solve an exponential equation when both sides can be written as a power of the same base.

122. Explain how to solve an exponential equation when both sides cannot be written as a power of the same base. Use $3^x = 140$ in your explanation.

123. Explain the differences between solving $\log_3(x - 1) = 4$ and $\log_3(x - 1) = \log_3 4$.

124. In many states, a 17% risk of a car accident with a blood alcohol concentration of 0.08 is the lowest level for charging a motorist with driving under the influence. Do you agree with the 17% risk as a cutoff percentage, or do you feel that the percentage should be lower or higher? Explain your answer. What blood alcohol concentration corresponds to what you believe is an appropriate percentage?

Technology Exercises

In Exercises 125–132, use your graphing utility to graph each side of the equation in the same viewing rectangle. Then use the x-coordinate of the intersection point to find the equation's solution set. Verify this value by direct substitution into the equation.

125. $2^{x+1} = 8$ {2} **126.** $3^{x+1} = 9$ {1}

127. $\log_3(4x - 7) = 2$ {4} **128.** $\log_3(3x - 2) = 2$ $\{\frac{11}{3}\}$

129. $\log(x + 3) + \log x = 1$ {2} **130.** $\log(x - 15) + \log x = 2$ *

131. $3^x = 2x + 3$ * **132.** $5^x = 3x + 4$ *

Hurricanes are one of nature's most destructive forces. These low-pressure areas often have diameters of over 500 miles. The function $f(x) = 0.48 \ln(x + 1) + 27$ models the barometric air pressure, $f(x)$, in inches of mercury, at a distance of x miles from the eye of a hurricane. Use this function to solve Exercises 133–134.

133. Graph the function in a [0, 500, 50] by [27, 30, 1] viewing rectangle. What does the shape of the graph indicate about barometric air pressure as the distance from the eye increases? *

134. Use an equation to answer this question: How far from the eye of a hurricane is the barometric air pressure 29 inches of mercury? Use the TRACE and ZOOM features or the intersect command of your graphing utility to verify your answer. about 63.5 miles

135. The function $P(t) = 145e^{-0.092t}$ models a runner's pulse, $P(t)$, in beats per minute, t minutes after a race, where $0 \le t \le 15$. Graph the function using a graphing utility. TRACE along the graph and determine after how many minutes the runner's pulse will be 70 beats per minute. Round to the nearest tenth of a minute. Verify your observation algebraically. about 7.9 min*

136. The function $W(t) = 2600(1 - 0.51e^{-0.075t})^3$ models the weight, $W(t)$, in kilograms, of a female African elephant at age t years. (1 kilogram \approx 2.2 pounds) Use a graphing utility to graph the function. Then TRACE along the curve to estimate the age of an adult female elephant weighing 1800 kilograms. about 20 years old*

Critical Thinking Exercises

Make Sense? *In Exercises 137–140, determine whether each statement makes sense or does not make sense, and explain your reasoning.*

137. Because the equations $2^x = 15$ and $2^x = 16$ are similar, I solved them using the same method. does not make sense

138. Because the equations does not make sense

$$\log(3x + 1) = 5 \quad \text{and} \quad \log(3x + 1) = \log 5$$

are similar, I solved them using the same method.

139. I can solve $4^x = 15$ by writing the equation in logarithmic form. makes sense

140. It's important for me to check that the proposed solution of an equation with logarithms gives only logarithms of positive numbers in the original equation. makes sense

In Exercises 141–144, determine whether each statement is true or false. If the statement is false, make the necessary change(s) to produce a true statement. Changes to false statements will vary.

141. If $\log(x + 3) = 2$, then $e^2 = x + 3$. false

142. If $\log(7x + 3) - \log(2x + 5) = 4$, then the equation in exponential form is $10^4 = (7x + 3) - (2x + 5)$. false

143. If $x = \dfrac{1}{k}\ln y$, then $y = e^{kx}$. true

144. Examples of exponential equations include $10^x = 5.71$, $e^x = 0.72$, and $x^{10} = 5.71$. false

145. If $4000 is deposited into an account paying 3% interest compounded annually and at the same time $2000 is deposited into an account paying 5% interest compounded annually, after how long will the two accounts have the same balance? Round to the nearest year. after 36 yr

Solve each equation in Exercises 146–148. Check each proposed solution by direct substitution or with a graphing utility.

146. $(\ln x)^2 = \ln x^2$ $\{1, e^2\}; e^2 \approx 7.389$

147. $(\log x)(2 \log x + 1) = 6$ $\{10^{-2}, 10^{3/2}\}; 10^{3/2} = 10\sqrt{10} \approx 31.62$

148. $\ln(\ln x) = 0$ $\{e\}; e \approx 2.718$

Group Exercise Answers will vary.

149. Research applications of logarithmic functions as mathematical models and plan a seminar based on your group's research. Each group member should research one of the following areas or any other area of interest: pH (acidity of solutions), intensity of sound (decibels), brightness of stars, human memory, progress over time in a sport, profit over time. For the area that you select, explain how logarithmic functions are used and provide examples.

Retaining the Concepts

150. Use the graph of f to determine each of the following. Where applicable, use interval notation.

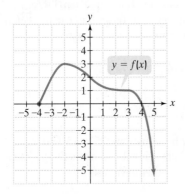

a. the domain of f $[-4, \infty)$
b. the range of f $(-\infty, 3]$
c. the x-intercepts -4 and 4
d. the y-intercept 2
e. interval(s) where f is increasing $(-4, -2)$
f. interval(s) where f is decreasing $(-2, \infty)$
g. the number at which f has a relative maximum $x = -2$
h. the relative maximum of f 3
i. $f(-3)$ 2

(Section 2.1, Examples 7 and 8; Section 2.2, Example 1, Figure 2.17)

151. Consider the quadratic function

$$f(x) = -4x^2 - 16x + 3.$$

a. Determine, without graphing, whether the function has a minimum value or a maximum value. maximum value

b. Find the minimum or maximum value and determine where it occurs. Maximum is 19 at $x = -2$.

c. Identify the function's domain and its range. (Section 3.1, Example 4) domain: $(-\infty, \infty)$; range: $(-\infty, 19]$

152. Solve the equation $x^3 - 9x^2 + 26x - 24 = 0$ given that 4 is a zero of $f(x) = x^3 - 9x^2 + 26x - 24$. (Section 3.3, Example 6) $\{2, 3, 4\}$

Preview Exercises

Exercises 153–155 will help you prepare for the material covered in the next section.

153. The formula $A = 10e^{-0.003t}$ models the population of Hungary, A, in millions, t years after 2006.

a. Find Hungary's population, in millions, for 2006, 2007, 2008, and 2009. Round to two decimal places. 10 million; 9.97 million; 9.94 million; 9.91 million

b. Is Hungary's population increasing or decreasing? decreasing

154. a. Simplify: $e^{\ln 3}$. 3

b. Use your simplification from part (a) to rewrite 3^x in terms of base e. $e^{(\ln 3)x}$

155. U.S. soldiers fight Russian troops who have invaded New York City. Incoming missiles from Russian submarines and warships ravage the Manhattan skyline. It's just another scenario for the multi-billion-dollar video games *Call of Duty*, which have sold more than 100 million games since the franchise's birth in 2003.

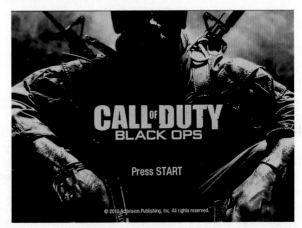

The table shows the annual retail sales for *Call of Duty* video games from 2004 through 2010. Create a scatter plot for the data. Based on the shape of the scatter plot, would a logarithmic function, an exponential function, or a linear function be the best choice for modeling the data?

Annual Retail Sales for *Call of Duty* Games

Year	Retail Sales (millions of dollars)
2004	56
2005	101
2006	196
2007	352
2008	436
2009	778
2010	980

Source: The NPD Group exponential function*

Section 4.5 Exponential Growth and Decay; Modeling Data

What am I supposed to learn?

After studying this section, you should be able to:

1. Model exponential growth and decay.

2. Use logistic growth models.

3. Choose an appropriate model for data.

4. Express an exponential model in base *e*.

On October 31, 2011, the world marked a major milestone: According to the United Nations, the number of people on Earth reached 7 billion—and counting. Since the dawn of humankind some 50,000 years ago, an estimated total of 108 billion people have lived on our planet, which means that about 6.5% of all humans ever born are alive today. That's a lot of bodies to feed, clothe, and shelter. Scientists, politicians, economists, and demographers have long disagreed when it comes to making predictions about the effects of the world's growing population. Debates about entities that are growing exponentially can be approached mathematically: We can create functions that model data and use these functions to make predictions. In this section, we will show you how this is done.

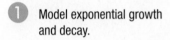

1. Model exponential growth and decay.

Exponential Growth and Decay

One of algebra's many applications is to predict the behavior of variables. This can be done with *exponential growth* and *decay models*. With exponential growth or decay, quantities grow or decay at a rate directly proportional to their size. Populations that are growing exponentially grow extremely rapidly as they get larger because there are more adults to have offspring. For example, world population is increasing at approximately 1.2% per year. This means that each year world population is 1.2% more than what it was in the previous year. In 2010, world population was 6.9 billion. Thus, we compute the world population in 2011 as follows:

6.9 billion + 1.2% of 6.9 billion = 6.9 + (0.012)(6.9) = 6.9828.

This computation indicates that 6.9828 billion people populated the world in 2011. The 0.0828 billion represents an increase of 82.8 million people from 2010 to 2011, the equivalent of the population of Germany. Using 1.2% as the annual rate of increase, world population for 2012 is found in a similar manner:

6.9828 + 1.2% of 6.9828 = 6.9828 + (0.012)(6.9828) ≈ 7.067.

This computation indicates that approximately 7.1 billion people populated the world in 2012.

The explosive growth of world population may remind you of the growth of money in an account subject to compound interest. Just as the growth rate for world population is multiplied by the population plus any increase in the population, a compound interest rate is multiplied by the original investment plus any accumulated interest. The balance in an account subject to continuous compounding and world population are special cases of *exponential growth models*.

Why does the formula for exponential growth look familiar?

You have seen the formula for exponential growth before, but with different letters. It is the formula for compound interest with continuous compounding.

$$A = Pe^{rt}$$

| Amount at time t | Principal is the original amount. | Interest rate is the growth rate. |

$$A = A_0 e^{kt}$$

Exponential Growth and Decay Models

The mathematical model for **exponential growth** or **decay** is given by

$$f(t) = A_0 e^{kt} \quad \text{or} \quad A = A_0 e^{kt}.$$

- **If $k > 0$, the function models the amount, or size, of a *growing* entity.** A_0 is the original amount, or size, of the growing entity at time $t = 0$, A is the amount at time t, and k is a constant representing the growth rate.

- **If $k < 0$, the function models the amount, or size, of a *decaying* entity.** A_0 is the original amount, or size, of the decaying entity at time $t = 0$, A is the amount at time t, and k is a constant representing the decay rate.

(a) Exponential growth (b) Exponential decay

Sometimes we need to use given data to determine k, the rate of growth or decay. After we compute the value of k, we can use the formula $A = A_0 e^{kt}$ to make predictions. This idea is illustrated in our first two examples.

EXAMPLE 1 Modeling the Growth of the U.S. Population

The graph in **Figure 4.22** shows the U.S. population, in millions, for five selected years from 1970 through 2010. In 1970, the U.S. population was 203.3 million. By 2010, it had grown to 308.7 million.

a. Find an exponential growth function that models the data for 1970 through 2010.

b. By which year will the U.S. population reach 335 million?

SOLUTION

a. We use the exponential growth model

$$A = A_0 e^{kt}$$

in which t is the number of years after 1970. This means that 1970 corresponds to $t = 0$. At that time the U.S. population was 203.3 million, so we substitute 203.3 for A_0 in the growth model:

$$A = 203.3 e^{kt}.$$

We are given that 308.7 million was the population in 2010. Because 2010 is 40 years after 1970, when $t = 40$ the value of A is 308.7. Substituting these numbers into the growth model will enable us to find k, the growth rate. We know that $k > 0$ because the problem involves growth.

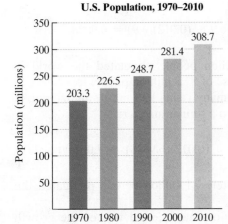

U.S. Population, 1970–2010

FIGURE 4.22
Source: U.S. Census Bureau

$$A = 203.3e^{kt}$$

Use the growth model with $A_0 = 203.3$.

$$308.7 = 203.3e^{k \cdot 40}$$

When $t = 40$, $A = 308.7$. Substitute these numbers into the model.

$$e^{40k} = \frac{308.7}{203.3}$$

Isolate the exponential factor by dividing both sides by 203.3. We also reversed the sides.

$$\ln e^{40k} = \ln\left(\frac{308.7}{203.3}\right)$$

Take the natural logarithm on both sides.

$$40k = \ln\left(\frac{308.7}{203.3}\right)$$

Simplify the left side using $\ln e^x = x$.

$$k = \frac{\ln\left(\dfrac{308.7}{203.3}\right)}{40} \approx 0.01$$

Divide both sides by 40 and solve for k. Then use a calculator.

The value of k, approximately 0.01, indicates a growth rate of about 1%. We substitute 0.01 for k in the growth model, $A = 203.3e^{kt}$, to obtain an exponential growth function for the U.S. population. It is

$$A = 203.3e^{0.01t},$$

where t is measured in years after 1970.

b. To find the year in which the U.S. population will reach 335 million, substitute 335 for A in the model from part (a) and solve for t.

$$A = 203.3e^{0.01t}$$

This is the model from part (a).

$$335 = 203.3e^{0.01t}$$

Substitute 335 for A.

$$e^{0.01t} = \frac{335}{203.3}$$

Divide both sides by 203.3. We also reversed the sides.

$$\ln e^{0.01t} = \ln\left(\frac{335}{203.3}\right)$$

Take the natural logarithm on both sides.

$$0.01t = \ln\left(\frac{335}{203.3}\right)$$

Simplify on the left using $\ln e^x = x$.

$$t = \frac{\ln\left(\dfrac{335}{203.3}\right)}{0.01} \approx 50$$

Divide both sides by 0.01 and solve for t. Then use a calculator.

Because t represents the number of years after 1970, the model indicates that the U.S. population will reach 335 million by $1970 + 50$, or in the year 2020. ● ● ●

In Example 1, we used only two data values, the population for 1970 and the population for 2010, to develop a model for U.S. population growth from 1970 through 2010. By not using data for any other years, have we created a model that inaccurately describes both the existing data and future population projections given by the U.S. Census Bureau? Something else to think about: Is an exponential model the best choice for describing U.S. population growth, or might a linear model provide a better description? We return to these issues in Exercises 66–70 in the Exercise Set.

⌀ Check Point **1** In 2000, the population of Africa was 807 million and by 2011 it had grown to 1052 million.

a. Use the exponential growth model $A = A_0 e^{kt}$, in which t is the number of years after 2000, to find the exponential growth function that models the data. $A = 807e^{0.024t}$

b. By which year will Africa's population reach 2000 million, or two billion? 2038

Our next example involves exponential decay and its use in determining the age of fossils and artifacts. The method is based on considering the percentage of carbon-14 remaining in the fossil or artifact. Carbon-14 decays exponentially with a *half-life* of approximately 5715 years. The **half-life** of a substance is the time required for half of a given sample to disintegrate. Thus, after 5715 years a given amount of carbon-14 will have decayed to half the original amount. Carbon dating is useful for artifacts or fossils up to 80,000 years old. Older objects do not have enough carbon-14 left to determine age accurately.

EXAMPLE 2 Carbon-14 Dating: The Dead Sea Scrolls

a. Use the fact that after 5715 years a given amount of carbon-14 will have decayed to half the original amount to find the exponential decay model for carbon-14.

b. In 1947, earthenware jars containing what are known as the Dead Sea Scrolls were found by an Arab Bedouin herdsman. Analysis indicated that the scroll wrappings contained 76% of their original carbon-14. Estimate the age of the Dead Sea Scrolls.

SOLUTION

a. We begin with the exponential decay model $A = A_0 e^{kt}$. We know that $k < 0$ because the problem involves the decay of carbon-14. After 5715 years ($t = 5715$), the amount of carbon-14 present, A, is half the original amount, A_0. Thus, we can substitute $\dfrac{A_0}{2}$ for A in the exponential decay model. This will enable us to find k, the decay rate.

$A = A_0 e^{kt}$	Begin with the exponential decay model.
$\dfrac{A_0}{2} = A_0 e^{k5715}$	After 5715 years ($t = 5715$), $A = \dfrac{A_0}{2}$ (because the amount present, A, is half the original amount, A_0).
$\dfrac{1}{2} = e^{5715k}$	Divide both sides of the equation by A_0.
$\ln\left(\dfrac{1}{2}\right) = \ln e^{5715k}$	Take the natural logarithm on both sides.
$\ln\left(\dfrac{1}{2}\right) = 5715k$	Simplify the right side using $\ln e^x = x$.
$k = \dfrac{\ln\left(\dfrac{1}{2}\right)}{5715} \approx -0.000121$	Divide both sides by 5715 and solve for k.

Substituting -0.000121 for k in the decay model, $A = A_0 e^{kt}$, the model for carbon-14 is

$$A = A_0 e^{-0.000121t}.$$

b. In 1947, the Dead Sea Scrolls contained 76% of their original carbon-14. To find their age in 1947, substitute $0.76A_0$ for A in the model from part (a) and solve for t.

$A = A_0 e^{-0.000121t}$	This is the decay model for carbon-14.
$0.76A_0 = A_0 e^{-0.000121t}$	A, the amount present, is 76% of the original amount, so $A = 0.76A_0$.
$0.76 = e^{-0.000121t}$	Divide both sides of the equation by A_0.
$\ln 0.76 = \ln e^{-0.000121t}$	Take the natural logarithm on both sides.
$\ln 0.76 = -0.000121t$	Simplify the right side using $\ln e^x = x$.
$t = \dfrac{\ln 0.76}{-0.000121} \approx 2268$	Divide both sides by -0.000121 and solve for t.

The Dead Sea Scrolls are approximately 2268 years old plus the number of years between 1947 and the current year. • • •

Blitzer Bonus ||

Carbon Dating and Artistic Development

The artistic community was electrified by the discovery in 1995 of spectacular cave paintings in a limestone cavern in France. Carbon dating of the charcoal from the site showed that the images, created by artists of remarkable talent, were 30,000 years old, making them the oldest cave paintings ever found. The artists seemed to have used the cavern's natural contours to heighten a sense of perspective. The quality of the painting suggests that the art of early humans did not mature steadily from primitive to sophisticated in any simple linear fashion.

⊘ Check Point **2** Strontium-90 is a waste product from nuclear reactors. As a consequence of fallout from atmospheric nuclear tests, we all have a measurable amount of strontium-90 in our bones.

 a. The half-life of strontium-90 is 28 years, meaning that after 28 years a given amount of the substance will have decayed to half the original amount. Find the exponential decay model for strontium-90. $A = A_0 e^{-0.0248t}$

 b. Suppose that a nuclear accident occurs and releases 60 grams of strontium-90 into the atmosphere. How long will it take for strontium-90 to decay to a level of 10 grams? about 72 years

② Use logistic growth models.

Logistic Growth Models

From population growth to the spread of an epidemic, nothing on Earth can grow exponentially indefinitely. Growth is always limited. This is shown in **Figure 4.23** by the horizontal asymptote. The *logistic growth model* is a function used to model situations of this type.

FIGURE 4.23 The logistic growth curve has a horizontal asymptote that identifies the limit of the growth of A over time.

Logistic Growth Model

The mathematical model for limited logistic growth is given by

$$f(t) = \frac{c}{1 + ae^{-bt}} \quad \text{or} \quad A = \frac{c}{1 + ae^{-bt}},$$

where a, b, and c are constants, with $c > 0$ and $b > 0$.

As time increases ($t \to \infty$), the expression ae^{-bt} in the model approaches 0, and A gets closer and closer to c. This means that $y = c$ is a horizontal asymptote for the graph of the function. Thus, the value of A can never exceed c and c represents the limiting size that A can attain.

 EXAMPLE 3 Modeling the Spread of the Flu

The function

$$f(t) = \frac{30,000}{1 + 20e^{-1.5t}}$$

describes the number of people, $f(t)$, who have become ill with influenza t weeks after its initial outbreak in a town with 30,000 inhabitants.

 a. How many people became ill with the flu when the epidemic began?

 b. How many people were ill by the end of the fourth week?

 c. What is the limiting size of $f(t)$, the population that becomes ill?

SOLUTION

 a. The time at the beginning of the flu epidemic is $t = 0$. Thus, we can find the number of people who were ill at the beginning of the epidemic by substituting 0 for t.

$$f(t) = \frac{30,000}{1 + 20e^{-1.5t}} \qquad \text{This is the given logistic growth function.}$$

$$f(0) = \frac{30,000}{1 + 20e^{-1.5(0)}} \qquad \text{When the epidemic began, } t = 0.$$

$$= \frac{30,000}{1 + 20} \qquad\qquad e^{-1.5(0)} = e^0 = 1$$

$$\approx 1429$$

Approximately 1429 people were ill when the epidemic began.

TECHNOLOGY

The graph of the logistic growth function for the flu epidemic

$$y = \frac{30,000}{1 + 20e^{-1.5x}}$$

can be obtained using a graphing utility. We started x at 0 and ended at 10. This takes us to week 10. (In Example 3, we found that by week 4 approximately 28,583 people were ill.) We also know that 30,000 is the limiting size, so we took values of y up to 30,000. Using a [0, 10, 1] by [0, 30,000, 3000] viewing rectangle, the graph of the logistic growth function is shown below.

b. We find the number of people who were ill at the end of the fourth week by substituting 4 for t in the logistic growth function.

$$f(t) = \frac{30,000}{1 + 20e^{-1.5t}} \qquad \text{Use the given logistic growth function.}$$

$$f(4) = \frac{30,000}{1 + 20e^{-1.5(4)}} \qquad \text{To find the number of people ill by the end of week four, let } t = 4.$$

$$\approx 28,583 \qquad \text{Use a calculator.}$$

Approximately 28,583 people were ill by the end of the fourth week. Compared with the number of people who were ill initially, 1429, this illustrates the virulence of the epidemic.

c. Recall that in the logistic growth model, $f(t) = \dfrac{c}{1 + ae^{-bt}}$, the constant c represents the limiting size that $f(t)$ can attain. Thus, the number in the numerator, 30,000, is the limiting size of the population that becomes ill. • • •

✓ **Check Point 3** In a learning theory project, psychologists discovered that

$$f(t) = \frac{0.8}{1 + e^{-0.2t}}$$

is a model for describing the proportion of correct responses, $f(t)$, after t learning trials.

a. Find the proportion of correct responses prior to learning trials taking place. 0.4 correct responses

b. Find the proportion of correct responses after 10 learning trials. 0.7 correct responses

c. What is the limiting size of $f(t)$, the proportion of correct responses, as continued learning trials take place? 0.8 correct responses

③ Choose an appropriate model for data.

Modeling Data

Throughout this chapter, we have been working primarily with models that were given. However, we can create functions that model data by observing patterns in scatter plots. **Figure 4.24** shows scatter plots for data that are exponential, logarithmic, and linear.

$$y = ab^x, \quad a > 0, b > 1$$
Exponential

$$y = ab^x, \quad a > 0, 0 < b < 1$$
Exponential

$$y = a + b \ln x, \quad a > 0, b > 0$$
Logarithmic

$$y = a + b \ln x, \quad a > 0, b < 0$$
Logarithmic

$$y = ax + b, \quad a > 0$$
Linear

$$y = ax + b, \quad a < 0$$
Linear

FIGURE 4.24 Scatter plots for exponential, logarithmic, and linear models

Table 4.6 Growth of the Human Brain

Age	Percentage of Adult Size Brain
1	30%
2	50%
4	78%
6	88%
8	92%
10	95%
11	99%

Source: Gerrig and Zimbardo, *Psychology and Life*, 18th Edition, Allyn and Bacon, 2008.

EXAMPLE 4 Choosing a Model for Data

The data in **Table 4.6** indicate that between the ages of 1 and 11, the human brain does not grow linearly, or steadily. A scatter plot for the data is shown in **Figure 4.25**. What type of function would be a good choice for modeling the data?

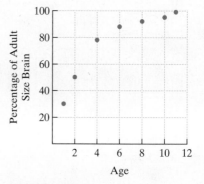

FIGURE 4.25

SOLUTION

Because the data in the scatter plot increase rapidly at first and then begin to level off a bit, the shape suggests that a logarithmic function is a good choice for modeling the data.

● ● ●

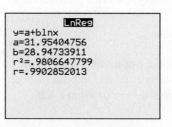

Check Point 4 **Table 4.7** shows the populations of various cities, in thousands, and the average walking speed, in feet per second, of a person living in the city. Create a scatter plot for the data. Based on the scatter plot, what type of function would be a good choice for modeling the data? logarithmic function*

Table 4.7 Population and Walking Speed

Population (thousands)	Walking Speed (feet per second)
5.5	0.6
14	1.0
71	1.6
138	1.9
342	2.2

Source: Mark H. Bornstein and Helen G. Bornstein, "The Pace of Life." *Nature*, 259, Feb. 19, 1976, pp. 557–559

Table 4.8

x, Age	y, Percentage of Adult Size Brain
1	30
2	50
4	78
6	88
8	92
10	95
11	99

FIGURE 4.26 A logarithmic model for the data in **Table 4.8**

How can we obtain a logarithmic function that models the data for the growth of the human brain? A graphing utility can be used to obtain a logarithmic model of the form $y = a + b \ln x$. **Because the domain of the logarithmic function is the set of positive numbers, zero must not be a value for x.** This is not a problem for the data giving the percentage of an adult size brain because the data begin at age 1. We will assign x to represent age and y to represent the percentage of an adult size brain. This gives us the data shown in **Table 4.8**. Using the logarithmic regression option, we obtain the equation in **Figure 4.26**.

From **Figure 4.26**, we see that the logarithmic model of the data, with numbers rounded to three decimal places, is

$$y = 31.954 + 28.947 \ln x.$$

The number r that appears in **Figure 4.26** is called the **correlation coefficient** and is a measure of how well the model fits the data. The value of r is such that $-1 \le r \le 1$. A positive r means that as the x-values increase, so do the y-values. A negative r means that as the x-values increase, the y-values decrease. **The closer that r is to -1 or 1, the better the model fits the data.** Because r is approximately 0.99, the model fits the data very well.

*See Answers to Selected Exercises.

EXAMPLE 5 Choosing a Model for Data

Figure 4.27(a) shows world population, in billions, for seven selected years from 1950 through 2010. A scatter plot is shown in **Figure 4.27(b)**. Suggest two types of functions that would be good choices for modeling the data.

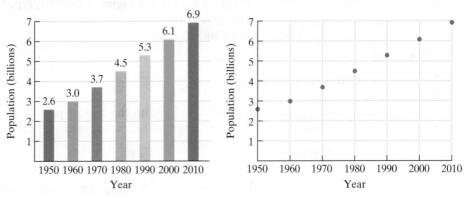

FIGURE 4.27(a)
Source: U.S. Census Bureau, International Database

FIGURE 4.27(b)

Table 4.9 Percentage of U.S. Men Who Are Married or Who Have Been Married, by Age

Age	Percent
18	2
20	7
25	36
30	61
35	75

Source: National Center for Health Statistics

SOLUTION

Because the data in the scatter plot appear to increase more and more rapidly, the shape suggests that an exponential model might be a good choice. Furthermore, we can probably draw a line that passes through or near the seven points. Thus, a linear function would also be a good choice for modeling the data. •••

 Check Point **5** **Table 4.9** shows the percentage of U.S. men who are married or who have been married, by age. Create a scatter plot for the data. Based on the scatter plot, what type of function would be a good choice for modeling the data?
exponential function, although answers may vary*

EXAMPLE 6 Comparing Linear and Exponential Models

The data for world population are shown in **Table 4.10**. Using a graphing utility's linear regression feature and exponential regression feature, we enter the data and obtain the models shown in **Figure 4.28**.

> Although the domain of $y = ab^x$ is the set of all real numbers, some graphing utilities only accept positive values for x. That's why we assigned x to represent the number of years after 1949.

Table 4.10

x, Number of Years after 1949	y, World Population (billions)
1 (1950)	2.6
11 (1960)	3.0
21 (1970)	3.7
31 (1980)	4.5
41 (1990)	5.3
51 (2000)	6.1
61 (2010)	6.9

FIGURE 4.28 A linear model and an exponential model for the data in **Table 4.10**

Because r, the correlation coefficient, is close to 1 in each screen in **Figure 4.28**, the models fit the data very well.

 a. Use **Figure 4.28** to express each model in function notation, with numbers rounded to three decimal places.
 b. How well do the functions model world population in 2000?
 c. By one projection, world population is expected to reach 8 billion in the year 2026. Which function serves as a better model for this prediction?

SOLUTION

a. Using **Figure 4.28** and rounding to three decimal places, the functions

$$f(x) = 0.074x + 2.294 \quad \text{and} \quad g(x) = 2.577(1.017)^x$$

model world population, in billions, x years after 1949. We named the linear function f and the exponential function g, although any letters can be used.

b. **Table 4.10** on the previous page shows that world population in 2000 was 6.1 billion. The year 2000 is 51 years after 1949. Thus, we substitute 51 for x in each function's equation and then evaluate the resulting expressions with a calculator to see how well the functions describe world population in 2000.

$f(x) = 0.074x + 2.294$	This is the linear model.
$f(51) = 0.074(51) + 2.294$	Substitute 51 for x.
≈ 6.1	Use a calculator.
$g(x) = 2.577(1.017)^x$	This is the exponential model.
$g(51) = 2.577(1.017)^{51}$	Substitute 51 for x.
≈ 6.1	Use a calculator:
	2.577 $\boxed{\times}$ 1.017 $\boxed{y^x}$ (or $\boxed{\wedge}$) 51 $\boxed{=}$.

Because 6.1 billion was the actual world population in 2000, both functions model world population in 2000 extremely well.

c. Let's see which model comes closer to projecting a world population of 8 billion in the year 2026. Because 2026 is 77 years after 1949 ($2026 - 1949 = 77$), we substitute 77 for x in each function's equation.

$f(x) = 0.074x + 2.294$	This is the linear model.
$f(77) = 0.074(77) + 2.294$	Substitute 77 for x.
≈ 8.0	Use a calculator.
$g(x) = 2.577(1.017)^x$	This is the exponential model.
$g(77) = 2.577(1.017)^{77}$	Substitute 77 for x.
≈ 9.4	Use a calculator:
	2.577 $\boxed{\times}$ 1.017 $\boxed{y^x}$ (or $\boxed{\wedge}$) 77 $\boxed{=}$.

The linear function $f(x) = 0.074x + 2.294$ serves as a better model for a projected world population of 8 billion by 2026. ● ● ●

GREAT QUESTION!

How can I use a graphing utility to see how well my models describe the data?

Once you have obtained one or more models for the data, you can use a graphing utility's $\boxed{\text{TABLE}}$ feature to numerically see how well each model describes the data. Enter the models as y_1, y_2, and so on. Create a table, scroll through the table, and compare the table values given by the models to the actual data.

∅ Check Point **6** Use the models $f(x) = 0.074x + 2.294$ and $g(x) = 2.577(1.017)^x$ to solve this problem.

a. World population in 1970 was 3.7 billion. Which function serves as a better model for this year? the exponential function g

b. By one projection, world population is expected to reach 9.3 billion by 2050. Which function serves as a better model for this projection? the linear function f

When using a graphing utility to model data, begin with a scatter plot, drawn either by hand or with the graphing utility, to obtain a general picture for the shape of the data. It might be difficult to determine which model best fits the data—linear, logarithmic, exponential, quadratic, or something else. If necessary, use your graphing utility to fit several models to the data. The best model is the one that yields the value r, the correlation coefficient, closest to 1 or -1. Finding a proper fit for data can be almost as much art as it is mathematics. In this era of technology, the process of creating models that best fit data is one that involves more decision making than computation.

 Express an exponential model in base *e*.

Expressing $y = ab^x$ in Base *e*

Graphing utilities display exponential models in the form $y = ab^x$. However, our discussion of exponential growth involved base *e*. Because of the inverse property $b = e^{\ln b}$, we can rewrite any model in the form $y = ab^x$ in terms of base *e*.

> **Expressing an Exponential Model in Base *e***
>
> $$y = ab^x \quad \text{is equivalent to} \quad y = ae^{(\ln b)\cdot x}.$$

Blitzer Bonus ||

Global Population Increase

By the time you finish reading Example 7 and working Check Point 7, more than 1000 people will have been added to our planet. By this time tomorrow, world population will have increased by more than 220,000.

Doubling of the World's Population

Year	World Population (millions)	Years to Double Population
480 B.C.	110	
800	220	1280
1330	440	530
1810	875	480
1910	1750	100
1967	3500	57
2012	7000	45

Source: U.S. Census Bureau

| EXAMPLE 7 | Rewriting the Model for World Population in Base *e*

We have seen that the function

$$g(x) = 2.577(1.017)^x$$

models world population, $g(x)$, in billions, *x* years after 1949. Rewrite the model in terms of base *e*.

SOLUTION

We use the two equivalent equations shown in the voice balloons to rewrite the model in terms of base *e*.

$$y = ab^x \qquad\qquad y = ae^{(\ln b)\cdot x}$$

$$g(x) = 2.577(1.017)^x \quad \text{is equivalent to} \quad g(x) = 2.577e^{(\ln 1.017)x}.$$

Using $\ln 1.017 \approx 0.017$, the exponential growth model for world population, $g(x)$, in billions, *x* years after 1949 is

$$g(x) = 2.577e^{0.017x}. \qquad \bullet\bullet\bullet$$

In Example 7, we can replace $g(x)$ with *A* and *x* with *t* so that the model has the same letters as those in the exponential growth model $A = A_0e^{kt}$.

$$A = \;A_0\;e^{kt} \qquad \text{This is the exponential growth model.}$$

$$A = 2.577e^{0.017t} \qquad \text{This is the model for world population.}$$

The value of *k*, 0.017, indicates a growth rate of 1.7%. Although this is an excellent model for the data, we must be careful about making projections about world population using this growth function. Why? World population growth rate is now 1.2%, not 1.7%, so our model will overestimate future populations.

✓ **Check Point 7** Rewrite $y = 4(7.8)^x$ in terms of base *e*. Express the answer in terms of a natural logarithm and then round to three decimal places.
$y = 4e^{(\ln 7.8)x}; y = 4e^{2.054x}$

CONCEPT AND VOCABULARY CHECK

Fill in each blank so that the resulting statement is true.

1. Consider the model for exponential growth or decay given by
$$A = A_0e^{kt}.$$
If $k \underline{\;>0\;}$, the function models the amount, or size, of a growing entity. If $k \underline{\;<0\;}$, the function models the amount, or size, of a decaying entity.

2. In the model for exponential growth or decay, the amount, or size, at $t = 0$ is represented by $\underline{\;A_0\;}$. The amount, or size, at time *t* is represented by $\underline{\;A\;}$.

3. Consider the model for limited logistic growth given by
$$A = \frac{c}{1 + ae^{-bt}}.$$
The amount, or size, at time *t* is represented by $\underline{\;A\;}$. This value can never exceed $\underline{\;c\;}$.

For each of the scatter plots in Exercises 4–6, determine whether an exponential function, a logarithmic function, or a linear function is the best choice for modeling the data.

4.
logarithmic

5.
exponential

6.
linear

7. $y = 3(5)^x$ can be written in terms of base e as $y = 3e^{(\ln 5) \cdot x}$.

EXERCISE SET 4.5

Practice Exercises and Application Exercises

The exponential models describe the population of the indicated country, A, in millions, t years after 2010. Use these models to solve Exercises 1–6.

India	$A = 1173.1e^{0.008t}$
Iraq	$A = 31.5e^{0.019t}$
Japan	$A = 127.3e^{-0.006t}$
Russia	$A = 141.9e^{-0.005t}$

1. What was the population of Japan in 2010? 127.3 million
2. What was the population of Iraq in 2010? 31.5 million
3. Which country has the greatest growth rate? By what percentage is the population of that country increasing each year? Iraq; 1.9%
4. Which countries have a decreasing population? By what percentage is the population of these countries decreasing each year? Japan: 0.6%; Russia: 0.5%
5. When will India's population be 1377 million? 2030
6. When will India's population be 1491 million? 2040

About the size of New Jersey, Israel has seen its population soar to more than 6 million since it was established. The graphs show that by 2050, Palestinians in the West Bank, Gaza Strip, and East Jerusalem will outnumber Israelis. Exercises 7–8 involve the projected growth of these two populations.

Source: *Newsweek*

*See Answers to Selected Exercises.

7. a. In 2000, the population of Israel was approximately 6.04 million and by 2050 it is projected to grow to 10 million. Use the exponential growth model $A = A_0e^{kt}$, in which t is the number of years after 2000, to find an exponential growth function that models the data. $A = 6.04e^{0.01t}$

b. In which year will Israel's population be 9 million? 2040

8. a. In 2000, the population of the Palestinians in the West Bank, Gaza Strip, and East Jerusalem was approximately 3.2 million and by 2050 it is projected to grow to 12 million. Use the exponential growth model $A = A_0e^{kt}$, in which t is the number of years after 2000, to find the exponential growth function that models the data. $A = 3.2e^{0.026t}$

b. In which year will the Palestinian population be 9 million? 2040

In Exercises 9–14, complete the table. Round projected populations to one decimal place and values of k to four decimal places.

	Country	2010 Population (millions)	Projected 2050 Population (millions)	Projected Growth Rate, k
9.	Philippines	99.9	146.1	0.0095
10.	Pakistan	184.4	334.7	0.0149
11.	Colombia	44.2	62.9	0.0088
12.	Madagascar	21.3	42.7	0.0174
13.	Germany	82.3	70.5	−0.0039
14.	Bulgaria	7.1	5.4	−0.0068

Source: International Programs Center, U.S. Census Bureau

An artifact originally had 16 grams of carbon-14 present. The decay model $A = 16e^{-0.000121t}$ describes the amount of carbon-14 present after t years. Use this model to solve Exercises 15–16.

15. How many grams of carbon-14 will be present in 5715 years? *

16. How many grams of carbon-14 will be present in 11,430 years? ≈ 4 grams

17. The half-life of the radioactive element krypton-91 is 10 seconds. If 16 grams of krypton-91 are initially present, how many grams are present after 10 seconds? 20 seconds? 30 seconds? 40 seconds? 50 seconds? *

18. The half-life of the radioactive element plutonium-239 is 25,000 years. If 16 grams of plutonium-239 are initially present, how many grams are present after 25,000 years? 50,000 years? 75,000 years? 100,000 years? 125,000 years? *

Use the exponential decay model for carbon-14, $A = A_0 e^{-0.000121t}$, to solve Exercises 19–20.

19. Prehistoric cave paintings were discovered in a cave in France. The paint contained 15% of the original carbon-14. Estimate the age of the paintings. ≈ 15,679 yr old

20. Skeletons were found at a construction site in San Francisco in 1989. The skeletons contained 88% of the expected amount of carbon-14 found in a living person. In 1989, how old were the skeletons? ≈ 1056 yr old

In Exercises 21–26, complete the table. Round half-lives to one decimal place and values of k to six decimal places.

	Radioactive Substance	**Half-Life**	**Decay Rate, k**
21.	Tritium	12.6 years	5.5% per year = −0.055
22.	Krypton-85	11.0 years	6.3% per year = −0.063
23.	Radium-226	1620 years	0.0428% per yr = −0.000428
24.	Uranium-238	4560 years	0.0152% per yr = −0.000152
25.	Arsenic-74	17.5 days	3.9608% per day = −0.039608
26.	Calcium-47	113 hours	0.6134% per hr = −0.006134

27. The August 1978 issue of *National Geographic* described the 1964 find of bones of a newly discovered dinosaur weighing 170 pounds, measuring 9 feet, with a 6-inch claw on one toe of each hind foot. The age of the dinosaur was estimated using potassium-40 dating of rocks surrounding the bones.

 a. Potassium-40 decays exponentially with a half-life of approximately 1.31 billion years. Use the fact that after 1.31 billion years a given amount of potassium-40 will have decayed to half the original amount to show that the decay model for potassium-40 is given by $A = A_0 e^{-0.52912t}$, where t is in billions of years. *

 b. Analysis of the rocks surrounding the dinosaur bones indicated that 94.5% of the original amount of potassium-40 was still present. Let $A = 0.945A_0$ in the model in part (a) and estimate the age of the bones of the dinosaur. ≈ 106,900,000 years old

Use the exponential decay model, $A = A_0 e^{kt}$, to solve Exercises 28–31. Round answers to one decimal place.

28. The half-life of thorium-229 is 7340 years. How long will it take for a sample of this substance to decay to 20% of its original amount? 17,043.0 yr

29. The half-life of lead is 22 years. How long will it take for a sample of this substance to decay to 80% of its original amount? 7.1 yr

30. The half-life of aspirin in your bloodstream is 12 hours. How long will it take for the aspirin to decay to 70% of the original dosage? 6.2 hr

31. Xanax is a tranquilizer used in the short-term relief of symptoms of anxiety. Its half-life in the bloodstream is 36 hours. How long will it take for Xanax to decay to 90% of the original dosage? 5.5 hr

32. A bird species in danger of extinction has a population that is decreasing exponentially ($A = A_0 e^{kt}$). Five years ago the population was at 1400 and today only 1000 of the birds are alive. Once the population drops below 100, the situation will be irreversible. When will this happen? *

33. Use the exponential growth model, $A = A_0 e^{kt}$, to show that the time it takes a population to double (to grow from A_0 to $2A_0$) is given by $t = \dfrac{\ln 2}{k}$. *

34. Use the exponential growth model, $A = A_0 e^{kt}$, to show that the time it takes a population to triple (to grow from A_0 to $3A_0$) is given by $t = \dfrac{\ln 3}{k}$. *

Use the formula $t = \dfrac{\ln 2}{k}$ that gives the time for a population with a growth rate k to double to solve Exercises 35–36. Express each answer to the nearest whole year.

35. The growth model $A = 4.3e^{0.01t}$ describes New Zealand's population, A, in millions, t years after 2010.

 a. What is New Zealand's growth rate? 1%

 b. How long will it take New Zealand to double its population? ≈ 69 yr

36. The growth model $A = 112.5e^{0.012t}$ describes Mexico's population, A, in millions, t years after 2010.

 a. What is Mexico's growth rate? 1.2%

 b. How long will it take Mexico to double its population? *

37. The logistic growth function

$$f(t) = \frac{100,000}{1 + 5000e^{-t}}$$

describes the number of people, $f(t)$, who have become ill with influenza t weeks after its initial outbreak in a particular community.

 a. How many people became ill with the flu when the epidemic began? ≈20 people

 b. How many people were ill by the end of the fourth week? ≈1080 people

 c. What is the limiting size of the population that becomes ill? 100,000 people

Shown, again, in the following table is world population, in billions, for seven selected years from 1950 through 2010. Using a graphing utility's logistic regression option, we obtain the equation shown on the screen.

x, Number of Years after 1949	**y, World Population (billions)**
1 (1950)	2.6
11 (1960)	3.0
21 (1970)	3.7
31 (1980)	4.5
41 (1990)	5.3
51 (2000)	6.1
61 (2010)	6.9

```
        Logistic
y=c/(1+ae^(-bx))
a=4.110936638
b=.0264832525
c=12.57374318
```

We see from the calculator screen at the bottom of the previous page that a logistic growth model for world population, f(x), in billions, x years after 1949 is

$$f(x) = \frac{12.57}{1 + 4.11e^{-0.026x}}.$$

Use this function to solve Exercises 38–42.

38. How well does the function model the data showing a world population of 6.1 billion for 2000? *

39. How well does the function model the data showing a world population of 6.9 billion for 2010? *

40. When did world population reach 7 billion? 2012

41. When will world population reach 8 billion? 2025

42. According to the model, what is the limiting size of the population that Earth will eventually sustain? 12.57 billion

The logistic growth function

$$P(x) = \frac{90}{1 + 271e^{-0.122x}}$$

models the percentage, P(x), of Americans who are x years old with some coronary heart disease. Use the function to solve Exercises 43–46.

43. What percentage of 20-year-olds have some coronary heart disease? ≈3.7%

44. What percentage of 80-year-olds have some coronary heart disease? ≈88.6%

45. At what age is the percentage of some coronary heart disease 50%? ≈48 years old

46. At what age is the percentage of some coronary heart disease 70%? ≈56 years old

Exercises 47–52 present data in the form of tables. For each data set shown by the table,

 a. *Create a scatter plot for the data.*

 b. *Use the scatter plot to determine whether an exponential function, a logarithmic function, or a linear function is the best choice for modeling the data. (If applicable, in Exercise 72, you will use your graphing utility to obtain these functions.)*

47. Percent of Miscarriages, by Age **b.** exponential function*

Woman's Age	Percent of Miscarriages
22	9%
27	10%
32	13%
37	20%
42	38%
47	52%

Source: *Time*

48. Savings Needed for Health-Care Expenses during Retirement *

Age at Death	Savings Needed
80	$219,000
85	$307,000
90	$409,000
95	$524,000
100	$656,000

Source: Employee Benefit Research Institute

49. Intensity and Loudness Level of Various Sounds *

Intensity (watts per meter²)	Loudness Level (decibels)
0.1 (loud thunder)	110
1 (rock concert, 2 yd from speakers)	120
10 (jackhammer)	130
100 (jet takeoff, 40 yd away)	140

50. Temperature Increase in an Enclosed Vehicle *

Minutes	Temperature Increase (°F)
10	19°
20	29°
30	34°
40	38°
50	41°
60	43°

51. Hamachiphobia logarithmic function*

Generation	Percentage Who Won't Try Sushi	Percentage Who Don't Approve of Marriage Equality
Millennials	42	36
Gen X	52	49
Boomers	60	59
Silent/Greatest Generation	72	66

Source: Pew Research Center

52. Teenage Drug Use linear function*

Country	Percentage Who Have Used	
	Marijuana	Other Illegal Drugs
Czech Republic	22	4
Denmark	17	3
England	40	21
Finland	5	1
Ireland	37	16
Italy	19	8
Northern Ireland	23	14
Norway	6	3
Portugal	7	3
Scotland	53	31
United States	34	24

Source: De Veaux et al., *Intro Stats*, Pearson, 2009.

In Exercises 53–56, rewrite the equation in terms of base e. Express the answer in terms of a natural logarithm and then round to three decimal places.

53. $y = 100(4.6)^x$ $y = 100e^{(\ln 4.6)x}; y = 100e^{1.526x}$

54. $y = 1000(7.3)^x$ $y = 1000e^{(\ln 7.3)x}; y = 1000e^{1.988x}$

55. $y = 2.5(0.7)^x$ $y = 2.5e^{(\ln 0.7)x}; y = 2.5e^{-0.357x}$

56. $y = 4.5(0.6)^x$ $y = 4.5e^{(\ln 0.6)x}; y = 4.5e^{-0.511x}$

Explaining the Concepts 57–65. Answers will vary.

57. Nigeria has a growth rate of 0.025 or 2.5%. Describe what this means.

58. How can you tell whether an exponential model describes exponential growth or exponential decay?

59. Suppose that a population that is growing exponentially increases from 800,000 people in 2010 to 1,000,000 people in 2013. Without showing the details, describe how to obtain the exponential growth function that models the data.

60. What is the half-life of a substance?

61. Describe a difference between exponential growth and logistic growth.

62. Describe the shape of a scatter plot that suggests modeling the data with an exponential function.

63. You take up weightlifting and record the maximum number of pounds you can lift at the end of each week. You start off with rapid growth in terms of the weight you can lift from week to week, but then the growth begins to level off. Describe how to obtain a function that models the number of pounds you can lift at the end of each week. How can you use this function to predict what might happen if you continue the sport?

64. Would you prefer that your salary be modeled exponentially or logarithmically? Explain your answer.

65. One problem with all exponential growth models is that nothing can grow exponentially forever. Describe factors that might limit the size of a population.

Technology Exercises

In Example 1 on page 520, we used two data points and an exponential function to model the population of the United States from 1970 through 2010. The data are shown again in the table. Use all five data points to solve Exercises 66–70.

x, Number of Years after 1969	y, U.S. Population (millions)
1 (1970)	203.3
11 (1980)	226.5
21 (1990)	248.7
31 (2000)	281.4
41 (2010)	308.7

66. **a.** Use your graphing utility's exponential regression option to obtain a model of the form $y = ab^x$ that fits the data. How well does the correlation coefficient, r, indicate that the model fits the data? $y = 201.2(1.011)^x; r \approx 0.999*$

b. Rewrite the model in terms of base e. By what percentage is the population of the United States increasing each year? $y = 201.2e^{(\ln 1.011)x}; y = 201.2e^{0.0109x}; \approx 1\%$

67. Use your graphing utility's logarithmic regression option to obtain a model of the form $y = a + b \ln x$ that fits the data. How well does the correlation coefficient, r, indicate that the model fits the data? $y = 191.9 + 24.569 \ln x; r \approx 0.870*$

68. Use your graphing utility's linear regression option to obtain a model of the form $y = ax + b$ that fits the data. How well does the correlation coefficient, r, indicate that the model fits the data? $y = 2.657x + 197.923; r \approx 0.997;$ well

69. Use your graphing utility's power regression option to obtain a model of the form $y = ax^b$ that fits the data. How well does the correlation coefficient, r, indicate that the model fits the data? $y = 195.05x^{0.100}; r \approx 0.896;$ fairly well

70. Use the values of r in Exercises 66–69 to select the two models of best fit. Use each of these models to predict by which year the U.S. population will reach 335 million. How do these answers compare to the year we found in Example 1, namely 2020? If you obtained different years, how do you account for this difference? *

71. The figure shows the number of people in the United States age 65 and over, with projected figures for the year 2020 and beyond.

United States Population Age 65 and Over

Source: U.S. Census Bureau

a. Let x represent the number of years after 1899 and let y represent the U.S. population age 65 and over, in millions. Use your graphing utility to find the model that best fits the data in the bar graph. $y = 3.46(1.02)^x$

b. Rewrite the model in terms of base e. By what percentage is the 65 and over population increasing each year? *

72. In Exercises 47–52, you determined the best choice for the kind of function that modeled the data in the table. For each of the exercises that you worked, use a graphing utility to find the actual function that best fits the data. Then use the model to make a reasonable prediction for a value that exceeds those shown in the table's first column. *

Critical Thinking Exercises

Make Sense? *In Exercises 73–76, determine whether each statement makes sense or does not make sense, and explain your reasoning.*

73. I used an exponential model with a positive growth rate to describe the depreciation in my car's value over four years. *

74. After 100 years, a population whose growth rate is 3% will have three times as many people as a population whose growth rate is 1%. does not make sense

75. When I used an exponential function to model Russia's declining population, the growth rate k was negative. *

76. Because carbon-14 decays exponentially, carbon dating can determine the ages of ancient fossils. makes sense

The exponential growth models describe the population of the indicated country, A, in millions, t years after 2006.

Canada $A = 33.1e^{0.009t}$

Uganda $A = 28.2e^{0.034t}$

In Exercises 77–80, use this information to determine whether each statement is true or false. If the statement is false, make the necessary change(s) to produce a true statement. Changes to false statements will vary.

77. In 2006, Canada's population exceeded Uganda's by 4.9 million. true

78. By 2009, the models indicate that Canada's population will exceed Uganda's by approximately 2.8 million. true

79. The models indicate that in 2013, Uganda's population will exceed Canada's. true

80. Uganda's growth rate is approximately 3.8 times that of Canada's. true

81. Over a period of time, a hot object cools to the temperature of the surrounding air. This is described mathematically by Newton's Law of Cooling:

$$T = C + (T_0 - C)e^{-kt},$$

where t is the time it takes for an object to cool from temperature T_0 to temperature T, C is the surrounding air temperature, and k is a positive constant that is associated with the cooling object. A cake removed from the oven has a temperature of 210°F and is left to cool in a room that has a temperature of 70°F. After 30 minutes, the temperature of the cake is 140°F. What is the temperature of the cake after 40 minutes? $\approx 126°F$

Group Exercise Answers will vary.

82. Each group member should consult an almanac, newspaper, magazine, or the Internet to find data that can be modeled by exponential or logarithmic functions. Group members should select the two sets of data that are most interesting and relevant. For each set selected, find a model that best fits the data. Each group member should make one prediction based on the model and then discuss a consequence of this prediction. What factors might change the accuracy of the prediction?

Retaining the Concepts

83. After a 60% price reduction, you purchase a computer for $440. What was the computer's price before the reduction? (Section 1.3, Example 4) $1100

84. Begin by graphing $y = |x|$. Then use this graph to obtain the graph of $y = |x - 2| + 1$. (Section 2.5, Example 8) *

85. Write an equation in point-slope form and slope-intercept form of the line passing through $(1, -4)$ and parallel to the line whose equation is $3x - y + 5 = 0$. (Section 2.4, Example 1) $y + 4 = 3(x - 1)$; $y = 3x - 7$

Preview Exercises

Exercises 86–88 will help you prepare for the material covered in the first section of the next chapter.

86. **a.** Does $(4, -1)$ satisfy $x + 2y = 2$? yes

 b. Does $(4, -1)$ satisfy $x - 2y = 6$? yes

87. Graph $x + 2y = 2$ and $x - 2y = 6$ in the same rectangular coordinate system. At what point do the graphs intersect? *

88. Solve: $5(2x - 3) - 4x = 9$. {4}

CHAPTER 4 Summary, Review, and Test

SUMMARY

DEFINITIONS AND CONCEPTS	**EXAMPLES**
4.1 Exponential Functions	
a. The exponential function with base b is defined by $f(x) = b^x$, where $b > 0$ and $b \neq 1$.	Ex. 1, p. 465
b. Characteristics of exponential functions and graphs for $0 < b < 1$ and $b > 1$ are shown in the box on page 467.	Ex. 2, p. 466; Ex. 3, p. 466
c. Transformations involving exponential functions are summarized in Table 4.1 on page 468.	Ex. 4, p. 469; Ex. 5, p. 469
d. The natural exponential function is $f(x) = e^x$. The irrational number e is called the natural base, where $e \approx 2.7183$. e is the value that $\left(1 + \dfrac{1}{n}\right)^n$ approaches as $n \to \infty$.	Ex. 6, p. 470

DEFINITIONS AND CONCEPTS	EXAMPLES

e. Formulas for compound interest: After t years, the balance, A, in an account with principal P and annual interest rate r (in decimal form) is given by one of the following formulas:

Ex. 7, p. 473

1. For n compoundings per year: $A = P\left(1 + \dfrac{r}{n}\right)^{nt}$

2. For continuous compounding: $A = Pe^{rt}$.

4.2 Logarithmic Functions

a. Definition of the logarithmic function: For $x > 0$ and $b > 0$, $b \neq 1$, $y = \log_b x$ is equivalent to $b^y = x$. The function $f(x) = \log_b x$ is the logarithmic function with base b. This function is the inverse function of the exponential function with base b.

Ex. 1, p. 479; Ex. 2, p. 480; Ex. 3, p. 480

b. Graphs of logarithmic functions for $b > 1$ and $0 < b < 1$ are shown in Figure 4.8 on page 482. Characteristics of the graphs are summarized in the box on page 483.

Ex. 6, p. 482

c. Transformations involving logarithmic functions are summarized in Table 4.5 on page 483.

Figures 4.9–4.12, pp. 483–484

d. The domain of a logarithmic function of the form $f(x) = \log_b x$ is the set of all positive real numbers. The domain of $f(x) = \log_b g(x)$ consists of all x for which $g(x) > 0$.

Ex. 7, p. 484; Ex. 10, p. 487

e. Common and natural logarithms: $f(x) = \log x$ means $f(x) = \log_{10} x$ and is the common logarithmic function. $f(x) = \ln x$ means $f(x) = \log_e x$ and is the natural logarithmic function.

Ex. 8, p. 485; Ex. 9, p. 486; Ex. 11, p. 487

f. Basic Logarithmic Properties

Base b ($b > 0, b \neq 1$)	Base 10 (Common Logarithms)	Base e (Natural Logarithms)	
$\log_b 1 = 0$	$\log 1 = 0$	$\ln 1 = 0$	
$\log_b b = 1$	$\log 10 = 1$	$\ln e = 1$	Ex. 4, p. 481
$\log_b b^x = x$	$\log 10^x = x$	$\ln e^x = x$	
$b^{\log_b x} = x$	$10^{\log x} = x$	$e^{\ln x} = x$	Ex. 5, p. 481

4.3 Properties of Logarithms

a. *The Product Rule:* $\log_b(MN) = \log_b M + \log_b N$

Ex. 1, p. 493

b. *The Quotient Rule:* $\log_b\left(\dfrac{M}{N}\right) = \log_b M - \log_b N$

Ex. 2, p. 494

c. *The Power Rule:* $\log_b M^p = p \log_b M$

Ex. 3, p. 495

d. *The Change-of-Base Property:*

The General Property	Introducing Common Logarithms	Introducing Natural Logarithms	
$\log_b M = \dfrac{\log_a M}{\log_a b}$	$\log_b M = \dfrac{\log M}{\log b}$	$\log_b M = \dfrac{\ln M}{\ln b}$	Ex. 7, p. 499; Ex. 8, p. 500

e. Properties for expanding logarithmic expressions are given in the box on page 496.

Ex. 4, p. 496

f. Properties for condensing logarithmic expressions are given in the box on page 497.

Ex. 5, p. 497; Ex. 6, p. 498

4.4 Exponential and Logarithmic Equations

a. An exponential equation is an equation containing a variable in an exponent. Some exponential equations can be solved by expressing each side as a power of the same base: If $b^M = b^N$, then $M = N$. Details are in the box on page 504.

Ex. 1, p. 504

DEFINITIONS AND CONCEPTS

b. The procedure for using logarithms to solve exponential equations is given in the box on page 505. The solution procedure involves isolating the exponential expression. Take the common logarithm on both sides for base 10. Take the natural logarithm on both sides for bases other than 10. Simplify using

$$\ln b^x = x \ln b \quad \text{or} \quad \ln e^x = x \quad \text{or} \quad \log 10^x = x.$$

Ex. 2, p. 505;
Ex. 3, p. 506;
Ex. 4, p. 507;
Ex. 5, p. 507

c. A logarithmic equation is an equation containing a variable in a logarithmic expression. Some logarithmic equations can be expressed in the form $\log_b M = c$. The definition of a logarithm is used to rewrite the equation in exponential form: $b^c = M$. See the box on page 508. When checking logarithmic equations, reject proposed solutions that produce the logarithm of a negative number or the logarithm of 0 in the original equation.

Ex. 6, p. 508;
Ex. 7, p. 510

d. Some logarithmic equations can be expressed in the form $\log_b M = \log_b N$. Use the one-to-one property to rewrite the equation without logarithms: $M = N$. See the box on page 510.

Ex. 8, p. 510

4.5 Exponential Growth and Decay; Modeling Data

a. Exponential growth and decay models are given by $A = A_0 e^{kt}$ in which t represents time, A_0 is the amount present at $t = 0$, and A is the amount present at time t. If $k > 0$, the model describes growth and k is the growth rate. If $k < 0$, the model describes decay and k is the decay rate.

Ex. 1, p. 520;
Ex. 2, p. 522

b. The logistic growth model, given by $A = \dfrac{c}{1 + ae^{-bt}}$, describes situations in which growth is limited. $y = c$ is a horizontal asymptote for the graph, and growth, A, can never exceed c.

Ex. 3, p. 523

c. Scatter plots for exponential, logarithmic, and linear models are shown in Figure 4.24 on page 524. When using a graphing utility to model data, the closer that the correlation coefficient, r, is to -1 or 1, the better the model fits the data.

Ex. 4, p. 525;
Ex. 5, p. 526;
Ex. 6, p. 526

d. Expressing an Exponential Model in Base e: $y = ab^x$ is equivalent to $y = ae^{(\ln b) \cdot x}$.

Ex. 7, p. 528

REVIEW EXERCISES

4.1

In Exercises 1–4, the graph of an exponential function is given. Select the function for each graph from the following options:

$$f(x) = 4^x, \quad g(x) = 4^{-x},$$
$$h(x) = -4^{-x}, \quad r(x) = -4^{-x} + 3.$$

1.

2.

3.

4.

In Exercises 5–9, graph f and g in the same rectangular coordinate system. Use transformations of the graph of f to obtain the graph of g. Graph and give equations of all asymptotes. Use the graphs to determine each function's domain and range.

5. $f(x) = 2^x$ and $g(x) = 2^{x-1}$ *

6. $f(x) = 3^x$ and $g(x) = 3^x - 1$ *

7. $f(x) = 3^x$ and $g(x) = -3^x$ *

8. $f(x) = \left(\frac{1}{2}\right)^x$ and $g(x) = \left(\frac{1}{2}\right)^{-x}$ *

9. $f(x) = e^x$ and $g(x) = 2e^{\frac{x}{2}}$ *

Use the compound interest formulas to solve Exercises 10–11.

10. Suppose that you have $5000 to invest. Which investment yields the greater return over 5 years: 5.5% compounded semiannually or 5.25% compounded monthly? 5.5% semiannually

11. Suppose that you have $14,000 to invest. Which investment yields the greater return over 10 years: 7% compounded monthly or 6.85% compounded continuously? 7% monthly

*See Answers to Selected Exercises.

12. A cup of coffee is taken out of a microwave oven and placed in a room. The temperature, T, in degrees Fahrenheit, of the coffee after t minutes is modeled by the function $T = 70 + 130e^{-0.04855t}$. The graph of the function is shown in the figure.

Use the graph to answer each of the following questions.

a. What was the temperature of the coffee when it was first taken out of the microwave? 200°

b. What is a reasonable estimate of the temperature of the coffee after 20 minutes? Use your calculator to verify this estimate. 120°; 119°

c. What is the limit of the temperature to which the coffee will cool? What does this tell you about the temperature of the room? 70°; room temperature: 70°

4.2

In Exercises 13–15, write each equation in its equivalent exponential form.

13. $\frac{1}{2} = \log_{49} 7$ ∗ **14.** $3 = \log_4 x$ ∗ **15.** $\log_3 81 = y$ ∗

In Exercises 16–18, write each equation in its equivalent logarithmic form.

16. $6^3 = 216$ ∗ **17.** $b^4 = 625$ ∗ **18.** $13^y = 874$ ∗

In Exercises 19–29, evaluate each expression without using a calculator. If evaluation is not possible, state the reason.

19. $\log_4 64$ 3 **20.** $\log_5 \frac{1}{25}$ −2 **21.** $\log_3(-9)$ ∗

22. $\log_{16} 4$ $\frac{1}{2}$ **23.** $\log_{17} 17$ 1 **24.** $\log_3 3^8$ 8

25. $\ln e^5$ 5 **26.** $\log_3 \frac{1}{\sqrt{3}}$ $-\frac{1}{2}$ **27.** $\ln \frac{1}{e^2}$ −2

28. $\log \frac{1}{1000}$ −3 **29.** $\log_3(\log_8 8)$ 0

30. Graph $f(x) = 2^x$ and $g(x) = \log_2 x$ in the same rectangular coordinate system. Use the graphs to determine each function's domain and range. ∗

31. Graph $f(x) = \left(\frac{1}{3}\right)^x$ and $g(x) = \log_{\frac{1}{3}} x$ in the same rectangular coordinate system. Use the graphs to determine each function's domain and range. ∗

In Exercises 32–35, the graph of a logarithmic function is given. Select the function for each graph from the following options:

$$f(x) = \log x, \quad g(x) = \log(-x),$$
$$h(x) = \log(2 - x), \quad r(x) = 1 + \log(2 - x).$$

32. **33.**

34. **35.**

In Exercises 36–38, begin by graphing $f(x) = \log_2 x$. Then use transformations of this graph to graph the given function. What is the graph's x-intercept? What is the vertical asymptote? Use the graphs to determine each function's domain and range.

36. $g(x) = \log_2(x - 2)$ ∗ **37.** $h(x) = -1 + \log_2 x$ ∗

38. $r(x) = \log_2(-x)$ ∗

In Exercises 39–40, graph f and g in the same rectangular coordinate system. Use transformations of the graph of f to obtain the graph of g. Graph and give equations of all asymptotes. Use the graphs to determine each function's domain and range.

39. $f(x) = \log x$ and $g(x) = -\log(x + 3)$ ∗

40. $f(x) = \ln x$ and $g(x) = -\ln(2x)$ ∗

In Exercises 41–43, find the domain of each logarithmic function.

41. $f(x) = \log_8(x + 5)$ (−5, ∞) **42.** $f(x) = \log(3 - x)$ (−∞, 3)

43. $f(x) = \ln(x - 1)^2$ (−∞, 1) ∪ (1, ∞)

In Exercises 44–46, use inverse properties of logarithms to simplify each expression.

44. $\ln e^{6x}$ 6x **45.** $e^{\ln \sqrt{x}}$ \sqrt{x} **46.** $10^{\log 4x^2}$ $4x^2$

47. On the Richter scale, the magnitude, R, of an earthquake of intensity I is given by $R = \log \dfrac{I}{I_0}$, where I_0 is the intensity of a barely felt zero-level earthquake. If the intensity of an earthquake is $1000I_0$, what is its magnitude on the Richter scale? 3.0

48. Students in a psychology class took a final examination. As part of an experiment to see how much of the course content they remembered over time, they took equivalent forms of the exam in monthly intervals thereafter. The average score, $f(t)$, for the group after t months is modeled by the function $f(t) = 76 - 18 \log(t + 1)$, where $0 \le t \le 12$.

a. What was the average score when the exam was first given? 76

b. What was the average score after 2 months? 4 months? 6 months? 8 months? one year? ∗

c. Use the results from parts (a) and (b) to graph f. Describe what the shape of the graph indicates in terms of the material retained by the students. ∗

49. The formula

$$t = \frac{1}{c}\ln\left(\frac{A}{A - N}\right)$$

describes the time, t, in weeks, that it takes to achieve mastery of a portion of a task. In the formula, A represents maximum learning possible, N is the portion of the learning that is to be achieved, and c is a constant used to measure an individual's learning style. A 50-year-old man decides to start running as a way to maintain good health. He feels that the maximum rate he could ever hope to achieve is 12 miles per hour. How many weeks will it take before the man can run 5 miles per hour if $c = 0.06$ for this person? ≈ 9 weeks

4.3

In Exercises 50–53, use properties of logarithms to expand each logarithmic expression as much as possible. Where possible, evaluate logarithmic expressions without using a calculator.

50. $\log_6(36x^3)$ $2 + 3\log_6 x$

51. $\log_4\left(\frac{\sqrt{x}}{64}\right)$ $\frac{1}{2}\log_4 x - 3$

52. $\log_2\left(\frac{xy^2}{64}\right)$ $\log_2 x + 2\log_2 y - 6$

53. $\ln\sqrt[3]{\frac{x}{e}}$ $\frac{1}{3}\ln x - \frac{1}{3}$

In Exercises 54–57, use properties of logarithms to condense each logarithmic expression. Write the expression as a single logarithm whose coefficient is 1.

54. $\log_b 7 + \log_b 3$ $\log_b 21$

55. $\log 3 - 3\log x$ $\log \frac{3}{x^3}$

56. $3\ln x + 4\ln y$ $\ln(x^3 y^4)$

57. $\frac{1}{2}\ln x - \ln y$ $\ln \frac{\sqrt{x}}{y}$

In Exercises 58–59, use common logarithms or natural logarithms and a calculator to evaluate to four decimal places.

58. $\log_6 72{,}348$ 6.2448

59. $\log_4 0.863$ −0.1063

In Exercises 60–63, determine whether each equation is true or false. Where possible, show work to support your conclusion. If the statement is false, make the necessary change(s) to produce a true statement.

60. $(\ln x)(\ln 1) = 0$ true

61. $\log(x + 9) - \log(x + 1) = \frac{\log(x + 9)}{\log(x + 1)}$ false*

62. $(\log_2 x)^4 = 4\log_2 x$ false; $4\log_2 x = \log_2 x^4$

63. $\ln e^x = x\ln e$ true

4.4

In Exercises 64–73, solve each exponential equation. Where necessary, express the solution set in terms of natural or common logarithms and use a calculator to obtain a decimal approximation, correct to two decimal places, for the solution.

64. $2^{4x-2} = 64$ $\{2\}$

65. $125^x = 25$ $\{\frac{2}{3}\}$

66. $10^x = 7000$ $\{\log 7000\}$; ≈3.85

67. $9^{x+2} = 27^{-x}$ $\{-\frac{4}{5}\}$

68. $8^x = 12{,}143$ $\{\frac{\ln 12{,}143}{\ln 8}\}$; ≈4.52

69. $9e^{5x} = 1269$ $\{\frac{1}{5}\ln 141\}$; ≈0.99

70. $e^{12-5x} - 7 = 123$ $\{\frac{12 - \ln 30}{5}\}$; ≈1.43

71. $5^{4x+2} = 37{,}500$ $\{\frac{\ln 37{,}500 - 2\ln 5}{4\ln 5}\}$ or $\{\frac{\ln 1500}{\ln 625}\}$; ≈1.14

72. $3^{x+4} = 7^{2x-1}$ $\{\frac{\ln 7 + 4\ln 3}{2\ln 7 - \ln 3}\}$; ≈2.27

73. $e^{2x} - e^x - 6 = 0$ $\{\ln 3\}$; ≈1.10

In Exercises 74–79, solve each logarithmic equation.

74. $\log_4(3x - 5) = 3$ $\{23\}$

75. $3 + 4\ln(2x) = 15$ $\{\frac{e^3}{2}\}$; ≈10.04

76. $\log_2(x + 3) + \log_2(x - 3) = 4$ $\{5\}$

77. $\log_3(x - 1) - \log_3(x + 2) = 2$ ∅

78. $\ln(x + 4) - \ln(x + 1) = \ln x$ $\{2\}$

79. $\log_4(2x + 1) = \log_4(x - 3) + \log_4(x + 5)$ $\{4\}$

80. The function $P(x) = 14.7e^{-0.21x}$ models the average atmospheric pressure, $P(x)$, in pounds per square inch, at an altitude of x miles above sea level. The atmospheric pressure at the peak of Mt. Everest, the world's highest mountain, is 4.6 pounds per square inch. How many miles above sea level, to the nearest tenth of a mile, is the peak of Mt. Everest? 5.5 mi

81. **Newest Dinosaur: The PC-osaurus?** For the period from 2009 through 2012, worldwide PC sales stayed relatively flat, while tablet sales skyrocketed. The bar graph shows worldwide PC and tablet sales, in millions, from 2009 through 2012.

Worldwide PC and Tablet Sales

Source: Canalys

The function

$$f(t) = 33.4(1.66)^t$$

models worldwide PC and tablet sales combined, $f(t)$, in millions, t years after 2009. When does this model project that 421 million PC and tablets were sold? Round to the nearest year. Based on the relatively flat PC sales over the four years shown by the graph, estimate the number of tablet sales for that year. 5 years after 2009, or 2014; 385 million

82. 60 + Years after "Brown v. Board" In 1954, the Supreme Court ended legal segregation in U.S. public schools. The bar graph shows the percentage of black and white U.S. adults ages 25 or older who completed high school since the end of segregation.

U.S. Adults Ages 25 or Older Who Completed High School

Source: Pew Research Center

The functions

$$B(x) = 16\ln x + 23 \text{ and } W(x) = 11\ln x + 49$$

model the percentage of black adults, $B(x)$, and white adults, $W(x)$, who had completed high school x years after 1961.

a. Use the appropriate model to determine the percentage of white adults who had completed high school in 2012. Round to the nearest percent. How well does this rounded value model the percents displayed by the graph? 92%; It's the same.

b. Use the appropriate model to determine when 90% of black adults will have completed high school. Round to the nearest year. 66 years after 1961, or 2027

83. Use the formula for compound interest with n compoundings per year to solve this problem. How long, to the nearest tenth of a year, will it take $12,500 to grow to $20,000 at 6.5% annual interest compounded quarterly? 7.3 yr

Use the formula for continuous compounding to solve Exercises 84–85.

84. How long, to the nearest tenth of a year, will it take $50,000 to triple in value at 7.5% annual interest compounded continuously? 14.6 yr

85. What interest rate, to the nearest percent, is required for an investment subject to continuous compounding to triple in 5 years? about 22%

4.5

86. According to the U.S. Bureau of the Census, in 2000 there were 35.3 million residents of Hispanic origin living in the United States. By 2010, the number had increased to 50.5 million. The exponential growth function $A = 35.3e^{kt}$ describes the U.S. Hispanic population, A, in millions, t years after 2000.

a. Find k, correct to three decimal places. 0.036

b. Use the resulting model to project the Hispanic resident population in 2015. 60.6 million

c. In which year will the Hispanic resident population reach 70 million? 2019

87. Use the exponential decay model, $A = A_0e^{kt}$, to solve this exercise. The half-life of polonium-210 is 140 days. How long will it take for a sample of this substance to decay to 20% of its original amount? 325 days

88. The function

$$f(t) = \frac{500,000}{1 + 2499e^{-0.92t}}$$

models the number of people, $f(t)$, in a city who have become ill with influenza t weeks after its initial outbreak.

a. How many people became ill with the flu when the epidemic began? 200 people

b. How many people were ill by the end of the sixth week? *

c. What is the limiting size of $f(t)$, the population that becomes ill? 500,000 people

Exercises 89–91 present data in the form of tables. For each data set shown by the table,

a. *Create a scatter plot for the data.*

b. *Use the scatter plot to determine whether an exponential function, a logarithmic function, or a linear function is the best choice for modeling the data.*

89. Number of Seriously Mentally Ill Adults in the United States linear function*

Year	Number of Seriously Mentally Ill Adults (millions)
2006	9.0
2008	9.2
2010	9.4
2012	9.6

Source: U.S. Census Bureau

90. Percentage of Moderate Alcohol Users in the United States (Not Binge or Heavy Drinkers) logarithmic function*

Age	Percentage of Moderate Drinkers
20	15%
25	24%
30	28%
35	32%
40	34%
45	35%

Source: Substance Abuse and Mental Health Services Administration

91. Number of Jobs in the U.S. Solar-Energy Industry exponential function*

Year	Number of Jobs (thousands)
2010	94
2011	100
2012	119
2013	143
2014	174
2015	209
2016	240

Source: The Solar Foundation

In Exercises 92–93, rewrite the equation in terms of base e. Express the answer in terms of a natural logarithm and then round to three decimal places.

92. $y = 73(2.6)^x$ $y = 73e^{(\ln 2.6)x}; y = 73e^{0.956x}$

93. $y = 6.5(0.43)^x$ $y = 6.5e^{(\ln 0.43)x}; y = 6.5e^{-0.844x}$

94. The figure on the right shows world population projections through the year 2150. The data are from the United Nations Family Planning Program and are based on optimistic or pessimistic expectations for successful control of human population growth. Suppose that you are interested in modeling these data using exponential, logarithmic, linear, and quadratic functions. Which function would you use to model each of the projections? Explain your choices. For the choice corresponding to a quadratic model, would your formula involve one with a positive or negative leading coefficient? Explain. Answers will vary.

Projections in World Population Growth

Source: United Nations

CHAPTER 4 TEST

1. Graph $f(x) = 2^x$ and $g(x) = 2^{x+1}$ in the same rectangular coordinate system. *

2. Graph $f(x) = \log_2 x$ and $g(x) = \log_2(x - 1)$ in the same rectangular coordinate system. *

3. Write in exponential form: $\log_5 125 = 3$. $5^3 = 125$

4. Write in logarithmic form: $\sqrt{36} = 6$. $\log_{36} 6 = \frac{1}{2}$

5. Find the domain: $f(x) = \ln(3 - x)$. $(-\infty, 3)$

In Exercises 6–7, use properties of logarithms to expand each logarithmic expression as much as possible. Where possible, evaluate logarithmic expressions without using a calculator.

6. $\log_4(64x^5)$ $3 + 5\log_4 x$

7. $\log_3\left(\dfrac{\sqrt[3]{x}}{81}\right)$ $\frac{1}{3}\log_3 x - 4$

In Exercises 8–9, write each expression as a single logarithm.

8. $6 \log x + 2 \log y$ $\log(x^6 y^2)$

9. $\ln 7 - 3 \ln x$ $\ln \dfrac{7}{x^3}$

10. Use a calculator to evaluate $\log_{15} 71$ to four decimal places. 1.5741

In Exercises 11–18, solve each equation.

11. $3^{x-2} = 9^{x+4}$ $\{-10\}$

12. $5^x = 1.4$ $\left\{\dfrac{\ln 1.4}{\ln 5}\right\}$

13. $400e^{0.005x} = 1600$ $\left\{\dfrac{\ln 4}{0.005}\right\}$

14. $e^{2x} - 6e^x + 5 = 0$ $\{0, \ln 5\}$

15. $\log_6(4x - 1) = 3$ $\{54.25\}$

16. $2 \ln(3x) = 8$ $\left\{\dfrac{e^4}{3}\right\}$

17. $\log x + \log(x + 15) = 2$ $\{5\}$

18. $\ln(x - 4) - \ln(x + 1) = \ln 6$ \varnothing

19. On the decibel scale, the loudness of a sound, D, in decibels, is given by $D = 10 \log \dfrac{I}{I_0}$, where I is the intensity of the sound, in watts per meter², and I_0 is the intensity of a sound barely audible to the human ear. If the intensity of a sound is $10^{12} I_0$, what is its loudness in decibels? (Such a sound is potentially damaging to the ear.) 120 db

In Exercises 20–22, simplify each expression.

20. $\ln e^{5x}$ $5x$

21. $\log_b b$ 1

22. $\log_6 1$ 0

Use the compound interest formulas to solve Exercises 23–25.

23. Suppose you have \$3000 to invest. Which investment yields the greater return over 10 years: 6.5% compounded semiannually or 6% compounded continuously? How much more (to the nearest dollar) is yielded by the better investment? 6.5% semiannually; \$221

24. How long, to the nearest tenth of a year, will it take \$4000 to grow to \$8000 at 5% annual interest compounded quarterly? 13.9 years

25. What interest rate, to the nearest tenth of a percent, is required for an investment subject to continuous compounding to double in 10 years? ≈ 6.9%

26. The function
$$A = 82.3e^{-0.004t}$$
models the population of Germany, A, in millions, t years after 2010.
 a. What was the population of Germany in 2010? 82.3 million
 b. Is the population of Germany increasing or decreasing? Explain. decreasing; The growth rate is negative.
 c. In which year will the population of Germany be 79.1 million? 2020

27. The 2010 population of Asia was 4121 million; in 2050, it is projected to be 5231 million. Write the exponential growth function that describes the population of Asia, in millions, t years after 2010. $A = 4121e^{0.006t}$

28. Use the exponential decay model, $A = A_0 e^{kt}$, to solve this exercise. The half-life of iodine-131 is 7.2 days. How long will it take for a sample of this substance to decay to 30% of its original amount? Round to one decimal place. 12.5 days

29. The logistic growth function

$$f(t) = \frac{140}{1 + 9e^{-0.165t}}$$

describes the population, $f(t)$, of an endangered species of elk t years after they were introduced to a nonthreatening habitat.

a. How many elk were initially introduced to the habitat? *

b. How many elk are expected in the habitat after 10 years? ≈ 51 elk

c. What is the limiting size of the elk population that the habitat will sustain? 140 elk

In Exercises 30–33, determine whether the values in each table belong to an exponential function, a logarithmic function, a linear function, or a quadratic function.

30.

x	y
0	3
1	1
2	−1
3	−3
4	−5

linear

31.

x	y
$\frac{1}{3}$	−1
1	0
3	1
9	2
27	3

logarithmic

32.

x	y
0	1
1	5
2	25
3	125
4	625

exponential

33.

x	y
0	12
1	3
2	0
3	3
4	12

quadratic

34. Rewrite $y = 96(0.38)^x$ in terms of base e. Express the answer in terms of a natural logarithm and then round to three decimal places. $y = 96e^{(\ln 0.38)x}$; $y = 96e^{-0.968x}$

CUMULATIVE REVIEW EXERCISES (CHAPTERS 1–4)

In Exercises 1–8, solve each equation or inequality.

1. $|3x - 4| = 2$ $\{\frac{2}{3}, 2\}$

2. $\sqrt{2x - 5} - \sqrt{x - 3} = 1$ {3, 7}

3. $x^4 + x^3 - 3x^2 - x + 2 = 0$ {−2, −1, 1}

4. $e^{5x} - 32 = 96$ $\left\{\frac{\ln 128}{5}\right\}$

5. $\log_2(x + 5) + \log_2(x - 1) = 4$ {3}

6. $\ln(x + 4) + \ln(x + 1) = 2\ln(x + 3)$ \varnothing

7. $14 - 5x \geq -6$ $(-\infty, 4]$

8. $|2x - 4| \leq 2$ [1, 3]

In Exercises 9–14, graph each equation in a rectangular coordinate system. If two functions are indicated, graph both in the same system.

9. $(x - 3)^2 + (y + 2)^2 = 4$ *

10. $f(x) = (x - 2)^2 - 1$ *

11. $f(x) = \dfrac{x^2 - 1}{x^2 - 4}$ *

12. $f(x) = (x - 2)^2(x + 1)$ *

13. $f(x) = 2x - 4$ and $f^{-1}(x)$ *

14. $f(x) = \ln x$ and $g(x) = \ln(x - 2) + 1$ *

15. Write the point-slope form and the slope-intercept form of the line passing through $(1, 3)$ and $(3, -3)$. *

16. If $f(x) = x^2$ and $g(x) = x + 2$, find $(f \circ g)(x)$ and $(g \circ f)(x)$.
$(f \circ g)(x) = (x + 2)^2$; $(g \circ f)(x) = x^2 + 2$

17. You discover that the number of hours you sleep each night varies inversely as the square of the number of cups of coffee consumed during the early evening. If 2 cups of coffee are consumed, you get 8 hours of sleep. If the number of cups of coffee is doubled, how many hours should you expect to sleep? 2 hours

A baseball player hits a pop fly into the air. The function

$$s(t) = -16t^2 + 64t + 5$$

models the ball's height above the ground, $s(t)$, in feet, t seconds after it is hit. Use the function to solve Exercises 18–19.

18. When does the baseball reach its maximum height? What is that height? after 2 sec; 69 ft

19. After how many seconds does the baseball hit the ground? Round to the nearest tenth of a second. 4.1 sec

20. You are paid time-and-a-half for each hour worked over 40 hours a week. Last week you worked 50 hours and earned $660. What is your normal hourly salary? $12 per hr

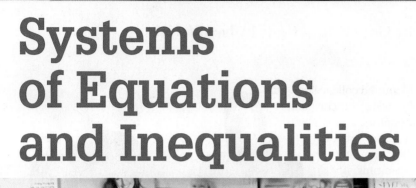

Systems of Equations and Inequalities

CHAPTER 5

Television, movies, and magazines place great emphasis on physical beauty. Our culture emphasizes physical appearance to such an extent that it is a central factor in the perception and judgment of others. The modern emphasis on thinness as the ideal body shape has been suggested as a major cause of eating disorders among adolescent women.

Cultural values of physical attractiveness change over time. During the 1950s, actress Jayne Mansfield embodied the postwar ideal: curvy, buxom, and big-hipped. Men, too, have been caught up in changes of how they "ought" to look. The 1960s ideal was the soft and scrawny hippie. Today's ideal man is tough and muscular.

Given the importance of culture in setting standards of attractiveness, how can you establish a healthy weight range for your age and height? In this chapter, we will use systems of inequalities to explore these skin-deep issues.

HERE'S WHERE YOU'LL FIND THESE APPLICATIONS:
You'll find a weight that fits you using the models (mathematical, not fashion) in Example 5 of Section 5.5 and Exercises 77–80 in Exercise Set 5.5. Exercises 85–86 use graphs and a formula for body-mass index to indicate whether you are obese, overweight, borderline overweight, normal weight, or underweight.

Section 5.1 Systems of Linear Equations in Two Variables

Researchers identified college students who generally were procrastinators or nonprocrastinators. The students were asked to report throughout the semester how many symptoms of physical illness they had experienced. **Figure 5.1** shows that by late in the semester, all students experienced increases in symptoms. Early in the semester, procrastinators reported fewer symptoms, but late in the semester, as work came due, they reported more symptoms than their nonprocrastinating peers.

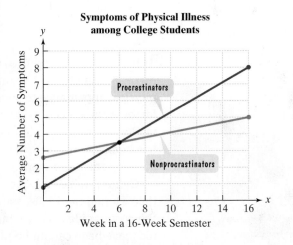

Symptoms of Physical Illness among College Students

Procrastinators

Nonprocrastinators

Week in a 16-Week Semester

FIGURE 5.1
Source: Gerrig and Zimbardo, *Psychology and Life*, 18th Edition, Allyn and Bacon, 2008.

The data in **Figure 5.1** can be analyzed using a pair of linear models in two variables. The figure shows that by week 6, both groups reported the same number of symptoms of illness, an average of approximately 3.5 symptoms per group. In this section, you will learn two algebraic methods, called *substitution* and *addition*, that will reinforce this graphic observation, verifying (6, 3.5) as the point of intersection.

1. Determine whether an ordered pair is a solution of a linear system.

Systems of Linear Equations and Their Solutions

All equations in the form $Ax + By = C$ are straight lines when graphed. Two such equations are called a **system of linear equations** or a **linear system**. **A solution to a system of linear equations in two variables** is an ordered pair that satisfies both equations in the system. For example, (3, 4) satisfies the system

$$\begin{cases} x + y = 7 & \text{(3 + 4 is, indeed, 7.)} \\ x - y = -1. & \text{(3 − 4 is, indeed, −1.)} \end{cases}$$

Thus, (3, 4) satisfies both equations and is a solution of the system. The solution can be described by saying that $x = 3$ and $y = 4$. The solution can also be described using set notation. The solution set to the system is {(3, 4)} — that is, the set consisting of the ordered pair (3, 4).

A system of linear equations can have exactly one solution, no solution, or infinitely many solutions. We begin with systems that have exactly one solution.

EXAMPLE 1 Determining Whether Ordered Pairs Are Solutions of a System

Consider the system:

$$\begin{cases} x + 2y = 2 \\ x - 2y = 6. \end{cases}$$

Determine if each ordered pair is a solution of the system:

a. $(4, -1)$ **b.** $(-4, 3)$.

SOLUTION

a. We begin by determining whether $(4, -1)$ is a solution. Because 4 is the x-coordinate and -1 is the y-coordinate of $(4, -1)$, we replace x with 4 and y with -1.

$$x + 2y = 2 \qquad\qquad\qquad x - 2y = 6$$
$$4 + 2(-1) \overset{?}{=} 2 \qquad\qquad 4 - 2(-1) \overset{?}{=} 6$$
$$4 + (-2) \overset{?}{=} 2 \qquad\qquad 4 - (-2) \overset{?}{=} 6$$
$$2 = 2, \quad \text{true} \qquad\qquad 4 + 2 \overset{?}{=} 6$$
$$6 = 6, \quad \text{true}$$

The pair $(4, -1)$ satisfies both equations: It makes each equation true. Thus, the ordered pair is a solution of the system.

b. To determine whether $(-4, 3)$ is a solution, we replace x with -4 and y with 3.

$$x + 2y = 2 \qquad\qquad\qquad x - 2y = 6$$
$$-4 + 2 \cdot 3 \overset{?}{=} 2 \qquad\qquad -4 - 2 \cdot 3 \overset{?}{=} 6$$
$$-4 + 6 \overset{?}{=} 2 \qquad\qquad -4 - 6 \overset{?}{=} 6$$
$$2 = 2, \quad \text{true} \qquad\qquad -10 = 6, \quad \text{false}$$

The pair $(-4, 3)$ fails to satisfy *both* equations: It does not make both equations true. Thus, the ordered pair is not a solution of the system. **• • •**

The solution of a system of linear equations can sometimes be found by graphing both of the equations in the same rectangular coordinate system. For a system with one solution, the **coordinates of the point of intersection give the system's solution**. For example, the system in Example 1,

$$\begin{cases} x + 2y = 2 \\ x - 2y = 6, \end{cases}$$

is graphed in **Figure 5.2**. The solution of the system, $(4, -1)$, corresponds to the point of intersection of the lines.

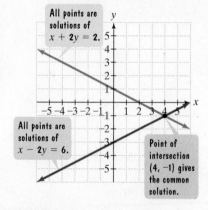

FIGURE 5.2 Visualizing a system's solution

☑ Check Point **1** Consider the system:

$$\begin{cases} 2x - 3y = -4 \\ 2x + y = 4. \end{cases}$$

Determine if each ordered pair is a solution of the system:

a. $(1, 2)$ solution **b.** $(7, 6)$. not a solution

② Solve linear systems by substitution.

Eliminating a Variable Using the Substitution Method

Finding the solution to a linear system by graphing equations may not be easy to do. For example, a solution of $\left(-\frac{2}{3}, \frac{157}{29}\right)$ would be difficult to "see" as an intersection point on a graph.

Let's consider a method that does not depend on finding a system's solution visually: the substitution method. This method involves converting the system to one equation in one variable by an appropriate substitution.

Solving Linear Systems by Substitution

1. Solve either of the equations for one variable in terms of the other. (If one of the equations is already in this form, you can skip this step.)
2. Substitute the expression found in step 1 into the *other* equation. This will result in an equation in one variable.
3. Solve the equation containing one variable.
4. Back-substitute the value found in step 3 into one of the original equations. Simplify and find the value of the remaining variable.
5. Check the proposed solution in both of the system's given equations.

EXAMPLE 2 Solving a System by Substitution

Solve by the substitution method:

$$\begin{cases} 5x - 4y = 9 \\ x - 2y = -3. \end{cases}$$

SOLUTION

Step 1 Solve either of the equations for one variable in terms of the other. We begin by isolating one of the variables in either of the equations. By solving for x in the second equation, which has a coefficient of 1, we can avoid fractions.

$x - 2y = -3$ This is the second equation in the given system.

$x = 2y - 3$ Solve for x by adding $2y$ to both sides.

Step 2 Substitute the expression from step 1 into the other equation. We substitute $2y - 3$ for x in the first equation.

$$x = \boxed{2y - 3} \qquad 5\boxed{x} - 4y = 9$$

This gives us an equation in one variable, namely

$$5(2y - 3) - 4y = 9.$$

The variable x has been eliminated.

Step 3 Solve the resulting equation containing one variable.

$5(2y - 3) - 4y = 9$	This is the equation containing one variable.
$10y - 15 - 4y = 9$	Apply the distributive property.
$6y - 15 = 9$	Combine like terms.
$6y = 24$	Add 15 to both sides.
$y = 4$	Divide both sides by 6.

Step 4 Back-substitute the obtained value into one of the original equations. Now that we have the y-coordinate of the solution, we back-substitute 4 for y into one of the original equations to find x. Let's use both equations to show that we obtain the same value for x in either case.

Using the first equation:	**Using the second equation:**
$5x - 4y = 9$	$x - 2y = -3$
$5x - 4(4) = 9$	$x - 2(4) = -3$
$5x - 16 = 9$	$x - 8 = -3$
$5x = 25$	$x = 5$
$x = 5$	

With $x = 5$ and $y = 4$, the proposed solution is $(5, 4)$.

Step 5 Check. Take a moment to show that $(5, 4)$ satisfies both given equations. The solution set is $\{(5, 4)\}$. • • •

TECHNOLOGY

Graphic Connections

A graphing utility can be used to solve the system in Example 2. Solve each equation for y, graph the equations, and use the intersection feature. The screen displays the solution $(5, 4)$ as $x = 5$, $y = 4$.

$[-10, 10, 1]$ by $[-10, 10, 1]$

✓ **Check Point 2** Solve by the substitution method:

$$\begin{cases} 3x + 2y = 4 \\ 2x + y = 1. \end{cases} \quad \{(-2, 5)\}$$

3 Solve linear systems by addition.

Eliminating a Variable Using the Addition Method

The substitution method is most useful if one of the given equations has an isolated variable. A second, and frequently the easier, method for solving a linear system is the addition method. Like the substitution method, the addition method involves eliminating a variable and ultimately solving an equation containing only one variable. However, this time we eliminate a variable by adding the equations.

For example, consider the following system of linear equations:

$$\begin{cases} 3x - 4y = 11 \\ -3x + 2y = -7. \end{cases}$$

When we add these two equations, the x-terms are eliminated. This occurs because the coefficients of the x-terms, 3 and -3, are opposites (additive inverses) of each other:

$$\begin{cases} 3x - 4y = 11 \\ \underline{-3x + 2y = -7} \\ \quad -2y = 4 \end{cases}$$ The sum is an equation in one variable.

$$y = -2.$$ Divide both sides by -2 and solve for y.

Now we can back-substitute -2 for y into one of the original equations to find x. It does not matter which equation you use; you will obtain the same value for x in either case. If we use either equation, we can show that $x = 1$ and the solution $(1, -2)$ satisfies both equations in the system.

When we use the addition method, we want to obtain two equations whose sum is an equation containing only one variable. The key step is to **obtain, for one of the variables, coefficients that differ only in sign**. To do this, we may need to multiply one or both equations by some nonzero number so that the coefficients of one of the variables, x or y, become opposites. Then when the two equations are added, this variable is eliminated.

Isn't the addition method also called the elimination method?

Although the addition method is also known as the elimination method, variables are eliminated when using both the substitution and addition methods. The name *addition method* specifically tells us that the elimination of a variable is accomplished by adding two equations.

Solving Linear Systems by Addition

1. If necessary, rewrite both equations in the form $Ax + By = C$.
2. If necessary, multiply either equation or both equations by appropriate nonzero numbers so that the sum of the x-coefficients or the sum of the y-coefficients is 0.
3. Add the equations in step 2. The sum is an equation in one variable.
4. Solve the equation in one variable.
5. Back-substitute the value obtained in step 4 into either of the given equations and solve for the other variable.
6. Check the solution in both of the original equations.

EXAMPLE 3 Solving a System by the Addition Method

Solve by the addition method:

$$\begin{cases} 3x + 2y = 48 \\ 9x - 8y = -24. \end{cases}$$

SOLUTION

Step 1 Rewrite both equations in the form $Ax + By = C$. Both equations are already in this form. Variable terms appear on the left and constants appear on the right.

Step 2 If necessary, multiply either equation or both equations by appropriate numbers so that the sum of the x-coefficients or the sum of the y-coefficients is 0. We can eliminate x or y. Let's eliminate x. Consider the terms in x in each equation, that is, $3x$ and $9x$. To eliminate x, we can multiply each term of the first equation by -3 and then add the equations.

$$\begin{cases} 3x + 2y = 48 \\ 9x - 8y = -24 \end{cases} \xrightarrow[\text{No change}]{\text{Multiply by } -3.} \begin{cases} -9x - 6y = -144 \\ \underline{9x - 8y = -24} \end{cases}$$

Step 3 Add the equations. Add: $-14y = -168$

Step 4 Solve the equation in one variable. We solve $-14y = -168$ by dividing both sides by -14.

$$\frac{-14y}{-14} = \frac{-168}{-14} \qquad \text{Divide both sides by } -14.$$

$$y = 12 \qquad \text{Simplify.}$$

Step 5 Back-substitute and find the value for the other variable. We can back-substitute 12 for y into either one of the given equations. We'll use the first one.

$$3x + 2y = 48 \qquad \text{This is the first equation in the given system.}$$
$$3x + 2(12) = 48 \qquad \text{Substitute 12 for } y.$$
$$3x + 24 = 48 \qquad \text{Multiply.}$$
$$3x = 24 \qquad \text{Subtract 24 from both sides.}$$
$$x = 8 \qquad \text{Divide both sides by 3.}$$

We found that $y = 12$ and $x = 8$. The proposed solution is $(8, 12)$.

Step 6 Check. Take a few minutes to show that $(8, 12)$ satisfies both of the original equations in the system. The solution set is $\{(8, 12)\}$. ● ● ●

✓ Check Point **3** Solve by the addition method:

$$\begin{cases} 4x + 5y = 3 \\ 2x - 3y = 7. \end{cases} \quad \{(2, -1)\}$$

Some linear systems have solutions that are not integers. If the value of one variable turns out to be a "messy" fraction, back-substitution might lead to cumbersome arithmetic. If this happens, you can return to the original system and use the addition method to find the value of the other variable.

EXAMPLE 4 Solving a System by the Addition Method

Solve by the addition method:

$$\begin{cases} 2x = 7y - 17 \\ 5y = 17 - 3x. \end{cases}$$

SOLUTION

Step 1 Rewrite both equations in the form $Ax + By = C$. We first arrange the system so that variable terms appear on the left and constants appear on the right. We obtain

$$\begin{cases} 2x - 7y = -17 \\ 3x + 5y = 17. \end{cases}$$ Subtract $7y$ from both sides of the first equation.
Add $3x$ to both sides of the second equation.

Step 2 If necessary, multiply either equation or both equations by appropriate numbers so that the sum of the x-coefficients or the sum of the y-coefficients is 0. We can eliminate x or y. Let's eliminate x by multiplying the first equation by 3 and the second equation by -2.

$$\begin{cases} 2x - 7y = -17 \\ 3x + 5y = 17 \end{cases} \xrightarrow[\text{Multiply by } -2.]{\text{Multiply by 3.}} \begin{cases} 6x - 21y = -51 \\ -6x - 10y = -34 \end{cases}$$

Step 3 Add the equations. Add: $-31y = -85$

Step 4 Solve the equation in one variable. We solve $-31y = -85$ by dividing both sides by -31.

$$\frac{-31y}{-31} = \frac{-85}{-31}$$ Divide both sides by -31.

$$y = \frac{85}{31}$$ Simplify.

Step 5 Back-substitute and find the value for the other variable. Back-substitution of $\frac{85}{31}$ for y into either of the given equations results in cumbersome arithmetic. Instead, let's use the addition method on the given system in the form $Ax + By = C$ to find the value for x. Thus, we eliminate y by multiplying the first equation by 5 and the second equation by 7.

$$\begin{cases} 2x - 7y = -17 \\ 3x + 5y = 17 \end{cases} \xrightarrow[\text{Multiply by 7.}]{\text{Multiply by 5.}} \begin{cases} 10x - 35y = -85 \\ 21x + 35y = 119 \end{cases}$$

Add: $31x = 34$

$$x = \frac{34}{31}$$ Divide both sides by 31.

We found that $y = \dfrac{85}{31}$ and $x = \dfrac{34}{31}$. The proposed solution is $\left(\dfrac{34}{31}, \dfrac{85}{31} \right)$.

Step 6 Check. For this system, a calculator is helpful in showing that $\left(\frac{34}{31}, \frac{85}{31} \right)$ satisfies both of the original equations in the system. The solution set is $\left\{ \left(\frac{34}{31}, \frac{85}{31} \right) \right\}$.

• • •

Check Point **4** Solve by the addition method:

$$\begin{cases} 2x = 9 + 3y \\ 4y = 8 - 3x. \end{cases} \quad \left\{ \left(\frac{60}{17}, -\frac{11}{17} \right) \right\}$$

④ Identify systems that do not have exactly one ordered-pair solution.

Linear Systems Having No Solution or Infinitely Many Solutions

We have seen that a system of linear equations in two variables represents a pair of lines. The lines either intersect at one point, are parallel, or are identical. Thus, there are three possibilities for the number of solutions to a system of two linear equations in two variables.

The Number of Solutions to a System of Two Linear Equations

The number of solutions to a system of two linear equations in two variables is given by one of the following. (See **Figure 5.3**.)

Number of Solutions	What This Means Graphically
Exactly one ordered-pair solution	The two lines intersect at one point.
No solution	The two lines are parallel.
Infinitely many solutions	The two lines are identical.

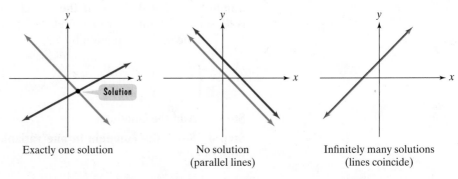

| Exactly one solution | No solution (parallel lines) | Infinitely many solutions (lines coincide) |

FIGURE 5.3 Possible graphs for a system of two linear equations in two variables

A linear system with no solution is called an **inconsistent system**. If you attempt to solve such a system by substitution or addition, you will eliminate both variables. A false statement, such as $0 = 12$, will be the result.

EXAMPLE 5 A System with No Solution

Solve the system:

$$\begin{cases} 4x + 6y = 12 \\ 6x + 9y = 12. \end{cases}$$

SOLUTION

Because no variable is isolated, we will use the addition method. To obtain coefficients of x that differ only in sign, we multiply the first equation by 3 and multiply the second equation by -2.

$$\begin{cases} 4x + 6y = 12 \xrightarrow{\text{Multiply by 3.}} \\ 6x + 9y = 12 \xrightarrow{\text{Multiply by } -2.} \end{cases} \begin{cases} 12x + 18y = 36 \\ -12x - 18y = -24 \end{cases}$$

$$\text{Add:} \qquad 0 = 12$$

> There are no values of x and y for which $0 = 12$. No values of x and y satisfy $0x + 0y = 12$.

The false statement $0 = 12$ indicates that the system is inconsistent and has no solution. The solution set is the empty set, \varnothing. • • •

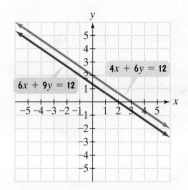

FIGURE 5.4 The graph of an inconsistent system

The lines corresponding to the two equations in Example 5 are shown in **Figure 5.4**. The lines are parallel and have no point of intersection.

DISCOVERY

Show that the graphs of $4x + 6y = 12$ and $6x + 9y = 12$ must be parallel lines by solving each equation for y. What are the slope and y-intercept for each line? What does this mean? If a linear system is inconsistent, what must be true about the slopes and y-intercepts for the system's graphs?

⊘ **Check Point 5** Solve the system:

$$\begin{cases} 5x - 2y = 4 \\ -10x + 4y = 7. \end{cases}$$ no solution or ∅

A linear system that has at least one solution is called a **consistent system**. Lines that intersect and lines that coincide both represent consistent systems. If the lines coincide, then the consistent system has infinitely many solutions, represented by every point on either line.

The equations in a linear system with infinitely many solutions are called **dependent**. If you attempt to solve such a system by substitution or addition, you will eliminate both variables. However, a true statement, such as $10 = 10$, will be the result.

EXAMPLE 6 A System with Infinitely Many Solutions

Solve the system:

$$\begin{cases} y = 3x - 2 \\ 15x - 5y = 10. \end{cases}$$

SOLUTION

Because the variable y is isolated in $y = 3x - 2$, the first equation, we choose to use the substitution method. We substitute the expression for y into the second equation.

$y = \boxed{3x - 2}$ $\qquad 15x - 5\boxed{y} = 10$ Substitute $3x - 2$ for y.

$\qquad 15x - 5(3x - 2) = 10$ The substitution results in an equation in one variable.

$\qquad 15x - 15x + 10 = 10$ Apply the distributive property.

This statement is true for all values of x and y. $\qquad 10 = 10$ Simplify.

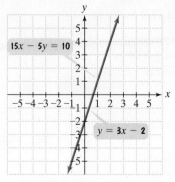

FIGURE 5.5 The graph of a system with infinitely many solutions

In our final step, both variables have been eliminated and the resulting statement, $10 = 10$, is true. This true statement indicates that the system has infinitely many solutions. The solution set consists of all points (x, y) lying on either of the coinciding lines, $y = 3x - 2$ or $15x - 5y = 10$, as shown in **Figure 5.5**.

We express the solution set for the system in one of two equivalent ways:

$$\{(x, y) \mid y = 3x - 2\} \quad \text{or} \quad \{(x, y) \mid 15x - 5y = 10\}.$$

The set of all ordered pairs (x, y) such that $y = 3x - 2$ The set of all ordered pairs (x, y) such that $15x - 5y = 10$

• • •

GREAT QUESTION!

The system in Example 6 has infinitely many solutions. Does that mean that any ordered pair of numbers is a solution?

No. Although the system in Example 6 has infinitely many solutions, this does not mean that any ordered pair of numbers you can form will be a solution. The ordered pair (x, y) must satisfy one of the system's equations, $y = 3x - 2$ or $15x - 5y = 10$, and there are infinitely many such ordered pairs. Because the graphs are coinciding lines, the ordered pairs that are solutions of one of the equations are also solutions of the other equation.

 Check Point **6** Solve the system:

$$\begin{cases} x = 4y - 8 & \{(x,y) \mid x = 4y - 8\} \text{ or} \\ 5x - 20y = -40. & \{(x,y) \mid 5x - 20y = -40\} \end{cases}$$

⑤ Solve problems using systems of linear equations.

Functions of Business: Break-Even Analysis

Suppose that a company produces and sells x units of a product. Its *revenue* is the money generated by selling x units of the product. Its *cost* is the cost of producing x units of the product.

Revenue and Cost Functions

A company produces and sells x units of a product.

Revenue Function

$$R(x) = (\text{price per unit sold})x$$

Cost Function

$$C(x) = \text{fixed cost} + (\text{cost per unit produced})x$$

The point of intersection of the graphs of the revenue and cost functions is called the **break-even point**. The x-coordinate of the point reveals the number of units that a company must produce and sell so that money coming in, the revenue, is equal to money going out, the cost. The y-coordinate of the break-even point gives the amount of money coming in and going out. Example 7 illustrates the use of the substitution method in determining a company's break-even point.

EXAMPLE 7 Finding a Break-Even Point

Technology is now promising to bring light, fast, and beautiful wheelchairs to millions of people with disabilities. A company is planning to manufacture these radically different wheelchairs. Fixed cost will be $500,000 and it will cost $400 to produce each wheelchair. Each wheelchair will be sold for $600.

a. Write the cost function, C, of producing x wheelchairs.

b. Write the revenue function, R, from the sale of x wheelchairs.

c. Determine the break-even point. Describe what this means.

SOLUTION

a. The cost function is the sum of the fixed cost and variable cost.

Fixed cost of $500,000 plus Variable cost: $400 for each chair produced

$$C(x) = 500,000 + 400x$$

b. The revenue function is the money generated from the sale of x wheelchairs.

Revenue per chair, $600, times | the number of chairs sold

$$R(x) = 600x$$

c. The break-even point occurs where the graphs of C and R intersect. Thus, we find this point by solving the system

$$\begin{cases} C(x) = 500{,}000 + 400x \\ R(x) = 600x \end{cases} \quad \text{or} \quad \begin{cases} y = 500{,}000 + 400x \\ y = 600x. \end{cases}$$

Using substitution, we can substitute $600x$ for y in the first equation:

$$600x = 500{,}000 + 400x \quad \text{Substitute } 600x \text{ for } y \text{ in } y = 500{,}000 + 400x.$$

$$200x = 500{,}000 \quad \text{Subtract } 400x \text{ from both sides.}$$

$$x = 2500. \quad \text{Divide both sides by } 200.$$

Back-substituting 2500 for x in either of the system's equations (or functions), we obtain

$$R(2500) = 600(2500) = 1{,}500{,}000.$$

We used $R(x) = 600x$.

The break-even point is $(2500, 1{,}500{,}000)$. This means that the company will break even if it produces and sells 2500 wheelchairs. At this level, the money coming in is equal to the money going out: $1,500,000. • • •

Figure 5.6 shows the graphs of the revenue and cost functions for the wheelchair business. Similar graphs and models apply no matter how small or large a business venture may be.

The intersection point confirms that the company breaks even by producing and selling 2500 wheelchairs. Can you see what happens for $x < 2500$? The red cost graph lies above the blue revenue graph. The cost is greater than the revenue and the business is losing money. Thus, if the company sells fewer than 2500 wheelchairs, the result is a *loss*. By contrast, look at what happens for $x > 2500$. The blue revenue graph lies above the red cost graph. The revenue is greater than the cost and the business is making money. Thus, if the company sells more than 2500 wheelchairs, the result is a *gain*.

FIGURE 5.6

✓ Check Point **7** A company that manufactures running shoes has a fixed cost of $300,000. Additionally, it costs $30 to produce each pair of shoes. They are sold at $80 per pair.

a. Write the cost function, C, of producing x pairs of running shoes.

b. Write the revenue function, R, from the sale of x pairs of running shoes.

c. Determine the break-even point. Describe what this means.

What does every entrepreneur, from a kid selling lemonade to Mark Zuckerberg, want to do? Generate profit, of course. The *profit* made is the money taken in, or the revenue, minus the money spent, or the cost. This relationship between revenue and cost allows us to define the *profit function, $P(x)$*.

The Profit Function

The profit, $P(x)$, generated after producing and selling x units of a product is given by the **profit function**

$$P(x) = R(x) - C(x),$$

where R and C are the revenue and cost functions, respectively.

FIGURE 5.7

The profit function for the wheelchair business in Example 7 is

$$P(x) = R(x) - C(x)$$
$$= 600x - (500,000 + 400x)$$
$$= 200x - 500,000.$$

The graph of this profit function is shown in **Figure 5.7**. The red portion lies below the x-axis and shows a loss when fewer than 2500 wheelchairs are sold. The business is "in the red." The black portion lies above the x-axis and shows a gain when more than 2500 wheelchairs are sold. The wheelchair business is "in the black."

Problems Involving Mixtures

Chemists and pharmacists often have to change the concentration of solutions and other mixtures. In these situations, the amount of a particular ingredient in the solution or mixture is expressed as a percentage of the total solution.

For example, if a 40-milliliter solution of acid in water contains 35% acid, the amount of acid in the solution is 35% of the total solution.

Amount of acid in the solution	is	35%	of	total number of milliliters in the solution.

$$= (0.35) \cdot (40)$$
$$= 14$$

There are 14 milliliters of acid in the solution.

When we solved problems in Chapter 1, we let x represent a quantity that was unknown. Mixture problems involve two unknown quantities. We will let x and y represent these quantities. We then translate from the verbal conditions of the problem into a *system* of linear equations.

EXAMPLE 8 Solving a Mixture Problem

A chemist working on a flu vaccine needs to mix a 10% sodium-iodine solution with a 60% sodium-iodine solution to obtain 50 milliliters of a 30% sodium-iodine solution. How many milliliters of the 10% solution and of the 60% solution should be mixed?

SOLUTION

Step 1 Use variables to represent unknown quantities.

Let $x =$ the number of milliliters of the 10% solution to be used in the mixture.

Let $y =$ the number of milliliters of the 60% solution to be used in the mixture.

Step 2 Write a system of equations that models the problem's conditions. The situation is illustrated in **Figure 5.8**. The chemist needs 50 milliliters of a 30% sodium-iodine solution. We form a table that shows the amount of sodium-iodine in each of the three solutions.

Volume	x	y	50
Strength	10% solution	60% solution	30% solution

FIGURE 5.8

Solution	Number of Milliliters	×	Percent of Sodium-Iodine	=	Amount of Sodium-Iodine
10% Solution	x		10% = 0.1		$0.1x$
60% Solution	y		60% = 0.6		$0.6y$
30% Mixture	50		30% = 0.3		$0.3(50) = 15$

The chemist needs to obtain a 50-milliliter mixture.

$$x \quad + \quad y \quad = \quad 50$$

The 50-milliliter mixture must be 30% sodium-iodine. The amount of sodium-iodine must be 30% of 50, or $(0.3)(50) = 15$ milliliters.

$$0.1x \quad + \quad 0.6y \quad = \quad 15$$

Step 3 Solve the system and answer the problem's question. The system

$$\begin{cases} x + y = 50 \\ 0.1x + 0.6y = 15 \end{cases}$$

can be solved by substitution or addition. Let's use substitution. The first equation can easily be solved for x or y. Solving for y, we obtain $y = 50 - x$.

$$y = \boxed{50 - x} \qquad 0.1x + 0.6\boxed{y} = 15$$

We substitute $50 - x$ for y in the second equation. This gives us an equation in one variable.

$$0.1x + 0.6(50 - x) = \quad 15 \quad \text{This equation contains one variable, } x.$$
$$0.1x + 30 - 0.6x = \quad 15 \quad \text{Apply the distributive property.}$$
$$-0.5x + 30 = \quad 15 \quad \text{Combine like terms.}$$
$$-0.5x = \quad -15 \quad \text{Subtract 30 from both sides.}$$
$$x = \frac{-15}{-0.5} = \quad 30 \quad \text{Divide both sides by } -0.5.$$

Back-substituting 30 for x in either of the system's equations ($x + y = 50$ is easier to use) gives $y = 20$. Because x represents the number of milliliters of the 10% solution and y the number of milliliters of the 60% solution, the chemist should mix 30 milliliters of the 10% solution with 20 milliliters of the 60% solution.

Step 4 Check the proposed solution in the original wording of the problem. The problem states that the chemist needs 50 milliliters of a 30% sodium-iodine solution. The amount of sodium-iodine in this mixture is 0.3(50), or 15 milliliters. The amount of sodium-iodine in 30 milliliters of the 10% solution is 0.1(30), or 3 milliliters. The amount of sodium-iodine in 20 milliliters of the 60% solution is 0.6(20) = 12 milliliters. The amount of sodium-iodine in the two solutions used in the mixture is 3 milliliters + 12 milliliters, or 15 milliliters, exactly as it should be.

• • •

 Check Point **8** A chemist needs to mix a 12% acid solution with a 20% acid solution to obtain 160 ounces of a 15% acid solution. How many ounces of each of the acid solutions must be used? 12% solution: 100 ounces; 20% solution: 60 ounces

ACHIEVING SUCCESS

Warm up your brain before starting the assigned homework. Researchers say the mind can be strengthened, just like your muscles, with regular training and rigorous practice. Think of the book's Exercise Sets as brain calisthenics. If you're feeling a bit sluggish before any of your mental workouts, try this warmup:

In the list below say the color the word is printed in, not the word itself. Once you can do this in 15 seconds without an error, the warmup is over and it's time to move on to the assigned exercises.

Blue Yellow Red Green **Yellow Green Blue** Red **Yellow Red**

CONCEPT AND VOCABULARY CHECK

Fill in each blank so that the resulting statement is true.

1. A solution to a system of linear equations in two variables is an ordered pair that ____satisfies both equations in the system____.

2. When solving a system of linear equations by graphing, the system's solution is determined by locating ___the intersection point___.

3. When solving

$$\begin{cases} 3x - 2y = 5 \\ y = 3x - 3 \end{cases}$$

by the substitution method, we obtain $x = \frac{1}{3}$, so the solution set is ___$\left\{ \left(\frac{1}{3}, -2 \right) \right\}$___.

4. When solving

$$\begin{cases} 2x + 10y = 9 \\ 8x + 5y = 7 \end{cases}$$

by the addition method, we can eliminate y by multiplying the second equation by ___−2___ and then adding the equations.

5. When solving

$$\begin{cases} 4x - 3y = 15 \\ 3x - 2y = 10 \end{cases}$$

by the addition method, we can eliminate y by multiplying the first equation by 2 and the second equation by ___−3___, and then adding the equations.

6. When solving

$$\begin{cases} 12x - 21y = 24 \\ 4x - 7y = 7 \end{cases}$$

by the addition method, we obtain 0 = 3, so the solution set is ___∅___. The linear system is a/an ___inconsistent___ system. If you attempt to solve such a system by graphing, you will obtain two lines that are ___parallel___.

7. When solving

$$\begin{cases} x = 3y + 2 \\ 5x - 15y = 10 \end{cases}$$

by the substitution method, we obtain $10 = 10$, so the solution set is __$\{(x, y) \mid x = 3y + 2\}$ or $\{(x, y) \mid 5x - 15y = 10\}$__. The equations in this system are called __dependent__. If you attempt to solve such a system by graphing, you will obtain two lines that __are identical or coincide__.

8. A company's __revenue__ function is the money generated by selling x units of its product. The difference between this function and the company's cost function is called its __profit__ function.

9. A company has a graph that shows the money it generates by selling x units of its product. It also has a graph that shows its cost of producing x units of its product. The point of intersection of these graphs is called the company's __break-even point__.

EXERCISE SET 5.1

Practice Exercises

In Exercises 1–4, determine whether the given ordered pair is a solution of the system.

1. $(2, 3)$ solution
$$\begin{cases} x + 3y = 11 \\ x - 5y = -13 \end{cases}$$

2. $(-3, 5)$ solution
$$\begin{cases} 9x + 7y = 8 \\ 8x - 9y = -69 \end{cases}$$

3. $(2, 5)$ not a solution
$$\begin{cases} 2x + 3y = 17 \\ x + 4y = 16 \end{cases}$$

4. $(8, 5)$ not a solution
$$\begin{cases} 5x - 4y = 20 \\ 3y = 2x + 1 \end{cases}$$

In Exercises 5–18, solve each system by the substitution method.

5. $\begin{cases} x + y = 4 \\ y = 3x \end{cases}$ $\{(1, 3)\}$

6. $\begin{cases} x + y = 6 \\ y = 2x \end{cases}$ $\{(2, 4)\}$

7. $\begin{cases} x + 3y = 8 \\ y = 2x - 9 \end{cases}$ $\{(5, 1)\}$

8. $\begin{cases} 2x - 3y = -13 \\ y = 2x + 7 \end{cases}$ $\{(-2, 3)\}$

9. $\begin{cases} x = 4y - 2 \\ x = 6y + 8 \end{cases}$ $\{(-22, -5)\}$

10. $\begin{cases} x = 3y + 7 \\ x = 2y - 1 \end{cases}$ $\{(-17, -8)\}$

11. $\begin{cases} 5x + 2y = 0 \\ x - 3y = 0 \end{cases}$ $\{(0, 0)\}$

12. $\begin{cases} 4x + 3y = 0 \\ 2x - y = 0 \end{cases}$ $\{(0, 0)\}$

13. $\begin{cases} 2x + 5y = -4 \\ 3x - y = 11 \end{cases}$ $\{(3, -2)\}$

14. $\begin{cases} 2x + 5y = 1 \\ -x + 6y = 8 \end{cases}$ $\{(-2, 1)\}$

15. $\begin{cases} 2x - 3y = 8 - 2x \\ 3x + 4y = x + 3y + 14 \end{cases}$ $\{(5, 4)\}$

16. $\begin{cases} 3x - 4y = x - y + 4 \\ 2x + 6y = 5y - 4 \end{cases}$ $\{(-1, -2)\}$

17. $\begin{cases} y = \dfrac{1}{3}x + \dfrac{2}{3} \\ y = \dfrac{5}{7}x - 2 \end{cases}$ $\{(7, 3)\}$

18. $\begin{cases} y = -\dfrac{1}{2}x + 2 \\ y = \dfrac{3}{4}x + 7 \end{cases}$ $\{(-4, 4)\}$

In Exercises 19–30, solve each system by the addition method.

19. $\begin{cases} x + y = 1 \\ x - y = 3 \end{cases}$ $\{(2, -1)\}$

20. $\begin{cases} x + y = 6 \\ x - y = -2 \end{cases}$ $\{(2, 4)\}$

21. $\begin{cases} 2x + 3y = 6 \\ 2x - 3y = 6 \end{cases}$ $\{(3, 0)\}$

22. $\begin{cases} 3x + 2y = 14 \\ 3x - 2y = 10 \end{cases}$ $\{(4, 1)\}$

23. $\begin{cases} x + 2y = 2 \\ -4x + 3y = 25 \end{cases}$ $\{(-4, 3)\}$

24. $\begin{cases} 2x - 7y = 2 \\ 3x + y = -20 \end{cases}$ $\{(-6, -2)\}$

25. $\begin{cases} 4x + 3y = 15 \\ 2x - 5y = 1 \end{cases}$ $\{(3, 1)\}$

26. $\begin{cases} 3x - 7y = 13 \\ 6x + 5y = 7 \end{cases}$ $\{(2, -1)\}$

27. $\begin{cases} 3x - 4y = 11 \\ 2x + 3y = -4 \end{cases}$ $\{(1, -2)\}$

28. $\begin{cases} 2x + 3y = -16 \\ 5x - 10y = 30 \end{cases}$ $\{(-2, -4)\}$

29. $\begin{cases} 3x = 4y + 1 \\ 3y = 1 - 4x \end{cases}$ $\{(\frac{7}{25}, -\frac{1}{25})\}$

30. $\begin{cases} 5x = 6y + 40 \\ 2y = 8 - 3x \end{cases}$ $\{(\frac{32}{7}, -\frac{20}{7})\}$

In Exercises 31–42, solve by the method of your choice. Identify systems with no solution and systems with infinitely many solutions, using set notation to express their solution sets.

31. $\begin{cases} x = 9 - 2y \\ x + 2y = 13 \end{cases}$ \varnothing

32. $\begin{cases} 6x + 2y = 7 \\ y = 2 - 3x \end{cases}$ \varnothing

33. $\begin{cases} y = 3x - 5 \\ 21x - 35 = 7y \end{cases}$ *

34. $\begin{cases} 9x - 3y = 12 \\ y = 3x - 4 \end{cases}$ *

35. $\begin{cases} 3x - 2y = -5 \\ 4x + y = 8 \end{cases}$ $\{(1, 4)\}$

36. $\begin{cases} 2x + 5y = -4 \\ 3x - y = 11 \end{cases}$ $\{(3, -2)\}$

37. $\begin{cases} x + 3y = 2 \\ 3x + 9y = 6 \end{cases}$ *

38. $\begin{cases} 4x - 2y = 2 \\ 2x - y = 1 \end{cases}$ *

39. $\begin{cases} \dfrac{x}{4} - \dfrac{y}{4} = -1 \\ x + 4y = -9 \end{cases}$ $\{(-5, -1)\}$

40. $\begin{cases} \dfrac{x}{6} - \dfrac{y}{2} = \dfrac{1}{3} \\ x + 2y = -3 \end{cases}$ $\{(-1, -1)\}$

41. $\begin{cases} 2x = 3y + 4 \\ 4x = 3 - 5y \end{cases}$ $\{(\frac{29}{22}, -\frac{5}{11})\}$

42. $\begin{cases} 4x = 3y + 8 \\ 2x = -14 + 5y \end{cases}$ $\{(\frac{41}{7}, \frac{36}{7})\}$

In Exercises 43–46, let x represent one number and let y represent the other number. Use the given conditions to write a system of equations. Solve the system and find the numbers.

43. The sum of two numbers is 7. If one number is subtracted from the other, their difference is -1. Find the numbers. 3 and 4*

44. The sum of two numbers is 2. If one number is subtracted from the other, their difference is 8. Find the numbers. *

45. Three times a first number decreased by a second number is 1. The first number increased by twice the second number is 12. Find the numbers. $3x - y = 1; x + 2y = 12; 2$ and 5

46. The sum of three times a first number and twice a second number is 8. If the second number is subtracted from twice the first number, the result is 3. Find the numbers. 2 and 1*

Practice Plus

In Exercises 47–48, solve each system by the method of your choice.

47. $\begin{cases} \dfrac{x + 2}{2} - \dfrac{y + 4}{3} = 3 \\ \dfrac{x + y}{5} = \dfrac{x - y}{2} - \dfrac{5}{2} \end{cases}$ $\{(6, -1)\}$

48. $\begin{cases} \dfrac{x - y}{3} = \dfrac{x + y}{2} - \dfrac{1}{2} \\ \dfrac{x + 2}{2} - 4 = \dfrac{y + 4}{3} \end{cases}$ $\{(8, -1)\}$

*See Answers to Selected Exercises.

In Exercises 49–50, solve each system for x and y, expressing either value in terms of a or b, if necessary. Assume that $a \neq 0$ and $b \neq 0$.

49. $\begin{cases} 5ax + 4y = 17 \\ ax + 7y = 22 \end{cases}$ $\left\{ \left(\frac{1}{a}, 3 \right) \right\}$ **50.** $\begin{cases} 4ax + by = 3 \\ 6ax + 5by = 8 \end{cases}$ $\left\{ \left(\frac{1}{2a}, \frac{1}{b} \right) \right\}$

51. For the linear function $f(x) = mx + b$, $f(-2) = 11$ and $f(3) = -9$. Find m and b. $m = -4; b = 3$

52. For the linear function $f(x) = mx + b$, $f(-3) = 23$ and $f(2) = -7$. Find m and b. $m = -6; b = 5$

Use the graphs of the linear functions to solve Exercises 53–54.

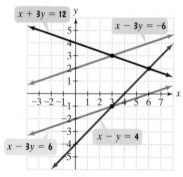

53. Write the linear system whose solution set is $\{(6, 2)\}$. Express each equation in the system in slope-intercept form. *

54. Write the linear system whose solution set is \emptyset. Express each equation in the system in slope-intercept form. *

Application Exercises

The figure shows the graphs of the cost and revenue functions for a company that manufactures and sells small radios. Use the information in the figure to solve Exercises 55–60.

Radios Produced and Sold

55. How many radios must be produced and sold for the company to break even? *500 radios*

56. More than how many radios must be produced and sold for the company to have a profit? *more than 500 radios*

57. Use the formulas shown in the voice balloons to find $R(200) - C(200)$. Describe what this means for the company. *

58. Use the formulas shown in the voice balloons to find $R(300) - C(300)$. Describe what this means for the company. *

59. a. Use the formulas shown in the voice balloons to write the company's profit function, P, from producing and selling x radios. $P(x) = 20x - 10,000$

b. Find the company's profit if 10,000 radios are produced and sold. $190,000

60. a. Use the formulas shown in the voice balloons to write the company's profit function, P, from producing and selling x radios. $P(x) = 20x - 10,000$

b. Find the company's profit if 20,000 radios are produced and sold. $390,000

Exercises 61–64 describe a number of business ventures. For each exercise,

 a. *Write the cost function, C.*
 b. *Write the revenue function, R.*
 c. *Determine the break-even point. Describe what this means.*

61. A company that manufactures small canoes has a fixed cost of $18,000. It costs $20 to produce each canoe. The selling price is $80 per canoe. (In solving this exercise, let x represent the number of canoes produced and sold.) *

62. A company that manufactures bicycles has a fixed cost of $100,000. It costs $100 to produce each bicycle. The selling price is $300 per bike. (In solving this exercise, let x represent the number of bicycles produced and sold.) *

63. You invest in a new play. The cost includes an overhead of $30,000, plus production costs of $2500 per performance. A sold-out performance brings in $3125. (In solving this exercise, let x represent the number of sold-out performances.) *

64. You invested $30,000 and started a business writing greeting cards. Supplies cost 2¢ per card and you are selling each card for 50¢. (In solving this exercise, let x represent the number of cards produced and sold.) *

An important application of systems of equations arises in connection with supply and demand. As the price of a product increases, the demand for that product decreases. However, at higher prices, suppliers are willing to produce greater quantities of the product. The price at which supply and demand are equal is called the **equilibrium price**. *The quantity supplied and demanded at that price is called the* **equilibrium quantity**. *Exercises 65–66 involve supply and demand.*

65. The table shows the price of a gallon of unleaded premium gasoline. For each price, the table lists the number of gallons per day that a gas station sells and the number of gallons per day that can be supplied.

Supply and Demand for Unleaded Premium Gasoline

Price per Gallon	Gallons Demanded per Day	Gallons Supplied per Day
$3.20	1400	200
$3.60	1200	600
$4.40	800	1400
$4.80	600	1800

The data in the table are described by the following demand and supply models:

Demand Model	**Supply Model**
$p = -0.002x + 6$	$p = 0.001x + 3.$

| Price per gallon | Number of gallons demanded per day | Price per gallon | Number of gallons supplied per day |

a. Solve the system and find the equilibrium quantity and the equilibrium price for a gallon of unleaded premium gasoline. 1000 gallons; $4

b. Use your answer from part (a) to complete this statement: If unleaded premium gasoline is sold for ___$4___ per gallon, there will be a demand for ___1000___ gallons per day and ___1000___ gallons will be supplied per day.

66. The table shows the price of a package of cookies. For each price, the table lists the number of packages that consumers are willing to buy and the number of packages that bakers are willing to supply.

Supply and Demand for Packages of Cookies

Price of a Package of Cookies	Quantity Demanded (millions of packages) per Week	Quantity Supplied (millions of packages) per Week
30¢	150	70
40¢	130	90
60¢	90	130
70¢	70	150

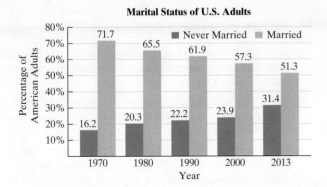

The data in the table can be described by the following demand and supply models:

Demand Model **Supply Model**

$p = -0.5x + 105$ $p = 0.5x - 5.$

Price per package (cents)	Number of packages demanded per week (millions)

Price per package (cents)	Number of packages supplied per week (millions)

a. Solve the system and find the equilibrium quantity and the equilibrium price for a package of cookies. *

b. Use your answer from part (a) to complete this statement: If cookies are sold for ___50¢___ per package, there will be a demand for ___110___ million packages per week and bakers will supply ___110___ million packages per week.

67. The bar graph indicates that fewer U.S. adults are getting married.

Marital Status of U.S. Adults

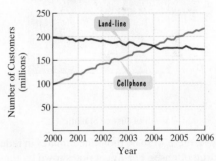

Source: U.S. Census Bureau

The data can be modeled by the following system of linear equations:

$$\begin{cases} -3x + 10y = 160 \\ x + 2y = 142. \end{cases}$$

Percentage of never-married American adults, y, x years after 1970

Percentage of married American adults, y, x years after 1970

a. Use these models to determine the year, rounded to the nearest year, when the percentage of never-married adults will be the same as the percentage of married adults. For that year, approximately what percentage of Americans, rounded to the nearest percent, will belong to each group? 2039; 37%

b. How is your approximate solution from part (a) shown by the following graphs? by the intersection point (69, 37)

68. The graph shows that from 2000 through 2006, Americans unplugged land lines and switched to cellphones.

Number of Cellphone and Land-Line Customers in the United States

Source: Federal Communications Commission

a. Use the graphs to estimate the point of intersection. In what year was the number of cellphone and land-line customers the same? How many millions of customers were there for each? *

b. The function $4.3x + y = 198$ models the number of land-line customers, in millions, x years after 2000. The function $y = 19.8x + 98$ models the number of cellphone customers, in millions, x years after 2000. Use these models to determine the year, rounded to the nearest year, when the number of cellphone and land-line customers was the same. According to the models, how many millions of customers, rounded to the nearest ten million, were there for each? 2004; 180 million

c. How well do the models in part (b) describe the point of intersection of the graphs that you estimated in part (a)? Answers will vary from "quite well" to "extremely well."

69. We opened this section with a study showing that late in the semester, procrastinating students reported more symptoms of physical illness than their nonprocrastinating peers.

 a. At the beginning of the semester, procrastinators reported an average of 0.8 symptoms, increasing at a rate of 0.45 symptoms per week. Write a function that models the average number of symptoms after x weeks. *

 b. At the beginning of the semester, nonprocrastinators reported an average of 2.6 symptoms, increasing at a rate of 0.15 symptoms per week. Write a function that models the average number of symptoms after x weeks. $y = 0.15x + 2.6$

 c. By which week in the semester did both groups report the same number of symptoms of physical illness? For that week, how many symptoms were reported by each group? How is this shown in **Figure 5.1** on page 542? *

70. Harsh, mandatory minimum sentences for drug offenses account for more than half the population in U.S. federal prisons. The bar graph shows the number of inmates in federal prisons, in thousands, for drug offenses and all other crimes in 1998 and 2010. (Other crimes include murder, robbery, fraud, burglary, weapons offenses, immigration offenses, racketeering, and perjury.)

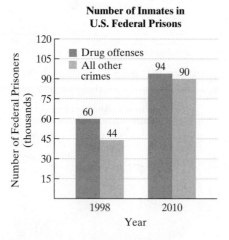

Number of Inmates in U.S. Federal Prisons

Source: Bureau of Justice Statistics

 a. In 1998, there were 60 thousand inmates in federal prisons for drug offenses. For the period shown by the graph, this number increased by approximately 2.8 thousand inmates per year. Write a function that models the number of inmates, y, in thousands, for drug offenses x years after 1998. $y = 2.8x + 60$

 b. In 1998, there were 44 thousand inmates in federal prisons for all crimes other than drug offenses. For the period shown by the graph, this number increased by approximately 3.8 thousand inmates per year. Write a function that models the number of inmates, y, in thousands, for all crimes other than drug offenses x years after 1998. $y = 3.8x + 44$

 c. Use the models from parts (a) and (b) to determine in which year the number of federal inmates for drug offenses was the same as the number of federal inmates for all other crimes. How many inmates were there for drug offenses and for all other crimes in that year? 2014; 104.8 thousand

71. The graphs show changing attitudes toward gay marriage for the period from 2001 through 2015.

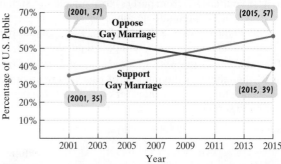

U.S. Public Opinion on Gay Marriage

Source: Pew Research Center

 a. Write the slope-intercept equation of the line that models the percentage of the U.S. public that supported gay marriage, y, x years after 2001. Round the value of the slope, m, to one decimal place. If necessary, round the value of the y-intercept to the nearest whole number. $y = 1.6x + 35$

 b. Write the slope-intercept equation of the line that models the percentage of the U.S. public that opposed gay marriage, y, x years after 2001. Round the value of the slope, m, to one decimal place. If necessary, round the value of the y-intercept to the nearest whole number. $y = -1.3x + 57$

 c. Use the models from parts (a) and (b) to determine the year, to the nearest whole year, during which the percentage who supported gay marriage was the same as the percentage who opposed gay marriage. 8 years after 2001, or during 2009

72. The graphs show per capita consumption of soda and bottled water in the United States, in gallons, from 2000 through 2015.

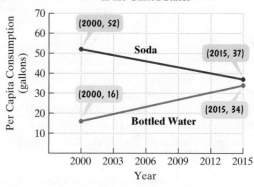

Consumption of Soda and Bottled Water in the United States

Source: Beverage Marketing Corporation

 a. Write the slope-intercept equation of the line that models soda consumption per capita, y, in gallons, x years after 2000. $y = -x + 52$

 b. Write the slope-intercept equation of the line that models bottled water consumption per capita, y, in gallons, x years after 2000. $y = 1.2x + 16$

 c. Use the models from parts (a) and (b) to determine the year, to the nearest whole year, during which per capita soda consumption was the same as per capita water consumption. 16 years after 2000, or during 2016

Use a system of linear equations to solve Exercises 73–84.

73. The current generation of college students grew up playing interactive online games, and many continue to play in college. The bar graph shows the percentage of U.S. college students playing online games, by gender.

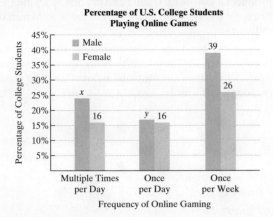

Percentage of U.S. College Students Playing Online Games

Source: statista.com

A total of 41% of college men play online games multiple times per day or once per day. The difference in the percentage who play multiple times per day and once per day is 7%. Find the percentage of college men who play online games multiple times per day and the percentage of college men who play online games once per day. *

74. A number of studies have emphasized the importance of sleep for students' success in their academic performance. The bar graph shows the actual sleep hours and the sleep hours to function best for U.S. college students.

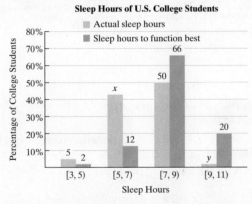

Sleep Hours of U.S. College Students

Source: Chiang et al., "The Effects of Sleep on Academic Performance and Job Performance," *College Student Journal*, Spring 2014

A total of 45% of college students sleep between 5 and 7 hours or between 9 and 11 hours. The difference in the percentage who sleep between 5 and 7 hours and between 9 and 11 hours is 41%. Find the percentage of college students who sleep between 5 and 7 hours and the percentage of college students who sleep between 9 and 11 hours. *

75. How many ounces of a 15% alcohol solution must be mixed with 4 ounces of a 20% alcohol solution to make a 17% alcohol solution? 6 ounces

76. How many ounces of a 50% alcohol solution must be mixed with 80 ounces of a 20% alcohol solution to make a 40% alcohol solution? 160 ounces

77. At the north campus of a performing arts school, 10% of the students are music majors. At the south campus, 90% of the students are music majors. The campuses are merged into one east campus. If 42% of the 1000 students at the east campus are music majors, how many students did each of the north and south campuses have before the merger? The north had 600 students and the south had 400 students.

78. At the north campus of a small liberal arts college, 10% of the students are women. At the south campus, 50% of the students are women. The campuses are merged into one east campus. If 40% of the 1200 students at the east campus are women, how many students did each of the north and south campuses have before the merger? *

79. A hotel has 200 rooms. Those with kitchen facilities rent for $100 per night and those without kitchen facilities rent for $80 per night. On a night when the hotel was completely occupied, revenues were $17,000. How many of each type of room does the hotel have? with: 50; without: 150

80. A new restaurant is to contain two-seat tables and four-seat tables. Fire codes limit the restaurant's maximum occupancy to 56 customers. If the owners have hired enough servers to handle 17 tables of customers, how many of each kind of table should they purchase? two: 6; four: 11

81. When a crew rows with the current, it travels 16 miles in 2 hours. Against the current, the crew rows 8 miles in 2 hours. Let x = the crew's rowing rate in still water and let y = the rate of the current. The following chart summarizes this information:

	Rate	×	Time	=	Distance
Rowing with current	$x + y$		2		16
Rowing against current	$x - y$		2		8

Find the rate of rowing in still water and the rate of the current. still water: 6 mph; current: 2 mph

82. When an airplane flies with the wind, it travels 800 miles in 4 hours. Against the wind, it takes 5 hours to cover the same distance. Find the plane's rate in still air and the rate of the wind. still air: 180 mph; wind: 20 mph

In Exercises 83–84, an isosceles triangle containing two angles with equal measure is shown. The degree measure of each triangle's three interior angles and an exterior angle is represented with variables. Find the measure of the three interior angles.

83. 80°, 50°, 50°

84. 30°, 75°, 75°

Explaining the Concepts 85–92. Answers will vary.

85. What is a system of linear equations? Provide an example with your description.

86. What is the solution of a system of linear equations?

87. Explain how to solve a system of equations using the substitution method. Use $y = 3 - 3x$ and $3x + 4y = 6$ to illustrate your explanation.

88. Explain how to solve a system of equations using the addition method. Use $3x + 5y = -2$ and $2x + 3y = 0$ to illustrate your explanation.

89. When is it easier to use the addition method rather than the substitution method to solve a system of equations?

90. When using the addition or substitution method, how can you tell if a system of linear equations has infinitely many solutions? What is the relationship between the graphs of the two equations?

91. When using the addition or substitution method, how can you tell if a system of linear equations has no solution? What is the relationship between the graphs of the two equations?

92. Describe the break-even point for a business.

Technology Exercise

93. Verify your solutions to any five exercises in Exercises 5–42 by using a graphing utility to graph the two equations in the system in the same viewing rectangle. Then use the intersection feature to display the solution. Answers will vary.

Critical Thinking Exercises

Make Sense? *In Exercises 94–97, determine whether each statement makes sense or does not make sense, and explain your reasoning.*

94. Even if a linear system has a solution set involving fractions, such as $\left\{\left(\frac{8}{11}, \frac{43}{11}\right)\right\}$, I can use graphs to determine if the solution set is reasonable. makes sense

95. Each equation in a system of linear equations has infinitely many ordered-pair solutions. makes sense

96. Every linear system has infinitely many ordered-pair solutions. does not make sense

97. If I know the perimeter of this rectangle and triangle, each in the same unit of measure, I can use a system of linear equations to determine values for x and y. makes sense

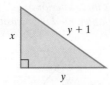

98. Write a system of equations having $\{(-2, 7)\}$ as a solution set. (More than one system is possible.) Answers will vary.

99. Solve the system for x and y in terms of $a_1, b_1, c_1, a_2, b_2,$ and c_2:
$$\begin{cases} a_1x + b_1y = c_1 \\ a_2x + b_2y = c_2. \end{cases}$$

100. Two identical twins can only be distinguished by the characteristic that one always tells the truth and the other always lies. One twin tells you of a lucky number pair: "When I multiply my first lucky number by 3 and my second lucky number by 6, the addition of the resulting numbers produces a sum of 12. When I add my first lucky number and twice my second lucky number, the sum is 5." Which twin is talking? the twin who always lies

101. A marching band has 52 members, and there are 24 in the pom-pom squad. They wish to form several hexagons and squares like those diagrammed below. Can it be done with no people left over? yes; hexagons: 8; squares: 4

B = Band Member
P = Pom-pom Person

Group Exercise Answers will vary.

102. The group should write four different word problems that can be solved using a system of linear equations in two variables. All of the problems should be on different topics. The group should turn in the four problems and their algebraic solutions.

Retaining the Concepts

In Exercises 103–104, find the domain of each function.

103. $f(x) = \ln(6 - x)$ (Section 4.2, Example 10) $(-\infty, 6)$

104. $g(x) = \dfrac{x - 6}{x^2 - 36}$
(Section 3.5, Example 1) $(-\infty, -6) \cup (-6, 6) \cup (6, \infty)$

105. Solve: $\log_3 x + \log_3(x + 6) = 3$
(Section 4.4, Example 7) $\{3\}$

Preview Exercises

Exercises 106–108 will help you prepare for the material covered in the next section.

106. If $x = 3, y = 2,$ and $z = -3,$ does the ordered triple (x, y, z) satisfy the equation $2x - y + 4z = -8$? yes

107. Consider the following equations: $11x + 4y = -3$
$$\begin{cases} 5x - 2y - 4z = 3 & \text{Equation 1} \\ 3x + 3y + 2z = -3. & \text{Equation 2} \end{cases}$$

Eliminate z by copying Equation 1, multiplying Equation 2 by 2, and then adding the equations.

108. Write an equation involving $a, b,$ and c based on the following description:

When the value of x in $y = ax^2 + bx + c$ is 4, the value of y is 1682. $1682 = 16a + 4b + c$

Section 5.2 | Systems of Linear Equations in Three Variables

What am I supposed to learn?

After studying this section, you should be able to:

① Verify the solution of a system of linear equations in three variables.

② Solve systems of linear equations in three variables.

③ Solve problems using systems in three variables.

All animals sleep, but the length of time they sleep varies widely: Cattle sleep for only a few minutes at a time. We humans seem to need more sleep than other animals, up to eight hours a day. Without enough sleep, we have difficulty concentrating, make mistakes in routine tasks, lose energy, and feel bad-tempered. There is a relationship between hours of sleep and death rate per year per 100,000 people. How many hours of sleep will put you in the group with the minimum death rate? In this section, we will answer this question by solving a system of linear equations with more than two variables.

Systems of Linear Equations in Three Variables and Their Solutions

An equation such as $x + 2y - 3z = 9$ is called a *linear equation in three variables*. In general, any equation of the form

$$Ax + By + Cz = D,$$

where A, B, C, and D are real numbers such that A, B, and C are not all 0, is a **linear equation in three variables: x, y, and z**. The graph of this linear equation in three variables is a plane in three-dimensional space.

The process of solving a system of three linear equations in three variables is geometrically equivalent to finding the point of intersection (assuming that there is one) of three planes in space. (See **Figure 5.9.**) A **solution** of a system of linear equations in three variables is an ordered triple of real numbers that satisfies all equations of the system. The **solution set** of the system is the set of all its solutions.

① Verify the solution of a system of linear equations in three variables.

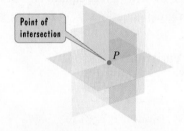

Point of intersection

P

FIGURE 5.9

EXAMPLE 1 | Determining Whether an Ordered Triple Satisfies a System

Show that the ordered triple $(-1, 2, -2)$ is a solution of the system:

$$\begin{cases} x + 2y - 3z = 9 \\ 2x - y + 2z = -8 \\ -x + 3y - 4z = 15. \end{cases}$$

SOLUTION

Because -1 is the x-coordinate, 2 is the y-coordinate, and -2 is the z-coordinate of $(-1, 2, -2)$, we replace x with -1, y with 2, and z with -2 in each of the three equations.

$x + 2y - 3z = 9$	$2x - y + 2z = -8$	$-x + 3y - 4z = 15$
$-1 + 2(2) - 3(-2) \stackrel{?}{=} 9$	$2(-1) - 2 + 2(-2) \stackrel{?}{=} -8$	$-(-1) + 3(2) - 4(-2) \stackrel{?}{=} 15$
$-1 + 4 + 6 \stackrel{?}{=} 9$	$-2 - 2 - 4 \stackrel{?}{=} -8$	$1 + 6 + 8 \stackrel{?}{=} 15$
$9 = 9$, true	$-8 = -8$, true	$15 = 15$, true

The ordered triple $(-1, 2, -2)$ satisfies the three equations: It makes each equation true. Thus, the ordered triple is a solution of the system. ● ● ●

 Check Point **1** Show that the ordered triple $(-1, -4, 5)$ is a solution of the system:

$$\begin{cases} x - 2y + 3z = 22 \\ 2x - 3y - z = 5 \\ 3x + y - 5z = -32. \end{cases}$$

$(-1) - 2(-4) + 3(5) = 22;$
$2(-1) - 3(-4) - 5 = 5;$
$3(-1) + (-4) - 5(5) = -32$

 Solve systems of linear equations in three variables.

Solving Systems of Linear Equations in Three Variables by Eliminating Variables

The method for solving a system of linear equations in three variables is similar to that used on systems of linear equations in two variables. We use addition to eliminate any variable, reducing the system to two equations in two variables. Once we obtain a system of two equations in two variables, we use addition or substitution to eliminate a variable. The result is a single equation in one variable. We solve this equation to get the value of the remaining variable. Other variable values are found by back-substitution.

GREAT QUESTION!

When solving a linear system in three variables, which variable should I eliminate first?

It does not matter which variable you eliminate first, as long as you eliminate the same variable in two different pairs of equations.

Solving Linear Systems in Three Variables by Eliminating Variables

1. Reduce the system to two equations in two variables. This is usually accomplished by taking two different pairs of equations and using the addition method to eliminate the same variable from both pairs.
2. Solve the resulting system of two equations in two variables using addition or substitution. The result is an equation in one variable that gives the value of that variable.
3. Back-substitute the value of the variable found in step 2 into either of the equations in two variables to find the value of the second variable.
4. Use the values of the two variables from steps 2 and 3 to find the value of the third variable by back-substituting into one of the original equations.
5. Check the proposed solution in each of the original equations.

EXAMPLE 2 Solving a System in Three Variables

Solve the system:

$$\begin{cases} 5x - 2y - 4z = 3 & \text{Equation 1} \\ 3x + 3y + 2z = -3 & \text{Equation 2} \\ -2x + 5y + 3z = 3. & \text{Equation 3} \end{cases}$$

SOLUTION

There are many ways to proceed. Because our initial goal is to reduce the system to two equations in two variables, **the central idea is to take two different pairs of equations and eliminate the same variable from both pairs.**

Step 1 Reduce the system to two equations in two variables. We choose any two equations and use the addition method to eliminate a variable. Let's eliminate z using Equations 1 and 2. We do so by multiplying Equation 2 by 2. Then we add equations.

(Equation 1) $\begin{cases} 5x - 2y - 4z = 3 \end{cases}$ — No change → $\begin{cases} 5x - 2y - 4z = 3 \end{cases}$
(Equation 2) $\begin{cases} 3x + 3y + 2z = -3 \end{cases}$ — Multiply by 2. → $\begin{cases} 6x + 6y + 4z = -6 \end{cases}$

Add: $\quad 11x + 4y \qquad = -3 \quad$ Equation 4

Now we must eliminate the *same* variable using another pair of equations. We can eliminate z from Equations 2 and 3. First, we multiply Equation 2 by -3. Next, we multiply Equation 3 by 2. Finally, we add equations.

(Equation 2) $\begin{cases} 3x + 3y + 2z = -3 \end{cases}$ — Multiply by -3. → $\begin{cases} -9x - 9y - 6z = 9 \end{cases}$
(Equation 3) $\begin{cases} -2x + 5y + 3z = 3 \end{cases}$ — Multiply by 2. → $\begin{cases} -4x + 10y + 6z = 6 \end{cases}$

Add: $\quad -13x + y \qquad = 15 \quad$ Equation 5

Equation 4, $11x + 4y = -3$, and Equation 5, $-13x + y = 15$, on the previous page, give us a system of two equations in two variables:

$$\begin{cases} 11x + 4y = -3 & \text{Equation 4} \\ -13x + y = 15. & \text{Equation 5} \end{cases}$$

Step 2 Solve the resulting system of two equations in two variables. We will use the addition method to solve Equations 4 and 5 for x and y. To do so, we multiply Equation 5 on both sides by -4 and add this to Equation 4.

(Equation 4) $\begin{cases} 11x + 4y = -3 \end{cases}$ $\xrightarrow{\text{No change}}$ $\begin{cases} 11x + 4y = -3 \end{cases}$

(Equation 5) $\begin{cases} -13x + y = 15 \end{cases}$ $\xrightarrow{\text{Multiply by } -4.}$ $\begin{cases} 52x - 4y = -60 \end{cases}$

$$\begin{aligned} \text{Add:} \quad 63x &= -63 \\ x &= -1 \quad \text{Divide both sides by 63.} \end{aligned}$$

Step 3 Use back-substitution in one of the equations in two variables to find the value of the second variable. We back-substitute -1 for x in either Equation 4 or 5 to find the value of y.

$$\begin{aligned} -13x + y &= 15 && \text{Equation 5} \\ -13(-1) + y &= 15 && \text{Substitute } -1 \text{ for } x. \\ 13 + y &= 15 && \text{Multiply.} \\ y &= 2 && \text{Subtract 13 from both sides.} \end{aligned}$$

Step 4 Back-substitute the values found for two variables into one of the original equations to find the value of the third variable. We can now use any one of the original equations and back-substitute the values of x and y to find the value for z. We will use Equation 2.

$$\begin{aligned} 3x + 3y + 2z &= -3 && \text{Equation 2} \\ 3(-1) + 3(2) + 2z &= -3 && \text{Substitute } -1 \text{ for } x \text{ and 2 for } y. \\ 3 + 2z &= -3 && \text{Multiply and then add:} \\ & && 3(-1) + 3(2) = -3 + 6 = 3. \\ 2z &= -6 && \text{Subtract 3 from both sides.} \\ z &= -3 && \text{Divide both sides by 2.} \end{aligned}$$

With $x = -1$, $y = 2$, and $z = -3$, the proposed solution is the ordered triple $(-1, 2, -3)$.

Step 5 Check. Check the proposed solution, $(-1, 2, -3)$, by substituting the values for x, y, and z into each of the three original equations. These substitutions yield three true statements. Thus, the solution set is $\{(-1, 2, -3)\}$. • • •

○ Check Point **2** Solve the system:

$$\begin{cases} x + 4y - z = 20 \\ 3x + 2y + z = 8 \qquad \{(1, 4, -3)\} \\ 2x - 3y + 2z = -16. \end{cases}$$

In some examples, the coefficient of one of the variables is zero in one of the equations. In this case, the missing variable should be eliminated from the other two equations, thereby making it possible to omit one of the elimination steps. We illustrate this idea in Example 3.

EXAMPLE 3 Solving a System of Equations with a Missing Term

Solve the system:

$$\begin{cases} x + z = 8 & \text{Equation 1} \\ x + y + 2z = 17 & \text{Equation 2} \\ x + 2y + z = 16. & \text{Equation 3} \end{cases}$$

$$\begin{cases} x + \qquad\ z = 8 \quad \text{Equation 1} \\ x +\ y + 2z = 17 \quad \text{Equation 2} \\ x + 2y +\ z = 16 \quad \text{Equation 3} \end{cases}$$

The given system (repeated)

SOLUTION

Step 1 Reduce the system to two equations in two variables. Because Equation 1 contains only x and z, we could omit one of the elimination steps by eliminating y using Equations 2 and 3. This will give us two equations in x and z. To eliminate y using Equations 2 and 3, we multiply Equation 2 by -2 and add Equation 3.

(Equation 2) $\begin{cases} x + y + 2z = 17 \end{cases}$ $\xrightarrow{\text{Multiply by } -2.}$ $\begin{cases} -2x - 2y - 4z = -34 \end{cases}$

(Equation 3) $\begin{cases} x + 2y + z = 16 \end{cases}$ $\xrightarrow{\text{No change}}$ $\begin{cases} \underline{x + 2y +\ z =\ \ \ 16} \end{cases}$

Add: $\qquad -x \qquad - 3z = -18$ Equation 4

Equation 4 and the given Equation 1 provide us with a system of two equations in two variables:

$$\begin{cases} x +\ z = 8 & \text{Equation 1} \\ -x - 3z = -18. & \text{Equation 4} \end{cases}$$

Step 2 Solve the resulting system of two equations in two variables. We will solve Equations 1 and 4 for x and z.

$$\begin{cases} x +\ z =\ \ \ 8 & \text{Equation 1} \\ \underline{-x - 3z = -18} & \text{Equation 4} \end{cases}$$

Add: $\qquad\qquad -2z = -10$

$\qquad\qquad\qquad\ z =\ \ \ 5$ Divide both sides by -2.

Step 3 Use back-substitution in one of the equations in two variables to find the value of the second variable. To find x, we back-substitute 5 for z in either Equation 1 or 4. We will use Equation 1.

$x + z = 8$ Equation 1

$x + 5 = 8$ Substitute 5 for z.

$x = 3$ Subtract 5 from both sides.

Step 4 Back-substitute the values found for two variables into one of the original equations to find the value of the third variable. To find y, we back-substitute 3 for x and 5 for z into Equation 2 or 3. We cannot use Equation 1 because y is missing in this equation. We will use Equation 2.

$x + y + 2z = 17$ Equation 2

$3 + y + 2(5) = 17$ Substitute 3 for x and 5 for z.

$y + 13 = 17$ Multiply and add.

$y = 4$ Subtract 13 from both sides.

We found that $z = 5, x = 3$, and $y = 4$. Thus, the proposed solution is the ordered triple $(3, 4, 5)$.

Step 5 Check. Substituting 3 for x, 4 for y, and 5 for z into each of the three original equations yields three true statements. Consequently, the solution set is $\{(3, 4, 5)\}$. • • •

⊘ **Check Point 3** Solve the system:

$$\begin{cases} 2y -\ z =\ \ \ 7 \\ x + 2y +\ z = 17 \qquad \{(4, 5, 3)\} \\ 2x - 3y + 2z = -1. \end{cases}$$

A system of linear equations in three variables represents three planes. The three planes may not always intersect at one point. The planes may have no common point of intersection and represent an inconsistent system with no solution. By contrast, the planes may coincide or intersect along a line. In these cases, the planes have infinitely many points in common and represent systems with infinitely many solutions. Systems of linear equations in three variables that are inconsistent or that contain dependent equations will be discussed in Chapter 6.

③ Solve problems using systems in three variables.

Applications

Systems of equations may allow us to find models for data without using a graphing utility. Three data points that do not lie on or near a line determine the graph of a quadratic function of the form $y = ax^2 + bx + c, a \neq 0$. Quadratic functions often model situations in which values of y are decreasing and then increasing, suggesting the bowl-like shape of a parabola.

EXAMPLE 4 Modeling Data Relating Sleep and Death Rate

In a study relating sleep and death rate, the following data were obtained. Use the function $y = ax^2 + bx + c$ to model the data.

x (Average Number of Hours of Sleep)	y (Death Rate per Year per 100,000 Males)
4	1682
7	626
9	967

SOLUTION

We need to find values for a, b, and c in $y = ax^2 + bx + c$. We can do so by solving a system of three linear equations in a, b, and c. We obtain the three equations by using the values of x and y from the data as follows:

$$y = ax^2 + bx + c \quad \text{Use the quadratic function to model the data.}$$

When $x = 4$, $y = 1682$:
When $x = 7$, $y = 626$:
When $x = 9$, $y = 967$:

$$\begin{cases} 1682 = a \cdot 4^2 + b \cdot 4 + c \\ 626 = a \cdot 7^2 + b \cdot 7 + c \\ 967 = a \cdot 9^2 + b \cdot 9 + c \end{cases} \quad \text{or} \quad \begin{cases} 16a + 4b + c = 1682 \\ 49a + 7b + c = 626 \\ 81a + 9b + c = 967. \end{cases}$$

The easiest way to solve this system is to eliminate c from two pairs of equations, obtaining two equations in a and b. Solving this system gives $a = 104.5$, $b = -1501.5$, and $c = 6016$. We now substitute the values for a, b, and c into $y = ax^2 + bx + c$. The function that models the given data is

$$y = 104.5x^2 - 1501.5x + 6016. \qquad \bullet\bullet\bullet$$

We can use the model that we obtained in Example 4 to find the death rate of males who average, say, 6 hours of sleep. First, write the model in function notation:

$$f(x) = 104.5x^2 - 1501.5x + 6016.$$

Substitute 6 for x:

$$f(6) = 104.5(6)^2 - 1501.5(6) + 6016 = 769.$$

According to the model, the death rate for males who average 6 hours of sleep is 769 deaths per 100,000 males.

TECHNOLOGY

The graph of

$$y = 104.5x^2 - 1501.5x + 6016$$

is displayed in a [3, 12, 1] by [500, 2000, 100] viewing rectangle. The minimum function feature shows that the lowest point on the graph, the vertex, is approximately $(7.2, 622.5)$. Men who average 7.2 hours of sleep are in the group with the lowest death rate, approximately 622.5 deaths per 100,000 males.

Minimum
X=7.1842088 Y=622.45395

☑ Check Point **4** Find the quadratic function $y = ax^2 + bx + c$ whose graph passes through the points $(1, 4)$, $(2, 1)$, and $(3, 4)$. $y = 3x^2 - 12x + 13$

ACHIEVING SUCCESS

Two Ways to Stay Sharp

- **Concentrate on one task at a time.** Do not multitask. Doing several things at once can cause confusion and can take longer to complete the tasks than tackling them sequentially.
- **Get enough sleep.** Fatigue impedes the ability to learn and do complex tasks.

CONCEPT AND VOCABULARY CHECK

Fill in each blank so that the resulting statement is true.

1. A solution of a system of linear equations in three variables is an ordered ___triple___ of real numbers that satisfies <u>all/some</u> of the equations in the system. ___all___

> **Circle the correct choice.**

2. Consider the following system:

$$\begin{cases} x + y - z = -1 & \text{Equation 1} \\ 2x - 2y - 5z = 7 & \text{Equation 2} \\ 4x + y - 2z = 7. & \text{Equation 3} \end{cases}$$

We can eliminate x from Equations 1 and 2 by multiplying Equation 1 by ___-2___ and adding equations. We can eliminate x from Equations 1 and 3 by multiplying Equation 1 by ___-4___ and adding equations.

3. Consider the following system:

$$\begin{cases} x + y + z = 2 & \text{Equation 1} \\ 2x - 3y = 3 & \text{Equation 2} \\ 10y - z = 12. & \text{Equation 3} \end{cases}$$

Equation 2 does not contain the variable ___z___. To obtain a second equation that does not contain this variable, we can ___add Equations 1 and 3___.

EXERCISE SET 5.2

Practice Exercises

In Exercises 1–4, determine if the given ordered triple is a solution of the system.

1. $(2, -1, 3)$ solution
$$\begin{cases} x + y + z = 4 \\ x - 2y - z = 1 \\ 2x - y - 2z = -1 \end{cases}$$

2. $(5, -3, -2)$ solution
$$\begin{cases} x + y + z = 0 \\ x + 2y - 3z = 5 \\ 3x + 4y + 2z = -1 \end{cases}$$

3. $(4, 1, 2)$ solution
$$\begin{cases} x - 2y = 2 \\ 2x + 3y = 11 \\ y - 4z = -7 \end{cases}$$

4. $(-1, 3, 2)$ solution
$$\begin{cases} x - 2z = -5 \\ y - 3z = -3 \\ 2x - z = -4 \end{cases}$$

Solve each system in Exercises 5–18.

5. $\begin{cases} x + y + 2z = 11 \\ x + y + 3z = 14 \\ x + 2y - z = 5 \end{cases}$
$\{(2, 3, 3)\}$

6. $\begin{cases} 2x + y - 2z = -1 \\ 3x - 3y - z = 5 \\ x - 2y + 3z = 6 \end{cases}$
$\{(1, -1, 1)\}$

7. $\begin{cases} 4x - y + 2z = 11 \\ x + 2y - z = -1 \\ 2x + 2y - 3z = -1 \end{cases}$
$\{(2, -1, 1)\}$

8. $\begin{cases} x - y + 3z = 8 \\ 3x + y - 2z = -2 \\ 2x + 4y + z = 0 \end{cases}$
$\{(1, -1, 2)\}$

9. $\begin{cases} 3x + 2y - 3z = -2 \\ 2x - 5y + 2z = -2 \\ 4x - 3y + 4z = 10 \end{cases}$
$\{(1, 2, 3)\}$

10. $\begin{cases} 2x + 3y + 7z = 13 \\ 3x + 2y - 5z = -22 \\ 5x + 7y - 3z = -28 \end{cases}$
$\{(-1, -2, 3)\}$

11. $\begin{cases} 2x - 4y + 3z = 17 \\ x + 2y - z = 0 \\ 4x - y - z = 6 \end{cases}$
$\{(3, 1, 5)\}$

12. $\begin{cases} x + z = 3 \\ x + 2y - z = 1 \\ 2x - y + z = 3 \end{cases}$
$\{(1, 1, 2)\}$

13. $\begin{cases} 2x + y = 2 \\ x + y - z = 4 \\ 3x + 2y + z = 0 \end{cases}$
$\{(1, 0, -3)\}$

14. $\begin{cases} x + 3y + 5z = 20 \\ y - 4z = -16 \\ 3x - 2y + 9z = 36 \end{cases}$
$\{(0, 0, 4)\}$

15. $\begin{cases} x + y = -4 \\ y - z = 1 \\ 2x + y + 3z = -21 \end{cases}$
$\{(1, -5, -6)\}$

16. $\begin{cases} x + y = 4 \\ x + z = 4 \\ y + z = 4 \end{cases}$
$\{(2, 2, 2)\}$

17. $\begin{cases} 3(2x + y) + 5z = -1 \\ 2(x - 3y + 4z) = -9 \\ 4(1 + x) = -3(z - 3y) \end{cases}$
$\left\{ \left(\frac{1}{2}, \frac{1}{3}, -1\right) \right\}$

18. $\begin{cases} 7z - 3 = 2(x - 3y) \\ 5y + 3z - 7 = 4x \\ 4 + 5z = 3(2x - y) \end{cases}$
$\left\{ \left(\frac{1}{2}, 3, -2\right) \right\}$

In Exercises 19–22, find the quadratic function $y = ax^2 + bx + c$ whose graph passes through the given points.

19. $(-1, 6), (1, 4), (2, 9)$ $y = 2x^2 - x + 3$
20. $(-2, 7), (1, -2), (2, 3)$ $y = 2x^2 - x - 3$
21. $(-1, -4), (1, -2), (2, 5)$ $y = 2x^2 + x - 5$
22. $(1, 3), (3, -1), (4, 0)$ $y = x^2 - 6x + 8$

In Exercises 23–24, let x represent the first number, y the second number, and z the third number. Use the given conditions to write a system of equations. Solve the system and find the numbers.

23. The sum of three numbers is 16. The sum of twice the first number, 3 times the second number, and 4 times the third number is 46. The difference between 5 times the first number and the second number is 31. Find the three numbers. 7, 4, and 5

24. The following is known about three numbers: Three times the first number plus the second number plus twice the third number is 5. If 3 times the second number is subtracted from the sum of the first number and 3 times the third number, the result is 2. If the third number is subtracted from 2 times the first number and 3 times the second number, the result is 1. Find the numbers. -1, 2, and 3

Practice Plus

Solve each system in Exercises 25–26.

25. $\begin{cases} \dfrac{x+2}{6} - \dfrac{y+4}{3} + \dfrac{z}{2} = 0 \\ \dfrac{x+1}{2} + \dfrac{y-1}{2} - \dfrac{z}{4} = \dfrac{9}{2} \\ \dfrac{x-5}{4} + \dfrac{y+1}{3} + \dfrac{z-2}{2} = \dfrac{19}{4} \end{cases}$ $\{(4, 8, 6)\}$

26. $\begin{cases} \dfrac{x+3}{2} - \dfrac{y-1}{2} + \dfrac{z+2}{4} = \dfrac{3}{2} \\ \dfrac{x-5}{2} + \dfrac{y+1}{3} - \dfrac{z}{4} = -\dfrac{25}{6} \\ \dfrac{x-3}{4} - \dfrac{y+1}{2} + \dfrac{z-3}{2} = -\dfrac{5}{2} \end{cases}$ $\{(-3, 0, 2)\}$

In Exercises 27–28, find the equation of the quadratic function $y = ax^2 + bx + c$ whose graph is shown. Select three points whose coordinates appear to be integers.

27. $y = -\frac{3}{4}x^2 + 6x - 11$

28. 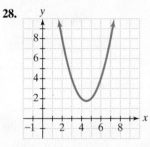 $y = x^2 - 9x + 22$

In Exercises 29–30, solve each system for (x, y, z) in terms of the nonzero constants a, b, and c.

29. $\begin{cases} ax - by - 2cz = 21 \\ ax + by + cz = 0 \\ 2ax - by + cz = 14 \end{cases}$ * **30.** $\begin{cases} ax - by + 2cz = -4 \\ ax + 3by - cz = 1 \\ 2ax + by + 3cz = 2 \end{cases}$ *

Application Exercises

31. You throw a ball straight up from a rooftop. The ball misses the rooftop on its way down and eventually strikes the ground. A mathematical model can be used to describe the relationship for the ball's height above the ground, y, after x seconds. Consider the following data:

x, seconds after the ball is thrown	y, ball's height, in feet, above the ground
1	224
3	176
4	104

*See Answers to Selected Exercises.

a. Find the quadratic function $y = ax^2 + bx + c$ whose graph passes through the given points. $y = -16x^2 + 40x + 200$

b. Use the function in part (a) to find the value for y when $x = 5$. Describe what this means. $y = 0$ when $x = 5$*

32. A mathematical model can be used to describe the relationship between the number of feet a car travels once the brakes are applied, y, and the number of seconds the car is in motion after the brakes are applied, x. A research firm collects the following data:

x, seconds in motion after brakes are applied	y, feet car travels once the brakes are applied
1	46
2	84
3	114

a. Find the quadratic function $y = ax^2 + bx + c$ whose graph passes through the given points. $y = -4x^2 + 50x$

b. Use the function in part (a) to find the value for y when $x = 6$. Describe what this means. 156*

In Exercises 33–41, use the four-step strategy to solve each problem. Use x, y, and z to represent unknown quantities. Then translate from the verbal conditions of the problem to a system of three equations in three variables.

33. The bar graph shows the average number of hours U.S. college students study per week for seven selected majors. Study times are rounded to the nearest hour.

Number of Hours U.S. College Students Study per Week, by Major

Source: statista.com

The combined weekly study time for students majoring in chemical engineering, mathematics, and psychology is 52 hours. The difference between weekly study time for chemical engineering majors and math majors is 6 hours. The difference between weekly study time for chemical engineering majors and psychology majors is 8 hours. Find the average number of hours per week that chemical engineering majors, mathematics majors, and psychology majors spend studying. *

34. The bar graph shows the average number of hours U.S. college students study per week for seven selected majors. Study times are rounded to the nearest hour.

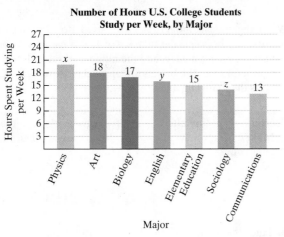

Number of Hours U.S. College Students Study per Week, by Major

Source: statista.com

The combined weekly study time for students majoring in physics, English, and sociology is 50 hours. The difference between weekly study time for physics majors and English majors is 4 hours. The difference between weekly study time for physics majors and sociology majors is 6 hours. Find the average number of hours per week that physics majors, English majors, and sociology majors spend studying. *

35. On a recent trip to the convenience store, you picked up 2 gallons of milk, 5 bottles of water, and 6 snack-size bags of chips. Your total bill (before tax) was $19.00. If a bottle of water costs twice as much as a bag of chips, and a gallon of milk costs $2.00 more than a bottle of water, how much does each item cost? milk: $3.50; water: $1.50; chips: $0.75

36. On a recent trip to the convenience store, you picked up 1 gallon of milk, 7 bottles of water, and 4 snack-size bags of chips. Your total bill (before tax) was $17.00. If a bottle of water costs twice as much as a bag of chips, and a gallon of milk costs $2.00 more than a bottle of water, how much does each item cost? milk: $3.50; water: $1.50; chips: $0.75

37. At a college production of *A Streetcar Named Desire*, 400 tickets were sold. The ticket prices were $8, $10, and $12, and the total income from ticket sales was $3700. How many tickets of each type were sold if the combined number of $8 and $10 tickets sold was 7 times the number of $12 tickets sold? $8: 200; $10: 150; $12: 50

38. A certain brand of razor blades comes in packages of 6, 12, and 24 blades, costing $2, $3, and $4 per package, respectively. A store sold 12 packages containing a total of 162 razor blades and took in $35. How many packages of each type were sold? six: 5; twelve: 3; twenty-four: 4

39. A person invested $6700 for one year, part at 8%, part at 10%, and the remainder at 12%. The total annual income from these investments was $716. The amount of money invested at 12% was $300 more than the amount invested at 8% and 10% combined. Find the amount invested at each rate. $1200 at 8%, $2000 at 10%, $3500 at 12%

40. A person invested $17,000 for one year, part at 10%, part at 12%, and the remainder at 15%. The total annual income from these investments was $2110. The amount of money invested at 12% was $1000 less than the amount invested at 10% and 15% combined. Find the amount invested at each rate. $4000 at 10%, $8000 at 12%, $5000 at 15%

41. In the following triangle, the degree measures of the three interior angles and two of the exterior angles are represented with variables. Find the measure of each interior angle. $x = 60, y = 55, z = 65$

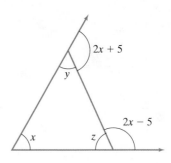

Explaining the Concepts 42–45. Answers will vary.

42. What is a system of linear equations in three variables?

43. How do you determine whether a given ordered triple is a solution of a system in three variables?

44. Describe in general terms how to solve a system in three variables.

45. AIDS is taking a deadly toll on southern Africa. Describe how to use the techniques that you learned in this section to obtain a model for African life span using projections with AIDS, shown by the red graph in the figure. Let x represent the number of years after 1985 and let y represent African life span in that year.

Source: United Nations

Technology Exercises 46–47. Answers will vary.

46. Does your graphing utility have a feature that allows you to solve linear systems by entering coefficients and constant terms? If so, use this feature to verify the solutions to any five exercises that you worked by hand from Exercises 5–16.

47. Verify your results in Exercises 19–22 by using a graphing utility to graph the resulting parabola. Trace along the curve and convince yourself that the three points given in the exercise lie on the parabola.

Critical Thinking Exercises

Make Sense? *In Exercises 48–51, determine whether each statement makes sense or does not make sense, and explain your reasoning.*

48. Solving a system in three variables, I found that $x = 3$ and $y = -1$. Because z represents a third variable, z cannot equal 3 or -1. does not make sense

49. A system of linear equations in three variables, x, y, and z, cannot contain an equation in the form $y = mx + b$. *

50. I'm solving a three-variable system in which one of the given equations has a missing term, so it will not be necessary to use any of the original equations twice when I reduce the system to two equations in two variables. makes sense

51. Because the percentage of the U.S. population that was foreign-born decreased from 1910 through 1970 and then increased after that, a quadratic function of the form $f(x) = ax^2 + bx + c$, rather than a linear function of the form $f(x) = mx + b$, should be used to model the data.

52. Describe how the system

$$\begin{cases} x + y - z - 2w = -8 \\ x - 2y + 3z + w = 18 \\ 2x + 2y + 2z - 2w = 10 \\ 2x + y - z + w = 3 \end{cases}$$

could be solved. Is it likely that in the near future a graphing utility will be available to provide a geometric solution (using intersecting graphs) to this system? Explain. Answers will vary.

53. A modernistic painting consists of triangles, rectangles, and pentagons, all drawn so as to not overlap or share sides. Within each rectangle are drawn 2 red roses and each pentagon contains 5 carnations. How many triangles, rectangles, and pentagons appear in the painting if the painting contains a total of 40 geometric figures, 153 sides of geometric figures, and 72 flowers? 13 triangles, 21 rectangles, and 6 pentagons

Group Exercise Answers will vary.

54. Group members should develop appropriate functions that model each of the projections shown in Exercise 45.

Retaining the Concepts

55. Solve: $\dfrac{x+3}{4} - \dfrac{x+1}{10} = \dfrac{x-2}{5} - 1$.

(Section 1.2, Example 3) {41}

56. Solve: $3x^2 + 1 = x^2 + x$. (Section 1.5, Example 8)

57. The height of a ball thrown across a field, $f(x)$, in feet, can be modeled by

$$f(x) = -0.005x^2 + x + 5,$$

where x is the ball's horizontal distance, in feet, from the point where it was thrown.

a. What is the maximum height of the ball and how far from where it was thrown does this occur? 55 ft; 100 ft

b. How far down the field will the ball travel before hitting the ground?

(Section 3.1, Example 5) ≈ 204.9 ft

Preview Exercises

Exercises 58–60 will help you prepare for the material covered in the next section.

58. Subtract: $\dfrac{3}{x-4} - \dfrac{2}{x+2}$. $\dfrac{x+14}{(x-4)(x+2)}$

59. Add: $\dfrac{5x-3}{x^2+1} + \dfrac{2x}{(x^2+1)^2}$. $\dfrac{5x^3 - 3x^2 + 7x - 3}{(x^2+1)^2}$

60. Solve:

$$\begin{cases} A + B = 3 \\ 2A - 2B + C = 17 \\ 4A - 2C = 14. \end{cases} \quad \{(5, -2, 3)\}$$

Section 5.3 Partial Fractions

What am I supposed to learn?

After studying this section, you should be able to:

① Decompose $\dfrac{P}{Q}$, where Q has only distinct linear factors.

② Decompose $\dfrac{P}{Q}$, where Q has repeated linear factors.

③ Decompose $\dfrac{P}{Q}$, where Q has a nonrepeated prime quadratic factor.

④ Decompose $\dfrac{P}{Q}$, where Q has a prime, repeated quadratic factor.

The rising and setting of the Sun suggest the obvious: Things change over time. Calculus is the study of rates of change, allowing the motion of the rising Sun to be measured by "freezing the frame" at one instant in time. If you are given a function, calculus reveals its rate of change at any "frozen" instant. In this section, you will learn an algebraic technique used in calculus to find a function if its rate of change is known. The technique involves expressing a given function in terms of simpler functions.

The Idea behind Partial Fraction Decomposition

We know how to use common denominators to write a sum or difference of rational expressions as a single rational expression. For example,

$$\frac{3}{x-4} - \frac{2}{x+2} = \frac{3}{(x-4)} \cdot \frac{(x+2)}{(x+2)} - \frac{2}{(x+2)} \cdot \frac{(x-4)}{(x-4)} = \frac{3(x+2) - 2(x-4)}{(x-4)(x+2)}$$

$$= \frac{3x + 6 - 2x + 8}{(x-4)(x+2)} = \frac{x+14}{(x-4)(x+2)}.$$

For solving the kind of calculus problem described in the section opener, we must reverse this process:

$$\underbrace{\frac{x + 14}{(x - 4)(x + 2)}}_{\substack{\text{is} \\ \text{expressed as the} \\ \text{sum of two} \\ \text{simpler fractions.}}} = \overbrace{\frac{3}{x - 4}}^{\text{Partial fraction}} + \overbrace{\frac{-2}{x + 2}}^{\text{Partial fraction}}.$$

This is the partial fraction decomposition of $\dfrac{x + 14}{(x - 4)(x + 2)}$.

Each of the two fractions on the right is called a **partial fraction**. The sum of these fractions is called the **partial fraction decomposition** of the rational expression on the left-hand side.

Partial fraction decompositions can be written for rational expressions of the form $\dfrac{P(x)}{Q(x)}$, where P and Q have no common factors and the highest power in the numerator is less than the highest power in the denominator. In this section, we will show you how to write the partial fraction decompositions for each of the following rational expressions:

$$\frac{9x^2 - 9x + 6}{(2x - 1)(x + 2)(x - 2)}$$

$P(x) = 9x^2 - 9x + 6$; highest power = 2

$Q(x) = (2x - 1)(x + 2)(x - 2)$; multiplying factors, highest power = 3.

$$\frac{5x^3 - 3x^2 + 7x - 3}{(x^2 + 1)^2}.$$

$P(x) = 5x^3 - 3x^2 + 7x - 3$; highest power = 3

$Q(x) = (x^2 + 1)^2$; squaring the expression, highest power = 4.

The partial fraction decomposition of a rational expression depends on the factors of the denominator. We consider four cases involving different kinds of factors in the denominator:

1. The denominator is a product of distinct linear factors.

2. The denominator is a product of linear factors, some of which are repeated.

3. The denominator has prime quadratic factors, none of which is repeated.

4. The denominator has a repeated prime quadratic factor.

❶ Decompose $\dfrac{P}{Q}$, where Q has only distinct linear factors.

The Partial Fraction Decomposition of a Rational Expression with Distinct Linear Factors in the Denominator

If the denominator of a rational expression has a linear factor of the form $ax + b$, then the partial fraction decomposition will contain a term of the form

$$\underbrace{\frac{\overbrace{A}^{\text{Constant}}}{ax + b}}_{\text{Linear factor}}.$$

Each distinct linear factor in the denominator produces a partial fraction of the form *constant over linear factor*. For example,

$$\frac{9x^2 - 9x + 6}{(2x - 1)(x + 2)(x - 2)} = \frac{A}{2x - 1} + \frac{B}{x + 2} + \frac{C}{x - 2}.$$

We write a constant over each linear factor in the denominator.

The Partial Fraction Decomposition of $\dfrac{P(x)}{Q(x)}$: $Q(x)$ Has Distinct Linear Factors

The form of the partial fraction decomposition for a rational expression with distinct linear factors in the denominator is

$$\frac{P(x)}{(a_1x + b_1)(a_2x + b_2)(a_3x + b_3) \cdots (a_nx + b_n)}$$

$$= \frac{A_1}{a_1x + b_1} + \frac{A_2}{a_2x + b_2} + \frac{A_3}{a_3x + b_3} + \cdots + \frac{A_n}{a_nx + b_n}.$$

EXAMPLE 1 Partial Fraction Decomposition with Distinct Linear Factors

Find the partial fraction decomposition of

$$\frac{x + 14}{(x - 4)(x + 2)}.$$

SOLUTION

We begin by setting up the partial fraction decomposition with the unknown constants. Write a constant over each of the two distinct linear factors in the denominator.

$$\frac{x + 14}{(x - 4)(x + 2)} = \frac{A}{x - 4} + \frac{B}{x + 2}$$

Our goal is to find A and B. We do this by multiplying both sides of the equation by the least common denominator, $(x - 4)(x + 2)$.

$$(x - 4)(x + 2)\frac{x + 14}{(x - 4)(x + 2)} = (x - 4)(x + 2)\left(\frac{A}{x - 4} + \frac{B}{x + 2}\right)$$

We use the distributive property on the right side.

$$(x - 4)(x + 2)\frac{x + 14}{(x - 4)(x + 2)}$$

$$= (x - 4)(x + 2)\frac{A}{(x - 4)} + (x - 4)(x + 2)\frac{B}{(x + 2)}$$

Dividing out common factors in numerators and denominators, we obtain

$$x + 14 = A(x + 2) + B(x - 4).$$

To find values for A and B that make both sides equal, we will express the sides in exactly the same form by writing the variable x-terms and then writing the constant terms. Apply the distributive property on the right side.

$x + 14 = Ax + 2A + Bx - 4B$ Distribute A and B over the parentheses.

$x + 14 = Ax + Bx + 2A - 4B$ Rearrange terms.

$1x + 14 = (A + B)x + (2A - 4B)$ Rewrite to identify the coefficient of x and the constant term.

As shown by the arrows, if two polynomials are equal, coefficients of like powers of x must be equal ($A + B = 1$) and their constant terms must be equal ($2A - 4B = 14$). Consequently, A and B satisfy the following two equations:

$$\begin{cases} A + B = 1 \\ 2A - 4B = 14. \end{cases}$$

We can use the addition method to solve this linear system in two variables. By multiplying the first equation by -2 and adding equations, we obtain $A = 3$ and $B = -2$. Thus,

$$\frac{x + 14}{(x - 4)(x + 2)} = \frac{A}{x - 4} + \frac{B}{x + 2} = \frac{3}{x - 4} + \frac{-2}{x + 2} \left(\text{or } \frac{3}{x - 4} - \frac{2}{x + 2} \right).$$

• • •

Steps in Partial Fraction Decomposition

1. Set up the partial fraction decomposition with the unknown constants A, B, C, etc., in the numerators of the decomposition.

2. Multiply both sides of the resulting equation by the least common denominator.

3. Simplify the right side of the equation.

4. Write both sides in descending powers, equate coefficients of like powers of x, and equate constant terms.

5. Solve the resulting linear system for A, B, C, etc.

6. Substitute the values for A, B, C, etc., into the equation in step 1 and write the partial fraction decomposition.

GREAT QUESTION!

Is there any way to speed up the process of finding partial fraction decompositions?

Sometimes. In certain situations, you can use the equation in step 2, the identity obtained after clearing fractions, to determine the constants A, B, C, etc., needed to write the partial fraction decomposition. In Example 1, the identity obtained after fractions were cleared was

$$x + 14 = A(x + 2) + B(x - 4).$$

Using suitable choices for x, we can quickly find A and B.

$$x + 14 = A(x + 2) + B(x - 4) \qquad x + 14 = A(x + 2) + B(x - 4)$$

If we let $x = -2$, the term containing A drops out.	If we let $x = 4$, the term containing B drops out.
$12 = -6B$	$18 = 6A$
$-2 = B$	$3 = A$

As in the solution to Example 1, $A = 3$ and $B = -2$. This shortcut sometimes makes things easier by letting you skip steps 3, 4, and 5 in the box. The downside: Using suitable choices for x to determine constants does not always work.

$\dfrac{2}{x - 3} + \dfrac{3}{x + 4}$

✓ Check Point **1** Find the partial fraction decomposition of $\dfrac{5x - 1}{(x - 3)(x + 4)}$.

② Decompose $\dfrac{P}{Q}$, where Q has repeated linear factors.

The Partial Fraction Decomposition of a Rational Expression with Linear Factors in the Denominator, Some of Which Are Repeated

Suppose that $(ax + b)^n$ is a factor of the denominator. This means that the linear factor $ax + b$ is repeated n times. When this occurs, the partial fraction decomposition will contain a sum of n fractions for this factor of the denominator.

The Partial Fraction Decomposition of $\dfrac{P(x)}{Q(x)}$: $Q(x)$ Has Repeated Linear Factors

The form of the partial fraction decomposition for a rational expression containing the linear factor $ax + b$ occurring n times as its denominator is

$$\frac{P(x)}{(ax + b)^n} = \frac{A_1}{ax + b} + \frac{A_2}{(ax + b)^2} + \frac{A_3}{(ax + b)^3} + \cdots + \frac{A_n}{(ax + b)^n}.$$

Include one fraction with a constant numerator for each power of $ax + b$.

EXAMPLE 2 Partial Fraction Decomposition with a Repeated Linear Factor

Find the partial fraction decomposition of $\dfrac{x - 18}{x(x - 3)^2}$.

SOLUTION

Step 1 Set up the partial fraction decomposition with the unknown constants. Because the linear factor $x - 3$ occurs twice, we must include one fraction with a constant numerator for each power of $x - 3$.

$$\frac{x - 18}{x(x - 3)^2} = \frac{A}{x} + \frac{B}{x - 3} + \frac{C}{(x - 3)^2}$$

Step 2 Multiply both sides of the resulting equation by the least common denominator. We clear fractions, multiplying both sides by $x(x - 3)^2$, the least common denominator.

$$x(x - 3)^2 \left(\frac{x - 18}{x(x - 3)^2} \right) = x(x - 3)^2 \left(\frac{A}{x} + \frac{B}{x - 3} + \frac{C}{(x - 3)^2} \right)$$

We use the distributive property on the right side.

$$x(x - 3)^2 \cdot \frac{x - 18}{x(x - 3)^2} = x(x - 3)^2 \cdot \frac{A}{x} + x(x - 3)^2 \cdot \frac{B}{(x - 3)} + x(x - 3)^2 \cdot \frac{C}{(x - 3)^2}$$

Dividing out common factors in numerators and denominators, we obtain

$$x - 18 = A(x - 3)^2 + Bx(x - 3) + Cx.$$

You can use this equation to find C (let $x = 3$) and to find A (let $x = 0$).

Step 3 Simplify the right side of the equation. Square $x - 3$. Then apply the distributive property.

$$x - 18 = A(x^2 - 6x + 9) + Bx(x - 3) + Cx \qquad \text{Square } x - 3 \text{ using } (A - B)^2 = A^2 - 2AB + B^2.$$

$$x - 18 = Ax^2 - 6Ax + 9A + Bx^2 - 3Bx + Cx \qquad \text{Apply the distributive property.}$$

GREAT QUESTION!

When setting up the partial fraction decomposition for

$$\frac{x - 18}{x(x - 3)^2},$$

can I just use a constant over $x - 3$ and another constant over $x - 3$? After all,

$$(x - 3)^2 = (x - 3)(x - 3).$$

No. Avoid this common error:

$$\frac{x - 18}{x(x - 3)^2} = \frac{A}{x} + \frac{B}{x - 3} + \frac{C}{x - 3}.$$

Listing $x - 3$ twice does not take into account $(x - 3)^2$.

Step 4 Write both sides in descending powers, equate coefficients of like powers of x, and equate constant terms. We use the equation from step 3:

$$x - 18 = Ax^2 - 6Ax + 9A + Bx^2 - 3Bx + Cx.$$

The left side, $x - 18$, is in descending powers of x: $x - 18x^0$. We will write the right side in descending powers of x.

$$x - 18 = Ax^2 + Bx^2 - 6Ax - 3Bx + Cx + 9A \quad \text{Rearrange terms on the right side.}$$

Express both sides in the same form.

$$0x^2 + 1x - 18 = (A + B)x^2 + (-6A - 3B + C)x + 9A \quad \text{Rewrite to identify coefficients and the constant term.}$$

Equating coefficients of like powers of x and equating constant terms results in the following system of linear equations:

$$\begin{cases} A + B = 0 \\ -6A - 3B + C = 1 \\ 9A = -18. \end{cases}$$

Step 5 Solve the resulting system for A, B, and C. Dividing both sides of the last equation by 9, we obtain $A = -2$. Substituting -2 for A in the first equation, $A + B = 0$, gives $-2 + B = 0$, so $B = 2$. We find C by substituting -2 for A and 2 for B in the middle equation, $-6A - 3B + C = 1$. We obtain $C = -5$.

Step 6 Substitute the values of A, B, and C, and write the partial fraction decomposition. With $A = -2$, $B = 2$, and $C = -5$, the required partial fraction decomposition is

$$\frac{x - 18}{x(x - 3)^2} = \frac{A}{x} + \frac{B}{x - 3} + \frac{C}{(x - 3)^2} = -\frac{2}{x} + \frac{2}{x - 3} - \frac{5}{(x - 3)^2}. \quad \bullet\bullet\bullet$$

$\frac{2}{x} - \frac{2}{x - 1} + \frac{3}{(x - 1)^2}$

✓ **Check Point 2** Find the partial fraction decomposition of $\dfrac{x + 2}{x(x - 1)^2}$.

GREAT QUESTION!

In Examples 1 and 2, the denominators of the given rational expressions were factored. What do I do when I have to determine a partial fraction decomposition in which the denominator of the given rational expression is not already factored?

Begin by factoring the denominator. Then apply the steps needed to obtain the partial fraction decomposition.

③ Decompose $\dfrac{P}{Q}$, where Q has a nonrepeated prime quadratic factor.

The Partial Fraction Decomposition of a Rational Expression with Prime, Nonrepeated Quadratic Factors in the Denominator

Our final two cases of partial fraction decomposition involve prime quadratic factors of the form $ax^2 + bx + c$. Based on our work with the discriminant, we know that $ax^2 + bx + c$ is prime and cannot be factored over the integers if $b^2 - 4ac < 0$ or if $b^2 - 4ac$ is not a perfect square.

The Partial Fraction Decomposition of $\dfrac{P(x)}{Q(x)}$: $Q(x)$ Has a Nonrepeated, Prime Quadratic Factor

If $ax^2 + bx + c$ is a prime quadratic factor of $Q(x)$, the partial fraction decomposition will contain a term of the form

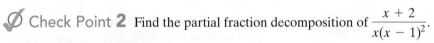

$$\frac{Ax + B}{ax^2 + bx + c}.$$

[Linear numerator] [Quadratic factor]

The voice balloons in the box show that each distinct prime quadratic factor in the denominator produces a partial fraction of the form *linear numerator over quadratic factor*.

Here's an example of a partial fraction decomposition containing a linear numerator over a quadratic factor:

$$\frac{3x^2 + 17x + 14}{(x - 2)(x^2 + 2x + 4)} = \frac{A}{x - 2} + \frac{Bx + C}{x^2 + 2x + 4}.$$

We write a constant over the linear factor in the denominator.

We write a linear numerator over the prime quadratic factor in the denominator.

Our next example illustrates how a linear system in three variables is used to determine values for A, B, and C.

EXAMPLE 3 Partial Fraction Decomposition

Find the partial fraction decomposition of

$$\frac{3x^2 + 17x + 14}{(x - 2)(x^2 + 2x + 4)}.$$

SOLUTION

Step 1 Set up the partial fraction decomposition with the unknown constants. We put a constant (A) over the linear factor and a linear expression $(Bx + C)$ over the prime quadratic factor.

$$\frac{3x^2 + 17x + 14}{(x - 2)(x^2 + 2x + 4)} = \frac{A}{x - 2} + \frac{Bx + C}{x^2 + 2x + 4}$$

Step 2 Multiply both sides of the resulting equation by the least common denominator. We clear fractions, multiplying both sides by $(x - 2)(x^2 + 2x + 4)$, the least common denominator.

$$(x - 2)(x^2 + 2x + 4)\left(\frac{3x^2 + 17x + 14}{(x - 2)(x^2 + 2x + 4)}\right) = (x - 2)(x^2 + 2x + 4)\left(\frac{A}{x - 2} + \frac{Bx + C}{x^2 + 2x + 4}\right)$$

We use the distributive property on the right side.

$$(x - 2)(x^2 + 2x + 4) \cdot \frac{3x^2 + 17x + 14}{(x - 2)(x^2 + 2x + 4)}$$

$$= (x - 2)(x^2 + 2x + 4) \cdot \frac{A}{x - 2} + (x - 2)(x^2 + 2x + 4) \cdot \frac{Bx + C}{x^2 + 2x + 4}$$

Dividing out common factors in numerators and denominators, we obtain

$$3x^2 + 17x + 14 = A(x^2 + 2x + 4) + (Bx + C)(x - 2).$$

Step 3 Simplify the right side of the equation. We simplify on the right side by distributing A over each term in parentheses and multiplying $(Bx + C)(x - 2)$ using the FOIL method.

$$3x^2 + 17x + 14 = Ax^2 + 2Ax + 4A + Bx^2 - 2Bx + Cx - 2C$$

Step 4 Write both sides in descending powers, equate coefficients of like powers of x, and equate constant terms. The left side, $3x^2 + 17x + 14$, is in descending powers of x. We write the right side in descending powers of x and express both sides in the same form.

$$3x^2 + 17x + 14 = Ax^2 + Bx^2 + 2Ax - 2Bx + Cx + 4A - 2C$$
$$3x^2 + 17x + 14 = (A + B)x^2 + (2A - 2B + C)x + (4A - 2C)$$

We use the equation from the previous page:

$$3x^2 + 17x + 14 = (A + B)x^2 + (2A - 2B + C)x + (4A - 2C).$$

Equating coefficients of like powers of x and equating constant terms results in the following system of linear equations:

$$\begin{cases} A + B = 3 \\ 2A - 2B + C = 17 \\ 4A - 2C = 14. \end{cases}$$

Step 5 Solve the resulting system for A, B, and C. Because the first equation involves A and B, we can obtain another equation in A and B by eliminating C from the second and third equations. Multiply the second equation by 2 and add equations. Solving in this manner, we obtain $A = 5$, $B = -2$, and $C = 3$.

Step 6 Substitute the values of A, B, and C, and write the partial fraction decomposition. With $A = 5$, $B = -2$, and $C = 3$, the required partial fraction decomposition is

$$\frac{3x^2 + 17x + 14}{(x - 2)(x^2 + 2x + 4)} = \frac{A}{x - 2} + \frac{Bx + C}{x^2 + 2x + 4} = \frac{5}{x - 2} + \frac{-2x + 3}{x^2 + 2x + 4}. \;\bullet\bullet\bullet$$

TECHNOLOGY

Numeric Connections

You can use the $\boxed{\text{TABLE}}$ feature of a graphing utility to check a partial fraction decomposition. To check the result of Example 3, enter the given rational function and its partial fraction decomposition:

$$y_1 = \frac{3x^2 + 17x + 14}{(x - 2)(x^2 + 2x + 4)}$$

$$y_2 = \frac{5}{x - 2} + \frac{-2x + 3}{x^2 + 2x + 4}.$$

X	Y₁	Y₂
-3	.28571	.28571
-2	.5	.5
-1	0	0
0	-1.75	-1.75
1	-4.857	-4.857
2	ERROR	ERROR
3	4.8421	4.8421
4	2.3214	2.3214
5	1.4872	1.4872
6	1.0769	1.0769
7	.83582	.83582

X= -3

No matter how far up or down we scroll, $y_1 = y_2$, so the decomposition appears to be correct.

✓ **Check Point 3** Find the partial fraction decomposition of

$$\frac{8x^2 + 12x - 20}{(x + 3)(x^2 + x + 2)}. \qquad \frac{2}{x+3} + \frac{6x - 8}{x^2 + x + 2}$$

④ Decompose $\dfrac{P}{Q}$, where Q has a prime, repeated quadratic factor.

The Partial Fraction Decomposition of a Rational Expression with a Prime, Repeated Quadratic Factor in the Denominator

Suppose that $(ax^2 + bx + c)^n$ is a factor of the denominator and that $ax^2 + bx + c$ cannot be factored further. This means that the quadratic factor $ax^2 + bx + c$ occurs n times. When this occurs, the partial fraction decomposition will contain a linear numerator for each power of $ax^2 + bx + c$.

The Partial Fraction Decomposition of $\dfrac{P(x)}{Q(x)}$: $Q(x)$ Has a Prime, Repeated Quadratic Factor

The form of the partial fraction decomposition for a rational expression containing the prime factor $ax^2 + bx + c$ occurring n times as its denominator is

$$\frac{P(x)}{(ax^2 + bx + c)^n} = \frac{A_1 x + B_1}{ax^2 + bx + c} + \frac{A_2 x + B_2}{(ax^2 + bx + c)^2} + \frac{A_3 x + B_3}{(ax^2 + bx + c)^3} + \cdots + \frac{A_n x + B_n}{(ax^2 + bx + c)^n}.$$

Include one fraction with a linear numerator for each power of $ax^2 + bx + c$.

EXAMPLE 4 Partial Fraction Decomposition with a Repeated Quadratic Factor

Find the partial fraction decomposition of

$$\frac{5x^3 - 3x^2 + 7x - 3}{(x^2 + 1)^2}.$$

SOLUTION

Step 1 Set up the partial fraction decomposition with the unknown constants. Because the quadratic factor $x^2 + 1$ occurs twice, we must include one fraction with a linear numerator for each power of $x^2 + 1$.

$$\frac{5x^3 - 3x^2 + 7x - 3}{(x^2 + 1)^2} = \frac{Ax + B}{x^2 + 1} + \frac{Cx + D}{(x^2 + 1)^2}$$

Step 2 Multiply both sides of the resulting equation by the least common denominator. We clear fractions, multiplying both sides by $(x^2 + 1)^2$, the least common denominator.

$$(x^2 + 1)^2 \left(\frac{5x^3 - 3x^2 + 7x - 3}{(x^2 + 1)^2} \right) = (x^2 + 1)^2 \left(\frac{Ax + B}{x^2 + 1} + \frac{Cx + D}{(x^2 + 1)^2} \right)$$

Now we multiply and simplify.

$$5x^3 - 3x^2 + 7x - 3 = (x^2 + 1)(Ax + B) + Cx + D$$

Step 3 Simplify the right side of the equation. We multiply $(x^2 + 1)(Ax + B)$ using the FOIL method.

$$5x^3 - 3x^2 + 7x - 3 = Ax^3 + Bx^2 + Ax + B + Cx + D$$

Step 4 Write both sides in descending powers, equate coefficients of like powers of x, and equate constant terms.

$$5x^3 - 3x^2 + 7x - 3 = Ax^3 + Bx^2 + Ax + Cx + B + D$$

$$5x^3 - 3x^2 + 7x - 3 = Ax^3 + Bx^2 + (A + C)x + (B + D)$$

Equating coefficients of like powers of x and equating constant terms results in the following system of linear equations:

$$\begin{cases} A = 5 \\ B = -3 \\ A + C = 7 \quad \text{With } A = 5, \text{ we immediately obtain } C = 2. \\ B + D = -3. \quad \text{With } B = -3, \text{ we immediately obtain } D = 0. \end{cases}$$

Step 5 Solve the resulting system for A, B, C, and D. Based on our observations in step 4, $A = 5$, $B = -3$, $C = 2$, and $D = 0$.

Step 6 Substitute the values of A, B, C, and D, and write the partial fraction decomposition.

$$\frac{5x^3 - 3x^2 + 7x - 3}{(x^2 + 1)^2} = \frac{Ax + B}{x^2 + 1} + \frac{Cx + D}{(x^2 + 1)^2} = \frac{5x - 3}{x^2 + 1} + \frac{2x}{(x^2 + 1)^2} \quad \bullet\bullet\bullet$$

GREAT QUESTION!

When setting up partial fraction decompositions, when should I use constant numerators and when should I use linear numerators?

When the denominator of a rational expression contains a **power of a linear factor**, set up the partial fraction decomposition with **constant numerators** (A, B, C, etc.). When the denominator of a rational expression contains a **power of a prime quadratic factor**, set up the partial fraction decomposition with **linear numerators** ($Ax + B$, $Cx + D$, etc.).

$\frac{2x}{x^2 + 1} + \frac{-x + 3}{(x^2 + 1)^2}$

✓ **Check Point 4** Find the partial fraction decomposition of $\dfrac{2x^3 + x + 3}{(x^2 + 1)^2}$.

GREAT QUESTION!

When a rational expression contains a power of a factor in the denominator, how many fractions do I have to include in the partial fraction decomposition?

When a rational expression contains a power of a factor in the denominator, be sure to set up the partial fraction decomposition to allow for every natural-number power of that factor less than or equal to the power. Example:

$$\frac{2x+1}{(x-5)^2 x^3} = \frac{A}{x-5} + \frac{B}{(x-5)^2} + \frac{C}{x} + \frac{D}{x^2} + \frac{E}{x^3}$$

Although $(x-5)^2$ and x^2 are quadratic, they are still expressed as powers of linear factors, $x-5$ and x. Thus, the numerators are constant.

CONCEPT AND VOCABULARY CHECK

Determine whether each partial fraction decomposition is set up correctly. If the setup is incorrect, make the necessary changes to produce the correct decomposition.

1. Correct or incorrect:

$$\frac{7x}{(x+2)(x-3)} = \frac{A}{x+2} + \frac{B}{x-3} \quad \underline{\text{correct}}$$

2. Correct or incorrect:

$$\frac{3x}{(x+5)(x-4)^2} = \frac{A}{x+5} + \frac{B}{x-4} + \frac{C}{x-4} \quad \underline{\text{incorrect}}$$

3. Correct or incorrect:

$$\frac{1}{(x+1)(x^2+4)} = \frac{A}{x+1} + \frac{B}{x^2+4} \quad \underline{\text{incorrect}}$$

4. Correct or incorrect:

$$\frac{7x-5}{(x^2+x+1)^2} = \frac{Ax+B}{x^2+x+1} + \frac{Cx+D}{(x^2+x+1)^2} \quad \underline{\text{correct}}$$

EXERCISE SET 5.3

Practice Exercises

In Exercises 1–8, write the form of the partial fraction decomposition of the rational expression. It is not necessary to solve for the constants.

1. $\dfrac{11x-10}{(x-2)(x+1)}$ *

2. $\dfrac{5x+7}{(x-1)(x+3)}$ *

3. $\dfrac{6x^2-14x-27}{(x+2)(x-3)^2}$ *

4. $\dfrac{3x+16}{(x+1)(x-2)^2}$

5. $\dfrac{5x^2-6x+7}{(x-1)(x^2+1)}$ *

6. $\dfrac{5x^2-9x+19}{(x-4)(x^2+5)}$ *

7. $\dfrac{x^3+x^2}{(x^2+4)^2}$ $\dfrac{Ax+B}{x^2+4} + \dfrac{Cx+D}{(x^2+4)^2}$

8. $\dfrac{7x^2-9x+3}{(x^2+7)^2}$ *

In Exercises 9–42, write the partial fraction decomposition of each rational expression.

9. $\dfrac{x}{(x-3)(x-2)}$ *

10. $\dfrac{1}{x(x-1)}$ $\dfrac{1}{x-1} - \dfrac{1}{x}$

11. $\dfrac{3x+50}{(x-9)(x+2)}$ *

12. $\dfrac{5x-1}{(x-2)(x+1)}$ *

13. $\dfrac{7x-4}{x^2-x-12}$ *

14. $\dfrac{9x+21}{x^2+2x-15}$ *

15. $\dfrac{4}{2x^2-5x-3}$ *

16. $\dfrac{x}{x^2+2x-3}$ *

17. $\dfrac{4x^2+13x-9}{x(x-1)(x+3)}$ *

18. $\dfrac{4x^2-5x-15}{x(x+1)(x-5)}$ *

19. $\dfrac{4x^2-7x-3}{x^3-x}$ *

20. $\dfrac{2x^2-18x-12}{x^3-4x}$ *

21. $\dfrac{6x-11}{(x-1)^2}$ $\dfrac{6}{x-1} - \dfrac{5}{(x-1)^2}$

22. $\dfrac{x}{(x+1)^2}$ $\dfrac{1}{x+1} - \dfrac{1}{(x+1)^2}$

23. $\dfrac{x^2-6x+3}{(x-2)^3}$ *

24. $\dfrac{2x^2+8x+3}{(x+1)^3}$ *

25. $\dfrac{x^2+2x+7}{x(x-1)^2}$ *

26. $\dfrac{3x^2+49}{x(x+7)^2}$ *

27. $\dfrac{x^2}{(x-1)^2(x+1)}$ *

28. $\dfrac{x^2}{(x-1)^2(x+1)^2}$ *

29. $\dfrac{5x^2-6x+7}{(x-1)(x^2+1)}$ *

30. $\dfrac{5x^2-9x+19}{(x-4)(x^2+5)}$ *

31. $\dfrac{5x^2+6x+3}{(x+1)(x^2+2x+2)}$ *

32. $\dfrac{9x+2}{(x-2)(x^2+2x+2)}$ *

33. $\dfrac{x+4}{x^2(x^2+4)}$ *

34. $\dfrac{10x^2+2x}{(x-1)^2(x^2+2)}$ *

35. $\dfrac{6x^2-x+1}{x^3+x^2+x+1}$ *

36. $\dfrac{3x^2-2x+8}{x^3+2x^2+4x+8}$ *

*See Answers to Selected Exercises.

37. $\dfrac{x^3 + x^2 + 2}{(x^2 + 2)^2}$ *

38. $\dfrac{x^2 + 2x + 3}{(x^2 + 4)^2}$ *

39. $\dfrac{x^3 - 4x^2 + 9x - 5}{(x^2 - 2x + 3)^2}$ *

40. $\dfrac{3x^3 - 6x^2 + 7x - 2}{(x^2 - 2x + 2)^2}$ *

41. $\dfrac{4x^2 + 3x + 14}{x^3 - 8}$ *

42. $\dfrac{3x - 5}{x^3 - 1}$ *

Practice Plus

In Exercises 43–46, perform each long division and write the partial fraction decomposition of the remainder term.

43. $\dfrac{x^5 + 2}{x^2 - 1}$ *

44. $\dfrac{x^5}{x^2 - 4x + 4}$ *

45. $\dfrac{x^4 - x^2 + 2}{x^3 - x^2}$ *

46. $\dfrac{x^4 + 2x^3 - 4x^2 + x - 3}{x^2 - x - 2}$ *

In Exercises 47–50, write the partial fraction decomposition of each rational expression.

47. $\dfrac{1}{x^2 - c^2}$ $(c \neq 0)$ *

48. $\dfrac{ax + b}{x^2 - c^2}$ $(c \neq 0)$ *

49. $\dfrac{ax + b}{(x - c)^2}$ $(c \neq 0)$ $\dfrac{a}{x - c} + \dfrac{ac + b}{(x - c)^2}$

50. $\dfrac{1}{x^2 - ax - bx + ab}$ $(a \neq b)$ $\dfrac{\frac{1}{a - b}}{x - a} + \dfrac{\frac{1}{b - a}}{x - b}$

Application Exercises

51. Find the partial fraction decomposition for $\dfrac{1}{x(x + 1)}$ and use the result to find the following sum:

$$\dfrac{1}{1 \cdot 2} + \dfrac{1}{2 \cdot 3} + \dfrac{1}{3 \cdot 4} + \cdots + \dfrac{1}{99 \cdot 100}.$$ $\dfrac{1}{x} - \dfrac{1}{x + 1}; \dfrac{99}{100}$

52. Find the partial fraction decomposition for $\dfrac{2}{x(x + 2)}$ and use the result to find the following sum:

$$\dfrac{2}{1 \cdot 3} + \dfrac{2}{3 \cdot 5} + \dfrac{2}{5 \cdot 7} + \cdots + \dfrac{2}{99 \cdot 101}.$$ $\dfrac{1}{x} - \dfrac{1}{x + 2}; \dfrac{100}{101}$

Explaining the Concepts 53–58. Answers will vary.

53. Explain what is meant by the partial fraction decomposition of a rational expression.

54. Explain how to find the partial fraction decomposition of a rational expression with distinct linear factors in the denominator.

55. Explain how to find the partial fraction decomposition of a rational expression with a repeated linear factor in the denominator.

56. Explain how to find the partial fraction decomposition of a rational expression with a prime quadratic factor in the denominator.

57. Explain how to find the partial fraction decomposition of a rational expression with a repeated, prime quadratic factor in the denominator.

58. How can you verify your result for the partial fraction decomposition for a given rational expression without using a graphing utility?

Technology Exercise

59. Use the ⎡TABLE⎤ feature of a graphing utility to verify any three of the decompositions that you obtained in Exercises 9–42. Answers will vary.

Critical Thinking Exercises

Make Sense? *In Exercises 60–63, determine whether each statement makes sense or does not make sense, and explain your reasoning.*

60. Partial fraction decomposition involves finding a single rational expression for a given sum or difference of rational expressions. does not make sense

61. I apply partial fraction decompositions for rational expressions of the form $\dfrac{P(x)}{Q(x)}$, where P and Q have no common factors and the degree of P is greater than the degree of Q. *

62. Because $x + 5$ is linear and $x^2 - 3x + 2$ is quadratic, I set up the following partial fraction decomposition:

$$\dfrac{7x^2 + 9x + 3}{(x + 5)(x^2 - 3x + 2)} = \dfrac{A}{x + 5} + \dfrac{Bx + C}{x^2 - 3x + 2}.$$ *

63. Because $(x + 3)^2$ consists of two factors of $x + 3$, I set up the following partial fraction decomposition: does not make sense

$$\dfrac{5x + 2}{(x + 3)^2} = \dfrac{A}{x + 3} + \dfrac{B}{x + 3}.$$

64. Use an extension of the Great Question! on page 577 to describe how to set up the partial fraction decomposition of a rational expression that contains powers of a prime cubic factor in the denominator. Give an example of such a decomposition. Answers will vary.

65. Find the partial fraction decomposition of

$$\dfrac{4x^2 + 5x - 9}{x^3 - 6x - 9}.$$ $\dfrac{2}{x - 3} + \dfrac{2x + 5}{x^2 + 3x + 3}$

Retaining the Concepts

66. Perform the operations and write the result in standard form:

$$\dfrac{-20 + \sqrt{-32}}{10}.$$ $-2 + i\dfrac{2\sqrt{2}}{5}$

(Section 1.4, Example 5)

67. Does $x^2 + y = 10$ define y as a function of x? (Section 2.1, Example 3) yes

68. Given $f(x) = 6x + 5$ and $g(x) = x^2 - 3x + 2$, find each of the following:

a. $(f \circ g)(x)$ * **b.** $(g \circ f)(x)$ * **c.** $(f \circ g)(-1)$. 41

(Section 2.6, Example 5)

Preview Exercises

Exercises 69–71 will help you prepare for the material covered in the next section.

69. Solve by the substitution method:

$$\begin{cases} 4x + 3y = 4 \\ y = 2x - 7. \end{cases}$$ $\{(2.5, -2)\}$

70. Solve by the addition method:

$$\begin{cases} 2x + 4y = -4 \\ 3x + 5y = -3. \end{cases}$$ $\{(4, -3)\}$

71. Graph $x - y = 3$ and $(x - 2)^2 + (y + 3)^2 = 4$ in the same rectangular coordinate system. What are the two intersection points? Show that each of these ordered pairs satisfies both equations. intersection points: $(0, -3)$, $(2, -1)$*

Section 5.4 · Systems of Nonlinear Equations in Two Variables

What am I supposed to learn?

After studying this section, you should be able to:

1. Recognize systems of nonlinear equations in two variables.
2. Solve nonlinear systems by substitution.
3. Solve nonlinear systems by addition.
4. Solve problems using systems of nonlinear equations.

Scientists debate the probability that a "doomsday rock" will collide with Earth. It has been estimated that an asteroid, a tiny planet that revolves around the Sun, crashes into Earth about once every 250,000 years, and that such a collision would have disastrous results. In 1908, a small fragment struck Siberia, leveling thousands of acres of trees. One theory about the extinction of dinosaurs 65 million years ago involves Earth's collision with a large asteroid and the resulting drastic changes in Earth's climate.

Understanding the path of Earth and the path of a comet is essential to detecting threatening space debris. Orbits about the Sun are not described by linear equations in the form $Ax + By = C$. The ability to solve systems that contain nonlinear equations provides NASA scientists watching for troublesome asteroids with a way to locate possible collision points with Earth's orbit.

① Recognize systems of nonlinear equations in two variables.

Systems of Nonlinear Equations and Their Solutions

A **system of** two **nonlinear equations** in two variables, also called a **nonlinear system**, contains at least one equation that cannot be expressed in the form $Ax + By = C$. Here are two examples:

$$\begin{cases} x^2 = 2y + 10 \\ 3x - y = 9 \end{cases}$$
Not in the form $Ax + By = C$. The term x^2 is not linear.

$$\begin{cases} y = x^2 + 3 \\ x^2 + y^2 = 9. \end{cases}$$
Neither equation is in the form $Ax + By = C$. The terms x^2 and y^2 are not linear.

A **solution** of a nonlinear system in two variables is an ordered pair of real numbers that satisfies both equations in the system. The **solution set** of the system is the set of all such ordered pairs. As with linear systems in two variables, the solution of a nonlinear system (if there is one) corresponds to the intersection point(s) of the graphs of the equations in the system. Unlike linear systems, the graphs can be circles, parabolas, or anything other than two lines. We will solve nonlinear systems using the substitution method and the addition method.

② Solve nonlinear systems by substitution.

Eliminating a Variable Using the Substitution Method

The substitution method involves converting a nonlinear system into one equation in one variable by an appropriate substitution. The steps in the solution process are exactly the same as those used to solve a linear system by substitution. However, when you obtain an equation in one variable, this equation may not be linear. In our first example, this equation is quadratic.

EXAMPLE 1 Solving a Nonlinear System by the Substitution Method

Solve by the substitution method:

$$\begin{cases} x^2 = 2y + 10 & \text{(The graph is a parabola.)} \\ 3x - y = 9. & \text{(The graph is a line.)} \end{cases}$$

SOLUTION

Step 1 Solve one of the equations for one variable in terms of the other. We begin by isolating one of the variables raised to the first power in either of the equations. By solving for y in the second equation, which has a coefficient of -1, we can avoid fractions.

$$3x - y = 9 \qquad \text{This is the second equation in the given system.}$$
$$3x = y + 9 \qquad \text{Add } y \text{ to both sides.}$$
$$3x - 9 = y \qquad \text{Subtract 9 from both sides.}$$

Step 2 Substitute the expression from step 1 into the other equation. We substitute $3x - 9$ for y in the first equation.

$$y = \boxed{3x - 9} \qquad x^2 = 2\,\boxed{y} + 10$$

This gives us an equation in one variable, namely,

$$x^2 = 2(3x - 9) + 10.$$

The variable y has been eliminated.

Step 3 Solve the resulting equation containing one variable.

$$x^2 = 2(3x - 9) + 10 \qquad \text{This is the equation containing one variable.}$$
$$x^2 = 6x - 18 + 10 \qquad \text{Use the distributive property.}$$
$$x^2 = 6x - 8 \qquad \text{Combine numerical terms on the right.}$$
$$x^2 - 6x + 8 = 0 \qquad \begin{array}{l}\text{Move all terms to one side and set the} \\ \text{quadratic equation equal to O.}\end{array}$$
$$(x - 4)(x - 2) = 0 \qquad \text{Factor.}$$
$$x - 4 = 0 \quad \text{or} \quad x - 2 = 0 \qquad \text{Set each factor equal to O.}$$
$$x = 4 \qquad\qquad x = 2 \qquad \text{Solve for } x.$$

Step 4 Back-substitute the obtained values into the equation from step 1. Now that we have the x-coordinates of the solutions, we back-substitute 4 for x and 2 for x into the equation $y = 3x - 9$.

If x is 4, $y = 3(4) - 9 = 3$, so $(4, 3)$ is a solution.
If x is 2, $y = 3(2) - 9 = -3$, so $(2, -3)$ is a solution.

Step 5 Check the proposed solutions in both of the system's given equations. We begin by checking $(4, 3)$. Replace x with 4 and y with 3.

$x^2 = 2y + 10$	$3x - y = 9$	These are the given equations.
$4^2 \overset{?}{=} 2(3) + 10$	$3(4) - 3 \overset{?}{=} 9$	Let $x = 4$ and $y = 3$.
$16 \overset{?}{=} 6 + 10$	$12 - 3 \overset{?}{=} 9$	Simplify.
$16 = 16$, true	$9 = 9$, true	True statements result.

The ordered pair $(4, 3)$ satisfies both equations. Thus, $(4, 3)$ is a solution of the system. Now let's check $(2, -3)$. Replace x with 2 and y with -3 in both given equations.

$x^2 = 2y + 10$	$3x - y = 9$	These are the given equations.
$2^2 \overset{?}{=} 2(-3) + 10$	$3(2) - (-3) \overset{?}{=} 9$	Let $x = 2$ and $y = -3$.
$4 \overset{?}{=} -6 + 10$	$6 + 3 \overset{?}{=} 9$	Simplify.
$4 = 4$, true	$9 = 9$, true	True statements result.

The ordered pair $(2, -3)$ also satisfies both equations and is a solution of the system. The solutions are $(4, 3)$ and $(2, -3)$, and the solution set is $\{(4, 3), (2, -3)\}$.

Figure 5.10 shows the graphs of the equations in the system and the solutions as intersection points. • • •

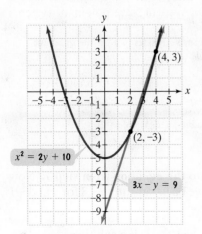

FIGURE 5.10 Points of intersection illustrate the nonlinear system's solutions.

✓ Check Point **1** Solve by the substitution method:

$$\begin{cases} x^2 = y - 1 \\ 4x - y = -1. \end{cases} \quad \{(0, 1), (4, 17)\}$$

A Brief Review •
The Standard Form of the Equation of a Circle
Recall that

$$(x - h)^2 + (y - k)^2 = r^2$$

describes a circle with center (h, k) and radius r. For more detail, see Section 2.8, Objectives 3 and 4.

EXAMPLE 2 Solving a Nonlinear System by the Substitution Method

Solve by the substitution method:

$$\begin{cases} x - y = 3 & \text{(The graph is a line.)} \\ (x - 2)^2 + (y + 3)^2 = 4. & \text{(The graph is a circle.)} \end{cases}$$

SOLUTION

Graphically, we are finding the intersection of a line and a circle with center $(2, -3)$ and radius 2.

Step 1 Solve one of the equations for one variable in terms of the other. We will solve for x in the linear equation—that is, the first equation. (We could also solve for y.)

$$x - y = 3 \qquad \text{This is the first equation in the given system.}$$
$$x = y + 3 \qquad \text{Add } y \text{ to both sides.}$$

Step 2 Substitute the expression from step 1 into the other equation. We substitute $y + 3$ for x in the second equation.

$$x = \boxed{y + 3} \qquad (\boxed{x} - 2)^2 + (y + 3)^2 = 4$$

This gives an equation in one variable, namely,

$$(y + 3 - 2)^2 + (y + 3)^2 = 4.$$

The variable x has been eliminated.

Step 3 Solve the resulting equation containing one variable.

$$(y + 3 - 2)^2 + (y + 3)^2 = 4 \qquad \text{This is the equation containing one variable.}$$
$$(y + 1)^2 + (y + 3)^2 = 4 \qquad \text{Combine numerical terms in the first parentheses.}$$
$$y^2 + 2y + 1 + y^2 + 6y + 9 = 4 \qquad \text{Use the formula } (A + B)^2 = A^2 + 2AB + B^2 \text{ to square } y + 1 \text{ and } y + 3.$$
$$2y^2 + 8y + 10 = 4 \qquad \text{Combine like terms on the left.}$$
$$2y^2 + 8y + 6 = 0 \qquad \text{Subtract 4 from both sides and set the quadratic equation equal to O.}$$
$$2(y^2 + 4y + 3) = 0 \qquad \text{Factor out 2.}$$
$$2(y + 3)(y + 1) = 0 \qquad \text{Factor completely.}$$
$$y + 3 = 0 \quad \text{or} \quad y + 1 = 0 \qquad \text{Set each variable factor equal to O.}$$
$$y = -3 \qquad\qquad y = -1 \quad \text{Solve for } y.$$

Step 4 Back-substitute the obtained values into the equation from step 1. Now that we have the y-coordinates of the solutions, we back-substitute -3 for y and -1 for y in the equation $x = y + 3$.

If $y = -3$: $x = -3 + 3 = 0$, so $(0, -3)$ is a solution.
If $y = -1$: $x = -1 + 3 = 2$, so $(2, -1)$ is a solution.

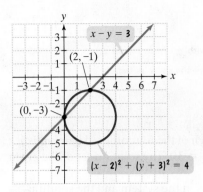

FIGURE 5.11 Points of intersection illustrate the nonlinear system's solutions.

③ Solve nonlinear systems by addition.

Step 5 Check the proposed solutions in both of the system's given equations. Take a moment to show that each ordered pair satisfies both given equations, $x - y = 3$ and $(x - 2)^2 + (y + 3)^2 = 4$. The solutions are $(0, -3)$ and $(2, -1)$, and the solution set of the given system is $\{(0, -3), (2, -1)\}$.

Figure 5.11 shows the graphs of the equations in the system and the solutions as intersection points. ● ● ●

☑ Check Point **2** Solve by the substitution method:

$$\begin{cases} x + 2y = 0 \\ (x - 1)^2 + (y - 1)^2 = 5. \end{cases} \quad \left\{\left(-\frac{6}{5}, \frac{3}{5}\right), (2, -1)\right\}$$

Eliminating a Variable Using the Addition Method

In solving linear systems with two variables, we learned that the addition method works well when each equation is in the form $Ax + By = C$. For nonlinear systems, the addition method can be used when each equation is in the form $Ax^2 + By^2 = C$. If necessary, we will multiply either equation or both equations by appropriate numbers so that the coefficients of x^2 or y^2 will have a sum of 0. We then add equations. The sum will be an equation in one variable.

EXAMPLE 3 Solving a Nonlinear System by the Addition Method

Solve the system:

$$\begin{cases} 4x^2 + y^2 = 13 & \text{Equation 1} \\ x^2 + y^2 = 10. & \text{Equation 2} \end{cases}$$

SOLUTION

We can use the same steps that we did when we solved linear systems by the addition method.

Step 1 Write both equations in the form $Ax^2 + By^2 = C$. Both equations are already in this form, so we can skip this step.

Step 2 If necessary, multiply either equation or both equations by appropriate numbers so that the sum of the x^2-coefficients or the sum of the y^2-coefficients is 0. We can eliminate y^2 by multiplying Equation 2 by -1.

$$\begin{cases} 4x^2 + y^2 = 13 \\ x^2 + y^2 = 10 \end{cases} \xrightarrow[\text{Multiply by } -1.]{\text{No change}} \begin{cases} 4x^2 + y^2 = 13 \\ -x^2 - y^2 = -10 \end{cases}$$

Steps 3 and 4 Add equations and solve for the remaining variable.

$$\begin{cases} 4x^2 + y^2 = 13 \\ -x^2 - y^2 = -10 \end{cases}$$
$$\begin{aligned} 3x^2 \quad\;\; &= \quad 3 \qquad \text{Add equations.} \\ x^2 &= 1 \qquad \text{Divide both sides by 3.} \\ x &= \pm 1 \qquad \text{Use the square root property:} \\ &\qquad\qquad\;\; \text{If } x^2 = c, \text{ then } x = \pm\sqrt{c}. \end{aligned}$$

Step 5 Back-substitute and find the values for the other variable. We must back-substitute each value of x into either one of the original equations. Let's use $x^2 + y^2 = 10$, Equation 2. If $x = 1$,

$$\begin{aligned} 1^2 + y^2 &= 10 \qquad \text{Replace } x \text{ with 1 in Equation 2.} \\ y^2 &= 9 \qquad \text{Subtract 1 from both sides.} \\ y &= \pm 3. \qquad \text{Apply the square root property.} \end{aligned}$$

$(1, 3)$ and $(1, -3)$ are solutions.

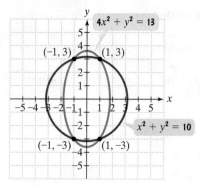

FIGURE 5.12 A system with four solutions

If $x = -1$,

$$(-1)^2 + y^2 = 10 \quad \text{Replace } x \text{ with } -1 \text{ in Equation 2, } x^2 + y^2 = 10.$$
$$y^2 = 9 \quad \text{The steps are the same as before.}$$
$$y = \pm 3.$$

$(-1, 3)$ and $(-1, -3)$ are solutions.

Step 6 Check. Take a moment to show that each of the four ordered pairs satisfies the given equations, $4x^2 + y^2 = 13$ and $x^2 + y^2 = 10$. The solution set of the given system is $\{(1, 3), (1, -3), (-1, 3), (-1, -3)\}$.

Figure 5.12 shows the graphs of the equations in the system and the solutions as intersection points. ● ● ●

⊘ Check Point **3** Solve the system:

$$\begin{cases} 3x^2 + 2y^2 = 35 \\ 4x^2 + 3y^2 = 48. \end{cases} \quad \{(3, 2), (3, -2), (-3, 2), (-3, -2)\}$$

In solving nonlinear systems, we include only ordered pairs with real numbers in the solution set. We have seen that each of these ordered pairs corresponds to a point of intersection of the system's graphs.

GREAT QUESTION!

When solving nonlinear systems, do I really have to go to the hassle of checking my solutions?

Yes. Extra solutions may be introduced that do not satisfy both equations in the system. Therefore, you should get into the habit of checking all proposed pairs in each of the system's two equations.

EXAMPLE 4 Solving a Nonlinear System by the Addition Method

Solve the system:

$$\begin{cases} y = x^2 + 3 & \text{Equation 1 (The graph is a parabola.)} \\ x^2 + y^2 = 9. & \text{Equation 2 (The graph is a circle.)} \end{cases}$$

SOLUTION

We could use substitution because Equation 1, $y = x^2 + 3$, has y expressed in terms of x, but substituting $x^2 + 3$ for y in $x^2 + y^2 = 9$ would result in a fourth-degree equation. However, we can rewrite Equation 1 by subtracting x^2 from both sides and adding the equations to eliminate the x^2-terms.

Notice how like terms are arranged in columns.

$$\begin{cases} -x^2 + y & = 3 & \text{Subtract } x^2 \text{ from both sides of Equation 1.} \\ x^2 \quad + y^2 & = 9 & \text{This is Equation 2.} \end{cases}$$
$$\overline{ y + y^2 = 12} \quad \text{Add the equations.}$$

We now solve this quadratic equation.

$$y + y^2 = 12 \quad \text{This is the equation containing one variable.}$$
$$y^2 + y - 12 = 0 \quad \text{Subtract 12 from both sides and set the quadratic equation equal to 0.}$$
$$(y + 4)(y - 3) = 0 \quad \text{Factor.}$$
$$y + 4 = 0 \quad \text{or} \quad y - 3 = 0 \quad \text{Set each factor equal to 0.}$$
$$y = -4 \qquad \qquad y = 3 \quad \text{Solve for } y.$$

To complete the solution, we must back-substitute each value of y into either one of the original equations. We will use $y = x^2 + 3$, Equation 1. First, we substitute -4 for y.

$$-4 = x^2 + 3$$
$$-7 = x^2 \quad \text{Subtract 3 from both sides.}$$

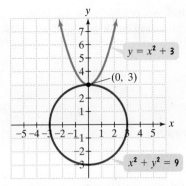

FIGURE 5.13 A system with one real solution

Because the square of a real number cannot be negative, the equation $x^2 = -7$ does not have real-number solutions. We will not include the imaginary solutions, $x = \pm\sqrt{-7}$, or $i\sqrt{7}$ and $-i\sqrt{7}$, in the ordered pairs that make up the solution set. Thus, we move on to our other value for y, 3, and substitute this value into Equation 1.

$$
\begin{aligned}
y &= x^2 + 3 &&\text{This is Equation 1.}\\
3 &= x^2 + 3 &&\text{Back-substitute 3 for } y.\\
0 &= x^2 &&\text{Subtract 3 from both sides.}\\
0 &= x &&\text{Solve for } x.
\end{aligned}
$$

We showed that if $y = 3$, then $x = 0$. Thus, $(0, 3)$ is the solution with a real ordered pair. Take a moment to show that $(0, 3)$ satisfies the given equations, $y = x^2 + 3$ and $x^2 + y^2 = 9$. The solution set of the system is $\{(0, 3)\}$. **Figure 5.13** shows the system's graphs and the solution as an intersection point. ● ● ●

✓ Check Point **4** Solve the system:

$$
\begin{cases}
y = x^2 + 5\\
x^2 + y^2 = 25.
\end{cases}
\qquad \{(0, 5)\}
$$

④ Solve problems using systems of nonlinear equations.

Applications

Many geometric problems can be modeled and solved by the use of systems of nonlinear equations. We will use our step-by-step strategy for solving problems using mathematical models that are created from verbal conditions.

EXAMPLE 5 An Application of a Nonlinear System

You have 36 yards of fencing to build the enclosure in **Figure 5.14**. Some of this fencing is to be used to build an internal divider. If you'd like to enclose 54 square yards, what are the dimensions of the enclosure?

FIGURE 5.14 Building an enclosure

SOLUTION

Step 1 Use variables to represent unknown quantities. Let $x =$ the enclosure's length and $y =$ the enclosure's width. These variables are shown in **Figure 5.14**.

Step 2 Write a system of equations modeling the problem's conditions. The first condition is that you have 36 yards of fencing.

Fencing along both lengths	plus	fencing along both widths	plus	fencing for the internal divider	equals	36 yards.
$2x$	$+$	$2y$	$+$	y	$=$	36

Adding like terms, we can express the equation that models the verbal conditions for the fencing as $2x + 3y = 36$.

The second condition is that you'd like to enclose 54 square yards. The rectangle's area, the product of its length and its width, must be 54 square yards.

Length	times	width	is	54 square yards.
x	\cdot	y	$=$	54

Step 3 Solve the system and answer the problem's question. We must solve the system

$$
\begin{cases}
2x + 3y = 36 &\text{Equation 1}\\
xy = 54. &\text{Equation 2}
\end{cases}
$$

$$\begin{cases} 2x + 3y = 36 & \text{Equation 1} \\ \quad xy = 54 & \text{Equation 2} \end{cases}$$

The system that models the problem's conditions (repeated)

We will use substitution. Because Equation 1 has no coefficients of 1 or -1, we will work with Equation 2 and solve for y. Dividing both sides of $xy = 54$ by x, we obtain

$$y = \frac{54}{x}.$$

Now we substitute $\frac{54}{x}$ for y in Equation 1 and solve for x.

$$2x + 3y = 36 \qquad \text{This is Equation 1.}$$

$$2x + 3 \cdot \frac{54}{x} = 36 \qquad \text{Substitute } \frac{54}{x} \text{ for } y.$$

$$2x + \frac{162}{x} = 36 \qquad \text{Multiply.}$$

$$x\left(2x + \frac{162}{x}\right) = 36 \cdot x \qquad \text{Clear fractions by multiplying both sides by } x.$$

$$2x^2 + 162 = 36x \qquad \text{Use the distributive property on the left side.}$$

$$2x^2 - 36x + 162 = 0 \qquad \text{Subtract } 36x \text{ from both sides and set the quadratic equation equal to O.}$$

$$2(x^2 - 18x + 81) = 0 \qquad \text{Factor out 2.}$$

$$2(x - 9)^2 = 0 \qquad \text{Factor completely using } A^2 - 2AB + B^2 = (A - B)^2.$$

$$x - 9 = 0 \qquad \text{Set the repeated factor equal to zero.}$$

$$x = 9 \qquad \text{Solve for x.}$$

We back-substitute this value of x into $y = \frac{54}{x}$.

$$\text{If } x = 9, \quad y = \frac{54}{9} = 6.$$

This means that the dimensions of the enclosure in **Figure 5.14** are 9 yards by 6 yards.

Step 4 Check the proposed solution in the original wording of the problem. Take a moment to check that a length of 9 yards and a width of 6 yards results in 36 yards of fencing and an area of 54 square yards. •••

FIGURE 5.14 (repeated)

 Check Point **5** Find the length and width of a rectangle whose perimeter is 20 feet and whose area is 21 square feet. length: 7 ft; width: 3 ft or length: 3 ft; width: 7 ft

ACHIEVING SUCCESS

As you continue taking college math courses, remember that you are expected to study and do homework for at least two hours for each hour of class time. In order to make education a top priority, be consistent in the time and location for studying. Arrange your schedule so that you can work in a quiet location without distractions and at a time when you are alert and best able to study. Do this on a regular basis until the routine of studying and doing homework becomes familiar and comfortable.

CONCEPT AND VOCABULARY CHECK

Fill in each blank so that the resulting statement is true.

1. A system of two equations in two variables that contains at least one equation that cannot be expressed in the form $Ax + By = C$ is called a system of ___nonlinear___ equations.

2. When solving

$$\begin{cases} x^2 - 4y = 4 \\ x + y = -1 \end{cases}$$

by the substitution method, we obtain $x = -4$ or $x = 0$, so the solution set is ___$\{(-4, 3), (0, -1)\}$___.

3. When solving

$$\begin{cases} 3x^2 + 2y^2 = 35 \\ 4x^2 + 3y^2 = 48 \end{cases}$$

by the addition method, we can eliminate x^2 by multiplying the first equation by -4 and the second equation by ___3___ and then adding the equations.

4. When solving

$$\begin{cases} x^2 + 4y^2 = 16 \\ x^2 - y^2 = 1 \end{cases}$$

by the addition method, we obtain $y^2 = 3$, so the solution set is ___$\{(2, \sqrt{3}), (2, -\sqrt{3}), (-2, \sqrt{3}), (-2, -\sqrt{3})\}$___.

5. When solving

$$\begin{cases} x^2 + y^2 = 13 \\ x^2 - y = 7 \end{cases}$$

by the addition method, we can eliminate x^2 by multiplying the second equation by ___-1___ and then adding the equations. We obtain ___$y^2 + y = 6$___, a quadratic equation in y.

6. When solving

$$\begin{cases} x^2 + 4y^2 = 20 \\ xy = 4 \end{cases}$$

by the substitution method, we can eliminate y by solving the second equation for y. We obtain $y = $ ___$\frac{4}{x}$___. Then we substitute ___$\frac{4}{x}$___ for ___y___ in the first equation.

EXERCISE SET 5.4

Practice Exercises

In Exercises 1–18, solve each system by the substitution method.

1. $\begin{cases} x + y = 2 \\ y = x^2 - 4 \end{cases}$ $\{(-3, 5), (2, 0)\}$

2. $\begin{cases} x - y = -1 \\ y = x^2 + 1 \end{cases}$ $\{(0, 1), (1, 2)\}$

3. $\begin{cases} x + y = 2 \\ y = x^2 - 4x + 4 \end{cases}$ $\{(1, 1), (2, 0)\}$

4. $\begin{cases} 2x + y = -5 \\ y = x^2 + 6x + 7 \end{cases}$ $\{(-6, 7), (-2, -1)\}$

5. $\begin{cases} y = x^2 - 4x - 10 \\ y = -x^2 - 2x + 14 \end{cases}$ *

6. $\begin{cases} y = x^2 + 4x + 5 \\ y = x^2 + 2x - 1 \end{cases}$ $\{(-3, 2)\}$

7. $\begin{cases} x^2 + y^2 = 25 \\ x - y = 1 \end{cases}$ $\{(4, 3), (-3, -4)\}$

8. $\begin{cases} x^2 + y^2 = 5 \\ 3x - y = 5 \end{cases}$ $\{(2, 1), (1, -2)\}$

9. $\begin{cases} xy = 6 \\ 2x - y = 1 \end{cases}$ $\{(-\frac{3}{2}, -4), (2, 3)\}$

10. $\begin{cases} xy = -12 \\ x - 2y + 14 = 0 \end{cases}$ $\{(-2, 6), (-12, 1)\}$

11. $\begin{cases} y^2 = x^2 - 9 \\ 2y = x - 3 \end{cases}$ $\{(-5, -4), (3, 0)\}$

12. $\begin{cases} x^2 + y = 4 \\ 2x + y = 1 \end{cases}$ $\{(3, -5), (-1, 3)\}$

13. $\begin{cases} xy = 3 \\ x^2 + y^2 = 10 \end{cases}$ *

14. $\begin{cases} xy = 4 \\ x^2 + y^2 = 8 \end{cases}$ $\{(2, 2), (-2, -2)\}$

15. $\begin{cases} x + y = 1 \\ x^2 + xy - y^2 = -5 \end{cases}$ $\{(4, -3), (-1, 2)\}$

16. $\begin{cases} x + y = -3 \\ x^2 + 2y^2 = 12y + 18 \end{cases}$ $\{(-6, 3), (-2, -1)\}$

17. $\begin{cases} x + y = 1 \\ (x - 1)^2 + (y + 2)^2 = 10 \end{cases}$ $\{(0, 1), (4, -3)\}$

18. $\begin{cases} 2x + y = 4 \\ (x + 1)^2 + (y - 2)^2 = 4 \end{cases}$ $\{(\frac{1}{5}, \frac{18}{5}), (1, 2)\}$

*See Answers to Selected Exercises.

In Exercises 19–28, solve each system by the addition method.

19. $\begin{cases} x^2 + y^2 = 13 \\ x^2 - y^2 = 5 \end{cases}$ *

20. $\begin{cases} 4x^2 - y^2 = 4 \\ 4x^2 + y^2 = 4 \end{cases}$ $\{(1, 0), (-1, 0)\}$

21. $\begin{cases} x^2 - 4y^2 = -7 \\ 3x^2 + y^2 = 31 \end{cases}$ *

22. $\begin{cases} 3x^2 - 2y^2 = -5 \\ 2x^2 - y^2 = -2 \end{cases}$ *

23. $\begin{cases} 3x^2 + 4y^2 - 16 = 0 \\ 2x^2 - 3y^2 - 5 = 0 \end{cases}$ *

24. $\begin{cases} 16x^2 - 4y^2 - 72 = 0 \\ x^2 - y^2 - 3 = 0 \end{cases}$ *

25. $\begin{cases} x^2 + y^2 = 25 \\ (x - 8)^2 + y^2 = 41 \end{cases}$ $\{(3, 4), (3, -4)\}$

26. $\begin{cases} x^2 + y^2 = 5 \\ x^2 + (y - 8)^2 = 41 \end{cases}$ *

27. $\begin{cases} y^2 - x = 4 \\ x^2 + y^2 = 4 \end{cases}$ *

28. $\begin{cases} x^2 - 2y = 8 \\ x^2 + y^2 = 16 \end{cases}$ *

In Exercises 29–42, solve each system by the method of your choice.

29. $\begin{cases} 3x^2 + 4y^2 = 16 \\ 2x^2 - 3y^2 = 5 \end{cases}$ *

30. $\begin{cases} x + y^2 = 4 \\ x^2 + y^2 = 16 \end{cases}$ *

31. $\begin{cases} 2x^2 + y^2 = 18 \\ xy = 4 \end{cases}$ *

32. $\begin{cases} x^2 + 4y^2 = 20 \\ xy = 4 \end{cases}$ *

33. $\begin{cases} x^2 + 4y^2 = 20 \\ x + 2y = 6 \end{cases}$ $\{(2, 2), (4, 1)\}$

34. $\begin{cases} 3x^2 - 2y^2 = 1 \\ 4x - y = 3 \end{cases}$ *

35. $\begin{cases} x^3 + y = 0 \\ x^2 - y = 0 \end{cases}$ $\{(0, 0), (-1, 1)\}$

36. $\begin{cases} x^3 + y = 0 \\ 2x^2 - y = 0 \end{cases}$ $\{(0, 0), (-2, 8)\}$

37. $\begin{cases} x^2 + (y - 2)^2 = 4 \\ x^2 - 2y = 0 \end{cases}$ $\{(0, 0), (-2, 2), (2, 2)\}$

38. $\begin{cases} x^2 - y^2 - 4x + 6y - 4 = 0 \\ x^2 + y^2 - 4x - 6y + 12 = 0 \end{cases}$ $\{(2, 2), (2, 4)\}$

39. $\begin{cases} y = (x + 3)^2 \\ x + 2y = -2 \end{cases}$ $\{(-4, 1), (-\frac{5}{2}, \frac{1}{4})\}$

40. $\begin{cases} (x - 1)^2 + (y + 1)^2 = 5 \\ 2x - y = 3 \end{cases}$ $\{(0, -3), (2, 1)\}$

41. $\begin{cases} x^2 + y^2 + 3y = 22 \\ 2x + y = -1 \end{cases}$ $\left\{\left(\frac{12}{5}, -\frac{29}{5}\right), (-2, 3)\right\}$

42. $\begin{cases} x - 3y = -5 \\ x^2 + y^2 - 25 = 0 \end{cases}$ $\{(-5, 0), (4, 3)\}$

In Exercises 43–46, let x represent one number and let y represent the other number. Use the given conditions to write a system of nonlinear equations. Solve the system and find the numbers.

43. The sum of two numbers is 10 and their product is 24. Find the numbers. 4 and 6

44. The sum of two numbers is 20 and their product is 96. Find the numbers. 8 and 12

45. The difference between the squares of two numbers is 3. Twice the square of the first number increased by the square of the second number is 9. Find the numbers. *

46. The difference between the squares of two numbers is 5. Twice the square of the second number subtracted from three times the square of the first number is 19. Find the numbers. *

Practice Plus

In Exercises 47–52, solve each system by the method of your choice.

47. $\begin{cases} 2x^2 + xy = 6 \\ x^2 + 2xy = 0 \end{cases}$ *

48. $\begin{cases} 4x^2 + xy = 30 \\ x^2 + 3xy = -9 \end{cases}$ *

49. $\begin{cases} -4x + y = 12 \\ y = x^3 + 3x^2 \end{cases}$ *

50. $\begin{cases} -9x + y = 45 \\ y = x^3 + 5x^2 \end{cases}$ *

51. $\begin{cases} \dfrac{3}{x^2} + \dfrac{1}{y^2} = 7 \\ \dfrac{5}{x^2} - \dfrac{2}{y^2} = -3 \end{cases}$ *

52. $\begin{cases} \dfrac{2}{x^2} + \dfrac{1}{y^2} = 11 \\ \dfrac{4}{x^2} - \dfrac{2}{y^2} = -14 \end{cases}$ *

In Exercises 53–54, make a rough sketch in a rectangular coordinate system of the graphs representing the equations in each system.

53. The system, whose graphs are a line with positive slope and a parabola whose equation has a positive leading coefficient, has two solutions. *

54. The system, whose graphs are a line with negative slope and a parabola whose equation has a negative leading coefficient, has one solution. *

Application Exercises

55. A planet's orbit follows a path described by $16x^2 + 4y^2 = 64$. A comet follows the parabolic path $y = x^2 - 4$. Where might the comet intersect the orbiting planet? $(0, -4), (-2, 0), (2, 0)$

56. A system for tracking ships indicates that a ship lies on a path described by $2y^2 - x^2 = 1$. The process is repeated and the ship is found to lie on a path described by $2x^2 - y^2 = 1$. If it is known that the ship is located in the first quadrant of the coordinate system, determine its exact location. $(1, 1)$

57. Find the length and width of a rectangle whose perimeter is 36 feet and whose area is 77 square feet. 11 ft and 7 ft

58. Find the length and width of a rectangle whose perimeter is 40 feet and whose area is 96 square feet. 12 ft and 8 ft

Use the formula for the area of a rectangle and the Pythagorean Theorem to solve Exercises 59–60.

59. A small television has a picture with a diagonal measure of 10 inches and a viewing area of 48 square inches. Find the length and width of the screen. width: 6 in.; length: 8 in.

60. The area of a rug is 108 square feet and the length of its diagonal is 15 feet. Find the length and width of the rug. length: 12 ft; width: 9 ft

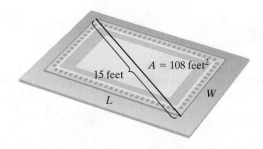

61. The figure shows a square floor plan with a smaller square area that will accommodate a combination fountain and pool. The floor with the fountain-pool area removed has an area of 21 square meters and a perimeter of 24 meters. Find the dimensions of the floor and the dimensions of the square that will accommodate the pool. $x = 5$ m, $y = 2$ m

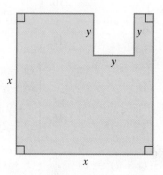

62. The area of the rectangular piece of cardboard shown below is 216 square inches. The cardboard is used to make an open box by cutting a 2-inch square from each corner and turning up the sides. If the box is to have a volume of 224 cubic inches, find the length and width of the cardboard that must be used. $W = 12$ in., $L = 18$ in

63. Between 1990 and 2013, there was a drop in violent crime and a spike in the prison population in the United States. The bar graph shows the number of violent crimes per 100,000 people and the number of imprisonments per 100,000 people for six selected years from 1990 through 2013.

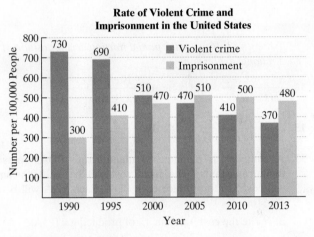

Rate of Violent Crime and Imprisonment in the United States

Sources: U.S. Justice Department; FBI

a. Based on the information in the graph, it appears that there was a year when the number of violent crimes per 100,000 Americans was the same as the number of imprisonments per 100,000 Americans. According to the graph, between which two years did this occur? between 2000 and 2005

b. The data can be modeled by quadratic and linear functions.

Violent crime rate $y = 0.6x^2 - 28x + 730$

Imprisonment rate $-15x + y = 300$

In each function, x represents the number of years after 1990 and y represents the number per 100,000 Americans. Solve a nonlinear system to determine the year described in part (a). Round to the nearest year. How many violent crimes per 100,000 Americans and how many imprisonments per 100,000 Americans were there in that year? 2002; 480 per 100,000 Americans; 480 per 100,000 Americans

Explaining the Concepts 64–66. Answers will vary.

64. What is a system of nonlinear equations? Provide an example with your description.

65. Explain how to solve a nonlinear system using the substitution method. Use $x^2 + y^2 = 9$ and $2x - y = 3$ to illustrate your explanation.

66. Explain how to solve a nonlinear system using the addition method. Use $x^2 - y^2 = 5$ and $3x^2 - 2y^2 = 19$ to illustrate your explanation.

Technology Exercises 67–68. Answers will vary.

67. Verify your solutions to any five exercises from Exercises 1–42 by using a graphing utility to graph the two equations in the system in the same viewing rectangle. Then use the intersection feature to verify the solutions.

68. Write a system of equations, one equation whose graph is a line and the other whose graph is a parabola, that has no ordered pairs that are real numbers in its solution set. Graph the equations using a graphing utility and verify that you are correct.

Critical Thinking Exercises

Make Sense? *In Exercises 69–72, determine whether each statement makes sense or does not make sense, and explain your reasoning.*

69. I use the same steps to solve nonlinear systems as I did to solve linear systems, although I don't obtain linear equations when a variable is eliminated. makes sense

70. I graphed a nonlinear system that modeled the orbits of Earth and Mars, and the graphs indicated the system had a solution with a real ordered pair. does not make sense

71. Without using any algebra, it's obvious that the nonlinear system consisting of $x^2 + y^2 = 4$ and $x^2 + y^2 = 25$ does not have real-number solutions. makes sense

72. I think that the nonlinear system consisting of $x^2 + y^2 = 36$ and $y = (x - 2)^2 - 3$ is easier to solve graphically than by using the substitution method or the addition method. *

In Exercises 73–76, determine whether each statement is true or false. If the statement is false, make the necessary change(s) to produce a true statement. Changes to false statements will vary.

73. A system of two equations in two variables whose graphs are a circle and a line can have four real ordered-pair solutions. false

74. A system of two equations in two variables whose graphs are a parabola and a circle can have four real ordered-pair solutions. true

75. A system of two equations in two variables whose graphs are two circles must have at least two real ordered-pair solutions. false

76. A system of two equations in two variables whose graphs are a parabola and a circle cannot have only one real ordered-pair solution. false

77. The points of intersection of the graphs of $xy = 20$ and $x^2 + y^2 = 41$ are joined to form a rectangle. Find the area of the rectangle. 18 sq units

78. Find a and b in this figure. $b = 6, a = 8$

Solve the systems in Exercises 79–80.

79. $\begin{cases} \log_y x = 3 \\ \log_y (4x) = 5 \end{cases}$ $\{(8, 2)\}$

80. $\begin{cases} \log x^2 = y + 3 \\ \log x = y - 1 \end{cases}$ $\{(10,000, 5)\}$

Retaining the Concepts

81. Solve: $x^4 + 2x^3 - x^2 - 4x - 2 = 0$.
(Section 3.4, Example 5) $\{-\sqrt{2}, -1, \sqrt{2}\}$

82. Expand: $\log_8\left(\dfrac{\sqrt[4]{x}}{64y^3}\right)$. (Section 4.3, Example 4) *

83. Use the exponential growth model, $A = A_0 e^{kt}$, to solve this exercise. In 1975, the population of Europe was 679 million. By 2015, the population had grown to 746 million.

a. Find an exponential growth function that models the data for 1975 through 2015. $A = 679e^{0.00235t}$

b. By which year, to the nearest year, will the European population reach 800 million? by 2045

(Section 4.5, Example 1)

Preview Exercises

Exercises 84–86 will help you prepare for the material covered in the next section. In each exercise, graph the linear function.

84. $2x - 3y = 6$ * **85.** $f(x) = -\tfrac{2}{3}x$ * **86.** $f(x) = -2$ *

CHAPTER 5 Mid-Chapter Check Point

WHAT YOU KNOW: We learned to solve systems of equations. We solved linear and nonlinear systems in two variables by the substitution method and by the addition method. We solved linear systems in three variables by eliminating a variable, reducing the system to two equations in two variables. We saw that some linear systems, called inconsistent systems, have no solution, whereas other linear systems, called dependent systems, have infinitely many solutions. We applied systems to a variety of situations, including finding the break-even point for a business, solving problems involving mixtures, finding a quadratic function from three points on its graph, and finding a rational function's partial fraction decomposition.

In Exercises 1–12, solve each system by the method of your choice.

1. $\begin{cases} x = 3y - 7 \\ 4x + 3y = 2 \end{cases}$ $\{(-1, 2)\}$

2. $\begin{cases} 3x + 4y = -5 \\ 2x - 3y = 8 \end{cases}$ $\{(1, -2)\}$

3. $\begin{cases} \dfrac{2x}{3} + \dfrac{y}{5} = 6 \\ \dfrac{x}{6} - \dfrac{y}{2} = -4 \end{cases}$ $\{(6, 10)\}$

4. $\begin{cases} y = 4x - 5 \\ 8x - 2y = 10 \end{cases}$ *

5. $\begin{cases} 2x + 5y = 3 \\ 3x - 2y = 1 \end{cases}$ $\{(\tfrac{11}{19}, \tfrac{7}{19})\}$

6. $\begin{cases} \dfrac{x}{12} - y = \dfrac{1}{4} \\ 4x - 48y = 16 \end{cases}$ \varnothing

7. $\begin{cases} 2x - y + 2z = -8 \\ x + 2y - 3z = 9 \\ 3x - y - 4z = 3 \end{cases}$ *

8. $\begin{cases} x - 3z = -5 \\ 2x - y + 2z = 16 \\ 7x - 3y - 5z = 19 \end{cases}$ *

9. $\begin{cases} x^2 + y^2 = 9 \\ x + 2y - 3 = 0 \end{cases}$ *

10. $\begin{cases} 3x^2 + 2y^2 = 14 \\ 2x^2 - y^2 = 7 \end{cases}$ *

11. $\begin{cases} y = x^2 - 6 \\ x^2 + y^2 = 8 \end{cases}$ *

12. $\begin{cases} x - 2y = 4 \\ 2y^2 + xy = 8 \end{cases}$ $\{(0, -2), (6, 1)\}$

In Exercises 13–16, write the partial fraction decomposition of each rational expression.

13. $\dfrac{x^2 - 6x + 3}{(x - 2)^3}$ *

14. $\dfrac{10x^2 + 9x - 7}{(x + 2)(x^2 - 1)}$ *

15. $\dfrac{x^2 + 4x - 23}{(x + 3)(x^2 + 4)}$ *

16. $\dfrac{x^3}{(x^2 + 4)^2}$ *

17. A company is planning to manufacture PDAs (personal digital assistants). The fixed cost will be $400,000 and it will cost $20 to produce each PDA. Each PDA will be sold for $100.

a. Write the cost function, C, of producing x PDAs. *

b. Write the revenue function, R, from the sale of x PDAs. *

c. Write the profit function, P, from producing and selling x PDAs. $P(x) = 80x - 400,000$

d. Determine the break-even point. Describe what this means. $(5000, 500,000)$*

18. The manager of a gardening center needs to mix a plant food that is 13% nitrogen with one that is 18% nitrogen to obtain 50 gallons of a plant food that is 16% nitrogen. How many gallons of each of the plant foods must be used? 20 gallons of 13% nitrogen and 30 gallons of 18% nitrogen

19. Find the measure of each angle whose degree measure is represented with a variable. $x = 55°, y = 35°$

20. Find the quadratic function $y = ax^2 + bx + c$ whose graph passes through the points $(-1, 0)$, $(1, 4)$, and $(2, 3)$. $y = -x^2 + 2x + 3$

21. Find the length and width of a rectangle whose perimeter is 21 meters and whose area is 20 square meters. *

Section 5.5

Systems of Inequalities

We opened the chapter noting that the modern emphasis on thinness as the ideal body shape has been suggested as a major cause of eating disorders. In this section (Example 5), as well as in the Exercise Set (Exercises 77–80), we use systems of linear inequalities in two variables that will enable you to establish a healthy weight range for your height and age.

What am I supposed to learn?

After studying this section, you should be able to:

1. Graph a linear inequality in two variables.
2. Graph a nonlinear inequality in two variables.
3. Use mathematical models involving linear inequalities.
4. Graph a system of inequalities.

Linear Inequalities in Two Variables and Their Solutions

We have seen that equations in the form $Ax + By = C$ are straight lines when graphed. If we change the symbol $=$ to $>$, $<$, \geq, or \leq, we obtain a **linear inequality in two variables**. Some examples of linear inequalities in two variables are $x + y > 2$, $3x - 5y \leq 15$, and $2x - y < 4$.

A **solution of an inequality in two variables**, x and y, is an ordered pair of real numbers with the following property: When the x-coordinate is substituted for x and the y-coordinate is substituted for y in the inequality, we obtain a true statement. For example, $(3, 2)$ is a solution of the inequality $x + y > 1$. When 3 is substituted for x and 2 is substituted for y, we obtain the true statement $3 + 2 > 1$, or $5 > 1$. Because there are infinitely many pairs of numbers that have a sum greater than 1, the inequality $x + y > 1$ has infinitely many solutions. Each ordered-pair solution is said to **satisfy** the inequality. Thus, $(3, 2)$ satisfies the inequality $x + y > 1$.

1. Graph a linear inequality in two variables.

The Graph of a Linear Inequality in Two Variables

We know that the graph of an equation in two variables is the set of all points whose coordinates satisfy the equation. Similarly, the **graph of an inequality in two variables** is the set of all points whose coordinates satisfy the inequality.

Let's use **Figure 5.15** to get an idea of what the graph of a linear inequality in two variables looks like. Part of the figure shows the graph of the linear equation $x + y = 2$. The line divides the points in the rectangular coordinate system into three sets. First, there is the set of points along the line, satisfying $x + y = 2$. Next, there is the set of points in the green region above the line. Points in the green region satisfy the linear inequality $x + y > 2$. Finally, there is the set of points in the purple region below the line. Points in the purple region satisfy the linear inequality $x + y < 2$.

A **half-plane** is the set of all the points on one side of a line. In **Figure 5.15**, the green region is a half-plane. The purple region is also a half-plane. A half-plane is the graph of a linear inequality that involves $>$ or $<$. The graph of a linear inequality that involves \geq or \leq is a half-plane and a line. A solid line is used to show that a line is part of a solution set. A dashed line is used to show that a line is not part of a solution set.

Half-plane
$x + y > 2$

Half-plane
$x + y < 2$

Line $x + y = 2$

FIGURE 5.15

Graphing a Linear Inequality in Two Variables

1. Replace the inequality symbol with an equal sign and graph the corresponding linear equation. Draw a solid line if the original inequality contains a \leq or \geq symbol. Draw a dashed line if the original inequality contains a $<$ or $>$ symbol.
2. Choose a test point from one of the half-planes. (Do not choose a point on the line.) Substitute the coordinates of the test point into the inequality.
3. If a true statement results, shade the half-plane containing this test point. If a false statement results, shade the half-plane not containing this test point.

EXAMPLE 1 Graphing a Linear Inequality in Two Variables

Graph: $2x - 3y \geq 6$.

SOLUTION

Step 1 Replace the inequality symbol by = and graph the linear equation. We need to graph $2x - 3y = 6$. We can use intercepts to graph this line.

We set $y = 0$ to find the x-intercept.	We set $x = 0$ to find the y-intercept.
$2x - 3y = 6$	$2x - 3y = 6$
$2x - 3 \cdot 0 = 6$	$2 \cdot 0 - 3y = 6$
$2x = 6$	$-3y = 6$
$x = 3$	$y = -2$

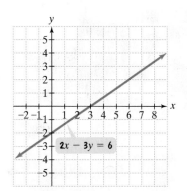

FIGURE 5.16 Preparing to graph $2x - 3y \geq 6$

The x-intercept is 3, so the line passes through $(3, 0)$. The y-intercept is -2, so the line passes through $(0, -2)$. Using the intercepts, the line is shown in **Figure 5.16** as a solid line. This is because the inequality $2x - 3y \geq 6$ contains a \geq symbol, in which equality is included.

Step 2 Choose a test point from one of the half-planes and not from the line. Substitute its coordinates into the inequality. The line $2x - 3y = 6$ divides the plane into three parts—the line itself and two half-planes. The points in one half-plane satisfy $2x - 3y > 6$. The points in the other half-plane satisfy $2x - 3y < 6$. We need to find which half-plane belongs to the solution of $2x - 3y \geq 6$. To do so, we test a point from either half-plane. The origin, $(0, 0)$, is the easiest point to test.

$2x - 3y \geq 6$ This is the given inequality.

$2 \cdot 0 - 3 \cdot 0 \overset{?}{\geq} 6$ Test (O, O) by substituting O for x and O for y.

$0 - 0 \overset{?}{\geq} 6$ Multiply.

$0 \geq 6$ This statement is false.

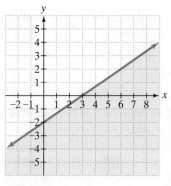

FIGURE 5.17 The graph of $2x - 3y \geq 6$

Step 3 If a false statement results, shade the half-plane not containing the test point. Because 0 is not greater than or equal to 6, the test point, $(0, 0)$, is not part of the solution set. Thus, the half-plane below the solid line $2x - 3y = 6$ is part of the solution set. The solution set is the line and the half-plane that does not contain the point $(0, 0)$, indicated by shading this half-plane. The graph is shown using green shading and a blue line in **Figure 5.17**. ● ● ●

✓ Check Point 1 Graph: $4x - 2y \geq 8$. *

When graphing a linear inequality, choose a test point that lies in one of the half-planes and *not on the line dividing the half-planes.* The test point $(0, 0)$ is convenient because it is easy to calculate when 0 is substituted for each variable. However, if $(0, 0)$ lies on the dividing line and not in a half-plane, a different test point must be selected.

*See Answers to Selected Exercises.

EXAMPLE 2 Graphing a Linear Inequality in Two Variables

Graph: $y > -\dfrac{2}{3}x$.

SOLUTION

Step 1 Replace the inequality symbol by = and graph the linear equation. Because we are interested in graphing $y > -\frac{2}{3}x$, we begin by graphing $y = -\frac{2}{3}x$. We can use the slope and the y-intercept to graph this linear function.

The y-intercept is 0, so the line passes through $(0, 0)$. Using the y-intercept and the slope, the line is shown in **Figure 5.18** as a dashed line. This is because the inequality $y > -\frac{2}{3}x$ contains a > symbol, in which equality is not included.

Step 2 Choose a test point from one of the half-planes and not from the line. Substitute its coordinates into the inequality. We cannot use $(0, 0)$ as a test point because it lies on the line and not in a half-plane. Let's use $(1, 1)$, which lies in the half-plane above the line.

$$y > -\frac{2}{3}x \qquad \text{This is the given inequality.}$$

$$1 \overset{?}{>} -\frac{2}{3}\cdot 1 \qquad \text{Test (1, 1) by substituting 1 for } x \text{ and 1 for } y.$$

$$1 > -\frac{2}{3} \qquad \text{This statement is true.}$$

Step 3 If a true statement results, shade the half-plane containing the test point. Because 1 is greater than $-\frac{2}{3}$, the test point $(1, 1)$ is part of the solution set. All the points on the same side of the line $y = -\frac{2}{3}x$ as the point $(1, 1)$ are members of the solution set. The solution set is the half-plane that contains the point $(1, 1)$, indicated by shading this half-plane. The graph is shown using green shading and a dashed blue line in **Figure 5.18**. ● ● ●

TECHNOLOGY

Most graphing utilities can graph inequalities in two variables. The procedure varies by model, so consult your manual. For most graphing utilities, you must first solve for y if it is not already isolated. The figure shows the graph of $y > -\frac{2}{3}x$. Most displays do not distinguish between dashed and solid boundary lines.

✓ **Check Point 2** Graph: $y > -\dfrac{3}{4}x$. *

Graphing Linear Inequalities without Using Test Points

You can graph inequalities in the form $y > mx + b$ or $y < mx + b$ without using test points. The inequality symbol indicates which half-plane to shade.

- If $y > mx + b$, shade the half-plane above the line $y = mx + b$.
- If $y < mx + b$, shade the half-plane below the line $y = mx + b$.

Observe how this is illustrated in **Figure 5.18**. The graph of $y > -\frac{2}{3}x$ is the half-plane above the line $y = -\frac{2}{3}x$.

It is also not necessary to use test points when graphing inequalities involving half-planes on one side of a vertical or a horizontal line.

*See Answers to Selected Exercises.

FIGURE 5.18 The graph of $y > -\frac{2}{3}x$

GREAT QUESTION!

When is it important to use test points to graph linear inequalities?

Continue using test points to graph inequalities in the form $Ax + By > C$ or $Ax + By < C$. The graph of $Ax + By > C$ can lie above or below the line given by $Ax + By = C$, depending on the values of A and B. The same comment applies to the graph of $Ax + By < C$.

For the Vertical Line $x = a$:	For the Horizontal Line $y = b$:
• If $x > a$, shade the half-plane to the right of $x = a$.	• If $y > b$, shade the half-plane above $y = b$.
• If $x < a$, shade the half-plane to the left of $x = a$.	• If $y < b$, shade the half-plane below $y = b$.

EXAMPLE 3 Graphing Inequalities without Using Test Points

Graph each inequality in a rectangular coordinate system:

a. $y \le -3$ **b.** $x > 2$.

SOLUTION

a. $y \le -3$

> Graph $y = -3$, a horizontal line with y-intercept -3. The line is solid because equality is included in $y \le -3$. Because of the less than part of \le, shade the half-plane below the horizontal line.

b. $x > 2$

> Graph $x = 2$, a vertical line with x-intercept **2**. The line is dashed because equality is not included in $x > 2$. Because of $>$, the greater than symbol, shade the half-plane to the right of the vertical line.

• • •

✓ **Check Point 3** Graph each inequality in a rectangular coordinate system:

a. $y > 1$ * **b.** $x \le -2$. *

② Graph a nonlinear inequality in two variables.

Graphing a Nonlinear Inequality in Two Variables

Example 4 illustrates that a nonlinear inequality in two variables is graphed in the same way that we graph a linear inequality.

EXAMPLE 4 Graphing a Nonlinear Inequality in Two Variables

Graph: $x^2 + y^2 \le 9$.

SOLUTION

Step 1 Replace the inequality symbol with = and graph the nonlinear equation. We need to graph $x^2 + y^2 = 9$. The graph is a circle of radius 3 with its center at the origin. The graph is shown in **Figure 5.19** as a solid circle because equality is included in the \le symbol.

*See Answers to Selected Exercises.

A Brief Review • Circles Centered at the Origin

The graph of

$$x^2 + y^2 = r^2$$

is a circle with center at the origin and radius r. For more detail, see Section 2.8, Objective 3.

Step 2 Choose a test point from one of the regions and not from the circle. Substitute its coordinates into the inequality. The circle divides the plane into three parts—the circle itself, the region inside the circle, and the region outside the circle. We need to determine whether the region inside or outside the circle is included in the solution. To do so, we will use the test point $(0, 0)$ from inside the circle.

$$x^2 + y^2 \leq 9 \qquad \text{This is the given inequality.}$$
$$0^2 + 0^2 \overset{?}{\leq} 9 \qquad \text{Test (0, 0) by substituting 0 for } x \text{ and 0 for } y.$$
$$0 + 0 \overset{?}{\leq} 9 \qquad \text{Square 0: } 0^2 = 0.$$
$$0 \leq 9 \qquad \text{Add. This statement is true.}$$

Step 3 If a true statement results, shade the region containing the test point. The true statement tells us that all the points inside the circle satisfy $x^2 + y^2 \leq 9$. The graph is shown using green shading and a solid blue circle in **Figure 5.20**.

FIGURE 5.19 Preparing to graph $x^2 + y^2 \leq 9$

FIGURE 5.20 The graph of $x^2 + y^2 \leq 9$

• • •

✓ Check Point **4** Graph: $x^2 + y^2 \geq 16$. *

③ Use mathematical models involving linear inequalities.

Modeling with Systems of Linear Inequalities

Just as two or more linear equations make up a system of linear equations, two or more linear inequalities make up a **system of linear inequalities**. A **solution of a system of linear inequalities** in two variables is an ordered pair that satisfies each inequality in the system.

EXAMPLE 5 Does Your Weight Fit You?

The latest guidelines, which apply to both men and women, give healthy weight ranges, rather than specific weights, for your height. **Figure 5.21** shows the healthy weight region for various heights for people between the ages of 19 and 34, inclusive.

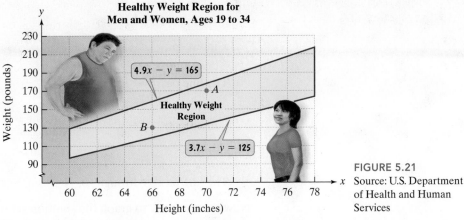

Healthy Weight Region for Men and Women, Ages 19 to 34

4.9x − y = 165

Healthy Weight Region

3.7x − y = 125

A

B

Weight (pounds)

Height (inches)

FIGURE 5.21
Source: U.S. Department of Health and Human Services

*See Answers to Selected Exercises.

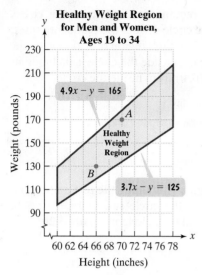

Healthy Weight Region for Men and Women, Ages 19 to 34

FIGURE 5.21 (repeated)

Point $B = (66, 130)$; $4.9(66) - 130 \geq 165$, or $193.4 \geq 165$, is true; $3.7(66) - 130 \leq 125$, or $114.2 \leq 125$, is true.

 4 Graph a system of inequalities.

If x represents height, in inches, and y represents weight, in pounds, the healthy weight region in **Figure 5.21** can be modeled by the following system of linear inequalities:

$$\begin{cases} 4.9x - y \geq 165 \\ 3.7x - y \leq 125. \end{cases}$$

Show that point A in **Figure 5.21** is a solution of the system of inequalities that describes healthy weight.

SOLUTION

Point A has coordinates $(70, 170)$. This means that if a person is 70 inches tall, or 5 feet 10 inches, and weighs 170 pounds, then that person's weight is within the healthy weight region. We can show that $(70, 170)$ satisfies the system of inequalities by substituting 70 for x and 170 for y in each inequality in the system.

$4.9x - y \geq 165$	$3.7x - y \leq 125$
$4.9(70) - 170 \overset{?}{\geq} 165$	$3.7(70) - 170 \overset{?}{\leq} 125$
$343 - 170 \overset{?}{\geq} 165$	$259 - 170 \overset{?}{\leq} 125$
$173 \overset{?}{\geq} 165,$ true	$89 \leq 125,$ true

The coordinates $(70, 170)$ make each inequality true. Thus, $(70, 170)$ satisfies the system for the healthy weight region and is a solution of the system. • • •

✓ Check Point **5** Show that point B in **Figure 5.21** is a solution of the system of inequalities that describes healthy weight.

Graphing Systems of Linear Inequalities

The **solution set of a system of linear inequalities in two variables** is the set of all ordered pairs that satisfy each inequality in the system. Thus, to graph a system of inequalities in two variables, begin by graphing each individual inequality in the same rectangular coordinate system. Then find the region, if there is one, that is common to every graph in the system. This region of intersection gives a picture of the system's solution set.

> **EXAMPLE 6** Graphing a System of Linear Inequalities

Graph the solution set of the system:

$$\begin{cases} x - y < 1 \\ 2x + 3y \geq 12. \end{cases}$$

SOLUTION

Replacing each inequality symbol with an equal sign indicates that we need to graph $x - y = 1$ and $2x + 3y = 12$. We can use intercepts to graph these lines.

$x - y = 1$		$2x + 3y = 12$
x-intercept: $x - 0 = 1$	Set $y = 0$ in each equation.	x-intercept: $2x + 3 \cdot 0 = 12$
$x = 1$		$2x = 12$
The line passes through $(1, 0)$.		$x = 6$
		The line passes through $(6, 0)$.
y-intercept: $0 - y = 1$	Set $x = 0$ in each equation.	y-intercept: $2 \cdot 0 + 3y = 12$
$-y = 1$		$3y = 12$
$y = -1$		$y = 4$
The line passes through $(0, -1)$.		The line passes through $(0, 4)$.

Now we are ready to graph the solution set of the system of linear inequalities.

Graph $x - y < 1$. The blue line, $x - y = 1$, is dashed: Equality is not included in $x - y < 1$. Because $(0, 0)$ makes the inequality true $(0 - 0 < 1$, or $0 < 1$, is true), shade the half-plane containing $(0, 0)$ in yellow.

Add the graph of $2x + 3y \geq 12$. The red line, $2x + 3y = 12$, is solid: Equality is included in $2x + 3y \geq 12$. Because $(0, 0)$ makes the inequality false $(2 \cdot 0 + 3 \cdot 0 \geq 12$, or $0 \geq 12$, is false), shade the half-plane not containing $(0, 0)$ using green vertical shading.

The solution set of the system is graphed as the intersection (the overlap) of the two half-planes. This is the region in which the yellow shading and the green vertical shading overlap.

$x - y = 1$: passes through $(1, 0)$ and $(0, -1)$

$2x + 3y = 12$: passes through $(6, 0)$ and $(0, 4)$

$x - y = 1$

This open dot shows $(3, 2)$ is not in the solution set. It does not satisfy $x - y < 1$.

The graph of $x - y < 1$

Adding the graph of $2x + 3y \geq 12$

The graph of $x - y < 1$ and $2x + 3y \geq 12$

⊘ Check Point **6** Graph the solution set of the system:

$$\begin{cases} x - 3y < 6 \\ 2x + 3y \geq -6. \end{cases}$$

EXAMPLE 7 Graphing a System of Inequalities

Graph the solution set of the system:

$$\begin{cases} y \geq x^2 - 4 \\ x - y \geq 2. \end{cases}$$

SOLUTION

We begin by graphing $y \geq x^2 - 4$. Because equality is included in \geq, we graph $y = x^2 - 4$ as a solid parabola. Because $(0, 0)$ makes the inequality $y \geq x^2 - 4$ true (we obtain $0 \geq -4$), we shade the interior portion of the parabola containing $(0, 0)$, shown in yellow in **Figure 5.22**.

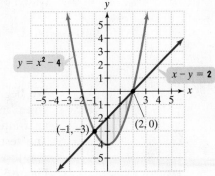

FIGURE 5.22 The graph of $y \geq x^2 - 4$

FIGURE 5.23 Adding the graph of $x - y \geq 2$

FIGURE 5.24 The graph of $y \geq x^2 - 4$ and $x - y \geq 2$

Now we graph $x - y \geq 2$ in the same rectangular coordinate system. First we graph the line $x - y = 2$ using its x-intercept, 2, and its y-intercept, -2. Because $(0, 0)$ makes the inequality $x - y \geq 2$ false (we obtain $0 \geq 2$), we shade the half-plane below the line. This is shown in **Figure 5.23** using green vertical shading.

The solution of the system is shown in **Figure 5.24** by the intersection (the overlap) of the solid yellow and green vertical shadings. The graph of the system's solution set consists of the region enclosed by the parabola and the line. To find the

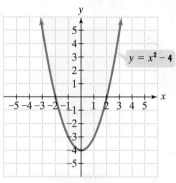

FIGURE 5.22 (repeated) The graph of $y \geq x^2 - 4$

points of intersection of the parabola and the line, use the substitution method to solve the nonlinear system containing $y = x^2 - 4$ and $x - y = 2$. Take a moment to show that the solutions are $(-1, -3)$ and $(2, 0)$, as shown in **Figure 5.24** on the previous page.

• • •

GREAT QUESTION!

In Example 7, how did you graph $y = x^2 - 4$?

We used the parabola's vertex and its x-intercepts.

$$y = 1x^2 + 0x - 4$$

$$a = 1 \quad b = 0 \quad c = -4$$

• $a > 0$: opens upward
• Vertex:

$$\left(-\frac{b}{2a}, f\left(-\frac{b}{2a}\right)\right) = (0, f(0)) = (0, -4)$$

• x-intercepts:

$$x^2 - 4 = 0$$

$$x^2 = 4$$

$$x = \pm 2$$

The graph of $y = x^2 - 4$, shown in blue in **Figure 5.22**, passes through $(-2, 0)$ and $(2, 0)$ with a vertex at $(0, -4)$.

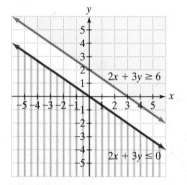

FIGURE 5.25 A system of inequalities with no solution

✓ Check Point **7** Graph the solution set of the system:

$$\begin{cases} y \geq x^2 - 4 \\ x + y \leq 2. \end{cases}$$

A system of inequalities has no solution if there are no points in the rectangular coordinate system that simultaneously satisfy each inequality in the system. For example, the system

$$\begin{cases} 2x + 3y \geq 6 \\ 2x + 3y \leq 0, \end{cases}$$

whose separate graphs are shown in **Figure 5.25**, has no overlapping region. Thus, the system has no solution. The solution set is \varnothing, the empty set.

EXAMPLE 8 Graphing a System of Inequalities

Graph the solution set of the system:

$$\begin{cases} x - y < 2 \\ -2 \leq x < 4 \\ y < 3. \end{cases}$$

SOLUTION

We begin by graphing $x - y < 2$, the first given inequality. The line $x - y = 2$ has an x-intercept of 2 and a y-intercept of -2. The test point $(0, 0)$ makes the inequality $x - y < 2$ true, and its graph is shown in **Figure 5.26**.

Now, let's consider the second given inequality, $-2 \leq x < 4$. Replacing the inequality symbols by $=$, we obtain $x = -2$ and $x = 4$, graphed as red vertical lines in **Figure 5.27**. The line of $x = 4$ is not included. Because x is between -2 and 4, we shade the region between the vertical lines. We must intersect this region with the yellow region in **Figure 5.26**. The resulting region is shown in yellow and green vertical shading in **Figure 5.27**.

FIGURE 5.27 The graph of $x - y < 2$ and $-2 \leq x < 4$

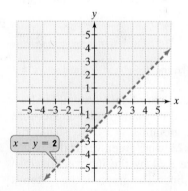

FIGURE 5.26 The graph of $x - y < 2$

*See Answers to Selected Exercises.

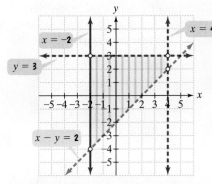

FIGURE 5.28 The graph of $x - y < 2$ and $-2 \le x < 4$ and $y < 3$

Finally, let's consider the third given inequality, $y < 3$. Replacing the inequality symbol by $=$, we obtain $y = 3$, which graphs as a horizontal line. Because of the less than symbol in $y < 3$, the graph consists of the half-plane below the line $y = 3$. We must intersect this half-plane with the region in **Figure 5.27**. The resulting region is shown in yellow and green vertical shading in **Figure 5.28**. This region represents the graph of the solution set of the given system.

In **Figure 5.28** it may be difficult to tell where the graph of $x - y = 2$ intersects the vertical line $x = 4$. Using the substitution method, it can be determined that this intersection point is $(4, 2)$. Take a moment to verify that the four intersection points on the border of the shaded region in **Figure 5.28** are, clockwise from upper left, $(-2, 3)$, $(4, 3)$, $(4, 2)$, and $(-2, -4)$. These points are shown as open dots because none satisfies all three of the system's inequalities. $\bullet\bullet\bullet$

Ⓧ **Check Point 8** Graph the solution set of the system:

$$\begin{cases} x + y < 2 \\ -2 \le x < 1 \\ y > -3. \end{cases}$$

ACHIEVING SUCCESS

Try to relax during a math test. Many of us have had the experience of feeling anxious and panicky during a test. If this happens to you, close your eyes and breathe deeply. Take a short break and think about something pleasant. This may help you to calm down and put the test in perspective.

CONCEPT AND VOCABULARY CHECK

Fill in each blank so that the resulting statement is true.

1. The ordered pair $(3, 2)$ is a/an __solution__ of the inequality $x + y > 1$ because when 3 is substituted for __x__ and 2 is substituted for __y__, the true statement __$5 > 1$__ is obtained.

2. The set of all points that satisfy a linear inequality in two variables is called the __graph__ of the inequality.

3. The set of all points on one side of a line is called a/an __half-plane__.

4. True or false: The graph of $2x - 3y > 6$ includes the line $2x - 3y = 6$. __false__

5. True or false: The graph of the linear equation $2x - 3y = 6$ is used to graph the linear inequality $2x - 3y > 6$. __true__

6. True or false: When graphing $4x - 2y \ge 8$, to determine which side of the line to shade, choose a test point on $4x - 2y = 8$. __false__

7. When graphing $x^2 + y^2 > 25$, to determine whether to shade the region inside the circle or the region outside the circle, we can use __$(0, 0)$__ as a test point. Answers may vary.

8. The solution set of the system
$$\begin{cases} x - y < 1 \\ 2x + 3y \ge 12 \end{cases}$$
is the set of ordered pairs that satisfies __$x - y < 1$__ and __$2x + 3y \ge 12$__.

9. True or false: The graph of the solution set of the system
$$\begin{cases} x - 3y < 6 \\ 2x + 3y \ge -6 \end{cases}$$
includes the intersection point of $x - 3y = 6$ and $2x + 3y = -6$. __false__

EXERCISE SET 5.5

Practice Exercises

In Exercises 1–26, graph each inequality.

1. $x + 2y \le 8$ *

2. $3x - 6y \le 12$ *

3. $x - 2y > 10$ *

4. $2x - y > 4$ *

5. $y \le \dfrac{1}{3}x$ *

6. $y \le \dfrac{1}{4}x$ *

7. $y > 2x - 1$ *

8. $y > 3x + 2$ *

9. $x \le 1$ *

10. $x \le -3$ *

*See Answers to Selected Exercises.

11. $y > 1$ *

12. $y > -3$ *

13. $x^2 + y^2 \leq 1$ *

14. $x^2 + y^2 \leq 4$ *

15. $x^2 + y^2 > 25$ *

16. $x^2 + y^2 > 36$ *

17. $(x - 2)^2 + (y + 1)^2 < 9$ *

18. $(x + 2)^2 + (y - 1)^2 < 16$ *

19. $y < x^2 - 1$ *

20. $y < x^2 - 9$ *

21. $y \geq x^2 - 9$ *

22. $y \geq x^2 - 1$ *

23. $y > 2^x$ *

24. $y \leq 3^x$ *

25. $y \geq \log_2(x + 1)$ *

26. $y \geq \log_3(x - 1)$ *

In Exercises 27–62, graph the solution set of each system of inequalities or indicate that the system has no solution.

27. $\begin{cases} 3x + 6y \leq 6 \\ 2x + y \leq 8 \end{cases}$ *

28. $\begin{cases} x - y \geq 4 \\ x + y \leq 6 \end{cases}$ *

29. $\begin{cases} 2x - 5y \leq 10 \\ 3x - 2y > 6 \end{cases}$ *

30. $\begin{cases} 2x - y \leq 4 \\ 3x + 2y > -6 \end{cases}$ *

31. $\begin{cases} y > 2x - 3 \\ y < -x + 6 \end{cases}$ *

32. $\begin{cases} y < -2x + 4 \\ y < x - 4 \end{cases}$ *

33. $\begin{cases} x + 2y \leq 4 \\ y \geq x - 3 \end{cases}$ *

34. $\begin{cases} x + y \leq 4 \\ y \geq 2x - 4 \end{cases}$ *

35. $\begin{cases} x \leq 2 \\ y \geq -1 \end{cases}$ *

36. $\begin{cases} x \leq 3 \\ y \leq -1 \end{cases}$ *

37. $-2 \leq x < 5$ *

38. $-2 < y \leq 5$ *

39. $\begin{cases} x - y \leq 1 \\ x \geq 2 \end{cases}$ *

40. $\begin{cases} 4x - 5y \geq -20 \\ x \geq -3 \end{cases}$ *

41. $\begin{cases} x + y > 4 \\ x + y < -1 \end{cases}$ \varnothing

42. $\begin{cases} x + y > 3 \\ x + y < -2 \end{cases}$ \varnothing

43. $\begin{cases} x + y > 4 \\ x + y > -1 \end{cases}$ *

44. $\begin{cases} x + y > 3 \\ x + y > -2 \end{cases}$ *

45. $\begin{cases} y \geq x^2 - 1 \\ x - y \geq -1 \end{cases}$ *

46. $\begin{cases} y \geq x^2 - 4 \\ x - y \geq 2 \end{cases}$ *

47. $\begin{cases} x^2 + y^2 \leq 16 \\ x + y > 2 \end{cases}$ *

48. $\begin{cases} x^2 + y^2 \leq 4 \\ x + y > 1 \end{cases}$ *

49. $\begin{cases} x^2 + y^2 > 1 \\ x^2 + y^2 < 16 \end{cases}$ *

50. $\begin{cases} x^2 + y^2 > 1 \\ x^2 + y^2 < 9 \end{cases}$ *

51. $\begin{cases} (x - 1)^2 + (y + 1)^2 < 25 \\ (x - 1)^2 + (y + 1)^2 \geq 16 \end{cases}$ *

52. $\begin{cases} (x + 1)^2 + (y - 1)^2 < 16 \\ (x + 1)^2 + (y - 1)^2 \geq 4 \end{cases}$ *

53. $\begin{cases} x^2 + y^2 \leq 1 \\ y - x^2 > 0 \end{cases}$ *

54. $\begin{cases} x^2 + y^2 < 4 \\ y - x^2 \geq 0 \end{cases}$ *

55. $\begin{cases} x^2 + y^2 < 16 \\ y \geq 2^x \end{cases}$ *

56. $\begin{cases} x^2 + y^2 \leq 16 \\ y < 2^x \end{cases}$ *

57. $\begin{cases} x - y \leq 2 \\ x > -2 \\ y \leq 3 \end{cases}$ *

58. $\begin{cases} 3x + y \leq 6 \\ x > -2 \\ y \leq 4 \end{cases}$ *

59. $\begin{cases} x \geq 0 \\ y \geq 0 \\ 2x + 5y < 10 \\ 3x + 4y \leq 12 \end{cases}$ *

60. $\begin{cases} x \geq 0 \\ y \geq 0 \\ 2x + y < 4 \\ 2x - 3y \leq 6 \end{cases}$ *

61. $\begin{cases} 3x + y \leq 6 \\ 2x - y \leq -1 \\ x > -2 \\ y < 4 \end{cases}$ *

62. $\begin{cases} 2x + y \leq 6 \\ x + y > 2 \\ 1 \leq x \leq 2 \\ y < 3 \end{cases}$ *

Practice Plus

In Exercises 63–64, write each sentence as an inequality in two variables. Then graph the inequality.

63. The y-variable is at least 4 more than the product of -2 and the x-variable. $y \geq -2x + 4$*

64. The y-variable is at least 2 more than the product of -3 and the x-variable. $y \geq -3x + 2$*

In Exercises 65–68, write the given sentences as a system of inequalities in two variables. Then graph the system.

65. The sum of the x-variable and the y-variable is at most 4. The y-variable added to the product of 3 and the x-variable does not exceed 6. $x + y \leq 4; 3x + y \leq 6$*

66. The sum of the x-variable and the y-variable is at most 3. The y-variable added to the product of 4 and the x-variable does not exceed 6. $x + y \leq 3; 4x + y \leq 6$*

67. The sum of the x-variable and the y-variable is no more than 2. The y-variable is no less than the difference between the square of the x-variable and 4. $x + y \leq 2; y \geq x^2 - 4$*

68. The sum of the squares of the x-variable and the y-variable is no more than 25. The sum of twice the y-variable and the x-variable is no less than 5. $x^2 + y^2 \leq 25; 2y + x \geq 5$*

In Exercises 69–70, rewrite each inequality in the system without absolute value bars. Then graph the rewritten system in rectangular coordinates.

69. $\begin{cases} |x| \leq 2 \\ |y| \leq 3 \end{cases}$ $\begin{cases} -2 \leq x \leq 2 \\ -3 \leq y \leq 3 \end{cases}$

70. $\begin{cases} |x| \leq 1 \\ |y| \leq 2 \end{cases}$ $\begin{cases} -1 \leq x \leq 1 \\ -2 \leq y \leq 2 \end{cases}$

*The graphs of solution sets of systems of inequalities involve finding the intersection of the solution sets of two or more inequalities. By contrast, in Exercises 71–72, you will be graphing the **union** of the solution sets of two inequalities.*

71. Graph the union of $y > \frac{3}{2}x - 2$ and $y < 4$. *

72. Graph the union of $x - y \geq -1$ and $5x - 2y \leq 10$. *

Without graphing, in Exercises 73–76, determine if each system has no solution or infinitely many solutions.

73. $\begin{cases} 3x + y < 9 \\ 3x + y > 9 \end{cases}$ no solution

74. $\begin{cases} 6x - y \leq 24 \\ 6x - y > 24 \end{cases}$ no solution

75. $\begin{cases} (x + 4)^2 + (y - 3)^2 \leq 9 \\ (x + 4)^2 + (y - 3)^2 \geq 9 \end{cases}$ infinitely many solutions

76. $\begin{cases} (x - 4)^2 + (y + 3)^2 \leq 24 \\ (x - 4)^2 + (y + 3)^2 \geq 24 \end{cases}$ infinitely many solutions

Application Exercises

The figure shows the healthy weight region for various heights for people ages 35 and older.

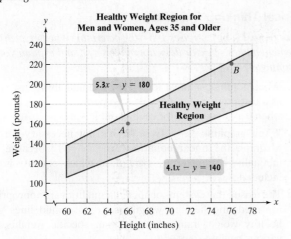

Healthy Weight Region for Men and Women, Ages 35 and Older

Source: U.S. Department of Health and Human Services

If x represents height, in inches, and y represents weight, in pounds, the healthy weight region can be modeled by the following system of linear inequalities:

$$\begin{cases} 5.3x - y \geq 180 \\ 4.1x - y \leq 140. \end{cases}$$

Use this information to solve Exercises 77–80.

77. Show that point A is a solution of the system of inequalities that describes healthy weight for this age group. *

78. Show that point B is a solution of the system of inequalities that describes healthy weight for this age group. *

79. Is a person in this age group who is 6 feet tall weighing 205 pounds within the healthy weight region? no

80. Is a person in this age group who is 5 feet 8 inches tall weighing 135 pounds within the healthy weight region? no

81. Many elevators have a capacity of 2000 pounds.

 a. If a child averages 50 pounds and an adult 150 pounds, write an inequality that describes when x children and y adults will cause the elevator to be overloaded. *

 b. Graph the inequality. Because x and y must be positive, limit the graph to quadrant I only. *

 c. Select an ordered pair satisfying the inequality. What are its coordinates and what do they represent in this situation? *

82. A patient is not allowed to have more than 330 milligrams of cholesterol per day from a diet of eggs and meat. Each egg provides 165 milligrams of cholesterol. Each ounce of meat provides 110 milligrams.

 a. Write an inequality that describes the patient's dietary restrictions for x eggs and y ounces of meat. *

 b. Graph the inequality. Because x and y must be positive, limit the graph to quadrant I only. *

 c. Select an ordered pair satisfying the inequality. What are its coordinates and what do they represent in this situation? *

83. On your next vacation, you will divide lodging between large resorts and small inns. Let x represent the number of nights spent in large resorts. Let y represent the number of nights spent in small inns.

 a. Write a system of inequalities that models the following conditions:

 You want to stay at least 5 nights. At least one night should be spent at a large resort. Large resorts average $200 per night and small inns average $100 per night. Your budget permits no more than $700 for lodging. $y \geq 0; x + y \geq 5; x \geq 1; 200x + 100y \leq 700$

 b. Graph the solution set of the system of inequalities in part (a). *

 c. Based on your graph in part (b), what is the greatest number of nights you could spend at a large resort and still stay within your budget? 2 nights

84. A person with no more than $15,000 to invest plans to place the money in two investments. One investment is high risk, high yield; the other is low risk, low yield. At least $2000 is to be placed in the high-risk investment. Furthermore, the amount invested at low risk should be at least three times the amount invested at high risk. Find and graph a system of inequalities that describes all possibilities for placing the money in the high- and low-risk investments. *

The graph of an inequality in two variables is a region in the rectangular coordinate system. Regions in coordinate systems have numerous applications. For example, the regions in the following two graphs indicate whether a person is obese, overweight, borderline overweight, normal weight, or underweight.

Females

Males

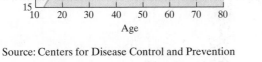

Source: Centers for Disease Control and Prevention

(Refer to the graphs at the bottom of the previous page.) The horizontal axis shows a person's age. The vertical axis shows that person's body-mass index (BMI), computed using the following formula:

$$\text{BMI} = \frac{703W}{H^2}.$$

The variable W represents weight, in pounds. The variable H represents height, in inches. Use this information to solve Exercises 85–86.

85. A man is 20 years old, 72 inches (6 feet) tall, and weighs 200 pounds.

 a. Compute the man's BMI. Round to the nearest tenth. *

 b. Use the man's age and his BMI to locate this information as a point in the coordinate system for males. Is this person obese, overweight, borderline overweight, normal weight, or underweight? overweight

86. A woman is 25 years old, 66 inches (5 feet, 6 inches) tall, and weighs 105 pounds.

 a. Compute the woman's BMI. Round to the nearest tenth. 16.9

 b. Use the woman's age and her BMI to locate this information as a point in the coordinate system for females. Is this person obese, overweight, borderline overweight, normal weight, or underweight? underweight

Explaining the Concepts 87–96. Answers will vary.

87. What is a linear inequality in two variables? Provide an example with your description.

88. How do you determine if an ordered pair is a solution of an inequality in two variables, x and y?

89. What is a half-plane?

90. What does a solid line mean in the graph of an inequality?

91. What does a dashed line mean in the graph of an inequality?

92. Compare the graphs of $3x - 2y > 6$ and $3x - 2y \leq 6$. Discuss similarities and differences between the graphs.

93. What is a system of linear inequalities?

94. What is a solution of a system of linear inequalities?

95. Explain how to graph the solution set of a system of inequalities.

96. What does it mean if a system of linear inequalities has no solution?

Technology Exercises

Graphing utilities can be used to shade regions in the rectangular coordinate system, thereby graphing an inequality in two variables. Read the section of the user's manual for your graphing utility that describes how to shade a region. Then use your graphing utility to graph the inequalities in Exercises 97–102.

97. $y \leq 4x + 4$ *

98. $y \geq \frac{2}{3}x - 2$ *

99. $y \geq x^2 - 4$ *

100. $y \geq \frac{1}{2}x^2 - 2$ *

101. $2x + y \leq 6$ *

102. $3x - 2y \geq 6$ *

103. Does your graphing utility have any limitations in terms of graphing inequalities? If so, what are they?

104. Use a graphing utility to verify any five of the graphs that you drew by hand in Exercises 1–26.

105. Use a graphing utility to verify any five of the graphs that you drew by hand for the systems in Exercises 27–62. 103–105. Answers will vary.

Critical Thinking Exercises

Make Sense? *In Exercises 106–109, determine whether each statement makes sense or does not make sense, and explain your reasoning.*

106. When graphing a linear inequality, I should always use $(0, 0)$ as a test point because it's easy to perform the calculations when 0 is substituted for each variable. does not make sense

107. When graphing $3x - 4y < 12$, it's not necessary for me to graph the linear equation $3x - 4y = 12$ because the inequality contains a $<$ symbol, in which equality is not included. does not make sense

108. The reason that systems of linear inequalities are appropriate for modeling healthy weight is because guidelines give healthy weight ranges, rather than specific weights, for various heights. makes sense

109. I graphed the solution set of $y \geq x + 2$ and $x \geq 1$ without using test points. makes sense

In Exercises 110–113, write a system of inequalities for each graph.

110. *

111. *

112. *

113. *
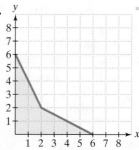

114. Write a system of inequalities whose solution set includes every point in the rectangular coordinate system.

115. Sketch the graph of the solution set for the following system of inequalities:

$$\begin{cases} y \geq nx + b \ (n < 0, b > 0) \\ y \leq mx + b \ (m > 0, b > 0). \end{cases}$$ *

Retaining the Concepts

116. Let $f(x) = \begin{cases} x + 3 & \text{if } x \geq 5 \\ 8 & \text{if } x < 5. \end{cases}$

 Find $f(12) - f(-12)$. (Section 2.2, Example 6) 7

117. Solve and graph the solution set on a number line:

$$\frac{x - 2}{x - 1} \geq 1.$$

 (Section 3.6, Example 4) $(-\infty, 1)$*

118. The function

$$f(t) = \frac{25.1}{1 + 2.7e^{-0.05t}}$$

models the population of Florida, $f(t)$, in millions, t years after 1970.

a. What was Florida's population in 1970? ≈6.8 million

b. According to this logistic growth model, what was Florida's population, to the nearest tenth of a million, in 2010? Does this underestimate or overestimate the actual 2010 population of 18.8 million? 18.4 million; underestimates by 0.4 million

c. What is the limiting size of the population of Florida? (Section 4.5, Example 3) 25.1 million

Preview Exercises

Exercises 119–121 will help you prepare for the material covered in the next section.

119. a. Graph the solution set of the system:

$$\begin{cases} x + y \geq 6 \\ x \leq 8 \\ y \geq 5. \end{cases}$$

b. List the points that form the corners of the graphed region in part (a). (1, 5), (8, 5), (8, −2)

c. Evaluate $3x + 2y$ at each of the points obtained in part (b). (1, 5): 13; (8, 5): 34; (8, −2): 20

120. a. Graph the solution set of the system:

$$\begin{cases} x \geq 0 \\ y \geq 0 \\ 3x - 2y \leq 6 \\ y \leq -x + 7. \end{cases}$$

b. List the points that form the corners of the graphed region in part (a). (0, 0), (2, 0), (4, 3), (0, 7)

c. Evaluate $2x + 5y$ at each of the points obtained in part (b). (0, 0): 0; (2, 0): 4; (4, 3): 23; (0, 7): 35

121. Bottled water and medical supplies are to be shipped to survivors of an earthquake by plane. The bottled water weighs 20 pounds per container and medical kits weigh 10 pounds per kit. Each plane can carry no more than 80,000 pounds. If x represents the number of bottles of water to be shipped per plane and y represents the number of medical kits per plane, write an inequality that models each plane's 80,000-pound weight restriction. $20x + 10y \leq 80{,}000$

Section 5.6 Linear Programming

What am I supposed to learn?

After studying this section, you should be able to:

1. Write an objective function describing a quantity that must be maximized or minimized.

2. Use inequalities to describe limitations in a situation.

3. Use linear programming to solve problems.

West Berlin children at Tempelhof airport watch fleets of U.S. airplanes bringing in supplies to circumvent the Soviet blockade. The airlift began June 28, 1948, and continued for 15 months.

The Berlin Airlift (1948–1949) was an operation by the United States and Great Britain in response to military action by the former Soviet Union: Soviet troops closed all roads and rail lines between West Germany and Berlin, cutting off supply routes to the city. The Allies used a mathematical technique developed during World War II to maximize the amount of supplies transported. During the 15-month airlift, 278,228 flights provided basic necessities to blockaded Berlin, saving one of the world's great cities.

In this section, we will look at an important application of systems of linear inequalities. Such systems arise in **linear programming**, a method for solving

problems in which a particular quantity that must be maximized or minimized is limited by other factors. Linear programming is one of the most widely used tools in management science. It helps businesses allocate resources to manufacture products in a way that will maximize profit. Linear programming accounts for more than 50% and perhaps as much as 90% of all computing time used for management decisions in business. The Allies used linear programming to save Berlin.

① Write an objective function describing a quantity that must be maximized or minimized.

Objective Functions in Linear Programming

Many problems involve quantities that must be maximized or minimized. Businesses are interested in maximizing profit. An operation in which bottled water and medical kits are shipped to earthquake survivors needs to maximize the number of survivors helped by this shipment. An **objective function** is an algebraic expression in two or more variables describing a quantity that must be maximized or minimized.

EXAMPLE 1 Writing an Objective Function

Bottled water and medical supplies are to be shipped to survivors of an earthquake by plane. Each container of bottled water will serve 10 people and each medical kit will aid 6 people. If x represents the number of bottles of water to be shipped and y represents the number of medical kits, write the objective function that models the number of people that can be helped.

SOLUTION

Because each bottle of water serves 10 people and each medical kit aids 6 people, we have

$$= \quad 10x \quad + \quad 6y.$$

Using z to represent the number of people helped, the objective function is

$$z = 10x + 6y.$$

Unlike the functions that we have seen so far, the objective function is an equation in three variables. For a value of x and a value of y, there is one and only one value of z. Thus, z is a function of x and y. •••

⊘ Check Point **1** A company manufactures bookshelves and desks for computers. Let x represent the number of bookshelves manufactured daily and y the number of desks manufactured daily. The company's profits are $25 per bookshelf and $55 per desk. Write the objective function that models the company's total daily profit, z, from x bookshelves and y desks. (Check Points 2 through 4 are related to this situation, so keep track of your answers.) $z = 25x + 55y$

② Use inequalities to describe limitations in a situation.

Constraints in Linear Programming

Ideally, the number of earthquake survivors helped in Example 1 should increase without restriction so that every survivor receives water and medical supplies. However, the planes that ship these supplies are subject to weight and volume restrictions. In linear programming problems, such restrictions are called **constraints**. Each constraint is expressed as a linear inequality. The list of constraints forms a system of linear inequalities.

EXAMPLE 2 Writing a Constraint

Each plane can carry no more than 80,000 pounds. The bottled water weighs 20 pounds per container and each medical kit weighs 10 pounds. Let x represent the number of bottles of water to be shipped and y the number of medical kits. Write an inequality that models this constraint.

SOLUTION

Because each plane can carry no more than 80,000 pounds, we have

| The total weight of the water bottles | plus | the total weight of the medical kits | must be less than or equal to | 80,000 pounds. |

$$20x \quad + \quad 10y \quad \leq \quad 80{,}000.$$

Each bottle weighs 20 pounds. Each kit weighs 10 pounds.

The plane's weight constraint is modeled by the inequality

$$20x + 10y \leq 80{,}000.$$

••••

✓ **Check Point 2** To maintain high quality, the company in Check Point 1 should not manufacture more than a total of 80 bookshelves and desks per day. Write an inequality that models this constraint. $x + y \leq 80$

In addition to a weight constraint on its cargo, each plane has a limited amount of space in which to carry supplies. Example 3 demonstrates how to express this constraint.

EXAMPLE 3 Writing a Constraint

Each plane can carry a total volume of supplies that does not exceed 6000 cubic feet. Each water bottle is 1 cubic foot and each medical kit also has a volume of 1 cubic foot. With x still representing the number of water bottles and y the number of medical kits, write an inequality that models this second constraint.

SOLUTION

Because each plane can carry a volume of supplies that does not exceed 6000 cubic feet, we have

| The total volume of the water | plus | the total volume of the medical kits | must be less than or equal to | 6000 cubic feet. |

$$1x \quad + \quad 1y \quad \leq \quad 6000.$$

Each bottle is 1 cubic foot. Each kit is 1 cubic foot.

The plane's volume constraint is modeled by the inequality $x + y \leq 6000$. ••••

In summary, here's what we have described so far in this aid-to-earthquake-survivors situation:

$$z = 10x + 6y \qquad \text{This is the objective function modeling the number of people helped with } x \text{ bottles of water and } y \text{ medical kits.}$$

$$\begin{cases} 20x + 10y \leq 80{,}000 \\ x + y \leq 6000. \end{cases} \qquad \text{These are the constraints based on each plane's weight and volume limitations.}$$

③ Use linear programming to solve problems.

Solving Problems with Linear Programming

The problem in the earthquake situation described previously is to maximize the number of survivors who can be helped, subject to each plane's weight and volume constraints. The process of solving this problem is called *linear programming*, based on a theorem that was proven during World War II.

Solving a Linear Programming Problem

Let $z = ax + by$ be an objective function that depends on x and y. Furthermore, z is subject to a number of constraints on x and y. If a maximum or minimum value of z exists, it can be determined as follows:

1. Graph the system of inequalities representing the constraints.
2. Find the value of the objective function at each corner, or **vertex**, of the graphed region. The maximum and minimum of the objective function occur at one or more of the corner points.

EXAMPLE 4 Solving a Linear Programming Problem

Determine how many bottles of water and how many medical kits should be sent on each plane to maximize the number of earthquake survivors who can be helped.

SOLUTION

We must maximize $z = 10x + 6y$ subject to the following constraints:

$$\begin{cases} 20x + 10y \le 80{,}000 \\ x + y \le 6000. \end{cases}$$

Step 1 Graph the system of inequalities representing the constraints. Because x (the number of bottles of water per plane) and y (the number of medical kits per plane) must be nonnegative, we need to graph the system of inequalities in quadrant I and its boundary only.

To graph the inequality $20x + 10y \le 80{,}000$, we graph the equation $20x + 10y = 80{,}000$ as a solid blue line (**Figure 5.29**). Setting $y = 0$, the x-intercept is 4000 and setting $x = 0$, the y-intercept is 8000. Using $(0, 0)$ as a test point, the inequality is satisfied, so we shade below the blue line, as shown in yellow in **Figure 5.29**.

Now we graph $x + y \le 6000$ by first graphing $x + y = 6000$ as a solid red line. Setting $y = 0$, the x-intercept is 6000. Setting $x = 0$, the y-intercept is 6000. Using $(0, 0)$ as a test point, the inequality is satisfied, so we shade below the red line, as shown using green vertical shading in **Figure 5.29**.

We use the addition method to find where the lines $20x + 10y = 80{,}000$ and $x + y = 6000$ intersect.

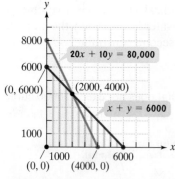

FIGURE 5.29 The region in quadrant I representing the constraints $20x + 10y \le 80{,}000$ and $x + y \le 6000$

$$\begin{cases} 20x + 10y = 80{,}000 \\ x + y = 6000 \end{cases} \quad \xrightarrow[\text{Multiply by } -10.}{\text{No change}} \quad \begin{cases} 20x + 10y = 80{,}000 \\ -10x - 10y = -60{,}000 \end{cases}$$

$$\begin{array}{rcl} \text{Add:} \quad 10x & = & 20{,}000 \\ x & = & 2000 \end{array}$$

FIGURE 5.30

Back-substituting 2000 for x in $x + y = 6000$, we find $y = 4000$, so the intersection point is (2000, 4000).

The system of inequalities representing the constraints is shown by the region in which the yellow shading and the green vertical shading overlap in **Figure 5.29**. The graph of the system of inequalities is shown again in **Figure 5.30**. The red and blue line segments are included in the graph.

Step 2 Find the value of the objective function at each corner of the graphed region. The maximum and minimum of the objective function occur at one or more of the corner points. We must evaluate the objective function, $z = 10x + 6y$, at the four corners, or vertices, of the region in **Figure 5.30**.

Corner (x, y)	Objective Function $z = 10x + 6y$
(0, 0)	$z = 10(0) + 6(0) = 0$
(4000, 0)	$z = 10(4000) + 6(0) = 40{,}000$
(2000, 4000)	$z = 10(2000) + 6(4000) = 44{,}000 \leftarrow$ maximum
(0, 6000)	$z = 10(0) + 6(6000) = 36{,}000$

Thus, the maximum value of z is 44,000 and this occurs when $x = 2000$ and $y = 4000$. In practical terms, this means that the maximum number of earthquake survivors who can be helped with each plane shipment is 44,000. This can be accomplished by sending 2000 water bottles and 4000 medical kits per plane. • • •

⌀ Check Point **4** For the company in Check Points 1–3, how many bookshelves and how many desks should be manufactured per day to obtain maximum profit? What is the maximum daily profit? 50 bookshelves and 30 desks; $2900

EXAMPLE 5 Solving a Linear Programming Problem

Find the maximum value of the objective function

$$z = 2x + y$$

subject to the following constraints:

$$\begin{cases} x \geq 0, \, y \geq 0 \\ x + 2y \leq 5 \\ x - y \leq 2. \end{cases}$$

SOLUTION

We begin by graphing the region in quadrant I ($x \geq 0$, $y \geq 0$) formed by the constraints. The graph is shown in **Figure 5.31**.

Now we evaluate the objective function at the four vertices of this region.

Objective function: $z = 2x + y$

At (0, 0): $z = 2 \cdot 0 + 0 = 0$

At (2, 0): $z = 2 \cdot 2 + 0 = 4$

At (3, 1): $z = 2 \cdot 3 + 1 = 7$ Maximum value of z

At (0, 2.5): $z = 2 \cdot 0 + 2.5 = 2.5$

Thus, the maximum value of z is 7, and this occurs when $x = 3$ and $y = 1$. • • •

We can see why the objective function in Example 5 has a maximum value that occurs at a vertex by solving the equation for y.

FIGURE 5.31 The graph of $x + 2y \leq 5$ and $x - y \leq 2$ in quadrant I

$z = 2x + y$ This is the objective function of Example 5.

$y = -2x + z$ Solve for y. Recall that the slope-intercept form of a line is $y = mx + b$.

Slope $= -2$ y-intercept $= z$

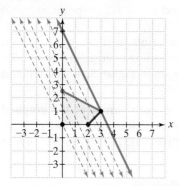

In the form $y = -2x + z$, z represents the y-intercept of the objective function. The equation describes infinitely many parallel lines, each with slope -2. The process in linear programming involves finding the maximum z-value for all lines that intersect the region determined by the constraints. Of all the lines whose slope is -2, we're looking for the one with the greatest y-intercept that intersects the given region. As we see in **Figure 5.32**, such a line will pass through one (or possibly more) of the vertices of the region.

FIGURE 5.32 The line with slope -2 with the greatest y-intercept that intersects the shaded region passes through one of its vertices.

☑ **Check Point 5** Find the maximum value of the objective function $z = 3x + 5y$ subject to the constraints $x \geq 0, y \geq 0, x + y \geq 1, x + y \leq 6$. 30

CONCEPT AND VOCABULARY CHECK

Fill in each blank so that the resulting statement is true.

1. A method for finding the maximum or minimum value of a quantity that is subject to various limitations is called __linear programming__.

2. An algebraic expression in two or more variables describing a quantity that must be maximized or minimized is called a/an __objective__ function.

3. A system of linear inequalities is used to represent restrictions, or __constraints__, on a function that must be maximized or minimized. Using the graph of such a system of inequalities, the maximum and minimum values of the function occur at one or more of the __corner__ points.

EXERCISE SET 5.6

Practice Exercises

In Exercises 1–4, find the value of the objective function at each corner of the graphed region. What is the maximum value of the objective function? What is the minimum value of the objective function?

1. Objective Function ∗
$z = 5x + 6y$

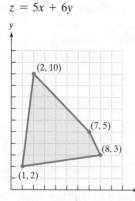

2. Objective Function ∗
$z = 3x + 2y$

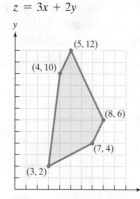

3. Objective Function ∗
$z = 40x + 50y$

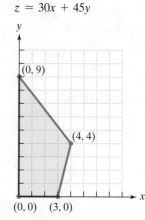

4. Objective Function ∗
$z = 30x + 45y$

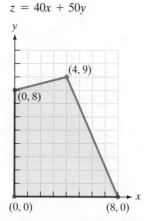

In Exercises 5–14, an objective function and a system of linear inequalities representing constraints are given.

a. *Graph the system of inequalities representing the constraints.*

b. *Find the value of the objective function at each corner of the graphed region.*

c. *Use the values in part (b) to determine the maximum value of the objective function and the values of x and y for which the maximum occurs.*

5. Objective Function
Constraints

c. Maximum value is 16 and occurs at
$x = 0$ and $y = 8$.*

$z = 3x + 2y$
$$\begin{cases} x \geq 0, y \geq 0 \\ 2x + y \leq 8 \\ x + y \geq 4 \end{cases}$$

6. Objective Function
Constraints

c. Maximum value is 12 and occurs at
all points on the line segment between
$(0, 4)$ and $(3, 2)$.*

$z = 2x + 3y$
$$\begin{cases} x \geq 0, y \geq 0 \\ 2x + y \leq 8 \\ 2x + 3y \leq 12 \end{cases}$$

7. Objective Function
Constraints

c. Maximum value is 24 and occurs at
$x = 6$ and $y = 0$.*

$z = 4x + y$
$$\begin{cases} x \geq 0, y \geq 0 \\ 2x + 3y \leq 12 \\ x + y \geq 3 \end{cases}$$

8. Objective Function
Constraints

c. Maximum value is 38 and occurs at
$x = 2$ and $y = 6$.*

$z = x + 6y$
$$\begin{cases} x \geq 0, y \geq 0 \\ 2x + y \leq 10 \\ x - 2y \geq -10 \end{cases}$$

9. Objective Function
Constraints

c. Maximum value is 11 and occurs at
$x = 5$ and $y = 2$.*

$z = 3x - 2y$
$$\begin{cases} 1 \leq x \leq 5 \\ y \geq 2 \\ x - y \geq -3 \end{cases}$$

10. Objective Function
Constraints

c. Maximum value is 25 and occurs at
$x = 5$ and $y = 0$.

$z = 5x - 2y$
$$\begin{cases} 0 \leq x \leq 5 \\ 0 \leq y \leq 3 \\ x + y \geq 2 \end{cases}$$

11. Objective Function
Constraints

c. Maximum value is 16 and occurs at
$x = 4$ and $y = 0$.

$z = 4x + 2y$
$$\begin{cases} x \geq 0, y \geq 0 \\ 2x + 3y \leq 12 \\ 3x + 2y \leq 12 \\ x + y \geq 2 \end{cases}$$

12. Objective Function
Constraints

c. Maximum value is 36 and occurs at
$x = 0$ and $y = 9$.*

$z = 2x + 4y$
$$\begin{cases} x \geq 0, y \geq 0 \\ x + 3y \geq 6 \\ x + y \geq 3 \\ x + y \leq 9 \end{cases}$$

13. Objective Function
Constraints

c. Maximum value is 78 and occurs at
$x = 3$ and $y = 4$.*

$z = 10x + 12y$
$$\begin{cases} x \geq 0, y \geq 0 \\ x + y \leq 7 \\ 2x + y \leq 10 \\ 2x + 3y \leq 18 \end{cases}$$

14. Objective Function
Constraints

c. Maximum value is 60 and occurs at
$x = 0$ and $y = 10$.*

$z = 5x + 6y$
$$\begin{cases} x \geq 0, y \geq 0 \\ 2x + y \geq 10 \\ x + 2y \geq 10 \\ x + y \leq 10 \end{cases}$$

Application Exercises

15. A television manufacturer makes rear-projection and plasma televisions. The profit per unit is $125 for the rear-projection televisions and $200 for the plasma televisions.

a. Let x = the number of rear-projection televisions manufactured in a month and let y = the number of plasma televisions manufactured in a month. Write the objective function that models the total monthly profit. *

b. The manufacturer is bound by the following constraints:

- Equipment in the factory allows for making at most 450 rear-projection televisions in one month.

- Equipment in the factory allows for making at most 200 plasma televisions in one month.

- The cost to the manufacturer per unit is $600 for the rear-projection televisions and $900 for the plasma televisions. Total monthly costs cannot exceed $360,000.

Write a system of three inequalities that models these constraints. *

c. Graph the system of inequalities in part (b). Use only the first quadrant and its boundary, because x and y must both be nonnegative. *

d. Evaluate the objective function for total monthly profit at each of the five vertices of the graphed region. [The vertices should occur at $(0, 0)$, $(0, 200)$, $(300, 200)$, $(450, 100)$, and $(450, 0)$.] *

e. Complete the missing portions of this statement: The television manufacturer will make the greatest profit by manufacturing ___300___ rear-projection televisions each month and ___200___ plasma televisions each month. The maximum monthly profit is $___77,500___.

16. a. A student earns $15 per hour for tutoring and $10 per hour as a teacher's aide. Let x = the number of hours each week spent tutoring and let y = the number of hours each week spent as a teacher's aide. Write the objective function that models total weekly earnings. $z = 15x + 10y$

b. The student is bound by the following constraints:

- To have enough time for studies, the student can work no more than 20 hours per week.

- The tutoring center requires that each tutor spend at least three hours per week tutoring.

- The tutoring center requires that each tutor spend no more than eight hours per week tutoring.

Write a system of three inequalities that models these constraints. $x + y \leq 20; x \geq 3; x \leq 8$

c. Graph the system of inequalities in part (b). Use only the first quadrant and its boundary, because x and y are nonnegative. *

d. Evaluate the objective function for total weekly earnings at each of the four vertices of the graphed region. [The vertices should occur at $(3, 0)$, $(8, 0)$, $(3, 17)$, and $(8, 12)$.] *

e. Complete the missing portions of this statement: The student can earn the maximum amount per week by tutoring for ___8___ hours per week and working as a teacher's aide for ___12___ hours per week. The maximum amount that the student can earn each week is $___240___.

Use the two steps for solving a linear programming problem, given in the box on page 606, to solve the problems in Exercises 17–23.

17. A manufacturer produces two models of mountain bicycles. The times (in hours) required for assembling and painting each model are given in the following table:

	Model *A*	Model *B*
Assembling	5	4
Painting	2	3

The maximum total weekly hours available in the assembly department and the paint department are 200 hours and 108 hours, respectively. The profits per unit are $25 for model *A* and $15 for model *B*. How many of each type should be produced to maximize profit? 40 model *A*; 0 model *B*

18. A large institution is preparing lunch menus containing foods A and B. The specifications for the two foods are given in the following table:

Food	Units of Fat per Ounce	Units of Carbohydrates per Ounce	Units of Protein per Ounce
A	1	2	1
B	1	1	1

Each lunch must provide at least 6 units of fat per serving, no more than 7 units of protein, and at least 10 units of carbohydrates. The institution can purchase food A for $0.12 per ounce and food B for $0.08 per ounce. How many ounces of each food should a serving contain to meet the dietary requirements at the least cost? food A: 4 oz; food B: 2 oz

19. Food and clothing are shipped to survivors of a natural disaster. Each carton of food will feed 12 people, while each carton of clothing will help 5 people. Each 20-cubic-foot box of food weighs 50 pounds and each 10-cubic-foot box of clothing weighs 20 pounds. The commercial carriers transporting food and clothing are bound by the following constraints:

- The total weight per carrier cannot exceed 19,000 pounds.
- The total volume must be less than 8000 cubic feet.

How many cartons of food and clothing should be sent with each plane shipment to maximize the number of people who can be helped? food: 300; clothing: 200

20. On June 24, 1948, the former Soviet Union blocked all land and water routes through East Germany to Berlin. A gigantic airlift was organized using American and British planes to bring food, clothing, and other supplies to the more than 2 million people in West Berlin. The cargo capacity was 30,000 cubic feet for an American plane and 20,000 cubic feet for a British plane. To break the Soviet blockade, the Western Allies had to maximize cargo capacity but were subject to the following restrictions:

- No more than 44 planes could be used.
- The larger American planes required 16 personnel per flight, double that of the requirement for the British planes. The total number of personnel available could not exceed 512.
- The cost of an American flight was $9000 and the cost of a British flight was $5000. Total weekly costs could not exceed $300,000.

Find the number of American and British planes that were used to maximize cargo capacity. 20 American; 24 British

21. A theater is presenting a program for students and their parents on drinking and driving. The proceeds will be donated to a local alcohol information center. Admission is $2.00 for parents and $1.00 for students. However, the situation has two constraints: The theater can hold no more than 150 people and every two parents must bring at least one student. How many parents and students should attend to raise the maximum amount of money? students: 50; parents: 100

22. You are about to take a test that contains computation problems worth 6 points each and word problems worth 10 points each. You can do a computation problem in 2 minutes and a word problem in 4 minutes. You have 40 minutes to take the test and may answer no more than 12 problems. Assuming you answer all the problems attempted correctly, how many of each type of problem must you answer to maximize your score? What is the maximum score? 4 computation; 8 word; 104

23. In 1978, a ruling by the Civil Aeronautics Board allowed Federal Express to purchase larger aircraft. Federal Express's options included 20 Boeing 727s that United Airlines was retiring and/or the French-built Dassault Fanjet Falcon 20. To aid in their decision, executives at Federal Express analyzed the following data:

	Boeing 727	Falcon 20
Direct Operating Cost	$1400 per hour	$500 per hour
Payload	42,000 pounds	6000 pounds

Federal Express was faced with the following constraints:

- Hourly operating cost was limited to $35,000.
- Total payload had to be at least 672,000 pounds.
- Only twenty 727s were available.

Given the constraints, how many of each kind of aircraft should Federal Express have purchased to maximize the number of aircraft? 10 Boeing 727s; 42 Falcon 20s

Explaining the Concepts 24–28. Answers will vary.

24. What kinds of problems are solved using the linear programming method?

25. What is an objective function in a linear programming problem?

26. What is a constraint in a linear programming problem? How is a constraint represented?

27. In your own words, describe how to solve a linear programming problem.

28. Describe a situation in your life in which you would really like to maximize something, but you are limited by at least two constraints. Can linear programming be used in this situation? Explain your answer.

Critical Thinking Exercises

Make Sense? *In Exercises 29–32, determine whether each statement makes sense or does not make sense, and explain your reasoning.*

29. In order to solve a linear programming problem, I use the graph representing the constraints and the graph of the objective function. does not make sense

30. I use the coordinates of each vertex from my graph representing the constraints to find the values that maximize or minimize an objective function. makes sense

31. I need to be able to graph systems of linear inequalities in order to solve linear programming problems. makes sense

32. An important application of linear programming for businesses involves maximizing profit. makes sense

33. Suppose that you inherit $10,000. The will states how you must invest the money. Some (or all) of the money must be invested in stocks and bonds. The requirements are that at least $3000 be invested in bonds, with expected returns of $0.08 per dollar, and at least $2000 be invested in stocks, with expected returns of $0.12 per dollar. Because the stocks are medium risk, the final stipulation requires that the investment in bonds should never be less than the investment in stocks. How should the money be invested so as to maximize your expected returns? $5000 in each

34. Consider the objective function $z = Ax + By$ ($A > 0$ and $B > 0$) subject to the following constraints: $2x + 3y \leq 9$, $x - y \leq 2, x \geq 0$, and $y \geq 0$. Prove that the objective function will have the same maximum value at the vertices $(3, 1)$ and $(0, 3)$ if $A = \frac{2}{3}B$. *

Group Exercises 35–36. Answers will vary.

35. Group members should choose a particular field of interest. Research how linear programming is used to solve problems in that field. If possible, investigate the solution of a specific practical problem. Present a report on your findings, including the contributions of George Dantzig, Narendra Karmarkar, and L. G. Khachion to linear programming.

36. Members of the group should interview a business executive who is in charge of deciding the product mix for a business. How are production policy decisions made? Are other methods used in conjunction with linear programming? What are these methods? What sort of academic background, particularly in mathematics, does this executive have? Present a group report addressing these questions, emphasizing the role of linear programming for the business.

Retaining the Concepts

37. Solve for x: $Ax + By = Cx + D$. (Section 1.3, Example 8)
$$x = \frac{D - By}{A - C}$$

38. Solve: $\sqrt{2x - 5} - \sqrt{x - 3} = 1$. (Section 1.6, Example 4)
{3, 7}

39. If $f(x) = 5x^2 - 6x + 1$, find $\dfrac{f(x + h) - f(x)}{h}$.
(Section 2.2, Example 8) $10x + 5h - 6$

Preview Exercises

Exercises 40–42 will help you prepare for the material covered in the first section of the next chapter.

40. Solve the system: {(6, 3, 5)}; Answers may vary.
$$\begin{cases} x + y + 2z = 19 \\ y + 2z = 13 \\ z = 5. \end{cases}$$

What makes it fairly easy to find the solution?

41. Solve the system: {(−2, 1, 4, 3)}; Answers may vary.
$$\begin{cases} w - x + 2y - 2z = -1 \\ x - \frac{1}{3}y + z = \frac{8}{3} \\ y - z = 1 \\ z = 3. \end{cases}$$

Express the solution set in the form $\{(w, x, y, z)\}$. What makes it fairly easy to find the solution?

42. Consider the following array of numbers:
$$\begin{bmatrix} 1 & 2 & -1 \\ 4 & -3 & -15 \end{bmatrix}.$$

Rewrite the array as follows: Multiply each number in the top row by −4 and add this product to the corresponding number in the bottom row. Do not change the numbers in the top row. *

CHAPTER 5 Summary, Review, and Test

SUMMARY

DEFINITIONS AND CONCEPTS	EXAMPLES
5.1 Systems of Linear Equations in Two Variables	
a. Two equations in the form $Ax + By = C$ are called a system of linear equations. A solution of the system is an ordered pair that satisfies both equations in the system.	Ex. 1, p. 542
b. Systems of linear equations in two variables can be solved by eliminating a variable, using the substitution method (see the box on page 544) or the addition method (see the box on page 546).	Ex. 2, p. 544; Ex. 3, p. 546; Ex. 4, p. 547

DEFINITIONS AND CONCEPTS	**EXAMPLES**

c. Some linear systems have no solution and are called inconsistent systems; others have infinitely many solutions. The equations in a linear system with infinitely many solutions are called dependent. For details, see the box on page 548.

d. Functions of Business

Revenue Function

$$R(x) = \text{(price per unit sold)}x$$

Cost Function

$$C(x) = \text{fixed cost} + \text{(cost per unit produced)}x$$

Profit Function

$$P(x) = R(x) - C(x)$$

The point of intersection of the graphs of R and C is the break-even point. The x-coordinate of the point reveals the number of units that a company must produce and sell so that the money coming in, the revenue, is equal to the money going out, the cost. The y-coordinate gives the amount of money coming in and going out.

e. Mixture problems involve two unknown quantities and can be solved using a system of linear equations in two variables.

5.2 Systems of Linear Equations in Three Variables

a. Three equations in the form $Ax + By + Cz = D$ are called a system of linear equations in three variables. A solution of the system is an ordered triple that satisfies all three equations in the system.

b. A system of linear equations in three variables can be solved by eliminating variables. Use the addition method to eliminate any variable, reducing the system to two equations in two variables. Use substitution or the addition method to solve the resulting system in two variables. Details are found in the box on page 562.

c. Three points that do not lie on a line determine the graph of a quadratic function $y = ax^2 + bx + c$. Use the three given points to create a system of three equations. Solve the system to find a, b, and c.

5.3 Partial Fractions

a. Partial fraction decomposition is used on rational expressions in which the numerator and denominator have no common factors and the highest power in the numerator is less than the highest power in the denominator. The steps in partial fraction decomposition are given in the box on page 572.

b. Include one partial fraction with a constant numerator for each distinct linear factor in the denominator. Include one partial fraction with a constant numerator for each power of a repeated linear factor in the denominator.

c. Include one partial fraction with a linear numerator for each distinct prime quadratic factor in the denominator. Include one partial fraction with a linear numerator for each power of a prime, repeated quadratic factor in the denominator.

5.4 Systems of Nonlinear Equations in Two Variables

a. A system of two nonlinear equations in two variables contains at least one equation that cannot be expressed as $Ax + By = C$.

b. Systems of nonlinear equations in two variables can be solved algebraically by eliminating all occurrences of one of the variables by the substitution or addition methods.

DEFINITIONS AND CONCEPTS **EXAMPLES**

5.5 Systems of Inequalities

 a. A linear inequality in two variables can be written in the form
 $Ax + By > C, Ax + By \geq C, Ax + By < C,$ or $Ax + By \leq C.$

 b. The procedure for graphing a linear inequality in two variables is given in the box on page 592. Ex. 1, p. 592;
 A nonlinear inequality in two variables is graphed using the same procedure. Ex. 2, p. 593;
 Ex. 3, p. 594;
 Ex. 4, p. 594

 c. To graph the solution set of a system of inequalities, graph each inequality in the system in the Ex. 6, p. 596;
 same rectangular coordinate system. Then find the region, if there is one, that is common to every Ex. 7, p. 597;
 graph in the system. Ex. 8, p. 598

5.6 Linear Programming

 a. An objective function is an algebraic expression in three variables describing a quantity that Ex. 1, p. 604
 must be maximized or minimized.

 b. Constraints are restrictions, expressed as linear inequalities. Ex. 2, p. 605;
 Ex. 3, p. 605

 c. Linear programming is a method for solving problems in which an objective function that must Ex. 4, p. 606;
 be maximized or minimized is limited by constraints. Steps for solving a linear programming Ex. 5, p. 607
 problem are given in the box on page 606.

REVIEW EXERCISES

5.1

In Exercises 1–5, solve by the method of your choice. Identify systems with no solution and systems with infinitely many solutions, using set notation to express their solution sets.

1. $\begin{cases} y = 4x + 1 \\ 3x + 2y = 13 \end{cases}$ {(1, 5)} **2.** $\begin{cases} x + 4y = 14 \\ 2x - y = 1 \end{cases}$ {(2, 3)}

3. $\begin{cases} 5x + 3y = 1 \\ 3x + 4y = -6 \end{cases}$ {(2, -3)} **4.** $\begin{cases} 2y - 6x = 7 \\ 3x - y = 9 \end{cases}$ ∅

5. $\begin{cases} 4x - 8y = 16 \\ 3x - 6y = 12 \end{cases}$ {(x, y) | 4x − 8y = 16} or {(x, y) | 3x − 6y = 12}

6. A company is planning to manufacture computer desks. The fixed cost will be $60,000 and it will cost $200 to produce each desk. Each desk will be sold for $450.

 a. Write the cost function, C, of producing x desks. *

 b. Write the revenue function, R, from the sale of x desks. *

 c. Determine the break-even point. Describe what this means. (240, 108,000)*

7. A chemist needs to mix a solution that is 34% silver nitrate with one that is 4% silver nitrate to obtain 100 milliliters of a mixture that is 7% silver nitrate. How many milliliters of each of the solutions must be used? 10 milliliters of 34% and 90 milliliters of 4% silver nitrate must be used.

8. The bar graph in the next column shows the percentage of Americans who used cigarettes, by ethnicity, in 1985 and 2005. For each of the groups shown, cigarette use has been linearly decreasing.

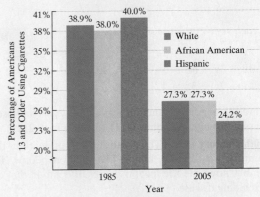

Cigarette Use in the United States

Source: Department of Health and Human Services

 a. In 1985, 38% of African Americans used cigarettes. For the period shown by the graph, this has decreased at an average rate of 0.54% per year. Write a function that models the percentage of African Americans who used cigarettes, y, x years after 1985. $y = -0.54x + 38$

 b. The function $0.79x + y = 40$ models the percentage of Hispanics who used cigarettes, y, x years after 1985. Use this model and the model you obtained in part (a) to determine the year during which cigarette use was the same for African Americans and Hispanics. What percentage of each group used cigarettes during that year? 1993; 33.68%

9. The perimeter of a table tennis top is 28 feet. The difference between 4 times the length and 3 times the width is 21 feet. Find the dimensions. \quad 9 feet by 5 feet

Width: y

Length: x

10. A travel agent offers two package vacation plans. The first plan costs \$360 and includes 3 days at a hotel and a rental car for 2 days. The second plan costs \$500 and includes 4 days at a hotel and a rental car for 3 days. The daily charge for the hotel is the same under each plan, as is the daily charge for the car. Find the cost per day for the hotel and for the car. room: \$80 per day; car: \$60 per day

11. The calorie-nutrient information for an apple and an avocado is given in the table. How many of each should be eaten to get exactly 1000 calories and 100 grams of carbohydrates? \quad 3 apples and 2 avocados

	One Apple	**One Avocado**
Calories	100	350
Carbohydrates (grams)	24	14

5.2

Solve each system in Exercises 12–13.

12. $\begin{cases} 2x - y + z = 1 \\ 3x - 3y + 4z = 5 \\ 4x - 2y + 3z = 4 \end{cases}$ $\{(0, 1, 2)\}$

13. $\begin{cases} x + 2y - z = 5 \\ 2x - y + 3z = 0 \\ 2y + z = 1 \end{cases}$ $\{(2, 1, -1)\}$

14. Find the quadratic function $y = ax^2 + bx + c$ whose graph passes through the points $(1, 4), (3, 20),$ and $(-2, 25).$ *

15. 20th Century Death The greatest cause of death in the 20th century was disease, killing 1390 million people. The bar graph shows the five leading causes of death in that century, excluding disease.

Number of Deaths in the 20th Century, by Cause

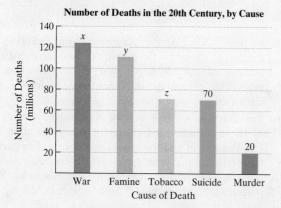

Source: Wikipedia

War, famine, and tobacco combined resulted in 306 million deaths. The difference between the number of deaths from war and famine was 13 million. The difference between the number of deaths from war and tobacco was 53 million. Find the number of 20th century deaths from war, famine, and tobacco. \quad war: 124 million; famine: 111 million; tobacco: 71 million

5.3

In Exercises 16–24, write the partial fraction decomposition of each rational expression.

16. $\dfrac{x}{(x - 3)(x + 2)}$ *

17. $\dfrac{11x - 2}{x^2 - x - 12}$ *

18. $\dfrac{4x^2 - 3x - 4}{x(x + 2)(x - 1)}$ *

19. $\dfrac{2x + 1}{(x - 2)^2}$ $\dfrac{2}{x - 2} + \dfrac{5}{(x - 2)^2}$

20. $\dfrac{2x - 6}{(x - 1)(x - 2)^2}$ *

21. $\dfrac{3x}{(x - 2)(x^2 + 1)}$ *

22. $\dfrac{7x^2 - 7x + 23}{(x - 3)(x^2 + 4)}$ *

23. $\dfrac{x^3}{(x^2 + 4)^2}$ *

24. $\dfrac{4x^3 + 5x^2 + 7x - 1}{(x^2 + x + 1)^2}$ $\dfrac{4x + 1}{x^2 + x + 1} + \dfrac{2x - 2}{(x^2 + x + 1)^2}$

5.4

In Exercises 25–35, solve each system by the method of your choice.

25. $\begin{cases} 5y = x^2 - 1 \\ x - y = 1 \end{cases}$ $\{(4, 3), (1, 0)\}$

26. $\begin{cases} y = x^2 + 2x + 1 \\ x + y = 1 \end{cases}$ *

27. $\begin{cases} x^2 + y^2 = 2 \\ x + y = 0 \end{cases}$ $\{(1, -1), (-1, 1)\}$

28. $\begin{cases} 2x^2 + y^2 = 24 \\ x^2 + y^2 = 15 \end{cases}$ *

29. $\begin{cases} xy - 4 = 0 \\ y - x = 0 \end{cases}$ $\{(2, 2), (-2, -2)\}$

30. $\begin{cases} y^2 = 4x \\ x - 2y + 3 = 0 \end{cases}$ $\{(9, 6), (1, 2)\}$

31. $\begin{cases} x^2 + y^2 = 10 \\ y = x + 2 \end{cases}$ $\{(-3, -1), (1, 3)\}$

32. $\begin{cases} xy = 1 \\ y = 2x + 1 \end{cases}$ $\{(\frac{1}{2}, 2), (-1, -1)\}$

33. $\begin{cases} x + y + 1 = 0 \\ x^2 + y^2 + 6y - x = -5 \end{cases}$ *

34. $\begin{cases} x^2 + y^2 = 13 \\ x^2 - y = 7 \end{cases}$ *

35. $\begin{cases} 2x^2 + 3y^2 = 21 \\ 3x^2 - 4y^2 = 23 \end{cases}$ $\{(3, 1), (3, -1), (-3, 1), (-3, -1)\}$

36. The perimeter of a rectangle is 26 meters and its area is 40 square meters. Find its dimensions. \quad 8 m and 5 m

37. Find the coordinates of all points (x, y) that lie on the line whose equation is $2x + y = 8,$ so that the area of the rectangle shown in the figure is 6 square units. \quad $(1, 6), (3, 2)$

$2x + y = 8$

(x, y)

38. Two adjoining square fields with an area of 2900 square feet are to be enclosed with 240 feet of fencing. The situation is represented in the figure. Find the length of each side where a variable appears. $x = 46, y = 28$ or $x = 50, y = 20$

5.5

In Exercises 39–45, graph each inequality.

39. $3x - 4y > 12$ *

40. $y \leq -\dfrac{1}{2}x + 2$ *

41. $x < -2$ *

42. $y \geq 3$ *

43. $x^2 + y^2 > 4$ *

44. $y \leq x^2 - 1$ *

45. $y \leq 2^x$ *

In Exercises 46–55, graph the solution set of each system of inequalities or indicate that the system has no solution.

46. $\begin{cases} 3x + 2y \geq 6 \\ 2x + y \geq 6 \end{cases}$ *

47. $\begin{cases} 2x - y \geq 4 \\ x + 2y < 2 \end{cases}$ *

48. $\begin{cases} y < x \\ y \leq 2 \end{cases}$ *

49. $\begin{cases} x + y \leq 6 \\ y \geq 2x - 3 \end{cases}$ *

50. $\begin{cases} 0 \leq x \leq 3 \\ y > 2 \end{cases}$ *

51. $\begin{cases} 2x + y < 4 \\ 2x + y > 6 \end{cases}$ no solution

52. $\begin{cases} x^2 + y^2 \leq 16 \\ x + y < 2 \end{cases}$ *

53. $\begin{cases} x^2 + y^2 \leq 9 \\ y < -3x + 1 \end{cases}$ *

54. $\begin{cases} y > x^2 \\ x + y < 6 \\ y < x + 6 \end{cases}$ *

55. $\begin{cases} y \geq 0 \\ 3x + 2y \geq 4 \\ x - y \leq 3 \end{cases}$ *

5.6

56. Find the value of the objective function $z = 2x + 3y$ at each corner of the graphed region shown. What is the maximum value of the objective function? What is the minimum value of the objective function? maximum value: 10; minimum value: 2*

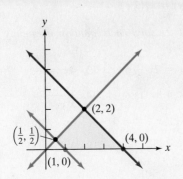

In Exercises 57–59, graph the region determined by the constraints. Then find the maximum value of the given objective function, subject to the constraints.

57. Objective Function
Constraints
Maximum is 24
and occurs at $x = 0, y = 8$.*

$z = 2x + 3y$
$\begin{cases} x \geq 0, y \geq 0 \\ x + y \leq 8 \\ 3x + 2y \geq 6 \end{cases}$

58. Objective Function
Constraints
Maximum is 33 and occurs at
$x = 5, y = 7$.*

$z = x + 4y$
$\begin{cases} 0 \leq x \leq 5, 0 \leq y \leq 7 \\ x + y \geq 3 \end{cases}$

59. Objective Function
Constraints
Maximum is 44
and occurs at $x = 4, y = 4$.*

$z = 5x + 6y$
$\begin{cases} x \geq 0, y \geq 0 \\ y \leq x \\ 2x + y \leq 12 \\ 2x + 3y \geq 6 \end{cases}$

60. A paper manufacturing company converts wood pulp to writing paper and newsprint. The profit on a unit of writing paper is $500 and the profit on a unit of newsprint is $350.

a. Let x represent the number of units of writing paper produced daily. Let y represent the number of units of newsprint produced daily. Write the objective function that models total daily profit. $z = 500x + 350y$

b. The manufacturer is bound by the following constraints:

• Equipment in the factory allows for making at most 200 units of paper (writing paper and newsprint) in a day.

• Regular customers require at least 10 units of writing paper and at least 80 units of newsprint daily.

Write a system of inequalities that models these constraints. *

c. Graph the inequalities in part (b). Use only the first quadrant, because x and y must both be positive. (*Suggestion:* Let each unit along the x- and y-axes represent 20.) *

d. Evaluate the objective function at each of the three vertices of the graphed region. *

e. Complete the missing portions of this statement: The company will make the greatest profit by producing ___120___ units of writing paper and ___80___ units of newsprint each day. The maximum daily profit is $___88,000___.

61. A manufacturer of lightweight tents makes two models whose specifications are given in the following table:

	Cutting Time per Tent	Assembly Time per Tent
Model A	0.9 hour	0.8 hour
Model B	1.8 hours	1.2 hours

Each month, the manufacturer has no more than 864 hours of labor available in the cutting department and at most 672 hours in the assembly division. The profits come to $25 per tent for model A and $40 per tent for model B. How many of each should be manufactured monthly to maximize the profit? model A: 480; model B: 240

CHAPTER 5 TEST

In Exercises 1–5, solve the system.

1. $\begin{cases} x = y + 4 \\ 3x + 7y = -18 \end{cases}$ $\{(1, -3)\}$

2. $\begin{cases} 2x + 5y = -2 \\ 3x - 4y = 20 \end{cases}$ $\{(4, -2)\}$

3. $\begin{cases} x + y + z = 6 \\ 3x + 4y - 7z = 1 \\ 2x - y + 3z = 5 \end{cases}$ $\{(1, 3, 2)\}$

4. $\begin{cases} x^2 + y^2 = 25 \\ x + y = 1 \end{cases}$ $\{(4, -3), (-3, 4)\}$

5. $\begin{cases} 2x^2 - 5y^2 = -2 \\ 3x^2 + 2y^2 = 35 \end{cases}$ $\{(3, 2), (3, -2), (-3, 2), (-3, -2)\}$

6. Find the partial fraction decomposition for $\dfrac{x}{(x + 1)(x^2 + 9)}$. *

In Exercises 7–10, graph the solution set of each inequality or system of inequalities.

7. $x - 2y < 8$ *

8. $\begin{cases} x \geq 0, y \geq 0 \\ 3x + y \leq 9 \\ 2x + 3y \geq 6 \end{cases}$ *

9. $\begin{cases} x^2 + y^2 > 1 \\ x^2 + y^2 < 4 \end{cases}$ *

10. $\begin{cases} y \leq 1 - x^2 \\ x^2 + y^2 \leq 9 \end{cases}$ *

11. Find the maximum value of the objective function $z = 3x + 5y$ subject to the following constraints: $x \geq 0, y \geq 0, x + y \leq 6, x \geq 2$. 26

12. You need to mix a 6% peroxide solution with a 9% peroxide solution to obtain 36 ounces of an 8% peroxide solution. How many ounces of each of the solutions must be used? 12 ounces of a 6% peroxide solution and 24 ounces of a 9% peroxide solution must be used.

13. A company is planning to produce and sell a new line of computers. The fixed cost will be $360,000 and it will cost $850 to produce each computer. Each computer will be sold for $1150.

a. Write the cost function, C, of producing x computers. *

b. Write the revenue function, R, from the sale of x computers. $R(x) = 1150x$

c. Determine the break-even point. Describe what this means. $(1200, 1,380,000)$*

14. Find the quadratic function whose graph passes through the points $(-1, -2)$, $(2, 1)$, and $(-2, 1)$. $y = x^2 - 3$

15. The rectangular plot of land shown in the figure is to be fenced along three sides using 39 feet of fencing. No fencing is to be placed along the river's edge. The area of the plot is 180 square feet. What are its dimensions? $x = 7.5$ ft, $y = 24$ ft or $x = 12$ ft, $y = 15$ ft

16. A manufacturer makes two types of jet skis, regular and deluxe. The profit on a regular jet ski is $200 and the profit on the deluxe model is $250. To meet customer demand, the company must manufacture at least 50 regular jet skis per week and at least 75 deluxe models. To maintain high quality, the total number of both models of jet skis manufactured by the company should not exceed 150 per week. How many jet skis of each type should be manufactured per week to obtain maximum profit? What is the maximum weekly profit? regular: 50; deluxe: 100; $35,000

CUMULATIVE REVIEW EXERCISES (CHAPTERS 1–5)

The figure shows the graph of $y = f(x)$ and its two vertical asymptotes. Use the graph to solve Exercises 1–10.

1. Find the domain and the range of f. *

2. Find the zeros. -1 and 1

3. What is the relative maximum and where does it occur? *

4. Find the interval(s) on which f is decreasing. $(0, 2)$

5. Is $f(-0.7)$ positive or negative? positive

6. Find $(f \circ f)(-1)$. 3

7. Use arrow notation to complete this statement:

$$f(x) \rightarrow -\infty \text{ as } \underline{x \rightarrow -2^+} \text{ or as } \underline{x \rightarrow 2^-}.$$

8. Does f appear to be even, odd, or neither? even

9. Graph $g(x) = f(x + 2) - 1$. *

10. Graph $h(x) = \frac{1}{2}f\left(\frac{1}{2}x\right)$. *

In Exercises 11–21, solve each equation, inequality, or system of equations.

11. $\sqrt{x^2 - 3x} = 2x - 6$ *

12. $4x^2 = 8x - 7$ *

13. $\left|\dfrac{x}{3} + 2\right| < 4$ $(-18, 6)$

14. $\dfrac{x + 5}{x - 1} > 2$ $(1, 7)$

15. $2x^3 + x^2 - 13x + 6 = 0$ *

16. $6x - 3(5x + 2) = 4(1 - x)$ *

17. $\log(x + 3) + \log x = 1$ *

18. $3^{x+2} = 11$ $\{-2 + \log_3 11\}$

19. $x^{\frac{1}{2}} - 2x^{\frac{1}{4}} - 15 = 0$ {625}

20. $\begin{cases} 3x - y = -2 \\ 2x^2 - y = 0 \end{cases}$ $\left\{\left(-\frac{1}{2}, \frac{1}{2}\right), (2, 8)\right\}$

21. $\begin{cases} x + 2y + 3z = -2 \\ 3x + 3y + 10z = -2 \\ 2y - 5z = 6 \end{cases}$ {(8, −2, −2)}

In Exercises 22–28, graph each equation, function, or inequality in a rectangular coordinate system. If two functions are indicated, graph both in the same system.

22. $f(x) = (x + 2)^2 - 4$ *

23. $2x - 3y \le 6$ *

24. $y = 3^{x-2}$ *

25. $f(x) = \dfrac{x^2 - x - 6}{x + 1}$ *

26. $f(x) = 2x - 4$ and f^{-1} *

27. $(x - 2)^2 + (y - 4)^2 > 9$ *

28. $f(x) = |x|$ and $g(x) = -|x - 2|$ *

In Exercises 29–30, let $f(x) = 2x^2 - x - 1$ and $g(x) = 1 - x$.

29. Find $(f \circ g)(x)$ and $(g \circ f)(x)$. *

30. Find $\dfrac{f(x + h) - f(x)}{h}$ and simplify. $4x + 2h - 1$

In Exercises 31–32, write the linear function in slope-intercept form satisfying the given conditions.

31. Graph of f passes through $(2, 4)$ and $(4, -2)$. $f(x) = -3x + 10$

32. Graph of g passes through $(-1, 0)$ and is perpendicular to the line whose equation is $x + 3y - 6 = 0$. $g(x) = 3x + 3$

33. You invested $4000 in two stocks paying 12% and 14% annual interest. At the end of the year, the total interest from these investments was $508. How much was invested at each rate? $2600 at 12%; $1400 at 14%

34. The length of a rectangle is 1 meter more than twice the width. If the rectangle's area is 36 square meters, find its dimensions. 4 m by 9 m

35. What interest rate is required for an investment of $6000 subject to continuous compounding to grow to $18,000 in 10 years? 10.99%

APPENDIX

Where Did That Come From?
Selected Proofs

The Product Rule

Let b, M, and N be positive real numbers with $b \neq 1$.

$$\log_b(MN) = \log_b M + \log_b N$$

Proof

We begin by letting $\log_b M = R$ and $\log_b N = S$.
Now we write each logarithm in exponential form.

$$\log_b M = R \quad \text{means} \quad b^R = M.$$
$$\log_b N = S \quad \text{means} \quad b^S = N.$$

By substituting and using a property of exponents, we see that

$$MN = b^R b^S = b^{R+S}.$$

Now we change $MN = b^{R+S}$ to logarithmic form.

$$MN = b^{R+S} \quad \text{means} \quad \log_b(MN) = R + S.$$

Finally, substituting $\log_b M$ for R and $\log_b N$ for S gives us

$$\log_b(MN) = \log_b M + \log_b N,$$

the property that we wanted to prove.

The quotient and power rules for logarithms are proved using similar procedures.

The Change-of-Base Property

For any logarithmic bases a and b, and any positive number M,

$$\log_b M = \frac{\log_a M}{\log_a b}.$$

Proof

To prove the change-of-base property, we let x equal the logarithm on the left side:

$$\log_b M = x.$$

Now we rewrite this logarithm in exponential form.

$$\log_b M = x \quad \text{means} \quad b^x = M.$$

Because b^x and M are equal, the logarithms with base a for each of these expressions must be equal. This means that

$$\log_a b^x = \log_a M$$
$$x \log_a b = \log_a M \qquad \text{Apply the power rule for logarithms on the left side.}$$
$$x = \frac{\log_a M}{\log_a b} \qquad \text{Solve for } x \text{ by dividing both sides by } \log_a b.$$

619

In our first step we let x equal $\log_b M$. Replacing x on the left side by $\log_b M$ gives us

$$\log_b M = \frac{\log_a M}{\log_a b},$$

which is the change-of-base property.

ANSWERS TO SELECTED EXERCISES

CHAPTER P

Exercise Set P.1

19. 60 ft **20.** 40 ft **24.** $\{r, e, a, l\}$ **30.** $\{1, 2, 3, 7, 8\}$ **31.** $\{1, 2, 3, 4, 5, 6, 7, 8, 10\}$ **32.** $\{0, 1, 2, 3, 4, 5, 6\}$ **34.** $\{e, m, p, t, y\}$

35. a. $\sqrt{100}$ **b.** $0, \sqrt{100}$ **c.** $-9, 0, \sqrt{100}$ **d.** $-9, -\dfrac{4}{5}, 0, 0.25, 9.2, \sqrt{100}$ **e.** $\sqrt{3}$ **f.** $-9, -\dfrac{4}{5}, 0, 0.25, \sqrt{3}, 9.2, \sqrt{100}$

36. a. $\sqrt{49}$ **b.** $0, \sqrt{49}$ **c.** $-7, 0, \sqrt{49}$ **d.** $-7, -0.\overline{6}, 0, \sqrt{49}$ **e.** $\sqrt{50}$ **f.** $-7, -0.\overline{6}, 0, \sqrt{49}, \sqrt{50}$

37. a. $\sqrt{64}$ **b.** $0, \sqrt{64}$ **c.** $-11, 0, \sqrt{64}$ **d.** $-11, -\dfrac{5}{6}, 0, 0.75, \sqrt{64}$ **e.** $\sqrt{5}, \pi$ **f.** $-11, -\dfrac{5}{6}, 0, 0.75, \sqrt{5}, \pi, \sqrt{64}$ **38. a.** $\sqrt{4}$

b. $0, \sqrt{4}$ **c.** $-5, 0, \sqrt{4}$ **d.** $-5, -0.\overline{3}, 0, \sqrt{4}$ **e.** $\sqrt{2}$ **f.** $-5, -0.\overline{3}, 0, \sqrt{2}, \sqrt{4}$ **40.** Answers may vary; an example is $\dfrac{1}{2}$.

72. $|-3 - (-26)|; 23$ **74.** $|-1.2 - (-5.4)|; 4.2$ **79.** commutative property of addition **88.** $24x - 2$ **100.** $-5x + 13y + 1$

121. $x - (x + 4); -4$ **123.** $6(-5x); -30x$ **124.** $10(-4x); -40x$ **129. a.** 140 beats per minute **b.** 160 beats per minute

130. a. 95 beats per minute **b.** 114 beats per minute **145.** does not make sense

Exercise Set P.2

48. $30x^{11}y^{12}$ **75.** $-60,000,100,000$ **76.** $-70,000,100,000$ **87.** 6.3×10^7 **88.** 8.2×10^7 **89.** 6.4×10^4 **90.** 4.2×10^4

91. 1.22×10^{-11} **92.** 1.53×10^{-11} **93.** 2.67×10^{13} **94.** 3.77×10^{13} **129.** does not make sense **131.** does not make sense

143. about 2.94×10^9 times

Exercise Set P.3

117. $\dfrac{\sqrt{5} + 1}{2}$; 1.62 to 1

Exercise Set P.4

4. yes; $x^4 - x^3 + x^2 - 5$ **9.** $11x^3 + 7x^2 - 12x - 4; 3$ **10.** $12x^3 - 5x^2 - 4x - 4; 3$ **11.** $12x^3 + 4x^2 + 12x - 14; 3$

12. $9x^4 + 4x^3 - 2x + 1; 4$ **13.** $6x^2 - 6x + 2; 2$ **14.** $6x^3 + 10x^2 + 11x - 8; 3$ **16.** $x^3 + 125$ **17.** $2x^3 - 9x^2 + 19x - 15$

18. $2x^3 - 9x^2 + 10x - 3$ **19.** $x^2 + 10x + 21$ **20.** $x^2 + 13x + 40$ **21.** $x^2 - 2x - 15$ **22.** $x^2 + x - 2$ **23.** $6x^2 + 13x + 5$

24. $21x^2 + 19x + 4$ **25.** $10x^2 - 9x - 9$ **26.** $14x^2 - 31x - 10$ **27.** $15x^4 - 47x^2 + 28$ **28.** $21x^4 - 41x^2 + 10$

29. $8x^5 - 40x^3 + 3x^2 - 15$ **30.** $7x^5 - 14x^3 + 5x^2 - 10$ **37.** $16x^4 - 25x^2$ **38.** $9x^4 - 16x^2$ **52.** $x^3 + 6x^2 + 12x + 8$

53. $8x^3 + 36x^2 + 54x + 27$ **54.** $27x^3 + 108x^2 + 144x + 64$ **57.** $27x^3 - 108x^2 + 144x - 64$ **58.** $8x^3 - 36x^2 + 54x - 27$

61. $2x^2y + 13xy + 13$ is of degree 3 **62.** $-11x^4y^2 - 11x^2y^2 + 2xy$ is of degree 6 **63.** $-5x^3 + 8xy - 9y^2$ is of degree 3

64. $-5x^4 - 4xy - 9y^3$ is of degree 4 **65.** $x^4y^2 + 8x^3y + y - 6x$ is of degree 6 **66.** $2x^4y^2 + 11x^3y - y - 8x$ is of degree 6

67. $7x^2 + 38xy + 15y^2$ **68.** $6x^2 + 61xy + 63y^2$ **69.** $2x^2 + xy - 21y^2$ **70.** $6x^2 + 13xy - 5y^2$ **71.** $15x^2y^2 + xy - 2$

72. $14x^4y^2 - 19x^2y - 3$ **77.** $x^3 - y^3$ **78.** $x^3 + y^3$ **79.** $9x^2 - 25y^2$ **80.** $49x^2 - 9y^2$ **81.** $49x^2y^4 - 100y^2$

82. $9x^2y^4 - 16y^2$ **106.** makes sense, although answers may vary

Mid-Chapter P Check Point

1. $12x^2 - x - 35$ **5.** $x + 45$ **10.** $2x^3 - 11x^2 + 17x - 5$ **12.** $18a^2 - 11ab - 10b^2$ **13.** $\{a, c, d, e, f, h\}$ **15.** $5x^2y^3 + 2xy - y^2$

Exercise Set P.5

5. $9x^2(x^2 - 2x + 3)$ **6.** $6x^2(x^2 - 3x + 2)$ **7.** $(x + 5)(x + 3)$ **8.** $(2x + 1)(x + 4)$ **9.** $(x - 3)(x^2 + 12)$ **10.** $(2x + 5)(x^2 + 17)$

11. $(x - 2)(x^2 + 5)$ **12.** $(x - 3)(x^2 + 4)$ **13.** $(x - 1)(x^2 + 2)$ **14.** $(x + 6)(x^2 - 2)$ **15.** $(3x - 2)(x^2 - 2)$ **16.** $(x - 1)(x^2 - 5)$

25. $(3x - 28)(x + 1)$ **27.** $(2x - 1)(3x - 4)$ **28.** $(2x - 3)(3x - 4)$ **29.** $(2x + 3)(2x + 5)$ **30.** $(8x + 1)(x + 4)$ **33.** $(5x + 8)(4x - 1)$

34. $(5x - 3)(3x - 2)$ **35.** $(2x + y)(x + y)$ **36.** $(3x + y)(x + y)$ **37.** $(3x + 2y)(2x - 3y)$ **38.** $(3x - 5y)(2x + y)$

43. $(3x + 5y)(3x - 5y)$ **44.** $(6x + 7y)(6x - 7y)$ **45.** $(x^2 + 4)(x + 2)(x - 2)$ **46.** $(x^2 + 1)(x + 1)(x - 1)$ **47.** $(4x^2 + 9)(2x + 3)(2x - 3)$

48. $(9x^2 + 1)(3x + 1)(3x - 1)$ **57.** $(x + 3)(x^2 - 3x + 9)$ **58.** $(x + 4)(x^2 - 4x + 16)$ **59.** $(x - 4)(x^2 + 4x + 16)$
60. $(x - 3)(x^2 + 3x + 9)$ **61.** $(2x - 1)(4x^2 + 2x + 1)$ **62.** $(3x - 1)(9x^2 + 3x + 1)$ **63.** $(4x + 3)(16x^2 - 12x + 9)$
64. $(2x + 5)(4x^2 - 10x + 25)$ **67.** $4(x + 2)(x - 3)$ **68.** $6(x + 2)(x - 5)$ **69.** $2(x^2 + 9)(x + 3)(x - 3)$ **70.** $7(x^2 + 1)(x + 1)(x - 1)$
71. $(x - 3)(x + 3)(x + 2)$ **72.** $(x - 5)(x + 5)(x + 3)$ **73.** $2(x - 8)(x + 7)$ **74.** $6(x - 2)(x + 1)$ **79.** $(x - 2)(x + 2)^2$
80. $(x - 1)(x + 1)(x + 2)$ **81.** $y(y^2 + 9)(y + 3)(y - 3)$ **82.** $y(y^2 + 4)(y + 2)(y - 2)$ **83.** $5y^2(2y + 3)(2y - 3)$
84. $3y^2(4y + 1)(4y - 1)$ **85.** $(x - 6 + 7y)(x - 6 - 7y)$ **86.** $(x - 5 + 6y)(x - 5 - 6y)$ **95.** $\dfrac{4(1 + 2x)}{x^{2/3}}$
96. $\dfrac{6(2 + x)}{x^{3/4}}$ **97.** $-(x + 3)^{1/2}(x + 2)$ **98.** $(x^2 + 4)^{3/2}(x^4 + 8x^2 + 17)$ **102.** $\dfrac{2(100x^2 + 95x + 11)}{(4x + 3)^2}$ **103.** $(x + 1)(5x - 6)(2x + 1)$
104. $(x - 1)(6x - 5)(2x + 1)$ **105.** $(x^2 + 6)(6x^2 - 1)$ **106.** $(x^2 + 5)(7x^2 - 1)$ **107.** $y(y^2 + 1)(y^4 - y^2 + 1)$ **108.** $(y + 2)(y^2 + y + 1)$
109. $(x + 2y)(x - 2y)(x + y)(x - y)$ **110.** $(x + 3y)(x - 3y)(x + y)(x - y)$ **113.** $(2x - y^2)(x - 3y^2)$ **114.** $(3x + 2y^2)(x + y^2)$
117. a. $9x^2 - 16$ **b.** $(3x + 4)(3x - 4)$ **118. a.** $49x^2 - 36$ **b.** $(7x + 6)(7x - 6)$ **119. a.** $x(x + y) - y(x + y)$ **b.** $(x + y)(x - y)$
131. makes sense **139.** $-(x + 5)(x - 1)$ **141.** $-\dfrac{10}{(x - 5)^{3/2}(x + 5)^{1/2}}$ **143.** $b = 0, 3, 4, -c(c + 4)$, where $c > 0$ is an integer

Exercise Set P.6

8. $\dfrac{4}{x - 2}, x \neq 2$ **10.** $\dfrac{x - 4}{3}, x \neq 4$ **12.** $\dfrac{y - 5}{y + 4}, y \neq -4, -1$ **14.** $\dfrac{x - 7}{x + 7}, x \neq 7, -7$ **16.** $\dfrac{1}{2}, x \neq 5, -\dfrac{3}{2}$ **17.** $\dfrac{(x - 3)(x + 3)}{x(x + 4)}, x \neq 0, -4, 3$

18. $2, x \neq 2, -2$ **21.** $\dfrac{x^2 + 2x + 4}{3x}, x \neq -2, 0, 2$ **22.** $\dfrac{1}{x^2 - 3x + 9}, x \neq -3$ **24.** $\dfrac{9}{28}, x \neq -5$ **25.** $\dfrac{(x - 2)^2}{x}, x \neq 0, -2, 2$

26. $4(x - 2), x \neq 2, -2$ **31.** $\dfrac{(x + 2)(x + 4)}{x - 5}, x \neq -6, -3, -1, 3, 5$ **32.** $\dfrac{7(x + 1)^2}{2x^2}, x \neq -5, -1, 0, 1, 5$ **34.** $2, x \neq -\dfrac{4}{3}$ **38.** $\dfrac{x}{x - 2}, x \neq 2$

41. $\dfrac{9x + 39}{(x + 4)(x + 5)}, x \neq -4, -5$ **42.** $\dfrac{10x - 28}{(x - 2)(x - 3)}, x \neq 2, 3$ **44.** $\dfrac{x + 12}{x(x + 3)}, x \neq -3, 0$ **45.** $\dfrac{3x^2 + 4}{(x + 2)(x - 2)}, x \neq -2, 2$

46. $\dfrac{2x^2 + 5x + 12}{(x - 3)(x + 2)}, x \neq 3, -2$ **47.** $\dfrac{2x^2 + 50}{(x - 5)(x + 5)}, x \neq -5, 5$ **48.** $\dfrac{2x^2 + 18}{(x - 3)(x + 3)}, x \neq -3, 3$ **49.** $\dfrac{13}{6(x + 2)}, x \neq -2$

50. $\dfrac{29}{6(x + 4)}, x \neq -4$ **51.** $\dfrac{4x + 16}{(x + 3)^2}, x \neq -3$ **52.** $\dfrac{20x - 6}{(5x - 2)(5x + 2)}, x \neq -\dfrac{2}{5}, \dfrac{2}{5}$ **53.** $\dfrac{x^2 - x}{(x + 5)(x - 2)(x + 3)}, x \neq -5, 2, -3$

54. $-\dfrac{5x}{(x - 6)(x - 1)(x + 4)}, x \neq 6, 1, -4$ **55.** $\dfrac{-x^2 - 2x + 1}{(x + 1)(x - 1)}, x \neq -1, 1$ **56.** $\dfrac{-x^2 - 2x + 3}{(x + 2)(x - 2)}, x \neq -2, 2$ **58.** $\dfrac{x^2 + 40x - 25}{(x + 5)(x - 4)}, x \neq -5, 4$

66. $\dfrac{x - 2}{x + 1}, x \neq 2, 3, -1$ **68.** $\dfrac{2(x + 2)}{x + 1}, x \neq -1, 2, -2, 1$ **71.** $-\dfrac{2x + h}{x^2(x + h)^2}, x \neq 0, h \neq 0, x \neq -h$

72. $\dfrac{1}{(x + 1)(x + h + 1)}, x \neq -1, h \neq 0, x \neq -h - 1$ **79.** $\dfrac{2d}{a^2 + ab + b^2}$ **80.** $\dfrac{a^2 + b^2}{a^2 + ab + b^2}$ **81. a.** $86.67, 520, 1170$; It costs \$86,670,000 to
inoculate 40% of the population against this strain of flu, \$520,000,000 to inoculate 80% of the population, and \$1,170,000,000 to inoculate 90% of the
population. **c.** The cost increases rapidly; it is impossible to inoculate 100% of the population.

82. $\dfrac{2r_1 r_2}{r_1 + r_2}; 34\dfrac{2}{7}$ mph **83. c.** $\dfrac{-33x^2 + 263x + 515}{-60x^2 + 499x + 295}$ **98.** does not make sense **101.** makes sense **108.** $\dfrac{x - y + 1}{(x - y)(x - y)}$

110. $-5; 0; 3; 4; 3; 0; -5$ **111.** $-8; -3; 0; 1; 0; -3; -8$

Chapter P Review Exercises

8. a. $\sqrt{81}$ **b.** $0, \sqrt{81}$ **c.** $-17, 0, \sqrt{81}$ **d.** $-17, -\dfrac{9}{13}, 0, 0.75, \sqrt{81}$ **e.** $\sqrt{2}, \pi$ **f.** $-17, -\dfrac{9}{13}, 0, 0.75, \sqrt{2}, \pi, \sqrt{81}$
12. $|4 - (-17)|; 21$ **17.** commutative property of multiplication **18.** commutative property of addition **36.** $390,000$ **46.** $4x\sqrt{3}$
73. $8x^3 + 10x^2 - 20x - 4$; degree 3 **74.** $8x^4 - 5x^3 + 6$; degree 4 **75.** $12x^3 + x^2 - 21x + 10$ **76.** $6x^2 - 7x - 5$
80. $8x^3 + 12x^2 + 6x + 1$ **82.** $-x^2 - 17xy - 3y^2$; degree 2 **83.** $24x^3y^2 + x^2y - 12x^2 + 4$; degree 5 **84.** $3x^2 + 16xy - 35y^2$
85. $9x^2 - 30xy + 25y^2$ **86.** $9x^4 + 12x^2y + 4y^2$ **87.** $49x^2 - 16y^2$ **90.** $(x - 4)(x - 7)$ **91.** $(3x + 1)(5x - 2)$
94. $3x^2(x - 5)(x + 2)$ **96.** $(x + 3)(x - 3)^2$ **98.** $(x^2 + 4)(x + 2)(x - 2)$ **101.** $3x^2(x - 2)(x + 2)$ **102.** $(3x - 5)(9x^2 + 15x + 25)$
103. $x(x - 1)(x + 1)(x^2 + 1)$ **104.** $(x^2 - 2)(x + 5)$ **107.** $(x + 2)(x - 2)(x^2 + 3)^{1/2}(-x^4 + x^2 + 13)$ **109.** $x^2, x \neq -2$
110. $\dfrac{x - 3}{x - 6}, x \neq -6, 6$ **111.** $\dfrac{x}{x + 2}, x \neq -2$ **112.** $\dfrac{(x + 3)^3}{(x - 2)^2(x + 2)}, x \neq 2, -2$ **113.** $\dfrac{2}{x(x + 1)}, x \neq 0, 1, -1, -\dfrac{1}{3}$
114. $\dfrac{x + 3}{x - 4}, x \neq -3, 4, 2, 8$ **115.** $\dfrac{1}{x - 3}, x \neq 3, -3$ **116.** $\dfrac{4x(x - 1)}{(x + 2)(x - 2)}, x \neq 2, -2$ **117.** $\dfrac{2x^2 - 3}{(x - 3)(x + 3)(x - 2)}, x \neq 3, -3, 2$
118. $\dfrac{11x^2 - x - 11}{(2x - 1)(x + 3)(3x + 2)}, x \neq \dfrac{1}{2}, -3, -\dfrac{2}{3}$ **119.** $\dfrac{3}{x}, x \neq 0, 2$ **120.** $\dfrac{3x}{x - 4}, x \neq 0, 4, -4$ **121.** $\dfrac{3x + 8}{3x + 10}, x \neq -3, -\dfrac{10}{3}$

Chapter P Test

5. $6x^2y^3 + 4xy + 2y^2$ **11.** $\dfrac{x+3}{x-2}, x \neq 2, 1$ **12.** 2.5×10^1 **13.** $2x^3 - 13x^2 + 26x - 15$ **14.** $25x^2 + 30xy + 9y^2$

15. $\dfrac{2(x+3)}{x+1}, x \neq 3, -1, -4, -3$ **16.** $\dfrac{x^2 + 2x + 15}{(x+3)(x-3)}, x \neq 3, -3$ **17.** $\dfrac{11}{(x-3)(x-4)}, x \neq 3, 4$ **20.** $(x^2 + 3)(x + 2)$

24. $(x + 5 + 3y)(x + 5 - 3y)$ **26.** $-7, -\dfrac{4}{5}, 0, 0.25, \sqrt{4}, \dfrac{22}{7}$ **31.** 1.32×10^{10}

CHAPTER 1

Section 1.1
Check Point Exercises

1. **2.** **3.**

$y = 4 - x$ $y = |x + 1|$

4. minimum x-value: -100; maximum x-value: 100; distance
between tick marks on x-axis: 50; minimum y-value: -100;
maximum y-value: 100; distance between tick marks on
y-axis: 10

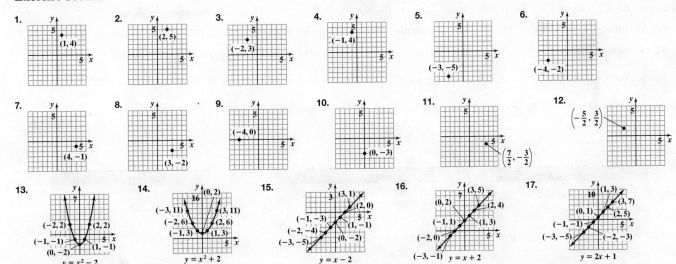

Exercise Set 1.1

1. $(1, 4)$ **2.** $(2, 5)$ **3.** $(-2, 3)$ **4.** $(-1, 4)$ **5.** $(-3, -5)$ **6.** $(-4, -2)$

7. $(4, -1)$ **8.** $(3, -2)$ **9.** $(-4, 0)$ **10.** $(0, -3)$ **11.** $\left(\dfrac{7}{2}, -\dfrac{3}{2}\right)$ **12.** $\left(-\dfrac{5}{2}, \dfrac{3}{2}\right)$

13. $y = x^2 - 2$; $(-2, 2), (-1, -1), (0, -2), (1, -1), (2, 2)$

14. $y = x^2 + 2$; $(0, 2), (-3, 11), (-2, 6), (-1, 3), (1, 3), (2, 6), (3, 11)$

15. $y = x - 2$; $(3, 1), (2, 0), (1, -1), (0, -2), (-1, -3), (-2, -4), (-3, -5)$

16. $y = x + 2$; $(3, 5), (2, 4), (1, 3), (0, 2), (-1, 1), (-2, 0), (-3, -1)$

17. $y = 2x + 1$; $(1, 3), (3, 7), (2, 5), (0, 1), (-1, -1), (-2, -3), (-3, -5)$

18. $y = 2x - 4$; $(2, 0), (3, 2), (1, -2), (-1, -6), (0, -4), (-2, -8), (-3, -10)$

19. $y = -\dfrac{1}{2}x$; $(-2, 1), \left(-3, \dfrac{3}{2}\right), (2, -1), \left(3, -\dfrac{3}{2}\right), \left(-1, \dfrac{1}{2}\right), (0, 0), \left(1, -\dfrac{1}{2}\right)$

20. $y = -\dfrac{1}{2}x + 2$; $\left(-3, \dfrac{7}{2}\right), (0, 2), \left(1, \dfrac{3}{2}\right), (2, 1), (-2, 3), \left(-1, \dfrac{5}{2}\right), \left(3, \dfrac{1}{2}\right)$

21. $y = 2|x|$; $(-3, 6), (3, 6), (-2, 4), (2, 4), (-1, 2), (1, 2), (0, 0)$

22. $y = -2|x|$; $(0, 0), (-1, -2), (1, -2), (-2, -4), (2, -4), (-3, -6), (3, -6)$

23.
$y = |x| + 1$

24.
$y = |x| - 1$

25.
$y = 9 - x^2$

26.
$y = -x^2$

27.
$y = x^3$

28.
$y = x^3 - 1$

39. $(-2, 4)$ and $(1, 1)$

47. $y = 2x + 4$

$y = 2x + 4$

48. $y = 4 - 2x$

$y = 4 - 2x$

49. $y = 3 - x^2$

$y = 3 - x^2$

50. $y = x^2 + 2$

$y = x^2 + 2$

51.
$y = 5$

52.
$y = -1$

53.

54.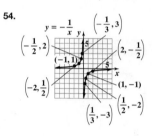

55. b. 45%; It's greater than the estimate, although answers may vary.
d. 70%; It's less than the estimate, although answers may vary.
e. 2000; ≈49% **56. b.** 66.4%; It's greater than the estimate, although answers may vary. **d.** 45.4%; It's greater than the estimate, although answers may vary. **e.** 1990; 90% **67.** makes sense
68. does not make sense

Exercise Set 1.2

4. $\left\{\dfrac{25}{3}\right\}$ **8.** $\{-19\}$ **17.** $\{12\}$ **18.** $\{30\}$ **20.** $\{15\}$ **21.** $\{-15\}$ **36. a.** 0 **b.** $\{3\}$ **37. a.** 0 **b.** $\{4\}$ **38. a.** 0 **b.** $\{8\}$

39. a. 1 **b.** $\{3\}$ **40. a.** -4 **b.** $\{-3\}$ **41. a.** -1 **b.** \varnothing **42. a.** 2 **b.** \varnothing **61.** $\{x|x$ is a real number$\}$; identity

62. $\{x|x$ is a real number$\}$; identity **63.** \varnothing; inconsistent equation **64.** \varnothing; inconsistent equation **65.** $\{0\}$; conditional equation

66. $\{0\}$; conditional equation **67.** \varnothing; inconsistent equation **68.** \varnothing; inconsistent equation **69.** $\{-7\}$; conditional equation

70. $\left\{\dfrac{46}{5}\right\}$; conditional equation **71.** \varnothing; inconsistent equation **72.** \varnothing; inconsistent equation **75.** $\{8\}$; conditional equation

76. $\{6\}$; conditional equation **77.** $\{-1\}$; conditional equation **78.** $\{3\}$; conditional equation **80.** $\left\{\dfrac{1}{7}\right\}$; conditional equation

106. a. $C = \dfrac{x + 0.35(200)}{x + 200}$

116. $\{3\}$

117. $\{5\}$

118. $\{-7\}$

119. $\{-5\}$

125. false

Exercise Set 1.3

7. c.

8. c.

21. $2000 at 6%; $5000 at 8%

22. $5000 at 5%; $6000 at 8% **45.** $p = \dfrac{T - D}{m}$

46. $M = \dfrac{P - C}{C}$ **47.** $a = \dfrac{2A}{h} - b$

48. $b = \dfrac{2A}{h} - a$ **49.** $r = \dfrac{S - P}{Pt}$

50. $t = \dfrac{S - P}{Pr}$ **51.** $S = \dfrac{F}{B} + V$ **52.** $r = -\dfrac{C}{S} + 1$

62. a. $F = 30 + 5x; F = 7.5x$

b.
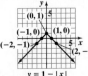

66. does not make sense

67. North campus had 600 students; South campus had 400 students.

71. $4000 for the mother; $8000 for the boy; $2000 for the girl

75. $-14 - 29x + 15x^2$ or $15x^2 - 29x - 14$

Exercise Set 1.4

57. $(5 + i\sqrt{15}) + (5 - i\sqrt{15}) = 10; (5 + i\sqrt{15})(5 - i\sqrt{15}) = 25 - 15i^2 = 25 + 15 = 40$

Exercise Set 1.5

31. $\left\{\dfrac{1 - \sqrt{7}}{5}, \dfrac{1 + \sqrt{7}}{5}\right\}$ **32.** $\left\{\dfrac{3 - \sqrt{5}}{8}, \dfrac{3 + \sqrt{5}}{8}\right\}$ **33.** $\left\{\dfrac{4 - 2\sqrt{2}}{3}, \dfrac{4 + 2\sqrt{2}}{3}\right\}$ **34.** $\left\{\dfrac{-8 - 3\sqrt{3}}{2}, \dfrac{-8 + 3\sqrt{3}}{2}\right\}$

35. $36; x^2 + 12x + 36 = (x + 6)^2$ **36.** $64; x^2 + 16x + 64 = (x + 8)^2$ **37.** $25; x^2 - 10x + 25 = (x - 5)^2$ **38.** $49; x^2 - 14x + 49 = (x - 7)^2$

39. $\dfrac{9}{4}; x^2 + 3x + \dfrac{9}{4} = \left(x + \dfrac{3}{2}\right)^2$ **40.** $\dfrac{25}{4}; x^2 + 5x + \dfrac{25}{4} = \left(x + \dfrac{5}{2}\right)^2$ **41.** $\dfrac{49}{4}; x^2 - 7x + \dfrac{49}{4} = \left(x - \dfrac{7}{2}\right)^2$ **42.** $\dfrac{81}{4}; x^2 - 9x + \dfrac{81}{4} = \left(x - \dfrac{9}{2}\right)^2$

43. $\dfrac{1}{9}; x^2 - \dfrac{2}{3}x + \dfrac{1}{9} = \left(x - \dfrac{1}{3}\right)^2$ **44.** $\dfrac{4}{25}; x^2 + \dfrac{4}{5}x + \dfrac{4}{25} = \left(x + \dfrac{2}{5}\right)^2$ **45.** $\dfrac{1}{36}; x^2 - \dfrac{1}{3}x + \dfrac{1}{36} = \left(x - \dfrac{1}{6}\right)^2$ **46.** $\dfrac{1}{64}; x^2 - \dfrac{1}{4}x + \dfrac{1}{64} = \left(x - \dfrac{1}{8}\right)^2$

57. $\left\{\dfrac{-3 + \sqrt{13}}{2}, \dfrac{-3 - \sqrt{13}}{2}\right\}$ **58.** $\left\{\dfrac{3 + \sqrt{29}}{2}, \dfrac{3 - \sqrt{29}}{2}\right\}$ **61.** $\left\{\dfrac{1 + \sqrt{2}}{2}, \dfrac{1 - \sqrt{2}}{2}\right\}$ **62.** $\left\{\dfrac{2 + \sqrt{6}}{2}, \dfrac{2 - \sqrt{6}}{2}\right\}$ **63.** $\left\{\dfrac{1 + \sqrt{7}}{3}, \dfrac{1 - \sqrt{7}}{3}\right\}$

64. $\left\{\dfrac{5 + \sqrt{145}}{6}, \dfrac{5 - \sqrt{145}}{6}\right\}$ **67.** $\left\{\dfrac{-5 + \sqrt{13}}{2}, \dfrac{-5 - \sqrt{13}}{2}\right\}$ **68.** $\left\{\dfrac{-5 + \sqrt{17}}{2}, \dfrac{-5 - \sqrt{17}}{2}\right\}$ **69.** $\left\{\dfrac{3 + \sqrt{57}}{6}, \dfrac{3 - \sqrt{57}}{6}\right\}$

70. $\left\{\dfrac{-1 + \sqrt{41}}{10}, \dfrac{-1 - \sqrt{41}}{10}\right\}$ **71.** $\left\{\dfrac{1 + \sqrt{29}}{4}, \dfrac{1 - \sqrt{29}}{4}\right\}$ **72.** $\left\{\dfrac{3 + \sqrt{6}}{3}, \dfrac{3 - \sqrt{6}}{3}\right\}$ **75.** $36; 2$ unequal real solutions that are rational

76. $-44; 2$ complex imaginary solutions that are complex conjugates **77.** $97; 2$ unequal real solutions that are irrational **78.** $169; 2$ unequal real solutions that are rational **79.** $0; 1$ real solution that is rational **80.** $-8; 2$ complex imaginary solutions that are complex conjugates

81. $37; 2$ unequal real solutions that are irrational **82.** $40; 2$ unequal real solutions that are irrational **90.** $\left\{\dfrac{-3 + \sqrt{17}}{4}, \dfrac{-3 - \sqrt{17}}{4}\right\}$

91. $\left\{\dfrac{-11 + \sqrt{33}}{4}, \dfrac{-11 - \sqrt{33}}{4}\right\}$ **92.** $\left\{\dfrac{3 + \sqrt{65}}{4}, \dfrac{3 - \sqrt{65}}{4}\right\}$ **102.** $\left\{\dfrac{1}{5} + i\dfrac{\sqrt{14}}{5}, \dfrac{1}{5} - i\dfrac{\sqrt{14}}{5}\right\}$ **106.** $\left\{\dfrac{5 + \sqrt{73}}{2}, \dfrac{5 - \sqrt{73}}{2}\right\}$

109. -1 and $5;$ d **110.** $3 - \sqrt{2}$ and $3 + \sqrt{2};$ a **111.** -3 and $1;$ f **112.** -4 and $-2;$ e **113.** no x-intercepts; b **141.** $20\sqrt{2}$ ft, or about 28.3 ft

142. $90\sqrt{2}$ ft, or about 127.3 ft **144. b.** The length of the hypotenuse is the length of a leg times $\sqrt{2}$. **170.** false

Mid-Chapter 1 Check Point

1. $\{6\}$ **4.** $\left\{\dfrac{3 \pm \sqrt{15}}{3}\right\}$ **7.** $\left\{\dfrac{3}{4} \pm i\dfrac{\sqrt{23}}{4}\right\}$ **20.** $\{-5 \pm 2\sqrt{7}\}$ **21.** two complex imaginary solutions that are complex conjugates

22. two unequal real solutions that are rational

23.

$y = 2x - 1$

24.

$y = 1 - |x|$

25.

$y = x^2 + 2$

Exercise Set 1.6

3. $\left\{-2, -\dfrac{2}{3}, 2\right\}$ **4.** $\left\{-\dfrac{3}{2}, \dfrac{3}{2}, 3\right\}$ **5.** $\left\{-\dfrac{1}{2}, \dfrac{1}{2}, \dfrac{3}{2}\right\}$ **6.** $\left\{-1, -\dfrac{1}{3}, \dfrac{1}{3}\right\}$ **8.** $\left\{-\dfrac{2}{3}, \dfrac{2}{3}, 2\right\}$ **10.** $\left\{0, 3, -\dfrac{3}{2} + i\dfrac{3\sqrt{3}}{2}, -\dfrac{3}{2} - i\dfrac{3\sqrt{3}}{2}\right\}$

23. $\{8\}$ **24.** $\{4\}$ **39.** $\{-4, 5\}$ **40.** $\{1, 2\}$ **41.** $\{-2, -1, 1, 2\}$ **42.** $\{-3, -2, 2, 3\}$ **43.** $\left\{-\dfrac{4}{3}, -1, 1, \dfrac{4}{3}\right\}$ **50.** $\left\{-125, \dfrac{27}{8}\right\}$

93. $-2, 0,$ and 2 **94.** -9 and $\dfrac{3}{4}$ **107.** $(5.4, 1.16)$ **108.** $(2.9, 0.85)$ **128.** makes sense

Section 1.7
Check Point Exercises

1. a. $\{x \mid -2 \le x < 5\}$

b. $\{x \mid 1 \le x \le 3.5\}$

c. $\{x \mid x < -1\}$

3. $[-1, \infty)$ or $\{x \mid x \ge -1\}$

4. $\{x \mid x < 4\}$ or $(-\infty, 4)$

5. $\{x \mid x \ge 13\}$ or $[13, \infty)$

7. $[-1, 4)$ or $\{x \mid -1 \le x < 4\}$

8. $(-3, 7)$ or $\{x \mid -3 < x < 7\}$

9. $\left\{x \,\middle|\, -\dfrac{11}{5} \le x \le 3\right\}$ or $\left[-\dfrac{11}{5}, 3\right]$

10. $\{x \mid x < -4 \text{ or } x > 8\}$ or $(-\infty, -4) \cup (8, \infty)$

Exercise Set 1.7

1. $\{x \mid 1 < x \le 6\}$

2. $\{x \mid -2 < x \le 4\}$

3. $\{x \mid -5 \le x < 2\}$

4. $\{x \mid -4 \le x < 3\}$

5. $\{x \mid -3 \le x \le 1\}$

6. $\{x \mid -2 \le x \le 5\}$

7. $\{x \mid x > 2\}$

8. $\{x \mid x > 3\}$

9. $\{x \mid x \ge -3\}$

10. $\{x \mid x \ge -5\}$

11. $\{x \mid x < 3\}$

12. $\{x \mid x < 2\}$

13. $\{x \mid x < 5.5\}$

14. $\{x \mid x \le 3.5\}$

27. $(-\infty, 3)$

28. $(-\infty, 6)$

29. $\left[\dfrac{20}{3}, \infty\right)$

30. $[2, \infty)$

31. $(-\infty, -4]$

32. $[-6, \infty)$

33. $\left(-\infty, -\dfrac{2}{5}\right]$

34. $\left(-\infty, -\dfrac{53}{6}\right]$

35. $[0, \infty)$

36. $(5, \infty)$

37. $(-\infty, 1)$

38. $(-\infty, -4)$

39. $[6, \infty)$

40. $[2, \infty)$

41. $[-10, \infty)$

42. $[-2, \infty)$

43. $(-\infty, -6)$

44. $(8, \infty)$

45. $[13, \infty)$

46. $\left[-\dfrac{19}{6}, \infty\right)$

47. $(-\infty, \infty)$

55. $(-5, -2]$ **56.** $\left[\dfrac{3}{2}, \dfrac{11}{2}\right)$

58. $[-4, 2)$ **66.** $\left[-\dfrac{19}{3}, 7\right]$

71. $(-\infty, -1] \cup [3, \infty)$ **72.** $(-\infty, -7] \cup [1, \infty)$ **73.** $\left(-\infty, \dfrac{1}{3}\right) \cup (5, \infty)$ **74.** $\left(-\infty, -\dfrac{11}{5}\right) \cup (3, \infty)$ **75.** $(-\infty, -5] \cup [3, \infty)$

76. $(-\infty, -2] \cup [4, \infty)$ **77.** $(-\infty, -3) \cup (12, \infty)$ **78.** $(-\infty, -8) \cup (16, \infty)$ **79.** $(-\infty, -1] \cup [3, \infty)$ **80.** $\left(-\infty, -\dfrac{17}{10}\right] \cup \left[\dfrac{7}{10}, \infty\right)$

83. $(-\infty, -3) \cup (5, \infty)$ **84.** $(-\infty, 2) \cup (8, \infty)$ **85.** $(-\infty, -1] \cup [2, \infty)$ **86.** $(-\infty, -4] \cup \left[\dfrac{1}{2}, \infty\right)$ **89.** $\left(-\infty, \dfrac{1}{3}\right) \cup (1, \infty)$

90. $(-\infty, -2) \cup (6, \infty)$ **91.** $\left(-\infty, -\dfrac{75}{14}\right) \cup \left(\dfrac{87}{14}, \infty\right)$ **92.** $(-\infty, \infty)$ **93.** $(-\infty, -6] \cup [24, \infty)$ **104.** $(-\infty, -1] \cup [9, \infty)$

107. $\left(-\infty, -\dfrac{1}{3}\right) \cup [3, \infty)$ **108.** $\left[-2, \dfrac{9}{2}\right]$ **111.** intimacy \geq passion or passion \leq intimacy **112.** commitment \geq intimacy or intimacy \leq commitment

113. commitment $>$ passion or passion $<$ commitment **114.** commitment $>$ passion or passion $<$ commitment
119. between 80 and 110 minutes, inclusive **122.** $50 + 0.20x < 20 + 0.50x$; more than 100 mi **123.** $15 + 0.08x < 3 + 0.12x$; more than 300 texts
124. $1800 + 0.03x < 200 + 0.08x$; greater than $32,000 **125.** $2 + 0.08x < 8 + 0.05x$; 199 checks or fewer **126.** $2x > 10,000 + 0.40x$;
more than 6250 tapes **127.** $5.50x > 3000 + 3x$; more than 1200 packages **128.** $265 + 65x \leq 2800$; at most 39 bags **129.** $245 + 95x \leq 3000$;
at most 29 bags **130.** $\dfrac{86 + 88 + 92 + 84 + x + x}{6} \geq 90$; at least 95% **131. a.** $\dfrac{86 + 88 + x}{3} \geq 90$; at least a 96 **b.** $\dfrac{86 + 88 + x}{3} < 80$;
a grade less than 66 **132.** 1.5 to 3.5 hours, inclusive

142. $\{x \mid x < 4\}$ or $(-\infty, 4)$

143. $\{x \mid x < -3\}$ or $(-\infty, -3)$

144. a.

b.

145. b.

154. Because $x > y$, $y - x$ represents a negative number, so when both sides are multiplied by $(y - x)$, the inequality must be reversed.

158.

159. c. all real numbers **d.** all real numbers greater than or equal to 1

Chapter 1 Review Exercises

1.

2.

3.

4.

5.

9. $(20, 8)$; 8% of college students anticipated a starting salary of $20 thousand. **22.** $\{x \mid x \text{ is a real number}\}$; identity

30. all real numbers except -1 and 1; identity **42.** $4500 at 2\%; $3500 at 5\%

46. $g = \dfrac{T}{r + vt}$ **63.** $\left\{\dfrac{4 - 3\sqrt{2}}{3}, \dfrac{4 + 3\sqrt{2}}{3}\right\}$ **69.** $\{1 + 3i\sqrt{2}, 1 - 3i\sqrt{2}\}$

71. 2 complex imaginary solutions **72.** 2 unequal real solutions that are rational **74.** $\left\{-2, \dfrac{10}{3}\right\}$

75. $\left\{\dfrac{7 + \sqrt{37}}{6}, \dfrac{7 - \sqrt{37}}{6}\right\}$ **78.** $\left\{\dfrac{1}{6} + i\dfrac{\sqrt{23}}{6}, \dfrac{1}{6} - i\dfrac{\sqrt{23}}{6}\right\}$ **80.** $\{-2 - 2i, -2 + 2i\}$ **87.** $\left\{-3, \dfrac{1}{2}, 3\right\}$ **89.** $\{8\}$ **92.** $\{-2, -1, 1, 2\}$

96. $\left\{-1, -\dfrac{2\sqrt{6}}{9}, \dfrac{2\sqrt{6}}{9}, 1\right\}$ **99.** $\{-3, -2, 3\}$ **101.** $\{-3, -\sqrt{2}, \sqrt{2}\}$

104. $\{x| {-3} \le x < 5\}$

105. $\{x| x > -2\}$

106. $\{x| x \le 0\}$

111. $[-2, \infty)$

112. $\left[\dfrac{3}{5}, \infty\right)$

113. $\left(-\infty, -\dfrac{21}{2}\right)$

114. $(-3, \infty)$

115. $(-\infty, -2]$

117. $(2, 3]$

118. $[-9, 6]$

119. $(-\infty, -6) \cup (0, \infty)$

120. $(-\infty, -3] \cup [-2, \infty)$

121. $(-\infty, -5] \cup [1, \infty)$

126. at least \$120,000

Chapter 1 Test

20. $(-\infty, 12]$

21. $\left[\dfrac{21}{8}, \infty\right)$

22. $\left[-7, \dfrac{13}{2}\right)$

23. $\left(-\infty, -\dfrac{5}{3}\right] \cup \left[\dfrac{1}{3}, \infty\right)$

30.

31.

CHAPTER 2

Section 2.1

Check Point Exercises

5. $f(x) = 2x$
$g(x) = 2x - 3$

Exercise Set 2.1

7. function; $\{-3, -2, -1, 0\}$; $\{-3, -2, -1, 0\}$ **8.** function; $\{-7, -5, -3, 0\}$; $\{-7, -5, -3, 0\}$ **27. c.** $-4x + 5$ **28. c.** $-3x + 7$

29. b. $x^2 + 12x + 38$ **c.** $x^2 - 2x + 3$ **30. b.** $x^2 - 6x - 19$ **c.** $x^2 + 10x - 3$ **33. c.** $\sqrt{x} + 3$ **34. c.** $\sqrt{2x} - 6$

36. c. $\dfrac{-4x^3 + 1}{-x^3}$ or $\dfrac{4x^3 - 1}{x^3}$ **38. c.** $\begin{cases} 1, & \text{if } x < -6 \\ -1, & \text{if } x > -6 \end{cases}$

39.

The graph of g is the graph of f shifted up by 3 units.

40.

The graph of g is the graph of f shifted down by 4 units.

41.

The graph of g is the graph of f shifted down by 1 unit.

42.

The graph of g is the graph of f shifted up by 3 units.

43.

The graph of g is the graph of f shifted up by 1 unit.

44.

The graph of g is the graph of f shifted down by 2 units.

45.

The graph of g is the graph of f shifted down by 2 units.

46.

The graph of g is the graph of f shifted up by 1 unit.

47.

The graph of g is the graph of f shifted up by 2 units.

48.

The graph of g is the graph of f shifted down by 1 unit.

49.

The graph of g is the graph of f shifted up by 2 units.

50.

The graph of g is the graph of f shifted up by 5 units.

51.

The graph of g is the graph of f shifted down by 1 unit.

52.

The graph of g is the graph of f shifted up by 2 units.

53.

The graph of g is the graph of f shifted to the right by 1 unit.

54.

The graph of g is the graph of f shifted to the left by 2 units.

77. a. $(-\infty, \infty)$ **b.** $[-4, \infty)$ **c.** -3 and 1 **d.** -3 **e.** $f(-2) = -3$ and $f(2) = 5$ **78. a.** $(-\infty, \infty)$ **b.** $(-\infty, 4]$ **c.** -3 and 1 **d.** 3 **e.** $f(-2) = 3$ and $f(2) = -5$ **79. a.** $(-\infty, \infty)$ **b.** $[1, \infty)$ **c.** none **d.** 1 **e.** $f(-1) = 2$ and $f(3) = 4$ **80. a.** $(-\infty, \infty)$ **b.** $[0, \infty)$ **c.** -1 **d.** 1 **e.** $f(-4) = 3$ and $f(3) = 4$ **81. a.** $[0, 5)$ **b.** $[-1, 5)$ **c.** 2 **d.** -1 **e.** $f(3) = 1$ **82. a.** $(-6, 0]$ **b.** $[-3, 4)$ **c.** -4 **d.** -3 **e.** $f(-4) = 0$ **83. a.** $[0, \infty)$ **b.** $[1, \infty)$ **c.** none **d.** 1 **e.** $f(4) = 3$ **84. a.** $[-1, \infty)$ **b.** $[0, \infty)$ **c.** -1 **d.** 1 **e.** $f(3) = 2$ **85. a.** $[-2, 6]$ **b.** $[-2, 6]$ **c.** 4 **d.** 4 **e.** $f(-1) = 5$ **86. a.** $[-3, 2]$ **b.** $[-5, 5]$ **c.** $-\dfrac{1}{2}$ **d.** 1 **e.** $f(-2) = -3$ **87. a.** $(-\infty, \infty)$ **b.** $(-\infty, -2]$ **c.** none **d.** -2 **e.** $f(-4) = -5$ and $f(4) = -2$ **88. a.** $(-\infty, \infty)$ **b.** $[0, \infty)$ **c.** $\{x | x \le 0\}$ **d.** 0 **e.** $f(-2) = 0$ and $f(2) = 4$ **91. a.** $\{-5, -2, 0, 1, 3\}$ **b.** $\{2\}$ **c.** none **d.** 2 **e.** $f(-5) + f(3) = 4$ **92. a.** $\{-5, -2, 0, 1, 4\}$ **b.** $\{-2\}$ **c.** none **d.** -2 **e.** $f(-5) + f(4) = -4$ **93.** $-2; 10$ **94.** $-8; 76$ **99. a.** $\{$(Iceland, 9.7), (Finland, 9.6), (New Zealand, 9.6), (Denmark, 9.5)$\}$ **b.** Yes; each country corresponds to exactly one corruption rating. **c.** $\{$(9.7, Iceland), (9.6, Finland), (9.6, New Zealand), (9.5, Denmark)$\}$ **d.** No; 9.6 in the domain corresponds to two countries in the range, Finland and New Zealand. **100. a.** $\{$(Bangladesh, 1.7), (Chad, 1.7), (Haiti, 1.8), (Myanmar, 1.8)$\}$ **b.** Yes; each country corresponds to exactly one corruption rating. **c.** $\{$(1.7, Bangladesh), (1.7, Chad), (1.8, Haiti), (1.8, Myanmar)$\}$ **d.** No; 1.7 in the domain corresponds to two countries in the range, Bangladesh and Chad, and 1.8 in the domain corresponds to two countries in the range, Haiti and Myanmar. **101. a.** 83; The chance that a 60-year-old will survive to age 70 is 83%. **b.** 76; The chance that a 60-year-old will survive to age 70 is 76%. **102. a.** 25; The chance that a 60-year-old will survive to age 90 is 25%. **b.** 10; The chance that a 60-year-old will survive to age 90 is 10%. **103. a.** 81; In 2010, the wage gap was 81%.; (30, 81) **104. a.** 69; In 1990, the wage gap was 69%.; (10, 69) **105.** $C = 100,000 + 100x$, where x is the number of bicycles produced; $C(90) = 109,000$; It costs \$109,000 to produce 90 bicycles. **106.** $V = 22,500 - 3200x$; $V(3) = 12,900$; The value of the car after 3 years is \$12,900.

107. $T = \dfrac{40}{x} + \dfrac{40}{x + 30}$, where x is the rate on the outgoing trip; $T(30) = 2$; It takes 2 hours, traveling 30 mph outgoing and 60 mph returning.
108. $S = 0.1x + 0.6(50 - x)$, where x is the number of milliliters of the 10% solution; $S(30) = 15$; There are 15 milliliters of sodium iodine in the mixture when 30 milliliters of the 10% solution are used. **118.** makes sense **120.** does not make sense **121.** does not make sense
127. Answers will vary; an example is $\{(1, 1), (2, 1)\}$. **128.** $f(2) = 6; f(3) = 9; f(4) = 12$; no **131.** snakes: 94 thousand deaths; mosquitoes: 755 thousand deaths; snails: 200 thousand deaths **132.** 36; For 100 calling minutes, the monthly cost is \$36.

133. $f(x) = x + 2, \ x \le 1$ **134.** $4xh + 2h^2 + 3h$

Section 2.2
Check Point Exercises

7.

$$f(x) = \begin{cases} 3 & \text{if } x \le -1 \\ x - 2 & \text{if } x > -1 \end{cases}$$

Exercise Set 2.2

1. a. $(-1, \infty)$ **b.** $(-\infty, -1)$ **c.** none **2. a.** $(-\infty, -1)$ **b.** $(-1, \infty)$ **c.** none **3. a.** $(0, \infty)$ **b.** none **c.** none
4. a. $(-1, \infty)$ **b.** none **c.** none **5. a.** none **b.** $(-2, 6)$ **c.** none **6. a.** $(-3, 2)$ **b.** none **c.** none
7. a. $(-\infty, -1)$ **b.** none **c.** $(-1, \infty)$ **8. a.** $(0, \infty)$ **b.** none **c.** $(-\infty, 0)$ **9. a.** $(-\infty, 0)$ or $(1.5, 3)$ **b.** $(0, 1.5)$ or $(3, \infty)$ **c.** none
10. a. $(-5, -4)$ or $(-2, 0)$ or $(2, 4)$ **b.** $(-4, -2)$ or $(0, 2)$ or $(4, 5)$ **c.** none **11. a.** $(-2, 4)$ **b.** none **c.** $(-\infty, -2)$ or $(4, \infty)$
12. a. none **b.** $(-4, 2)$ **c.** $(-\infty, -4)$ or $(2, \infty)$ **37.** odd; symmetric with respect to the origin **38.** odd; symmetric with respect to the origin
39. neither; symmetric to neither the y-axis nor the origin **40.** neither; symmetric to neither the y-axis nor the origin **41.** even; symmetric with
respect to the y-axis **42.** even; symmetric with respect to the y-axis **45.** even; symmetric with respect to the y-axis **46.** odd; symmetric with
respect to the origin **47.** odd; symmetric with respect to the origin **48.** even; symmetric with respect to the y-axis

59. a.

$$f(x) = \begin{cases} -x & \text{if } x < 0 \\ x & \text{if } x \ge 0 \end{cases}$$

60. a.

$$f(x) = \begin{cases} x & \text{if } x < 0 \\ -x & \text{if } x \ge 0 \end{cases}$$

61. a.

$$f(x) = \begin{cases} 2x & \text{if } x \le 0 \\ 2 & \text{if } x > 0 \end{cases}$$

62. a.

$$f(x) = \begin{cases} \frac{1}{2}x & \text{if } x \le 0 \\ 3 & \text{if } x > 0 \end{cases}$$

63. a.

$$f(x) = \begin{cases} x + 3 & \text{if } x < -2 \\ x - 3 & \text{if } x \ge -2 \end{cases}$$

64. a.

$$f(x) = \begin{cases} x + 2 & \text{if } x < -3 \\ x - 2 & \text{if } x \ge -3 \end{cases}$$

65. a.

$$f(x) = \begin{cases} 3 & \text{if } x \le -1 \\ -3 & \text{if } x > -1 \end{cases}$$

66. a.

$$f(x) = \begin{cases} 4 & \text{if } x \le -1 \\ -4 & \text{if } x > -1 \end{cases}$$

67. a.

$$f(x) = \begin{cases} \frac{1}{2}x^2 & \text{if } x < 1 \\ 2x - 1 & \text{if } x \ge 1 \end{cases}$$

68. a.

$$f(x) = \begin{cases} -\frac{1}{2}x^2 & \text{if } x < 1 \\ 2x + 1 & \text{if } x \ge 1 \end{cases}$$

69. a.

$$f(x) = \begin{cases} 0 & \text{if } x < -4 \\ -x & \text{if } -4 \le x < 0 \\ x^2 & \text{if } x \ge 0 \end{cases}$$

70. a.

$$f(x) = \begin{cases} 0 & \text{if } x < -3 \\ -x & \text{if } -3 \le x < 0 \\ x^2 - 1 & \text{if } x \ge 0 \end{cases}$$

77. $2x + h - 4, h \ne 0$ **84.** $-6x - 3h + 2, h \ne 0$
78. $2x + h - 5, h \ne 0$ **85.** $-4x - 2h - 1, h \ne 0$
79. $4x + 2h + 1, h \ne 0$ **86.** $-6x - 3h + 1, h \ne 0$
80. $6x + 3h + 1, h \ne 0$
81. $-2x - h + 2, h \ne 0$ **91.** $\dfrac{1}{\sqrt{x + h} + \sqrt{x}}, h \ne 0$
82. $-2x - h - 3, h \ne 0$ **92.** $\dfrac{1}{\sqrt{x + h - 1} + \sqrt{x - 1}}, h \ne 0$
83. $-4x - 2h + 5, h \ne 0$

95. $C(t)$

96. $C(t)$

97. $C(t) = \begin{cases} 50 & \text{if } 0 \le t \le 400 \\ 50 + 0.30(t - 400) & \text{if } t > 400 \end{cases}$

98. $C(t) = \begin{cases} 60 & \text{if } 0 \le t \le 450 \\ 60 + 0.35(t - 450) & \text{if } t > 450 \end{cases}$

99. increasing: $(25, 55)$; decreasing: $(55, 75)$ **100.** increasing: $(25, 65)$; decreasing: $(65, 75)$

101. 55 years old; 38% **102.** 65 years old; 26% **103.** domain: $[25, 75]$; range: $[34, 38]$

104. domain: $[25, 75]$; range: $[23, 26]$ **107.** 2575; A single taxpayer with taxable income of $20,000 owes $2575. **108.** 8625; A single taxpayer with taxable income of $50,000 owes $8625. **109.** $42,449 + 0.33(x - 174,400)$ **110.** $110,016.50 + 0.35(x - 379,150)$ **111.** 0.93; It costs $0.93 to mail a 3-ounce first-class letter. **112.** 1.15; It costs $1.15 to mail a 3.5-ounce first-class letter. **113.** $0.71 **114.** $0.71

115.

125.

The number of doctor visits decreases during childhood and then increases as you get older. The minimum is $(20.29, 3.99)$, which means that the minimum number of annual doctor visits, about 4, occurs at around age 20.

126.

increasing: $(-\infty, 1)$ or $(3, \infty)$
decreasing: $(1, 3)$

127.

increasing: $(-2, 0)$ or $(2, \infty)$
decreasing: $(-\infty, -2)$ or $(0, 2)$

128.

increasing: $(2, \infty)$
decreasing: $(-\infty, -2)$
constant: $(-2, 2)$

129.

increasing: $(1, \infty)$
decreasing: $(-\infty, 1)$

130.

increasing: $(0, \infty)$
decreasing: $(-\infty, 0)$

131.

increasing: $(-\infty, 0)$
decreasing: $(0, \infty)$

132. a.

b.

c. increasing: $(0, \infty)$; decreasing: $(-\infty, 0)$

d. $f(x) = x^n$ is increasing for $(-\infty, \infty)$ when n is positive and odd.

e.

As n increases the steepness increases.

139. a. h is even if both f and g are even or if both f and g are odd. **b.** h is odd if f is odd and g is even or if f is even and g is odd.

142. $\left\{ \dfrac{7}{10} + i\dfrac{\sqrt{11}}{10}, \dfrac{7}{10} - i\dfrac{\sqrt{11}}{10} \right\}$ **143.** 3

Section 2.3

Check Point Exercises

4.

$f(x) = \frac{3}{5}x + 1$

5.

6.

7.

$3x + 6y - 12 = 0$

8.

$3x - 2y - 6 = 0$

Exercise Set 2.3

5. 0; horizontal **6.** 0; horizontal **9.** undefined; vertical **10.** undefined; vertical **11.** $y - 5 = 2(x - 3); y = 2x - 1$

12. $y - 3 = 4(x - 1); y = 4x - 1$ **13.** $y - 5 = 6(x + 2); y = 6x + 17$ **14.** $y + 1 = 8(x - 4); y = 8x - 33$

15. $y + 3 = -3(x + 2); y = -3x - 9$ **16.** $y + 2 = -5(x + 4); y = -5x - 22$ **17.** $y - 0 = -4(x + 4); y = -4x - 16$

18. $y + 3 = -2(x - 0); y = -2x - 3$ **19.** $y + 2 = -1\left(x + \frac{1}{2}\right); y = -x - \frac{5}{2}$ **20.** $y + \frac{1}{4} = -1(x + 4); y = -x - \frac{17}{4}$

21. $y - 0 = \frac{1}{2}(x - 0); y = \frac{1}{2}x$ **22.** $y - 0 = \frac{1}{3}(x - 0); y = \frac{1}{3}x$ **23.** $y + 2 = -\frac{2}{3}(x - 6); y = -\frac{2}{3}x + 2$

24. $y + 4 = -\frac{3}{5}(x - 10); y = -\frac{3}{5}x + 2$ **25.** using $(1, 2), y - 2 = 2(x - 1); y = 2x$ **26.** using $(3, 5), y - 5 = 2(x - 3); y = 2x - 1$

27. using $(-3, 0), y - 0 = 1(x + 3); y = x + 3$ **28.** using $(-2, 0), y - 0 = 1(x + 2); y = x + 2$

29. using $(-3, -1), y + 1 = 1(x + 3); y = x + 2$ **30.** using $(-2, -4), y + 4 = 1(x + 2); y = x - 2$

31. using $(-3, -2), y + 2 = \frac{4}{3}(x + 3); y = \frac{4}{3}x + 2$ **32.** using $(-3, 6), y - 6 = -\frac{4}{3}(x + 3); y = -\frac{4}{3}x + 2$

33. using $(-3, -1), y + 1 = 0(x + 3); y = -1$ **34.** using $(-2, -5), y + 5 = 0(x + 2); y = -5$ **35.** using $(2, 4), y - 4 = 1(x - 2); y = x + 2$

36. using $(1, -3), y + 3 = -\frac{3}{2}(x - 1); y = -\frac{3}{2}x - \frac{3}{2}$ **37.** using $(0, 4), y - 4 = 8(x - 0); y = 8x + 4$

38. using $(4, 0), y - 0 = \frac{1}{2}(x - 4); y = \frac{1}{2}x - 2$

39. $m = 2; b = 1$ **40.** $m = 3; b = 2$ **41.** $m = -2; b = 1$ **42.** $m = -3; b = 2$

$y = 2x + 1$

$y = 3x + 2$

$f(x) = -2x + 1$

$f(x) = -3x + 2$

43. $m = \frac{3}{4}; b = -2$ **44.** $m = \frac{3}{4}; b = -3$ **45.** $m = -\frac{3}{5}; b = 7$ **46.** $m = -\frac{2}{5}; b = 6$

$f(x) = \frac{3}{4}x - 2$

$f(x) = \frac{3}{4}x - 3$

$y = -\frac{3}{5}x + 7$

$y = -\frac{2}{5}x + 6$

47. $m = -\frac{1}{2}; b = 0$ **48.** $m = -\frac{1}{3}; b = 0$ **49.** $y = -2$ **50.** $y = 4$ **51.** $x = -3$

$g(x) = -\frac{1}{2}x$

$g(x) = -\frac{1}{3}x$

52. $x = 5$

53. $y = 0$

54. $x = 0$

55. $f(x) = 1$

56. $f(x) = 3$

57. $3x - 18 = 0$, $(6, 0)$

58. $3x + 12 = 0$, $(-4, 0)$

59. a. $y = -3x + 5$
b. $m = -3; b = 5$
c. $(0, 5)$, $(1, 2)$, $3x + y - 5 = 0$

60. a. $y = -4x + 6$
b. $m = -4; b = 6$
c. $(0, 6)$, $(1, 2)$, $4x + y - 6 = 0$

61. a. $y = -\dfrac{2}{3}x + 6$
b. $m = -\dfrac{2}{3}; b = 6$
c. $(0, 6)$, $(3, 4)$, $2x + 3y - 18 = 0$

62. a. $y = -\dfrac{2}{3}x - 2$
b. $m = -\dfrac{2}{3}; b = -2$
c. $(0, -2)$, $(3, -4)$, $4x + 6y + 12 = 0$

63. a. $y = 2x - 3$
b. $m = 2; b = -3$
c. $(2, 1)$, $(0, -3)$, $8x - 4y - 12 = 0$

64. a. $y = \dfrac{6}{5}x - 4$
b. $m = \dfrac{6}{5}; b = -4$
c. $(5, 2)$, $(0, -4)$, $6x - 5y - 20 = 0$

65. a. $y = 3$
b. $m = 0; b = 3$
c. $3y - 9 = 0$, $(0, 3)$

66. a. $y = -7$
b. $m = 0; b = -7$
c. $(0, -7)$, $4y + 28 = 0$

67. $(2, 0)$, $(0, -6)$, $6x - 2y - 12 = 0$

68. $(3, 0)$, $(0, -2)$, $6x - 9y - 18 = 0$

69. $2x + 3y + 6 = 0$, $(-3, 0)$, $(0, -2)$

70. $3x + 5y + 15 = 0$, $(-5, 0)$, $(0, -3)$

71. $(0, 6)$, $\left(-\dfrac{3}{2}, 0\right)$, $8x - 2y + 12 = 0$

72. 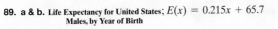 $(0, 5)$, $\left(-\dfrac{5}{2}, 0\right)$, $6x - 3y + 15 = 0$

74. $m = -\dfrac{b}{a}$; falls **75.** undefined slope; vertical **76.** $m = \dfrac{a}{b}$; rises **77.** $m = -\dfrac{A}{B}; b = \dfrac{C}{B}$ **78.** $m = \dfrac{A}{B}; b = \dfrac{C}{B}$

81. $3x - 4f(x) - 6 = 0$, $\left(4, \dfrac{3}{2}\right)$, $\left(0, -\dfrac{3}{2}\right)$

82. $(5, 2)$, $(0, -4)$, $6x - 5f(x) - 20 = 0$

87. a. $y - 38.9 = 0.89(x - 20)$ or $y - 47.8 = 0.89(x - 30)$
88. a. $y - 51.7 = 1.09(x - 20)$ or $y - 62.6 = 1.09(x - 30)$

89. a & b. Life Expectancy for United States; $E(x) = 0.215x + 65.7$
Males, by Year of Birth

Life Expectancy; $(20, 70)$, $(40, 74.3)$; Birth Years after 1960

90. Answers will vary due to rounding.
a & b. Life Expectancy for United States; $E(x) = 0.17x + 72.9$ or $E(x) = 0.17x + 73$
Females, by Year of Birth

Life Expectancy; $(40, 79.7)$, $(10, 74.7)$; Birth Years after 1960

c. 83.1 yr or 83.2 yr

91. Answers will vary; an example is $y = -2.3x + 255$, where x is the percentage of adult females who are literate and y is under-five mortality per thousand.; Predictions will vary.

100.

101. NORMAL FLOAT AUTO REAL RADIAN MP

102.

103.

104. a.

x	y
2.5	211
3.9	167
⋮	⋮
2.7	172

b & d

c. $a = -22.96876741$
$b = 260.5633751$
$r = -0.8428126855$

106. does not make sense **108.** makes sense **110.** false **113.** coefficient of x: -6; coefficient of y: 3 **114.** coefficient of x: 1; coefficient of y: -2

119. $(-\infty, 5]$; **120.** $(-7, 1)$;

Exercise Set 2.4

9. $y - 2 = \dfrac{2}{3}(x + 2)$; $2x - 3y + 10 = 0$ **10.** $y - 3 = \dfrac{3}{2}(x + 1)$; $3x - 2y + 9 = 0$ **11.** $y + 7 = -2(x - 4)$; $2x + y - 1 = 0$

12. $y + 9 = 7(x - 5)$; $7x - y - 44 = 0$ **27.** 137; There was an average increase of approximately 137 discharges per year.

28. -130; There was an average decrease of approximately 130 discharges per year. **29. b.** overestimates by 5 discharges per year

30. b. underestimates by 34 discharges per year

37. b. ; no **c.** ; The lines now appear to be perpendicular. **42.** $\dfrac{B}{A}$

47. a. $f(x) = |x|$ **b.** **c.** The graph in part (b) is the graph in part (a) shifted down 4 units.

48. a. $f(x) = x^2$ **b.** **c.** The graph in part (b) is the graph in part (a) shifted to the right 2 units.

49. a. **b.** **c.** The graph in part (b) is the graph in part (a) reflected across the y-axis.

Mid-Chapter 2 Check Point

3. function; domain: $\{x \mid -2 \le x < 2\}$ or $[-2, 2)$; range: $\{y \mid 0 \le y \le 3\}$ or $[0, 3]$ **4.** not a function; domain: $\{x \mid -3 < x \le 4\}$ or $(-3, 4]$; range: $\{y \mid -1 \le y \le 2\}$ or $[-1, 2]$ **5.** not a function; domain: $\{-2, -1, 0, 1, 2\}$; range: $\{-2, -1, 1, 3\}$
6. function; domain: $\{x \mid x \le 1\}$ or $(-\infty, 1]$; range: $\{y \mid y \ge -1\}$ or $[-1, \infty)$ **9.** No vertical line intersects the graph in more than one point.

27. $y = -2x$ **28.** $y = -2$ **29.** $x + y = -2$ **30.** $y = \dfrac{1}{3}x - 2$ **31.** $x = 3.5$

32. $4x - 2y = 8$ **33.** $f(x) = x^2 - 4$ **34.** $f(x) = x - 4$ **35.** $f(x) = |x| - 4$ **36.** $5y = -3x$

37. $5y = 20$ **38.** $f(x) = \begin{cases} -1 \text{ if } x \le 0 \\ 2x + 1 \text{ if } x > 0 \end{cases}$ **39. a.** $f(-x) = -2x^2 - x - 5$; neither

Section 2.5
Check Point Exercises

1.

2.

3.

4.

5.

6.

7. a.
$g(x) = f(2x)$

b.
$h(x) = f\left(\frac{1}{2}x\right)$

8.
$y = -\frac{1}{3}f(x+1) - 2$

9.
$g(x) = 2(x-1)^2 + 3$

Exercise Set 2.5

1.
$g(x) = f(x) + 1$

2.
$g(x) = f(x) - 1$

3.
$g(x) = f(x+1)$

4.
$g(x) = f(x-1)$

5.
$g(x) = f(x-1) - 2$

6.
$g(x) = f(x+1) + 2$

7.
$g(x) = f(-x)$

8.
$g(x) = -f(x)$

9.
$g(x) = -f(x) + 3$

10.
$g(x) = f(-x) + 3$

11.
$g(x) = \frac{1}{2}f(x)$

12.
$g(x) = 2f(x)$

13. $g(x) = f\left(\frac{1}{2}x\right)$

14. $g(x) = f(2x)$

15.
$g(x) = -f\left(\frac{1}{2}x\right) + 1$

16.
$g(x) = -f(2x) - 1$

17.
$g(x) = f(x) - 1$

18.
$g(x) = f(x) + 1$

19.
$g(x) = f(x-1)$

20.
$g(x) = f(x+1)$

21.
$g(x) = f(x-1) + 2$

22.
$g(x) = f(x+1) - 2$

23.
$g(x) = -f(x)$

24.
$g(x) = f(-x)$

25.
$g(x) = f(-x) + 1$

26.
$g(x) = -f(x) + 1$

27.
$g(x) = 2f(x)$

28.
$g(x) = \frac{1}{2}f(x)$

29.
$g(x) = f(2x)$

30.
$g(x) = f\left(\frac{1}{2}x\right)$

31.
$g(x) = 2f(x+2) + 1$

32.
$g(x) = 2f(x+2) - 1$

33.
$g(x) = f(x) + 2$

34.
$g(x) = f(x) - 2$

35.
$g(x) = f(x+2)$

81. **82.** **83.** **84.** **85.**

86. **87.** **88.** **89.** **90.**

91. **92.** **93.** **94.** **95.**

96. **97.** **98.** **99.** **100.**

101. **102.** **103.** **104.** **105.**

106. **107.** **108.** **109.** **110.**

111. **112.** **113.** **114.** **115.**

116. **117.** **118.** **119.** **120.**

121. $h(x) = \text{int}\,(-x) + 1$

122. $h(x) = \text{int}\,(-x) - 1$

123. $y = \sqrt{x - 2}$ **124.** $y = -x^3 + 2$ **125.** $y = (x + 1)^2 - 4$

126. $y = \sqrt{x - 2} + 1$ **127. a.** First, vertically stretch the graph of $f(x) = \sqrt{x}$ by the factor 2.9; then, shift the result up 20.1 units. **d.** 0.2 in. per month; This is a much smaller rate of change; The graph is not as steep between 50 and 60 as it is between 0 and 10.

128. a. First, vertically stretch the graph of $f(x) = \sqrt{x}$ by the factor 3.1; then, shift the result up 19 units. **d.** 0.2 in. per month; This is a much smaller rate of change; The graph is not as steep between 50 and 60 as it is between 0 and 10.

135. a. **b.**

136. a. **b.** **137.** makes sense **138.** makes sense

Exercise Set 2.6

3. $(-\infty, 4) \cup (4, \infty)$ **4.** $(-\infty, -5) \cup (-5, \infty)$ **5.** $(-\infty, \infty)$ **6.** $(-\infty, \infty)$ **7.** $(-\infty, -3) \cup (-3, 5) \cup (5, \infty)$ **8.** $(-\infty, -4) \cup (-4, 3) \cup (3, \infty)$

9. $(-\infty, -7) \cup (-7, 9) \cup (9, \infty)$ **10.** $(-\infty, -8) \cup (-8, 10) \cup (10, \infty)$ **13.** $(-\infty, 0) \cup (0, 3) \cup (3, \infty)$ **14.** $(-\infty, 0) \cup (0, 4) \cup (4, \infty)$

15. $(-\infty, 1) \cup (1, 3) \cup (3, \infty)$ **16.** $(-\infty, 2) \cup \left(2, \dfrac{10}{3}\right) \cup \left(\dfrac{10}{3}, \infty\right)$ **27.** $[2, 5) \cup (5, \infty)$ **28.** $[3, 6) \cup (6, \infty)$

31. $(f + g)(x) = 3x + 2$; domain: $(-\infty, \infty)$; $(f - g)(x) = x + 4$; domain: $(-\infty, \infty)$; $(fg)(x) = 2x^2 + x - 3$; domain: $(-\infty, \infty)$; $\left(\dfrac{f}{g}\right)(x) = \dfrac{2x + 3}{x - 1}$;

domain: $(-\infty, 1) \cup (1, \infty)$ **32.** $(f + g)(x) = 4x - 2$; domain: $(-\infty, \infty)$; $(f - g)(x) = 2x - 6$; domain: $(-\infty, \infty)$; $(fg)(x) = 3x^2 + 2x - 8$;

domain: $(-\infty, \infty)$; $\left(\dfrac{f}{g}\right)(x) = \dfrac{3x - 4}{x + 2}$; domain: $(-\infty, -2) \cup (-2, \infty)$ **33.** $(f + g)(x) = 3x^2 + x - 5$; domain: $(-\infty, \infty)$; $(f - g)(x) = -3x^2 + x - 5$;

domain: $(-\infty, \infty)$; $(fg)(x) = 3x^3 - 15x^2$; domain: $(-\infty, \infty)$; $\left(\dfrac{f}{g}\right)(x) = \dfrac{x - 5}{3x^2}$; domain: $(-\infty, 0) \cup (0, \infty)$ **34.** $(f + g)(x) = 5x^2 + x - 6$;

domain: $(-\infty, \infty)$; $(f - g)(x) = -5x^2 + x - 6$; domain: $(-\infty, \infty)$; $(fg)(x) = 5x^3 - 30x^2$; domain: $(-\infty, \infty)$; $\left(\dfrac{f}{g}\right)(x) = \dfrac{x - 6}{5x^2}$; domain: $(-\infty, 0) \cup (0, \infty)$

35. $(f + g)(x) = 2x^2 - 2$; domain: $(-\infty, \infty)$; $(f - g)(x) = 2x^2 - 2x - 4$; domain: $(-\infty, \infty)$; $(fg)(x) = 2x^3 + x^2 - 4x - 3$;

domain: $(-\infty, \infty)$; $\left(\dfrac{f}{g}\right)(x) = 2x - 3$; domain: $(-\infty, -1) \cup (-1, \infty)$ **36.** $(f + g)(x) = 6x^2 - 2$; domain: $(-\infty, \infty)$; $(f - g)(x) = 6x^2 - 2x$;

domain: $(-\infty, \infty)$; $(fg)(x) = 6x^3 - 7x^2 + 1$; domain: $(-\infty, \infty)$; $\left(\dfrac{f}{g}\right)(x) = \dfrac{6x^2 - x - 1}{x - 1}$; domain: $(-\infty, 1) \cup (1, \infty)$ **37.** $(f + g)(x) = 2x - 12$;

domain: $(-\infty, \infty)$; $(f - g)(x) = -2x^2 - 2x + 18$; domain: $(-\infty, \infty)$; $(fg)(x) = -x^4 - 2x^3 + 18x^2 + 6x - 45$;

domain: $(-\infty, \infty)$; $\left(\dfrac{f}{g}\right)(x) = \dfrac{3 - x^2}{x^2 + 2x - 15}$; domain: $(-\infty, -5) \cup (-5, 3) \cup (3, \infty)$ **38.** $(f + g)(x) = 4x - 7$;

domain: $(-\infty, \infty)$; $(f - g)(x) = -2x^2 - 4x + 17$; domain: $(-\infty, \infty)$; $(fg)(x) = -x^4 - 4x^3 + 17x^2 + 20x - 60$;

domain: $(-\infty, \infty)$; $\left(\dfrac{f}{g}\right)(x) = \dfrac{5 - x^2}{x^2 + 4x - 12}$; domain: $(-\infty, -6) \cup (-6, 2) \cup (2, \infty)$ **39.** $(f + g)(x) = \sqrt{x} + x - 4$;

domain: $[0, \infty)$; $(f - g)(x) = \sqrt{x} - x + 4$; domain: $[0, \infty)$; $(fg)(x) = \sqrt{x}(x - 4)$; domain: $[0, \infty)$; $\left(\dfrac{f}{g}\right)(x) = \dfrac{\sqrt{x}}{x - 4}$;

domain: $[0, 4) \cup (4, \infty)$ **40.** $(f + g)(x) = \sqrt{x} + x - 5$; domain: $[0, \infty)$; $(f - g)(x) = \sqrt{x} - x + 5$; domain: $[0, \infty)$; $(fg)(x) = \sqrt{x}(x - 5)$;

domain: $[0, \infty)$; $\left(\dfrac{f}{g}\right)(x) = \dfrac{\sqrt{x}}{x - 5}$; domain: $[0, 5) \cup (5, \infty)$ **41.** $(f + g)(x) = \dfrac{2x + 2}{x}$; domain: $(-\infty, 0) \cup (0, \infty)$; $(f - g)(x) = 2$;

domain: $(-\infty, 0) \cup (0, \infty)$; $(fg)(x) = \dfrac{2x + 1}{x^2}$; domain: $(-\infty, 0) \cup (0, \infty)$; $\left(\dfrac{f}{g}\right)(x) = 2x + 1$; domain: $(-\infty, 0) \cup (0, \infty)$ **42.** $(f + g)(x) = 6$;

domain: $(-\infty, 0) \cup (0, \infty)$; $(f - g)(x) = \dfrac{6x - 2}{x}$; domain: $(-\infty, 0) \cup (0, \infty)$; $(fg)(x) = \dfrac{6x - 1}{x^2}$; domain: $(-\infty, 0) \cup (0, \infty)$; $\left(\dfrac{f}{g}\right)(x) = 6x - 1$;

domain: $(-\infty, 0) \cup (0, \infty)$ **43.** $(f + g)(x) = \dfrac{9x - 1}{x^2 - 9}$; domain: $(-\infty, -3) \cup (-3, 3) \cup (3, \infty)$; $(f - g)(x) = \dfrac{x + 3}{x^2 - 9} = \dfrac{1}{x - 3}$;

domain: $(-\infty, -3) \cup (-3, 3) \cup (3, \infty)$; $(fg)(x) = \dfrac{20x^2 - 6x - 2}{(x^2 - 9)^2}$; domain: $(-\infty, -3) \cup (-3, 3) \cup (3, \infty)$; $\left(\dfrac{f}{g}\right)(x) = \dfrac{5x + 1}{4x - 2}$;

domain: $(-\infty, -3) \cup \left(-3, \dfrac{1}{2}\right) \cup \left(\dfrac{1}{2}, 3\right) \cup (3, \infty)$ **44.** $(f + g)(x) = \dfrac{5x - 3}{x^2 - 25}$; domain: $(-\infty, -5) \cup (-5, 5) \cup (5, \infty)$; $(f - g)(x) = \dfrac{x + 5}{x^2 - 25} = \dfrac{1}{x - 5}$;

domain: $(-\infty, -5) \cup (-5, 5) \cup (5, \infty)$; $(fg)(x) = \dfrac{6x^2 - 10x - 4}{(x^2 - 25)^2}$; domain: $(-\infty, -5) \cup (-5, 5) \cup (5, \infty)$; $\left(\dfrac{f}{g}\right)(x) = \dfrac{3x + 1}{2x - 4}$;

domain: $(-\infty, -5) \cup (-5, 2) \cup (2, 5) \cup (5, \infty)$ **45.** $(f + g)(x) = \dfrac{8x^2 + 30x - 12}{(x - 2)(x + 3)}$; domain: $(-\infty, -3) \cup (-3, 2) \cup (2, \infty)$; $(f - g)(x) = \dfrac{8x^2 + 18x + 12}{(x - 2)(x + 3)}$;

domain: $(-\infty, -3) \cup (-3, 2) \cup (2, \infty)$; $(fg)(x) = \dfrac{48x}{(x - 2)(x + 3)}$; domain: $(-\infty, -3) \cup (-3, 2) \cup (2, \infty)$; $\left(\dfrac{f}{g}\right)(x) = \dfrac{4x(x + 3)}{3(x - 2)}$;

domain: $(-\infty, -3) \cup (-3, 2) \cup (2, \infty)$ **46.** $(f + g)(x) = \dfrac{9x^2 + 79x - 28}{(x - 4)(x + 8)}$; domain: $(-\infty, -8) \cup (-8, 4) \cup (4, \infty)$; $(f - g)(x) = \dfrac{9x^2 + 65x + 28}{(x - 4)(x + 8)}$;

domain: $(-\infty, -8) \cup (-8, 4) \cup (4, \infty)$; $(fg)(x) = \dfrac{63x}{(x - 4)(x + 8)}$; domain: $(-\infty, -8) \cup (-8, 4) \cup (4, \infty)$; $\left(\dfrac{f}{g}\right)(x) = \dfrac{9x(x + 8)}{7(x - 4)}$;

domain: $(-\infty, -8) \cup (-8, 4) \cup (4, \infty)$ **47.** $(f + g)(x) = \sqrt{x + 4} + \sqrt{x - 1}$; domain: $[1, \infty)$; $(f - g)(x) = \sqrt{x + 4} - \sqrt{x - 1}$;

domain: $[1, \infty)$; $(fg)(x) = \sqrt{x^2 + 3x - 4}$; domain: $[1, \infty)$; $\left(\dfrac{f}{g}\right)(x) = \dfrac{\sqrt{x + 4}}{\sqrt{x - 1}}$; domain: $(1, \infty)$ **48.** $(f + g)(x) = \sqrt{x + 6} + \sqrt{x - 3}$;

domain: $[3, \infty)$; $(f - g)(x) = \sqrt{x + 6} - \sqrt{x - 3}$; domain: $[3, \infty)$; $(fg)(x) = \sqrt{x^2 + 3x - 18}$; domain: $[3, \infty)$; $\left(\dfrac{f}{g}\right)(x) = \dfrac{\sqrt{x + 6}}{\sqrt{x - 3}}$;

domain: $(3, \infty)$ **49.** $(f + g)(x) = \sqrt{x - 2} + \sqrt{2 - x}$; domain: $\{2\}$; $(f - g)(x) = \sqrt{x - 2} - \sqrt{2 - x}$; domain: $\{2\}$; $(fg)(x) = \sqrt{x - 2} \cdot \sqrt{2 - x}$;

domain: $\{2\}$; $\left(\dfrac{f}{g}\right)(x) = \dfrac{\sqrt{x - 2}}{\sqrt{2 - x}}$; domain: \varnothing **50.** $(f + g)(x) = \sqrt{x - 5} + \sqrt{5 - x}$; domain: $\{5\}$; $(f - g)(x) = \sqrt{x - 5} - \sqrt{5 - x}$;

domain: $\{5\}$; $(fg)(x) = \sqrt{x - 5} \cdot \sqrt{5 - x}$; $\{5\}$; $\left(\dfrac{f}{g}\right)(x) = \dfrac{\sqrt{x - 5}}{\sqrt{5 - x}}$; domain: \varnothing **55. a.** $(f \circ g)(x) = 20x^2 - 11$ **b.** $(g \circ f)(x) = 80x^2 - 120x + 43$

56. a. $(f \circ g)(x) = 14x^2 - 62$ **b.** $(g \circ f)(x) = 98x^2 + 28x - 7$ **57. a.** $(f \circ g)(x) = x^4 - 4x^2 + 6$ **b.** $(g \circ f)(x) = x^4 + 4x^2 + 2$

58. a. $(f \circ g)(x) = x^4 - 6x^2 + 10$ **b.** $(g \circ f)(x) = x^4 + 2x^2 - 2$ **59. a.** $(f \circ g)(x) = -2x^2 - x - 1$ **b.** $(g \circ f)(x) = 2x^2 - 17x + 41$

60. a. $(f \circ g)(x) = -5x^2 + 20x - 7$ **b.** $(g \circ f)(x) = -25x^2 + 40x - 13$ **61. a.** $(f \circ g)(x) = \sqrt{x - 1}$ **b.** $(g \circ f)(x) = \sqrt{x} - 1$

62. a. $(f \circ g)(x) = \sqrt{x + 2}$ **b.** $(g \circ f)(x) = \sqrt{x} + 2$ **67. b.** $\left(-\infty, -\dfrac{1}{3}\right) \cup \left(-\dfrac{1}{3}, 0\right) \cup (0, \infty)$ **68. b.** $\left(-\infty, -\dfrac{1}{4}\right) \cup \left(-\dfrac{1}{4}, 0\right) \cup (0, \infty)$

69. b. $(-\infty, -4) \cup (-4, 0) \cup (0, \infty)$ **70. b.** $\left(-\infty, -\dfrac{6}{5}\right) \cup \left(-\dfrac{6}{5}, 0\right) \cup (0, \infty)$ **73. b.** $(-\infty, 1]$ **74. b.** $(-\infty, 2]$ **75.** $f(x) = x^4, g(x) = 3x - 1$

76. $f(x) = x^3, g(x) = 2x - 5$ **77.** $f(x) = \sqrt[3]{x}, g(x) = x^2 - 9$ **78.** $f(x) = \sqrt{x}, g(x) = 5x^2 + 3$ **79.** $f(x) = |x|, g(x) = 2x - 5$

80. $f(x) = |x|, g(x) = 3x - 4$ **81.** $f(x) = \dfrac{1}{x}, g(x) = 2x - 3$ **82.** $f(x) = \dfrac{1}{x}, g(x) = 4x + 5$ **87.** $[-4, 3]$ **88.** $(-4, 3)$

89. **90.** **95.** 1 and 2 **96.** $-\dfrac{4}{3}$ and 1

99. $(R - C)(20,000) = -200,000$; The company loses $200,000 when 20,000 radios are sold.; $(R - C)(30,000) = 0$; The company breaks even when 30,000 radios are sold.; $(R - C)(40,000) = 200,000$; The company makes a profit of $200,000 when 40,000 radios are sold.
100. a. $m = -0.44$; Profit is decreasing. **b.** $m = 0.51$; Profit is increasing. **c.** $0.07x + 24.76; m = 0.07$; Profit is increasing.
101. a. f gives the price of the computer after a $400 discount. g gives the price of the computer after a 25% discount.
b. $(f \circ g)(x) = 0.75x - 400$; This models the price of a computer after first a 25% discount and then a $400 discount. **c.** $(g \circ f)(x) = 0.75(x - 400)$; This models the price of a computer after first a $400 discount and then a 25% discount. **d.** $f \circ g$ because $0.75x - 400 < 0.75(x - 400)$
102. a. f gives the cost of a pair of jeans after a $5 discount. g gives the cost of a pair of jeans that has been discounted 40%.
b. $(f \circ g)(x) = 0.6x - 5$; The cost of a pair of jeans is 60% of the regular price minus $5 rebate. **c.** $(g \circ f)(x) = 0.6(x - 5) = 0.6x - 3$; The cost of a pair of jeans is 60% of the price that has been reduced by $5. **d.** $f \circ g$ because $0.6x - 5 < 0.6x - 3$

108. NORMAL FLOAT AUTO REAL RADIAN MP

y_1 y_2 y_3

At $x = 0$, there is no y-value since the function is not defined there.

109. NORMAL FLOAT AUTO REAL RADIAN MP

Domain: $[0, 4]$

117. true **118.** Assume f and g are even; then $f(-x) = f(x)$ and $g(-x) = g(x)$. $(fg)(-x) = f(-x)g(-x) = f(x)g(x) = (fg)(x)$, so fg is even.
119. Answers will vary; One possible answer is $f(x) = x + 1$ and $g(x) = x - 1$. **123.** $\{(4, -2), (1, -1), (1, 1), (4, 2)\}$; no

Section 2.7
Check Point Exercises

6.

7. $f^{-1}(x) = \sqrt{x-1}$

Exercise Set 2.7

3. $f(g(x)) = x; g(f(x)) = x; f$ and g are inverses. **4.** $f(g(x)) = x; g(f(x)) = x; f$ and g are inverses.

5. $f(g(x)) = \dfrac{5x - 56}{9}; g(f(x)) = \dfrac{5x - 4}{9}; f$ and g are not inverses. **6.** $f(g(x)) = \dfrac{3x - 40}{7}; g(f(x)) = \dfrac{3x - 4}{7}; f$ and g are not inverses.

7. $f(g(x)) = x; g(f(x)) = x; f$ and g are inverses. **8.** $f(g(x)) = x; g(f(x)) = x; f$ and g are inverses. **10.** $f(g(x)) = x; g(f(x)) = x; f$ and g are inverses.

27. $f^{-1}(x) = \dfrac{3x + 1}{x - 2}; x \neq 2$ **28.** $f^{-1}(x) = \dfrac{-x - 3}{x - 2}; x \neq 2$

29. The function is not one-to-one, so it does not have an inverse function. **30.** The function is one-to-one, so it does have an inverse function.

31. The function is not one-to-one, so it does not have an inverse function. **32.** The function is not one-to-one, so it does not have an inverse function.

33. The function is one-to-one, so it does have an inverse function. **34.** The function is one-to-one, so it does have an inverse function.

35.

36.

37.

38.

39. a. $f^{-1}(x) = \dfrac{x + 1}{2}$

b.

c. domain of f = range of $f^{-1} = (-\infty, \infty)$;
range of f = domain of $f^{-1} = (-\infty, \infty)$

40. a. $f^{-1}(x) = \dfrac{x + 3}{2}$

b.

c. domain of f = range of $f^{-1} = (-\infty, \infty)$;
range of f = domain of $f^{-1} = (-\infty, \infty)$

41. a. $f^{-1}(x) = \sqrt{x + 4}$

b.

c. domain of f = range of $f^{-1} = [0, \infty)$;
range of f = domain of $f^{-1} = [-4, \infty)$

42. a. $f^{-1}(x) = -\sqrt{x + 1}$

b.

c. domain of f = range of $f^{-1} = (-\infty, 0]$;
range of f = domain of $f^{-1} = [-1, \infty)$

43. a. $f^{-1}(x) = -\sqrt{x} + 1$

b.

c. domain of f = range of $f^{-1} = (-\infty, 1]$;
range of f = domain of $f^{-1} = [0, \infty)$

44. a. $f^{-1}(x) = \sqrt{x} + 1$

b.

c. domain of f = range of $f^{-1} = [1, \infty)$;
range of f = domain of $f^{-1} = [0, \infty)$

45. a. $f^{-1}(x) = \sqrt[3]{x + 1}$

b.

c. domain of f = range of $f^{-1} = (-\infty, \infty)$;
range of f = domain of $f^{-1} = (-\infty, \infty)$

46. a. $f^{-1}(x) = \sqrt[3]{x - 1}$

b.

c. domain of f = range of $f^{-1} = (-\infty, \infty)$;
range of f = domain of $f^{-1} = (-\infty, \infty)$

47. a. $f^{-1}(x) = \sqrt[3]{x} - 2$

b.

c. domain of f = range of $f^{-1} = (-\infty, \infty)$;
range of f = domain of $f^{-1} = (-\infty, \infty)$

48. a. $f^{-1}(x) = \sqrt[3]{x} + 2$

b.

c. domain of f = range of $f^{-1} = (-\infty, \infty)$;
range of f = domain of $f^{-1} = (-\infty, \infty)$

49. a. $f^{-1}(x) = x^2 + 1, x \geq 0$

b.

c. domain of f = range of $f^{-1} = [1, \infty)$;
range of f = domain of $f^{-1} = [0, \infty)$

50. a. $f^{-1}(x) = (x - 2)^2, x \geq 2$

b.

c. domain of f = range of $f^{-1} = [0, \infty)$;
range of f = domain of $f^{-1} = [2, \infty)$

51. a. $f^{-1}(x) = (x - 1)^3$

b.

c. domain of f = range of $f^{-1} = (-\infty, \infty)$;
range of f = domain of $f^{-1} = (-\infty, \infty)$

52. a. $f^{-1}(x) = x^3 + 1$

b.

c. domain of f = range of $f^{-1} = (-\infty, \infty)$;
range of f = domain of $f^{-1} = (-\infty, \infty)$

65. a. {(Zambia, 4.2), (Colombia, 4.5), (Poland, 3.3), (Italy, 3.3), (United States, 2.5)}
b. {(4.2, Zambia), (4.5, Colombia), (3.3, Poland), (3.3, Italy), (2.5, United States)}; no; The inverse of f is not a function.
66. a. {(Zambia, −7.3), (Colombia, −4.5), (Poland, −2.8), (Italy, −2.8), (United States, −1.9)}
b. {(−7.3, Zambia), (−4.5, Colombia), (−2.8, Poland), (−2.8, Italy), (−1.9, United States)}; no; The inverse of g is not a function.
67. a. f is a one-to-one function. **b.** $f^{-1}(0.25)$ is the number of people in a room for a 25% probability of two people sharing a birthday. $f^{-1}(0.5)$ is the number of people in a room for a 50% probability of two people sharing a birthday. $f^{-1}(0.7)$ is the number of people in a room for a 70% probability of two people sharing a birthday. **68. a.** No; the horizontal line $y = 3$ intersects the graph in more than one point.
c. No, the graph does not represent a one-to-one function; the points (12, 3) and (19, 3) have the same second coordinate but different first coordinates.

69. $f(g(x)) = \dfrac{9}{5}\left[\dfrac{5}{9}(x - 32)\right] + 32 = x$ and $g(f(x)) = \dfrac{5}{9}\left[\left(\dfrac{9}{5}x + 32\right) - 32\right] = x$

76.

not one-to-one

77.

one-to-one

78.
one-to-one

79.
not one-to-one

80.
not one-to-one

81.
not one-to-one

82.

one-to-one

83.

not one-to-one

84.

f and g are inverses.

85.

86.

f and *g* are inverses. *f* and *g* are inverses.

95. $(f \circ g)^{-1}(x) = \dfrac{x - 15}{3}$; $(g^{-1} \circ f^{-1})(x) = \dfrac{x}{3} - 5 = \dfrac{x - 15}{3}$

97. No; The spacecraft was at the same height, $s(t)$, for two different values of t—once during the ascent and once again during the descent.

102. $(-\infty, -3] \cup [4, \infty)$; **104.**

Section 2.8

Check Point Exercises

5. b. **6.**

$(x + 3)^2 + (y - 1)^2 = 4$ $x^2 + y^2 + 4x - 4y - 1 = 0$

Exercise Set 2.8

3. $2\sqrt{29} \approx 10.77$ **4.** $\sqrt{73} \approx 8.54$ **7.** $\sqrt{29} \approx 5.39$ **8.** $2\sqrt{10} \approx 6.32$ **10.** $\sqrt{41} \approx 6.40$ **11.** $2\sqrt{5} \approx 4.47$ **12.** $5\sqrt{2} \approx 7.07$

18. $\sqrt{2} \approx 1.41$ **21.** $(-4, -5)$ **22.** $\left(-\dfrac{5}{2}, -5\right)$ **31.** $x^2 + y^2 = 49$ **32.** $x^2 + y^2 = 64$ **33.** $(x - 3)^2 + (y - 2)^2 = 25$

34. $(x - 2)^2 + (y + 1)^2 = 16$ **35.** $(x + 1)^2 + (y - 4)^2 = 4$ **36.** $(x + 3)^2 + (y - 5)^2 = 9$ **37.** $(x + 3)^2 + (y + 1)^2 = 3$

38. $(x + 5)^2 + (y + 3)^2 = 5$ **39.** $(x + 4)^2 + y^2 = 100$ **40.** $(x + 2)^2 + y^2 = 36$

41. center: $(0, 0)$
radius: 4
domain: $[-4, 4]$
range: $[-4, 4]$

$x^2 + y^2 = 16$

42. center: $(0, 0)$
radius: 7
domain: $[-7, 7]$
range: $[-7, 7]$

$x^2 + y^2 = 49$

43. center: $(3, 1)$
radius: 6
domain: $[-3, 9]$
range: $[-5, 7]$

$(x - 3)^2 + (y - 1)^2 = 36$

44. center: $(2, 3)$
radius: 4
domain: $[-2, 6]$
range: $[-1, 7]$

$(x - 2)^2 + (y - 3)^2 = 16$

45. center: $(-3, 2)$
radius: 2
domain: $[-5, -1]$
range: $[0, 4]$

$(x + 3)^2 + (y - 2)^2 = 4$

46. center: $(-1, 4)$
radius: 5
domain: $[-6, 4]$
range: $[-1, 9]$

$(x + 1)^2 + (y - 4)^2 = 25$

47. center: $(-2, -2)$
radius: 2
domain: $[-4, 0]$
range: $[-4, 0]$

$(x + 2)^2 + (y + 2)^2 = 4$

48. center: $(-4, -5)$
radius: 6
domain: $[-10, 2]$
range: $[-11, 1]$

$(x + 4)^2 + (y + 5)^2 = 36$

49. center: $(0, 1)$
radius: 1
domain: $[-1, 1]$
range: $[0, 2]$

$x^2 + (y - 1)^2 = 1$

50. center: $(0, 2)$
radius: 2
domain: $[-2, 2]$
range: $[0, 4]$

$x^2 + (y - 2)^2 = 4$

51. center: $(-1, 0)$
radius: 5
domain: $[-6, 4]$
range: $[-5, 5]$

$(x + 1)^2 + y^2 = 25$

52. center: $(-2, 0)$
radius: 4
domain: $[-6, 2]$
range: $[-4, 4]$

$(x + 2)^2 + y^2 = 16$

53. $(x + 3)^2 + (y + 1)^2 = 4$
center: $(-3, -1)$
radius: 2

$x^2 + y^2 + 6x + 2y + 6 = 0$

54. $(x + 4)^2 + (y + 2)^2 = 4$
center: $(-4, -2)$
radius: 2

$x^2 + y^2 + 8x + 4y + 16 = 0$

55. $(x - 5)^2 + (y - 3)^2 = 64$
center: $(5, 3)$
radius: 8

$x^2 + y^2 - 10x - 6y - 30 = 0$

56. $(x - 2)^2 + (y - 6)^2 = 49$
center: $(2, 6)$
radius: 7

$x^2 + y^2 - 4x - 12y - 9 = 0$

57. $(x + 4)^2 + (y - 1)^2 = 25$
center: $(-4, 1)$
radius: 5

$x^2 + y^2 + 8x - 2y - 8 = 0$

58. $(x + 6)^2 + (y - 3)^2 = 49$
center: $(-6, 3)$
radius: 7

$x^2 + y^2 + 12x - 6y - 4 = 0$

59. $(x - 1)^2 + y^2 = 16$

center: $(1, 0)$

radius: 4

$x^2 - 2x + y^2 - 15 = 0$

60. $x^2 + (y - 3)^2 = 16$

center: $(0, 3)$

radius: 4

$x^2 + y^2 - 6y - 7 = 0$

61. $\left(x - \dfrac{1}{2}\right)^2 + (y + 1)^2 = \dfrac{1}{4}$

center: $\left(\dfrac{1}{2}, -1\right)$

radius: $\dfrac{1}{2}$

$x^2 + y^2 - x + 2y + 1 = 0$

62. $\left(x + \dfrac{1}{2}\right)^2 + \left(y + \dfrac{1}{2}\right)^2 = 1$

center: $\left(-\dfrac{1}{2}, -\dfrac{1}{2}\right)$

radius: 1

$x^2 + y^2 + x + y - \dfrac{1}{2} = 0$

63. $\left(x + \dfrac{3}{2}\right)^2 + (y - 1)^2 = \dfrac{17}{4}$

center: $\left(-\dfrac{3}{2}, 1\right)$

radius: $\dfrac{\sqrt{17}}{2}$

$x^2 + y^2 + 3x - 2y - 1 = 0$

64. $\left(x + \dfrac{3}{2}\right)^2 + \left(y + \dfrac{5}{2}\right)^2 = \dfrac{25}{4}$

center: $\left(-\dfrac{3}{2}, -\dfrac{5}{2}\right)$

radius: $\dfrac{5}{2}$

$x^2 + y^2 + 3x + 5y + \dfrac{9}{4} = 0$

65. a. $(5, 10)$ **b.** $\sqrt{5}$ **c.** $(x - 5)^2 + (y - 10)^2 = 5$

67. $x^2 + y^2 = 16$
$x - y = 4$

68. $x^2 + y^2 = 9$
$x - y = 3$

69. $(x - 2)^2 + (y + 3)^2 = 4$
$y = x - 3$

70. $(x - 3)^2 + (y + 1)^2 = 9$
$y = x - 1$

83.

84.

85.

89. makes sense

95. a. Distance between (x_1, y_1) and $\left(\dfrac{x_1 + x_2}{2}, \dfrac{y_1 + y_2}{2}\right)$

$$= \sqrt{\left(\frac{x_1 + x_2}{2} - x_1\right)^2 + \left(\frac{y_1 + y_2}{2} - y_1\right)^2}$$

$$= \sqrt{\left(\frac{x_1 + x_2 - 2x_1}{2}\right)^2 + \left(\frac{y_1 + y_2 - 2y_1}{2}\right)^2}$$

$$= \sqrt{\left(\frac{x_2 - x_1}{2}\right)^2 + \left(\frac{y_2 - y_1}{2}\right)^2}$$

$$= \sqrt{\frac{x_2^2 - 2x_1x_2 + x_1^2}{4} + \frac{y_2^2 - 2y_1y_2 + y_1^2}{4}}$$

$$= \sqrt{\frac{x_1^2 - 2x_1x_2 + x_2^2}{4} + \frac{y_1^2 - 2y_1y_2 + y_2^2}{4}}$$

$$= \sqrt{\left(\frac{x_1 - x_2}{2}\right)^2 + \left(\frac{y_1 - y_2}{2}\right)^2}$$

$$= \sqrt{\left(\frac{x_1 + x_2 - 2x_2}{2}\right)^2 + \left(\frac{y_1 + y_2 - 2y_2}{2}\right)^2}$$

$$= \sqrt{\left(\frac{x_1 + x_2}{2} - x_2\right)^2 + \left(\frac{y_1 + y_2}{2} - y_2\right)^2}$$

$$= \text{Distance between } (x_2, y_2) \text{ and } \left(\frac{x_1 + x_2}{2}, \frac{y_1 + y_2}{2}\right)$$

b. $\sqrt{\left(\dfrac{x_2 - x_1}{2}\right)^2 + \left(\dfrac{y_2 - y_1}{2}\right)^2} + \sqrt{\left(\dfrac{x_2 - x_1}{2}\right)^2 + \left(\dfrac{y_2 - y_1}{2}\right)^2}$

$$= 2\sqrt{\left(\frac{x_2 - x_1}{2}\right)^2 + \left(\frac{y_2 - y_1}{2}\right)^2}$$

$$= 2\sqrt{\frac{(x_2 - x_1)^2 + (y_2 - y_1)^2}{4}}$$

$$= \sqrt{(x_2 - x_1)^2 + (y_2 - y_1)^2}$$

$$= \text{Distance from } (x_1, y_1) \text{ to } (x_2, y_2)$$

98. $\{x \mid x$ is a real number$\}$; identity

100. $[-2, 4)$;

103.

$f(x) = x^2$
$g(x) = (x + 3)^2 + 1$

Chapter 2 Review Exercises

1. function; domain: $\{2, 3, 5\}$; range: $\{7\}$　　**2.** function; domain: $\{1, 2, 13\}$; range: $\{10, 500, \pi\}$　　**20. b.** $-2, 3$; $f(-2) = -3, f(3) = -5$

31.

$$f(x) = \begin{cases} 5 \text{ if } x \le -1 \\ -3 \text{ if } x > -1 \end{cases}$$

32.

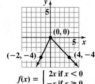

$$f(x) = \begin{cases} 2x \text{ if } x < 0 \\ -x \text{ if } x \ge 0 \end{cases}$$

34. $-4x - 2h + 1$　　**35. a.** yes; The graph passes the vertical line test.
b. $(3, 12)$; The eagle descended.　　**c.** $(0, 3)$ and $(12, 17)$; The eagle's height held steady during the first 3 seconds and the eagle was on the ground for 5 seconds.
d. $(17, 30)$; The eagle was ascending.

36.

41. $y - 2 = -6(x + 3)$; $y = -6x - 16$
42. using $(1, 6)$, $y - 6 = 2(x - 1)$; $y = 2x + 4$

46. Slope: $\dfrac{2}{5}$; y-intercept: -1

47. Slope: -4; y-intercept: 5

48. Slope: $-\dfrac{2}{3}$; y-intercept: -2

49. Slope: 0; y-intercept: 4

$y = \dfrac{2}{5}x - 1$　　$f(x) = -4x + 5$　　$2x + 3y + 6 = 0$　　$2y - 8 = 0$

50.

$2x - 5y - 10 = 0$

51. $2x - 10 = 0$

55. $y = g(x)$

56. $y = g(x)$

57.

$y = g(x)$

58.

$y = g(x)$

59.

$y = g(x)$

60.

61.

62.

63.

64.

65.

66.

67.

68.

69.

70.

71.

72.

73.

74. $\left(-\frac{1}{2}, 1\right)$

75.

77. $(-\infty, 7) \cup (7, \infty)$

82. $(f + g)(x) = 4x - 6$; domain: $(-\infty, \infty)$; $(f - g)(x) = 2x + 4$; domain: $(-\infty, \infty)$; $(fg)(x) = 3x^2 - 16x + 5$; domain: $(-\infty, \infty)$;
$\left(\dfrac{f}{g}\right)(x) = \dfrac{3x - 1}{x - 5}$; domain: $(-\infty, 5) \cup (5, \infty)$ **83.** $(f + g)(x) = 2x^2 + x$; domain: $(-\infty, \infty)$; $(f - g)(x) = x + 2$;
domain: $(-\infty, \infty)$; $(fg)(x) = x^4 + x^3 - x - 1$; domain: $(-\infty, \infty)$; $\left(\dfrac{f}{g}\right)(x) = \dfrac{x^2 + x + 1}{x^2 - 1}$; domain: $(-\infty, -1) \cup (-1, 1) \cup (1, \infty)$

84. $(f + g)(x) = \sqrt{x + 7} + \sqrt{x - 2}$; domain: $[2, \infty)$; $(f - g)(x) = \sqrt{x + 7} - \sqrt{x - 2}$; domain: $[2, \infty)$; $(fg)(x) = \sqrt{x^2 + 5x - 14}$;
domain: $[2, \infty)$; $\left(\dfrac{f}{g}\right)(x) = \dfrac{\sqrt{x + 7}}{\sqrt{x - 2}}$; domain: $(2, \infty)$ **85. a.** $(f \circ g)(x) = 16x^2 - 8x + 4$ **b.** $(g \circ f)(x) = 4x^2 + 11$

86. a. $(f \circ g)(x) = \sqrt{x + 1}$ **b.** $(g \circ f)(x) = \sqrt{x} + 1$ **87. b.** $(-\infty, 0) \cup \left(0, \dfrac{1}{2}\right) \cup \left(\dfrac{1}{2}, \infty\right)$ **88. b.** $[-2, \infty)$

91. $f(g(x)) = x - \dfrac{7}{10}$; $g(f(x)) = x - \dfrac{7}{6}$; f and g are not inverses of each other.

92. $f(g(x)) = x$; $g(f(x)) = x$; f and g are inverses of each other.

100.

101. $f^{-1}(x) = \sqrt{1 - x}$ **102.** $f^{-1}(x) = (x - 1)^2, x \geq 1$

109. center: $(0, 0)$; radius: 1
domain: $[-1, 1]$; range: $[-1, 1]$

110. center: $(-2, 3)$; radius: 3
domain: $[-5, 1]$; range: $[0, 6]$

111. center: $(2, -1)$; radius: 3
domain: $[-1, 5]$; range: $[-4, 2]$

Chapter 2 Test

2. e. $(-5, -1)$ or $(2, 6)$

3. g.

h.

i.

4. $x + y = 4$

domain: $(-\infty, \infty)$;
range: $(-\infty, \infty)$

5. $x^2 + y^2 = 4$

domain: $[-2, 2]$;
range: $[-2, 2]$

6. $f(x) = 4$

domain: $(-\infty, \infty)$;
range: $\{4\}$

7.

$f(x) = -\dfrac{1}{3}x + 2$

domain: $(-\infty, \infty)$;
range: $(-\infty, \infty)$

8.

$(x + 2)^2 + (y - 1)^2 = 9$

domain: $[-5, 1]$;
range: $[-2, 4]$

9.

$f(x) = \begin{cases} 2 \text{ if } x \le 0 \\ -1 \text{ if } x > 0 \end{cases}$

domain: $(-\infty, \infty)$;
range: $\{-1, 2\}$

10.

$x^2 + y^2 + 4x - 6y - 3 = 0$

domain: $[-6, 2]$;
range: $[-1, 7]$

11.

domain of f = domain of $g = (-\infty, \infty)$;
range of $f = [0, \infty)$; range of $g = [-2, \infty)$

12.

domain of f = domain of $g = (-\infty, \infty)$;
range of $f = [0, \infty)$; range of $g = (-\infty, 4]$

13.

domain of f = range of $f^{-1} = (-\infty, \infty)$;
range of f = domain of $f^{-1} = (-\infty, \infty)$

14.

domain of f = range of $f^{-1} = (-\infty, \infty)$;
range of f = domain of $f^{-1} = (-\infty, \infty)$

15.

domain of f = range of $f^{-1} = [0, \infty)$;
range of f = domain of $f^{-1} = [-1, \infty)$

16. $f(x - 1) = x^2 - 3x - 2$ **17.** $2x + h - 1$ **19.** $\left(\dfrac{f}{g}\right)(x) = \dfrac{x^2 - x - 4}{2x - 6}; (-\infty, 3) \cup (3, \infty)$

20. $(f \circ g)(x) = 4x^2 - 26x + 38$ **21.** $(g \circ f)(x) = 2x^2 - 2x - 14$

23. $f(-x) = x^2 + x - 4$; neither **25.** using $(2, 1)$, $y - 1 = 3(x - 2)$; $y = 3x - 5$

Cumulative Review Exercises (Chapters 1–2)

4.

$g(x) = f(x - 1) + 1$

5.

$h(x) = -2f\left(\dfrac{1}{2}x\right)$

11.

$3x - 6y - 12 = 0$

domain: $(-\infty, \infty)$;
range: $(-\infty, \infty)$

12.

$(x - 2)^2 + (y + 1)^2 = 4$

domain: $[0, 4]$;
range: $[-3, 1]$

13.

domain of f = domain of $g = (-\infty, \infty)$;
range of f = range of $g = (-\infty, \infty)$

14.

domain of f = range of $f^{-1} = [3, \infty)$;
range of f = domain of $f^{-1} = [2, \infty)$

17. $y - 5 = 4(x + 2)$; $y = 4x + 13$; $4x - y + 13 = 0$

CHAPTER 3

Section 3.1
Check Point Exercises

1.
(0, 3)
(−1, 0) 5 (1, 4) (3, 0) 5 x
$f(x) = -(x-1)^2 + 4$

2.
(0, 5) 7
(2, 1) 5 x
$f(x) = (x-2)^2 + 1$

3.
(2, 5)
(0, 1) (2+√5, 0)
(2−√5, 0) 5 x
$f(x) = -x^2 + 4x + 1$

5. c.
$f(x) = -0.005x^2 + 2x + 5$
(200, 205)
(0, 5) (402, 0)
100 300 500 x
Arrow's Horizontal Distance (feet)

Exercise Set 3.1

10. $(2, 12)$ **11.** $(-1, 5)$ **12.** $(-4, -8)$ **14.** $(2, -11)$ **16.** $(2, 7)$

17. domain: $(-\infty, \infty)$
range: $[-1, \infty)$
axis of symmetry: $x = 4$

(0, 15)
(3, 0) (5, 0)
(4, −1) 9 x
$f(x) = (x-4)^2 - 1$

18. domain: $(-\infty, \infty)$
range: $[-2, \infty)$
axis of symmetry: $x = 1$

$(1-\sqrt{2}, 0)$ $(1+\sqrt{2}, 0)$
(0, −1) 5 x
(1, −2)
$f(x) = (x-1)^2 - 2$

19. domain: $(-\infty, \infty)$
range: $[2, \infty)$
axis of symmetry: $x = 1$

(0, 3)
(1, 2) 5 x
$f(x) = (x-1)^2 + 2$

20. domain: $(-\infty, \infty)$
range: $[2, \infty)$
axis of symmetry: $x = 3$

(0, 11)
(3, 2) 16 x
$f(x) = (x-3)^2 + 2$

21. domain: $(-\infty, \infty)$
range: $[1, \infty)$
axis of symmetry: $x = 3$

(0, 10)
(3, 1) 16 x
$y - 1 = (x-3)^2$

22. domain: $(-\infty, \infty)$
range: $[3, \infty)$
axis of symmetry: $x = 1$

(0, 4) (1, 3)
5 x
$y - 3 = (x-1)^2$

23. domain: $(-\infty, \infty)$
range: $[-1, \infty)$
axis of symmetry: $x = -2$

$\left(-2-\frac{\sqrt{2}}{2}, 0\right)$ $\left(-2+\frac{\sqrt{2}}{2}, 0\right)$
(−2, −1) 2 x
$f(x) = 2(x+2)^2 - 1$

24. domain: $(-\infty, \infty)$
range: $\left(-\infty, \frac{5}{4}\right]$
axis of symmetry: $x = \frac{1}{2}$

$\left(\frac{1}{2}-\frac{\sqrt{5}}{2}, 0\right)$ $\left(\frac{1}{2}, \frac{5}{4}\right)$ $\left(\frac{1}{2}+\frac{\sqrt{5}}{2}, 0\right)$
5 x
$f(x) = \frac{5}{4} - \left(x - \frac{1}{2}\right)^2$

25. domain: $(-\infty, \infty)$
range: $(-\infty, 4]$
axis of symmetry: $x = 1$

(0, 3) (1, 4)
(−1, 0) (3, 0) 5 x
$f(x) = 4 - (x-1)^2$

26. domain: $(-\infty, \infty)$
range: $(-\infty, 1]$
axis of symmetry: $x = 3$

(3, 1)
(2, 0) (4, 0) 9 x
$f(x) = 1 - (x-3)^2$

27. domain: $(-\infty, \infty)$
range: $[-4, \infty)$
axis of symmetry: $x = 1$

(3, 0)
(−1, 0) 5 x
(0, −3) (1, −4)
$f(x) = x^2 - 2x - 3$

28. domain: $(-\infty, \infty)$
range: $[-16, \infty)$
axis of symmetry: $x = 1$

(5, 0)
(−3, 0) 6 x
(0, −15) (1, −16)
$f(x) = x^2 - 2x - 15$

29. domain: $(-\infty, \infty)$
range: $\left[-\frac{49}{4}, \infty\right)$
axis of symmetry: $x = -\frac{3}{2}$

(−5, 0) (2, 0)
4 x
$\left(-\frac{3}{2}, -\frac{49}{4}\right)$ (0, −10)
$f(x) = x^2 + 3x - 10$

30. domain: $(-\infty, \infty)$
range: $\left[-\frac{81}{8}, \infty\right)$
axis of symmetry: $x = \frac{7}{4}$

(4, 0)
$\left(-\frac{1}{2}, 0\right)$ 6 x
(0, −4) $\left(\frac{7}{4}, -\frac{81}{8}\right)$
$f(x) = 2x^2 - 7x - 4$

31. domain: $(-\infty, \infty)$
range: $(-\infty, 4]$
axis of symmetry: $x = 1$

(0, 3) (1, 4)
(3, 0)
(−1, 0) 5 x
$f(x) = 2x - x^2 + 3$

32. domain: $(-\infty, \infty)$
range: $(-\infty, 9]$
axis of symmetry: $x = -2$

(−2, 9)
(0, 5)
(−5, 0) (1, 0) 4 x
$f(x) = 5 - 4x - x^2$

33. domain: $(-\infty, \infty)$
range: $[-6, \infty)$
axis of symmetry: $x = -3$

$f(x) = x^2 + 6x + 3$

34. domain: $(-\infty, \infty)$
range: $[-5, \infty)$
axis of symmetry: $x = -2$

$f(x) = x^2 + 4x - 1$

35. domain: $(-\infty, \infty)$
range: $[-5, \infty)$
axis of symmetry: $x = -1$

$f(x) = 2x^2 + 4x - 3$

36. domain: $(-\infty, \infty)$

range: $\left[-\dfrac{13}{3}, \infty\right)$

axis of symmetry: $x = \dfrac{1}{3}$

$f(x) = 3x^2 - 2x - 4$

37. domain: $(-\infty, \infty)$
range: $(-\infty, -1]$
axis of symmetry: $x = 1$

$f(x) = 2x - x^2 - 2$

38. domain: $(-\infty, \infty)$
range: $[2, \infty)$
axis of symmetry: $x = 2$

$f(x) = 6 - 4x + x^2$

39. a. minimum **b.** Minimum is -13 at $x = 2$.
 c. domain: $(-\infty, \infty)$; range: $[-13, \infty)$
40. a. minimum **b.** Minimum is -11 at $x = 2$.
 c. domain: $(-\infty, \infty)$; range: $[-11, \infty)$
41. a. maximum **b.** Maximum is 1 at $x = 1$.
 c. domain: $(-\infty, \infty)$; range: $(-\infty, 1]$
42. a. maximum **b.** Maximum is 21 at $x = -3$.
 c. domain: $(-\infty, \infty)$; range: $(-\infty, 21]$

43. a. minimum **b.** Minimum is $-\dfrac{5}{4}$ at $x = \dfrac{1}{2}$. **c.** domain: $(-\infty, \infty)$; range: $\left[-\dfrac{5}{4}, \infty\right)$

44. a. minimum **b.** Minimum is $-\dfrac{3}{2}$ at $x = \dfrac{1}{2}$. **c.** domain: $(-\infty, \infty)$; range: $\left[-\dfrac{3}{2}, \infty\right)$

45. domain: $(-\infty, \infty)$; range: $[-2, \infty)$ **46.** domain: $(-\infty, \infty)$; range: $(-\infty, -4]$ **51.** $f(x) = 2(x + 10)^2 - 5$ **53.** $f(x) = -3(x + 2)^2 + 4$
54. $f(x) = -3(x - 5)^2 - 7$ **55.** $f(x) = 3(x - 11)^2$ **56.** $f(x) = 3(x - 9)^2$

59. c.

60. c.

67. 12.5 yd by 12.5 yd; 156.25 sq yd
68. 20 yd by 20 yd; 400 sq yd
72. 3 in.; 18 sq in.

81. a.

You cannot see the parabola.

d. You can choose Xmin and Xmax so the x-value of the vertex is in the center of the graph. Choose Ymin to include the y-value of the vertex.

82. $(80, 1600)$

83. $(2.5, 185)$

84. $(-4, 520)$

85. $(-30, 91)$

86. a & e.

d. The greatest number of viewers actually occurred in Season 5, not Season 6, and the model underestimates the greatest number by 1.1 million.

106. $f(2) = -1$; $f(3) = 16$; The graph passes through $(2, -1)$, which is below the x-axis, and $(3, 16)$, which is above the x-axis. Since the graph of f is continuous, it must cross the x-axis somewhere between 2 and 3 to get from one of these points to the other.

Section 3.2

Check Point Exercises

9.

$$f(x) = x^3 - 3x^2$$

10.

$$f(x) = 2(x + 2)^2 (x - 3)$$

Exercise Set 3.2

3. polynomial function; degree: 5 **4.** polynomial function; degree: 7

25. $x = 5$ has multiplicity 1; The graph crosses the x-axis; $x = -4$ has multiplicity 2; The graph touches the x-axis and turns around.

26. $x = -5$ has multiplicity 1; The graph crosses the x-axis; $x = -2$ has multiplicity 2; The graph touches the x-axis and turns around.

27. $x = 3$ has multiplicity 1; The graph crosses the x-axis; $x = -6$ has multiplicity 3; The graph crosses the x-axis.

28. $x = -\dfrac{1}{2}$ has multiplicity 1; The graph crosses the x-axis; $x = 4$ has multiplicity 3; The graph crosses the x-axis.

29. $x = 0$ has multiplicity 1; The graph crosses the x-axis; $x = 1$ has multiplicity 2; The graph touches the x-axis and turns around.

30. $x = 0$ has multiplicity 1; The graph crosses the x-axis; $x = -2$ has multiplicity 2; The graph touches the x-axis and turns around.

31. $x = 2, x = -2$ and $x = -7$ have multiplicity 1; The graph crosses the x-axis.

32. $x = 3, x = -3$, and $x = -5$ have multiplicity 1; The graph crosses the x-axis.

37. $f(-3) = -11; f(-2) = 1$ **39.** $f(-3) = -42; f(-2) = 5$ **40.** $f(2) = -4; f(3) = 14$

41. a. $f(x)$ rises to the right and falls to the left.
 b. $x = -2, x = 1, x = -1$;
 $f(x)$ crosses the x-axis at each.
 c. The y-intercept is -2.
 d. neither
 e.

$$f(x) = x^3 + 2x^2 - x - 2$$

42. a. $f(x)$ rises to the right and falls to the left.
 b. $x = -1, x = 2, x = -2$;
 $f(x)$ crosses the x-axis at each.
 c. The y-intercept is -4.
 d. neither
 e.

$$f(x) = x^3 + x^2 - 4x - 4$$

43. a. $f(x)$ rises to the left and the right.
 b. $x = 0, x = 3, x = -3$;
 $f(x)$ crosses the x-axis at -3 and 3;
 $f(x)$ touches the x-axis at 0.
 c. The y-intercept is 0.
 d. y-axis symmetry
 e.

$$f(x) = x^4 - 9x^2$$

44. a. $f(x)$ rises to the left and the right.
 b. $x = 0, x = 1, x = -1$;
 $f(x)$ touches but does not cross the x-axis at 0;
 $f(x)$ crosses the x-axis at -1 and 1.
 c. The y-intercept is 0.
 d. y-axis symmetry
 e.

$$f(x) = x^4 - x^2$$

45. a. $f(x)$ falls to the left and the right.
 b. $x = 0, x = 4, x = -4$;
 $f(x)$ crosses the x-axis at -4 and 4;
 $f(x)$ touches the x-axis at 0.
 c. The y-intercept is 0.
 d. y-axis symmetry
 e.

$$f(x) = -x^4 + 16x^2$$

46. a. $f(x)$ falls to the left and the right.
 b. $x = 0, x = 2, x = -2$;
 $f(x)$ touches but does not cross the x-axis at 0;
 $f(x)$ crosses the x-axis at -2 and 2.
 c. The y-intercept is 0.
 d. y-axis symmetry
 e.

$$f(x) = -x^4 + 4x^2$$

47. a. $f(x)$ rises to the left and the right.
 b. $x = 0, x = 1$;
 $f(x)$ touches the x-axis at 0 and 1.
 c. The y-intercept is 0.
 d. neither
 e.

$f(x) = x^4 - 2x^3 + x^2$

48. a. $f(x)$ rises to the left and the right.
 b. $x = 0, x = 3$;
 $f(x)$ touches the x-axis at 3 and 0.
 c. The y-intercept is 0.
 d. neither
 e.

$f(x) = x^4 - 6x^3 + 9x^2$

49. a. $f(x)$ falls to the left and the right.
 b. $x = 0, x = 2$;
 $f(x)$ crosses the x-axis at 0 and 2.
 c. The y-intercept is 0.
 d. neither
 e.

$f(x) = -2x^4 + 4x^3$

50. a. $f(x)$ falls to the left and the right.
 b. $x = 0, x = 1$;
 $f(x)$ crosses the x-axis at 0 and 1.
 c. The y-intercept is 0.
 d. neither
 e.

$f(x) = -2x^4 + 2x^3$

51. a. $f(x)$ rises to the left and falls to the right.
 b. $x = 0, x = \pm\sqrt{3}$;
 $f(x)$ crosses the x-axis at 0;
 $f(x)$ touches the x-axis at $\sqrt{3}$ and $-\sqrt{3}$.
 c. The y-intercept is 0.
 d. origin symmetry
 e.

$f(x) = 6x^3 - 9x - x^5$

52. a. $f(x)$ rises to the left and falls to the right.
 b. $x = 0, x = \pm\sqrt{2}$;
 $f(x)$ crosses the x-axis at $-\sqrt{2}$, 0, and $\sqrt{2}$.
 c. The y-intercept is 0.
 d. origin symmetry
 e.

$f(x) = 6x - x^3 - x^5$

53. a. $f(x)$ rises to the left and falls to the right.
 b. $x = 0, x = 3$;
 $f(x)$ crosses the x-axis at 3;
 $f(x)$ touches the x-axis at 0.
 c. The y-intercept is 0.
 d. neither
 e.

$f(x) = 3x^2 - x^3$

54. a. $f(x)$ falls to the left and the right.
 b. $x = \pm 1$;
 $f(x)$ crosses the x-axis at -1 and 1.
 c. The y-intercept is $\dfrac{1}{2}$.
 d. y-axis symmetry
 e.

$f(x) = \dfrac{1}{2} - \dfrac{1}{2}x^4$

55. a. $f(x)$ falls to the left and the right.
 b. $x = 1, x = -2, x = 2$;
 $f(x)$ crosses the x-axis at -2 and 2;
 $f(x)$ touches the x-axis at 1.
 c. The y-intercept is 12.
 d. neither
 e.

$f(x) = -3(x - 1)^2 (x^2 - 4)$

56. a. $f(x)$ falls to the left and the right.
 b. $x = 4, x = -5, x = 5$;
 $f(x)$ crosses the x-axis at -5 and 5;
 $f(x)$ touches the x-axis at 4.
 c. The y-intercept is 800.
 d. neither
 e.

$f(x) = -2(x - 4)^2 (x^2 - 25)$

57. a. $f(x)$ rises to the left and the right.
b. $x = -2, x = 0, x = 1$;
$f(x)$ crosses the x-axis at -2 and 1;
$f(x)$ touches the x-axis at 0.
c. The y-intercept is 0.
d. neither
e.

$f(x) = x^2 (x - 1)^3 (x + 2)$

58. a. $f(x)$ rises to the left and the right.
b. $x = -2, x = -1, x = 0$;
$f(x)$ crosses the x-axis at -1 and 0;
$f(x)$ touches the x-axis at -2.
c. The y-intercept is 0.
d. neither
e.

$f(x) = x^3 (x + 2)^2 (x + 1)$

59. a. $f(x)$ falls to the left and the right.
b. $x = -3, x = 0, x = 1$;
$f(x)$ crosses the x-axis at -3 and 1;
$f(x)$ touches the x-axis at 0.
c. The y-intercept is 0.
d. neither
e.

$f(x) = -x^2 (x - 1)(x + 3)$

60. a. $f(x)$ falls to the left and the right.
b. $x = -2, x = 0, x = 2$;
$f(x)$ crosses the x-axis at -2 and 2;
$f(x)$ touches the x-axis at 0.
c. The y-intercept is 0.
d. y-axis symmetry
e.

$f(x) = -x^2 (x + 2)(x - 2)$

61. a. $f(x)$ falls to the left and the right.
b. $x = -5, x = 0, x = 1$;
$f(x)$ crosses the x-axis at -5 and 0;
$f(x)$ touches the x-axis at 1.
c. The y-intercept is 0.
d. neither
e.

$f(x) = -2x^3 (x - 1)^2 (x + 5)$

62. a. $f(x)$ falls to the left and the right.
b. $x = -3, x = 0, x = 1$;
$f(x)$ crosses the x-axis at -3 and 0;
$f(x)$ touches the x-axis at 1.
c. The y-intercept is 0.
d. neither
e.

$f(x) = -3x^3 (x - 1)^2 (x + 3)$

63. a. $f(x)$ rises to the left and the right.
b. $x = -4, x = 1, x = 2$;
$f(x)$ crosses the x-axis at -4 and 1;
$f(x)$ touches the x-axis at 2.
c. The y-intercept is -16.
d. neither
e.

$f(x) = (x - 2)^2 (x + 4)(x - 1)$

64. a. $f(x)$ rises to the right and falls to the left.
b. $x = -4, x = -3, x = -1$;
$f(x)$ crosses the x-axis at each.
c. The y-intercept is 12.
d. neither
e.

$f(x) = (x + 3)(x + 1)^3 (x + 4)$

65. a. $-2,$ odd; $1,$ odd; $4,$ odd **b.** $f(x) = (x + 2)(x - 1)(x - 4)$ **c.** 8 **66. a.** $-3,$ odd; $2,$ odd; $5,$ odd **b.** $f(x) = (x + 3)(x - 2)(x - 5)$
c. 30 **67. a.** $-1,$ odd; $3,$ even **b.** $f(x) = (x + 1)(x - 3)^2$ **c.** 9 **68. a.** $-2,$ odd; $1,$ even **b.** $f(x) = (x + 2)(x - 1)^2$ **c.** 2
69. a. $-3,$ even; $2,$ even **b.** $f(x) = -(x + 3)^2 (x - 2)^2$ **c.** -36 **70. a.** $-1,$ even; $4,$ even **b.** $f(x) = -(x + 1)^2 (x - 4)^2$ **c.** -16
71. a. $-2,$ even; $-1,$ odd; $1,$ odd **b.** $f(x) = (x + 2)^2 (x + 1)(x - 1)^3$ **c.** -4 **72. a.** $-2,$ odd; $-1,$ even; $1,$ odd
b. $f(x) = (x + 2)(x + 1)^2 (x - 1)^3$ **c.** -2 **73. a.** 3167; The world tiger population in 2010 was approximately 3167.; $(40, 3167)$
b. underestimates by 33 **c.** rises to the right; no; The model indicates an increasing world tiger population that will actually decrease without
conservation efforts. **74. a.** $26{,}747$; The world tiger population in 1980 was approximately $26{,}747$.; $(10, 26{,}747)$ **b.** underestimates by 1253
c. rises to the right; yes; The model indicates an increasing world tiger population that might actually increase with conservation efforts.
75. a. from 1 through 4 min and from 8 through 10 min **b.** from 4 through 8 min and from 10 through 12 min **e.** negative; The graph falls to the
left and falls to the right. **f.** 116 ± 1 beats per min; 10 min **g.** 64 ± 1 beats per min; 8 min **76. e.** positive; The graph falls to the left and rises
to the right. **f.** $\$3.15 \pm \0.05; 2011 **g.** $\$1.85 \pm \0.05; 2009

94.

95.

96.

97.

98.

99.

111. $(-\infty, -5]$; **112.** $y - 3 = -1(x + 10)$ or $y + 5 = -1(x + 2)$; $y = -x - 7$ **114.** $6x^3 - x^2 - 5x + 4$

Exercise Set 3.3

9. $2x^2 + x + 6 - \dfrac{38}{x + 3}$ **10.** $3x + 7 + \dfrac{26}{x - 3}$ **11.** $4x^3 + 16x^2 + 60x + 246 + \dfrac{984}{x - 4}$ **13.** $2x + 5$ **14.** $x^2 + x - 3 - \dfrac{12}{x^2 + x - 2}$

15. $6x^2 + 3x - 1 - \dfrac{3x - 1}{3x^2 + 1}$ **16.** $x^2 - 4x + 1 + \dfrac{4x - 1}{2x^3 + 1}$ **17.** $2x + 5$ **18.** $x + 2$ **23.** $6x^4 + 12x^3 + 22x^2 + 48x + 93 + \dfrac{187}{x - 2}$

24. $x^4 + 7x^3 + 21x^2 + 60x + 182 + \dfrac{549}{x - 3}$ **27.** $x^4 + x^3 + 2x^2 + 2x + 2$ **28.** $x^6 - 2x^5 + 5x^4 - 10x^3 + 10x^2 - 20x + 40 - \dfrac{68}{x + 2}$

29. $x^3 + 4x^2 + 16x + 64$ **30.** $x^6 + 2x^5 + 4x^4 + 8x^3 + 16x^2 + 32x + 64$ **31.** $2x^4 - 7x^3 + 15x^2 - 31x + 64 - \dfrac{129}{x + 2}$ **55. a.** 70; When the

tax rate is 30%, tax revenue is \$700 billion.; (30, 70) **b.** $80 + \dfrac{800}{x - 110}$; $f(30) = 70$; yes **c.** No, f is a rational function because it is a quotient of

two polynomials. **56. a.** $68\dfrac{4}{7}$; When the tax rate is 40%, tax revenue is $68\dfrac{4}{7}$ tens of billions of dollars, or approximately \$685.7 billion.; $\left(40, 68\dfrac{4}{7}\right)$

b. $80 + \dfrac{800}{x - 110}$; $f(40) \approx 68.57$; yes **c.** No; f is a rational function because it is a quotient of two polynomials. **79.**

Exercise Set 3.4

4. $\pm 1, \pm 3, \pm 5, \pm 15, \pm\dfrac{1}{2}, \pm\dfrac{3}{2}, \pm\dfrac{5}{2}, \pm\dfrac{15}{2}$ **5.** $\pm 1, \pm 2, \pm 3, \pm 6, \pm\dfrac{1}{2}, \pm\dfrac{1}{4}, \pm\dfrac{3}{2}, \pm\dfrac{3}{4}$ **6.** $\pm 1, \pm 2, \pm 4, \pm 8, \pm\dfrac{1}{3}, \pm\dfrac{2}{3}, \pm\dfrac{4}{3}, \pm\dfrac{8}{3}$

7. $\pm 1, \pm 2, \pm 3, \pm 4, \pm 6, \pm 12$ **9. b.** $-2, -1,$ or 2 **10. a.** $\pm 1, \pm 2, \pm 3, \pm 4, \pm 6, \pm 12$ **b.** $-3, 1,$ or 4 **11. a.** $\pm 1, \pm 2, \pm 3, \pm 6, \pm\dfrac{1}{2}, \pm\dfrac{3}{2}$

b. $-2, \dfrac{1}{2},$ or 3 **12. b.** $-\dfrac{1}{2}, 1,$ or 2 **13. c.** $-1, \dfrac{-3 - \sqrt{33}}{2},$ and $\dfrac{-3 + \sqrt{33}}{2}$ **14. c.** $\dfrac{1}{2}, \dfrac{-1 - \sqrt{5}}{2},$ and $\dfrac{-1 + \sqrt{5}}{2}$ **15. c.** $-2, \dfrac{-1 + i}{2},$ and $\dfrac{-1 - i}{2}$

16. c. $1, \dfrac{3 + i\sqrt{11}}{2},$ and $\dfrac{3 - i\sqrt{11}}{2}$ **17. a.** $\pm 1, \pm 2, \pm 3, \pm 4, \pm 6, \pm 12$ **b.** $-3, 1,$ or 4 **18. b.** -1 or 4 **19. a.** $\pm 1, \pm 2, \pm 3, \pm 4, \pm 6, \pm 12$

21. a. $\pm 1, \pm 5, \pm\dfrac{1}{2}, \pm\dfrac{5}{2}, \pm\dfrac{1}{3}, \pm\dfrac{5}{3}, \pm\dfrac{1}{6}, \pm\dfrac{5}{6}$ **b.** $-5, \dfrac{1}{3},$ or $\dfrac{1}{2}$ **22. a.** $\pm 1, \pm 2, \pm 4, \pm\dfrac{1}{2}$ **23. a.** $\pm 1, \pm 2, \pm 4$ **b.** -2 or 2

24. a. $\pm 1, \pm 3, \pm 5, \pm 15$ **b.** -1 or 3 **25.** $f(x) = 2x^3 - 2x^2 + 50x - 50$ **26.** $f(x) = 2x^3 - 8x^2 + 8x - 32$

27. $f(x) = x^3 - 3x^2 - 15x + 125$ **28.** $f(x) = 3x^3 + 12x^2 - 93x - 522$ **30.** $f(x) = 2x^4 + 5x^3 + 4x^2 + 5x + 2$

31. $f(x) = x^4 - 9x^3 + 21x^2 + 21x - 130$ **32.** $f(x) = 3x^4 - x^3 - 9x^2 + 159x - 52$

53. b.

$f(x) = -x^3 + x^2 + 16x - 16$

54. b.

$f(x) = -x^3 + 3x^2 - 4$

55. b.

$f(x) = 4x^3 - 8x^2 - 3x + 9$

56. b.

$f(x) = 3x^3 + 2x^2 + 2x - 1$

57. b.

$f(x) = 2x^4 - 3x^3 - 7x^2 - 8x + 6$

58. b.

$f(x) = 2x^4 + 2x^3 - 22x^2 - 18x + 36$

59. b.

$f(x) = 3x^5 + 2x^4 - 15x^3 - 10x^2 + 12x + 8$

60. b.

$f(x) = -5x^4 + 4x^3 - 19x^2 + 16x + 4$

72. $\left\{-\dfrac{1}{2}, 3, 5\right\}$ **77.** 5, 3, or 1 positive real zeros; no negative real zeros

79.

1 real zero, 2 nonreal complex zeros

80.

3 real zeros, 2 nonreal complex zeros

81.

2 real zeros, 2 nonreal complex zeros

82.

2 real zeros, 4 nonreal complex zeros

83. makes sense
85. makes sense

Mid-Chapter 3 Check Point

1. $f(x) = (x - 3)^2 - 4$
domain: $(-\infty, \infty)$
range: $[-4, \infty)$

2. $f(x) = 5 - (x + 2)^2$
domain: $(-\infty, \infty)$
range: $(-\infty, 5]$

3. $f(x) = -x^2 - 4x + 5$
domain: $(-\infty, \infty)$
range: $(-\infty, 9]$

4. $f(x) = 3x^2 - 6x + 1$
domain: $(-\infty, \infty)$
range: $[-2, \infty)$

5. $f(x) = (x - 2)^2 (x + 1)^3$

6. $f(x) = -(x - 2)^2 (x + 1)^2$

7. $f(x) = x^3 - x^2 - 4x + 4$

8. $f(x) = x^4 - 5x^2 + 4$

9. $f(x) = -(x + 1)^6$

10. $f(x) = -6x^3 + 7x^2 - 1$

11. $f(x) = 2x^3 - 2x$

12. $f(x) = x^3 - 2x^2 + 26x$

13. $f(x) = -x^3 + 5x^2 - 5x - 3$

23. $2x^2 - x - 3 + \dfrac{x + 1}{3x^2 - 1}$

24. $2x^3 - 5x^2 - 3x + 6$
25. $f(x) = -2x^3 + 2x^2 - 2x + 2$
26. $f(x) = x^4 - 4x^3 + 13x^2 - 36x + 36$
27. yes

Section 3.5

Check Point Exercises

4.

$g(x) = \dfrac{1}{x + 2} - 1$

5.

$f(x) = \dfrac{3x - 3}{x - 2}$

6.

$f(x) = \dfrac{2x^2}{x^2 - 9}$

7.

$f(x) = \dfrac{x^4}{x^2 + 2}$

Exercise Set 3.5

21. vertical asymptote: $x = -4$; no holes **22.** vertical asymptote: $x = 3$; no holes **23.** vertical asymptotes: $x = -4, x = 0$; no holes
24. vertical asymptotes: $x = 3, x = 0$; no holes **25.** vertical asymptote: $x = -4$; hole at $x = 0$ **26.** vertical asymptote: $x = 3$; hole at $x = 0$
27. no vertical asymptotes; no holes **28.** no vertical asymptotes; no holes **29.** no vertical asymptotes: hole at $x = 3$
30. no vertical asymptotes: hole at $x = 5$ **31.** vertical asymptote: $x = -3$; hole at $x = 3$ **32.** vertical asymptote: $x = -5$; hole at $x = 5$
33. vertical asymptote: $x = 3$; hole at $x = -7$ **34.** vertical asymptote: $x = 4$; hole at $x = -6$ **35.** no vertical asymptotes; hole at $x = -7$
36. no vertical asymptotes; hole at $x = -6$

45.
$x = 1$
$(2, 1)$
$(0, -1)$
$g(x) = \dfrac{1}{x - 1}$

46.
$x = 2$
$\left(0, -\dfrac{1}{2}\right)$
$(3, 1)$
$(1, -1)$
$g(x) = \dfrac{1}{x - 2}$

47.
$(1, 3)$
$(-1, 1)$
$y = 2$
$\left(-\dfrac{1}{2}, 0\right)$
$h(x) = \dfrac{1}{x} + 2$

48.
$(1, 2)$
$y = 1$
$(-1, 0)$
$h(x) = \dfrac{1}{x} + 1$

49.
$x = -1$
$(0, -1)$
$y = -2$
$(-2, -3)$
$g(x) = \dfrac{1}{x + 1} - 2$

50.
$x = -2$
$\left(-\dfrac{3}{2}, 0\right)$
$\left(0, -\dfrac{3}{2}\right)$
$(-1, -1)$
$y = -2$
$(-3, -3)$
$g(x) = \dfrac{1}{x + 2} - 2$

51.
$x = -2$
$(-1, 1)$
$\left(0, \dfrac{1}{4}\right)$
$(-3, 1)$
$g(x) = \dfrac{1}{(x + 2)^2}$

52.
$x = -1$
$(-2, 1)$
$(0, 1)$
$g(x) = \dfrac{1}{(x + 1)^2}$

53.
$\left(-\dfrac{1}{2}, 0\right)$
$\left(\dfrac{1}{2}, 0\right)$
$(-1, -3)$
$(1, -3)$
$y = -4$
$h(x) = \dfrac{1}{x^2} - 4$

54.
$\left(-\dfrac{\sqrt{3}}{3}, 0\right)$
$\left(\dfrac{\sqrt{3}}{3}, 0\right)$
$(-1, -2)$
$(1, -2)$
$y = -3$
$h(x) = \dfrac{1}{x^2} - 3$

55.
$x = 3$
$(2, 2)$
$(4, 2)$
$y = 1$
$h(x) = \dfrac{1}{(x - 3)^2} + 1$

56.
$x = 3$
$(2, 3)$
$(4, 3)$
$y = 2$
$h(x) = \dfrac{1}{(x - 3)^2} + 2$

57.
$x = 2$
$(4, 8)$
$(6, 6)$
$y = 4$
$(-2, 2)$
$(0, 0)$
$f(x) = \dfrac{4x}{x - 2}$

58.
$x = 1$
$(2, 6)$
$(4, 4)$
$y = 3$
$(-2, 2)$
$(0, 0)$
$f(x) = \dfrac{3x}{x - 1}$

59.
$x = 2$
$(0, 0)$
$\left(-1, \dfrac{2}{3}\right)$
$\left(3, \dfrac{6}{5}\right)$
$\left(-3, -\dfrac{6}{5}\right)$
$\left(1, -\dfrac{2}{3}\right)$
$x = -2$
$f(x) = \dfrac{2x}{x^2 - 4}$

60.
$x = 1$
$\left(-\dfrac{1}{2}, \dfrac{8}{3}\right)$
$(0, 0)$
$\left(3, \dfrac{3}{2}\right)$
$\left(-3, -\dfrac{3}{2}\right)$
$\left(\dfrac{1}{2}, -\dfrac{8}{3}\right)$
$x = -1$
$f(x) = \dfrac{4x}{x^2 - 1}$

61.
$\left(-2, \dfrac{8}{3}\right)$
$x = 1$
$y = 2$
$\left(2, \dfrac{8}{3}\right)$
$(0, 0)$
$x = -1$
$f(x) = \dfrac{2x^2}{x^2 - 1}$

62.
$x = 3$
$\left(-5, \dfrac{25}{4}\right)$
$\left(5, \dfrac{25}{4}\right)$
$y = 4$
$(0, 0)$
$x = -3$
$f(x) = \dfrac{4x^2}{x^2 - 9}$

63.
$x = -1$
$(0, 0)$
$y = -1$
$(-2, -2)$
$f(x) = \dfrac{-x}{x + 1}$

64.
$x = -2$
$(0, 0)$
$y = -3$
$(-4, -6)$
$(4, -2)$
$f(x) = \dfrac{-3x}{x + 2}$

65.
$x = -2$
$\left(0, \dfrac{1}{4}\right)$
$\left(-3, -\dfrac{1}{5}\right)$
$\left(3, -\dfrac{1}{5}\right)$
$x = 2$
$f(x) = -\dfrac{1}{x^2 - 4}$

66.
$x = -1$
$(0, 2)$
$\left(-2, -\dfrac{2}{3}\right)$
$\left(2, -\dfrac{2}{3}\right)$
$x = 1$
$f(x) = -\dfrac{2}{x^2 - 1}$

67.
$x = -2$
$\left(-3, \dfrac{1}{2}\right)$
$\left(2, \dfrac{1}{2}\right)$
$(-1, -1)$
$(0, -1)$
$x = 1$
$f(x) = \dfrac{2}{x^2 + x - 2}$

68.
$x = -1$
$(0, 1)$
$(1, 1)$
$\left(-2, -\dfrac{1}{2}\right)$
$\left(3, -\dfrac{1}{2}\right)$
$x = 2$
$f(x) = \dfrac{-2}{x^2 - x - 2}$

69.
$(-2, 1)$
$y = 2$
$(0, 0)$
$(2, 1)$
$f(x) = \dfrac{2x^2}{x^2 + 4}$

70.
$y = 4$
$(-1, 2)$
$(1, 2)$
$(0, 0)$
$f(x) = \dfrac{4x^2}{x^2 + 1}$

71.
$x = 2$
$(-2, 0)$
$\left(3, \dfrac{5}{6}\right)$
$\left(-4, -\dfrac{1}{3}\right)$
$\left(0, -\dfrac{1}{3}\right)$
$x = -3$
$f(x) = \dfrac{x + 2}{x^2 + x - 6}$

72.
$\left(2, \dfrac{1}{2}\right)$
$x = -2$
$\left(0, \dfrac{2}{3}\right)$
$(4, 0)$
$\left(-3, -\dfrac{7}{6}\right)$
$x = 3$
$f(x) = \dfrac{x - 4}{x^2 - x - 6}$

73.
$\left(0, \dfrac{1}{2}\right)$
$x = -2$
$(-1, 1)$
$\left(2, \dfrac{1}{4}\right)$
$(-3, -1)$
$f(x) = \dfrac{x - 2}{x^2 - 4}$

74.
$x = -3$
$\left(0, \dfrac{1}{3}\right)$
$(-2, 1)$
$\left(3, \dfrac{1}{6}\right)$
$(-4, -1)$
$f(x) = \dfrac{x - 3}{x^2 - 9}$

75.
$\left(-2, \dfrac{8}{3}\right)$
$\left(2, \dfrac{8}{3}\right)$
$(0, 0)$
$f(x) = \dfrac{x^4}{x^2 + 2}$

76.
$(-1, 1)$
$(1, 1)$
$(0, 0)$
$f(x) = \dfrac{2x^4}{x^2 + 1}$

77.
$$f(x) = \frac{x^2 + x - 12}{x^2 - 4}$$

78.
$$f(x) = \frac{x^2}{x^2 + x - 6}$$

79.
$$f(x) = \frac{3x^2 + x - 4}{2x^2 - 5x}$$

80.
$$f(x) = \frac{x^2 - 4x + 3}{(x + 1)^2}$$

81. b.
$$f(x) = \frac{x^2 - 1}{x}$$

82. b.
$$f(x) = \frac{x^2 - 4}{x}$$

83. b.
$$f(x) = \frac{x^2 + 1}{x}$$

84. b.
$$f(x) = \frac{x^2 + 4}{x}$$

85. a. Slant asymptote: $y = x + 4$
b.
$$f(x) = \frac{x^2 + x - 6}{x - 3}$$

86. a. Slant asymptote: $y = x$
b.
$$f(x) = \frac{x^2 - x + 1}{x - 1}$$

87. b.
$$f(x) = \frac{x^3 + 1}{x^2 + 2x}$$

88. b.
$$f(x) = \frac{x^3 - 1}{x^2 - 9}$$

89.
$$f(x) = \frac{x + 2}{2x(x - 2)}$$

90.
$$f(x) = \frac{5x + 1}{2(x - 5)}$$

91.
$$f(x) = \frac{x - 6}{2(x - 3)}$$

92.
$$f(x) = \frac{-2}{(x + 2)(x + 3)}$$

93.
$$f(x) = \frac{x - 2}{x + 2}$$

94.
$$f(x) = \frac{(x - 1)(x + 1)}{x^2 + 1}$$

95. $g(x) = \dfrac{1}{x + 3} + 2$

$$g(x) = \frac{1}{x + 3} + 2$$

96. $g(x) = \dfrac{1}{x + 2} + 3$

$$g(x) = \frac{1}{x + 2} + 3$$

97. $g(x) = \dfrac{-1}{x - 2} + 3$

$$g(x) = \frac{-1}{x - 2} + 3$$

98. $g(x) = \dfrac{-1}{x - 4} + 2$

$$g(x) = \frac{-1}{x - 4} + 2$$

99. b. $\overline{C}(x) = \dfrac{100x + 100,000}{x}$

c. $\overline{C}(500) = 300$, when 500 bicycles are produced, it costs $300 to produce each bicycle; $\overline{C}(1000) = 200$, when 1000 bicycles are produced, it costs $200 to produce each bicycle; $\overline{C}(2000) = 150$, when 2000 bicycles are produced, it costs $150 to produce each bicycle; $\overline{C}(4000) = 125$, when 4000 bicycles are produced, it costs $125 to produce each bicycle.
d. $y = 100$; The cost per bicycle approaches $100 as more bicycles are produced.

100. b. $\overline{C}(x) = \dfrac{30x + 300,000}{x}$

c. $\overline{C}(1000) = 330$, when 1000 pairs of shoes are produced, it costs $330 to produce each pair; $\overline{C}(10,000) = 60$, when 10,000 pairs of shoes are produced, its costs $60 to produce each pair; $\overline{C}(100,000) = 33$, when 100,000 pairs of shoes are produced, it costs $33 to produce each pair.
d. $y = 30$; The cost per pair of shoes approaches $30 as more shoes are produced.
101. d. $y = 6.5$; Over time, the pH level rises back to normal. **e.** It quickly drops below normal and then slowly begins to approach the normal level. **102. c.** $y = 0$; Over time, the drug's concentration will approach 0 mg per liter. **103.** 90; An incidence ratio of 10 means 90% of the deaths are smoking related. **104.** 89; An incidence ratio of 9 means about 89% of the deaths are smoking related. **105.** $y = 100$; The percentage of deaths cannot exceed 100% as the incidence ratios increase. **106.** No; according to the model and its graph, there is no incidence ratio that will produce 100 percentage of deaths. **107. a.** $f(x) = \dfrac{1.75x^2 - 15.9x + 160}{2.1x^2 - 3.5x + 296}$ **c.** 66%; underestimates by 1%

118.

The graph approaches the horizontal asymptote faster and the vertical asymptote more slowly as n increases.

119.

The graph approaches the horizontal asymptote faster and the vertical asymptote more slowly as n increases.

120.

$g(x)$ is the graph of a line whereas $f(x)$ is the graph of a rational function with a slant asymptote; In $g(x)$, $x - 2$ is a factor of $x^2 - 5x + 6$.

121. a.

b. The graph increases and reaches a maximum of about 356 arrests per 100,000 drivers at age 25.

122. does not make sense

Section 3.6

Check Point Exercises

1. *(number line with open circles at −4 and 5)*

2. *(number line with marks at $\dfrac{-3-\sqrt{7}}{2}$ and $\dfrac{-3+\sqrt{7}}{2}$)*

3. *(number line with marks at −3, −1, 1)*

4. *(number line with marks at −1, 1)*

Exercise Set 3.6

1. $(-\infty, -2) \cup (4, \infty)$
(−2, 4)

2. $(-\infty, -3) \cup (5, \infty)$
(−3, 5)

3. $[-3, 7]$
(−3, 7)

4. $[-1, 7]$
(−1, 7)

5. $(-\infty, 1) \cup (4, \infty)$
(1, 4)

6. $(1, 3)$
(1, 3)

7. $(-\infty, -4) \cup (-1, \infty)$
(−4, −1)

8. $(-\infty, -3) \cup (2, \infty)$
(−3, 2)

9. \varnothing

10. $(-\infty, 1) \cup (1, \infty)$
(1)

11. $\left[-4, \dfrac{2}{3}\right]$
(−4, 2/3)

12. $\left(-\infty, -\dfrac{2}{3}\right] \cup \left[\dfrac{1}{3}, \infty\right)$
(−2/3, 1/3)

13. $\left(-3, \dfrac{5}{2}\right)$
(−3, 5/2)

14. $\left(-\infty, -\dfrac{1}{2}\right) \cup \left(\dfrac{1}{3}, \infty\right)$
(−1/2, 1/3)

15. $\left(-1, -\dfrac{3}{4}\right)$
(−1, −3/4)

16. $\left(-5, -\dfrac{1}{3}\right)$
(−5, −1/3)

17. $\left[-2, \dfrac{1}{3}\right]$
(−2, 1/3)

18. $(-\infty, \infty)$

19. $(-\infty, 0] \cup [4, \infty)$
(0, 4)

20. $(-2, 0)$
(−2, 0)

21. $\left(-\infty, -\dfrac{3}{2}\right) \cup (0, \infty)$
(−3/2, 0)

22. $\left[0, \dfrac{5}{3}\right]$
(0, 5/3)

23. $[0, 1]$
(0, 1)

24. $[0, 2]$
(0, 2)

25. $[2 - \sqrt{2}, 2 + \sqrt{2}]$
(2 − √2, 2 + √2)

26. $[1 - \sqrt{3}, 1 + \sqrt{3}]$
(1 − √3, 1 + √3)

27. \varnothing

28. $(-\infty, \infty)$

29. $[1, 2] \cup [3, \infty)$
(1, 2, 3)

30. $[-3, -2] \cup [-1, \infty)$
(−3, −2, −1)

31. $[0, 3] \cup [5, \infty)$
(0, 3, 5)

32. $[0, 4] \cup [6, \infty)$
(0, 4, 6)

33. $(-\infty, 2) \cup \left(2, \frac{7}{2}\right)$

34. $(-\infty, 5) \cup \left(5, \frac{13}{2}\right)$

35. $[-2, -1] \cup [1, \infty)$

36. $\{-2\} \cup [2, \infty)$

37. $(-\infty, -3)$

38. $(-\infty, -7) \cup (-1, 1)$

39. $(-1, \infty)$

40. $(1, \infty)$

41. $\{0\} \cup [9, \infty)$

42. $(-\infty, 4]$

43. $(-\infty, -3) \cup (4, \infty)$

44. $(-\infty, -5) \cup (2, \infty)$

45. $(-4, -3)$

46. $(-5, -2)$

47. $[2, 4)$

48. $(-\infty, -3] \cup (-2, \infty)$

49. $\left(-\infty, -\frac{4}{3}\right) \cup [2, \infty)$

50. $\left[-\frac{5}{3}, 3\right)$

51. $(-\infty, 0) \cup (3, \infty)$

52. $(-\infty, -4) \cup (0, \infty)$

53. $(-\infty, -4] \cup (-2, 1]$

54. $(-\infty, -3] \cup (-1, 2]$

55. $(-\infty, -5) \cup (-3, \infty)$

56. $(1, 2)$

57. $\left(-\infty, \frac{1}{2}\right) \cup \left[\frac{7}{5}, \infty\right)$

58. $(-\infty, 3) \cup (4, \infty)$

59. $(-\infty, -6] \cup (-2, \infty)$

60. $[-4, -2)$

61. $\left(-\infty, \frac{1}{2}\right] \cup [2, \infty)$

62. $\left(-\infty, \frac{1}{4}\right) \cup (2, \infty)$

63. $(-\infty, -1) \cup [1, \infty)$

64. $\left(\frac{1}{2}, 1\right]$

65. $(-\infty, -8) \cup (-6, 4) \cup (6, \infty)$

66. $(-\infty, -7) \cup (1, \infty)$

67. $(-3, 2)$

68. $(-\infty, -3) \cup (-1, 1)$

69. $(-\infty, -1) \cup (1, 2) \cup (3, \infty)$

70. $(-\infty, -1) \cup (1, 2) \cup (3, \infty)$

71. $\left[-6, -\frac{1}{2}\right] \cup [1, \infty)$

72. $(-\infty, -6) \cup \left(-\frac{1}{2}, 1\right)$

73. $(-\infty, -2) \cup [-1, 2)$

74. $(-2, -1) \cup (2, \infty)$

76. between 0 and 3 seconds **77. c.** extremely well; Function values and data are identical. **d.** speeds exceeding 76 miles per hour; points on (b) to the right of (76, 540) **78. c.** extremely well **d.** speeds exceeding 68 miles per hour; points on (a) to the right of (68, 540)

86. $(-\infty, -5) \cup (2, \infty)$ **87.** $\left[-3, \frac{1}{2}\right]$ **88.** $(-2, -1) \cup (2, \infty)$ **91.** $(-4, -1) \cup [2, \infty)$ **92. b.** speeds exceeding 60 miles per hour

93. b. speeds exceeding 52 miles per hour **95.** does not make sense **96.** makes sense **97.** does not make sense **104.** $(-\infty, 2) \cup (2, \infty)$

107. $(-\infty, 2) \cup (2, \infty)$

109. $27 - 3x^2 \geq 0$
$3x^2 \leq 27$
$x^2 \leq 9$
$-3 \leq x \leq 3$

112.

113. a. 16 **c.** 400

114. a. 96 **c.** 32

Exercise Set 3.7

13. $x = \dfrac{kz^3}{y}$; $y = \dfrac{kz^3}{x}$ **14.** $x = \dfrac{k\sqrt[3]{z}}{y}$; $y = \dfrac{k\sqrt[3]{z}}{x}$ **15.** $x = \dfrac{kyz}{\sqrt{w}}$; $y = \dfrac{x\sqrt{w}}{kz}$ **16.** $x = \dfrac{kyz}{w^2}$; $y = \dfrac{xw^2}{kz}$ **17.** $x = kz(y + w)$; $y = \dfrac{x - kzw}{kz}$

18. $x = kz(y - w)$; $y = \dfrac{x + kzw}{kz}$ **19.** $x = \dfrac{kz}{y - w}$; $y = \dfrac{xw + kz}{x}$ **20.** $x = \dfrac{kz}{y + w}$; $y = \dfrac{kz - xw}{x}$ **36.** $\dfrac{1}{3}$ as long as it seems now

37. a. $C = \dfrac{kP_1P_2}{d^2}$ **38.** Yes, the wind will exert a force of 360 pounds.

39. a.

b. Current varies inversely as resistance.

51. makes sense **55.** Reduce the resistance by a factor of $\frac{1}{3}$.

61.

$f(x) = 2^x$

62.

$g(x) = f(-x) = 2^{-x}$

63.

$h(x) = f(x) + 1 = 2^x + 1$

Chapter 3 Review Exercises

1.

$f(x) = -(x + 1)^2 + 4$

axis of symmetry: $x = -1$
domain: $(-\infty, \infty)$; range: $(-\infty, 4]$

2.

$f(x) = (x + 4)^2 - 2$

axis of symmetry: $x = -4$
domain: $(-\infty, \infty)$; range: $[-2, \infty)$

3.

$f(x) = -x^2 + 2x + 3$

axis of symmetry: $x = 1$
domain: $(-\infty, \infty)$; range: $(-\infty, 4]$

4.

$f(x) = 2x^2 - 4x - 6$

axis of symmetry: $x = 1$
domain: $(-\infty, \infty)$; range: $[-8, \infty)$

5. a. maximum is -57 at $x = 7$ **b.** domain: $(-\infty, \infty)$; range: $(-\infty, -57]$ **6. a.** minimum is 685 at $x = -3$ **b.** domain: $(-\infty, \infty)$; range: $[685, \infty)$

7. b. 6 ft **d.**

$f(x) = -0.025x^2 + x + 6$

Ball's Height (feet)

Ball's Horizontal Distance (yards)

14. b. no; The model indicates increasing deforestation despite a declining rate in which the forest is being cut down. **d.** no; The model indicates the amount of forest cleared, in square kilometers, will eventually be negative, which is not possible. **15.** No; the graph falls to the right, so eventually there would be a negative number of thefts, which is not possible. **16.** $x = 1$, multiplicity 1, crosses; $x = -2$, multiplicity 2, touches; $x = -5$, multiplicity 3, crosses **17.** $x = -5$, multiplicity 1, crosses; $x = 5$, multiplicity 2, touches **18.** $f(1)$ is negative and $f(2)$ is positive, so by the Intermediate Value Theorem, f has a real zero between 1 and 2.

19. a. The graph falls to the left and rises to the right.
b. no symmetry
c.

$f(x) = x^3 - x^2 - 9x + 9$

20. a. The graph rises to the left and falls to the right.
b. origin symmetry
c.

$f(x) = 4x - x^3$

21. a. The graph falls to the left and rises to the right.
b. no symmetry
c.

$f(x) = 2x^3 + 3x^2 - 8x - 12$

22. a. The graph falls to the left and to the right.
b. y-axis symmetry
c.

$f(x) = -x^4 + 25x^2$

23. a. The graph falls to the left and to the right.
b. no symmetry
c.

$f(x) = -x^4 + 6x^3 - 9x^2$

24. a. The graph rises to the left and to the right.
b. no symmetry
c.

$f(x) = 3x^4 - 15x^3$

25.

$f(x) = 2x^2(x - 1)^3(x + 2)$

26.

$f(x) = -x^3(x + 4)^2(x - 1)$

28. $2x^2 - 4x + 1 - \dfrac{10}{5x - 3}$ **36.** $\pm 1, \pm 2, \pm 4, \pm 8, \pm\dfrac{8}{3}, \pm\dfrac{4}{3}, \pm\dfrac{2}{3}, \pm\dfrac{1}{3}$ **38.** 3 or 1 positive real zeros; 2 or 0 negative real zeros

39. No sign variations exist for either $f(x)$ or $f(-x)$, so no real roots exist. **40. b.** 1 positive real zero; 2 or no negative real zeros **c.** -2 or 1

41. b. 2 or 0 positive real zeros; 1 negative real zero **c.** $-1, \dfrac{1}{3}$, or $\dfrac{1}{2}$

42. a. $\pm 1, \pm 3, \pm 5, \pm 15, \pm\dfrac{1}{2}, \pm\dfrac{1}{4}, \pm\dfrac{1}{8}, \pm\dfrac{3}{2}, \pm\dfrac{3}{4}, \pm\dfrac{3}{8}, \pm\dfrac{5}{2}, \pm\dfrac{5}{4}, \pm\dfrac{5}{8}, \pm\dfrac{15}{2}, \pm\dfrac{15}{4}, \pm\dfrac{15}{8}$ **b.** 3 or 1 positive real solutions; no negative real solutions

c. $\dfrac{1}{2}, \dfrac{3}{2}$, or $\dfrac{5}{2}$ **43. b.** 2 or 0 positive real solutions; 1 negative solution **d.** $\left\{\dfrac{1}{2}, \dfrac{-5 - \sqrt{29}}{2}, \dfrac{-5 + \sqrt{29}}{2}\right\}$

44. b. 2 or zero positive real solutions; 2 or zero negative real solutions **c.** $-2, -1, 1$, or 3

45. b. 1 positive real root; 1 negative real root **c.** $-\dfrac{1}{2}$ or $\dfrac{1}{2}$ **46. b.** 2 or no positive zeros; 2 or no negative zeros **c.** $-2, -1, \dfrac{1}{2}$, or 2

47. $f(x) = x^3 - 6x^2 + 21x - 26$ **48.** $f(x) = 2x^4 + 12x^3 + 20x^2 + 12x + 18$ **49.** $f(x) = (x - i)(x + i)(x + 2)(2x - 1)$
50. $g(x) = (x + 1)^2(x - 4)^2$ **51.** 4 real zeros, one with multiplicity two **52.** 3 real zeros; 2 nonreal complex zeros
53. 2 real zeros, one with multiplicity two; 2 nonreal complex zeros **54.** 1 real zero; 4 nonreal complex zeros

55.

56.

57. Vertical asymptotes: $x = 3$ and $x = -3$
horizontal asymptote: $y = 0$

58. Vertical asymptote: $x = -3$
horizontal asymptote: $y = 2$

59. Vertical asymptotes: $x = 3$ and $x = -2$
horizontal asymptote: $y = 1$

60. Vertical asymptote: $x = -2$
horizontal asymptote: $y = 1$

61. Vertical asymptote: $x = -1$
no horizontal asymptote
slant asymptote: $y = x - 1$

62. Vertical asymptote: $x = 3$
no horizontal asymptote
slant asymptote: $y = x + 5$

63. No vertical asymptote
no horizontal asymptote
slant asymptote: $y = -2x$

64. Vertical asymptote: $x = \dfrac{3}{2}$
no horizontal asymptote
slant asymptote: $y = 2x - 5$

65. b. $\overline{C}(x) = \dfrac{25x + 50,000}{x}$

c. $\overline{C}(50) = 1025$, when 50 calculators are manufactured, it costs \$1025 to manufacture each; $\overline{C}(100) = 525$, when 100 calculators are manufactured, it costs \$525 to manufacture each; $\overline{C}(1000) = 75$, when 1000 calculators are manufactured, it costs \$75 to manufacture each; $\overline{C}(100,000) = 25.5$, when 100,000 calculators are manufactured, it costs \$25.50 to manufacture each. **d.** $y = 25$; costs will approach \$25. **66.** $y = 3000$; The number of fish in the pond approaches 3000. **67.** $y = 0$; As the number of years of education increases the percentage rate of unemployment approaches zero. **68. b.** $R(x) = \dfrac{1.48x + 115.1}{2.92x + 236}$ **c.** $y = 0.51$; Over time the percentage of men in the U.S. population will approach 51%.

69. $\left(-3, \dfrac{1}{2}\right)$

70. $\left(-\infty, -4\right] \cup \left[-\dfrac{1}{2}, \infty\right)$

71. $(-3, 0) \cup (1, \infty)$

72. $(-\infty, -2) \cup (6, \infty)$

73. $[-1, 1) \cup [2, \infty)$

74. $(-\infty, 4) \cup \left[\dfrac{23}{4}, \infty\right)$

75. a. 261 ft; overestimates by 1 ft **b.** speeds exceeding 40 miles per hour

Chapter 3 Test

1.

$f(x) = (x + 1)^2 + 4$

axis of symmetry: $x = -1$
domain: $(-\infty, \infty)$; range: $[4, \infty)$

2.

$f(x) = x^2 - 2x - 3$

axis of symmetry: $x = 1$
domain: $(-\infty, \infty)$; range: $[-4, \infty)$

3. a. maximum of 2 at $x = 3$;
b. domain: $(-\infty, \infty)$; range:
$(-\infty, 2]$

6. b.

$f(x) = x^3 - 5x^2 - 4x + 20$

7. Since the degree of the polynomial is odd and the leading coefficient is positive, the graph of f should fall to the left and rise to the right. The x-intercepts should be $-1, 0,$ and 1.

10. 3 or 1 positive real zeros; no negative real zeros

12. a. $\pm 1, \pm 3, \pm 5, \pm 15, \pm\dfrac{1}{2}, \pm\dfrac{3}{2}, \pm\dfrac{5}{2}, \pm\dfrac{15}{2}$

15.

$f(x) = -3x^3 - 4x^2 + x + 2$

16. domain: $\{x \mid x \neq -3\}$

$f(x) = \dfrac{1}{(x + 3)^2}$

17. domain: $\{x \mid x \neq 1\}$

$f(x) = \dfrac{1}{x - 1} + 2$

18. domain: $\{x \mid x \neq 4, x \neq -4\}$

$f(x) = \dfrac{x}{x^2 - 16}$

19. domain: $\{x \mid x \neq 2\}$

$f(x) = \dfrac{x^2 - 9}{x - 2}$

20. domain: $\{x \mid x \neq -3, x \neq 1\}$

$f(x) = \dfrac{x + 1}{x^2 + 2x - 3}$

21. domain: all real numbers

$f(x) = \dfrac{4x^2}{x^2 + 3}$

22. a. $\overline{C}(x) = \dfrac{300{,}000 + 10x}{x}$

b. $y = 10$; As the number of satellite radio players increases, the average cost approaches \$10.

23. $(-3, 4)$

24. $(-\infty, 3) \cup [10, \infty)$

Cumulative Review Exercises (Chapters 1–3)

1. domain: $(-2, 2)$; range: $[0, \infty)$

6.

$g(x) = f(x + 2) + 1$

13.

$f(x) = x^3 - 4x^2 - x + 4$

14.

$f(x) = x^2 + 2x - 8$

15.

$f(x) = x^2(x - 3)$

16.

$f(x) = \dfrac{x - 1}{x - 2}$

17.

18.

$x^2 + y^2 - 2x + 4y - 4 = 0$

CHAPTER 4

Section 4.1
Check Point Exercises

2.
$f(x) = 3^x$

3.
$f(x) = \left(\frac{1}{3}\right)^x$

4.
$f(x) = 3^x$
$g(x) = 3^{x-1}$

5.
$f(x) = 2^x$
$g(x) = 2^x + 1$

Exercise Set 4.1

1. 10.556　**2.** 13.967　**3.** 11.665　**4.** 16.242　**5.** 0.125　**6.** 0.116　**7.** 9.974　**8.** 29.964　**9.** 0.387　**10.** 0.472

11.
$f(x) = 4^x$

12.
$f(x) = 5^x$

13.
$g(x) = \left(\frac{3}{2}\right)^x$

14.
$g(x) = \left(\frac{4}{3}\right)^x$

15.
$h(x) = \left(\frac{1}{2}\right)^x$

16.
$h(x) = \left(\frac{1}{3}\right)^x$

17.
$f(x) = (0.6)^x$

18.
$f(x) = (0.8)^x$

19. $H(x) = -3^{-x}$
20. $g(x) = 3^{x-1}$
21. $F(x) = -3^x$
22. $f(x) = 3^x$
23. $h(x) = 3^x - 1$
24. $G(x) = 3^{-x}$

25.
$f(x) = 2^x$
$g(x) = 2^{x+1}$
asymptote: $y = 0$
domain: $(-\infty, \infty)$
range: $(0, \infty)$

26.
$f(x) = 2^x$
$g(x) = 2^{x+2}$
asymptote: $y = 0$
domain: $(-\infty, \infty)$
range: $(0, \infty)$

27.
$f(x) = 2^x$
$g(x) = 2^x - 1$
asymptote: $y = -1$
domain: $(-\infty, \infty)$
range: $(-1, \infty)$

28.
$f(x) = 2^x$
$g(x) = 2^x + 2$
asymptote: $y = 2$
domain: $(-\infty, \infty)$
range: $(2, \infty)$

29.
$f(x) = 2^x$
$h(x) = 2^{x+1} - 1$
asymptote: $y = -1$
domain: $(-\infty, \infty)$
range: $(-1, \infty)$

30.
$f(x) = 2^x$
$h(x) = 2^{x+2} - 1$
asymptote: $y = -1$
domain: $(-\infty, \infty)$
range: $(-1, \infty)$

31.
$f(x) = 2^x$
$g(x) = -2^x$
asymptote: $y = 0$
domain: $(-\infty, \infty)$
range: $(-\infty, 0)$

32.
$f(x) = 2^x$
$g(x) = 2^{-x}$
asymptote: $y = 0$
domain: $(-\infty, \infty)$
range: $(0, \infty)$

33.
$f(x) = 2^x$
$g(x) = 2 \cdot 2^x$
asymptote: $y = 0$
domain: $(-\infty, \infty)$
range: $(0, \infty)$

34.
$f(x) = 2^x$
$g(x) = \frac{1}{2} \cdot 2^x$
asymptote: $y = 0$
domain: $(-\infty, \infty)$
range: $(0, \infty)$

35.
$g(x) = e^{x-1}$
asymptote: $y = 0$
domain: $(-\infty, \infty)$
range: $(0, \infty)$

36.
$g(x) = e^{x+1}$
asymptote: $y = 0$
domain: $(-\infty, \infty)$
range: $(0, \infty)$

37.
$g(x) = e^x + 2$
asymptote: $y = 2$
domain: $(-\infty, \infty)$
range: $(2, \infty)$

38.
$g(x) = e^x - 1$
asymptote: $y = -1$
domain: $(-\infty, \infty)$
range: $(-1, \infty)$

39.
$h(x) = e^{x-1} + 2$
asymptote: $y = 2$
domain: $(-\infty, \infty)$
range: $(2, \infty)$

40.

$h(x) = e^{x+1} - 1$

asymptote: $y = -1$
domain: $(-\infty, \infty)$
range: $(-1, \infty)$

41.

$h(x) = e^{-x}$

asymptote: $y = 0$
domain: $(-\infty, \infty)$
range: $(0, \infty)$

42.

$h(x) = -e^x$

asymptote: $y = 0$
domain: $(-\infty, \infty)$
range: $(-\infty, 0)$

43.

$g(x) = 2e^x$

asymptote: $y = 0$
domain: $(-\infty, \infty)$
range: $(0, \infty)$

44.

$g(x) = \frac{1}{2} e^x$

asymptote: $y = 0$
domain: $(-\infty, \infty)$
range: $(0, \infty)$

45.

$h(x) = e^{2x} + 1$

asymptote: $y = 1$
domain: $(-\infty, \infty)$
range: $(1, \infty)$

46.

$h(x) = e^{x/2} + 2$

asymptote: $y = 2$
domain: $(-\infty, \infty)$
range: $(2, \infty)$

47.

$f(x) = 3^x$
$g(x) = 3^{-x}$

asymptote of f: $y = 0$
asymptote of g: $y = 0$

48.

$f(x) = 3^x$
$g(x) = -3^x$

asymptote of f: $y = 0$
asymptote of g: $y = 0$

49.

$f(x) = 3^x$
$g(x) = \frac{1}{3} \cdot 3^x$

asymptote of f: $y = 0$
asymptote of g: $y = 0$

50.

$f(x) = 3^x$
$g(x) = 3 \cdot 3^x$

asymptote of f: $y = 0$
asymptote of g: $y = 0$

51.

$f(x) = \left(\frac{1}{2}\right)^x$
$g(x) = \left(\frac{1}{2}\right)^{x-1} + 1$

asymptote of f: $y = 0$
asymptote of g: $y = 1$

52.

$f(x) = \left(\frac{1}{2}\right)^x$
$g(x) = \left(\frac{1}{2}\right)^{x-1} + 2$

asymptote of f: $y = 0$
asymptote of g: $y = 2$

53. a. $13,116.51
b. $13,140.67
c. $13,157.04
d. $13,165.31
54. a. $9479.19
b. $9527.79
c. $9560.92
d. $9577.70

57.

$g(x) = 2^{-x}$ $f(x) = 2^x$

58.

$g(x) = 2^{-x+1}$ $f(x) = 2^{x+1}$

59.

$y = 2^x$
$x = 2^y$

60.

$y = 3^x$
$x = 3^y$

69. 3.249009585; 3.317278183; 3.321880096; 3.321995226; 3.321997068; $2^{\sqrt{3}} \approx 3.321997085$; The closer the exponent is to $\sqrt{3}$, the closer the value is to $2^{\sqrt{3}}$.
70. 8; 8.5741877; 8.815240927; 8.821353305; 8.824411082; 8.824961595; 8.824979946; $2^\pi \approx 8.824977827$; The closer the exponent gets to π, the closer the value is to 2^π. **74. a.** approximately $5,027,378,919 **b.** approximately $5,231,970,592

81. a. $A = 10,000\left(1 + \dfrac{0.05}{4}\right)^{4t}; A = 10,000\left(1 + \dfrac{0.045}{12}\right)^{12t}$

b.

5% interest compounded quarterly

82. a.

b.

c.

84. makes sense

92.

$f(x) = 2^x$

93. a. $\cosh(-x) = \dfrac{e^{-x} + e^{-(-x)}}{2} = \dfrac{e^{-x} + e^{x}}{2} = \dfrac{e^{x} + e^{-x}}{2} = \cosh x$

b. $\sinh(-x) = \dfrac{e^{-x} - e^{-(-x)}}{2} = \dfrac{e^{-x} - e^{x}}{2} = -\dfrac{e^{x} - e^{-x}}{2} = -\sinh x$

c. $\left(\dfrac{e^x + e^{-x}}{2}\right)^2 - \left(\dfrac{e^x - e^{-x}}{2}\right)^2 \stackrel{?}{=} 1$

$$\dfrac{e^{2x} + 2 + e^{-2x}}{4} - \dfrac{e^{2x} - 2 + e^{-2x}}{4} \stackrel{?}{=} 1$$

$$\dfrac{e^{2x} + 2 + e^{-2x} - e^{2x} + 2 - e^{-2x}}{4} \stackrel{?}{=} 1$$

$$\dfrac{4}{4} \stackrel{?}{=} 1$$

$$1 = 1$$

94. $y = \dfrac{18 - 7x}{3}$ or $y = -\dfrac{7}{3}x + 6$ **96.** $\left(-4, \dfrac{3}{2}\right);$

97. We don't know how to solve $x = 2^y$ for y.

Section 4.2
Check Point Exercises

6.

$f(x) = 3^x$

$g(x) = \log_3 x$

Exercise Set 4.2

9. $\log_2 8 = 3$ **10.** $\log_5 625 = 4$ **11.** $\log_2 \dfrac{1}{16} = -4$ **12.** $\log_5 \dfrac{1}{125} = -3$ **13.** $\log_8 2 = \dfrac{1}{3}$ **14.** $\log_{64} 4 = \dfrac{1}{3}$ **15.** $\log_{13} x = 2$

16. $\log_{15} x = 2$ **17.** $\log_b 1000 = 3$ **18.** $\log_b 343 = 3$ **19.** $\log_7 200 = y$ **20.** $\log_8 300 = y$

43.

$f(x) = 4^x$

$g(x) = \log_4 x$

44.

$f(x) = 5^x$

$g(x) = \log_5 x$

45.

$f(x) = \left(\dfrac{1}{2}\right)^x$

$g(x) = \log_{1/2} x$

46.

$f(x) = \left(\dfrac{1}{4}\right)^x$

$g(x) = \log_{1/4} x$

47. $H(x) = 1 - \log_3 x$ **48.** $G(x) = \log_3(-x)$ **49.** $h(x) = \log_3 x - 1$ **50.** $F(x) = -\log_3 x$ **51.** $g(x) = \log_3(x - 1)$ **52.** $f(x) = \log_3 x$

53.

$f(x) = \log_2 x$

$g(x) = \log_2(x+1)$

vertical asymptote: $x = -1$
domain: $(-1, \infty)$
range: $(-\infty, \infty)$

54.

$f(x) = \log_2 x$

$g(x) = \log_2(x+2)$

vertical asymptote: $x = -2$
domain: $(-2, \infty)$
range: $(-\infty, \infty)$

55.

$f(x) = \log_2 x$

$h(x) = 1 + \log_2 x$

vertical asymptote: $x = 0$
domain: $(0, \infty)$
range: $(-\infty, \infty)$

56.

$f(x) = \log_2 x$

$h(x) = 2 + \log_2 x$

vertical asymptote: $x = 0$
domain: $(0, \infty)$
range: $(-\infty, \infty)$

57.

$f(x) = \log_2 x$

$g(x) = \dfrac{1}{2} \log_2 x$

vertical asymptote: $x = 0$
domain: $(0, \infty)$
range: $(-\infty, \infty)$

58.

$f(x) = \log_2 x$

$g(x) = -2 \log_2 x$

vertical asymptote: $x = 0$
domain: $(0, \infty)$
range: $(-\infty, \infty)$

59.

$g(x) = \log(x - 1)$

asymptote: $x = 1$
domain: $(1, \infty)$
range: $(-\infty, \infty)$

60.

$g(x) = \log(x - 2)$

asymptote: $x = 2$
domain: $(2, \infty)$
range: $(-\infty, \infty)$

61.

asymptote: $x = 0$
domain: $(0, \infty)$
range: $(-\infty, \infty)$

62.

asymptote: $x = 0$
domain: $(0, \infty)$
range: $(-\infty, \infty)$

63.

asymptote: $x = 0$
domain: $(0, \infty)$
range: $(-\infty, \infty)$

64.

asymptote: $x = 0$
domain: $(0, \infty)$
range: $(-\infty, \infty)$

65.

asymptote: $x = -2$
domain: $(-2, \infty)$
range: $(-\infty, \infty)$

66.

asymptote: $x = -1$
domain: $(-1, \infty)$
range: $(-\infty, \infty)$

67.

asymptote: $x = 0$
domain: $(0, \infty)$
range: $(-\infty, \infty)$

68.

asymptote: $x = 0$
domain: $(0, \infty)$
range: $(-\infty, \infty)$

69.

asymptote: $x = 0$
domain: $(0, \infty)$
range: $(-\infty, \infty)$

70.

asymptote: $x = 0$
domain: $(0, \infty)$
range: $(-\infty, \infty)$

71.

asymptote: $x = 0$
domain: $(0, \infty)$
range: $(-\infty, \infty)$

72.

asymptote: $x = 0$
domain: $(-\infty, 0)$
range: $(-\infty, \infty)$

73.

asymptote: $x = 0$
domain: $(0, \infty)$
range: $(-\infty, \infty)$

74.

asymptote: $x = 0$
domain: $(0, \infty)$
range: $(-\infty, \infty)$

119. b. 71.5; 63.9; 58.8; 55; 52; 49.5
c.

Material retention decreases as
time passes.

128.

$g(x)$ is $f(x)$ shifted left 3 units.

129.

$g(x)$ is $f(x)$ shifted upward 3 units.

130.

$g(x)$ is $f(x)$ reflected about the x-axis.

131.

$g(x)$ is $f(x)$ shifted right 2 units and upward 1 unit.

132.

The score falls below 65 after 9 months.

133. a. **b.** **c.**

d. They are the same.
$\log_b M + \log_b N$

134. $y = \ln x, y = \sqrt{x}, y = x, y = x^2, y = e^x, y = x^x$

135. makes sense **138.** does not make sense

142. true **150. c.** $\log_2(8 \cdot 4) = \log_2 8 + \log_2 4$

Exercise Set 4.3

1. $\log_5 7 + \log_5 3$ **2.** $\log_8 13 + \log_8 7$ **3.** $1 + \log_7 x$ **4.** $1 + \log_9 x$ **5.** $3 + \log x$ **6.** $4 + \log x$ **7.** $1 - \log_7 x$ **8.** $1 - \log_9 x$

9. $\log x - 2$ **10.** $\log x - 3$ **11.** $3 - \log_4 y$ **12.** $3 - \log_5 y$ **13.** $2 - \ln 5$ **14.** $4 - \ln 8$ **21.** $2\log_b x + \log_b y$ **22.** $\log_b x + 3\log_b y$

23. $\frac{1}{2}\log_4 x - 3$ **24.** $\frac{1}{2}\log_5 x - 2$ **25.** $2 - \frac{1}{2}\log_6(x+1)$ **26.** $2 - \frac{1}{2}\log_8(x+1)$ **27.** $2\log_b x + \log_b y - 2\log_b z$

28. $3\log_b x + \log_b y - 2\log_b z$ **29.** $1 + \frac{1}{2}\log x$ **30.** $\frac{1}{2} + \frac{1}{2}\ln x$ **31.** $\frac{1}{3}\log x - \frac{1}{3}\log y$ **32.** $\frac{1}{5}\log x - \frac{1}{5}\log y$

33. $\frac{1}{2}\log_b x + 3\log_b y - 3\log_b z$ **34.** $\frac{1}{3}\log_b x + 4\log_b y - 5\log_b z$ **35.** $\frac{2}{3}\log_5 x + \frac{1}{3}\log_5 y - \frac{2}{3}$ **36.** $\frac{1}{5}\log_2 x + \frac{4}{5}\log_2 y - \frac{4}{5}$

37. $3\ln x + \frac{1}{2}\ln(x^2+1) - 4\ln(x+1)$ **38.** $4\ln x + \frac{1}{2}\ln(x^2+3) - 5\ln(x+3)$ **39.** $1 + 2\log x + \frac{1}{3}\log(1-x) - \log 7 - 2\log(x+1)$

40. $2 + 3\log x + \frac{1}{3}\log(5-x) - \log 3 - 2\log(x+7)$ **47.** $\log\left(\dfrac{2x+5}{x}\right)$ **48.** $\log\left(\dfrac{3x+7}{x}\right)$ **51.** $\ln(x^{1/2}y)$ or $\ln(y\sqrt{x})$

52. $\ln(x^{1/3}y)$ or $\ln(y\sqrt[3]{x})$ **55.** $\ln\left(\dfrac{x^5}{y^2}\right)$ **56.** $\ln\left(\dfrac{x^7}{y^3}\right)$ **57.** $\ln\left(\dfrac{x^3}{y^{1/3}}\right)$ or $\ln\left(\dfrac{x^3}{\sqrt[3]{y}}\right)$ **58.** $\ln\left(\dfrac{x^2}{y^{1/2}}\right)$ or $\ln\left(\dfrac{x^2}{\sqrt{y}}\right)$ **59.** $\ln\dfrac{(x+6)^4}{x^3}$

60. $\ln\dfrac{(x+9)^8}{x^4}$ **61.** $\ln\left(\dfrac{x^3y^5}{z^6}\right)$ **62.** $\ln\left(\dfrac{x^4y^7}{z^3}\right)$ **65.** $\log_5\left(\dfrac{\sqrt{xy}}{(x+1)^2}\right)$ **66.** $\log_4\left((x+1)^2\sqrt[3]{\dfrac{x}{y}}\right)$ **67.** $\ln\sqrt[3]{\dfrac{(x+5)^2}{x(x^2-4)}}$

68. $\ln\sqrt[3]{\dfrac{(x+6)^5}{x(x^2-25)}}$ **69.** $\log\dfrac{x(x^2-1)}{7(x+1)} = \log\dfrac{x(x-1)}{7}$ **70.** $\log\dfrac{x(x^2-4)}{15(x+2)} = \log\dfrac{x(x-2)}{15}$

79. **80.** **81.**

82.

87. $\dfrac{1}{2}A - \dfrac{3}{2}C$ **88.** $\dfrac{1}{2}C - 2A$ **90.** false; ln 0 is undefined. **91.** false; $\log_4(2x)^3 = 3\log_4(2x)$

94. false; $\ln(x \cdot 1) = \ln x + \ln 1$ **95.** true **96.** false; $\ln x + \ln(2x) = \ln(2x^2)$

97. false; $\log(x + 3) - \log(2x) = \log\left(\dfrac{x + 3}{2x}\right)$ **98.** false; $\log\left(\dfrac{x + 2}{x - 1}\right) = \log(x + 2) - \log(x - 1)$

102. false; $e^x = \ln e^{e^x}$

113. a & b

$y = 2 + \log_3 x$ shifts the graph
of $y = \log_3 x$ two units upward;
$y = \log_3(x + 2)$ shifts the
graph of $y = \log_3 x$ two units left;
$y = -\log_3 x$ reflects the graph of
$y = \log_3 x$ about the x-axis.

114.

$y = \log(10x) = 1 + \log x$
shifts $y = \log x$ one unit upward;
$y = \log(0.1x) = -1 + \log x$
shifts $y = \log x$ one unit downward;
the product rule

115.

c. The graph of the equation with the largest b
will be on the top in the interval $(0, 1)$ and on
the bottom in the interval $(1, \infty)$.

121. makes sense **123.** makes sense **129.** $\log e = \log_{10} e = \dfrac{\ln e}{\ln 10} = \dfrac{1}{\ln 10}$ **130.** $\dfrac{2A}{B}$

132. $\dfrac{f(x + h) - f(x)}{h} = \dfrac{\log_b(x + h) - \log_b x}{h}$

$= \dfrac{\log_b\left(\dfrac{x + h}{x}\right)}{h}$

$= \dfrac{1}{h}\log_b\left(1 + \dfrac{h}{x}\right)$

$= \log_b\left(1 + \dfrac{h}{x}\right)^{\frac{1}{h}}$

133. Let $\log_b M = R$ and $\log_b N = S$.

Then $\log_b M = R$ means $b^R = M$ and $\log_b N = S$ means $b^S = N$.

$\dfrac{M}{N} = \dfrac{b^R}{b^S} = b^{R-S}$

$\log_b \dfrac{M}{N} = \log_b b^{R-S} = R - S = \log_b M - \log_b N$

136.

Mid-Chapter 4 Check Point

1.

$f(x) = 2^x$
$g(x) = 2^x - 3$

asymptote of f: $y = 0$
asymptote of g: $y = -3$
domain of f = domain of g = $(-\infty, \infty)$
range of f = $(0, \infty)$; range of g = $(-3, \infty)$

2.

$f(x) = \left(\dfrac{1}{2}\right)^x$

$g(x) = \left(\dfrac{1}{2}\right)^{x-1}$

asymptote of f: $y = 0$
asymptote of g: $y = 0$
domain of f = domain of g = $(-\infty, \infty)$
range of f = range of g = $(0, \infty)$

3.

$f(x) = e^x$
$g(x) = \ln x$

asymptote of f: $y = 0$
asymptote of g: $x = 0$
domain of f = range of g = $(-\infty, \infty)$
range of f = domain of g = $(0, \infty)$

4.

$f(x) = \log_2 x$
$g(x) = \log_2 (x - 1) + 1$

asymptote of f: $x = 0$
asymptote of g: $x = 1$
domain of $f = (0, \infty)$; domain of $g = (1, \infty)$
range of $f =$ range of $g = (-\infty, \infty)$

5.

$f(x) = \log_{1/2} x$
$g(x) = -2 \log_{1/2} x$

asymptote of f: $x = 0$
asymptote of g: $x = 0$
domain of $f =$ domain of $g = (0, \infty)$
range of $f =$ range of $g = (-\infty, \infty)$

8. $(-\infty, -6) \cup (-6, \infty)$

15. Evaluation not possible; $\log_2 \dfrac{1}{8} = -3$ and $\log_3(-3)$ is undefined.

Exercise Set 4.4

23. $\{\log 3.91\}$ or $\left\{\dfrac{\ln 3.91}{\ln 10}\right\}$; ≈ 0.59 **24.** $\{\log 8.07\}$ or $\left\{\dfrac{\ln 8.07}{\ln 10}\right\}$; ≈ 0.91 **27.** $\left\{\dfrac{\ln 17}{\ln 5}\right\}$; ≈ 1.76 **28.** $\left\{\dfrac{\ln 143}{\ln 19}\right\}$; ≈ 1.69

31. $\left\{\dfrac{\ln 659}{5}\right\}$; ≈ 1.30 **32.** $\left\{\dfrac{\ln 2568.25}{7}\right\}$; ≈ 1.12 **33.** $\left\{\dfrac{1 - \ln 793}{5}\right\}$ or $\left\{\dfrac{\ln 793 - 1}{-5}\right\}$; ≈ -1.14

34. $\left\{\dfrac{1 - \ln 7957}{8}\right\}$ or $\left\{\dfrac{\ln 7957 - 1}{-8}\right\}$; ≈ -1.00 **35.** $\left\{\dfrac{\ln 10{,}478 + 3}{5}\right\}$; ≈ 2.45 **36.** $\left\{\dfrac{\ln 11{,}250 + 5}{4}\right\}$; ≈ 3.58 **37.** $\left\{\dfrac{\ln 410}{\ln 7} - 2\right\}$; ≈ 1.09

38. $\left\{\dfrac{\ln 137}{\ln 5} + 3\right\}$; ≈ 6.06 **39.** $\left\{\dfrac{\ln 813}{0.3 \ln 7}\right\}$; ≈ 11.48 **40.** $\left\{\dfrac{7 \ln 0.2}{\ln 3}\right\}$; ≈ -10.25 **41.** $\left\{\dfrac{3 \ln 5 + \ln 3}{\ln 3 - 2 \ln 5}\right\}$; ≈ -2.80

42. $\left\{\dfrac{2 \ln 3 - \ln 7}{2 \ln 7 - \ln 3}\right\}$; ≈ 0.09 **43.** $\{0, \ln 2\}$; $\ln 2 \approx 0.69$ **44.** $\{\ln 3\}$; ≈ 1.10 **45.** $\left\{\dfrac{\ln 3}{2}\right\}$; ≈ 0.55 **46.** $\left\{\dfrac{\ln 6}{2}\right\}$; ≈ 0.90

48. $\left\{\dfrac{\ln 3}{\ln 2}\right\}$; ≈ 1.58 **61.** $\left\{\dfrac{e^4}{2}\right\}$; ≈ 27.30 **62.** $\left\{\dfrac{e^5}{2}\right\}$; ≈ 74.21 **63.** $\{e^{-1/2}\}$; ≈ 0.61 **64.** $\{e^{-1/3}\}$; ≈ 0.72 **65.** $\{e^2 - 3\}$; ≈ 4.39

66. $\{e^2 - 4\}$; ≈ 3.39 **97.** $\left\{\pm \sqrt{\dfrac{\ln 45}{\ln 3}}\right\}$ **98.** $\left\{\pm \sqrt{\dfrac{\ln 50}{\ln 5}}\right\}$ **103. a.** 37.3 million **104. b.** 2016 **119. b.** $10^{-2.4}$ mole per liter

130. $\{20\}$ **131.** $\{-1.391606, 1.6855579\}$ **132.** $\{-1.291641, 1.2793139\}$

133.

As distance from eye increases, barometric
air pressure increases.

135.

about 7.9 min

136.

about 20 years old

155.

Section 4.5
Check Point Exercises

4.

5.

Exercise Set 4.5

15. approximately 8 grams **17.** 8 grams after 10 seconds; 4 grams after 20 seconds; 2 grams after 30 seconds; 1 gram after 40 seconds; 0.5 gram after 50 seconds **18.** 8 grams after 25,000 years; 4 grams after 50,000 years; 2 grams after 75,000 years; 1 gram after 100,000 years; 0.5 gram after

125,000 years **27. a.** $\dfrac{1}{2} = e^{1.31k}$ yields $k = \dfrac{\ln\left(\dfrac{1}{2}\right)}{1.31} \approx -0.52912$. **32.** about 39 years after the year when the population was 1400

33. $2A_0 = A_0 e^{kt}; 2 = e^{kt}; \ln 2 = \ln e^{kt}; \ln 2 = kt; \dfrac{\ln 2}{k} = t$ **34.** $3A_0 = A_0 e^{kt}; 3 = e^{kt}; \ln 3 = \ln e^{kt}; \ln 3 = kt; \dfrac{\ln 3}{k} = t$

36. b. about 58 yr **38.** quite well **39.** quite well

47. a.

48. a.

b. exponential function

49. a.

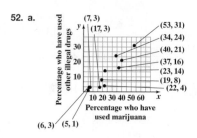

b. logarithmic function

50. a.

b. logarithmic function

51. a.

52. a.

66. a. $r \approx 0.999$; Since r is close to 1, the model fits the data well. **67.** $r \approx 0.870$; Fit is ok, but not great.

70. $y = 201.2(1.011)^x$; $y = 2.657x + 197.923$; using exponential; by 2016; using linear: by 2021; Answers will vary.

71. b. $y = 3.46e^{(\ln 1.02)x}$; $y = 3.46e^{0.02x}$; by approximately 2% **72.** Models will vary. Examples are given. Predictions will vary. For Exercise 47: $y = 1.402(1.078)^x$; For Exercise 48: $y = 2896.7(1.056)^x$; For Exercise 49: $y = 120 + 4.343 \ln x$; For Exercise 50: $y = -11.629 + 13.424 \ln x$; For Exercise 51: $y = -175.58 + 56.81 \ln x$; For Exercise 52: $y = 0.615x - 3.068$ **73.** does not make sense **75.** makes sense

84.

87.

Chapter 4 Review Exercises

1. $g(x) = 4^{-x}$ **2.** $h(x) = -4^{-x}$ **3.** $r(x) = -4^{-x} + 3$ **4.** $f(x) = 4^x$

5.

$f(x) = 2^x$
$g(x) = 2^{x-1}$

asymptote of f: $y = 0$
asymptote of g: $y = 0$
domain of f = domain of $g = (-\infty, \infty)$
range of f = range of $g = (0, \infty)$

6.

$f(x) = 3^x$
$g(x) = 3^x - 1$

asymptote of f: $y = 0$
asymptote of g: $y = -1$
domain of f = domain of $g = (-\infty, \infty)$
range of $f = (0, \infty)$; range of $g = (-1, \infty)$

7.

$f(x) = 3^x$
$g(x) = -3^x$

asymptote of f: $y = 0$
asymptote of g: $y = 0$
domain of f = domain of $g = (-\infty, \infty)$
range of $f = (0, \infty)$; range of $g = (-\infty, 0)$

8.

$f(x) = \left(\dfrac{1}{2}\right)^x$

$g(x) = \left(\dfrac{1}{2}\right)^{-x}$

asymptote of f: $y = 0$
asymptote of g: $y = 0$
domain of f = domain of g = $(-\infty, \infty)$
range of f = range of g = $(0, \infty)$

9.

$f(x) = e^x$
$g(x) = 2e^{x/2}$

asymptote of f: $y = 0$
asymptote of g: $y = 0$
domain of f = domain of g = $(-\infty, \infty)$
range of f = range of g = $(0, \infty)$

13. $49^{1/2} = 7$ **14.** $4^3 = x$ **15.** $3^y = 81$
16. $\log_6 216 = 3$ **17.** $\log_b 625 = 4$
18. $\log_{13} 874 = y$
21. undefined; $\log_b x$ is defined only for $x > 0$.

30.

$f(x) = 2^x$
$g(x) = \log_2 x$

domain of f = range of g = $(-\infty, \infty)$
range of f = domain of g = $(0, \infty)$

31.

$f(x) = \left(\dfrac{1}{3}\right)^x$

$g(x) = \log_{1/3} x$

domain of f = range of g = $(-\infty, \infty)$
range of f = domain of g = $(0, \infty)$

32. $g(x) = \log(-x)$
33. $r(x) = 1 + \log(2 - x)$
34. $h(x) = \log(2 - x)$
35. $f(x) = \log x$

36.

$f(x) = \log_2 x$
$g(x) = \log_2(x - 2)$

x-intercept: $(3, 0)$
vertical asymptote: $x = 2$
domain: $(2, \infty)$; range: $(-\infty, \infty)$

37.

$f(x) = \log_2 x$
$h(x) = -1 + \log_2 x$

x-intercept: $(2, 0)$
vertical asymptote: $x = 0$
domain: $(0, \infty)$; range: $(-\infty, \infty)$

38.

$f(x) = \log_2 x$
$r(x) = \log_2(-x)$

x-intercept: $(-1, 0)$
vertical asymptote: $x = 0$
domain: $(-\infty, 0)$; range: $(-\infty, \infty)$

39.

$f(x) = \log x$
$g(x) = -\log(x + 3)$

asymptote of f: $x = 0$
asymptote of g: $x = -3$
domain of f = $(0, \infty)$;
domain of g = $(-3, \infty)$
range of f = range of g = $(-\infty, \infty)$

40.

$f(x) = \ln x$
$g(x) = -\ln(2x)$

asymptote of f: $x = 0$
asymptote of g: $x = 0$
domain of f = domain of g = $(0, \infty)$
range of f = range of g = $(-\infty, \infty)$

48. b. $\approx 67, \approx 63, \approx 61, \approx 59, \approx 56$
c.

$f(t) = 76 - 18\log(t + 1)$

61. false; $\log(x + 9) - \log(x + 1) = \log\left(\dfrac{x + 9}{x + 1}\right)$ **88. b.** about 45,411 people

89. a.

90. a.

91. a.

Chapter 4 Test

1.
$f(x) = 2^x$
$g(x) = 2^{x+1}$

2.
$f(x) = \log_2 x$
$g(x) = \log_2(x - 1)$

29. a. 14 elk

Cumulative Review Exercises (Chapters 1–4)

9.
$(x - 3)^2 + (y + 2)^2 = 4$

10.
$f(x) = (x - 2)^2 - 1$

11.
$f(x) = \dfrac{x^2 - 1}{x^2 - 4}$

12.
$f(x) = (x - 2)^2(x + 1)$

13.
$f(x) = 2x - 4$
$f^{-1}(x) = \dfrac{x + 4}{2}$

14.
$f(x) = \ln x$
$g(x) = \ln(x - 2) + 1$

15. using $(1, 3)$, $y - 3 = -3(x - 1)$; $y = -3x + 6$

CHAPTER 5

Exercise Set 5.1

33. $\{(x, y) \mid y = 3x - 5\}$ **34.** $\{(x, y) \mid y = 3x - 4\}$ **37.** $\{(x, y) \mid x + 3y = 2\}$ **38.** $\{(x, y) \mid 2x - y = 1\}$ **43.** $x + y = 7$; $x - y = -1$; 3 and 4
44. $x + y = 2$; $x - y = 8$; 5 and -3 **46.** $3x + 2y = 8$; $2x - y = 3$; 2 and 1 **53.** $y = x - 4$; $y = -\dfrac{1}{3}x + 4$ **54.** $y = \dfrac{1}{3}x + 2$, $y = \dfrac{1}{3}x - 2$
57. -6000; When the company produces and sells 200 radios, the loss is $6000. **58.** -4000; When the company produces and sells 300 radios, the loss is $4000. **61. a.** $C(x) = 18{,}000 + 20x$ **b.** $R(x) = 80x$ **c.** $(300, 24{,}000)$; When 300 canoes are produced and sold, both revenue and cost are $24,000. **62. a.** $C(x) = 100{,}000 + 100x$ **b.** $R(x) = 300x$ **c.** $(500, 150{,}000)$; When 500 bicycles are produced and sold, both cost and revenue are $150,000. **63. a.** $C(x) = 30{,}000 + 2500x$ **b.** $R(x) = 3125x$ **c.** $(48, 150{,}000)$; For 48 sold-out performances, both cost and revenue are $150,000. **64. a.** $C(x) = 30{,}000 + 0.02x$ **b.** $R(x) = 0.5x$ **c.** $(62{,}500, 31{,}250)$; For 62,500 cards, both cost and revenue are $31,250. **66. a.** 110 million packages; 50¢ **68. a.** Answers will vary; approximately $(2004, 180)$; 2004; 180 million **69. a.** $y = 0.45x + 0.8$
c. week 6; 3.5 symptoms; by the intersection point $(6, 3.5)$ **73.** multiple times per day: 24%; once per day: 17% **74.** between 5 and 7 hours: 43%; between 9 and 11 hours: 2% **78.** The north had 300 students and the south had 900 students. **99.** $y = \dfrac{a_1c_2 - a_2c_1}{a_1b_2 - a_2b_1}$; $x = \dfrac{b_2c_1 - b_1c_2}{a_1b_2 - a_2b_1}$

Exercise Set 5.2

29. $\left\{\left(\dfrac{8}{a}, -\dfrac{3}{b}, -\dfrac{5}{c}\right)\right\}$ **30.** $\left\{\left(-\dfrac{9}{a}, \dfrac{5}{b}, \dfrac{5}{c}\right)\right\}$ **31. b.** $y = 0$ when $x = 5$; The ball hits the ground after 5 seconds. **32. b.** 156; When a car is in motion for 6 seconds after the brakes are applied, it travels 156 feet. **33.** chemical engineering: 22 hours; mathematics: 16 hours; psychology: 14 hours
34. physics: 20 hours; English: 16 hours; sociology: 14 hours **49.** does not make sense **51.** makes sense **56.** $\left\{\dfrac{1}{4} \pm i\dfrac{\sqrt{7}}{4}\right\}$

Exercise Set 5.3

1. $\dfrac{A}{x - 2} + \dfrac{B}{x + 1}$ **2.** $\dfrac{A}{x - 1} + \dfrac{B}{x + 3}$ **3.** $\dfrac{A}{x + 2} + \dfrac{B}{x - 3} + \dfrac{C}{(x - 3)^2}$ **4.** $\dfrac{A}{x + 1} + \dfrac{B}{x - 2} + \dfrac{C}{(x - 2)^2}$ **5.** $\dfrac{A}{x - 1} + \dfrac{Bx + C}{x^2 + 1}$
6. $\dfrac{A}{x - 4} + \dfrac{Bx + C}{x^2 + 5}$ **8.** $\dfrac{Ax + B}{x^2 + 7} + \dfrac{Cx + D}{(x^2 + 7)^2}$ **9.** $\dfrac{3}{x - 3} - \dfrac{2}{x - 2}$ **11.** $\dfrac{7}{x - 9} - \dfrac{4}{x + 2}$ **12.** $\dfrac{3}{x - 2} + \dfrac{2}{x + 1}$ **13.** $\dfrac{24}{7(x - 4)} + \dfrac{25}{7(x + 3)}$
14. $\dfrac{6}{x - 3} + \dfrac{3}{x + 5}$ **15.** $\dfrac{4}{7(x - 3)} - \dfrac{8}{7(2x + 1)}$ **16.** $\dfrac{3}{4(x + 3)} + \dfrac{1}{4(x - 1)}$ **17.** $\dfrac{3}{x} + \dfrac{2}{x - 1} - \dfrac{1}{x + 3}$ **18.** $\dfrac{3}{x} - \dfrac{1}{x + 1} + \dfrac{2}{x - 5}$
19. $\dfrac{3}{x} + \dfrac{4}{x + 1} - \dfrac{3}{x - 1}$ **20.** $\dfrac{3}{x} - \dfrac{5}{x - 2} + \dfrac{4}{x + 2}$ **23.** $\dfrac{1}{x - 2} - \dfrac{2}{(x - 2)^2} - \dfrac{5}{(x - 2)^3}$ **24.** $\dfrac{2}{x + 1} + \dfrac{4}{(x + 1)^2} - \dfrac{3}{(x + 1)^3}$

25. $\dfrac{7}{x} - \dfrac{6}{x-1} + \dfrac{10}{(x-1)^2}$ **26.** $\dfrac{1}{x} + \dfrac{2}{x+7} - \dfrac{28}{(x+7)^2}$ **27.** $\dfrac{1}{4(x+1)} + \dfrac{3}{4(x-1)} + \dfrac{1}{2(x-1)^2}$

28. $-\dfrac{1}{4(x+1)} + \dfrac{1}{4(x-1)} + \dfrac{1}{4(x+1)^2} + \dfrac{1}{4(x-1)^2}$ **29.** $\dfrac{3}{x-1} + \dfrac{2x-4}{x^2+1}$ **30.** $\dfrac{3}{x-4} + \dfrac{2x-1}{x^2+5}$ **31.** $\dfrac{2}{x+1} + \dfrac{3x-1}{x^2+2x+2}$

32. $\dfrac{2}{x-2} - \dfrac{2x-1}{x^2+2x+2}$ **33.** $\dfrac{1}{4x} + \dfrac{1}{x^2} - \dfrac{x+4}{4(x^2+4)}$ **34.** $\dfrac{14}{3(x-1)} + \dfrac{4}{(x-1)^2} - \dfrac{14x-4}{3(x^2+2)}$ **35.** $\dfrac{4}{x+1} + \dfrac{2x-3}{x^2+1}$ **36.** $\dfrac{3}{x+2} - \dfrac{2}{x^2+4}$

37. $\dfrac{x+1}{x^2+2} - \dfrac{2x}{(x^2+2)^2}$ **38.** $\dfrac{1}{x^2+4} + \dfrac{2x-1}{(x^2+4)^2}$ **39.** $\dfrac{x-2}{x^2-2x+3} + \dfrac{2x+1}{(x^2-2x+3)^2}$ **40.** $\dfrac{3x}{x^2-2x+2} + \dfrac{x-2}{(x^2-2x+2)^2}$

41. $\dfrac{3}{x-2} + \dfrac{x-1}{x^2+2x+4}$ **42.** $-\dfrac{2}{3(x-1)} + \dfrac{2x+13}{3(x^2+x+1)}$ **43.** $x^3 + x - \dfrac{1}{2(x+1)} + \dfrac{3}{2(x-1)}$

44. $x^3 + 4x^2 + 12x + 32 + \dfrac{80}{x-2} + \dfrac{32}{(x-2)^2}$ **45.** $x + 1 - \dfrac{2}{x} - \dfrac{2}{x^2} + \dfrac{2}{x-1}$ **46.** $x^2 + 3x + 1 + \dfrac{5}{x-2} + \dfrac{3}{x+1}$ **47.** $\dfrac{\frac{1}{2c}}{x-c} - \dfrac{\frac{1}{2c}}{x+c}$

48. $\dfrac{\frac{ac+b}{2c}}{x-c} + \dfrac{\frac{ac-b}{2c}}{x+c}$ **61.** does not make sense **62.** does not make sense **68. a.** $(f \circ g)(x) = 6x^2 - 18x + 17$ **b.** $(g \circ f)(x) = 36x^2 + 42x + 12$

71. ; $(0, -3)$ and $(2, -1)$; $0 - (-3) = 3$ and $(0-2)^2 + (-3+3)^2 = 4$ are true;
$2 - (-1) = 3$ and $(2-2)^2 + (-1+3)^2 = 4$ are true.

Exercise Set 5.4

5. $\{(4, -10), (-3, 11)\}$ **13.** $\{(3, 1), (-3, -1), (1, 3), (-1, -3)\}$ **19.** $\{(3, 2), (3, -2), (-3, 2), (-3, -2)\}$ **21.** $\{(3, 2), (3, -2), (-3, 2), (-3, -2)\}$

22. $\{(1, 2), (1, -2), (-1, 2), (-1, -2)\}$ **23.** $\{(2, 1), (2, -1), (-2, 1), (-2, -1)\}$ **24.** $\{(\sqrt{5}, \sqrt{2}), (-\sqrt{5}, \sqrt{2}), (-\sqrt{5}, -\sqrt{2}), (\sqrt{5}, -\sqrt{2})\}$

26. $\left\{\left(\dfrac{\sqrt{31}}{4}, \dfrac{7}{4}\right), \left(-\dfrac{\sqrt{31}}{4}, \dfrac{7}{4}\right)\right\}$ **27.** $\{(0, 2), (0, -2), (-1, \sqrt{3}), (-1, -\sqrt{3})\}$ **28.** $\{(0, -4), (2\sqrt{3}, 2), (-2\sqrt{3}, 2)\}$

29. $\{(2, 1), (2, -1), (-2, 1), (-2, -1)\}$ **30.** $\{(4, 0), (-3, \sqrt{7}), (-3, -\sqrt{7})\}$ **31.** $\{(-2\sqrt{2}, -\sqrt{2}), (-1, -4), (1, 4), (2\sqrt{2}, \sqrt{2})\}$

32. $\{(-2, -2), (2, 2), (4, 1), (-4, -1)\}$ **34.** $\left\{(1, 1), \left(\dfrac{19}{29}, -\dfrac{11}{29}\right)\right\}$ **39.** $\left\{(-4, 1), \left(-\dfrac{5}{2}, \dfrac{1}{4}\right)\right\}$ **45.** 2 and 1, 2 and -1, -2 and 1, or -2 and -1

46. 3 and 2, 3 and -2, -3 and 2, or -3 and -2 **47.** $\{(2, -1), (-2, 1)\}$ **48.** $\{(-3, 2), (3, -2)\}$ **49.** $\{(2, 20), (-2, 4), (-3, 0)\}$

50. $\{(-5, 0), (-3, 18), (3, 72)\}$ **51.** $\left\{\left(-1, -\dfrac{1}{2}\right), \left(-1, \dfrac{1}{2}\right), \left(1, -\dfrac{1}{2}\right), \left(1, \dfrac{1}{2}\right)\right\}$ **52.** $\left\{\left(-1, -\dfrac{1}{3}\right), \left(-1, \dfrac{1}{3}\right), \left(1, -\dfrac{1}{3}\right), \left(1, \dfrac{1}{3}\right)\right\}$

53. **54.** **72.** makes sense
$y = x^2 - 4$ $y = 4 - x^2$ **82.** $\dfrac{1}{4} \log_8 x - 2 - 3 \log_8 y$
$y = 2x - 4$ $y = -4x + 8$

84. **85.** **86.**
$2x - 3y = 6$ $f(x) = -\dfrac{2}{3} x$ $f(x) = -2$

Mid-Chapter 5 Check Point

4. $\{(x, y) \mid y = 4x - 5\}$ or $\{(x, y) \mid 8x - 2y = 10\}$ **7.** $\{(-1, 2, -2)\}$ **8.** $\{(4, -2, 3)\}$ **9.** $\left\{\left(-\dfrac{9}{5}, \dfrac{12}{5}\right), (3, 0)\right\}$

10. $\{(-2, -1), (-2, 1), (2, -1), (2, 1)\}$ **11.** $\{(-\sqrt{7}, 1), (-2, -2), (2, -2), (\sqrt{7}, 1)\}$ **13.** $\dfrac{1}{x-2} - \dfrac{2}{(x-2)^2} - \dfrac{5}{(x-2)^3}$

14. $\dfrac{5}{x+2} + \dfrac{3}{x+1} + \dfrac{2}{x-1}$ **15.** $-\dfrac{2}{x+3} + \dfrac{3x-5}{x^2+4}$ **16.** $\dfrac{x}{x^2+4} - \dfrac{4x}{(x^2+4)^2}$ **17. a.** $C(x) = 400{,}000 + 20x$ **b.** $R(x) = 100x$

d. $(5000, 500{,}000)$; The company will break even when it produces and sells 5000 PDAs. At this level, both revenue and cost are \$500,000.
18. 13%: 20 gallons; 18%: 30 gallons **21.** length: 8 m; width: 2.5 m

Section 5.5

Check Point Exercises

1. $4x - 2y \geq 8$

2. $y > -\frac{3}{4}x$

3. a. $y > 1$

b. $x \leq -2$

4.

$x^2 + y^2 \geq 16$

6. $\begin{cases} x - 3y < 6 \\ 2x + 3y \geq -6 \end{cases}$

7.

$\begin{cases} y \geq x^2 - 4 \\ x + y \leq 2 \end{cases}$

8. $\begin{cases} x + y < 2 \\ -2 \leq x < 1 \\ y > -3 \end{cases}$

Exercise Set 5.5

1.

$x + 2y \leq 8$

2.

$3x - 6y \leq 12$

3.

$x - 2y > 10$

4.

$2x - y > 4$

5.

$y \leq \frac{1}{3}x$

6.

$y \leq \frac{1}{4}x$

7.

$y > 2x - 1$

8.

$y > 3x + 2$

9.

$x \leq 1$

10.

$x \leq -3$

11.

$y > 1$

12.

$y > -3$

13.

$x^2 + y^2 \leq 1$

14.

$x^2 + y^2 \leq 4$

15.

$x^2 + y^2 > 25$

16.

$x^2 + y^2 > 36$

17.

$(x - 2)^2 + (y + 1)^2 < 9$

18.

$(x + 2)^2 + (y - 1)^2 < 16$

19.

$y < x^2 - 1$

20.

$y < x^2 - 9$

21.

$y \geq x^2 - 9$

22.

$y \geq x^2 - 1$

23.

$y > 2^x$

24.

$y \leq 3^x$

25. $x = -1$

$y \geq \log_2 (x + 1)$

26. $x = 1$

$y \geq \log_3 (x - 1)$

27. $\begin{cases} 3x + 6y \leq 6 \\ 2x + y \leq 8 \end{cases}$

28. $\begin{cases} x - y \geq 4 \\ x + y \leq 6 \end{cases}$

29. $\begin{cases} 2x - 5y \leq 10 \\ 3x - 2y > 6 \end{cases}$

30. $\begin{cases} 2x - y \leq 4 \\ 3x + 2y > -6 \end{cases}$

31. $\begin{cases} y > 2x - 3 \\ y < -x + 6 \end{cases}$

32. $\begin{cases} y < -2x + 4 \\ y < x - 4 \end{cases}$

33. $\begin{cases} x + 2y \leq 4 \\ y \geq x - 3 \end{cases}$

34. $\begin{cases} x + y \leq 4 \\ y \geq 2x - 4 \end{cases}$

35. $\begin{cases} x \leq 2 \\ y \geq -1 \end{cases}$

36. $\begin{cases} x \le 3 \\ y \le -1 \end{cases}$

37. $-2 \le x < 5$

38. $-2 < y \le 5$

39. $\begin{cases} x - y \le 1 \\ x \ge 2 \end{cases}$

40. $\begin{cases} 4x - 5y \ge -20 \\ x \ge -3 \end{cases}$

43. $\begin{cases} x + y > 4 \\ x + y > -1 \end{cases}$

44. $\begin{cases} x + y > 3 \\ x + y > -2 \end{cases}$

45. $\begin{cases} y \ge x^2 - 1 \\ x - y \ge -1 \end{cases}$

46. $\begin{cases} y \ge x^2 - 4 \\ x - y \ge 2 \end{cases}$

47. $\begin{cases} x^2 + y^2 \le 16 \\ x + y > 2 \end{cases}$

48. $\begin{cases} x^2 + y^2 \le 4 \\ x + y > 1 \end{cases}$

49. $\begin{cases} x^2 + y^2 > 1 \\ x^2 + y^2 < 16 \end{cases}$

50. $\begin{cases} x^2 + y^2 > 1 \\ x^2 + y^2 < 9 \end{cases}$

51. $\begin{cases} (x - 1)^2 + (y + 1)^2 < 25 \\ (x - 1)^2 + (y + 1)^2 \ge 16 \end{cases}$

52. $\begin{cases} (x + 1)^2 + (y - 1)^2 < 16 \\ (x + 1)^2 + (y - 1)^2 \ge 4 \end{cases}$

53. $\begin{cases} x^2 + y^2 \le 1 \\ y - x^2 > 0 \end{cases}$

54. $\begin{cases} x^2 + y^2 < 4 \\ y - x^2 \ge 0 \end{cases}$

55. $\begin{cases} x^2 + y^2 < 16 \\ y \ge 2^x \end{cases}$

56. $\begin{cases} x^2 + y^2 \le 16 \\ y < 2^x \end{cases}$

57. $\begin{cases} x - y \le 2 \\ x > -2 \\ y \le 3 \end{cases}$

58. $\begin{cases} 3x + y \le 6 \\ x > -2 \\ y \le 4 \end{cases}$

59. $\begin{cases} x \ge 0 \\ y \ge 0 \\ 2x + 5y < 10 \\ 3x + 4y \le 12 \end{cases}$

60. $\begin{cases} x \ge 0 \\ y \ge 0 \\ 2x + y < 4 \\ 2x - 3y \le 6 \end{cases}$

61. $\begin{cases} 3x + y \le 6 \\ 2x - y \le -1 \\ x > -2 \\ y < 4 \end{cases}$

62. $\begin{cases} 2x + y \le 6 \\ x + y > 2 \\ 1 \le x \le 2 \\ y < 3 \end{cases}$

63. $y \ge -2x + 4$

64. $y \ge -3x + 2$

65. $\begin{cases} x + y \le 4 \\ 3x + y \le 6 \end{cases}$

66. $\begin{cases} x + y \le 3 \\ 4x + y \le 6 \end{cases}$

67. $\begin{cases} x + y \le 2 \\ y \ge x^2 - 4 \end{cases}$

68. $\begin{cases} x^2 + y^2 \le 25 \\ 2y + x \ge 5 \end{cases}$

69. $\begin{cases} -2 \le x \le 2 \\ -3 \le y \le 3 \end{cases}$

70. $\begin{cases} -1 \le x \le 1 \\ -2 \le y \le 2 \end{cases}$

71. $y > \frac{3}{2}x - 2$ or $y < 4$

72. $x - y \ge -1$ or $5x - 2y \le 10$

77. Point $A = (66, 160)$; $5.3(66) - 160 \ge 180$, or $189.8 \ge 180$, is true; $4.1(66) - 160 \le 140$, or $110.6 \le 140$, is true.

78. Point $B = (76, 220)$; $5.3(76) - 220 \ge 180$, or $182.8 \ge 180$, is true; $4.1(76) - 220 \le 140$, or $91.6 \le 140$, is true.

81. a. $50x + 150y > 2000$

b.
$50x + 150y > 2000$

c. Answers may vary. Example: $(20, 20)$: 20 children and 20 adults will cause the elevator to be overloaded.

82. a. $165x + 110y \le 330$

b.
$165x + 110y \le 330$

c. Answers may vary. Example: $(1, 1)$: One egg and one ounce of meat are within the patient's dietary restrictions.

83. b. $\begin{cases} y \ge 0 \\ x + y \ge 5 \\ x \ge 1 \\ 200x + 100y \le 700 \end{cases}$

84. $\begin{cases} x + y \leq 15,000 \\ x \geq 2000 \\ y \geq 3x \\ x \geq 0 \\ y \geq 0 \end{cases}$ **85. a.** 27.1

$\begin{cases} x + y \leq 15,000 \\ x \geq 2000, \ y \geq 3x \\ x \geq 0, \ y \geq 0 \end{cases}$

97.

98.

99.

100.

101.

102.

110. $\begin{cases} x \geq -2 \\ y > -1 \end{cases}$ **111.** $\begin{cases} y > x - 3 \\ y \leq x \end{cases}$ **112.** $\begin{cases} x^2 + y^2 \leq 9 \\ y < x^2 \end{cases}$ **113.** $x + 2y \leq 6$ or $2x + y \leq 6$

115. $\begin{array}{l} y \geq nx + b \\ y \leq mx + b; \end{array}$ **117.** $(-\infty, 1)$ **119. a.** **120. a.**

Exercise Set 5.6

1. $(1, 2)$: 17; $(2, 10)$: 70; $(7, 5)$: 65; $(8, 3)$: 58; maximum: $z = 70$; minimum: $z = 17$

2. $(3, 2)$: 13; $(4, 10)$: 32; $(5, 12)$: 39; $(8, 6)$: 36; $(7, 4)$: 29; maximum: $z = 39$; minimum: $z = 13$

3. $(0, 0)$: 0; $(0, 8)$: 400; $(4, 9)$: 610; $(8, 0)$: 320; maximum: $z = 610$; minimum: $z = 0$

4. $(0, 0)$: 0; $(0, 9)$: 405; $(4, 4)$: 300; $(3, 0)$: 90; maximum: $z = 405$; minimum: $z = 0$

5. a.

6. a.

7. a.

b. $(0, 8)$: 16; $(0, 4)$: 8; $(4, 0)$: 12

b. $(0, 0)$: 0; $(0, 4)$: 12; $(3, 2)$: 12; $(4, 0)$: 8

b. $(0, 4)$: 4; $(0, 3)$: 3; $(3, 0)$: 12; $(6, 0)$: 24

8. a.

b. $(0, 5): 30; (0, 0): 0; (2, 6): 38; (5, 0): 5$

9. a.

b. $(1, 2): -1; (1, 4): -5; (5, 8): -1; (5, 2): 11$

10. a.

b. $(0, 3): -6; (0, 2): -4; (2, 0): 10;$
$(5, 0): 25; (5, 3): 19$

11. a.

b. $(0, 4): 8; (0, 2): 4; (2, 0): 8; (4, 0): 16;$
$\left(\dfrac{12}{5}, \dfrac{12}{5}\right): \dfrac{72}{5}$

12. a.

b. $(0, 9): 36; (0, 3): 12; (6, 0): 12; (9, 0): 18;$
$\left(\dfrac{3}{2}, \dfrac{3}{2}\right): 9$

13. a.

b. $(0, 6): 72; (0, 0): 0; (5, 0): 50; (3, 4): 78$

14. a.

b. $(0, 10): 60; (10, 0): 50; \left(\dfrac{10}{3}, \dfrac{10}{3}\right): \dfrac{110}{3}$

15. a. $z = 125x + 200y$ **b.** $\begin{cases} x \le 450 \\ y \le 200 \\ 600x + 900y \le 360,000 \end{cases}$

c.

d. $(0, 0): 0; (0, 200): 40,000;$
$(300, 200): 77,500; (450, 100): 76,250;$
$(450, 0): 56,250$

16. c.

d. $(3, 0): 45; (8, 0): 120; (3, 17): 215; (8, 12): 240$

34. $(0, 0): 0; (0, 3): 3B; (3, 1): 3B; (2, 0): \dfrac{4}{3}B;$ maximum value: $3B$ at $(0, 3)$ and $(3, 1)$

42. $\begin{bmatrix} 1 & 2 & -1 \\ 0 & -11 & -11 \end{bmatrix}$

Chapter 5 Review Exercises

6. a. $C(x) = 60,000 + 200x$ **b.** $R(x) = 450x$ **c.** $(240, 108,000);$ This means the company will break even if it produces and sells 240 desks.

14. $y = 3x^2 - 4x + 5$ **16.** $\dfrac{3}{5(x-3)} + \dfrac{2}{5(x+2)}$ **17.** $\dfrac{6}{x-4} + \dfrac{5}{x+3}$ **18.** $\dfrac{2}{x} + \dfrac{3}{x+2} - \dfrac{1}{x-1}$

20. $-\dfrac{4}{x-1} + \dfrac{4}{x-2} - \dfrac{2}{(x-2)^2}$ **21.** $\dfrac{6}{5(x-2)} + \dfrac{-6x+3}{5(x^2+1)}$ **22.** $\dfrac{5}{x-3} + \dfrac{2x-1}{x^2+4}$ **23.** $\dfrac{x}{x^2+4} - \dfrac{4x}{(x^2+4)^2}$ **26.** $\{(0, 1), (-3, 4)\}$

28. $\{(3, \sqrt{6}), (3, -\sqrt{6}), (-3, \sqrt{6}), (-3, -\sqrt{6})\}$ **33.** $\left\{\left(\dfrac{5}{2}, -\dfrac{7}{2}\right), (0, -1)\right\}$ **34.** $\{(2, -3), (-2, -3), (3, 2), (-3, 2)\}$

39.

$3x - 4y > 12$

40.

$y \le -\dfrac{1}{2}x + 2$

41.

$x < -2$

42.

$y \ge 3$

43.

$x^2 + y^2 > 4$

44.

$y \le x^2 - 1$

45.

$y \le 2^x$

46.

$\begin{cases} 3x + 2y \ge 6 \\ 2x + y \ge 6 \end{cases}$

47.

$\begin{cases} 2x - y \ge 4 \\ x + 2y < 2 \end{cases}$

48.

$\begin{cases} y < x \\ y \le 2 \end{cases}$

49.
$$\begin{cases} x + y \le 6 \\ y \ge 2x - 3 \end{cases}$$

50.
$$\begin{cases} 0 \le x \le 3 \\ y > 2 \end{cases}$$

52.
$$\begin{cases} x^2 + y^2 \le 16 \\ x + y < 2 \end{cases}$$

53.
$$\begin{cases} x^2 + y^2 \le 9 \\ y < -3x + 1 \end{cases}$$

54.
$$\begin{cases} y > x^2 \\ x + y < 6 \\ y < x + 6 \end{cases}$$

55.
$$\begin{cases} y \ge 0 \\ 3x + 2y \ge 4 \\ x - y \le 3 \end{cases}$$

56. $(2, 2)$: 10; $(4, 0)$: 8; $\left(\dfrac{1}{2}, \dfrac{1}{2}\right)$: $\dfrac{5}{2}$; $(1, 0)$: 2; maximum value: 10; minimum value: 2

57.
$(0, 8)$ $(0, 3)$ $(2, 0)$ $(8, 0)$
$$x \ge 0, y \ge 0$$
$$x + y \le 8$$
$$3x + 2y \ge 6$$
Maximum is 24 at $x = 0, y = 8$.

58.
$(0, 7)$ $(5, 7)$ $(0, 3)$ $(3, 0)$ $(5, 0)$
$$0 \le x \le 5$$
$$0 \le y \le 7$$
$$x + y \ge 3$$
Maximum is 33 at $x = 5, y = 7$.

59.
$\left(\dfrac{6}{5}, \dfrac{6}{5}\right)$ $(4, 4)$ $(3, 0)$ $(6, 0)$
$$x \ge 0, y \ge 0$$
$$y \le x$$
$$2x + y \le 12$$
$$2x + 3y \ge 6$$
Maximum is 44 at $x = y = 4$.

60. b. $\begin{cases} x + y \le 200 \\ x \ge 10 \\ y \ge 80 \end{cases}$

c.
$(10, 190)$ 200 $(120, 80)$ $(10, 80)$ 200
$$x + y \le 200$$
$$x \ge 10, y \ge 80$$

d. $(10, 80)$: 33,000; $(10, 190)$: 71,500; $(120, 80)$: 88,000

Chapter 5 Test

6. $\dfrac{-1}{10(x + 1)} + \dfrac{x + 9}{10(x^2 + 9)}$

7.
$$x - 2y < 8$$

8.
$$\begin{cases} x \ge 0, y \ge 0 \\ 3x + y \le 9 \\ 2x + 3y \ge 6 \end{cases}$$

9.
$$\begin{cases} x^2 + y^2 > 1 \\ x^2 + y^2 < 4 \end{cases}$$

10.
$$\begin{cases} y \le 1 - x^2 \\ x^2 + y^2 \le 9 \end{cases}$$

13. a. $C(x) = 360,000 + 850x$ **c.** $(1200, 1,380,000)$; The company will break even if it produces and sells 1200 computers.

Cumulative Review Exercises (Chapters 1–5)

1. domain: $(-2, 2)$; range: $(-\infty, 3]$ **3.** maximum of 3 at $x = 0$

9.
$x = -4$ $(-2, 2)$ $(-3, -1)$ $(-1, -1)$
$$g(x) = f(x + 2) - 1$$

10.
$x = -4$ $x = 4$ $\left(0, \dfrac{3}{2}\right)$ $(-2, -0)$ $(2, 0)$
$$h(x) = \dfrac{1}{2}f\left(\dfrac{1}{2}x\right)$$

11. $\{3, 4\}$ **12.** $\left\{\dfrac{2 + i\sqrt{3}}{2}, \dfrac{2 - i\sqrt{3}}{2}\right\}$

15. $\left\{-3, \dfrac{1}{2}, 2\right\}$ **16.** $\{-2\}$

17. $\{2\}$ **20.** $\left\{\left(-\dfrac{1}{2}, \dfrac{1}{2}\right), (2, 8)\right\}$

22.
$(-4, 0)$ $(0, 0)$ $(-2, -4)$
$$f(x) = (x + 2)^2 - 4$$

23.
$$2x - 3y \le 6$$

24.
$\left(0, \dfrac{1}{9}\right)$ $(3, 3)$ $(2, 1)$
$$y = 3^{x-2}$$

25.
$x = -1$ $y = x - 2$ $(-2, 0)$ $(-3, -3)$ $(3, 0)$ $(0, -6)$
$$f(x) = \dfrac{x^2 - x - 6}{x + 1}$$

26.
$(0, 2)$ f f^{-1} $(-4, 0)$ $(2, 0)$ $(0, -4)$
$$f(x) = 2x - 4$$
$$f^{-1}(x) = \dfrac{x + 4}{2}$$

27.
$$(x - 2)^2 + (y - 4)^2 > 9$$

28.
$(-2, 2)$ $(2, 2)$ $(0, 0)$ $(2, 0)$ $(0, -2)$ $(4, -2)$ g
$$f(x) = |x|$$
$$g(x) = -|x - 2|$$

29. $(f \circ g)(x) = 2x^2 - 3x$;
$(g \circ f)(x) = -2x^2 + x + 2$

SUBJECT INDEX

HORRORS OF WAR

OF WAR

HISTORICAL REALITY AND PHILOSOPHY

Franjo Tudjman

Translated from Croatian by Katarina Mijatovic

REVISED EDITION

M. EVANS AND COMPANY, INC.
New York

M. Evans and Company, Inc.
216 East 49th Street
New York, NY 10017

ISBN: 0-87131-838-5

Library of Congress Number: 96-61794

Manufactured in the United States of America

10 9 8 7 6 5 4 3 2

In memory of
my father,
my brother,
and my stepmother,
who fell victims
in the wastelands of history.

in a mock communist trial in 1946 he was found guilty and sentenced to sixteen years of forced labor for "war crimes." Most western governments joined the U.S. and the Vatican in calling the trial a travesty of justice.

After being released from prison by Tito in 1951, the Archbishop went into "internal" exile, house-arrest rather than leave the country. Shortly after that, Pope Pius XII elevated Archbishop Stepinac to the rank of Cardinal. Tito, furious with the Pope's actions, broke diplomatic relations with the Holy See in 1952. Cardinal Stepinac, who died in 1960, was one of the many victims of communist repression. He was in the thoughts of Pope John Paul II when he visited Zagreb in September 1994, more than 1,000 years after the first Papal contact with Croatia. At this time, Pope Jean X was in communication with the first Croatian king Tomislov (910-930).

Pope John Paul II, in his visit, escorted by Cardinal Kuharic, prayed at the tomb of Cardinal Stepinac. It was apparent to all observers that the Holy Father wished to indicate to the world that Cardinal Stepinac was held in high regard by the Church and that he, like so many priests, bishops and cardinals in the communist controlled areas, was the victim of communist misinformation.

After praising various clergy and lay persons in Croatia who "through many centuries . . . left an extraordinary spiritual heritage," Pope John Paul II went on to say that the "most prestigious figure, is however, Cardinal Stepinac."

On that late summer day, on September 10, 1994, the Pope declared that Cardinal Stepinac "in his life, work, courage and patience, in his silence and finally in his death, he showed himself to be a true man of the church."

In his examination of the historical record, Tudjman examines the phenomenon of aggression and violence throughout the sweep of history, explores the record of wartime Yugoslavia, and traces the purposes and processes behind the creation of the Croatian "black legend." At the same time, the author does not lose site of the fact that each statistic was a human life. Having experienced the hor-

rors of the Second World War himself, and having lost several members of his own family in the fighting, Tudjman understands like few other scholars that the horror of the Holocaust cannot be reduced to a statistical exercise, but must be understood in terms of the individual lives ended out of hate and ignorance. Tudjman's exploration of historical violence serves to point up the unique character of this tragedy. Indeed, given his insights into the suffering throughout history of peoples who have been deprived for protracted periods of time of a state of their own, President Tudjman deals with particular sympathy with various matters pertaining to the war-time persecution of the Jewish people. Tudjman's book also strongly condemns the unique and terrible tragedy of the Holocaust against the Jewish people, which is qualitatively different from all other genocides perpetuated throughout human history.

The discussion of war crimes and atrocities is only one aspect of this book and of Franjo Tudjman's life. He is now the leader of a free and independent Croatia. The transition from Tito's dictatorship to a socialist state and then to a sovereign democratic republic has not been easy. And, in addition, there was the conflict over Bosnia-Hercegovina that raged until 1996.

The people of Croatia have been served well by President Tudjman's experience and intellect. This book is many things: a political memoir, intellectual tour de force, a political document of an important period in the country's history and a broad historical philosophical survey. The book gives many insights into the mind of the statesmen who guided the Croatian people to the establishment of a state, fully accepted by the international community. Both President Tudjman and the Republic of Croatia will be prominent participants in the new Europe and in the world community of the third millennium.

The readers of this book will have a candid insight into the mind of one of the most important statesmen in contemporary Europe. *Horrors of War* is a contribution to the efforts of all people of good will to learn about the alienations of the past so that they will not be repeated. Once this is done, we can build a new creative framework of peace-oriented relationships for the new century.

Horrors of War

PART ONE

■

History and Life's Destiny

I. JASENOVAC — TRAGEDY AND DISTORTION

Well, heaven forgive him! And forgive us all! Some rise by sin and some by virtue fall:
Some run from brakes of vice, and answer none; And some condemned for a fault alone.

—W. Shakespeare

In History There Are No Coincidences

IT IS AN OLD AND FAMILIAR NOTION that in history there are no accidents and that, in essence, almost nothing is new.

From its very beginning, philosophical thought has arrived at two fundamental conclusions. Even what seems to be sheer coincidence is conditioned by deeper reasons and linked by unseen threads to historical existence. This also applies to experience in the deepest sense: ever since man has noted and considered history, he has done so because of his future, which is to say—how history will judge him.

3

These concepts—like all concepts—are subject to various, and even contrary, interpretations. Such interpretations require intelligent reflection upon the historical existence of man, as well as that of a nation, as social components of the human community. That reflection should result in a better understanding and even a harmonious coexistence among contradictory self- or national interests, and in the ever more possible, but never fully attainable, universality of the world. And such interpretations come into conflict with all that opposes them.

Historical insights on this level can serve as points of departure for the consideration of contemporary history of the multi-ethnic community of Yugoslavia. I will bring into this the examination of my own personal experience, which might otherwise belong to a different sort of literary work. I believe they can facilitate a better understanding of the controversy over Jasenovac and contemporary Yugoslav history in general. Thomas Hobbes concluded that in history there exists an eternal war as if among wolves, not only because of conflicting material needs, but also because of irrational impulses. And he also claimed that "there is properly no history; only biography," because all ideological and real historical events of a given period, movement, or nation are reflected through people.

Theory of the Croatian "Genocidal Nature"

Since the beginning of the 1980s—after Tito's departure from the historical scene, and the acceleration of general sociopolitical, economic, inter-ethnic, and moral crises—the interpretations of controversial, muddled, and "delicate" historical questions became more pressing. This occurred first in the literary and journalistic sphere, but also within the narrower field of historiography. The tendencies toward a complete re-evaluation of historical events and personalities became more prevalent. Among such interpretations of weighty and controversial issues of contemporary history, a special significance was attached to a distortion of the historical facts concerning the victims of the Jasenovac camp. Further, an attempt has been made to substantiate the distortion

by ascribing genocidal tendencies to every manifestation of Croatiandom.

The death camp at Jasenovac was a horrible tragedy. Thousands of innocent people were slaughtered by the Nazis and their collaborators. But in the nearly half century after these terrible events, the historical record has been repeatedly and intentionally altered for political purposes. The numbers of non-Jews killed at Jasenovac have been greatly exaggerated in an attempt to create a burden of historical guilt against the Croatian people.

First, it must be pointed out that while many Croatians did collaborate with the Nazis, and the quisling Croatian government took part in the slaughter of innocents, the Croatian people as a whole are no more to blame for the terror than those of other occupied countries whose governments collaborated with the Third Reich. The blame for the death camp at Jasenovac and other wartime atrocities should not be ascribed to the Croatian people as a whole, but rather to those who collaborated. However, the enemies of the Croatian people have found it politically effective to accuse them of ghastly crimes. Such charges are both inaccurate and undeserved.

Distortions of the number of deaths in Jasenovac are not only inaccurate and politically motivated, but also morally repugnant. In my examination of the historical data, you will notice that only the numbers of non-Jewish victims have escalated. This is a brazen attempt to downplay the unique and unprecedented tragedy of the Jewish people during World War II. The Jewish people were singled out for destruction by the Nazi regime. While the war claimed many victims, no other people suffered as much as the Jews. Any effort to distort the historical record concerning war victims not only demeans the memory of all who died, but also risks obscuring the singular tragedy suffered by the Jewish people. The Jasenovac distortion is a slander against the Croatian people. But it is a sacrilege against the Jewish people.

Several decades ago, during a general evaluation of fundamental issues in contemporary national history, I stated my opposition to any political manipulation of the historical record concerning

Jasenovac, contrived by hitherto virtually unheard-of exaggerations of the nevertheless grievous numbers of victims of the fascist and Ustasha terror. Because of that I have found myself—for a long time—in the center of what was unfortunately not so much an intellectual debate as a mutual rivalry and even retaliation among various groups in the Yugoslav community.

This was no coincidence. As a scholar and politician, a man of ideas and a man of action, throughout my life I have been directly involved, in both history and historiography, with the problems, hardships, and the confusion of the recent Croatian history in monarchist and socialist Yugoslavia. My experience traverses the crucial period of the Occupation, the Independent State of Croatia (NDH), and the revolution. In my younger days, like many others all over the world at the time, I nurtured the illusion that the Marxist and socialist doctrines alone offered a solid foundation for ideal solutions to all national and class differences. But a more rational understanding of historical movements began to develop within me, as much from a deeper scholarly immersion into the mysteries of history, as from my own experiences.

From this immersion into the mysteries of history, and my own reflections on personal experience, I have changed some views that I had previously held, and become more sensitive to certain issues. I first began writing this book while a political prisoner in Tito's Yugoslavia. Since then I have seen the disintegration of that country and the emergence of independent nations, including my own Croatia. This process has been difficult, and unfortunately, bloody. And in that process I have become more aware of the terrible tragedy of history. Not just the history of my own people, but the tragedy that has befallen so many other races and nations.

When the Croatian people were under the yoke of Tito's dictatorship and a Serb-dominated Yugoslavia, my primary concern was to protect the Croatian nation and culture, particularly from unprincipled attacks by its enemies. Sometimes my patriotism got the better of me. Focused so intently on the sufferings of my own people, I was not sufficiently aware of the sufferings of others. Some of my scholarly attempts to shine the light of truth on crimes

against the Croatian people have led critics to believe that I was attempting to deny or obscure crimes committed by the Ustasha government. Others felt that my scholarly investigations into war crimes diminished or trivialized the memories of war victims, both at Jasenovac and elsewhere.

I realize now that in my efforts to correct the historical record I might have caused pain to those who suffered, their families, friends, fellow countrymen and co-religionists. Whenever one uses statistical or demographic data in historical investigations, it is easy to think of them as just numbers, and not individual human lives. Of course, throughout my studies, I have never lost sight of the fact that these war victims were indvidual human beings who suffered and died horribly. I myself had lost a father, brother and stepmother in the war, and their memories are with me always. And I never wished to trivialize the experiences of any victims of war by my scholarly attentions.

In this light, the experience of Europe's Jewish population deserves special mention. The Jewish people have suffered immensely throughout their history, but most tragically in this century. Later on in this book, I will describe many of the barbarous acts of political, religious, and ethnic violence that have been committed throughout the years. Among them, Hitler's Final Solution stands alone as the single most devastating act of attempted genocide in all of history. The Holocaust was the greatest evil that the world has ever known. It was not just mindless barbarism but an intentional and organized strategy of total genocide, carried out by a mechanized bureaucracy. Never before has an entire race of people been targeted for complete destruction by a state. And one can only hope, pray, and fight to make sure that it never happens again.

In all of my investigations, I meant to clarify the historical record so that differing interpretations could, if not be reconciled, at least be explained and understood. Differences in the interpretation of historical events are nothing new, nor are they characteristic of any particular time or place. One of the earliest European historians, Hecateus of Miletus (sixth century B.C.),

states that he has merely recounted what he supposed to be the truth, and that the oral tradition of the Greeks contained so many contradictions as to be comical. Even Don Quixote's attacks upon the windmills have their justification because wartime adventures are subject to all manner of diverse interpretations after the fact. (For he could instruct Sancho, "All depends on our will.") Arthur Schopenhauer had ample reason to declare that "the muse of history, Cleo, is as completely and totally infected with lies, as a street prostitute with syphilis." Very few participants in history are capable of objective examination and impartial judgment. The differences in the conclusions of historians—insofar as they do not stem from national origin—commonly depend on differences in their philosophical understanding of history. This manifests itself even in the works of the greatest historians. Thus "Mommsen looks on a period in a different way than Gibbon; Taine shrinks in horror from a political occurrence which Michelet applauds."[1] The founders of Marxism cautioned that one should not believe "what each age claims and boasts about itself."[2] Of course this must apply also to Marxist historians.

This brief retrospection is necessary in order to show that differences and oppositions in the understanding of contemporary Yugoslav history are not an anomaly. They should not be surprising even when we take into consideration that they are often most vehement among followers of the same world view. Even in international Marxism and in the Socialist Federative Republic of Yugoslavia, for example, the interests of national subjects outweigh all forms of supranational aims.

Jasenovac Distortions Play a Chief Role against Croatiandom

Since I realized that the Jasenovac controversy was crucial to contemporary understandings of the recent history of the Croatian people a full quarter century ago, I encouraged historical research on the subject.[3] I also conducted my own studies and wrote my own memoirs.[4] In these works I argued that the stories about war victims, and those of Jasenovac in particular, had been used methodically to

create a distorted historical consciousness, into which many well-intentioned public activists and even historians have fallen, or embraced willingly for their own opportunistic reasons.

Since the time of my original studies, the arguments I made about the exaggeration of war victims at Jasenovac have been corroborated by other scholars. In his book *Serbia's Secret War*, Dr. Cohen not only corroborates my estimate of the number of deaths at the camp, but also agrees that the falsifications were intentional and politically motivated.[5]

Dr. Cohen writes:

> The Belgrade government was aware from the outset that the original claims of 1.7 million war victims were false. However, since inflated figures served Yugoslavia's interest in anticipated war reparations from Germany, the fiction was perpetuated. In 1964, when census data, which excluded foreigners and collaborators, demonstrated one third of the original claimed war victims, the resulting figures were made a state secret and remain so at the time of this writing [1996]. For many Serbs, inflated war losses, especially the alleged 700,000 to 800,000 at the Jasenovac concentration camp in Croatia, remain integral to the mythology surrounding the Serbian war record. The continued falsification of these losses, however, can no longer be uncritically accepted.[6]

However, back when I originally aired my views I did not have such support. My ideas were sharply opposed to the maintenance of a greater Serbia—they did not conform with the views of political pragmatism, so that I found myself in an increasingly difficult position. It was a time when the historical understanding of the revolution and contemporary national history in general was under the influence, not only of the official ideology, but also of direct political influence. In standing by the principle that no kind of political or ideological interest can justify an unscholarly approach to national history, I suddenly found myself in the midst of all manner of gyrations and power plays, set up as a target for political exile, and later, even for legal persecution. This first occurred during the time of Zanko's march (1967) against the

"focuses of Croatian nationalism," which were accused of writing the "Declaration of the Name and Position of the Croatian Literary Language," then later during the crackdown on the "mass movement counterrevolution" in Croatia (1971), and finally in the "first great political show trial after Tito" (1981).

Opposition to the Jasenovac Distortion Used as a Pretext for Persecution

Throughout this time, no attempt was made by means of serious critical analysis to disprove my scholarly views or my interpretations of controversial historical issues. There was absolute silence about all my essays and books—as if they had never existed. One of my theses, on the magnification of the distortions concerning Jasenovac, served as a pretext for a campaign of persecution, which included several years of personal attacks against me in almost all the daily and periodical journals. Because I have argued against the inflated numbers of Jasenovac victims, all sorts of grievous sins have been ascribed to me—a maneuver well known in the arsenal of Stalinist retribution. I was accused not just of the "crime" of Croatian nationalism, but also of association with a Fascist and terrorist Ustasha emigration and foreign intelligence center.[7] Some of my attackers didn't mention me by name. Instead, they used expressions such as "a falsifier of history," "Fascist elements," and agent of "genocidal-maniacal ideas." A certain "Professor H. M." from Bosnia called me a follower of Khomeini, claiming that I was, among other things, the author of the "Islamic Declaration" of the Moslem fundamentalists.[8]

I was submitted to a campaign of persecution. Fortunately, this did not end in physical liquidation, as often happened to those out of favor in a Stalinist state. I was allowed to remain alive even though such officials as Dr. Dusan Dragosavac and Marinko Grujic "theoretically" explained that I fell into the category of counterrevolutionaries, who should be struck down and thrown into the "rubbish pile of history."[9]

In time, this frenzied persecution relaxed considerably, but did not completely disappear. On the contrary, it will probably never

go away, only take on new and more profound dimensions. Indeed, the problems surrounding the Jasenovic controversy are of such significance and gravity that they cannot be permanently removed from the agenda, no matter how many are persecuted because of their views on the subject. Ultimately, these questions require scholarly examination.

In recent times we have witnessed the reexamination of old contributions and the presentation of new ones, which, instead of being researched and objectively scrutinized, are being judged from a biased standpoint.

The 1986 reprint of Viktor Novak's book, *Magnum Crimen* (1948), which from the scholarly standpoint is a pure anachronism (reducing the question of Croatian autonomy to "clericalism" or even a notorious "clerical nationalism" and "clerical Fascism"), is of particular significance. Its reprint signified the revival of forces opposed to any form of Croatiandom which did not conform to their narrow conception of Yugoslav unity. Among other examples is Velimir Terzic, who in his texts (*The Fall of the Kingdom of Yugoslavia* and others) places the blame for the fall solely upon the Croatian nation, and links this accusation with Jasenovac. According to Terzic, a lieutenant general and former director of the Military Historical Institute, "at least a million Serbs were killed" at the camp.[10] Another is *New Contributions* by Vladimir Dedijer, who inclines toward the demythologization of the National Liberation War and Tito, and is particularly obsessed with the Jasenovac victims.[11]

Along the same lines, the scholarly reflections of Veselin Djuretic put forward such theses as the minimization of the "Serbian tragedy," that is, the unfounded "equalization of crimes," the minimization of the necessity to reject the anti-Serbian "hypotheses" of the Comintern on the "Serbian imperialist hegemony," and the need to take into account "existential motives" in considering the orientation of the Chetnik movement in the Second World War. At the same time, he doesn't even remotely consider the possibility of similar incentives on the part of other participants in the Yugoslav conflicts.[12]

Exaggerations of the Historical Guilt
of the Croatian Nation

Those mentioned and others, whose numbers are constantly increasing, make many claims about the historical guilt of the Croatian nation. They blame all of Croatiandom's popular movements and most prominent historical figures—from Josip Strossmayer's Yugoslavianism and Stjepan Radic's Croatian Peasant Party's confederalism, to Tito's basis of equality according to AVNOJ (the Anti-fascist Council for the National Liberation of Yugoslavia). And they reflect, if not a state of hopelessness, then at least a profound crisis of historical understanding. The crisis has manifested itself most clearly in the shaping of the Jasenovac controversy, and in the theory of Professor Vasilije Krestic's *On the Genesis of the Genocide Against Serbs in the Independent State of Croatia*.[13] Meanwhile, the persistent effort to support the Jasenovac exaggerations theoretically, and even scientifically—despite compelling evidence to the contrary—makes for an atmosphere in which it is neither possible nor permitted to discuss the controversy rationally.

Research projects, discussions among experts, and even international academic conferences are organized concerning Jasenovic. In these forums speakers repeat "authenticated" truths about the "at least 700,000" executed, while the historians and other scholars present, who do not dare to oppose openly the manipulation of facts and erroneous interpretations from various sources, point out the necessity for further investigation. By such intervention on behalf of the truth, they ease their consciences. But further biased investigations result only in an even higher number of victims, not in an accurate record of the historical truth.

In 1986 a long announced and impatiently expected historiographical work was issued which was finally to establish the disputed facts about Jasenovac. The book—two volumes of 1,124 pages prepared by the military historian, Colonel Dr. Antun Miletic—contained documents, a foreword by Lieutenant General Jefto Sasic, and an introduction in which the author himself summarized the results of his research.[14] It will remain the most

damning proof of the confusion on this most serious issue.[15] As far as the materials collected are concerned, the book undoubtedly is the most comprehensive contribution to the research on these problems, regardless of the fact that it contains documentation of differing origins, and that it omits any critical analysis of its credibility.

This enormous two-volume document does not offer factual evidence—neither real, nor even something which could be debated —yet it has served not only to further the distortions about Jasenovic, but also to continue the persecution of those who advocate unbiased research and objective judgment of that controversial issue, precisely because of its sensitivity and historical significance.[16]

In the foreword to Miletic's book, Jefto Sasic (retired lieutenant general and member of the Council of the Commemoration of the Jasenovac region) brings out some accurate ideas. But he also says certain things that left me dumbfounded. I have known Sasic ever since we worked for several years together in the Federal Secretariat of National Defense. I knew him as a man who always strove to judge situations and people from the standpoint of Marxist impartiality. Since I too was a follower of that world view at the time, this was the basis for a certain personal esteem and even trust. Our paths diverged long ago, in both the physical and intellectual sense, but the fact that we would later find ourselves in such opposing positions cannot be explained by mere personal changes, but only by the complexity of Yugoslav inter-ethnic relations, and the corruption of historical consciousness in whose exclusivity there is no place for other subjectivities and, consequently, no place for objectivity.

Jefto Sasic is conscious of what he says and writes. He knows the subject matter well. He even reminds us of Tito's outstretched "hand of reconciliation to all the righteous, that such a [WWII] hell never be repeated."[17] Even so, he resolutely chooses the distorted view of history. The essence of all his arguments in the foreword is aimed against those who question that distorted view. He speaks of "hardened reactionaries" and of "reactionism and

counterrevolution." To him: "The same master as during the Occupation is waging a new fratricidal war." And "at issue is a poorly camouflaged anti-Communism with familiar centers, both in the country and abroad." According to Sasic, "the deceivers are trying by means of falsification—as they are not dealing merely with some sort of supplemental research—to gain corrupt political capital," and "our young generation should be aware of this, to recognize those who prolong a horrible crime by these forms and actions!"[18]

Lest there be any doubt as to the noble inspirations that move him, and to whom he directs the attention of the young generation, Jefto Sasic writes at the end of his foreword:

> Jasenovac demands to speak of the victims only in noble terms, truthfully and with dignity. It is always *the chauvinistic and criminal extremists*, unworthy of man, who do not allow in the name of "historical truth" that any more than 50,000 people were killed in Jasenovac, primarily Croatians, and all other claims they classify as Great Serbian [italics added].[19]

However, Sasic also wants to be objective. As it is, to him, not an exaggeration, but a historical truth that more than 700,000 people, predominantly Serbs, were put to death in Jasenovac, he also condemns the "other side," which "doesn't admit less than a million killed, exclusively Serbian." Sasic concludes: "In fact, *the criminals claim that this crime is not what it is* . . . a crime of special human determinant—genocide" (italics added).[20]

Sasic categorized me as among the "falsifiers" of history, and as one of the "chauvinistic and criminal elements," and along with me such people as the scholar, humanist writer, and professor Ivo Supek, Colonel General Nikola Kajic, and even Dr. Bogoljub Kocovic, a Serb (because Kocovic's scholarly results are akin to mine).[21]

Virtually everyone, without exception, grasps and interprets history in a way acceptable and beneficial to himself. And Sasic did this precisely in his conflict with my historical views. Sasic, as

a formed, educated Marxist, states his position with a ruthless imprudence that surpasses all his predecessors.

Since the results of the completed census of war victims (1964-1966) shattered the Jasenovac distortions, Sasic simply castigates the Federal Statistical Institute for not refuting that census to which "anti-communists and enemies of NOB [People's Liberation Struggle] refers," and by which "in fact, a green light is actually given to the *criminals*" to turn to the path of "deceit" (italics added).[22]

Can there be a more convincing proof as to the degree of the corruption of historical consciousness and common sense than the labeling of pre-war Marxists, war heroes and revolutionaries, writers and humanists, as "anti-Communists and enemies of the NOB" and "criminals"? This is an explicit warning that disregard of principle in historical issues has given too much credence to currents which my friend the Croatian novelist Miroslav Krleza spoke of when he said that, after Tito, we would arrive at a time during which he would be lucky not to be alive.

It might be helpful to repeat here what I have said many times about the Jasenovac deaths: even if the crimes were many times fewer than they are alleged to be, they would still be unspeakably horrible and enormous. Had they killed only six people and not some 60,000—the number of those who died in all the camps and prisons, according to the statistical data of the Federative Republic of Croatia—it would have been an appalling crime! So to whose benefit is it to multiply that crime only in the case of Jasenovac, and to multiply the Serbian victims, ten or twenty, or thirty or fifty times? Doesn't the systematic exaggeration of the Jasenovac terror serve the purpose of creating a black legend of the historical guilt of the entire Croatian nation, a guilt which should be expiated?

I did not have to read Sasic's quotation of Borges—"Only one man is borne,/ Only one man died on Earth./ To claim otherwise, is pure statistics;/ Adding up is impossible"—to understand that death, crime, and the horrors of suffering are immeasurable by numbers. I have lived through, and I carry with me, all the tragedy

of individual and mass death, all kinds of victimizations, and hor-
rible sufferings which we have had to endure during the
Occupation and NOB, and also in the post-revolutionary days.
Those who themselves have not felt this do not approach the level
and the shock of a personal tragedy. I, unfortunately, have experi-
enced such a tragedy within the context of the historical events
which we are exploring, a tragedy of such dimensions and absurd
contrasts, that it gives me not only the right, but also the obliga-
tion, to uncover the truth.

In 1941 I could not return to high school (although I longed for
knowledge), because I was a Marxist anti-Fascist, and a Croatian
nationalist. That is why I decided in 1961 to take off my general's
uniform in order to devote myself completely to intellectual work
and why I did the same after the events of 1971, 1981, and 1987.[23]
For as far back as I can remember, I have been carrying the fate-
ful impressions of a persecuted Croatiandom, through the exam-
ple of my father and many others, and particularly during the dic-
tatorship of King Alexander. As a young man I followed a radical
path that promised a free socialist Croatia and its equality in the
new union of South Slavic states. But on that path there were also
arrest warrants by the Ustasha authorities because of my illegal
work before leaving to join the Partisan forces; the imprisonment
of my father Stjepan by the Ustashas; the death of my youngest
brother Stjepan in 1943 in the Partisan forces at the young age of
seventeen; the deportation of my other younger brother Ivan to
the camp at Sisak; and finally, the tragic loss of both my father and
stepmother Olga almost a full year after the end of the war. My
father's name, along with my brother's, was especially prominent
on the monument to the fallen fighters and victims of Fascism.[24]
But my father's death was shrouded in the postwar settling of
accounts, which made it all the more tragic for me when added to
the fact that only after forty years did I learn the names of his
murderers.

As a member in the front ranks of the revolution, as a victim of
personal tragedy, and finally as a historian of the very events that
I lived through, I have been inspired to research and reflect

upon all aspects of our recent history. Thus my opposition to the Jasenovac distortions has stemmed directly from my understanding that it distorts and corrupts historical consciousness.

The champions of the Jasenovac exaggerations, and their calculating followers, should explain why the "cohesive tissue" of the Yugoslav community can be strong only if it is woven by the deliberate distortion of historical reality and the deception of the public. This distortion takes the shape of claims that in Jasenovac alone, as many, or even more, perished than in all of Yugoslavia from all of the warring factions. Why cannot the cohesive strength of this community be built on a respect for historical truth and scientifically substantiated facts about real victims of all Yugoslav nationalities?

Historical judiciousness warns us that *no* manipulation can last long, for it is historically and scientifically untenable, politically damaging, and morally dangerous. At least on that point there is ample experience even from our most recent history. One should take the time to consider this in the spirit of Aldous Huxley's wise reflection that "experience is not what happens to a man. It is what a man does with what happens to him."

II. THE ORIGINS OF THE CONTROVERSY

After all, immortal gods, I know not whether to speak or remain silent! But why keep still when that which I must say is truer than truth?

—Desiderius Erasmus (1509)

Dogmatic Unitarism and the People's Defense

In 1955 I finished my comprehensive manuscript, *War Against War,* a treatise on the "Partisan war in the past and future." Opponents of my views were so influential that they blocked the printing of the book in Belgrade. But they were unable to keep it from being printed in Zagreb and subsequently translated into Slovenian.[25] The book was very well received by the public. Some twenty commentators wrote about it, including military leaders, historians, literary critics, and publicists. However, a critical review appeared from the camp of the opposition, which was intended to discredit the author, accusing me of an insufficient level of Marxism, which at that time was the greatest of deadly sins.[26] This was obviously a matter of disagreement with my historical judgment, for no one refuted me by argument. Rather, evaluations were presented from an arbitrary and distorted interpretation of the text: that I approached "the partisan war through history" from a "quite vague sociopolitical and military basis," and "that the writer doesn't always stand firmly on positions of historical materialistic understanding of history and dialectic methodology."[27] Such criticism wouldn't deserve a special review, had those who were behind it not made certain that the editors of *Nasa stvarnost* refuse to print the response of Andro Gabelic, a pre-war Marxist, and then one of the most prominent military publicists. That response[28] was printed only after its credibility had been judged on the level of Milentije Popovic (member of the CK SKJ [Central Committee of the League of Communists of Yugoslavia] and Editor-in-Chief of *Nasa stvarnost*) and Otmar Kreacic (Lieutenant General, Under-Secretary of National Defense

and Plenipotentiary in the Central Committee for the Yugoslav National Army). In a discussion over this, Popovic said that we must consider as Marxist that which is historically true. But Gabelic, in his conclusion about the book and in his criticism, stated: "Tudjman undoubtedly succeeded in his undertaking," and, "If a discussion on some of his conclusions is possible and desirable, its scholarly viability, that is from the Marxist standpoint . . . cannot be brought into question. The opposite conclusions reached in Moraca's criticism have no basis in the book, *A War Against War,* whose spirit and letter that criticism not only did not respect, but simply ignored—for reasons unclear to the reader."[29]

This, however, is not the end of the story of the book. It served as a point of departure from my theoretical work, *On the Fundamental Specific Characteristics and Essential Experiences of Military Expertise of the National Liberation War,* which was my personal contribution to the development of the theory of total national defense.[30] In 1959 it was awarded first place for the best work in the field of history and military-theoretical literature published in the past five years. Meanwhile, *A War Against War* entered all our military libraries and excerpts were translated in several foreign countries. The strategic concept of total national defense, upon being made official, became a generally accepted achievement.

I was not only in a small group of those who worked out all the details, but my contribution to the theoretical discussion of the experiences of the National Liberation War also had substantial weight. In fact, it had such weight that because of it, I was entrusted by the highest-ranking members of the General Headquarters to lecture to classes of high-level military and diplomatic officers who had come to visit the Yugoslav National Army. This duty would typically fall to the regular instructors of the military college.

In the "Outline of the Program of the SKJ" (League of Communists of Yugoslavia), which was prepared by a special Commission of the Central Committee, I prepared some comments on the section relating to the National Liberation War and the Yugoslav National Army. Those comments, in both content

and form, were accepted virtually in their entirety and entered into the program.

So when today many, even those from the Partisan revolutionary and Marxist current, attack "Tudjmanism," it should remind one, and particularly those who would completely refute it, that it can be found in the historical SKJ Program of 1958. The various igniters of inquisitional pyres for "Tudjmanism" should stop to consider this. It is not a question of insignificant, minor, personal trifles, but of issues concerning the fate of Yugoslav inter-ethnic relations.

Key Disputes Over the Fall of Monarchist Yugoslavia

In the background of the campaign against "Tudjmanism"—yesterday and today—is more than mere Marxist theoretical disputes. It is the irreconcilability between the opponents of "Tudjmanism" and the normal expression of the national being of the Croatian people.

The crucial event that testifies to this irreconcilability is the controversy concerning the historical record of the National Liberation War in Croatia for the *Military Encyclopedia* which occured toward the end of 1960.

The historical account of the NOR (People's Liberation War) in Croatia, as in the case of the other Yugoslav republics, was prepared in the *Military Historical Institute* by a contributor who was most qualified for the task. At that time General Terzic was no longer head of the Institute; General P. Brajovic was. And as far as the spirit of the Institute is concerned, it certainly never expressed his personal standpoint. The Institute's account was so biased and inaccurate that the *Military Encyclopedia* Council rejected it as totally unacceptable. Its rejection was so unprecedented that the Council concluded that it provided no foundation even for additional work or correction. It was not, as was customary, returned with comments. Instead, I was called on to write a completely new text, even though printing dates were already under contract. I was able to accept the assignment, having already completed, in addition to my previous works, *Making of*

a Socialist Yugoslavia, a book which was published in the summer of that year. I found myself in a very delicate position, because I knew that my judgments on controversial issues, which differed entirely from those of the rejected text, would not only be subjected to intense debate, but would also be battered by the criticism of the *Military Encyclopedia* Council itself, because of the first text they had commissioned and rejected.

My text on the history of the NOR in Croatia was accepted by the *Military Encyclopedia* Council virtually without significant comments. Then it was handed over for consideration to the top political and military leaders of the National Liberation War and Revolution in Croatia, in particular to those who were in office in the federal government agencies in Belgrade, and to Dr. Vladimir Bakaric on behalf of the Central Committee of the Communist Party of Croatia (CK SKH).

In the accompanying statement, the editorial staff of the *Military Encyclopedia* did not speak (for understandable reasons) of the events preceding the article's preparation, but did air the reason for the conflict, which was, in any case, a "public secret" among the concerned top military and political circles. It stated: "We ask that attention be drawn foremost to the political assessments and correct treatment of the development of the liberation war and revolution in Croatia as a whole (and in connection with that, to specifics in particular regions)."[31]

The chief editor of the *Military Encyclopedia,* Bosko Siljegovic (Lieutenant General and member of the CK SKJ), wrote to Vladimir Bakaric alongside the text, "The material has been reviewed and remarks have been made by permanent members of the Military Encyclopedia Council: Generals Kreacic, Rukavina, Jaksic, Manola, P. Brajovic, and Tomac. Outside of the Council, M. Spiljak, J. Sasic, I. Kukoc, and E. Santini participated in consultations. After these consultations and editing, the material was given for inspection to Army General Ivan Gosnjak and Vlado Popovic. Their minor comments have not been brought into the text, but I will present them orally at the meeting in Zagreb when we will read the text."[32] Announcing that he would come to

Zagreb with me as the author of the text, Siljegovic turned his attention to Bakaric and said that it would be sufficient to appoint "three or four other comrades" with whom the text could be examined in detail in the CK SKH. Bakaric invited twice as many.

Before the meeting in Zagreb, we received comments from General Gosnjak, the war commander of the Main Headquarters of the NOV i PO (People's Liberation Party and Partisan Units) of Croatia and the Secretary of National Defense at that time, and from Vlado Popovic, member of the CK SKJ, who was a delegate of the CK KPJ and Secretary of the Operational leadership of the CK KPH during the war. Gosnjak submitted his comments in writing, while Popovic wrote his in the margins of the text and stated them in conversations with Siljegovic.

At the meeting of the Central Committee of Croatia (December 23, 1960), apart from Bakaric, Jakov Blazevic, Dr. Pavle Gregoric, Anka Berus, Karlo Mrazovic, Zvonko Brkic, and others participated in the discussion. Of course, they had some comments (in accord with their "duty") which were worth hearing as an expression of personal experience and viewpoint, but their content did not touch upon the text. Moreover, as far as the political assessments of the National Liberation War in Croatia were concerned, they were more in opposition to the rejected text.

My text was finally accepted without having suffered any significant changes. One comment challenging my views was formal in nature, namely, that the text was twice as long (forty-two typed pages) as the others completed by that time—Bosnia-Hercegovina and Montenegro. The *Military Encyclopedia* Council, meanwhile, permitted such a large amount of space precisely because of the need for the historical and political assessment of all the circumstances surrounding the fall of Yugoslavia, the establishment and disappearance of the NDH (Independent State of Croatia) in the process of the Occupation, and the development of the NOB (People's Liberation Struggle), and, likewise, the formation of the federal state of Croatia in the AVNOJ's Yugoslavia.

Considering all that happened around the preparation of the historical account of the NOR in Croatia, it was to be expected

that passions would now be pacified and the opposing views rec-
onciled, since all those with any authority in the matter had spoken
on the controversial issues. Indeed, reconciliation had to be facili-
tated by the fact that fourteen Croatian revolutionary (military and
political) leaders were added as participants in the discussion.

For the sake of comparison, it is not uninteresting to note that
along with the account of the NOR in Bosnia-Hercegovina, a total
of six names are mentioned, and from Montenegro, only five
names of those who participated in producing the articles, while
in the case of Macedonia, Slovenia, and Serbia, no one was need-
ed for that purpose.[33] There the writers alone sufficed, without any
political arbitrage.

I had barely caught my breath after the hasty completion of my
writing followed by hurried procedural preparations of the text,
when I received a call from General Gosnjak. "You and your
Croatia have high feelings to such an extent that they—even after
everything—will not desist with their criticism, that is to say, with
their intentions!" He informed Tito of the situation and asked him
to read it and to judge for himself. Even though he was preparing
for an important trip abroad, the Marshal agreed. I held off send-
ing the text to the printers (at that time I was working as assistant
to the chief editor of the *Military Encyclopedia*), and delivered the
text to Gosnjak who immediately sent it to Tito.

I knew quite well of Tito's adherence to principle on key issues,
and also the statesman's skill of remaining above any conflict
between opposing views, striving to make them harmonious, or at
least to pacify them. The possibility of compromise was pressed
upon me even by Gosnjak's benign reminder—when I handed him
the manuscript—that particularly as a young general, I must learn
to accept the fact that in every war there are retreats as well.[34]

Alongside a paragraph of mine on the development of the NOB
at the end of 1941 in Croatia, Bakaric wrote in the margin: "This
is quite good!" Tito accepted my treatment of the history of the
NOR in Croatia, without giving any concession to opposing
views. On the returned copy of my text, Tito himself wrote by
hand, "I have no remarks. T."[35]

And thus that peace-time (though hardly peace-loving in intent) battle over the NOR in Croatia ended in a complete victory by those views which Sasic, a quarter of a century later, would call "Tudjmanism." The war, however, was not over; that was merely the end of the first struggle. It continued with even more severe methods. And it continues still.

Since I could no longer be directly attacked over this text, which the leadership of the JNA (the *Military Encyclopedia* Council and the Secretary of State, Gosnjak himself), the CK SKH, and even the Commander in Chief and President Tito stood behind, now "Tudjmanism" was attacked indirectly. The next assault was aimed at my book, *Stvaranje Socijalisticke Jugoslavije* (*Making of a Socialist Yugoslavia*).[36] This work described the sociopolitical causes of the fall of the Kingdom of Yugoslavia and the victory of the NOR. In a sharp critical attack on the book, all those questions around which the controversy was directed, and which are still breaking lances today, were again set in motion.[37] My opponents were influential enough to prevent the translation of my book into Macedonian and English, though not into Slovenian.[38]

Gap Between Political Voluntarism and Scientific Efforts

In the summer of 1961, I decided to leave the army, whatever the costs. Until then, my requests had always been denied, but this time I had personal reasons (I wished to devote myself to intellectual work), family reasons (my children's schooling), and a request from the CK SKH. Since at that time Croatia was the only Yugoslavian republic without an institute for the history of the workers' movement, Zvonko Brkic, then the organizational secretary of the CK SKH, proposed that I begin work as the director to found such an institute. This helped bring about my discharge from the army, even though Gosnjak himself tried to dissuade me, warning me that "conditions in civilian life are even more envenomed." To me, however, it was most important that I obtain civilian status, hoping that I would thus be able to pursue my scholarship without the interference of politics. Unfortunately, that was not to be.

After the founding of the Institute in the fall of 1961, the controversy concerning my opinions on substantial issues of recent history did not subside, but rather became more intense. This was first seen at the Third Congress of Yugoslav Historians in Ljubljana, December 1961, and then also in *A Survey of the History of the SKJ,* and especially in Terzic's book on the fall of Yugoslavia (1963).

In the main paper at the historians' congress in Ljubljana, such critical judgments concerning Croatia emerged that Vaso Bogdanov was impelled to impart a word of caution in Miroslav Krleza's periodical, *Forum:* "These claims are only a step away from those malicious, slanderous, chauvinistic, reactionary generalizations that all Croatians are Ustashas."[39]

Such generalizations had been in existence for quite a while, but only in extremist emigrant literature. In Yugoslav scholarly and periodical literature, the ground was systematically being prepared for such charges, particularly by the magnification of the Jasenovac deaths.

Finding myself facing old adversaries, but now in quite a different environment, I strove to create conditions at the Institute that were as favorable as I could make them for the gathering of both young intellectuals and mature, distinguished revolutionaries and Marxists. This effort caused suspicions, as did my friendship with Krleza, which brought about the formation of an intellectual circle (the likes of which—in the words of Krleza himself—had not existed in Zagreb since the war).[40] Some held that we were creating a political leadership with an excessively independent center which might create problems. Some people from the CK, for example, took particular offense that Vaso Bogdanov collaborated in the Institute, even though he was an academic and chairman of the Department of Contemporary National History. They objected to Bogdanov because, even before the war, he had dropped out of the party, due to his defense of Vladko Macek and the HSS (Croatian Peasant Party). They completely ignored his work at that time, and especially that he was one of very few Marxist Serbs who defended the historical truth about the Croatian nation.

In 1963 *A Survey of the History of the SKJ* appeared.[41] This work was a history of the Party, revolution, and building of the socialist state written by collaborators of the Belgrade Institute and edited by R. Colakovic, Dr. D. Jankovic, and P. Moraca. It was written without the assistance of anyone from Croatia. Our Institute was given the task of preparing critical comments and suggestions for work on the second edition. For that purpose, a symposium was organized on which about fifty scholars participated.[42] Virtually all of them, as well as the politicians and revolutionaries present, came out with quite sharp criticisms of the *Survey*. The work was found to contain too many of those same tendencies that surrounded the history of the NOR in Croatia at the Ljubljana Congress of Historians. With the *Survey* and the Zagreb Symposium, the conflict in historiography became even more intense. The Slovenians were also quite dissatisfied with the *Survey*, but they were allowed to make corrections in the Slovenian translation, i. e., to change everything they felt was necessary. Meanwhile, virtually all comments from the Croatian side were rejected as unacceptable. Because of this, we at the Institute prepared a summary of controversial issues for internal discussion by the CK SKH.[43] We agreed that controversial historical questions exceeded the Institute's mandate. Thus, at the CK SKH meeting, we found it necessary to discuss how those questions should be posed. The members of the Croatian leadership should debate the question at the highest levels because the second edition would also be prepared inside the Commission for the History of the CK SKH whose president, Rodoljub Colakovic, was ill disposed to our arguments.

In March of 1964 a meeting was held in the CK SKH. Instead of an internal discussion, we were presented with the written statement that the Croatian CK could not support the views of the Institute because of its "bourgeois nationalistic deviation in its approach to national question."[44] Thus in the name of Croatian "adherence to principle" I was branded a "nationalist." And a path was cleared for all those actions taken against me up to Sasic's declaration that "Tudjmanism" was equivalent to "criminal."

Considering the fact that many open and controversial questions in historiography have been reflected both in textbooks and in history classes at all levels, the organizers of a seminar for history teachers of the SRH (Socialist Republic of Croatia) asked me to present an introductory paper on the topic. At first I declined, wishing to avoid a further deepening of the conflict. But I finally agreed, because refusing the invitation would have meant submitting to political pressure, and a retreat from the scholarly discussion of issues, however sensitive. Precisely for those reasons, I chose the most controversial issues for the subject of my lecture at the seminar in Split of October 9, 1964: the causes of the fall of the Kingdom of Yugoslavia, and reasons for the development of the NOB in Croatia.[45] The lecture was given such approval that I received invitations to repeat it in many other cities (Varazdin, Osijek, Karlovac, Pula, Rijeka, Zadar, Dubrovnik, and even Mostar). I was not, however, able to accept those invitations, first, because I was too busy with my own work, and second, because that very lecture served as an inducement for a new campaign against me. I repeated the lecture only in Karlovac on March 2, 1965, because of the overtures of the mayor of the city, Josip Boljkovac, and many local teachers.

After the lectures in Split and Karlovac, Bakaric and the CK SKH received two anonymous letters in which I was accused of chauvinistic views as well as nationalism. The letters were supported with "information" from the Documentation and Information Service of the SKH Committees of Labin and Karlovac, stating that my views had provoked "different interpretations and reactions among those present." That information was also anonymous.

It did not surprise me that I was seen in such a light by those who do not give their *Black-hand* (Serbian terrorist organization) threats under their own names. However, I was intensely embittered by the fact that Bakaric and the CK SKH would accept *anonymous letters* as a reason for forming a party commission in the Central Committee to investigate my fault. This commission was obviously considered important, since the organizational secretary of the CK SKH, M. Cvetkovic, was placed at its head.

(Other members were Josip Hrncevic, Bogdan Lasic, and Frane Boko.)[46] My revulsion at such actions was so extreme that I wanted to resign from my duty as the Institute's director, to cease my membership in the SK, and not even appear before the Commission. I considered the whole procedure to be beneath human dignity. But both Krleza and other close friends convinced me that such a move would only make my opponents' job easier; that it was precisely what they wanted to achieve.

Since my lecture had been tape recorded, I presented it for inspection to the Commission to judge for itself whether there was any basis for the accusations. I am not certain, however, what the final decision of the Central Committee and Bakaric would have been, had I not delivered a copy of the lecture to General Gosnjak and to Tito himself. I also felt that it was necessary to express my opinion in a separate letter to the Executive Committee of the CK SKH so that they would have "an all-sided insight into the complete problem, and of my positions on various open questions."[47]

Of course, there was no close examination of my scholarly views, nor any public discussion by experts and scholars of my works, but the power relations in the political struggle behind the scenes were such that they could not remove me.

But they were able to prevent my election to the Yugoslav Academy of the Arts and Sciences (JAZU or YAAS). The initiative for my nomination originated in 1964—I don't know whether from Krleza himself—though certainly from people predominantly of Krlez's intellectual circle at the Academy and others who were with me on the editorial staff of the JAZU *Forum* journal since its founding.[48] A formal proposal for my nomination as JAZU associate member was signed by the academicians Vaso Bogdanov and Ferdo Culinovic. The secretary of the Social Sciences Department, I. Krbek, requested that I inform him in writing as to whether I accept the nomination, with which there was complete agreement in the Department and in the presidency of the Academy.[49] In that part of the election procedure which is executed outside of the Academy, in addition to its President Grga Novak, even Krleza's participated; he considered it necessary due to my conflict with

Bakaric and Bakaric's interest in YAAS. Bakaric himself said: "We have nothing against the nomination of Tudjman; our disputes over the Party's history are one thing, and his general academic activity is another." Meanwhile, on the very day before the electoral conference, the Chief of the Cabinet of Yugoslav Vice President, Aleksandar Rankovic, telephoned Grga Novak, the YAAS president, and informed him that he must drop my name from the election. Novak replied that it was impossible, the whole procedure had already been completed, and I had been unanimously chosen by secret ballot.

Novak became so agitated over this demand that he suffered a heart attack. Meanwhile, I was summoned to see the President of the Croatian Parliament, I. Krajacic. He told me that my election could not be carried out until the completion of the CK Commission's work, and asked me to write a letter to the Academy, withdrawing of my own accord. If I would not do so, he said, I could no longer remain at the Institute. At first, I refused. But then I vacillated. Perhaps he spoke more as a follower and friend of Tito's, than from his longtime official link with Rankovic. Besides, I thought: Didn't I relinquish my general's epaulets, because I was more interested in scholarly work? Now I could be thwarted in these efforts. And as for my academic title, I had already completed the work for my dissertation.

So I wrote a letter to the President of the Academy in which I requested that he notify the Assembly that my name be withdrawn from the election because I had supposedly discovered that, on the occasion of my election by unanimous secret ballot there had been no formal quorum in the Department.[50]

The nomination was informally "postponed." The act of preventing my entry into the Academy by such a crude method was by no means aimed only at me personally. It was intended to prevent the fortification of "Tudjmanism" on the floor of the Academy. It was also directed against Krleza himself, who had stood behind my nomination. His circle could not be permitted to gain an even more conspicuous position.

While writing of my "voluntary" withdrawal from the nomi-

nation for YAAS membership—it became clear to me that there was a connection between the anonymous letters and the prevention of my entrance into that organization. Only later would I be informed that attempts had been made to prevent my obtaining a doctorate in history as well. Marijan Cvetkovic, in his capacity as organizational secretary of the Central Committee of the Communist Party of Croatia, informed me that the response given to that attempt was that the University should respect the law.

The appearance of Terzic's book about the collapse of the Yugoslav monarchy evoked among the Croatian public such great dissatisfaction that Anka Berus, the president of the Historical Commission of the CK SKH, took offense because the Institute did not react immediately by criticizing the book. I explained that we didn't do so for two reasons. We were occupied with the symposium on *A Survey of the History of the SKJ*. And Terzic's work had more of a journalistic than a scholarly-historical character.[51] For these reasons, I did not personally write about Terzic's book, but I participated as a reviewer of critical contributions prepared by collaborators in the Institute, and in professional journals and daily newspapers.[52] Apart from those works, criticisms of Terzic's book—as far as I know—appeared only in the official papers of the Yugoslav National Army, the weekly *Narodna Armija* (The People's Army). Given those negative evaluations, the publication of the second edition of Terzic's book was prevented at that time. During the discussion among the Institute's Council members and the CK SKH, there were quite serious proposals that Terzic's book be banned. I said, however, that banning the book would be wrong. I explained that the problem was not in Terzic himself but in a recrudescence of the past. And this cannot be overcome by any kind of ban, but only by open, principled discussion of all controversial questions.

Since I came to realize, long ago, that the problem of inter-ethnic relations in the Socialist Federative Republic of Yugoslavia, and in the world, is blurred by interpretations stemming from class ideology, or else avoided altogether by escapism, I have sought other, more effective methods to shed scholarly light on the problem. To

that end, during 1963 and 1964, in encounters with scholars from similar institutes and universities in Warsaw and Cracow, Prague and Bratislava, Rome and Milan, I reached an agreement on the launching of an international conference. All involved were enthusiastic about the idea, considering us to be the most competent for its organization because the national question was central both to the victory of the revolution in Yugoslavia, and to its opposition to Stalin. The subject of the symposium was originally planned to be "Marxism and the National Question." But we agreed to broaden it to the "The National Question in the Contemporary World."

In the outline for the conference, which I completed in December 1964, the goal was established—to conduct as many debates of and contributions as possible to the origins and development of nations and the evolution of historical, philosophical, and scholarly theories on the national question up to the present day, with the purpose of clarifying, from all sides, the development of Marxist theories and, particularly, present-day views on the role of the nation in the contemporary world. Invitations to the conference were to be sent to scholars and public workers and institutions all over the world, who had been involved in these issues, or who were believed to have something to offer to Marxist theory and scholarly examination of the national question. The conference was planned for May or September of 1966.[53] It was expected that interest would be quite widespread, and that we would find publishers for its materials in several countries.

For such an ambitious program, I needed the agreement not only of the Institute's Council, but also of the highest political officials. In two conversations with Bakaric, he gave me his principled agreement. Moreover, he joined me in the opinion that such a conference was necessary and useful. However, when I sent him an outline of the program, I received no response.[54] It remains a mystery to me why he changed his mind. Not only did I have to abandon plans for the conference, but my survival at the Institute was in question.

Because of this situation I accepted the suggestion of my friends, Professor V. B. and Academician M. M. to use a manu-

script which I had prepared for publication as a dissertation, and to apply for my doctorate. I had not previously considered this, particularly because I didn't think it necessary since I had completed a historical-theoretical text.[55] I was a general and had already been named professor of contemporary history at the Political Science School of the University of Zagreb. But I agreed to the suggestion, above all because it would enable me to discuss the most essential controversial issues on the highest scholarly level. Thus I did not take a narrow monographic issue from the material which I had been working on intensively and unceasingly for ten years, even though it would have been simpler. Instead I decided upon *The Causes of the Crisis in Monarchist Yugoslavia from the Unification in 1918 to Its Fall in 1941.*[56] I had come to the conclusion that the major part of our differences of opinion was the result of conflicting views on 1918 and 1941 as fateful turning points in the history of the South Slavic peoples. Without an understanding of these conditions and their particular significance for each individual nation, there could be no correct interpretation of either the NOR or the revolution, or an understanding of the main problems in relations among the nations of the SFR of Yugoslavia. I dealt with these issues in my above-mentioned work, and directly in individual texts: *The Fall of Monarchist Yugoslavia and the Sociopolitical Origins of the Socialist Revolution* (1959),[57] and *An Introduction to the History of Socialist Yugoslavia,* which I published in installments in the *Forum* (1963 and 1964).[58]

The defended dissertation was sent for printing that same year by the company which had already published my book on Yugoslavia; it negotiated a contract with me after conducting the normal review procedure. In early 1966, after the manuscript had passed all the customary technical preparations, the printing was halted. One could hardly ascribe to me the sins of Croatian nationalism because the dissertation had been defended before a committee panel in which two out of three members were Serbs.[59] It was nevertheless "recommended" to the publisher that the manuscript be subjected to another review. For that purpose, one

reviewer was found from the ranks of political science educators and another from among academic professionals (I do not mention their names because they were not officially revealed to me, nor were the reviews that were given to me signed). These reviewers strove conscientiously to execute the assigned task, but they didn't recommend the rejection of the manuscript. Notifying me that the manuscript had been handed over for yet another review because of the warning that "certain chapters in it contained statements and evaluations which at this time might provoke, as much for the author as for the publisher, quite undesirable scholarly and political repercussions," the director of the publishing company wrote to me that "both reviews confirm, more or less," that warning. Therefore, he asked me to perform "the necessary emendations so that the book can be published as soon as possible."[60]

Unwilling to change my historical conclusions, but in order to avoid indecisive haggling, I suggested breaking the contract and banning the manuscript, even though it had already been accepted by a committee of three university professors and the entire school council. Later I published excerpts, including the final chapter.[61] It is interesting to note that no one has ever undertaken a critical discussion of them.

In 1965 I raised the problem of war victims and the Jasenovac controversy. It was not Bruno Busic, in his 1969 writings, but I myself who had first broached the subject, despite what J. Sasic claims. Behind Sasic's claims hides the transparent motive to dump me into the category of anti-communist and nationalist.[62] It all began when I charged General N. Kajic, my assistant for the documentation and archival sector, with procuring from the Statistical Institute for the Republic of Croatia the results of the "List of War Victims 1941-1945," which had already been tabulated and was to be made public in its entirety for the SFRJ (Socialist Federative Republic of Yugoslavia). The statistics state that the total number of victims of the NOP (People's Liberation Movement) "slain, killed in action, dead and missing" in the Socialist Republic of Croatia were 185,327. Of that number, in the NOR and allied forces—63,336; in "direct terror"—50,806;

in camps, prisons, deportation, and forced labor—59,639 (of that number, 51,534 in the camps); and by unknown causes—11,546.[63]

Experts at the Croatian Statistical Institute judged that the data might—as in all similar statistics—vary by plus or minus 10 percent. However, taking into consideration that the statistics gathered in this case were to be used for negotiations for reparation with the Federal Republic of Germany, and that all were interested in having their slain, killed, or missing included among the "victims" and not "enemies," it is realistic to assume that the statistics are more likely to be inflated.

These statistical data, supplemented with other demographic data, were supposed to serve as a final estimate of the number of war victims in our research on the NDH, and particularly for a lengthy monograph on Croatia in the NOB and the revolution. The fact that these data were the *most authoritative source* until that time confirmed the results of my own research. With them in hand I was able to oppose more decisively the political manipulations of historical arguments over war victims, and particularly the Jasenovac distortions.

At that same time, a particularly urgent need for action appeared. It was proposed that the principal celebration of the twenty-fifth anniversary of the beginning of the NOR and Revolution should be held in Jasenovac. At the celebration there was to be a speech, if not by Tito himself, then by the vice president. And on that occasion, the number of victims at Jasenovic was to be carved into granite. It was merely a question of resolving whether the number was closer to 600,000 or 900,000.

Because of my firm conviction that every distortion of historical truth, and this one in particular, can only create an even greater animosity and evil, I made my opposition to that idea quite clear. I contended that there was no sense in celebrating the twenty-fifth anniversary of the revolution at a mass death site, which is a place for commemoration. Instead it should be held at the site of a great victory. And on the question of the number of Ustasha victims at Jasenovac I asked: Since even tens of thousands would make a

horrible and appalling crime, why do you need to multiply that number to nearly a million? Moreover, these were not the only victims of the war and revolution. We have in our hands the official statistics and reliable demographic data—we have no right to falsify them. Our public, and the rest of the world, would never forgive us for that. Nor would history.

Marijan Cvetkovic, the organizational secretary of the CK SKH, said to me in his office, "On the subject of Jasenovac, you're right, but for political reasons we cannot support you." My ideas were shared, however, by prominent members of the Institution's Council and the majority of reasonable people at the meetings where this was discussed, even by those who were reluctant to express their views publicly.

I substantiated my position in a detailed written report submitted to the Historical Commission of the CK SKH on October 20, 1965. A summary of the report even reached Tito and Gosnjak.

It stated:

> It is a general characteristic in the presentation of data on the victims in the course of the NOR that the number who died in the camps in the NDH is exaggerated many times over, particularly in the case of Jasenovac, where, in publications and on various commemorative ceremonies and annual observances, it is usually claimed that between six and eight hundred thousand people were killed there. These numbers are treated as facts even on occasions of the raising of monuments, etc.
>
> Until now, the only exact data are those which we have, based on the census of war victims from 1941-1945, completed by the Statistical Institute of the Socialist Republic of Croatia in order to present data for reparational negotiations with the Federal Republic of Germany. According to those, till now the only reliable data, the total number of victims in the territory of the Socialist Republic of Croatia (killed in the NOB and which succumbed to Fascist terror), amounts to 185,327, and of that number, deaths in camps, prisons, and during deportation totalled 59,639.
>
> However, on the question of the total number of victims in

the NDH and Jasenovac, it is necessary to add the territory of
Bosnia and Hercegovina which entered into the Independent
State of Croatia, so that the number of all the victims in all the
camps may be at most three to four times greater, which would
mean 180 to 240,000. The insistence that the number of vic-
tims is six to eight hundred thousand only in Jasenovac not
only does not comply with historical truth, but is also so inflat-
ed, that it would make Jasenovac the second largest camp in
Nazi Europe. Such exaggeration is completely unnecessary
and detrimental because it results in long-lasting historical,
moral, and political implications, as if the objective numbers of
the victims in Jasenovac and other camps of the quisling
Ustasha Independent State of Croatia were in themselves not
horrible and appalling enough.[64]

The proposal for the planned main celebration at Jasenovac
was not accepted. The proposal and its rejection were mere power
struggles.

Why did Josip Broz Tito never go to Jasenovac, even though he
visited every other important site in all the republics? Some deep
reasons must exist, since one can certainly not attribute to him any
partiality or deference toward Croatian sins. As a historical figure,
knowing the historical facts, and with the instinct of a great politi-
cian and statesman, he must have had a premonition that the
monstrous Jasenovac untruth was ultimately headed toward the
undermining of the foundation of his work and bringing into
question the very purpose of the truly grievous number of victims
of the war and revolution.

In 1965 and 1966, however, I survived at the Institute, thanks
only to the direct intervention of Tito and Edvard Kardelj.
Rodoljub Colakovic, as president of the Historical Commission of
the CK SKJ, made it known that Tudjman, with his opinions, was
not suitable as a collaborator on the Party's history, so that the CK
SKH was already searching for a replacement. I discovered eight
years later that on one occasion, when Tito was staying in Zagreb
with his wife Jovanka (who also participated in the talks), he
declared that there was no fault on my part. And on another occa-

sion, Kardelj, in Tito's presence, said to the Croatian leadership: "Leave Tudjman alone because it is he who in the field of historiography offers the most decisive resistance to centralist and hegemonist tendencies!"

However, political circumstances were quickly changing only on the surface. At the Eighth Congress of the SKJ in 1964, an ideological victory had been carried out over the bureaucratic, centralistic and unitarist current which opposed socio-economic reform and general national equality. Some political factions felt it would be appropriate to strike down upon the "nationalist deviationists." The "Declaration of the Name and Position of the Croatian Literary Language" (March 1967) was used as an inducement for the final settling of accounts with "nationalistic centers of activity" because "the matter of the Declaration need not be considered separately," but rather from the angle of the "atmosphere that was created by it" and the institutions and people who created it.[65]

I was mentioned by Milos Zanko among the foremost culprits of the "Declaration of the Language." In fact, I did not participate in bringing it about, except in helping create the pro-Declaration "atmosphere." When, at the Chief Council of the SSRNH, in a discussion about the Declaration, Josip Sentija reproached Zanko for behaving as an "inquisitor during a witch hunt" (forgetting that he, too, bore a part of the responsibility for the condition which had led up to it), Zanko, full of maliciousness, actually uncovered the incentives behind the persecution of the Declaration. Revealing that he had been receiving letters in which "people were requesting" that not the Declaration, but its supporters, be condemned, Zanko said, "Yes, comrades, I am guilty for the present condition of that problem, as I am in other matters concerning comrade Tudjman in connection with the Institute for the History of the Workers' Movement. I am guilty of not having driven out those problems to the finish."[66]

We are left to imagine why Zanko neglected the opportunity to settle the matter, as he could and should have, in order to clear himself of that "guilt." He did not have any connection with the

Institute or with me, either by his official duties or in any other way known to me.

A month later, at a meeting of the CK SKH, Bakaric gave his theoretical explanations for the "atmosphere" in which the Matica Hrvatska, the Society of Croatian Writers, the Emigration Office (because of Holjevac), and the Institute (because of me) had been branded as "centers of nationalist deviations." He spoke of "unacceptable positions" in *A Survey of the History of the SKJ*, and said that the CK had sought "a discussion of the problems in historiography because there had been criticism (of a very reasonable nature) in which Tudjman was correct." On the other hand, his disagreement "was not with the discussion, but with the *political evaluation of the material given to our Commission*," and it was apparent that "comrade Tudjman had differing views on this matter."[67]

It can be concluded from Bakaric's report that the issue was essentially not the difference in scholarly assessments of historical events. But even though he was correct in saying that I had "differing views" from his and that I was not prepared to accept his as scholarly opinions, these were not sufficient reasons for classifying me with the "centers of nationalist devationists." It was no accident, after all, that in the course of that witch hunt against the "Declaration," I was the only one in Croatia forced to leave my job, and to be expelled from the SKH.[68] Krleza had to resign from the CK SKH, V. Holjevac was completely driven out politically, and some of those who had signed the Declaration were thrown out of the SK or received milder chastisement from the Party.

III. SETTLING OF ACCOUNTS: THE FORCES OF DOGMATISM VERSUS THE DEMOCRATIC PROGRESSIVES

You, white bird, are guilty
even if you do not do anything wrong

—Ivan Aralica

And thus the Institute's "era of Tudjmanism" came to an end. The consequences of my exchanging a military career for a civilian one could not have been more contrary to my expectations. Rather than extricating me from the circumstances in which my historical works and views were taken into consideration only after the arbitration of a political hierarchy, politics had systematically shackled me and finally removed me from the official chairs of scholarly work.

Since the dawn of history every human idea or written word, and particularly historical literature, has been the subject of political power. But, in my case, the price paid to power was not a small one. And for a decade and a half it only increased. In 1971, with the removal of political leadership in Croatia, there was a final settling of accounts with "Tudjmanism."

During that time I was a member of the intellectual group concerned with the reform of the socialist system. We felt that it must escape the prolonged stagnation which was about to become a general crisis. Because of this position I was grouped with the "anti-socialist forces." But in actuality I belonged to one of those reformist Marxist movements which recurrently appeared in socialist countries from Poland, Hungary, and Czechoslovakia to China and the Soviet Union, in addition to Yugoslavia.

I believed that in order to surmount the crises of socialism, it was necessary to abandon Stalinist absolutism. History has shown that the essential premise for the development of every society— but middle-class society in particular—has been the free flow of human thoughts and ideas. Contrary to this, totalitarianism of

any kind usually acts as a constraint upon a society's development, expressed by a crisis which no society has ever been able to sustain for very long. Because of this understanding, I was among the proponents of the need for pluralism in socialist thinking. In Yugoslavia, pluralism was even more important because of the fact that we were a multi-ethnic community.

In its conflict with Stalinism, Yugoslav socialist society sought an escape from bureaucratic sclerosis by means of self-management and socio-economic reforms. At the Eighth Congress of the SKJ in 1964, the principle of self-management was raised to the national level. The principle of the "pluralism of different interests," which assumes a pluralism of ideas as well, was also adopted. However, all of this was carried out only halfway. Meanwhile, the crisis progressed relentlessly. Dogmatic Marxist views, deeply rooted in their theory, and grown oppressive within the highly developed bureaucratic structures of state and political power, were proving themselves, once again, to be dominant enough to be able at least to beat back more progressive reforms, if not to render them powerless altogether.

In the course of their continuous retaliations against "enemy" and "counterrevolutionary forces" of all kinds, "Tudjmanism" too was targeted. In 1972 I was arrested and tried with a group consisting chiefly of highly influential people from the Emigration Office. We were accused of being those "most responsible" for the supposed "nationalistic counterrevolution" in Croatia. My attorneys, Dr. Ivo Glowatsky and Mate Tafra, informed me that there had been talk of as much as twelve to fifteen years of hard labor. Having discovered what my opponents intended, I agreed with my wife that she should ask Krleza to speak with Tito about my case. Previously I had rejected such action because I had believed that, in the course of the investigative work, the charges against me would be proved groundless. My accusers had even contrived a false charge that the "Declaration of the Language" had been written in my office at the Institute, even though Bakaric himself had declared that I was not involved in its preparation. Despite the fact that I had engaged in no other activities aside from my writ-

ings and lectures on historical issues (which were predominantly about the revolution), I was the first defendant in a group charged with espionage. When Krleza read the charge, his comment was, "This is evidently a smokescreen!" After that, he spoke with Tito, who sent word to "stop the trumped up charges against Tudjman."[69] I was not accused of espionage but "of connections with foreign countries" and "with emigrant elements," which I had brought about in the course of my official duties, including those with the Emigration Office.[70]

In his concluding address during the trial, my attorney Mate Tafra said that he had prosecuted many postwar trials of enemy ministers and generals, but that in no instance had there been such a construction of false accusations as in my case. The Court Council accepted the proposal of my attorneys to call Miroslav Krleza, Army General Ivan Rukavina, and Lieutenant General Nikola Kajic as defense witnesses. However, no summons were delivered to them because their testimony on my behalf would have reflected upon the trial itself. In 1972 I was sentenced to "only" two years, having already spent ten months in prison while under investigation. Thanks to Tito's involvement, I managed to avoid more than ten years of imprisonment.[71]

For the entire remainder of the 1970s I was effectively silenced. But my ideas were being represented by others in their own ways, even by those who in their time had been included among my opponents. Despite a new wave of hounding terrorism and persecution, I felt that it would be intellectual cowardice, unworthy of my life, if I were to surrender in silence. I kept in mind the ancient Epicurean philosophical insight that man's fate is never completely in his own hands, but also never completely independent of his own will. I was indeed aware that I was now in a state such as Socrates remarked upon so wisely: "Whatever you do, you will regret it!" Similarly, Anaxarchus, who upon Nicocreon's decision that he should be killed, answered that the body can be killed, but not the spiritual work, which will eventually prevail if it is just.

At a time when thousands of people in Croatia had been black-listed, once again, a very exceptional and singular role was intended

for me. It would have been impossible for me to defend myself in print. So I decided to send my manuscript, *The National Question in Contemporary Europe,* to a foreign publisher and to speak up on controversial issues.[72] I would offer my case to be judged by the world's public, in an environment where, contrary to the principles of the recent Helsinki Accord, my personal rights were being trampled even more viciously.

My statements, mostly on historical issues, were used as "evidence" of "enemy activities." Concerning those statements, two members of the political and state leadership (the Presidium of the CK SKJ and SFRJ), said at internal meetings that there was nothing in them that was hostile and that persecution on that basis was a political mistake. Still, my statements served as the inducement for my sentence of three years of hard labor and a five-year ban on any kind of public activity.[73] It is easy to see why I was under attack immediately after Tito's departure from the historical scene. The first great political show trial after his death was that of "Tito's general" and the "Croatian historian," as I was called in the world media.

The only "factual evidence" of my "false" and "malicious" accounts was precisely my views of the Jasenovac controversy. On the basis of such "evidence," certain mindless buffoons tried to convince the public that I belonged not only to the Macek nationalists but also to the Fascist Ustasha terrorist group. A commentator for the major Zagreb newspaper wrote the following attack against me:

> Wicked lies... the fabrications and provocations of Nazi nostalgists, Tudjman associates himself with them by his "investigations," but without any perception of authenticity: it is well known that the Ustasha and pro-Ustasha elements abroad are the actual authors of the black legend of Jasenovac... Let us recall that Franjo Tudjman in the early 1970s was often quoted in extremist emigrant periodicals... history is repeating itself... which is not surprising at all because this is clearly a man who has not changed his position, and particularly not his ideas because they are—as the prose-

cutor said in his concluding speech—precisely the ideas of the
Ustasha and pro-Ustasha terrorist organizations abroad.[74]

Since I was extremely concerned, not so much about myself
personally as about the danger of spreading historical nonsense, I
addressed a letter to the highest officials of the SFRJ and the SRH,
and also to such sharers in my historical "guilt" as Krleza,
Generals Rukavina, Kajic and Sibl, and to those of the removed
leadership of Croatia. I stated to the first group, even at the risk
of causing great harm to myself, that distortion of facts and the
loss of historical common sense threatened to bring on new cata-
clysms or tragic consequences for a good many years. I called
upon the others as witnesses before the inescapable judgment of
history.

Here I will cite some sections of that missive, in order that the
present-day supporters of the Jasenovac distortions, from
Vladimir Dedijer to Vuk Draskovic, may ruminate over them:

To all eighteen addressees:
 In order to prevail upon you to read this letter attentively,
 as well as my enclosed response to the prosecution and closing
 remarks at the trial, may I remind you of the horrible histori-
 cal consequences—upon great numbers of people, upon all
 socialist countries and upon the socialist movement in the
 world—brought about by the Stalinist suppression of human
 thought and the trampling of the conscience and dignity of
 men, revolutionaries, intellectuals, and workers.

To those at the head of the government (Cvijetin Mijatovic,
Lazar Mojsov, Vladimir Bakaric, Jakov Blazevic, Nikola Ljubicic,
Milos Minic, Aleksandar Grlickov, Mika Spiljak, and Milka
Planinc):
 I address you with this letter, not so much as a participant
 in the socialist movement, but as a historian of the revolution
 and the most recent national history, and not so much because,
 as such, I have found myself standing before the abyss of incar-

ceration, but because I wish, on the basis of complete life experience and scholarly historical insights (not only my own), to place before you some conclusions concerning certain present-day political movements and tendencies which, in the opinions of many reasonable people, can have far reaching and even fatal consequences.

As I laid out in the enclosed response to the charges against me and to the concluding statement at the trial, for decades, many stipulations pertaining to material and spiritual life have had a negative effect upon the shaping of the political, moral, and psychological condition in Croatia. Through the most recent political and historical "assessments" by the Tanjug agency or Milutin Baltic or Jure Bilic, and the ways in which they have been carried out, according to which virtually all components of contemporary Croatian history, without exception, are identified with the defeated powers in the struggle for national liberation, that is, with Fascism and Ustashaism—this has reached most drastic and extremely disturbing dimensions.

All my guilt is contained only and exclusively in the fact that in practice and in scholarly activity, I worked in favor of unveiling and removal of causes which in any way influence the Croatian issue in the Yugoslav socialist community.

I write this letter, too, from the dictates of conscience—even at the price of the danger of worsening my situation further—but with the hope that it will, in discussing these problems in the highest political and state bodies, impel you to keep in mind the thoughts and understandings which, you may be certain, are not only mine, but also those of a great majority of Croatian revolutionaries and intellectuals.

To Generals Ivan Rukavina and Nikola Kajic:

By a coincidence, you are still the living "crown" witnesses of that sequence of events which will in these days bring not only a court sentence against me, but also the appalling regeneration of those "political-historical evaluations," in the contemplation of which we all participated together.

We are, in fact, representatives of three generations of

Croatian revolutionaries (born in 1902, 1912, and 1922, in three different parts of Croatia), but it is indeed no coincidence that long ago we became like-minded intellectually, on the basis of our individual experiences and insights, at which we arrived in the course of the revolutionary movement and in the construction of socialism...

All three of us were in that small gathering of military experts that formulated the strategic concept of total national defense, contrary to various other views...

Both of you are likewise aware—one as the assistant to the director, and the other as a member of the Council of the Institute of History of the Workers' Movement in Croatia— that I, as director of the Institute (and professor of Contemporary History and the Revolution at the University's School of Political Science), did not make of that institution any sort of "nationalist center," as was proclaimed by Milos Zanko (on the occasion of the Declaration of the Language), who managed, in 1967, by means of persecution, to bring about my replacement. It was my only intention, along with that of several others, that we should make of the Institute a serious institution, which through its work would provide scholarly contributions to the history of the workers' movement and to contemporary national history in general.

However, each of you is a special "historical" witness in my case to the essential issues on whose account I was again brought to the defendant's stand and am now sentenced to three years of imprisonment...

You, Ivo, were witness to the historical controversy which broke out more than two decades ago over the history of the NOR in Croatia...

The controversy was that from the perspective not only of Croatian national but also of revolutionary thinking, that is, from the viewpoint of historical science and everyday education, one cannot accept the theses drawn from the following assertions: that the Croatian nation was virtually the exclusive culprit of the collapse of monarchist Yugoslavia; that the Croatian national movement, under the leadership of the

Croatian Peasant Party [HSS] between the two World Wars was reactionary and virtually the same as the Ustasha movement, i.e., separatist and Fascist, that is, that Vladko Macek was just like Ante Pavelic; that the Home Guard [domobranstvo] philosophy was identical to that of the Ustashas; that the entire Croatian Catholic Church was Fascist and Ustasha in orientation; in a word, that the vast majority or, so to speak, the entire Croatian nation, because of dissatisfaction with their situation under the Yugoslav monarchy, were Ustasha Fascists, and that they did not participate in the NOB, which was waged in Croatia supposedly only by Serbs, along with an unsignificant number of Communists. To make such an argument more easily acceptable, the Croatian regions of Dalmatia and Slavonia are systematically separated from the whole of Croatia.

That for some persons this controversy still continues, has been made evident in these days in the issuing of certain political-historical judgments and in the rationalizing of the court sentence against me...

And you, dear Kole, played a special role in my "historical" guilt concerning the war victims... you yourself, in the Croatian Statistical Institute, wrote out the data about the actual war victims, which we needed for the monograph, "Croatia in the Revolution." With those original facts in hand, we opposed the proliferation of the black legend according to which the number of people put to death in Jasenovac is several times greater than in the notorious Nazi camps of Mauthausen and Dachau, by which, then, the Fascist character and the collective historical guilt of the Croatian people was outlined. We did it not merely to fight for the correct judgments about Croatian history in general, but above all for the affirmation of the truth about the National Liberation Movement and the revolution in Croatia.

But of what use is argumentation, even when based on original, official documents, if it is implemented by a Croatian lieutenant general, and historical assessments justified by another general and Doctor of historical science if they do not conform to certain police views of Croatian and Yugoslav

history and, naturally, their present...

Because in the trial farce in which I was sentenced to three years imprisonment, there was no room even for evidence in the form of official statistics, I have had to call you forth as witnesses before the court of history!

Allow me, by the way, to pose a question to the scenarists and directors of such a rationalized [political] sentence. Are they aware that by this action, they condemn not only me, but a great multitude of those who think the way I do, friends and acquaintances, Croatian generals, revolutionaries and intellectuals, as well as ordinary thinking people? It is a terrible thing if they are not aware of this, and even more terrible if they are.

To Krleza:

...Even as a high school student, I had committed myself to socialism chiefly through the magic of your literary word, but we met when I was an adult with broad experience... In our many lively discussions on history, the life experiences and views of such an intellectual progenitor of the revolutionary movement among us were, for me, of the highest value, but not as a substitute for—rather as a supplement to the experiences of a younger generation and my own views.

Because you have dealt, since the beginning of your work, with the historical destiny of our men and nation, and as director of the Yugoslav Lexicographical Institute, you have attained such insights (as probably no one else in the country) into the present state of all disciplines related to history, as well as the events and motives that affect these disciplines. Likewise, you were familiar with my efforts as well as my difficulties in the field of the history of the workers' movement and recent national history, which was also tightly linked with the sociopolitical movements of that time. You interceded on my behalf at the highest levels when, in 1972, I was arrested in connection with the suppression of the alleged nationalistic counterrevolution in Croatia.

Upon observing that the accusations were "clear smokescreen," you spoke with President Tito...

In order to inform you about the act we have reached in this historical drama—in which, of course, we do not appear in the same scene coincidentally, and in which we have been assigned roles that are as similar as they are different—I am moved to address you, also, in this letter.

But believe me, dear Krleza, if this matter concerned me alone, I would not be addressing you.

You yourself have been able to conclude that I have never solicited anyone on my own behalf.

I write to you because I am... firmly convinced that at this moment in history... a noble-minded personage like Krleza would and should have something to say! This is expected of you by the anxious consciences and spirits of our public which nurtures respect for your work. And your word is feared by all those who refuse to recognize your work.

And your words should resonate from the mountain peaks of life! And so they will, both today and in History!

To the removed leadership of the SR Croatia (Savka Dabcevic-Kucar, Mika Tripalo, Ivan Sibl, Dragutin Haramija, Jure Saric, and Srecko Bijelic):

As I am writing this letter in the capacity of a participant in and chronicler of history, but also as a sentenced person, and since I am summoning witnesses before the court of history because of its present, it is understandable that I should address you as well.

Historically viewed, you were the "typical" or exemplary children of the revolution and the Party; you arrived on the historical stage of political life at a time when Croatia was covered with bitter scars from wrestling with the unitaristic centralist tendencies which were then present as manifestations of Rankovicism, but which did not disappear...

History has assigned to you the role of the new wave in the socialist movement which strove to walk in step with the general democratic trends of contemporary society and the world.

Because of this, harkening to the pulse of all the strata within the Croatian nation, from the Partisan fighters, the workers

and peasants, to the intellectuals, and, following the tendencies of progressive streams in all the republics of the SFRJ [Socialist Federative Republic of Yugoslavia] and in all of Europe, which, from Warsaw to Budapest, from Prague to Rome and Madrid, have striven toward the discovery of new, more democratic, and more humanist "Euro-Communist" paths in the building of socialism, you have, in Croatia, brought about a "renaissance" wave even within the communist movement, from which then sprang the unity of a "mass" Croatian national movement in the socialist spirit.

Historically viewed, it was entirely of secondary importance that there appeared, on the margins of that movement, non-socialist, and even anti-socialist elements—such an occurrence is inevitable at any time and in any movement—to which elements an unjustified primary significance would later be assigned as a justification for a retaliation against the "counterrevolution." My "historical" guilt—in the words of a petty thinker whose judgments will remain a permanent testimony in the treasury of Croatian culture—was in the fact that I was the first among the "high-placed first-fighters of the revolution" who dared to think and act in the historical field, in a consistent Marxist and Croatian way (amazed that such a "free thinker" could ever have reached the rank of general). Similarly, your "guilt" was in the fact that you—by the same judgment—had become the first Croatian leadership of the SR Croatia, just as the leaderships of the other republics are respectively Serbian, Slovenian, Macedonian, and Montenegrin.

Your collective and my individual fate serve as proof not only to the above-mentioned lofty personage, that this [Croatian position] is not only infeasible, but intolerable.

My response was and still is: if one loses faith in human reason in general, and in the nation to whom one belongs, then nothing is left in life but hopelessness and despair. I am nevertheless inclined to believe in human reason to my last breath.

Today all of you, individually and collectively, and the entire democratic socialist movement, with all the social strata which

had aligned with it by plebiscite, according to Jure Bilic and those who before him (toward the end of 1979) pledged themselves to Zivorad Mihailovic-Silja—are categorized along with me in the "defeated powers of the NOB," "imbued with the ideology of neo-Fascism and the Ustasha movement"!...

Hitherto you have remained silent.

Silence too is sometimes the most eloquent form of speech—but only for those who can hear even the stifled palpitations of the despairing and restless population.

Despite the fact that your silence was devised upon politically very well-intentioned considerations, it was also taken as an inducement for the ominous passage of appalling judgments on the unity of the defeated Ustasha Fascist powers in the NOR and originally socialist movement in 1971.

And finally:

That which has been written and said lately about the political events and personages from the most recent history of the Croatian people, and which serves as the ideological political explanation of my sentencing and for renewed persecution, can be historically compared with those from the period of Pribicevic's unitarist rampage in Croatia in the 1920s and with Pero Zivkovic's consecration of Aleksandar's dictatorship in 1929.

For history it remains an open question why it happens repeatedly that there is a preponderance of such irrational police-state views, anti-historical, anti-political, and arch-Stalinist interpretations of historical and political events. This finds the aims and interests of a social democracy to be in the multitudinous creations and incarcerations of ever newer sorts of enemies, counterrevolutionaries, and now, also, fifth columnists, even from among the founders of the existing political order.

The preponderance of such "political" judgments, which in Croatia leads to the state of absurdity in which everyone and everything is equated with the defeated powers in the anti-Fascist war and battle for National Liberation, i.e., Fascism

and Ustashaism, simultaneously leads off into an abyss in which all historical circles are closed and out of which there is scarcely any view or exit...

It is understandable that those who, blinded from maintaining such warped mirrors, initiate even further such dangerous "historical" opinions, are unable to foresee all the consequences of completing such circles, which certainly cannot be permanently held in check, neither by police terror, nor by political psychological terror. But for precisely that reason, a great deal more judiciousness is sought at all other levels, upon which all necessary consequences should be drawn before the events assume excessively undesired or unforeseeable dimensions.

I do not believe that in the present time such injudicious politics can be even temporarily justified, nor that anyone who does not agree with the views of Bilic or Baltic is a counter-revolutionary who should be lynched as a traitor and rendered powerless.

On the question of myself or the essence of the socialist movement of 1971, the waving of a red rag and stirring up of hatred between peoples is one of the vilest, and also most dangerous deceptions of the public...

From a historical perspective, it is indubitable that there can be no positive changes in the political and psychological atmosphere as long as a Croatian person feels that moral and psychic violence is being done to his being and conscience, principally of all by an interpretation of the history of his forbears and of his present life that is distorted and unacceptable to him, or rather as long as he feels fettered and insulted in his personal and national dignity.

This is because man, both as an individual, people, and national community, is an entity which cannot be separated from its history, nor can it spring out of it at their or anyone else's will. Historicity—with all its objective determinants and irrational components—is one of the most essential characteristics of entire human and national existence and survival. Because of that, independent of man's will, all that is not in accord with his beliefs and his understandings of his own his-

> toricity, continuously operates upon his psyche in a negative
> way, causing him to feel every alien and unnatural interpreta-
> tion of his history as a crude violence against his very being.[75]

Legal norms were still being transgressed in my treatment, espe-
cially the disregard for my seriously declining health. The entire
proceeding—with periodic interruptions of the sentence and
repeated returns to prison—became a systematic course of physi-
cal and psychological terror which lasted a full four years, while
my actual incarceration, since the initial hearing, lasted three
times longer. It became clear that the directors of this ordeal were
not about to desist from persecution, but wanted to broaden the
charges. So my only alternative was to protest with my own life.
In addition to giving verbal notice, I wrote about this in a pre-
pared protest to the presidents of the top agencies of the SFRJ and
SRH (Socialist Republic of Croatia). I wrote that instead of my
being released, the charges were being broadened.

This follows from a "Tanjug" release that announced on
October 6, 1984 in its telexes (which the Republic's Information
Committees then put out in its bulletins) concerning an article by
the Swedish journalist Sune Olofson in the Stockholm newspaper
Svenska Dagbladet. In the summary of that article, most of the
space was devoted to my case; it said, in part, "...the government
was irritated most by the fact that Tudjman accused the regime of
falsifying history. Tudjman demonstrated, among other things,
that the Serbs in 1945, after the war, had killed at least 200,000
Croatians. He also spoke about a ritual mass killing of 70
Croatian priests at Siroki Brijeg..."

The new charge for which I should be punished, according to
my accusers, should carry a prison sentence of fifteen years!

To me it is of secondary importance who was the source of this
report. For me it is all the same whether it came from someone in
Yugoslavia or from someone abroad. According to Tanjug's news,
Olofson might even have heard that from me personally.

But the factual truth is:

1. I have never spoken a word with Sune Olofson.

2. I have never made any such claims in any of my printed or handwritten texts.

3. Contrary to these and similar claims, even in private conversations on historical topics, I have denied their credibility, defining them as similar irrational exaggerations, as are the ones made on the other side in the case of Jasenovac.

This gives me grounds for the following conclusions:

1. Olofson did not appear in Zagreb by accident, and, moreover, with such a text about me, at a time when my case was to be decided due to my illness.

2. The accusation against me by the Tanjug, with the aid of Olofson's writing in November 1984, is a logical continuation of those charges which Z. Mihailovic-Silja presented in his columns in *Politika Ekspres* and *Nedjeljna Dalmacija* in 1979. Then Mihailovic identified me with the Ustasha movement. And in order to prosecute me in 1981 because of Jasenovac, I was charged with having my alleged interview printed in the most extremist emigrant publication, the *Hrvatska drzava* (Croatian State).

3. This publishing of monstrous insinuations through the Tanjug service and other confidential bulletins intended for a narrower political circle was obviously used to defame and disqualify me.

In view of these conclusions and the fact that all the judicial investigations, in spite of medical recommendations, did not free me from serving my sentence further, but thrust me again into prison, my only alternative was to express my protest by going on a hunger strike without regard to the possible consequences.

Both my actions and words reiterated Mesa Selimovic's demand: "I wish to be a man; confront me in a human way as a man!" But it was in vain, like someone crying in the wilderness. I am left with the consolation of the purposefulness of history, as others are, but also with a responsibility.[76] History will, in the end, testify about us—individuals and nations—and about the

movements to which we belonged. As an individual, I was only one of many examples of a game of probability in historical necessity, which are—even according to the wise Solon—ready by words and deeds, that is, even by their lives, to testify to the truth about their nation, even when that seems to some historically shortsighted people, to be unwise.

Even to contemplate historical events too deeply, let alone to act upon them according to personal convictions, has been perilous since ancient times, as can be seen in Shakespeare's historical plays and especially in the chained environments of contemporary ideological systems. Indeed, the historical circumstances have not been essentially altered from those at the beginning of the post-Christian ascent of human thought when the French poet Francois Villon (1431-1463) called out from under the gallows, "Ravens plucked out the eyes of him who has looked too much in life!" The English statesman and historian, Sir Walter Raleigh (1559-1618), paid with his head for alleged treason only because he "approached the truth too closely" which was displeasing to those in power.

Despite such experience, throughout time man has responded as Aeschylus' chained Prometheus did to Hermes: "For your servant position I would never exchange my own ill fortune!" There have always been and always will be "triumphant defeats" which are envied even by the most celebrated victories (Montaigne), because the honor and virtue of the very battle for truth is eternal, regardless of its temporary outcome.

Authenticated historical experience must not be forgotten in view of the fact that it originates in the self-awareness of every individual's share in ill-fated historical events. And indeed, history would not be what it is if there did not appear in it, since time immemorial, as a necessary and lawful phenomenon, courageous protest, as was said by Ante Starcevic, against all forms of imposition by foreign patronage and one's own loss of autonomy, because that inevitably causes people to feel bereft of their rights, like strangers in their own homeland. Nor can they ever find this acceptable—not as individuals, nor as a nation.

PART TWO

■

History as Crime and Mythical Appearance

I. ON THE HISTORY OF THE EXAGGERATION OF WAR VICTIMS AND THE CREATION OF THE JASENOVAC AND BLEIBURG CONTROVERSIES

Qui quae vult dicit, quae non vult audit. (He who says whatever he wishes will hear what he does not wish.)
—Terence (2nd c. B.C.)

Bridle your tongues, you who never consider the consequences [of your words].
—Enguerrand Coucy (14th c.)

THE EXAGGERATION OF THE numbers of war victims began with two Memoranda from the Serbian Orthodox Church to the German commanders of Serbia. The first was on July 9, 1941, to General Ludwig von Schroeder. The second, and more detailed, was delivered in August 1941 to General H. Danckelmann. The final text of the Memoranda was written by Professor Pero Sli-

jepcevic, and delivered in September 1941 to the exiled Yugoslav government in London by Dr. Milos Sekulic, a prominent member of the Serbian Agrarian Party who had had contact with both General Nedic and Draza Mihailovic. He added his own detailed account to the Memoranda.[1]

In these documents, the heads of the Serbian Orthodox Church request that the leadership of Hitler's Reich intercede on behalf of the persecuted Serbian population in the NDH. They claim that during the first four months of the Ustasha government more than 180,000 Serbs had been killed.

Since the Memoranda were written with the belief that the Croatian nation as a whole bore the responsibility for the Ustasha crimes, their appearance in London provoked a bitter dispute among the monarchist government. The Memoranda essentially deepened the discord between its Serbian and Croatian members, which had already been exacerbated during their first meetings in exile, due to attempts by Simovic and others to ascribe the whole blame for the kingdom's collapse in the April 1941 War to the Croatian nation. The royal leadership concluded that the Memoranda should not be publicized until their claims could be confirmed and their background investigated. It was obvious that the documents had been prepared and delivered in collaboration with the Germans just before the establishment of Nedic's government in Serbia. They were nonetheless immediately presented to the British government and the Anglican Church and then published in the United States. There they were used by the *Americki Srbobran* [American Serbian Shield], as well as by the Yugoslav envoy to Washington, Konstantin Fotic, and by the diplomat and poet Jovan Ducic, to kindle not only anti-Croatian, but also anti-Yugoslav sentiments.[2] During this same time, the number of alleged victims doubled to 360,000 slain Serbs.

Dr. Rudolf Bicanic prepared an answer to these charges on behalf of the Croatian ministers in the royal government. He pointed out the absurdity of such a figure through an obviously convincing comparison: the number of alleged Serb victims was equal to the number of victims at approximately the same time in

the German-Soviet war where battles were waged with the most modern weapons and with some of the largest forces the world had ever seen. But comparisons aside, even from an analysis of the Memoranda themselves, it could be deduced how greatly the number of victims had been overestimated. In the documents are listed the names of 593 slain persons (of whom thirty-nine were priests), while from references to confirmed and estimated numbers the figure is 28,979.

In Bicanic's response, the crimes of Pavelic's Ustashas were condemned, along with the plan to resolve the Serbian question by liquidating a third of the Serbs, resettling another third, and converting the rest to Catholicism. And it was emphasized that for that plan, as well as the persecution of the Jews, Pavelic did not have the approval of the Croatian people. The victims mentioned in the Memoranda were not the only casualties because "the war was combined with a civil war and guerrilla warfare on the part of various groups." Even Slovenian reports confirmed "that the Serbs themselves estimate that 50,000 Croatians were killed in eastern Bosnia, the majority of whom were Moslems." Thus in a situation where "one atrocity overshadows another," to separate the crimes of only one group "represents a biased and tendentious account of events." If the Ustashas presented the execution of their plan as justifiable retaliation for the suffering of the Croatian nation from 1918 to 1941, it was a preparation for new mutual atrocities. In connection with this, Bicanic turned his attention to the "distortion of history" in the Memoranda, with the assertion that the "Serbian people created Yugoslavia by their blood" which means "the continuation of the 'megalomaniacal' opinions of particular Serbian circles which have poisoned the relations between Serbs and Croatians" and repelled Croatians from the idea of Yugoslavia because "it negates the entire politics of the Yugoslav Committee." In response to the claim that "the Serbian nation suffered at the hands of its Croatian and Moslem brethren graver and more painful blows than from its enemies in the period from 1914 to 1918," Bicanic responded that this claim represented "a falsehood which now for the first time, as far as we know, has

appeared in an official document."

Bicanic argued that the Memoranda ignored historical facts, attributing the "perpetrated crimes to Croatians indiscriminately," and accusing the Catholic Church as a whole. He made the point that "the regional phenomena of various Ustasha fanatics cannot be considered a criterion for the work of the Catholic Church in general, especially when the position of the Catholic Episcopate is well known, especially that of the Zagreb archbishop Stepinac, who protested resolutely against bloodshed and persecution." Calling these anti-Croatian slanders "characteristic of the mentality of the composers of the Memoranda," Bicanic recalled that the Croatians in the kingdom of Yugoslavia had experienced deep disillusionment and that "solemn promises and obligations," such as the Corfu declaration, the Geneva Treaty, and the decision of the Croatian Sabor (parliament) of Oct. 29, 1918, had been trampled. "The Croatian name and symbols were outlawed... while Croatians were persecuted, imprisoned and tortured." Still, despite all these injustices, "at no time did Croatians through their legitimate leadership, even in their most difficult moments, blame the Serbian nation for all this: not for the iniquities, nor for the violation of agreements, nor for the persecution." Because of this, "no 'absolution of sins' is necessary, as is stated in the Memoranda; however, the offensive language concerning 'deviant Croatian and Moslem brethren' is out of place, and Croatians do not accept such lessons."

Bicanic points out that "the Memoranda, directed against the Croatian nation, had the political aim of embittering Serbian public opinion against Croatians and thus facilitating its acceptance of Nedic's [Serbian quisling] government" as well as "inciting a conflict between Serbs and Croatians" within the London-based government. In his conclusion, Bicanic addresses the claim of Dr. Sekulic, that "the repercussions of the lamentable events in Croatia among the Serbian people are unique: ... *an extreme thirst for revenge*, which seeks that all those who are not Serbs *vanish* precisely from those [Croatian] areas." Can responsible and reasonable people allow themselves "such passions"? For the

"politics of revenge will generate among Croatians who do not wish to identify themselves with Ustashas a desire for retaliation against the Serbs, and where do such policies then lead?... Will it be possible for people to live in a common state under such conditions?"[3]

All this was said in 1941, long before the birth of the Jasenovac controversy. In our time, the Slovenian historian Dr. Dusan Biber confirmed Bicanic's judgment concerning the Germans' part in the writing of the Memoranda. The documents did, in fact, deepen the conflict within the emigrant government, so that even its president, General Simovic, was harshly attacked for not believing the data and supposedly being on the side of the Croatians. Biber reproached Dr. V. Duretic for engaging in polemics with Dr. Bicanic, "rather than undertaking a sober documented investigation of the problem."[4]

Duretic questioned the necessity for a critical assessment of the Memoranda. According to him, this would only set up an "equilibrium of evil" and prove that "the issue was not a Croatian attack on the Serbs, but a civil war." This would ignore the fact that the Ustashas did not kill only Serbs, but Croatians and Communists as well, and, at the same time, that Serbs were killing Croatians. Duretic concluded that "critics of the Memoranda... skillfully capitalizing on certain imprecisions and errors in the documents are bringing into question not only the stated numbers of those slain, but also the Memoranda as a whole." The doubling in number of those who were liquidated was not in Duretic's mind a confirmation of "propagandist exaggeration of the offenses." But the request for investigation into the truth of these figures "carries an accompanying ring of the extreme heartlessness... of this cynical nationalistic commentary."[5]

New reports appeared abroad in September 1941, which told of 350,000 and 500,000 slain Serbs, and concluding that "mass killings are an integral part of the ideology of the NDH" and "the Magna Carta of the Croatian State." *A Call to Prayer* in the *Americki Srbobran* [American Serbian Shield] spoke of "the slaying of 360,000 Serbs." Here is an excerpt: "Alongside all the

many instances of treason, today in our afflicted old homeland there is a fratricidal and religious war being waged. The Croatian nation has risen in battle against Serbs in Croatia, Bosnia, Hercegovina, Lika, Banija and Slavonia, in short, in all Serbian lands and areas, and has placed itself in the service of our old enemies: the Germans and Italians, resolving to exterminate the Serbs completely from all those areas… In Croatia there is no heart that would beat or cry for a Serb. There is no longer any Serbian heart that would shed a tear over the innumerable Serbian graves, because there are no longer any living Serbs there. Some have been killed, others have resettled in their Serbian motherland, and still others have scattered into the forests and mountains resisting the German, Italian and Croatian enemy."[6]

Not all Serbs went along with these exaggerations. Sava Kosanovic, a member of the government in exile and head of the Independent Democratic Party in America, criticized the publication of the Memoranda and protested its misuse by certain members of the leadership, including the *Srbobran*. In his opinion, the Memoranda were "one of the most successful acts of Goebbelsian propaganda."[7] Kosanovic sought the replacement of his envoy Fotic, who had played the part of "a representative of democracy," while his chief task was to systematically spread anti-Croatian news and to rouse chauvinistic hatred in every way possible, together with the *Srbobran* and Jovan Ducic, who, as an envoy to Madrid, went to America for health reasons. "Ducic invested his authority as ambassador and poet into the vilest and most unintelligent rousing of chauvinism, which in no way differed from an incitement to Fascism. In this operation, he was also assisted by Bishop Dionisije along with a large part of the priests, the military attaches and almost the entire diplomatic apparatus."[8]

Dr. Branko Cubrilovic, on the other hand, testified that in the beginning he believed Dr. Sekulic's report and the Memoranda. However, he was soon disillusioned when Sekulic "began to accuse all Croatians for the slayings, by which he aroused suspicion that he had been concealing his intent, presenting himself as a humble and honest witness to the occurrences in the country,

while it is probable that he actually came here to bring about confusion and conflict between Serbs and Croatians," assigning the blame for the "calamities and fratricidal war in the land in the main to the Croatian leaders and the entire Croatian nation."[9]

Here it is important to keep in mind that Kosanovic and Cubrilovic—as well as the public—were at that time being confronted with only the initial stage of the manipulation of statistics. The figure of 180,000 had merely been doubled to 360,000. The figure of a half million had not yet been publicized, and even larger numbers were still to come in the following months, years, and even decades.

The next step in the multiplication process was undertaken at the meeting between Draza Mihailovic and the representative of the German commanding general of Serbia on November 11, 1941, in the village of Divci. There, Mihailovic's assistant, Colonel Branislav J. Panic, explained why Mihailovic remained on the sidelines of the battle that had been initiated in Serbia by the Communists: "he had been preparing to strike out in the direction of Bosnia and Sandzak" because "he wanted to retaliate against the Ustashas who had torturously slaughtered *hundreds of thousands of Serbs.*"[10]

Then the zenith was reached in such a frenzied incitement of vengeful hatred that even the numerical exaggeration and multiplication of war victims will remain in the shadow of the discovery of these horrors. In the month of May of 1942, an incredible piece of news was spread from Rio de Janiero through diplomatic and other channels, with the evident intention of provoking a soul-stirring reaction to the monstrous deformity and "sadism" of Croatian "criminals." In order to be even more impressive, the following testimony was placed in the mouth of a Croat "witness" who reported the following: "On the streets of Croatian cities they have eyes for sale, collected from corpses, the eyes of Serbs: 30 and 40 eyes at once." And then he asks: "When the civilized world finds out about that, the revolt and indignation will be such that I don't know what will become of us Croats."[11] It is only strange that the informant himself forgot about the language of the testi-

mony, which was in Serbian.

This was only one of many similar displays of "horror propaganda," as the Germans called it. According to the assessment of Dr. Prvislav Grisogono, a former minister and staunch defender of Yugoslav unitarism, its aim was "to fire up public opinion to the highest degree in such a way that it would no longer be able to resist any suggestion as to the origin of the evil and where its remedy was to be found."[12]

In 1942 a letter was released which was allegedly written by P. Grisogono to the archbishop of Zagreb, Alojzije Stepinac, that Grisogono might "save his [Grisogono's] soul." The letter described horrible atrocities in the NDH: "They impaled Serbs on spits, lit fires upon their bare chests, roasted them alive, set fire to them in their houses and churches... they gouged out their eyes while they were still alive, chopped off their noses, ears, tongues... castrated them and put their organs in their mouths... drove nails into their heads... dismembered their children and bashed their heads against walls." In a small boat on the Sava River there was "a pile of children's heads" with a sign reading "Meat for Jovan's Market" and Serbs were "forced to drink the still warm blood of their slaughtered brothers." According to the letter, there were "rapes of innumerable women, girls and small children," and even "in the district of Petrinja, a son was forced to rape his own mother." "The Italians photographed a container with (number unclear) and a half kilos of the eyes of Serbs and one (man) who had come from Dubrovnik girded with two ropes of severed Serbian tongues." Of course, "after every great crime in history, there has been a great punishment as recompensation."[13]

P. Grisogono himself, in a letter to A. Stepinac, denied that he would have written such a letter. Meanwhile S. Jovanovic, the president of the royal government, attached a significance to it even after this, claiming that it had already been made public by the domestic radio station of General Mihailovic "Karadorde," although that station operated from the Middle East under the direction of the government.[14]

In his report to the Yugoslav Government, which reached

London in the summer of 1942, Grisogono himself said, "Today, entire circles of people are working, according to a plan, to convince the Serbian people here that all Croatians (apart from such a small number of exceptions that it is not worthwhile to consider them) are traitors in the army, that all Croatians consciously worked on the establishment of the NDH, and that they are participating, if not actively, then at least through their approval, in the persecution of the Serbs. To that end, atrocities are needlessly magnified... specific occurrences are generalized... the persecution of Croatians is passed over in silence... In this unfortunate and systematic effort, some people are not reluctant to resort to lies and falsification." As an example of such lies and falsifications, Grisogono mentions rumors which were circulating in Belgrade "concerning a sealed railroad car bearing the sign *Lard for Belgrade* and full of severed Serbian heads," and "concerning certain large trunks filled with the extracted eyes of Serbs." There was also talk of a letter from the "Ustasha headquarters" of Belgrade in which hopes were expressed "that the future Catholic Belgrade will be incorporated into the Croatian Homeland."[15]

These Satanic hyperbolics, directed not so much at the ghastliness of Ustasha crimes as they were at Croatia in general, remained—as far as I know—unsurpassed throughout the war. Only after four decades would a comparable imagination manifest itself in the form of certain literary works and films.

Don't the horrors of war thus depicted call for a conclusive investigation into the extent of their veracity? The actual truth of the crimes must be uncovered completely, but so should the nature of the diabolical imaginations that fabricated such monstrous "testimonies."[16]

There was a similar intent in Draza Mihailovic's report concerning the participation of the Croatian Army, along with the Germans and Bulgarians, in the clearing of the area of Kopaonik Serbia in November of 1942. Along with the description of the burning of several villages, the killing and burning of many innocent people, and the rape of fifteen-year-old girls, this was also mentioned: "A church was burned down, and in it 120 people

were roasted alive. Dogs and eagles carried off their remains because the Germans had banned funerals... After these slaughters, the Germans, Croatians and Bulgarians retreated into their garrisons." The minister of foreign affairs of the royal government, Dr. Nincic, reported on this in a letter to the British minister, Anthony Eden, with his own statement that: "Therefore, Croatians are slaughtering Serbs in Serbia too!" In the Foreign Office, however, Sir Orme Sargent wrote across Nincic's letter: "I do not believe a word of this letter." Sargent's doubts were based on an intercepted message to Draza Mihailovic, telling him to accuse the Croatians with as much as possible.[17]

In June 1942 reports with the latest statistics from the homeland reached the royal government in London. According to these statistics, 30 percent of the Serbs had already converted to Catholicism, 300,000 had fled over to Serbia, and "at least 700,000 Serbs have been slain" in the span of one year.[18] These reports triggered the growth of British pessimism regarding Yugoslavia. But Duretic was not concerned with their content, that is, with exaggerated numbers of victims. Rather, he was worred about Gavrilovic's conviction that "the British government is anti-Serbian."[19]

The number of "at least 700,000 slain Serbs" in the NDH was "confirmed" in mid-1942. But since the very beginning of its presentation to the public, this figure's credibility has been under fundamental suspicion. To all those who strove to deal more seriously with the Yugoslav situation, that figure surpassed all the limits, for all wars, of the customary exaggeration between opposing sides.

In March of 1942, the Serb Sava Kosanovic turned the attention of Slobodan Jovanovic, president of the Yugoslav government, to "the firing up of the most repulsive campaign against the Croatians" by the circulation of the Memoranda. He said that it represented something that no opponent of Yugoslavia, not Hitler, nor Mussolini, nor the Habsburgs "could have worked out better." The campaign was not aimed "against Pavelic, but against all Croatians," not only in their native land, but also in America, in

the interests of Nedic's Serbia, and against the very idea of a Yugoslav state.[20]

American officials arrived at the same conclusion as the British. In January 1943, Blair Bolls wrote that the Serbian National Committee in the United States was spreading "inflammatory propaganda, not against the obvious enemies of Yugoslavia, the Germans and the Italians, but against Croatians, Catholic constituents, together with the Orthodox Serbs, of the Kingdom of Yugoslavia, and against Croatian Americans." Bolls wrote that Fotic, as "a man of the ruling Serbian nation," used his position as emissary for a "campaign against... the Croatians and the Slovenians."[21] An official U.S. government communication on June 1, 1943, stated that "the American government has been observing the politics of the *Americki Srbobran* with some concern. Its wild attacks against all population of Croatian descent and their clergy, its strongly anti-Catholic articles and concealed efforts at defending the quisling Nedic... serve as an adjunct to the Nazi campaign of racial intolerance and hatred, and are detrimental to the American war efforts."[22]

One should recall that Jasenovac—at that time—was never cited separately. Moreover, it is not even mentioned as a prominent execution site. This role was to be assigned to it only after the war. Then the monstrosity in the description of the methods used in executing those crimes would be replaced progressively by the no less monstrous multiplication of the numbers of victims at Jasenovac alone. The reasons for this are rather transparent. Instead of establishing the total number of war victims in all the camps and execution sites throughout Yugoslavia, the crimes were to be centered in only one place.

It is interesting to look at what the purported letter by Grisogono to Archbishop Stepinac says about Jasenovac. In that missive, Ustasha crimes are generally presented in a way that Grisogono himself designated as a recourse to "lies and falsification." Therefore, it can rightly be assumed that the letter's composers described the suffering in the camps in the same way. This is what was said in the document concerning the camps: "The fear

of being taken to the camp in which thousands of Serbs have been killed or left to die of maltreatment, hunger and cold, is horrible. The Germans speak of a certain camp in Lika in which there were thousands of Serbs, but when they got there, they found it empty, drenched with blood... There, they say, a Serbian bishop was executed. In the camp of Jasenovac, thousands of Serbs are tortured and have disappeared. They have been taken there in this terrible winter weather to stay in pseudo-barracks (and have been treated worse than) Gypsies..."[23]

Consequently, when this apocryphal document mentions Ustasha crimes in the first year of the war—when they were the most extreme and most brutal—Jasenovac is mentioned after a camp in Lika. And Jasenovac is given no special significance, neither with respect to numbers, nor the methods used there. The fact that Jasenovac was no more notorious than the other camps throughout Yugoslavia can be concluded from other war documents, as well as from the alleged letter of Grisogono. In one of the copies of that letter, which is quoted by A. Miletic in his book on Jasenovac, the camp in Jasenovac is not even mentioned. Instead, the writer merely described in general the "camps in which thousands of Serbs have been killed." The camp in Lika was discovered, not by the Germans, but by the Italians, and there "the corpses were thrown into an underground stream near the village of Jadnikovo" and it is "estimated that about 80,000 Serbs were *slain here.*" To which is immediately added: "In the history of Europe there have been no such cases."[24]

It was to be only in later reports that Jasenovac would be assigned the significance of an example. It would be interesting to substantiate by whom, when, and for what purpose this was done. One can suppose that a part was played in this phenomenon by the patriarchal folklore tradition of reducing historical occurrences to mythical legends, but it is more probable that in the formation of the Jasenovac deception, a certain common purpose was found, or at least a silent concordance of political interests.

It is significant that even those from the ranks of the exiled royal government—who formulated their politics upon the mag-

nification of wartime horrors for the sake of revenge upon Croatians—did not argue in particular with Jasenovac, nor was Jasenovac the object of particular attention by the Commission for the Investigation of War Crimes at the Presidium of the Ministerial Council which worked until 1942 when its report was released to the Yugoslav State Commission in Belgrade.[25]

In its report on November 15, 1945, the Republic Commission of Croatia for the Establishment of Crimes of the occupiers and their collaborators stated that because of a lack of statistical data "it is, therefore, not possible to answer the question as to how many victims lost their lives in Jasenovac." Based on accounts by surviving witnesses, the Commission concluded that the most intense Ustasha terror was in 1941 and 1942, while "all of 1943 and half of 1944 experienced a relative lull," but that at the end of the war there were more mass liquidations. In all, it mentions "some 50 mass crimes" in which, according to the accounts of the witnesses, in some cases as many as 800 prisoners were killed. Some witnesses reported that there were as many as 3,000 killed, while some mentioned numbers as high as 7,000 and 8,000 victims. Adding to these mass liquidations the internees who were executed individually, "one arrives at a number between 500,000 and 600,000."[26] At the same time, the Regional Commission for Vojvodina claimed 800,000 deaths in Jasenovac,[27] while the Federal State Commission reported that in Jasenovac "more than 600,000 victims were slain."[28]

In the report by the Reparation Commission of the FNRJ (Federative People's Republic of Yugoslavia) on December 13, 1945, at the Allied Conference for Reparation in Paris, it is stated that the total number of Yugoslav losses amounts to 1,706,000 victims. In the report of the Federal State Commission for the Establishment of War Crimes, issued on December 26, 1945, at the International Military Court in Nuremberg, signed by Dr. Dusan Nedeljkovic, president of the State Commission, and Dr. Ivan Grgic, secretary, Jasenovac is still listed third in order among the largest concentration camps in occupied Yugoslavia.[29] Later, not only would all the other camps be forgotten but so would all

the various crimes committed in them.

Due to the lack of scholarly study, and because of the neglect in publishing official statistical data concerning war victims as a whole, it has come about that in encyclopedia articles about the Jasenovac concentration camp, differing data are presented, even within the same encyclopedia editions. Thus in the general Encyclopedia of the Lexicographical Institute, 1959, one can read under "Jasenovac" that "about 500-600,000 were killed" in this camp,[30] but also that "about 350,000 persons suffered" (under "Concentration Camps," in the same edition).[31] Under "Jasenovac" in the 1978 edition, "several hundreds of thousands of Serbs, Croatians, Jews and Gypsies were killed."[32] In the encyclopedia *Iugoslavica*, it is stated that "the exact number can not be confirmed," but according to estimates based on the reports of survivors and on confessions, "the number of victims exceeds 700,000."[33] This number is mentioned in the Lexicon of the Miroslav Krleza Institute.[34] The *Small Prosveta's Enciklopedia* states that in Jasenovac "several hundred thousand patriots were killed."[35] According to the *Military Encyclopedia,* in Jasenovac "about 600,000" and in Stara Gradiska, 75,000 were killed.[36]

In some of the above-mentioned encyclopedia accounts there are reservations regarding the accuracy of the numbers or qualifying remarks that the numbers are unconfirmed. However, that makes it all the more interesting to ask why the numbers of victims at Jasenovac continued to multiply and how it acquired the significance of the central historical event in Croatian-Serbian relations during the period of the occupation and revolution.

In 1980 the question regarding the war victims, particularly and foremost those of Jasenovac, was to serve as the basis for literary, dramatic, and even cinematographic creations about the war in which the Croatian Catholic and Moslem peoples are given homicidal roles. At the same time, the whole problem of the war victims was almost completely reduced to Jasenovac in the form of a progressive exaggeration of the number of Serbs slain. The obvious goal was the establishment of hitherto "unexpiated" special responsibility of the Croatian nation.

In the field of historiography, Vladimir Dedijer would offer particular support toward that end with his *New Contributions.* Striving, as always, to maintain the appearance of scholarly objectivity, he announced in 1981 that "the exact number of Serbs killed in Croatia has not yet been established. Some authors claim that the number of victims amounts to 200,000, while others assess the loss to be at least 600,000."[37] However, in the same book Dedijer forgets those uncertainties regarding the total Serbian losses in the NDH, and writes that through the Jasenovac camp "there passed over a million people, while between 480,000 and 800,000 were killed."[38] Here it does not dawn on him to consider the logic of a claim that far more people were slain in Jasenovac than in all of the NDH. Rather, he quotes "the historian Miletic," in order to impute a scholarly significance to his own historical views.

It is precisely this kind of uncritical approach to the issue of war victims that served Dedijer both for his intemperate and untenable judgments concerning Croatia's Catholic Church, Archbishop A. Stepinac, and the role of Catholicism and the Vatican in general. Because Pope Pius XII had an audience with Ante Pavelic, there was created a "justified suspicion that a collusion existed between certain circles in the Vatican and Archbishop Stepinac and Ante Pavelic."[39] Moreover, Archbishop Stepinac "was his [Pavelic's] spiritual father, as well as his political mentor."[40] Without much regard for accuracy, Dedijer stamps Archbishop Stepinac as a representative of "extreme Croatian nationalism," who does not shrink from "giving the initiative for war crimes." And Dedijer juxtaposes Stepinac, not Ante Pavelic, with Draza Mihailovic when he says of Mihailovic that "he acquired the reputation of the ideological originator and executor of mass crimes against the Croatians and Moslems."[41]

It remains unanswered whether Dedijer, by this comparison, wished to attribute a higher level of guilt to the Croatians and the Catholic Church or to shed doubt upon the crimes of D. Mihailovic precisely by comparing him to Stepinac. Or was he striving for both effects? Dedijer's own letter to Pope John Paul II

at the end of 1986 protests the fact that the Pope, supposedly "according to the advice of the local Catholic hierarchy," refused to visit Jasenovac in which "the crime of genocide was committed" and "several hundreds of thousands of Serbs were slain" while "several Catholic priests were in command at this camp for mass extermination."[42]

In the beginning of 1983, Vuk Draskovic, a young Communist Serb, explained that he wrote his novel *The Knife*, because "about 800,000 Serbs were killed in Jasenovac at the hands of their brothers," while "those who wish that this soon be forgotten are prepared to begin it again."[43] Apart from Jasenovac, Draskovic also mentioned Borac, Gracanica, and Kula Fazlagica, which as a "fortresses of hatred and evil... and by brutality and cruelty of crimes against Serbs... are not in the least secondary to Jasenovac, Jadovno, Gradiska, and other Ustasha slaughterhouses." He added that "the chief, and often exclusive executors of the crimes were Ustashas of the Moslem faith—not of the Moslem nationality, because those murderers presented themselves as 'the flower of Croatiandom,'or as Moslem Croatians."[44]

In the spring of 1983, Terzic's two-volume book, *The Fall of the Kingdom of Yugoslavia in 1941*, appeared—a full two decades after that first controversial edition. The work was not merely a repetition of old viewpoints, but brought them to their ultimate conclusions. In his youth, Terzic was a member of the "Royal Guard," an officer in the April 1941 War, then one of the highest ranking generals in the NOR and, finally, a chief of the Military Historical Institute for many years. He based his historical judgments on a broad life experience and on knowledge of archival documents which in part are not accessible to many historians. Terzic reached the profound conclusion that only through a genuine familiarity with all the recent history of Croatia can one "more easily get at Jasenovac where, in that one place alone (according to the latest statistics) there were killed (between 1941 and 1945) at least a million Serbs, not counting victims of other nationalities."[45]

The tragedy of Jasenovac is that many innocent people were

brutally killed. But Terzic and others are more interested in distorting the historical record for their own political advantage than simply confronting the historical truth — which is awful enough without any distortions. As we can see in the above quotation, Terzic is only concerned with the Serbian victims of Jasenovac. By exaggerating the number of Serbs victims Terzic blasphemes the memories of other victims, particularly the Jews who suffered so horribly at the camp.

Thus the problem of the war victims was explicitly established as the central issue in recent Croatian history, and the Jasenovac casualty figure was elevated to "at least a million." This was done in the spirit of Lenin's message to historians: "Truth, truth, and only truth; otherwise a bloody punishment threatens." As Terzic himself explains, "deception and self-deception in dealing with the past will later be paid for dearly." In his own words, "I have presented the crucial characteristics of the April 1941 War based on years of research by scholarly methods and a critical approach to all events and occurrences." And Milo Gligorijevic concludes that Terzic presented "completely new historical facts, regardless of whom they might offend, now or in the future."[46]

Historical facts are flexible material. All manner of things can be constructed from them, just as every architect or mason can turn a pile of bricks into differing structures, from majestic edifices to ugly shacks. Aristotle taught that a correct principle can be realized only by a reasonable, wise, and noble person, whose judgment is in harmony with, and not in opposition to, the factual world. The entire problem of all historical disagreements boils down to how prepared one is to respect the individuality of others. Is one willing to recognize—alongside one's own truth—the truths of one's neighbors? History is either an uncertain battle for the imposition of one's own truth, or the attainment of a more-or-less harmonious coexistence.

One wonders whether those who multiply war victims are aware of the consequences of such actions. There are both the deceivers and the deceived. And there is also a third group, the historically powerless or ignorant.

In order to see all the dimensions of the Jasenovac controversy, it is necessary to remember the relentless frenzy of publicity that accompanied it. In discussing the literature of Bosnia-Hercegovina, the Serbian writer Vojislav Lubarda said: "In Jasenovac, on Mt. Velebit, in Hercegovinian pits, and in hollows all over the Ustasha Croatia, there were killed, mostly slaughtered... more than a million Serbs and almost all the Jews from the territories of Bosnia-Hercegovina and Croatia."[47]

In 1984, when the creation of the Jasenovac deception was complete, Miodrag Dukic wrote an article in the newsletter of the Serbian Orthodox Church, *Pravoslavlje*. The distortion of the historical record served the writer as a sacrosanct argument under the appropriate heading "Saint Sava—Determinant of the Serbian Nation."[48] Dukic claims that Jasenovac stands out as the greatest crime which "in relentless attempts at genocide" has been brought about against the Serbian nation by "papist and caesarist appetites on one hand and Moslem imperialism on the other."[49] According to Dukic, "a million Serbian men, women, and children were tortured and killed" at Jasenovac.[50]

The purpose of the Jasenovac distortion is to create the "historical guilt" of the Croatian nation. Because of this, it was not enough simply to raise the number of victims to one million. Even more ardent mythmakers raised it to 1,200,000 "Serbs only, who were killed in Jasenovac," as it has been read and heard at various conferences.[51]

However, this still did not suffice.

When Vuk Draskovic first publicized the number of 800,000 victims at Jasenovac, I was inclined to believe he had just been misled but, in 1985, he published his new "discovery" that the total number of Serbian victims in Croatia and Bosnia-Hercegovina "is at least one and a half million, and perhaps even more than two million slain."[52]

The above-mentioned book of A. Miletic on Jasenovac states that the estimates range from 480,000 to 900,000, and even over a million, but in most cases remain at about 700,000.[53] Each of these figures has since been taken as evidence, not only by various

journalists and propagandists, but also by those such as V. Dedijer or J. Sasic to develop their own theses. In so doing, they pay little heed to the fact that in dispensing these figures Miletic himself offered qualifications. On the next page he writes: "Therefore, the total number of those killed in the territory of the so-called NDH according to various sources, can, for now, be approximated only at several hundred thousand."[54] But that doesn't keep him from using the initial figure just a few pages later, at the end of his introduction to the documentation: "In the Ustasha concentration camp of Jasenovac-Stara Gradiska, from August 1941 to May 1945, according to an estimate, several hundred thousand men, women and children were killed."[55]

In a footnote Miletic emphasizes the statistical data concerning the real number of Serbs in the territory of the NDH before and after the war. These data indicate that some 275,000 to 356,000 Serbs were killed total, and that my estimate of about 50,000 slain in all the camps in Croatia is supported by the Serbian writer Dr. Bogoljub Kocovic in his book.[56] Despite all of this, Miletic does not reach the obvious conclusions. His enumeration of completely unreliable figures from all kinds of sources—personal views, propaganda statements, and biased superficial reports—is taken by Dedijer, Sasic, and their followers as facts confirmed through long years of investigation. Virtually all of those who accept such bogus calculations, because they are deluded by them or for hypocritical ill-intentioned reasons, regularly appeal to their authenticity. Thus, for example, Milan Basta speaks out on behalf of objectivity in the study of history and against "nationalists on both sides" who "falsify the facts according to their own needs." However, for him the problem lies with those who claim that "some tens of thousands of people" were liquidated, while they seem to forget "that only in Jasenovac more than 700,000 lives were destroyed." For greater credibility he adds that "there were more than 100,000 Croatians" among them, a figure which is likewise completely arbitrary.

Basta forgot what he once wrote on that topic himself. Twenty years earlier when memories of the war were much fresher,

Jasenovac was just one in a list of camps as far as Basta was concerned. In 1963 he wrote in his book that in the NDH, "following the pattern of the concentration camps in Germany, horrible death camps were organized: in Slano on the island of Pag, in Jadovno on Mt. Velebit, Jasenovac, Stara Gradiska, Lepoglava, Sisak, Slavonska Pozega, Koprivnica, Dakovo, Zemun, numerous torture sites and ditches in Bosnia, Hercegovina, Lika, Banija, Kordun, Dalmatia and other areas. In the territory of the NDH, about 800,000 people were killed."[57] After two decades of serious research in the field, Basta offers no foundation for the inflated numbers, and his claims remain the most eloquent examples of even those of former participants in the NOR yielding to the Jasenovac deception.

There is no essential difference between those who, presenting allegedly official statistics, reproach all who reportedly or actually falsify them, and those who deliberately multiply the numbers of war victims. This was made particularly clear by the repetition of approximately the same figures in special Tanjug news agency editions. On the occasion of the trial of Andrija Artukovic, in its "Testimonies" series, Tanjug published (in 1985, in English) a pamphlet by Branimir Stanojevic, entitled "Under the Wings of Fascism."[58] Stanojevic writes: "Jasenovac can be singled out, if for no other reason than for the fact that in that one place between 700,000 and 900,000 men, women and children were killed or driven to death in the most bestial ways."[59] Yet, just as all others before him (similar, however different they may appear), the Tanjug writer was unable to escape the illogic of his own assertions concerning the total number of victims in the NDH and in Jasenovac.

Among the ranks of the Croatian emigrant population abroad, data were presented about the "crimes" of the "Serbian" opponents of the NDH ("Chetniks" and "Communists") against the Croatian, Catholic, and Moslem populations. And just as the multiplication of Serbian war victims led to the Jasenovac controversy and a total number larger than the entire Serbian population of the NDH, similarly, the exaggeration of Croatian victims led to

the so-called *Bleiburg myth*.[60] According to this argument: "hundreds of thousands of Croatians were killed" who were being extradited by the English allies of the Yugoslav Partisan Army. Like the Jasenovac controversy, the controversy of the "Bleiburg tragedy" or the "Croatian super-Katyn" is supported by some foreign journalists and historians in addition to those from among the emigrant circles.

In the summer of 1942, the Ministry of Foreign Affairs of the NDH printed its so-called *Gray Book* (in Croatian, German, Italian, and French) on the "seditious atrocities and destruction in the NDH."[61] "Based on material evidence," this publication claimed "tens of thousands of Croatians lost their lives" as early as 1941. "The atrocities by Chetnik and Communist bandits" were very similar to the reports of Ustasha horrors. (Apart from the slaughter, incineration of villages, and all manner of brutality, only the street sale of eyes was missing.)[62]

In 1955 "the news from the homeland" reached Ivan Mestrovic that after the Bleiburg events, "the Serbs, disguised in the mask of Communism... had killed 300,000 Croatian soldiers" near Dravograd, Maribor, and in other places (in Samobor—6,000; in the forest of Macelj—30,000; in Sarajevo—15,000). "At various locations in Slavonia, over 100,000 civilians were killed." So it "can be concluded that the Serbs, after the collapse of the monarchy and during the war, killed 500,000 Croatians—more than the Turks did in 100 years."[63]

The German historian Rudolf Kiszling, in his book about the Croatians, wrote in 1956 that the number of "slain Croatian officers and soldiers, after their surrender at Bleiburg, amounted to 100,000 to 150,000 people."[64] Two other German historians, L. Hory and M. Broszat, relate in their 1964 book that in the middle of May 1945, "more than 100,000 Croatian soldiers" were handed over to the Partizans, and more than 10,000 met their deaths.[65]

The English historian, Lord Nicholas Bethell, in his 1974 book on the extradition of the Cossacks to the Russians, writes that in Bleiburg there were two groups of Croatian soldiers, a total of

some 200,000 accompanying about 500,000 civilians. According to the British whom Bethell used as a source, "the conditions of the extradition were relatively fair," but he did not know their fate, and "wouldn't be surprised if they were killed." But Bethell brings in an account "from a strongly anti-Titoist source," according to which "tens of thousands of Croatians" were grouped into many columns and only "a very few" of them survived the "death marches."[66]

The French writer, Andre Ott claims in his book on Serbian-Croatian relations that after the war "more than 100,000 former Croatian fighters were put to death."[67]

A separate collection of papers concerning *The Bleiburg Tragedy* published in 1976 states that the "exact number of those who found themselves in this tragic exodus will never be established," but probably "that number was between 300 and 400 thousand people, partly soldiers, partly civilians." In their foreword, V. Nikolic and F. Nevistic conclude that after Bleiburg there were "200,000 Croatian victims."[68] The same number is mentioned by Bruno Busic for the victims "on the Slovenian-Austrian border" and on "death marches." Additionally, Busic cites M. Djilas as saying that "they had to die in order for Yugoslavia to exist."[69] Djilas mentioned the figure of 30,000 victims at Bleiburg. My own studies indicate some 35,000 to 40,000.

Colonel Josip Aleksic, a member of the main headquarters of the NDH, cites the estimates that "over 300,000 refugees left the homeland," and that in the mass of refugees near Bleiburg, there were 200,000 people.[70] At the end, referring to the book *Kriegsschauplatz Kroatien* by Colonel F. Schraml, he quotes the estimates of Vice Admiral Joachim Lietzmann that after Bleiburg "no fewer than 150,000 victims fell at one blow," that "in the course of the war about 300,000 Croatian fighters fell" and "apart from that, about half a million Croatian citizens were massacred."[71] According to these quite arbitrary numbers, the total of Croatian victims should amount to about 950,000 people.

In the same book, J. Petricevic agrees with the estimate that in

May of 1945, about 500,000 "Croatians left their homeland," of which number 200,000 to 300,000 were "extradited to Yugoslavia and killed," while about 200,000 were scattered all over the world.[72]

In similar assessments, George J. Prpic, an American professor of Croatian descent, went even further in his examination of the *Tragedies and Migrations in Croatian History* from the Turkish conquest to the present day. He writes of the fate of the Bleiburg prisoners as an example of "horrifying genocide" in "the bloodiest orgy in the history of the Balkans." He concludes: "Of all the tragedies, Bleiburg was the worst in Croatian history. The result of that tragedy was the death and exodus of over a million men, women and children."[73]

Ante Beljo, in his bilingual (Croatian-English)[74] "documented analysis" estimates that "the columns of Croatian military and civilians" on the border of Yugoslavia and Austria numbered more than 300,000 people.[75] However, he also brings in all the various hitherto mentioned data and opinions from which it clearly follows that even the participants themselves give quite contradictory numbers. For instance, D. Crljen states that a Croatian delegation (in which he took part, along with Generals Herencic and Servatzy) submitted a request to the English general T. P. Scott on behalf of the head of the NDH and the "half million Croatians" who find themselves "in flight from Bolshevism" that they might surrender and "place themselves under the protection of the Western Allies."[76] Colonel Milan Basta, a participant in the same negotiations, quotes D. Crljen as having reproached the English commander "that history would never forgive him if he were to allow 300,000 Croatians to fall into the hands of their enemy that would annihilate them."[77]

M. Basta writes that near Bleiburg about 100,000 soldiers of the NDH and "many civilians as well" fell into the hands of the Yugoslav Partisans.[78] Basta's commanding general, Kosta Nadj, states that in that area, 105,000 prisoners fell into the hands of the Yugoslav Army, of whom about 40,000 were members of the German Army, about 60,000 members of the army of the NDH,

and 5,000 Chetniks from Montenegro and Hercegovina. At the same time, Nadj asserts that from May 9, 1945, when World War II ended in Europe, to May 15, when the military operations of the Yugoslav Army at the upper branch of the Drava River came to an end, of the "enemy formations, over 130,000 soldiers strong," a total of "25,000 enemy soldiers died."[79]

The settling of wartime accounts was not yet over. Nor was the number of victims complete. West of Bleiburg, in the area of Klagenfurt in a camp near Viktring, there were about 70,000 soldiers of Yugoslav quisling military organizations (Croatian Ustashas and Domobrans, Slovenian Domobrans, Serbian and Montenegrin Chetniks) who had penetrated through to the West and surrendered into English custody.[80] By a decision on the part of the English, they were extradited for the most part to the Yugoslav Army as late as the end of May, when about 45,000 Cossacks, Ukrainians, and Russian soldiers of Gen. Andrei Vlasov were handed over to the Soviet Army.

The members of the Yugoslav quisling formations, having been led to believe that they were being sent to Italy, were transported by trucks and trains to Yugoslavia between May 23 and May 31. By June 1, due to protests on the part of the Red Cross, the decision of the Allied Supreme Command concerning the curtailment of forced mass repatriation was finally implemented. The total number of prisoners who had been extradited amounted to 26,339, out of whom there were 12,196 Croatians, 8,263 Slovenians, 5,480 Serbs, and 400 Montenegrins. Meanwhile, to a lesser extent, the extraditions were still continuing, and the official allied reports state that by May 29, 29,972 prisoners had been repatriated.[81]

This forcible repatriation, as well as the fatal destiny of the prisoners, long remained an unknown fact. Perhaps it was not considered to be of significance despite the personal testimonies of the survivors. Only after Alexander Solzhenitsyn pointed out in his famous *Gulag Archipelago* that one must not forget the responsibility of the Allies for the fate of those returned by force to the Soviet Union and Yugoslavia, did this event draw the interest of

historians and the public. British historian Lord Bethell's *The Last Secret of the Second World War* provoked lively discussions and new historical research.[82] In the works of the English historian (of Russian descent) Count Nikolai Tolstoy[83] and then in the contributions of participants at the first international symposium on forcible repatriation after the Second World War at Oxford University (in March 1987), it was asserted that the prisoners who were surrendered to the Yugoslav and Soviet authorities "proceeded to certain death."[84]

Nikolai Tolstoy starts with those documents which contain the exact numbers of prisoners extradited after the end of the war, convinced that all of them—about 30,000—met with a tragic fate.[85] Tolstoy concludes that in mid-May of 1945, "about 200,000 Croatians" were forced by the British into Yugoslav Partisan hands.[86] In his opinion, the British soldiers witnessed the machine gun execution of a great many of them on the very spot where they were handed over.[87] And the "majority of them" were "killed without any kind of differentiation under the barbaric conditions of Tito's infamous death marches," which extended over weeks and months throughout the whole of Yugoslavia.[88]

Concerning the extradition of the soldiers and civilian refugees near Bleiburg on May 15 to the Yugoslav Army, Tolstoy finds that this was done in accordance with the British policy toward Tito's government, as well as with the actual conditions on the battlefield. At the time of the crisis on the battlefield the day before, "when it seemed that the sudden congregation of about half a million Croatians" would exceed the strength and capability of the Fifth British Corps, Field Marshal Alexander approved their repatriation. But as soon as that "danger was over," by May 17, he revoked those instructions, ordering that no more Yugoslav citizens be extradited.[89]

However, despite such an order, some 30,000 Croatians, Slovenians, and Serbs were forcibly repatriated to Yugoslavia. The British government argued that the members of the Chetnik formations should be considered prisoners of war but were "not to be extradited to Tito's forces." They also held that members of

"the Croatian Army are in a quite different position from the Chetniks." For, "although the Croatian puppet state was established by the Germans... the Croatian formations were, in fact, a regular army... under German direction." Because of that, "it would have been far less justified than in the case of the Chetniks, that they should be considered irregular forces which participated in the Yugoslav civil war." Besides that, their extradition would "certainly please Tito, proving to him in a specific way that we are willing to relate to him as a real and responsible ally." If, however, "the government of the USA should be unwilling to accept such a drastic action, which could, as Field Marshal Alexander warns, be fatal to the Croatian military formations," then the Foreign Office consented to "deal with the Croatians in the same way as with the Chetniks."[90]

The Americans did not agree with the English recommendations. Despite the protests of the American advisor in Field Marshal Alexander's Command, A. Kirk, nearly all of those who still remained in the British zone in Austria were extradited. Only a small number was later removed to Italy.[91]

While near Bleiburg, fate had overtaken the majority of the soldiers and refugees from the NDH, the majority of those who were withdrawing from Slovenia met their fate later in the area of Viktring near Klagenfurt. There, about 17,000 Slovenians were found (12,000 to 13,000 Home Guards, about 3,000 women, and 550 children), the majority of whom were extradited despite the fact that Dr. Joze Basaj, president of the Slovenian National Council, and General Franz Krenner, commander of the Domobrans, protested to the British general, C. Keightley, who denied the report that the transports were directed toward Yugoslavia rather than Italy and warned that any resistance would be subdued by force. The extradited group of about 30,000 prisoners apparently consisted of Slovenian Domobrans (about 12,000), no fewer Croatian Ustashas and Domobrans (12,000 to 15,000), and somewhat more than 5,000 Serbian and Montenegrin Chetniks.[92]

In the Slovenian *White Book* (printed in 1985 in the United

States) the following data are presented: At the end of the war, 30,000 Slovenians fled to Austria and 20,000 to Italy; of the extradited Slovenian Domobrans (who were killed along with others at Kocevski Rog, near Teharje, in Podutik, and other places) only three survived; the names of 13,222 "victims of the revolution" are presented (of whom 3,103 were "slain civilians," 1,072 were "fallen fighters," and 9,047 were "extradited fighters"), while "within Slovenia, fewer than a thousand Italians and as many Germans were killed."[93]

In the summer of 1945, after the forcible repatriation from Austria of the above-mentioned groups, in the Allied refugee camps there were left about 70,000 more Yugoslav citizens with the status of "expatriates unable to return to their country."[94] Individual "war criminals" were extradited from this group upon the specific requests of the Yugoslav government.

The Serb emigré historian Borivoje Karapandzic, in a book based on non-Yugoslav archival documentation and printed in Serbian and English, contends that after the war 250,000 military and civilian refugees from Serbia, Croatia, and Slovenia were killed.[95] He, too, like all the aforementioned supporters of the Bleiburg controversy, points the accusatory finger at the World War II Allies and particularly at the English for their role in the tragedy.

This discussion of the war victims is by no means complete. However, it is certainly sufficient to convince one of the extreme seriousness and complexity of the whole problem. These examples of the darkest sides of our recent national and general history confirm the permanent presence of a historical evil and of an irremediable tendency toward exaggeration by humankind in general, but particularly in matters concerning criminal acts and atrocities. This consideration led the American historian Barbara Tuchman to posit the following: "It is a fact that a sad event made public increases its apparent scope from five to ten times (or by however much the reader wishes to add)."[96] Let us add: by however much is needed to serve the purposes of some new devotee or interpreter of old events.

As proof of this law, one can bring in a multitude of historical examples, but for the sake of comparison, two examples from Tuchman's account of fourteenth-century Europe will suffice. Concerning mercenary soldiers in the Guegler War in Alsace in 1375, various estimates set the number at 50,000, 60,000, and even 100,000, while there were probably only about 10,000 of them.[97] In the chaos of war, unrestrained bloodshed and crimes were perpetrated equally by both sides, not only by the enemy but by the home troops as well, as punishment for disobedience, but also as deliberate provocation. ("And those that fled were killed, while those that remained were burned to death." A ballad testifies to the "chivalric" behavior of the victorious knights: "... You will not escape from me! / I shall strike you down, slaughter you and reduce you to ashes.")[98]

A second and in many ways more didactic example is found in the battle at Nikopolje in 1396, in which the Turkish sultan Bayazid I crushed the army of the Hungarian-Croatian King Sigismund, who was being aided by the Crusaders' Army from all of Christian Europe. It is written that this army numbered 100,000 soldiers, while some reports estimate as many as 400,000. Chroniclers from both sides regularly wrote that the enemy outnumbered them twofold. Only one young participant in the battle, the Bavarian nobleman Schiltberger, presented upon his return from thirty years' captivity what best approximated the truth: that the entire Christian Army may have had 16,000 warriors. The nineteenth-century historians have concluded that 7,500 to 9,000 is a realistic estimate for the army of the Crusaders, while the Turks had between 12,000 and 20,000 soldiers.[99] "The flower of European nobility" fell in that war. Only a few managed to survive, and only at the price of a copious ransom. The treatment of the prisoners, however, could not have been more bestial. In front of the sultan, the executioners decapitated, castrated, or butchered the prisoners. However, while more moderate assessments of that slaughter range from 300 to 3,000 slain, there are also exaggerations which multiply the already exaggerated number tenfold.[100]

This battle sealed the fate of the Byzantine Empire. Christian Europe was defeated as it offered assistance to the Emperor Manuel II Paleologue, and Turkish power over the Balkans was ensured for the next five centuries, which in its way was also fateful for the history of the rest of Europe. But one more significant aspect is worth mentioning. It is the accepted view that the Turks won this victory because the Christian Army was trapped between the Turkish Army and a Serbian Army of 1,500 cavalry, which tipped the scales at the decisive moment. "As a vassal of the Sultan, the Serbian Despot, Stephen Lazarevich, might have chosen passive neutrality like the Bulgarians on whose soil the struggle was being fought, but he hated the Hungarians more than the Turks, and chose active fidelity to his Moslem overlord. His intervention was decisive. Sigismund's forces were overwhelmed."[101]

Is it not curious why the Bulgarians, who were in the same vassalic relationship toward the Sultan as the Serbs, and territorially even more firmly controlled, remained neutral, and why the Serbs wagered on the side of the Turks? Was it just the opposition of the Serbs and Bulgarians? Or does the fact that Stevan Lazarevich "hated the Hungarians more than the Turks" have its deeper cause in distrust of Catholic Europe, that is to say, in the oppositions of Orthodoxy and Catholicism, which the Islamic Ottoman conquerors would use to their advantage at the expense of both? And have the evil roots of that discord not spread out into the tragic events of later history?

To get back to the original point of the unreliability of historical testimonies, let me add a few more observations.

It is an ancient Thucydidian insight that to those who wield the power of the state, nothing is unlawful which they deem useful, even deceit. But every historical untruth, delusion, or lie—however purposeful it may momentarily seem—not only resembles a snowball, but inevitably also experiences the same fate. In the beginning, the more it rolls, the bigger it becomes. For a certain period it freezes the truth. But it cannot evade historical inevitability. Sooner or later, the snowball of deceit will catch upon a shard of

truth, or it will gradually thaw and in the end melt in the light of that truth.

That many no doubt well-intentioned people, along with conscious delusionists, have participated in this absurd game with the numbers of war victims is a result of the condition which has led to the reign of either "learned ignorance" (N. Cusanus's *docta ignorantia*) or, as Pierre Vilar would say, "bartering with history." In the first case, presumable truths formed as accidental or conscious delusions are reiterated. In the latter everyone introduces and elevates his own truths. Vilar reached the conclusion that there are times and circumstances "when anyone may call himself a historian."

All this only proves that everything in history remains essentially the same. Even when everything is changing, nothing is actually lost from the nature of man, nation, or humanity. And if history never repeats itself in exactly the same form, man in his actions does, and far too often. He can never escape his own shadow which binds him by invisible threads to the soil of his own land, even if he is not chasing after false stars, but only striving in a human way after the real but unattainable.

II. HISTORICAL SUBSTANCE, AND IRRATIONALITY

Because history is experience, it is the present. What the historian does when he imagines that only he knows the events of the past as they actually occurred in reality is, indeed, merely the manifestation of his present consciousness.

—Michael B. Oakeshott

1. Mythomaniacal Tales

To be sure, supporters of the Jasenovac distortions (as well as the Bleiburg distortions) do not belong in the main to the ranks of professional historians. However, by a convergence of circumstances, they have succeeded in persuading the general public that the maintenance of those distortions is important, or more accurately, necessary "for the present as well as for the future." Convincing evidence for this can be seen not only in the unusually ruthless settling of accounts with a historian and revolutionary, as in my case, but also in the branding of others who dared in any way to vent their opposition to those distortions (from writers Vjekoslav Kaleb and Vlado Gotovac to Ivan Supek and General Nikola Kajic), and even by condemning an encyclopedia on Croatian history and culture.[102] Of course, it is significant that the Jasenovac distortion has remained a taboo subject, while all other taboos have been subjected to necessary discussion. Historians as well as others who are familiar with the problem are forced to remain silent, demonstrating by their unspoken exchanges the need for deeper investigations. Meanwhile, an ever increasing circle of people is emerging from scientific and cultural fields who have succumbed to the influence of those distortions, or use particular facts or testimonies outside of their historical contexts. The great humanist poet, Francesco Petrarch, understood that "when gray-haired children begin to vomit their syllogisms, it's time to

run." In our case, we should not run from the battlefield, but at least wait for the "gray-haired children" to be replaced by mature adults, though many of them would no doubt display a similar lack of historical judgment.

One should recall the consequences which have sprung from the tremendously irresponsible and one-sided presentation of "data" concerning the war victims. These consequences are both direct and indirect.

It is not necessary to have any extraordinary wisdom based on historical experience, but only common sense, in order to comprehend the direct effects in the wartime environment. During the Second World War, the factions in the conflict were continually adding fuel to the fire of an already flaming frenzy through their one-sided portrayals and exaggerations. Accusations from both sides necessarily and inevitably deepened the chasm of discord, firing up hatred and spurring the blind instincts of an imperiled people for vengeance. The progenitors and instigators of mutual destruction on both sides have found, both in the past and present, reasons not only for the protection of truly threatened vital national interests, but also, using the excuse of wrongs suffered in that past, the attainment of their long-term plans at the expense of their national and ideological adversaries.

Pavelic's Ustasha movement, as an expression of that faction of Croatian political life which could not see any possibility of coexistence with the Serbs, obtained power in the NDH with the help of the Axis powers and did not shrink from any means under the aegis of Fascist ideology and Hitler's "New European Order" to achieve its final goals of the ethnically homogeneized Croatian state. In furtherance of that goal, the program of the Ustasha movement regarded the Moslems as an integral part of the Croatian nation and ventured to resolve the Serbian question by the radical means of liquidation, expulsion, and conversion as described previously.[103]

The very establishment of the NDH, and especially of a program such as Pavelic's, served as new and final proof to the Chetnik movement and Great Serbian circles that there could be

no reconciliation with the Croats. They, in turn, created their own no-less-radical program which involved purging Serbian lands of non-Serbs, and the formation of a mutual border between a homogeneous Great Serbia and Slovenia, after the defeat of Hitler by the allied coalition.[104]

Under such circumstances, which had arisen as much from the impact of World War II as from a clash between two unreconcilable nation-states, a situation was created which had many of the characteristics of a fratricidal and religious war. However, the situation also created the possibility for a new Yugoslav union of free and equal peoples.

Although the Croatian people accepted the collapse of monarchist Yugoslavia as the end of evil and an exit from the "prison of nations," the Croats remained for the most part indifferent towards Pavelic's NDH, both because he handed Dalmatia over to Mussolini and because of their disapproval of the Ustasha-Fascist regime. Many Croatians had also some reservation toward the NOP under the leadership of the KPJ (Communist Party of Yugoslavia), fearing that it might lead to the return of a Great Serbian Yugoslavia. They remained reserved until Croatian Communists and other anti-Fascists (mostly leftist radicals) convinced them that this was a movement for a Free Croatia within a new democratic and federative Yugoslavia, in which there would be no subjugation because each nation would be able to determine its own fate. In other words, the chief prerequisite for the winning over of the Croatian people to the side of the revolutionary NOP and the new Yugoslavia was the guarantee against the return of a Great Serbian hegemony. The same was true in Slovenia, Macedonia, and, in part, Montenegro. This fact, however, is inconceivable or unacceptable to all those such as V. Duretic, who deny the existence of the Serb hegemony in the Yugoslavia of the Versailles Treaty, and who maintain that such judgments and such terms were imposed upon the KPJ by the Comintern.

Serbian distrust of all things Croatian was demonstrated by the Serbs' readiness to make pacts with the Italian and German occupiers against the NDH as a Croatian state. They had even more

distrust of the Croatian communists, who stood from the begin-
ning on the side of the Serbs against Ustasha persecutions. To take
but one example: The fate of Marko Oreskovic, killed while car-
rying out negotiations with the Serbian rebellion leadership in the
vicinity of Drvar, would be exalted in a Partisan song. One of the
most convincing testimonies to the grisly dimensions of the hys-
terical anti-Croatian attitude on the part of the Serbian rebels was
left by Rade Zigic. As commissar of the 2nd Army, he testified in
a speech in 1945 in Zagreb's Jelacic Square that "mortal danger
threatened the courageous sons of the Croatian nation" whom the
"KP had sent... among the Serbs in order to organize them... All
of them set out, prepared to meet their doom, and many paid with
their lives simply because they were Croatians... Marko
Oreskovic was bound and imprisoned several times until he was
finally killed by Chetnik criminals. I watched Jakov Blazevic as he
awaited a bullet to the forehead for an entire hour, while I, as a
Serb, was unable to help him. Only when other people confirmed
the fact that he was a Croatian Communist did an enraged mob
set him free."[105]

Among the ranks of the Serbian populace tormented by perse-
cution and ready for revenge, not many people—except, of course,
the few prewar Communists—took a more sober look at things.
Vasilije Gacesa, for one, had the fortitude—even after the Ustasha
slaughter in Glina—to discourage his compatriots from their
revenge by reminding them that they were the cause of their own
suffering by their injudicious actions between the two World Wars
when they desecrated Catholic holidays with gunfire and affronted
Croatian dignity in other ways.[106] While researching that crucial
period, I myself came across other similarly impressive memories
from the first insurgent days and the first Partisan mandates of
such Croatian revolutionaries as Generals Ivan Rukavina, Srecko
Manola, Veco Holjevac, Ivan Sibl, Nikola Kajic, Vlado Mutak,
Janko Bobetko, and others. In the beginning, the lives of some of
them were in danger. And during the war it was at times impossi-
ble to prevent unreasonable measures against innocent individuals
and even entire Croatian settlements.

This legacy of national and religious distrust, which during the war and among the ranks of the NOP was repressed by the revolutionary goals concerning the new union, manifested itself again more strongly at the end of the war. At that time, brute force often took irrational and vengeful actions. In the final operations of the war, many Partisan units did not differentiate between those who had been forced to join the Domobran ranks and the Ustashas. Not only did they not take into consideration the suffering of the supporters, they even ignored that of the active collaborators of the NOP, who allegedly were undercover Ustashas. There is, of course, hardly any documentation on this subject, but the imprints remain in the minds of the population. Such actions took place all over Yugoslavia, and they were often inescapable when the revolution settled its accounts. Only this can clarify the circumstances under which such actions were possible, but it cannot remove their consequences—neither in reality, nor in historical memory. Because of this, the historical responsibility is all the greater to expose the true causes of these events as objectively and comprehensively as possible, and to determine the actual conditions under which they occurred. Every biased approach serves only to deepen heinous old wounds and to produce new misunderstandings.

The extent to which the problem concerning the causes and dimensions of the war victims, particularly at Jasenovac, has become a serious and central issue, can be seen in the fact that it is used as a symbol of our entire contemporary history. This issue has placed on opposing sides not only coincidental associates from the anti-Fascist war, but also prewar Marxists, former like-minded ideologues, revolutionary fighters and collaborators in managing postwar crises. Their descendants, that is, the younger generation in political, academic, and cultural life, also found themselves on opposing sides.

In addition to the causes of that phenomenon, no less interesting are all the socio-historical factors—national, political-ideological, and psychological—that have led us to this situation. Some of these causes were the main reason why we suddenly found that it was not possible to conduct a normal investigation into a factual

historical condition, and why those who advocated debating the historical truth were denounced and forced into silence.

Is the chief role here played by the ideology of a monopolistic and sectist perversion of historical consciousness? Did it choose this direction because of a feeling of omnipotence, or perhaps because of the uncertainty of its own historical position? Was the predominant factor the inter-ethnic (in this case, Serbo-Croatian) relations in the Yugoslav community? If both factors were at work here, as they obviously were, which of them was the more decisive? Wasn't their role on both sides quite different, depending on the subjects and circumstances of particular periods?

There is also the systematic creation of the black legend of historical guilt of the entire Croatian nation. For if the dimensions of the Ustasha crimes are stretched to hundreds of thousands and even millions of victims, and if, by contrast, there are no commensurate crimes on the opposite side, then the responsibility for the crimes does not fall upon a mere handful of Pavelic's fanatical followers, blinded by vengeful impulses, but rather upon the entire Croatian nation. From this point, it follows logically that (implicitly but also explicitly) Croatianness can be equated with Ustashism, which is branded as worse than Fascism or Nazism.

The effect of such a historical approach can be deduced from some more than significant examples from the 1960s when the Jasenovac distortion had not yet even nearly achieved its present dimensions. There were instances in which a child, after a history lesson in primary school, would announce to his parents that he no longer wanted to be a Croatian. The police of a municipality submitted a written report that some young hooligans were summoned for taking part in "singing an Ustasha song—*Our Beautiful Homeland* [the Croatian national anthem]." At the train station and at certain newspaper stands in Sarajevo, the Zagreb *Vjesnik,* the newspaper of the Socialist Alliance of Croatia, was not displayed along with all the other newspapers, but was hidden under the counter, and thus were these "Ustasha" newspapers secretly purchased.[107]

But the legend of the historical guilt of the Croatian nation did

not remain limited only to the primary-school level and within domestic boundaries. Ernst Bloch, the well-known philosopher who enjoyed international recognition as an unorthodox Marxist, referred to Croatians as "Fascists" or at least "semi-Fascists" in discussions on the problems of socialism and in the company of certain philosophers from Yugoslavia. He later repeated the labels in an interview in the German magazine *Spiegel*. There Bloch himself reveals the philosophical connection between historical guilt and the present. Namely, he admits to wondering whether "all Croatians in Croatia are Fascists" (in his opinion, those hundreds of thousands of Croatians temporarily working in Germany certainly are). Because of this, he similarly questions their Marxism because it "could be, after all, merely a cloak."[108] In his late years, Bloch could certainly not have arrived at such conclusions through his own research on the matter, but only on the basis of the opinions of his Yugoslav acquaintances and followers. And while Bloch was written of in Croatia as "the greatest mind of our time," my response to him could not even be published.[109]

The Jasenovac distortion served as the basis for an entire theory of the genocidal character of all Croatians. That myth rejects the fact of the Great Serbian hegemony in the Versailles Treaty-Yugoslavia. And the mythical presentation of history has brought about the accumulation of historical guilt. In addition to Jasenovac, the Croatian nation should be on its knees for all the other historical "evils." We should be on our knees even for what the Vatican and the Comintern did to the South Slavs.

Writing about the creation of a belligerent atmosphere prior to World War I, Stefan Zweig reveals some of its signs which, in his opinion, were characteristic of all countries. Since it is "characteristic of human nature," wrote Zweig, "to be unable to prolong strong passions indefinitely, whether in regard to an individual or a nation," it is therefore necessary "to stimulate them artificially, constantly spurring the emotions." And thus, "that task of keeping the fires blazing must be performed—with either a clear or unclear conscience, sincerely or routinely—by intellectuals, poets, writers, and journalists. It was up to them to beat the drums of

hatred, and they did so exuberantly, until every impartial person's ears were ringing and every heart grew cold." Everywhere "those people, in Germany, in France, in Italy, Russia, and Belgium served obediently" that war psychosis based on someone's great guilt, contributing "thereby, to the mass hysteria and mass hatred which incites war instead of working against it." "The worst of rumors was instantly transformed into truth, the most absurd slander was given credence," and thus, "the fairy tales of extracted eyes and severed hands, which appear in every war, were filling newspaper columns" while "the harmless people who were spreading such lies" were not aware of the devastating consequences. Masses of people, entire cities and settlements "yielded to the ghastly hysteria of hate."[110]

Regarding the behavior of those in the Second World War—even more bloodied by nationalism and ideological fanaticism—we are clearly reminded of the ever-present horrors of war. Foremost among them is the reality of the war victims of Jasenovac, but along with it Banjica (the concentration camp in Serbia) and all the other death sites.

We have seen the methods used by politicians and religious leaders, revolutionaries, scholars, journalists, and movie-makers in transforming rumor into the truth about Jasenovac. However, as an introduction to a discussion of the remaining issues, it is necessary to note the culmination of the Jasenovac distortion as expounded by the book of A. Miletic. How his book fulfilled that objective has for the most part been examined on previous pages, but the task of revealing its full-scale confrontation with historical reality and scholarly data still lies ahead.

The reaction among Yugoslavian historians, judged by their review in their major journal, was exactly what one might have expected. Miomir Dasic, in his review, does not see even those doubts that Miletic himself points out in his book. In Dasic's opinion, "the latest anthropometric research indicates that in Jasenovac, as many as 700,000 people were liquidated." He continues: "virtually all documents state unambiguously that the number of victims is at least ten times greater than that which

Ustasha immigrant organizations launch abroad, distorting historical truth and falsifying documents, in which endeavor they are sometimes aided by the *scientific calculations* and statistics of certain domestic quasi-historians and blind nationalist publicists."

According to Dasic, the "two-volume collection of Dr. Miletic... will be complemented by a third book of documents concerning the Ustasha camps," along with the reprint of *Magnum Crimen* by Viktor Novak, books by Simo Simic on the Vatican and Yugoslavia, and a third book by Jakov Blazevic. Together these works will "provide enough opportunities for the general public to learn more about the horrors of the war and the issue of genocide." However, "the fact that Yugoslavia has no written history of genocide in World War II... is due primarily to the disintegration and excessive decentralization of the scholarly research institutions." For, "the territorialization of scholarly work has caused a territorialization of the historical way of thinking." Therefore, there exists "an increased distancing from broader historical themes such as the history of genocide, which is inescapable and urgently necessary if we are to arrive at the desired synthesis of the history of the nations and nationalities of Yugoslavia."[111]

There you have the "scholarly expert" view of the problem of Jasenovac. If the situation stands thus in the scholarly field, then one cannot expect that it should be any different in schools, in daily newspapers, on radio and television, and in all the mass media which participate in the shaping of public opinion and historical consciousness.

One example of such views can be seen in the newsletter of the League of Socialist Youth of Yugoslavia. There, Miletic's book and the related papers of Jefto Sasic and Radovan Trivuncic, statements by witnesses, and Lordan Zafranovic's films (*The Blood and Ashes of Jasenovac, The Occupation in 26 Pictures*) were used as the basis of an article by Braka Culjic, the creator of "A Dossier of *Mladost*." To Culjic it is peculiar that we are not given the whole truth about Jasenovac, that "most horrible human slaughterhouse of the Second World War." In Jasenovac there is to be found, "all that is negative, pathological, and criminal about

the entire Ustasha movement... surpassing in its perfidy, blood-thirstiness, and repugnance all the most horrible sites of torture in the hell that was Europe under Hitler and Mussolini." And since "history has no recollection of any crime of such magnitude ever having been recorded, the insanity of Jasenovac cannot and must not be forgotten."[112] Culjic rightly asks: Why was there no resistance of the camp inmates? Did the Partisans attempt to liberate the camp? Why is there no list of Jasenovac victims?[113]

It is necessary and possible to answer these questions, but only by recognizing the facts concerning historical reality, such as it was. But similar provocative questions which arouse confusion and feelings of vengeance will remain unanswered as long as mythical images are forged upon rumors and lies are deliberately spread. Culjic chose to overlook my investigations of the false accounts, as he did the similar conclusions of the Serbian scholar, Dr. Bogoljub Kocovic. If Culjic had taken them into account, he never would have been able to come up with his assertions. This youngster, with the help of Sasic, includes all of us who strive toward the truth, among the "forces that wish to reduce the crimes, to falsify and manipulate the facts."[114]

Culjic pays attention to the "main question" of Radovan Trivuncic, who was the director of the Jasenovac Memorial Area: "Who decided in 1946 and 1947 that the buildings of the camp should be torn down?... Who ordered that all traces of the Jasenovac camp be destroyed?"[115]

In these questions and answers, and particularly in their implied accusation, the essence of the Jasenovac controversy is made manifest—how it was created, its factual basis, and finally and above all, its purpose. For in presenting such a question and such an answer before the public, but above all before the state and experts and scholars, R. Trivuncic must have known the game he was already playing. He must have been aware of what he wanted to achieve with this gross deception and building up of the Jasenovac distortion. As the director of the Jasenovac Memorial Area, as an expert personally engaged in research on the subject, and as the author of a published work called *The Drama of*

Jasenovac, Trivuncic could not have been unfamiliar with the following facts:[116]

■ The remains of the Jasenovac camp of the NDH were for the most part destroyed by the Ustashas in the days before their withdrawal.

■ The area was used as a camp under the new authorities, first by the units of the Yugoslav Army and then by the civilian authorities of the NR Croatia.

■ In 1946 and 1947 regardless of whose orders they were and what effect they had, not only were the traces of the Ustasha camp destroyed, but also of that later camp.

If Trivuncic did not come across these facts in archival documents, it is difficult to imagine that he wouldn't have learned them from local residents. I myself am personally familiar with these facts, both from conversations with high-ranking officials, as well as through other witnesses and people who observed the truth about Jasenovac first-hand.

The use of the Jasenovac camp by military units of the NOVJ (People's Liberation Pary of Yugoslavia) and by security agencies of the new revolutionary authorities was no exception. In final operations and immediately after the war, former camps were commonly used (for example, the Banjica camp in Serbia) because new ones were needed for prisoners.

These facts lead to other conclusions which are contrary to those of the propagators of the Jasenovac distortion. They should at least serve to restrain the mythmakers each time a new grave site is discovered. However, they pay no heed to the meaning of the wise doctrines of biblical times: "Condemn not, lest ye be condemned!" For "as you judge, so shall you be judged," and "with the same measure that you mete withal, it shall be measured to you again." And woe unto you Pharisees, scribes, hypocrites, and keepers of false laws "who build sepulchers," while you accept the blameful deeds of your fathers, for if you seek revenge "for bloodshed from the beginning of the world" and "of this generation,"

that cannot be solely by your judgment.[117] And about four centuries before these teachings of Christ, it was none other than Plato who said that those horrors which stem from vengeance "should not be recounted in front of innocent young children, but should be kept secret even if they are true."[118]

The Jasenovac distortion has proven that the Machiavellian effectiveness of great deception does not begin or end with Hitler. Both before and after him, many have discovered that great lies are more effective because the "broader masses" of people, in the "primitive simplicity of their minds," fall prey to big lies sooner than to lesser ones.

2. War and Genocide in Earlier History

And Moab shall be destroyed from being a people...
—(Jerermiah 48:42)

Cut off the sower from Babylon and him that handleth the sickle in time of harvest...
For fear of the oppressing sword they shall turn every one to his people and they shall flee every one to his own land... saith the Lord
—(Jeremiah 50:16)

No historical phenomenon can be understood in itself nor by itself nor taken separately, but only in connection with other phenomena. Wise minds understood this from the very beginning of history. In spite of this wisdom, human judgment remains incapable of removing from historical events the horror of any sort of crime, including that of genocide. In this respect, humanity is caught in a vicious cycle, never relinquishing the hope that it will at some point, by reason or by force, find an escape.

In our age of general spiritual and material progress, humanity has experienced the most widespread and most destructive war devastation with the horrors of the Holocaust and Gulag, Katyn,

the Jasenovac and Banjica camps, the apocalypse of Hiroshima, the plight in Cambodia, Afghanistan, Lebanon, the Palestine-Israeli war and the Iran-Iraq holy war. Yet, as A. Heller rightfully concludes, there are no more nor fewer atrocities going on today than there were, proportionately, in the past.[119] Humanity has not been liberated from the beastly impulses of its nature.

As far back as our historical memory can reach, it reminds us that genocidal crimes were not merely an accompanying phenomenon of total historical upheaval. There are unambiguous testimonies to this in the Bible, that original "Book of the books" of world literature, which has become a treasure and holy scripture of human wisdom because it is one of the first sources of religious and moral doctrine as well as historical experience. The historical value of the Old Testament data consists, above all, in providing a link with the fates of the people of vanished civilizations (from the Phoenicians, Egyptians, Assyro-Babylonians, Sumerians, Hittites, and Persians to the Greeks and Romans). However, it also tells us of the transition from a tribal polydemonism into the national monotheism of Judaism. The New Testament history, bequeathing to us all the wisdom of ancient cultures, leads to the sources of our present-day civilization.

The Old Testament testifies, in quite a dramatic way, how violence, hatred, crime, and revenge are inseparable components in the life of man as an individual and in society. All those instances of violence, from back in primordial times, are just the other side of human existence which arises from love and strives for peaceful coexistence. Because humankind is forever unable to extricate itself from destructive evil, that evil is not merely incidental or secondary, but an essential component of human nature. Even though human life has always been assigned the highest value, the fundamental commandment "Thou shalt not kill"[120] has never become the only determinant of any regime, either human or divine. In Cain's murder of Abel, one finds a proto-pattern of human nature which leads to fratricidal disputes even among the closest of relatives and fellow tribesmen.

Since the times of the first known ethnic state-communities, and especially since the confrontation of the first "chosen people" with the "pagans," and all subsequent "civilized" nations with "barbarians," one sees unceasing wars of conquest and defense throughout history, with both sides attempting genocidal extermination.

Since Abel's case, the blood of the righteous man crieth out for revenge upon the murderer.[121] And while customary law legalized the act of blood revenge,[122] by the principle of "an eye for an eye, a tooth for a tooth," the law endeavored at least to control revenge. The Covenant between Jahweh and his people was founded upon a blood rite so that God himself became the executor or patron of revenge.

Babylon was first "the whip of God" in the execution of the judgment upon Nineveh and Judah, then the hammer used by Jahweh who handed Israel and the neighboring kingdoms to the Babylonian king, Nebuchadnezzar.[123] But when their songs were drowned by cries "by the rivers of Babylon," Babylon felt Jahweh's revenge against its gods, and not a stone of that city remained standing.[124]

Moses was a Hebrew prophet, a leader, "a chief and liberator" of the Israelites, a law-giver who spoke in God's name in order to extricate his people from oppression and to make them the "chosen ones," strengthening the covenant with God.[125] By his prophecy he preserved that covenant, and for the sake of victory over the enemy, he nurtured his nation in a rebellious and belligerent spirit.[126] And he protected it from the wrath of God for the sins they had committed.[127] For the enemies of the chosen people were the enemies of Jahweh, while from the justified hatred of such enemies of God, there arose the necessity of holy war.[128] Thus the Sinai covenant was contracted, not for the sake of peace, but for battle.

God gave his people a homeland, but they had to conquer it.[129] This justifies a war of aggression and makes it holy: all wars of Israel were to be "Jahweh's wars." He was to punish or even uproot and annihilate Canaan, the lands of Kings Sihonand and

Og and others who stood in his people's way.[130] Thus Israel, in both aggressive and defensive wars, acted as the executor of the will of God in history, which induced religious fervor and military heroism, but also mercilessness.[131] Through Moses, Jahweh tells the Israelites how He will "pass through the land of Egypt this night and will smite all the firstborn in the land of Egypt, both man and beast" and punish "all the gods of Egypt."[132]

In those days the aim of most wars was extermination. When the Israelites conquered the people of Amalek, Moses proclaimed Jahweh's decision that He will "utterly put out the remembrance of Amalek from under heaven."[133] And in battles against one's oppressor, it was not enough merely to "strike down the foe, destroy the enemy." Provoked, one sought that "a wild fire devour [all] survivors."[134] Nor were only the warriors destroyed. Both the Israelites and their adversaries slaughtered the populace of the enemy nation and threw them into pits and even cisterns, "man and woman, child and suckling" were eradicated, with "none to remain."[135] The biblical books are full of accounts of real occurrences and prophesies.

In the New Testament, Christ introduced essential changes to the Old Testament understanding of man and nation. Despite the incorporation of all nations of the world into "God's people," discord, hatred, and violence did not vanish. In the beginning violence was aimed against the Christians. Later it persisted, though with a different intensity, both among the Christians themselves, and by Christians against other peoples. And thus the mystery of blind hate and bloodshed continued even after the death of Jesus Christ.

Confronted with the horrors brought about by the application of the law of revenge, the Old Testament arrived at the need to find justness in revenge. Christ's New Testament idea was that violence be tamed by forgiveness and repentance, and that evil be "overcome with good" by the surrender of retribution to the Supreme Judge, God.[136] Today we have reached the insight: that every new act of violence or revenge only provoked future horrors.

If one were to examine today's historical climate, one would be dumfounded by the continual recurrence of human evil. All the development of spiritual and creative forces of human society have not resulted in the removal, or even the decrease, of that evil. Even after the transition from savage prehistory into history, savage violence remains. Nor did the periods of civilization, marked by the exchange of pagan polytheism for pluralistic monotheism, remove barbarian savagery from the Greek and Roman civilizations. In much the same way, the spread of universalistic monotheistic religions like Christianity and Islam, and after them, atheist ideologies (from democracy to totalitarianism) did not contribute to the removal but rather to the increase of crimes, including those of genocide.

The Roman Empire arose from the Hellenic civilization of Europe, which was also influenced by the old civilizations of the Middle East. Of course, its establishment over wide territories was preceded by centuries of conquest with all its characteristic violence and destruction, devastation of cities and annihilation of states, repression of populations and even extermination of entire tribes and peoples. To be sure, history has marked the destruction of Carthage, and to an extent, the life-and-death battles of whole populations and entire peoples, from the Armenians, Pergamum, and Parthia in Asia Minor, through the Illyrian and Celtic tribes in the Balkans and Germanic tribes in Central Europe, to the Gallic and Iberian tribes in the West and the Briton and Scotts in the North. But history was to bestow far more of its attentions to the advantages of the Roman civilization. Usually, it is forgotten that the *Pax Romana* lasted a far shorter time than its despotic encroachment upon the "barbaric" peoples. In the description of the decline of that first empire, too much attention is given to internal oligarchic bickering and too little to the impossibility of the survival of such a tyrannically created, ethnically diverse state organization. Its political vicissitudes are of entirely secondary significance. After the destruction of Roman civilization, barbarian peoples committed not fewer, but scarcely more, crimes (including genocide) than their Roman predecessors.

Every national history is concerned predominantly with the creative side of its people, neglecting or leaving totally in the shadows their side of destruction and despotic assimilation. These histories do not comprehend, or are reluctant to admit that only both together form the whole of a nation. The nonviolent forms of historical life were able to reign only when they relied on the superiority of their power.

Under the eternal law of self-preservation, both in nature and in society, the situation with ethnic groups is no exception. Yet the majority of historical reflections overlook this fact, often for moralistic reasons. This is the reason for the incomplete, and even incorrect views on the causes of individual historical phenomena and events, as well as for the one-sided assessments of their consequences. And it of course produces mistaken judgments of fateful events even in the lives of individual nations and of historical existence in general.

Closing the eyes of reason before the facts of reality occurs, understandably, as a result of moral and ethical principles opposed to all forms of violence and which call for the condemnation of all, and especially genocidal, crimes. Of course, such escapism cannot change or even significantly influence reality. However, a one-sided illumination of historical events—whether it entails the embellishment of the bright side or the exaggeration of the atrocities—will only blur the real causes of historical reality, however crude or frightful they may be. Meanwhile it will not contribute anything to illuminating murky historical horizons, regardless of desires or intentions.

To substantiate these remarks, a few more examples from other historical periods will be useful.

In Croatian national history, for instance, one embarks from the biased assumption that the Croatians, in the period of great migrations of nations, took over the area from the Drava and the Danube rivers to the Adriatic Sea, and that there they began their history by establishing their state. Neglecting the fate of the previous population of the area (more or less Romanized Illyrian and Celtic tribes and Avar conquerors), particular importance is

placed on the peace-loving, nonviolent, and nonaggressive nature of the Croatians. As proof of this, one is told of a treaty which the Croatians made with Pope Agathon (678-681), promising to refrain from attacks upon the neighboring peoples. However, it is often ignored that the main reason for not waging aggressive wars of conquest was the need for Croatia to defend against neighboring peoples, that is, from foreign attackers. War in the Middle Ages had the same genocidal characteristics as in the ancient period or in biblical times. This is evident from examples of hostilities between the Croatians and the Franks at the beginning of the ninth century. Attempting to force their rule upon the Croatians, "the Franks were so brutal that they killed babies and fed them to the dogs." Unable to withstand such barbarity, the Croatians rebelled against the Franks, killing all the Frankish governors. Then in a seven-year-long war they defeated with much difficulty the great Frankish Army, killing "all the Franks and their archon, Kocel."[137]

By means of many similar wars, Charlemagne established the European Roman Empire in the West as a counterbalance to the Byzantine Empire in the East. In the Western Empire the Roman popes wanted to turn Christianity into a secular European, and then a worldwide empire. For the sake of attaining eternal peace on earth, they waged holy wars, not only to annihilate the Moslem "infidels," but also to reunify eastern Christianity with western Christianity. It is important to recall some facts about the Crusades and other religious wars, and particularly about chivalric warfare, and of the genocidal crimes, particularly those against the Jewish people.

A united Europe, civilized by Christianity, waged the Crusades from the eleventh to the thirteenth centuries against the Turkish Seljuks and Mamelukes for the liberation of the "Holy Land" and "Christ's sepulcher." The religiously fanatical crusaders, joined by troops from all the lands of Europe under the leadership of renowned kings and feudal nobles, performed acts of violence no less cruel and extensive than those of the "Asian barbarians" and Moslem infidels. Having captured Jerusalem in 1099 during the

First Crusade, the crusaders committed horrible acts of plunder and carnage. They killed all the Jewish and Moslem inhabitants of the city. In the Second Crusade in 1145, Jews were killed on the way in German territory. In the Third Crusade campaign, King Richard I (the Lion-hearted) became famous for having given orders to kill 2,700 prisoners of war to whom he had promised freedom if they surrendered the city of Akka. In the Fourth Crusade, the Croatian city of Zadar, which in 1202 was taken by the Venetians with the help of the Crusaders, fell victim to slaughter and pillaging. In 1204 the same fate befell Constantinople, the capital city of Byzantium. The frenzied violence, plundering, and destruction of the sacred relics and the cultural riches of the wealthiest city in the world lasted three days. The Crusaders founded a Latin Empire in the captured city, the first emperor of which was the Flemish Count Balduin I. The Greeks and Bulgarians soon defeated Balduin, but the empire still managed to last until 1261, when the Nicaean emperor Michael VIII Palaeologus captured Constantinople and renewed the Byzantine Empire.

The level of perversity that these wars reached can be seen in the "Children's Crusade." In 1212, several groups of fanatical children of both sexes set out from France and Germany, via Marseilles, Genoa, Pisa, and Ancona in the naive belief that the sea would part before them and open up a path to the Holy Land. Their numbers dwindled as they met with various hardships. Disenchanted by the lack of miracles, many of them drifted off, and some decided to return home. Many children were lured by merchants onto ships to be sold as slaves in the Moslem lands of North Africa, though the usual arrangement was for slaves from the East to be sold in the West. In the sixth campaign, the French chivalric order of Templars violated a cease-fire, captured Nablus in 1242, and killed, among a large portion of the inhabitants, indigenous Christians along with the Moslems. In the Seventh and final Crusade, Alexandria suffered immensely after its conquest in 1365. It was pillaged to the core, while in addition to Moslems, large numbers of Christians and Jews fell in the

slaughter. In their brutality, these acts of violent outrage and criminality were equal to those of previous centuries in Jerusalem and Constantinople, and reportedly even worse than those perpetrated by the Moslems in their conquests of Antioch, Akka, and other places.[138]

In the Crusades, Christianity as a whole confronted Islam as a whole. Therefore the causes of all sorts of crimes were rooted, above all, in the oppositions of different civilizations, specifically the irreconcilability of their religions. Both sides encourged and even blessed such crimes by raising the murder of mortal enemies to the level of absolution for sins and divine merits for "eternal life." However, this century-long war did not remove the ethnic and social conflicts within the warring civilizations. Nor was there any less internal violence. In fact, it was quite the opposite. The actions of various religious power figures and secular tyrants toward heretics, renegades, and adversaries of all kinds among their own ranks was even more brutal. A broken penitent was of little value to anyone. From this context come all the horrors of inquisitional torture chambers, and the burning alive of "witches" and of those obsessed by satanic ideas, particularly when the heretics and false prophets of hell were learned people and national leaders like Jan Hus, or entire ethnic groups.

While the Crusades against Islam continued, religious wars were beginning in Europe. These wars were being waged by the Pope and individual rulers against various "heretical" currents. They were to continue over even more centuries and in virtually all parts of Europe: against the Albigenses in France (1208-29), against the Bogumils in Bosnia (1234-39), against the Hussites in Bohemia (1414-38), and against the Huguenots in France (1562-98 and 1702-05). Luther's reformation led to armed conflicts of the Counter-Reformation, which reached their peak in the Thirty Years War (1618-48). Not only all of Germany, but also every other European state took part in that conflict.

Although all the Crusades and religious wars were set off by religious inspirations and schisms, they had a deeper sociopolitical and national basis, namely liberation from foreign religious

and political domination. Although waged predominantly by mercenary armies, the ones who suffered the most were the civilian populations of individual cities and entire regions, in catastrophic devastation, savagery, and slaughter of all kinds, followed by epidemics of disease and starvation. In these wars, the population of Germany fell from approximately 16 million to 6 million. In West Phalz, where the greatest casualties occurred, up to 90 percent of the population lost their lives.[139] Yet when one speaks of the atrocities of these wars, one usually loses sight of the grievous suffering of the population, exemplified by the massacre on St. Bartholomew's Night, when in Paris on August 23-24, 1572, about 2,000 Huguenot leaders were killed by trickery. In the days that followed, all over France some 20,000 to 25,000 French Protestants were slaughtered.[140]

In connection with the brutality of the Crusaders, and generally in the Middle Ages and Renaissance, it is interesting to look at the military order of knighthood. In the broad sense, neither noblemen nor lords belonged to the order of knights, who were selected by a solemn ceremony in which they took an oath to adhere to certain ethical principles: justice, magnanimity toward the adversary, the protection of the weak, and the cult of women. (They were raised in that spirit in the courts of noblemen and rulers as pages until the age of fourteen, and as squires until twenty-one.) In the narrower sense, that order was made up of knights who had agreed to military service for individual sovereigns or feudal noblemen. The equestrian orders of knights were the basis of the European Christian armies (German, French, English, and others) in the Crusades against Islam, but also in mutual wars within Europe. (They were to be replaced gradually by hired military formations.)

This was the military order of knights which respected the rules of announcing the place and time of the beginning and end of a battle, and of truce, so that each side could collect its wounded, or in order to attempt a solution by negotiations. The highest form of combat upon such principles undoubtedly was in the agreement that a conflict be settled by a personal duel between army leaders,

or among a set number of warriors from both sides. But examples
of such conduct were few. Among those examples mentioned from
the Crusades, the following ones will demonstrate how this
humane chivalric warfare degenerated or how it was overcome by
the blind impulses of carnage and madness.

In the conflict between the French-Breton and Anglo-Breton
sides (1351), during the agreed-upon chivalric combat of thirty
knights, the bloodied and exhausted French nobleman
Beaumanoir asked for water. He received the famous answer of
Bramborough on the English side, which was defeated in the end:
"Drink your own blood, Beaumanoir, and your thirst will be
quenched!"[141] That battle became celebrated in verse and art, but
also as an event of particular challenge to the justifiability of bru-
tality, undoubtedly because of the bloodthirsty answer given to the
enemy knight. However, it seems there was no similar ambiva-
lence regarding the unchivalrous revenge against the populace
which participated in the "Jacquerie" rebellion of 1358. To be
sure, the chronicles recorded that the revolting mobs of peasants
and townspeople in many French regions (Ile-de-France, Picardie,
Champagne, Orlanais) commited crimes of such magnitude
against the landowners that they "confound anyone's imagina-
tion." But the army of knights, too, sank to a similar level of bes-
tial savagery. Having ravaged and set fire to hundreds of the
landowners' castles and properties in the first wave, the enraged
peasants continued mercilessly to perform acts the likes of which
"have never been committed either among Christians or
Saracens." Among the lootings and horrors, a knight was "roasted
on a spit before the eyes of his children and wife," whom they
"forced to eat the flesh of her husband." After "ten or twelve men
had raped her," she was killed. In the mad rage of collective
hatred, plundering, and devastation, not only the nobility's prop-
erty, but the Church's dominions were also at risk. Fear of incin-
eration, rape, and death magnified the horror of revenge in the sti-
fling of the Jacquerie. Disorderly and poorly armed insurgents fell
before the armored horsemen. In the decisive battle of Meaux,
"Knights were assailing and madly hacking, killing the common-

ers like animals, until they were exhausted from the slaughter." All who were caught trying to escape were killed, while "some were burned alive in their homes." In addition to unrestrained force, the army of knights also used trickery and guile. Thus one of the leaders of the rebellion, Guillaume Cale, agreed to negotiate with Carl of Navarre in the belief that his royal opponent would adhere to his knightly word. Instead the latter had him chained and executed after first crowning him king of the Jacquerie with a red hot crown. Carl then killed 3,000 peasants in the chase, including 300 who were burned alive in a monastery where they were hiding. In all the areas of the uprising, the army of the nobility proceeded in full force to burn down villages and hamlets, while captured rebels were hanged from the trees along the roadsides or from the doorways of their own huts.[142] Thus the chivalrous methods of waging war had disappeared both from the homeland and from all the roads to the "Holy Land."

Similar acts of wanton violence occurred during the Renaissance on the Apennine Peninsula as well, in wars between city-states and the papal state, and in conflicts between opposing factions. For example, in the second half of the fourteenth century, Bernabo of the house of Visconti, one of the governors of Milan and Lombardy, achieved notoriety as a particularly wrathful, uncontrollably greedy, cruel, and irresponsible tyrant. When the Pope, during a war, excommunicated him for his violent actions, Bernabo forced the envoy who delivered the bull of excommunication to eat it, along with the lead seal that fastened it. His *quaresima* is also well known: a forty-day program of torture for traitors and enemies in which they were subjected to stretching on the wheel, the stripping of skin, gouging out of eyes, ears, nose, and castration. Within their fortified cities and walled fortresses, the Viscontis also built extravagant palaces. The one in Pavia, built by Bernabo's brother, Galeazzo, was called "the greatest palace in the entire world" and "the most beautiful residence in Europe," and was to be admired by Petrarch whose presence adorned Viscontis' palace for eight years.[143]

The Hundred Years War, which continued intermittently from

1337 to 1453 between France and England over the right of succession to the French throne, also devastated many European lands. The military orders of knights almost completely disappeared and were replaced by mercenary detachments. Thus the suffering of the civilian population rose increasingly. In Limoges horrifying bloodshed was perpetrated by the English under the command of the "Black Prince," who ordered the closing of all exits from the city and the slaughter of all the inhabitants regardless of age or sex. About forty villages were destroyed in Alsace, where hundreds were killed and such atrocities were committed in the name of vengeance that the military commanders hanged mercenaries guilty for the rebellion almost daily. In the war that took place in Switzerland between the detachments of Leopold of Habsburg and the Anglo-Breton detachments under the command of Coucy, the local Swiss populace, frenzied by the violence and destruction perpetrated by the undisciplined foreign soldiers, lay in ambush and rose against them, killing and burning hundreds in their sleep. "And those who tried to escape were killed, while those that remained were burned," and "in England and France the wails of women rang out."[144] Similar events took place in Flanders. As the war enveloped wider territories and greater multitudes of people, the violence on both sides grew ever more ruthless and brutal.

Historical events of all times and from all corners of the earth testify to the fact that violence and bloodshed have always taken on the greatest and most widespread proportions when their causes have been rooted in racial and ethnic differences, even when they could have been provoked by other reasons, above all, reasons of religious or class conflicts.

All of the nations along the path of the Ottoman Turkish conquests felt the genocidal characteristics and consequences of their military campaigns and their reign of terror in the Balkans and Southeastern Europe—particularly in those lands which fell under Turkish slavery for any longer period. Thus, for instance, in Croatia large central areas were seized and their further development within the Mediterranean and Central European cultural

sphere was discontinued. A large number of Croatian noblemen perished in an extended war. Entire regions were deserted, both due to the effects of the war and Turkish rule, as well as from mass migrations of the Croatian population. Hundreds of thousands of people fled from Hercegovina, the Lika, Adriatic coast, Bosnia, Slavonia, Banija, and Kordun, to the remaining parts of Croatia, as well as to Italy, Hungary, and Austria, all the way to Slovakia (to the foothills of the Carpathians, around Bratislava), and even to Romania.[145] The Turkish occupation brought in its own settlers, mercenaries of Wallachian descent, to take their place. The occupiers gave special privileges to the Orthodox Church as opposed to the Catholic, because the Roman Church had instigated the Crusades against the Islamic conquerors. A significant portion of the population converted from the Catholic to the Moslem faith, in order to remain on their own land. However, one of the most ominous and treacherous forms of genocide was unquestionably the collecting of tribute in blood—the seizure and removal to Turkey of the children of the oppressed Christian subjects in order to turn these Moslem converts of various nationalities into janissaries for the elite military units and administrative officials, raised to be the most fanatical warriors for the Moslem faith and the most merciless assassins of their own people.

What should one mention from the horrible array of colonial conquests? Which European nation through its share in the creation of colonial empires and dominions—from Spain, Portugal, England, France, Russia, Italy, and Germany, to Holland and Belgium—did not bloody its hands in the tyrannical imposition of "high" European civilization upon the lands of North and South America, Africa, Asia, and Australia? Didn't this initially require the total subjugation of the people of the conquered lands and even of entire nations—to the point of genocidal extermination and the eradication of their original cultures? As is well known, these actions in the "new worlds" did not involve only "primitive" American "Indian" tribes, or uncivilized black African natives, or culturally "backward" nations of Asia, but also states of a high degree of civilization like the Aztecs and the Incas, old

cultures which disappeared under the ruins of their own cities, while even their people were barely able to survive in their scattered and assimilated form under the merciless rule of the Spanish and Portuguese conquistadors. Similar events occurred in other areas under the hand of other colonial rulers. It was under such circumstances, for example, that the capturing, shackling, and transporting of black slaves took place, probably in an environment of greater bloodshed than during their mass lynchings up until recent times.

Napoleon in his conquest of European lands ceremoniously proclaimed far and wide the realization of the great ideas of the French Revolution, of the liberties and equality of the citizens and nations, of an end to feudal and class privileges, of the prohibition of corporal punishment, and of judicial order. Those changes, nonetheless, remained for the most part in the shadow of the violence and lawlessness perpetrated both by vassal rulers by whom the foreign lands were subjugated into his empire, and by his conquering army, from the pillaging of property and the raping of women, to the various brutal actions against the local populations. When such actions stripped away national rights they often provoked armed resistance by the local populations. In response, Napoleon ordered hostages taken from Swiss, Austrian, and German lands to Russia, and chastized entire cities and settlements in certain areas which were suspected of aiding the liberation movements against France.[146]

In the Franco-German War (1870-71), one finds typical examples of the violence of conquering armies against the inhabitants of vanquished lands, but also violence against foreigners. In that war, the French army of Napoleon III, whose imperialistic aim was to prevent the unification of German lands under Bismarck, suffered a shameful defeat. The Prussian Army captured, along with the French emperor and marshall, the entire regular army. When it seemed that the war was over, the Prussians engaged in wanton violence, plundering and killing civilians, and burning cities and villages. Their request that two French provinces be surrendered roused the anger of the French people who rose to a war

to the finish in which both sides used all means available to destroy their opponents.[147]

In the Balkan Wars (1912-13) for the liberation of the Balkan lands still under Turkish rule (Kosovo and Metohija, Albania, Macedonia, Epirus, Thrace), and afterwards, for the division of the spoils among the Balkan states (Bulgaria, Rumania, Greece, Serbia, Montenegro), one finds a picture of the dismal consequences of centuries of foreign rule as well as the genesis of the causes of new and even fiercer opposition between nations. It is an uncontestable historical fact that under the several-centuries-old Ottoman rule, the ethnic composition of Kosovo was changed to the disadvantage of the Serbs. In the territory of Raska, which was the cradle of the Serbian state, it is also true that Serbia attempted to achieve its imperialistic goals with regard to Albania (to clear a way to the Adriatic) by such actions against the Albanian populace that it caused a deepening of inter-ethnic animosity and the beginning of a new source of irreconcilable oppositions. The Serbian Army carried out the occupation of Albanian territories with such brutal force that it created "a state of havoc which thrust the Albanian people into a desperate battle for survival," as described by the Serbian Marxist Dimitrije Tucovic. "Embittered by the savagery of the Serbian Army," the Albanian people similarly exhibited extreme mercilessness. In order to stifle the uprising, Serbia had to mobilize "nearly three divisions." But the "vengeful burning of villages and the massacre of the Albanian populace" was neither "reparation for their [Serbian] unnecessary losses" nor could it calm passions. Remaining "blind and deaf to the brutal practice of colonial practice," Serbian propaganda raised an infernal uproar against the "Albanian barbarism," calling for extermination without mercy. Albanian villages were "transformed into ashes" while "the Serbian soldiery spared neither the children, nor the women, nor the sick people." By such actions, claims Tucovic, the Serbian bourgeoisie entered the colonial ranks of the English, Dutch, French, Germans, Italians, and Russians.[148]

Immediately after the First World War, S. Zweig would write:

"I believe I know history rather fundamentally, but it seems to me it has never yet produced such a mad genocide on such a monstrous scale."[149] Yet taken as a whole, World War I was undoubtedly one of the more regular wars. For the most part, there were not such genocidal retaliations as those we have seen throughout history. But of course such violence was not altogether absent. The most pronounced example of genocide in that period is certainly the extermination of Armenians in Turkey. In the Middle Ages that nation was the battlefield for the struggles between the Arabs and the Byzantines. Then in the eleventh century it was divided between the Seljuk Turks and the Byzantines. Later it became the object of partitioning among Persia, Turkey, and Russia, as well as the colonial politics of England and France in the Middle East. In the nineteenth century, the Armenian Question became an important part of the Eastern Question. In its endeavor to break a path to the Bosporus and the Dardanelles, Russia took the land under the pretext of "battling for the liberation of Christianity from Moslem bondage." However, by a decision of the Berlin Congress in 1878, Russia had to withdraw from the occupied areas of Turkey, which in turn was obligated to carry out reforms in the Armenian regions. However, instead of obeying, Turkey used the Kurds and their religious fanaticism to initiate a pogrom against the Armenians in 1890. After Armenian resistance, the Turkish government resorted to the politics of even more forceful extermination. In 1894, the massacre of Armenians in Sasun and throughout all of Asian Turkey was carried out. In 1896, 8,000 Armenian villages were destroyed and burned down, 50,000 Armenians were killed, and about 100,000 were mutilated (in Istanbul alone, 6,000 were massacred), and 300,000 of those who were left homeless fled to Russia. A new massacre was effected in 1909 in Kiliki, where about 20,000 Armenians were killed. On the eve of World War I, the Turks signed an agreement that the Armenians would be given broad autonomy, but they took advantage of the wartime environment to enact a "final solution" to the Armenian question. By 1916, the Armenians in Turkey had been almost entirely eliminated: a million to a million and a half people had

been killed or scattered, while about 400,000 were converted to Islam. This was one of the most horrible cases of genocide in recent history. The idea of a united and free Armenia, the bearers of which were precisely those Armenians who were under the Turks, was to materialize only partially after the October Revolution in Russia in the form of the Soviet Republic of Armenia (in the USSR, there are about 2.5 million Armenians, not even half of the total of Armenians, while the other half is scattered over the world).[150]

From the period of World War I, it is necessary to mention the radical way in which the ethnic mingling of the Greek and Turkish populations was resolved. Turkey, which was on the side of the vanquished Central Powers, lost at the end of the war, along with all its European possessions, as well as those in the Middle East. Greece, however, attempted the conquest of extensive portions of Anatolia itself, where there was a sizeable Greek minority. In 1919 the weakened Turkey was unable to resist the Greek invasion of its Asia Minor territory. The Greeks had been facilitated by aid from France. However, the danger of partitioning awoke a Turkish national resistance and a revolutionary democratic overthrow under the leadership of Mustafa Kemal Pasha (Ataturk) who, casting off the sultan's disordered regime, defeated the Greek Army (1921-22) and threw it out of Asia Minor, regaining Istanbul and Eastern Thrace. The Peace Treaty of Lausanne confirmed Turkish sovereignty over the Bosporus and Dardanelles, and a mutual exchange of citizens was worked out. About a million and a half Greeks from Asia Minor crossed over into Greece, while about half a million Turks left Greece for Asia Minor. By this radical exchange of populations which had been mixed during the course of a long history, the prerequisites were realized for the more normal national development of both countries.

On Cyprus, a Greco-Turkish separation was carried out several decades later, when an invasion of the Turkish Army brought about a territorial and administrative partition of the island into a Turkish and a Greek part, actually into two separate states, each bound to its respective mainland.[151] However, the Greek side still

strongly opposes these conditions. Not having agreed, before the division, to a settlement which would have brought about inter-ethnic relations based on equal rights, Greece still insists upon a unified Cypriot state, under Greek control, of course.

3. In the Shadow of the Holocaust

The more a tree strives toward the skies and light, the more firmly its roots grip the earth and strive downward into darkness, into the depths, into evil.

—Friedrich Nietzsche

During the Second World War, the madness of violence, partic-ularly against entire nations and with distinctly genocidal aims and consequences, reached dimensions which were unquestion-ably the greatest in all history, in terms of the number dead and the manner in which they were exterminated.

Most terrible among these manifestations of violence was, of course, the Holocaust. Never before had a race or nation been singled out for total destruction by another nation as a matter of state policy. The previous historical examples of genocide that we have examined were acts of mindless barbarism, not an an explic-it "final solution" meant to destroy an entire race of people.

The name "Holocaust" is used to describe the horrible suffer-ing and extermination of Europe's Jewish population by the Nazis. The term was introduced by writer and Nobel prize-winner Elie Wiesel according to the Greek translation of the biblical Hebrew word *shoah*, meaning burnt offering.[152] This designation for the persecution of the Jewish people could lead one to believe that the extermination was carried out only or predominantly in the gas chambers and ovens in the notorious camps. However, in this as in previous periods of history, the Jewish people were sub-jected to three forms of pogrom: forcible conversion to Christianity, banishment, and extermination.

When Christianity became the dominant religion in the Roman Empire in the fourth century and was later accepted by all European nations, it was assumed that the conversion of Jews to Catholicism would be achieved, whether willingly or by force. But over twelve long centuries the policy of Christianization brought little success. So from the thirteenth to the sixteenth centuries, in all the Western European lands, extremely brutal methods were undertaken against the Jews.[153] The command to convert or be banished was always followed by mass bloodshed. The rationalism of the Enlightenment and the nineteenth-century Romanticism removed the religious basis for anti-Semitism, but gave it instead a foundation in various theories and ideologies.

Beginning in fifteenth-century Spain and continuing into this century, racial theories became the justification for imperialistic conquests. However, they would experience their fullest development in Germany. This manifested itself to the extreme in Roseberg's racist doctrine and Hitler's pan-Germanic program. The idea of the world mission of the German *Herrenvolk* as the supreme race was founded also on the premise of a "final solution" of the "Jewish Question," that is, their absolute elimination from Germany and Europe.

From its advent to power in 1933 until the end of its first successful period of the Second World War at the end of 1941, Hitler's leadership of the Third Reich planned to realize the "final solution" territorially, by facilitating their emigration, and later by their forced resettlement. Until 1938, Hitler encouraged the emigration of Jews to Palestine, and also their founding a homeland there, which was in keeping with the pre-Weimar politics of German diplomacy. But since Palestine was under British mandate, the British limited the immigration of Jews, because Arabs were rising in protest against it. Only later did Hitler, because of the war with the British Empire and the persecution of the Jewish people, win the undivided sympathies of the Arab world.[154] From 1933 to the middle of 1938, of the 150,000 Jews who emigrated from Germany, 50,000 settled in Palestine. By the time war broke out in 1939, the total number of Jewish immigrants into Palestine

(from Germany and other countries) had reached 215,232, and by the end of the war in 1945, about 270,000.[155] In order to spur on the emigration of Jews into Palestine, Hitler's government offered them special concessions for the transfer of their capital. According to the so-called Haavara Treaty between the government of the Third Reich and the Jewish representatives of Palestine, Jews were compensated for their blocked capital in such a way that Germany, in a trade exchange with Palestine, gave its equivalent in industrial products, while the representatives of the Jewish "Haavara society" paid individual Jewish refugees off in local currency upon their entry into Palestine. All sides were equally satisfied with this agreement, except the "Palestinian Germans, who protested that their interests were being gravely neglected to the benefit of the Jews."[156]

The publication of the British Peel Report, which recommended the formation of a Jewish state in a partitioned Palestine, motivated Hitler to change his mind concerning the efficacy of the Haavara Treaty. Although Peel's proposal was not accepted at that time, Hitler arrived at the conclusion that the transfer of "Jewish-owned German capital into Palestine" would facilitate the founding of a Jewish state, which in the long run would go against German interests. Such a state, rather than merely drawing the Jews from the rest of the world, "would one day bring about a substantial increase in the political power of Jewry in the world." Because of this, Germany was not interested in facilitating the emigration of wealthy individuals who would take along their capital, but rather a mass exodus of the entire Jewish population.[157]

In searching out the possibilities for a territorial solution to the "Jewish Question" in 1938, Hitler's leadership accepted the well-known plan of relocating the Jews to Madagascar, which was under French colonial rule. However, that idea was not an invention of Hitler's. After the first Zionist congress in Basel in 1897, when negotiations with the Turkish sultan on the resettlement of Jews to Palestine proved unfruitful, the British government offered a territory in its colony of Uganda. At that time, even Theodor Herzl, the originator of the Zionist movement, was inclined to

that proposal. But the majority was for a revival of the Jewish state through resettlement in Palestine.

In 1934, Himmler allegedly advocated Hitler's idea concerning Madagascar. But it was actually a proposal by Poland's government. The Polish government sent a special commission to Madagascar in 1937 to investigate the possibilities of a settlement, and even the French colonial office did a special study on it. Despite strong opposition by the original population, the Polish minister of foreign affairs, Joseph Beck, upheld the resettlement of the Jewish people to Madagascar (which was almost twice the size of Poland in territory and had almost one-tenth of Poland's population).[158] Because of this, Madagascar became the subject of broader diplomatic negotiations. Besides French-Polish discussions, the German government and Hitler himself negotiated with the representatives of France and England about that plan, while its particular advocates were South Africa and Italy. In discussing territorial solutions, all of these governments were opposed to the Palestinian solution, some in order not to lose, and others to gain the friendship of the Arabs. In February of 1939, Alfred Rosenberg publicly presented the plan for the settlement of 15 million Jews in Madagascar or Guiana (in northeast South America).[159]

The crisis surrounding the outbreak of World War II pushed aside international debates, but also caused a wave of emigration. Particularly after the occupation of Czechoslovakia, the number of Jews wishing to escape Nazism was much greater than the number which the rest of the world was willing to accept. The Jews were fleeing to independent European countries and abroad: from Scandinavia to Italy, France, and England, from Palestine to Shanghai (which was under Japanese occupation, but to which Jews were able to immigrate without limit), to Cuba and both Americas.

According to the available statistics for 1939, 1940, and 1941, the approximate number of Jews that immigrated were: 155,000 to the United States; 70,000 to Palestine; 130,000 to other countries outside of Germany's reach; and about 100,000 to countries

occupied by Germany. (It is necessary to note that the immigration of Jews into the United States was limited by a general quota restricting all foreign immigration to an annual total of 150,000 people.)[160]

At the very beginning of the war, Hitler's government, in its peace proposal submitted on December 1, 1939, through the intercession of Pope Pius XII, included a plan for the emigration of all German Jews to Palestine, Ethiopia, and Madagascar.[161] Although the British, French, and American governments felt that the final victory of the Allies would eliminate the necessity for implementing a program of a broad emigration, that problem was nonetheless discussed in detail among them, spurred on, in particular, by Roosevelt. From these discussions, there resulted a plan for the settlement or colonization of 100,000 Jews in the Dominican Republic, which began to be implemented in part. It was addressed more seriously after the occupation of France.[162]

Resettling the Jews in Madagascar was now out of the question, due to the proposal's rejection by other powers. The leadership of Hitler's Germany, having gotten hold of the majority of Europe's Jewish population through the partition of Poland, accepted a territorial solution by creating the "Jewish autonomous region" of Lublin. There they intended to collect about 1.4 million Polish Jews and about 600,000 from Germany and other areas. However, as early as May of 1940, the project was abandoned. There were insurmountable difficulties in gathering such a multitude in an inadequate area without enough lodgings, even though regulation works and labor camps were foreseen. Later, the territorial solution was abandoned in that area, which was to be part of the German ethnic corridor from the Baltic to the Carpathians. But the idea of an autonomous Jewish district ("a reservation") in that area continued throughout the war. The German government would use it to justify its demands for sending Jews from other countries to the East. Among the Polish and other Jewish populations, the opinion was widespread that the Germans were trying to set up a Jewish region in the east of Poland. Repercussions of that proposal can be found in the out-

line drawn up by Zygmunt Kaczynski of the Polish emigrant government in the United States in 1942, concerning the postwar "solution" to the "Jewish Question" by their settlement in east Poland or in Bessarabia. All things considered, this proposal seems to have been supported even by the Polish resistance movement in Warsaw.[163]

Germany's victory over France unlocked visions of a final victory, and thereby, the possibility of a "final territorial" solution in one of the African or American colonies, with Madagascar heading the list. It is interesting to note that all of Hitler's collaborators were in agreement with such a solution, from Goering, von Ribbentrop, and Goebbels, to Himmler, Rosenberg, Frank, and Heydrich. In a conversation concerning the division of the French colonies (on June 18, 1940), Hitler told Mussolini that a "Jewish state could be established in Madagascar"; this was also told to Galeazzo Ciano by von Ribbentrop. Dr. F. Rademacher was assigned the task of working out that plan from the standpoint of international law. According to his memorandum, which was approved by von Ribbentrop, Petain's French government was to surrender Madagascar to Germany and transfer 25,000 Frenchmen from the island in order to create conditions for the "solution to the Jewish Question." Europe's Jewish population were to be moved to Madagascar where they would enjoy complete autonomy, which would ensure "good behavior" of American Jews towards Germany. Under the direction of Eichmann and Rademacher, all necessary arrangements were made for the implementation of the plan to transport 4 million Jews by sea to Madagascar, as well as a study of the island in collaboration with the French Ministry for the colonies. Of course the transfer could not commence until the war was over.[164]

But Hitler's war machine became stuck in the mud in the Soviet Union. And along with the myth of German military invincibility, the possibility of a territorial "solution to the Jewish Question" outside of Europe also vanished. From the beginning of the German-Soviet war in the summer of 1941, the persecution of the Jewish people was escalated by fanatical propaganda about the

necessity for a merciless uprooting of all members and adherents of "Jewish" Bolshevism. The possibility of deporting Europe's Jewish population to Madagascar was now out of the question. The conquest of large Polish and Soviet expanses in the East opened up new possibilities for a territorial solution. In the beginning of 1942, Hitler decided to expel or "deport" the Jews to the East. During the Berlin conference on January 21, 1942, in which Heydrich gave instructions to high level Nazi officials to carry out Hitler's new directive, they spoke only of the "evacuation" of Jews from European countries to the East. Still, it was obvious that the "final solution to the Jewish Question" in this manner aimed at their gradual destruction. H. Frank spoke before the German officials in Crakow more clearly about the "final solution." Stating that in Poland there were now about 2.5 million Jews, and with those who were *Mischling* (part Jewish), possibly even 3.5 million, he said that they "cannot all be executed or poisoned," but that it was necessary to "undertake measures which would lead to their destruction" because the war would not be a complete success if the Jews were to survive it.[165]

Thus in the fourth year of the Second World War (1942), the period began in which the leadership of the Third Reich set about the execution of the diabolical "final solution," i.e., the exclusion of Jews from the life of Germany and other European nations by their gradual extermination. However, since such a goal could not be expressed openly before the entire world, it was communicated in the form of secret instructions only to a small circle of Nazi confidants. Therefore, it remained hidden even from the majority of Germans who understood the deportation of Jews to the East to mean their resettlement into Poland and Russia, and assumed that the concentration camps were labor and not death camps. Apart from that, neither Mussolini, nor Antonescu, nor Horthy, nor Petain, nor the other quisling governments, and in particular the population, from Norway and the Netherlands to Greece and Bulgaria, were in favor of following Hitler's policy of the exclusion and eradication of the Jewish people. With respect to the Jews, the situation in Tiso's Slovakia was decidedly different from

that of Pavelic's Croatia or Nedic's Serbia. Because of this, in that period there were instances of individual and group rescues, and even emigration of Jews from "Hitler's fortress."

4. The Timeless Universality of Genocidal Acts

Deep calleth unto deep...
—(Psalm 42:7)

Or, according to Leszek Kollakowski:
Deep calleth unto deep by its echo...

Though not on the scale of the Nazis, the Allies themselves killed many innocent citizens. In dozens of German cities ruined by systematic air attacks, a million people, mostly unfit for military service, were killed. However much the American atom bombs dropped on Hiroshima and Nagasaki may have shortened the war, so also did their burnt offerings—in the most literal sense—remind and warn of the mindless evil of all military retaliation.

Common military violence is transformed into genocidal extermination and deportation of populations when the conqueror has permanent designs upon the territory and ethnic assimilation of the opposing nation, as both Hitler's Germany and Stalin's Soviet Russia had toward Poland. While the world had long been familiar with the German annihilation of the Poles as well as the Jews, the discussion concerning the fate of the Poles under Soviet occupation was obstructed by political considerations.

Another illustration of the distortions and lies that often surround historical occurrences and their subsequent debates is the controversy surrounding the murder of the Polish officers in Katyn Forest. As early as 1943, the Germans disclosed that they had discovered in the woods of Katyn the graves of thousands of Polish officers, but the Russians attributed the crime to the

Germans, rejecting an international investigation by the Red Cross. In May 1943, the Germans entrusted an international commission made up of medical and criminological experts from eleven European countries to confirm their findings. From the pits of Katyn, 4,143 slain Poles were exhumed, of which 2,914 were identified. Further investigation organized by the exiled Polish government and other international agencies determined that of 14,920 active and reserve Polish officers and Polish intellectuals who had been held in three Polish prison camps (near Starobielsk, Kozielsk, and Ostaszkov) only 448 remained alive—those who had been taken into the Polish military corps in the USSR. Aside from those found near Katyn, it remains unknown where the rest were killed. The international Nuremberg trials, despite all Soviet efforts, were unable to find any basis for accusing the Germans of the Katyn massacre, even though they were charged with all of the crimes of the war.

Finally, the Soviet Union acknowledged guilt in the Katyn massacre, when Mikhail Gorbachev sent a communique to Pope John Paul II memorializing the tragedy and formally admitting that the officers had been killed by the Red Army. After years of claiming that the murders were committed by the Nazis, Gorbachev and the Soviets came clean about historic crimes that had been long denied. Gorbachev's glastnost policy applied even to the Soviet's notorious rewriting of history. Today one can see increasingly normalized discussions in the field of historical study. Soviet sources maintain that about 230,000 Polish soldiers were imprisoned and interned. Of this number, 10,000 were officers. In addition, both Polish and Soviet historians have revealed that in 1940, from the eastern part of Poland, more than a million people (600,000 to 700,000 Poles, and the rest mostly Jewish) were deported to various parts of the Soviet Union, of which only a few returned to their homeland.[166] If one adds to this the fact that before the war, by a decision of the Comintern, the Polish Communist Party had been dissolved and that a large number of its members also disappeared into Stalin's camps, one comes to the harrowing conclusion that this was all part of a plan to

remove the intellectual and political leadership of the Polish nation.

Furthermore, should one forget the horrible fate that befell entire nations or millions of people after the end of World War II? If the fate of the German nation was predetermined because it had to atone for all the sins of Hitler's lost war, the destiny of many nations was determined by agreement of the Big Three— Roosevelt, Stalin, and Churchill— concerning the establishment of a new international order. The order from the start was burdened with the earlier inconsistent Treaty of Versailles, upon the weaknesses of which Hitler's "New European Order" was able to arise. Winston Churchill was indeed correct when, speaking of the Yalta conference of the Big Three, attested that never in history had the fate of so many depended on so few. However, this was no new phenomenon. Throughout history, a small number of victorious men have arbitrarily decided the fates of multitudes, and chiefly of small nations, which are scarcely ever able to escape their tragic doom under the millstone of historical dramas.

The consequences of the Yalta-Potsdam Order must, of course, be viewed from the perspective of those nations over which colonial rule had actually been established through their division by the interest spheres in Europe, at a time when colonialism was being abolished both in Asia and Africa. But apart from that lengthy historical era of ruthless domination, after the disappearance of Hitler's Germany as the major power in Europe and the rise of the USSR as a new European and world superpower, the forced migrations of an even greater multitude of people than during the war occurred in Eastern Europe. The forces of the anti-Hitler coalition emphasized, among the proclaimed goals of their postwar policy, the inalienable right of each nation and national minority to self-determination, and opposed all violence, and particularly forced expulsion, of people. Nevertheless, at the Potsdam Conference, the victorious powers agreed to the transfer of the German minority from Poland, Czechoslovakia, and Hungary to Germany. That was the basis of their expulsion from Yugoslavia as well. The addendum that this should be done in a fair and humane way was of course *contradictio in adiecto,* just a pretext

for a totally contrary result. This decision by the Allies formally sanctioned the right to revenge. Its implementation resulted in banishment and the fleeing of a populace in virtually unprecedented numbers. Immediately after the Second World War, about 12 million Germans were forced to abandon their residences. Of that number about 3.3 million were from Czechoslovakia, more than 3 million from Silesia, about 2 million from Eastern Prussia, about 1.5 million from Eastern Pomerania, about 400,000 from Eastern Brandenburg, and about 400,000 to 500,000 each from Rumania, Hungary, and Yugoslavia. To that number German sources add about 3 million refugees from the Soviet-occupied zone, which eventually became the German Democratic Republic.

After the above-mentioned pogrom and dispersal of Poles by the German and Russian occupation authorities during the war, further migrations continued even after the war was over. From the Soviet Union, those who had been exiled during the war were returned, as well as those who had remained in areas annexed to the Soviet republics, and they were resettled in those Western regions allocated to Poland which had been taken from Germany as recompense for the territory that Poland was forced to give up in the East. Of the other large migrations caused by the war's outcome, there were about 400,000 Finnish refugees from areas which fell to the Soviet Union after the Finnish-Russian War and World War II and, likewise, a considerable number of Italians and Turks from Yugoslavia.[167]

All of these predominantly forcible deportations, but partly spontaneous migrations (though brought on by fear), changed the ethnic characteristics of entire regions, and even some countries. These were the consequences of the changes in the relations among the powers on the European continent, but also in essence the result of conflicting national interests and, therefore, of rivalry not only between empires but between states.

Such radical changes as were made after World War II always bring two-sided consequences. On the one hand, they inevitably deepen historical chasms, incite inter-ethnic hatred, and arouse vengeance, contributing to continued international tensions and

outbreaks of new conflicts. On the other hand, they lead to the ethnic homogenization of certain nations, to a greater harmony between the ethnic composition of the population and the state borders of certain countries. That can minimize reasons for new acts of violence and provocation.

After World War II, the leaders of Hitler's Germany and the satellite states of the "New European Order," as well as all the members of their Nazi-Fascist organizations, in addition to being charged with war crimes and crimes against peace (military aggression), were also accused of crimes against humanity. Thus the idea of genocidal crime—that which leads to the total or partial annihilation of a national, ethnic, racial, or religious group— now includes the crime of inflicting physical suffering, and also psychological torment (genocide would later be defined in this way by the UN).[168] This means that a crime against humanity, or genocidal crime, apart from the direct destruction of a population or destruction of cities and devastation of territories, is understood to include all forms of forced migration or forced labor. In addition to bodily abuse it includes such mental mistreatment as forcible conversion from national and religious affiliations. All this came to pass, as we have seen, in a wartime setting, though it hardly dissipated after the war.

Here is an example that falls in the category of genocide of spiritual believers. In the first years after World War II, by a decision of the Stalinist regimes in Eastern European countries, in the USSR, Romania, and Czechoslovakia, between 7 and 10 million Catholics (of the Eastern Rite) were converted to Orthodox Christianity. This was clearly done by force due to the widespread distrust of the Catholic Church under the Pope and of the need to destroy it as the chief obstacle to atheistic socialism. Moscow viewed Orthodoxy, rooted in Russian tradition and under the direct influence of state authorities, as posing not nearly so serious a problem as Catholicism.[169] In 1987, the Rumanian Patriarch Teoctist announced in Vienna: "The problem of the former Romanian Greek Catholics is solved for good because they have been 100 percent integrated into Romanian Orthodoxy." This only confirmed

that there was no difference between the Russian (Stalinist) and Rumanian (Ceausescuan) Orthodox approach. Every viewpoint—including that of the Church, especially when it is subjected to state totalitarianism—originates, if not exclusively from its own self-interest, then certainly from its own experience and the prospect of survival. Around the same time, an official publication of the Croatian Catholic Church rightfully turned its attention to the difference between the above-mentioned viewpoints of the Rumanian patriarch and those in Yugoslavia which, in the name of the Serbian Orthodox Church and supposedly impartial Marxist science, repeatedly raised the question of the responsibility of Cardinal A. Stepinac and the entire Croatian nation for the war crime of converting 200,000 Orthodox Christians to Catholicism in the first year of the NDH. "Five million Ukrainian and 2 million Romanian Catholics of the Eastern Rite... have been converted to Orthodoxy. And now the patriarch says there are no more problems!" And no one was prosecuted, nor has anyone identified that forced conversion as a war crime or genocide.[170]

In order to complete the picture of the scope of the violence and suffering in the contemporary world, much must be kept in mind. In 1947 the Indian subcontinent was liberated from the British colonial authorities and partitioned into India and Pakistan. During that period about half a million people died in religious conflicts, while about 6.5 million Moslems fled from India to Pakistan, and about 5.5 million Hindus from Pakistan to India. The cruelty of the West Pakistan Army toward the Bengali population in East Pakistan incited disturbances and the flight of several million Bengalis to India. This in turn provoked the Indo-Pakistani War (1971), the establishment of Bangladesh, and the vengeful reprisals of the Bengalis. Today, differences between the Hindus and Sikhs break out increasingly in more open confrontations, while the conflict between the Singhalese and the Tamils in Sri Lanka takes on the form of war and territorial partition. After the Soviet suppression of the Hungarian national revolution of 1956, about 100,000 Hungarians were thrown into concentration camps, while a far greater number fled to the West. A stream of

refugees flowed also from Czechoslovakia after the Soviet military intervention in 1968, as well as from Poland in the early 1980s. The Soviet occupation of Afghanistan provoked an anti-Soviet and civil war and caused the exodus of several million Afghan refugees to Pakistan. And there is virtually no African state, formerly colonial and now independent, in which there has been no internal inter-tribal violence and repeated oustings of both despotic and more or less democratic regimes. Inevitable outbreaks of genocidal savagery break out especially in cases of racial differences, along with ethnic and class differences.

Based on all these observations about genocidal crimes, one may draw several conclusions.

First, throughout history there have always been attempts at a "final solution" of foreign or dissimilar racial and ethnic or religious groups by expulsion, extermination, or conversion to the "true faith."

Second, there are many examples which indicate that such crimes which can be classified as genocidal have occurred to a much greater extent, both in terms of numbers and kinds, after wars rather than during wars because then the victors are able to execute their desires unimpeded. During war these goals are limited by the very uncertainty concerning the final outcome of military conflict.

Third, every attempt at establishing the appearance of all forms of genocidal acts as characteristic of one historical period alone is a futile task. Genocidal acts have existed since time immemorial in some form or other, regardless of all differences in appearance and dimensions.

Fourth, reasoning that ascribes genocidal inclinations to only some nations or racial-ethnic communities, to only some cultures and social movements, to only some individual religions and ideologies is completely mistaken and lacks historical validity.

Genocidal violence is a phenomenon present at all times and in all civilizations. However it does occur on an increasingly smaller scale and sometimes even virtually completely disappears in certain parts of the world. Genocide diminishes or disappears where the causes of its outbreak have been removed, for example, in Scandinavia and Switzerland. Or when and where the causes for

rivalry between countries disappear, as was the case in the earlier history of England and France, and most recently, between Germany and France. This occurs always in a place and time where no particular socio-ethnic community is endangered, and when the conditions favorable to benefitting from a crime are removed. Such a situation creates conditions that overcome historical irrationality and a judicious search for ways and means to a harmonious life in the peace of an organized social and international order. However, even in an environment of the highest degree of development, it is extremely difficult if not impossible to remove completely the bloodthirsty impulses of man's nature and the blind forces of destruction from the history of human society. One is reminded of this by such phenomena as the madness of the various terrorist acts in Italy, Germany, even Sweden, and everywhere else in the world, or the outbreak of repressed rage and aggressive instincts, if nowhere else, then in the field of sports. (This was demonstrated in a ghastly way in the bloodshed of English and Italian fans at a soccer game in Brussels.)

The human mind has never flagged in the noble effort to master the destructive powers in man. The tragedy and grandeur of these attempts is rooted in the contradiction of being itself: the elimination of all kinds of violence cannot be achieved without the use of some kind of violence. Just as an innocuous cold and a fatally incurable disease do not threaten man's life equally, thus it is also with violence in history.

PART THREE

■

Philosophy and Historical Evil

I. THE ANGUISH OF HUMAN REASON OVER THE ESSENCE OF VIOLENCE

From whichever angle one views world history, one sees a tapestry of brutality, madness and catastrophe.

—Oliver Goldsmith (1728-1774)

The best history is the kind that has happened (because there is no other).

—G. W. Leibniz (1646-1716)

1. Mythological and Religious Reflection

Arouse your anger and pour out your wrath. Destroy the adversary and wipe out the enemy. Let him that would save himself be consumed in furious fire. And let those who harm your people meet with destruction.

—(Sir 36:8-11)

FOR AS LONG AS IT HAS been aware of its own existence, the human mind has been occupied with the essence of violence. Acts that in the mid-twentieth century have been labeled genocide

have existed since the dawn of our primitive prehistory. In the historical procession of man and nations, violence has been a constant phenomenon, the cause and precondition of his survival and disappearance. One is faced with its inescapable presence in the eternal struggle of contradictions which guides the destiny of man and nation—which is to say, life and death, good and evil, love and hate, harmony and conflict, peace and war, the bliss of all that is beautiful and the horror of all atrocity.

Man's historical thought has been in a state of such contradictions since the very beginning of time. We are aware of the interdependence between the instincts of creation and destruction, but unable to eliminate them. Whenever human thought has striven to go beyond good and evil, it loses contact with firm ground, fleeing from historical reality into heavenly rapture, into other worlds, or into some peaceful future that remains forever at the unattainable horizon.

Every act of violence is the fruit of hatred, which is in opposition to love. However, hatred is also very close to love. In many ways it represents love's other inseparable facet. According to the Bible, God created mankind as brethren who should live in love. But hatred, born out of envy, led to murder in the very first generation of mankind.[1] Since then, the world has been prey to hatred and violence.[2] And just as Cain rose against Abel, Esau against Jacob, and Jacob's sons against Joseph, so, too, did the Egyptians rise against Israel, and outlanders against Jerusalem.[3] The enemies of the chosen people are God's enemies, equated with criminals, which justifies hatred and Israel's holy wars.[4] The principle "Thou shalt love thy neighbour and hate thy enemy,"[5] supported by the doctrine that every enemy of the people of Israel should also be seen as God's enemy, necessarily led to the justification of violence. Thus the God of love is ambiguously portrayed also as a God of hatred who hates the enemy, the tyrant, and every sinner, whether it be an individual or entire nations.[6] Thus Jewish biblical thought made God a frightful warrior who, for the sake of Israel, killed the Egyptian first-born children.[7] He demands utter annihilation; that is, complete destruction of the enemy, both of

living beings and material goods. Thus the slaughter of the van-
quished enemy becomes one of the religious rules of holy war, and
failure to annihilate the enemy can bring on defeat as punishment
for sacrilege.[8]

Biblical thought, born in the midst of a violent age, sees the sal-
vation of an oppressed and imperiled people only in the power of
the counterstroke: "May disaster pursue the violent man with
blow upon blow!"[9] And it regards holy war as the means to erad-
icate the evil in the world. Despite its approval of vengeful,
destructive force, biblical thought condemns unjustified violence,
the scattering of populations, slaughter of pregnant women,
immolation, the dashing of infants against rocks, and the oppres-
sion of newcomers.[10]

The Bible recognizes that every war of conquest, including the
holy war of Israel, provokes new wars. It also realizes that the
enemy who cannot be defeated by the righteous side will eventu-
ally perish of its own wickedness.[11] War and violence cause moral
anxiety and spiritual doubt from which there is no escape.
Although peace is God's intent, experience shows that every
nation achieves it only at the cost of victory obtained through
fighting and violence.

Since genocidal violence can arise not only out of the desire for
conquest but also out of vengefulness, biblical thought permits
revenge to the extent that it re-establishes justice and victory over
evil. The duty is to avenge trampled rights, but it is forbidden to
take revenge on the evildoers out of hatred. In order to restrain
vengeful instincts and the crimes they incite, the duty and the
right to revenge were gradually transferred from the individual to
the social community, and finally to God, who was characterized
as the only impartial avenger. As an omniscient being, only He
can be a just judge and lawful executor of revenge. The concept
of vengance develops from the idea that spilled blood cries out
for revenge, to the renunciation of retaliation, because "the man
who takes vengeance will have vengeance taken on him by the
Lord."[12]

The New Testament does not completely reject violence either.

It acknowledges the necessity for struggle—Christ, for example, would defend His holy city against the heathen—but it also seeks to give it meaning.[13] Despite the fact that Christ himself fell victim to blind, senseless hatred, the New Testament develops the principle of abhorring evil, not men and nations, and preaches forgiveness and love even toward the enemy.[14] The kingdom of God was established not by earthly, but by divine power, and "all they that take the sword shall perish with the sword."[15] Hostility is a satanic mystery rather than a mere fact, but it is the spirit of love, not revenge, which turns an enemy into a friend. This is true in the lives of individuals, and particularly in that of nations.

Yet the anguish remains. Even while forgiving, those who have suffered and whose blood has been spilled cry out to God: "How long, O Lord, holy and true, dost thou not judge and avenge our blood on them that dwell on the earth?"[16] History is full of examples in which the failure of self-sacrifice or forced reconciliation resulted in even worse outbursts of hatred and violence.

In the early thought of the great Eastern religions one finds reflections on the essence of sacrifice and violence similar to those of the Bible despite the profound differences between those religious worldviews and the tenets of Christianity.

In Brahmanism's oldest period, from the tenth century B.C. to the sixth century B.C., Indra, the first of its four main deities, is the god of thunder, fertility, and war. Of the four vast collections of the *Veda*, the oldest literary historical monument of the Indo-European period which contains all of Brahman wisdom, one entire collection explains the rituals of sacrifice, with the purpose of furthering the cult of the "thunderer and warrior." It also justifies human sacrifice offered for the "purging" of the state, and for the health and welfare of the community. In Brahmanism, too, one finds ambiguity in the teaching that every act of good or evil supports the "life cycle," that is, the ceaseless interchange of life and death.

This idea was taken over by Buddhism around the sixth century B.C., which adopted very different characteristics in various Asian countries, depending on their individual religious traditions. It is

frequently pointed out that Buddhism teaches that man's fate is predetermined and that redemption from all evil can be achieved only by abandoning mundane activity in Nirvana, the divine primeval nothingness. This, however, is only one side of Buddhism. There are several very different traditions. Among the mythological images of Tibetan Tantric Buddhism, one sees a supreme sorcerer, with the ritual dagger in one hand and a vessel full of human blood in the other, challenging Shiva the bull who carries on his head a crown made of human skulls. His blood-thirsty character is also represented through an erotic act with his mate Shakti, performed over a sacrificed human being and a bull—a symbol of his power. The sorcerer's disciples observe this act in amazement while another sorcerer, a wise man, is immersed in anguished contemplation.[17]

It is interesting to note that in the mythology of Hinduism, Shiva became the main deity with essentially the same traits—the creator and destroyer of the world (forming a supreme trinity with Brahma and Vishnu). More than a religion, Hinduism is an all-encompassing philosophical worldview, whose origin can be found in the old Indian heroic epic poem, *Mahabharata*. The main theme of this epic poem is the struggle for domination between two branches of the royal house of Bharata, which turned into a war of nations. Even where there was room for the Buddhist passive surrender to fate, there was never a reconciliation with the violence inevitably bred by oppression, nor a renunciation of its use. A particularly poignant speech from that epic is the "Mother's Curse" upon her faltering people: "My poor nation, joy of our enemies, you are not a child of mine nor of your father's. Whose are you? How can you consider yourself a part of humanity when you lack any wrath, like a eunuch!... Why do you lie like a corpse, as if struck by lightening? Arise, you coward; do not sleep after the defeat! Do not recoil miserably; show your bravery! Arise and thunder!" For, if you "wish to preserve your life," struggle is inevitable. Therefore, you "blaze like a torch."

Of course, such a moving call to arms for the sake of self-preservation can also be found in the Old Testament psalms, in the

"Song of Songs" and other biblical messages, in Isaiah's boosting of national conscience and consciousness, in David's lament over the heroes fallen on the hills of Gilboa, and especially in Jeremiah's laments that mourn the downfall of his people.[18] This biblical philosophy, as well as all of ancient mythological thought, usually does not conceal its intention to evidence the existence of one's own people through the pitiless elimination of others.

A certain exception in this regard can be found in the original Confucianism that arose on Chinese soil, out of K'ung Fu-tse's philosophical thought in the sixth century B.C., whose aim was fundamental social reform. In the midst of the evils of war and violence brought about by conflicts between the small feudal states and the tyrannies of corrupt governments, K'ung Fu-tse preached moral principles which were to usher in peace, justice, and nonviolent universal order. He believed in the original goodness of man and in the historical role of rulers, but he also called attention to the fact that "grass is not bent by the wind as forcefully as people by the wills of their rulers," regardless of who those rulers may be. In striving to achieve his ideal for humanity K'ung Fu-tse advocated the avoidance of violence and of any extremes. The "golden rule" of his teaching—"Do not do unto others what you do not want done unto you"—is essentially the same as Christ's principle. And initially it shared a similar fate: cruel persecution and the burning of books, contradictory interpretations, transformation of philosophical ideas into a popular, dogmatic religion, and even the transformation of K'ung Fu-tse himself from a philosopher to a deity to whom even sacrifices were offered. However, his teaching has also been subjected to repeated elaboration, which has made Confucianism one of the most fundamental components of the spiritual being of the Chinese people to this day.[19] This, in a way, also proves the inability of the oldest and most sagacious thoughts of the human mind to eliminate violence.

In the part of the world where Islam, the youngest of the great world religions, was established, the role of violence is even greater. The Qur'an teaches that "all people constitute one

nation"[20] and that the division of mankind into tribes and nations is not an obstacle to their fraternity.[21] But it also teaches that both individuals and nations will be called to render an account of their deeds.[22] And the "holy war" against the "infidel" was the best way for the followers of Islam to achieve not only the benefit of earthly domination, but also to attain eternal life in heaven. Since the Qur'an ordered its followers to fight with "all fortitude" against the "infidels" who "dwell in hell" and "to be harsh toward them,"[23] war campaigns of Moslem conquerors usually involved merciless violence toward entire populations of the invaded and conquered countries.

Although rationalist thought has emerged in Moslem countries, too, philosophical reflection has remained, for the most part, within the confines of theology. It is significant that Islam, the latest of the three great monotheistic religions, ever since its appearance in the seventh century, has shown enmity and intolerance toward Judaism and Christianity—out of which it has emerged, and with which it has the most in common.

2. In Ancient Philosophical Reasoning

Everything exists on the basis of strife... War is the father of everything and the king of all.

—Heraclitus

Vae victis! (Woe to the vanquished!)

—Roman proverb

In ancient history one can perceive man's doubts about the sense and purpose of violence. In Hellenic and Roman antiquity one can find the foundations of all the subsequent depth of human thought. In this way the historical insights of vanished civilizations have reached us and continue to exist.

The Hellenic goddess of revenge, Nemesis, had a very important place in ancient Greek mythology. She was the mistress of man's fate and of the communities in which man lives and shares his fate with others. The goddess Nemesis was the embodiment of

the Greek notion that only through the vindication of injustice and crime can law and order among men and nations be established.[24] In his epic poems *The Iliad* and *The Odyssey*, Homer offers a picture of Greek mythological thought and historical reality at the time of the Trojan War (twelfth century B.C.). The *Iliad* recounts only a brief period of that ten-year war between allied Greek city-states and the Trojans. At the conclusion of this war, the Greeks performed an indisputably genocidal crime which is glorified as a splendid victory. After capturing Troy through wily deception, the Greeks completely burned down the city and killed all its inhabitants except for a few individuals who had fled. Although Homer ascribes these historical events to destiny and to the will and actions of the Moirai—three sisters who spin and cut the threads of human lives and historical events—human activity is shown as part of a whole connected by fate. Even the gods themselves are an inseparable part of that fate as they, too, participate in the war, separated into two hostile camps. Here, the right of the stronger is accepted as a natural necessity. And just as in nature the "wind scatters the leaves" so new ones may grow, men also suffer the same fate. "Some succeed; others perish," says Homer.

In the *Iliad* one also finds the glorification of excessive retaliation, as when the angry Achilles swears over the body of the dead Patroclus that he will properly avenge him by "slaughtering twelve sons of Troy." Nevertheless, Homer's world strives for the triumph of justice in the struggle between good and evil, for the sake of achieving progress and greater harmony in the world.[25]

About a century later, Hesiod understood the cosmos as a battle of opposites, and saw the way toward reconciliation. The power of the stronger serves to introduce order, and Zeus's "men of heroes" are a "divine order," who in the fervor of war "mistakenly kill each other." However, for humankind there is no salvation from evil and misfortune, judging by Hesiod's ageless counsel that "he who brings evil to others brings evil to himself."[26]

According to Diogenes Laertius, the "greatest deed" of the wise Solon—thinker, poet, and great politician—was when he induced

the Athenians to go to war against the Megarians because of his birthplace, Salamis. When the Athenians, after several defeats, decided that whoever suggested a new war would be sentenced to death, Solon provoked the Athenians to start yet another war.[27] Solon, the wisest man of his time, would not have resorted to such a peculiar ruse if he had not regarded military conquests as normal and suited to Athenian interests. One finds such conceptions shared by essentially all the great men of Hellenic culture.

Heraclitus taught that "all things are created by fate," and can be brought "into harmony [only] through a struggle between opposite currents," for the world is simultaneously the harmony and the struggle between opposites. Moreover, everything can be recognized only by its opposite. And, "of those opposites, the one which strives toward birth and creation is called war and struggle, and the one which strives toward destruction through fire is called harmony and peace." Heraclitus certainly knew the true nature of war, and yet he glorified it as a supreme act of creation and a socio-ethical principle: "War is the father of everything and the king of everybody; it has made some into gods, others into men, some into slaves, and others into free citizens."[28]

The nature of warfare at the time of Heraclitus can be concluded from the fate of the city of Miletus, which was completely destroyed in the uprising against the Persians in 494 B.C. All its inhabitants were either killed or deported to Persia. Aeschylus himself fought at Marathon, Salamis, and Plataea, and celebrated the Greek victories over the Persians in enraptured verses. In his numerous dramas and tragedies, the gods are protectors of justice and avengers of crime. Descendants, too, are subject to retaliation for the crimes of their ancestors. His gods and heroes are extreme in their noble actions, but also in passion and crime.

Sophocles, another great Greek dramatist, also participated in the war against Samos. For him, too, conflict and war are mysteries of dark fate, the solutions to which he does not question. Instead he takes these mysteries as a model for the historical characters in his dramas, whose essence is revealed precisely through contradictions and violence. It is not surprising to hear him say in

his play *Antigone* that "there are many strange things, but none is stranger than man."

Without Pericles' triumph as a military commander, the flourishing of Athens would not have come about. It must not be forgotten that the greatness of Athens did not rest only on defensive wars, but also on violent wars of conquest. Pericles himself urged the Athenians to go to war even after he had learned that both of his sons had perished.[29]

Empedocles, who suffered violence as an exile, and who was regarded by Aristotle as the founder of rhetoric, concluded that love and hatred reign as inseparable principles, as the cause and origins of everything. They rule alternately, as elements and forces, and their cycle of harmony and hostility never ceases.[30]

Socrates also excelled in courage as a young Athenian soldier. But he was equally courageous later in his life when he was condemned to death. The trial of Socrates is the model for the persecution of thinking people in any age. Reflecting on justice and injustice, Socrates arrived at the insight that truth and morality should be identical concepts, that nobody sins intentionally, so that sin is only an error. But his rational ethics also remained subject to the mystical inner voice of the demon which directs all actions in one's life.

Similarly, Democritus realized that lust and passions can dominate men.[31] He nevertheless had faith in the power of reason and good over evil and violence. Life's ultimate goal of happiness can be achieved only through spiritual balance and peace of soul. This was also reflected in Protagoras' (481-411 B.C.) great principle: "Man is the measure of all things."[32] According to Protagoras, "there are two statements about everything, one contrary to the other." This means that truth is not something in itself, independent and objective, but rather a subjective perception. The truth is what seems to be true to someone. Thus all opinions and ideas may be true from the viewpoint of individual perception.[33]

The circumstances that Plato experienced—the breakdown of the power of the Athenian state after the Peloponnesian War, political turmoil, terror by individuals and the mob, and, in par-

ticular, the sad fate of Socrates himself—inspired him to work out a system for the ideal state and society. He presented this in his *Republic* and *Laws*. Plato finds an explanation for everything in the necessary conflict of opposites: good and evil, justice and injustice, reason and irrationality, law and violence. In Plato's philosophical system of ideas, the highest is good, because everything that comes into being and changes strives toward the realization of good. Yet Plato's ethical idealism retains the dualistic principle. He does not ignore life's horrors and cruelties. Man cannot escape from the eternal cosmic struggle between good and evil because divine harmony in general and social harmony in the state can be achieved only through this struggle. Consequently, Plato, in his ideal republic, gives an especially high and honorable place to the special class of soldiers and warriors, second only to the philosopher-rulers. The warriors' only duty is to protect the republic from the enemy, and for this purpose they are specially educated and trained in the skills of warfare. Even Plato regarded force and war as necessary and inevitable phenomena. And to him a war could be a true war only if it is between the Greeks and the barbarians, "who are enemies by nature," while wars among the Greeks themselves, however bloody, represent only conflicts or friction, because mere rivals are not "natural" enemies but only adversaries. Consequently, in his utopian *Republic*, Plato explicitly states that "justice is nothing but the interests of the stronger."

While Plato focused on the ideal state, Aristotle described the reality of the state-political system of the Greek world in his *Politics*. According to Aristotle, the love of oneself, one's family, and homeland are not evils but deeply rooted human needs. Such love becomes evil only in the case of its exaggeration, when it is accommodated at the expense of others. As a man who had witnessed all the evils and misfortunes brought upon the citizens of Greece by unrestrained passions, violent upheavals, and conflicts between the city-states for dominance, Aristotle became an advocate of stability and moderation in political life and of the principle of the "golden mean" between intolerant extremes. Like Plato, he too regarded the four basic human virtues—courage, modera-

tion, justice, and reason—equally necessary in both peace and war. According to Aristotle, life consists of struggle and peace, work and rest, useful labor and exertion. Peace, leisure, and the arts are the blessings of life, while war, work, and labor are necessary evils. Therefore, peaceful leisure must be secured even by means of struggle. That is why citizens must be trained for valiant warfare as well as for noble leisure.

Man is by nature a "social animal" (*zoon politikon*), and only within a community does he become a "social being." However, it is natural that there be inequality, freedom, and slavery among people. As a social being, man can achieve his bliss only within the framework of a community. A state is the highest and most powerful community because it embraces all the other communities, and strives toward the highest good from everyone and for everyone. Since, according to Aristotle, all communities have their purpose and strive toward some good, he regards the freedom of some and the slavery of others as natural phenomena. He quotes poets who say that "the Greeks rightfully rule over the barbarians." Apart from the fact that he believed that the Greeks were determined by nature to be free and to rule, and that all barbarians (non-Greeks) were determined to be slaves, Aristotle also felt it was justified that prisoners of war become slaves by law. This shows that he accepted war violence as being normal, and even purposeful.

Aristotle developed the old Hellenic conviction of the exceptionality of the Greeks and of the natural subordination of the "barbarian" nations into a theoretical justification for imperial expansionism. It is no accident that Philip of Macedonia invited him to become the tutor of his son Alexander. Aristotle was still Alexander's teacher at the time when the latter took part in the battle at Heronea (338 B.C.) in which Athens and Thebes were defeated.

One of the Athenian participants in that battle was Demosthenes, who became famous for his "philippics"—speeches against the Macedonian king, Philip II, branding him a conqueror and destroyer of Greek liberty. Demosthenes's efforts to prompt

the Athenian alliance to offer strong resistance to Macedonian conquests was futile. The old Greek *polis* was crushed, and Macedonia became the greater power. In fact, Alexander the Great united the Greek world and then, with his conquests in the Middle East and Asia, expanded Macedonian-Greek domination temporarily, and Hellenic culture more permanently.[34] The fact that Aristotle did not accompany Alexander on his military expedition in the East, but returned to Athens, and that later, after Alexander's death, he had to flee Athens, are examples both of his personal duality and the fact that there cannot be lasting harmony between philosophical thought about the world and statesmen's operations of conquest. It also shows that history does not forgive anyone who serves his people's enemies.

Epicurus belongs to the list of great Greek philosophers who exercised particular influence on the subsequent development of European philosophical thought. Lucretius Carus (96-55 B.C.), who continued Epicurus' enlightenment philosophy in Rome two centuries later, said that "his mind surpassed all humankind." Even twenty centuries later he has great influence. According to Epicurus, the aim and final goal of everything must be man's happiness, which can be achieved by knowing and respecting the laws of nature. Everything in nature, life, and the world has its model. It follows that events are the result of necessity—which acts as an uncompromising force. But some events take place only by chance, and still others are prompted by us ourselves. Therefore, "the future is not entirely in our hands, nor is it completely independent of our will." Epicurus believed that "Justice has never existed in an absolute sense," but only as an agreement concerning mutual relationships. People who cannot agree "not to harm others or to tolerate harm, know neither justice nor injustice." He concluded that this applies "also to nations that are not able to reach an agreement not to cause or tolerate harm." The essence of the problem is embodied in the fact that "injustice by itself is no evil, but the evil is realized only in the consequences" as well as in that every community observes the rule that "just is that which is useful to the community."[35]

Epicurus advocated searching for happiness through separation from the community, through peace of mind and personal satisfaction in life, and by striving to overcome pain and evil through wisdom. He realized the powerlessness of man's will and the impossibility of changing historical and social reality in which the agreement of men and nations about mutual restraint from doing violence and harm remains an almost unattainable goal, a rare and usually short-lived phenomenon, a fleeting moment in the whole parade of history.

The victory of Philip of Macedonia over the Athens-Thebes alliance at Heronea (338 B.C.) marked the end of the era of the Greek *polis*, but not the end of Hellenic culture. By assuming the leadership of the association of Greek states, not as a conqueror, but as the leader in the war against the Persians, Philip helped create conditions for the spread of Hellenism, which was furthered by the military expeditions of his son. Alexander's world state could not have the same purpose, nor could it be the same in structure and operation as the earlier Hellenic states. The larger and more ecumenical the world was becoming, the more powerful man felt over nature and space, and the more alone and helpless in relation to the ever more awful armed forces of the state and increase of violence. This was manifested in the philosophical views of Stoicism, which was founded by the Cypriot Zeno (366-264 B.C.) and developed further by Roman thinkers. Confronted with a world upon which he had an ever smaller influence, the Stoic thinker sought solutions in the strength of his own spiritual being to rise above existing conditions that he could not change.

With the establishment of Alexander's empire, barbarians were given the right to become Greeks. In this way, Hellenism, by the imposition of state-political dominance and by the expansion of the Greek culture and language, essentially meant the assimilation of non-Greeks.[36] Hellenism can undoubtedly serve as one more convincing example of the tacit acceptance of violence as a normal component of history. But Hellenism can also be taken as proof that it is futile to attempt to assimilate numerous historic and ethnic peoples into a single nation. This was the main reason for the

breakdown of Alexander's empire soon after his death. And when Greece became the booty of the powerful Roman Empire, the same elements were to determine its fate as well.

The imperial eras of Hellenism and the Roman Empire produced fewer original philosophical and historical thinkers than the previous periods, except in the areas of military strategy and state administration. Roman thought began in Greece, but it was more preoccupied with practical than theoretical questions.

One of the best known mottos from Roman history is *vae victis!* (woe to the vanquished). It was actually spoken by the Gallic leader, Breno (387 B.C.), but within its meaning the entire history of the great Roman Empire unfolds—its wars of conquest to enslave various foreign peoples, its frequent civil wars, and the uprisings of subjugated nations and of slaves.[37] Roman history is, in fact, a state of continuous conflict, armed assaults, and bitter perils.

This moved the Roman poet and philosopher, Lucretius Carus, to invoke the divine powers of Venus that "the rumble and clangor be quelled" and that "the cries of the raging war on both land and sea subside" in an appeal for an attractive and coveted peace which, for the Romans, was to remain unattainable. To Lucretius, war had such a significance in social life that he used the image of war to explain and evoke the atomistic picture of the creation of the world in the eternal motion of matter, and collision and fusion of atoms in the infinity of empty space.[38]

At that time, conflict and war were the normal condition. And long-coveted peace was attained, just as in the Greek era, only through war (*pax paritur bello*), as Cornelius Nepos (100-24 B.C.) noted in his description of world events and famous rulers.[39] The hell created by force and violence, symbolized by the expression "iron and blood" (*ferrum et sanguis*), neither originated nor ended with Rome, but we have inherited the expression from the glory that was Rome.

The first true Roman historian, Polybius (200-120 B.C.), testified with his life and work to the continuity between Greek and Roman history. He spoke of the Roman Empire as the best possi-

ble regime, but at the same time he was far from the opinion that the mistakes and evils of the past could be overcome or escaped in history. Polybius's views were based on his own personal experience. Having found himself among the thousand hostages that the Achaeans had to hand over to Rome, he joined in the Roman cultural life and became the tutor and friend of Publius Cornelius Scipio, Jr. With Scipio he went to war against the Carthaginians and became a witness to the destruction of Carthage in 146 B.C. When the Romans in the same year crushed the Achaean League and conquered Corinth, Polybius tried to assuage the fate of his own people.[40]

The philosopher, statesman, and greatest of Roman orators, Marcus Tullius Cicero (106-43 B.C.) was a participant in and witness to internecine political violence as well as wars. He uncovered and sentenced Catiline's conspirators to death, and was celebrated as the savior of the state, only to experience the bitterness of exile followed by the glory of his return to power. In civil battles and wars, he was with Pompey and then with Caesar. After the assassination of Caesar, he became the leader of the senate, leading a decisive fight for the defense of the republic against Marcus Antonius's plans to become dictator. When Antonius succeeded, Cicero, along with his son and his brother were persecuted. Cicero's head and right hand were cut off and exhibited on the platform where he had delivered his fiery speeches against the very tyrants who eventually killed him. Cicero contributed to the spread and latinization of the Hellenic culture throughout the vast territories of the Roman Empire. And his ideas are woven into the most valuable elements of Greco-Roman culture, which, with the advent of Christianity, were to become the foundation of European and Western civilization.

Like the other Roman thinkers, Cicero did not recognize an equality between the Romans and other "barbarian" nations. In his view, "other nations can bear slavery, but to the Romans, freedom is innate." Cicero saw the use of violence as a necessary evil. "There are two ways of solving disputes, one by discussion, the other by force. Since the former is appropriate to man and the lat-

ter to animals, the latter is used only when the former cannot be applied. It is therefore permissible to wage war if its purpose is to live in peace, without injustice." Cicero reduced the relations between nations and states to their observance of the laws of warfare. In this context, he also advocated the limitation of violence. "When victory is achieved, those who were not inhumane in combat should be spared." When a "war is waged for the sake of glory," the motives for the war should be just. In this way Cicero expressed the already established rule that a war is just (*bellum justum*) if it has a just cause and if it is started in the proper way. Regarding violence in general, Cicero pointed out that "people desecrate the very essence of nature when they make a distinction between profit and morality," and that pillaging and doing harm to others for the sake of one's own bounty is more contrary to nature than death itself. However, according to the understanding of the laws of warfare at the time, everything belonged to the winner: personal property was taken as booty, land was appropriated by the state, people were taken into slavery, and truces arranged by military commanders were frequently guaranteed by taking hostages. From his own experience, Cicero was aware of the powerlessness of moral principles and said that it was deplorable to fear the evil that cannot be avoided. He also knew that no matter how improbable, base, repugnant, or inconceivable an action might be, hypocritical rhetoric could always make it probable, attractive, or worthy of admiration. He testified to his conception of liberty with his own life, and his view that "death is terrible to those who lose everything with their life, but not to those whose glory will never die" contains an understanding of the essence of force and violence since primordial times, but also an insight into the necessity and justification of an unremitting instinctual resistance to evil, and of the nobility of sacrifice for the sake of freedom.[41]

The philosophical historical approach of Livy (*Titus Livius*), expressed much more strongly than Cicero the idea of the exceptionality of the Romans, their superiority over other nations, and their right to dominate the world. He was self-confidently biased

against the opponents of Rome, even when they proved their supe-
riority, as in the case of Hannibal.[42]

The poetic works of Ovid (43 B.C.-18 A.D.) exerted great influ-
ence on later European literature. His bitter recognition that
"brotherly love is rare" (*fratrum gratia rara est*) seems to reflect
the relations between very nations rather than among a family. In
his collections of elegiac poems written in exile (*Tristia* and *Ex
Ponto*), he describes the abhorrent barbarian surroundings and
longs for the blessings of Rome.

The thought and the fate of Lucretius Aeneas Seneca (4 B.C.-
65 A.D.) likewise describe the perpetual contradictions within
Roman society. This follower of Stoic philosophy became tutor and
counselor to Nero. At first, under Seneca's influence, Nero ruled
sensibly and moderately. But the intoxicating effect of power soon
prevailed over Seneca's humanistic ideas. Seneca himself was dis-
missed, accused of conspiring against the emperor, and was forced
to take his own life. Nero committed such reckless violence and
such horrible crimes—from the assassinations of his mother, his
wife, conspirators, and innocents, to the burning of Rome and
bloody persecution of Christians—that he has remained a symbol
of terrorism and bloodthirstiness. His violence led to rebellions in
the provinces (Gallia, Hispania, and Germania) and in Rome itself,
so that the Senate proclaimed him the enemy of the Roman people
and sentenced him to death, after which he committed suicide.
Since Nero is not a unique or exceptional villain in history, but only
an extremely exaggerated case of savagery, the example of his age
is an illustration of how the most outrageous crimes can happen in
the shadow and presence of suppressed humanistic ideas.

This is why Seneca asked, "Of what use is philosophy if a god
rules the world?" In looking for a solution, he remained firmly
rooted in reality, not in dreams, because he knew that "any
attempt to defy nature is a futile labor." He preached a doctrine of
humane relations built on wisdom and wholesome human virtues.
Seneca urged those who could not imagine liberating slaves to rec-
ollect all those who had fallen into slavery and even to imagine
themselves slaves. Regarding the prevalence of violence among

nations, one recalls Seneca's wise comment that "No one loves his homeland because of its greatness, but because it is his."[43]

One finds a confirmation of all of this in the works of the great Roman historian Cornelius Tacitus (55-120 A.D.). His contribution to political life was considerable (he held the post of consul for the emperor Nerva), but he was spared the fate of Seneca. Tacitus attributed the main reasons for historical events to the characters of individual people, which are predetermined. In his opinion, a good man is incapable of evil deeds, but he who proves himself to be bad must always have been so, while his true nature was concealed by pretense. Here, Tacitus followed Aristotle's view that no force can change the character of a man, but merely serves to reveal his true nature. Tacitus was an advocate of republican freedoms, but he was also a supporter of violent means, expressing contempt toward peaceful government. He glorified the Roman conquests in Britain and Germany, and praised military successes, even though he was aware of what war campaigns brought with them. He said of the victors: "they propagate devastation and call it peace." In his depiction, the crude but uncorrupted Germanic tribes present a striking contrast to the moral corruption of Roman society at that time.[44]

The Roman emperor Marcus Antonius Aurelius (121-180 A.D.) was a great leader and also a philosopher. Historians note that from his youth he was an ardent follower of Stoicism, and that he remained faithful to those philosophical principles even as emperor. A confirmation of this can be found in his work *Conversations with Myself*, but is it also to be found in his imperial rule? In fact, his reign was characterized disproportionately more by the sword than by any philosophy. His armies waged war on the borders of the empire and suppressed uprisings in Britain and the Middle East. They crushed attacks by Germanic tribes that attempted to invade Roman territories. In the East, they battled against the Parthians, in Armenia, Syria, and Mesopotamia. Marcus Aurelius, through his "extraordinary courage and perseverance" proved himself to be "as great a warrior as he was a philosopher."

The philosopher's reign was filled with war and, of course, with all the unavoidable forms of violence that come with it. Nevertheless, even in that chaos Marcus Aurelius was able to institute paid teaching positions in schools of philosophy in Athens, Alexandria, Smyrna, and Syria. According to him, war and philosophy were equal natural phenomena.[45]

3. In Christian Scholastics

The fate of the weaker hangs on the desire of the stronger within whose power it lies.

—Saint Augustine

Internal social frictions and an increasing disarray within the Roman Empire was seen in the violence of its civil and foreign wars, its ever more obvious inability to maintain rule over the vast territories of three continents against the uprisings of conquered barbarian nations and invasions by others. These conflicts all created the conditions that deepened not only the social crisis in the empire, but the entire spiritual worldview of the ancient world. It also helped elevate persecuted Christianity to the level of a world religion.

Although Christ said, "I have not come to destroy the Law, but to fulfill it,"[46] Christianity advanced as a revolutionary movement of oppressed peoples. In a time of unrestrained rule of force and violence, Christianity conquered much of the world with its new ethics, by preaching the commandments of love and brotherhood, peace and nonviolence, equality and reverence, forgiveness and mercy. Aware that such a state of bliss cannot be realized here on Earth, Christianity promises it in the other world. On the Day of Judgment all will receive just punishments for the evil they have done, and rewards for their good deeds. In this way, Christianity disrupted the internal harmony between body and soul which was the ideal of ancient philosophical thought. Its supranational, universalist ideology likewise contradicted reality.

Christianity teaches that "there is neither Jew nor Greek, there

is neither bond nor free" but rather that "all are one in Christ Jesus."[47] While Christianity was still a small, secret, and persecuted sect, such a view was possible. However, as soon as Christianity became dominant this proved to be a delusion. There was proof of this even at the very beginning when Christianity was cast out from the bosom of Judaism, where it had originated.[48]

Aurelius Augustinus (Saint Augustine, 354-430 A.D.) was the first great Christian philosopher. He turned to Christianity at the mature age of thirty-three, having previously been a pagan. According to Augustine, one arrives at truth through self-cognition, while rational cognition rests upon knowledge (*scientia*) and wisdom (*sapientia*). God exists as the creative Perfection of truth, good and beauty, as the Purpose, Ideal, and Source of all. Nature strives toward order and harmony. However, the struggle never ends and there is no final peace in this world. Although the meaning of life is in love and good, the presence of evil is constant. For this reason, a moral life presupposes a free will. He who disturbs the desire for harmony and violates the order and laws of nature must bear the responsibility and the penalty.

For Augustine, the state rests upon the love and interdependence of its members, rationally united for their collective benefit. But authority comes from God, and apart from the temporal state (*civitas terrena*), there exists an eternal, divine (*civitas Dei*) or celestial state (*civitas coelestis*). The Church as the representative of God's Kingdom on Earth must prevail over the state in all matters regarding religion and morality, while in mundane matters, the Church recognizes the authority of the state.

This Augustinian view concerning the separation of church and state became a permanent element and even an essential determinant of Western culture. To be sure, there were times when the papal and secular authorities vied for power or clashed in armed confrontation. But in the West, this principle was to attain permanence by the explanation that combining secular and spiritual power in one person often results in despotism. Contrary to Augustinian views, the caesaro-papism of the Byzantine Empire prevailed in the East. There, the efforts of the Eastern Roman

emperors toward bringing the Christian Church under their control brought about a schism with incalculable historical consequences, resulting in the development of the Eastern Roman or Orthodox Christian Church. In this century, it manifested itself in the Western European Marxist ideology, taking the form of Leninist-Stalinist totalitarianism.

Augustine preached that the greatest good on Earth is achieved in peaceful order. He opposed slavery and advocated freedom and equality among people. Violence, armed warfare, unrest, and disaster are brought about through man's attempts to impose his will and power in order to gain control over others, rather than yielding to the will of God. The creation of great empires was stained with blood, and their survival depended on continual conflict and war, which does not bring people closer to God, but to the demons of evil. For this reason the worldly state is proclaimed the state of Satan (*civitas diaboli*). Man can find his deliverance and his realization only in the kingdom of heaven.

However, Augustine did not unconditionally oppose all war. A war can be just if it is waged in defense against injustice, for the protection of the sacred rights and properties of man, or if its true aim is to establish or ensure peace. Augustine advocates the moral virtues to curb the impulses of passions and greed, gluttony and vanity, and to honor the principles of justice in society and in the state. Augustine's cry of protest—"What is a state without justice if it is a great band of outlaws?"—was powerful in his own time and is still relevant today.[49]

In appraising and interpreting historical events, Augustine always embarks from the standpoint that man—as the only good and the being of highest value—is also a measure of all occurrences.[50,51] He judges that the mobility and migration of religions and cultures is a phenomenon which leads to a more civilized world. Meetings and clashes between them bring about acclimatization and transformation, gradual assimilation, or even extinction.[52] Augustine sees free will as the independent cause for man's actions, despite God's foreknowledge. And fate is an unavoidable necessity independent of man's will, "the fate of the weaker is the

will of the stronger, who has power over him, rather than allow an order of the ideal [which the Stoics in their own way call fate] to annihilate the freedom of our will."

Augustine points out the difficulties in trying to extract from many various historical events one universal law, or a common suprahistorical purpose, as a foundation for the understanding of events.[53] With regard to the distorted and erroneous interpretations of certain historical phenomena, Augustine points out that the blame does not lie with the ignorant, but with those "who are educated in the liberal arts, love history and know all that quite well, but in order to stir up the uneducated mobs against us, they pretend that they do not know and they do their utmost to convince the masses that for all the misery (which has plagued humankind in certain times and places), the very Christian name is to blame."[54]

Although he wrote in defense of Rome's wars of conquest, Augustine took advantage of every possible opportunity to reveal the destructive and inhuman side of war. In writing of the Second Punic War, instead of the victories he points to the "defeat of both nations." He mentions that even those whose motive was to "glorify the Roman leadership" admit "that the conqueror seemed more like the conquered." Augustine was to present examples of ruthless violence as evidence of the need to accept Christian views. He mentions the case of "the king of Asia, Mithridates, who ordered that in a single day all the Roman citizens found anywhere in Asia be killed," an order which was carried out in such a way that "every one of them was struck down no matter where he was found: at home, on the street, at the market, in the temple, in bed, or at a feast!" and the local inhabitants were forced not only "to watch this grisly bloodshed in their own homes," but they also had to participate in it. From this, Augustine draws a profound message: "the wounds were reciprocal: those who were assaulted were physically wounded, while their assailants incurred wounds to the soul!"[55]

With even greater fervor, Augustine described the irrationality of civil wars which "are more bitter than any war against a for-

eign enemy." In the civil war between the Marians and the fol-
lowers of Sulla, when the actual war came to an end, "it turned
into an even more brutal peace." First those on the Marian side,
having lost faith "not only in victory, but in life itself... continued
the slaughter of their own people and their enemy," but when
"Sulla later entered the town as the victor... a frenzy of peace
broke out." On the Field of Mars, "he slaughtered seven thousand
of those who had surrendered and were, of course, not armed,"
after which two thousand high-ranking military officials were sen-
tenced to death, confiscation of property, or banishment. The tor-
mentors performed more savage acts of carnage "than animals
when they tear apart an abandoned carcass." Augustine rightful-
ly asks: "Which rage of foreign tribes, which barbarian rancor can
compare with that victory of one citizen over another? What did
Rome see as more ruinous, more horrible, more bitter: the past
invasion of the Gauls and the recent attack by the Goths, or the
savagery of Marius and Sulla and the finest men of their leagues,
pitted against each other just as one's own eyes against one's own
limbs?" But apart from the fact that many distinguished men lost
their lives in civil wars and in the reign of terror that accompanied
them, those wars also brought on the corruption and trampling of
all moral values.[56]

In the entire medieval period after Augustine, one finds no new
or more profound reflections upon violence. Quite the contrary.
Along with taking up the position of the ruling religion, the
Christian Church was to transform philosophy into *ancilla the-
ologiae,* so that virtually all scholastic ideas were to be directed
toward justifying the battle against the nonbelievers and toward
proving the necessity for their extermination. One finds support
for this even in the views of Thomas Aquinas (1225-1274), the
greatest Christian theologian and philosopher of the Middle Ages.
His belief that it was "more just to put heretics to death than fal-
sifiers of money" reflects the spiritual situation of a time which
exhibited the brutality of the inquisition and religious wars. Harsh
and primitive incentives to brutal actions for the sake of victory
were to be increasingly justified by loftier goals: religious, imperi-

alistic, national, and ideological.[57] The Middle Ages were in this way no "dark" exception.

Niccolo Machiavelli (1469-1527), is one of the best known thinkers of the Renaissance. His ideas concerning government, violence, and war have caused many to flinch. But in *The Prince*, Machiavelli was simply presenting the truth of his times. He explained how to gain authority by utilizing the chaos of violence from the standpoint of pure exploitation and self-interest on behalf of his imperiled land. He was familiar with both Plato's and Dante's ideal states, but also with the reality of history in which such utopian orders have never existed. For this reason, Machiavelli consecrates the principle of force to such a degree that he gives the absolute ruler not only the unlimited right, but also the duty to wage war with all available means in order to achieve power and impose his authority. He did not advise criminal violence at all costs, but "really there is no more reliable method than destruction if you wish to maintain your control over conquered cities." Therefore, "one should condemn the destructive violence, but not the violence that renews." He supported peaceful measures and philanthropic magnanimity as necessary components of a good reign, but he also advocates caution and unity, for "man is inclined more towards evil than towards good. Moreover, people will always be wicked unless some necessity forces them to be good."

However, to ascribe to Machiavelli the historical blame for all later abuses of the notorious principle that the end justifies the means is a mistake. In the name of false virtue, Machiavelli is blamed for the crimes which he wanted to oppose by the only effective means according to his understanding of history.[58]

II. CONTEMPLATING THE MEANING
OF INEVITABLE EVIL

4. In Humanism and the Renaissance

Where the greatest atrocities originate, from there the greatest virtues arise.

—M. A. Muret

Desiderius Erasmus of Rotterdam (1466-1536), the great humanist of his age, rose against the utilitarian acceptance of war and violence, and particularly against the scholastic teaching that "every heretic is a criminal" who should be destroyed from the face of the Earth. Erasmus warned that "one war stems from another and from one, two wars ensue," so that this most uncivil and violent way of resolving mutual antagonism can never achieve human harmony.[59]

In the sixteenth century, German humanist M. Neander proclaimed his opinion against violence: "One must not do evil in order to achieve good." But he also said, "It is permissible to deflect force with force." The French humanist writer M. A. Muret (1526-1585) commented that "Where the greatest atrocities originate, from there the greatest virtues arise." More skeptical and better known than Muret, was his teacher, Michel de Montaigne (1533-1592). Montaigne remarked that "There is no animal in the world a man must fear more than another man." In his opinion, the cost of struggle is greater than the victory. "Are there not triumphant defeats of which even the most glorious victories are envious?" Montaigne was a skeptic, but not a nihilist. Although he understood that eternal evil is inevitably present, he remained an advocate of the continual struggle for freedom and a more civilized world.[60]

In a very distinctive and deeply moving way, the life and fate of

Giordano Bruno (1548-1600) relate to us the character of violence in the Renaissance. His heretical opinions forced him to leave the clerical profession and Italy. He wandered all over Europe teaching as a professor in Switzerland, France, England, and Germany. In Geneva, he converted to Calvinism, but his attacks against Christian dogma were aimed at Protestantism as well as Catholicism. In virtually all of his arguments, he raised strong accusations against the Jews, whom he despised. When he returned to Venice, Bruno was extradited to face the Roman inquisition, before which he tried not to reveal his heretical opinions, convincing his inquisitors that he had been misunderstood. Still, as he did not want to retract his fundamental belief about "the infinity of the universe and the plurality of worlds," he was, after seven years of imprisonment as a heretic and apostate, condemned to death. His executioner was given orders to "treat him kindly and, if possible, without bloodshed," so that Bruno was merely burned at the stake.

Since his youth, Bruno had sensed the contradictions of his age. This feeling of division is revealed in his shrewd remark: "joyful in sorrow, sorrowful in joy." Bruno resolutely rejected the value of majority opinion, the authority of tradition, and in particular, religious dogma. For the scientist, "the only authority should be reason and free investigation." To him, God is "nature which works." With his interpretation that events in nature are the manifestations of God as a cosmic power, Bruno was a prominent advocate of naturalistic pantheism. God as the primary cause unites in himself all the opposites, although his qualities are power, wisdom, and love. According to Bruno the world is a unified whole, yet it is not only ruled by love and friendship, but also by hatred and hostility. "The principles of decline and emergence are one and the same," for "decline is nothing other than emergence, and emergence is decline." And ultimately, "Love is hate, and hate is love."[61]

Bruno endures as a lasting component of modern civilization. The same cannot be said for his executioners who disappeared anonymously into the wilderness of history, only to reaffirm that

violence can destroy the physical, but not the spiritual, life.

The Croatian philosophers and writers of this period should not be ignored. Their thinking reflects, on the one hand, the circumstances of the direct war atrocities caused by the hostile penetration of the Ottoman conquerors into Croatian lands and by the Venetian menace. On the other hand, Croatian thought reflects the universalist bond among the European intelligentsia who wrote their works and communicated primarily in Latin at that time.

The erudite philosopher Franjo Petric (Franciscus Petricius, 1529-1597), was one of the greatest names in world philosophy of his time. The German philosopher F. Ûberweg, author of the most thorough and objective philosophical survey in the nineteenth century, considers him to be Giordano Bruno's predecessor. Renowned experts in the Latin tradition in Croatia and Europe say that Petric was, by virtue of his early work, *The Fortunate City* (*La Citta Felice*), a precursor to Campanella and his *City of the Sun* (*Civitas Solis*), which appeared seventy years later; with his *New Universal Philosophy,* Petric also preceded certain later philosophical ideas of Descartes and Kant.

As a young man, Petric fought on a galley against the Turks in Croatian and Greek waters. Because of his anti-Venetian views, he was banished from his native island of Cres. He was at first self-educated, but he later completed philosophical studies and became one of the most prominent members of the Italian and Spanish intellectual circles. He finally became a professor of Platonic philosophy in Rome and the dean of the St. Jerome College. Petric discussed every scientific field from astronomy and geometry to medicine and meteorology. In his *Dialogues on History,* he allowed many representatives of the contemporary cultural life to speak, making efforts to expound upon the course and purpose of history from a philosophical standpoint. In his treatises on the military and warfare (*Milizia Romana* and *Paralleli Militari*), he brought up critical observations concerning the military of that period. He bluntly attacked the use of foreign mercenaries, who were always hired for aggressive military under-

takings and plundering, and maintained that the military should be national in its structure, which implied democratic freedom and equality among the population. In his philosophical interpretation of the world, Petric explicated a theory that all of nature is a reflection of divine primeval radiance. Everything that exists is immanent to the harmony of the world: everything is "one and all" (*unomnia*), undistinguished and undivided in its original element. The soul, too, is among the primary fundamental qualities of all the material phenomena (light, warmth, spatiality, and fluidity). Everything in nature is permeated by the soul, and the universe as a whole has its rational "universal soul," of which every human soul is a part.

Due to its naturalistic and pantheistic elements, Petric's *New Universal Philosophy* was subjected to the criticism of orthodox theologians, and was later banned. His friendship with Pope Clement VIII saved him from persecution, but he was no longer permitted to publish. In a letter of Petric to Clement's predecessor, Gregory XIV and "the future Popes of Rome," in which he explained his philosophical views, he said, among other things, since "every kind of violence and deception causes fear" for "the human intellect conforms only to intellect, voluntarily follows intellect, and allows itself, whether willingly or unwillingly, to be drawn by intellect," it is therefore only "by the intellect that people should be brought towards God." So concludes Petric in order to win the popes over to the idea of spreading Christianity, not by force, but by religious instruction.[62] This is certainly a profound reflection upon the pointlessness of violence against human nature and on faith in the power of the human intellect.

The poet, Latinist, and humanist Ivan Cesmicki (Ianus Pannonius, 1434-1472) was also a participant in warfare. He was sent to school in Italy where he received a doctorate in canon law. After his return, Cesmicki was appointed bishop of the city of Pecs, and in compliance with Matthias Corvinus's wish, came to the royal court. In the year 1463, he took part in the king's war campaign in Bosnia and fought in the small town of Jajce, and again the following year when he became seriously ill. In the year

1465, as head of the king's mission to Rome, he asked Pope Paul II for assistance against the Turks. Three years later he accompanied the King in a war against the Czech King, Podebrad. In 1469-1470, he was appointed, together with Ivan Thuz, to the position of Lord of Medvedgrad and the viceroy of all of Croatia. But soon they were relieved of the post, and Thuz was punished for disregarding the defense of the country. After that, Cesmicki took part in a plot against the King in Hungary, and when it was stifled, he took refuge in Medvedgrad, where he later died. Even during his short life, Cesmicki was celebrated as a distinguished poet in Latin. He left behind an impassioned testimony about the bloody wars against the Turks and about the indifference of Christian Europe. "Frenzied slaughter, everywhere flames, smoldering embers, and dust," and help comes from nowhere: "France is slumbering, Spain does not bother with Christ, England is preoccupied with the quarrels of her nobility, neighboring Germany wastes her time in meaningless discussions, and Italy only wants to do business." In his poems he ridiculed superstition and cried out for peace, cursing the god of war, the propagator of fear and suppressor of the human spirit.[63]

One comes across similar reflections in the person of Antun Vrancic (1504-1573). Like Cesmicki, Vrancic is one of the most distinguished representatives of Croatian Renaissance humanist thought, the difference being that he was from Dalmatia while Cesmicki was from Pannonian Croatia. As Church dignitary (archbishop and cardinal), diplomat, statesman, historian, poet, and accomplished writer, Vrancic became most prominent in European humanistic circles, but always remained proud to be a Croatian. King John (Janos) Zapolya entrusted him, as his secretary, with a series of diplomatic missions to the chief European courts. Along the way, he befriended some of the most distinguished personalities of European intellectual life and particularly those from Croatia.

After Zapolya's death, Vrancic went into in the service of Ferdinand I of Habsburg who sent him in 1533 to the Sultan Suleiman in Constantinople to investigate the possibility of mak-

ing peace. After a truce was broken, Maximilian II again dispatched Vrancic to Turkey to negotiate a peace settlement with Selim II, which, after long discussions with Mehmed-pasha Sokolovic (Muhammad Sokolli), ended in the signing of an eight-year peace treaty. For his success, Vrancic was appointed primate, as well as viceroy and ruler of Hungary. His authority did not reach as far as Croatia, but after the Croatian Peasant Rebellion was crushed in 1573, he intervened to Maximilian on behalf of the peasants. While traveling through Turkey, Vrancic himself had written in his journal about the violence of the Turks: first they would collect their "tax in blood," by taking away little boys every five years, then every two or three years. And they treated the impoverished peasantry most mercilessly "not immediately after the victory, but after they had eliminated the borders of the newly conquered territory" and "after they had consolidated their authority over it." Vrancic warned that "they who long for Turkish rule and believe that the Turks are indulgent with our people should consider this." This was probably in reference to the fact that such views existed because of the widespread despair felt in a Croatia wearied by Turkish advances, pointing in particular to the numerous examples of people of Croatian origin who had been educated into Islam since their youth, and had been incorporated into the Turkish military and state administration.

The poet and philosopher Marko Marulic (1450-1524) composed both poetry and prose in the fifteenth century in his mother tongue of Croatian, because he wanted to elevate the morale of his people in the struggle for self-preservation in the face of foreign conquerors. His philosophical works, however, were written mostly in Latin, and they underwent numerous editions in Latin and all the major European languages. The theme of the danger of the Turkish advance through Croatian lands is present in all of Marulic's works. In his impassioned poem, *A Prayer Against the Turks*, he described all those terrors, and in "Laments of the City of Jerusalem," he depicted the imperilled city of Split and the fields "white with bones," evoking the cry for the lost heritage. In his chief work, *Judith,* Marulic spoke through biblical themes

about the horrors of the Turkish encroachment and encouraged faith in the deliverance from the dreadful enemy.

In his letter to Pope Hadrian VI in 1522, Marulic appealed for aid to his homeland, imploring the Pope to unite the European rulers in the struggle against the common enemy. The tribulations are here described lucidly: "The infidel Turks with their daily advances... torture us constantly; some slaughter, others enslave; possessions are being carried off, the cattle are dragged away, hamlets and villages are being burned down, and the fields... left with no one to cultivate them, bear thorns instead of grain." Reminding him of how many provinces had already been enslaved, towns and churches ruined, alters desecrated, girls and women raped, boys circumcised and converted to heresy by the "Mohammedan infidels," Marulic warned that, if help should not arrive on time, "the enemy would have an open route by which to attack Germany and Italy, to conquer all of Illyria, and finally, to submit to its rule the rest of the Christian world." To be able to survive these hardships, assistance in the form of arms, money, and necessary supplies would be needed, and such "generosity will be extolled to the heavens by the still remaining towns in Croatia.... "As there was no immediate and effective response to such requests, it is remarkable that Croatia endured.

In his poem, *Davidijadi*, Marulic asked: "What is the cause of your hatred? You promise me peace, and then make me to raise hateful arms against you. You break your vow and then try deceitfully to pledge that broken promise again. You pretend to be my friend in public while in your heart you are my fiercest enemy. Oh King, what kind of madness is this?" Crimes cannot be justified in any way, the one who stands in defense of justice should not have to fear "any misfortune or violence." To validate his Christian humanism, Marulic looked for examples of the triumph of the intellect over the instincts of hatred and passion in ancient literature and philosophy, particularly in Stoicism and Platonism.[64]

A historian and Benedictine from Dubrovnik who finished his studies in Paris, Ludovik Crijevic Tuberon (Ludovicus Cerva Tubero, 1459-1527) also spent some time in Buda at the court of

the Croatian-Hungarian King, Matthias Corvinus, together with other well-known humanists, most of whom came from Croatian regions. In *Comments on the Events of the Time* (*Commentaria de temporibus suis*) he described the infiltration of the West by the Turks, in particular, the battles between the Croatians and Turks and between the Venetians and Turks, and also the sociopolitical and religious turmoil in southeastern Europe. Because of his objective account of national and state relations, and other religions (Hussite, Orthodox Christianity, and even Islam), and especially because of his unbridled criticism of several Roman popes, his masterly *Comments* was banned two hundred years later (in 1734). As a contemporary eyewitness, Tuberon felt all the historical tragedy which overtook Croatia after the defeat of the Croatian forces led by Viceroy Derencin in 1493 in the battle against the Turkish forces under the command of Jacob-pasha on the Field of Krbava. In the defeat on the Field of Krbava, wrote Tuberon, "the Croatian nobility was almost completely annihilated... and the peasants and cattle having been chased away, the fields were almost entirely deserted." Under the very walls of the city of Zadar, 5,000 people and 70,000 head of cattle were seized; in the regions of Carniola, Lika, and Istria, one-third of the population was exterminated.

Tuberon sharply criticized other nations, but also his own. He disparaged the Turks for their wild brutality, but when writing about their military fanaticism and discipline, he sometimes emphasized their morale which he felt should be a model to the Christians. However, he was disgusted that religious differences "lead people to hatred of the kind that is found between wild animals of different species."

Tuberon acknowledged the Croatian-Hungarian king, but, along with other Croatians, he looked upon the Croatian-Hungarian union as a personal union (with one crown). In his observation that "the Hungarians and Croatians constantly quarrel about who is braver" and that "the Dalmatians, most of whom are Croatian, did not acquiesce to Hungarian rule by force or arms, but by hereditary right," he was describing a centuries-old

feud between the Croatians and Hungarians. With his powerful depiction of the serf rebellion led by Juraj Doza (Georgius Dozsa), against the brutality of their landlords, Tuberon, unlike anyone in world literature of that time, upheld the side of the oppressed and persecuted. He found fault with the policy of the corrupted Roman popes, especially with that of Alexander VI, the Spaniard Rodrigo Borgia, and his son, Caesar. About the latter, known for his evil, Tuberon said that he had been "a debauched fellow" and "the seller of sacred relics," and that "he should rather be called a bloodthirsty person than a pontifex." Concerning the acts of violence and terror which Caesar Borgia's mercenary troops had committed in Italian towns, Tuberon wrote: "Everything was polluted with blood, desertion, lust, cupidity, and every manner of crime." But the very fate of the Borgias led him to conclude that every "reign that is established through crime carries out in the vilest way," and that "every bad example—and not only crime—falls back upon the head of its perpetrator."[65]

Simun Kozicic (1460-1536), bishop of the Modrus and Senj, spoke resolutely and indignantly to Christian Europe concerning the terrible suffering in Croatia during the wars against the Turks, when Croatian lands either fell under Ottoman domination or were subjected to their looting invasions. This ecclesiastical prelate and active participant in cultural life, writer and founder of the Glagolitic printing house in Rijeka, published his *Croatian Missal* (1531) and other "Booklets About the Croatian Land" in praise of his homeland. In his speeches at the Sixth Lateran Council in 1513, and then before Pope Leo X in 1516, he presented the condition of "Desolate Croatia" and in making his appeal, he did not plead for, but demanded help. With regard to the strategic importance of Croatia as the bulwark of Christianity for the whole of Europe, he drew the Pope's attention to the fact that the latter would have to assume responsibility before the entire Christian world if the depleted Croatia were not given the necessary assistance. But if "our miserable homeland should be spurned," our people would be "forced to form an alliance or peace treaty with the Turks or to pay tribute to them" but at that time, "bearing in mind their for-

mer glory" and "the fact that it was befitting of the most stalwart defenders of Christianity," they "did not want to agree to it, although the Turks had on several occasions" proposed it. Having thus brought up the suffering of his people before the Lord who knew "that we can no longer endure against the enemy," Simun Kozicic concluded his historical speech with the words: "Finally, with no feelings of guilt, I shall also say this: I fear, Holy Father, that our poor fellow countrymen might, because of poverty, indigence and despair, be forced into warfare and the looting of other Christians together with the Turks. Whoever shall be guilty for that, must account for it before almighty God. This I say to you— if you are not inclined to believe me now—so that when it happens, you may recall my announcement."[66]

Since that time, all Croatian poets and writers related in their works the terrible sufferings of Croatia against the conquering Ottoman force. They also wrote about the ingratitude and insufficient aid on the part of Christian Europe in whose defense Croatia was bleeding. At the end of the seventeenth century, Pavao Ritter Vitezovic marked that tragic period in Croatian history in the titles of his historical works: *The Revived Croatia* (*Croatia Rediviva*) (Zagreb, 1700), and *Two Centuries of Weeping Croatia* (*Plorantis Croatiae Saecula Duo*) (Zagreb, 1703). Besides the already mentioned reprimands addressed to Europe, stands the poetic rebuke by Vladislav Mencetic in the mid-seventeenth century in his *Slavic Trumpet*:

> *Into the waves would Italy*
> *Long since have drowned from slavery,*
> *Had not the Ottoman tide before*
> *Crashed upon a Croatian shore!*

In addition to protest here, there is undoubtedly also a certain amount of pride, despite the tragic position of its people in defense of the West. It also offers evidence that the Croatians had always understood the concept of Slavdom (Slavic Trumpet), not so as to drown the imperilled Croatiandom in the Slavic community, but to strengthen it.

5. On the Paths of Rationalism

The war of all against all is a natural condition.

—Thomas Hobbes

Let us return from Croatia to the larger Europe. Hugo Grotius (1538-1645), a Dutch humanist who was imprisoned because of his participation in political-religious conflicts, laid down the foundations of modern philosophy of natural and international law in his famous work *On the Law of War and Peace* (*De iure belli ac pacis*). He started from the fact that killing during wartime had been justifiable ever since the beginning of armed combat. It was a matter of course, not codified, but customary law. Since, in earlier history, wars were waged between nations, no distinction was drawn between warriors and peaceful inhabitants. So Grotius, too, looked upon the citizens of the enemy state as being completely without rights. In that respect, changes were to come with the emergence of the feudal, knightly, and mercenary troops, so that Frederick the Great was heard to say that "a citizen needn't take heed when his king is at war." In reality, however, wars always advanced at the expense of the population.

In his theoretical commentaries on war, E. Vattel (1714-1767), a Swiss jurist, wrote in the middle of the eighteenth century: "When one ruler declares war on another, it is viewed as if the entire nation has declared war on the other nation because the ruler represents the nation and acts on behalf of the whole community," and "therefore, all state subjects of one nation are mutual enemies to those of the other."[67] Such an understanding was to remain even after the endeavor to restrict warfare to armies and their military targets by codifying international law.

Jakob Boehme (1575-1624), a self-educated German philosopher, was persecuted officially by his enemies, held in contempt and derided as the "cobbler anti-Christ." For him the basis and origin of all good and love, as well as evil and hatred, lay within

the eternal and immense being of God. All of nature consisted of two opposing qualities: one celestial and holy, the other infernal and wrathful. "The dark inferno and the clear light originate from one heart according to the words in the Holy Writ: I create both light and true darkness; I give peace and deal out evil." Thus, divine kindness is the source of heaven, and divine anger is the foundation of hell. In the dialectics of eternal nature, life and all its appearances are the result of the everlasting struggle between the principles of good and evil, in which one or the other alternately has the upper hand.[68]

The various struggles that rocked Europe gave Thomas Hobbes (1588-1679) reason to judge that there is no order and peace in the natural social condition, but only disorder and a struggle of all against all. The saying "man is a wolf [enemy] to man" (*homo homini lupus*) is particularly true when it becomes "man is a god to man" (*homo homini deus*). The transition from such a natural condition of disorder and uncertainty to the condition of peace and organized society is achieved through the creation of a state and the realization of order and sovereign will. "Without a common power," wrote Hobbes in his *Leviathan*, "to keep them in fear, people find themselves in the condition which is called war, the war of all against all." This *bellum omnium contra omnes* is actually a natural condition from which one can extricate oneself only by wise state systems. And "the concept of permanence should also be included within the essence of war... just as the essence of bad weather is not in just one or two drops of rain, but in its inclination toward continual rain... and so the essence of war is not in the real battle, but in the familiar tendency toward fighting all the time, when there is no certainty to the opposite."

War is, according to Hobbes, something innate to man, a requirement for his existence, so that the wise state must force him out of this condition and mandate peace, which must be maintained by authority. Since this kind of power is possible only in a specific political community, the state, international life is dominated by the "natural condition" of war.

It should be added that Hobbes' judgment is shared by other

seventeenth-century philosophers (Spinoza, Pufendorf, et al.). Spinoza pointed out that armed conflict could be avoided only if the oppressors would satisfy the needs and aspirations of the oppressed, thereby enabling the realization of freedom. But as the oppressors customarily refuse to do so, tyranny itself causes the battle in which tyranny must desist because, in this power struggle, the human aspiration for freedom wins.[69] But what freedom has not unlocked new paths to violence and even crueler tyranny?

Rene Descartes's (1596-1650) rationalistic philosophical world view contained in his saying, *Cogito ergo sum* (I think; therefore I am) has often been repeated. This implies that all problems of human existence can be understood and solved by the human intellect. Thus the triumph of intellect provides the arrangement of the world according to human standards. However, such an interpretation of Descartes's and of subsequent Cartesian philosophy in general, is oversimplified. His essential thought (which in French is: *Je pense, donc je suis*) might more correctly be interpreted as: "I think and nevertheless I exist," which would respond better to his skepticism concerning the imperfect world. Descartes himself experienced wartime devastation. He fought in the Thirty Years War, joining up as a volunteer in 1617 when he was twenty-one and spending four years in service. He was later criticized for his philosophical views, by both Jesuits and Protestants, as a heretic and atheist.[70] At that time, in which the methods of the inquisition had not yet disappeared, Descartes was able to go to Sweden, where he became the queen's adviser in philosophy and continued his philosophical deliberations.

Around the same time, in England, the Puritan leader Oliver Cromwell (1599-1658) defeated the army of Charles I in a civil war and then had him executed by a decision of the court. Even after the regicide, Cromwell never ceased with the internal violence and wars abroad. He put down in blood the rebellions not only of the Royalists, but also of those who demanded equality (the so-called Levellers, in 1647); he conquered Ireland in 1649, and then Scotland in 1651 after he had defeated Charles II. Then he began spreading English authority beyond the British Isles. He

wrested Jamaica from Spain, and New Amsterdam (later New York) from the Netherlands, and in an alliance with France, he obtained Dunkerque. In comparing the English with the French Revolution, F. Engels said that "Cromwell united in himself the personalities of Robespierre and Napoleon." He ruled as a dictator, and yet he was tolerant of the Jews and non-Anglican Protestants, though he fanatically hated the Catholics.

Cromwell found a deeper philosophical explanation for his own irrepressible hostility. Inciting war against Catholic Spain, Cromwell said in a speech in 1656, "The very first matter that I am going to discuss is what the first task of nature is: existence and survival... We have to keep in mind the maintenance of our national being in view of those who try to oust and destroy it." Peace can be made with France, concludes Cromwell, but not with Spain because it is a Catholic state under the Roman Pope.[71]

6. In the Reason of Enlightenment

To give up one's own freedom is to renounce the characteristics of a human being...

—J. J. Rousseau

Cromwell's understanding of the hostility between the opponents at war was, of course, by no means isolated. Numerous examples can be cited to prove that it appeared similarly in the following centuries, in the reasoning of great statesmen and military leaders, but also of great thinkers. One of the more pronounced examples is a fervent message from Heinrich von Kleist to the German soldiers in the liberation war against Napoleon's conquering army: "Finish them off; the court of the world does not ask you for reasons!"[72]

The freedom-loving philosophical spirit of Montesquieu (1689-1755), who, with his analysis of political despotism, slavery, and spiritual intolerance, inspired the initiators of the revolution and

the authors of the *Declaration of the Rights of Man,* was unquestionably mistaken when attributing specific characteristics only to particular societies, states, or nations. According to him, conquest was the goal of Rome, and war of the Spartans; religion was the goal of Jewish laws, and public order of Chinese laws; trade was specific to one and sea-faring to another city, pleasure a feature of tyrants and glory of monarchists. Just "one nation" in the world—the English—"have political freedom as the immediate aim of their constitution."[73] The partial accuracy of such notions certainly cannot justify their remarkable one-sidedness.

J. J. Rousseau (1712-1778) is the best known and most influential thinker of the Enlightenment, having provided the theoretical basis for the programs of the French Revolution with his *The Social Contract.* Rousseau opened his work stating: "Man is born free, yet everywhere he is in chains." But there are problems with his claim that man is by nature good and not prone toward evil and sin, and that civilization itself has corrupted him. Rousseau's views about the necessity for a return to nature were not accepted by his contemporaries. "When reading your book, one feels like walking on all fours. However, since I abandoned the habit more than sixty years ago, I feel that it's fortunately impossible for me to take it up again," Voltaire answered him. Rousseau believed that a return to nature would result in a higher level of civilization and transform man "from a dull and witless animal into an intelligent being and a human." No less mistaken was his reasoning that private ownership had been the cause of all disparity, violence, and wars and that its eradication would wipe all evil from the face of the Earth. "How many crimes, murders and wars could have been prevented, how much misery and terror could humankind have been spared if someone were to have plucked up the signs and covered the hole, calling to his fellows, 'Do not listen to the lies; you are doomed if you forget that the fruits of the Earth belong to all and the Earth to no one!'"

The failure of Rousseau's judgment was not so much due to his ignorance of the new terrors revolution and ultimately communism would cause, as to his misunderstanding of the very nature

of human history. This, however, does not lessen the importance of his views, which pleaded for a civic society based on justice and intellect in which the principle of freedom for all people and the sovereignty of nations would be guaranteed. His thoughts justified the struggle for freedom as essential to the "natural rights which resound in every human being."[74]

The ingenious Encyclopedist Denis Diderot (1713-1784) was one of the most daring thinkers of his time. His works were burned, and he was imprisoned, though ultimately he found a sponsor in the Russian empress Catharina II. The history of humanity is, according to Diderot, "the history of its being oppressed by a handful of rogues." But this is only a small part of the truth. If the problem were merely in the "handful of rogues," humanity would certainly already have rid itself of them. Instead "oppressed humanity" has sunk even deeper into pain and suffering.

It has always been difficult to determine the one true way out of this historical entanglement in theory, let alone in practice. All experience says that this is beyond the human faculties. As the main author of the *Declaration of Independence*, Thomas Jefferson (1743-1826) wrote Enlightenment ideas about the freedom and equality of people and nations into law. He was also a predecessor to the French Revolution. Jefferson condemned slavery, yet he himself died the owner of two hundred slaves. His ideas, of course, did not therefore have lesser, but greater significance.

7. On the Horizon of European "Classical" Philosophy

"Is war detrimental or beneficial?" That depends on the circumstances.

—G. W. F. Hegel

How Immanuel Kant (1724-1804) regarded the problem of force, and especially war, within the scope of his comprehensive philosophical system, is of vital importance. With Kant one reaches the "final expression" of rationalistic enlightened thought.

Kant's views on this matter are crucial also because he was the founder of German Classical Philosophy and the predecessor to all its Hegelian and Marxist thinkers.

Kant observed war as necessary and unavoidable and, moreover, a determinant of historical reality. He therefore concluded: "War is an unavoidable means which enables the culture to advance further," so that "eternal peace will only be beneficial when the advancement of culture is completed."[75] He stated that war "develops talents which serve the culture to the highest degree." Such convictions led Kant to consider war as something generally "exalted if conducted according to order and with regard for the sanctity of civil rights."[76]

Frederick the Great, who certainly understood "the exalted side" of real war, seemed to answer Kant directly: "It is futile to boast of exalted feelings. Every war which does not lead to conquest weakens the victor and deprives the state of its forces. One should never engage in hostilities if the prospects for conquest are not certain."

Kant based his idea of the possibility for humanity to attain eternal peace—which was the title of his famous work (*Zum ewigen Frieden*, 1795)—not on historical experience, but on his own moral principles, and this as a categorical imperative. Demanding that one should act in such a way "that the principles of one's will may at any time become universal law," Kant wanted to convince himself and others of the possibility of doing so, and consequently of the realization of eternal peace. As later history showed, Kant's noble aspirations remained unrealized. His reflections on war should have been based on the insight that "a human being is made of a too gnarled wood that could never be carved into something different." And as for his vision of eternal peace, he was neither the first nor the last dreamer to visualize it.[77]

In Kant's time others too emphasized the necessity of legal, righteous, and controlled warfare. The middle class, which had personally experienced the terrors of revolution and war, attempted to humanize future warfare through international laws. However, in the periods of civil war and ideological and imperial-

istic conflicts, such agreements were abandoned, just as chivalric principles were in the madness of religious wars, and racial and colonial pogroms.

These experiences suggest that the philosophical, legal, and military interpretations of war are historically and socially determined. When the European governments of the *ancien regime,* alarmed by the French Revolution, decided to bring back the "legitimate regime" in France by military force, they did not conduct a war against France, but ordered a punitive campaign against the usurpers of the power. In the manifesto of the commander of the allied (Austro-Prussian) army before the invasion of France in 1792, the duke of Braunschweig denied the army of revolutionary France the status of combatant, and excluded it from the international war rules and conventions. On the other hand, the National Convention, as the government of revolutionary France, called on all the populace for a national war: "From this moment until the day when our enemies are expelled from the territory of the Republic, the French people will be in continuous military service. Young men will join in battle. Married men will construct arms and transport supplies. Women will sew uniforms and work in hospitals. Children will make bandages from old linens, and old people will go out into the squares to encourage the soldiers, preaching hatred toward kings and the unity of the Republic." Owing to their eagerness for freedom, such a national army succeeded in defending the Republic and the Revolution.

When Napoleon started a war of conquest against Europe his conduct toward the armed national units was in no way different from that of Braunschweig. In fact, it was even more cruel. Napoleon himself ordered his soldiers to treat mercilessly those who belonged to the liberation movements. The urgency of war has always been an excuse for the most violent actions. And the despotism of all conquerors has trampled upon not only the rights of others, but also their own principles. To every aggressor, war has always been just, and all possible forms of violence and evil deeds have been justified. And usually, the more exalted and holy the aim, the crueller and more demented the violence.

Thinkers often personally shared the fate of their national and social community. J. G. Fichte (1762-1814), the great philosopher of German Idealism, joined the ranks of the German resistance movement against Napoleon as a volunteer. Like Kant, Fichte was inspired with the idea of eternal peace achieved by the reign of intellect and enlightenment. But the German authorities forbade him to lecture, due to charges of his being an atheist. Although Fichte was a zealous follower of the French Enlightenment, in his ardent *Speeches to the German People* from occupied Berlin, he called for a struggle against the French subjugators. Regarding the defense of one's own homeland the first duty of a citizen, Fichte said, "everyone is responsible for the condition he is in," and "both the individual and the whole nation can extricate themselves from that condition only through their own actions and strength."

Fichte was the first European thinker to expound systematically the theory of the nation as a singular ethnic being and state. To Fichte, the life of a nation represented eternity on Earth. The national spirit is an expression of the nation's faith in its particular task. The national language is the first, but not only, expression of that original life strength. And the state is a means to ensure the identity and existence of the national being. Fichte completed his nationalistic thoughts during the days of the disastrous collapse in 1806 of Prussia, as the greatest German state.

Fichte's further thoughts on this subject run as follows:

1. If your neighbor does not see in you his natural ally against another power dangerous to both of you, he will always be ready to turn against you as soon as the first opportunity is offered to him whereby he can do it with certainty. He would have to do so if he is wise, and could not hold back, even if you were his own brother.

2. It is by no means enough to defend only your own territory, but you should be cautious of everything that could have an influence upon your position; do not allow anything within your con-

trol to change at your expense; do not hesitate for a moment if you can change something to your benefit, for you can be certain that others would do the same as soon as they had the chance; and if you miss your opportunity, you will fall behind. The one who does not become stronger loses when the others do.

Having concluded that an individual can protect his life even if enslaved, while a nation as a separate individuality in such conditions must perish, Fichte attached special importance to patriotism. At the moment when national survival is endangered, love for the homeland must be ready to make any sacrifice in order to preserve its eternal mission. Such a manifestation of love for the homeland is not national egoism. A nation that defends its individual identity performs its duty toward humanity.[78]

The philosophical system of G. W. F. Hegel (1778-1831) marked a further step toward the comprehension of the essense and purpose of human history. The highest stage of philosophical contemplations on history was reached with Hegel, and all future reasoning had to start with him. Hegel's premise was that the task of philosophy is to "comprehend that which is," for only "that which is actual" is rational too; and vice versa, "that which is rational is actual." This provides philosophy with a solid basis, so that it may not be misled by one-sided speculation or empty analysis.

For Hegel, the world is a dynamic, dialectic process, the struggle of opposites. The process of historical events, as well as of their understanding, passes along in the conflict of opposites. And the essence of that dialectic flux is expressed in thesis, antithesis, and synthesis. Naturally, synthesis is by no means an end, but a new beginning of the three-part succession.

For Hegel, the power of the spirit is the subject, and the life of the spirit fears no destruction or death, but endures death and maintains itself in it. Thus Hegel arrives at the understanding that the life and death struggle is an inevitable necessity. Everything is found in the struggle: death and life, personality and freedom, the whole of history. Accordingly, a human being, just like a nation,

is mortal, but also an infinite, free historical subject.

The most powerful factor of history is the state. It is not, however, a mere sum of individuals, but an elevated entirety which manifests itself as a determined national spirit. Every nation is a separate expression of the spirit of mankind, just as every individual is an expression of national spirit. Thus the struggle for the self-preservation of the state, which expresses one particular national spirit, has the deepest historical justification. Moreover, the very essence of history is expressed in this struggle because a state, insofar as it possesses the will and the power, which is the result of its particular national spirit, becomes a fateful factor in the creation of history. It is only in a time of war that a state can prove to be such a spiritual whole. For Hegel, world history is a judgment with verdicts against which no appeal can be petitioned. Within the historical destiny, justice is revealed, before which only the worthy and the divine can endure. There is neither purpose nor justification for any feeling of shock or despair over historical events, regardless of their cruelties, injustices, and violence. Not only can such feelings change nothing, but the act of historical justice should be accepted as the highest expression of historical reason.

Hegel thought that an enemy is a stranger who is denied totality in every aspect of his living. But an enemy can only be an enemy to the nation. A war is not a war of an individual or of a family, "but of a nation against a nation." In this way hatred becomes generalized, "freed from any personal admixture." Still, Hegel said: "In world history, neither formal bravery, nor so-called merits, but the value of the cause determines glory."[79] According to that same criterion, Hegel condemned many wars in history in which kings were involved, who had "extended their state powers" setting "demands of all sorts towards outside." The purpose of such wars had always been conquest, and "one such particular object of conquest was Italy," which was to the French, the Spaniards, and later the Austrians, to serve as an object of plundering. The wars that Germany waged were not "especially honorable" in this respect, either.[80]

Carl von Clausewitz entered the war against France as a twelve-

year-old boy, and at the age of thirteen, he rose from the ranks. From then on, he fought in the Prussian and then Russian armies against Napoleon. He studied and taught war skills (he was tutor to the Prussian heir to the throne), gaining glory, not as a Prussian general, but as a military writer. Both Engels and Lenin regarded Clausewitz as such a great writer that every thinking person should be acquainted with his work. Yet Stalin regarded him as an ideologue and representative of Prussian militarism and backwardness.

Clausewitz was a participant in the tumult of contemporary military and political events, but he sought the basis for their better understanding in history. Ancient history as well as Machiavelli and Frederick the Great directed him to realism. Kant gave him philosophical depth, and Fichte inspired him with the ideas of homeland and national honor. Like Hegel, he began with historical reality and wrote that "examples in history explain everything" and "have the greatest power of evidence." This refers in particular to combat.

And in the area of combat, no one has ever delved so deeply as Clausewitz. His idea that "war is nothing other than a state policy carried on through other means" was far ahead of his time. War is not merely one of the arms of politics, but also the *ultima ratio* (the final argument) against the enemy, which changes the relationship between the two sides. Politics determines aims and, therefore, war takes on the characteristics of politics. The more "magnificent and powerful the politics, the more so the war, and this can mount to such heights that war can reach its absolute form." That insight brought him to the conclusion that the essence of every battle is the ambition of each side to impose its will upon the other by the use of all available force, with no limits to the violence. That is the quintessence of his theory of *absolute war,* which in this century was to be called total war. But in reality, armed struggle usually remained a more or less *limited war.* However, although a war may not reach its absolute form, it is never an abstract act. Therefore, it must be led with all forces. According to Clausewitz, Napoleon brought war to absolute perfection. Still,

a war always rests upon possibility and probability, good and ill fortune. The logical order of things is often disturbed, and blind luck can predominate over the certainty of reason.

Clausewitz attached great importance to the role of military leaders. Only leaders of firm disposition, strong reasoning, and piercing spirit can achieve success in war. Clausewitz is the first theoretician to realize the purpose of national war, of calling out the nation and waging guerrilla warfare, and the participation of the whole population. Laying aside the political reasons, Clausewitz said, "In the majority of cases, a nation that would use such a means rationally would prevail comparatively over those who reject it with contempt. If this is true, there still remains the question of whether this new strengthening of the war element is useful to humanity or not—a question which one should answer in the same way as the question about war itself. We leave both to the philosophers."

To the criticism that the energy which a national war expends could be better used, Clausewitz answered that experience shows that, "for the most part, those energies are not at one's disposal to be used according to one's own will," and that "their essential part, namely the moral element, comes into being through that same way of use."[81] Clausewitz found justification for the increased sacrifice of a national war in the possibility of saving the future of the state and nation, which is imperiled if the regular army is defeated. "And just as it is a natural instinct for a drowning man to grab at a straw, so also, in the natural order of the moral world, one nation tries to use even the last possible means for its salvation when it realizes that it has come to the brink of its downfall."[82]

Thus Clausewitz gave a theoretical justification for national guerrilla warfare which has been used by the liberation and revolutionary movements all over the world in this century. And with those movements, war was to have neared its absolute form as hardly any war between established states had ever done. Since both sides in guerrilla wars seek not only military victory, but also the complete destruction of the enemy and of everything that

could be of use to them, this kind of war, through the mutual use of violence without limitations, assumes the form of almost total, unlimited war.

III. MAKING SENSE OF THE AIM AND AIMLESSNESS OF FORCE

8. Agreements with and Oppositions to Philosophical Legacy

Every nation has its own language of good and evil: it doesn't understand its neighbor.

—F. Nietzsche

From the historical turning point in which the aforementioned great thinkers lived, philosophical thought concerning war and violence in human society headed in two directions. Those who took the more traditional path continued to interpret war as an inevitable historical necessity, while the others denied the necessity of both social and military conflict, or at least preached the need for their final overcoming from various and contrasting standpoints: pacifist, anarchistic, or revolutionary.

Joseph de Maistre (1753-1821) declared war to be divine, although he was otherwise a fierce opponent of the French Revolution precisely because of its violence, and regarded man as a rational, free, and noble being. Friedrich Schlegel (1772-1829) shared this opinion as well, calling religious wars "the most beautiful flower of humanity." V. Cousin (in his *A Course in the History of Modern Philosophy,* published in the middle of the nineteenth century) wrote that war was "an empirically necessary form of struggle for survival for the purpose of achieving progress." And in his book *The Fact of War and the Idea of Peace,* Giorgio del Vecchio viewed the very thought of war as healthy, and judged that "the beneficial consequences of war outweigh the detrimental ones."

For the Prussian historian, Heinrich von Treitschke (1834-1894), all means are permitted in attaining the unity and power of the homeland, even those of "the most intolerable despotism," and war is actually a political necessity, for without war there would be no state. Precisely for this reason, the rule of war is an eternal rule of human history.

Treitschke also contributed to the theory of racial inequality. The world had always been the scene of opposition between the races. In contrast to the others, only the Whites, who were divided into Aryan and Semite nations, were capable of establishing states and freedom. (The Red race was decreasing in number, the Blacks had always been slaves, and the Asians had never risen to the greatness of a state.) And "one should not object to this" because "the world would be intolerably empty if all the races were equal." Yet, in that respect, not all Whites, and indeed, not even all Aryan nations are equal. Although the Aryans are generally characterized by bravery and physical and moral endurance, there are two proto-nations among them that stand out: the Greeks and the Germans, "whose ingeniousness has not been surpassed." On the other hand, those who do not have the aforementioned attributes and capabilities, and who do not comprehend life and history, advocate "that nonsense called eternal peace."[83]

While Treitschke is rightly considered to be a contributor to the racial and state-totalitarian political theories in German thinking, the same cannot be said for the English historian and statesman James Bryce (1838-1922). In his works he gave evidence about how racial intolerance and violence has always existed in history, even among the most civilized and democratic nations. Among the essential shortcomings of democracy, which, in theory, he otherwise justified, Bryce cited the violence on the part of the "democratic nations" against the "subdued countries." From the Athenian democracy and the Roman civilization, through various Swiss nationalities, down to all the examples of colonial conquest, he illustrated how throughout history the democratic principles of freedom and equality were not applied to others, that is to say,

how they "were of no value" when it was not in the interest of the superior powers.[84]

Friedrich Nietzsche (1844-1900) took part as a volunteer in the Franco-Prussian War in 1870. However, that personal experience was not as crucial to his philosophical insights as his contemplation of the entirety of philosophical thought. Nietzsche was not afraid to tell the world quite plainly that "this life is immoral." It is not a path toward divine salvation, but a wild and merciless struggle in which "God is dead," and those who are stronger and more capable, braver and more heartless, are the ones who survive and prevail. In this struggle for self-preservation, only those who have a will to live can endure and succeed.

Nietzsche's remarkable work *Thus Spake Zarathustra* remains a profound testimony to the despair of the human intellect which is aware that man is stretched "between the beast and the superman" (*Uebermensch*) and doomed to dangle above the eternal historical abyss. In his despair over the reality of historical life, he reveals the Utopian wish to flee from this to another world, one in which man, having transcended himself and all the evil and atrocities of the existing world, would rise to the state of superman, who did not need a "God." Nietzsche found no other way to achieve that end than by struggle and war. Wisdom teaches that life and history want only "the brave, the carefree, the mocking, and the forceful" to subsist in them, for "wisdom is a woman and always loves only a warrior." And just as there is always "something insane" in love, so, too, is there "always some amount of reason in insanity" (par. 306). The more a man strives upward "into the heights and toward the light, the more firmly his roots strive for the ground, downward, into the darkness, the depth— into evil," while "the invisible hands bend and torture" him "the most cruelly" (par. 307).

No one spoke of war and warriors as did Nietzsche: "My brothers at war! I love you with all my heart; I was and I am of your kind." ... "You should love peace as a means for new wars. Even a short-lived peace is more than a long one." "Let your peace be victory!" A good cause has not to justify a war, for "it is

the good war that sanctifies every cause." What is more, "War and bravery have performed more great deeds than the love of neighbor." A miserable person is not saved by pity but by courage. And it is good "to have only such enemies that should be hated, but not such that should be scorned."

True to his convictions, and in his other works as well, Nietzsche regarded struggle and war as the law of life. Thus he was not unnerved by the fact that "the world of the future will call our era the classical period of war." Nor can one deny the inevitability of war because "if one renounces war, one renounces the great life." It is precisely "war that educates one for freedom." Therefore, "no peace at all, but war," nor the illusions and self-deceptions about the possibility of peace, but the need "of the consciousness for the most severe and most necessary wars." He even prophesied enthusiastically that "such wars are to take place as never yet on Earth."[85]

In his last work, The Will to Power (1890), Nietzsche wrote: "Life is a consequence of war; society is a means for it," and he was therefore "delighted at the militaristic progress of Europe" because it was "a confirmation of the barbarian... indeed, of the beastly in all of us," and "the society that rejects war and conquest is on the road to decline."[86] By preaching such a philosophy, Nietzsche had as an aim the overthrow of all the usual notions about the world as it seemed to be, or should be according to religious and ideological perceptions. History does not give anyone the right to reshape people or the world according to one's own fancy, or to deceive them with any kind of model or morality which has no backing in reality. Nietzsche attacked ruthlessly the promotion of such delusions, exposing the futility of bowing to the false creations of some ideal and moral world without evil. The depiction of ideal morality, society, and world peace actually denies real life, which with all its inner instincts and impulses carries within itself all the negative as well as the positive, the destructive as well as the creative; it is "beyond good and evil." Nietzsche longed for the beauty and resplendence of life to come true in the real, not an imagined, world. "There are

no moral phenomena; there are only moral explanations of the phenomena."

All in all, Nietzsche remained the greatest scorner of historical delusions and misconceptions, but he became, in fact, the very spirit of that historical reality which he wanted to transcend. However, this offers no foundation to those who include Nietzsche among the progenitors of Nazi-fascist totalitarianism. Nietzsche broke with Richard Wagner because of his Christianity and glorification of the ancient Germanic myth. And Nietzsche was even ashamed of writing his works in German, for it might "seem to serve the strengthening of any kind of aspirations of the German empire." He wrote that "Germans, too, had on their conscience all the great cultural crimes of the last four centuries," and also that anti-Semites were in his view generally "ill-fated persons" and "orphans." And, after all, did he not foresee prophetically in his *Zarathustra:* "My enemies have become powerful and they have given my teachings such a monstrous shape that even my dearest must be ashamed of the gifts that I have given them."[87]

Like Nietzsche, Oswald Spengler (1880-1936), a German historian and philosopher who gained fame particularly with his work *The Decline of the West*, also rejected every delusion about peace, surpassing Nietzsche in this respect. Viewing "history as existing only through people who have become races," and who have succeeded in surviving the unsparing struggle, he ridicules every idea of peaceful cooperation in history. Spengler says, "The road from Alexander to Caesar is unambiguous and inevitable, and even the most powerful nation has to walk it, whether it wants to or not." Historical facts cannot be ignored. "The peace conference, held at the Hague, was a prologue to a World War," and "the one in Washington D.C., in 1921, a prologue to new wars." The only lesson from this was: "remain alive or be destroyed—there is no third alternative."[88]

Spengler regarded the history of mankind as the history of cultures, which are independent, separate structures. All cultures are predetermined by organic laws: they have a childhood, a maturity, and an old age. The last stage of every culture is civilization, in

which struggle and war emerge in full force, necessarily and inevitably, and everything else is merely "empty philosophy." In struggles like this, one should strive to be the first, no matter how cruel and difficult the contest might be. Spengler therefore stressed the motto of Frederick the Great as a principle of permanent value: "To be a servant to the state," to possess "honor, virtue, race, blood and nobility," and "he will be the master who possesses these qualities."

To Spengler war is an ideal. In fact, it is an end which is worthy of every culture. In this respect it is "all the same" to Spengler "which slogans are being tossed around while doors and brains are being smashed." Spengler believed that "World history is a history of states and it will forever remain such." The state's responsibility is to be prepared "for external struggle, whether it be of the military, diplomatic, or economic kind."[89]

Considering that Spengler's major work was finished right before the First World War and published shortly after its end, his remarks about conflict are of particular interest. "Great conflicts between states are everywhere imbued with internal revolutions of the most dreadful kind, but these—whether willingly or not—with no exception, benefit non-state as well as purely personal issues of power." Wars and the violence they entail were becoming more cruel and more horrible. And although he spoke with abhorrence of the destruction of culturally valuable goods, war remained for Spengler "a magnificent event." "Life is hard and cruel—when it should be great."

In keeping with his understanding of history, Spengler outlined an even more frightful picture of the future. According to him, in two generations' time the will of those who want war would overwhelm that of those who need peace. And "in those wars for the inheritance of the whole world, continents will take part, India, China, Africa, Russia, and Islam will rise up. There will be a clash of new technologies and tactics. Great centers of world power will have at their disposal the smaller states, their economies and their people: they are still mere provinces, objects, a means to an end; their fate will be insignificant to the great pace of events. In the

span of a few years, we have learned to pay hardly any attention to events which before the war would have paralyzed the world with fear."[90]

Regarding war as a natural and social phenomenon established through a thousand-year experience, Max Scheler (1874-1928) also dealt with the ethical relationship between man as an individual and war as the crudest form of violence against human beings. Taking this as a starting point, in his work *The Genius of War and the German War* (1915), he argues that "wars are not waged against individuals but against states." Their basic purpose is the disarmament of the enemy "and not the killing of people." Therefore the enemy does not stand as an individual, "but as a united enemy force, as a weapon of a foreign government, whose will is active in that force as a whole." And so "this very fact eliminates completely the ethical fact of murder."[91] Accordingly, Scheler affirmed that "killing at war is killing without hatred... hatred toward the enemy is an element completely alien to real war." According to Scheler, war does not divide, but unites people: "If viewed factually and more thoroughly, it becomes the greatest force of human unification, so that its genius can actually be called the strongest factor of unification." In war, "the differences between individuals, classes, higher and lower ranks, the poor and the rich, suddenly decrease."[92] Therefore, and in that sense, "war is indeed the storm of the moral world," for it corrects social and class oppositions.

As his predecessors before him, Scheler also rejected pacifist protests against war as being empty phrases behind which other intentions and values were hidden. "Here then lies the quintessence of the great ethical contradiction of war: in the name of some sort of *universal human love*, in the name of *humanity*, the objections of the modern liberals are directed at war," but, in fact, "the noble name of love is misused here." For the epitome of true love toward man and humanity is found "precisely within the war itself." Scheler sensed the one-sidedness of his own interpretations and continually spoke of the abstract "real war" in his reflections. He was aware that there had been very few such "real wars" in

taken goods, and the victor must even offer gifts to the defeated. In all cases, warfare and fatal retaliations are more frequent and more cruel between racially and ethnically different communities than between those that differ only in their levels of civilization.[96]

War and violence, like human communities, are conditioned by biological, geographic, economic, political, and cultural factors. They experience changes in space and time, but, in a particular sense, they remain to some degree constant in the history of social existence, which has kept its primeval characteristics even in the midst of universal change.

9. In Pacifist and Marxist Contemplation

They have healed also the hurt of my people slightly, saying, Peace, peace; when there is no peace.

—Jeremiah (6:14)

It is not up to us to prove the moral nonsense of war, but to understand its historical sense.

—F. Mehring

Since time immemorial man has longed for peace. The thought of peace conjures up all the delights of life, both real and imagined: the abundance of pleasure, creativity and repose, wealth and glory, the joy of bearing offspring, and the serenity of death. Peace is the security of a nation and of the international order. But peace is never a permanent natural condition. In fact, it must be fought for; either defended or established through victory over an enemy.

In the very beginning, humanity used prayer and sacrifice to attempt to resolve this contradiction. Later it used pacifist ideas and anti-war movements, or, conversely, armed revolutionary coups. Since the condition of peace often means an attempt to preserve a state of inequality against particular social classes and entire nations, peace itself generates turmoil and war.

As far back as the Bible, endeavors to maintain peace were fruitless. In vain did Isaiah dream of "the Prince of peace" and of

the arrival of the great "Counsellor," who was to bring "peace with no end."[97] Jahweh promised his people peace in which "truth shall spring out of the earth, and righteousness shall look down from heaven."[98] But there was to be no peace, and the less there was of it, the more of a spiritual good it would become.

Despite Christ's teaching of peace among people and nations, He pointed out the contradictions of that unattainable goal. "Suppose ye that I am come to give peace on earth? I tell you nay; but rather division."[99] Jesus' peace was the peace of the Resurrection, after the final victory. The spreading of His teaching to all parts of the world would mean the metamorphosis of that which was called *Pax Romana* into a *Pax Christiana*.

In Christianity peace was raised to the level of the highest moral value. But it would only be when the doctrine of Christianity prevailed that eternal peace could be established throughout the world. Every attempt to impose its idea of peace on the rest of the world would result in its exact opposite. This also includes the idea of *pacifism*, which, since the time of the Enlightenment, has sought to eradicate war.

But pacifism is not exclusively Christian or European. In the foundation myth of the Iroquois alliance, its leader went unarmed among the nations to spread in that way the idea of "the great peace" throughout the world, such as the Iroquois imagined it. However, the actual alliance of the five related Iroquois tribes (Cayuga, Mohawk, Oneida, Onondaga, and Seneca, which later merged with the Tuscarora tribe) did not come about through peaceful means, but by arms. Afterward, the Iroquois alliance wanted to expand its "great peace" to other tribes. Wars occurred when one or another tribe refused to accept its incorporation into the alliance. The "great peace" was then imposed by war, and for the subjugated group, it meant the acceptance of domination and assimilation into the extended tribe of their more powerful neighbor, or in the case of war defeat, dispersion, and extermination.[100]

Demands for the prohibition of arms and the abrogation of the army were present in a number of documents of the sixteenth century mystical spiritualists, as well as in the doctrines of certain reli-

gious sects (e.g., Quakers and Mennonites). The French writer, Abbé de Saint-Pierre (1658-1743), presented a detailed program for achieving eternal peace in Europe, in which he recommended setting up a permanent union of European States. In order to secure peace, the Union was to have its own international court as well as its own army. Frederick the Great wrote in response to Saint-Pierre: "Whoever wants to establish permanent peace [escapes] into the world of ideas where there is no 'mine' or 'yours' and where all people are devoid of passion."[101]

Saint-Pierre's point of view, regardless of its utopianism, can at least be considered progressive. His union was a precursor to the League of Nations, the United Nations, and even to the European Community which is currently taking shape. And of course these supreme international institutions are and have been equally unable to ensure permanent world peace.

From the beginning of the nineteenth century, special pacifist organizations were founded: the Peace Society in England in 1816, the American Peace Society in the USA in 1828, the International Peace and Freedom League in France in 1846, and the Austrian Peace Society in 1846, in Austria. Many European writers and philosophers took up the cause. The Austrian writer Bertha von Suttner (1843-1914) aroused so much attention with her novel *Down with Arms* and her support of the peace movement, that she was awarded the Nobel Prize for peace in 1905. The French socialist spokesman Jean Jaures (1859-1914), a historian and a philosopher, fought for the unity of all the socialist organizations and for the peaceful transformation of society, believing that revolution should not be carried out through violence but by democratic means. A member of the anti-war movement on the eve of the First World War, Jaures was assassinated by a supporter of violence. The newspaper he founded, *L'Humanite,* became a pulpit for the promotion of the struggle against imperialistic wars, but also for the support of revolutionary civil wars, and then later socialist wars.

Moved by the horrors of the battle of Solferino in 1859, the Swiss peacemaker, J. H. Dunant (1828-1910), established an asso-

ciation for aid to the victims of war. In 1863, the International Red Cross was formed in Geneva, and in the next year an international convention took place establishing the neutral position of army hospitals and medical staff and the obligatory care of enemy wounded. For this, Dunant won the Nobel Prize for peace in 1901. All that did not help to eliminate war, but did, to an extent, help to alleviate its consequences.

Among German pacifists, F. W. Foerster (1869-1966) was a prominent figure. An educator and philosopher, Foerster was forced to relinquish his professorship in 1920 because of his views which he expounded in the book *A Pacifist: My Battle against Militarist and Nationalist Germany.* With peace conferences held at the Hague from 1899 to 1907, the European pacifist movement gained considerable momentum. Expressions of the embittered European spirit against the evil of war were to be found in particular in literary works like H. Barbusse's *Under Fire* and E. M. Remarque's *All Quiet on the Western Front.*

After World War I, more peace-making organizations were operating in Germany than in other European countries. They were all incorporated into the German League for Peace established in 1921, but already banned by 1929. And it promoters were scattered all over the world after 1933. Various Catholic organizations, such as the Peace Union of German Catholics, and Protestant organizations represented conciliatory pacifism, which acknowledged the right to "a nation's armed defense." One can assume that the more radical ones with their unrealistic utopian dreams of peace directly contributed to Hitler's rise.

Pacifism is contrary to the nature of life and to the experience of history. Moreover, it is harmful to the community or the nation in which it prevails, and the more consistent it is, the more it contradicts itself.

That contradiction cuts two ways. On the one hand, pacifism leads to the weakening of the will to life and the defensive capability of a nation. It could therefore be identified as a harmful and even traitorous phenomenon. This, after all, is the reason why pacifist thought remained fruitless at all crucial moments in

history despite the nobleness of its aspirations. On the other hand, when the pacifist opposition to war was so intense that it led the pacifists into an armed struggle against the supporters of war, then pacifism itself prompted hatred toward its opponents and preached the necessity of eliminating the enemy. Thus, peace-making confirms the very sense of war. This became a "particularly interesting kind of justification for wars," as C. Schmitt concluded in the early 1930s. He also called attention to the fact that such wars necessarily took on particularly cruel and callous characteristics. And precisely because a war of this kind assumed the form of "the ultimate and final war of humanity," in it, enemies were divested of all moral and any other value, and reduced to inhuman freaks "who must finally be finished off." [102]

Such reflections are affirmed even in the title of my book, *A War Against War*, which I wrote in the 1950s. I was compelled by a personal need to resolve the causes and sense of national liberation and revolutionary war as opposed to war in general from the viewpoint of history, ethical philosophy, law, and military strategy. I was not familiar then with the above-mentioned ideas, and I wrote in the preface that the title reflected my personal motivation, based on pacifist and revolutionary convictions, for having joined in the "war against war" at the age of nineteen. But I also stated that I had chosen such a title "with the profound conviction that it expresses exactly the quintessence and the essential idea of every national liberation and revolutionary partisan war" because it "embraces and carries with it the specific idea" of "war for the elimination of war both in our land and in every other, so that it should not return." [103] I was soon brought to my senses by the harsh reality of war, which mocked those noble ideals. And all my later historical perceptions, certainly not without despair, pointed out that reflection upon the reality of the world is the only meditation worthy of the human intellect.

Marx's materialist dialectic gave a theoretical basis for proving the inadequacy of the pacifist will to abolish war. According to the Marxist understanding of history, force need not always be immoral; on the contrary, it can have a progressive revolutionary

The fall of I
central Eui
monarchy a
the countric
October Bo
March on F
the menacin
beginning t
movement,
Maurras, w

Benito M
tor of the s
and founde
Journal of I
Fascism, Be
negation of
not mean a
called" the
else. "Politi
doctrine mu
will to pow

Fascism
of the huma
thy of him.
nation or to
newfangled
self outside
"that *happr*
advocated f
the highest
a spiritual o
the value o
of warfare
human ene:
nations wh
As a "R

role. The use of violence is socially necessary and ethically justified when employed by progressive anti-war forces that aim to eradicate violence and the insanity of war.

One should recall Marx's conflict with Proudhon, who until then had long been his friend and fellow analyst of such issues. Proudhon (1809-1865) was a theoretician of socialism and anarchism (a rebel during the revolutions of 1848 who was sentenced to three years of exile) who mocked not only bourgeois morality but also bourgeois rationality. Nevertheless, he advocated the idea of a nonviolent, peaceful transition from a capitalist to a socialist society. Understandably, this led him into a dispute with Marx, who answered Proudhon's *The Philosophy of Poverty* (1846) with his own *The Poverty of Philosophy* (1847), in which he called Proudhon a side-show comedian and sophist.

The notion was made firm in Marxism that force has "the role of a midwife" in the birth of the socialist order. Engels openly justified every use of force for a purpose, rejecting as absurd the statement that force morally diminishes everyone that uses it. He proved that theories about the absolute unjustifiability of the use of force were untenable, citing the "obviously high moral (spiritual) surge which was a consequence of every victorious revolution."[104] In all its contemplation over the complicated problem of force and violence, Marxism adopted the ethical and legal classification of wars into just and unjust. The use of force and violence acquired the stamp not only of legitimacy but even of high ethical value when it facilitated the realization of Marxist goals. Thus, contrary to pacifist theories, Marx and Engels justified the use of force and violence in conducting a revolution, in defensive, national liberation and revolutionary wars against foreign aggressors, and even in wars of conquest when they serve to expand a more progressive social system.

Engels himself advocated the destruction of the "historical" "trash-nations," and even of all the "tribes and nations, large or small," which were not capable of becoming revolutionary, such as Germans, Poles, or Hungarians. In the name of revolution, Engels even advocated the total extermination of certain nations.

He c
Spair
Gaeli
adde
preta
natio
Brezł

Le
make
viole
Inde
prem
were
eradi
histo
feste

Fasci
nal
hum

All
insti
eithe
ardly

D
litica
brou

and the will to power. "The craving for an empire, i.e., for the expansion of the nation, is a reflection of the vital force." According to *il Duce*, Fascism was the most appropriate doctrine for the renewed spiritual elevation of the Italian nation in the twentieth century, in contrast to the obsolete ideologies of the preceding century. However, as a doctrine of this century, Fascism also had a general cosmopolitan significance because, Mussolini said, "from now on, throughout the world, it has the universality of all doctrines which in their realization have represented one moment in the history of the human spirit."[109]

It is significant that probably in the emergence of no other Fascist movement did prominent representatives of the spiritual life take so much part as they did in the case of Italy. Gabriele D'Annunzio (1863-1938), a writer of great artistic power, put himself in command of a military unit which carried out the forcible annexation of the Croatian city of Rijeka, and he later became the people's representative and national poet of Fascist Italy. The philosopher Giovanni Gentile (1875-1944) joined Mussolini from the beginning (he was a member of the Great Fascist Council and Minister of Education) and shared his fate (both were killed by the Partisans). Even the philosopher and historian Benedetto Croce (1866-1952) became Mussolini's minister, although he did not mask his opposition to Fascism. Giovanni Papini (1881-1951) also belonged to this circle. He was one of the most controversial literary and philosophical personalities, who had gained fame as a "consummate, total atheist," a "universal revolutionary and an oppositionist." Giuseppe Prezzolini became a devotee of Fascism, having previously been a follower of socialism, but he was critical of Mussolini, and so ended up in the United States in the 1930s where he continued his career as a university professor.

The Fascist doctrine also displayed academic restraint and tolerance, and in its politics there was no such explicit terrorist violence or ruthless anti-Semitism as was found in other Fascist movements, especially the Nazis. Nevertheless, the essential component of Fascism—totalitarianism—predominated over any

regard for historical perspective and reality. The Fascist structure of the totalitarian state meant the establishment of totalitarian violence (spiritual and physical) over its population and determined its policy of conquest, with all the accompanying phenomena, from Ethiopia and North Africa to France and the Balkan countries along the Adriatic.

The German version of Fascism was a consequence of the same spiritual crisis in European and world society from which Italian Fascism emerged. (The same crisis created the New Deal policy in the United States.) This can be seen in the very name of Hitler's movement: National Socialism (the full name was *Nationalsozialistische Deutsche Arbeiterpartei (NDSAP)* or German National Socialist Laborers' Party). Hitler's party had from the start the characteristics of those social democratic and social Christian currents that were apparent in the Austrian and German labor and socialist movements. This socialism also had as its philosophical starting point the criticism of capitalistic production relations, but it sought a solution in emulating the old Germanic cooperative communes and the class unions.[110]

Although there were not as many distinguished personalities of German intellectual life directly within Hitler's circle as there were in Mussolini's, Hitler did not lack support. His movement as a whole was more an expression of the German spirit—both in the defense of Germanism against its rival European powers and in his expansionist and imperialistic aspirations—than of German leadership in the spreading of Fascist views throughout Europe. The emergence and the victory of Hitlerism in Germany was primarily a consequence of the German spiritual legacy and the general crisis in which Germany found itself after World War I, but undeniably also of the spiritual climate in Europe.

Up until the nineteenth century, the main problem had been the uniting of the scattered and divided German people. During that time, the idea took shape that anyone who could unite the German people, even if by fanaticism and force, would be glorified as a hero, and his sins would all be forgiven because nothing could stand in the way of the powerful and great Germany. In the

tion" where the German element had once helped in the construction of states before "Jewry" encroached upon the Slavs. And now, that "giant empire is ripe for collapse," and the end of "Jewish rule" there would coincide with the end of the Russian state and the Bolshevik revolution.

Such aims could not be achieved through an ordinary war. Hitler, therefore, spoke openly about an aggressive, desolating, and destructive war of conquest. According to Hitler, all those kinds of war were regarded as the highest form of life, in which boys became men and the populace would rise to the greatness of its racial duty. Hitler said that a period of peace which lasted longer than twenty-five years was harmful to a nation. Therefore, he did not consider even German world rule as a permanent, peaceful order, but as a continual state of war. He predicted that farther into the future, "perhaps in a hundred years," Germany would also have to fight against Italy after the completion of Mussolini's program.[114]

Life is a struggle, he wrote in *Mein Kampf,* and the natural law of the stronger indicates the domination of the higher race. Within it, only one Aryan nation is a bearer of the totality of cultural and creative strength: "the Prometheus of humankind, from whose astute forehead the divine ray of genius has radiated in all ages." Hitler believed that the misunderstanding and failure to resolve the racial problem can lead to national catastrophe. Therefore, he felt that the awareness of race and of the purity of blood is a necessity for national survival.[115]

His views on the Jewish people were vile, irrational, and hate-filled. To Hitler, Jews were the bearers of destruction, without any idealistic convictions, filled with hatred and egoism, and tightly congregated and isolated within their religious community in order to protect their race, cult, and customs. To him they were the eternal strangers, parasites, and mischief-makers, destroyers of the national culture and will, of the very national being. Hitler wrote that the Jewish people stood out as a secret, dark, and destructive force since the beginning of history, as a general "Bolshevism" of humanity with its false humanism and pacifism,

but actual revolutionary and conspiratorial tendencies against all genuine values.[116]

Hitler ascribed all failures of capitalist society and democracy to the Jewish people, but its greatest sin is the leadership of the international Marxist movement. Moreover, according to his evil doctrine, Bolshevism is the final and highest historical limit of Jewish influence, which, with the help of the dictatorship of the proletariat, deprived the people of democracy and subjugated them with brutal force. As part of a consistent application of the principle of racial purity, "the task of maintaining those racial proto-elements" which "create the beauty and dignity of a higher form of humanity" was set as the state's highest aim. Hitler's doctrine could "grasp the idea of state only as a living organism of one nationality, which provides for the continuation of that nation, and leads it to the greatest freedom by the further development of its spiritual and ideal abilities."[117] To this was added, as "the greatest task in the mission of the German people," that the German Reich, "as a state, should include all Germans," not only to gather and preserve them, but "to lead them slowly but surely" down the path to "world domination." Then, perhaps it would even be possible to have "that which many blind pacifists hoped to obtain by begging, lamenting and moaning: peace, protected not by the palm branch of the teary-eyed pacifist wailer, but by the victorious sword of the ruling nation (*Herrenvolk*) which overpowered the world in favor of one higher culture."[118]

The horrible racial basis for the inequality of people and nations is also at the heart of Hitler's *Lebensraum* principle. It meant, essentially, the need for a balance between the number of people of a nation and the amount of territory necessary for vital functions as well as strategic reasons. Since "only a large enough space guarantees freedom of life for a nation," the German nation in particular had the right to such space for the purpose of fulfilling its world task. And considering its geopolitical position, Hitler felt that Germany had no other alternative but to spread within Europe. From there, the natural direction would be a spread toward the East, and the conquest, colonization, and Germaniza-

tion of the Slavic lands east of the Elbe. And that Germanization of the Eastern agricultural, i.e., Polish, Russian, and Ukrainian lands, was to be achieved by replacing the Polish and Jewish population with German settlers, that is, through "the biological weakening" of the local population by repression and extermination (especially of the Jews) and by turning the rest of the population into slaves (*Sklavenvolk*) under German masters.

The nefarious doctrine of racial purity and of living space was actually centered around Hitler's Eastern Politics (*Ostpolitik*). Its implementation meant the "solution" to almost all key issues in Hitler's doctrine for turning Germany into a world power: the gathering together of the Germans, the expansion of the living space through the Germanization of Slavic territories, the annihilation of the "Russian Empire," and the threat of Communism, with which the "destruction of the Jewish-Bolshevik head" could be simultaneously achieved, along with the "solution" to the "Jewish question" in Eastern Europe and proximity to the Middle East and Asia in order to overthrow the English colonial empire.

The old German idea (from Hitler's *Mein Kampf*) that "England does not want Germany as a world power, and France wants no power at all called Germany" probably influenced Germany's orientation toward *Eastern politics*. And since it was inconceivable that the obstacles in the West (especially England), could be removed, an outlet into the world scene was sought in the East.[119] Only there could Germany expect to conquer the large territory necessary to every world power, of which Hitler argued Germany had been deprived in the past.

In the doctrine of Hitlerism, racial hatred was brought to the utmost limits of ruthlessness. The consequences of Hitler's insanity did grave harm to his victims, most tragically the Jewish people, who suffered more than anyone else under the Nazis. The Holocaust was not some isolated and irrational act of violence but part of an overall plan for the extermination of the Jewish people and culture. This plan had its basis in anti-Semitic ideology and was orchestrated by Hitler, and carried out by his Nazi regime, along with collaborators all across Europe. In the end Hitler's evil

doctrines and policies brought Europe to the brink of destruction and almost destroyed the Jewish people. By the application of the Nazi racial doctrine, he caused unprecedented violence in one of the darkest periods of human history.

11. Contemporary Reflection on Violence

History has always given the same answer to meticulous questions: if you are not the oppressed, then you are the oppressor!
—Carlo Bini

Civilization is a terrible plant which grows and blossoms only when it is doused with tears and blood.
—Arturo Graf

What an ethical phenomenon! What a stupefying revelation!
—Romain Rolland

If there is no God, everything is permitted, wrote Fyodor Mikhailovich Dostoyevsky, that master of Russian and world literature of the nineteenth century, who delved into the secret depths of man's soul and into the infinite spaces of human nature.

In this century new movements appeared on the world stage, movements that committed horrible crimes in order to rearrange the wicked world and establish their new order under the supremacy of their earthly gods and with their supposedly scientific minds. Such historical vortices, which in this century swept first over Europe and then over the whole world, stimulated further meditations and debates on war atrocities and crimes in general, particularly on racism and anti-Semitism, as well as on revolutionary violence.

As we have seen, throughout the course of history, struggle and war have been seen as inseparable components of historical existence. "War is a proto-historical phenomenon for all living

things," O. Spengler said in this age, "and to such an extent that struggle and life are identical in the deepest sense, for when the will to fight has reached its term, then so has life." And it is indeed pointless and even unreasonable to brand Spengler, or Nietzsche before him, as war-mongers, as if their reflections were the cause of any war. No matter how much moral abhorrence he may induce in us, one must admit that Spengler only brought to our attention the harsh and unyielding historical reality that "Life is hard and cruel—when it should be great. It offers only the choice between victory and defeat and not between war and peace, and the sacrifices of the victory are a part of the victory as well."[120]

When someone ventures to expose the naked truth about historical existence in such a crude way, then the truth seems to us even more dreadful than the reality which it reflects. Civilized man is unwilling to accept the horror of such a truth but instead identifies those who speak it as creators of war atrocities and violence—undoubtedly in order to placate his own conscience.

One should confront the fact that on each side of every conflict, there has always occurred an extreme, willful exertion of physical, spiritual, and intellectual force toward identification with one's own side against the opponent. Whether the struggle was one of defense or conquest, enslavement or national liberation, revolution or counterrevolution was irrelevant. Every side has always found justification for its conduct, for the justness of its war, and the unjustness of the enemy's, for the divine mission of its religion and the heresy of the enemy's, for the progressiveness, righteousness, and future of its ideology and the backwardness and historical doom of the enemy's—in short, for the holiness of its struggle and the damnation of the enemy.

In this discussion, we have already seen examples of mythological and biblical invocations of historical reality, as well as those from classical literature which have become the eternal and universal model for future contemplations. But there are many more examples of literary and intellectual endeavors that can shed light on our subject. Let us recall the great cultural and artistic monuments left in Spain from the golden age of its empire (fifteenth and

sixteenth centuries) when, as a leading European power, it controlled territories from the Netherlands to Mexico through the brutality of conquest. And on the other hand, it was the spirit of the subjugated Flanders that created the *Legend of Till Eulenspiegel and Lamme Goedzak*. That legend was immortalized first in the imagination of national tradition, and later by the writings of Charles de Coster (1827-1879).

Till Eulenspiegel became a character uniquely representative of national vengeance in world literature, and Coster's work came to be known as a "book of the homeland" and the "folk bible" of Flanders. R. Rolland placed that "unique epic of courage" in the same category as *Don Quixote*, and Maxim Gorki classed it among "the most momentous works of world literature." Coster himself, in the last century, wrote:

"Eulenspiegel - Uliedenspiegel. A mirror for you—for those who are ruled and for those who rule—a mirror of the madness, absurdity and crime of one century." (And one could certainly add: of every century—for there has not been one century without insanity and evil violence.)

"What has nourished your politics since you have been ruling the world? Massacre and murder?... "

The innocent victims of the subjugated nation call out for vengence: "Ashes burn upon my chest... the flame of revenge which burns in my heart was kindled by God, and it will become a sword, a fire, a rope... war and the destruction of the butcher... an eye for an eye, a tooth for a tooth... seize the bludgeon without mercy: these are our joys. Fight the enemy Roman Spaniard wherever you find him... Rip out the duke's intestines and smash them into his face... But when will peace come?... It will come when in Flemish orchards, from every branch of the apple, plum and cherry trees, instead of fruit, a Spaniard would hang!"

That epic poem about the struggle for freedom, but also about the merciless and unsparing lust for revenge, led even such a humanist and pacifist as R. Rolland to write with greatest enthusiasm, though also with resignation:

"It is a whip which takes vengeance, a spirit which thrashes

with its sword, it is a step of freedom, a battle without sympathy,
where blood flows in rivulets ... and where even hatred is a kind
of wine. It is a poem of the independence and revenge of an
oppressed nation." Rolland was deeply aware of the notion that
all wars were "cruel wars without sympathy, without compassion,
both on the part of the oppressor and the oppressed. Wars which
obviously cannot be repented, for the poet has forgiven nothing
even after three centuries!... Such hatred which lasts three cen-
turies! And upon that hatred a new homeland is built! What an
ethical phenomenon! What a stupefying revelation!"[121]

We are reminded of all the fruits of the human intellect and
heart, with all their achievements and their principles, destruc-
tions and contradictions, aspiration and despair: all the living leg-
ends of all communities, both the mythological worlds and the
Old Testament psalmists, the anonymous plebeian poets, as well
as Homer, Sophocles, Euripides, Virgil and Dante, Shakespeare
and Byron, Milton and Shaw, Breughel and Goya, Schiller and
Goethe, Hemingway and Lorca, Mazzini and Verdi, Manzoni and
Papini, von Kleist and Korner, Hugo and A. France, Beethoven
and Wagner, Dostoyevsky and Tolstoy, Erenberg and
Shostakovich, Petar Petrovic Njegos and Dobrica Cosic, and (to
mention a few more names from Croatia) Marko Marulic and
Ivan Gundulic, Ivan Mestrovic and Vanja Radaus, Anton Gustav
Matos and Tin Ujevic, Miroslav Krleza and Petar Segedin, Ranko
Marinkovic and Ivan Goran Kovacic, Jure Kastelan, Jozo Lausic
and Ivan Aralica, and likewise, Ivo Andric and Mesa Selimovic;
Delacroix and Rude, Maupassant and Eluard, Mauriac, Sartre,
Camus and Boll, Mailer and Baldwin, etc. All spiritual represen-
tatives of their nations and times, and the greatest masters of
human culture, both in peace and in war, they glorified in their
works the battle for freedom and justice for their nation and called
for revenge and the destruction of the cursed enemy.

All of that is to say: they actually encouraged, if not openly
preached, hatred and destruction, no matter how much they oth-
erwise pleaded for peace and brotherhood among nations or
pledged themselves to universal human ideals. Even Goethe, who

was cold toward the national liberation movement of his own German nation, adhering to his philosophical stance that "a poet must rise above all national disputes" in order to serve all of humanity, did not remain consistent with that precept. Moreover, he eventually admitted his error by celebrating in verse the character of Goetz von Berlichingen, the leader of a vengeful band of robbers and a peasant rebellion as an example of a freedom fighter. Goethe publicly praised the participants in the national struggle, expressing regret that he himself had not joined in it, in his play *The Awakening of Epimenides* with the words: "It would have been a privilege to suffer with you because you were greater than I—for the pain that you felt!"[122]

Goethe's example, however, is not isolated. The person who resembled him the most closely in Croatian history was undoubtedly Miroslav Krleza. Krezla's detached and reserved attitude toward both the original national movement and the revolutionary movement was a consequence of his being torn between his own roots in the Croatian nationalism and his predilection toward Marxist internationalism. He, too, has left us with more than enough acknowledgment of his error.

In examining the manifestations of violence in history, one encounters the interesting fact that they appear in ever new and unpredictable forms. Thus, before the French Revolution, the ruling aristocratic class was enthusiastic over, and the leaders of the Enlightenment were inspired by, the notion of "the naturally good man" and "the virtuous nation." In his description of the *ancient regime*, Tocqueville spoke of how there were no indications which could have suggested the impending revolution. There was relaxed and optimistic talk of the virtue, benevolence, and harmlessness of the people, while secretly, a *spectacle ridicule et terrible was already ablaze underground, which no rational person could have imagined would erupt into such a frenzy in the most enlightened European society.*[123]

A similar situation existed in France at the beginning of World War I, after a comparably long period of peace. Suddenly, it was evident that the possibility for eternal peace and friendship among

European nations, particularly between France and Germany, was a delusion. We have convincing testimony about this period left to us by Stefan Zweig, the great defender of peace and the spiritual unity of Europe, who was in France in 1914. "Only one second revealed to me how in a moment of serious crisis, a nation on this or that side could easily be incited to revolt, in spite of all our attempts to come to an agreement, in spite of all our advocacy of peace." The most ordinary film report about the arrival of the German emperor Wilhelm to Vienna caused an outburst of such virulent hatred on the part of "the good-natured inhabitants of Tours," that Zweig was astonished and frightened. "I was unable to sleep," he wrote. "If this had taken place in Paris, I would have been equally uneasy, but not as shocked. I could not believe that hatred had pervaded so deeply into the provinces, into the good-natured, naive people, and I was overcome with horror."

Zweig's contemporary, Rolland, judged things a little differently. Having arrived at the understanding "the more naive a nation is, the easier it is to subdue it," he concluded that "we have come to the age of mass feelings and mass hysteria, whose potency in the case of war cannot yet be fully fathomed."[124] The accuracy of such a profound comprehension of the condition of European society and the relations between nations would be confirmed at the end of World War I, but certainly not in its full sense until the Second World War.

Every war in itself amounted to pouring oil on the fire of such instinctive tendencies of people and nations toward hatred and inflamed passions. War retributions—where there were special reasons for inciting mobs to fanaticism and bloodthirstiness, even if they had no rational basis—resulted in outbursts of endless violence and crime. Leo Tolstoy called such phenomena the corruption of people at war. "Every war," wrote the great author of *War and Peace*, "even the mildest sort, with all the usual consequences: destruction, uprisings, plundering, kidnapping, debauchery, murder, pretenses of necessity and justifiability, the exaggerated praise of military undertakings, the love of the national flag and homeland, the pretense of caring for the wounded—corrupted more

people in a year than thousands of robberies, arson and murders committed by individuals incited by passion could do in a hundred years." In general, Tolstoy was right, especially with regard to the comparison in the proportion of war and peacetime crimes. However, it would probably be more accurate to conclude that this was not so much a question of the "corruption" of people, because virtues that even Tolstoy himself endorsed (the intuition of military leaders, courage, devotion) become more pronounced in times of war. Instead, the impulses of passion—which in times of peace were suppressed by customs and laws—become chaotic and enraged in war by the very nature of an apparent life and death struggle against the enemy.

The notion that man, both at peace and at war, "is tantamount to an animal"—even worse than an animal, for it kills its own kind—probably led Nietzsche to cry out in pain, "Oh, my friends! He who has comprehended says: shame, shame, shame - that is the history of man!"[125] This idea was expressed about half a century before the Second World War.[126] In the meantime Maxim Gorki said: "Man—how proud that sounds!" We also find this in Shakespeare's reflection through the character of Hamlet: "What a piece of work is man! How noble in reason! How infinite in faculties!" Shakespeare also reminds us that man is in apprehension "like a god" and "in action how like an angel," according to which he is "the beauty the earth" and "the paragon of all living beings." But along with all those characteristics, he added: "And yet... ," drawing attention to the fact that man's unbounded skill appears not only in good, but in evil as well.

However, one usually forgets that insight, so only one side of human nature is considered and judged. Thus, in this century, a great deal of emphasis was placed on just that notion that man, through his elevated consciousness and material progress, had finally reached the point where he could pass out of inhuman history. But the primeval historical plays continued to make their successive appearances, which meant a drama for all the new actors, for some a short rhapsody or even a tragicomedy, but for most an ill-fated tragedy.

In the early twentieth century, the world seemed to have succumbed to oblivion: the Russo-Japanese War, the Balkan Wars, all the known but scarcely recorded retributions on the part of the imperialistic powers in the colonies from Africa to Asia and both South and North America, and even World War I with its triumph in the form of the Treaty of Versailles. The world is still very much preoccupied with the Russian Revolution, the gulags of Stalin, Hitler's Final Solution and New European Order from World War II, virtually ignoring what in fact happened during and after that war in the process of ending the Chinese and other revolutions and of liberating the colonial lands.

Now, human intellect has allowed itself to be deceived about a supposed world peace only because World War II has not yet been followed by a World War III. It doesn't recognize that just because World War III has not come about due to the balance between two great nuclear powers, the world still has not come any closer to a peace without war. Actually, soon after the end of World War II and the Yalta Conference, the world entered a period which could rightly be called a sort of new world retribution. The so-called cold war, which emerged in the early 1950s, exhibited all the traits of the various forms of violence, especially psychological, and even the characteristics of acts of genocide. And with the partial decline of the cold war—that is, with the understanding that no single power could have its own will in world relations—international crises, armed intervention, and local wars quickly multiplied on all sides of the world. Not one of the conflicts was without the direct or indirect participation of one, if not of both, nuclear superpowers. The people of those countries which once were plundered by the colonial forces continue to be at war or suffer from internal or external conflicts. All in all, one can scarcely find more convincing reasons for the view that in the present world—despite the highest degree of cultural, technological, and economic integration—we are no closer in eliminating racial, ethnic, social, political, religious, and ideological conflicts.

The ability to use nuclear power has marked a new historical phenomenon, in which the notion of all destructive omnipotence

brings with it the threat of self-destruction. But this has only led to the exclusion of a nuclear world war between the nuclear super-powers. It is human to hope that exactly such "godlike" power in the hands of man could help preserve the civilization which mastered it, instead of creating the possibility for its destruction. However, even if this civilization could evade the doom of vanished cultures in human history, it is hard to imagine that the battle of opposites, between good and evil, darkness and light, peace and war, could ever disappear.

Back in early historical times, the human intellect, through the Old Testament prophets, envisioned the idyllic condition of permanent-lasting peace this way: "The wolf also shall dwell with the lamb ... the calf and the young lion and the fatling together ... they shall not hurt nor destroy ... the earth shall be full of the knowledge of the Lord, as the waters cover the sea ... they shall beat their swords into plowshares and their spears into pruning hooks; nation shall not lift up sword against nation, neither shall they learn war any more."[127]

If the human intellect was able to imagine such a condition, then it undoubtedly had enough sense to realize that such a world could no longer be this world, nor have anything in common with it. None of the other conceptions of peace that followed this biblical vision have given pacifist ideas any meaningful content. Instead they relied on two mistaken visions—either longing for the creation of such a social and international order which would be able to restrain opposing forces, or seeking a new god who would create a completely new world. History has taught us that such aspirations, instead of bringing the bliss of peace, have only given rise to greater evil than that which had already existed.

Thus an attempt to escape into another world, outside of and contrary to the historical one, has proven not only fruitless, but tragic: it has brought much greater evil than that which it tried to eliminate. Definitely, the human mind must soberly face the revelation that, in reality, "life lives through life," as Petar Segedin quaintly put it, no matter how much (according to another great humanist, R. Rolland) that revelation stupefies. Therefore, one

can and must look for reasonable explanations only in history itself, for "it is only through the comparative history of nations that we may, to a certain extent, locate that vein," wrote Franz Kafka, perceiving that "the battles from our most ancient history are being fought only now."[128] Or to be more precise: they are being fought still.

12. Despair in Doubts

Since the other world of truth has vanished, it is the task of history to establish truth in this world.

—Karl Marx

To be sure, the demand for understanding of historical events in no case means an attempt to justify any sort of historical crimes. Instead that demand comes from the realization that simplistic and one-sided moralistic condemnations, from whichever standpoint, do nothing more than contribute to the worsening of the circumstances which caused the tragic events in the first place. Ascribing historical misdeeds and evil only to certain nations, or only to particular social movements, is a sign both of historical illiteracy and unreasonableness.

Historical reality always takes care to reveal quickly the errors and lapses of memory, making amends for all the delusions and deceptions. And it always shows that the truth is never found solely in the beliefs of the victorious side, nor of the defeated. Rather, to a certain extent, it resides in both. History is filled with abundant examples that prove this point.

The dark side of the great French Revolution is hidden behind the glory of the victory of its ideals. There one finds not only the horrors of revolutionary violence and the Thermidorian Terror, but also the primitiveness and the imperialism of Bonapartism. And even contemporaries of the Revolution expressed some accurate judgments on it, which would be confirmed by all subsequent revolutions. Thus Joseph de Maistre (1753-1821) said, in keeping with his Christian philosophical conviction, "There is a kind of satanical character of the French revolution which distinguishes it

from everything that has been seen and perhaps from everything that will be seen." In the name "of the coronation of the Goddess of Reason" there was a myriad of evil deeds "in which the provinces tried to outdo Paris," to such an extent that "it all departs from the ordinary realm of crime and seems to belong to another world." Even when the Revolution was pacified, "major breaches of law vanished, but the principles still remained."[129]

The opinion of Edmund Burke (1729-1797) is even more interesting because of his comparison between the English and French revolutions. In the English Revolution, the ancient laws, freedoms and the constitution of the state were preserved: "Everything that we have can be traced back to our country's past, as a legacy left to us by our ancestors' testament." In France, everything was overturned and destroyed—with disdain for their entire legacy. Therefore, "they were punished by their own success, which was thus: laws were abandoned and courts demolished, industry was weakened and trade obliterated, taxes were not levied, while the people were nevertheless reduced to poverty, the church was robbed, but that did not make matters easier for the state." The civil and military anarchy established a new, uncertain, and precarious rule, the currency was devaluated and the principle of property destroyed, a condition of poverty and plundering emerged, everything was sacrificed to the idol, and "the consequence was—national bankruptcy."[130]

The judgments of de Maistre were rejected as the reactionary views of a French ultramontanist ideologue. Burke's views were rejected as those of an English conservative. Of course, all other similar judgments were trampled with scorn in the name of omniscient reason and the necessity of social progress.

Indeed, they were trampled also in the name of Hegel's *historical world spirit*, which is "the substantial"—quintessential and fundamental—in world history. Certainly Hegel's historical ideas signify one of the greatest reaches of human thought in order to comprehend "that which eternal wisdom longs for." Yet everyone must wonder about this, and also about Hegel's judgments about revolution and violence in view of new wisdom with which histo-

ry has endowed us. For Hegel, the Jacobin reign of terror was, in fact, "the pure horror of the negative." In a time of great passions, history has no time to deal with "the triviality of individual human destinies." In the French Revolution, as in every other "world-historical event," the use of force "proved to be an absolute need." Through the transfer of government to the people, "the abstract principle of freedom" took hold, and "doubt gained tremendous power."

Robespierre himself established "the principle of morality as the ultimate" principle. But now what reigned were "*virtue and terror* because subjective virtue, which rules purely according to disposition, brings with itself the most dreadful tyranny. It executes all its power without forms of court," so its punishment is "simple—death." The instincts of passion in satisfying special interests or vindictive impulses "are the most powerful" in actions which "have no regard for any of the limitations that law or ethics want to set for them because these natural forces are much closer to man than any artificial and prolonged training for order and art, for law and morality."

Hegel also said that revolutionary (and any other) tyranny must collapse under the necessity of the reestablishment of the reign of order "because all inclinations, all interests, and wisdom itself" rise "against that terrible consequential freedom" of violence and terror.[131] Considering the fact that the instincts of passion lead to evil deeds most due to vindictive impulses, Hegel felt that revenge was justified in its content only if it was retribution. However, "revenge, insofar as it is a positive action on the part of one particular will, becomes a new offense." And thus, "as a contradiction, it becomes caught up in a progression, in the endlessness, and it is conveyed from generation to generation into infinity."[132]

Marx and Engels rejected Hegel's "world spirit" and "historical mind." In their place the two philosophers focused on materialism and social relations. In their theory of the inevitability of social changes and the necessity of revolutions, they removed any possible obstacles to the justification of violent means for the sake

of envisioned progress. It was precisely this perception of world history which found its clear realization first in the Marxist ideology of Leninism and Stalinism and then even more in the reality of socialist societies based on Marxist doctrine.

Not one of these societies succeeded in freeing the shackled human spirit from the delusions of philosophical ideology transformed into practice, whose inviolable divinity was to be the only correct and most progressive doctrine of the world—in the name of which everything was permitted. All attempts at finding a way out of the repression of the human spirit and state control of productive power have so far remained fruitless.

Even after four decades, Titoist self-management remained unsuccessful. A stormy tumult crushed the "hundred different flowers" in Mao's garden. With ruthless force the Polish national movement was repeatedly suppressed with the use of tanks. The Hungarian revolution was similarly stamped out. The "Prague Spring" froze over in the Siberian winter of Soviet military intervention. And the "Croatian Spring" was destroyed by the violence of combined dogmatism and hegemony. The use of force and the incitement of hatred between nations deepened the crisis in Kosova. And the military occupation of Cambodia and Afghanistan revealed their notorious order, inciting armed national resistance.

Quite contrary to Marxist expectations, the socialist world fell into an ever deepening crisis, because of its excessive dissociation from spiritual and material life, and, even more, its spiritual violence and reckless use of armed force.

Just as all similar historical events, regardless of time and space, bear some identical characteristics, so, too, similar revolutions are in many ways different depending on their historical and national circumstances. Such a conclusion was drawn by Mikhail Bakunin, who, half a century before it occurred, predicted the revolution in tsarist Russia. "Let him who wishes to preserve the purity and innocence of his virtue remain in his office; let him dream there, meditate, compose essays and verses. But he who wishes to be a

real revolutionary fighter in Russia must take off his gloves because there are no gloves that can save him from the endless, multiform Russian mud. The Russian world—that state, privileged and national world—is a terrible world. And the Russian revolution will certainly be terrible."[133]

And such it was. And so were all its leaders and those who carried it out, from Lenin and Trotsky onward. Trotsky wrote even before the overthrow of tsarism: "The revolution asks of the revolutionary class to achieve its aim by *any* means available: if necessary, through armed rebellion, and if there is no other way—through terror. The revolutionary class, which gained power with arms in hand, is obligated to, and certainly will, again with arms, suppress all attempts to take that power away from it."[134]

Lenin himself did not hesitate to apply extreme and inhumane actions in order to accomplish his aims. He approved of and himself initiated all forms of terror: execution by firing squad, banishment, taking hostages, conducting show trials, decimating, and killing the underage members of families. He even left testimonies as to why he had done this. "I decide rationally and categorically: which is better—to catch tens or hundreds of instigators, whether guilty or innocent, aware or unaware, or to lose thousands of Red Army soldiers and workers?— The first choice is better. And let them accuse me of all the mortal sins and violations of freedom— I will plead guilty, but the interests of workers will benefit."[135] However, this was not only a case of brutal retribution against the "class enemy" (the representatives of the feudal and urban classes and of the overthrown empire) but it referred also to the equally brutal measures taken against the Mensheviks, members of the same socialist movement who happened to oppose Bolshevik methods. Lenin explicitly directed, "Our revolutionary courts must deal with the public demonstration of Menshevism by execution; otherwise, they would not be our courts, but God knows what!" Lenin presented an even more harrowing response to objections from the social democrats: "Let us place you up against a wall. Or take care to refrain from expressing your opinions, or

if you wish to express them ... we shall treat you as the worst and the most harmful elements of the White Army."[136]

In the days of such revolutionary insanity there were those, even among the followers of revolution, who had the courage to raise the voice of reason against it. In 1920 the chief Soviet newspapers *Izvestia* and *Pravda,* published the decision of the Soviet government that the members of the social revolutionary groups, the nationalist White Army, and Wrangel's officers should be taken hostage and executed by firing squad. This prompted P. A. Kropotkin, one of the best known Russian and European anarchist thinkers, to address Lenin in a letter thus: "...is there no person in your circle who would suggest that such measures signify a return to the darkest days of the Middle Ages and the religious wars, that the measures are not worthy of the people who took it upon themselves to build the future society on Communist principles!" And Maxim Gorki, the most original proletarian writer and defender of the revolution, wrote at that time: "For almost 100 years, the best among the Russian people have lived for the idea of a constituent assembly—a political forum where all of Russian democracy would be able to express its will freely. For that idea, thousands of intellectuals, tens of thousands of workers and peasants have died in prison, in exile and in hard labor, died on gallows or were executed by firing squads. Rivers of blood have flowed as a sacrifice to that sacred idea—and now national commissars give orders to fire upon the crowds that demonstrate in support of that idea."[137] It is a testimonial to Lenin's magnanimity that upon such a protest, he recommended that his friend Gorki go "for therapy" abroad. This he did—living mostly in Mussolini's Italy. When Gorki returned to Russia in 1931, a new wave of violence had begun. It came first in the form of Stalin's liquidation of the wealthy *kulak* peasantry, followed by new purges at the top Soviet ranks and also among members of the international *Comintern.*

It is ridiculous to seek an explanation for all the violence and the vast number of victims of Communist revolution (which

according to various estimates is between 80 and 139 million).[138] This violence is not the result of the deformations of Stalinism; instead, it exists in every revolution, and the Russian revolution in particular. It is clearly in view of such an understanding that one should judge all other evil deeds in the dark maelstrom of history. All the perpetrators of even the most dreadful deeds have done them out of love for that which they held sacred and hatred against their demonized opponents.

Both great minds and the uneducated masses succumb to such Manichaeist delusions. "When someone directs oneself to (his/her) God" there is no sacrifice that he will not bring to the altar in order to achieve his paradise. Moreover, for such, even "God's punishment" is a "gift of God!"—explains Jean Ellenstein, one of the leading Marxist ideologues of the French Communist Party, who, only after the Soviet occupation of Afghanistan, became disenchanted with the infallibility of the divine idea of Communism. Along with his correct remark that one should judge Stalin within the framework of history, for "Caesar killed more Bretons and Gauls than Stalin did Russians,"[139] one mustn't neglect the fact that Stalinism destroyed more Marxists than Fascism did Communists.

However, one may very well ask whether a more complete understanding of history has ever had a decisive influence upon the course of history. Even back in biblical times, a wise man came to understand that common sense, which has embraced the principle of love thy neighbour, but hate thine enemy, does not lead one to salvation from the hell of enmity. Thence came Christ's remarkable principle: "Love your enemies, bless them that curse you, do good to them that hate you, and pray for them which use you and persecute you."[140] That selfless sentiment has remained throughout history a cry of the weak, or an expression of those who had accepted their doom. They have relinquished punishment of their enemies to fate, or believed that causing evil to others cannot bring lasting good to anyone.

No matter how many examples can be found in life and history

to support such renunciation, it has never overcome the passions of hatred and the desire to dominate or to take revenge. On the contrary, hatred has repeatedly assumed the form of mass insanity of social, class, or political movements, or even of entire nations. And insanity—as Miguel de Unamuno noted—ceases to be insanity at the moment when it becomes a collective phenomenon, when it turns into the insanity of an entire community, and especially of an entire nation. At that point even the ordinary man loses his judgment and begins to act in a way described by G. Papini: "No crime was strange to me, no baseness unfamiliar; I didn't yield to any fear. I killed innocent old men through the most ingenious tortures; I poisoned the water supply of entire towns; I burned the hair of many women; I tore up with my own teeth, wild with the desire to destroy, the children that I came across on my way." And in so doing, he became *conscious* of his own *self*, "aware that I will *never* stop being me myself."[141]

The humanist and Nobel Prize winner Hermann Hesse (1877-1962) presented the same insights into human nature in his work. The "way to innocence" leads deeper and deeper into sin on the road to becoming human. Nothing can substitute "the ecstasy of experiencing a celebration within a community, the secret of the individual's disappearance in the crowd, the experience of the joy of unifying." The man-wolf is astonished "that shooting gave him such satisfaction" because he was previously opposed to war, but he also observes with distress the tamed Steppenwolf, forced to "reject his own nature." He is aware that "what we are doing is probably insane," but he maintains that it is also probably "good and necessary." And "it is not good when humanity strains the brain and with the help of reason attempts to put things in order" which are not at all within the realm of reason.[142]

In the gyrations of history neither defeats nor victories are indications of a permanent condition. Regardless of an historically temporary predominance of force, which imposes its "justice," it is entirely uncertain whether the victor will establish an even more sinister state or induce even more evil. From the fertile womb of

inspired, jubilant victory, "a loathsome beast" may emerge, as Bertolt Brecht described it.

Besides, one cannot abstractly judge the evil in history. For despite the fact that every crime is in itself evil, each is essentially different according to its motivation. In psychology it is necessary to distinguish between defensive (benign) aggression, which is "phyllogenetically programmed" through the instinct of self-preservation, and instrumentally adaptable (malignant) aggression, which is determined by belligerent impulses and the desire to conquer. In his reflections on human destructiveness, Erich Fromm concluded that man—as an individual and as a community—is biologically "programmed": "with his phylogenetic readiness for defensive aggression, when his life, health, or freedom are in danger." And "since defensive aggression is a phyllogenetically prepared reaction to the state of imperilled vital interests, it is not possible to change its biological basis, although it can be controlled and modified as impulses rooted in other instinctive dispositions." And "the main condition for reducing defensive aggression" is the reduction of those actual factors that initiate and provoke rational and irrational passions, which, according to Fromm, are always part of the "life-advancement syndrome." Lacking that natural, phyllogenetic basis, instrumental (malignant) aggression, even when it seems momentarily inevitable and necessary, always belongs to the "life-deterioration syndrome." Thus the more terroristic violence it does in the name of "absolute state sovereignty," the more destructiveness it causes in each new act of defensive aggression.[143]

And with regard to an assessment of crimes of one and the other form of aggression and destructiveness, Fromm is unambiguous: "no crime that a man performs for his freedom can be as great as the crime of those who deny him his freedom."[144]

Another contemporary philosopher, Leszek Kolakowski, takes a similar position. "We need a combative, aggressive defense of freedom. Freedom cannot be maintained by clemency... it demands watchfulness, pugnacity, and a constant combative spirit." This is

irrefutable, although "one can hardly decide rationally between the uncertain evil of the present and the certain evil of the future." From ancient times to the present, there has been no other key to salvation and liberation for man than "to expel the devil by means of the devil... to destroy evil with evil." Since man strives for good, "everyone wants to believe that his work is motivated by noble will," so that in history "every necessity becomes a virtue." One always finds a theory and a preacher to glorify every evil, even slavery. We live in a world in which "every form of evil must mercilessly generate a new and greater evil," where even "love is a sin, and the sin as a fiery tumult spreads of itself and rekindles itself endlessly." Where "destructive force" is present "in defeat and delusion," or "in unfulfilled yearning," it is there that cruelty and evil are not only the means but the aim itself. Therefore: "You can remove the sense from every evil that has sense; you can arrange the world differently. But you cannot eliminate vitality from that kind of evil that is its own sense, simply as evil."[145]

And so, having reached our times, in seeking deeper explanations to genocidal violence in the course of history, we have certainly come across no new wisdom. Instead, the quest serves only to convince one of how accurate Lao Tse, the founder of Taoism, was when he said some 2,500 years ago: "to go farther and farther means to come back again." On that path, it is inevitable that one would meet not only with rational and wise people, but also with criminals and fools—in all eras and in all places.

Of course, it is true that *tempora mutantur* ("times change"). However, the fact that *nihil novi sub sole* ("there is nothing new under the sun") remains no less despairing. The world today is still torn by essentially the same contrariness as in the beginning, and the same battles are still being fought, only on a higher technological level. The more one has tried to change and adjust the world by force, the more anguish man has put himself through, plunging all the deeper into the evils of the past. He transgresses the present through his headlong leap toward some other, imagined future.

In spite of everything, the longing to bridge contradictions in order to realize the greatest possible harmony has never vanished from the world. There is no doubt that there has, after all, been some progress in that respect in certain periods and in certain parts of the world. For that reason, there is no room for hopeless despair. Whenever and wherever nations in mutual competition are convinced that they cannot defeat each other or impose their domination, they finally in one way or another, reconcile themselves to peaceful coexistence. The recent history of the Scandinavian nations offers a more than convincing example of this phenomenon. Their differences do not disappear, but are resolved in equality and respect by mutually recognized subjects.

It seems that the undertaking to create the contemporary Western European Community is a similar effort. The realization of a union of independent European countries in a *united states of Europe*, based on the recognition of the identity of the historical European nations, could mean the elimination of such military retributions as have racked their history for centuries. However, Western Europe is still burdened with intense and turbulent conflicts between nations. The minority and stateless nations seek their sovereignty here, too, as everywhere else in the world.

But a new crisis is also emerging. The economic migration of populations from Southeastern Europe as well as from non-European countries is creating unrest resulting from differences in culture and religion. Those who oppose any restriction on immigration forget that it can cause great disturbances in ethnic and political communities resulting in conflicts with far-reaching consequences. Even here on European soil, there is no peace in regard to overcoming or eliminating contradictions and violence, although a higher degree of civilization has been reached here than on some other continents.

And what of the world as a whole? The human intellect in general, man as an individual, and the historical wisdom of nations as subjects of the international order are faced with the same problems they have faced throughout history. However, there is one

important difference: the impossibility of using their nuclear weapons because it would endanger all of civilization. One has to believe that, hidden in this evil news (as the opposite to the Gospel "good news"), there is the urgency—and also hope—for at least a bit more harmony in future.

■

Historical Facts and Unscientific Theories

I. THE DISTORTION OF HISTORICAL FACTS

It's diabolically murky here on our sentry duty of the human intellect.

—Miroslav Krleza

On the river of history, the totality lies behind and the rough waters ahead, but people only see that which is directly in front of their eyes.

—Barbara Tuchman

1. On the Credibility of Sources and Testimonies Concerning the War Victims

A mere fact, separated from other facts, is something monstrous... but nothing real.

—R.G. Collingwood

As WE HAVE SEEN THROUGH many examples, the verification of numerical data in historic events is dubious at best. As a rule, every side involved has a tendency to exaggerate, supporting its view with more or less credible evidence. But with the passage of time, as blinding passions are subdued, we are usually able to achieve a more accurate picture.

In the case of the war victims in Yugoslavia, however, we are witnesses to an interesting and unusual phenomenon. With each passing decade since the war, instead of coming closer to establishing the truth, we have actually moved farther away from it. Moreover, it is not only impossible now to consider the facts, but to attempt to do so provokes persecution. As is evident in my own case, as well as that of the Encyclopedia of Croatian History and Culture, scholarly endeavor was frustrated, while mythomania and other chimeras were encouraged.

People ask why the "truth" about Jasenovac hasn't long since been confirmed. At the same time, the excuse is made that the real situation can hardly be established, considering the conditions of war and revolution at the time. However, one can see that this is only partially true. The revolution did not spare any energy or means in order to make its victory eternal. Where the names of "fallen fighters and victims of fascist terror" are not carved into the granite of innumerable monuments, they are found on the lists of all local and municipal committees of the SUBNOR Veterans organization of the National Liberation War. Besides that, a special state statistical census of war victims was carried out, which have been concealed from the public.

In order to really delve into the basis and the aim of the Jasenovac distortion, it is necessary to consider the sources upon which it was built. Or to put it more accurately, how the truth was obscured, the past was re-shaped, and how, with the help of a biased selection and interpretation of sources the distortion was created.

All the previously mentioned creators of the Jasenovac distortion base their claims upon personal estimates and the testimonies

of individuals, and not on any proven facts. This begs several questions. How is it possible to make a judgment about such an important issue without subjecting every available source to critical investigation? Why were documents from the NOR disregarded as if they didn't exist, and the documents from the enemy side completely ignored? Why is it that out of the personal "testimonies" available, only those are taken as being reliable which distort the already monstrous crimes to the point of implausibility, while others which are at least rational are totally overlooked? Finally, how can one speak of a Jasenovac out of context, as if it were the only such phenomenon in its time?

The NOR in Croatia had a well organized intelligence service within the NDH, its army and police, and it was familiar with the actual conditions in all areas, including the camps. Moreover, a party organization was operating within the Jasenovac camp itself. Thus, the leadership of the NOR was able to make judgments based on information from its own organizations as well as from reliable sources in the NDH. It is therefore difficult to reject the number reported in an injunction of the Supreme Headquarters of the NOV and POJ, dated March 31, 1942, which was personally signed by the commander in chief, Josip Broz Tito. This injunction ordered the Headquarters at Bosanska Krajina to examine "the possibility of a potential attack on the concentration camp at Jasenovac, where about 10,000 of our people were imprisoned, and about 1500 are still alive." That this number of camp inmates and dead is considered accurate can be deduced from the fact that Tito, shortly thereafter (on April 4, 1942), repeated it in a report to the Executive Committee of the Comintern. And it is evident from the additions to the figures that, for political reasons, the aim was not to reduce the number, but to emphasize the enormity of the losses. Namely, it was stated that: "The most horrible concentration camp in Croatia is located in Jasenovac. Into that camp, Hitler's henchman, Pavelic, has thrown more than 10,000 of the best sons of Croatia. At this time only 1500 are still alive, while all the others were subjected to horrible torture and killed. The Ustashas themselves admit that since their

coming to power, as many as 500,000 people have been killed, most of the Serbs."[1]

Consequently, after the first, most horrible wave of terror, during the initial year of the Occupation, the leadership of the NOP estimated that 8,500 people had been put to death in Jasenovac. Meanwhile, the total of victims in the entire NDH a figure is cited to be "as many as 500,000" which certain NDH madmen boasted in order to magnify their accomplishment.

However, by the end of 1942 (as until the end of the war) Jasenovac was not nearly the kind of death site that the distortion depicts. This can be concluded from the report that the CK KPH (Central Committee of Communist Party of Croatia) sent on November 14, 1942, to the CK KPJ (Central Committee of Communist Party of Yugoslavia). In an exhaustive account concerning the general military and political situation and conditions in specific parts of Croatia, Jasenovac is not mentioned at all. However, a great deal is said about the Chetniks and their crimes. In the last Italian offensive, the Chetniks "forcibly mobilized 300 Serbian peasants" in Gorski Kotar, about 1000 people in the Otocac district, and about 500 peasants in Plasko. "In increasing offensives in the districts of Primorje and Gorski Kotar, the Italians, with the help of the Chetniks, are aiming to destroy the people of this area completely. Their frequent attacks, the large number of completely destroyed and plundered villages, and the appalling hunger which prevails in these areas curtail the further Partisan mobilization of the masses." Because of this, "Lately one notices among Croatians in all regions an ever increasing fear of Chetniks."[2]

In comparison to the Chetnik atrocities, Chetnik genocidal crimes against the Croatian populace in other parts of the NDH were even more horrible and widespread. They were by no means any less terrible than the crimes of the Ustashas. Historical sources record that in some places Ustasha elements initiated the crimes, and in other places it was the Chetniks. Thus, even though acts of revenge were mutually provoked, the essence of this fratricidal and religious war was not as much in the initiation of terror or

reprisals as in the climate which arose with the fall of the Yugoslav monarchy and the establishment of the NDH.

In addition to confirmations from the ranks of the NDH concerning Chetnik genocidal crimes, there are quite enough testimonies from the ranks of the NOP. One even finds tenable evidence about them in the books of Vladimir Dedijer, who despite this, was to become an increasingly biased and obdurate champion of the Jasenovac distortion. Dedijer "heard, even in Serbia, that Draza Mihailovic was mercilessly killing every Moslem and Croatian he could get hold of." On December 23, 1941, "for the first time" Dedijer saw "slaughtered Moslems," recording in his war journal en route from Rudo near Lim: "Yesterday the Chetniks slaughtered 30 people and strangled 20 in the Lim river." And in the Bosnian town of Foca, on January 30, 1942, he encountered "an even more horrifying picture": there, "Moslem families are burying the victims of the Chetniks" who, according to the inhabitants, "have killed ... over 3000 people."

On this subject, Tito wrote (in his 1952 autobiography for *Life* magazine): "I will never forget the sight at Foca in January of 1942. That lovely town, hidden in the mountains of Bosnia in the deep valley of the clear and swift Drina river, was under the control of the Chetniks at that time. When we entered the town, we heard that they were killing off the Moslem populace. Every second morning in an ordinary ritual, a group of prisoners from among those unfortunate people was taken to a bridge, their throats were cut, and their bodies thrown into the river."

While Tito was crossing the river, a Moslem woman cried out, "My husband is there!" and begged Tito to take her 12-year-old son along with the Partisans, for "if they return they will kill him as well." Tito continued, "I looked down, and there in the torrent, I saw one of the ghastliest sights in my memory: there in the clear blue-green water, five or six bodies stood upright as if they were wax statues." He ordered them removed from the water and had them buried "in accordance with Moslem rite." Tito took the woman's son with him. He became a courier in one of the Partisan brigades and later an officer in the Yugoslav National Army. And

his mother was right, for in May when the Partisans "were thrown out of Foca by the Chetniks and Italians, and the massacres began anew, she, too, was killed."[3]

Tito, unlike many others, never forgot the horror of the occurrences at Foca and similar sights. He said much more about this to Krleza. Those bodies which he had seen in the river Drina, had been standing upright because before they had been slaughtered, bags of stones had been tied to their feet so that the current would not take them away and they could remain as a reminder and a testimony. This is probably one of the reasons why Tito never visited the Jasenovac death site, thus demonstrating that it was neither the only, nor could it be the symbol of all the victims.

Dedijer forgets this when he sums up the problem of the war victims as being mainly centered in Jasenovac. In 1985 Jasenovac was the only site to which he led "a working team of the Committee of the SANU [Serbian Academy of Arts and Sciences] for collecting material concerning genocide against Serbian and other nations in the twentieth century."[4] And even the JAZU announced in 1987 the need for an investigation into the victims of Jasenovac, though not a comprehensive one into all the victims.

Dedijer even mentions the "instructions" that the Chetnik commander-in-chief Draza Mihailovic gave to his commanders in which some of the chief goals established are:

> 2) To form a Greater Yugoslavia, and within it a Greater Serbia, ethnically pure, within the borders of Serbia, Montenegro, Bosnia, Hercegovina, Srem, Banat and Backa.
>
> 4) To purge the state's territories of all national minorities and unnational elements.
>
> 5) To form a direct common border between Serbia and Montenegro, as well as between Serbia and Slovenia, by means of purging the Sandzak area of Moslem inhabitants, and Bosnia of Moslem and Croatian inhabitants.[5]

How, in the context of such "instructions" can one speak of genocide against the Serbian nation and not against the Croatian

nation as well? The common border "between Serbia and Slovenia" implies that Croatia must be destroyed or incorporated in Serbia. Croatia is the only Yugoslav republic which has a common border with Slovenia.

There are plenty of documents in the archives of the NOR, as well as in those of the NDH, in the German archives, and also in Dedijer's *New Contributions* which show that Mihailovic's "instructions" were carried out in the literal sense.

Major Pavle Durisic, commander of the Lim-Sandzak Chetnik regiment, states in his report of February 13, 1943, to the Supreme Command of Draza Mihailovic: "Operations were executed exactly according to orders and the issued directive." After the "enemy resistance" was subdued, the battles were over, for the most part, and "then the cleansing of the liberated territory was begun. All the Moslem villages in the three mentioned districts were completely burned down, so that not one of their houses was left standing."

"All their possessions were destroyed except livestock, grain and hay...

"During the operation, the total annihilation of Moslem inhabitants regardless of sex and age was carried out."

"Casualties—our total was 22 dead, of which 2 deaths were accidental, and 32 injured.

"The Moslem casualties totaled about 1200 fighters, and up to 8000 others: women, old men and children."

Dedijer himself comments on this: "A similar report of genocide against Moslems and Croatians in other parts of Yugoslavia, as for example in Dalmatia in December, 1942, are deposited in the archives of Draza Mihailovic." The "Dinara Chetnik Division, under the command of the priest Momcilo Dujic, issued daily order on the cleansing of Croatians and Moslems from Lika and northern Dalmatia [two regions of Croatia]."[6]

Among the ranks of the revolutionary Partisan movement in Croatia and other parts of Yugoslavia, the Chetniks were considered no less heinous adversaries than were the Ustashas. In addition to military and political documents of the NOP which point

this out, one can see this in the term Ustaches (meaning "Ustasho-Chetniks"), which was coined by the Montenegrin poet Radovan Zogovic, a prewar Marxist.

Just as it opposed Ustasha genocide against the Serbian nation, Tito's NOP opposed with equal resolution Chetnik genocide against the Croatian nation. However, since both the NOP and the Chetniks were striving to destroy Pavelic's NDH, the NOP was causing difficulties for that segment of the Croatian population which because of its national and political ideologies did not join in the revolutionary Partisan movement.

The consequences were large numbers of victims in the war operations of the occupiers and of both adversaries of the NDH, and even greater numbers of refugees from all parts of Croatia. As early as the summer of 1942, for example, there were about 12,000 Croatian refugees in Banja Luka alone. By the end of the war there were even hundreds of thousands in Zagreb, which brought on an increasingly unhealthy atmosphere of endangerment, hatred, and vengeance.[7]

There are some interesting statistics collected by the Germans from an investigation carried out during the first two years of retaliations, while the war was still in progress in the NDH. In his report dated June, 25, 1943, Siegfried Kasche, the German envoy to the NDH, stated that the claims that Ustashas had killed 400,000 "Orthodox Christians" were dangerously exaggerated. To show that this could not be true, he presented the data from the investigation. In six east Bosnian and four central Bosnian districts—regions where unrest was most prevalent—the number of fatalities amounted to: 8,400 Catholic Croatians; 24,400 Moslem Croatians; or a total of 32,000 Croatians, as well as 37,300 "Orthodox Christians," i.e., Serbs. Kasche claimed that the disturbances had been initiated by violence on the part of the "Orthodox Christians" or members of the former Yugoslav Royal Army, and that the counterattacks of the Ustashas were radical.[8]

The statistical data collected by the war agencies of the NOP point to similar findings concerning total numbers of victims. These data were published in 1944 as "documents on fascist

crimes" in Naprijed, the publication of the CK KPH. The exhaustive statistics were compiled according to singular instructions for the organizing the gathering of documentation concerning the crimes of the occupiers and their helpers, which the CK KPJ put out in 1942. Consequently they provide evidence that there actually aren't any great mysteries concerning the victims.

In all of the Croatian region of Lika, the total of victims amounted to 11,290 persons, of which number 6,734 had been killed by Ustashas, Chetniks, Italians and Germans, 2,702 had died of typhoid, and 1,744 had been taken away to camps. In the Kordun region, 16,704 had died. Of that number, 8,140 had been killed, 6,814 had died of typhoid, and 1,750 had fallen in battle. An additional 2,415 had been taken to camps, 3,173 had "surrendered," and 2,346 had moved away from the area. In the Banija region, 4,040 had been killed, 3,965 had been taken to camps, and 1,551 had resettled in Serbia. (Data on the district of Kostajnica and on those who had died from epidemics was not included.) For the regions of Slavonia and Baranja, in addition to the material losses cited, there are only data about camps and emigration. From 13 Slavonian districts, 10,937 persons were taken to forced labor camps, while from 4 districts of Baranja, 2,033 were taken to camps, and 9,202 resettled in Serbia.[9]

According to these statistics, from the major part of Croatia, the total number of people who had been taken to camps comes to 21,147. This does not include the last year of the war, when the practice of sending people into the camps had declined. Nor does it include statistics from the Zagreb district, from Gorski Kotar, Dalmatia and Bosnia-Hercegovina. However,)f one assumes that the number of people sent to camps from these places was proportional to the statistics we have from Lika, Kordun, Banija, and Slavonia, where reprisals were the most intense, this leads us to the conclusion that the fabled numbers of hundreds of thousands of slayings at Jasenovac are utter nonsense. One should also remember that of the documented numbers, a considerable number of prisoners were sent to camps in Germany.

According to the falsifiers of the historical record—in Jasenovac

alone, ten to twenty times more people were killed than had been
sent to all the camps in Croatia. Considering this discrepancy, it is
quite understandable why, according to them, no one was able to
leave that "death factory" alive. However, among the membership
of the Croatian SUBNOR (Union of Veterans of the People's
Liberation War), every fifth war veteran is a former prisoner or an
internee (of 329,570 members, 60,192 had been in prison, includ-
ing 9,138 prisoners of war, 26,209 internees, which means that
every ninth member of the SUBNOR was a former inmate since
prisoners of war were also sent to the existing camps).[10]

Even more significant are the official statistical data concerning
the total of war victims in the SR Croatia, which include even
those who had already died as well as those who perhaps hadn't
achieved the status of members in the SUBNOR veterans' organi-
zation. From the total of 163,126 persecuted, there were 103,387
survivors (from prisons, camps, and "deportation") or 63.4
percent.[11]

All this shows the necessity for objective research into all "col-
lecting" and "labor" camps, and above all, the Jasenovac camp.
The crimes of the Ustashas—like those of the Chetniks or any
other group in any other place and time—will lose none of their
monstrosity if they are shown to be as they really were. In fact, the
outrageous exaggeration of these crimes produces precisely the
opposite effect—on the one hand because it is unconvincing, and
on the other because it provokes an even more irrational hatred.

The promoters of the Jasenovac distortion insist that the camp
was organized with the express purpose of liquidating all the
inmates, and that daily executions were conducted of hundreds
and even thousands of Serbs, Jews, Gypsies, and Communists.
The fact is, however, that the camp had been set up as a "labor
camp" with a large number of agricultural and industrial labor
units. Thousands, and even tens of thousands of unfortunate
people were sent to the camp. But many were set free, while
others were deported to camps and forced labor in Germany. The
inmates were continually overworked and maltreated under
incredibly difficult and unsanitary work conditions. They were

individually tortured and killed for the most trivial reasons, particularly the exhausted and the old. Occasionally, usually under the pretext of revenge for the fallen Ustashas or as punishment for attempted escape, they were brutally killed in groups of tens or hundreds. At the Jasenovac camp, between 30,000 and 40,000 inmates actually died, most of them Gypsies, Jews, and Serbs, and even Croatians. I am firmly convinced that this number can be accurately established without dispute—except perhaps for the Gypsies—by means of further systematic investigation.

In assessing the conditions in the camp, taking into consideration its purpose and the crimes committed therein, we must be aware of its internal organization, which was based in part on the inmates' "self" management. I have long since come to the understanding that in the magnification of Ustasha crimes, certain individuals stand out who have had specific, clandestine reasons for their actions. A report by Vojislav Prnjatovic, a Jasenovac inmate from Sarajevo, which was given to the commissariat for the exiles of Nedic's government in Belgrade on March 11, 1942, after Prnjatovic's release from Jasenovac, provides a relatively accurate picture. The inmates were used for labor on a number of large agricultural estates in the area between Jasenovac and Stara Gradiska, the owners of which were Serbs who had been "resettled, in part voluntarily and in part by force." And since "industrial plants ... which had been confiscated for the benefit of the Croatian state" were located in this area it was the intention of the Ustashas to turn Jasenovac into not only a collecting camp, but also an industrial labor camp.

The area already contained good industrial plants for the manufacture of chains, locks, tin, upholstery, guns, and carts, as well as a stone-masonry shop, an auto repair shop, a carpentry shop, a sawmill, a brickyard, and an electrical power plant, which also provided power for Jasenovac itself. Inmates were divided into three categories: "freemen," "semi-freemen," and "common inmates." In addition to the Ustashas' management of the camp, there was also internal self-management by the inmates. The Ustasha management consisted of a command-giving supervisory

government, while the "self-management on the part of the inmates consists of their own regulation of camp life and work in accordance with the orders and plans of the Ustashas. Thus, the inmates themselves managed the foodstuffs, lodgings, maintenance and sanitation in the camp." Those inmates categorized as freemen were allowed not only to move about freely within the confines of the camp and the town of Jasenovac, "but also a 'freeman' goes alone without an escort on authorized trips for the purpose of procuring supplies to all areas of Croatia." Semi-freemen were allowed to move freely within the camp and the town of Jasenovac where they "go at any time to procure or receive supplies." Common inmates were permitted to move about the camp during the day as long as they stayed five meters away from the wire fence. "At the top of the hierarchy of prisoner," there was a camp manager who had a replacement, an economic director and clerk. Next, there were commanders of the barracks and work groups. All those in the internal administration were freemen. Each barracks had a sergeant with four assistants. "All the inmates were immediately employed according to their civilian occupations," while "the best off were the skilled workers and tradesmen because they are used for labor in closed workshops," and for outdoor work only when the trade positions were filled.

"Freemen do not reside in the camp itself, but in the town of Jasenovac in private housing, and they are permitted to bring along their families. Thus, they live together freely with their families in Jasenovac."

"The inmates who belong to work groups receive special food which is good and palatable and they each get a whole 90 gram loaf of bread daily." But this only applied to members of work groups, while "common inmates are given simple food." Poor nourishment, hard labor and unsanitary conditions were the cause of weakness and disease among the inmates, "which leads directly to death, for when they take ill and are unable to work, they are routinely liquidated." Only in January of 1942, prior to the arrival of an international commission, a clinic and hospital were constructed. But the inmates were sent there only in the most dire

circumstances, since the Ustashas "raided the clinic and hospital every two to three weeks," removing the patients and liquidating them in front of the hospital or in the cemetery. All of the inmates, except the incapacitated ones in the hospital, had to work. An Ustasha "had the right to kill an inmate without any accountability," and "not a day passes when at least one Ustasha exercises this right." In addition to such individual killings, there were occasional "mass killings," prior to the arrival of new inmates or whenever the total number of inmates rose "above 1,000." Then the Ustashas would order an "exhibition" and would choose "a specified number of inmates for the purpose of liquidation." As far as numbers are concerned, Prnjatovic indicates that from December 1941 to January 1942, "between three and four thousand newly arrived inmates were killed." There were 5 to 8 times more Jews in the camp than Serbs and they were used more by the prison hierarchy.[12]

This and similar testimonies are supplemented in particular by Ante Ciliga, a former member of the Politburo of the Central Committee of the Communist Party of Yugoslavia. He spent more than a year in an Ustasha prison and then in the camp at Jasenovac, from December 14, 1941 to the last day of 1942, after which he observed further developments, residing in Zagreb until the summer of 1944. In his memoirs, he makes a comparison between that most notorious of Pavelic's camps and those of Stalin in Siberia, at which he had also sojourned.[13]

In comparing the Jasenovac camp with the Siberian gulag, Ciliga sees the main difference in that the "chief concern" at the Siberian camps was "to destroy the dignity and moral integrity of its victims." This was something the Ustashas, as well as the Chetniks and Partisans, were not concerned with.

Ciliga agrees with the remark Engels made in 1892: "With regard to these nations—Serbs and Croats—religion still precedes nationality, and each religion wishes to rule. While there is no progress in this area which could at least make religious tolerance possible, a Greater Serbia can only mean civil war." This is true with regard to religious tolerance, and likewise with regard to

utopianism. The reprisals between the Ustashas and the Chetniks, in all their forms, were indeed an expression both of the irreconcilability and of the impossibility of the domination of one national objective over the other.

From the war period, there are data concerning war crimes and Jasenovac in reports which were sent by opponents of Pavelic's regime to members of the exiled Yugoslav royal government in London. They deserve attention as they were usually sent by well informed people from both official and opposition ranks. Even in cases in which their information is not reliable, their tone at least is worth noting.

Monsignor Dr. Augustin Juretic wrote of Jasenovac in his exhaustive and critical report to Krnjevic, the vice president of the government in exile, dated June 10, 1942. The report comes from a man who is politically experienced—he was formerly a leader of the Croatian Populist Party and was on good terms with Anton Korosec, and with the royal court in Belgrade, and in Zagreb with the Roman Catholic administration center and the leadership of the Croatian Peasant Party. He maintained ties with Rome throughout the war, and remained in Switzerland from 1943.[14] Juretic writes in his report: "The camp Jasenovac is a true death camp ... People are tortured in the most ghastly ways." About Kvaternik he says that the man is such a "beast" that his mother not only disowned him, but committed suicide because of him. "Jasenovac is Ustasha's darkest side. Thousands of people have been killed there."[15]

Similarly, in a detailed report to London from an unknown person, concerning the situation in Yugoslavia in mid-1942, it is stated that: "Despite the Pope's intervention, the regime in the camps has not yielded. About 10,000 individuals are in the camps, and it is estimated that about 15,000 have been executed so far. About 1,000 people have died of typhoid which is rampant in the camps." After the inclusion of a list of prominent Croatian leftists and the mention of 500 students, the report continues: "One can estimate that in all of Croatia about 200,000 people have perished. Another 150,000 have emigrated. About 120,000

are working in Germany. About 50,000 Slovenians have immigrated."[16]

A report written by the Franciscan Dr. Dominik Mandic on March 24, 1943, and sent to the assistant of the minister Dr. M. Martinovic, states: "It is estimated that between 350 and 400,000 Croatians have been killed by Chetniks, Partisans and local friends [the Ustashas]. As many, and perhaps somewhat more Serbs have been perished in the territory of the NDH."[17]

In the main document of Tito's NOP and socialist revolution, the declaration of the second session of the AVNOJ, there is specific mention of the killings and crimes on the part of the Chetniks of Draza Mihailovic, who "under the flag of the King and the monarchy" are committing "the most loathsome acts of treachery" and "most horrible crimes."[18]

Considering the multiplication of the Jasenovac victims—from 500,000 to 1,500,000—it might be useful to look into what was being said on this subject toward the end and after the war.

In the spring of 1944 the *Naprijed* newspaper, within a systematic account of war crimes, stated that of about 30,000 Jews who had been taken to Jasenovac, less 1000 remained alive, that the Gypsies there had completely been exterminated, and that "the number of Serbs killed is over 150,000."[19]

V. Dedijer, in his first book about Tito (1953), writes that "during the war, more than 200,000 persons were put to death" at Jasenovac.[20] In his *New Contributions* (1981), one can read: "The exact number of Serbs killed in Croatia has not yet been determined. Some authors claim that the number amounts to 200,000, while others assess the loss to be at least 600,000." These numbers are assigned to total of Serbian victims in the NDH. Then, in the same book, Dedijer writes about Jasenovac alone: "Through that camp, according to the investigations of the historian Antun Miletic, more than a million people passed, while between 480,000 and 800,000 were killed. It is difficult to carry out an exact reconstruction as not all of the lists of those arrested and killed have been preserved."[21]

Dedijer is not at all concerned with the contradiction of one

claim by the other. He established his alleged scientific objectivity by stating that "some authors claim" these numbers of total losses, while the "investigations" of Miletic, in Dedijer's opinion, leave no doubt that in Jasenovac between 480,000 and 800,000 were killed, and here it is commonly understood that this refers only to Serbs.

Dedijer simply accepts Miletic's figures, not only without any scholarly criticism, but also with the aim of adding to their impact. Dedijer dedicates his latest book *The Vatican and Jasenovac* (1987) to the "historian Antun Miletic, author of the three truth-loving volume monograph about Jasenovac, through which the suffering of the hundreds of thousands killed there speaks out... "[22] It was upon these numbers that he was to construct his theses about genocide, not merely in that book, but also in the international promotion of the Jasenovac distortion.

Towards the end of the 1950s, an outline of the chronicles of the NOR—in the writing of which, as I have mentioned, not one Croatian historian took part—describes the camp victims with Marxist objectivity. To support assertions concerning "reprisals against the people" committed by the occupiers and "local traitors," it is stated that "... tens of thousands of men, women and children were sent off to concentration camps in Jasenovac, Stara Gradiska, Banjica, Sajmiste, Nis and Kostolac, the islands of Rab and Molat and other places. A large number of Yugoslavs have been interned in concentration camps outside of Yugoslavia— Buchenwald, Dachau, Auschwitz, Mauthausen, Korca and others. Thousands of men and women have been transported by the German occupier to forced labor in Germany."[23]

The authors certainly knew about the "findings" of the Commission for the Establishment of War Crimes which asserted that 500,000 to 600,000 persons were killed in Jasenovac. But they obviously didn't wish to include such inflated figures for war reparations in the historical text. Perhaps they believed—for reasons we will yet discuss—in the figure of 1.7 million total victims from "all the Yugoslav nations in the war from 1941 to 1945," both those perished and killed.[24]

One finds the same numbers regarding total victims as well as "tens of thousands" transported to "death camps" in the printed *Survey of the History of the SKJ* (1963), with the omission of the names of the camps in Yugoslavia and Germany.[25] (One should be reminded that this is that same Survey which was so sharply criticized both by Croatian historians and leaders of the revolution in Croatia.)

In the early 1950s, the UDB (the Yugoslav secret service) for the SR Croatia, made a detailed four volume study concerning the Ustasha movement. The study was produced for "internal use" i.e., to instruct and brief UDB functionaries, as well as "for lectures in the [Communist] party's school." In the study's description of Pavelic's false Christianity, it is stated that in Jasenovac "thousands and thousands of innocent people were executed," savagely and villainously, on the grounds of certain "legal provisions" regarding retaliation, hostage-taking, and the like. The total number of victims in the NDH are denoted by "all that unspeakable horror of the slaughter and burning of hundreds of thousands of corpses which marked the path of that Balkan Genghis Khan of the 20th century," while "the church at Glina, Jasenovac, hostages and reprisals" are cited as examples.[26]

In his multi-volume work, Antun Miletic presented two documents which are extremely important to this consideration. Under Appendix No. 2 one finds an "Excerpt from the Trial of Ljubo Milos, one of the functionaries of the UNS [Ustasha security service] and Commander of the Jasenovac camp."[27] (Appendix No. 1 contains an excerpt from the memoirs of Vladko Macek concerning his five-month internment at Jasenovac, where he was "handed over to the authority" of Ljubo Milos, to whom Macek does not state "any objection," although he later found that Milos "had become one of the worst butchers."[28]) That excerpt from Milos's trial contains questions and answers ranging from general information about the organization of the camp and terror in the NDH to cases of individual and mass slaughters in Jasenovac. But there are no questions about the total number of victims at that camp.

However, one finds in the same chapter the Appendix No. 7: Ljubo Milos about the Jasenovac concentration camp.[29] This appendix is three times the length of the Appendix No. 2, and in content was clearly meant to have a greater significance. However, it remains a mystery why Miletic gives no explanation as to where the information contained in this section comes from, that is, how, when, and where it came about, and where the original document is deposited. According to the title of the appendix, this should be the truth about Jasenovac according to Ljubo Milos. Namely, it is written as if all the information in it had been provided by Milos himself. Thus, the concluding statement to the discussion on "Liquidations" (one of the subtitles), reads: "I would only like to add for the sake of truthfulness, that we are dealing here not with an exception, with accidental occurrences, but with a system which in this way certainly liquidated a couple of hundred thousand people."[30]

Milos's penitent evocation of the truth was obviously meant to carry the greatest possible weight of his alleged testimony that "certainly... a couple of hundred thousand people" had been killed. Regardless of the historical fact that people standing before the gallows are apt to tell both incredible truths and most incredible untruths—this claim by such a key witness could well serve as a support for the continuance of the Jasenovac distortion.

In Ljubo Milos one does not see a case of a guilt-ridden soul bidding its last farewell to the world, nor a testimony to the actual (to him unfavorable) truth, nor a defiant insistence upon some fabrication of his own in order to beguile and mislead his mortal enemy. Namely, from the stenographic record of his trial an altogether different truth follows.

The president of the judicial council asked Milos how many people had been liquidated in Jasenovac, and whether the number stated in the indictment "can be said to be accurate," and that "bill of indictment specifies 40,000 to 60,000." Milos first answered that he wasn't able to say, and then he agreed that "it can."

Upon repeated questioning concerning the total number of victims at Jasenovac, Milos later repeated that neither he "nor any-

one else" could say how many had been killed. Concerning his statement in the records of the investigation "that the number was enormous and quite incomprehensible," he said "that means it was an immense figure." Upon the continued insistence that he express this figure numerically, Milos repeats that he cannot say how many there were because he does not know, and he was not there at all times. And when his interrogators presented the figure of eighty thousand, Milos replied: "Maybe more, but maybe less."

However, that "enormous" and "quite incomprehensible" number was, for some, neither large enough nor incomprehensible enough. One sentence from Milos's responses was twice—most probably at a later time—rewritten in ink. Thus one finds in the record that Milos said "I think that all together in Jasenovac some hundreds of thousands of people were killed." The first correction said "a couple of hundred thousand," but a stronger phrase was obviously needed.[31] Such hand-written alterations of judicial minutes are not permitted. By law, each error or change in a deposition must be separately registered and entered in an addendum by the same means, i.e., typewriter, as the statement which is being amended. So much for the legal formalities of this incident.

This reveals not only the extreme legal and historical irresponsibility of the document's falsifiers, but also their arrogant unscrupulousness. This "correction" was thrown in after the first statement by the judge referring to "tens" of thousands, and before the other statement, also by the judge, and followed by that of the presiding moderator, concerning "eighty thousand." Therefore, all of this even at first glance seems quite illogical, for what sense might there be in pressing Milos to accept the figure of "eighty thousand" after he had just admitted to "some hundreds of thousands?"

Mesmerized by their short-sighted omnipotence some people assume that they can get away with anything, never considering the far-reaching consequences of their actions. Such malicious deeds must be judged from both a political and a moralpsychological standpoint. What are the motivating factors involved? What aspect of human nature can explain the depravity of those

to whom the enormity of the horrors and madness of war are not enough, but they must exaggerate them? And who for the sake of petty interests are prepared to willingly comply with such hellish plans?

Because of the total illogic of the corrections, it was "scientifically" inappropriate that this fundamental and most important part of Milos's interrogation be published. For this reason it was excluded from Appendix No. 2. But because of the need to strengthen the Jasenovac distortion, it was published in Appendix No. 7 as a reliable testimony by a man who was supposed to have known the truth.

Thus, instead of striving to investigate scientifically and to answer objectively the many open questions concerning Jasenovac, the infernal game is perpetuated. How, for instance, does the number of total victims of Jasenovac keep climbing from indictment to indictment? Against Pavelic, it stood at "thousands" of victims; against Stepinac, at 40,000; against Milos 40-60,000; and I myself answered to the law for my disavowal of the 700,000 figure. And how did it come about that during Milos's trial even the court itself—at beginning of the hearing and without any new facts—raised the figure in the charges from 40,000 to 60,000 towards 80,000, which in the end was to miraculously become "some hundreds of thousands?" How also does it happen that the indictment against Andrija Artukovic refers to "hundreds of thousands" for all the victims in the Independent State of Croatia, which, naturally, must also include Jasenovac?

Considering the repeated assertions that a vast number of Jasenovac victims were thrown into the Sava River—which is used as another reason why the number cannot be established—one notices that the documents presented in all three of Miletic's books deal mainly with corpses found and exhumed near Belgrade and Sabac [in Serbia] and in Jasenovac itself, in the period between May 13 and June 9, 1945 (totaling 287 corpses).[32] It is a well-known fact that Nedic's government had been collecting data about crimes in the NDH since 1941 to use them in its protests to the Germans.

In addition to all the above considerations, original wartime documents, and testimonies concerning the Jasenovac camp and the total number of war victims, I base my knowledge also upon personal conversations with a large number of very serious and responsible persons who were well acquainted with this controversial issue. I speak, namely, of people who were either themselves persecuted and incarcerated during the time of the NDH, or who, as members of the high courts and judicial bureaus, had access to all the materials used as evidence concerning the crimes committed.

Based on all the above, I maintain that the following conclusions can be drawn without any doubt:

■ The war crimes of the Ustasha authorities of the NDH were immense and appalling in number, in the manner in which they were carried out, and in their genocidal intention. However, they cannot be judged separately from the no less genocidal crimes of the Chetnik movement. And one should keep in mind the general circumstances of the Second World War, the fall of monarchist Yugoslavia, the establishment and fall of the NDH, and the revolution.

■ The original documents from the war provide no basis for the promotion of a Jasenovac distortion, nor for the supposedly conservative estimate of about 700,000 victims, let alone for the hysteria which claims that over a million were killed. For this reason, its propagandists have had to resort to crude falsification of testimonies and vicious obstruction of truth, while their followers have found it necessary to use unscientific methods in their selection, analysis, and interpretation of historical material.

2. On Scientific Findings

The truth is a torch, but a terrible one; thus, squinting, we all endeavor to pass by it as if we feared being burned.

—Goethe

Politics need not be honest; history must.

—Sartre

The acceptance and perpetuation of the Jasenovac distortion was possible as part of the officially verified and constantly repeated fact that the total of direct war losses in Yugoslavia amounted to more than 1,700,000 victims. This figure is accurate, but only in the sense of total demographic losses, which include the decrease in population due to all the consequences of the war. In other words, apart from immediate victims of the war, population losses due to lower birth rate, higher natural death rate, emigration, and deportation are incorporated into this figure.

The number of 1,700,000 victims has become sacrosanct; no one has ever been permitted to doubt it. It has been presented in official documents of the government of the SFRJ in international conferences as a basis for requests for reparation, and throughout the world for the sake of relating Yugoslavia's contribution to the victory of the Allied Forces, and as an example of its anti-Fascist Partisan movement as compared with those of other countries. In Yugoslavia itself, through the constant repetition at state and political gatherings, in journalism and in textbooks, and even in academic and informational publications, that number has permeated into public opinion as proof of the magnitude of the NOB and the cost of the socialist revolution.

Since it is the case in the area of all the social sciences, and particularly in the field of history, that judgments based on the allegedly impeccable Marxist method are viewed as law, this figure was uncontestable. In 1960, I myself defined that number in terms of demographic losses. In my book, *The Creation of the Socialist Yugoslavia*, I wrote, "Yugoslavia lost a total of

1,706,000 people in the war (in addition to those killed in battle, also through the brutality of the Fascist occupiers and their servants, bombing and other consequences of war)... "[33] However, it seems that this fine distinction was not perceived by my persecutors so that I was not indicted and tried for reducing the total number of victims, but only those of Jasenovac.

The reasons why the number of 1,700,000 victims was enthroned and consecrated also served to justify the decision not to make public the results of the official statistical Census of War Victims (1941-1945), which was conducted toward the end of 1964. Such was the explanation I was given as the director of the Institute by the members of the Federal government. This, in a certain sense, influenced my not having received support from within the political leadership of Croatia (in 1965) in clearing up the problem of the Jasenovac victims from the standpoint of the Institute. But it would seem that the predominant reasons were to be found beyond all that and within the political and inter-ethnic relations of that time.

There were those who needed the Jasenovac controversy in order to keep Croatianness in shackles, as well as those to whom that legend of Croatian historical guilt served not only to instigate Serbdom against Croatianness in general, but also against founders of the federal socialist Yugoslavia. However, by tolerating and even encouraging the Jasenovac distortion, both within and apart from such calculated consequences, another objective was achieved. By appealing to the horrors of Jasenovac, the dogmatic bureaucratic forces consented to Serbian collaboration (particularly within Croatia). Of course, the majority of participants in all these more or less hidden currents were aware neither of their intentions nor of their consequences.

The Jasenovac controversy, as well as the obstruction of the scholarly community, developed despite the fact that in the self-managing socialism in the SFRJ, the independence of scholarly work has long been guaranteed. Tito himself had high regard for historical truth, emphasizing that in historical documents "there may be no alterations" or re-wording. "Something is either a fact

or it is not." Tito conceded to the decision "not to publicize" a fact, but never to modify it.[34]

Nevertheless, as usually happens with delusions and hoaxes, the Jasenovac distortion, too, showed itself to be ineffectual. The victory of the revolution and the establishment of an AVNOJ Yugoslavia does not gain in historical value if too high a price was paid for it. Quite the contrary.

We must establish how, through whose fault or decision, this number came about. In 1947, Vladeta Vuckovic worked in the Federal Bureau of Statistics in Belgrade on the assessment of losses among the Yugoslav population caused by the war, and reached the conclusion that "the population loss in Yugoslavia between 1941 and 1945 was 1,700,000 people." However, Edvard Kardelj transformed that demographic loss, "probably through some sort of teachers' semantics" into war victims, presenting it as an officially established truth at a conference in Paris. And thus, "this number has become the official number of war victims, and not the numerical loss in population." And "This is how myths are created"—concludes Professor Vuckovic, now a professor of mathematics in the United States. He writes of that estimate as if it were his, but it is clear that it had to reach Kardelj as a result of the Federal Office of Statistics, regardless of the fact that the professional statisticians, Vogelnik, Debevc, and Tasic had "unloaded" that task on him, a mere novice, while "the only instructions he was given were that the number should be significant, but based on solid statistics."[35]

This is certainly an invaluable testimony as to how the distortion came about. If it had been clarified at that time that those were demographic losses, then it would have been clear that the total number of war victims would have to amount to less than one million. This very fact would have rendered impossible the creation of a second Jasenovac distortion, with figures even greater than those of the total war victims. However, it is difficult to imagine that the official blame rests only with Kardelj whether because he made no semantic distinction between demographic losses and war victims, or because of a calculated distortion of the

truth on his part. On such an important issue, he must, in any case, have received approval from the state political leadership.

In the beginning of 1950 (from January 22 to February 22), a census of war victims was conducted in Yugoslavia. However, the results were never made public. Nor were they made available for examination by the scholarly community and institutional researchers until more than a decade later, when I founded the institute in Zagreb. The decision concerning the census was made on January 6, 1950, by the Chief Committee of the Yugoslav SUB-NOR organization, which issued at the same time instructions to the commissions for gathering data concerning war victims. Those instructions contained detailed directions concerning the method by which the census was to be undertaken, and they were communicated by all the newspapers in the country.[36]

The census was to include all who had lost their lives "because of wartime causes and conditions," but "all those... on the side of the occupiers or local traitors... are not considered victims of war." In order to avoid sectarianism and superficiality, and an obviously smaller number of victims, the instructions specify that the census-takers go from house to house according to a strictly systematized plan, while "In case of doubt... it is better to include the person than to leave him out," and later to verify the information and "settle" the matter. These guidelines were certainly intended to insure that the census be executed thoroughly, nor is there any reason to believe that it was not carried out in this manner. Why then were the results not publicized? And why, to this very day, has no researcher of war victims taken that census into account? In the context of all the considerations hitherto mentioned, one cannot escape the conclusion that all this secrecy was in the interest of facilitating the continuance of the Jasenovac distortion.

In the beginning of the 1950s, the first systematic discussion concerning demographic losses took place. In his 1951 work Ivo Lah concluded that the demographic losses caused by the war totaled 1,999,000 by the middle of 1945 (2,177,000 by the middle of 1950).[37] Using a different method of calculation, Dolfe

Vogelnik asserted that the total demographical loss was 2,854,000.[38] However, Lah refuted Vogelnik's estimate, pointing out the untenableness of the analytical method which he used, as well as those figures attained by using the Malthusian geometric progression, according to which the losses should have totaled 2,699,000. Calculated by Verhulst's logistic curve, they come to 2,165,000, and by linear extrapolation, 2,073,000, while the Princeton University analytical method gives us only 983,000. Lah accepted as most accurate the method of linear extrapolation, and he concluded that the demographic losses amounted to 2,073,000, even though he maintained that demographic losses due to emigration were much greater than even Vogelnik conjectured.[39] Since Vogelnik did not enter into any further discussion on the subject, Lah's estimate may be considered to be scientifically valid.

According to the 1954 estimation òf German demographer Professor Gunther Ipsen, the total of demographic losses of the Yugoslav populace came to 1,650,000, of which number 1,010,000 were male and 640,000 were female. He estimated wartime losses of soldiers on all sides to be 410,000, while demographic losses among just the Yugoslav nationalities, he judged to be approximately 1,000,000 in the dead, unborn, and refugees, which amounts to 6 percent of the prewar population and is similar to the losses experienced in World War I.[40]

In his 1953 discussion on "The Population of the FNRJ According to National Affiliation," Rudolf Bicanic didn't deal separately with the problem of war losses. But his comparison of the prewar and postwar national composition indicate the extent of these losses. In comparing the census of 1931 and that of 1948, in Yugoslavia the following groups increased in ratio: Serbs by .6 percent (from 40 percent to 40.6 percent), Moslems by .3 percent (from 5.3 percent to 5.6 percent), Slovenians by .9 percent (from 8.1 percent to 9 percent), and Macedonians by .5 percent (from 4.6 percent to 5.1 percent), while there was a decrease in the ratios of Croatians by .2 percent (from 23.1 percent to 22.9 percent), and Montenegrins also by .2 percent (from 2.9 percent to 2.7 percent). In absolute numbers, all nationalities increased, while the

demographic changes, caused not only by the war, but also by differences in natural growth and migration, did not appreciably change the relative ratio among the nationalities of Yugoslavia even though there was a bit of a decrease in Croatians and Montenegrins.

It was quite a different matter with national minorities. From 498,000 Germans in 1931, the 1948 census counted only 55,000. The number of Jews fell from 68,000 to only 7,000, though about 9,000 of them had emigrated to Israel before, during, and immediately after the war. The total number of Gypsies during that time increased from 64,909 to 72,671, but the fact that in 1931 there had been 14,499 of them in the Sava and Primorje provinces (approximately, the current Croatia), and only 405 in the entire SR Croatia in 1948, indicates the magnitude of the Ustasha racist crimes.

The number of Serbs within the territory of the SR Croatia fell by 49,000 (from 583,000 in 1931 to 534,000 in 1948), but that entire sum cannot be attributed to war victims because a considerable number had emigrated to Serbia. In Bosnia-Hercegovina, all three nationalities experienced a growth in absolute numbers, but the ratio of Serbs dropped by 2.4 percent (from 43.8 percent to 41.4 percent), while the percentage of Moslems rose by 3.4 percent (from 31 percent to 34.5 percent) and Croatians by .8 percent (from 22.1 percent to 22.9 percent).

Such a factual situation, that is, an essentially unchanged national structure, can be interpreted—according to Bicanic—both as a serious and exemplary lesson "that the relations between our [Yugoslav] nationalities cannot be resolved by force and violence on the part of one nation against another," and as a most dire verdict against "the Ustasha-Chetnik methods and mentality."[41]

Another thought concerning the increase in population in the republics, which was presented by Bruno Busic, according to Simeunovic's book (1964), points out that the demographic losses in Croatia were comparatively the greatest, as the 1948 census showed a decrease in population by 8,713, while all the other republics showed an increase: Macedonia by 22.9 percent, Serbia

by 14.2 percent (or 802,054), Bosnia-Hercegovina by 10.8 percent, Slovenia by 4.2 percent, Montenegro by 3.7 percent, and the whole of Yugoslavia by 9 percent.[42]

On June 10, 1964, the government of the SFRJ mandated that a census of war victims be conducted by the state organization because official documentation was needed for negotiations with Germany, which, according to the peace treaty, was obligated to provide reparation for war damages and human losses. The census was conducted in October and November of that year. The provisions and guidelines were the same as for the 1950 census, only with even stricter obligations of the commission not to overlook any victim and to complete the data using the previous censuses. General Kajic in the Statistical Institute of Croatia discovered that some victims had been entered into the census forms in more than one place. The case with those statistical data for the SR Croatia is quite well known. The same is said of the reasons why the data for all of Yugoslavia were not publicized. It is not known whether and how the data were used in negotiations with the German government.

The state census of 1964, just as the one done by the SUBNOR Veteran organization in 1950, remained inaccessible even to researchers, while the doors remained wide open to those who would distort the historical truth.

In 1967, Dr. Ivan Klauzer, in a study of the migration of populations, prepared at the Institute for Social Research of Zagreb University, came to the following assessments concerning demographic losses:[43]

Socialist Republic	War Losses	PERCENT of Population (1948)
Bosnia-Hercegovina	580,767	22.65
Croatia	661,543	17.5
Macedonia	25,621	2.22
Montenegro	62,700	16.62
Slovenia	136,000	9.45
Serbia	575,741	8.82
Serbia proper	471,552	11.35
Kosova	38,616	5.27
Vojvodina	65,573	4
Yugoslavia - total	2,042,373	12.89

According to these figures, it is clear that the relative demographic losses were greatest in those republics in whose territory the battles against the occupier and quisling forces took place in the broadest scope, that is, where mutual retaliations between the warring factions in the civil war were the most intense. And this was the case in the territory of the NDH and Montenegro, while losses in Serbia (proportionally half when compared with those of Bosnia-Hercegovina and by about 2/3 compared with those of Croatia and Montenegro) were greater in the final period of the war.

Based on Klauzer's estimated demographic losses and the available statistics on war victims from the SR Croatia, according to Busic, the composition of those losses for the SR Croatia and the SR Bosnia-Hercegovina—from which information we can get an approximate picture of the NDH—would be as follows:[44]

Type of Loss	SR Croatia	SR B&H	Total
Members of NOP	185,327	171,000	356,327
Opponents of NOP	219,216	235,767	454,983
Decrease in Population	107,000	94,000	201,000
Emigration	150,000	80,000	230,000
Totals	661,543	580,767	1,242,310

In his 1973 study about the increase in population during one century, Vladimir Stipetic states: "Direct losses during the Second World War were 1,700,000 victims, while demographic losses exceed 2.2 million persons."[45] He does not present any evidence for this claim, but is obviously using the official figures, while he refers to Lah's estimate for total demographic losses in a footnote. It is interesting, however, that he does not use the figure for losses up to the middle of 1945 which Lah assesses at 1,999,000, but rather losses to the middle of 1950, amounting to 2,177,000, and even these he increases without any explanation as to why the number exceeds 2.2 million. This was probably done in order to attain at least an idea of the relationship between demographic losses and war victims. Otherwise, at first glance the number of victims would seem untenable (1.7 million out of 1.99 million). However, this study offers the interesting data that the demographic losses in the territory of Yugoslavia as a result of the Balkan Wars and World War I were very close to those brought about by World War II—"about two million." Moreover, the former are proportionally even greater since they amounted to "about a sixth of the population of that time," while the demographic loss in World War II comprised about a seventh of the prewar population. Likewise, it took about the same length of time for the population to return to former levels. According to Stipetic, after World War I, the population level of 1910 was

reached by the end of 1923. As we have seen, after World War II the total population was higher in 1948 than it had been in 1931, with the exception of the SR Croatia.

The Academician V. Stipetic also writes on the problem of demographic war losses in his foreword to M. Korencic's book on settlements and population of the SR Croatia.[46] Croatia's demographic loss during the First World War amounted to nearly 380,000 inhabitants or 11.1 percent of the total population of 1921, while he judges that the Second World War reduced the population by 590,000, or 15.6 percent of the population of 1948. In World War I, the demographic loss experienced by Croatia was considerably less than in the other areas of Yugoslavia: in Croatia the population decreased by 1.2 percent, while the other areas lost 5 percent. During the Second World War "the population of the SR Croatia decreased somewhat more than .1 percent between the census of 1931 and that of 1948, while the rest of Yugoslavia experienced a 12.2 percent population increase during the same period."[47]

It should be noted that Stipetic's estimate of demographic losses in Croatia in World War II (590,000) is a full 11 percent lower than Busic's (661,543), even though Stipetic, raising Lah's estimate, starts out with a larger total of demographic losses in Yugoslavia (by about 8 percent) than does Busic, who uses Klauzer's figures.

Since reasonable estimates by demographic statisticians concerning total demographic losses range, as we have seen, from 1.69 million to 2.2 million, I myself have arrived at the conclusion that it is justified, in contesting the Jasenovac distortion, to begin from the higher figure. Thus, I asserted that when one goes by the assumption of 2.1 million demographic losses, one arrives at the conclusion that the number of direct war losses must be between 700,000 and 800,000 deaths. This is approximately the same number that the propagators of the distortion claimed for Jasenovac alone.[48] I have no reason to retract such an assertion today, even after certain new findings. Perhaps further statistical demographic investigation will reveal that the decreased birthrate

at that time did not amount to a demographic loss of 450,000, but probably closer to 350,000 people. It has also already been established that the members of the sizeable German minority who emigrated during that period numbered 450,000 rather than 500,000, although one should add the figure of 200,000 for the emigrated Italian and Turkish and diminished Hungarian minorities, as well as 200,000 for political emigration during and immediately after the war.[49] If one adds to this those who died of natural causes, the number of deaths can be raised somewhat. Nevertheless, my guess is that it will not exceed 800,000 to 900,000.

I am also directed to such a judgment by the most scientific study on this problem. I am referring to the book *Victims of the Second World War in Yugoslavia* by Dr. Bogoljub Kocovic, a Serbian writer who received his education in the area of law and economic statistics in Paris, the United States, and Canada. His book was published in London in 1985.[50]

Kocovic analyzed the problem of demographic losses and war victims from all sides using an unbiased method of demographic statistics, such that his work might be viewed as the final scientific word. Some secondary questions at least partially affect the final result. The solution to such problems (e.g., the method by which one calculates the population of the annexed areas which were under Italian occupation; the estimated growth from the previous census and the decreased birthrate during the war, which varied according to republic; unknowns concerning refugees and emigration, as well as calculations according to republic and nationality), cannot essentially change his results. Kocovic himself expresses doubt in the whole and in certain details, submitting reasons for his choice in cases of possible variants, where the approach itself stipulates different results. This will facilitate further deliberation by demographic statisticians, which depart from the scope of historical considerations.

According to Kocovic's calculations, the total demographic losses in Yugoslavia number 1,985,000, but he concedes that this number may range between 1,900,000 and 2,100,000.

From Kocovic's examination of earlier statistical calculations,

it is worth noting that he concludes that the smaller figure for demographic losses calculated by German professor Gunther Ipsen results from his having taken a smaller "pure accretion factor" after 1939, but that "in any case his number" (1,690,000) "is closer to the truth than are the assertions of D. Tasic or D. Vogelnik." (According to Tasic, the demographic losses were 2.4 million, and according to Vogelnik, between 2.8 and 3.2 million.) At the same time, Kocovic emphasizes that the differences between his own calculations and Lah's are "quite small," and "at approximately the same level regarding the total number of losses." So, too, are his results in relation to Klauzer's, which were used also by Busic. However, differences among figures for individual republics and provinces are considerable, a fact which Kocovic does not discuss because he doesn't know upon what basis Klauzer established his numbers.

There is an approximate correspondence in the whole but also very significant differences in their figures:

Demographic losses in thousands according to:

Socialist Republic or Province	Klauzer	Kocovic	difference
Bosnia-Hercegovina	581	668	+87
Croatia	661	605	-56
Macedonia	26	101	+75
Montenegro	63	112	+49
Slovenia	136	97	-39
Serbia	(575)	(402)	(-173)
Serbia proper	471	260	-211
Kosova	39	23	-16
Vojvodina	65	119	+54
SFR Yugoslavia (total)	2,042	1,985	-57

Kocovic explains that the differences appear because the difficulties are greater in estimating demographic losses in the individ-

ual republics and provinces than in the country as a whole. Those are the already mentioned problems: the "formerly Italian" territories, the unborn, factors of natural growth, refugees, migrations, "assimilations," as well as the "problem of the relativization of district factors of natural growth."[51]

Even if one assumes that the other elements which affect the calculation of losses—the number of unborn and the migration of inhabitants, which are also controversial—retain the same premise, by taking different factors of natural accretion, one arrives at very different results. Those used in Kocovic's calculations yield the following figures for war victims: under premise A - 862,000, B - 824,000, C - 727,000, D - 1,084,000, E - 1,171,000, F - 1,429,000.

Kocovic judges that it is "difficult to accept" the figures generated under premises C and E, even though they remain "within the realm of possibility," while the number attained by using F "goes beyond the scope of possible estimates," even though it is "still very far" from the propaganda figure of 1,700,000 victims. Consequently, he maintains that it can be asserted that the war victims numbered between 824,000 and 1,084,000. He himself favors the premise which results in the number of 1,014,000 real war losses. However, in this assessment, he concedes the possibility of error in both the premise of accretion and in other calculations. He says, "it is not excluded that the number of real victims may be higher or lower by 100,000 to 150,000. In other words, the number of actual victims may vary between 900,000 and 1,150,000."[52]

Kocovic describes these results, presented in a multitude of reasonably explained tables and graphs, in this way:

"As difficult as it may be to establish with complete accuracy the number of demographic, let alone real victims for the entire territory of Yugoslavia… , one can still assert with complete certainty that some numbers which appear both in some emigrant circles abroad, as well as in Yugoslavia, are, to put it mildly, pure fantasy. Claims that there were from 1,700,000 to 3,000,000 real victims cannot be supported by any possible premises. Likewise,

to claim, for example, that 700,000 people were killed in the concentration camp Jasenovac is pure nonsense." Kocovic continues, expressing amazement "that Dr. F. Tudjman was tried and even sentenced for having declared that there were not that many victims." However, Kocovic does not support his criticism of my alleged "underestimation of the number of Serbian victims and overestimation of victims within Croatia [which] has no material basis," because there is indeed no material basis for such a criticism. It is probably his realization of having so unprincipledly departed from scientific principledness which led him to write that my views are "much closer to reality than are many official and emigrant numbers of real victims," adding: "Ultimately, even Dr. Tudjman has the same right as others do to make a mistake."[53]

After all, even Kocovic, whose scientific integrity and honesty I salute, has remained in the shadows of national partiality, which clouds other perspectives. For him, "the word genocide corresponds only" to the Ustashas, as if there had been no genocide by the Chetniks, as if the Chetnik crimes were "only acts of defense and retaliation," and not part of a plan for creating an "ethnically pure Great Serbia" which was formulated before the onset of Ustasha crimes. His view that "the stain made by one group of Croatians, the Ustashas, will unfortunately remain the shame of several Croatian generations, just as the crimes of the Nazis have tainted the image of the German nation,"[54] is as biased as it is reasonable.

Kocovic arrives at the following results concerning demographic and real losses according to republics and provinces:

Republic or Province	Probable Number of Inhabitants (in thousands)	Total Losses (in thousands)		Relative Losses (in percents) in 1948	
		demogr.	real	demogr.	real
Slovenia	1,312	97	35	7.4	2.7
Croatia	4,045	605	295	15	7.3
B-H	3,233	668	382	20.7	11.8
Montenegro	489	112	50	22.9	10.2
Serbia proper	4,397	260	141	5.9	3.2
Vojvodina	1,782	119	76	6.7	4.3
Kosova	751	23	10	3.1	1.3
Macedonia	1,254	101	25	8.1	2
SFR Yugo.	17,263	1,985	1,014	11.5	5.9

From this it is evident that three republics suffered major losses: Montenegro, Bosnia-Hercegovina, and Croatia. Slovenia and Macedonia had losses at about a third, Vojvodina three fifths, Serbia proper about half, and Kosova three fourths of the Yugoslav average.

It is a similar case with real war victims. In comparison to the Yugoslav ratio of victims to population, the ratio of victims in Bosnia-Hercegovina were 200 percent of the Yugoslav figure, in Montenegro 173 percent, and in Croatia 124 percent. In Vojvodina, the ratio of victims to population was 73 percent of the Yugoslav ratio, in Serbia proper 54 percent, in Slovenia 46 percent, in Macedonia 34 percent, and in Kosova 22 percent.[55]

Both the total number of demographic and actual losses as well as their respective ratios to the population reveal that these losses were greatest where the population was involved to a greater extent in the war and in reciprocal retaliations. But the numbers also show that Croatia is far behind Montenegro and Bosnia-Hercegovina, whereas according to the Jasenovac distortion, Croatia's figures should vastly surpass them. This means that

along with losses generated by the struggle against the fascist occupation, Montenegro actually suffered more casualties from mutual Partisan and Chetnik reprisals, between revolutionaries and counterrevolutionaries, which was, in essence, the battle between the Great Serbian and Montenegrin national federalist programs. The fact that the greatest relative number of real war losses was in the territory of Bosnia-Hercegovina shows that these losses are above all the consequence of extremely intense national and religious conflicts. This is also seen to a considerable (though lesser) extent in the example of Croatia, while the examples of Serbia proper, Slovenia, Macedonia, and Kosova show that losses were fewer where, due to ethnic homogeneity, there was no inter-ethnic and religious retaliation, regardless of the fact that the setting was the same: the struggle between the revolution and the counterrevolution in the war against the occupiers.

Presenting the results of his calculations of real war victims according to nationality, i.e., looking at each nation and national minority separately, Kocovic lists the figures resulting from his premises, but also the possible variants for minimum and maximum losses within the given range of possibility (with percentages in relation to the probable number of inhabitants in 1948):

Nationality	Real losses			Possible Variants of Losses			
	Population (in thousands)		Absolute Victims	Minimal Percent of Victims (in thous.)		Maximal Percent of Victims (in thous.)	
Serbs	487	6.9	48	452	6.4	528	7.5
Montenegrins	50	10.4	4.9	48	10	52	10.9
Croats	207	5.4	20.4	190	5	240	6.3
Moslems	86	6.8	8.5	80	6.3	100	7.9
Slovenes	32	2.5	3.2	28	2.2	36	2.8
Macedonians	7	.9	.7	5	.6	10	1.3
All Yugoslavs	869	5.9	85.7	803	5.5	966	6.6
Other Slavic nationalities*	12	3.9	1.2	6	1.9	18	5.8
Other Balkan nationalities**	13	1.3	1.3	8	.8	19	1.8
Germans	26	4.8	2.6	20	3.7	50	9.3
Hungarians	5	1	.5	3	.6	10	2
Italians	1			2			
Gypsies	27	31.4	2.7	20	23.3	35	40.7
Jews	60	77.9	5.9	57	74	63	81
Other people	1			2			
Other nationalities	120	9.9	11.8	100	8.2	162	13.3
Total	1,014	5.9	100	917	5.3	1,165	6.8

*Other Slavic nationalities include Bulgarians, Russians, Ukrainians, Czechs, Slovaks, and Poles.

**Other Balkan nationalities include Vlachs, Rumanians, Greeks, Albanians, and Turks. For each of these, Kocovic lists separate figures. It is interesting to note that the figures for Albanians are; 6,000 [1 percent]; min.: 4,000 [.6 percent]; max.: 8,000 [1.1 percent].

In order to achieve a more accurate picture of the numbers of real war victims, Kocovic's estimates of relative losses according to republics and provinces compared with their probable population in 1948 are of particular interest. The losses are as follows:

Nationality	percentages
Serbs from Croatia	16.3
Serbs from Bosnia-Hercegovina	14.6
Serbs from Montenegro	11.6
Croatians from Bosnia-Hercegovina	11.4
Serbs from Macedonia	10.9
Moslems from Bosnia-Hercegovina	7.4
Moslems from Montenegro	6.0
Moslems from Serbia proper	5.2
Serbs from Vojvodina	4.9
Croatians from Croatia	4.2
Serbs from Serbia proper	2.9
Slovenians from Slovenia	2.3
Croatians from Vojvodina	2.1
Serbs from Kosova	2.0
Macedonians from Macedonia	0.8

From all these data one can conclude, both in terms of demographic losses as well as in real victims, that the Yugoslav nationalities did not suffer similar absolute nor relative losses. The differences among them are considerable for reasons we have already mentioned in part, while other reasons follow from facts that are outside the scope of this discussion. However, it is worthwhile to mention some comments of Kocovic on the figures he obtained. For the total actual victims among Serbs (487,000) and Montenegrins (50,000)—which he accounts for as a single group and assesses together as being between 500,000 and 580,000 maximum—he says "this is a huge number, far too great by any standard." But for many that is a "scandalously small number"

because "Many search for spiritual food for their positions in their anti-Croatianism. There is a deeply rooted opinion, that there were at least a million, if not more, Serbs killed," as well as a conviction "that the Serbs were practically the only ones who had real victims." Along with many figures which are "part of the myth which has seized all circles," Kocovic points out in particular the ridiculous claims of Terzic's that in Jasenovac "at least a million Serbs were killed" and in all, "more than two million people, and not 1,706,000, as is officially purported to both the Yugoslav and the world public opinion." Considering the indisputable fact that the lowest number of losses was experienced by the Macedonians, Kocovic finds: "Both the war and the civil war, and even the Bulgarian occupation did not affect that part of Yugoslavia very gravely."

Concerning the problem of Gypsy victims, Kocovic offers his figure of 27,000, pointing out that there may have been between 20,000 and 35,000. However, the claim that the highest number of Gypsies were killed in the NDH must be viewed in light of the fact that there were 14,000 of them in Croatia and only 2000 in Bosnia-Hercegovina, out of a total of 70,000 (according to the 1931 census). Concerning the German minority, he deduces that 26,000 died, 335,000 ended up in West Germany with a few going to East Germany, 51,000 were taken away to the USSR, and about 10,000 were placed in two camps in Vojvodina, but all these numbers are unreliable.

In the conclusion of his study, Kocovic addresses the question of whether he will succeed in destroying "certain myths which are deeply rooted among nations, among politicians, and even among scholars," for "myths have a resistant life" and furthermore, "the myths that the victims were, in the main, only of one nation, or that victimization was brought about by only one." In addressing this issue, Kocovic expresses the hope that there will be people who will "at least doubt" some of these myths.[56] The Slovenian historian Dr. Dusan Biber disputed the theses of Dr. V. Duretic, citing Kocovic's estimate of 487,000 Serbs who had been killed (or had died of natural causes) in all the various circumstances of war.

Biber expressed the hope that with regard to establishing the number of victims, scholarly initiatives will finally bear fruit, or else we shall have to rely, just as we do on foreign debts, so too on foreign experts.[57]

But we continue to acquire new testimonies about the people and the times which reveal how firmly anchored the distortions are and how a spiritual climate still predominates in which even scholars find it more advantageous to contribute to the glorification of these untruths.

One of the most remarkable examples of this is a book about two centuries of demographic changes in Croatia, which appeared in 1987. This monograph was Jakov Gelo's 1983 doctoral dissertation.[58] One may suppose that this is a scholarly work corroborated by a university, a work which was then judged for the purpose of publication by a panel of esteemed professors to be an "undoubtedly valuable original scholarly work based on Marxist theories of contemporary demography." The book's structure, style and graphics appear at first glance to be the epitome of scholarly writing of highest level. Let us take a look, however, at how its author deals with the problem of demographic and war losses. One should expect to find in such a work answers to at least the disparity and inconclusiveness in demographic discussions until now, if not of all the controversial issues concerning demographic changes in Croatia in the period of the Second World War.

Instead, Jakov Gelo not only simply accepts the lie that the war victims totaled 1,706,000, but he even adds to that figure considerably. Calculating that "the war in Yugoslavia lasted 1,500 days (from April 6, 1941 to May 15, 1945)," and that it took an average of "1,137 human lives daily," he adds to this figure another "approximately 440 Yugoslavs" daily, who died of natural causes. The result is "that the total number of Yugoslav citizens who died from 1941 to the end of 1945 (sic) was 2,509,000, or 31.4 per 1000 inhabitants per year (during five years of the war)."[59] And thus the inflated number of war victims was for the first time "scientifically" expanded to 2.5 million. Who cares that in that careful calculation of victims per day World War II was first short-

ened to "four full years," and immediately afterward lengthened by almost seven months "to the end of 1945," and then (in the same sentence) by a whole year—"in the course of five years"?

It is almost unbelievable that in such a book, no attempt was made at a systematic analysis of the problem concerning demographic and war losses. Instead, one is given the simple statement: "During World War II in Yugoslavia, 305,000 fighters of the NOP were killed, as well as about 150,000 collaborators and 1,251,000 victims of fascist terror." Gelo concludes "that Croatia had a total of 91,500 dead fighters of the NOR, 45,000 collaborators with the occupiers, and 298,500 victims of fascist terror." This comes to 435,000 fatalities (or 11.5 percent of the population), and together with the 231,000 who died naturally, totals 666,100 (or 17.6 percent of the population according to the census of 1948).[60]

Some fifty pages later, on the subject of "estimating the emigration figures," Gelo writes that "due to consequences of World War II," the population of Croatia between 1940 and 1948 "dropped by 315,142 people." But he immediately mentions the number of the born (631,744) and says that "because of the war, Croatia lost considerably more than these 315,000." Then he repeats the figures just mentioned, without any explanation as to the difference. In this discussion, it even remains ambiguous whether the estimated "emigration figure" for Croatia was in the negative by 123,085 persons, since it was arrived at by subtracting the "negative accretion" which amounted to 192,057 "from the established decline in population" (315,142), which is simultaneously "confirmed" and "exceeded."[61]

An attempt to approach the problem of assessing the number of war victims scientifically was made by Dr. Fikreta Jelic-Butic in her monograph concerning the Ustasha movement and the Independent State of Croatia. Still, she did not dare venture out from the shadows of the Jasenovac distortion.[62] The results of her many years of investigative work (I myself assigned her that topic at the Institute some fifteen years before the appearance of her book) led her to write: "Through the camp of Jasenovac tens of thousands of people passed, most of whom were killed in the

numerous massacres that occurred there."[63] That scholarly state-
ment is supported by a source quoted in a footnote (214). The first
sentence of the source reads: "In Jasenovac, by far the largest
number of people were killed in the NDH—several hundred thou-
sand." Continuing, she cites data from the Land Commission for
the Establishment of War Crimes, stating that "the number is
between 500 and 600 thousand." Then she cites some new "evi-
dence" that "Investigations of the gravesite complex being under-
taken by the Jasenovac Memorial Area have confirmed that in the
57,000 square meter space which has been examined to date, over
360,000 dead inmates were buried." Butic quotes R. Trivuncic's
conclusion: "Based on space indicators and claims by surviving
inmates, the figure of 700,000 destroyed persons is very realistic."

What are these "space indicators" on the basis of which some
can arrive at—according to Trivuncic—"very realistic" numbers?
The Jasenovac Memorial Area Council explains this by its own
authority in its 1972 "Open Letter" to the Croatian author and
academician Vjekoslav Kaleb. The letter reprimanded Kaleb for
having dared to write that there are those "who multiply tenfold
the number of the killed." The Council reminds Kaleb: Jasenovac
is "the largest gravesite in Yugoslavia"; "it is one of the largest
camps in Europe, and that it is, without exaggeration, the largest
in Central Europe and the Balkans"; furthermore, "up to the pre-
sent time, in the area of the entire complex... a vast amount of
investigative work has been conducted which has shown" that
what Kaleb writes about the victims in the *Hrvatski tjednik*
(Croatian Weekly) is "irresponsible, without argumentation, and
completely improvised."

Here is the evidence the Council offers: "Namely, in the entire
area of the Jasenovac Memorial Area, there are seven sites of
which only one—Gradina—has been scientifically examined. By
the simple mathematical operation of addition, it has been calcu-
lated that in a 57,000 square meter gravesite, there are 366,000
killed inmates." So that when "our scientists investigate" all the
rest, "Will you need more evidence to convince you that you des-
ecrate the blood of 700,000 who have died in Jasenovac?"[64]

In the *Encyclopedia of Croatian History and Culture*, which came out in 1980, Dr. Ivan Jelic presents quite a reasonable judgment concerning the camps in general and Jasenovac in particular. Mentioning that in the beginning, groups of Communists and participants in the NOP were brought to that camp, "and then increasingly larger groups of Serbs, Jews, Croatians, Moslems, Gypsies," Jelic says: "Jasenovac was turned into the largest center for torture and liquidation of all who were considered undesirable persons to the Ustasha regime. Tens of thousands of people passed through that camp, and the majority of them were executed in the many massacres conducted there."[65] This is not merely taken from Fikreta Jelic-Butic's work, but Jelic himself had to have been convinced of its validity since he was researching that same period at the Institute, though his specific area of study was the NOB and the KPH, so that they were both directed to the same sources, including those from the enemy side.

The Encyclopedia of Croatian History and Culture appeared when the legal persecution against me had begun, above all because of my stance on the Jasenovac controversy. In such circumstances the entire Encyclopedia came under attack because of the mentioned text of Ivan Jelic. Defenders of the distortion contrived a course of action in which the Encyclopedia was taken from the bookstores, and Jelic was forced to withdraw.

A large number of members of the Croatian and Serbian emigrants have also dealt with the problem of the war victims, as well as foreign scholars who study the contemporary history of this area. Thus, American professor Pedro Ramet, in *Nationalism and Federalism in Yugoslavia 1963-1983*, reports that 350,000 Serbs were killed in Pavelic's NDH.[66] The English journalist Nora Beloff, in her book, *Tito's Flawed Legacy*, maintains that the claimed figures for Jasenovac "of 600,000 to 800,000" require the removal of one zero.[67]

Jure Parsic was the parish priest at Jasenovac during the war. As a supporter of the NDH, he accepted the rank of Ustasha captain, but he was against the crimes of the Ustashas and he assisted in the rescue of victims. He was personally familiar with the condi-

tions in the camp and its officials, and in 1985 he cited information he received from Ivica Matkovic, one of the commanding officers of the camp. Among his recollections, he tells of how he heard the confessions of Pavelic's government ministers, Mladen Lorkovic and Ante Vokic. Parsic, a well-informed and reasonable man writes: "In the Jasenovac prison camp, it is possible that from the beginning to the end (1941-45) between 30 and 40 thousand people lost their lives."[68] This estimate is within the range of my investigations as well.

The estimates of three of the most informed and most responsible people in Croatia, who had access to all the available documentation, were considerably lower than Parsic's estimate. Their response to the observation that this contradicts even the disclosures of the Ustashas themselves was that the Ustashas had been boasting of their crimes, believing that by exaggerating them, they better advanced their aims.

At the time when an effort was being initiated in America to raise a monument to Draza Mihailovic, which even reached a Congressional committee, the newspaper of the Croatian Fraternal Union, that emigrant organization which has always remained independent from emigrant political currents, also spoke out. In a detailed account, Ante Beljo, presented all of those above-mentioned arguments with which I had been contesting the Jasenovac distortion since 1965. In addition, the editorial staff, in a separate commentary, felt it necessary to state that Beljo's article was "our final response" to those "who ascribe to the Croatian nation the guilt ... for the lives of some 700,000 to 1,200,000 Serbs."[69]

The problem has long since exceeded the limits of "petty" internal squabbles and conjectures, manifesting itself as one of the major issues in the contemporary history of Croatian-Serbian relations, from both national and international perspectives.

Finally, in 1986-1987, there appeared the three-volume *Documents* concerning the Jasenovac camp by A. Miletic (published by Narodna Knjiga, Belgrade and the Jasenovac Memorial Area). As a collection of documentation, this is undoubtedly a

serious contribution to research, regardless of the lack of critical evaluations of the published documents and of material as a whole. However, from the historiographical viewpoint, the undertaking is a total scholarly failure. The work's conclusions are contrary to the real facts in that very documentation.

What can be concluded from this documentation as being factually undisputable is only the following:

1. The register containing the names of the people who were in the camp, listing 25,740 names including victims, survivors, and those who comprised the administration of the camp.

2. The total of recorded statistics concerning those who had been taken to camps from various parts of the NDH, who numbered 22,219.

3. The total of properly recorded statistics concerning the victimized (i.e., killed) persons in various parts of the NDH, who numbered 27,951.

All the rest are unsubstantiated hypotheses, unreliable testimonies, or negligently drawn estimates.

By this, of course, I do not wish to claim that the number of victims does not exceed the above-mentioned statistical figures. If it is logical to assume that all of the Jewish victims were probably registered by name, that was obviously not the case with the Gypsies. But one must first establish how many of them there were in the NDH. In part, the Serbs were also brought in groups, but their number and names can also be established according to their places of residence. The problem of establishing the number of Jewish prisoners as well as others who were taken to German camps (from Jasenovac, as well) is left out of the scope of this documentation, as well as the fact that tens of thousands of "internees" and "deportees" survived the horrors of wartime persecution. Accordingly, both within and outside of the published material there are unanswered questions which should be investigated. As it has been presented to the public, this material gives no basis for the conclusive judgments it hopes to pass.

There is no evidence for Miletic's estimate that "several hundred thousand men, women and children were killed in the

Jasenovac camp." It is substantiated only by the fact that "in publications, scholarly works and the press, the number ranges from 480,000 to 900,000, and even over a million," but in "most cases," it is around 700,000. In support of this statement, he can cite not one, let alone several, truly scholarly works. He does not discuss opposing assessments; he merely mentions them in a footnote.

It is significant that all of Miletic's estimates are found in the first volume, as the summary of the introduction.[70] This is significant because it demonstrates the method and the intent of this publication. At the end of this undertaking, in the Epilogue of the third volume, Miletic did not find it necessary to change or add to the estimates in his introduction, only to prove his "scholarly principledness" by disparaging those who do not agree with his ridiculous figures.[71]

Dragoje Lukic wrote the foreword to Miletic's completed work. Scorning the belief "that we will sooner and more easily overcome evil if we do not mention it," Lukic raises his prophetic voice: "A painful past lives the longest. Jasenovac is a grievous wound, and as long as this land exists, that wound must hurt." According to Lukic, the book "through its authenticity, demystifies the various representations of Jasenovac." The book with its "strength of incontestable arguments" replies to "all those who, by audaciously reducing the number of victims and the magnitude of the Jasenovac tragedy, attempt to justify those dark forces which led to the fratricidal massacres," and it also "strikes down any argument by those who through Jasenovac have tried to blame an entire nation for that which was committed by its degenerates."[72]

In his introduction Miletic presents the following: "According to some statistics, there were in 1941, in the territory of the so-called NDH, between 1,885,943 and 1,966,000 Serbs, and according to the census of March 15, 1948, there were 1,610,000 Serbs in the territories of Croatia, Bosnia-Hercegovina."[73]

What do these statistics tell us? Do they not reveal the mathematically indisputable truth that in Croatia and Bosnia-Hercegovina, there were between 275,943 and 356,000 fewer

Serbs after the war? That number includes those "hundreds of thousands" or even "over a million" Jasenovac victims, and all those who died in the Partisan and Chetnik conflicts, the occupiers and the NDH, and even those who died of natural causes, as well as those forcibly and voluntarily resettled in Serbia, of whom there were also from "tens of thousands" to "two hundred thousand," according to Miletic.

According to Kocovic's estimates that Serbian wartime losses in Croatia amounted to 16.3 percent, and in Bosnia-Hercegovina 14.6 percent of their total numbers in those regions, one gets the result (125 + 209 thousand) of 334,000 Serbs (including Montenegrins) which is in the range of those mentioned above and calculated by a simple comparison of prewar and postwar figures.[74]

II. THE UNFOUNDED BASIS FOR THE THEORIES OF GENOCIDAL CHARACTER OF CROATIANNESS

Many very learned people have no mind.
—Democritus (460-370 B.C.)

Destiny has bestowed upon us the mercy to have been born and to have lived in a time which has deceived us.
—Petar Segedin

3. From Myth to Theories

The eyes and ears are evil witnesses to people who have barbarian souls.
—Heraclitus (535-475 B.C.)

Having traced the genesis of the Jasenovac controversy, we should now turn to a general outline of the consequences which logically stemmed from the various distortions and untruths, reaching their culmination in ideas concerning the genocidal nature of Croatianness.

These ideas were always present, even among the ranks of the Partisan movement. I have already mentioned the display of distrust on the part of the Serbian fighters toward Croatian revolutionaries. I have mentioned the necessity of restraining vengeance against Croatian villages and populations during and at the end of the war. Likewise, immediately after the war, the magnification of only Ustasha crimes led to public demands for executing vengeance upon the captive Home Guards.

Judgments concerning the historical guilt of the Croatian people for the collapse of the Monarchy began to reappear in socialist Yugoslavia in the early 1960s. They were made public in the most severe form by V. Terzic in his previously mentioned book in 1963. What it says, in brief, is that Croatian leaders shifted from their 1920's federalistic program to separatism, that the presentation of the Croatian question before the international public caused the weakening of the "unity and resistive strength of the country," and finally, that Macek's leadership of the Croatian Peasant Party which had represented virtually all of the Croatian people, had committed "high treason" which "our generations will eternally condemn," while the decline of the royal state is blamed upon the entire Croatian nation.[75]

In Terzic's book, the legend of the historical guilt of the Croatian nation for the collapse of the Kingdom of Yugoslavia and for all the war suffering was systematically explained. The legend has not been officially accepted, but it was firmly maintained, and with time, gained an ever greater number of followers, and new characteristics. In the end—as we shall see—the Croatian nation would become the primary culprit for both Fascism and Communism in this country.

In his 1983 book, *The Fall of the Kingdom of Yugoslavia in 1941*, which was actually a revision of his book from the 1960s, Terzic recalled his old conclusions, explained them, and added new dimensions to them. Since Milo Gligorijevic had reminded him that Vuk Vinaver was of the opinion "that Jasenovac was the first consequence of the fall of the Monarchy," Terzic commented that in the issue of genocide, "there is a fear of the truth" or "an

evasion of truth, an evasion of reality." The truth, according to Terzic, is "that Croatia before Jasenovac had a long history of clericalism, at least fifty years of indoctrination from Ante Starcevic to Ante Pavelic: in historical continuity, a consistent political doctrine against Serbs and the Serbian nation was carefully set in motion."[76]

Terzic never considers asking for the reasons why "in Serbian, Montenegrin, Slovenian, and even Dalmatian regions, the aggressor was received very coldly, antagonistically and with hatred. However, in some areas of northern Croatia, in many sections of Bosnia with Catholic and Moslem populations, parts of eastern Macedonia, as well as in areas chiefly inhabited by minority nationalities [in Kosova, Banat, and Backa], the aggressors were awaited with enthusiasm and joy, and with exceptional honor, as liberators." Such facts do not motivate him to consider this matter more deeply; instead, he simply ascribes this "enthusiasm toward the aggressor" to the "combined efforts of foreign and domestic enemies," and to the "international conspiracy against Yugoslavia."[77]

Terzic might have discovered at least a part of the answer, had he given Starcevic a fraction of the consideration such a great Croatian thinker deserves. Namely, he could have read that a nation which "has reached the state in which it has nothing to defend, nor to lose ... will sacrifice nothing" for the benefit of its enemies nor for a state which it does not consider its own. "An oppressed nation cannot be the friend of its oppressor," while "no one has ever found it easy to unite or preserve an illegal state; to preserve the state against the will of those who should be its citizens." Starcevic said this over a hundred years ago about "Austria, which has always been the same," for "dictatorships do not improve, but collapse," warning the Viennese hegemonists that "we refuse any responsibility for the consequences which may result from those illegal actions" taken against the Croatian nation, which desires to "free itself from yoke of anyone" and not merely to "change yokes." For "the Croatian nation needs and wants to be free and happy, and it has that right as does every other nation."[78]

Terzic gives himself the sovereign right to make conclusions and to pass judgment upon Croatian history from a standpoint that is contrary to the interests of the Croatian people. It is likely that he finds Starcevic so odious because the latter said about just such a standpoint, "To be unwilling to admit that a nation knows best what is of true benefit to it, and to claim that someone outside that nation has that knowledge, means blindly taking on an immense responsibility of the sort that prudent rulers avoid; it means scorning a nation... It is foolish and dangerous to claim that the right can be altered... There isn't a fallacy more groundless or a doctrine more harmful than this conviction."

Starcevic saw "Every man's nationality as the most sacred thing." He said openly, "We are concerned about the good or bad fortune of Austria only insofar as it serves Croatia." No less significant was his statement that "The Croatian nation will respect the freedom of every nation in the same way that other nations respect the freedom of the Croatians."[79] When one speaks of Starcevic as the founder of modern Croatian national consciousness, it is understood that his negation of Serbdom cannot be forgotten. However, it is historically inadmissible to judge this one-sidedly, for it is a historical fact that Starcevic's renunciation of Serbian domination came about in defense of Croatianness against Vuk Karadzic and his followers who by his text "Serbians All and Everywhere" (and his later "Serbs and Croats") founded modern Great Serbian ideology, denying Croatians their history, their name, and their future.[80]

Karadzic wrote that "if [Croatians] do not wish to be called Serbs, then they have no national name." Starcevic's philippics against the Croatian "Slavoserbs" (pro-pan-Slavist Croatians) were aimed at those politicians who were ready to assimilate Croatianness into the idea of Yugoslavism, and who were prepared to serve both Austria and Hungary. Starcevic's renunciation of Serbdom was the same as Vuk Karadzic's renunciation of Croatianness. However, in the 1870s, he wrote affirmatively on "The Principality of Serbia." He spoke negatively only of those

Serbs who lived in Croatia and were working against Croatian interests. That Starcevic's views were based on historical facts and were democratic in the spirit of the times. This can be concluded from the fact that they had a significant effect upon the Croatian intelligentsia. His ideas were accepted by a substantial number of Orthodox Christians, some of whom even became prominent followers of Starcevic's Party of Rights.[81] Associating Starcevic with the Jasenovac victims, as Terzic does, is a sign either of historical ignorance, or the denial of a Croatian national being, of which Starcevic's views were the most prominent expression in the past century.

In the fall of 1986, Vasilije Krestic published in the Belgrade *Knjizevne novine* (Literary Gazette) his essay "On the Genesis of the Genocide against the Serbs in the NDH."[82] In the history of the genesis of Croatian-Serbian discord Krestic's essay will probably take its place right next to Nikola Stojanovic's 1902 thesis "Until Your Extermination or Ours."

Krestic wrote the essay because of the "enthusiastic interest on the part of broad strata of our society regarding the question of genocide." That interest was awakened by the reprint of Viktor Novak's *Magnum Crimen*, as well as by the emergence of Miletic's collection of material on Jasenovac, Stanojevic's books on Stepinac and by Artukovic's trial. Since the years "not only have failed to erase from our memories the horrors of the war, particularly those which took the lives of Serbs" but have "revived the picture of those horrors," Krestic takes on the task of answering the "irresistible" question: "How was this made possible and why did it happen?" (emphasized in the Serbian original). The fact that this question "remains unanswered to this day" as well as the polemics about *Magnum Crimen*, "have shown the necessity for strictly objective, scholarly discussion of such delicate ... issues as are those in connection with the genocide against Serbs in the Ustasha NDH." This is, with no doubt, the right approach. So, too, is his introductory remark that an answer is possible only if one traces the history of relations between the Croatians and Serbs "in the continuum of several centuries." However, the author

looks only for genocidal intentions of Croatians against Serbs and not of Serbs against Croatians.

Krestic was not at all moved by the views of Vasa Bogdanov who dealt with the history of Croatian-Serbian conflict in five of his works from 1957 to 1964.[83] These works present the historical causes of the conflict between the Croatians and Serbs in what is to this day the most detailed account. It is strange that Krestic does not take these works into consideration, all the more since they were written by a Yugoslav and Croatian historian who was a Serb by national origin. It is even more unusual that Krestic should judge that Bogdanov's text which was published in *Izraz* ("The beginning of misunderstanding between Croatians and Serbs") "merely broached the subject," even though Krestic was well aware that Bogdanov reprinted it after the war. Kresic should presumably be aware of the works by Bogdanov in which the latter completed his thesis. According to Krestic, Bogdanov "relied exclusively on the biased opinions of Ante Starcevic, which he accepted without criticism." This claim is not supported by any evidence, and it is also significant that there is no reference to any of Bogdanov's other discussions so that the reader could examine accuracy of this himself.

It is no surprise that Krestic should consider Viktor Novak's book, *Magnum Crimen*, "the first and greatest" work of its kind, regretting only that it is "still a solitary breakthrough in the disclosure of the truth concerning the causes of genocide in the Ustasha NDH." Krestic writes: "It is certain that the genesis of the genocidal acts against the Serbs in Croatia must be sought in that era when the so-called Orthodox Christian Vlachs [Vlasi], i.e., Serbs, under the pressure of the Turks, began to settle in Croatian lands in the 16th and 17th centuries."

This claim is valid only insofar as it explains the beginning of the conflict. However, it is untenable to equate Orthodox Vlachs with Serbs because, at that time, the majority of them were not Serbs either by ethnic origin, or, certainly, by national consciousness. Not only the Croatians, but all the southern Slavs, and even the Serbs themselves used the term Vlachs in reference to all

Romans, Romanized Illyrians and Thracians in the Balkans. Among these, only a part were ethnic Serbians, while the others were regarded as Serbs through the identification of Orthodoxy with Serbdom. That practice began when Vienna granted the Orthodox Christians special privileges in Krajina, and has continued since the 19th century.

Having been given the status of freemen, the Vlachs not only disturbed the feudal order, but they also favored the separation of border zones of the Military Frontier from the control of the Croatian Parliament and Governor [Ban]. The Vlachs served to violate the sovereignty of Croatian authorities, and aided in the parceling and militarizing of Croatian lands, thereby strengthening Viennese centralism in relation to the government of the Ban, weakened by the Military Frontier. Ill will toward the Vlachs for these reasons was later to increase after the establishment of dualism and demilitarization of Krajina, particularly because of the collaboration of Orthodox Serbs with the Hungarian hegemonic politics of limiting and weakening Croatian autonomy. In this context, one must not forget that in Dalmatia, Serbian politicians, together with the pro-Italians, opposed the unification of Croatian Dalmatia with the rest of Croatia ruled by the Ban, not to mention the issue of Bosnia-Hercegovina. Therefore, the problem is much more complex, and it could hardly be historically deduced to a class "antagonism between the Krajina Serbs and the Croatian feudal lords."

Krestic does have certain grounds for the claim that in the 17th and 18th centuries, "feudal circles of Croatia... were ready for genocide against the Serbian Orthodox populace which had settled in their dominions under special conditions and against their will, violating their rights as feudal lords." But he is wrong when he ascribes this to the fact that "the Croatian upper class" were supposedly "known for their religious exclusivism." The 1608 decision of the Croatian Parliament, according to which public rights are recognized only to those of the Catholic religion, gives no basis for such a claim. It was not directed primarily against Orthodoxy, but was decreed for the purpose of guarding Croatian

sovereignty because the military administration, aided by the spread of Protestantism in Croatian regions, was attempting to weaken Croatian unity and strengthen Viennese centralism. It never occurs to him to mention the extermination of Moslems from Serbian territories. In their case it was not genocide, but an act worthy of exaltation [sic] in the greatest of "Serbian epics," *The Mountain Wreath* (Gorski vijenac), which remains above reproach as the most distinguished poetic work and loftiest moral guide.

The central thesis upon which Krestic's entire theory concerning the genocidal predisposition of Croatianness rests is denial of the existence of a Croatian "political" nation. According to Krestic, the discord "between the Croatian and Serbian political leadership, as well as among broader strata of people," throughout the entire period of bourgeois development, arose "from Croatian national-political ideology, which was entirely based on the idea of the existence of a political or diplomatic nation." It was this idea which made "the differences between the views of Croatians and Serbians difficult to bridge." Consequently, "there resulted fierce mutual confrontations and the growth of an intolerance which, within certain Croatian bourgeois and especially petty-bourgeois circles, gradually developed into ever more apparent genocidal aspects." It is a pity that we are left wondering why, under the same conditions, there wasn't any similar phenomenon among Serbian circles.

Krestic argues that the Croatian nation, "was taken from Hungarian state-feudal political ideology." Thus Krestic establishes the groundwork for a new political and international legal explanation of all historical occurrence. In keeping with this notion, according to Krestic, the underlying basis for the programs of all Croatian civic parties was the Croatian state and historical rights since Croatia had "maintained a semi-feudal character."

The Croatians did not acquire the idea of the existence of a nation from Hungarian state-feudal ideology. In fact, the Croatians used this idea to oppose ideology which sought to aggrandize Hungary by means of eliminating Croatia. It was pre-

cisely because of the inextinguishable living consciousness of their historical and state right that the Croatians resisted hegemonic encroachment and attempts at assimilation by Budapest, Vienna, and later, Belgrade. The origins of this idea are to be found nowhere else but in the medieval Croatian state. Preserved by the Croatian nobility, and then by feudal lords who died in battles and dungeons, it was the "underlying basis" for the programs of all Croatian civic parties, not only "until the beginning of the 20th century," but even further. It was with this idea (supplemented with the principles of natural rights) that Radic's party became the representative of the all-national movement. This idea is contained—against Pavelic's NDH—in that deeply moving message from 44 Croatian Marxist revolutionaries who in 1941, on their way to meet their deaths, carved into a wall their pledge to a free socialist Croatia.

According to Krestic the "reactionary" Croatian feudal aristocracy had no historical rights to insist upon the Croatian political nation. At the very appearance of socially progressive Vlach population communities the Croatians should have surrendered. Let us ask him whether according to this same logic, the Croatian "secular and feudal lords" should not have abandoned their Croatian rights even before the arrival of the Orthodox Vlachs and laid it at the feet of the repeated Frankish and Byzantine, Venetian and Ottoman, Hungarian and German onslaughts and invasions of Croatian lands? Naturally, Croatian noblemen couldn't—shouldn't have, and didn't wish to—"fulfill" such a "historical" demand.

Krestic attacks the school system which the Croatian Parliament legally adopted under the rule of Ban Ivan Mazuranic because he strove for "Croatization." It would be interesting to find even one historical example in which a sovereign government of a nation did not strive toward national homogenization. And why is such an endeavor only in Croatia called genocide? It turns out that because of this, even Mazuranic was a genocidal reactionary because he would not have permitted "Serbization" or perhaps "Hungarization" or "Germanization" of Croatia.

In his treatise, Krestic is unquestionably correct when he claims that after the revolution of 1848-1849, all Croatian politics turned toward Croatian integration. However, he is incorrect when he claims that this was "contrary to the Illyrian movement," which was a Croatian revival movement. It is interesting to note here that Krestic calls those politics of Croatian integration Great Croatian, which necessarily had to clash with Serbian and Great Serbian politics.

Why is such a policy seen to be negative? It is nothing other than a demand that all the inhabitants see themselves as citizens of the Croatian state, regardless of religious or ethnic differences. In other words, that sovereignty over Croatian territories should belong to the Croatian nation and that it cannot be shared with anyone beyond its borders. Insistence upon that principle was a defense against the weakening of the right to sovereignty through Hungarian, Venetian, German, and, naturally, Serbian encroachment as well, when their time came. Krestic does not see that Croatian policy and the Croatian Parliament were, in fact, quite obliging toward the requests of the Serbian minority in order to win them over when they fought together with the Serbs of Vojvodina in a common battle against Vienna and Budapest.

Croatian politics found itself faced with the historical task of uniting the lands of the Croatian Triune Kingdom (Dalmatia, Croatia, and Slavonia). When Slavonia was completely incorporated into the Croatian union under the Ban, what remained to be accomplished was its unification with Dalmatian Croatia, as well as the liberation of lands that had been seized by foreigners. Here, on the matter of Bosnia-Hercegovina, one encounters the heart of the Croatian-Serbian conflict. On this subject, certain facts should be noted. Jajce and Bihac belonged to the Croatian Kingdom. The entire area all the way to the Vrbas river was part of Croatia even before the designation of the Habsburgs by the Croatian Parliament (in 1527) as ruler of Croatia. Even on Serbian maps from the 19th century this area was called "Turkish Croatia." As far as Croatian politics toward the Serbs regarding Bosnia-

Hercegovina is concerned, along with the desire to annex this area to Croatia, there was always the recognition of the necessity for a united effort in the area's liberation.

Krestic's considerations about the relations of the Croatians and Serbs with Austria and Austro-Hungary are extremely over-simplified. It is historically incorrect to claim that Croatians, contrary to Serbs, were predominantly pro-Austrian and "based all their politics on a collaboration with Austria." Exactly the opposite was true. All Croatian politics strove toward complete autonomy from Vienna, but it had to deal with the reality of the Habsburg monarchy as the greatest power in Europe at that time.

In his interpretation of the intensification of the Croatian-Serbian conflict which occurred in 1902, Krestic does not see that the cause of this escalated tension was in the fact that the Belgrade *Srpski knjizevni glasnik* [Serbian Literary Messenger] published the previously mentioned article, "Serbs and Croats" in which a battle is openly prophesied "until our extermination or yours," while the *Zagreb Srbobran* [Serbian Shield] had the audacity to reprint that publicly proclaimed program of genocide against Croatians in the capital city of Croatia.

In keeping with his one-sided approach to the Croatian-Serbian conflict, Krestic has arrived at the conclusion that the prerequisites for the elimination of this conflict had begun to emerge because "advanced Croatian youth and progressive strata of the Croatian middle class" had begun to relinquish the "traditional paths" of Croatian politics based on the historical and political right. However, the very logic behind the historical view of the genesis of the conflict as it is presented by Krestic, and especially the entirety of historical experience—tell us just the opposite. Namely, that "progressive stratum" which was enraptured by Yugoslav integrationist illusions was quite thin, and it sobered in its confrontation with the unrenounced Great Serbian ideas. Mistrust only increased and the conflict between two national-state entities deepened. Krestic tells that by 1914 a program was already completed "according to which Serbs were to be dealt with most radically" and "their massacre" was to be undertaken "in 1914." Krestic doesn't ask why no genocidal crimes

occurred during the First World War. Is it not perhaps significant that the fact that there was no genocide in World War I rests virtually exclusively with the Croatian side? Is it not significant that the responsibility for the fact that it did occur in World War II, rests primarily with the inter-war situation, marked first by the murder of Stjepan Radic and then of King Aleksandar, and during the war by both the actions of the Ustashas, Serbs and Chetniks, under the general circumstances of that war?

Krestic's theory was founded in a pure vacuum, as if among Serbs there was no genocidal ideas against the Croatians, let alone acts of genocide. This theory is opposite to historical reality, which shows that genocidal ideas sprang from Vuk Karadzic's denial of Croatianness, and that the Serbian genocidal ideas were farther reaching than the denial of Serbdom by the Party of Rights or by Frankists. It is difficult to understand how Krestic does not see the fact that his concluding remark applies not only to the Croatians but also to the Serbs when he asserts that genocidal crimes "are not the result of only one system, of one party or another, of one society or another, of one individual or another, but rather from the convergence of a sequence of circumstances over a long period of time." He is likewise correct when he says: "The persistence of genocidal ideas... has rooted itself deeply into the consciousness of many generations," but he is entirely wrong when he ascribes this only to "some centers of Croatian society."[84] He is even more mistaken if he thinks that by labeling Croatianness as genocidal, and attempting to conceal genocidal ideas on the part of the Serbs, he is not providing new incitements for genocide. Such one-sided views on the problem of genocidal phenomena in Croatian-Serbian relations are given the right to exist in the heart of the highest Serbian scholarly institution—the Serbian Academy of Arts and Sciences [SANU]. It is significant that the Committee for the Investigation of Genocide was founded at that institution, which could certainly be a sign of a positive approach toward elucidating the problem. But to this day, there have been no other results except for the participation of its members in the endorsement of the Jasenovac distortion.

After the trial of Andrija Artukovic in Zagreb, the Committee deliberated upon the report by their observer to the proceedings, Dr. Milan Bulajic, with a critical retrospect on "why Artukovic was not tried for genocide" and why the court did not establish "the site and number" of victims, along with the criticism of his address to the Croatian people in his closing statement.[85] These are signs of a political, not a scholarly approach to this problem. And it can be muddled even further by such "scholarship" as is represented by V. Dedijer, the president of the Committee of the SANU, and also president of the Russell Court, which in its time was a form of political protest against crimes in the contemporary world. Not only from Dedijer's books, but also from his presentations, especially that on Kurt Waldheim, one can easily ascertain the aim of his approach to genocide.

First, Dedijer emphasized the fact that "it is to Yugoslavia's great misfortune" that "each Republic and province has its own stance toward Dr. Waldheim." And the alleged concealment of the truth about Waldheim in Yugoslavia came about "because of pertial interests," since the main archives concerning Dr. Waldheim's units in genocide are to be found in Croatia." He claims that "certain Croatian officials, led by Andrija Hebrang and Stevo Krajacic, felt that it was unnecessary to save those archives. It was also their view that the remains of the concentration camp in Jasenovac should be destroyed." As an addendum, Dedijer supports his discoveries with the fact that he searched in vain for original documents for his book, *The Vatican and Jasenovac*, since "a large part of the Ustasha archives have been destroyed."[86]

Dedijer is all too familiar with the structure of the Communist authorities to be ignorant of the fact that Hebrang, after being replaced as head of the KPH (Communist Party of Croatia) in 1944, and upon his withdrawal to Belgrade into the economic sector, was unable in the years right after the war, to execute political decisions in Croatia. Yet Dedijer holds him responsible, along with Krajacic, for the alleged destruction both of archival material and the remains of the Jasenovac camp. If Dedijer is perhaps unfamiliar with the fact that in the first days after the war, the area

of Jasenovac and Stara Gradiska served also as a camp to the new socialist authorities, then he certainly could not have been unaware or could not deny the fact that archival material exists which is contrary to his claims.

Namely, after the war, the main archival documentation from the NDH was transferred to Belgrade. The fate of the more sensitive portion of the party records, which was not handed over into the Archives of the CK KPH in Zagreb, remains unknown. For this reason, after the establishment of the Institute for the History of the Workers' Movement, I undertook measures to recover at least copies from the federal archival institutions. I learned that the Slovenians hadn't allowed the removal of any Party documents or "enemy" documents. It remains to be established, by whose decision the material was transferred to Belgrade (at that time, Mosa Pijade dealt with those issues, and they were within Rankovic's scope of official duties involving party organization and internal affairs). The fate of those documents was perhaps the same as that of other material at the time of its removal from Belgrade during the danger of Soviet aggression after the Cominform Resolution of 1948.

Croatia can indeed be blamed as the primary guilty party for republican federalism. The only historical controversy here might be whether the restoration and survival of a Yugoslavia would have been at all possible without the recognition of national republics. However, it seems that this is not at all controversial to the creators of the Jasenovac distortion and theories of genocide. It is remarkable only that one finds such conjectures even in more serious and more methodical considerations than those of Terzic and Dedijer.

The 1986 Memorandum of the SANU argues that the Serbian nation was "in less than fifty years, in two consecutive generations, twice exposed to physical annihilation, forced assimilation, religious conversion, cultural genocide—ideological indoctrination, the devaluation and abdication of their traditions under an imposed guilt complex, intellectually and politically incapacitated." Except for the NDH period, "Serbs in Croatia have never in the past been as threatened as they are today." Because of this, "the dominating

ideological and political symmetry in historical guilt cannot be accepted."[87]

Such views are possible only under the following premises: Either a total ignorance of those historical facts, the unwillingness to perceive them, or the insistence that all forms of Croatianness can be vanquished.

When a distorted picture of historical occurrences acquires a pseudo-scientific aura, and is even expressed in the inspired wrath of literature, an attempt should be made to clear up the historical accounts. Every nation "recognizes the outermost limits to its humiliation and tolerance!" as Vuk Draskovic says of the Serbian nation in his *Petition* to the highest government agencies in the SR Croatia. The petition contains the message: "On a bitter wound—a bitter poultice. Open afflictions require open dialogue."[88]

I endeavored to enter into open discussions, presenting certain facts to counter perilous fallacies, even before a judicial tribunal. But instead of dialogue, the alleged genocidal monster has grown more menacing.

"Any crimes committed by anyone in these spaces of ours cannot and must not be compared with the extent of the horrors numbering 1.5 million slain Serbs, at least, in Pavelic's Croatia."

Thus Vuk Draskovic openly unveiled my own magnum crimen, magnifying it with his own figures. To those unfamiliar with history, Draskovic's thinking may even seem to have some basis, but of course, only if viewed from his side. "The distance between Ante Pavelic and General Draza Mihailovic, Andrija Artukovic and Professor Slobodan Jovanovic, Alojzije Stepinac and Bishop Nikolaj Velimirovic, Mile Budak and Jovan Ducic, is no smaller than the abyss between cause and effect, crime and defense from the crime, genocide and the infrequent retaliatory cases of counterattacks, which the Serbian people have so many times since the war repented and condemned."

As far as cause and effect are concerned—from the Vuk Karadzic of the previous century to the present-day Vuk Draskovic—one should take certain things into consideration. To the provocatory threats addressed to Croatians as a nation let us

add: about 30 thousand Croatian peasants imprisoned and beaten after the 1918 unification because of their dissatisfaction with the exchange of currency and the imposition of the Vidovdan-constitution centralism; the murder of the Croatian leadership (Radic, et al.) in the midst of the Belgrade Parliament; thousands persecuted and imprisoned during the January 6th 1929 dictatorship, when there wasn't a family which because of its Croatiandom did not in some way experience the terror and tumult; the murder of the Croatian Professor of history Milan Sufflay on a Zagreb street and a similar attempt on the life of the writer Mile Budak; the victims of Sibinj, Senj, and others. One shouldn't forget that Croatians went off in desperation into economic and political emigration, in which Pavelic formed the core of his Ustasha movement. The people saw no other exit than the painful departure from the homeland or acceptance of a battle to extermination. Thus, the Croatians—and not only the Serbs—had reason indeed for "retaliatory counterattacks." In scrutinizing all these circumstances, it is necessary to consider the victims of the Serbian nation as well as the Croatian nation, that is, to consider in its entirety the fate which history has dealt out to them, for which they also hold a share of the responsibility.

If Vuk Draskovic had even a partial understanding of these ideas, he probably would not have been able to blame everything on the "cosmic crimes" of the Ustashas, who "unintentionally inflicted shame and insult upon their Croatian nation, as well as upon the Moslem nation," which "cannot be absolved," so that "the genocide will be hushed up" and "the same degree of shame will be symmetrically the same crimes committed by the supposed Serbian degenerates." These, according to him, are the "fabrications ... to secure an alibi for the genocidal wedding feast of the Ustashas."[89]

From its very inception, the Chetnik movement took decisive steps toward the realization of the Great Serbian program of a homogeneous Serbian state, which was first worked out by Dr. Stevan Moljevic in his Memorandum in June of 1941, in Belgrade, and then supplemented by Dragisa Vasic, both of whom were to

become chief advisors to Draza Mihailovic. In order for this program to materialize, at the end of 1941 and the beginning of 1942, Mihailovic gave his subordinates specific directives as to the goals of their activities:

"To cleanse the state's territories of all national minorities and foreign elements" and "To create a direct mutual border between Serbia and Montenegro, as well as between Serbia and Slovenia, by means of cleansing the Sandzak region of Moslem inhabitants, and Bosnia of Moslem and Croatian inhabitants."[90]

Not only was this an explicitly genocidal program, but it proved to be more radical than Moljevic's memorandum, which had anticipated the resettlement of populations, and more radical even than the program of Pavelic's government, which had in its plan to liquidate one third of the Serbian populace, to resettle another third, and to convert the rest to Catholicism or into a Croatian Orthodox Church. This rash political course was directed to Pavelic by Hitler himself, who wanted an orderly NDH in his New European Order, so he proposed that policies be merciless in its first years.[91]

As we have seen, the estimates of actual war victims show that losses among both the Croatian and Serbian nations as a whole are similar, with the Serbs in the Independent State of Croatia suffering somewhat greater losses, and the Montenegrins relatively even more, but not from the Ustasha genocide. Along with the Chetnik genocide against the Croatian Catholic and Moslem populations within the NDH, it is important to remember the Serbian victims who suffered at the hands of the Chetnik ranks because they sided with the NOP. For example, in the Knin Krajina[92] and in Bosnia, the Chetniks participated in the German-Ustasha attack on Mt. Kozara,[93] and committed terrorist crimes within Serbia itself. Inside the territory of the NDH, the Chetniks, particularly in collaboration with the Italians, conducted their own "cleansing" of populations unfriendly to them, which included all Croatians regardless of orientation, and those Serbs who were collaborating with the NDH or the Partisan movement. An example of the kind of role they played can be seen in their proclamation

"To the People of Bosnia, Lika, and Dalmatia" on February 10, 1943 by the commanders of the Chetnik detachments: "Having cleansed Serbia, Montenegro and Hercegovina, we have come to your assistance to destroy the pitiful remains of the international band of Communist criminals... in this historic onslaught, fifteen thousand Chetniks armed to the teeth are attacking from all sides and are destroying anyone who shows any resistance." At the same time, in his directives to other Chetnik leaders Draza Mihailovic said: "This is an opportune moment for settling accounts with the Communists once and for all." And the Chetnik commanders again issued directives to kill the political commissars in the Partisan units, and that "all the Communists" in Hercegovina, as well as the Partisan prisoners be killed.[94]

The political and military strengthening of the NOP, in connection with the ever more certain outcome of World War II, had manifold consequences. After the initial collaboration between the Partisan and Chetnik movements, relations between them became increasingly strained. They were in conflict predominantly because of their fundamentally different revolutionary aims. Contrary to this, the Chetnik movement was expanding its collaboration with the Italians which had already been established early in 1941, as well as with the Germans, and even with the authorities of the NDH. The NDH itself, increasingly imperilled by the Partisan and Chetnik movements, not having succeeded in suppressing by force the resistance and the Partisan movement, was forced to relax its violence and terror and, in general, to seek political solutions to issues such as the Serbian question. The more moderate leaders of the Ustasha movement advocated undertaking political measures, and even the leadership of the Third Reich, who wanted peace in the NDH, made such a request. The Croatian Peasant Party and the Catholic Church, which were quite influential among the people, totally opposed violence and terror. This helped relax the politics of Pavelic's government. From the beginning of 1942, after a visit by a commission of the International Red Cross, the conditions in the Jasenovac camp improved. Political concessions were given to the Serbs with the

recognition of the status of "Orthodox [Christian] citizens" of the NDH; the Croatian Orthodox Church was established; Serbs were recruited into military service, and were even called upon to enter into the Ustasha movement. The government of the NDH began to give protection and aid to Serbian villages in defense against the Partisans, while the Chetniks recognized by oath the sovereignty of the NDH.

From the very beginning of the creation of the Croatian Home Guard as the armed force of NDH, there were Orthodox officers who declared themselves to be Croats in it, and some in very high positions. From 1942, special Home Guard work regiments (the so-called dora regiments) were set up, into which mainly Serbs were recruited. When the Partisans in northern Croatia captured a dora regiment, not even 10 percent of the Serbian Home Guards chose to remain with the Partisans, but rather returned to their homes, not fearing that they would be held accountable for their capture, as indeed they were not (nor were Croatians from other units).

In January 1942, Pavelic summoned the Croatian State Assembly, which usually began sessions in February, and in which there were representatives of the Serbs. Together with the restoration of the work of the Orthodox Church (though under a Croatian name) in April, 1942, this was a clear sign of softening of the NDH positions. The Ustasha regime in this way was giving significance of a recognition of the full rights of Orthodoxy in the NDH, stressing that neither in doctrine nor in actuality did it have anything against Orthodoxy as a faith, but merely did not wish it to become an instrument of foreign, Serbian policy in the Croatian state. The Constitution of the Croatian Orthodox Church (April, 1942) was, in the religious sense, equal in content to the Constitution of the Serbian Orthodox Church (1931), except for the difference in the name of the state, its leader, and national emblems. M. Obrknezevic, lawyer of the Serbian Orthodox Church in Srijemski Karlovci and captain in the former Yugoslav royal army during World War I, led negotiations concerning the establishment of the Croatian Orthodox Church and was the

author of its Constitution and then the secretary to the metropolitan of the Croatian Orthodox Church, Germogen. He gave the following view: "The Croatian Orthodox Church was for the Orthodox Serbs in Croatia a protector and a temporary religious solution in an authoritarian regime which did not tolerate the Serbian Orthodox Church as the national church of the neighboring state of Serbia. To Orthodox Croatians, it was a long-desired and repeatedly defeated solution to their predicament of being torn between feelings of Croatian national affiliation and loyalty to the Croatian fatherland on the one hand, and Serbian national exclusionism with a single Orthodox Church in Croatia, the Serbian Orthodox Church."

Among other reform measures, E. Kvaternik, the person most responsible for the policy of violence, was removed from office, as were some notorious Ustashas.[95] As the war came closer to an end, and with it, the NDH, its entire organization was less inclined to resort to terror. But toward the very end there was a repetition of mass killings out of a desperate retaliation by those who were losing the war.

Without taking into consideration all these circumstances, one cannot make accurate judgments about the genocidal actions taken on one side by the Ustasha movement and on the other by the Chetniks. Those actions can be considered purely criminal, both legally and morally, only when viewed in terms of their execution. In considering them from a historical perspective, it is necessary to uncover all the causes which gave them an impetus, and all the objectives in whose name they were conducted.

4. On the Historical Guilt of the Catholic Church

The fall of clericalism... means the fall of Croatianness.

—Nikola Stojanovic

O Jerusalem, wash thine heart from wickedness that thou mayest be saved.

—Jeremiah 4:14

At the very beginning of this century, Dr. Nikola Stojanovic revealed that from the standpoint of exclusivist Serbdom, a complete distrust of Croatianness and the Catholic Church arises. At that time Stojanovic was on the threshold of his legal and political career.[96] He was later to change his views, in part influenced by his participation as the representative for Serbs in Trumbic's Yugoslav Committee. However, the views he held in his earlier years are now circulated by many people. These views are not only appearing in increasing numbers of newspapers and journals, but also in ever more numerous, more exhaustive, and murkier books with scholarly and literary ambitions.

From Stojanovic's theses one learns that the historical guilt of the Croatians lies in the fact that they "rejected" the Slavic and "accepted the Latin liturgy." This is taken as evidence of their lack of national consciousness, even though the general acceptance of the Latin language was beneficial to the unification of the lands of the Kingdom of Croatia. Apart from that, worship in the local vernacular in Croatia had long been the custom, and was recognized by Rome. This criticism of Croatianness implies its supposed guilt for its religious separation from Serbdom, even though the causes of the schism into the Eastern and Western Churches are much deeper and more historical and geopolitical than religious in nature, transcending Croatian-Serbian relations. Besides that, there wasn't any unity, either in the national or religious sense, between the Serbs and Croatians even in the 10th century, or later, until the unification in 1918.

Stojanovic emphasizes Orthodoxy as the faith of the Slavs,

ascribing to Serb Orthodoxy the characteristics of "democratic nationalism," and to Croatian Catholicism the characteristics of "reactionary clericalism" and "ultramontanist cosmopolitanism." According to him, these traits predetermined the characteristics of the struggle between the Serbs and Croatians. Stojanovic and his followers are not interested in the positive contributions of Catholicism to the history of the Croatian nation, both as a foundation to the entire culture and in the defense of Croatian national survival against foreign encroachment. Instead, he is bothered by the fact that Croatian Catholicism created a great barrier to the penetration of Serbdom. Catholicism indeed played a chief role in this respect, but the identification of Croatianness with Catholicism was never as prevalent as was the case of Orthodoxy and Serbdom. Despite all its Catholicism, Croatianess had the spiritual broadmindedness to recognize the Moslem faith. And Croatianness did not have a negative relationship either toward Orthodoxy or toward those of Jewish faith. This was evident in the fact that there were eminent members of those faiths in the major Croatian parties.

Stojanovic's equating Croatiandom with backward clericalism is not only exaggerated, but also essentially incorrect. A backward clerical current never played a decisive role in Croatian political life. Croatia is, after all, one of very few European Catholic lands in which a strong Catholic party never developed. Thus, the Croatian Populist Catholic Party remained of secondary importance, and as such, quickly died out. A significant number of clergymen participated in Starcevic's Party of Rights, at a time when Croatian national consciousness was taking shape in the last century. This is evidence of a progressive national democratic orientation among the clergy and that party's national liberal-democratic program.

When Stojanovic announced the need for the destruction of clericalism as a prerequisite to the fall of Croatianness, he was expressing the conviction that the Orthodox Great Serbian ideology would be capable of dominating Croatianness if it could succeed at crushing Catholicism. This delusion continues to the present day.

In most recent times, V. Dedijer stands out as one of the main propagandists against the Roman Catholic Church. To Stojanovic's thesis on the interdependence of "clericalism" and Croatianness he adds new arguments concerning the culpability of the Catholic Church and the Vatican for the fall of the Yugoslav monarchy, for the genocide perpetrated by the Ustasha authorities, and the threat it poses not only to Serbdom, but to all Orthodoxy in the Balkans. In attacks upon the Catholic Church, Zagreb's archbishop Alojzije Stepinac is presented as the personification of the Church's guilt, in the capacity of a "Croatian metropolitan." The accusations against Stepinac have even less of a foundation than does the Jasenovac distortion.

In his *New Contributions* (1981), Dedijer mentions that there are conflicting views concerning the role of Stepinac "both as to his participation in direct genocide—the slaughter of the Serbian populace in Croatia—as well as concerning the degree of his responsibility for the forced re-christening of the Serbian Orthodox" populace in Croatia into the Roman Catholic faith, as well as inhabitants of the "Old Catholic" faith, and also "for the persecution of Protestants." Equating Stepinac with Draza Mihailovic, as representatives of "the extremist wings of Croatian and Serbian chauvinism," respectively, he calls Stepinac: "spiritual and political mentor to Ante Pavelic." He also invokes "Stepinac's responsibility for the killing in the camp of Jasenovac," and "the role of Stepinac in the extermination of the Jews."

For evidence of these charges, Dedijer quotes passages from Stepinac's own diary. He presents one entry from January, 1940: "thus, in the interest of Roman Catholicism we must do everything to insure that the Croatian nation remain healthy and culturally strong." It "would be ideal if the Serbs would return to the faith of their fathers," i.e., under the auspices of the Pope—for "then we, too, would finally be able to heave a sigh of relief in this part of Europe, for Byzantism has played a terrible role in the history of this part of the world in connection with the Turks."

According to Dedijer, Stepinac expressed this kind of "intention

to convert Serbs" several times in his diaries. However, Dedijer does not cite any other examples, and he doesn't think of the historical reasons which might have moved Stepinac to his statement. In arguing that Stepinac's views were little different from Hitler's regarding "the need to destroy Yugoslavia," Dedijer cites what Stepinac wrote on the occasion of the Belgrade military coup of March 27, 1941: "All in all, Croatians and Serbs are two different worlds, north and south poles, which will never, short of a divine miracle, move closer to each other. The schism (i.e., Orthodoxy) is the greatest curse upon Europe, almost greater than that of Protestantism. It has no morals, no principles, no truth, no justice, no honesty."[97]

These words of the highest Croatian prelate certainly deserve particular attention, but they cannot be judged outside of the historical context of the Croatian-Serbian disparity. This conflict manifested itself so dramatically in 1902, and again in 1914, when the assassination of Franz Ferdinand in Sarajevo had an anti-Croatian appearance because he was a supporter of the triune solution to the Croatian question. Similarly it cropped up in 1928, 1934, and finally with the elimination of the Yugoslav concordat with the Vatican and the Croatian-Serbian agreement on the very eve of the war. After the state political crisis, which emerged after the assassination of Radic in the Parliament and then of King Alexander in Marseilles, conditions had begun to improve. Then the concordat which Stojadinovic had arranged with the Vatican was rejected, followed by the overthrow of the Cvetkovic-Macek government. These steps were taken as proof that the Great Serbian hegemonists did not accept any sort of compromise with the Croatians. This Croatian view was present, in essence, even after Macek's concession to call for loyalty to the new authorities of the NDH as proclaimed by Kvaternik on April 10, 1941. Macek himself had refused to do so when urged by the Germans because he was convinced that they would lose the war. Dedijer takes none of this into account.

Dedijer bases his claim that Stepinac was a "spiritual father" and even "political mentor" to Ante Pavelic upon the fact that he

discovered through the diary that Stepinac was "far more intelligent" than Pavelic, and that in pre-war Yugoslavia, he developed a wide range of operations in the defense of "the interests of the Croatian right-wing middle class."

His judgement that "the greatest responsibility for the conversion of Serbs to Catholicism falls upon Archbishop Stepinac," who used Ante Pavelic as a mere "instrument" is made without any support. It is a completely incorrect and arbitrary claim on Dedijer's part when he says: "It is significant that Archbishop Stepinac never condemned the genocide against the Serbs, Jews, Gypsies, and Croatians in Jasenovac, nor did he condemn individual priests who carried out their duties there as functionaries and murderers." It is equally untrue that "only once" did Stepinac speak in the Cathedral "against the evil of revenge," and that it was at the beginning of 1943, when Stepinac's brother was among those executed by the Ustashas.[98]

Dedijer makes such allegations despite the fact that there are several published books all over the world containing documentation which testifies to the contrary.[99]

Rejecting the false accusations against Stepinac, Cardinal Kuharic cited 33 documents which tell of the position of the Catholic Church and Stepinac toward the persecution of Orthodox Christians and "non-Aryans," which were published in the Vatican archives six years prior to Dedijer's book.[100]

In the first circular letter which Stepinac issued to his clergy after the declaration of the NDH, he was quite reserved toward the newly established regime. He said that circumstances had led the Croatian people "to an encounter with the long dreamed of and coveted ideal," but he reminds them that he speaks "not only as a son of the Croatian nation, but even more as the representative of the Church," which must be "the pillar and ground of the truth" (1 Tim 3:15). Stepinac was already in disagreement with the Ustasha regime when he reminds the clergy that "kingdoms pass from one nation to another because of iniquity and injustice, scandals and deceptions," and therefore "we must consider it our greatest duty" to "spiritualize our entire national being," to cau-

tion and teach everyone "that divine rapture and noble elation in the building of the foundations of the young Croatian State should be inspired with the fear of God and love of God's laws and His commandments." Stepinac calls upon his clergy not to cease urging their congregations to pray to the Father "that He fill the chief of state with the spirit of wisdom to execute his duties in an exalted and responsible way to the veneration of God and salvation of his people in justice and in truth."[101]

Stepinac protested to Pavelic's government against the execution of Serbian hostages, which was done in retaliation in the first days of the NDH. And upon receiving the news that Orthodox Christians had been killed in Glina on May 13, 1941, he sent Pavelic the following letter:

"I just received word that the Ustashas have executed 260 Serbs in Glina without trial or investigation. I know that the Serbs have committed grievous crimes in our homeland in the last twenty years of their rule. However, I feel that it is my duty as Archbishop to raise my voice in protest and to say that this is not allowable according to Catholic morality, and I beseech you to undertake most urgent measures throughout the entire territory of the NDH that not one more Serb be killed unless it is proven that he is guilty of a crime for which he deserves to die. Otherwise we cannot count on blessing of the heavens, without which we will surely meet our ruin. I hope that you will not take offense at my open word."[102]

Naturally, Stepinac's interfering in the affairs of the "Ustasha revolution" was criticized. Pavelic himself, within the circle of his close associates, was often heard to say, "How can a bishop defend Serbs and Jews?" Some Ustasha "bosses" even considered the idea of removing Stepinac by force. Siegfried Kasche, the German envoy to the government of the NDH, also felt this way because of Stepinac's aid in the saving the Jewish population.

In July of 1941, Stepinac protested to Pavelic of the inhuman treatment of the Jews, reminding him of the Christian principle of "seeing a man in every man." Shortly thereafter, when the Zagreb canon, Dr. Pavao Loncar, was sentenced to death for offending

Pavelic and for anti-Ustasha propaganda (his sentence was reduced through the intervention of a Vatican representative), it was not an isolated instance of an anti-Fascist sentiment in the top ranks of the Church.

After two attempts at intervening directly with Pavelic, Stepinac asked Artukovic in 1942, as minister of justice and religion, that he arrange for a priest to have access to the camps of Jasenovac, Gradiska, and Lobor. "Is it any wonder that many decent people are asking what difference there is between a Bolshevik camp and one of our camps?"Stepinac wrote, "And can the Ustasha movement count on the blessing of the Heaven when it denies dying people that which no civilized state has ever refused?" In September, 1942, at the request of the catechist of the town of Pakrac, Tomo Sruk, Stepinac successfully negotiated with Pavelic for the release of about 200 Orthodox Christians from Stara Gradiska, who had been taken as hostages for the Ustasha functionaries, Dr. Karlo Wagner and Mirko Vutuc. The Archbishop continually intervened on behalf of the Jewish population, and assisted the Jewish community of Zagreb. Stepinac sought from the Ministry of Public Order and Security in April, 1942 that it be arranged for Jewish children to go to Turkey via Hungary. Dr. Weltman, an emissary from the Commission for Aid the European Jews in Istanbul, wrote in June, 1943, to the apostolic delegate, Monsignor Roncalli (later Pope John XXIII): "We know that Monsignor Dr. Stepinac has done all he can to ease and alleviate the dreadful fate of the Jews in Croatia." Weltman asked Monsignor Roncalli to convey to Stepinac "our deepest thanks" and the entreaty that "he carry on his distinguished work of saving our unfortunate brothers, sisters and children."

In answer to Pavelic's excuse that he was compelled to carry out anti-Jewish laws, Stepinac wrote in his diary: "Chief, pound your fist upon the table and tell the Germans and Italians that this cannot go on." He also wrote, in a letter to Pavelic on March 6, 1943:

"I implore you in the name of humanity, which our people have always valued so highly, that you not permit any of the remaining citizens of our state to suffer unjustly. In the collection camps

there are many who are innocent or who do not deserve so severe a punishment.

"If the meddling of some foreign government in our internal national and political affairs is at issue, then I would not fear if this message and protest should reach that foreign government. The Catholic Church recognizes no fear of any earthly power when it comes to the defense of the most basic human rights...

"Do not permit irresponsible and uninvited elements to sin against the true good of our nation. Violation of the national law in the name of the nation and state strikes back at that very nation and state: a bitterness is growing in the country which strives for revenge."

The imprisonment of seven Slovenian Catholic priests and their execution in Jasenovac served as an occasion for Stepinac to raise an embittered voice against the crimes. In a letter to Pavelic dated February 24, 1943 Stepinac says, "This is a shameful stain and a crime which cries out for revenge, just as the whole camp of Jasenovac is a shameful stain upon the NDH. As a priest and bishop I say together with Christ on the cross: Father forgive them, for they know not what they do! But the whole public, and especially the relatives of the murdered priests, is asking for compensation and atonement, that these murderers, who are Croatia's greatest misfortune, be placed on trial. Be assured that it is not hatred, but the love of truth and of the Croatian nation which compels me to write this letter."[103]

In his public appearances Stepinac expressed his disagreement with the violent methods of the Ustasha rule. He was to speak out on behalf of a Croatian state, but at the same time express with increasing decisiveness the Church's disagreement with the violent and racist politics of the NDH government.

In the autumn of 1941, from the pulpit of the Zagreb Cathedral, Stepinac spoke out against the spirit of hatred and vengeance which was being roused in the wartime confusion. Stepinac wished to warn them that "true subjects of Christ the King" mustn't forget that "love of one's neighbor, love of all men regardless of how they be called" is the highest Christian law. Thus it is

"impossible to unite love of God with hatred of man."[104]

In a public sermon of May 1942, Stepinac spoke against the misconception of those who "think peace can be achieved only with weapons." Instead peace is nothing other—as St. Augustine said—than constancy of order. In history, every great earthly potentate from King Balthazar to Napoleon who has had the audacity to trifle with the supreme laws has been punished. Stepinac said: "a good relationship with one's neighbor requires that we see in him not a beast, but a man, a child of God, as we ourselves are; a brother whom it is our duty to love... It would be frivolous to speak of some new order in the world, regardless of which side it is coming from, if in that order there is no respect for the human person, the immortal soul, which transcends all systems, which cannot be substituted by anything, which has its inalienable rights, which cannot and must not be restricted by any form of man's authorities. It would be wrong to think that perhaps the Catholic Church, in defense of the basic rights of the human person and freedom of consciousness would acknowledge a fear of any human force."[105]

In another public sermon, Stepinac repudiated racism.[106]

And in a similar speech on the Holy Father's Day, March 14, 1943, he warned that every violation of human rights "can amount to nothing but adverse consequences" and that it is "erroneous to think that the Catholic Church will ever condone measures which conflict with the basic human rights."[107]

In the beginning of 1943, Stepinac requested that the parish offices send him data on the crimes committed by the fascist authorities, and he himself took the documentation to the Vatican on May 26, 1943. There he told Ivan Mestrovic how the situation in Croatia was becoming ever more lamentable, which was due for the most part to clashes between the Chetniks and the Ustashas. He also said "that hatred has become the general rule on both sides," but that "it isn't true that all Serbs hate the Croatians." In Rome, Stepinac was cautioned by the General of the Jesuit order to be careful because the Nazis were planning to liquidate him.[108]

In his sermon on the occasion of a pilgrimage to Marija Bistrica in the summer of 1944, Stepinac spoke of all the atrocities of war that Croatia was experiencing. "My heart bleeds and weeps over all these horrors which have befallen our homeland and our weary Croatian nation." Mentioning the ruins and ashes of cities and villages, the thousands of innocent people killed, Stepinac asks: "While it is battering our land with such horrors, does the warring faction perhaps view as a crime the fact that for centuries the Croatian nation has been struggling for freedom with all the strength of its being... ? If so, then all other nations that maintain an unwavering aspiration for freedom and independence are also criminals." And "let no one doubt that Croatians will ever renounce their right."[109] Just as from 1941 on, he persistently protested against the reign of violence in the NDH, now, as the war was coming to a close, he raised his voice against the madness of the victorious side, affirming the right of the Croatian people to freedom and autonomy.

Just before the end of the war, Stepinac expressed his anxiety that the victors over fascism and racism relied on violence in the establishment of an international order. There is no doubt that in Stepinac's worldview "the minority nations are more secure in their freedom and independence in the order based on the doctrine of the Church and its highest representatives than in the order based on the principle of brute force."[110] Many times he repeated his conviction that such a peace brings no good when larger states "impose their will upon smaller and weaker ones by the power of their weapons," and "one social stratum seizes the government into its own hands by fire and sword," leaving no rights to the other classes, even though the latter are in the majority. Such a peace actually means "a latent war which is being prolonged in people's hearts and souls, until, one day, it acquires its external characteristics as well, in the form of killing and destruction, which we are witnessing today, and in even worse forms." On that occasion, he made it clear to the representatives of the new order that he would "say what he had to say" without seeking permission from anyone, just as he "did not have to solicit the Croatian government."[111]

This survey of Stepinac's views and activities leaves no doubt that Dedijer's claims are completely contrary to the truth. Stepinac was not a "spiritual and political mentor to Ante Pavelic," but was decisively opposed to the ideas and methods of Pavelic's rule. Nor can one ascribe to him "responsibility for the killings at Jasenovac" or that he gave "the initiative for war crimes." The truth is that he did all that was within his power to prevent them. The facts incontestably show that the claim that Stepinac did not publicly express his disagreement with the regime of violence in the NDH until 1943 when his brother was killed, is not true. In fact, he protested since the first violent acts in 1941 until the end of the war. There was probably no high-ranking ecclesiastic in the lands of Hitler's New European Order throughout the entire war, who opposed Nazi-fascist and racist ideology and practice so resolutely, consistently, and systematically.

Then why do Dedijer and similarly-minded individuals classify Stepinac among the spiritual originators of Ustasha crimes? As the highest Croatian Catholic dignitary, Stepinac stood unwaveringly in defense of the right of Croatian national freedom and independence. Contrary to Pavelic, who equated the Independent State of Croatia with Ustashism, Stepinac's stance was determined by the historical understanding that ideologies change, and party regimes come and go, while nations remain with their need and their right to an independent existence.

A further reason for the negative appraisal of Stepinac's role in the war may be, of course, of an ideological nature. In keeping with the duties that come with the high office which he held in the Church hierarchy, and even more, with his conviction in the Christian worldview, Stepinac was a resolute and uncompromising opponent to every totalitarian ideology and doctrine, whether it was that of Fascism and Nazism, or Marxist Communism. Nazi-fascism was alien and unacceptable to him, particularly because of its racism and its earthly "gods," and so was Communism, especially because of its atheism and its Stalinist forms.

It is hard to suppose that the facts presented above are unfa-

miliar to Dedijer. Indeed, in the same book (*New Contributions*), he even relates that in a postwar encounter, Marshal Tito personally suggested to Archbishop Stepinac "that it would be best for the Catholic Church in Yugoslavia that it separate from the Vatican, to be an independent, national church," but that Stepinac didn't wish to discuss the matter at all.[112] From this it follows, and it was explicitly confirmed by J. Blazevic, that the legal persecution of Stepinac did not come about because of his alleged war crimes, but because of his refusal to abandon his principles and his disagreement with the new socialist order.

Despite all this, Dedijer persists with his monstrous judgement that "the supreme representative of the Roman Catholic Church in Yugoslavia, Archbishop Alojzije Stepinac, crucifix in hand, waged a Crusade against those who were of a different religion and nationality."[113] In 1984 he sent a letter to Pope John Paul II, in which he criticized him for not having come to visit Yugoslavia because he did not wish to see Jasenovac. Dedijer asks to what extent was the Vatican itself involved "in the massacres of the Fascist state, in the forced re-christening of 200,000 Orthodox Serbs"?[114] Two years later Dedijer again sent a letter to the Pope, claiming that the trial of Andrija Artukovic "revealed the relationship between Pavelic's government and the Catholic Church." Dedijer wrote that despite the fact that what was "revealed" was exactly the opposite of what he claims. Artukovic himself testified that relations between Pavelic and Stepinac "were obviously unfriendly."

In section VII of his Foreword, Dedijer acquaints us with the fact that "there are many examples in history of forced conversion by all Christian churches," and that religious conversion "was one of the ways in which conquered nations were subdued." Here we find that Vuk Karadzic noted that after the liberation of Belgrade "Turkish women and children were converted to Christianity by force." Likewise, in Gusinje and Plav in Montenegro in 1912, at the end of the First Balkan War, "the Montenegrin king Nikola and some Orthodox priests forcibly converted the Moslem populace. Those who did not want to be christened were executed."[115] It

should be mentioned, however, that this phenomenon is presented in only three examples, as if it did not reach its farthest dimensions since the last century, in both Montenegro and Serbia, and that there is absent the label of genocide, the kind of which we have had and will have occasion to read.

In the next section, Dedijer reminds us that the Roman Catholic Church, since its very inception, "has organized and preached a particular hatred" toward Jews, and that in the 19th century, Christian Churches were the "chief instruments" in Jewish persecution. Then he makes two judgements of "historical" significance. First, "Adolf Hitler developed his anti-Semitism and racism (particularly against Serbs and Slovenians) in his native Austria under the influence of Dr. Liger (Lueger), the leader of the Roman Catholic political party in Austria, and Archduke Francis Ferdinand, the most Catholic of European rulers in the 20th century. Both of them preached the physical extermination of Jews and Serbs." Second, "One mustn't forget that the Vatican on July 20, 1933, enacted a Concordat which made it the first state to bestow its blessing upon Hitler's regime. We will later see that Hitler and the Vatican united in genocide against the Serbs, Slovenians and Jewish population in Yugoslavia."

While it is true that Dr. Karl Lueger, a dominant Austrian Christian socialist and mayor of Vienna, was an anti-Semite. But he was certainly not an originator of anti-Semitic and racist theory, the roots of which (as we have seen in previous chapters) were much deeper and were found both all across Europe and in Germany. Furthermore, Lueger's anti-Semitism surfaced around 1895 when Hitler was still a small child. In 1909, when Hitler arrived in Vienna, Lueger had little to do with political propaganda. Moreover, as mayor, Lueger collaborated successfully with Jewish colleagues in public affairs, explaining openly that anti-Semitism can be beneficial only as "a means of political agitation by which a path is cleared, but when one reaches the top, such sentiments are unnecessary. That is sports for the lower types."[116]

Dedijer refers to Franz Ferdinand as the most Catholic of rulers in Europe (forgetting that in another part of the text he called him

an anti-Catholic Freemason), although he wasn't a ruler at all, nor, as a mere heir to the throne, had he been given any authority to govern. How then did Hitler develop his racism, and, moreover, "particularly against Serbs and Slovenians" through the influence of Lueger and Franz Ferdinand, and where and when did those two "preach the extermination" not only of Jews, but also of Serbs? Dedijer doesn't even attempt to document his claim.

The other above-mentioned "historical" judgment is an even clearer example of his method of thoughtlessly piling even contradictory pieces of historical material to formulate total misconceptions about a historical situation. From the isolated fact that the Vatican enacted a Concordat with the government of Hitler's Germany on July 20, 1933, which suggests that it was prepared before Hitler came to power, he draws a conclusion about the hasty desire to "bestow its blessing upon Hitler's regime," and about the later collaboration between Hitler and the Vatican in the genocide against Serbs, Slovenians, and Jews. This goes beyond historical responsibility and coherence. Setting aside the various circumstances and purposes to signing such agreements between the Vatican and many states in the world throughout history (the first Concordat was signed in the 12th century), one is compelled to ask Dedijer why the charge that the Vatican "blessed" the regimes with which it signed such agreements should not be applied also to the one with the Principality of Montenegro in 1886, or with the Kingdom of Serbia on the eve of World War I on June 24, 1914? Didn't the Vatican sign a Concordat after lengthy negotiations from 1920 with the government of the Kingdom of Yugoslavia on July 25, 1935, though Stojadinovic had to refrain from ratifying it since a united Serbian right and left wing were against it? Furthermore, wasn't the Protocol signed by the Vatican and the SFRJ likewise the blessing of the socialist regime in Yugoslavia?

The fact that in its world politics, the Vatican believed that Fascism and National Socialism were lesser evils than Marxism and Bolshevik Communism does not offer any grounds for identifying the Vatican with Nazi-fascism, and even less with its politics of plundering and genocide. This was not the case even in its rela-

tions toward Mussolini's Italy, let alone toward Hitler's Germany. Nazi ideology had an obvious anti-Catholic tendency, holding the Protestant Church to be the Germanic national church. At the time when Mussolini led Italy down the path of military conquest, Pope Pius XI sent him an angry warning that God's power scatters the nations that desire war. Likewise, relations between the Vatican and Germany became increasingly strained as Hitler, while carrying out his racist politics, also more openly encroached upon the authority of the Catholic Church. Because of this, but also in order to instruct the Catholics, a large number of whom were also blinded by Hitler's program, Pope Pius XI presented an encyclical in 1937, in which he clearly laid out the stance of the Catholic Church against the racist politics of Nazism. The Pope unambiguously declared: "The swastika is not the cross of Christ." His successor Pope Pius XII was unwilling to accept proposals that the war of Hitler's Germany and his New Europe against the USSR be proclaimed a crusade for the defense of the Christian world. It was no secret that the Vatican was officially unsympathetic toward Nazism and Hitler, who at the peak of his power threatened to "destroy the Church like a toad."

Such are the facts regarding any similarity of interest between the Vatican and Hitler. In order to achieve a general framework for his ideas, Dedijer simply twists and distorts the historical reality, and figures that no one will care if the facts are not in agreement with him.

Dedijer's final concluding remark is as follows: "Dignitaries of the Roman Catholic Church and the Ustasha state organized the mass re-christening of the Serbian Orthodox populace into Catholicism. Ustashas were killing off Serbs, while Catholic priests were forcing the Serbian populace to change its faith. Hundreds of Serbian Orthodox churches were pillaged and destroyed, three of the highest church dignitaries and 200 priests were murdered in cold blood, while the remainder of the clergy were persecuted. Several Catholic priests commanded the Jasenovac concentration camp where several hundred thousands of Serbs were killed. A special papal emissary, Marcone, was in

Croatia at that time and he did not attempt to curtail the crimes."[117]

Dedijer's claim that "several" Catholic priests were commanders at the Jasenovac camp is pure nonsense. There was not even one Catholic priest in such a position, let alone several. To cite as evidence that "Brother Miroslav Majstorovic," an Ustasha major was one of them, or "priest Ivica Brkljacic," another Ustasha major, would be like calling il Duce "the socialist and Marxist Mussolini," according to what he was before he became a fascist leader. Tomislav Filipovic was discharged from the Franciscan order by a canonic trial on April 28, 1942, and in June 1942 was suspended from the priesthood by a decision of the papal envoy due to the fact that as a minister he followed an Ustasha battalion which liquidated a Chetnik village. Because of this, he was placed under judicial investigation. But after the investigation, under his assumed name (Miroslav Filipovic Majstorovic) he was accepted into the Ustashi troops and sent to Jasenovac, where he then functioned in the capacity of an Ustasha officer, not a priest. Ivica Brkljacic is mentioned in the documents on Jasenovac as a "priest" or "former clergyman," even though he never was such, for he withdrew from his theological studies before becoming an Ustasha officer.[118]

After the genocide carried out by means of "mass re-christening" and killings, it would seem that in the Independent State of Croatia, not one Orthodox priest was left. According to Dedijer 203 were "murdered in cold blood, while the remainder of the clergy were persecuted." However, this claim is exaggerated to such a degree that it gives a completely distorted picture of what actually occurred. With such portrayals of "re-christening" and genocidal crimes, the intent is to show that the Catholic Church and the Ustasha movement were actually waging a religious war of Catholicism against Orthodoxy. But the Catholic Church did not instigate a pogrom against the Orthodox populace. And from 1941 on, Pavelic and the other Ustasha leaders publicly announced that they did not oppose Orthodoxy as a religion, but that they would not allow the existence of a Serbian Orthodox

Church which acts as a foreign "political organization" on
Croatian soil. From the very beginning, it was possible for the
people of the Orthodox faith to attain even the highest military
positions in the Independent State of Croatia. In the spring of
1942 the establishment of a Croatian Orthodox Church was per-
mitted, which in essence meant the legal recognition of Orthodoxy
as a religion.

On the subject of the 203 victimized Orthodox priests—during
that same time the Catholic Church also suffered no less of a loss.
On September 22, 1945 Archbishop Stepinac addressed a memo-
rial letter to Marshal Tito, in which he stated that in the course of
four years: 243 Catholic clergymen had been killed, 89 had dis-
appeared, and 169 had been incarcerated in camps and prisons,
among them a bishop, while the fate of another bishop was
unknown. In a pastoral letter from the Episcopal conference (held
on that same day), he added to that number 26 seminarists, monks,
and nuns, and he stated that "in just the Franciscan monastery at
Siroki Brijeg, 28 friars had been killed, of whom not one was ever
armed." In his final statement at his trial (on October 3, 1946),
Stepinac said, "Again I declare before all of you that between 260
and 270 members of the (Catholic) clergy were killed" during the
course of the war.[119] Such facts are of no interest to Dedijer

In February 1942 Rusinovic, a representative of the NDH,
reported to Cardinal Maglione, that the Chetniks in the NDH had
begun burning villages and killing people, and that "in these
attacks, many Catholic priests have paid with their lives." One of
these was "a parish priest whom the Chetniks first robbed, and
then set fire to him in the rectory."[120]

And what of the facts concerning the conversion of Orthodox
Christians to Catholicism? The negligence and irresponsibility of
that charge is evident in the fact that Dedijer and the others speak
of "christening" or "re-christening." Both of these are incorrect
terms; what took place was a religious transition. Whoever is once
baptized into one of the Christian churches, whether Catholic or
Orthodox, does not, in converting to another, need to be chris-
tened or re-christened.

In 1981, Dedijer wrote that "many documents also confirm the involvement of the Vatican in the conversion of over 200,000 Orthodox inhabitants into the Catholic faith," which was conducted amid the persecution of Serbs.[121] There are no documents to support either the number he claimed or the actual participation of the Vatican.

Dedijer quotes a memorandum, which, according to the judicial indictment, Stepinac had allegedly given to Pope Pius XII on May 18, 1943. The memorandum says that with the termination of the NDH "not only would the approximately 240,000 converts from Serbian Orthodoxy be lost, but the entire Catholic populace as well." Dedijer then states: "In this way we discovered that by May of 1943, the number of re-christened Serbs in the NDH numbered about 240,00."

He says this means that "the number of re-christened Serbs in the year 1942-43 had increased by an additional 140,000." Here he makes reference to a statement which Archbishop Stepinac gave Rusinovic, but he cites only a portion of it, so as to make it appear as if Stepinac had been portending further conversions, while he had actually said that there could have been "up to 400,000 conversions, but there were only 100,000" for canonical reasons. However, somewhere along the way, Dedijer forgets two facts: first, that after the establishment of the Croatian Orthodox Church in the spring of 1942, there were no more conversions; second, that Stepinac's defense attorney, Dr. Ivo Politeo, presented "incontestable evidence against the authenticity of the copy" of this allegedly plausible memorandum from Stepinac to the Pope, which makes it clear that it is an "Ustasha falsification or at least an Ustasha product." Dedijer must have been familiar with the established truth concerning the document which he cites, as he uses Dr. Politeo's defense in the Appendices of his book.[122]

However, if Dedijer assigned even an ounce of importance to such truths, how then would he build his historical towers of genocide, which are so one-sidedly biased that Vuk Draskovic and others find in them a pretext for their monstrous attacks against Archbishop Stepinac as "the spiritual father of Ustasha geno-

cide,"[123] and for their revulsion over masses held in his honor.[124]

Clearly, it is indisputable that the transition of a considerable number of Orthodox Christian believers to Catholicism (and the figure of 100,000 seems to be the most accurate) came about under anomalous circumstances. Since that very act meant the possibility of extrication from the threat of genocidal annihilation or deportation to Serbia, there are actually no justifiable grounds for labeling mere transition as genocide under the given circumstances. It is particularly true since transition did not come about as a result of orders from the Ustasha authorities, but by the request of the Orthodox adherents themselves, precisely in order to avoid the actual genocidal crimes that the Ustasha authorities were committing against the Serbs.

It is significant that very few have attempted to contemplate the problem of this religious transition from an objective and reasonable viewpoint, except for M. Obrknezevic. According to him, the very declaration of Pavelic's NDH induced among Orthodox populace, on one hand a need for acquiescence and loyalty to the Croatian state, but on the other hand a psychological condition of uncertainty and fear. In such a condition, some fled (even before any persecution) to Serbia or to the Italian occupation zone, while others attempted to escape possible persecution by undertaking measures on their own toward transition to Catholicism, and a few to Protestantism. There were even cases of "transition of some Orthodox Christians to Catholicism in the belief that they were returning to the faith of their forefathers," which was stated in memorandums to Dr. Pavelic and Dr. Hefer.

Thus, Obrknezevic concludes that "neither was transition forced nor was it the intention of the authorities of the NDH, but rather, it came about because the Orthodox populace itself had arrived at the idea and underwent it voluntarily and through its own initiative even though this was done out of fear and in avoidance of eventual violence." Explaining his statement, Obrknezevic notes: (1) that the Catholic Church authorities opposed transition which was not sought by personal request, and only in cases of acute persecution was transition accepted "as

a quick means of protection for Orthodox Christians"; and (2) that "the leading personalities" of the Ustasha movement "were not particularly pro-Catholic, nor particularly anti-Orthodox," but rather "anti-Serbian and only through this and in connection with this were they anti-Serbian Orthodox, but not anti-Orthodox."[125]

Such a line of reasoning is confirmed also by Archbishop Stepinac's statement that he had to relocate certain parish priests who were in danger of losing their lives at the hands of Orthodox Christians because, abiding by religious laws, the priests were taking too long with the transitions. "It is a fact that in the course of the war, the Church had to drag itself through adversities like a snake, but it ministered to the Serbian nation with the intent of providing assistance, insofar as it was possible."[126]

There are also credible testimonies from the Serbs themselves. Dr. Marko Vidakovic, the first doctor of technological sciences at the University of Belgrade, spent the entire wartime period in Zagreb, developing widespread charitable activities with the support of Kaptol and the personal backing of Archbishop Stepinac, for protection and aid to persecuted Serbs and others. Since he was not allowed to testify at Dr. Stepinac's trial, Vidakovic handed over to reverend Dr. Josip Ladika, "in gratitude to Cardinal Stepinac," his documents showing that transition meant deliverance for Orthodox Christians. In a letter to the Synod of the Serbian Orthodox Church, Vidakovic describes the manner in which operations were developed for the aid and rescue of condemned and persecuted people and war orphans, for release from camps (he mentions mass releases from Jasenovac), for assistance to prisoners and returned emigrants, all of which unfolded with support not only from a multitude of citizens as individuals, but also from institutions, such as the Red Cross, the Catholic Caritas, the Union of Banks, the Health Department, and others. All of this was organized by "a small group of people ... only through the intervention of Archbishop Stepinac." Vidakovic himself "received authorization from Dr. Stepinac to act in his name in all ventures."

In the final days of the war, news spread through Zagreb that 500 Serbs were in danger of being killed. "The archbishop allowed me personally," Vidakovic wrote to the Synod, "if the need arose, to provide refuge for all the Serbs in Zagreb in the cellars of Kaptol."[127]

How can one overlook the testimonies of such participants? Why is there not even the slightest mention of them in Dedijer's and other similar books? The reason for this is not an unfamiliarity with the facts, but rather their misinterpretation, distortion and concealment. This can be seen in the example of another well-intentioned Serbian view of Stepinac. Priest Andreja D. Dimitrijevic, Serbian Orthodox rector and "senior priest of the Cathedral Church of St. Sava" in New York (who became a priest in 1953 in Pozarevac, Serbia), writes in his declaration in 1975: "I have in my possession the names and documentation of how many His Eminence aided during the war not only of Croats, but also Jews and Serbs. He was the first founder of Caritas in Europe... The late Cardinal Dr. ALOJZIJE STEPINAC, in the opinion of many spiritual figures from whom I have heard it personally, is the greatest apologist of the twentieth century, because he defended the sacred Christian faith in the most difficult period, when it was threatened, as it still is."[128]

However, considering the fact that Dedijer assigns no importance to those documents and testimonies which do not fit in with his version of history, it comes as no surprise that he overlooks similar views of Stepinac as recorded elsewhere. He seems to have forgotten the knowledge which he must have attained as a member of the close circle of political propagandists surrounding the highest state political leadership of the SFRJ in the post-war period. Thus, he, too, must have been implied in the statement by Milovan Djilas to Ivan Mestrovic in New York in the presence of the Yugoslav ambassador, Sava Kosanovic: "Every intelligent Communist knows that Stepinac is innocent and that he is being tried for political reasons." The statement received a great deal of publicity throughout the world.[129] Nor is it likely that Dedijer is unfamiliar with the whole background of the statement made by

President Tito in the 1950s to United States Senator Blair Moody and C. L. Sulzberger, correspondent with *The New York Times*, in 1950, and to E. Korry, correspondent with United Press International, in 1951. In those statements, Tito said that Stepinac had to be tried because he refused to leave the country, that the Orthodox populace viewed him as a war criminal, and his not being released from prison was due to their dissatisfaction and to the fact that "Russian agents" could take advantage of Stepinac's release "for an attempt at creating an internal rift." Tito stressed the problem of "internal unity," and on that subject openly said, "In the world there have been cases in which even innocent people, in the interests of the state, have sat in prison, and have even lost their lives."[130]

In his book Sulzberger was to note very laudatory remarks on the part of Pope Pius XII about Stepinac as a noble and courageous man and church dignitary, showing that even the Russian Orthodox Tsars were more merciful toward the Catholics than were the Bolsheviks and Communists of our era. In 1960, in the basilica of St. Peter, Pope John XXIII expressed extraordinary respect, love, and admiration toward the enthusiastic apostolic work of Cardinal Stepinac, who, in his testament treats nobly "those who inflicted upon him such unjust sufferings," repeating Christ's words: "Forgive them Lord, for they know not what they do." Likewise, Cardinal Montini—later Pope Paul VI—said (also on the occasion of Stepinac's death in 1960) that Stepinac had been "the Archbishop of one of the most respected and largest dioceses in the world, the archpriest and spiritual representative" of his nation. His trial, which poses the question: "criminal or martyr?" can be likened to those of Socrates and Christ—which "is a credit to the condemned, while, upon the one who dares to pronounce the sentence, it brings the condemnation of all humanity."[131]

Even among the Serbs, there are those who regard Stepinac as praiseworthy. On the other hand, there are also those who continue to distort the truth about him. They are now waging their attacks more forcefully and thoughtlessly—because of Stepinac's Catholicism and Croatianness—than was done in the postwar

period. It is clear that this phenomenon, too, is not accidental, but conditioned by history. And it takes on extremely distressing dimensions when it is clothed in scientific and artistic attire, squelching every utterance of judicious and reasonable reflection.

In his book, *In Defense of Faith and Nation* (1986), Dr. Zarko Gavrilovic writes, "it is chilling when Christians must fear would-be Christians, when Orthodox Christians must fear Roman Catholics."[132] Dr. Dragoljub Zivojinovic, professor at the School of Philosophy of the University of Belgrade, reduces Catholicism in Croatia to what he calls clerical fascism. According to him, "The prolonged and constant political indoctrination of non-Catholics in our regions culminated in the most drastic way in 1941." He, too, makes reference to "documents" which "show that Stepinac was on the side of the Ustasha movement, heart and soul." Zivojinovic maintains that "when one considers the phenomenon of clerical fascism and Ustashism, one can speak of a broader base of evil and crime. This was not created only by the clergy and by the extreme right-wing forces gathered around Pavelic."

The conclusion is self-evident—after all, we've already heard it from Terzic. However, Terzic, in his search for the origins of Jasenovac only went as far back as the 19th century and Ante Starcevic. He didn't identify Croatianness to such an extent with clerical fascism. Dr. Zivojinovic goes beyond Terzic and even beyond the religious and national apostate, Viktor Novak. To him, Stepinac, as the spiritus movens behind the collaboration between the Catholic clergy and the Ustasha movement, was even "somewhat more controlled than the other Catholic prelates." Among the "more aggressive" and "extremists" of that time, one finds bishops Ivan Saric in Sarajevo, Janko Simrak in Zagreb, and Jozo Garic in Banja Luka, while today "that more aggressive part of the Catholic clergy are concentrated around the Zagreb metropolitan Cardinal Franjo Kuharic."

Zivojinovic notes that in the 16th, 17th, and 18th centuries "the effort to Uniate Orthodox Christians was ever present." The Vatican politics of "re-Christening the Orthodox population" was

constant and consistent. Regardless of changes in the circum-
stances of "various epochs," that goal "was never abandoned for
any significant length of time." To Viktor Novak's idea about
"constant pressure from Vienna and the Vatican upon our lands,"
Zivojinovic's "new sources" give a new content—that of clerical
fascism. "The general conclusion about the intense times of cleri-
cal fascist crime is undeniable and irrevocable."[133]

It is significant that Dr. Zivojinovic gives his foreword that title
"Murderers in the Name of God" with the addition of "Some
Thoughts." There one certainly finds "some thoughts" which
amount to a deepening of the theses which we have traced from
Nikola Stojanovic to Terzic, Dedijer and Vuk Draskovic.
Classified among the promoters of "the philosophy of assimila-
tion of Serbs and Orthodox Christians," along with Ante
Starcevic, "father of the Croatian nation," one finds Dr. Ante
Trumbic, and it is revealed that "the clergy exerted extremely
powerful influence" upon the peasantry, which "was apparent on
the eve of the Second World War in the Croatian Peasant Party
movement." Later we read "a large part of the Catholic clergy
openly endorsed or secretly supported the Ustasha movement."
His previously mentioned judgment, by which he extended cleri-
cal fascism and Ustashism to include all strata of the Croatian
nation, is raised here to an even higher level and he calls this
phenomenon clerical Nazism. In presenting examples of collabo-
ration between the Catholic clergy and the Ustashas, he goes
"from the Vrh-Bosnian Archbishop, Ivan Saric and his protege,
Bozidar Brala, to that most sadistic murderer, Father Miroslav
Filipovic-Majstorovic, the hyphenated addendum to whose name
[Majstorovic = Master] refers to his mastery in killing and
slaughter."

Certainly, this is a new contribution to the Jasenovac contro-
versy. In 1987 Belgrade University professor Dr. Zivojinovic
supplemented the already rigidly one-sided Paris essay of 1951.
Regarding Jasenovac, he even improved it. Namely, the alleged
Frenchman, "Laurière" (alias Branko Miljus) wrote his text at a
time when the Jasenovac "phenomenon" had not evolved to its

present mythical dimensions, so he said that "in Jasenovac at least 50,000 people were killed." And this, of course, according to Zivojinovic, is an "inaccurate claim" for it is not unlikely that the number is even greater than 700,000, despite "attempts by certain individuals including Zagreb's Cardinal Kuharic to reduce the number to 40,000 people."[134] Zivojinovic happens to forget that Cardinal Kuharic gets this number from the indictment against his predecessor, Cardinal Stepinac.

Zivojinovic took it upon himself that the number of deaths in Jasenovac should "be even greater than 700,000." He presents us with new assertions for the sake of solidifying and magnifying the Jasenovac distortion. In 1988, together with a certain Dejan V. Lucic, Zivojinovic published *Barbarism in the Name of Christ*. Its very title shows its bias. In this work he presents the claim that "840,000 people were killed in Jasenovac." It is clear that Zivojinovic and Lucic view this figure as reliable from the fact that they present it in a subtitle of the twenty-eighth chapter of their book (which deals with "The Technology of Genocide against the Serbian Nation"). However, according to them, this figure could be even greater because in the preceding text, also authenticated by a separate subtitle, it says that "500-800 prisoners were killed daily." Considering the fact that the Jasenovac camp was in operation for 1,337 days (from August 20, 1941, to April 21, 1945), the claim indicates that the number of deaths should be between 668,500 to 1,069,600, or an average of 869,050 (650 a day).

On the basis of such historical "testimonies," it is not difficult to draw the conclusion (again a subtitle of his) that Jasenovac is "the greatest torture-chamber in the history of humankind." But it also renders historical judgments with far-reaching objectives, aimed at establishing the historical guilt and responsibility of the entire Croatian nation for war crimes in the NDH, not just by the one-sided depiction of Pavelic's Ustasha movement, but also of Macek and the Croatian Peasant Party, and particularly Stepinac and the Catholic Church. In this respect, the Zivojinovic-Lucic book is just a new and more explicit contribution to similar views about which I have written. The authors' aim with this work was

resentative of a certain [...]
paid a visit to Stepinac [...]
have sentenced him to [...]
tice." These are the wor[...]
opposed the death sente[...]
not comprehend or acc[...]
one can clearly deduce [...]
the simple reason that [...]
situation of every natio[...]
to its history and that [...]
Serbian nations. Finally[...]

"As is well known, [...]
Yugoslavia! He was sen[...]
He served 6, and was [...]
was born. Jurists felt a[...]
tence for initiating bar[...]
victims agree?"[135]

It is impossible, how[...]
ical research, let alone [...]
answer from the victim[...]
premise to any judiciou[...]
desirable to delve into [...]
spring the demonic for[...]
authors of *Barbarism*, [...]
of serious discussion a[...]
alone their conclusions.[...]

On the list of similar[...]
same time (1988) the m[...]
The Ustasha Crime [...]
Artukovic in 1986. Th[...]
thousands of... victims [...]
a foreword by V. Dedije[...]
its enormity and amplit[...]
standings regarding the [...]
only its relevant special[...]
crimes, first of all incon[...]

to supplement Novak's *Magnum Crimen* "with new material and documents," which would make it possible to pinpoint the issues and expose new problems, and to "survey the dimensions of the monstrousness and crimes which were inspired and guided by the top ranks of the Catholic Church." The Church has been involved with "preparations for a religious war against Serbia" since even before World War I. Thus, "the monstrousness and scope of the Ustasha crimes were not the product of an incident, but of an extensive indoctrination which was conducted through the religious organizations." In this respect, the authors don't spare Slovenia's Catholic Church, either, recalling that in a poem in one of its publications in 1914, the battle cry "Hang the Serbs from Willow Trees" was expressed for the fist time. But they forget the Yugoslav, which is to say, pro-Serbian orientation of high-level Catholic prelates, from Strossmayer and Racki to Ritig and D. Mandic, and in particular the politics of Korosec's Slovenia and even that of the Croatian Clerical Party after the 1918 unification. The role of Macek's leadership of the Croatian Peasant Party appears as an endeavor to "realize the Great Croatian chauvinist plan." Zivojinovic and Lucic refer to the testimony of Glaise von Horstenau saying that the leaders of the Croatian Peasant Party were "in a certain sense Yugoslavists" and "complete" anglophiles, but this does not change these historians' impression of them, clearly because "they did not want to withdraw one step from their claim on historical lands, both in Bosnia and in Srijem." Nor are they impressed with Horstenau's attestation to Stepinac's sharper criticism of Ustasha terror and German racism than of the Partisans and of Communism itself, which he had always opposed. To these writers, it is a "falsification of history" that Orthodox Christians came into Croatian lands along with the Turks, although in one place even they write that the Drina river "was the boundary of Catholicism." They conclude "that the chauvinist Croatian plan was exactly the same as that of the Vatican concerning the Croatization and religious cleansing of the territory up to the Drina river." And they assign equal blame to Pavelic's NDH, the Vatican, and Kaptol for genocide. They do not

recall the "Serb
west of the D
Catholic and M

It is remarka
demic caliber b
and apocrypha
war, they desce
tionary politica
radically one-si
ciousness. This
"Syphilis and C
where as a sync
Extremist"; "T
Against the Ser
Catholicized c
Concordat Trap
of the People
Catholic Chu
Gendarmes"; "
under the Prote
Program of Gre
Lament of the
Vatican, the Ep
condemned th
Bloodshed!" be
obstruct the ser
clusion, "The P

As proof tha
ers find it neces
with an unan
Sulzberger's de
Lepoglava, Ziv
Montenegrin C
killed." From J
cite the recolle
judicial council

scientific method. On one hand, the author as an attorney appears to be impartial, showing that Artukovic was accused of acts of mass murder for which there is no evidence in the history of the NOB, solely on the false testimony of Bajro Avdic, the only "crown witness." Because of his past (sentenced to 20 years imprisonment) and contradictory statements, Advic cannot be considered reliable or trustworthy.

Bulajic directs sharp criticism at the judiciary of the SRH for not having tried Artukovic for genocide against the Serbs (and others). In that way, "in the perspective of a number of decades, the possibility for a calm denial of the genocidal character of the Independent State of Croatia from 1941 to 1945, was legally sanctioned." This is the reason why Bulajic considers genocidal not only all the crimes, but the totality of the existence of the NDH. However, since he does not see any other genocidal crimes at the same time, either in the territory of the NDH or in Yugoslavia, this is clear evidence that as a lawyer, and especially as a historical writer, he does not remain on the level of scientific objectivity.

Bulajic equates the Ustasha government with the Catholic Church to such a degree that he devotes the entire last part of his book exclusively to the Church's accountability for Ustasha crimes in the NDH and its relations with the Vatican. In this way the affront of historical name-calling retains its spiked edge. In the final chapter of his book, which he entitles, "In Place of a Conclusion: an Invitation to Dialogue and the Establishment of Historical Truth," Bulajic leaves no room for doubt. There, in the words of Cardinal Montini, later Pope Paul VI, he reminds one that Stepinac was the "archpriest and spiritual representative of a Catholic nation" who was "respected by the whole nation." From a speech about Stepinac by Pope John XXIII, Bulajic stresses the Pope's blessing "to all the believers in devoted and God-loving Croatia." Bulajic uses such statements to support his viewpoint, as well as Artukovic's assertion that Stepinac was a "holy man." And to this he juxtaposes the judgment made by J. Blazevic that Stepinac was a "war criminal" fit for the ninth circle of hell. Despite statements that the Catholic Church and Stepinac dis-

resentative of a certain English Quaker organization who had also paid a visit to Stepinac had announced that "the English would have sentenced him to death. He would have been brought to justice." These are the words of a Quaker who, otherwise, personally opposed the death sentence. According to him, the English "could not comprehend or accept a prison term" in such a case. Nor, as one can clearly deduce from this, can the authors of this book, for the simple reason that they do not fathom the differences in the situation of every nation and the way in which its stance is linked to its history and that this is the case with the Croatian and Serbian nations. Finally, they end their book with these words:

"As is well known, Stepinac did not receive such a penalty in Yugoslavia! He was sentenced to 16 years deprivation of freedom. He served 6, and was then permitted to live in Krasic where he was born. Jurists felt at that time that this was enough of a sentence for initiating barbarism in the name of Christ. Would the victims agree?"[135]

It is impossible, however, to carry on any sort of serious historical research, let alone eliminate historical disputes, if one seeks an answer from the victims on only one side. This fact should be a premise to any judicious historical approach, insofar as it is even desirable to delve into the obscure circumstances out of which spring the demonic forces of mutual destruction. Concerning the authors of *Barbarism*, it is quite impossible to enter into any kind of serious discussion about the analysis of causes and facts, let alone their conclusions.

On the list of similar historiographic literature, one finds at the same time (1988) the massive two-volume work by Milan Bulajic, The Ustasha Crime of Genocide and the Trial of Andrija Artukovic in 1986. This work is dedicated "to the hundreds of thousands of... victims of the Ustasha crime of genocide," and has a foreword by V. Dedijer and an epilogue by L. Kljakic. In spite of its enormity and amplitude, it presents no new or different understandings regarding the subject at hand. Thus one should examine only its relevant special characteristics in its consideration of war crimes, first of all inconsistencies in the application of an objective

scientific method. On one hand, the author as an attorney appears to be impartial, showing that Artukovic was accused of acts of mass murder for which there is no evidence in the history of the NOB, solely on the false testimony of Bajro Avdic, the only "crown witness." Because of his past (sentenced to 20 years imprisonment) and contradictory statements, Advic cannot be considered reliable or trustworthy.

Bulajic directs sharp criticism at the judiciary of the SRH for not having tried Artukovic for genocide against the Serbs (and others). In that way, "in the perspective of a number of decades, the possibility for a calm denial of the genocidal character of the Independent State of Croatia from 1941 to 1945, was legally sanctioned." This is the reason why Bulajic considers genocidal not only all the crimes, but the totality of the existence of the NDH. However, since he does not see any other genocidal crimes at the same time, either in the territory of the NDH or in Yugoslavia, this is clear evidence that as a lawyer, and especially as a historical writer, he does not remain on the level of scientific objectivity.

Bulajic equates the Ustasha government with the Catholic Church to such a degree that he devotes the entire last part of his book exclusively to the Church's accountability for Ustasha crimes in the NDH and its relations with the Vatican. In this way the affront of historical name-calling retains its spiked edge. In the final chapter of his book, which he entitles, "In Place of a Conclusion: an Invitation to Dialogue and the Establishment of Historical Truth," Bulajic leaves no room for doubt. There, in the words of Cardinal Montini, later Pope Paul VI, he reminds one that Stepinac was the "archpriest and spiritual representative of a Catholic nation" who was "respected by the whole nation." From a speech about Stepinac by Pope John XXIII, Bulajic stresses the Pope's blessing "to all the believers in devoted and God-loving Croatia." Bulajic uses such statements to support his viewpoint, as well as Artukovic's assertion that Stepinac was a "holy man." And to this he juxtaposes the judgment made by J. Blazevic that Stepinac was a "war criminal" fit for the ninth circle of hell. Despite statements that the Catholic Church and Stepinac dis-

agreed with the Ustasha government, Bulajic states that "it cannot be denied that the prelates of the Catholic Church were responsible for backing the Ustasha regime," for Catholicization, and for all the crimes. To support this claim, Bulajic cites none other than Dedijer's letter to Pope John Paul II (dated December 6, 1986), in which he wrote that the "trial of Artukovic... has exposed the connection between Pavelic's government and the Catholic Church" and that "all the crimes in Croatia during the war have not yet been uncovered."

Bulajic extends "An Invitation to Dialogue" for the sake of "establishment of man's and God's truth, lest the prophesy of St. John be realized, that a time cometh that whosoever killeth you will think that he doeth God service." This is Bulajic's concluding sentence, encouraging because of his call to dialogue, but even more disheartening because of its warning that there may again come a time when the Biblical prophecy will come true if the dialogue does not conform to his one-sided views. For what is the purpose of citing "facts and documents," if we are to believe only and exclusively in those which direct us to the Gospel according to Jovan (John, in Serbian), while rejecting those which interpret it according to Ivan (John, in Croatian)?

In connection with Jasenovac and genocide, Bulajic mentions my name in two places. In the first instance, Miletic's text is quoted, saying that the stance that there were "40,000 victims" in Jasenovac "is supported by Tudjman, based on the suspended census of inhabitants." (Vol. 2, p. 227) In the other instance, having condemned the liquidation of "testimonies concerning the crimes" in Jasenovac itself, Bulajic says, "All this makes it possible to falsify the genocidal crimes that were committed, through the public assertions of Franjo Tudjman and others that only 60,000 persons were liquidated in the Jasenovac camps. Those are the obvious negative consequences of the attitude that the genocide camp should be turned into a horticultural park in which the criminal bloodshed will be forgotten." (Vol. 2, p. 244)

Setting aside the "triviality" that my figure of almost 60,000 did not refer to Jasenovac alone, I ask Bulajic: On the basis of

what principles, academic or moral, can he charge me with having falsified historical facts concerning war crimes—which in my opinion can be ascribed to every one of those who is engaged in these claims of genocide—even if I arrive at my conclusions only "based on suspended census" counts? If anyone who is involved with historical investigations does not understand that the census was suspended by a voluntary unacademic decision by the state party magnates on the grounds that the suspension served the interests of everyday politics, and that it would eventually prove detrimental to those who wished to use it for the Jasenovac distortion and the genocidal theory, then he has found himself in the wrong field. Or is he affiliated (surreptitiously) with those who advocate the "suspension" of factual statistics in order to construct their own theses and theories? In this respect, Bulajic's book is based on extremely biased assumptions and offers no contribution to real dialogue.[136]

In laying the blame upon the Catholic Church, these writers use its alleged Ustashism as evidence for historical sin. The Church's very existence is to blame for the existence of Croatianness and presents a threat to Orthodox Serbdom, or at the very least, is the chief obstacle to its advancement. Along the lines of such thinking, one finds all the interpretations of Croatian-Serbian relations in terms of both nationality and religion in the publications of the Serbian Orthodox Church.

Thus, in the official publication of the Serbian Patriarchate, Pravoslavlje [Orthodoxy], Svetozar Dusanic, an answer to a speech by Hrvoje Istuk presents the thesis of the irreconcilable division in which one finds the "Balkan, Byzantine, Orthodox" world on one side and the "Western, Roman Catholic" world on the other. "This division is a reality." The Ustashas arose from the "Western Roman sphere" because they were "adherents of the Roman Catholic Church," and so did all their crimes.[137] Thus, from the thesis of genocidal Croatianness, one arrives at this one of genocidal Catholicism. Of course, it is logical to expect a comprehensive theory as well. But it never occurs to the creators of these theories to consider from which sphere the Chetniks came

since their historical "actions" have been covered over by the Jasenovac distortion.

This theory concerning the genocidal character of Catholicism is not new. More than half a century ago, it was contained in a missive of the Serbian patriarch Varnava and in the interpretation of that missive. There it is written that "the Roman Catholic Church and faith is a parasitism which was and has remained the greatest misfortune in the history of Christianity." It is a "wild and infertile offshoot" which "has given birth to Freemasonry and all the philosophical, humanistic and social movements in the West: Protestantism, Marxism and atheism, movements which are merely justified reactions to the tyranny of Catholicism." The "inner life of the Western man and Western humanity is organized around hatred, on which grounds the Western Church is also based." The Catholic Church is "an old serpent in the terrestrial paradise," and "Catholics in Serbia are desert locusts, antipodes of the gospel, moths in the fabric of modern times, perverted doctrines."[138] This was written in 1931, during the time of the dictatorship, while today we are experiencing merely the further development of such ideas.

A more complete view is presented by M. Milovanovic in his explanation as to why the Serbian Orthodox Church rejected the Concordat so resolutely a half century ago. According to him, "The Roman Church, ever since the great schism" in 1054 has never ceased to dream of "regaining the schismatics," and even "little Serbia" was not spared from the Vatican's plans to bring the Orthodox East back into the "bosom of the Mother Church." This deliberation, which certainly has a rational basis, is followed, however, by this statement: "Gradually and logically bringing its plans to life, in the next three centuries, with the help of great powers, the Vatican converted entire Serbian regions in our western territories to Catholicism and Eastern [Uniate] Catholicism. This was often conducted by force, and sometimes with the help of deviant Serbian Orthodox renegades. Thus, it came about that many Serbs were designated Croatians, according to the rule that whoever is a Roman Catholic cannot be a Serb."[139]

At which point in history were the Croatian lands ever "entire Serbian regions" in Serbia's "western territories"—except in the Great Serbian plans of having a common border with Slovenia? When were the Serbian Orthodox inhabitants of these places converted to Catholicism and Eastern Catholicism? Where does one find conditions for any sort of serious discussion when after such an introduction all the blame for the events of 1941-45 is laid only upon the Vatican and the Catholic Church, the Croatians and Stepinac, and from that period only Jasenovac is seen? Why isn't all that transpired taken into consideration and why doesn't it provoke more reasonable consideration?

Those who accuse the Catholic Church, and particularly Stepinac, of the crime of genocide because of the religious conversion of Orthodox Christians to Catholicism ignore the given wartime circumstances and the role of the Church. There are other factors which should be taken into consideration as well if one is to make a responsible assessment of this case. What first comes to mind is the negative legacy from the Yugoslav monarchy in which the St. Sava ideology of Serbian Orthodoxy was the basis for hegemonical policy which was reflected in the realm of religion as well. As a consequence of the policy of the regime, cession of rights, instances of blackmail, and particularly through mixed marriages with Serbian military and government personnel, in two decades, about 200,000 Catholics converted to Orthodoxy. In order for the state to create an even greater Orthodox identity, it was the policy to build Orthodox churches as a symbol of "national statehood" even in the most Catholic areas, so that an Orthodox church was erected even on the Croatian island of Vis.[140] It is understandable that such religious conversions induced an even greater resentment toward the "foreign" predominantly Serbian composition of the military and administrative-police castes because that was the most overt expression of the policy of assimilatory unitarism. Because of this, one can assume that a part of the Catholic populace spitefully awaited the changes of 1941, just as it is probable that the memory of those circumstances prompted a part of the Orthodox populace to seek an outlet,

whether in conversion to Catholicism or in emigration to Serbia.

The fact that it is erroneous to view religious conversion under those wartime circumstances as genocide can best be seen in the case of the Slovenians who—expelled from their homeland—found refuge in Serbia but were forced to convert to Orthodoxy. We are told this by such witnesses as the Slovenian writer Anton Ingolic who spent four years in Serbian exile. He says, "I have written how the Serbian nation as a whole has accepted the Slovenians. My own misfortune was in the fact that I was situated in a pronouncedly Chetnik region." After the Germans succeeded in beating back the Partisan movement in the Fall of 1941, "conditions were difficult not only for the Slovenians, but also for the Serbs. Despite the general terror, we Slovenians managed to survive, thanks to the Serbs, even though as Slovenians in Cuprija we had to convert to Orthodoxy."[141] There also followed in the postwar period in eastern European countries the conversion of millions of Roman Catholics and Eastern Catholics to Orthodoxy which was organized by the socialist governments.

Taking into consideration all the facts, how can one reproach Cardinal Kuharic for conducting allegedly "aggressive" and "clerical nationalist" policy when he espoused probing into the truth concerning the role of the Catholic Church during the Second World War and Cardinal Stepinac as its most prominent representative? Anyone may disagree with the judgment reached by a large number of clerical and secular people in the world that "not one grievance against the Croatian clergy in general, and Cardinal Stepinac in particular is as unjustified as this one," as was stated by Dr. V. Macek regarding accusations of "forced re-christening" of Orthodox Serbs.[142] However, no one who wishes to engage in such serious historical issues has any right to accuse Cardinal Kuharic of conducting the allegedly Ustasha and clerical fascist policy of Stepinac because Kuharic persistently and resolutely upheld the historical truth. "If one advocates truth without the fear that someone will subvert the truth, then one allows the other side to be heard as well."[143] This "ethical principle," Stepinac's successor rightly reminds one, should constrain the conscience

of all public officials, particularly of those who wield the far-reaching historical sword.

In view of all the accusations concerning the historical guilt of the Catholic Church in general, and in Croatia in particular, one should, finally, address certain essential facts. No matter how many such accusations are made about the Vatican and the Roman Catholic Church as a whole, they are aimed at Catholicism and the Church on Croatian soil. Regardless of how much religious conflict there may be in the Catholic-Orthodox rivalry, their background is of a national nature. More clearly stated: they stem from the existence of two historically different individual national entities and the contrariety of their respective national interests. The state-political ideas of a Croatian and of a Serbian nation have taken on their severest form in the very designation of their respective religions. Despite the fact that these national-religious differences were expressed even before the unification of the two nations into a common state, they grew deeper within that state, to the point where they became explosive when that state collapsed. Despite attempts to overcome these differences through the idea of Marxism and a socialist community, the differences are apparent today in an even more envenomed form than ever before.

The reasons for this phenomenon are manifold. First of all, they each have their own historical legacy which cannot be overcome by ideological premises since even they contribute to the old discord. The socialist order itself, founded upon militant atheism, discriminated in favor of the unrestrained attacks against the Catholic Church due to the concatenation of various events. Marxist ideology made use of the blame placed upon the Catholic Church for the "fascist" sins of the war to misrepresent and denounce religion as "the opium of the people." Just as the revolution fought the "bourgeois counterrevolution" during the war and expropriated the bourgeoisie after the war because of its "national betrayal," so also did it deal with the Church as its chief ideological opponent and the only organized remainder of the "old society" under the pretext that it collaborated with the

Fascist occupation order. In this respect, the NDH presented quite a welcome opportunity for getting even with the Catholic Church, whose very nature made it less reconcilable with the new socialist order than was the Orthodox Church. Thus, a certain silent "alliance" between the old and new anti-Catholic elements was formed. The fact that behind all this there was a backdrop of historical Croatian-Serbian contrariety was to make itself increasingly evident as time went on. It was to reach its fullest expression in the Jasenovac distortion.

Around this distortion, as the most promising model for establishing the historical responsibility of the Catholic Church as well as that of the entire Croatian nation, otherwise contrary ideological and political currents became united. Both the old views

(quelled by the revolution, but historically extant) and the new (neo-hegemonic and socialist-integralist) currents needed the Jasenovac controversy and the theories concerning the genocidal character of Catholicism and Croatianness that it is based on. These theories provide a horrifying means for the obstruction of any attempts to realize equality and sovereignty of Yugoslav nations, and even for the removal both of the individuals and the principles of those historically responsible for federalization.

Thus one arrives at the objective of all the above-mentioned phenomena and the broader dominant issue in the context of which even the role of the historical responsibility of the Catholic Church is assigned the significance of the crown testimony for the genocidal character of Croatianness. One can conclude more about the depth of its historical roots and about the spiritual climate in which it emerged simply by observing its public repercussions than through a discussion of controversial issues, the likes of which has not yet taken place.

As far as the Croatian as well as the international Catholic public is concerned, historical continuity was manifested in the most impressive way in 1984, when at the closing part of the celebration of Thirteen Centuries of Christianity Among Croatians, at the National Eucharistic Congress at Marija Bistrica (September 8 and 9) a crowd of about 400,000 devotees congregated—along

with 1100 priests, 35 bishops and arch-Bishops, 5 Cardinals, as well as 200 reporters. Pope John Paul II was unable to attend that "greatest Croatian mass in history" due to the atmosphere created by the abuse of these untruths and theories. In his distinguished message for that occasion, Pope John Paul II emphasized that "the Croatian nation was the first of the Slavic nations" to enter into the world Christian community out of paganism. He added that the great anniversary of "thirteen centuries of Christianity among Croatians" should have been "celebrated in 1941, but because of the circumstances at that time" it had to be cancelled. "Be proud that you belong to an esteemed Church... and to a nation with a history and Christian heritage so ancient and noble."

The emissary of the Holy Father, Cardinal Konig, didn't neglect to mention Stepinac explicitly, and Cardinal F. Kuharic quoted Stepinac's words as well. Speaking of the embodiment of the Catholic Church in the thirteen century history of the Croatian nation "often doused with blood and tears," Kuharic stressed that bishops and priests are sent out and called upon to deal out both good and evil to their people, that they show their "evangelical patriotism most genuinely when by Redemption they strive to heal the soul of their nation from the mortal evils which destroy man, family and nation." Voicing his condemnation of all forms of spiritual and physical genocide in contemporary times, Kuharic repeatedly calls for the discovery of a path toward a peaceful and harmonious mutual existence with adherents of different religions, as well as with those of no religious affiliation.[144]

The nine-year-long celebration of thirteen centuries of Croatian Christianity, and especially the final segment at Marija Bistrica, was the historical expression of the spiritual state of the Croatian nation. It was also the answer, in its own way, to the exaggeration-based accusations and blame. From the historical viewpoint it deserves all the more attention since it took place in the realm of religious life, alongside, and to a large extent, in opposition to the public and official life of society in general.

Thus, one could actually conclude that the closing festivities of the celebration of thirteen centuries of Christianity among the

Croatians was a specific and indubitably convincing answer of Croatian Catholicism—with the full agreement of European Catholicism—to all those monstrous accusations and ideas which label the Catholic Church because of Croatianness and Croatianness because of the Catholic Church with historical guilt and even genocidal character.

In connection with this we should consider how, at the same time, Serbian Orthodox public opinion about Jasenovac manifested itself. A week prior to the festivity in Jasenovac itself, a celebration was held in honor of a new Orthodox parish church which was even attended by the Patriarch of the Serbian Orthodox Church, German, along with representatives of the government and of other religious communities, and about 10,000 devotees from all over the country. Such a small number of participants, even though they did not have to surmount all manner of difficulties as did the pilgrims to Marija Bistrica, tells us in itself that in the consciences of Orthodox adherents and the Serbian nation, the exaggeration of the Jasenovac victims is not alive in such dimensions as its creators and propagators wish to give it. Their tossing about of inflated numbers of victims has no analogous reverberations among the majority of the Serbian Orthodox population. On the other hand, the Croatian Catholic population through its behavior stands in defense of its history and its present.

In his speech on Jasenovac, Patriarch German mentioned the Jasenovac camp after the German camps, along with Jadovno and Glina, characterizing the war victims through biblical images. It was a time, he said, "when the Father of sin, evil and eternal darkness spread his death-bearing wings and spilled blood and tears, crime and death," when "Christened people were the executors of his will, enticed and misled, convinced that by doing harm to others, they themselves would benefit." He referred to Christ's warning to the apostles: "there is nothing covered, that shall not be revealed; and hid, that shall not be known" (Matt. 10:26).

However, it seems that even Patriarch German forgot that the evil was not all on one side, and good all on the other side, when

he said, "Brothers, forgive—we must, but forget—we cannot." All the more because a concrete flower on the Jasenovac field would remind the great grandsons of the grandsons of the victims of that place.[145] But historical wisdom teaches us that monuments no less permanent will remain in the hearts and minds of the descendants of those other victims, no fewer in number, who similarly will be unable to forget, particularly if the former are to be multiplied while the latter are cursed and branded. For, just as the personal representative of the Croatian Cardinal Kuharic, Bishop Dr. D. Koksa said at that assembly, "Things happened that should never have been allowed to happen."[146] They, however, did not just fall from the sky, but were precipitated by other events and were mutually conditioned, so that the problem of forgiving and forgetting has equal historical significance for all the participants. Besides that, such a call to not forget "sounds like, and is understood as, a threat," we are rightly reminded by writer P. Segedin, "and it brings under suspicion the very act of forgiveness. This may be a political gesture, but not a Christian one! He who is forgiven in a Christian way does not need to be forgotten; quite the contrary." For "forgiveness is a constructive act, connected to a particular notion, according to which a crime among people" is not the product of only one side.[147]

Every other approach leads to the inescapable cycle of revenge. Accordingly, every one-sided insistence upon one's right not to forget will never lead to the removal of old accumulations of destructive animosity, only new accretions of vengeful hatred. There can be no solution either in forgiving or in forgetting— because every act of forgetting is in its own way a form of remembering. The solution lies in a recognition of the truth.

5. *Views of Nationalism and Separatism, Hegemony and Federalism*

Different theories of history choose different histories of the past.
—A. Heller

Nations that have subjugated and crushed others have always acted foolishly and to their own detriment and destruction, but those that act the most foolishly are the ones that believe they can still do this today.

—Ivan Mazuranic

All around us we sense the snares and dark deeds of deceit, the abysmal perils beyond the hedges of words.

—Tin Ujevic

In a revolution there are two kinds of people: those who prepare it and those who benefit from it. As Zweig concluded: Whether classist, state-political, or national revolutions—they never belong to those who start them, but to those who finish them and take the spoils. Similarly, A. Toynbee showed that in all ages and all places, the social stratum or political movement, idea or historical personage that responds successfully to one historical challenge rarely succeeds in responding to the next historical challenge. Exhilaration over the victory in one historical battle is quite often the cause of failure and decline in subsequent attempts.[148] As a rule, that decline is all the swifter when the elation of victory unwisely and ruthlessly imposes its will, denouncing all that came before.

In the case of Yugoslavia, all this was clearly visible even in the first couple of decades after the victory of the revolution. From the beginning of the 1970s, it shifted into a critical decline that became no less of a historical challenge than the one to which the

revolution had successfully responded. The causes of this decline were twofold: one stemmed from the very foundations of socialist order which had never succeeded either in developing production capabilities, or in realizing a free and prosperous society. The other reason for this decline was in the multi-national and multi-ethnic composition of socialist Yugoslavia.

A treatment of the general problems of socialism would be outside the scope of this discussion. However, it is necessary to deal with a very significant phenomenon. It is uncontestable that the revolution in Yugoslavia was able to succeed and to establish a socialist order above all—and it would be no exaggeration to say exclusively—because it responded successfully to the national question. Nonetheless, many forget this essential fact when seeking a response to the present historical challenge. Moreover, in their disagreement with the AVNOJ principles of building the state community, they ascribe to those very principles all the causes of the entire socio-economic, political, and ethical crisis. In so doing, they forget one small fact—that without those principles, a renewal of that state community would have been impossible. Since its establishment as a socialist community of equal nations actually meant the realization of socialist ideas on Southern Slavic and Balkan soil, those who lay the historical blame for those AVNOJ principles upon Croatian and Slovenian representatives forget and quite openly refute the ideas and activity of those Serbian socialists and Marxists from Svetozar Markovic, Dimitrije Tucovic, Kosta Novakovic, and Filip Filipovic, to all those who participated in the leadership of the Yugoslav communist movement during Tito's period (M. Pijade and R. Colakovic, S. Zujovic and B. Neskovic, M. Djilas and A. Rankovic, M. Popovic and M. Todorovic, P. Stambolic and K. Popovic, et al.).

The historical perceptions and pragmatism of the Serbian socialists and Marxists were directed toward finding solutions that were acceptable to other, neighboring nations as well. This was contrary to the politics of Serbian imperialist expansionism, from the time of Garasanin, then Pasic, to Alexander Karadordevic. The theses presented by Nikola Stojanovic in 1902

expressed clearly their pretenses. As we have seen, the origin of the Jasenovac distortion as well as of the theory of the genocidal character of Croatiandom was the continuation of Stojanovic's directions or, to put it more precisely, the resurrection of those ideas and viewpoints that had been suppressed by the victory of the socialist revolution.

As in the past, today such views express a misunderstanding of the national question in general, and in this case the Croatian-Serbian situation. The misconception rests in the belief that a solution by way of Serbian expansionism can be achieved by imposing Great Serbian ideas and glorifying Serbdom and Yugoslavism, while defaming and condemning Croatiandom.

Of course, in Serbian, just as in Croatian, Slovenian, or any other intellectual life, there are those who distance themselves from a "strictly national," "conservative," and "historically traditional" approach to the problem of national and inter-ethnic relations, emphasizing their supra-national, pro-Yugoslav, socialist-integralistic, or common democratic cosmopolitan views. Even when such views are not merely a pretext for pragmatic problem solving and historical ignorance, but an expression of a utopian belief in one's own illusions, they only serve to obscure even further those already excessively clouded historical problems.

In this respect, particularly interesting are those viewpoints which, free from ideological bias, rightly stress that "all attempts to petrify the propagandist dimensions of the revolutionary premises" go beyond the scope of scholarly objectivity and that in interpreting historical events and "determining motives in the behavior of people" and nations, one must take into account those incentives and factors "which have an existential power." One should agree with the principled approach Djuretic advocates, for there certainly exists in history, always and "everywhere the presence of elements that transcend class and ideology," and thus, outside of "the functional designation of this phenomenon, Serbian-Croatian war conflicts are without any substance."[149]

The controversy commences, however, upon closer inspection into those existential factors, for one also remains outside the

realm of scholarly objectivity if one takes into consideration their presence in the Serbian case, and ignores them in the Croatian case and ascribes to all such manifestations a pathological genocidal predeterminance. We have seen many examples of blaming Croatian and Catholicism for genocide, or in linking the two. The one who went farthest in such "synthesizing" was Vuk Draskovic, who in his madness sees in Croatiandom nothing other than "Kaptolian, Starcevician and Frankian Serb-devouring" stemming from "Vatican-Frankian-Comintern propaganda," while in Kuharic's appeal for the truth about Stepinac, Draskovic sees nothing other than a demand "for another Jasenovac and another Jadovno."[150] If such unrestrained outpourings of literary malversation were not in harmony with an entire orchestra of assorted "scholarly" interpretations, one wouldn't need to pay them any heed. But since they are the expression of a specially created psychosis, it would be unreasonable not to investigate them.

All the various manifestations of such attacks upon Croatiandom have their primary origin in a misunderstanding of the national question in general and that on Yugoslav soil in particular. Duretic thinks that a Yugoslav nation could have been formed along the lines of a "European state-legal constitution," in the form that had been "tested in the Italian Risorgimento and in German Bismarckism." But he argues that "misunderstandings between Serbian and Croatian representatives" led "to a condition in which national patriotisms repressed Yugoslav patriotism" to "megalomaniacal national ambitions" and "old illusions... with projections of their own independent states" to conflict between strictly national and Yugoslav determinations."[151] In this one can clearly see a basis for all misunderstanding, all the theories of historical guilt, and even for the inadmissibility of any manifestation of Croatiandom.

This illusion has certainly been remarkable, starting with the romanticism of the Illyrian Movement in the last century, Austro-Slavism, and Yugoslavism. Let us limit ourselves to the last case, which was shattered by the separate identity of both the Croatian and Serbian national being and their distinct, and to a great

extent, contrary state-political ideas and interests. The determinants by which this was conditioned are a subject for separate consideration. However, the inability to see that history has proven that fact bluntly and unequivocally (the assassinations in the Belgrade Assembly and in Marseilles, genocide by the Ustashas and the Chetniks) amounts to the resurrection of a twice-shattered illusion, even in the form of socialist integrationism.

When the separate identity of a nation is not recognized, then the natural aspiration for national liberty and state independence is declared a megalomaniacal and even genocidal ambition. It would be interesting to hear from the promoters of such theses (Terzic, Duretic, Krestic, or Bulajic) why the Serbian nation has the right to its statehood (and even an expanded one), while the Croatian nation, with its eleven-hundred-year political tradition, does not—particularly in a century when new, until recently unheard-of nations are appearing in droves on the world stage.

In order to lay the blame upon the Croatian nation for the crimes of treason, separatism, and genocide, an effort is made to forget that every nation in history governs its own interests. And for this very reason, history attests that on existential grounds, an artificial patriotism cannot be imposed upon any nation toward a multinational state creation in which that nation is not afforded the right to develop its national identity. In modern history, the best confirmation of this are the breakdown of the Habsburg monarchy and tsarist Russia, as well as prewar Czechoslovakia and the Kingdom of Yugoslavia, and, in even more recent history, not only all the colonial empires, but also their multi-ethnic successor states—that is, artificial, colonially established state organizations.

In keeping with this willful historical ignorance is the remarkable denial of the existence of Great Serbian hegemony in monarchist and centralist Yugoslavia. According to Duretic, there in fact was no Great Serbian hegemony in Yugoslavia between the wars. He just barely acknowledges a certain "hegemony of the Serbian monarchist elite," which, however, was "without economic and cultural protectionism, and without any organizational elements

directed against any other nation in particular." Moreover, he reproaches that same elite for having been too lenient and misdirecting national politics, especially with regard to taking into service after the unification a contingent of officers and functionaries from the former Austro-Hungarian (i.e., Croatians and Slovenians). In the Kingdom of Serbs, Croats and Slovenes, and in particular in Yugoslavia after the January 6, 1929 dictatorship, when Serbia had renounced its separate, nation-state name, there was no ruling nation and no suppressed nation, so how was such a "myth about Serbian hegemony" able to arise? Those responsible for the genesis, the general pervasiveness, and the consequences of that myth, which its creators used "to carry their anti-Serbism and separatism to the point of absurdity," are both the Croatian right and left wings, as well as other nations across the political spectrum and around the world. In the "thick forest of anti-hegemony" there was often a mixture of "red and black colors." We have already learned from Dedijer and Zivojinovic that the Vatican even preceded Fascism in its anti-Serbism, giving "the color black" a more permanent mark, and now Dedijer's attempts to uncover connections between the CK KPH and Pavelic's Ustasha headquarters are historically unveiled.

According to Duretic, after the unification in 1918, "the new entity was not only ethnically related, but it also had a higher level of political and economic democracy." But "the constant confrontations between Serbian and Croatian bourgeois circles" prompted an "anational yearning for the good old days." This "turning to the past" was accompanied by "outpourings of the wrath" of inherited anti-Serbian imperialist policy with "taunting conceptual terminology": "Great Serbianism," "Serbian barbarism," "Serbian tyranny," "Serbian primitivism," "Byzantinism." And "in this way, not only was an atmosphere of national and religious intolerance created, but also the appearance of a clash of civilizations." This "condition of insinuations and prejudices," by means of a "calculated internationalization of Serbian-Croatian contradictions," took on the "form of a severe problem in Europe," whereby a special role is assigned to

Comintern Stalinist voluntarism and its simplified interpretation of all the evils of the Yugoslav reality with its "so-called Great Serbian hegemony." It is a fact "that not only in propaganda, but also in literature, ideas with scholarly pretensions were emerging concerning oppressed Yugoslav nations and the Great Serbian dictatorship, ideas which were quite reminiscent of... Austro-Hungarian anti-Serbian intrigues."[152]

What should be said of such a denial of Great Serbian hegemony? If it had not existed, would the "Croatian question" have become the chief cause of Yugoslavia's state political crisis during the interwar period, and even a severe European problem, as Duretic himself is compelled to admit? Any serious deliberation should begin with the assumption that this could not have come about accidentally, due to political and ideological whim. In other words, the entire world accepted the idea of the oppression of the Croatian and other non-Serbian nations by the Serbian hegemony, not on the basis of superficial propaganda, but only after careful examination of the issue from a political, diplomatic, and historical perspective.

Yugoslavia was an important link in the European order established by the Treaty of Versailles, to counter the threat of German revanchism and possible Soviet expansionism in the Balkans. The powers defending that Versailles order, France and Great Britain, first offered their full support to the Serbian hegemony so that it could secure its rule by all possible means. Only when it failed to do so did they advise King Aleksandar Karadordevic, and later, even more decisively, Prince Paul, that a solution to the Croatian question could only be brought about by arranging a Yugoslav state. Italy was opposed to a Yugoslav state, and because of that, it was to offer asylum to Pavelic. But in its relationship with him, Italy never overlooked the possibility of compromise with Belgrade. Hitler, because of the Belgrade putsch, was to declare war on Serbdom and allow the establishment of an independent Croatia in his New European Order according to earlier plans. Accusations that the Vatican was allegedly involved in the break-up of Yugoslavia and that it backed Croatian separatism can cer-

tainly find no substantiation except in its principled position that
the Church—because of its mission—must be on the side of rights
of every man and the rights of every nation "of God."

Duretic resolutely ascribes the stance of the Comintern against
Great Serbian hegemony and for the rights of the subjugated
Croatian and other non-Serbian nations to national independence
and state sovereignty to mere "Stalinist voluntarism and extrem-
ism," and to the Soviet misapprehension of "the essence of the
Yugoslav situation" and the true "nature of Croatian, Serbian,
and Slovenian nationalism." This shows either a complete unfa-
miliarity with this matter or a distorted, totally inaccurate judg-
ment. When the Comintern, in the mid-1920s, took the position
that it would be necessary to break Yugoslavia apart, it was not
because "after Lenin's death, the loudest voices in Yugoslavia were
those of individuals of doctrinarian character."[153] In fact, the con-
trary was true. Within the leadership of the Comintern at that
time were all the top names of the European Marxist movement
and the October Revolution (later eliminated by Stalin), working
in cooperation with Stalin in discussions on the national question
in Yugoslavia, contrary to the unitaristic ideas of the leader of the
KPJ (Communist Party of Yugoslavia), Sima Markovic. The deci-
sion concerning the right to self-determination of the Croatian,
Slovenian, and other nations was not made along the lines of a
dogmatic departure from Lenin's views, but by the consistent
application of those views. It was not imposed from outside on the
Yugoslav communist movement, but rather, it served to support
the views of the "left wing" of the KPJ, radicalized by their own
consistency, and which had already, contrary to Sima Markovic's
"right wing," arrived at the conclusion that there could be no rev-
olution in Yugoslavia without a solution to the national question.
Of course, it was no accident of history that the chief contenders
for the recognition of a nation's right to self-determination were
Croatian Marxists, while they were the minority among the
Serbian Marxists, since the majority were followers of Sima
Markovic, and maintained that the national question had been
solved by the 1918 unification. Similar divisions are not coinci-

dental in present-day discussions on similar issues. Indeed, then as now, in both ideological-theoretical discussions and state-political viewpoints, a predominance of views are conditioned by the origins and "existential" interests of one's own national being.

Duretic's claim that after the unification of the Kingdom of Serbs, Croats, and Slovenes, the new state—together with its ethnic kinship—achieved a higher degree of political and economic democracy is an example of generalizations and value judgments which are based on a subjective view. Here we see an opinion on a Croatian and Slovenian issue from the Serbian-Montenegrin standpoint. And the incongruity of these realities has been felt ever since. Even the Montenegrin ethnic kinship with the Serbs—which is incontestably closer than with the others—was not close enough to bridge Serbian-Montenegrin divisions. Montenegrin resistance to the imposition of Serbian hegemony had to be overcome by the ruthless force of arms.

But however much that judgment about a higher degree of democracy may relate to that particular part of the whole, it has no bearing at all in the case of Croatia, Slovenia, and other lands that were under Austro-Hungarian rule. And in this judgment, one sees the reflection of a broader and deeply rooted prejudice that democracy in Serbia—with all its dynastic and "Black Hand" terrorist vindictiveness, which appalled Europe—was supposedly on a higher level than in the "backward" Habsburg monarchy. When Serbian historians admit to a higher degree of economic development of Croatia and Slovenia, they forget that in the lawfully ordered state of Austria-Hungary, a high level of civilized society, parliamentary democracy and political freedom was reached. Even the European Marxist movement reached its zenith in Austria-Hungary. For no matter how much Austro-Marxism was being rejected by later Leninist views, history has clearly attested to its reformist path toward a socialist society, along with and even contrary to that of the Bolshevik revolution.

In the sphere of this European cultural and political climate, Croatia had a considerable degree of national-state autonomy with its Croatian Sabor (parliament) and Banus's government. But

in the new Serbian-dominated state it was suddenly left without political and economic rights. This uncovers the entire error in Duretic's accusations concerning Croatia's fault in intensifying Serbian-Croatian relations and even in his "anational longing for the good old days." The dissatisfaction of the Croatian nation with conditions in the Habsburg monarchy was due to the fragmentation of Croatian lands and the lack of complete Croatian sovereignty. In the new state, the dissatisfaction became even greater due to betrayed expectations concerning the attainment of a higher level of national liberty and better conditions for economic and cultural development. In setting demands for national political and economic equality, it was not a case of any sort of intensified relations with the Serbs or anational longing for times past. To the contrary, it was the expression of natural aspirations on the part of a national entity to secure a higher level of its development, never once giving in to circumstances that would endanger it.

Furthermore, the political views contained in Duretic's "fact" that the formation of the Kingdom of Serbs, Croats, and Slovenes meant a supposedly higher level of political and economic democracy, amount not only to a totally erroneous interpretation of actual facts, but also create a basis for virtually all misunderstandings in examining the entirety of later historical occurrences—particularly the causes of the collapse of the state in 1941, and the conditions that brought about the revolution, all the way to the present controversy about the constitutional order.

Unification into a single state came about under completely different historical circumstances and conditions in different South Slavic lands. It had very different, and in some ways completely opposite consequences, both with regard to the resulting spiritual state, and the future development of individual nations, and especially the Croatian and Serbian nations. Serbia emerged from World War I in triumph, experiencing the proclamation of the Kingdom of Serbs, Croats, and Slovenes in the euphoria of a redeemer. The expansion of the nation-state, the unification of all Serbdom under the national dynasty of Karadordevic, and the

liberation of the other brethren served to strengthen Serbian national consciousness and unity even further, establishing firmly in the Serbian nation the authority of the dynasty, the army, the state authorities, and the ruling parties.

The situation in Croatia as well as in other non-Serbian lands was quite different. The Croatian nation awaited the fall of the monarchy in revolutionary ferment, anticipating that the decision of the Croatian parliament concerning the separation from Austro-Hungary would result in national independence and a republic—in the spirit of the principles proclaimed in the name of democracy by Wilson and in the name of the proletarian revolution by Lenin. However, the creation of a common state betrayed all those expectations. In Croatia, the revolutionary movement of the peasant class took the form of an armed, so-called "Green Cadre" insurrection, both spontaneously and under the influence of the October revolution and other European revolutions. The old state system was brought down and the authority of the politicians of the old government was shaken, as was that of those who had submitted Croatia, through the unification, to a new foreign government.

The anxious Croatian populace had striven for complete national and social liberation. But in the new state, they experienced a counterrevolution in the form of betrayal on the part of the local bourgeoisie and the intervention of the foreign, Serbian, and French armies. The brutal armed introduction of order together with the application of sadistic "Balkan" methods unfamiliar to these regions, as well as economic impoverishment (the exchange rate of four crowns for one dinar amounted to "official robbery"); unrest in the villages due to the branding of livestock, and in the cities due to loss of jobs; and difficulties brought on by the breaking of business ties with the rest of Europe; all led to a state of general dissatisfaction. Because of this, neither the common people nor the intelligentsia of Croatia experienced the establishment of the common state as a "higher level" of political and economic democracy, but rather as the stifling of a revolutionary democratic movement, as foreign occupation, and as the recur-

rence of national subjugation.

These are the reasons why the Croatian nation as a whole turned resolutely against the new centralist system, which had cut off Croatian state autonomy, and against all those civic political parties and organizations that had betrayed national and democratic expectations, placing themselves at the service of the Great Serbian hegemony. Only in this way can one explain the fact that among the Croatian people, those parties and currents which had become compromised in the eyes of the people through their shared responsibility in the unification and formation of the Kingdom of Serbs, Croats, and Slovenes completely disappeared from the political stage. This was particularly the case with all promoters of the idea of integralist Yugoslavism and collaborators with the hegemonic centralist regime.

On the other hand, the renown of Stjepan Radic's, whose opposition in the National Council to the reckless and unconditional unification of the state of Slovenes, Croatians, and Serbs with the Kingdom of Serbia had been supported only by Dr. D. Hrvoj out of all the delegates to the Croatian parliament, grew rapidly. Radic's Croatian Populist Peasant Party, which, until the unification, had played an altogether minor role in political life, became a central political influence almost overnight, particularly because Radic was so determined in his opposition to the methods of the December 1, 1918, unification. This was also due to the fact that along with the populist, democratic program of the Peasant Party, which was contrary to unitarist centralism, Radic presented a program of demands for a Croatian Republic and for a common state structured as a confederacy. Although a stalwart defender of parliamentary democracy and an opponent of revolutionary methods in political struggles, Radic saw that he had no influence regarding the centralist Vidovdan Constitution, and listened to the voice of the Croatian people. As a result, not only did he change the name of his party from Populist to Republican, but also, in the name of the elected representation to the Constitutional Assembly of the Croatian people, he brought forth in March 1921 the Constitution of a Neutral Peasant Republic of Croatia. With that,

Radic's leadership of the Croatian Republican Peasant Party had sharpened to the limit its programmatical demand for Croatia's self-determination and sovereignty, as well as for a republican and confederative arrangement of the union of Southern Slavs.

However, one must bear in mind that as far as its political program was concerned, considering the existing order, this act of revolution did not satisfy the people because it remained within a declarative, amicable framework. Namely, conditions were such that it was expected that Radic, as the uncontested representative of the Croatian nation, would proclaim a republic in Zagreb, which the Croatian populace had declared itself in favor of, through its many uprisings in the period from 1918 to 1920, as well as through the constitutional elections.

The people were prepared even for a revolution for the sake of a Croatian Republic. But Radic, a pacifist, opposed revolution, knowing that Belgrade wouldn't hesitate to attempt to carry out the threat that in such a case, "no stone would be left standing" in Zagreb. Reminding them that the situation had reached "the point of revolution or compromise," Radic calmed down the Croatian populace by saying "every revolution is evil." But he also directed at Belgrade the historical warning: "If it were to come to a revolution, you would collapse and we would be ruined."

All his endeavors at bringing about a solution to the Croatian question by means of negotiation remained fruitless. In 1925, compelled by circumstances, he abandoned the republican program for tactical reasons. The Croatian public showed its great dissatisfaction with this concession to Serbian hegemony. In the 1927 election, Radic's party lost about a third of its votes and six representative mandates, of which two were in Zagreb, where Dr. A. Trumbic and Dr. A. Pavelic, with their more radical national program, were elected. (After Radic's assassination, Pavelic emigrated as a representative of Zagreb.)

Such was the status of Croatian nationalism in the period of the "Vidovdan" Constitution when Stjepan Radic's Croatian Peasant Party represented the peasant and middle classes, while the representatives of integralist Yugoslavism and collaborators of the cen-

tralistic order, regardless of factional changes in the government, came from among the ranks of the upper class and bureaucrats and, in general, from among the Serbian populace in Croatia. The assassination in the Assembly was to trigger a greater interest in the late Radic's program for a free Croatia among both left- and right-wing Croatian political currents. The very idea of an integralist Yugoslavism was totally undermined by Aleksandar's pro-Fascist dictatorship, which, as the most outward vehicle of Great Serbian hegemony, directed its energies toward merciless retaliation, first against Croatian nationalism, but also against revolutionary communism.

When one takes into consideration the conditions in Croatia, the dissatisfaction in Slovenia, as well as in Montenegro—where, alongside the strengthening of a pro-Serbian orientation, dissension due to the loss of Montenegrin statehood also existed—and especially the poisonous situation in Macedonia and in Kosova, it is clear that the Yugoslav state of the Treaty of Versailles was burdened from the onset with virtually irreconcilable problems which transcended the physical power and political wisdom of the ruling Serbian bourgeois class, with its dynasty and army at its head, which was trying to arrange Yugoslavia into an expanded Great Serbia.

The creation of a mutual state after the First World War did not have the same consequences for all the nations in that union. The idea of Yugoslavism did not achieve a spiritual basis even among the Serbian people, let alone the non-Serbs. Because of the undemocratic way in which unification was carried out and the forcibly imposed centralist hegemonical order, monarchist Yugoslavia, from the very beginning, was caught up in a grievous internal crisis. In the general circumstances at that time domestically, as well as in Europe, the ruling Serbian stratum had little difficulty in surmounting the initial state crisis with a very narrow stratum from Croatia and the other lands, but not when it came to the removal of its causes. The crisis was to grow deeper and become increasingly connected to the intensification of the European crisis which was to lead to World War II.[154]

Among those in Croatia who advocated the reconciliation of the Croatian-Serbian disparity, contrary to Croatian opposition parties, were representatives of the Catholic Church. For example, in the early 1920s, the obscure Croatian Populist Party (Catholic) joined with Korosec's Slovenian Populist Party in the Yugoslav Club. And at the time of the January 6, 1929, declaration of dictatorship, the parish priest of the St. Blaise Church in Zagreb, Fr. Svetozar Ritig, together with Ivan Mestrovic, organized a "deputation to pay homage" to King Aleksandar in Belgrade, and later, Aleksandar's visit to Zagreb, in order to ease relations between Croatian nationalism and Serbian hegemony.

As this had no effect in bringing about a change in the dictatorial regime, and seeing the futility of such endeavors, Ritig rejected his nomination for senator, and resigned from his position as representative of the city of Zagreb, explaining his move in a statement which he sent both to the king and the president of the government, General P. Zivkovic. Its deeper significance, however, did not reach those who, like Duretic, ignored the reality of international relations, declaring all criticisms of the hegemony to be invented myths. In that historically significant report (August 3, 1932), Dr. Ritig said, among other things:

"We are broken under the burden of a cross which a number of us have taken upon our shoulders in order to create a link between the Croatian nation, its history and its interests and the idea of a Yugoslav state. We have not succeeded, nor have those who are stronger and greater than we. The way of the cross continues for the Croatian nation, as it does for Yugoslavia. He will be its Savior who finds a bridge into the heart and soul of the nation which remains far away from the present-day system, and who establishes justice, brotherhood and equality as the founding basis of national and political life in Yugoslavia."[155]

In connection with the problem of hegemony, one meets with a very interesting phenomenon. All those who deny the existence of a Serbian hegemony in the Yugoslavia between the wars, do so, as a rule, with the explanation that it is a pure fabrication by the combined forces of separatist nationalism and Comintern dogma-

tism. Such an approach actually rejects any justification for solving the national question from the Marxist standpoint because that would be nothing more than the manipulation of a nation's self-determination in the interests of socialist revolution. To be sure, there is some truth to that comment. However, the intent, and the very act of manipulation for the purpose of achieving the goals of the revolution do not lessen the historical effect of the realization of a nation's right to freedom and independent growth. Some nations have achieved that goal in an evolutionary historical development, others through civil or socialist revolutions, while still others are having that right challenged in the name of none other than the lofty ideals of those revolutions.

Those who, claiming the Comintern myth of Serbian hegemony, wish to reject all that has been achieved through Marxism in relations between nations and in the battle against nationalism, are well versed at invoking the Marxist classist view of history. An outline of a memorandum of the Serbian Academy of Arts and Sciences reads, for example, that "Mimicking the thesis of a confrontation between the Serbian oppressor nation and the oppressed nations has led to a departure from Marxist doctrine concerning the classist division within every nation due to political pragmatism, which, in its endeavor to make use of friction between nations, has pushed aside classist internationalism." The Comintern, because of its strategic aim of dissolving Yugoslavia, took on a view of Serbian hegemony "actually from civic parties with a separatist orientation," while the KPJ has conformed to the Comintern's authority rather than arriving at the truth about relations between nations on its own.[156]

This claim is erroneous because even Yugoslav Marxists, and not only civic "separatists," expressed their opinions about the Great Serbian hegemony and oppressor and oppressed nations prior to the decision by the Comintern concerning the dissolution of Yugoslavia. In addition, the *a priori* condemnation of every form of nationalism and separatism is advocated by undemocratic, unitarist, or imperialistic standpoints, by means of which those nations which have found themselves in various state configura-

tions are denied the right to self-determination, a right which is a generally accepted principle in the modern world. Is it necessary to mention that in the four decades since the founding of the United Nations, the number of its member nations has quadrupled? These are independent states, which, according to such standards concerning separatism and hegemony, would never have found a place among the sovereign participants in the international order, but would have had to remain for all time content with patriotism toward some colonial, imperialistic, or other kind of expansionist political creation.

Casting away the "mythomaniacal lie" about Great Serbian hegemony with the same fervor with which he magnifies the "truth" about the Jasenovac distortion, Vuk Draskovic is, of course, only partially correct in his defense of Great Serbian ideology. Even if one is to accept his thesis that "only such" Serbs as are "desired and prescribed" by adversaries of Serbdom, whom he finds "in the Vatican, in Zagreb or in Sarajevo," are not true Serbs because "a true Serb is only one who is a Great Serb," it still remains unclear whether Serbdom survives only by Great Serbian ideology. He writes: "Only they, those Great Serbs, created Serbian culture and Serbian history—from the Nemanjics and Saint Sava to Njegos and Karadorde." It doesn't occur to him that a Great Serbia might for some be undesirable and threatening. He writes, "The idea of Great Serbia never grew out of hatred toward another nation, nor did it lead to anyone's downfall, collective hysteria, or genocide." According to Draskovic, the sails of "Great Serbia" were always "blown by the wind of the national aspiration to protect and gain back its own, but never to build its freedom and its fortune upon the subjugation and misfortune of others." As proof of this, he claims that the Serbian Church had "proclaimed as saints" all Serbian rulers except Dusan the Conqueror. To that idealized picture of Great Serbian ideology, unlike that of any other nation in world history, he finally adds a little substance: "Wherever there are Serbian pits, Serbian execution sites and Serbian graves, there was, is and will be Serbia." Thus, in just the history of this century, as we know, that means

everything from the ravines of Albania and from Kajmakcalan in Macedonia, across Croatia, to Koruska in Slovenia, and Trieste in Italy. There are, of course, graves and execution sites of other nations in these places as well, which should have no less a claim on their own lands. From this crude reality, it can be deduced that the alleged "mythomaniacal lie" concerning Serbian imperialist expansionism, oppression, and hegemony—"which bring about distrust and open animosity toward the Serbian nation"[157]—did not originate with the Vatican and the Comintern, but within the historical being of the nations which felt the oppression of Great Serbian activity.

Attempts to deny the existence of Serbian hegemony, and in so doing, not to allow the "equalization" of Croatian and Serbian "historical guilt" are completely incorrect, and therefore, not only futile, but also deleterious.

The idea of Great Serbia indubitably also played a positive role in the history of the Serbian nation. The sin is not in its mere existence nor in all the consequences of its appearances. There is no historical nation on earth which has not had expansionist ideas or the conception of a great state. The problem is in the sober consideration of when and in which cases such an idea, in this case that of Great Serbia, came into conflict with the survival of other nations, resulting in iniquity to them and to its own national being by overstepping the boundaries of its own capabilities and the given historical circumstances.

The one-sided understanding of national being in general, and particularly the refusal to recognize the historical identity of the Croatian nation and its right to realize its full national sovereignty are maintained, with particular duplicity, in the interpretation not only of the causes of the collapse of Yugoslavia in 1941, but also of the consequences—especially the Serbian and Croatian nations as the two primary factors in the reshaping of Yugoslavia's historical fate. Here one meets with the very phenomenon—which manifested itself most systematically in Duretic's books—that all significant issues are viewed, if not exclusively from the standpoint of the Serbian nation, then from the Serbian perception of a

Yugoslavism that in reality hardly existed anywhere.

According to Duretic, only the attitude of Serbian factors had to be "beyond all ideologies and politics," which is also true "to a somewhat lesser extent" for Slovenians, and, thus, they should be judged accordingly. He finds justification for both Draza Mihailovic's Chetnik movement as well as the government of General Nedic in its service to the German occupier for reasons of "existential realism" or "national realism." He judges their politics as "existential self-defensive dialectics" and "national maneuvering," whether in seeking a "modus vivendi with the occupier" in order to avoid unnecessary casualties among the Serbian people, or in opposition to the revolutionary Machiavellianism of Tito's Communist NOP. To Duretic, the Chetnik movement was one pole of "Serbian anti-Fascism," while the Partisan movement was its other pole, which succeeded by means of the "revolutionary ploy" to discredit the Chetniks as the "only anti-Fascist rivals." Duretic realizes that the Chetnik movement did not limit itself only to retaliation (mentioning again the genocide of "hundreds of thousands of Serbs"), but that it also has designs of Great Serbian expansionism and anti-Communist fanaticism.

However, not even this convinces him of the need for national existentialism on the Croatian side. He labels equally Pavelic's Ustasha movement and Macek's Croatian Peasant Party for their "anti-national foreign dependence" or for their "attack upon Yugoslav patriotism" in connection with the establishment of the NDH. He likewise condemns "vacillation among the ranks of Croatian revolutionaries," seeing in it "an alliance of all the opposition's forms of antihegemony" which "gave rise to a substratum of such animosity toward the Serbian nation that it bordered on insanity."[158]

Beyond or apart from an exclusively ideological interpretation of historical occurrences, there is a basis for finding reasons for Mihailovic's and Nedic's politics in the existential interests of the Serbian nation. But how from such a standpoint can one not see that the Croatian nation had no fewer existential interests for both Pavelic and Macek? After all, events did take place, and one ought

to consider why they happened.

One should recall that the fall of monarchist Yugoslavia took place at a time when Hitler was sovereign ruler of all Europe, when he was not yet at war, but in alliance with the USSR, "a great socialist country." Thus there were not yet any realistic assumptions upon which to assess the possible duration of his New European Order. It should not be forgotten that the Versailles order, destroyed by Hitler, had barely lasted two decades, while many strive to defend it as though it had existed forever, and as if it hadn't carried the seeds of its own distruction.

With the downfall of monarchist Yugoslavia, Serbian hegemony also experienced a historical decline. Mihailovic's Chetnik movement was, in the beginning, an expression not so much of existential realism as of a longing to establish a Greater Serbia in order to secure its hegemony over a shattered Croatia in a restored Yugoslavia. Scarcely having confronted the growth and objectives of the revolutionary movement, it became the expression of the "Realpolitik" of those Serbian currents which were resolutely opposed to Tito's Partisan movement because of its equally Communist and AVNOJ foundations. Because of this, the Chetnik movement entered a tactical collaboration with the Italian and German occupiers, while remaining, however, strategically linked to the anti-Hitler coalition. The Chetniks were the only real force upon which the Great Serbian representatives in the exiled government could rely in the re-establishment of a Versailles Treaty Yugoslavia, a hegemony for all non-Serbian nations within it.

The quisling government of General Milan Nedic in Serbia was from the onset an expression of the realistic existential politics of those Serbian currents which had concluded that it was necessary to take into account the reality in which Serbia found itself in Hitler's Europe, as well as of those currents who were for an independent Serbia, and not for the renewal of Yugoslavia, even in the case of an Allied victory. Both Mihailovic's and Nedic's politics of the existential and national realism of "guarding Serbdom" were the foundation for collaboration between the Chetnik movement and Nedic's quisling government, and particularly against the rev-

olutionary movement in Serbia. In this endeavor, both Mihailovic and Nedic had support from among the adherents of virtually all Serbian civic parties and currents, while Ljotic's ideological Serbian Fascist movement remained throughout the war on the fringes and of secondary importance in the organization of Serbian political life.

As a representative of that right-wing Croatian current which was resolutely in favor of separating Croatia from Yugoslavia, Pavelic was clearly more justified in adopting his politics of national existential realism than was Nedic. The NDH as part of Hitler's New European Order was set up to include its maximum historical and ethnic territory, while Serbia was diminished and territorially crippled. The same holds for Macek in relation to Mihailovic. Rejecting the German offer that he himself proclaim the NDH, both because of his democratic and anti-Fascist views, as well as his conviction that the Axis Powers were going to lose the war, Macek, in calling for loyalty to the Croatian state pro-claimed by Kvaternik, unambiguously pronounced his support of its establishment, but not of its Ustashan, pro-Fascist regime. Politically equidistant from both Pavelic's Ustashism and Tito's revolutionary movement, Macek's middle-of-the-road Croatian Peasant party was to remain the chief political force opposed to the revolutionary NOP on Croatian soil, just as Mihailovic's Chetnik movement was in Serbia. Macek's "politics of waiting" was to experience its political downfall when broad strata of Croatians perceived that it would not secure the existential inter-ests of Croatia. For this reason, they became increasingly oriented toward the program of the NOP. And for similar reasons, caused by the military and then political decline of Mihailovic's Chetnik movement, the orientation of a considerable portion of the Serbian populace of Croatia and Bosnia-Hercegovina was pro-NOP because its program was better suited to the existential inter-ests of the Serbs "on the other side" than was the Chetnik pro-gram, in the mere fact that it offered a basis for the restoration of a Yugoslav state.

Instead of a sober consideration of all the factors of the exis-

tential or national realism of individual nations—which I myself had attempted three decades ago, clashing with both hegemonical and ideological views[159]—Terzic, Dedijer, Duretic, and others engage in the superficial labeling of the Croatian nation because of its anti-Yugoslav attitude and "national treason." They find affirmation for their views in the welcome given to the German army in Zagreb, in the alleged announcement by Hebrang that the NDH meant the realization of "centuries-old dreams of the Croatian nation," and in the "vacillation" of the leadership of the KPH and the alleged underhanded dealings between the Croatian communists and the Ustashas.

It cannot be argued that the Croatian nation, taken as a whole, welcomed the fall of the Yugoslav hegemony as their extrication from "prison," and that this was publicly visible. Nor is it disputable that the declaration of an independent and free Croatian state at first meant the realization of a "centuries-old dream," dreamed by national, but also class, revolutionaries. However, it is even less arguable—both from the standpoint of the participants as well as from the objective investigations of the historical events of that time—that the Croatian people not only did not identify themselves with the Ustasha regime of the NDH, but that they also increasingly and more resolutely distanced themselves from that regime, having become suddenly sobered by its pro-Fascist, pogrom methods of rule and the handing of Dalmatia over to Italy. Thus even Hebrang's possible declaration should be taken as a sign of the spiritual state conditioned by Croatian-Serbian relations, and not as evidence of his endorsement of Pavelic's NDH.

Likewise, one should have a clear understanding of the historical background of the relationship of the Ustasha leaders towards Croatian communists in the first days of the NDH. Since many Croatian national revolutionaries had found themselves alongside Croatian communists in Yugoslav prisons, and since the programs of one and the other involved a free Croatia, in the first days of the NDH there was no persecution of communists. Rather, there were even promises that no harm would come to them as long as they didn't work against the Croatian state. (The afore-

mentioned invitation by the Ustasha representative, Wagner, to a member of the CK KPH, Vlado Janic in Sisak, was not an isolated instance in this respect.) However, as the revolutionary and anti-Fascist activities of Croatian communists became concretely aimed against the NDH, Ustasha terror soon came crashing down upon any communist it could find.

The fact that there was no mass uprising in Croatia in 1941, as there was in Serbia, was not the result of any sort of vacillation of the CK KPH, but of historical circumstances. Indeed, the Croatian people perceived themselves to be in a different position than were the Serbs, so that for the same "existential" reasons which in Serbia gave rise to a "general national uprising," such a thing could not occur in Croatia. Nor was there in Croatia such a crushing defeat of the NOP as there was in Serbia. The reasons for this, indeed, were again of an existential nature.

The Croatian communist leadership is also ascribed with separatist intentions (by Dedijer, et al.) in connection with attempts on the part of the Comintern to meddle in internal relations within the KPJ in 1941. At that time, through the interference of a "representative" of the Comintern, J. Kopinic, through the party committee in Zagreb, an attempt was made to supplant the CK KPH and appoint a new one, resulting in the tragic loss of a large number of prominent Croatian communists, after their sloppily organized liberation from the camp at Kerestinec.[160]

This interference on the part of the Comintern has not yet been completely elucidated historically, but there is a basis to the assumption that it was probably an attempt to choose a new, separate leadership for the Communist Party of the NDH. The following facts indicate this: the Soviet government accepted the collapse of Yugoslavia as a fait accompli so soon (May 8, 1941) after the fall and partition of the Yugoslav state that it broke off diplomatic relations with the monarchy's government in exile. That very act should be interpreted as a recognition of the newly established situation. However, the leadership of the KPJ, which under Tito had become independent of Moscow, did not accept this. Therefore, the Comintern's endeavors through its "representa-

tive" in Zagreb, did not succeed, while it did succeed in carrying out its directive to bring about the separation of the Provincial Macedonian communist organization from the KPJ and join it with the Bulgarian Communist Party (a move which was later retracted at Tito's request). One Croatian who was in Moscow in April 1941, was invited to remain there as a representative of the NDH, but he declined because he didn't have the consent of the leadership of the Croatian Peasant Party.[161] Certain diplomatic functionaries of the NDH have told of their recollections that Soviet representatives (in Berlin and Bratislava) announced that Moscow was prepared to recognize the NDH and to set up diplomatic ties with it (as Moscow recognozed Slovakia). Pavelic, however, did not hasten to accept the Soviet offer before the war between the USSR and Germany. And the fact that Comintern steps taken through Kopinic in Zagreb did not bring about their intended effect is certainly ascribable both to Tito's leadership of the KPJ and to the leadership of the KPH. Thus it is unfounded to make any claims about separatism on the part of the KPH itself. Indeed, in these historical examples, both the leaders of the Croatian right wing as well as the left wing gave priority to the demands of their ideological politics over their national politics, which was the cause of their irreconcilable conflict. However, the opponents of one and the other clearly did not recognize this as positive, but rather, in spite of this, they judged them to be "separatist" as well as "genocidal."

There is a general unwillingness to understand that hegemony was one of the chief causes of the instability of monarchist Yugoslavia. The blame is simply placed upon "separatism" on the part of all elements of Croatian politics (i.e., upon the entire Croatian nation, without questioning wherein lay the essence of the problem and how it could have been resolved). Along such a line of thinking, one either completely overlooks the fact that conditions in Croatia and Slovenia in the days of the monarchy's decline were similar, or, on the other hand, one does not perceive the reasons why their respective situations developed differently later.

In considering the problem of the occupation and revolution, a Slovenian journalist told me a number of decades ago that it is awkward to seek an answer to the question: Why did not only the residents of Maribor, but also the miners of Trbovlje, greet the German army with flowers? One of the most eminent Slovenian proletarian revolutionaries and secretary of the CK KP Slovenia, Miha Marinko, had the courage to ask that question immediately after the war. But as it becomes farther removed in time, it has become an uncomfortable subject even to serious authors.

Similar is the case with the historical assessment of the Slovenian National Council, which was created in Ljubljana on April 6, 1941. This fact is overlooked by the persecutors of Croatian separatism, as if this Slovenian event didn't happen before the proclamation of the NDH on April 10, or they (but not only they) incorrectly judge it as national treason. The Slovenian National Council—as a sovereign body representing the Slovenian people—was founded by representatives from all currents of Slovenian political life. On the very day when aggression against Yugoslavia began, in attendance at the founding conference were representatives of the Slovenian Humanist Party, the Yugoslav National Party, the National Radical Party, the Yugoslav Democratic Party, the Independent Peasant Party, and the Socialist Party of Yugoslavia. The CK KP Slovenia presented the request that its representative also be included in the National Council, but the request was not accepted with the principled explanation that only representatives of recognized political parties should form the Council.

As can be deduced from the composition of that council, it came about at a critical turning point in history, as the result of an existential necessity on the part of the Slovenian nation in order to have a favorable effect on the nation's fate. In the days of the Yugoslav collapse, the Council requested on April 7 that the Yugoslav military authorities prevent the destruction of manufacturing and transport industries, to which the Yugoslav Supreme Command bluntly replied that the Council's members would be arrested (April 9), but there was no one to carry out the order. On

April 10, when the German army began to invade Slovenia and
Croatia, the Slovenian National Council presented a resolution
and publicly announced to the Slovenian people that the "supreme
state administration of Yugoslavia over Slovenia" had ended,
and that the Council would take over the sovereign power of the
territory of Slovenia and "the protection of peace, order, and the
fate of the Slovenian nation," calling on the state administrative
authorities to remain at their posts. In the evening of the same day,
a delegation from the Council requested of Yugoslav generals
Pandurovic and Lavadinovic that the Yugoslav army on Slovenian
soil recognize the sovereign government of independent Slovenia,
which they did not accept, but it is significant of the state of dis-
solution of the monarchy that both sides were powerless. Upon
the arrival of the German army at Celje, a delegation from the
Council (Dr. Natlacen, I. Pucelj, Dr. Gosar) made a proposal on
April 12 to the German military representatives to establish the
groundwork for negotiations with the German government: that
the Germans accept the "sovereign and independent state" of
Slovenia as a new co-signatory to the tripartite treaty and that
Slovenia be recognized within Germany's New European Order.
Negotiations with the German government never materialized
because Hitler did not wish to change his plans for dividing up
Slovenia. The Council attempted to negotiate on the same points
with Italian representatives, but for the same reasons, efforts were
fruitless.[162]

From a Yugoslav viewpoint, the emergence of the Slovenian
National Council can indeed be labeled treason, while from a
class-ideological standpoint it can be viewed as an attempt by the
"Slovenian bourgeoisie" to protect the capitalist order and the
class interests of its rule. But from a scholarly historical viewpoint,
this is as biased as it is absurd. The endeavors of the Slovenian
Council had as their objective the prevention of the division of
Slovenian national territory, and thereby also the threat of dena-
tionalization and physical endangerment, which would be the
result of partitioning the Slovenian nation among three foreign
powers (Germany, Italy, Hungary). Herein lies the essence of the

Council's emergence in the days of the overthrow, while the failure of its efforts and the genocidal anti-Slovenian politics of the occupiers created a basis upon which, in the first days of the occupation, the KP Slovenia proposed the program of the Liberation Front. For this very reason, because the program expressed the existential interests of an imperiled Slovenian nation, broader political currents became involved in it than was the case in the other lands of a partitioned Yugoslavia.

While, at turning-points in history, individual national elements, political parties, currents, and movements orient themselves in accordance with their worldview and their understanding of historical development, nations as a whole behave more or less equally in a psychological sense, according to the historically acquired instinct for self-preservation, which can vary greatly, regardless of similar external factors.

The significantly different positions of the Croatian and Slovenian nations in the Second World War most clearly manifested themselves in the long-term plans of the architects of the New European Order. Hitler himself had offered to give an independent Croatia a place in the new Europe. Not even such an astute thinker in the field of international relations as Josip Vidmar was able to grasp the result of that quite different historical position. In his otherwise sober reflections, on the occasion of the publication of his book *On Slovenism and Yugoslavism* (in a conversation with J. Sentija), he concluded superficially that the Slovenian nation in 1941, "was left completely isolated" because "the Serbs had gone their way" and "the Croatians were going the way of Pavelic."[163] Certainly, the Croatians were not going "the way of Pavelic," but they had merely gotten that which the Slovenians had wanted, in order to secure their existential interests in the given circumstances. And had the Croatians not gone resolutely their anti-Pavelic way as well (since Pavelic was failing), also in order to secure their national state sovereignty, as had the Slovenians, it is quite clear that there would never had been a state community on the principles of the AVNOJ. This should not be enigmatic to a man who shared Krleza's "Croatian

stance: the Croatians are their own nation," and who correctly feels that Croatian-Serbian relations are essential "in the complex of Yugoslavism,"[164] and who understands (this point was made clear to me in a personal conversation with Vidmar) that a resolution to those relations cannot be achieved through attempts at mutual denial and extermination, but through sober understanding.

Those who did not understand the essence of the relations between the nations, or the causes of the collapse of Yugoslavia and the situation during the war, also misunderstand the essence of the NOP, i.e., the need for a revolutionary change in the structure of the state community of nations of Yugoslavia, and for both a victory in the anti-Fascist war and revolution, as well as for the realization of premises for its further survival and a viable co-existence of Yugoslav nations.

One of the most widespread prejudices along these lines is the one about a "Partisan Yugoslavism," as interpreted by Veselin Duretic and Ljubomir Tadic. In the concluding reflections in his "Yugoslav drama," Duretic presents judgments that are in total contradiction to the historical reality. According to him, among "Yugoslav specific traits... up until AVNOJ, international goals had top priority," while later, national problems "were taken up by the political determinant of the struggle for brotherhood and unity and national equality (without any clear statelegal statement)." This was "in the name of Yugoslav and not individual national goals," for "the majority of old communists had experienced the national folklore within the unfolding of the international course of growth." According to Duretic, the leadership of the revolution based its politics and its victory upon the myth of "Great Serbian hegemony and oppression" as well as upon the myth of the "Croatian Partisans." The "social and national unconstitutionality of Serbdom" became evident, while "Croatian and Moslem society demonstrated the strength of denationalizing factors and the predominance of centrifugal development over the ethnic Yugoslav development." And in his last sentence, as the crowning glory: "The indubitable merit of the leadership of

the Communist Party of Yugoslavia lies in the fact that its revolutionary activities" were "the sole negation of the existing national and religious contradictions, from which resulted the common skeleton of supranational strength, which was the historical chance for the Yugoslav idea to be regenerated upon a broad democratic and class basis and to be realized as part of the great process of social reform, that is, socialist process."[165]

Professor L. Tadic, who is quite familiar with the history of the debate within the communist movement concerning the national issue, first surprises us with the claim that Serbian Marxists had long been "making concession after concession," and, subordinating themselves "virtually masochistically" along party lines, they "sadistically persecuted real and imagined Serbian nationalism." If one allows that political-psychological reasons can condition even such individual phenomena, how can one explain the differences, which even among the ranks of the Serbian Marxists were great, ranging from the adherents of Leninist principles concerning the necessity of recognizing a nation's right to self-determination to the theoretical advocates of supranational integralism and state unitarism?

L. Tadic sees in "Partisan Yugoslavism" the "foundation for unity" while he defines it as "a living Yugoslav patriotism which has not yet been eroded by nationalistic particularism." Its origins are clearly the places where the KPJ was able successfully to organize a "Partisan uprising," and this was "primarily in areas in which the tradition of liberation battles still lived on after past wars." However, upon later constitutional solutions "it would seem... that in the destruction of monarchist Yugoslavia, the Serbian Partisans did not make a proportional decisive contribution, i.e., in numerical terms."[166]

The views expressed by Duretic and Tadic were present in part during the NOP and revolution; they drew their roots from the prewar legacy. Judging by the fact that many would give them a general and conclusive significance, they are to a great extent even more present today than they were in the prewar and revolutionary periods.

The fact that there were "old communists" who felt that the national question was, if not a "capitalist fabrication," then merely a tactical appendage in carrying out the proletarian revolution was manifested—after the debates at the beginning of the 1920s—again on the occasion of the establishment of the Communist Parties of Slovenia and Croatia. However, it is quite certain that such people did not make up the "majority," at least not in the Croatian and Slovenian workers' movement. In the period of the revolution, they did not play a major, but a secondary role. With their "two-stage" perception of the revolution, in new circumstances they were thrown to its fringes. After accepting the program for the defense of Yugoslavia from the dangers of Fascist aggression, the leadership of the KPJ did not abandon its program of struggle for the national liberation of oppressed nations. In the summer of 1935, the CK KPJ stated in explanation of the decision by the Fourth Land Conference of the KPJ concerning the founding of the Communist Parties of Croatia and Slovenia: "We are for freedom and the right to self-determination on the part of oppressed nations. But we do not seek a break-up of Yugoslavia at any price. In this respect our objective is freedom, equality, and the brotherhood of all nations; the right and freedom of every nation to determine for itself with whom and how it will create a state community. If Croatians, Slovenians and other nations, having won full liberty and the right to self-determination, decide freely and without any coercion or deception to remain within the federation together with the Serbian nation, we are not opposed to this, as we are likewise not opposed to secession, if that is the only path toward the liberation of those nations. The name of the Communist Party of Yugoslavia does not in any sense imply the existence of some impossible 'Yugoslav' nationality. It simply signifies that our party operates in the territory of a state which is called Yugoslavia."

Why was no Communist Party of Serbia established? The "first and main" reason was that "the working masses in Serbia do not seek the formation of a Communist Party of Serbia." And "this is understandable. The Serbian nation is not nationally oppressed.

The Serbs are the ruling nation and they do not feel national oppression on the part of another nation. Therefore, the class struggle [in Serbia] does not take on the shape of a battle for national liberation."[167] In the declaration of the KPJ of January 1937, it is stated: "In order to reorganize the state of Yugoslavia, it is necessary to summon a Constitutional Assembly, which will resolve the mutual relations among all the nations and component parts of Yugoslavia, free nations without any sort of supremacy of one nation over another," and thus, "individual national assemblies will be elected (the Croatian Parliament, etc.) which will determine the internal organization of those regions."[168]

The Central Committee of the Communist Party of Croatia, in its Declaration from the inaugural Congress addressed "To the Croatian nation" (1937), says: "The Croatian nation is unanimous in the struggle to break the bonds of national slavery"; "The working class of Croatia is doubly enslaved"; as a part of the Croatian nation, it is "nationally and economically oppressed by the Serbian imperialist hegemony," as well as "mercilessly exploited" by Croatian capitalists. In the struggle for "freedom and democracy," Croatian communists declare their support of concordance and collaboration with "all national parties" (the Croatian Peasant Party and the Independent Democratic Party) and groups, and for an agreement with the Democratic Peasant Coalition and the Serbian United Opposition. "Without national equality there will be no survival for the present state community... Only free, equal, concordant and satisfied nations of Yugoslavia will be willing and able to thwart every foreign attempt at Fascist invasion."[169]

But it should be recognized that certain differences existed in judging the importance and significance of the formation of national parties, between the Slovenian and Croatian communists on one side and the Serbian communists on the other. "Serbian communists," it was written in *Proleter* [*The Proletariat*], "have always been... for the self-determination of the Slovenian, Croatian and other nations of Yugoslavia and for their equality with the Serbian nation." But it is also stressed that "the creation

of the Communist Party of Slovenia and the Communist Party of Croatia does not amount to a victory for separatism and nationalism over proletarian internationalism within the ranks of the Communist Party of Yugoslavia and the working class." The formation of those parties is welcomed "as a step forward in the struggle for the unity of the working class of Yugoslavia" as well as "the concordance of all nations of Yugoslavia."[170]

The difference is clearly in the emphasis on the unity of the working class and one's own nation on the one hand, and the "unity of the working class of Yugoslavia" on the other. Herein is the probable reason for criticism of the CK KPH by the CK KPJ (in 1939) due to alleged submission to the influence of bourgeois nationalism. But in essence, the program viewpoints remained unchanged.

In the conclusions of the Symposium of the KPJ, held at the end of April 1941 in Zagreb, the judgment was made that twenty years of "the politics of horrible national oppression" was the reason for Yugoslavia's demise. Within the framework of the general goals for the battle against the occupiers and for the freedom of all the nations of Yugoslavia, it stands out that it is "simply untrue and slanderous" that the communists are "against the independence of the Croatian nation." It is particularly emphasized as an "urgent task" of the KPH that it organize a battle "for the true independence of the Croatian nation" and for its joining the other nations in the battle for national liberation.[171]

These were the understandings with which the revolutionary communist movement entered the war and revolution. The fact that among the ranks of all non-Serbian nations there was even less room for Duretic's unitarist "supranational" Yugoslavism as opposed to supposedly "strictly national objectives" can be seen throughout the entire development of the NOP both before and after the AVNOJ. Even in the 1958 Program of the SKJ, it is stated that "Great Serbian hegemony and centralism have caused a situation in which throughout the existence of the former Yugoslavia, the disparity between the Yugoslav nations grew ever deeper," and that "Socialist Yugoslavia came about, and could only have been

formed as a federal state made up of equal and sovereign nations."[172]

One can assume that the above-mentioned image of Duretic, as well as that of Tadic, concerning "Partisan Yugoslavism" undiminished by "nationalistic particularism" was present in the Serbian perception, but certainly not in that of the Croatians, Slovenians, and Macedonians. In Croatia, for example (and only to a slightly lesser extent in Slovenia), the very idea of "Yugoslavism" was so unpopular, and, moreover, detested among the general public because of its identification with Aleksandar's dictatorship and Chetnikism, that it was to be avoided and was ascribed with a completely new significance within the framework of the NOP. The extent of the distrust of anything Yugoslav can be seen in the fact that the agreement between Tito and Subasic provoked dissatisfaction and anxiety, both among the people and the Partisan units, and great political efforts had to be undertaken to explain that this in no sense meant a return to the conditions of the former Yugoslavia. Even the hardy slogan, "No return to the old way," contained this idea, and with it and because of it, also its socialist content. After all, the most appealing aspect of the NOP's program was that every nation by means of arms was to take its fate into its own hands.

In Duretic's denouncement of the myth of "Croatian Partisans," in Tadic's reference to uprisings in areas with a tradition of fighting for independence, and in the admonishment (on the part of both of them) to forget who gave the "decisive contribution," there lies hidden a deeper misunderstanding about the factors that determined the victory of the revolution.

The historical significance of the uprising in Serbia and Montenegro in the summer of 1941 is indubitable. But also the defeat of that uprising followed by the waning of the Partisan movement and the increase in Chetnikism in those areas, and the steady increase in the growth of the NOP in Croatia and Slovenia, followed by Macedonia, were conditioned by deeper national-political reasons. A mass uprising in 1941 was possible because of the general dissatisfaction of the Serbian nation and its expecta-

tion that when the "powerful" Soviet Union and "Orthodox mother Russia" entered the war, it would soon be over and an uprising would bring about a quicker end to the occupation and the restoration of the lost position of Serbdom in Yugoslavia. When this did not happen, the Serbian nation lacked the historical, national reasons for following the revolutionary program of the NOP that the non-Serbian nations had, in order to secure their futures by preventing a return to the old order. Because of this, in Serbia (and similarly in Montenegro) until just before the final operations for liberation, in the summer and fall of 1944, the influence of the Chetnik movement of Draza Mihailovic prevailed—that is, the influence of the exiled royal government in London. From this viewpoint, there is a basis even to Duretic's judgment that this was in accordance with the existential interests of the Serbian nation. Serbia was able to wait; its future was not endangered. It was otherwise with Serbdom in the NDH, and because of that, the Serbs rebelled and aligned with both the Partisan and Chetnik movements.

Since the position of the Croatian nation after the declaration of the NDH was fundamentally different from that of the Serbs there could be no existential motives for an uprising in Croatia in 1941. Because of that, the NOP was able to grow only gradually, upon the basis of the anti-Fascist activities of the communists (which began earlier in Croatia than anywhere else—the Sisak detachment). The spread of the Partisan Movement in Croatia came about as broad national strata realized that the NDH did not mean a solution to the Croatian question because the NDH would share the future of Fascism and Hitler's European order, and that the program of the NOP offered the prospect of securing the nation's own future against a return even to the exacerbated former state.

For the aforementioned reasons, conditions developed in such a way—after the failure of the uprising and the withdrawal of several hundred Partisans from Serbia—that in Croatia, there remained at the end of 1941 about twenty Partisan detachments and smaller military units with a total of about 7,000 Partisans.

When, in the course of 1942, the crisis among the rebel ranks in Serbia grew deeper, and Partisan forces were driven out of eastern Bosnia and Montenegro, very difficult conditions ensued for the NOP in the entire eastern part of Yugoslavia, so that east of the rivers Bosnia and Neretva, there remained only a few hundred Partisans.

At that same time, the NOP became more firmly established in Croatia and widespread both politically and militarily. In the summer of 1941, there were about 12,000 fighters in the Partisan units, and by 1942, eighteen Partisan brigades had been created in Croatia, out of a total of thirty-seven for all of Yugoslavia. Toward the end of 1943, the Partisan army of Croatia numbered three army corps, eleven divisions, forty brigades, and an even greater number of detachments, which made up virtually half of all the Partisan forces in Yugoslavia. And in the summer of 1944, prior to military operations for the liberation of Serbia and Macedonia, under the Chief Command in Croatia five corps were in operation out of a total of nine in the National Liberation Army of Yugoslavia. According to nationality, nearly two-thirds of the Partisans were Croatians and about one-third were Serbs. There were considerably more Serbs in the Partisan forces in proportion to their population in Croatia, but claims that they made up the majority of the Partisan army in Croatia have no foundation. Those who make such claims usually forget the Dalmatian Croatians and their contribution to the battles of Neretva and Sutjeska.

In addition, the NOP in Croatia achieved the highest degree of political organization of an anti-Fascist front and new revolutionary authorities: at the end of 1941, in Croatia there were 677 various committees of the NOP; a year later, there were already 1,609 NOP committees, and by the end of 1943, all of Croatia was covered with a network of 4,596 NOP committees, from local and district to regional committees.

Preparations for the creation of ZAVNOH, the Anti-Fascist Council for the National Liberation of Croatia, as the supreme representative body, began at the end of 1942. From March of

1943, the Initiative Committee worked. At its First (June 13-16, 1943) and Second (October 12-15, 1943) Convention, the ZAVNOH rendered decisions concerning the establishment of a legal means of political representation for the NOP and the Federal State of Croatia. Such developments in Croatia, and similar ones in Slovenia, provided a real basis for the decisions of the Second AVNOJ Convention.[173]

Thus there is no question of any sort of "myth of the Croatian Partisans"—not even as far as their military contribution is concerned. And in the political sense, neither the world in general, nor any of the Allies, and least of all Churchill himself were deluded by Tito's adeptness when they regarded Serbia as being under Mihailovic's control and Croatia as belonging to the Partisans. Throughout the war, the English had military missions on one side and the other, first with Mihailovic, and then both in the Partisan headquarters (in the Supreme Command, in the Main Headquarters of Croatia, with Croatian corps, and elsewhere) in which even Churchill's son Randolph participated. In those political speculations, of primary importance was the sober assessment of the enemy's operations, and the opinion that only Tito's program, even if it was undesirable due to its Communism, offered a basis for a restoration of Yugoslavia.

Another thing that contributed to ascribing Croatian attributes to the Partisan Movement, as opposed to Serbian Chetnik ones, was the fact that Tito was a Croatian. And that historical fact, even if taken as "Tito's direction in the game of symbols," according to Duretic, cannot be attributed to any sort of "mystification," or to the propagandist creation of a myth, or to a game of "remarkable camouflage" in presenting the "common Yugoslav" NOP "as a Croatian, and partially Slovenian counterbalace to the old Serbian hegemony and tyranny."

There are a number of indications that even today Duretic does not reach the correct conclusion as to wherein lay the essence of the program and the strength of the Titoist revolutionary movement which ensured its victory, although this was understood by world figures even during the war. Judging correctly that Tito,

based on the theoretical foundations of "the Russian revolution and Marxism," gave the NOP "clearly defined national objectives," Duretic nonetheless maintains that Comintern and Soviet "patronage" made it possible for his "chiefly Serbian movement to progress along the lines of pan-Slavic and Orthodox solidarity," and that through "an ideological common denominator of national positions," and uncompromising political stance toward the "Serbian factor" and "compromises" with Subasic, he was able to reach a settlement with his "only serious adversary."[174]

In such oversimplifications, we can certainly not find an explanation as to why, at the head of the AVNOJ structure of the new Yugoslav community together with Tito as its chief political figure, one again finds a Croatian (regardless of his pro-Yugoslav orientation), Dr. Ivan Ribar. Nor is it clear why not one Serbian politician from the government in exile was to be found—even at the insistence of their mentor, the British government—who was prepared to negotiate with Tito. Subasic, however, was found. Communism, of course, was no dearer to him, but as a Croatian, he was sooner able to accept Tito's federalist basis for a state community.

In connection with this, it is necessary to consider why "Serbia during the war was not in a position to participate completely equally in making the decisions that would determine future relations between the nations and social organization of Yugoslavia." This claim is presented in a draft memorandum of the Serbian Academy of Arts and Sciences.[175] It is in one sense a censure of Tito and Kardelj as the chief creators of the AVNOJ arrangement, which since their departure has been increasingly challenged. The claim itself has a basis in factual historical events, but the reasons for this are not to be found outside of, but within the Serbian communist movement itself and, moreover, within the Serbian nation. However, such a judgment does not mean an emphasis on some historical guilt on the part of that movement and nation, but rather points out the necessity of seeking the historical and national reasons for its origin.

It happened first of all, because those who found themselves at

the forefront of the Communist Party of Yugoslavia were representatives not of the Serbian revolutionary movement and of the ruling Serbian nation, but of the Croatia, and Slovenian revolutionary movement and nations. Secondly, it happened because the revolutionary program of the NOP "concerning future relations between nations and the social organization of Yugoslavia" clearly was not suited to the interests of the Serbian nation to the same degree that it was to the non-Serbian nations, above all, Croatians and Slovenians.

This was evident from the very beginning when, in the summer of 1941, concerned about the well-being of the Serbian people, not only the Holy Archiepiscopal Synod of the Serbian Orthodox Church, but also distinguished representatives of Serbian public life, called for loyalty to the German military command of Serbia and to Acimovic's quisling Commissariat (which preceded Nedic's government). At the behest of the minister of education, Velibor Jonic, from an assembly in Belgrade, 413 representatives of political and public life, several of them distinguished members of former civic parties and top names of various professions, signed an Appeal to the Serbian Nation in which he makes an appeal to embark on the road to national salvation and renewal against the mindless operations of criminal Bolshevism.[176] (That appeal was allegedly signed by all the university professors except the philosopher M. Duric.)

In the course of the war and revolution the vast majority of Serbian parties' representatives and the Serbian Cultural Club—which before the war had been a central gathering place for Serbdom—backed the Chetnik movement, the king, and the government in exile, and were resolutely opposed to the NOP. It is significant, for instance, that the leading personage of the Serbian Cultural Club, Dragisa Vasic, who otherwise had pan-Slavic and pro-Soviet sentiments, became also the leading ideologist of the Chetnik movement. Here, there was a clear difference between the circumstances in Serbia and those in Slovenia where the Liberation Front had gathered a considerable number of representatives of civic political life.

Circumstances in Croatia were similar. Macek's leadership of the Croatian Peasant Party, which before the war was backed by the majority of Croatians, remained equally distanced from both the Ustasha regime and from the revolutionary Partisan movement, which facilitated the mass shift of its membership over to the ranks of the NOP, where the dissident Executive Council of that party was formed. A consequence of this general state, and of the slowness of the development of the NOP in Serbia, was that at the AVNOJ conventions there were neither the numbers nor the distinguished personages that there were from Croatia and Slovenia. In addition, it was only after the liberation of Belgrade at the end of 1944 that the national representative body of the NOP and top agencies of the government of federal Serbia were formed. In Croatia this took place in the summer, and in Slovenia in the autumn of 1943, in the midst of war, upon liberated territory.

Part of Duretic's interpretation—according to which the "specifically Yugoslav trait" (that of supposedly setting "in its primary plan" international objectives with the "negation" of "narrow national objectives") was reflected in its emphasis upon brotherhood, unity, and equality—could just barely be justified from the viewpoint of formal legality if it were not a question of revolution. However, in this way, he managed to cloud the indisputably clear—in principle at least—national objectives of that revolution. This followed Duretic's support of the idea of a Yugoslavia (even in Comintern socialism) which should replace once and for all "narrow national" goals. The extent to which such a view was contrary to the objectives with which the KPJ entered the revolution has been shown in this deliberation. And the fact that during the revolution these goals were not changed, but realized, cannot be denied without gross distortions.

Even in 1941, the leadership of the KPJ, having no other political options, presented its revolutionary anti-Fascist mission in national form, transforming before the public its party leadership (even where there were no national Communist parties) into factional national Main Headquarters of the Partizan movement of

Serbia, Croatia, Slovenia, etc., with a Supreme Headquarters for all Yugoslavia (which actually made up the Politburo of the CK KPJ). There was certainly no "general Yugoslav" objective as opposed to "national" objectives. As far as the allegation that there were no specific objectives regarding state-national objectives, here are a few facts: In its proclamation of July 25, 1941, the CK KPJ calls upon all the nations of Yugoslavia to fight in the name of their "freedom and national independence," while it reminds the Croatian nation in particular of the tradition of struggle "for justice and liberty for the Croatian nation."[177] And in the proclamation concerning the anniversary of the October Revolution of 1941, which was completely permeated with Comintern internationalism, the CK KPJ calls nations, workers, youth, and communists "to the battle for a better future for you and freedom for all nations of Yugoslavia."[178]

In these positions of the KPJ, one does sense the presence of class internationalism, but even more so in that they are written so as to be acceptable to both the Serbian and non-Serbian nations, and they always emphasized the equal right of all nations to national freedom. However, there is no ambiguity in principled standpoints and decisions of the representative revolutionary bodies of the NOP.

In the resolution concerning the creation of the AVNOJ in November of 1942, the "further development of the united efforts of all the nations of Yugoslavia to gain final liberation for all of them and for creating conditions for their complete freedom and equality... "was established as a basic task.[179] And in the proclamation of the first convention, it was even emphasized that the Serbs "have had to bear the consequences of the Great Serbian oppressive politics of all the Yugoslav regimes," such that "they have felt most grievously the Fascist politics of extermination." Along with a condemnation of the Ustasha and Chetnik crimes, Croatians are called to the National Liberation Battle as the only path toward "the happiness and independence of the Croatian nation," while Serbs are called to arms "for freedom and for the fraternal community of Serbia, Montenegro, Croatia, Slovenia,

Bosnia and Hercegovina, and Macedonia."[180]

There were incorrect interpretations of the national politics of the NOP—opponents saw it as a deception, as means by which the KPJ misled the ignorant. Even within its own ranks there were dogmatic perceptions (on the part of certain individuals, predominantly "old communists") that this was just a tactic in the "first phase" of the revolution.

Therefore, in December 1942, Tito published his historic article, "The National Question in Yugoslavia in Light of the Struggle for National Liberation." The article was published in *Proleter*, the journal of the CK KPJ, whereby it was given the significance of an obligatory directive for all levels of the party's leadership, although it was intended for the "broadest masses" as well. In it, Tito takes as his starting point the opinion that the Versailles-created Yugoslavia "had become the most typical example of national oppression in Europe," that national oppression had facilitated the penetration of Fascism, that the government in exile and the Chetnik movement were "forging plans to expand the borders and the Great Serbian hegemony," that "Croatians as the strongest national identity among the oppressed nations" were offering "the fiercest resistance to the Great Serbian national policy," but that this resistance did not produce the results "which the Croatian nation had anticipated" because the Croatian upper class was prepared to share control of the government, was directing the battle against the entire Serbian nation, and was "ignoring a solution to the national question of other nations."

Tito referred to the Atlantic Charter, which stated that after the defeat of Fascism, "nations will be able to decide their fates on their own." He explained the essence of the revolutionary content of the NOP with these words: "Today's struggle for National Liberation is tied inseparably to the national question. Our struggle for national liberation would not be so resolute and so successful if the nations of Yugoslavia did not see in it, apart from victory over Fascism, also a victory over former regimes, a victory over those who have oppressed and strive toward further

oppression of the nations of Yugoslavia. The words National
Liberation War would be a mere phrase, and even a deception, if
they did not have, apart from a general Yugoslav significance, also
a national significance for each nation individually, i.e., if, in addi-
tion to the liberation of Yugoslavia, they did not also mean the lib-
eration of Croatians, Slovenians, Serbs, Macedonians, Albanians,
Moslems, etc.—if the National Liberation War did not embrace
true freedom, equality and brotherhood for all the nations of
Yugoslavia. Herein lies the essence of the National Liberation
War."

In the end, Tito repeats: "The flag of the National Liberation
War against the occupiers, which the KPJ raised in 1941, is the
same flag of the battle for national freedom and the equality of
each individual nation," for which the Communist Party of
Yugoslavia "uncompromisingly" has fought "for as long as
Yugoslavia has existed." It is significant that after all this, he
found it necessary to add: "Never has the Communist Party aban-
doned nor will it abandon its principles, which were established
by our great teachers Marx, Engels, and Lenin—the principles
that every nation has the right of self-determination to the point
of secession." The KPJ will fight in the name of that right equally
against a nation's imposing "colonial slavery" upon another, as in
the case of Pavelic's NDH, as well as against attempts at renewing
Great Serbian hegemony. All questions concerning nations will be
resolved to the general satisfaction, and "each nation will secure,
gun in hand" the right to make its own decisions.[181]

In the proclamation of the First ZAVNOH Convention on June
14, 1943, it is stated that upon the "ruins" of Pavelic's "shame-
ful" NDH, "we will build a truly democratic Free Croatia." Later
there is mention of "an independent Croatia, in which the people
will decide their own fate."[182] The ZAVNOH, from its very begin-
ning, acted as a representative body (the Croatian Parliament)
with an elected Presidium, and its Executive Council as a govern-
ment with departments in the form of ministries. The president of
the Presidium and Executive Committee was Vladimir Nazor and
the secretary was Pavle Gregoric.

ZAVNOH prescribed regulations for the operation of the entire system of the new national authorities, publicized official bonds of tender for the national liberation, and presented the historic Decision to Annex Istria, Rijeka, Zadar, the Adriatic islands, and the rest of the Italian-annexed Croatian territories to "the heartland—to Croatia, and through her to the new... community of Yugoslav nations."[183] At the Second ZAVNOH Convention (October 12-15, 1943), Andrija Hebrang said that the new Yugoslavia as a "democratic republic must be organized in such a way that it guarantees national rights, national freedom and autonomy for each of its nations." That "Democratic Republic of Yugoslavia must be founded upon a federal basis," and the Federative Republic of Croatia, the Federative Republic of Slovenia, the Federative Republic of Serbia—"will be integral parts of the Republic of Yugoslavia," and "national minorities... shall likewise have their guaranteed rights: equality and freedom."[184]

The situation in Croatia was such at that time that Dr. Milos Zanko, speaking as vice president of the National Liberation Committee of Dalmatia, said that "all of Dalmatia is unanimous on all the issues which Andrija Hebrang presented in his report," while Bozidar Magovac (editor of *Slobodni Dom*), representative of the dissident Executive Committee of the Croatian Peasant Party, recalled that in the past, many had regarded Stjepan Radic's struggle "for a republic within Yugoslavia" and a union with Bulgaria as "Radic's absurdity," but today it was being realized. In the name of "old Radic followers," Filip Lakus ardently upheld Hebrang's report, while Stanko Opacic said that Serbian "blood was spilled" for brotherhood and unity, for a "free republic of Croatia... in a federative Yugoslavia." Dr. Ivan Ribar, as president of the AVNOJ, "wholly accepted" Hebrang's report, emphasizing that together with Tito, "such a leadership," and the ZAVNOH at the head with Nazor, who is "one of the most distinguished sons of Croatia" we can march "bravely forward" in the battle for "freedom for the republic, for our independent and free state," affirming that "the Serbian worker and peasant is just as loyal to

a truly democratic and independent Republic of Serbia" as Croatian peasants and workers are to Croatia.[185]

The ideas presented there were also emphasized in the Resolution of that ZAVNOH convention, in which it is stated that the Croatian and Serbian nations, together with the others, are fighting "for a new democratic Yugoslavia, consisting of free and equal nations, in which—on the basis of self-determination—a free and democratic Croatia would be formed."[186]

In the same month (October 1943), a meeting of representatives of the Slovenian nation affirmed the politics of the Liberation Front "in terms of the self-determination of the Slovenian nation," declaring themselves "for a free and unified Slovenia in a free and united Yugoslavia." At that time, meetings were being held to establish ZAVNO assemblies for Bosnia and Hercegovina, ZAVNO of the Sandzak region, ZAVNO of Montenegro and Boka Kotorska, and in Macedonia, a Preparatory Committee was set up to establish the Initiative Committee of the ASNOM (the Macedonian counterpart to the ZAVNOH).[187]

From such political and military circumstances, it clearly follows that Croatia was at that time the firmest bastion of the NOP and revolution in Yugoslavia. The ZAVNOH had the power to speak in the name of the Croatian people, and its decisions—as was also the case with the Slovenian delegation—preceded in terms of demands those of the Second AVNOJ Convention in Jajce (on November 29, 1943). As far as Croatia and Slovenia were concerned, the AVNOJ's decisions concerning the organization of Yugoslavia as a federation of equal nations was simply the legalization of an already created condition. In this respect, the historical significance of those decisions was greater for the other nations, as well as in terms of proclaiming republics, and in the revolutionary act of seizing from the royal government in exile the right to rule Yugoslavia, of prohibiting the return of King Peter Karadordevic II, and of turning AVNOJ into the supreme legislative and executive representative body with a National Committee as the temporary government with Tito at its head.[188]

At the Third ZAVNOH Convention (at Otocac, May 8-9,

1944), the (revolutionary) establishment of the Federal State of Croatia within a Democratic Federative Yugoslavia was formally executed.[189]

This is a historical fact, in contradistinction to Duretic's neo-scholarly claim that in the Yugoslav NOP and revolution there were no "clearly stated legal statements," even with regard to national equality, let alone the right to self-determination and full national freedom.

Equally subjective and historically unfounded is Duretic's inter-pretation of central issues concerning relations between nations; i.e., his claim that "after the Second AVNOJ Convention... it was apparent that the revolutionary concept had significantly tran-scended the old demands of the Croatian civic opposition."[190]

Duretic, first of all, does not perceive that the establishment of an Independent State of Croatia in Hitler's European Order came about because of "the old demands of the Croatian civic opposi-tion." Since such demands were based on the natural, historical aspirations of the entire Croatian nation, they had to be taken into account, and advanced, as we have seen—even in the Croatian revolutionary communist movement. Accordingly, if those "old demands" had not existed, or the establishment of an NDH, there certainly would not have been such resolute demands on the part of Croatian communists from 1920s onward for the recognition of the right of the Croatian nation to self-determination, nor would there have been a decision on the part of ZAVNOH to that effect. After all, the totality of Duretic's understanding of Croatian-Serbian relations boils down to the thesis that the Croatians are to blame for both Fascism and Communism in Yugoslavia. Thus remaining on the level of political and ideologi-cal accusations—one actually espouses the old bourgeois Great Serbian views, attempting to institute a neo-hegemony upon alleged narrow national or limitations in the name of socialist integralism, even when socialism is rejected as unfeasible.

Would it not be worthwhile to ask Duretic and his allies: how is it that the two great ideologies of the twentieth century were both on the side of the self-determination of the Croatian nation,

and that even the capitalist world of pluralistic democracy stood against Serbian hegemony?

Until they find an answer to this question, which they find so peculiar and perplexing, here are a few points which might bring about more accurate comparisons of AVNOJ decisions with old Croatian demands.

At the turning point of World War I, all the major Croatian political currents were focused on ways to preserve Croatian political individuality and to create the best possible conditions for further development by achieving the unification of Croatian lands and winning back the independence of the old Croatian state. In order to achieve that goal, whether within the Habsburg monarchy or outside of it, both the Parties of Rights and Radic's Populist Peasant Party (which was linked to the Frank's Party of Rights in the Croatian Democratic Bloc of the State Rights) embarked upon the program of an independent Croatian or Croatian-Slovenian state. The ruling Croatian-Serbian coalition strove toward that goal as well, with the exception of Svetozar Pribicevic and his Serbian Autonomous Party. Pribicevic skillfully carried out Pasic's recommendation that Serbian, and not Croatian, politics be the dominant force on Yugoslav soil.

Dr. A. Korosec, in the so-called May Declaration (1917) at the Vienna parliament, presented a demand "on the basis of the national principle and the Croatian state right" for the unification of all lands inhabited by Slovenians, Croatians, and Serbs "into a single independent state body" under the scepter of the Austrian monarchy. Following that demand, Starcevic's Party of Rights, in a declaration in the Croatian parliament and in its publication *Hrvatska drzava* [*The Croatian State*], revealed the conditions under which unification into a South Slavic state would be acceptable. While embracing the idea of "national unity," they did not abandon the "historical right of the Kingdom of Croatia." They rose against the notion of Yugoslavism as a political idea and national program, regarding it as an illusion and a fantasy. "In this monarchy, there is a Croatian nation, there is a Slovenian nation, there is a Serbian nation, but there is no such thing as a

Yugoslav nation." To them, "Yugoslavism" can refer only to the "scholarly and geographic designation" of the ethnic and linguistic union of Croatians, Slovenians, and Serbs and of the geographic location of the area they inhabit.

After the Pasic-Trumbic Corfu Declaration of unification, Dr. Trumbic received the message from Zagreb that "Croatians, because of their historical state-right, wish, at least initially, to see the preservation of the Croatian state within that common state." In the Croatian Parliament in May 1918, representative Dr. Hrvoj expressed not just the views of his Party of Right, but also the general Croatian views of Yugoslavism, saying that it "means only the aspiration for the emancipation of the Croatian, Serbian and Slovenian nations from the German, Hungarian, and every foreign yoke," and "there is not, never has been, nor will there ever be a Yugoslav nation." There are only "a Croatian, a Slovenian, and a Serbian nation and their individual nationalisms, but there is no Yugoslav nationalism." Resolutely opposing every idea which would lead to the melting of Croatians and their statehood into a Great Serbia or into "some sort of unspecified Yugoslavia," Hrvoj presented a basis for a state community that was acceptable to Croatians: "We want our independent Croatian state together and in connection with our free brothers, the Slovenians and Serbs; thus the Serbs should have their own Serbian state, the Slovenians their Slovenian state, and we Croatians our own Croatian state, all within a fraternal federative union." According to Hrvoj, the Croatian state would stretch in a continuous geographic span from the Drava River to the sea—i.e., it would include Bosnia-Hercegovina—with the right of every individual to his own nationality, providing that he recognize Croatian state sovereignty.[191]

This was the general drift of Croatian views and the "old" Croatian demands on the eve of unification in 1918—that the nature of demands at the beginning of coexistence in the mutual state should not be judged according to a handful of integralist utopian or political careerists, without any support from the Croatian nation, but rather according to Radic's program for the establishment of a Neutral Croatian Republic, which was con-

trary to the institution of Serbian Vidovdan centralism.

The Croatian nation's response to the assassination in the Assembly, which was an introduction to Aleksandar's dictatorship should be noted. The Croatian representatives (together with Pribicevic's Serbian Autonomous Party in the Democratic Peasant Coalition) arrived at the decision on June 21, 1928, just one day after the assassination of Stjepan Radic, to leave the Assembly and Belgrade and not participate in its operation until it got a "guarantee of complete equality and justice."

In Zagreb, the Croatian Academic Youth organization in its proclamation said that shots fired at Radic and his colleagues "were actually meant for the entire Croatian nation," whom it called to unity regardless of differences in political ideology. The local communist committee in Zagreb called the working people and citizens to demonstrations and a general strike in protest of the crimes of the regime which was preparing a dictatorship. General Croatian public opinion was best expressed in an editorial in *Novosti [News]*, an otherwise quite moderate civic paper, in which it was written: "the real murderer" is that "corrupt Belgrade hegemony," whose crime touches the very foundations of the state, while "from the blood of Croatian martyrs, there must spring either a Croatian, Serbian and Slovenian state," in which all nations are completely equal, "or that blood will melt the foundations of this state so that it will crumble and disappear."[192]

During the course of the January-6-1929 dictatorship, and later, at the time of negotiations, Macek expressed the essence of the demand for a Croato-Serbian settlement with the proverb: my gun on my own shoulder and my wallet in my own pocket. Dr. Macek was a conservative, moderate, and judicious politician, who was bound to consistency and even inflexibility in Croatian demands by unanimity in the leadership, and especially by the radical disposition among wide strata of the Croatian nation.

From these facts one can clearly see that the claim that the AVNOJ "revolutionary concept" of the arrangement of the Yugoslav community surpassed the "old demands of the Croatian civic opposition" was totally unfounded.

If, along with the "old" Croatian demands, one keeps in mind the proclamation of the right of nations to self-determination among the objectives of the NOP's program, both from the Leninist revolutionary viewpoint and the universality of the Atlantic Charter—upon which the AVNOJ principles are based— then it is clear that there is no justification to the claims that only with the 1974 Constitution was the idea of confederation brought into relations among the Yugoslav nations. They have no founda- tion because that principle is implied in the very spirit of the accepted principle concerning the inalienable right to the self- determination of nations, which, as incontrovertible historical subjects, have the right to sovereign decision-making. It is quite another question if certain essential premises upon which the AVNOJ principles were to have rested—that is, upon which the practical efficacy of a confederative union is possible—are dis- turbed or betrayed by their application in concrete organizational decisions and in actual execution.

However, in that respect, any sort of polemic containing such views as are currently voiced by Vuk Draskovic is neither possible nor worthwhile. According to him (and those who are of like mind) it is necessary to "eliminate that catastrophic principle of the equality of nations and replace it with the principle of the equality of citizens."[193] Ignoring the entire history of relations between nations, he seeks solutions to the open national problems of the Serbian nation on the basis of unitarism under the old pretext of civil equality and democracy for all, except, of course, non-Serbs. Such a perception clearly preserves the idea from the previous century that Serbian supremacy is possible on South Slavic soil.

Kosta Cavoski's deliberation on the AVNOJ organization of Yugoslavia is of a completely different disposition, as is Svetozar Stojanovic's opinion on the national question in the Yugoslav cri- sis, as well as Dobrica Cosic's view on the historical instability and uncertain future of Yugoslavia.[194] Although, in their concern with a change of the 1974 Constitution, they lose sight of certain essentials, they do reveal problems to be dealt with. Stojanovic's inclusion in utopia of a "federation by consensus," and Cosic's

one-sided designation of Ustasha genocide against Serbs, while failing to perceive the same (in both program and actions) on the part of the Chetniks against the Croatians, his rejection of the "burden of guilt" imposed by the Serbian "ruling oppressor nation" and the "symmetry of war crimes," his oblivion of the contradiction between the condemnation of nationalism and separatism (from the Great Serbian viewpoint) and the recognition of the right of nations to self-determination to the point of secession (from the principles of democracy)—reveal more of the unyielding nature of Serbian views on Yugoslav community than of a reasonable scholarly approach.

Cavoski is basically correct when he claims that the chief decision-makers in the establishment of the AVNOJ Yugoslavia (Tito, "whose word certainly carried the greatest weight," followed, in order, by Kardelj, Rankovic, and Djilas) regarded all the proclaimed principles from the standpoint of pure revolutionary pragmatism. That, however, does not render valueless the principles contained in the decisions themselves, and especially their objective historical significance. From the formal side, his comment that through the 1974 Constitution, the AVNOJ principles could be more fully realized than by the 1946 Constitution which came about directly from the NOP and the revolution, is not unwarranted. In such an approach, one certainly loses sight of the fact that the leadership of the KPJ, in the war and revolution, was more consistent in emphasizing self-determination to the point of secession. Thus, after taking over power—in keeping with its Marxist doctrine and the Soviet example—it would lean more toward a centralist and integralist system, only to return again to the spirit of AVNOJ after thirty years of experience, and not at all coincidentally at a time of renewed intensification of relations between Yugoslavia's nations. This, however, does not mean that all solutions in it were compatible with the spirit of AVNOJ.

Cavoski is likewise correct only right from the formal point of view when he claims that the principle of the unanimity of the federal units is not contained in the decisions of AVNOJ. However,

considering the historical and ideological foundation of the pro-
claimed principles and the meaning of the decisions made, there is
a basis to the interpretation that there is contained within them
the principle of unanimity. This stems from the recognition of the
right to self-determination: if one has the right to separate from
the community, then one logically also has the right not to agree
with the decisions which go against one's essential interests.
Cavoski saw clearly that there "remains a great gain" from the
Second AVNOJ Convention which "raised to the highest level of
merit" the full national equality of nations (Serbs, Croatians,
Slovenians, Macedonians, and Montenegrins). He rightly pointed
out that the substitution or supplementation of the national prin-
ciple with republican and regional ones (with the fixing of bound-
aries) brought misunderstandings and opened up new problems
which still seek solutions.

The debate over the constitutional order, the problem of rights
of nations to self-determination one-sidedly viewed and from an
expressly legalistic standpoint, was joined by Radomir Lukic and
Jovan Dordevic. They advocated the theory of the consumption of
the right of nations to self-determination. Although they knew
that the right of a nation to self-determination has become a uni-
versal right—i.e., a generally acknowledged right in the world
community, and that it was the starting point of the Constitution
of the SFRJ—they nevertheless felt that once used it would remain
forever spent. This "using up" of the right refers (according to
many) to the unification of 1918, regardless of the fact that it
came about in an undemocratic way, and that the state collapsed,
and therefore, the "spent" right could not have been the basis for
the revolutionary act of forming an AVNOJ Yugoslavia. Such a
viewpoint, above all, expresses a disregard for the causes of the
collapse of Royal Yugoslavia and the premises for a renewal of the
Yugoslav state. It is legalistically but not historically relevant.
Such a legitimist viewpoint has always been advocated by repre-
sentatives of imperialistic multinational structures, since it
amounts to a defense of the existing order and denial of the right
of the unequal, oppressed nations to separation and existence as

an independent state. Thus, there has always been, and there remains, a strictly legal basis for the violent repression of national movements on the part of imperialistic hegemonies.

Lukic also supports his views with the fact that in the 1974 Constitution there is no explicit legal provision which would allow for the exercising of self-determination to the point of secession. However—along with Dordevic and throughout the entire development of his theory that from the constitutional standpoint, that right has been exercised and used up—he thus interprets the Constitution in a way contrary not only to its AVNOJ spirit, but also to its language. It is significant that the Constitution of the SFRJ begins with precisely that right in its first sentence: "The nations of Yugoslavia, starting from the right of each nation to self-determination, including the right to secession... have joined into a federative republic of free and equal nations and nationalities" and have formed the SFRJ in order to create and ensure the "national freedom and independence... of each nation and nationality individually and of all of them together."

No impartial person, and especially not anyone who knows the historical background to the origin of such Fundamental Principles, can deduce from this a constitutional denial of the right of a nation to self-determination. He who does this is not an advocate of the protection of these principles from hegemonical centralistic endangerment. Dr. J. Pleterski pointed out through the Slovenian example the inadequacy and untenability of the contention regarding the one-time use of the right to self-determination. And Dr. Z. Loretic rightly points out its incoherence by making the analogy that this is just as logical as saying that a worker's right to strike is used up when he becomes employed. It is understood that the thesis concerning the use of the right to self-determination by previous unification can be summed up by the theory of impermissibility of secession from the standpoint of protecting the interests of imperialist hegemony. As such, it is contrary to the entire movement toward national independence throughout world history.[195]

A surprising unfamiliarity with the history of relations between

nations and of the Yugoslav state is evident in the philosophical reflections of Mihailo Markovic. He labels the advocacy of confederalism on the part of Kermauner and Slovenian intellectuals as "the worst possible solution," which leads to the further disunion of Yugoslavia. Markovic does not feel that the stability of multi-national states cannot be achieved by any sort of general democratization if that democratization includes any form of guardianship over other nations, as well as the decision whether they want federative or confederative relations. However, that his viewpoint reflects an even deeper misapprehension of the existence of national entities as inalienable sovereign actors in the international order and in world culture can be deduced from his belief that all solutions should be sought in the recognition of general human rights and not the national rights, for the latter is allegedly the cause of "all that is happening in our reality," that is to say, of the total crisis. "This happened when we began to transform the federation into a confederation and when we raised the national principle above the classist principle."[196]

Regarding Markovic's preference of federation over confederation, two things should be mentioned: a federation is a more appropriate form for a state that is predominantly nationally unified with regional differences (such as Germany and even Austria), while a confederation is virtually the only solution for the survival of a state that is a multi-national conglomeration of historically separate nations. Second, we are living in an age of the integration of Western and Eastern Europe, though not upon federative, but confederative principles. Taking into consideration Markovic's philosophical criticism of renunciation of the Marxist view of history due to the disregard of the "class principle," one should note that to this day, history has not reached any sort of classless society, while nationally based societies in one form or another are to be found from time immemorial in all forms of real socialism. It is surprising, however, that these historical understandings have not penetrated Markovic's philosophical circle. Although they designated their Marxist or neo-Marxist direction in the Praxis journal, even the above-mentioned views of Markovic are an example of

how little heed some thinkers pay to historical reality.

A lack of understanding of essential issues about the history of relations between Yugoslav nationalities was also displayed by another eminent representative of the Praxis circle, Gajo Petrovic. Speaking about the constitutional definition of the name of the language, he would take upon himself—as a philosopher—first of all the judgment that the thesis of the Croatian literary language, which is supported by all Croatian linguistic scholars and almost all Croatian writers, and with them the entire Croatian thinking public, "is scientifically unfounded." He calls the usage of Croatian words (clearly loathsome to him) "the forcing of a linguistic grotesque" for the sake of having "their own" variants, "that is, their eventual separate language." Blind to the grotesque he displays in such a line of reasoning (not merely with respect to philosophy but to common sense), Gajo Petrovic clearly places himself in the category of those who claim that the Croatians do not have their own language. Moreover, he brings into question "their variant." However, he does not speak (as Nikola Stojanovic did in 1902) of only one Serbian language. In the name of "over a million Yugoslavs" he in fact seeks a "Yugoslav literary language," in order to avoid the undesirable choice between Serbian and Croatian.[197]

Since he sees the separateness of the Croatian language as a threat to the future, it is unfortunate that he doesn't use his philosophical logic to explain the language in which Marko Marulic, "the father of Croatian literature," wrote five centuries ago, or the engraver of the Baska tablet eight centuries ago. In the resurrection of such unitarist theses, one should—at least on a philosophical and historical level—be reminded of the fate of Svetozar Pribicevic, one of the most capable of Serbian politicians from the "other side" (Croatian), who was transformed from the fiercest advocate and enforcer of unitarist ideas into their repudiator, having become convinced of their harmful effect upon Croatian-Serbian relations, especially in Croatia.

However, socialist internationalism is not the only ideology, nor is imperialist hegemony or bureaucratism the only challenge to

national self-determination. Also, various and sundry followers of universalist internationalism (Coca-Cola internationalism, in the words of an American professor) as well as certain advocates of the economic, informatic, or democratic civilizational integration of the world see the specter of nationalism which should be subdued through enlightened supranational centralism. There is reflected in such an approach an unfamiliarity with the very essence of every form of nationalism and with the causes of crises and the disintegration of multi-national states throughout history.

In the views of the esteemed Zagreb economist Branko Horvat, one clearly sees a misapprehension of the central Croatian-Serbian problem. He knows that "history does not behave according to economic accounts" (or, let us add, even according to certain fixed economic theories and laws). As an economist, he also shows a remarkable familiarity with Balkan inter-ethnic contrariety, so he has found it necessary to enter into a fairly rational consideration of the very sensitive issue of the Albanians and Kosova. However, in his rejection of national economies—which hardly anyone in the world would seriously contest—there is reflected a deeper misapprehension of the essence of the self-determination of a nation. Moreover, contrary to Croatian linguists and scholars, he accepts the term "Serbo-Croatian" as a designation for language and even for nationality, which only Great Serbian unitarists (in addition to Jovan Cvijic, to whom he makes reference) have supported, but not for the most part, Croatian and South-Slavic integralists. This leads one to conclude that he regards the problems of Croatian-Serbian relations from past-century viewpoints which history has clearly refuted, and which are found today only in utopian integralist and neo-hegemonic tendencies.

The fact that he finds himself in that early period of national disorientation is seen in his support of the establishment of a new nation in Bosnia-Hercegovina, to be called the Bosnjaks,[198] probably because even in his economist reasoning he senses the insufficiency of turning the Moslem faith into a nation, as did others who opposed such an idea and its proclamation, foreseeing that a separate Moslemhood was not the solution either for Bosnia-

Hercegovina or for Croatian-Serbian relations as a whole.

Suggesting a Bosnjak instead of a Moslem nation, Horvat forgets—as do those whom he corrects with that proposal—that nations are not formed upon political, constitutional, or theoretical decisions. A simple geographic or religious designation is not sufficient for their creation. An entire set of various factors and determinants dependent upon the totality of historical evolution and conditioned by the intricacies of the geopolitical environment is required. If it were not so, then in that attempt at creating a separate "Bosnian political nation," contrary to the Serbian and Croatian, at least some measure of success would have been achieved in the last century by Benjamin Kallay, whose regime strove to accomplish just that, though it was then in the interests of different imperialist politics (for the sake of weakening both the Croatian and the Serbian).

Unhistorical assumptions concerning the relations between nations in Yugoslavia, above all Croatian-Serbian, are visible in the aforementioned views of L. Tadic. To him, the Yugoslav federation was "founded upon Partisan Yugoslavism," and only through the 1974 Constitution did it acquire "the form of a confederation according to the ideological paradigms of prewar party politics." His remark that the constitutional reorganization of the state "reflects the preserved continuity of distrust toward the Serbian nation and expresses dread of its hegemony" is correct. But the fact that there is reason for that distrust is confirmed, along with other points, by Tadic himself through these very views. Namely, he does not see that the "confederative configuration" was already contained in the AVNOJ decisions, but that in the course of three decades it was suppressed, most often in the name of the "ideological paradigms" which it challenges, and for that reason, had to be constitutionally more explicitly restated. That distrust and fear of Serbian hegemony continued cannot be ascribed to any particular entity, nor, accordingly, can any particular blame be placed on Serbia or other nations.

Herein is simply expressed the history of all multinational state creations. The numerically stronger nation, or the one that in par-

ticular circumstances has achieved domination, always strives to impose its hegemony. And every ruling structure aids it in this effort in accordance with the inescapable logic of imperialist bureaucracy, which typically finds its firmest support in one nation. Because of this, there is no multinational state in history without a "ruling" nation, at least in some respect.

In the SFRJ, similarly in the USSR, that general historical rule was supported by the ideology of socialist internationalism and class integralism, expressed here in the idea of "socialist Yugoslavism" or a unified "Yugoslav working class," the defenders of which were predominantly Serbs, while those of other nations were dogmatic careerists. Thus although the SFRJ was founded from the very beginning upon confederative principles which were to protect the new union from the old hegemony, the manifestation of hegemonical tendencies was unavoidable. Due to attempts at averting the deepening of already present crises, there arose the historical need to guarantee the AVNOJ spirit through constitutional provisions, particularly in view of the impending departure of Tito as well as other historical figures.

There is a certain justification to Tadic's criticism that the present constitutional order is a "formula for Balkanization," in the sense of P. Togliatti's criticism of the "fragmentation" and the "aspiration for complete independence of the smallest parts." But one must not forget that even Togliatti supported the implementation of the principle of self-determination of the nations of Yugoslavia. Thus, in reasoning out the decisions of the AVNOJ and the constitutional provisions of 1974, one should use as a starting point the extent to which that principle is consistently applied in them, and what their rejection due to pragmatic voluntaristic reasons would mean. Those who start with the idea that the present solution is only against the unity of the Serbian nation forget that as far as the Croatian nation is concerned regarding territory, it is far below that which, by the eve of the war, had been achieved through the Serbian-Croatian agreement as a temporary solution.

Tadic's entire line of reasoning is an example of how one's

obsession with seeing only the problems of one's own nation makes one totally blind to the problems of other nations, let alone to their right to self-determination. For instance, he labels the 1971 "mass movement" in Croatia ("and the similar one in Kosovo") as separatist. And he ascribes to it a hidden chauvinism under the communist flag, although that movement in its "confederalism" merely presented a demand for a "clearing of accounts" (of both investments and hard currency). As a whole, that Croatian movement was an expression both of Euro-communist and anti-dogmatic as well as general reformist social-democratic trends toward socialism with a human face. It is more than significant that Tadic judges that "the prewar history of the KPJ... provided enough ideological food" for that supposed separatism and chauvinism, while neither in the Party's nor in the national history does he find any sort of "food" either for the manifestation of hegemony or for its negation, i.e., for the national movement for self-determination. Because of this, he arrives at the completely distorted judgment that "the crisis of the Yugoslav federation began with the battle of the ruling hierarchy against an imaginary unitarism and bureaucratism which are either [tacite or *expressis verbis*] imputed to Belgrade and Serbia." And all this was along the lines of "prejudice and accusations contained in both chauvinist anti-Serbian propaganda and Party history, i.e., ideology." Contrary to this view, the truth is that a crisis in every respect (economic, political, and moral) started precisely with the removal of the supposedly nationalist leadership of Croatia, Slovenia, and Macedonia, and even the ("liberalist" leadership) of Serbia.

Contrary to those who viewed the problems of relations between the nations in the SFRJ from an essentially imperialist and unitarist, or socialist and integralist standpoint, there also appeared views from the Serbian ranks supporting the historical need to separate Serbdom from the Yugoslav union.

This was expressed in its most succinct and historically and programmatically elucidated form, at the end of 1987 in *Pravoslavlje*, the official publication of the Patriarchate of the Serbian Orthodox Church. There, in the previously mentioned

article by Svetozar S. Dusanic, along with a comment by the editors that it was being printed "because of its interesting presentation of a political conception of the distant future," Dusanic completes his theory of the two worlds: of the Byzantine Eastern and Roman Western world, and of the irreconcilability of Orthodox Serbdom and Catholic Croatianness, which is equated with both the Ustashas and the communists. Embarking from the idea that in Yugoslavia, "tailored" in Versailles, "there are indeed two worlds" and that out of the "reality" of their division and irreconcilability sprang the Ustasha-Catholic genocide in World War II, the author sees no other way out than their separation. The Byzantine world differs from the West "not only in religion" but in the totality of its culture as well as in its "ethics, psyche, and mentality," and, thus:

"This world doesn't see a future of coexistence in the same state with those of Western-Roman orientation, particularly not after the Second World War. This world's path does not lead toward the Karavankas and the Julian Alps [in Slovenia and Austria]. It does not wish to set its pace according to Northwestern Europe in order to compete with 80 million Germans and 60 million Italians. It has no intention of renouncing its blessed land, which was left to it by its forefathers, nor is it likely to abandon the historical path which was determined by the immortal Nemanjics eight centuries ago. The Nemanjic route leads to the warm seas which have always attracted the nations of the North. For that reason, a union of Balkan nations is an absolute imperative for the future." And "that future state is a natural, logical and necessary exit" out of this "Versailles-Treaty Yugoslavia," which, "in fact, is not Yugoslavia in the full sense of the term because it does not include the South Slavic state of Bulgaria."

In the creation of a Versailles Treaty Europe, he says, its architects "did not take into account the interests of the Balkan nations, and especially the needs of the Serbs." They intentionally fashioned a state "conglomerate with a strong Western component, as a counter-balance to the Eastern." And "that conglomerate" was doomed even "in its conception, because of its

composition, to be continuously torn by its contrasts. Thus instead of the "Sick man on the Bosporus" that the West had artificially kept alive for centuries, the West created in 1919 a "Sick man on the Danube." And "both sick men have the same task: to prevent the penetration of Slavs to the Mediterranean." The future task, according to the characteristics of the related Balkan nations, would be the deprival of "combinations of great powers"; "it would make fratricidal wars impossible," and moreover, would "eliminate the domination of one nation over another."

All in all, it would be the complete opposite of the Versailles "conglomerate" of incompatible worlds. Since there already existed "a Union of Western European states and a Moslem Arab Bloc, why should the Balkans be deprived of their political unity?"[199]

Such a vision of the position of Serbdom in the Balkans is nothing new. It had its origin in Dusan's Balkan kingdom, and it was present in evem more recent Serbian history, from the renewal of the Serbian state in the last century to the formation and continuance of the Kingdom of Yugoslavia. The Balkan aim of Serbian politics stemmed from the whole conception examined in Ilija Garasanin's Program "Outline" of 1844.

The greatest success of Garasanin's politics in the past century is considered to be the creation of the Union of Balkan Nations (1861-1867). At that time, even the most progressive Serbian minds (S. Miletic and M. Polit-Desancic) were for a confederation of Balkan lands, which would take the place of European Turkey. Later, even Serbian socialists, from Svetozar Markovic on, supported a union of federative republics in the Balkans, which could spread all the way to the Danube.

Controversies were to ensue over this issue between Serbian Marxist social democrats and those of Croatia and Slovenia. At the Yugoslav Socialist Conference in Ljubljana (1909), D. Tucovic disagreed with the Tivoli Resolution that the workers' movement seek a social and national solution in the democratization and federalization of the Dual Monarchy of Austria-Hungary and in an alliance with the developed socialist movement of Central Europe. Instead, at the Balkan Socialist Conference in Belgrade in 1910,

he proposed resolving the South Slavic question in the Balkans, seeing it as the next step in economic development.

After the October Revolution, those contrasts were magnified by the decision of the Tenth Congress of the Yugoslav Socialist Democratic Party (Ljubljana, December 25-26, 1917) by which, in the name of the international union of the proletariat and the self-determination of nations, an appeal was made for the formation of federated divisions in an Adriatic-Danubian-Sudete Union. At the time of the First World War, Nikola Pasic was more inclined to favor the expansion of the Serbian State into a Greater Serbia in the Balkans over a unification with all the Croatians and Slovenians. Even the government of the Russian Empire advised the Serbian government to be careful in its expansion and not to incorporate too much of the "Catholic element," which could challenge the domination of Serbia.[200]

Soon after the unification in 1918, it was established that the stabilization of the new state could not be achieved upon the centralistic Vidovdan constitution and that the Serbian concept of state was incapable of permeating the total organization of the new state, while the national concept founded upon Orthodoxy was rejected by all non-Serbian nations.

At that time, even in very influential Serbian circles, there appeared the notion that secession (the so-called amputation) of Croatia would be necessary. This idea was supported also by certain distinguished leaders of the Radical Party with the reasoning that it was necessary to "break with the Croatians because we can no longer remain together," while Greater Serbia would be "more at peace and freer to conduct an orderly political life." This idea was allegedly backed by Duke Misic, and an anonymous leaflet was printed showing the borders of the amputation. Since the idea was not accepted either by Yugoslav unitarists or, because of the borders, by Croatian confederationists, it was then abandoned. But it was repeated by Alexander himself to Svetozar Pribicevic after Radic's assassination, in the form of a message to Radic: "tomorrow in Zagreb, you may declare secession" because "we can no longer remain together." In so doing, Alexander made ref-

erence to the opinion of the late King Peter, resolutely rejecting the idea of solving the state crisis by implementing a federation. ("I do not want a federation; I shall never agree to a federation; I am rather for a partition than a federation.") Allegedly, diplomatic communiques had even been prepared concerning the notification of foreign powers as to the partition of the country.[201]

The rejection of Prince Paul's and Dragisa Cvetkovic's agreement with Macek concerning the formation of Banovina Croatia in 1939 by virtually all representatives of Serbian political life (including the democratic United Opposition), as well as the coup of March 1941, expressed the irreconcilable views toward any idea of federalization. During the Second World War, the vast majority of Serbian civic politicians was more inclined toward the creation of a Greater Serbia than the renewal of a Yugoslavia: from those centered around General Nedic and Mihailovic's Chetnik movement to those in the government in exile in London (associated with S. Jovanovic and M. Ninic) and in the embassy in Washington (K. Fotic). The decisions of AVNOJ concerning the establishment of a federative Yugoslavia, as well as the politics of the Western powers—above all Britain—forced rather than prompted acceptance of the renewal of a common state. But they certainly did not force agreement with the AVNOJ federative and, moreover, socialist order.

Thus, in the example of the historical reasoning of *Pravoslavlje* concerning the "conception of the distant future," there is manifested the ancient understanding that many old ideas are sometimes resurrected, if not in unchanged forms, then as new ideas with the same meaning.

CONCLUSION

Premises for the Realization of Historical Reasonableness

The first law of history is: one must fear to speak untruths; the second is that one must not fear to speak the truth!

—Cicero

People should behave so as not to make enemies of friends, but rather to turn enemies into friends.

—Pythagoras

People and nations will behave rationally when they have exhausted all other possibilities.

—A. Bloch: Katz's Law

And thus we have reached the end of this investigation, profoundly convinced of the general accuracy of Katz's Law. We arrive at this thought from the despairing question: Have all irrationalities been exhausted in relations between nations, in our case, between Croatians and Serbs?

Although I was prompted to make this inquiry for personal reasons, I did not write this only because of those reasons. It was the awareness that there is, after all, a great difference between the vainglory and malevolence of the incorrigible, and the destructiveness of the well-intentioned but uninformed, that inspired me to try to discover the origins of the historical Croatian-Serbian conflict in reality and in historiographic examination, and to seek a basis for its sober assessment.

Although I have already had considerable opportunity to experience the "benefit" of such labors, I do not perform them even

now only for that motive which we find in biblical wisdom: I said this and saved my soul—but from the understanding that every betrayal of past or present history, prettifying or soiling its face, is harmful. For just as no one can ruin the present so much as the person who wants to correct it by force, so too no one can harm it so much as the person who distorts its past. False foundations cannot bridge even the chasms between the past and the present, never mind serve to build lasting vaults for the future. Even historical truth is scarcely sufficient for such an undertaking. Silence about it is not allowed. For on the soil of all trampled and stifled truths, there spreads the poisonous weed of lies and deceit, and the historical wilderness becomes even more thorny. Misplaced truths will take a longer time and greater effort to uncover, but their value is more profound.

In all historical disputes—that is, in any historical arena where divergent historical subjects confront each other—one side is never completely right and the other completely wrong. The same is also true in the case of the conflict in the Croatian-Serbian historical wilderness. Since it is never the case in history that right and wrong keep court only in opposite camp, it is impossible in a historical discussion to divide them into one or the other camps, nor in such a way that each of them would be satisfied with the components of the other.

Among the essential insights brought out in this work is the fact that no historical reflection or interpretation—whether scholarly based or intentionally distorted—can change historical reality, although it may strongly influence the way it is perceived, and even change the manner in which one views what, how, and why events occurred.

As we have seen, the entire Jasenovac controversy was orchestrated in such a way that it appeared that genocidal crimes of equal proportions had not occurred during World War II nor throughout history. This controversy was promulgated not only in yellow journalism or short-sighted, pragmatic petty-politics, but also in books with scholarly ambitions by authors with titles

and high academic degrees. This is why it was necessary to consider the problem of war victims and genocide as part of the totality of historical occurrence—precisely in order to determine its actual dimensions, the causes of its rise, and the ensuing consequences in comparison with similar phenomena.

There can be no doubt that the distorted image of the Jasenovac victims—in both Yugoslav and world perspective—is so unfounded and distorted that its absurd incoherence would not deserve serious attention, were it not the basis for the theory of the genocidal character of every manifestation of Croatiandom, from that of Catholic origin to that which chose the path of self-determination. But the magnification of the number notwithstanding (no matter how many tenfold multiplied)—a crime took place. It was horrible and immense, both in its dimensions and in the way it was carried out. It even had genocidal characteristics.

Nevertheless, the problem of war victims on Yugoslav soil in the Second World War cannot be summed up solely in Jasenovac, and not only because less than 5 percent of the total of Yugoslav war victims (and less than 5 percent of the total of Serbian war victims) lost their lives in the Jasenovac camp. Indeed, that very fact leads us to the conclusion that the Jasenovac distortion came about through the symbiosis of two opposite currents: one—to muddy the actual losses of a supposedly altered land; and another —to redirect that same sludge into an old and murky river bed.

It is a historical fact that the Ustasha regime of the Independent State of Croatia, in carrying out its plans of reducing the enemy Serbian Orthodox population in Croatian lands, committed a genocidal crime against the Serbs, and a proportionally even greater one against the Gypsies and the Jews, in carrying out the Nazi racial policy.

However, it is an equally incontestible historical fact that the Chetnik movement also committed a genocidal crime in carrying out its plans to "purge Serbian lands of inimical Croatian, Catholic, and Moslem populations," in order to achieve an ethnically pure Greater Serbia and a mutual border with Slovenia.

Taken as a whole, the wartime losses on the part of both the

Croatian and Serbian nations in World War II are proportionally close. And even the total Serbian losses in all of Yugoslavia—in absolute numbers—are about two to three times fewer than those that the myth-makers claim for Jasenovac alone. Contrary to the Jasenovac distortions, which is aimed at placing the blame for all the evils of the war upon the Croatian nation, the Bleiburg myth ascribes all the evils to the Serbian nation because, to its promoters, apart from the Serbian Chetnik movement, the communist Partisan movement was also predominantly Serbian.

On no grounds can the Ustasha and Chetnik movements be ascribed to the entire Croatian and the entire Serbian nations. For crimes, just like every other act of a system or a movement, are attributable to a particular ruling structure or ideological and political movement or group, including personal responsibility for each deed and misdeed.

But from the historical standpoint, neither the Ustasha nor the Chetnik movement can be separated from the history of its respective nation. The same is true of the Nazi movement in the German nation and the Bolshevik movement in the Russian nation. Nor can the nation of any imperialist country be freed from the historical guilt for the crimes committed by the English, French, Spanish, and Portuguese imperialists in colonies all over the globe, or just yesterday by the Americans in Vietnam.

The Ustasha movement originated from within the Croatian nation, just as did the Chetnik movement from within the Serbian nation. Both the Ustasha and Chetnik movements were equally the expression of mutually opposing ideas concerning nation and state and of the programs for their implementation, both stemming from the judgment that coexistence in a common state was impossible. This means that Dr. A. Pavelic and General D. Mihailovic, in the circumstances of the Second World War, found themselves at the forefront of nationally exclusive and irreconcilable movements, which sought equally to exploit those circumstances for the realization of their respective national programs. From this, one can certainly deduce that neither the Ustasha nor the Chetnik movement can be viewed only through the prism of

the crimes that were primarily the result of reckless attempts at achieving the aims of their programs, and only secondarily the expression of an impulse for revenge and of the pathological degeneration of individuals and groups, a phenomenon that is characteristic of all wars and inevitable among the ranks of all warring factions.

Because of this, it is of far-reaching importance and historical significance that the problem of the war victims be presented impartially within the framework of an objective interpretation of the origin, position, and role of the Ustasha movement in the history of the Croatian nation and the Chetnik movement in that of the Serbian nation, both in terms of their mutual conflicts and of their individual confrontation with the NOP and the socialist revolution.

Obscuring and even distorting the truth about the war victims has given rise to extremely undesirable consequences. If the 1950s statistics of the SUBNOR veterans organization and the results of the official census of war victims compiled in 1964 had been made public, and if on the basis of these figures and demographic research, a normal scholarly investigation had been made possible (both for myself and others), the monstrous growth of the Jasenovac distortion would not have been possible, nor would the problem of war victims or "equalization of guilt" between the Croatian "genocide" and Serbian "hegemony" have been able to facilitate various theories on the genocidal tendencies of the Croatians *sui generis* and even Catholicism, and of a Vatican-Croatian-Comintern conspiracy against Yugoslavia. In that incongruous alliance, Croatia is assigned the role of scapegoat for all the historical consequences of both fascism and communism on Yugoslav soil. Indeed, this was just an introduction to a general attack against all the premises upon which the Titoist revolutionary NOP won its victory, and against the AVNOJ principles according to which it endeavored to establish a new union of equal nations.

However, the reduction of the revolutionary period of history to the Jasenovac distortions and to the specters of genocidal

Croatiandom and Catholicism that arose from it has, in addition to its aforementioned direct aims, a farther reaching and even more ominous historical significance. Such interpretations of history, which persist with theses that cite the historical guilt of the Croatian nation for all the misfortunes of the Yugoslav state and Serbia, not only maintain the climate of distrust in the spiritual lives of both the Croatian and Serbian nations, but also directly add to the buildup of hatred upon an already dubious legacy from the past of both sides. Such a one-sided acknowledgment of a sinful past undermines the possibilities for mutual existence in any acceptable form.

The nurturing of lies and deception is an extremely irresponsible manipulation of irrational passions. It stores in the historical memory of nations a devastating bomb which can explode in any situation or at any provocation, at which point reasonable efforts will prove impotent before its destructive force. For, as the historical wisdom of the ancient Arab proverb cautions, the ears of the mob are a wondrous wild field: sow promises and faith will grow; sow lies and blood will flow.

Therefore, in the interpretation of historical events, it is extremely irresponsible to go by the fantasy of one's own part of the truth. Historical thought founded on the exclusiveness of the interests of one's own nation inevitably carries the seeds of misunderstanding and discord from the past, burdened with new accretions of hatred, into a perhaps more tragic future. Besides, one should be mindful of the fact that in every military confrontation, no warring side can escape the madness of hatred toward the enemy, but that hatred is of a transitory nature and is easily transformed into a feigned and even amicable alliance. On the other hand, historically kindled hatred is transformed into the lasting madness of mutual intolerance. And nations, for which historical fate has preordained that they at least coexist as neighbors should take this into consideration.

The eighteenth-century English statesman and writer Philip Chesterfield left a wise philosophical message to posterity, applicable to all the nations of the world, on the inevitability of

historical permeation and conflict:

"In both friendship and enmity, one must establish certain boundaries to trust and hatred, so that trust may not become perilous and hatred may not exclude any possibility of reconciliation."

Historically viewed, the friction and distrust in Croato-Serbian relations have stemmed from:

(1) a misunderstanding of the historical nature of a national being, specifically, the misconception about the possibility of creating a single Yugoslav nation on South Slavic soil, or a single Croatian or Serbian nation from the two of them, or from all three of the supposed "tribes" (including the Slovenians);

(2) disrespect for the distinct historical identity of the Croatian as well as the Serbian nation, from which followed an ideological misconception that in the name of Yugoslavism or socialism, and then their symbiosis, the right to the self-determination and sovereignty of a nation may be restricted.

The conception of Yugoslavism was inspired by the idea of a supposed Illyrian proto-community, as well as by the romantic archetype of the Italian and the German unifications, but without an understanding of the historically determined differences. Its material basis, however, was in the desire to unite all Croatians, and similarly, unify all Serbs.

Until the creation of a Versailles Treaty Yugoslavia, the Serbs did not wish—and rightly so—to reject their idea of statehood for the sake of some sort of ambiguous Yugoslavism, which, as a mask for Catholic Croatiandom could imperil Orthodox Serbdom. When Aleksandar, by introducing the January 6, 1929, dictatorship, proclaimed a unitarist Yugoslav nation, in order to preserve Serbian hegemony, the idea of Yugoslavism was totally dead in Croatian social life.

In Croatian national thought, back in the last century, the Great Croatian idea of establishing a single identity with the Slovenians and Serbs or the possibility of spreading Croatiandom to the Slovenian and Serbian nations was abandoned, even

by Starcevic himself. In Yugoslavism, Croatia sought its self-determination upon confederationist grounds, agreeing of necessity to a federation.

In the Serbian national view, however, the expansionist idea of forming a unified nation upon a linguistic and political foundation remained. The Kingdom of Yugoslavia was essentially perceived as an expanded Serbia. Virtually all Serbian currents presented their fiercest resistance to the federalization of the state, a resistance which was triggered by an agreement on the formation of Banovina Croatia, with the objective of either preventing the agreement or ensuring, even in a federal Yugoslavia, both the territorial and political dominance of a Great Serbia.

In the Croatian national being, a federation upon AVNOJ principles might have succeeded in gaining acceptance even as an incomplete form of self-determination and the attainment of statehood for the Croatian nation.

In the Serbian national being, the establishment of an AVNOJ federation had a substantially different significance. The renewal of a Yugoslav state was clearly not enough compensation for the loss of the prestige which Serbdom enjoyed through the role of its dynasty, and in absolute superiority in the central government and state administration, particularly in the army and diplomacy. Thus the AVNOJ federative solution was always considered from the side of its primarily Yugoslav significance, while the significance for the self-determination of the non-Serbian nations was explicity neglected. And at the same time, on the other side, there was an increased feeling that the federation had divided Serbdom, and that supposedly it alone had not achieved self-determination and statehood.

Since Tito, the charismatic leader of both the revolution and the newly created Yugoslavia, stood behind both the AVNOJ decisions and the 1974 Constitution, opponents of the federative structure had to look for other routes gradually to bring into question both the constitutional order itself and the authority of its chief architect. In so doing, the opponents of federalism, coming as they did from totally opposite ideological camps, found them-

selves allied in the historical task of systematically discrediting and ousting both AVNOJ federalism and the authoritative government of the historic leader.

Because of the historical legacy and the extent of the essentially anti-federalist perception of the Yugoslav state, the attack on the AVNOJ principles and upon Tito himself was necessarily preceded by an attack against the Croatian nation and the revolutionary movement of which he was a part and which he symbolized, despite his ideological orientation toward a solution to the national question within the framework of the Yugoslav state.

From that arose both the distorted view of Jasenovac, as proof of the historical guilt and genocidal character of the Croatian nation, and that nation's association with the Vatican and the Comintern in their anti-Yugoslav conspiracy.

Just as the last century's idea of a Croatian-Serbian national linguistic unity had grown out of mutual interests concerning the threat of foreign domination and to achieve full national freedom, so too did the Croatian-Serbian discord of this century grow out of their differing perceptions of Yugoslavism, and even more out of the differing consequences of the common Yugoslav state upon their respective national beings.

The Croatian-Serbian understanding of the past century, first forged in the battle of the Banovina Croatia and the Serbs of Vojvodina against Hungarian hegemony, was perhaps best expressed in the lines of the Serbian poet Zmaj Jovan Jovanovic (1833-1904):

> *The Croat does not the battle's spoils desire*
> *On his own hearth he tends the holy fire.*

And all the distrust in this century, the contrasts, the disunion, and even the mutual destruction, are the consequence of aggressive pretensions and the defensive impulses of one side against the other. Croatia senses the threat of Serbian endeavors to centralism and unitarism, while Serbdom sees in Croatianness its chief obstacle to the realization of its idea of a national state.

And since this has not only persisted through history, but has shown itself to be stronger than all integralist and consensus-based ideological political currents, and even all spiritual and political contributions by personages of both nations, the only possible inference one can draw is that there are real grounds to this phenomenon. Accordingly, both sides should look historical reality in the eye without prejudice and deception, and draw the necessary conclusions.

From discussions on the creation of a common state, during the First World War, Frano Supilo arrived at the conviction that on the Serbian side, neither in Pasic's government, nor in the opposition, was there an understanding, let alone a guarantee of equality in a unified state. Therefore, though personally one of the most assertive advocates of unification, he refrained from participating in its final preparations. On the other side, the representatives of Serbia had their own strong reasons not to enter into an uncertain arrangement, for their primary task was to strengthen and expand Serbia.

Dr. Ante Trumbic was ready even to compromise with N. Pasic to prevent the partition of Croatian lands according to the London Pact, and he assumed that Croato-Serbian relations would later be settled by mutual agreement. Pasic agreed to the Corfu Agreement only after he was left without his primary support, that of tsarist Russia, knowing that the execution of any agreement in principle depended on actual power relations. When Pasic later ensured Serbian domination by Vidovdan-constitution centralism, Trumbic—a very moderate and judicious politician and, as president of the Yugoslav Committee, one of the chief creators of the common state—resigned from the government, went over to the opposition, and became a supporter of a federalistic reorganization of the Kingdom of Serbs, Croats, and Slovenes. After the institution of the January 6, 1929, dictatorship, he had more sympathy for Pavelic's movement for Croatian independence than for the few followers of Yugoslav unitarism in Croatia, who were scorned by the Croatian people.

While Pasic represented the old current of politicians of the

Kingdom of Serbia, who felt that Serbian politics must not renounce its name and its statehood regardless of any desire for a union, King Aleksandar and his court symbolized the new current. Recognizing that Serbian hegemony in the Kingdom of Serbs, Croats, and Slovenes could not be maintained under the sign of Serbdom, he was to attempt to bridge the old frictions between nations by proclaiming a Yugoslave state and nation.

Also headed in this direction, was the undisputed leader of the Croatian nation, Stjepan Radic, a politician prepared for any tactical shift that circumstances might demand, save abandoning his aim of a free and equal Croatia; he was assassinated in the Assembly.

Aleksandar gave Yugoslavism an explicitly Great Serbian and anti-Croatian meaning. In addition to pushing aside old Serbian parties from the political scene, he entered into considerable conflict with traditional Serbian political thought, which did not want to follow him into Yugoslavism, which had been predominantly considered an unnecessary Croatian import.

During Aleksandar's pro-Fascist January 6, 1929, dictatorship, there came about in Croatia a collaboration between national revolutionaries, Ustashas, and proletarian communist revolutionaries, followed by an attempt on the part of the communists to form a united Croatian national revolutionary movement. This occurred, above all, due to their having the same programmatical objectives, for both movements aimed at destroying Yugoslavia and creating an independent Croatia. Therefore, to seek an explanation for that collaboration in allegedly similar—that is, illegal and Ustasha-style methods of combat—means to be blind to, or to evade, the essence of these historical occurrences.

The communists in 1932 inspired the publication of a journal of that movement, *Hrvatski Put*, which was printed abroad, but the creation of a united national revolutionary movement never took place for two reasons. The Croatian people had merged in a unified national movement under the leadership of the Croatian Peasant Party, and the ideological differences between the Ustashas and the communists were of an irreconcilable nature—

to such a degree that some communists felt that it was permissible to collaborate with the police against the "Frankists" (Ustashas), and even the CK KPH had to take a stand.[202]

Aleksandar himself arrived at the conclusion that the proclamation of a Yugoslav nation did not contribute to the anticipated stabilization of the country, and that he would have to seek a solution to the Croatian question through negotiations. Prince Paul, his actual heir in the Regent's Council, was advised of such a need by the Western Allies in the interest of averting a state crisis and preserving Yugoslavia. However, not even for the sake of loftier state interests were the Serbian civic parties prepared to heed the politics of compromise with the Croatians. Stojadinovic, one of the most competent Serbian politicians after Pasic, felt that on the basis of his radical, Serbian-Slovenian-Moslem coalition (JRZ), and his good relations with Rome and Berlin, an inflexible stance toward the Croatians was permitted. The Serbian democratic United Opposition supported negotiations with them, but only for the purpose of achieving general democratization, and not a territorial solution to the Croatian question.

Dr. V. Macek, a conservative politician, but without doubt a consistent one when it came to actualizing Radic's national program, took account of Croatian interests, but not any more than of mutual state interests. Defenders of centralism and hegemony blamed him personally for the "Croatian treason," ignorant or unmindful of the fact that the dissatisfaction of the Croatian people toward conditions under the monarchy was much more extreme than Macek's leadership in negotiations with Belgrade.

From historically shortsighted, Serbian, and exclusively ideological viewpoints, the reasonable state politics (internal as well as international) of Prince Paul were incorrectly judged. On the other hand, there was blindness to the devastating consequences of the Serbian *putschist* adventure against the Serbian politics it was to serve.

Pavelic's and Mihailovic's attempts, respectively, at a radical solution to the Serbian and Croatian problem proved to be impossible to achieve and equally perilous to both nations.

Contrary to the Ustasha and Chetnik programs, the revolutionary AVNOJ program of the NOP was acceptable to the nations of Yugoslavia and to international politics.

Not even all the experiences of World War II were able to bring about agreement among the majority of Croatian and Serbian civic politicians: neither in the form of membership in a mutual government in exile, nor animosity toward the communist order in the new federative Yugoslavia. Only an insignificant number of them was to continue the politics of Cvetkovic and Macek's Croatian-Serbian accord. However, in this context, it is interesting that Stojadinovic and Pavelic allegedly came to an agreement in Argentina concerning the possibility of a permanent solution to Croatian-Serbian controversy through a territorial redrawing of boundaries in Bosnia that was acceptable to both sides.[203]

In this sequence of events and personages essential to understanding the history of Croatian-Serbian relations, one should recognize the fact that even the "cases" of A. Hebrang and A. Rankovic cannot be explained merely as "deviations," "internal party" politicking, or a battle for personal power.

Andrija Hebrang represented that current in the Croatian revolutionary movement which from the 1920s to 1941, when forty-four Croatian Marxists were sent to their death and left behind the staggering legacy written on the wall of their cell (A. Cesarec, et al.), consistently sought to realize a free socialist Croatia in the spirit of the Leninist principle of the right of nations to self-determination. In keeping with this and with the disposition of the Croatian people, Hebrang assigned to the first and second ZAVNOH conventions (prior to the AVNOJ decisions) the purpose of a supreme representational and legislative body of the Federal State of Croatia, which was not quite understood by those who wanted to give the NOP predominantly classist or Yugoslav integralist characteristics.

What gave the Croatian Partisan movement its power and attraction was precisely the idea of a free Croatia. Hebrang, as secretary of the CK KPH, symbolized the determination of the Croatian communist revolutionaries in the National Liberation

Movement that a Croatian socialist state be created in an equal union with other states. And Vladimir Nazor represented that part of the Croatian nation which joined the NOP, not for socialist reasons, but exclusively for national reasons. As a poet, he wrote the "Partisan Flagpole" in May 1944, because it was navigating the ship of Partisan Croatia to freedom and glory as had King Zvonimir in one of his earlier poems.[204] And as a "Croatian and the President of the Croatian State," he ordered the representative of the unified federal government's Committee for Information (NKOJ) to ensure that the bulletin issued in the territory of Croatia "be written in Croatian" and not in the "language of the Belgrade market-place." The Politburo of the Central Committee of the Communist Party of Yugoslavia, upon receipt of a report to this effect, felt it was necessary "to make an end to these out-pourings and to locate the person who inspired them." This says as much about their unfamiliarity with the conditions in Croatia as of their different perception of the aims of the revolution.[205]

Regardless of how much the removal of A. Rankovic, Tito's closest collaborator and vice president of the SFRJ, was based on "the deformations of the secret police" (which is characteristic of all single-party or totalitarian regimes), there were, in essence, much deeper reasons. The matter at hand was a clash between two currents: the dogmatic centralists, on the one hand, and the feder-alist supporters of self-management, on the other. The fact that Rankovic became a representative of the former was not so much due to his personal character as to the fact that the strengthening of centralism was in keeping with traditional Serbian views. Tito, though he authoritatively held the power in his own hands, was on the side of self-management federalism in the spirit of AVNOJ.

Similarly, the reasons for removing the leadership of the SR Croatia at the end of 1971, and the crackdown on the movement for socialist reforms are not to be found in a supposed counter-revolution, but in a complexity of contrasts between nations and ideologies.

In connection with the 1971 events, it is interesting that the younger, anti-dogmatic and reform-minded liberal leaderships of

the SR Croatia and the SR Serbia were likewise unable to find paths to a bilateral solution to their conflicts in order to achieve a mutual "settling of accounts," which made it easier for the dogmatic powers to force Tito, who was clearly reluctant, to strike back, with all the disastrous consequences that followed.

Along with this cross-section of events in political life, one should keep in mind that opinions and phenomena in the spiritual life of the Croatian and the Serbian nations paralleled the political. One should recall the fate of Milan Sufflay, Mile Budak, and Dragisa Vasic (who in his time was also a friend of Miroslav Krleza), the paths of Tin Ujevic and Ivan Mestrovic, etc., and the virtually opposite views of Croatian linguists and writers on one side, and Serbian ones on the other, on the issue of the Croatian literary language.

In all these and similar manifestations, it is necessary to consider their meaning not only within time and circumstances, but also in their historical succession. For behind superficial changes (even after revolutionary changes in political ideology and economic systems), historical programs return to their sources and continue toward their goals, predetermined by geopolitical facts and historical and national factors.

The evolution of Croatian-Serbian relations has created such a historical climate that in all segments of the Croatian national being, there is a built-in defense mechanism against Serbian hegemony, and in the Serbian national being, against Croatian nationalism or separatism. These defense mechanisms, as a result of instincts of irreconcilability, contrast and adopt aggressive forms in the presence of real or merely imagined manifestations. An essential difference between them is, however, the fact that the entire Croatian national being does in fact seem to be imperiled by Serbian hegemony or Yugoslav integralism, while the entire Serbian national being does not feel threatened by Croatian nationalism, which it always equates with separatism, except as it exists in Croatia. However, because of that, Croatian nationalism is accused of being a threat to the entire Yugoslav union. For the

other nationalities in Yugoslavia, the disparity between Croatian and Serbian views sometimes gained the appearance of mutual rivalry over their positions on hegemony. Regardless of the fact that the Croatian national movement with its federalism had always been the chief support of the other non-Serbian nations, their particular national interests have sometimes caused them to side with hegemonic centralism, which has also found justification for accommodating them to the detriment of its chief rival.

In this respect, historically most significant is the role of Slovenian politics in Croato-Serbian relations. Anton Korosec, the most distinguished Slovenian politician in the period between World War I and World War II, was to have a clear effect on those relations. In the Viennese Imperial Council in 1917, Korosec referred to the Croatian historical right to statehood and the unity of the Southern Slavs in the Austrian-Hungarian monarchy as the basis to the framework for the unification of Slovenian regions and their extrication from Viennese hegemony and even from the threat of German assimilation within the framework of the Habsburg monarchy. At the time of its disintegration, as president of the National Council of Slovenians, Croatians, and Serbs in Zagreb, he contributed (along with Pribicevic and some other Serbian and Croatian politicians) to the implementation of an unconditional unification with Serbia. He was opposed to Vidovdan centralism and unionism because of its proclamation of a unified—although threefold, Serbian-Croatian-Slovenian— nation and language, and he answered Radic's invitation to form a Federalist Bloc from the main Slovenian and Croatian parties together with Spaho's Moslem party.

Both before and after this, he was part of several Belgrade governments, and in the critical days of the crisis of the Vidovdan constitution order in the coalition government of Veljo Vukicevic—made up of Serbian radicals and democrats and Slovenian clerics whose task was to overcome the opposition in the Assembly and in the country—Korosec was minister of internal affairs. After the assassination in the Assembly ended Radic's Croatian leadership, Korosec even answered Aleksandar's invita-

tion to become president of the interim government and thus, before the whole world, lessened the impact of the crimes of the Serbian Black Hand "in the bloody Belgrade Assembly." This was the first and only case in which a non-Serb was president of the government, and because, in addition to being a Slovenian, he was also a Catholic priest, it appeared to the domestic and international public that Slovenians, too, were opponents of Croatian "tribal separatism."

S. Pribicevic called Korosec's government a "provocation to Croatians," and those who labeled Croatian Catholic clericalism a vipers' nest of anti-Yugoslavism should be reminded of its presence in Korosec's government. Namely, Korosec installed two Croatians in his government: the unitarist ideologist Grga Andelinovic and a leader of the Croatian clerical party, Stjepan Baric, who, along with Korosec, had been in the Yugoslav club back at the Constitutional Assembly.[206] Since imperialist hegemony during Aleksandar's dictatorship threatened even the Slovenian nation, Korosec joined in the Croatian federalist demands, and so he was interned on the island of Hvar, but he was not tried and incarcerated as were Macek and others. Later, he entered Stojadinovic's government, but he was on the side of a Serbo-Croatian agreement on the creation of Banovina Croatia, aware that this in itself meant the right to a Banovina Slovenia. Through his pragmatic politics, he succeeded in procuring considerable favors for Slovenia, and he based his politics on the opinion that even the worst Yugoslavia is the best solution for Slovenia, considering the physical threat to the Slovenian nation from German and Italian imperialism.[207] In fact, the politics of all Slovenian currents was conditioned by this notion, up until the completely changed international conditions in the contemporary world, in which European nations (including Slovenia) are no longer threatened by those forms of imperialism. Those geopolit-ical and historical factors conditioned even the unique form of Slovenian arbitration in Croatian-Serbian relations in the revolutionary and postwar period. Or, in other words, from the unique position of the Slovenian nation, the stance of Slovenian representatives

toward Croatian confederalism and Serbian imperialism resulted.

Edvard Kardelj, who was indubitably one of the best Marxist theoreticians of the national question and Tito's chief collaborator in structuring an AVNOJ plan for building the SFRJ, never fully comprehended the Croatian problem, both because he remained preoccupied with ideological doctrine and because he judged the problem predominantly from the viewpoint of Slovenian interests. Kardelj did not approve Hebrang's policy in Croatia because he did not understand the Croatian conditions, but also out of fear that too great an emphasis on the self-determination and independence of Croatia would threaten Yugoslavia as a whole, without which, in the circumstances at that time, a solution to the Slovenian national question would have been impossible. In his views, the right of a nation to self-determination was left in the mire of socialist integration. In 1971, he leaned, not for theoretical but pragmatic reasons, toward those who were seeking from Tito the restriction of the Croatian movement pushing for actual and not merely theoretical equality. In the discussions held at that time, he actually presented the fundamentally correct proposition that Serbs and Croatians, as the two most numerous nations, should not be allowed to make agreements at the expense of the smaller nations, but he clearly did not contribute to discouraging the unscrupulousness of other proposals concerning inter-republican relations in the federation.

An even sharper politically pragmatic, and non-historical stance toward this Croatian movement was demonstrated by Stane Dolanc. A full decade and a half later, when, not only all the catastrophic consequences of the anti-reform changes at the end of 1971 and in 1972 had become evident, but also when in Slovenia essentially the same developments were taking place as had in Croatia, only with more resolute demands for a confederation, Dolanc approved the removal of the leadership of the Croatian "mass movement" (i.e., the leadership of SR Croatia) because of its alleged ties with "Ustasha emigrants" abroad.[208] Since Dolanc probably does not fall into the category of those postwar communists whom T. Kermauner labeled Lumpen-naiven ("the naive

rabble"), and is otherwise a supporter of AVNOJ federalism, such a Stalinist police-state interpretation of historical events comes as a surprise, particularly because, based on experience both from our own history and that of all other socialist lands, one should come to the understanding that even in the Yugoslav reality, there are historically untenable, inaccurate political judgments, ousters, and condemnations of entire movements and individuals without whose "rehabilitation" there is no freedom from past errors. We shall repent these sins all the more the longer they are disregarded and historically unevaluated.

It is an incontestable historical experience that every society, every nation, or every consolidated state, after experiencing a historical rift and surviving a turning-point, make a rational assessment of its being in order to determine as realistic a place for itself as possible in the contemporary world.

If the formula at the beginning of this chapter—that people and nations will learn to behave rationally only when they have exhausted all other possibilities—is to be believed, then there are many reasons for the conclusion, based on all that we have covered in this work, that the time for rationality has come. For what sort of insanity could yet occur in material and spiritual life after all that has already come to pass?

The fundamental preconditions for an escape from the historical wilderness of Serbian-Croatian relations are very simple:

(1) To assess rationally and soberly the causality of all historical occurrences;

(2) To accept the national identity of the Croatian and the Serbian national beings, such as have been formed over the course of historical development with all their elements and determining factors;

(3) To recognize that each of the two nationalities has the right to self-determination and to their own statehood, i.e., to inviolable sovereignty in deciding their own fate; in other words, their lives and their pursuits; and

(4) To resolve questions in dispute by negotiation, in order to achieve harmonious coexistence in a community of peoples with

equal rights in both Yugoslavia and Europe.

Since these are generally accepted principles in the modern world, it is clear that any disagreement with them amounts to a desire for hegemony and an endeavor to impose one's own will by force. And thereby, it makes no difference whether that will is the expression of an unfulfilled desire for expansionism or an incoherent utopianism.

History has already given us its answers concerning the possibilities and consequences of such endeavors. Unfortunately, their true meaning has not been transformed into the understanding that is historically necessary, in order to overcome the cause-and-effect chain of irrationality.

At the end of this discussion on the sources and the purpose of historiographic and theoretical attempts to reduce the essential problems of the revolutionary period to the distorted view of Jasenovac, and on the relations between nations to the thesis of the alleged genocidal character of Croatiandom, one must bear in mind that it was by no means accidental that such ideas should appear in the wake of the departure from the historical stage of Josip Broz Tito as the founder of the SFRJ. As part of a historically inescapable scholarly reexamination of the revolutionary past, such views are above all an expression of those tendencies which are incapable of accepting the historical achievements of the revolution. Their aim is to render worthless its unquestionable historical value: the Titoist AVNOJ basis for relations between nations in the SFRJ. In connection with that one should not, after all, neglect certain essential lessons from historical experience gleaned from this laborious confrontation with the past. In spite of everything, we are an integral part of that world which, from Scandinavia to the Mediterranean, has for the most part exhausted irrationalities, and has reached the understanding that every attempt at stifling the national sovereignty of historical nations is an untenable form of tyranny in the contemporary world, since every threat to the unhampered development of individual nations threatens the international order itself.

The ever increasing integration of the world in terms of culture

and civilization, economy and trade, technology and information does not lead to the removal of national individualities, but just the contrary—to their development, though with ever increasing imperatives that they find a way toward a more harmonious, peaceful coexistence.

Since a considerable portion of the world has already arrived at such understandings on the road to overcoming its historical irrationalities, we hope that historical judiciousness will reign once and for all in the arena of the Croatian-Serbian wilderness, for the very reason that historical insanity on this transitional and unstable soil is becoming ever more ominous. In light of the mass historical contradictions between nations, and the continuing and deceitful rivalry of foreign powers in this important historical crossroads, the task of finding paths for the possibility of the coexistence of historically distinct, neighboring nations is all the more urgent.

There can be no doubt that an understanding of the necessity of transcending historical misfortunes and malicious defamation reveals the degree of development of an individual national entity. Moreover, it reveals the level of maturity in the primordial yearning to achieve, as an unsubstitutable objective, the status of a recognized subject with full rights within the state community, in the European community as well as the international community.

NOTES

■

History and Life's Destiny

PART ONE

[1] Sidney Hook, *The Hero in History* (Boston: 1965), 56.

[2] Karl Marx - Friedrich Engels, *Rani radovi* [Early Works] (Zagreb: 1973), 397.

[3] In the *Orijentacioni program djelatnosti Instituta za historiju radnickog pokreta* [Orientation Program of Activities at the Institute for the History of the Workers' Movement], which I prepared and which was accepted by the Council on December 25, 1961, apart from a monographic treatise on the history of the Communist Party of Croatia, the People's Liberation Struggle and the Revolution, the Ustasha movement and the NDH, there is a forecast of research into the terror and crimes of the occupiers, the Ustashas, and the Chetniks. See *Putovi revolucije* [Path of Revolution] Zagreb, nos. 1-2 (1963), 519-536.

[4] See my works: *Okupacija i revolucija* [The Occupation and the Revolution] (Zagreb: 1963); "Raspre o uzrocima sloma monarhisticke Jugoslavije i o pretpostavkama razvitka NOB-a u Hrvatskoj" [Discussions on the Causes of the Decline of Monarchist Yugoslavia and on the Premises for the Evolution of the People's Liberation Struggle in Croatia], 1964; "O opcim uvjetima i znacajkama razvitka revolucionarno-demokratskog poretka u Hrvatskoj" [On the General Conditions and Characteristics of the Development of the Revolutionary Democratic Movement in Croatia], 1966; the encyclopedia entry on the history of the Socialist Federal Republic of Yugoslavia (1967); "Ideja o slavenskoj uzajamnosti i narodi Jugoslavije u Drugom svjetskom ratu" [The Idea of the

Slavic Reciprocity and the Nations of Yugoslavia in WWII], 1968. All of these works were published in *Velike ideje i mali narodi* [Great Ideas and Small Nations], 1969, Second edition, 1970; dealing directly with the problem of war victims in *Nationalism in Contemporary Europe* (Columbia University Press, 1981); particularly, "Odgovor na optuzbu" [A Response to Allegations] (1981), as well as letters in response to high-level state and political agencies and public figures (from 1981 to 1986) in the manuscript *Usudbene povjestice* [Fated History].

⁵ Cohen, Philip J. *Serbia's Secret War: Propaganda and the Deceit of History.* Texas A&M Press, College Station, Texas, 1996.

⁶ Cohen, op. cit. pp. 111-112.

⁷ The far-reaching intent of those who orchestrated the retaliation against me is evident from the fact that they chose as a framework for exposing me as a falsifier of history and a collaborator with the Ustashas a news series about the liquidation of a terrorist Ustasha group in 1972, in which, apart from myself (most often mentioned), M. Djilas, and the leaders of the "Croatian Spring," M. Tripalo and S. Dabcevic-Kucar were included in the list. The series first appeared in the Belgrade *Politika Ekspres* under the title "Kako ubiti slobodu" [How to Kill Freedom] in twenty-nine installments (August 13 to September 10, 1979), then in "Operacija Radusa" [Radusa Operation], subtitled *Svi su povezani* ["All are Connected"] and *Veze ca spijunskim centrima* ["Ties with Espionage Centers,"] in *Nedjeljna Dalmacija*, Split, October 21 to December 23, 1979; no. 441-450. Mihailovic also published this in books and repeated his accusations in other newspapers, e.g., in *Reporter,* Belgrade, April 2-9, 1981, No. 771.

⁸ It is interesting that he did this in an otherwise rational review of those who are "hampered by the Constitution" (*Svijet,* Sarajevo, 1985, No. 1393).

⁹ This prompted me to address an open letter (July 30, 1981) to Dr. Dusan Dragosavac, a member of the Presidium of the Central Committee of the League of Communists of Croatia, and to Marinko Gruic, a member of the Central Committee of the League of Communists of Croatia and the presidency of the Republic Conference of the Socialist League of Working People of Croatia. I also sent that reply (nine pages) to the Presidiums of the Central Committee of the League of Communists of Yugoslavia and the Central Committee of the League of Communists of Croatia, as well as to the Presidiums of the federal and republican Socialist Alliance and to the editors of *Borba, Vjesnik,* and *Politika.* I harbored no illusions that they would publish it, but I firmly believed that this would take place at some future time. Six years later, during my stay in Canada in June 1987, I received a surprising letter

from an unknown countryman in California who had my response to Dragosavac and Gruic translated into English and had made 7,000 copies to send to prominent Americans.

[10] Interview with the pensioned lieutenant general V. Terzic, author of the book *Slom Kraljevine Jugoslavije 1941* [The Fall of the Kingdom of Yugoslavia 1941], *Intervju*, Belgrade, August 5, 1983, No. 57, 4-9.

[11] Dedijer's distortion of propagandized statistics, his "Russell Court" and his attempts to strengthen the Jasenovac myth by accusing the Vatican.

[12] Veselin Djuretic, *Saveznici i jugoslovenska ratna drama* (Allies and the Yugoslav War Drama), (Belgrade: 1985), I, 72-90, 111-150; II, 239-253, et al.

[13] *Knjizevne novine*, XXXII, No. 716, September 1986, 1, 4-5.

[14] Antun Miletic, *Koncentracioni logor Jasenovac 1941-1945: Dokumenti, I, II* [The Jasenovac Concentration Camp 1941-1945. Documents, I, II] (Belgrade; Narodna knjiga, Jasenovac Memorial Area, 1986-1987).

[15] Reviews in all the daily and weekly newspapers; comments by participants at a meeting concerning the book, organized by the journal *Nase teme* [Our Themes] (see No. 9, 1986, 1265-1320); a criticism of the book in the historical journal *JIC* (*Jugoslovenski istorijski casopis*— Yugoslav Historical Review), Nos. 1-4. 1986; 234-241.

[16] In a discussion of the book, which, by the way, came out in a second edition in the same year, A. Miletic announced its upcoming completion with a third volume. Indeed, this can be beneficial to further research in details, but presented in such a way it cannot shed new light upon an assessment of the total number of victims already presented by Miletic.

[17] A. Miletic, *op. cit.*, I, 12.

[18] Ibid., 12-13.

[19] Ibid., 14. (Italics mine - F.T.)

[20] Ibid. (Italics mine - F.T.)

[21] Jeffo Sasic, *Pregled istrazivanja genocida u Jasenovsu* (A Survey of Research into the Genocide in Jasenovac), *Nase teme*, no. 9, 1986, 1294.

[22] Ibid.

[23] As I did not leave the army at their request but my own, I also received a civilian salary. Unaware of this, my colleagues were envious: "It's easy for you to adhere to principles with a general's pension." This testifies to the spiritual condition of even this area.

[24] A monument marks my father's birthplace in Veliko Trgovisce (in the Zagorje region of Croatia), where, before the war, from 1936, he was twice elected president of the county seat; a memorial plaque marks the house of his birth, in which he was murdered along with his wife.

[25] F. Tudjman, *Rat protiv rata: partizanski rat u proslosti i buducnosti* [A War Against War: Partisan War in Past and Future], Zagreb, 1957; Slovenian translation: Vojna proti vojni, Ljubljana, 1964.

[26] Pero Moraca, "Rat protiv rata: Povodom Tudjmanove knjige" [A War Against War: On Tudjman's Book], *Nasa stvarnost,* Belgrade, 1958, XII, No. 1, 82-94.

[27] Ibid., 87 and 94.

[28] Andro Gabelic, "Rat protiv rata: Osvrt na knjigu F. Tudjmana i napis Pere Morace" [A War against War: A Review of F. Tudjman's Book and Pero Moraca's Essay], *Nasa stvarnost,* Belgrade, 1958, XII, 7-8; 109-123.

[29] Ibid., 123.

[30] The discussion was printed in the journal *Vojno delo,* Belgrade, 1985, X, 9; and reprinted in *Zbornici izabranih radova: Razmatranja o savremenom ratu* [Collection of Selected Works: Discussions on Contemporary War—Belgrade, 1959] and *O vojnoj vestini* [About Military Deeds], II, Belgrade, 1960.

[31] "Note from the editorial staff of the *Vojna enciklopedija (VE)* [*Military Encyclopedia*] accompanying the article on Croatia in the National Liberation War." A copy of the official communication is in my possession.

[32] Letter from the editor-in-chief of the *Vojna enciklopedija,* Bosko Siljegovic, to Dr. Vladimir Bakaric, dated December 15, 1960. A copy is in my possession.

[33] *VE,* vol. 2, 1959, 267.

[34] I was promoted to the rank of major general just that year. Although since 1945 I had continuously performed high-level functions in the Defense Ministry and in the General Command of the Yugoslav National Army, I was not promoted earlier because of my "youth" and the "principledness" that I applied to Croatian cadres.

[35] Original copy in my archives.

[36] *Franjo Tudjman, Stvaranje Socijalisticke Jugoslavije* [The Creation of Socialist Yugoslavia], Zagreb, 1960.

[37] F. Tudjman, "O sustini i metodu jedne kritike" [On the Substance and Method of a Criticism], *Vojnoistorijski glasnik,* Belgrade, No. 5, 1961, 67-107.

[38] Franjo Tudjman, *Rojstvo socijalisticne Jugoslavije,* Ljubljana, 1961.

[39] Vaso Bogdanov, "O godini 1941. u Hrvatskoj (povodom Moracina referata na Kongresu historicara)" [About the Year 1941 in Croatia (inspired by Moraca's report to the Congress of Historians)], *Forum,* I, 1962, 3.

[40] Ivan Sibl wrote of that intellectual circle in his memoirs, *Sjecanja III: poslijeratni dnevnik* [Memoirs III: A Postwar Journal], Zagreb, 1986, 147-149).

[41] *Pregled istorije SKJ* [A Survey of the History of the League of Communists of Yugoslavia], ed. R. Colakovic, Dr. D. Jankovic, and P. Moraca (Belgrade, 1963).

[42] Material for the symposium on *Pregled istorije SKJ* and on the problem of scholarly research into the history of the workers movement, the revolution, and the socialist building (December 12-14, 1963), *Putovi revolucije*, Zagreb, II, 1964: nos. 3-4.

[43] A commentary on *Pregled istorije SKJ*, IHRPH [Institute for the History of the Workers' Movement in Croatia]. Edited by I. Babic, F. Bikar, D. Bilandzic, J. Buljan, Z. Cepo, I. Jelic, Z. Lisinski, V. Stopar, H. Sarinic, D. Scukanec, I. Skiljan, F. Tudjman, Zagreb, February 1964, 91 (mimeographed).

[44] *Marinko Gruic, Burzoasko-nacionalisticka skretanja u pristupu "Nacionalnom pitanju"* [Bourgeois-nationalistic Detours in Approaches to the "National Question"], *Komunist*, XXII, March 26, 1964, No. 360, 4.

[45] The text of the lecture was published under the title *"Raspre o uzrocima sloma monarhisticke Jugoslavije i o pretpostavkama razvitka Narodno-oslobodilacke borbe u Hrvatskoj"* [Discussions on the Causes of the Decline of Monarchist Yugoslavia and on the Conditions for the Development of the National Liberation War in Croatia], Kolo, Zagreb, Matica Hrvatska, VI (CXXVI), 1968, 8-9; also in *Velike ideje i mali narodi* op. cit. On the conference in Split (October 5-10, 1964), held on the occasion of the twentieth anniversary of the first government of the Socialist Republic of Croatia and the Third Congress of the Anti-fascist ZAVNOH Council, see *Putovi revolucije*, Zagreb, III, 1965, 5, 283-284.

[46] As I recall, Frane Boko was then Cvetkovic's Office Chief.

[47] Handwritten copy of the letter to the Executive Committee of the Central Committee of the League of Communists of Croatia, dated July 22, 1965, in the author's archives.

[48] Among a large number of university professors, certain members of the Yugoslav Academy of the Arts and Sciences were also permanent or occasional collaborators with the institute. My collaboration was not limited only to the Academy journal *Forum;* I also published a lengthy work in *Zbornik radova JAZU* [Collection of Works of the Yugoslav Academy of the Arts and Sciences], "Jugoslavenski odbor u Londonu: U povodu 50-godisnjice osnivanja" [Yugoslav Committee in London: On the Occasion of the 50th Anniversary of its Founding], Zagreb, 1966. At

the meeting of Yugoslav Academy of Arts and Sciences, the University of Zagreb, and the Croatian Council for Scholarly Work, I did the introductory presentation: "O stanju i osnovnim problemima institucionalnog razvitka drustvenih znanosti u Hrvatskoj" [On the Condition and Fundamental Problems of the Institutional Development of the Social Sciences in Croatia], *Telegram,* Zagreb, March 1964.

[49] A copy of the proposal by Academicians Vaso Bogdanov and Ferdo Culinovic and my reply to Academician Ivo Krbek are in my archives.

[50] A copy of the letter to the President of the Yugoslav Academy of the Arts and Sciences, Academician Grga Novak, dated June 23, 1965, is in my archives, while the original should be at the academy.

[51] Velimir Terzic, *Jugoslavija u aprilskom ratu 1941* [Yugoslavia in the April War of 1941] (Titograd, 1963).

[52] Dragutin Scukanec, Povodom knjige V. Terzica: Jugoslavija u aprilskom ratu (On the Occasion of V. Terzic's book: Yugoslavia in the April War), *Socijalizam*, 10, 1964.; Stjepan Scetaric, and colleagues., O politickim i vojnim uzrocima sloma Jugoslavije (On the Political and Military Causes of the Fall of Yugoslavia); Fikreta Butic and Ivan Jelic, O takozvanom ustaskom ustanku u Hrvatskoj 1941 (On the So-called Ustasha Uprising in Croatia in 1941.); *Putovi revolucije*; Zagreb, II., 1964., 3-4, 491-520. The actuality of the problems of Terzic's book is attested to by the fact that Sehovic's (pre-war publicist, member of ZAVNOH, Yugoslav ambassador, and at that time my colleague at the Institute) review was published by the Zagreb *Vjesnik* on two pages under the title: Traganje za krivcima (Search for the Guilty) with the subtitles: O metodi istrazivanja istine (On the Method of Investigating the Truth); Stav prema nacionalnom pitanju (Position on the National Question); Separatizam i pravo samoodredenja (Separatism and the Right to Self-determination); Teza o ustanku u Hrvatskoj kao uzroku aprilskog sloma (The Thesis on the Uprising in Croatia as a a Cause of the Fall in April); Cemu to sluzi (Whom Does This Serve); Iluzija o mogucnosti spasavanja monarhisticke Jugoslavije (Illusions on the Possibility of Preserving Monarchist Yugoslavia); and others. Along with the expanded text reports were also provided. *Vjesnik*, Zagreb, XXV., July 26, 1964., No. 6263., 6-7. This was judged as the most radical historical-political text which the director of Vjesnik Bozo Novak took to V. Bakaric before its publication. Bakaric himself told Sehovic: "I read it, I congratulate you but don't tell anyone I read it before it was printed!" This tells you something about the conditions under which we worked in the Institute, but also about Bakaric!! It also tells something about all of us who participated in the creation of such conditions!

⁵³ Outline for the program of the conference, "Nacionalno pitanje u suvremenom suijetu" ("The National Question in the Contemporary World,") or Marksizam i nacionalno pitanje ("Marxism and the National Question,") December 22, 1964, published "Vjesnik" HAZU, II/1993., Nos. 1-3, 3-7 under the title "Nacionalno pitanje u suvremenom svijetu u svjetlosti teorije i prakse socijalistickih i drugih pokreta. Nacrt programa za simpozij, prosinac 1964." (The National Issue in the Contemporary World in Light of Theory and Practice in Socialist and Other Movements. Outline program for the Symposium, December 1964.)

⁵⁴ A copy of my letter to Dr. Vladimir Bakaric, Secretary of the Central Committee of the League of Communists of Croatia, dated January 19, 1965, is in my archives.

⁵⁵ Franjo Tudjman, "Evolucija nacina ratovanja: Pregled povijesti ratne vjestine i tendencije razvoja u suvremenom, nuklearnom dobu" [The Evolution of Warfare: a Survey of the History of Military Practice and Tendencies in the Modern Nuclear Age] (thesis for promotion to rank of general), Belgrade, 1960

⁵⁶ Franjo Tudjman, *"Uzroci krize monarhisticke Jugoslavije od ujedinjenja 1918 do sloma 1941"* [The Causes of the Crisis in Monarchist Yugoslavia from the Unification in 1918 to Its Fall in 1941], doctoral dissertation I, II, 1965, 675 pages.

⁵⁷ *Republika,* Zagreb, XV, 1959, 4-5.

⁵⁸ *Forum,* Zagreb, II, 1963, Nos. 2, 3, 4, 5, 6, 7-8; III, 1964, Nos. 1-2.

⁵⁹ Members of the review committee for my doctoral dissertation were Academician Dr. Vaso Bogdanov, head of the Department of History of the Nations of the Socialist Federal Republic of Yugoslavia, University of Zagreb; Dr. Kosta Milutinovic, professor, University of Zadar; and Dr. Dinko Foretic, interim professor of the University of Zadar.

⁶⁰ The letter from the director of the *Naprijed* publishing house, Kalman Vajs, dated January 6, 1967, addressed to me (as the director of the Institute) in its entirety:

> Having negotiated an agreement concerning the publication of your work Monarchist Yugoslavia, based on the positive review by Dr. Vaso Bogdanov and on your position as a historian and distinguished political and public figure, not anticipating any problems, we immediately began preparing the manuscript for publication.
>
> Unfortunately, during the course of the preparation of the manuscript, certain members of the editorial staff notified us

is my handwriting; that these are the statistics which I brought
from the official statistical survey of the Republican Statistical
Institute for the Socialist Republic of Croatia.

Zagreb, March 13, 1981.

Nikola Kajic

Retired Lieutenant General

[64] Copy of the paper I provided to the Central Committee of the
League of Communists of Croatia (Commission for history) in my letter
of October 20, 1965. From the author's archives.

[65] A presentation by Dr. Vladimir Bakaric at the seventh plenary ses-
sion of the Central Committee of the League of Communists of Croatia,
Vecernji list, Zagreb, IX, 2398, April 20, 1967, 1, 3-5.

[66] Fifth plenary session of the Chief Committee of the Socialist League
of the Working People of Croatia, "Politicka odgovornost pokretaca i
politickih nosilaca deklaracije" [The Political Responsibility of the Ini-
tiators and the Political Bearers of the Declaration], *Vjesnik,* Zagreb,
XXVII, 7217, March 23, 1967, 3.

[67] A presentation by Dr. Vladimir Bakaric at the 7th plenary session of
the Central Committee of the League of Communists of Croatia, op. cit.;
emphasized in the original, referring to material for discussion in the Cen-
tral Committee of the League of Communists of Croatia after which Gruic
published in *Komunist* his critique of the "bourgeois nationalist deviation."

[68] I was forced to resign from the directorship of the institute, which
had been founded by the Central Committee of the League of Commu-
nists of Croatia. At the Faculty of Political Science, where I was an hon-
orary lecturer, I was simply no longer needed, as I hadn't accepted an ear-
lier offer of permanent employment. At a meeting at the institute's party
organization after Zanko's attack, at which two were present (Pero Pirk-
er and Duje Katic), no one proposed any sort of punishment for me.
However, after I relinquished my post they expelled me from member-
ship in the League of Communists of Croatia in my absence at the explic-
it request of Marinko Gruic, who certainly had either an order or autho-
rization to take such action.

[69] I found this out from Krleza just as I came out of prison at the end
of 1972. To make a historical testimony of that fact, I mentioned it at the
occasion of my second judicial persecution in a letter dated Mar. 4, 1981,
after Tito had died, to eighteen members of the political leadership of
Socialist Federal Republic of Yugoslavia and the Socialist Republic of
Croatia as well as to Krleza and certain others. Krleza was still alive, and
I was able to mention him as a witness, while I could not mention two
others because of their official positions.

[70] I entered the *Matica Iseljenika* [Emigration Office] at the explicit request of Veco Holjevac, and out of indebtedness to the organizational secretary of the Central Committee of the League of Communists of Croatia, against my will.

[71] By the verdict of the District Court (October 11, 1972), I was sentenced to two years in prison, but I was immediately released from prison. The Supreme Court of the Socialist Republic of Croatia later reduced my sentence to the time I had spent in prison pending trial.

[72] My manuscript *Nacionalno pitanje u suvremenoj Europi* was printed in English (in somewhat shortened form) under the title *Nationalism in Contemporary Europe* (New York: Columbia University Press, 1981), 293 pages. The book came out just at the time of my second trial, so it was most unfavorable for me that it stated "Copyright by Matthew Mestrovic." At that time Mestrovic was being attacked in the Yugoslav press as the president of the expatriot Croatian National Council. It is a fact that I had given no authorization and that the copyright was determined without my knowledge.

At the same time, I had a contract with a European publisher for a German and a Croatian edition. While these editions were in preparation, an illegal edition appeared, without my knowledge, under the title *Nacionalno pitanje u suvremenoj Europi* (Munich-Barcelona: Hrvatska Revija, 1981, 373 pages; followed by a second edition, 1982, 429 pages).

Because of this, my publisher canceled a Croatian edition, printing only its concluding chapter (which was not contained in the above-mentioned Croatian edition), entitled *Drzavnost nacija kljuc mira Europe* [Statehood of Nations—the Key to Peace in Europe] (Lidingo, 1982), 53 pages, followed by the entire manuscript in German, *Die Nationalitaetenfrage im heutigen Europa* (Lidingo, 1986), 302 pages.

[73] By the verdict of the District Court in Zagreb on February 20, 1981, I was sentenced to 3 years of prison (in accordance with art. 133, par. 1 and 2, Yugoslav Criminal Code), and at the same time (in accordance with art. 57, par. 1 and 2, Yugoslav Criminal Code) to a ban on any public activity (public presentation in print, at meetings, and on radio and television) for a period of five years. The sentence was confirmed by the Supreme Court of the Socialist Republic of Croatia, while the Yugoslav Federal Court and the Yugoslav Prosecution remained silent on all evidence presented by the defense.

[74] Ivan Lugar, "Povijesne istine po Franji Tudjmanu" (Historical Truths According to Franjo Tudjman), *Vjesnik, Sedam dana*, Zagreb, XLII, No. 12029, February 21, 1981, 19.

[75] *My "Letter to the 18," dated March 4, 1981.*

[76] A Letter to the Presidiums of the Socialist Federal Republic of Yugoslavia and the Socialist Republic of Croatia, the Central Committee of the League of Communists of Yugoslavia and the Central Committee of the League of Communists of Croatia, the Socialist League of Working People of Yugoslavia and the Socialist League of Working People of Croatia, dated November 24, 1984. On the same day I had to return (for the third time) to complete my prison sentence (of which I had already completed over a year and a half, but had to interrupt twice because of illness). Along with this letter I also prepared an "Open Letter to the Public." The letters were not mailed because I received the decision (by a special courier) concerning my conditional release on the day before I was to return to prison, on November 23, 1984. I wrote in a letter to Mika Spiljak, as president of the Presidium of the Central Committee of the League of Communists of Croatia and to Josip Vrhovec, as a member of the Presidium of the Socialist Federal Republic of Yugoslavia, dated May 28, 1985, about Tanjug's issuance of new incriminations against me from the Swedish press, as one of the examples of the continuation of the witch hunt which was not to end until a retrial was allowed or until my replies were allowed to be published.

NOTES

■

History as Crime and Mythical Appearance

PART TWO

[1] Jozo Tomasevic, *Cetnici u drugom svjetskom ratu* 1941-1945 [The Chetniks in the Second World War 1941-1945], Zagreb, 1979, 130, 240. (The author of the memorandum, Pero Slijepcevic, was a Hercegovinian Serb who, during the 1930s, was a professor of philosophy in Skopje; 240.)

[2] Ibid., 340-341.

[3] Opinion on the Memorandum of the Serbian Orthodox Church to General Danckelmann (copies from the bequest of Dr. Bicanic are in my archives). Italics in the original.

[4] Dusan Biber, *Kome sluzi prebrojavanje* [Whom Does the Counting Serve?], Danas, Zagreb, no. 209, February 12, 1986, 71.

[5] Veselin Djuretic, *Vlada na bespucu* [Government at an Impasse], Belgrade, 1982, 197-199.

[6] Ibid., 166-167.

[7] Sava N. Kosanovic, *Sta se moglo videti iz emigracije (nekoliko dokumenata)* [What One Could See from Emigration: Several Documents], Belgrade, 1945, 9.

[8] Ibid., 8.

[9] Dr. Branko Cubrilovic, *Zapisi iz tudine* [Writing from emigration], Sarajevo, 1946, 72-76.

[10] J. Tomasevic, op. cit., 131.

[11] That Croatian who "disembarked" in the capital of Brazil was said to have been Nikola Pavelic. The importance attributed to his stories is

evident from the fact that a Yugoslav emissary immediately (May 15, 1942) reported the matter to I. Subasic in the United States (V. Djuretic, *Vlada na bespucu*, op. cit., 208).

¹² The February 1942 report by Dr. P. Grisogono from Belgrade was received by the president of the government of the King of Yugoslavia in London, top secret V. K. no. 724, August 13, 1942 (Ljubo Boban, *Hrvatska u arhivima izbjeglicke vlade, 1941-1943* [Croatia in the Archives of the Government in Exile, 1941-1943], Zagreb, 1985, document 4, Report by P. Grisogono, 293.

¹³ The alleged letter from P. Grisogono to Archbishop Stepinac, dated Febuary 8, 1942, sent to the president of the royal government in London, S. Jovanovic, by Ambassador Ilija Sumenkovic from Ankara via telegram July 24, 1942 (Lj. Boban, op. cit., document 2, 284-286).

¹⁴Grisogono denies the letter's authenticity; see Lj. Boban, op. cit., document 3, 288-292.

¹⁵ See Lj. Boban, op. cit., 292-296.

¹⁶ Curzio Malaparte (Kurt Suckert), an Italian journalist and writer of exceedingly contradictory and radical views. In the 1930s he was a fanatical adherent of Fascism, in which he saw the birth of a new civilization (he walked around Rome in boots and a Fascist uniform), was infatuated with Lenin and Mussolini and in the effectiveness of their totalitarianisms, but in the 1940s he decided that communism was the movement of the future (B. Radica, *Zivjeti nedozivjeti* [Living and Not Experiencing], Munich-Barcelona, 1982-1984, vol. I, 148-9; vol. II, 373. The fantasy of Pavelic's little basket of Serbian eyes was most likely meant to authenticate the sincerity of his change of sides, since Malaparte had before that written hymns of praise to the *Poglavnik* of the NDH [Ante Pavelic] as well as in the Italian papers, which were reprinted by Pavelic's publications (see *Spremnost*, 1942, no. 17, 2; *Hrvatski narod*, no. 455 and others). Raffaele Casertano, the Italian ambassador to the NDH government in Zagreb, who brought Malaparte to Pavelic, announced that Malaparte had written that sentence "out of sadism, sinning against the truth." And Victor Alexandrov wrote in 1960 that Malaparte himself admitted that he had "wanted to create an effect, to symbolize horror." He had seen currants, but "eyes and currants: is there any difference?" "You cán't wound the imagination with currants" (compare M. Rojnica's *Napukla Titova ostavstina u knjizi Nore Belof* [Tito's Fragile Heritage in Nora Beloff's book], *Hrvatska revija*, no. 3, 1986, 518.

¹⁷ Report by D. Mihailovic picked up by the presidency of the government under top secret V.K. no. 935, dated November 5, 1942. (Ilija Jukic, *Pogledi na proslost, sadasnjost i buducnost hrvatskog naroda*

[Views on the Past, Present and Future of the Croatian Nation], London, 1965, 130-131).

¹⁸ V. Djuretic, *Vlada nu bespucu*, op. cit., 208.

¹⁹ Ibid., 208-209.

²⁰ S. Kosanovic, op. cit., 18-25.

²¹ *Harper's*, January 1943, pages 183-185; compare P. D. Ostovic, *The Truth About Yugoslavia*, New York, 1952, pages 204-205.

²² In the name of the U.S. government, the report was presented by Elmer Davis, Director of the Office for Military Information [P. D. Ostovic, op. cit., page 205].

²³ Lj. Boban, op. cit., vol. I, pages 284-285.

²⁴ A. Miletic, op. cit., vol. I, document no. 152, pages 398-400. [Italics are mine, FT. It is clear that this is the same apocryphal letter based on the content and the marks on the letter; compare Lj. Boban, op. cit., pages 284-288.]

²⁵ V. Djuretic, *Vlada na bespucu*, op. cit., 93-96, 208-209, et al.

²⁶ Croatian Commission for the establishment of Crimes of the Occupiers and Their Collaborators, Zagreb, 1946; An excerpt from the report by that commission, dated November 15, 1945, concerning the crimes in the Jasenovac camp in: A. Miletic, op. cit., vol. II, 1090-1120; citations concerning the victims, 1100ff.

²⁷Compare Antun Miletic-Branko Obucina, Jasenovacki Igor "Gradina" 1949-1945 [Jasenovac Camp "Gradina" 1949-1945], *Vojnoistorijski glasnik*, Vol. XXXI, 1,1980.

²⁸ FNRJ, *Drzavna Komisija za utvrdjivanje zlocina okupatora i njihovih pomagaca: Saopcenje o talijanskim zlocinima protiv Jugoslavije i njenih naroda* [Federal People's Republic of Yugoslavia, State Commission to Establish Crimes of the Occupiers and Their Collaborators: The Report on Italian Crime Against Yugoslavia and Its Nations], Belgrade, 1946.

²⁹ A. Vojinovic, "Jugoslavenski dosje u Nurnbergu" [The Yugoslav Dossier in Nuremberg], *Danas*, Zagreb, no. 214, March 25, 1986, 73-75.

³⁰ *Enciklopedija leksikografskog zavoda* (ELZ) [Encyclopedia of the Lexicographical Institute], vol. 3 (1958), 648.

³¹ *ELZ*, vol. 4 (1959), 322.

³² *ELZ*, vol. 4 (1978), 504.

³³ *Enciklopedija Jugoslavije* [Encyclopedia of Yugoslavia], vol. 4 (1969), 467.

³⁴ *Leksikon Jugoslavenskog leksikografskog zavoda* [Lexicon of the Yugoslav Lexicographical Institute], 1974, 418.

[35] *Mala enciklopedija Prosveta* [Prosveta's Little Encyclopedia], vol. I (1959), 567.

[36] *VE*, vol. 4 (1961), 629.

[37] Vladimir Dedijer, *Novi prilozi za biografiju J. B. Tita* [New Contributions to the Biography of J. B. Tito] Rijeka and Zagreb, 1981, vol. 2, 381.

[38] Ibid., 489.

[39] Ibid., 539.

[40] Ibid., 535.

[41] Ibid., 546.

[42] Open letter to "His Holiness the Pope," signed by both Vladimir Dedijer and Christopher Farley in the name of the *"Boro Dedijer* Independent Foundation for Peace" centered in London. It was transmitted by Tanjug, the Yugoslav press agency, in a special supplement, under the title of "Pije XII je blagoslovio Pavelica i Artukovica" [Pius XII Gave His Blessing to Pavelic and Artukovic], *Tanjug*, November 3, 1986, special supplement 1.

[43] Tanjug, *Dnevni informativni bilten* [Daily Information Bullentin], March 30, 1983.

[44] Vuk Draskovic, *Otvoreno pismo predsjedniku opstinskog odbora SUBNOR Gackog* [Open Letter to the President of the Municipal Committee of SUBNOR Gacko), Nasa rec, London, May 1983.

[45] Milo Gligorijevic, interview with V. Terzic, entitled "Bili smo razbijeni pre izlaska na front" (We Were Crushed Before We Came to the Front), *Intervju*, no. 57, August 5, 1983, 4-9.

[46] Ibid.

[47] Compare Vojislav Lubarda, *Takodjer polemicki o knjizevnosti BiH* [Also Polemically on the Literature of Bosnia and Herzegovina], NIN, no. 1758, September 9, 1984, 33.

[48] *Pravoslavlje,* Belgrade, February 15, 1984.

[49] Ibid. See also *AKSA*, no. 8 (719), February 24, 1984, 1a.

[50] Ibid.

[51] This is discussed in "Petorica na drugom tasu a Zec na kojem?" [Five on the Other Platter, But Where's Zec?], March 20, 1986, my reply to Stevan Zec's writings in Duga (no. 314, March 8-21, 1986), who, on the occasion of Andrija Artukovic's trial, charged me with "acts against the Socialist Federal Republic of Yugoslavia" from an "Ustasha position." However neither the editors of *Duga* nor *Knjizevna rec* published my reply. I also mention that in "Pro Memoria," along with a letter to Mika Spiljak and Josip Vrhovec, in explaining the need to renew the process of establishing, among other things, the truth about the victims, and to do it in a judicial setting.

⁵² Draskovic's "Demand" to the Assembly of the Socialist Republic of Croatia and the presidencies of the Central Committee of the League of Communists of Croatia and the Socialist League of Working People of Croatia: Belgrade, December 16, 1985. (A photocopy from public circulation is in my possession.)

⁵³ A. Miletic, op. cit., vol. I, 37.

⁵⁴ Ibid., 38.

⁵⁵ Ibid., 42.

⁵⁶ Ibid., 38.

⁵⁷ Compare statement by Milan Basta in *Vjesnik's* supplement *Sedam dana*, no. 346, December 21, 1984, and M. Basta, *Rat posle rata* [A War After the War], Zagreb, 1963, 146.

⁵⁸ Branimir Stanojevic, "Under the Wing of Fascism: Andrija Artukovic and the 'Ustasa' Regime" (in English), Belgrade, Tanjug, Yugoslav Press Agency, 1985, Testimonies 2, 38.

⁵⁹ Ibid., 18.

⁶⁰ The Bleiburg controversy is named after the town of Bleiburg (Pliberk) in the Carinthian part of Austria. There the remainder of the army of the NDH surrendered to the English military command, who, in turn, handed them over, along with a multitude of civilian refugees, to the National Liberation Army of Yugoslavia.

⁶¹ *Odmetnicka zvjerstva i pustosenja u Nezavisnoj Drzavi Hrvatskoj u prvim mjesecima hrvatske narodne drzave* [Seditious Atrocities and Destruction in the Indepentent State of Croatia in the First Months of the Croatian National State], written and published by order of the Ministry of Foreign Affairs, based on evidence, Zagreb, June 1942.

⁶² Ibid., 20-22. Compare *Bleiburska tragedija hrvatskog naroda* [The Bleiburg Tragedy of the Croatian Nation]. Foreword by Frano Nevistic and Vinko Nikolic, edited by Vinko Nikolic, Munich-Barcelona, 1976.

⁶³ Pismo iz Hrvatske Ivanu Mestrovicu (A Letter From Croatia to Ivan Mestrovic), May 1955, *Hrvatska Revija*, vol. XXXIII, no. 131, September 1983, 489-492.

⁶⁴ Rudolf Kiszling, *Die Kroaten: Der Schicksalsweg eines Sudslawenvolkes* [The Croatians: The Path of the Destiny of a South Slavic People] (Graz-Cologne, 1956), 222-223.

⁶⁵ Ladislaus Hory and Martin Broszat, *Der Kroatische Ustasha-Staat 1941-1945* [The Croatian Ustasha State 1941-1945], Stuttgart, Deutsche Verlags Anstalt, 1964, 173.

⁶⁶ Nicholas Bethell, *The Last Secret - Forcible Repatriation to Russia, 1944-1947* (London, 1974), Chapter 4, "The Croats and the Cossacks," 82-88; compare *Bleiburska tragedija*, op. cit., 403-406; Ante Beljo, Jugoslavija, *Genocid, dokomentarna analiza* [Yugoslavia, Genocide: A

Documented Analysis] (Sudbury, 1985), 173-179.

⁶⁷ Andre Ott, *Dangers Serbes sur la Croatie* [The Serbian Danger Looming over Croatia], Conclusion by Mirko Vidovic, Paris, 1982, 26.

⁶⁸ Compare *Bleiburska tragedija,* op.cit., 3-10, 15, 46, 87, 117, 159, 167, 173, 263, 296-268, 403-406.

⁶⁹ Bruno Busic, *Jedino Hrvatska! Sabrani spisi* [Only Croatia! Collected Works] Toronto-Zürich-Chicago: ZIRAL, 1983, 134.

⁷⁰ Josip Aleksic, "Vojnicki aspekti hrvatskog povlacenja u svibnju 1945" [Military Aspects of the Croatian Withdrawal in May 1945], in: *Na prekretnici* [At the Turning Point], ed. J. Petricevic, Brugg, 1986, 24-49.

⁷¹ Ibid., 48-49 (cited as source: F. Schraml, *Kriegsschauplatz Kroatien* [The Croatian War Theater]) (Kurt Vowinkel Verlag), 294.

⁷² Jure Petricevic, "Razmatranja u povodu 40. godisnjice hrvatske narodne tragedije kod Bleiburga" [Some Considerations on the Occasion of the Fortieth Anniversary of the Croatian People's Tragedy in Bleiburg], in: *Na prekretnici,* op. cit., 50-54.

⁷³ George J. Prpic, *Tragedies and Migrations in Croation History,* Toronto, 1973, 5-24; compare A. Beljo, *Jugoslavija, Genocid,* op. cit., 268-269.

⁷⁴ A. Beljo, *Yugoslavia, Genocide,* op. cit.

⁷⁵ Ibid., 65.

⁷⁶ Ibid., 147-150.

⁷⁷ Milan Basta, *Rat je zavrsen sedam dana kasnije* [The War Ended Seven Days Later] Zagreb, 1977, 362; A. Beljo, *Jugoslavija, Genocid.,* op. cit., 170-171.

⁷⁸ M. Basta, *Rat je zavrsen sedam dana kasnije,* op. cit., 355-365; A. Beljo, *Jugoslavija, Genocid.,* op. cit., 164-173.

⁷⁹ Kosta Nadj, *Pobeda* [Victory] (Zagreb, 1980); A. Beljo, *Jugoslavija, Genocid.,* op. cit., 179.

⁸⁰ Nikolai Tolstoy, *The Minister and the Massacres,* London, 1986, 123-125, 175.

⁸¹ Ibid., 204, 411. et al.

⁸² N. Bethell, op. cit., 403-406.

⁸³ Nikolai Tolstoy, "The Klagenfurt conspiracy," *Encounter,* May 1983, LX, 5, 24-37; Nikolai Tolstoy, *The Minister and the Massacres,* op. cit.

⁸⁴ Participants in the first international symposium on forcible repatriation after World War II were: Prof. Michael McAdams (San Francisco), "Forcible Repatriation, Cause and Effect: Yalta and Bleiburg"; Dr. Tony Arko (Chicago), "Forcible Repatriation of Slovenes"; Dr. B. Harrell-Bond (Oxford), "The Continuing Relevance of the Subject"; Prof. L.

Kosinski (Alberta, Edmonton), "European Migration During and After WWII"; Prof. Mark Elliott (Illinois), "The U.S., Yalta and Forcible Repatriation"; Prof. Col. C. G. Draper (Sussex), "The Yalta Agreement and International Law"; Dr. Robert Knight (Vienna), "Repatriation and Its Prelude: The Von Pannwitza Cossacks in Croatia"; Count Nikolay Tolstoy (Berkshire), "Forcible Repatriation from British Occupied Austria in May-June 1945"; Dr. Katya Andreyev (Oxford), "The Vlasovites and the West"; Dr. Lubomyr Luciuk (Toronto), "Ukrainian Displaced Persons, Repatriation, and Soviet 'Refugees' in the British Zone of Germany, 1945-47"; Prof. D.G. MacRae (London), "Political Causes and Sociological Consequences of Population Movements in Postwar Central Europe"; Dr. Ante Cuvalo (Ohio), "Forcible Repatriation of Croatians"; Prof. P.J. Potchnyj (Hamilton), "Action Vistula"; Prof. Ron Vastokas (Peterborough, Ontario), "Forcible Repatriation as a Crime Against Humanity: Perceptions and Reality." Prof. P. R. Magosci, John Corellis, Dr. Joseph Feldner, Dr Hugh Macdonald, and others also took part in the work of the symposium.

[85] N. Tolstoy, *The Minister and the Massacres*, op. cit., 95-129, 130-175, 176-207.

[86] Ibid., 99-111, 387-399.

[87] Ibid., 103-109, 130.

[88] Ibid., 387-388. On the subject of the fate of the prisoners and on the "death marches," there is an abundance of printed material among the Croatian emigrees; see, e.g., *Bleiburska tragedija*, op. cit.; Theodore Benkovic, *The Tragedy of a Nation* (Chicago, 1947); Joseph Hecimovic, *In Tito's Death Marches and Extermination Camps* (New York, 1962); Borivoje M. Karapandzic, *Jugoslovensko krvavo prolece 1945: Titovi Katini i Gulazi* [The Bloody Yugoslav Spring of 1945 - Tito's Katyns and Gulags] (Cleveland, 1976); J. Prcela and S. Guldescu, op. cit., V. Nikolic, *Tragedija se dogodila u svibnju* [The Tragedy Happened in May] (Munich-Barcelona, 1984); A. Beljo, *Jugoslavija, Genocid*, op. cit.; *Bela knjiga Slovenskoga protikomunisticnega upora 1941-1945* [The White Book of Slovenian Anticommunist Resistance 1941-1945] published by TABOR ZDSPB, printed in the USA, 1985.

[89] N. Tolstoy, *The Minister and the Massacre*, op. cit., 128.

[90] Ibid., 125-126.

[91] Ibid., 8, 126, 422.

[92] Ibid., 10-14, 18-34, 79-97, 122-128, 135-147, 150-154, 165-176, 204, 287-306, 339-340.

[93] *Bela knjiga Slovenskega protikomunisticnega upora 1941-1945*, op. cit., 4, 154, 515, 518, 524-532, 546-549.

[94] Ibid., 530.

[95] Borivoje Karapandzic, *The Bloodiest Yugoslav Spring 1945-Tito's Katyns and Gulags* (New York: Carlton Press, 1980).

[96] Barbara W. Tuchman, *A Distant Mirror: The Calamitous 14th Century* (New York: Alfred A. Knopf, 1978). Cited from the Croatian translation of Tuchman's book, *Daleko zrcalo: Zlosretno XIV. stoljece*, vol. I, 18.

[97] Ibid., 360.

[98] Ibid., 362-370.

[99] Ibid., II, 270-271.

[100] Ibid., 279-280.

[101] Ibid., 278.

[102] *Enciklopedija hrvatske povijesti i kulure* [Encyclopedia of Croatian History and Culture], ed. J. Adamcek, A. Badurina, J. Bratulic, Z. Dadic, B. Draskovic, Z. Haramija, I. Jelic, I. Karaman, chief ed., J. Lucic, M. Prelog, N. Stancic, I. Supicic (Zagreb, Skolska knjiga, 1980).

[103] Fikreta Jelic-Butic, *Ustase i NDH* [The Ustashas and the Independent State of Croatia] (Zagreb, 1977), 162-178: J. Tomasevic, op. cit., 105.

[104] Jovan Marjanovic, "Prilozi istoriji sukoba Narodnooslobodilackog pokreta i cetnika Draze Mihailovica u Srbiji 1941. godine" [Contributions to the History of the Conflict Between the National Liberation Movement and the Chetniks of Draza Mihailovic in Serbia in 1941], *Istorija XX veka* [History of the Twentieth Century], I (Belgrade, 1959), 180; J. Tomasevic, op. cit., 155-165; Fikreta Jelic-Butic, *Cetnici u Hrvatskoj 1941-1945* [The Chetniks in Croatia from 1941 to 1945], Zagreb, 1986, 17-22.

[105] *Vjesnik,* September 28, 1945, no. 137, 1.

[106] Personal testimony of General Niloka Kajic who, through his service (as principal at a school) and participation in the People's Liberation Struggle in that area, knew the circumstances and people in Banija, and knew Vasilije Gacesa as well. General Kajic was with the 7th Banija Division (as the division's chief-of-staff) in the fourth and fifth enemy offensives and, though wounded, survived the battle of Sutjeska. At the end of the war, he was the chief-of-staff and deputy commander of the 10th Zagreb Corps of the National Liberation Army of Yugoslavia.

[107] We discussed these cases within the Ideological Commission of the Central Committee of the League of Communists of Croatia and before the Brijuni Plenum of the Central Committee of the League of Communists of Yugoslavia (1966), but it was "too sensitive," i. e., politically not recommended to place these issues directly on the agenda for public debate. We saw the trail left by the "evasive" presentation of such issues when we wanted to research them in a scholarly way within the framework of the Institute which I headed.

[108] Ernst Bloch, "Das Unabgegoltene als Problem des Gegenwartigen" [The Unresolved Issue as a Contemporary Problem] in: *Sozialismus in Osteuropa, Jugoslawien* [Socialism in Eastern Europe, Yugoslavia] (Goettingen 1973), 92-124; *Der Spiegel,* February 3, 1985, 80.

[109] I sent my reply under the title "Ernst Bloch and the Croatians, or (perhaps) Marxism and the National Question" to the editors of *Der Spiegel* as well. They apologized for having mistakenly printed Bloch's "imprecise statement," but they did not print my letter, as its contents went beyond the scope of the topics of the article. Here are a few excerpts from my reply:

By which principle of Marxist philosophy, Mr. Bloch, can the Croatian, or any other entire nation, be labeled Fascist?

In analyzing the contemporary state of Yugoslavia, you said, "Die Kroaten als Faschisten waren gefährlich" [The Croatians as Fascists were dangerous], and added "das kann ich mir gut vorstellen" [I can well imagine this]!

But why is it that Croatians are "Fascists" and why can you so "well imagine" this? And then you speak of Croatians in the socialist order, and, moreover, of their communist leadership!

Would you, distinguished professor, have pronounced such a judgment if you had been familiar with these historical facts:

First, until the Fascist Occupation in 1941, the working class in Croatia was considerably more well developed than in the other lands of Yugoslavia (except Slovenia); that in Croatia there was no Fascist party, nor even a religous party as in other Catholic countries of Europe, due to the strength and character of Radic's national democratic peasant movement, so that pro-Fascist ideology was politically unknown, while, to the contrary, there was a strong Croatian cultural left wing with predominantly liberal, anti-dogmatic tendencies in the areas of literature, art, and philosophy...

Second, during the anti-Fascist Partisan war in Yugoslavia (1941-45)—despite the fact that the Independent Croatian State had been established within the framework of Hitler's New European Order—the socialists revolution in Croatia was well under way... The nature of the political and military conditions in Croatia at that time is evident from the judgment given in 1943 by the creators and protectors of Hitler's new European order: "By far the weakest of all states in the Europe vanquished by the Axis powers was still Croatia." *Kriegstagebuch des Oberkommandos der Wehrmacht 1940-1945* (Frankfurt A/M, 1961), III, 1936.

However, let there be no misunderstanding about what I have cited. By my historical, not to say Marxist, judgment, such an unfavorable sit-

uation for Nazi-Fascism in Croatia, and thus favorable for an anti-Fascist partisan movement, existed not because the Croatians in themselves were more revolutionary than the other Yugoslav or European nations, but because their very existence was threatened in the bourgeois Kingdom of Yugoslavia as well as in the Independent State of Croatia, which was on the side of the Axis powers. Since they did not wish to return to their former bondage, a greater proportion of Croatians saw a way out in Tito's program of socialist society and in a Croatian Republic within a federative Yugoslavia.

[110]Stefan Zweig, *Jucerasnji svijet* [Yesterday's World], Rijeka, 1965, 500-505.

[111] Miomir Dasic on A. Miletic's book in *JIC*, XXI, 1-4, 1986, 234-241.

[112] Branka Culjic, "Dosije *Mladosti*: Konc-logor Jasenovac: [The *Mladost* Dossier: the Concentration Camp of Jasenovac], *Mladost*, published by SSOJ [League of Socialist Youth of Yugoslavia], vol. LXVIII, New Series, no. 30, December 8-21, 1986, 26. Culjic inaccurately cites the name of his crown witness; i. e., Trivuncic is misspelled "Trivundzic."

[113] Ibid.

[114] Ibid., 28.

[115] Ibid.

[116] *Radovan Trivuncic, Jasenovacka drama* [The Drama of Jasenovac], Zagreb, Skolske novine, 1986.

[117] Luke 6:37-8; Luke 11:37-53.

[118] Plato, *Drzava* [The Republic], Zagreb, Liber (Croatian translation), 1977, 65-66.

[119] Agnes Heller, *A Theory of History*, London-Boston, 1982, 307-309.

[120] Exodus 20:13.

[121] Genesis 4:10.

[122] Genesis 9:6

[123] Nahum 2:2-3, 19; Jeremiah 21: 3-7, Jeremiah 50:23, 51:20; Jeremiah 27:1-28, 17.

[124] Psalm 137; Jeremiah 51: 44-57; Isaiah 24: 7-18; 25:1-5.

[125] Exodus 24:3-8; Acts 7:21-35; Exodus 19:4-6; 24:3-8; Hebrews 9:18ff.

[126] Hosea 12:14; Exodus 17:9-13.

[127] Psalms 106:23; Exodus 32:31-35.

[128] Numbers 10:35; Psalms 83:3; Deuteronomy 7:1-6; et al.

[129] Exodus 23:27-33.

[130] Deuteronomy 3:6-9; Joshua 6.

[131] Numbers 21:21-35; Deuteronomy 2:26; 3:17; Joshua 6-12; Numbers 31; 1 Samuel 11-17; 28-30; 2 Samuel 5; 8; 10.

[132] Exodus 12:12.

[133] Exodus 17:14.

[134] Sirach 36:6-9.

[135] Jeremiah 41: 4-10; 44:7.

[136] Matthew 6:12, 14: Mark 11:25; Romans 12:21; Revelations 6:10, 22:3.

[137] Nada Klaic, *Povijest Hrvata u ranom srednjem vijeku* [The History of the Croatians in the Early Middle Ages] Zagreb, 1971, 209.

[138] *ELZ,* 4 (1959), 436-437; *VE,* 5 (1962), 52-61.

[139] *VE,* 10 (1967), 181-186; Franjo Tudjman, *Rat protiv rata* [War Against War], Zagreb, 970, 116-118.

[140] *Hrvatska Enciklopedija* [Croatian Encyclopedia], II (1941), 259; *Leksikon Jugoslavenskoga leksikografskoga zavoda* (1974), 84.

[141] B. Tuchman, op cit., 183-184.

[142] Ibid., 237-244.

[143] Ibid., 321-323.

[144] Ibid., 350, 362, 367-372.

[145] Veceslav Holjevac, *Hrvati izvan domovine* [Croatians Outside Their Homeland], 2nd ed., 9-16. Some scholars estimate that the number of Croatians that migrated or were taken captive came to about 1.6 million. See Ivan Muzic, *Razmatranja o povijesti Hrvata* [Contemplations About History of the Croats], (Split 1967), 15.

[146] F. Tudjman, *Rat protiv rata,* op. cit. 128-132, 133-137, 142-156.

[147] Ibid., 156-163.

[148] Dimitrije Tucovic, *Srbija i Albanija* (Serbia and Albania), Belgrade, 1914, 124-136; F. Tudjman, *Rat protiv rata,* op. cit., 109-112.

[149] S. Zweig, op. cit., 585.

[150] *Hrvatska enciklopedija,* I (1941), 631-637; *ELZ,* 1 (1955), 232-234; Michael Braga, *Völker zur Freiheit* [People to Freedom], Kiel, Arndt-Verlag, 221-229.

[151] F. Tudjman, *Nacionalno pitanje u suvremenoj Europi,* 2nd ed., op. cit., 25, 116-118.

[152] In biblical texts we find the motif of sacrifice by fire in pagan nations and in the Jewish nation. There were blood and non-blood sacrifices offered to God: food, various animals, but also people. Gradually, human sacrifices were rejected in temple sacrifices, but animal sacrifices were accepted in the name of God. However, in the New Testament there is reference to offering oneself in the service of God as a sacrific of reconciliation. The highest expression of the continuity of the Old and New Testment concept of sacrifice is when Jesus offers himself as a sacrifice on the Cross, which is simultaneously a sacrifice by burning, repentance,

and sacrifice of the unity (compare *Rjecnik biblijske teologije* [Dictionary of Bible Theology], Zagreb, KS, 1969, 1571-9). In classical Greek *holokaust* (holos-whole+kaustos-burnt), sacrifice by fire by which an entire animal is set afire (B. Klaic, *Rjecnik stranih rijeci* [Dictionary of foreign words], 1980, 551). On Wiesel's resurrection of the name, see *Glas Koncila,* XXVI, 36 (691), September 6, 1987, 1, 4.

153 Raul Hilberg, *The Destruction of the European Jews,* New York, New Viewpoints (1973), 1-3.

154 Support of Hitler's Germany on the part of the Arab world was so great that, e. g., G.A. Nasser (1918-1970), one of the most outstanding statesmen of Egypt, showed his respect for Hitler by keeping his photo even after he had begun to espouse "Arab socialism" and had become a "Third World" non-aligned leader.

155 Nora Levin, *The Holocaust: The Destruction of European Jewry,* 1933-1945, New York, Schocken Books, 1973, 129-130, 724.

156 R. Hilberg, op. cit., 95.

157 Memo to the German minister of Foreign Affairs, March 10, 1938, concerning the "Haavara Treaty," compare: N. Levin, op. cit., 130-132.

158 At that time, Poland was 388,635 sq. kilometers in size and had a population of 33 million (of which 3.25 million were Jews), and Madagascar was 627.327 sq. kilometers with only 3.6 million inhabitants.

159 N. Levin, op. cit., 199.

160 Compare: R. Hilberg, op. cit., 715-717.

161 C. M. Cianfarra, *The Vatican and the War,* New York, 1944, 145-146; compare: N. Levin, op. cit, 199-200.

162 N. Levin, op. cit., 143-144.

163 Compare: R. Hilberg, op. cit., 137-149, 165, 286-287, et al.; N. Levin, op. cit., 121, 152, 178-184, 188-191, 200, 230.

164 Compare: N. Levin, op. cit., 200-201.

165 Ibid., 203, 288-289, 293-294.

166 Compare: *The Columbia-Viking Desk Encyclopedia,* 1964, 526; Louis Fitzgibbon, *Katyn Massacre,* London, 1977, 3, 17-19, 29-39, 124-145, 155-158, 230-249, et al.; N. Levin, op. cit., 276-279, N. Tolstoy, *The Minister and the Massacres,* op. cit., 281-282, 387; "Poljsko-sovjetski odnosi: Istina o interniranima" [Polish-Soviet Relations: "The Truth About Internment"], *Vjesnik,* XLVII, 14378, September 22, 1987, 3.

167 Compare: F. Tudjman, *Nacionalno pitajie u suvremenoj Europi,* op. cit., 77-92.

168 The agreement concerning the prosecution of war, like the one concerning the division of Germany, was designed to prevent a renewal of "German militarism," particularly of Germany as a European power. That agreement of the three powers of the anti-Hitler coalition was

arranged through: the Moscow Declaration (October 30, 1943), at the Yalta Conference in Crimea (February 1945), and at the Potsdam Conference (August 1945), while several other nations joined in the decision (August 1945, in London) concerning the establishment of an international military court at Nuremberg. A United Nations Convention concerning the punishment for and repression of genocide (December 1948) binds the signatory members to its implementation within their own jurisdictions as well.

[169] In the 1960s, I discussed this issue with certain Slovakian historians, who, among other things, told me that the numbers of Slovakian Greek Catholics (according to some, several tens of thousands, and according to others, as many as 200,000) were forcibly converted to Orthodoxy and that this was done by specially chosen security personnel who had been quickly trained for Orthodox priesthood through special courses.

[170] "Dedijerov Artukovic" [Dedijer's Artukovic] *Glas Koncila* XXVI, 28, July 12, 1987, 8; "Prekrstavanje i genocid danas" [Forced Conversion and Genocide Today], Ibid., no. 36, September 6, 1987, 8.

NOTES

■

Philosophy and Historical Evil

PART THREE

[1] Gen 4:2-8
[2] Wisd 2:24; Tit 3:3
[3] Ps 105:25; Is 60:15
[4] Deut 7:1-6
[5] Mt 5:43
[6] Ps 11:5; 31:7; 5:6f
[7] Ex 12
[8] Num 21:2f; Josh 6; 1 Sam 15; Josh 7
[9] Ps 140:12
[10] Amos 1:1-2:8; Ex 3:9; 21-23; Deut 24:20
[11] Ps 7:13-17
[12] Sir 28:1
[13] Rev 20:8f
[14] Lk. 6:27
[15] Mt. 26:52
[16] Rev. 6:10; 16:6; 19:2
[17] Compare: *Hrvatska enciklopedija,* 3, 218-222, 444-448.
[18] Compare: F. Tudjman, *Rat protiv rata,* op. cit., 438-439.
[19] Compare: K. Ploetz, Auszug aus der Geschichte, Würzburg, Ploetz-Verlag, 1968, 402-403, 807, 1190; *The Columbia-Viking Desk Encyclopedia,* 1964, 226; *ELZ,* 4, 1959, 326-327.
[20] *Kuran casni* [Holy Qur'an], Croatian translation, Zagreb, 1974, 2:213; 10:19.

[21] Ibid., 49:13.

[22] Ibid., 45:28.

[23] Ibid, 9:74; 66:9.

[24] Compare: Robin George Collingwood, *The Idea of History*, Oxford, University Press, 1970, 56-57.

[25] Compare: Branko Bosnjak, *Grcka filozofija* [Greek Philosophy], 1956, 16-20, 143-148; K. Ploetz, op. cit., 113-115, 121; Nerkez Smailagic, "Politicka misao helenskog svijeta" [Political Thought in the Hellenic World] in *Historija politickih doktrina* [The History of Political Doctrine], I, 13-18; *VE*, 10, 1967, 198.

[26] Compare: B. Bosnjak, *Grcka filozofija*, op. cit., 21-23, 149-156; K. Ploetz, op. cit., 125.

[27] Compare: D. Laertius, *Zivot i misljenje istaknutih filozofa* (Life and Thoughts of Famous Philosophers), Belgrade, BIGZ, 1985, 14-15; B. Bosnjak, *Grcka filozofija*, op. cit., 30-31.

[28] Heraclitus, fragments, 53, 80, 82 (compare: F. Tudjman, *Rat protiv rata*, op. cit., 483); D. Laertius, op. cit., 296; B. Bosnjak, *Grcka filozofija*, op cit., 40-46, 168-174.

[29] F. Tudjman, *Rat protiv rata,* op cit., 480-481; K. Ploetz., op cit., 53-59, 106-127, 133-161, et al.

[30] Compare: B. Bosnjak, *Grcka filozofija*, op. cit., 55-58, 180-185; Hermann Diels, *Predsokratovci* [The Presocratic Philosophers], Zagreb, 1983, 247.

[31] H. Diels, op. cit., 203.

[32] D. Laertius, op. cit., 310; B. Bosnjak, *Grcka filozofija*, op. cit., 76-80, 198-200.

[33] Compare: H. Diels, op. cit., 244.

[34] Compare: Aristotle, "Priroda drzave" [The Nature of the State] in: N. Smailagic, op. cit., 60-66; Julije Makanec, *Razvog drzavne misli ad Platona do Hegela* [Development of the State Thought from Plato to Hegel], Zagreb, 1943, op. cit., 30-42; B. Bosnjak, op. cit., 114-139, 227-242.

[35] Compare: D. Laertius, op. cit., 333-396; Branko Bosnjak, *Filozofija od Aristotela do renesanse* [Philosophy from Aristotle to the Renaissance], Zagreb, Matica hrvatska, 1957, 32-40.

[36] Compare: R. G. Collingwood, op. cit., 42.

[37] T. Livius, *Ab Urbe Condita* [From the Founding of the City], V, 48, 9 (compare: Zvonimir Doroghy, *Blago latinskoq jezika* [Treasures of the Latin Language], Zagreb, Matica Hrvatska, 1966, 364).

[38] Compare: B. Bosnjak, *Filozofija od Aristotela do renesanse*, op. cit., 40-43.

[39] Cornelius Nepos, Vitae: Atticus, Chabrias, Epaminondas, Hannibal, Zagreb, Knjizara Jugoslavenske akademije, *bez. god.* (no year).

⁴⁰ Compare: Karl Hildebrand, *Geschichte und System der Rechts-und Staatsphilosophie* [The History and System of the Philosophy of Justice and the State], Leipzig, 1860, I, 511 et al.; Mason Hammond, *City-State and World-State in Greek and Roman Political Theory until Augustus*, Cambridge, HUP, 1951, 54, 161 et al.; N. Smailagic, op. cit., I, 45; R. G. Collingwood, op. cit., 43-46.

⁴¹ Compare: Cicero, *Filozofski Spisi* [Philosophical Essays], Belgrade, "Kultura" 1955, 63-64, 1148, 161, 195; Vinko Prostenik, "Ciceron" [Cicero], *Hrvatska Enciklopedija*, III, 1942, 744-747; F. Tudjman, *Rat protiv rata*, op. cit., 483-484, 506, 541, 578; Josip Andreis, *Tragovi Mudrosti* [Trails of Wisdom] Zagreb, 1944, 24, 37, 129, 179, 215.

⁴² Compare: K. Ploetz, op cit., 186-189, 308; R. G. Collingwood, op. cit., 47-49.

⁴³ Compare: Seneka, "Filozofija i zivot i druga pisma" [Philosophy and Life and Other Letters] translation in: B. Bosnjak, *Filozofija od Aristotela do renesanse*, op. cit., 135-143; K. Ploetz, op. cit., 290-292, 310-312, 340-356.

⁴⁴ Compare: K. Ploetz, op. cit., 314, 363-369, 377; R. G. Collingwood, op. cit. 49-57.

⁴⁵ Compare: K. Ploetz, op. cit., 282, 316-319, 341, 352-354, 361, 372; Grga Novak, "Aurelije, Marko" [Marcus Aurelius], *Hrvatska enciklopedija*, I, 749; N. Smailagic, op. cit., I, 45-46.

⁴⁶ (Mt 5:17).

⁴⁷ Gal 3:28

⁴⁸ Compare: B. Bosnjak, *Filozofija od Aristotela do renesanse*, op. cit., 67-81, 191-201; Boris Kalin, *Povijest filozofije* [The History of Philosophy], Zagreb, Skolska knjiga, 60-65.

⁴⁹ Compare: Aurelius Augustinus, *O drzavi Bozijoj* (De civitate Dei), Zagreb, 1982, XC-XCVIII, CXIX, CXXX, CXLII, 25-27, 41, 51, 99, 113-117, 127, 133-137, 169, 177, 181-189, 193, 209, 215-231, 235-239, 253-277, 323, 343-359, 367, 375-377, 387-389, 415, 451, 471, 547-551, 589, 613, 621; Karlo Balic, "Aurelius Augustinus," "Augustinianizam," *Hrvatska enciklopedija*, I, 742-745; J. Makanec, op. cit., 43-54; N. Smailagic, op. cit., I, 69-82. 91-112; Mladen Jurcic, "Borba za smisao" [Struggle for Sense], Zagreb, Marulic, XX, 1987, no. 5, 570-576.

⁵⁰ Referring to text of St. Augustine's work; all quotes according to the above mentioned edition (footnote 49), Latin original translated by Tomislav Ladan.

⁵¹ *De Civitate Dei*, 3:1-12.

⁵² Ibid., 3:12-13.

⁵³ Ibid., 5:21.

[54] Ibid., 5:21.

[55] Ibid., 3:22.

[56] Ibid., 3:27-31.

[57] Compare: K. Ploetz, op. cit., 295-309; J. Makanec, op. cit., 55-62; N. Smailagic, op. cit., I, 122-130; F. Tudjman, *Rat protiv rata*, op. cit., 484, 516-517, et al.

[58] Compare: Niccolo Machiavelli, *Vladar* [The Prince], Zagreb, 1983; Carl Schmitt, *Pojam politike*, [*Der Begnff des Politishen*] [Concept of Politics], Zagreb, 1943, 43; Vladimir Filipovic, *Filozofija renesanse (The Philosophy of the Renaissance)*, Zagreb, Matica hrvatska, 1956, 58-64; F. Tudjman, *Rat protiv rata*, op. cit., 229, 484; N. Smailagic, op. cit., I, 169-174, 247-256.

[59] Desiderius Erasmus, *Pohvala qluposti* [In Praise of Folly], 6, 77, 262-263 (compare: F. Tudjman, *Rat protiv rata*, op. cit., 484; V. Filipovic, op. cit., 45-46, 135-154).

[60] Compare: Michel de Montaigne, Essays in: V. Filipovic, op. cit., 250-255.

[61] Compare: Giordano Bruno, "O uzroku. principu i jednome" [On the Cause, the Principle and One] in: V. Filipovic, op. cit., 256-290; Albert Bazala, *Povijest filozofije*, (The History of Philosophy), Zagreb, Matica hrvatska, 1909, II, 210-212; Vilim Keilbach, "Giordano Bruno," *Hrvatska enciklopedija*, III, 408-409.

[62] Veljko Gortan, *Hrvatski latinisti* [Croatian Writers in Latin], I, 709-725; Kruno Krstic, "Petrisevic" (Petris, Petric) "Franjo" (Franciscus Patricius), *Enciklopedija Jugoslavije*, 6, 479-481; V. Filipovic, op. cit., 115-120; Franjo Petric, *Sretan Grad* (La citta felice) [The Happy City], Zagreb, 1975, (translated from the Italian, with accompanying analysis by V. Premec); Franjo Petric, *Nova sveopca filozofija* (Nova de universis philosophia) [New Universal Philosophy], Zagreb, 1979, (translated from Latin by T. Ladan, foreword by V. Filipovic, K. Kristic, editor).

[63] Compare: V. Gortan, *Hrvatski latinisti*, op. cit., I, 152-222; Veljko Gortan, "Janus Pannonius" (Ivan Cesmicki, Joanes de Chezmicze), *Enciklopedija Jugoslavije*, 4, 461-463.

[64] Compare: V. Gortan, *Hrvatski latinisti*, op. cit., I, 224-317; *Povijest hrvatske knjizevnosti*, Zagreb, Liber, 1978, 3, 28-38; V. Filipovic, op. cit., 122-124.

[65] Compare: V. Gortan, *Hrvatski latinisti*, op. cit., I, 319-351; Kruno Krstic, "Crijevic Ludovik Tuberon," *Enciklopedija Jugoslavije*, 2, 391-392.

[66] V. Gortan, *Hrvatski latinisti*, op. cit., I, 503-513.

[67] Compare: F. Tudjman, *Rat protiv rata*, op. cit., 484, 540-542 et al.

[68] Compare: Jakob Boehme, "Aurora ili jutarnje rumenilo na izlazu"

[Aurora, or the Morning Sunrise] - translation in: V. Filipovic, op. cit., 195-203; Albert Bazala, "Jakob Boehme," *Hrvatska enciklopedija*, 2, 724; V. Filipovic, op. cit., 74-79.

69 Thomas Hobbes, "Nastanak Drzave" [The Creation of the State] in N. Smailagic, op. cit., I, 256-266; C. Schmitt, op. cit., 37-38, 205-207; F. Tudjman, *Rat protiv rata*, op. cit., 486, 515.

70 Compare: K. Krstic, "Descartes (Cartesius) Rene," *Hrvatska enciklopedija*, 4, 672-673; N. Smailagic, op. cit., I, 133-138; R. G. Collingwood, op. cit., 74-80.

71 Compare: Grga Novak, "Oliver Cromwell," *Hrvatska Enciklopedija*, IV, 137-138; C. Schmitt, op. cit., 45-46; original from "Letters" cited according to the translation of the German text.

72 Compare: C. Schmitt, op. cit., 45; F. Tudjman, *Rat protiv rata*, op. cit., 449.

73 Compare: Charles Montesquieu, "Duh zakona" [The Spirit of the Law]—translation in: N. Smailagic, op. cit., I, 270-275.

74 Compare: Jean Jacques Rousseau, "Drustveni ugovor" [The Social Contract] - translated excerpts in: N. Smailagic, op. cit., I, 276-285, and contemplations 179-188; J. Makanec, op. cit., 132-143; B. Kalin, op. cit., 114-116; F. Tudjman, *Rat protiv rata*, op. cit., 443, 503.

75 Immanuel Kant, *Werke*, VI, 327.

76 Immanuel Kant, *Kritika Rasudne Moci* [Critique of Judgement], Zagreb, Kultura, 1957, 270.

77 Immanuel Kant, *Zum ewigen Frieden* [Toward Eternal Peace], Koenigsberg, 1795; Immanuel Kant, *Kritika prakticnog uma* [Critique of Practical Reason], Zagreb, 1956; Max Scheler, *Covjek i povijest* [Man and History], Sarajevo, 1987, 82; J. Makanec, op. cit., 144-155; F. Tudjman, *Rat protiv rata*, op. cit., 486-487, 500, 503, 542; N. Smailagic, op. cit., I, 138-148.

78 Compare: J. Makanec, op. cit., 156-168; F. Tudjman, *Rat protiv rata*, op. cit., 450, 487-488, 503.

79 Georg Wilhelm Friedrich Hegel, *Filozofija Povijesti* [Philosophy of History], Zagreb, Naprijed, 1966, 238.

80 Ibid. 387-388.

81 Karl von Clausewitz, *O ratu* [On War], Belgrade, 1951, 440-441.

82 Ibid., 445. Compare C. Schmitt, op. cit., 11; Petar Tomac, "Karl Klauzevic" *VE* 4, 476-478; F. Tudjman, *Rat Protiv rata*, op. cit., 32-35, 47, 482, 613, 676.

83 Compare: F. Tudjman, *Rat protiv rata*, op. cit., 489 et al.; N. Smailagic, op. cit., I, 221-226, 355-360.

84 Compare: James Bryce, *Modern Democracies*, 1921—translated excerpts in: N. Smailagic, op. cit., I, 294-302.

[85] Friedrich Nietzsche, *Goetterdaemmerung* [Twilight of the Idols], 32, 103, 160, 351, 400.

[86] Friedrich Nietzsche, *Der Wille zur Macht* [Will to Power], Leipzig, 1906, 41, 92, 490.

[87] Compare: Friedrich Nietzsche, *Also sprach Zarathustra* [Thus Spake Zarathustra] (1881-91), citations from Croatian translation: *Tako je govorio Zaratustra*, Zagreb, 1983; *Jenseits von Gut und Boese* [Beyond Good and Evil], Leipzig, 1886; *Zur Genealogie der Moral* [Genealogy of Morals], Leipzig 1887; *Der Wille zur Macht* [The Will to Power], Leipzig, 1895; F. Tudjman, *Rat protiv rata*, op. cit., 489-93; D. Grlic, "Smisao i sudbina filozofije Friedricha Nietzchea" [The Sense and Fate of Friedrich Nietzsche's Philosophy] (epilogue to *Tako je govorio Zaratustra*, 307-358.

[88] Compare: Oswald Spengler, *Propast Zapada* [*Der Untergang des Abendlandes*] [Decline of the West], Belgrade, 1936, 516.

[89] Ibid., 512, 603, 436.

[90] Compare: O. Spengler, op. cit., 514; F. Tudjman, *Rat protiv rata*, op. cit., 493-496.

[91] Max Scheler, *Der Genius des Krieges und der deutsche Krieg* [The Genius of War and the German War], Leipzig, 1915, 75, 78.

[92] Ibid., 81; 97-98.

[93] Compare: F. Tudjman, *Rat protiv rata*, op. cit., 497-498.

[94] Compare: Wilhelm Emil Muehlmann, *Rat i Mir* [*Krieg und Frieden*] [War and peace], Zagreb, Matica Hrvatska, 1943, 12, 42, 61, 100, 180, 190.

[95] Ibid., 180-184.

[96] Compare: Ibid, 62-65; 70-74; 98-114; 186-189; 197-214.

[97] Isa. 9:5,6.

[98] Ps. 85: 9-14.

[99] Luke 12:51.

[100] Compare: W. E. Muehlmann, op. cit., 151-154, 193-194.

[101] Ibid., 195.

[102] C. Schmitt, op. cit., 13-14.

[103] F. Tudjman, *Rat protiv rata*, 13-15.

[104] Friedrich Engels, *Anti-Duehring*, Zagreb, Naprijed, 1945, 190-191.

[105] Karl Marx - Friedrich Engles, *Werke* [Works], VI, 168-176; Compare: F. Tudjman, *Velike ideje i mali narodi*, op. cit., 27-36; F. Tudjman, *Rat protiv rata*, op. cit., 501-503, 516.

[106] Benito Mussolini, *Doktrina fasizma*, Rome, Novissima (no year), 17-18; Ernest Nolte, *Der Faschismus in seiner Epoche* [Fascism in Its Era], Munich, 1963.

[107] B. Mussolini, *Doktrina fasizma*, op. cit., 17-18.

[108] Ibid., 31-32.

[109] Ibid., 8, 15, 22-23, 27-30, 37-39, 48-49.

[110] Compare: E. Nolte, op. cit., 197-200, 208, 222, 312, 315-316, 367, 595.

[111] Ibid., 343-55, 408-409, et al.

[112] R. Hilberg, op. cit., 9-11.

[113] Compare: F. Tudjman, *Rat protiv rata*, op. cit., 498-500.

[114] Compare: E. Nolte, op. cit., 494-495.

[115] Adolph Hitler, *Mein Kampf* [My Struggle], Munich, Zentralverlag der NSDAP, 1941, 310, 369.

[116] Ibid., 329-347.

[117] Ibid, 431-434.

[118] Ibid., 436-439.

[119] Compare: Ibid., 153, 312-317, 329-347, 372, 425-35, 436-439, 461-493, 690-700, 726-733; E. Nolte, op. cit., 23-58, 61-89, 345-55, 366-369, 385-398, 403-409, 432-440, 494-500, 505-512, 541-545.

[120] O. Spengler, op. cit., 504, 545-546; Compare F. Tudjman, *Rat protiv rata*, op. cit., 493-496; R. Hilberg, op. cit., 662.

[121] Compare: F. Tudjman, *Rat protiv rata*, op. cit., 441-442. It is significant that August Cesarec translated the Charles de Coster work into Croatian, which adds to the evidence that his Marxist world view determined his choice of a revolutionary path in the battle for the national liberation of Croatia, his unfortunate "Czar's Kingdom." That he, too—as he said in 1941 (!) in the foreword to Coster's "Ulenspigel," considered it of primary importance to be "at least for national freedom, as well as for social justice, for the victory of well-being on earth" (Compare: Charles de Coster, *Legenda o Ulenspiegelu i Lamu Gudzaku* [Legend of Til Eulenspiegel and Lamme Goedzak], I-II, Zagreb, 1945, Cesarec's foreword, I, vii-xi).

[122] Compare: F. Tudjman, *Rat protiv rata*, op. cit., 443-450.

[123] Compare: C. Schmitt, op. cit., 46.

[124] S. Zweig, op. cit., 479-480.

[125] F. Nietzsche, *Tako je govorio Zaratustra*, op. cit., 80.

[126] Nietzsche published *Also Sprach Zarathustra* from 1883 to 1891.

[127] Isa, 11:6; 2:4.

[128] Franz Kafka, *Jastreb* [The Hawk], Zagreb, 1982, 87, 91.

[129] Joseph de Maistre, "Considerations sur la France" [Reflections on France] (*Oeuvres*, III); translations in: N. Smailagic, op. cit., I, 312.

[130] Edmund Burke, selections from his political writings and speeches; translation in: N. Smailagic, op. cit., I, 308-310.

[131] Compare: G. W. F. Hegel, *Filozofija povijesti*, op. cit., 32-40, 401-404.

132 G. W. F. Hegel, "Osnovne crte filozofije prava" [The basic framework of the Philosphy of Right], Sarajevo, 1964, par. 102.

133 Violence dans la violence, le debat Bakounine, Necaev, Paris, 1973, 107 (Compare: Nikola Milosevic, Dostojevski kao mislilac [Dostoevski as a Philosopher], Belgrade, 1981, 237).

134 Leo Trotsky, Iz revolucije ("Terorizam i komunizam") [From the Revolution (Terrorism and Communism)], Rijeka, 1971, 179.

135 Vladimir Ilic Lenin, Djela [Works], Belgrade, 1975, 30, 341.

136 Ibid., 35, 162-163 (Political report of the Central Committee of the Russian Communist Party/b. 1922).

137 Compare: Raphael R. Abramovitch, The Soviet Revolution, London, 1962, 128.

138 Based on its investigations, the Foreign Affairs Research Insitute claims that "under communism" 83.5 million people were killed: researcher Stewart Smith concludes that the "communist experiment" has cost between 85 and 125 million lives; a French study states that the number is 139.9 million, while Solzhenitsyn's estimate of the "victims of communism" is 110 million people (Compare: Neue Bildpost, Jan., 11, 1987; Hrvatsko pravo, Valencia, XXVIII, 305, May 1987, 12).

139 Compare: Intervju, no. 196, February 1, 1985, 23.

140 Matt 5:43-44.

141 Giovanni Papini, Zrcalo Koje Izmice [The Elusive Mirror], Zagreb, GZH, 1982, 56-57, 60.

142 Hermann Hesse, Stepski vuk (Steppenwolf), Belgrade, 1960, 6, 71, 165, 182, 188.

143 Erich Fromm, The Anatomy of Human Destructiveness, 1973; Croatian transaltion, Anatomija Ljudske Destruktivnosti, I-II, Zagreb, 1987 (3rd edition); II, 21, 40, et al.

144 Ibid.

145 Leszek Kolakowski, Kljuc nebeski & Razgovor s djavlom [The Key to Heaven & A Conversation with the Devil], Zagreb, 1981, 38, 63-64, 97, 117-118, 135-136.

■

Historical Facts and Unscientific Theories

PART FOUR

[1] Compare Josip Broz Tito, *Sabrana Djela* [Collected Works], vol. IX, Zagreb, Kultura, 172, 187; A. Miletic, op. cit., vol. I, 11, 193-194; vol. III, 706.

[2] Institute for the History of the Workers' Movement—Central Committee of the Communist Party of Croatia, copy in my possession.

[3] Compare V. Dedijer, *Novi prilozi za biografiju J.B.Tita,* op. cit., vol. 2, 547.

[4] The visit by that working group from the Council of the Serbian Academy of Arts and Sciences to Jasenovac on Oct. 11 and 12, 1985 was also documented by a photograph in Miletic's *Jasenovac* (III, 573) with the following caption: Milos Macura, scholar, Mile Dragic, director of the Museum of the National Liberation Battle in Novska, Dr. Djuro Mesterovic, retired general, Vladimir Dedijer, scholar, Antun Miletic, Yugoslav Army colonel, and Dr. Milan Bulajic.

[5] V. Dedijer, *Novi prilozi za biografiju J. B. Tita,* op. cit., vol. 2, 547-548.

[6] Ibid., 548-549.

[7] In a Memorandum of the government of the Independent State of Croatia, which was prepared for the commander of the Allied Forces, Field Marshal Alexander, that is, for the governments of the U.S. and Great Britain, it is stated that the number of refugees from all parts of Croatia, as well as from Serbia, Montenegro, Albania, and Bulgaria

447

numbered over a million (Compare F. Jelic-Butic, *Ustase i NDH*, op. cit., 296, 308; A. Miletic, op. cit., vol. III, 200-203).

[8] National Archives, Washington, D.C., T-501, roll 265 and T-120, roll 212 (Compare Bogdan Krizman, *Ustase i Treci Reich* [The Ustashas and the Third Reich], Zagreb, Globus, 1983, vol. I, 38-39); Bogdan Krizman, *Pavelic izmedju Hitlera i Mussolinija* [Pavelic between Hitler and Mussolini], Zagreb, Globus, 1980, 581-585).

[9] *Naprijed* [Forward (a Partisan newspaper)], no. 42, March 5, 1944; no. 44, March 26, 1944; no. 46, April 6, 1944; no. 49-50, 1944 (see also A. Miletic, op. cit., vol. III, 337-339, 344-346, 348-350, 448-450).

[10] Compare Report on the work of the veterans' organizations 1962-1965, by the presidency of the Croatian chapter of the Organization of Veterans of the National Liberation War, Zagreb, June 1965, 32-36.

[11] The Croatian Statistical Institute, initial results from the census of war victims 1941-1945, table ZR-9. The original, handwritten copy written by General N. Kajic at the Croatian Statistical Institute is in my possession. A copy is on file at the Institute for the History of the Worker's Movement in Croatia.

[12] Compare A. Miletic, op. cit., III, 106-119.

[13] Ante Ciliga, *Sam kroz Europu u ratu*, (Alone Through Europe in the War), Rome, 1978, 217-346, et al.

[14] Lj. Boban, op. cit., 145-151.

[15] Ibid, 155-158.

[16] Ibid, 477.

[17] Ibid, 248.

[18] Compare *Prvo i Drugo zasjedanje AVNOJ-a* [the First and Second Sessions of the Anti-Fascist Council for the People's Liberation of Yugoslavia (AVNOJ)], Zagreb, Stvarnost, 1963, 205-206.

[19] *Naprijed* [Forward], organ of the Central Committee of the communist Party of Croatia, no. 47, April 13, 1944 (compare: A. Miletic, op. cit., vol. III, document no. 185, 351-355).

[20] Vladimir Dedijer, *Josip Broz Tito, prilozi za biografiju*, [Josip Broz Tito: Contributions to the Biography], Belgrade, Kultura, 1953, 505.

[21] Vladimir Dedijer, *Novi Prilozi za biografiju J. B. Tita*, vol. 2, op. cit., 381, 489.

[22] Vladimir Dedijer, *Vatikan i Jasenovac, Dokumenti* [The Vatican and Jasenovac, Documents], Belgrade, "Rad," 1987, 5.

[23] Outline of *Chapter VI* (History of the Communist Party of Yugoslavia), 62, bound manuscript which had been sent to me in 1960 for my review, currently in my possession.

[24] Ibid., 129.

[25] *Pregled istorije SKJ* [A Survey of the History of the League of Com-

munists of Yugoslavia], ed. R. Colakovic, D. Jankovic, P. Moraca, Belgrade, 1963, 355, 421.

26 *Ustase*, book 3, copy 2, 385, 395.

27 A. Miletic, op. cit., vol. II, 1010-1023.

28 Ibid., 1003, 1005.

29 Ibid., 1051-1089.

30 Ibid., 1080.

31 Photocopy of a stenographic record of the trial against the accused (1948) Ljubo Milos, et al., (Miroslav Filipovic, Ante Vrban, Nikola Pehar, Adam Milicevic, Mate Vasilj, Jacob Martinovic, Mimo Rosandic, Josip Tomljenovic, Bozidar Petracic, and Ivica Grzeta) from the court records (several hundred pages of testimony by the accused and witnesses are in my archives).

32 Compare A. Miletic, op. cit., vol. III, document numbers 220, 221, 222, 240.

33 F. Tudjman, *Stvaranje socijalisticke Jugoslavije*, op. cit., 127; compare also: Slovenian translation, *Rojstvo socijalistice Jugoslavije*, op. cit., 138.

34 Compare V. Dedijer, *Novi Prilozi za biografiju J. B. Tita*, op. cit., vol. III, 104 and elsewhere; *NIN*, no. 1751, July 22, 1984.

35 Vladeta Vuckovic, "Zrtve rata, sahrana jednog mita" [Victims of the War, The Burial of a Myth], *Nasa Rec*, London, no. 368, October 1985, 2-3; compare Ante Beljo, "The Challenge to Our Dignity," *CIS* [Croatian Information Service], no. 9, March, 1987, 2-3.

36 *Vjesnik*, Zagreb, January 14, 1950, 2 (see also B. Busic, *Jedino Hrvatska*, op. cit., 543).

37 Ivo Lah, "Metode izracunavanja buduceg stanovnistva i njihova primena na stanovnistvo predratne Jugoslavije" [Methods of Calculating Future Population and Their Application to Prewar Yugoslavia], *Statisticka revija*, Belgrade, vol. I, 1951, no. 2, 246.

38 Dolfe Vogelnik, *Demografski gubici Jugoslavije u Drugom svjetskom ratu* [Demographic Losses in Yugoslavia in the Second World War]; *Statisticka revija*, vol. II, 1952, no. 1, 32.

39 Ivo Lah, *Istinski demografski gubici Jugoslavije u Drugom svjetskom ratu* [True Demographic Losses in Yugoslavia in the Second World War], *Statisticka revija*, vol. II, 1952, no. 2-3, 214-215 (compare B. Busic, *Jedino Hrvastka*, op. cit., 534-535).

40 Guenther Ipsen, "Wachstum und Gliederung der Bevoelkerung Jugoslawiens" [Growth and Categorization of the Population of Yugoslavia] in *Ost-Europa Handbuch*, Koeln-Graz, Boehlan-Verlag, 1954, vol. I, 37-63.

41 Rudolf Bicanic, "Stanovnistvo FNRJ po nacionalnoj pripadnosti"

[The Population of Yugoslavia According to Nationality], *Hrvatsko kolo* [Croatian Circle], Matica hrvatska, vol. VI, 1953, 74-84.

[42] Compare Vladimir Simeunovic, *Stanovnistvo Jugoslavije i socijalistickih republika od 1921-1961* [The Population of Yugoslavia and the Socialist Republics From 1921 to 1961], Belgrade, 1964, 31; Bruno Busic "Zrtve rata" [Victims of the War], *Hrvatski knjizevni list*, vol. II, 1969, no. 15, 2-3.

[43] Ivan Klauzer, "Stanovnistvo i domacinstva u Hrvatskoj" [Population and Households in Croatia], Institut za drustvena istrazivanja Sveucilista u Zagrebu, Zagreb, 1967; the manuscript is in the Institute's archives (see also Mladen Friganovic, "Socijalni aspekti demografskih kretanja u SR Hrvatskoj" [Social Aspects of Demographic Movement in the Socialist Republic of Croatia], *Revija za sociologiju*, vol. V, 1975, no. 1-2, 37; B. Busic, *Jedino Hrvatska*, op. cit. 599-600.

[44] Compare B. Busic, *Jedino Hrvatska*, op. cit., 560-561.

[45] Vladimir Stipetic, "Jedno stoljece u brojcanom razvoju stanovnistva na danasnjem podrucju Jugoslavije" [A Century in the Numerical Growth of Populations in the Area of the Present-Day Yugoslavia], *Forum*, vol. XII, 1973, no. 12, 888-889.

[46] Mirko Korencic, *Naselja i stanovnistvo SR Hrvatske 1847-1971* [Settlements and Population in the Socialist Republic of Croatia 1847-1971], Zagreb, JAZU (Yugoslav Academy of Arts and Sciences), 1979.

[47] Ibid., forword by V. Stipetic.

[48] F. Tudjman, *Nacionalno pitanje u suvremenoj Europi*, op. cit., chapter "O demografskim i ratnim gubicima" [On Demographic and War Losses], 192, 331, 339-340, 344-347, 368-370. At the same time that I was so concerned with historical truth, aware of the dangers therein, virtually all of the Yugoslav publications were writing in praise of the Jasenovac distortion in the way described above, seeking (before I had been packed off to prison) that my "words, life and hands" be put in chains, in keeping with the "Ustasha beasts" and "Stepinac's satanic priests" (*Arena*, no. 1080, September 2, 1981, 9).

[49] As late as 1948, there were about 130,000 displaced persons from Yugoslavia in European camps; in addition, more than 210,000 former Yugoslav citizens of German nationality were registered in Austria and West Germany (United Nations, *The Refugee in the Postwar World*, Genova, 1951, 58, 78). Undoubtedly, neither the first nor the second figure is complete.

[50] Dr. Bogoljub Kocovic, *Zrtve Drugog svetskog rata u Jugoslaviji* [Victims of the Second World War in Yugoslavia], forword by Desimir Tosic, London, 1985.

⁵¹ Ibid., 28-34, 41-58.

⁵² Ibid., 46-55.

⁵³ Ibid., 54.

⁵⁴ Ibid., 62-63.

⁵⁵ Ibid., 94-95, 183.

⁵⁶ Compare ibid., 97-109, 110-112, 119-122, 124-126, 129-134.

⁵⁷ *Danas*, no. 209, February 18, 1986, 70-72.

⁵⁸ Jakov Gelo, *Demografske promjene u Hrvatskoj: Od 1780. do 1981. godine* [Demographic Changes in Croatia: From 1780 to 1981]. Zagreb, 1987.

⁵⁹ Ibid., 152.

⁶⁰ Ibid., 152-154.

⁶¹ Ibid., 192-193.

⁶² F. Jelic-Butic, *Ustase i Nezavisna Drzava Hrvatska 1941-1945*, op. cit.

⁶³ Ibid., 186-187.

⁶⁴ "U kakve vrapce puca topovnik Vjekoslav Kaleb?" [What Manner of Sparrow Is Artilleryman Vjekoslav Kalebs Shelling?], an open letter by the Jasenovac Commemoration Committee to the renowned Croatian author, *Politika*, Belgrade, January 14, 1972, 6.

⁶⁵ *Enciklopedija hrvatske povijesti i kulture* Zagreb, 1980, 304-305 ("concentration camps").

⁶⁶ Pedro Ramet, *Nationalism and Federalism in Yugoslavia 1963-1983,* Bloomington, 1984 (compare: letter by the author in *Vjesnik*, January 12, 1986).

⁶⁷ Nora Beloff, *Tito's Flawed Legacy: Yugoslavia and the West: 1939-1984*, London, Victor Gollenz, 1985.

⁶⁸ Jure Parsic, *Sjecanje nekadasnjeg jasenovackog zupnika na dogadjaje onoga vremena* [Memoirs of a Former Jasenovac Priest on the Events of Those Times), *Hrvatska revija*, vol. XXXV, 1985, no. 3, 428.

⁶⁹ *Zajednicar* [Fraternalist], official newspaper of the Croatian Fraternal Union of America, vol. 80, no. 3, January 16, 1985; "World War II Victims," 2; Ante Beljo, *Croatia and the Victims of World War II*, 8-9; Ante Beljo, *Hrvatska i zrtve drugog svjetskog rata*, 16-17.

⁷⁰ A. Miletic, op. cit., vol. I, 37-42.

⁷¹ Ibid., vol. III, 717.

⁷² Ibid., vol. III, 5-6.

⁷³ Ibid., vol. I, 38. According to Dr. Zvonimir Dugacki's ethnographic map, the number of Serbs or those of the Orthodox faith in the Independent State of Croatia was somewhat lower: 1,847,000 or 30.56% of the population compared with 3,069,000 (50.57%) Croatian Catholics and 717,000 (11.86%) Muslims, as well as 410,000 (6,80%) others, of

a total population of 6,043,000 people. These figures were used by the Germans in 1941 (compare: B. Krizman, *Pavelic izmedju Hitlera i Mussolinija*, op. cit., 129).

⁷⁴ Compare B. Kocovic, op. cit., Appendix 35, 184.

⁷⁵ Compare V. Terzic, *Jugoslavija u aprilskom ratu*, op. cit., 5, 51-52, 157, 506, 644, 665, and elsewhere. See also D. Scukanec, op. cit.; S. Scetaric, op. cit.; F. Butic and I. Jelic, op. cit.

⁷⁶ Velimir Terzic, *Bili smo razbijeni pre izlaska na front*, op. cit.

⁷⁷ Ibid.

⁷⁸ Ante Starcevic, *Djela* [Works], Zagreb, 1893, vol. I, 25, 63; vol. II, 65; *Izabrani Spisi* [Selected Writings], Zagreb, 1943, 53, 79-80, 276, 382, 416, 427, 432; *Misli i pogledi* [Thoughts and Views], Zagreb, 1971, 59, 211, 244, 259.

⁷⁹ A. Starcevic, *Djela*, op. cit., vol. II, 47; *Izabrani Spisi*, op. cit., 137, 469, 472, 476-477; *Misli i pogledi*, op. cit., 207-208.

⁸⁰ The text was allegedly written back in 1836 and published in 1849 in the book: *Vuk Stefanovic Karadzic, Kovcezic za istoriju jezika i obicaja Srba sva tri zakona* [A Treasury of the History, Language, and Customs of the Serbs of all Three Creeds], Vienna, 1849.

⁸¹ Compare: Mile Starcevic, *Ante Starcevic i Srbi* [Ante Starcevic and the Serbs], Zagreb, 1936, 22-32, 87-120. Of the more prominent Orthodox Christians who acknowledged being Croatians (some were even members of the Croatian Party of Rights) are mentioned: Nikola Kokotovic, Dr. Ivo Malin, Stjepan Miletic, Dr. Dimitrije Dematar, Spiro Dimitrijevic Kotoranin, Dusan Kotur, Vladislav Nikolic, Danijel Medic, Makso Prica, Nikola Krestic, Dane Stanisavljevic, and others.

⁸² Vasilije Krestic, *O genocidu nad Srbima u NDH u periodu do prvog svetskog rata* [On Genocide Against the Serbs in the Independent State of Croatia: to the Period up to the First World War]. *Knjizevne novine*, Belgrade, vol. XXXII, no. 716, September 15, 1986, 1, 4-5.

⁸³ Compare Vaso Bogdanov, "Historijski uzroci sukoba izmedju Hrvata i Srba" [Historical Causes of the Conflict Between the Croatians and the Serbs], *Rad JAZU*, vol. 311, 1957, 353-477; *Ziva Proslost* [A Living Past] Zagreb, 1957; "Porijeklo i ciljevi sovinistickih teza o drzanju Hrvata 1941" [The Origins and Goals of the Chauvinistic Theses Concerning the Conduct of the Croatians in 1941], *Zbornik Historijskog instituta JAZU*, no. 4, 1961, 5-92; *Sest decenija Regierungsfaehig Politike* [Six Decades of Regierungsfaehig Politics], inspired by Veceslav Vilder's book, *Bika za rogove; Gde je izvor spora srpsko hrvatskog? Gde je resenje?* [A Bull by the Horns: What is the Source of the Serbian-Croatian Clash?], London, 1957, *Historijski zbornik JAZU*, no. 5, 1964, 456-530.

84 Compare V. Krestic, op. cit.

85 Compare *Vjesnik*, vol. XLVI, 13900, May 23, 1986, article: "Providne smicalice na akademskom nivou" [Transparent Subterfuge at the Academic Level].

86 *NIN*, no. 1937, February 14, 1988, 9-10 (A conversation with the scholar V. Dedijer, conducted by Milo Gligorijevic).

87 *Memorandum SANU Nacrt ili Radni tekst* [Memorandum of the Serbian Academy of Arts and Sciences: Outline or the Working Text] 1986, 1-74 (a photocopy of the typed text in the Cyrillic and Latin alphabets is in my possession). According to a May 23, 1985 decision by the Academy, on June 12, 1985, the Presidency of the Academy selected a commission of 16 members: Pavle Ilic, Antonije Isakovic, Dusan Kanazir, Mihailo Markovic, Milos Macura, Dejan Medakovic, Miroslav Pantic, Nikola Pantic, Ljubisa Rakic, Radovan Samardzic, Miomir Vukobratovic, Vasilije Krestic, Ivan Maksimovic, Kosta Mihailovic, Stojan Celic, and Nikola Cobeljic, with Dusan Kanazir as president. The work of the commission was directed by Antonije Isakovic. Compare Velizar Zecevic, *Memorandum jedne akademije* [Memorandum of an Academy], *NIN*, no. 1869, October 26, 1986, 14-15; Dragan Vlahovic, *Dosije tzv Memoranduma* [Dossier of the So-called Memorandum], *Intervju*, November 7, 1986, cit., 6-11.

88 Draskovic's letter to the government and political agencies of the Socialist Republic of Croatia, dated December 16, 1985. One of the photocopies, which was distributed all over the country is in my possession.

89 All citations from Vuk Draskovic's reply to Vjesnik, Belgrade, September 2, 1986, distributed by photocopy.

90 Compare J. Tomasevic, op. cit., 155-165.

91 Compare B. Krizman, *Pavelic izmedju Hitlera i Mussolinija*, op. cit., 49, et al.

92 Compare J. Tomasevic, op. cit., 212, 232-239; F. Jelic-Butic, *Cetnici u Hrvatkoj*, op. cit., 160-164. The local council of the veterans' organizations for the municipality of Knin, in its request for the extradition of the Chetnik leader, the priest Momcilo Djujic, commander of the Dinara Division, states that he was directly responsible for the execution, hanging, and slaughter of 1,372 members of the National Liberation Movement, of which 918 were of Serbian nationality.

93 In the "Kozara Offensive," which had as its purpose the destruction of the 1st and 2nd Krajina Partisan Detachments (starting June 10, 1942) along with a strong German force under the command of General Stahl and Ustasha and Domobran units, Chetnik detachments of 2,000 men also participated, *VE*, vol. 4, 793-796.

⁹⁴ *Zbornik dokumenata NOR-a*, (Collection of National Liberation War documents), vol. XIV/2, Belgrade, 1952, 175; *Dokumenti o izdajstvu Draze Mihailovica*, (Documents About the Treason of Draza Mihailovic), vol. I, 109-126, 433-436, 447-452; see also J. Tomasevic, op. cit., 217-218.

⁹⁵ Compare F. Jelic-Butic's *Ustase i NDH*, op. cit., 175-178, 238, 247-256; Milos Obrknezevic, *Razvoj pravoslavlja u Hrvatskoj i Hrvatska pravoslavna crkva* [The Development of Orthodoxy in Croatia and the Croatian Orthodox Church], offprint from *Hrvatska Revija*, vol. XXIX, 1979, no. 2 (114), 229, 248-249, 250-252, 259-262; Ivan Sibl, *Ratni dnevnik* [War Journal], 191-194. (Home Guard [Domobran] Labor Regiments were armed with old rifles and three bullets. Sibl spoke to a "Serbian domobran regiment" captured towards the end of 1942 on Bilogora: "And what do you think? How many of these young men from Kordun, Banija and Bosnian Krajina stated their readiness to stay with the Partisans? *Only thirty*, and after several hours of cajoling. The only thing we didn't do was to fall on our knees and, with clasped hands, beg them to stay with us. The rest of them were released but commander Captain L., also Serbian, was shot!)

⁹⁶ Dr. Nikola Stojanovic (1880-1964) was a lawyer and journalist, one of the foremost Serbian politicians from Bosnia, a member of the Yugoslav Committee on Corfu Island. He published *Srbija juce i sutra; Bosna i Hercegovina; Jugoslovenski odbor; Srbija i jugoslovenko ujedinjenje; Nasa politika; Bosanska kriza 1908-1928* [Serbia, Today and Tomorrow; Bosnia and Herzegovina, Yugoslav Council, Serbia and the Yugoslav Unification, Our Politics, Bosnian Crisis 1908-1928], and other works, according to the *Mala enciklopedija*, *Prosveta*, Belgrade, 1949, 630.

⁹⁷ V. Dedijer, *Novi prilozi za biografiju J. B. Tita*, op. cit., vol. 2, 1981, 531-534.

⁹⁸ Ibid., 534-535, 544.

⁹⁹ From the literature about Stepinac, I mention only more important works. O'Brien, A.H. *Archbishop Stepinac: The Man and his Case*, Westminster, Maryland, USA, 1947; R. Pattee, *The Case of Cardinal Aloysius Stepinac*, Milwaukee, 1953; T. Dragoun, *Le Dossier of Cardinal Stepinac*, Paris, 1958; E. Pezet, *Stepinac-Tito*, Paris, 1959; E. Beluhan Kostelic, *Stepinac govori* [Stepinac Speaks], Valencia, 1967; Ivan Mestrovic, *Uspomene na politicke ljude i dogadaje* [Memoirs of Politicians and Events], Buenos Aires, 1969; O. A. Benigar, *Alojzije Stepinac, hrvatski Kardinal* [Aloysius Stepinac, Croatian Cardinal], Rome, 1974; *Le saint seige et la guerre en Europe* [The Holy See and the War in Europe], Vatican Library edition, vol. 4 (covering June 1940 - June 1941), 1967; vol.

5 (covering July 1941 - 1942), 1969; vol. 8 (covering January 1941 - December 1942), 1974; vol. 9, (covering January 1943 - December 1943), 1975; E. Bauer, *Kardinal Aloisius Stepinac*, Recklinghausen, 1979. *Stepinac mu je ime: Zbornik uspomena, svjedocanstva i dokumenata* [Stepinac is His Name: Anthology of Memoirs, Testimonies and Documents], edited by V. Nikolic, Munich-Barcelona, vol. I, 1978, vol. II, 1980; J. Batelja, *Rendere testimonia alla fede, vita e programma pastorale del Cardinal Alozije Stepinac* [Testimonies About the Faith, Life, and Pastoral Program of Cardinal Aloysius Stepinac], Rome, 1984; S. Alexander, Church and State in Yugoslavia since 1945, Cambridge UP, 1979; M. Landercy, *Le cardinal Stepinac martyr des droits de l'homme* [Cardinal Stepinac. Martyr for Human Rights], Paris, 1981; Ivan Cvitkovic, *Ko je bio Alozije Stepinac* (Who Was Alozije Stepinac), Sarajevo, 1986; S. Alexander, *The Triple Myth, A Life of Archbishop Alojzije Stepinac*, Eastern European Monographs, Boulder (Columbia University Press), New York, 1987; L. Kordic, *Mucenistvo crkve u Hrvatskoj* [Martyrdom of the Church in Croatia], Chicago, 1988.

100 Sermon of Archbishop Franjo Kuharic in the Zagreb Cathedral, February 10, 1981. On the occasion of the 21st anniversary of the death of Cardinal Alozije Stepinac, supplement to *Sluzbeni vjesnik* [Official Messenger] of the Zagreb Diocese, no. 2, 1981, 3-10; see also the statement by Cardinal F. Seper and the statement by Dr. R. Peric, Ibid., 1-2, 11-12.

101 Official letter to the honorable clergy of Zagreb Archdiocese, April 28, 1941, Katolicki List, Zagreb, 92, April 29, 1941; O. A. Benigar, op. cit., 386-389; r. Pattee, op. cit. 258-260.

102 O. A. Benigar, op. cit., 403-404; Archbishop Franjo Kuharic's previously mentioned sermon.

103 O. A. Benigar, op. cit., 406-419.

104 *Katolicki List*, 92, October, 26, 1941. Compare O. A. Benigar, op. cit. 472-475.

105 O. A. Benigar, op. cit., 436-439; R. Pattee, op. cit., 267-271.

106 Since it could not be printed, this sermon was copied by and distributed among the parishioners (See O. A. Benigar, op. cit., 442-446).

107 O. A. Benigar, op. cit., 439-441; R. Pattee, op. cit., 271-276.

108 Compare I. Mestrovic's op. cit., 1961, 363, 377-378; O. A. Benigar, op. cit., 485-487.

109 Archbishop's sermon at the Shrine of Marija Bistrica on the 9th of July, 1944, *Katolicki list*, 95, 1944, 344; compare O. A. Benigar, op. cit., 453.

110 Compare O. A. Benigar, op. cit., 452-454.

111 Remarks at the Easter Conferences of Croatian University Students

on March 18, 1945, *Katolicki list*, 96, 1945, 95. Compare O. A. Benigar, op. cit., 467-468.

112 V. Dedijer, *Novi prilozi za biografiju J. B. Tita*, op. cit., 2, 949.

113 Ibid., 585.

114 Compare "Pismo papi" [A letter to the Pope], *NIN*, no. 1741, May 13, 1984.

115 V. Dedijer, *Vatikan i Jasenovac*, op. cit., 17-18.

116 Ibid., 18-19, 32-33.

117 Ibid., 38-39.

118 Compare A. Miletic, op. cit., vol. II, 1024-1025; A. Ciliga, op. cit., 315-316; V. Nikolic, *Stepinac mu je ime*, op. cit., vol. I, 283-284.

119 Compare O. A. Benigar, op. cit., 519-520, 542, 585.

120 V. Dedijer, *Vatikan i Jasenovac*, op. cit. 463-464.

121 V. Dedijer, *Novi prilozi za biografiju J. B. Tita*, op. cit., 2, 539.

122 V. Dedijer, *Vatikan i Jasenovac*, op. cit., 464, 534, 709-712.

123 Vuk Draskovic in his "Demand" to the Parliament of the Socialist Republic of Croatia, on December 16, 1985.

124 Vuk Draskovic's reply to *Vjesnik*, September 2, 1986.

125 Compare M. Obrknezevic, op. cit., 246-250.

126 Compare O. A. Benigar, op. cit., 538.

127 Dr. Marko Vidakovic, *Predmet: Ratno doba u savremenom secanju, Svetom Arhijerejskom Sinodu SPC Beograd* [Subject: War Period in Contemporary Recollection, to the Holy Synod of the Serbian Orthodox Church in Belgrade], Zagreb, August 25, 1973 (a copy of this statement is in my possession - F.T.).

128 Handwritten statement dated "New York, February 10, 1975" and signed "Prota Andreja Dimitrijevic, Serbian Orthodox Parish Priest and Guardian of the St. Sava Cathedral Church in New York" (photocopy in my possession - F.T.).

129 Compare V. Nikolic, *Stepinac mu je ime*, vol. II, 54.

130 Ibid., vol. I, 286-290, 316-320, 382-385.

131 C. L. Sulzberger, *A Long Row of Candles—Memoirs and Diaries, 1934-1954*, Macmillan 1969. (Compare V. Nikolic, *Stepinac mu je ime*, op. cit., vol. II, 397); *Discorsi, messagi, collogqui del Santo Padre Giovanni XXIII*, Vatican, 1961, vol. II, 202-205; *Rivista diocesana milanese*, 196-, 110-112. Compare AKSA, February 12, 1988, 6 (1925) Supplement II; Vjekoslav Cvrlje, *Vatikan u suvremenom svijetu* [The Vatican in the Contemporary World], Zagreb, 1980, 103-114, 346, et al.

132 Dr. Zarko Gavrilovic, *Na braniku vere i nacije* [On the Rampart of Faith and Nation], Belgrade, 1986 (compare *Glas Koncila*, vol. XXVI, no. 30, July 26, 1987).

133 *Duboki koreni zlodela* [Deep Roots of Misdeeds], Interview by

Miodrag Kujundzic with Professor Dr. Dragoljub Zivojinovic, *Nedeljni dnevnik*, Novi Sad, vol. XLVI, no. 14502, March 15, 1987, 6-7. In that interview it is important to stress among other things that Dr. Zivojinovic, "Professor of modern history" at the Philosophy Faculty in Belgrade, did his post-graduate work in the United States and scientific research in Great Britain and Italy. His 1966 dissertation was "The USA, Italy and the emergence of Yugoslavia 1917-1919" (in Serbian). He published a "unique" trilogy - *SAD ivatikanska politika 1914-1918* [The USA and the Vatican Policy, 1914-1918], *Vatikan i prvi svetski rat* [The Vatican and WWI], and *Vatikan i stvaranje jugoslovenske drzave 1914-1920* [The Vatican and the Creation of the Yugoslav State, 1914-1920]."

[134] Herve Lauriere, *Assassins au nom de Dieu* [Murderers in the Name of God], Paris 1951, also in Serbian translation by Mihailo B. Pavlovic; forward by Dr. Dragoljub Zivojinovic, Belgrade, 1987, 5-11.

[135]Dragoljub R. Zivojinovic and Dejan V. Lucic, *Varvarstvo u ime Hristovo: Prilozi za Magnum Crimen* [Barbarism in the Name of Christ: Contributions to Magnum Crimen], Belgrade, 1988; see especially 7, 44-52, 75, 126, 198, 317, 366, 437, 522, 545-547, 583, 667-669, 673-675, 731, 752-763, 774-782, 796-799, 807, 816, 853-856, 887-898.

[136] Dr. Milan Bulajic, *Ustaski zlocini genocida i sudjenje Andriji Artukovicu 1986. godine* [Ustasha Crimes of Genocide and the Trial of Andrija Arkutovic in 1986], Belgrade, 1988; forword by Dr. Vladimir J. Dedijer, academic, President of the Russell Court, President of the SANU Committee for the Collection of Materials About Genocide Against the Serbian People and Other Peoples of Yugoslavia in the 20th Century; "Afterward - the Topicality of the Subject and the Writer of This Book" by Ljubomir Kljakic; see: vol. I, 6-10, 11-105; II, 227, 244, 388-390, 391-407, 676-685, 943-948.

[137] Svetozar Dusanic, "Osvrt na jedan govor" [Comment on a Speech], *Pravoslavlje*, October 1, 1987, 8.

[138] From *Glasnik*, the official newspaper of the Serbian Orthodox Patriarchate, no. 11-12, 1931, and *Vesnik srpske crkve*, the organ of the Association of the Serbian Orthodox Priests, 1931 (Compare Filip Lukas, *Licnosti, stvaranja, pokreti* [Personalities, Creations, Movements], Zagreb, 1944, 64-65).

[139] M. Milovanovic, "Srpska crkva i Konkordat" [Serbian Church and the Concordat], *Glas crkve*, no. 4, 1987.

[140] Compare F. Lukas, op. cit., 241-242; *Nova Revija*, vol. V, 1926, no. 1, 90-91.

[141] Anton Ingolic, "Krenimo od suradnje" [Let's Start With Cooperation], interview with Stane Pucko, *Vjesnik's supplement Sedam dana*, no. 451, January 10, 1987, 5.

[142] Compare O. A. Benigar, op. cit., 216-233, et al.

[143] Zagreb Archbishop's sermon on the anniversary of the death of Cardinal Alojzije Stepinac, in the Zagreb Cathedral on February 10, 1977; supplement to *Sluzbeni Vjesnik Nadbiskupije zagrebacke*, no. 2, 1977, 1-6.

[144] Compare Glas Koncila, Zagreb, vol. XXIII, no. 18 (538) September 16, 1984, 1-6, 10-17.

[145] Ibid., 7.

[146] Compare *Glas Koncila*, vol. XXIII, no. 21 (541), October 7, 1984, 2.

[147] Petar Segedin. "Prica iz bolnickog vrta" [A Story From the Hospital Garden], *Forum*, vol. XXV, no. 11-12, 1986, 538-542.

[148] Compare Arnold Joseph Toynbee, *A Study of History*, New York, London, 1947, op. cit., 927-928.

[149] Compare V. Djuretic, *Saveznici i jugoslovenska ratna drama*, op. cit., vol. I, 6-7.

[150] See previously mentioned letters addressed to the leadership of the Socialist Republic of Croatia and in the reply to *Vjesnik*.

[151] V. Djuretic, *Saveznici i jugoslovenska ratna drama*, op. cit., vol. I, 13, 15, 32-33.

[152] Ibid., vol. I, 15-19, 23; as an example of the Marxist scholarly literature which allegedly unjustifiably wrote about the great Serbian hegemony, Djuretic quotes; I. Levin, *Nacionalnii vopros v poslevoenoj Europe* [The National Question in Postwar Europe], Moscow, 1934, 341-383.

[153] Ibid., vol. I, 23.

[154] I especially dealt with this problem in my doctoral dissertation (1965), *Uzroci drzavno-politicke krize monarhisticke Jugoslavije od ujedinjanja 1918. do sloma 1941,* op. cit. For citations about this consideration compare the texts in my published works: *Drustveni aspekt narodnooslobodilackog pokreta u Jugoslaviji* [Social Aspect of the National Liberation Movement in Yugoslavia], in: *Putovi revolucije*, Zagreb, vol. IV, 1966, 7-8, 10-12; *Hrvatska politika u prvim godinama borbe protiv vidovdanskog centralisticko-hegemonistickog poretka*, op. cit. 575-585 et al. *Slom vidovdanskog parlamentarnog centralizma i drzavno-politicka kriza Kraljevine SHS*, op. cit., 51-53, 73-106.

[155] The bequest of Ivan Persic, Historical Archives of Zagreb (Compare Ljubomir Antic, "Svetozar Ritig u biljeskama Ivana Persica" [Svetozar Ritig in the Notes by Ivan Persic], *Vjesnik's supplement Sedam dana*, December 4, 1982).

[156] Outline of the SANU Memorandum, op. cit., 41-42.

[157] All citations from already mentioned Vuk Draskovic's reply to "Vjesnik."

¹⁵⁸ V. Djuretic, *Saveznici i jugoslovenska ratna drama*, op. cit., vol. I, 6, 32-33, 97-98, 111, 135, 160, 213; vol. II, 8, 184, 214, 229, 241-250.

¹⁵⁹ In this work, see the part concerning the controversy about my book *Rat protiv rata* (1957), my encyclopedia entry entitled "NOR u Hrvatskoj" (1960), my doctoral dissertation "Uzroci drzavno - politicke krize...," op. cit., and especially "Raspre o uzrocima sloma monarhisticke Jugoslavije i o pretpostavkama razvitka NOB-a u Hrvatskoj," as published in: *Kolo*, Matica hrvatska, no. 8-9, 1968 and in *Velike ideje i mali narodi*, op. cit.

¹⁶⁰ For full details see Ivan Jelic, *Tragedija u Kerestincu: Zabrebacko ljeto 1941* [Tragedy in Kerestinac: Zagreb Summer of 1941] Zagreb, 1986.

¹⁶¹ I heard this information personally from Professor R. Bicanic, corroborated later by Franjo Gazi, to whom that was offered.

¹⁶² Compare F. Tudjman, *Uzroci drzavno-politicke krize...*, op. cit., (dissertation), 452-455.

¹⁶³ Josip Vidmar, *O slovenstvu i jugoslovenstvu* [On Slovenianism and Yugoslavism], Zagreb, 1986, 468.

¹⁶⁴ Ibid., 483.

¹⁶⁵ Djuretic, *Saveznica i jugoslovenska ratna drama*, op. cit., vol. II, 252-253.

¹⁶⁶ Ljubomir Tadic, "Balkanizacija partizanskog jugoslovenstva" [Balkanization of the Partisan Yugoslavism], *Knjizevne novine*, Belgrade, vol. XL, no. 749 and 750, March 1-15, 1988.

¹⁶⁷ On the creation of the Communist Party of Croatia and of Slovenia, *Proleter*, organ of the Central Committee of the Communist Party of Yugoslavia, vol. XI, July-August 1935, 7-8; Also in *Srp i Cekic*, Belgrade vol. II, 1935, 4. Compare *Osnivacki kongres KPH* [The Founding Congress of the Communist Party of Croatia], Zagreb, 1958, 71-78.

¹⁶⁸ Compare *Osnivacki kongres KPH*, op. cit., 115.

¹⁶⁹ Ibid., 81-90. It is indicative that from the Founding Congress of 1937, this party had the name Komunisticka *stranka* Hrvatske (Croatian Communist Party)—"stranka" is the Croatian word for "party"—led by its *Glavni odbor* (Chief Committee). When and on the basis of whose decision it took the name of the Komunisticka *Partija* hrvatke (KPH, or Communist Party of Croatia) led by its *Centralni komitet* (Central Committee)—surrendering the name CK KPH—is not clear from the monograph by I. Jelic, *Komunisticka partija Hrvatke 1937-1941* [Communist Party of Croatia 1937-1941], Zagreb, 1972. There was some "evolution;" first *Partija* replaced *Stranka* and then *Centralni komitet* replaced *Glavni odbor* (the Chief Committee) and *Sredisnji odbor* (the Central Committee). Allegedly because of internationism, the first two issues of

Srp i cekic, vol. VI, 1940, had as a subtitle "Organ sredisnjeg odbora" [the Organ of the Central Committee]; no. 3 carried only a slogan about the unity of the proletariat; and from no. 4 onward it was called the organ of the CK KPH.

[170] "Srpski komunisti i osnivanje Komunistickih stranaka Slovenije i Hrvatske" [Serbian Communists and the Founding of the Communist Parties of Slovenia and Croatia], *Proleter*, organ of the Central Committee of the Communist Party of Yugoslavia, vol. XIII October, 1937, 11. (See *Osnivacki kongres KPH*, op. cit., 124-126.)

[171] Conclusions reached at the session of the Communist Party of Yugoslavia held in Zagreb in April of 1941 were published in *Proleter* no. 3-4-5 (March-April-May), 1941. (Compare *Zbornik dokumenata i podataka o NOR-u* [Collection of Documents and Data About the National Liberation War], vol. II, Belgrade, 1952, 2, 7-23).

[172] "Program SKJ" [Program of the League of Communists of Yugoslavia], adopted at the 7th Congress of League of Communists of Yugoslavia, 1958, Belgrade, 1962, 70, 144.

[173] Compare F. Tudjman, *Velike ideje i mali narodi*, op. cit., 206-215.

[174] V. Djuretic, *Saveznici i jugoslavenska ratna drama*, op. cit., vol. 2, 246-251.

[175] Draft of the SANU Memorandum, op. cit., 45.

[176] *Novo vreme*, Belgrade, August 13, 1941 (Compare F. Tudjman, *Okupacija i revolucija*, op. cit., 197-198).

[177] Proclamation by the Central Committee of the Communist Party of Yugoslavia to the peoples of Yugoslavia on July 25, 1941 (*Zbornik dokumenata i podataka o NOR-u*, op. cit., vol. I, Belgrade, 1949, 22-26).

[178] "Narodima Jugoslavije" [To the Peoples of Yugoslavia], Central committee of the Communist Party of Yugoslavia, November 7, 1941. *Borba*, organ of the CPY, Belgrade, vol. VI, no. 8 and 9, Friday, November 7, 1941. (Compare Historical Archives of the Communist Party of Yugoslavia, vol. I, Belgrade, 1949, no. 1, 133-139).

[179] *Prvo i drugo zasjedanje AVNOJ-a* [First and second session of Anti-Fascist Council for the People's Liberation of Yugoslavia], (Zagreb, 1963), 61-63.

[180] Ibid., 67-69.

[181] J. B. Tito, "Nacionalno pitanje u Jugoslaviji u svjetlosti Narodnooslobodilacke borbe" [The National Question in Yugoslavia in Light of the People's Liberation Struggle], *Proleter*, Organ of Central Committee of Communist Party of Yugoslavia, vol. XVII, December 16, 1942.

[182] *ZAVNOH (Zemaljsko antifasisticko vijece narodnog aslobodjenja Hruatske): Zbornik dokumenata 1943* [Territorial Anti-Fascist Council for the Peoples' Liberation of Croatia] 208-214.

[183] Ibid., 397, et al.

[184] Ibid., 458.

[185] Ibid., 463-469.

[186] Ibid., 478-481.

[187] Compare F. Tudjman, *Okupacija i revolucija*, op. cit., 224-225; *Nacionalno pitanje u djelima klasika marksizma i u dokumentima i praksi KPJ/SKJ* [The National Question in the Works of the Marxist Classics and in the Documents and Practice of the Communist Party of Yugoslavia/League of Communists of Yugoslavia], Belgrade, 1967, 308-316.

[188] F. Tudjman, *Okupacija i revolucija*, op. cit., 225-228.

[189] Ibid., 154. (Compare *ZAVNOH: Zbornik dokumenata 1944*, op. cit., vol. III, from May 10 to December 31.)

[190] V. Djuretic, *Saveznici i jugoslovenska ratna drama*, op. cit., vol. II, 63.

[191] Compare my works: *Jugoslavenski odbor i stvaranje zajednicke drzave juznoslavenkih naroda* [The Yugoslav Commitee and the Creation of the Common State of South Slavic Peoples], in the collection, *Jugoslavenski odbor: u povodu 50-godisnjice osnivanja* [The Yugoslav Committee: On the Occasion of the 50th Anniversary of its Founding], Zagreb, The Yugoslav Academy of Arts and Sciences, 1966, 369-449; *Hrvatska politika na prekretnici prvog svjetskog rata* [Croatian Politics at the Crossroads of the First World War], *Forum*, The Yugoslav Academy of Arts and Sciences, vol. 9, 1970, no. 6, 895-945.

[192] F. Tudjman, *Slom vidovdanskog parlamentarnog centralizma...*, op. cit., 74-75.

[193] Statement of Vuk Draskovic given to German radio (Radio Deutsche Welle, Cologne), Febraury 28, 1987, from the Vuk Stefanovic Karadzic Club. Sources: my own records, Compare *Vjesnik's supplement Sedam dana*, December 5, 1987.

[194] Kosta Cavoski, "Iz istorije stvaranja nove Jugoslavije" [From the History of the Creation of the New Yugoslavia], *Knjizevna rec*, no. 291, December 25, 1986, 6-8; Svetozar Stojanovic, "Jugoslawiens Krise und die nationale Frage" [Yugoslav Crisis and the National Question], *Europaeische Rundschau*, vol. 16, 1988, no. 3, 43-58; Dobrica Cosic, "Jugoslavija—drzava izneverenih ocekivanja i neizvesne buducnosti" [Yugoslavia—a State of Betrayed Expectations and Uncertain Future], *Knjizevne novine*, no. 765, December 1, 1988, 4-5; Stojanovic's and Cosic's texts appeared while the manuscript for my book, *Bespuca* [the original Croatian version of this book], was in process of being printed.

[195] Compare *Delo*, Ljubljana, February 6, 1988; *NIN*, Belgrade, Feb-

ruary 21, 1988; *Ustav SFRJ* [The Constitution of Yugoslavia] Zagreb, 1974, 3-4.

¹⁹⁶ Mihailo Markovic, "Benigni i maligni nacionalizam: O angazmanu filozofa, o raslojavanju grupe oko *Praxisa*, o odnosu politike i nauke, o Kermaunerovim pismima" [Benign and Malignant Nationalism: On the Engagement of Philosophers, on the Differentiation Within the *Praxis* Group, on the Relations Between Politics and Science, and on Kermauner's Letters], *Danas*, September 8, 1987, 36-39.

¹⁹⁷ Gajo Petrovic, "Tko i zasto dijeli jedan jezik?' [Who is Splitting One Language, and Why?] *Vjesnik*, Panorama subotom, February 6, 1988.

¹⁹⁸ Compare Branko Horvat, *Kosovsko pitanje* [The Kosovo Question], Zagreb, 1988, 21-23, 105-108, 124-128, et al; some chapters from the book were published in *Start* magazine, Zagreb, no. 492, 493, 494 (1987) and no. 495 (1988).

¹⁹⁹ *Pravoslavlje*, October 1, 1987, 8.

²⁰⁰ Compare F. Tudjman, *Jugoslavenski odbor i stvaranje zajednicke drzave juznoslavenskih naroda*, op. cit.

²⁰¹ Compare F. Tudjman, *Hrvatska politika u prvim godinama...*, op. cit., 607-608; F. Tudjman, *Slom vidovdanskog parlamentarnog centralizma...*, op. cit., 77-79.

²⁰² See "Borba protiv frankovaca" [Struggle Against the Followers of Josip Frank], published in *Srp i Cekic*, organ of the Communist Party of Croatia, no. 1-2 (January-February), 1941. This article states, among other things, that all Party decisions and resolutions have emphasized "the need to strengthen the struggle against the followers of Josip Frank," but that "it is not correct to simply catagorize the followers of Frank as agents of imperialism." At that time, the Communist Party of Croatia and of Yugoslavia favored the defense of Yugoslavia. In the name of "the international brotherhood of the suppressed," the Party condemned "the most simple ideology in the world, that of state's rights, that the Croatian people has its own independent state," as enunciated by both the followers of Frank and the leadership of the Croatian Peasant Party. The Central Committee of the Communist Party of Croatia wrote about the collaboration between communists and the police: "The incorrect attitude of certain comrades who began to collaborate with the police in the persecution of Frank's followers has now been for the most overcome." (Compare *Srp i Cekic, 1940-1941: Dokumenti historije KPH* [Hammer and Sickle 1940-1941 Documents of the History of the Communist Party of Croatia] Zagreb, 1951, 132-136).

²⁰³ Vinko Nikolic, *Pred vratima domovine* [Before the Gates of the Homeland], Paris - Munich, 1967, vol. 2, 217-220; *Osvrt na sporazu-*

mijevanje Pavelic-Stojadinovic [A Comment on the Pavelic-Stojadinovic Agreement], *Nezavisna drzava Hrvatska*, no. 10 (123), Toronto, October 1970, 5-6; Bogdan Krizman, *Pavelic u bjegstvu* [Pavelic in Flight], Zagreb, 1986, 261-272, 278-288, 342-349, 403-410; *Glasnik HSS*, Vancouver, 1988, no. 4, 12.

204 Photocopy of the original manuscript of Vladimir Nazor's poem, "Mornarska" and the musical composition by Ivo Tijardovic, both dated May 28, 1944, in the town of Rakovica (in my possession).

205 Confidential report by Prvoslav Vasiljevic to Milovan Djilas sent also to Aleksandar Rankovic and Edvard Kardelj on September 27, 1944, as well as Djilas's remarks to Rankovic, the chief of OZNA of NKOJ (secret police) (Compare V. Dedijer, *Novi prilozi za biografiju J. B. Tita*, op. cit., vol. 2, 842-843).

206 Compare F. Tudjman, "Slom vidovdanskog parlamentarnog centralizma...," op. cit., 79-81. et al.

207 Compare F. Tudjman, *Nacionalno pitanje u suvremenoj Europi*, op. cit., 34, 134, 194-195.

208 Stane Dolanc, member of the Yugoslav presidency, interviewed by Mirko Galic, the editor-in-chief of *Danas*, Zagreb, no. 287-288, August 18 and 25, 1987, 8.

List of Acronyms

AVNOJ: Antifasisticko vijece narodnog oslobodenja Jugoslavije (Antifascist Council of People's Liberation of Yugoslavia)

CK SKH: Centralni komitet Saveza Komunista Hrvatske (Central Committee of the League of Communists of Croatia)

CK SKJ: Centralni komitet Saveza Komunista Jugoslavije (Central Committee of the League of Communists of Yugoslavia)

FNRJ: Federativna Narodna Republika Jugoslavija (Federative People's Republic of Yugoslavia)

JAZU (YAAS): Jugoslavenska akademija znanosti i umjetnosti (Yugoslav Academy of Arts and Science)—now Hrvatska akademija znanosti i umjetnosti (HAZU) / Croatian Academy of Arts and Science / - founded in 1867; although always Croatian, the name "Yugoslav" originates from its founder, Josip Juraj Stross-mayer, as a result of his idea of South-Slavic unity

JNA: Jugoslavenska narodna armija (Yugoslav People's Army)

KPH: Komunisticka partija Hrvatske (Communist Party of Croatia)

KPJ: Komunisticka partija Jugoslavije (Communist Party of Yugoslavia)

LCC: League of Communists of Croatia

LCY: League of Communists of Yugoslavia

NDH: Nezavisna Drzava Hrvatska (Independent State of Croatia) - state created in 1941 within the framework of the Axis powers

NOB: Narodno - oslobodilacka borba (People's Liberation Struggle)

NOP: Narodna - oslobodilacki pokret (People's Liberation Movement)

NOR: Narodno - oslobodilacki rat (People's Liberation War)

NOV (i PO): Narodno - oslobodilaca vojska / i partizanski odredi / (People's Liberation Party / and Partisan Units /)

NR (Hrvatska): Narodna Republika / Hrvatska / (People's Republic / of Croatia /)

SANU: Srpska akademija nauka i umetnosti (Serbian Academy of Arts and Science)

SFRJ: Socijalisticka Federativna Republika Jugoslavija (Socialist Federative Republic of Yugoslavia)

SKH: Savez komunista Hrvatske (League of Communists of Croatia)

SKJ: Savez komunista Jugoslavija (League of Communists of Yugoslavia)

SRH: Socijalisticka Republika Hrvatska (Socialist Republic of Croatia)

SUBNOR: Savez udruzenja boraca narodno - oslobodilackog rata (Union of Veterans of the People's Liberation War)

ZAVNOH: Zemaljsko antifasisticko vijece narodnog oslobodenja Hrvatske (Territorial Anti-fascist Council of People's Liberation of Croatia)

Index

465